A Treatise on Limnology

VOLUME I

GEOGRAPHY, PHYSICS, AND CHEMISTRY

G. Evelyn Hutchinson

DEPARTMENT OF ZOOLOGY

YALE UNIVERSITY

NEW YORK · JOHN WILEY & SONS, INC.

London · Chapman & Hall, Ltd.

6 7 8 9 10

Library of Congress Catalog Card Number: 57-8888

PRINTED IN THE UNITED STATES OF AMERICA

ISBN 0 471 42570 2

WASTWATER

FRONTISPIECE. *Above,* Wastwater; *below,* Esthwaite Water (from Harriet Martineau, *A Complete Guide to the English Lakes,* 1855).

ESTHWAITE WATER.
From the Ulverston Road.

VICTORIO LIVIAE

TONOLLI

VERBANENSIBUS

AVCTOR

D D

Quant Theseus, Hercules et Jazon
Pour accroistre leur pris et leur renon
Cercherent tout, et terre et mer profonde
Et pour veoir bien tout l'estat dou monde,
Moult furent dignes d'onnour

Guillaume de Machault

+ = −

Je mehr
die Sonne scheint
desto mehr
Wasser verdunstet
es entstehen
mehr Wolken
und
die Sonne
scheint weniger

Je weniger
die Sonne scheint
desto weniger
Wasser verdunstet
es entstehen
weniger Wolken
und
die Sonne
scheint mehr

da capo

Josef Albers

Preface

The aim of this book is to give as complete an account as is possible of the events characteristically occurring in lakes. The author, by training a biologist, is by inclination a naturalist who has tried to examine the whole sequence of geological, physical, chemical, and biological events that operate together in a lake basin and are dependent one on another. The book is addressed to all who are professionally concerned with limnology, but also to biologists who may wish to know something of the physicochemical environment, mode of life, and evolutionary significance of such fresh-water organisms as they may study from quite different points of view; to geologists who are desirous of learning something of modern lakes in order that they may better interpret the record of inland waters in past times; and to oceanographers who wish to compare the results of their own science with what has been learned of the small but very individual bodies of water which make up the nonmarine part of the hydrosphere.

In order to make the work as useful as possible to these and other scientists directly or indirectly concerned with lakes, a number of ancillary matters have been considered in detail. In choosing what collateral matter to elaborate, the author has been guided by the availability of other modern accounts, and has tried to provide discussions of subjects that most biologists or geologists would find difficult to obtain elsewhere. In developing such accounts he has endeavored to follow the principle that what he has himself found perplexing has probably perplexed other investigators.

In order that the research worker may discover what he needs as a background to his investigations, the text has been written in a very detailed manner, but at the end of each chapter an undocumented factual summary of all the main conclusions has been given. It is hoped that the elementary student may find that these summaries, taken together, provide a simple account of what is actually known about lakes.

The first volume is intended to cover geographical and physicochemical limnology, the second will deal with limnobiology and the ecological, typological, and stratigraphic problems of lake development. In dealing with some of these subjects it will be necessary to revert to certain chemical matters; this will provide an opportunity to mention works published too late to consider in this volume.

Although the present work is longer than any of its predecessors, it has proved impractical to refer to all the significant earlier investigations. Wherever large amounts of work have been summarized in a condensed form, references with an asterisk in the bibliography will indicate the most convenient source of the older material. Only in this way has it been possible to keep the length of the bibliography within reasonable bounds. As sources for the older work, W. Halbfass *Grundzüge einer vergleichenden Seenkunde* (Berlin, 1923), A. Thienemann *Die Binnengewässer Mitteleuropas* (Stuttgart, 1925), P. S. Welch *Limnology* (New York, 1952, 2nd ed.), and F. Ruttner *Fundamentals of Limnology* (trans. D. G. Frey and F. E. J. Fry, Toronto, 1953) are still indispensable. My debt in particular to Halbfass will be apparent throughout the present work, and I have neither the ability nor the desire to supersede a work the rarity of which is its only defect. The extraordinary bibliography prepared by J. Chumley for the first volume of J. Murray and L. Pullar *Bathymetric Survey of the Scottish Fresh-water Lochs* (Edinburgh, 1910) will be found invaluable to anyone wishing to explore the older literature. The review of Lauscher (1955) on lake optics unfortunately became available too late to use.

Any value that the work may possess is largely due to the help that I have received from many institutions and friends. To Dr. Ross G. Harrison, the late Dr. L. L. Woodruff, and Dr. J. S. Nicholas, Directors of the Osborn Zoological Laboratory of Yale University, I am indebted for their understanding of what I have tried to do, so that the most unzoological work has not seemed inappropriate in the laboratory under their charge. I would also express my gratitude to another Yale institution, the Sterling Memorial Library; without its enormous resources the book would have been quite an impossible task. The members of its reference department, particularly Miss Jean C. Smith, herself a limnologist, have been of great help on countless occasions.

To that admirable institution, the John Simon Guggenheim Foundation, I am deeply indebted for the opportunity to travel in Scandinavia and Italy, greatly increasing my acquaintance with both lakes and limnologists. I would particularly remember for their kindness and hospitality Dr. Wilhelm Rodhe and Dr. Gunnar Lohammar of Uppsala; Dr. H. C. Gilson, Dr. C. H. Mortimer, and the whole staff of the Freshwater Biological Association, then at Wray Castle; the late Dr. Edgaro Baldi and my good friends Dr. Vittorio and Dr. Livia Pirocchi Tonolli of the Instituto italiano di Idrobiologia at Pallanza.

I am most grateful to Dr. A. D. Hasler of the University of Wisconsin for free access to the great mass of unpublished data left by the pioneer limnologists of Madison, E. A. Birge and C. Juday, and at an earlier date to Dr. Birge and Dr. Juday themselves for enabling me to spend a week at their laboratory at Trout Lake.

A number of investigators have communicated to me data that had not been published; I would specially remember in connection with the preparation of this volume Dr. W. T. Edmondson and his associates, Dr. R. B. Benoit, Dr. D. Frey, Dr. R. F. Flint, Dr. Eville Gorham, Dr. Joyce C. Lewin, Dr. K. Sugawara, Dr. Joseph Shapiro, and most particularly Dr. Heinz Löffler and Dr. C. H. Mortimer. I have had the great pleasure and benefit of numerous discussions with Dr. Mortimer both in England and in America, and he has always made his knowledge and insight freely available. My immense debt to Dr. Löffler will be apparent from numerous footnotes. A considerable part of the manuscript has been read by Dr. E. S. Deevey, Dr. W. H. Edmondson, Dr. J. T. Edsall, Dr. G. A. Riley, Dr. J. Neumann, Dr. J. R. Vallentyne, Dr. Eva Low Verplanck, and Dr. T. H. Waterman, and in every case some improvement has resulted from their criticisms. Dr. J. H. Zumberge and Dr. Löffler read the entire work in its penultimate state, making many useful suggestions. My debt to Dr. Deevey and Dr. Riley is particularly great, as we have discussed limnological and cognate problems continually for a number of years. The numerous discussions that I have had with Dr. Ruth Patrick have also been exceedingly fruitful. Mr. Parry Larsen has carried through with great patience and skill the enormous task of making the geographical index, and Mr. Donald Martin has done much in the preparation of the bibliography.

Many individuals have helped in a variety of ways in preparing the text. I would especially thank my assistants, Miss Martha M. Dimock, whose linguistic skill has made possible the addition of recent material from the Russian, and whose willing, untroubled help in the assembling of the initial draft was invaluable, and Mrs. Nancy Kimball, who performed similar services in the later stages of the preparation of the manuscript.

The text figures, which have almost all been redrawn from original sources, are the work of Miss Dimock and Mrs. Kimball. The very few unmodified drawings copied directly from earlier publications are indicated by the omission of the word "after" before the acknowledgement of source.

I am greatly indebted to Mons. B. Dussart, Dr. W. T. Edmondson, Dr. Heinz Löffler, Dr. C. H. Mortimer, Dr. C. E. Prouty, Professor K. Strøm, Dr. J. H. Zumberge, the Geological Society of America, the *Lady* magazine of London, the National Geographic Society, the Royal Canadian Air Force, the South Africa Air Force, the United States Department of Agriculture, the United States Geological Survey, the United States Navy, and Widerøes Flyveselskap og Polarfly of Oslo for help in obtaining photographs or for permission to publish them.

To my wife I am, as always, more indebted than I can tell. I have had from her the support of common faith; *alitur enim liquantibus ceris, quas in substantiam pretiosae hujus lampadis apis mater eduxit.*

G. E. HUTCHINSON

New Haven, Connecticut, U.S.A.
July, 1957

Contents

CHAPTER 1

The Origin of Lake Basins

Lakes seem, on the scale of years or of human life spans, permanent features of landscapes, but they are geologically transitory, usually born of catastrophes, to mature and to die quietly and imperceptibly. The catastrophic origin of lakes, in ice ages or periods of intense tectonic or volcanic activity, implies a localized distribution of lake basins over the land masses of the earth, for the events, however grandiose, that have produced the basins have never acted on all the lands simultaneously and equally. Lakes therefore tend to be grouped together in *lake districts*; within each district the various basins may resemble each other in certain general characters yet differ markedly in size and depth and so in their rate of maturation and senescence. It is this diversity in unity that gives the peculiar fascination to limnology. A group of lakes confronts the investigator as a series of very complex physicochemical and biological systems, each member of which has its own characteristics and yet also has much in common with the other members of the group. Moreover, the whole group of lakes of a given lake district may be compared with another group, often under widely different geographical and climatic conditions. In this way it is possible to identify the causal factors producing the differences between lake and lake or lake district and lake district. Much of the present work will be devoted to this type of analysis. Lakes, moreover, form more or less closed systems, so that they provide a series of varying possible ecological worlds which permit a truly comparative approach to the mechanics of nature. It is in the study of the lake as a microcosm, as Forbes (1887) put it many years ago, that much of the intellectual significance of limnology lies, and this significance grows geometrically as a whole series of comparable yet different microcosms become available.

Throughout the succeeding pages, relations between the shape and size of a lake and the nature of the physicochemical and biological events taking place in its water will become apparent. The size and shape of a lake are, however, largely dependent on the forces that have produced the lake basin. Some discussion of the geomorphological aspects of lake basins and of the ways in which lakes have come into existence is therefore an appropriate point of departure.

THE ORIGIN OF LAKES

Some geomorphologists, notably W. M. Davis (1882) in his earlier work, have adopted a formal classification of the agencies which may produce basins, grouping them as *constructive*, *destructive*, or *obstructive*. Penck (1882) and Supan (1896) used similar primary divisions. This procedure is somewhat artificial and tends to obscure the regional grouping of lakes. Glacial action may form a basin by destruction, as in the case of the excavation of a cirque lake, or by obstruction, as when a valley is dammed by a moraine. Similarly, volcanic action may produce a basin by destruction when an explosion or volcanic subsidence occurs, by construction when a crater rim is built, or by obstruction when a valley is dammed by a lava flow. In general, glacial basins, whether formed destructively or by obstruction, will tend to occur in the same regions, and volcanic lakes, however formed, in other regions. The limnologist is usually more interested in keeping together the regionally grouped lakes than in separating them by formal intellectual barriers. Within the regional group, Davis's threefold classification of agencies producing basins is often most useful.

It is also possible to consider the origin of lakes and their geomorphological classification primarily in temporal terms, as Davis (1887) did in a later brief paper which emphasizes the contrast between the immature landscape rich in lakes and the relatively lakeless mature landscape. Such a temporal approach certainly calls attention to an important aspect of lake basins, namely their impermanence, but it is no more satisfactory than is the threefold formal classification of Davis's earlier paper in providing a classification of the lakes of the world.

In the present work, it will be more convenient to follow those authors who have considered the agencies producing lakes more empirically and in such a way that a regional point of view is easily achieved. Admittedly such a point of view easily emerges from Davis's earlier paper, which lies behind all later treatments of the subject, and which must still be read by the serious limnologist. The classification to be adopted is based primarily on those used by Penck (1894) in the chapters on lakes and basins of his classical *Morphologie der Erdoberfläche* and by Russell (1895) in his admirable book *Lakes of North America.* Penck gives an excellent discussion of the early history of the concepts that he employs. A comparable, if somewhat less natural, arrangement was employed by Delebecque (1898) in his masterly treatise on the lakes of France. Similar schemes were used by Forel (1901a), by Sir John Murray (1910) in the introductory volume of Murray and Pullar's monumental *Bathymetric Survey of the Scottish Freshwater Lochs*, and by W. Halbfass (1923) in his *Grundzüge einer vergleichenden Seenkunde*, the most learned and generally satisfactory work on the physical aspects of limnology that has yet appeared. In slightly different forms the same type of treatment has been used by Hobbs (1912); by Davis (1933) himself in his last contribution to limnology; following these authors, in the recent textbook by Cotton (1941), to name but one of several modern introductions to geomorphology; and, in so far as it applies to Minnesota, by Zumberge (1952) in his admirable treatise on the lakes of that state. The presentation to be made in the following pages owes much to all these writers and to their nineteenth-century predecessors. Where the treatment differs from earlier expositions, it is mainly to permit the addition of minor details not known to former writers and to provide examples from as many parts of the world as is possible in accordance with the international outlook that the present work attempts to achieve.

In discussing the origin of lake basins, it is obviously necessary to set some sort of limit to the inquiry, otherwise the consideration of any particular lake will rapidly become the study of the entire structural geology and geomorphology of the region in which the lake is situated. Since it will be necessary to consider lakes all over the world, the necessity for some limitation becomes all the more apparent. In the following pages only those events that actually determine that the basin can hold water will be considered. In particular, when obstructive phenomena are being treated, the valleys will be taken as given, and only the processes that turn an open valley into a lake basin will be discussed. Geologically, a landslide that dams a valley may be of quite minor and purely local importance, while the valley itself may reflect major structural features of great tectonic interest. To the geologist the dam therefore is an epiphenomenal afterthought, while to the limnologist it makes all the difference in the world, determining whether his science can or cannot be pursued in the valley. Constructive or destructive processes producing basins not necessarily dependent on the pre-existing drainage pattern will often have to be treated in somewhat greater detail than the obstructive processes building dams.

TECTONIC BASINS

The term tectonic basin will be used to include all lake basins formed by movements of the deeper parts of the earth's crust, except those in which manifest volcanic activity has played a major part. A few cases in which it is difficult to distinguish between the effects of volcanism and of other types of earth movement, and which may be most reasonably designated as *volcano-tectonic*, are reserved for consideration on page 35.

Relict lakes isolated from the sea by epeirogenetic earth movements (Type 1). The Ponto-Caspian region is of exceptional interest and importance, as within it lie the two great inland seas of the Eurasiatic continent, the Caspian Sea and the Sea of Aral. In discussing these two bodies of water, the first of which is the largest lake in the world, the history of the Euxine Basin, which now contains the Black Sea but has often in its past been occupied by more or less fresh bodies of water, must also be considered. This assemblage of large lakes, together with a few smaller lakes in the Aral and Caspian Basins, forms a unique group, for here alone can the history of a considerable body of water isolated from the ocean by earth movements be adequately studied. The biological implications of this history will be apparent in Volume II.

During the early Tertiary, almost the whole of the present Ponto-Caspian region was submerged as part of the Tethyan Sea, a marine basin that originally communicated with the Atlantic Ocean to the west and with the Indo-Pacific to the east. The eastern end of this sea was subsequently closed, and in the late Miocene communication with the ocean to the west was also interrupted by the extensive elevation and mountain-building that produced the Alps. The old Tethyan Sea then became a vast inland basin, the Sarmatian Sea, the arms of which ran northwestward, north of the Carpathians as far as Cracow and westward at least to the Iron Gate, with a connected basin higher up the Danube in the region of the present Hungarian plain. The approximate outlines of the western part of the Sarmatian Sea at the time of its maximum transgression are given in Fig. 1, mainly taken from Stahl (1923). Other closed basins existed in other parts of southern Europe during the Miocene and Pliocene. The early Sarmatian fauna was marine but depauperated, lacking echinoderms and other notably stenohaline organisms. During the very late Miocene (Maeotian) and the subsequent Pliocene (Pontian and Dacian) stages, there is evidence of the development of a special fauna consisting of hydrobiid snails,[1] particularly of the genera *Caspia* and *Micromelania*, and of lamellibranchs such as *Dreissena*, *Monodacna*, and *Didacna*, certainly of Sarmatian marine origin but adapted to low salinities. Along with this special fauna, which will be considered in greater detail in Volume II, the truly marine Sarmatian fauna also survived in parts of the area, and it seems to have been reinforced by a few marine species from

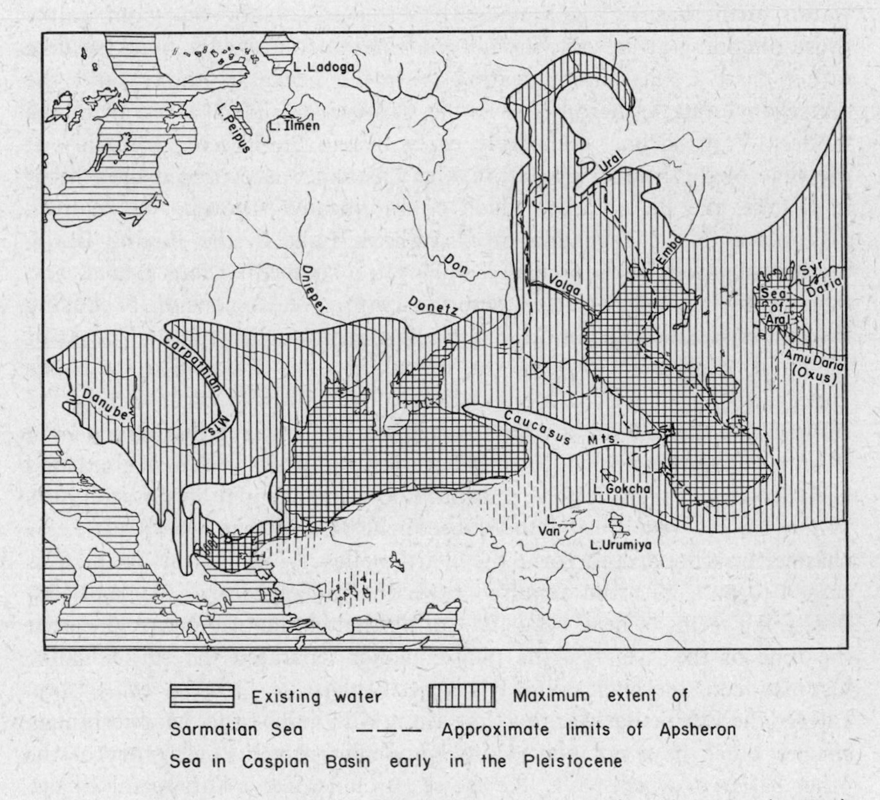

FIGURE 1. Relation of the modern Black Sea or Pontic Basin and the Caspian Basin to the coast line of the Sarmatian Sea (after Stahl and other sources).

outside, such as *Acicularia italica*, known from the Pliocene Aktschagyl beds. The very considerable variations in salinity indicated by these different faunas suggest that during the Pliocene the various basins into

[1] The former wide distribution of these and of allied genera in the waters of southern Europe is apparent from Wenz's (1925) catalog of the Tertiary *Hydrobiidae*.

which the Sarmatian Sea became more or less divided developed great salinity gradients, comparable to those exhibited today by the Baltic, the Sea of Azov, and the modern Caspian. Though temporary lowering and isolation of the various basins certainly occurred, there was a connection between the Euxine and Caspian at least during Maeotian and again in Aktschagyl times. As geographical changes occurred, the areas of maximum and minimum salinity presumably varied.

The latest Pliocene or early Pleistocene saw some contraction of the waters of the basins, which evidently retreated to shore lines not unlike those obtaining at present, though the Euxine contained a brackish lake rather than a sea communicating with the ocean as today, and the Aktschagyl and Apsheron seas in the Caspian extended far north of the modern Volga delta. Relatively early in the Pleistocene, probably at the time of the Mindel glaciation, the Caspian was occupied by a high-level lake, the Baku Lake, which communicated through the Manych Channel with the Tschauda or Chaudinsk Lake in the Euxine Basin. Somewhat lower levels prevailed during the succeeding interglacial, and then during the Riss glaciation there was again a great expansion, producing the Palaeo-Euxine Lake in the Euxine Basin connected with the Chosar or Khararsk high-level lake in the Caspian. The Palaeo-Euxine Lake appears to have received a great deal of glacial-melt water during the summers at the time of the Riss glaciation and to have discharged by a river which, cutting down toward base level at a time when the sea had been lowered by glaciation, produced a deep channel (Ramsay 1931). When the sea rose during the subsequent Riss-Würm interglacial, this channel was flooded to form the Dardanelles. The Euxine Basin thus ceased to be a lake and acquired a Mediterranean fauna. A transitory freshening seems to have occurred during the middle of the interglacial at the time of the Meso-Euxine phase, which separated the earlier saline Usunlar from the later saline Karangatsk phases. *Cardium edule* Linn. entered the Euxine Basin at this time along with many other Mediterranean species which have not played any important part in the history of the other basins or which have, like the sea urchins of the Karangatsk phase, been unable to survive subsequent changes in the Euxine. *C. edule* penetrated the Manych at this time but did not appear in the Caspian, as far as the fossil record goes, until postglacial time. Grahmann (1937), whose summary is the basis of the present account, supposes that the species survived the high-level fresh Khvalynsk stage of the Caspian, during the Würm glaciation, in isolated saline bays, only to appear in the main basin when the level fell and the salinity increased during post-glacial post-Khvalynsk time when there was no more possibility of direct introduction of the species from the Euxine. Kovalevsky (1933) appears,[2] however, to consider that the Manych Channel was still open in the middle of the second millennium B.C. and that it was navigated by the Argonauts. Another modern Mediterranean lamellibranch, *Mytilaster lineatum* (Gm.), also known from the Caspian seems to have had a history similar to that of *C. edule* (Bogatschew 1928).

The problem of the extent and causes of the changes in level of the Caspian during historic times is not fully solved. Three postglacial high-level beaches bearing shells of *C. edule* are known, but according to Berg (1934, [1938] 1950) none lie at altitudes greater than 5 m. above the modern level. Historic evidence considered extensively by many writers, notably Brückner (1890), Huntington (1907) and Kovalevsky (1933), has suggested that the Caspian had a high but declining level during classical antiquity, falling to a minimum 20 m. below the modern level during the seventh century A.D. A rise to a level 4 to 10 m. above the present appears then to have set in, the maximum being achieved about 900 A.D. There was then a slight fall and another rise to a maximum 12 to 14 m. above the present level (page 238) about 1300 A.D. Berg (1934), however, appears to regard the archeological and historic data on which these conclusions are based as due rather to local earth movements changing the position of artificial structures of known age on the shores. He points out that in the Euxine Basin subsidence in quite recent times has submerged Karangatsk and even Neo-Euxine littoral deposits to depths of up to 1500 m.

The history of the Sea of Aral is probably comparable to that of the Caspian but is less well known. It is evident, since *C. edule* occurs in both bodies of water, that there has been some hydrographic connection by way of saline channels in post-Khvalynsk time, but the nature of this connection is not entirely clear. The present level of the Sea of Aral stands at 52 m. above sea level, while the highest beach deposit containing *C. edule* in the Caspian basin lies 21 m. below sea level. During the highest glacial stages the Sea of Aral drained into the Caspian through the Uzboy Channel. Much more recently, during classical and again during medieval times, there was an indirect hydrographic connection, for a branch of the Amu-Darya or Oxus entered the Uzboy Channel and so drained into the Caspian, while the rest of the waters of the river reached the Sea of Aral as today. The Sarykamysh Basin further to the southwest, now containing two shallow salt lakes, appears to have been connected with the Caspio-Aral system since the entry of *C. edule*, but the shells of this mollusk are not found in the Sarykamysh Basin at such high levels as

---

[2] Kovalevsky's work is known to the writer only through the informative but naive review of Zavoico (1935). In this work *C. edule* is called *C. rusticum* (= *edule*), though the two species appear to be different.

in the Sea of Aral (Berg 1950). It is obvious that much remains to be done to elucidate the history of the earth movements and hydrographic connections involved in the formation and population of the Ponto-Caspian region.

The only analogy to the history of the basins just considered is provided by the Baltic (Fig. 2), which like the Euxine has alternated between a fresh and salt condition, and which, when saline, exhibits very marked salinity gradients, like the modern Black Sea. The earliest postglacial Baltic was an ice lake; when the ice that dammed the Billinger Gap in South Sweden retreated, the lake discharged, supposedly in 7912 B.C., and was invaded by ocean water to form the Yoldia Sea. This was converted into a lake about 6500 B.C. by the rebound of the southern part of the recently deglaciated Scandinavia, which rose isostatically more rapidly than the postglacial eustatic rise in sea level. Lake Vener, which lies in the region at the mouth of the Yoldia Sea, was formed as elevation continued. Later, about 5000 B.C., as deglaciation became more complete, the rising ocean broke through again, this time by way of the Danish Sounds rather than through the Vener Basin, to form the Litorina Sea, the ancestor of the modern Baltic. As far as information is available, the dating of the events, based primarily on the varve chronology, is in fair accord with the radiocarbon chronology (Flint and Deevey 1951).

Gentle epeirogenetic uplift of irregular marine surfaces (Type 2). A newly elevated land surface may inherit an irregular topography from the time that it was covered by the sea, such irregularity being due to uneven sedimentation determined by currents. If the irregularities form closed basins, lakes may collect in them after their elevation. Hobbs (1912) speaks of such lakes as *newland lakes*. It is generally believed that such of the shallow lakes of Florida as are not solution basins have been formed in this way. The largest of these lakes, Lake Okeechobee, though only about 3 m. deep, has an area[3] of 1840 km.^{2} and thus has a greater water surface than any other lake except Michigan wholly within the United States. According to Parker and Cooke (1944), the lake certainly represents a depression formed at the bottom of the Pliocene sea, though it has probably been modified in shape by solution, erosion, and sedimentation. That such a shallow lake has not drained naturally is largely due to the very small gradient that can develop in the region and the ease with which vegetation can block the effluent. A few smaller lakes in the neighborhood of Okeechobee have doubtless been formed in the same way.

Reversal of hydrographic pattern by tilting or folding (Type 3). Slight tilting or upwarping of a continental surface may produce lakes by inter-

[3] Somewhat variable according to lake level.

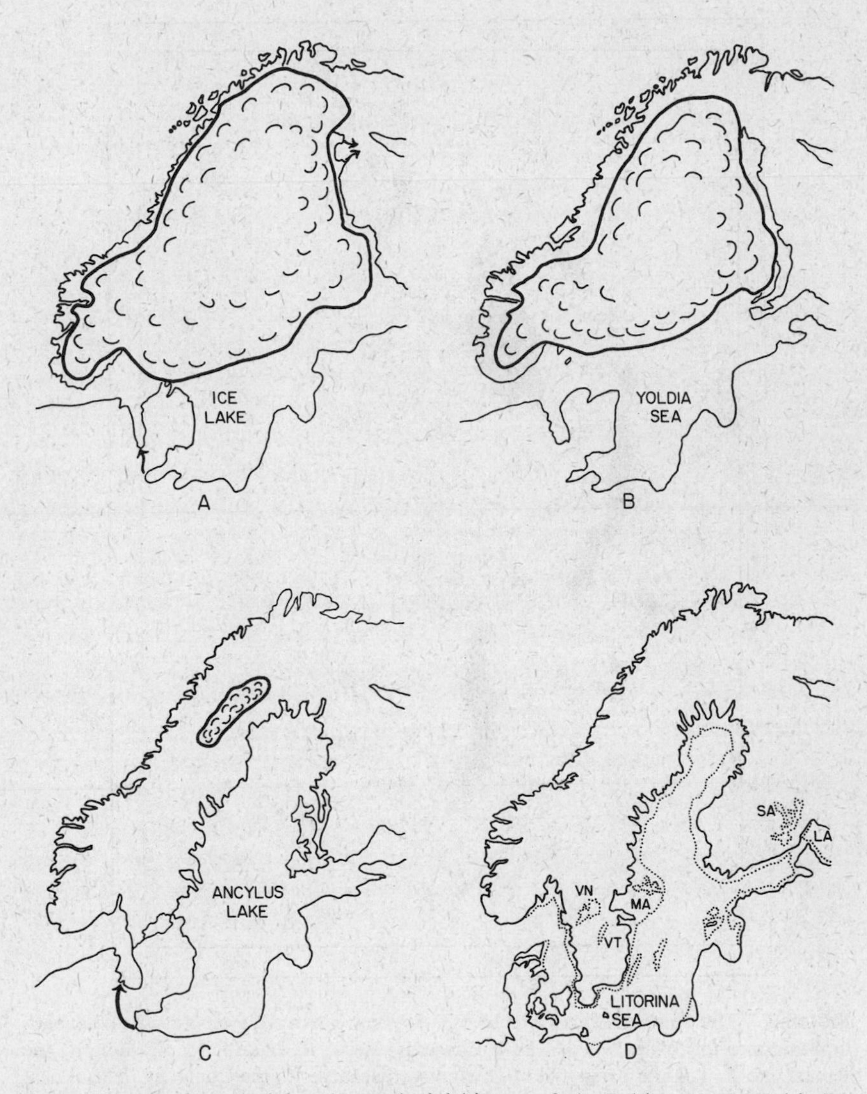

FIGURE 2. The late glacial and postglacial history of the Baltic. *A*, the Baltic Ice Lake, about 8000 B.C.; *B*, the Yoldia Sea, about 7900 B.C.; *C*, the Ancylus Lake, 6500–6000 B.C.; *D*, the early Litorina Sea, about 5000 B.C., with the modern coast line of the Baltic Sea and the positions of certain lakes indicated by dotted lines: VT, Lake Vetter; VN, Lake Vener; MA, Lake Mälar; SA, Lake Saimaa; LA, Lake Ladoga. (After Sauramo, Zeuner, and other sources.)

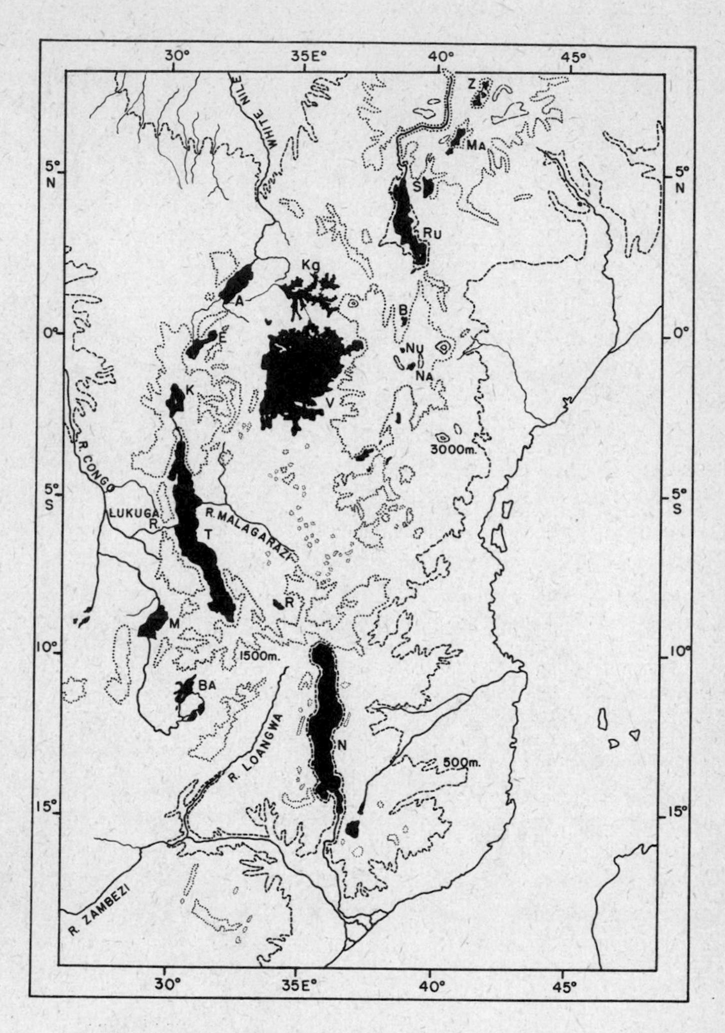

FIGURE 3. The Central African Rift lakes. In the western rift system: A, Lake Albert, draining into the White Nile; E, Lake Edward, closed; K, Lake Kivu, draining by the Ruzizi into T; T, Lake Tanganyika, which now discharges to the Congo by the Lukuga. The western rift converges on the eastern rift southeast of R, Lake Rukwa, and then diverges further south as the Loangwa Valley. In the eastern system: Z, Lake Zwai, and the closed lakes Hora Abjata, Oitu, and Hora Shala immediately to the south; MA, Lake Margherita, and Lake Chamo immediately to its south; S, Lake Stefanie; RU, Lake Rudolf, southern portion only; a number of small lakes of which B, Lake Baringo, NU, Lake Nakuru, and NA, Lake Naivasha, are of limnological importance;

fering with the hydrographic pattern. In central Africa (Fig. 3), the region north of Lake Victoria has been tilted in such a way as to reverse the flow in the upper part of the Kafu River. The valleys at the head of the river therefore became flooded, to form the practically continuous pair of Lakes Kioga and Kwania (Wayland 1934). There is evidence that the reversal which formed Lake Kioga first occurred during late Tertiary times; the old pattern was apparently re-established in the Pleistocene, but later was reversed again to give the modern Kioga. The older Kioga, before the temporary restoration of the original drainage in the Pleistocene, was a larger lake than the present body of water (O'Brien 1939 and earlier work cited therein). The northern part of Lake Victoria, with its numerous drowned valleys, has certainly been modified by processes that produced Kioga, and several other small lakes were doubtless formed in the same way at the same time (see Solomon in O'Brien 1939 for a critical summary of the earlier work).

The upwarping that formed the Great Basin of South Australia, which was occupied during the wet phases of the Pleistocene by Lake Dieri, now represented by the normally dry Lake Eyre, and several other like basins, is perhaps comparable to the process just described, but on an even greater scale. The same episode formed the basin of the Pleistocene Lake Nawait by partial reversal of drainage in the lower part of the Murray-Darling river system (David and Browne 1950). Lake George, near Canberra, probably provides another Australian example (Garretty 1937).

The origin of Lake Champlain, New York and Vermont, as a lake, involved the detachment of an arm of the sea, owing to the uplift of the northern end of the basin as the North American continent rebounded after deglaciation (Fairchild 1918, 1919; Goldring 1922). This process, in which a valley was slightly tilted, has much in common with the type of event just described; such a basin is clearly a composite type combining features of Type 1 and Type 3. Lake Mälar in Sweden is presumably another example of a basin of similar composite character.

Large-scale basins formed by warping (Type 4). In central Africa the movements responsible for the flooding of the valley of the upper Kafu River appear to be part of a process that has produced Lake Victoria as a whole, for it is evident that there has been some uptilting all around the margins of the plateau bounded by the Albert-Edward-Tanganyika

---

and N, Lake Nyasa, at the south end of the rift. Upwarping along the edges of the rifts has caused the flooding of the river valley now occupied by Kg, Lakes Kioga and Kwania, and has formed the immense flat basin of V, Lake Victoria, between the two rift systems. Outside the rifts, the large swamp lake BA, Lake Bangweolo, and M, Lake Mweru, are of some importance. (After Worthington, Willis, Brooks, and other sources.)

system of rifts on the west and the Gregory or Kenya rift valley on the east, and that Lake Victoria occupies a slightly depressed area in the middle of this plateau. The details of the process have been the subject of much discussion, summarized by Willis (1936), Solomon (1939) and other writers mentioned below in connection with the rifts themselves.

Earth movements producing local subsidence (Type 5). Depressions are sometimes produced by local subsidence during earthquakes, and water may then collect in the depressions to produce lakes. Lyell (1837) collected a number of cases from the earlier literature. During an earthquake that convulsed the island of Jamaica in 1692, much local subsidence seems to have occurred (Sloane 1694, p. 94), and in the northern part of the island several plantations are said to have been swallowed up and replaced by a lake which subsequently dried up, leaving nothing but a bed of sand and gravel. Very many depressions were also generated during the Calabrian earthquake of 1783. Vivenzio (1788) enumerates 50, and in the official report, drawn up by the authorities of the Kingdom of Naples, it is said that 215 new lakes and ponds were formed, though many of these were very small and some were due to landslides produced by the earthquake. One case, however, the Lago del Tolfilo, said to have been formed near Seminara, involved the opening of a chasm from the bottom of which water issued, producing a lake 545 m. long, 286 m. wide, and 15.9 m. deep.

Earth movements associated with the New Madrid earthquake of 1811 produced a series of shallow, irregular lakes in the "sunk" area of the states of Tennessee and Missouri and Arkansas. A low dome across the Mississippi River was elevated, while areas on either side of this dome were depressed. The river is reported to have flowed backwards for a short time after the earthquake, but soon established a new course across the obstruction. The areas of subsidence (Fig. 4), however, still contain lakes, notably Reelfoot Lake on the Tennessee side and several smaller lakes on the Missouri side of the river. The recent sediments in the raised area across the river are very clearly upwarped, but it is not apparent exactly what may have happened below them. McGee's (1893) study of the region provides one of the earliest cases of the use of tree-ring counts in dating events of physiographic interest.

Lakes in basins in tectonically dammed synclines (Type 6). A few examples are known where a basin lies in the axis of a syncline and may be reasonably attributed to folding. At least the northern part of Lake Rudolf in central Africa appears to belong in this category (Fuchs 1939). Heim (1905) gives an excellent Swiss example, the Fählensee in the Säntis massif, which lies in a synclinal valley blocked by a tectonic dam due to the thrusting of an anticlinal dome across its lower end (Fig. 5). Collet (1925) states that several lakes in the Jura, notably the Lac de Joux, are of like origin.

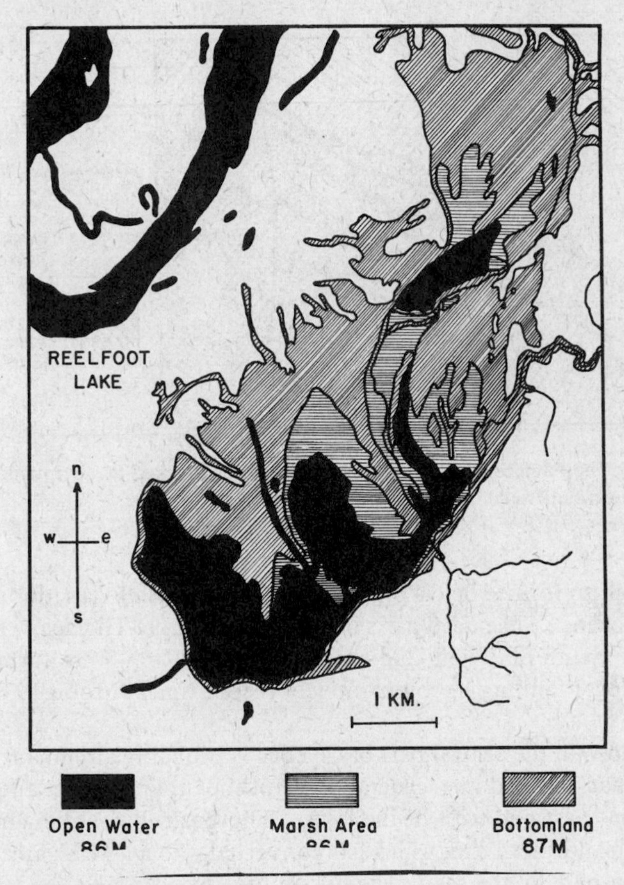

FIGURE 4. Reelfoot Lake, Tennessee, a very shallow lake formed as the result of an earthquake (after Baker).

Old peneplain surfaces as intermontane basins (Type 7). A few cases are known in which lake basins have been formed by the uplifting of practically undisturbed sections of ancient peneplains to form intermontane basins during the process of mountain building. Subsequent local block faulting may deepen parts of the intermontane basin so produced. The most remarkable of such basins is the Altiplano on the Peru-Bolivia frontier, on which lie Titicaca and its associated lakes. The Altiplano represents a fragment of a late Mesozoic land surface that was elevated nearly 4000 m. in the Tertiary, during the building of the Andes, and which now forms a broad intermontane valley delimited by overthrusts.

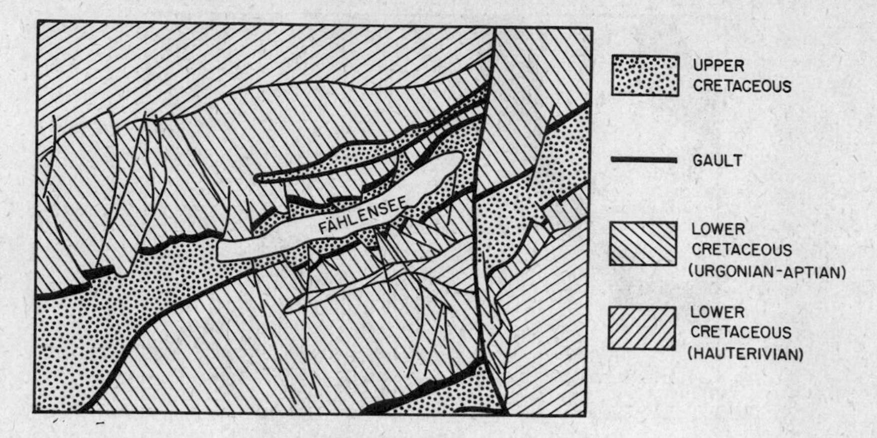

FIGURE 5. The Fählensee, a lake in a synclinal valley blocked by an anticlinal tectonic dam thrust across its end (after Collet, from Heim).

The trough so formed held a high-level lake, Lake Ballivian, during part of the Pleistocene. The deep parts of the modern Lake Titicaca, which has a maximum depth of 281 m. and an area of 7600 km.^{2} were formed, however, by local faulting in comparatively recent times (Moon 1939, Newell 1949).

Lake Poso in the central part of Celebes is probably a remnant of an old surface that escaped the general Pleistocene uplift which affected two geanticlines east and west of the lake. The basin has undergone a little uplift at the northern end, which has converted into a lake an area formerly covered in part by the sea. The possibility that the lake lies in a graben or fault trough is regarded by Van Bemmelen (1940) as very unlikely. It evidently originated by processes not unlike those forming the basin of Titicaca, but the elevation of the old surface has been mainly confined to a little tilting, formally effecting a transition to Type 4.

Basins associated with fault scarps (Types 8 and 9). The most important type of tectonic basin is due to faulting. In the American literature, for example in Hobbs (1912) or to a less extent in Davis (1933), a distinction is often made between lakes associated with single faults and lakes lying in down-faulted troughs or grabens. In the first case, Type 8, a region is supposed to be broken into fault blocks which may become slightly and irregularly tilted. Water then collects on the most depressed parts of the blocks, forming lakes. In the second case, Type 9, a more or less elongate area is depressed and the lake lies at the bottom of this depression. It is difficult to preserve this distinction, however, though it is perhaps useful enough to retain for the extreme cases. The Sierra Nevada and Great Basin region of western North America exhibits block faulting or basin-and-range topography on a scale that is probably unparalleled elsewhere in the world. A general account of the geology is given by Blackwelder (1948), and all the closed drainage systems of the region are enumerated in the admirable summary of Hubbs and Miller (1948) on the physiography and zoogeography of the area. Basin-and-range topography, moreover, extends under the sea well to the west of the southern California coast, producing "closed" basins which if elevated would contain isolated lakes. There is a good deal of evidence that the existing basins were so elevated during the early Pleistocene.

Abert Lake (Fig. 6) in Oregon is usually considered (Russell 1895) as the finest example in the basin-and-range area of a lake lying on a tilted fault block; Walker and Winnemucca lakes in the Lahontan Basin certainly belong to this type. Some other cases widely quoted in the literature are probably actually situated in grabens. Among these may be noted the three shallow lakes of the Surprise Valley in northeastern California. Though R. J. Russell (1927, 1928) points out that the presence of the lakes up against a fault scarp on their western shores argues in favor of recent rotation of the fault blocks involved, he also indicates that the modern

FIGURE 6. Abert Lake, a beautiful example of a lake against a fault scarp (Russell 1895).

lakes are but shallow and impermanent remnants of a large Pleistocene lake, Lake Surprise, which was a typical graben lake, though the eastern fault is less impressive and less well known than the western. Lake Surprise laid down at least 250 m. of sediments, all apparently of shallow-water origin, so that its basin must have been undergoing subsidence during the whole of the history of this ancient body of water.

Farther west, the lakes of the Modoc lava field are frequently regarded as lying on tilted block faults. For the modern Tule Lake and Lower Klamath Lake this is no doubt correct, but both these lakes are surrounded by areas of lake sediments which were deposited in typical graben lakes of the Pleistocene.

In the Sierra Nevada, Lake Tahoe (Fig. 7) is a magnificent example of a graben lake. Lying well above the lakes of the Great Basin to the east, in relatively well-watered mountains, it differs greatly from the foregoing examples in its great depth (501 m.). The lakes of southern Oregon and northern California already mentioned are shallow playas, though they retain their elongate form and obvious relation to at least one scarp.

Lake Tahoe, moreover, is at present somewhat deepened by the presence of a lava dam that is being eroded down to the old rock lip by the effluent of the lake. A much smaller graben to the east contains the shallow Big Washoe Lake, but this basin has been modified by outwash and wind action, and the modern lake is a typical pan of the kind so often found in semiarid regions.

The huge basin of inland drainage which contained the Pleistocene Lake Lahontan is almost entirely a complex of fault troughs. The accompanying map (Fig. 8), based mainly on those of Russell (1885) and of Hubbs and Miller (1948), showing the maximum extent of the ancient lake, the position of the principal faults that determined its outlines, and the presence of a few remaining water-filled basins, will give a better idea of this extraordinary region than could any detailed description. Lake Lahontan at its height had an area of 21,860 km.^{2} and a maximum depth of 270 m. Of the much smaller modern lakes occupying the deepest depressions, Pyramid is a typical graben lake. Honey Lake, now probably dry, and the shallow lakes of the Carson Sink are playas situated in sediment-filled grabens, while Walker Lake appears to lie on a tilted fault block with a scarp on its western shore.

To the west of Lake Lahontan an even larger Pleistocene lake, Lake Bonneville, formerly existed, though it is now reduced to three salt lakes, Great Salt Lake being the most important. Bonneville had an area of 51,300 km.^{2} at its maximum extent, and a maximum depth of 320 m. Unlike Lahontan, it was not quite a pure tectonic rock basin, for the water was held at the highest level by an alluvial cone deposited by a stream prior to the rise in level. But this cone was rapidly cut down to the rock lip. Both lakes resemble each other in their great area and extremely complex shore line; no modern tectonic lakes are known which combine these features to the same degree. During the Eocene, however, similar lakes,

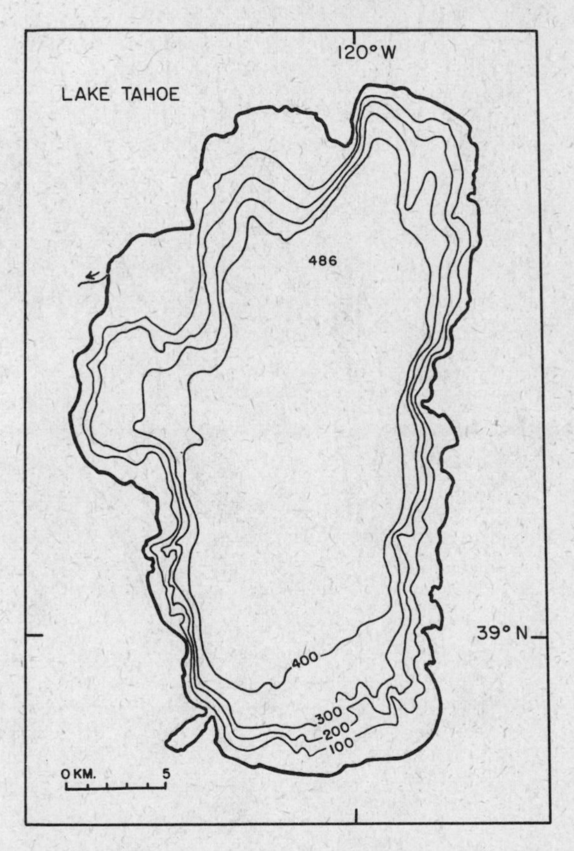

FIGURE 7. Bathymetric map of Lake Tahoe, a lake in a graben, with an extremely flat bottom. (Redrawn with metric contours from U. S. Coast and Geodetic Survey 5001; other investigators have found a maximum depth of 502 m.)

notably that in which the sediments of the Green River Formation were deposited, existed in western North America. In addition to the two immense lakes, about seventy other Pleistocene lakes of much smaller size, nearly all of tectonic origin, are known in the basin-and-range area (Meinzer 1922, Hubbs and Miller 1948).

It is desirable, before leaving the tectonic lakes of North America, to

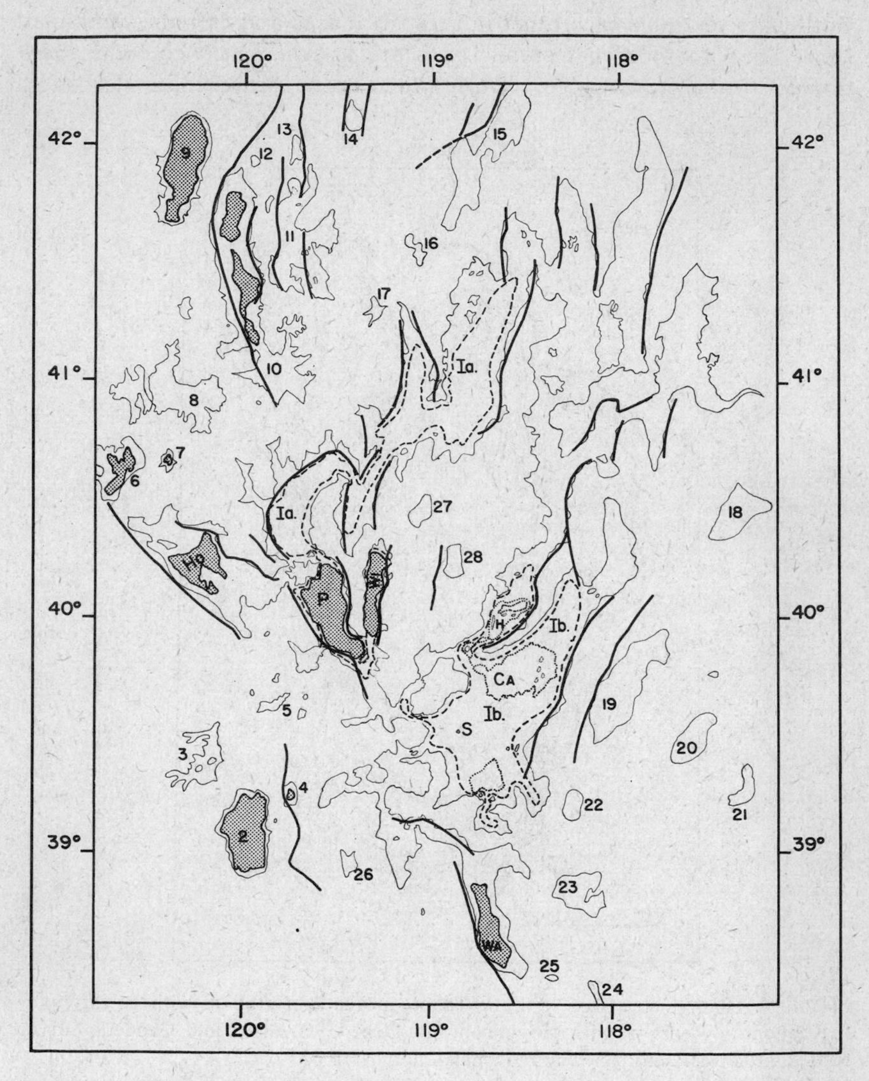

FIGURE 8. The lakes, ancient and modern, of the Lahontan Basin. I*a*, I*b*, pluvial Lake Lahontan, later divided, probably in a pre-Cary interstadial of the Wisconsin glaciation, into the east (I*a*) and west (I*b*) thinolite lakes and now reduced to Pyramid (P), Winnemucca (WI), Honey (HO), Walker (WA), and Soda (S) lakes, with a little water in the Carson Sink (CA) and until recently in Humboldt Lake (H). 2, Lake Yahoe. 3, pluvial Lake Truckee. 4, Washoe Lake. 5, pluvial Lake Lemmon, with several small basins of internal drainage in the vicinity. 6, Eagle Lake. 7, Horse Lake. 8, pluvial Lake Madeline. 9, Goose Lake, sometimes draining into the Sacramento system. 10, pluvial Lake Surprise, now reduced to three smaller lakes.

notice the presence of small "sag ponds" (Davis 1933) along some of the fault lines of California, notably the San Andreas and Elsinore faults. These small lakes appear to be due to minor local subsidence, which may in a small way perhaps be comparable to the more dramatic subsidences in which large tectonic lakes are found.

In Europe, the finest and most interesting lakes in grabens are those of the Balkans, namely, Lake Ohrid and Lake Prespa; in common with the other large lakes of the region (page 106), both show some karstic features. Lake Ohrid, at the southern end of the frontier between Yugoslavia and Albania, is the deeper and the more perfect example. The lake is rectangular in shape, elongated in a north and south direction, and has strikingly parallel eastern and western sides, corresponding to the main faults determining the basin. The area is 270 km.^{2} and the maximum depth 286 m.; in places the slope of the sides is so steep that it is possible to sound 230 m. only 300 m. from the shore. Lake Prespa, to the southeast, has a slightly greater area (298 km.^{2}) but is much shallower, having a maximum depth of 54 m., and is much less regular in outline. The date of the formation of the lakes has been supposed to be either Pliocene or early Pleistocene, but the region has had a long limnetic history and has contained large tectonic lakes since the middle Tertiary. A full summary of the relevant but often conflicting geological results of the study of the basin is given by Stanković (1932).

The rather large and numerous lakes of the Taurus region of Asia Minor are, according to Lahn (1945), situated in tectonic basins which contained lakes during the Tertiary. The older lake sediments are faulted, and the modern lakes lie in basins defined by these more recent faults. It seems unlikely that these basins have contained water continuously since the Tertiary, but further studies of the region would be of interest.

Farther eastward the great lakes of central Asia, notably Balkhash and Issyk-kul, lie in grabens (Leuchs 1937), as do many other smaller examples in the Altai region between these two lakes and the southern end of the Baikal grabens. Of these lakes, Issyk-kul, with a maximum depth of 702 m. is the fifth deepest lake known (Halbfass 1937).

11, pluvial Lake Meinzer. 12, Cowhead pluvial lake. 13, pluvial Lake Warner. 14, Guano pluvial lake. 15, pluvial Lake Alvord. 16, Summit Lake. 17, High Rock Lake, still with intermittent water on occasions. 18, Buffalo pluvial lake. 19, pluvial Lake Dixie. 20, pluvial Lake Edwards. 21, pluvial Lake Smith. 22, pluvial Lake Labou (doubtful). 23, pluvial Lake Bagg. 24, pluvial Lake Luning. 25, pluvial Lake Acme. 26, pluvial Lake Wellington. 27, pluvial Lake Kumiva. 28, Granite Springs pluvial lake. Principal pre-Lahontan faults are indicated by heavy lines. Lake basins now containing water are stippled; basins which probably would contain some water but for artificial diversion are dotted; Thinolite Lakes are indicated by broken line. (After Russell, Hubbs and Miller.)

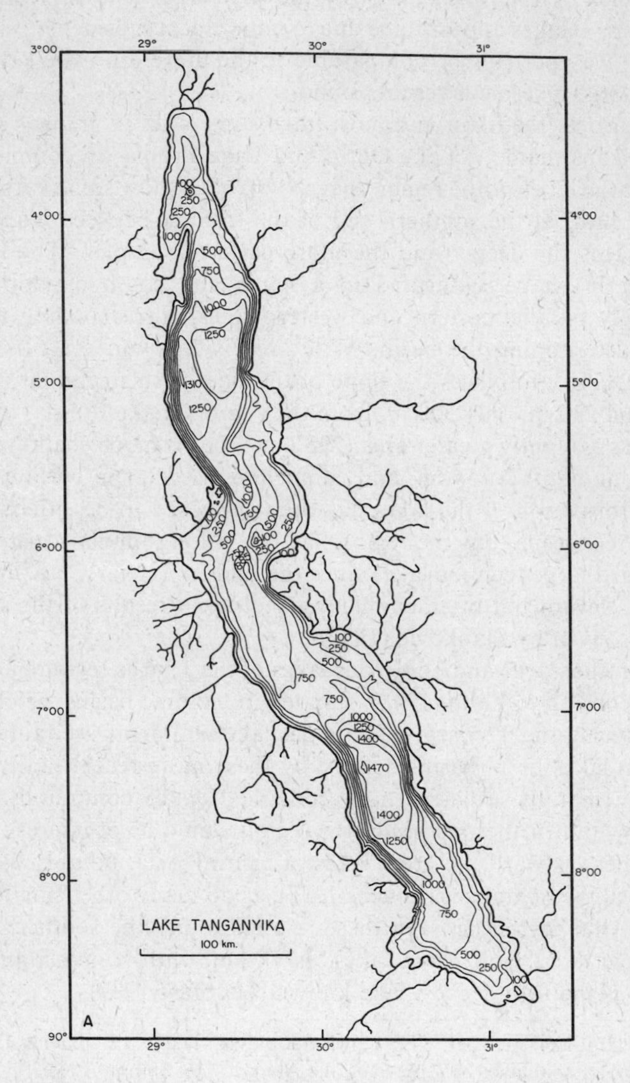

FIGURE 9. Bathymetric maps of the two deepest lakes, both of which lie in multiple grabens. *A*, Lake Tanganyika; *B*, portion of eastern shore of Lake Tanganyika, to show sublacustrine valleys at the mouths of the Lugufu and Malagarasi rivers (after Capart). *C*, Lake Baikal (after Suslov).

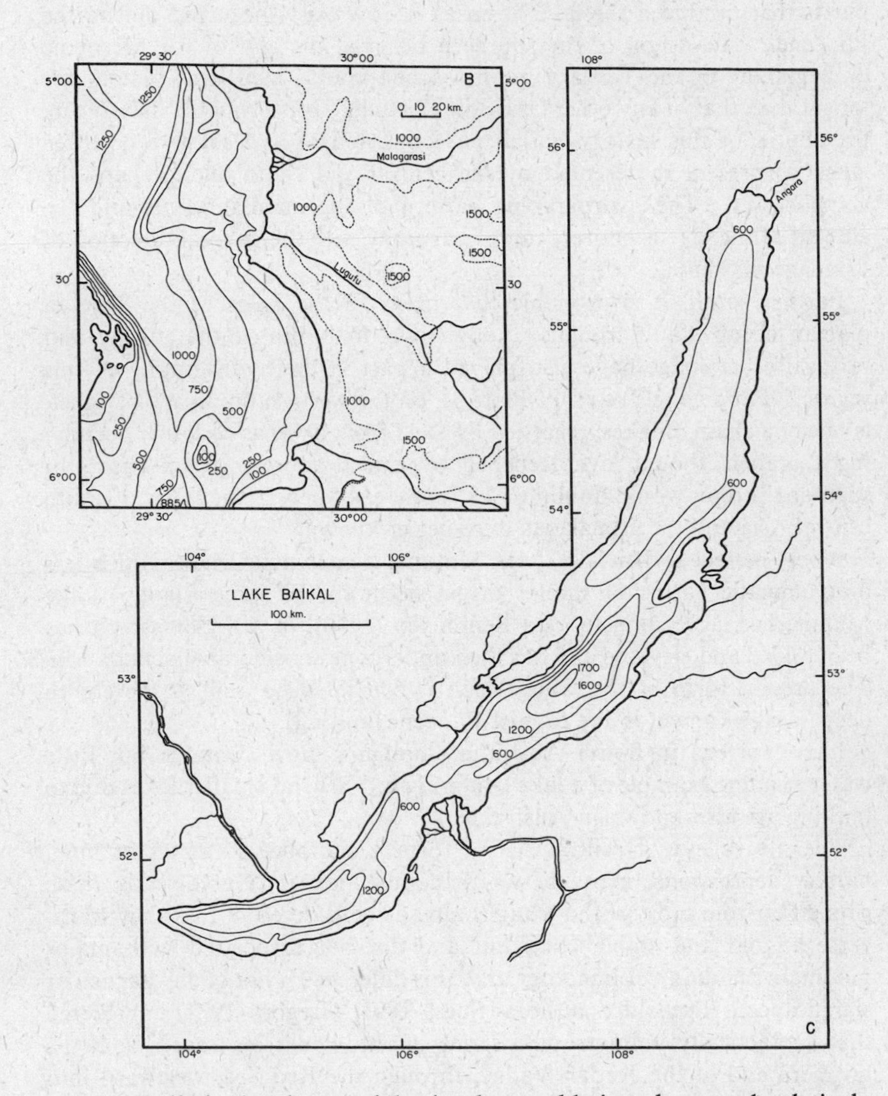

Lake Baikal, the deepest lake in the world, is a long and relatively narrow body of water (Fig. 9), being 674 km. in length and having a maximum breadth of 74 km. The area of the lake is 31,500 km.[2] It consists of three basins, the northern being the shallowest, the central having a known maximum depth of 1741 m., and the southern basin a maximum depth of 1441 m. These basins are separated by ridges, which are a little more than 530 m. deep. The southern basin is believed to date from the Paleocene or even the late Cretaceous, but most of the movements that produced the lake basin as it now exists occurred during the Pliocene. The origin of the southern basin at the end of the Mesozoic or beginning of the Tertiary implies a continuous lacustrine history far longer than that of any other lake now existing. It is probable that during the course of this history Baikal has consisted of at least two or three lakes, at present represented by the central and south and perhaps the north basins. The hydrographic pattern of the main influents and the effluent of the lake is entirely out of harmony with the general direction of drainage in Siberia.

Farther south, it is probable that many of the lakes of the Tibetan plateau occupy fault troughs, though it is to be noted that tilting and reversal of drainage have also played a part in the hydrography of this region. Pang-gong Tso (*C* of Plate 3), on the western boundary of Tibet, is one of a chain of lakes, regarded by De Terra (1934) as certainly occupying a graben, though overdeepened by glacial action. Rock basins of such dual origin would doubtless be found elsewhere if the lakes of south central Asia and the Himalayas were better known.

A very perfect graben lake, Lake Matana, is known in Celebes; other less diagrammatic basins of similar origin occur in the same island. Lake Matana lies, according to Van Bemmelen (1940), in a region which has been folded and elevated and has then undergone considerable denudation. The present form of the lake, which is 590 m. deep and so the eighth deepest lake known, is due to post-Pliocene faulting.

Lake Torrens in South Australia, though it now contains but little water, is a fine example of a lake in a graben. Several smaller lakes due to faulting are also known in Australia.

The rift valleys of central Africa form a complicated series of long, narrow depressions, many of which contain lakes. Geologically these rifts differ from most of the troughs already considered in that they transverse an old and stable continent and are not associated with major mountain building. Limnologically, this difference is not of any particular significance. The older authors (Suess 1891, Gregory 1921) considered that a practically uninterrupted series of rifts could be traced from the northern end of the Jordan Valley, through the Red Sea, westward into central Africa, and southward to Beira on the sea just south of Lake Nyasa. More recent studies, particularly by Willis (1936), indicate that the system is not as continuous as had been thought, though Solomon (1939) indicates that perhaps Willis somewhat overstated his case.

The rift system is not a single straight line but, in central Africa, consists of two branches, an eastern and a western, which as conceived by Willis lie symmetrically on either side of the Tanganyika central plateau, the northern part of which bears Lake Victoria. Northward, the western rift axis points in the direction of the Nile Valley, while the eastern rift axis continues just east of Lake Rudolf (Fuchs 1939) into Ethiopia, towards the Red Sea. South of the central plateau, the axes converge on Mount Rungwe, but diverge again at the valley of the Luanga River on the west and the trough of Lake Nyasa on the east. Considering the whole complex system from the Jordan Valley southward, there is evidence of an antimeric pattern of rifts on the floor of the Persian Gulf, Arabian Sea, and western Indian Ocean (Wiseman and Sewell 1937).

The western axis consists, from south to north, of the Luangwa Trough, containing a tributary of the Zambezi; the southern and northern Tanganyika troughs, which Willis regards as separate tectonic units; the Edward-Kivu Trough, separated from the north Tanganyika Trough by a gap; and the Albert Trough, separated from the Edward-Kivu by Mount Ruwenzori. As is indicated below, a recent chain of volcanoes has been thrown up across the Edward-Kivu Trough, separating Lake Edward from Lake Kivu. The main arc of western rifts therefore contains four major lakes, Tanganyika, Kivu, Edward, and Albert; of these, Kivu is due to a volcanic dam but the other three may be regarded as purely tectonic.

Lake Tanganyika is surpassed in depth by Lake Baikal only, the maximum known depth being 1435 m. The length of 650 km. and the area of 31,900 km.^{2} are comparable to those of Lake Baikal, and like the Siberian lake, Tanganyika is divided into three basins by ridges, the mean depth of which is 500 m. (Capart 1949). Willis (1936) points out that the ratio of maximum width of the whole trough to its maximum depth is 17 : 1 for the south and 20 : 1 for the north trough, values slightly exceeding those of the Dead Sea (16 : 1) and certain submarine basins, such as the Tonga Deep (14 : 1) and the Philippine Deep (15 : 1). The Baikal trough has a ratio of 32 : 1 obviously of the same order of magnitude as the Tanganyika values. For the Black Sea the ratio is 114 : 1 and for the Caspian Basin 300 : 1, implying a quite different type of depression.

Three peneplains are recognizable in the part of central Africa around the lake. The earliest is of Jurassic age, and the second of Miocene; the third, termed by Willis the Malagarasi, is early post-Miocene. When this last peneplain was developed, the area to the east of Tanganyika drained into the Congo, though at the extreme southern end of the present lake there is evidence of a river flowing eastward, the drowned valley of which forms Cameron Bay. During the Pliocene and Pleistocene the rifting movements began. The sides of the troughs were uplifted and two depressions were formed by subsidence, the present north and south basins of the lake. The ridge separating the two basins appears to represent the original Malagarasi peneplain level, left in place between the northern and southern Tanganyika troughs. As a result of this movement the Malagarasi River was beheaded, becoming the main eastern influent of the new lake. For most of its Pleistocene history, Tanganyika was probably a basin of internal drainage. After the ponding of Lake Kivu by volcanic action, a new important effluent from that lake increased the water running into Tanganyika above that needed to compensate for evaporation. The lake then acquired an effluent, the Lukuga River, draining westward into the Congo. Tanganyika therefore lies transverse to the main hydrographic pattern of the neighboring parts of central Africa, in an entirely disharmonic fashion. It is desirable to point out that southeast of Tankanyika, the Rukwa Trough contains a small shallow lake, Lake Rukwa, which may have had hydrographic connection with Tanganyika in the past (see Brooks 1950 for a critical summary of the evidence).

The other lakes of the western rift system are all much smaller and shallower than Tanganyika. Kivu, as has been indicated, is due to recent volcanic damming; Edward is a basin of internal drainage, but its eastern influents originate in the region of great hydrographic uncertainty to the west of Lake Victoria. Albert now drains into the Nile. It is reasonably certain that none of these lakes has a continuous history as long as that of Lake Tanganyika, and the history of Tanganyika is obviously very much shorter than that of Lake Baikal. Tanganyika presumably ranks third or fourth in age of the lakes of the world; Baikal, the Caspian, and perhaps the Sea of Aral, alone have histories extending well into the pre-Pliocene past.

The eastern series of rifts contains at its southern end Lake Nyasa. This lake occupies a trough which was, according to Dixey (1939, 1941), a valley as early as the Cretaceous, draining southward into what is now the Zambezi. The lake started to form as the result of faulting in the middle Pleistocene, as a small body of water in the northern end of the modern Nyasa Trough. The major earth movements associated with its production were downthrusts at the southern end of the lake.

In the central part of the eastern system, the Kenya or Gregory Rift, there are a number of smaller tectonic lakes, some of which are of interest to the chemical or biological limnologist and will therefore be discussed again on appropriate later pages of this work. These lakes are, from south to north, Lake Eyasi and Lake Manyara, the Natron Lake, Lake Magadi, Lake Naivasha, Lake Elmenteita, Lake Nakuru, now said to be dry, and Lake Bolossat and Lake Baringo. A much larger Pleistocene lake, Lake Kamasia, held by a volcanic dam to the north, formerly occupied the part of the Kenya Rift now containing Lake Baringo (Fuchs 1950).

Other than Nyasa, the only large modern lake in Africa associated with the eastern axis is Lake Rudolf (Willis 1936, Fuchs 1939), but only the southern end of the lake lies in a true rift, the larger northern section occupying a synclinal fold formed during rifting. The axis of the rift system runs northeastward; Lake Stephanie is presumably a true rift lake. The rift continues to form the Jordan Valley with its lakes of great historic and scientific interest.

LAKES ASSOCIATED WITH VOLCANIC ACTIVITY

Volcanic activity may give rise to lake basins in several different ways, so that regions of present or relatively recent volcanism are, unless they are too arid, characteristically lake districts. Remarkable examples of volcanic lakes of various kinds are found in Iceland, the Eifel district of Germany, the Auvergne district of France, the Roman Campagna and the Phlegrean Fields in Italy, most of Indonesia, northward through the Philippines into Japan, parts of central Africa, New Zealand and parts of Australia, the northwestern United States, much of Central America, and the Andes.

Crater lakes, maars, and calderas. Volcanoes are elevations of various shapes and sizes composed of material which has welled up or been violently ejected from some depth below the surface of the earth. The material is supposed to have originally occupied a magma reservoir or chamber usually situated under the volcano. The summit of the volcano usually has a well-defined crater from which ejecta escape to continue the building of the cone. The material produced by a volcano can very broadly be regarded as of two kinds, *lava* and *pumice*. Lava may be produced by the crater and continue building a dome or cone, but it may also appear as a flow breaking through the side of the volcano. It is possible to distinguish two extreme types of volcanic construction or positive activity, the formation of lava domes and the formation of what are commonly called cinder cones. The typical large volcano with a well-developed crater is an intermediate or composite structure, usually built of alternating layers of lava and cinders, and therefore called a stratovolcano. It is also possible to distinguish two kinds of destructive or negative volcanic activity: *explosion*, which may initiate volcanic activity and also frequently occurs late in eruptions of volcanoes primarily producing pumice, and *collapse*, which may take place whenever a considerable amount of material has been ejected from the magma reservoir, the roof of which is left bearing a considerable unsupported load. The reader wishing to learn more of the geomorphology of volcanic regions may profitably consult Cotton (1944) or Van Bemmelen (1940).

Any cavity produced by volcanic action, provided it is undrained, may contain a lake. In a good many cases, well-formed craters are dry owing to the porosity of their walls. The water level in the little calderas of

Mount Gambier, South Australia, fluctuates with that of the local ground water (Fenner 1921). At Big Soda Lake, Fallon, Nevada, the present lake level also represents the current water table of the surrounding terrain, which water table has been raised as the result of irrigation. In this case, much salt has been washed out of the lake by the passage of ground water through the basin (page 484; also Hutchinson 1937c). In many cases, however, the decomposition of the finest material produced by eruptions provides an impermeable clay seal, so that once the volcano has become extinct a permanent lake can occupy the depression in its summit.

Unmodified craters in cinder cones (*Type* 10). Crater lakes occupying unmodified constructional cones are probably rare, are certainly all small, and are hard to identify in the literature. The lake in Crater Butte, Mount Lassen National Park, California, a region described in great detail by a number of geologists, of whom Williams (1932) is the most recent, seems to provide an unquestionable example.

Explosion craters (*Type* 12) *and maars* (*Type* 11). The important crater lakes of the world all lie in depressions formed either by the explosive eruption of abortive volcanoes or of pre-existing cones, or by collapse. It is usually now believed that all the largest basins or *calderas* are the result of the collapse of the roof of a partially emptied magma chamber, while the smaller modified cavities may be due to either collapse or explosion. Williams (1941) considers that the upper limit of the diameter of an explosion crater is about 1 mile or 1.6 km.; subsequent slumping, while decreasing the depth, may of course somewhat enlarge the diameter at the top of the rim.

A considerable number of basins apparently represent abortive embryonic volcanoes. These basins were evidently formed by explosions at some depth below the surface, which produced low rims built of fragments of the superficial rock but with little or no volcanic material. Such embryonic volcanoes, if they develop no further, frequently fill with water and become small deep lakes. There are a number of excellent examples in the Eifel district of Germany, where they are locally designated *Maare*; the term maar is convenient for the earth form in question wherever it may be found. The deepest German example, the Pulvermaar, is 74 m. deep but has an area of only 0.35 km.^{2}, a great depth coupled with a very small water surface being characteristic of the lakes of this type. The maars of the Eifel appear to have originated during the closing stages of the Pleistocene glaciation, for volcanic ash attributed to the explosions producing them has been found in organic deposits of Alleröd age (Firbas 1949).

In France, several remarkable examples are found in the Auvergne.

Of these, the Lac d'Issarlès (Scrope 1858, Delebecque and Ritter 1892), with an area of 0.92 km.[2] and a depth of 108.6 m., again demonstrates the great depth possible in a quite small lake of this sort (Fig. 10).

Cotton (1944), following Suess (1888), regards most of the smaller lakes of the Roman Campagna as belonging to the same class. Among them

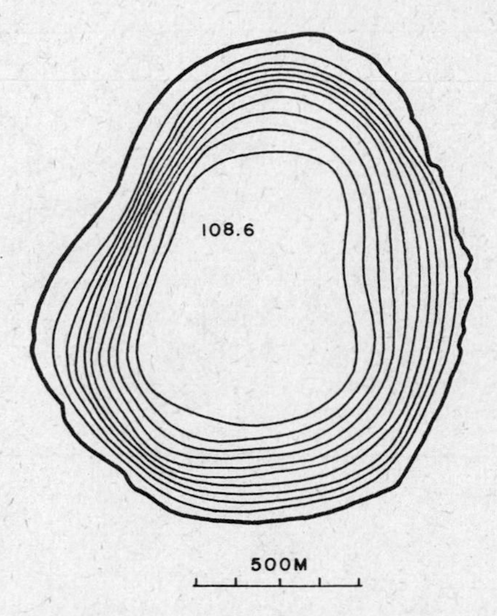

FIGURE 10. Bathymetric map of Lac d'Issarlès (after Delebeque).

the Lago di Nemi, famous in antiquity no less than today for its associations with the Golden Bough, is also a lake of considerable limnological interest. Another important Italian example is the Lago d'Averno in the Phlegrean Fields, also rich in mythological and poetic associations,[4] which Suess regarded as an explosion crater parasitic on the great Phlegrean Volcano, an immense structure now mostly collapsed and submerged.

Lake Viti in Iceland, a small lake 315 m. in diameter, is a maar in a basin which was formed by a sudden explosion on May 17, 1724, and is of

[4] According to Vergil, this lake at the entrance to the infernal regions was also a locality where the Golden Bough could be found:

"inde ubi venere ad fauces grave olentis Averni
tollunt se celeres liquidumque per aëra lapsae
sedibus optatis gemina super arbore sidunt,
discolor unde auri per ramos aura refulsit."

The Aeneid, VI, 201–204

Later folklorists seem to have neglected the association of the *ramus aureus* with these circular water-filled explosion craters.

interest as being one of the few examples known to have been produced in historic times. Other Icelandic lakes, such as Graenavatn and Gest-astadavatn, provide further examples from this volcanic region. Lake Tikitapu, N.I., New Zealand (Cotton 1944), is an excellent antipodean maar. Several fine examples occur around the base of Mount Ruwenzori

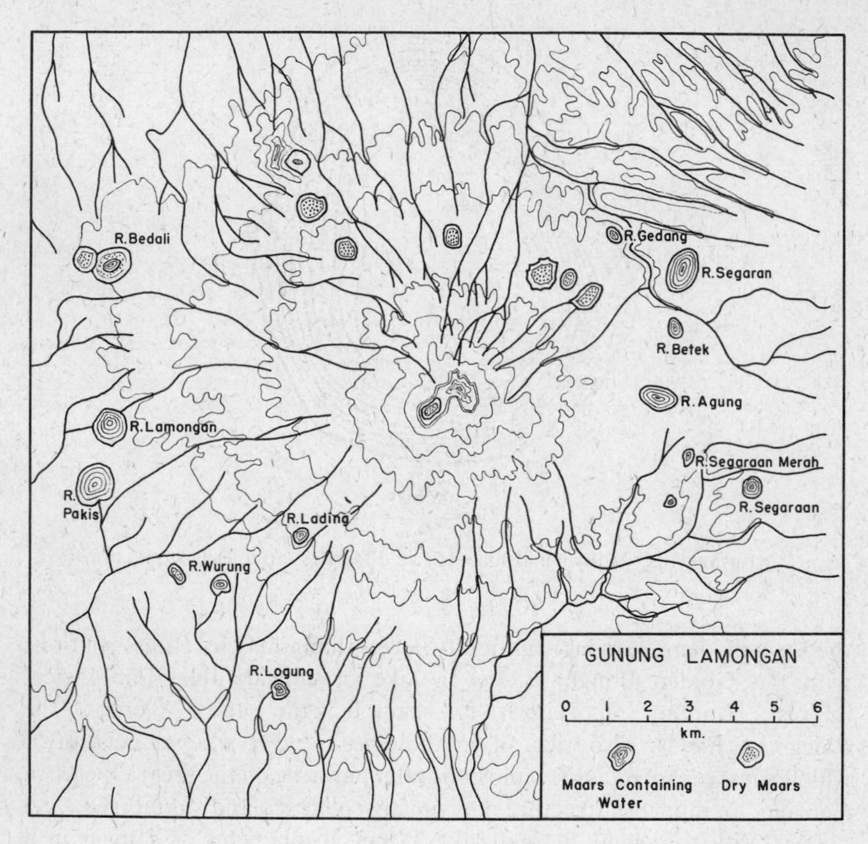

FIGURE 11. Parasitic maars on Gunung Lamongan (Ruttner 1931).

in central Africa (Willis 1936). Numerous other maars could be enumerated from the various volcanic areas of the world. Not infrequently, maars are situated, like the Lago d'Averno, as parasitic craters on the flanks of large volcanic peaks. Chanmico, on the side of the Volcan San Salvador in Central America, appears to be such a parasitic maar. Lake Grati, a lake 125 m. deep at the foot of Tennger in Java, provides another striking case, while a very remarkable example of a whole group of such parasitic maars (Fig. 11) is provided by the small lakes on the slopes of the volcano Gunung Lamongan in the eastern part of the same island (Ruttner 1931).

Certain maar-like depressions surrounded by rings of tuff apparently represent basins from which rather more volcanic matter has been ejected than in the previous cases. Cotton (1944) cites as examples of lakes in such tuff rings Pupuke Lake and a marshy lake in Crater Hill, Auckland,

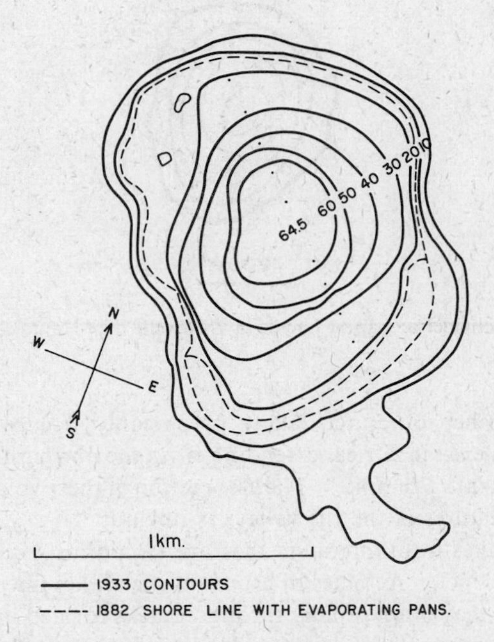

FIGURE 12. Bathymetric map of Big Soda Lake, Nevada.

New Zealand. He also considers the lakes of Mount Gambier, South Australia, to lie in comparable basins, though Fenner (1921), who described these lakes, thought that they occupied small basins due to subsidence, being in fact very minute calderas.

The typical small crater lake (Type 12a) found in the summit of an extinct stratovolcano occupies a crater that has generally undergone some enlargement during the terminal phases of an eruption of the vulcanian type in which a large amount of finely divided solid matter has been blown out. Big Soda Lake (Fig. 12), Fallon, Nevada, appears to occupy such an enlarged crater built on the floor of the Pleistocene Lake Lahontan but last active after the water had retreated from this part of the lake bed.

Numerous further examples can be found in other parts of the world. The Lac de la Godivelle d'en Haut (Fig. 13), maximum depth 43.7 m., the Lac de Servière, maximum depth 26.5 m., and perhaps the Gour de Tazanat, maximum depth 66.6 m., are well-known examples from the Auvergne (Glangeaud 1919, Delebecque and Ritter 1892), though the last named is sometimes regarded (Collet 1925) as a maar.

FIGURE 13. Bathymetric map of Lac de la Godivelle d'en Haut (after Delebeque).

A great number of crater lakes, presumably occupying somewhat modified cones, exist in Africa, extending from the northern part of Nyasaland northward into Ethiopia. The association of these volcanic structures with tectonic features of the rift valleys is obvious.

Sometimes lakes form in craters that are temporarily quiescent though by no means extinct. At times of eruption such lakes may be thrown out of the crater or may drain rapidly through cracks formed in the side of the cone, washing unconsolidated material down the slopes of the volcano as a mudflow or lahar. A lake that collects in the crater of the intermittently active volcano Kelut in Java has been discharged in this way in 1771, 1811, 1826, 1835, 1848, 1851, 1859, 1864, 1901, and 1919. In addition, partial collapse of the crater in 1875 reduced the volume from about 76×10^6 m.^{3} to 38.5×10^6 m.^{3} In order to avoid further destruction by these catastrophic discharges, water is now syphoned off from the lake through a tunnel that maintains its volume at 1.8×10^6 m.^{3} (Van Bemmelen 1940).

A disaster caused by a crater lake probably occurred at the eruption of the Guatemalan volcano Agua in 1541 (Russell 1897). A lake which filled the crater of La Soufrière, St. Vincent, was thrown out of the crater during the first phase of the eruption of 1902, producing extensive hot mudflows that rushed down the mountain into the sea (Anderson and Flett 1903).

The most curious and at the same time most devastating phenomenon of this kind is recorded in Iceland. A craterlike depression about 7.5 km. long and 5 km. wide, with an eccentrically situated volcanic vent, lies under a part of the great Vatnajökull icecap in the southeastern part of the island. This depression is presumably a caldera due to subsidence rather than an explosion crater, but it is conveniently noted in the present context. For a considerable part of its history between eruptions, the depression has been filled with an ice-covered lake kept melted by volcanic heat. At eruptions the lake is blown out of the depression, and up to 10 $km.^3$ of melt water may be produced and discharged beyond the margin of the icecap. The resulting flood, locally termed a *jökulhlaup* or glacier burst, may carry away whole parishes and leave the landscape in an unrecognizable condition (Wadell 1920, Nielsen 1937).

Mention must be made of one special case, a lake which occupies instead of a single crater a series of irregular, more or less confluent explosion craters (Type 12b), namely Lake Rotomahana, below Mount Tarawera on the North Island of New Zealand. The present lake was formed in 1886 as the result of intense local volcanic activity along a fissure. It is reasonably certain that the nature of the eruption at the site of the present lake was due to the presence of much water, in the form of three pre-existing smaller lakes lying on water-rotted lava and lake sediments, so that quantities of superheated steam were formed when the fissure became active. Such steam liberated under very weak rock produced a series of explosion craters, which now contain the modern lake.

Calderas (*Type* 13). Though in the past it has often been supposed that calderas were generally produced by explosive processes, most recent investigators conclude, as has already been indicated, that practically every caldera of importance has been formed by subsidence of the superstructure over a magmatic reservoir, when part of the magma has been extruded. The material formerly occupying the space may have been discharged in an explosive eruption as vast quantities of lapilli and ash (Krakatau type), or it may have been liberated on the slopes of the volcano as lava flows (Kilauea type), or may even have been intruded as dykes into the rock of the surrounding country. In any of these cases the central area over the reservoir is liable to collapse, so producing the caldera. The whole process has been discussed in detail by Williams (1941), whose paper may be consulted for the history of the subject and for the purely geological details.

Usually the collapse involves the central part of a volcanic cone that has undergone a catastrophic eruption. When the width of the magma reservoir is great compared with the diameter of the volcanic vent, the central part of the peak will be unsupported after a large amount of magma has been discharged, and will tend to fall into the formerly filled space. The descent usually involves a single more or less circular block, which becomes fractured in the process. A large steep-sided cavity with a fairly uniform floor results, and this when filled with water becomes a more or less circular lake. Where the collapse of a well-formed volcanic peak is involved, the resulting caldera occupies the summit of a truncated, if often steep and quite high, mountain. The most diagrammatic case of a lake in such a basin is certainly Crater Lake, Oregon, apparently the second deepest lake in the New World.[5] The basin of this lake was formed by collapse (Williams 1942) of the central part of a high volcanic peak, now called Mount Mazama, after a cataclysmic eruption about 4500 B.C., as is shown by application of radiocarbon dating to wood from a tree killed at the time (Arnold and Libby 1950). Crater Lake has an approximately circular form, is surrounded by a rim that rises about 600 m. above the present water level, has an area of 64.4 km.^{2}, and is 608.4 m. deep. There is a small eccentric secondary cone, forming Wizard Island.

Scarcely less remarkable caldera lakes surrounded by high rims are known in Japan. Tazawako, Honsyu, area 25.65 km.^{2} and depth 425 m., is surrounded by a rim rising at an angle of about 35 degrees to about 250 m. above the water surface, while Masyuko in southwestern Hokkaido has an area of 19.77 km.^{2} a maximum depth of 211.5 m., and a nearly vertical rim rising almost 300 m. above the level of the lake. Where volcanic activity has resulted in collapse with very little building of a cone, as may happen when large lateral lava flows occur, a rather different type of caldera on a domelike mound and without a high rim may result. Such more or less rimless calderas containing large lakes are well known in Japan and have been discussed by Tanakadate (1930; morphometric data in Yoshimura 1938b). Of these, Kuttyaroko in Hokkaido, in a depression of area 430 km.^{2}, has a water surface of 79.89 km.^{2} and a maximum depth of 120 m.; Tôyako, in the same island, has an area of 69.60 km.^{2} and a maximum depth of 179.3 m.

Conche (*Type* 14). A modified type of caldera is exemplified by the basin known as the Conca di Bolsena, in the Roman Campagna, in which the Lago di Bolsena lies. This depression has been formed by a number of eruptions, at first probably from craters occupying the central part of the Conca and later from new craters to the west. As material was thrown out of these craters, the central region collapsed in stages to form a basin with gently sloping sides interrupted by step faults, which produced terraces.

---

[5] Great Slave Lake, in north western Canada, the maximum depth of which is given as 614 m. as the result of rather restricted bathymetric studies, was proved recently to be the deepest (Rawson 1950).

Tanakadate (1930), who has given the most informative account of the basin, suggests that the term *conca* (plural *conche*) be used generically for calderas with gentle gradients formed in this way. Another Italian example is provided by the depression, south of the Conca di Bolsena, in which the Lago di Bracciano lies.

In the Pilomasin Basin, South Java (Van Bemmelen 1940, Williams 1941), the center of the depression is occupied by lake sediments interstratified with ash. In this case it is clear that intermittent volcanic activity, largely lateral to the conca, repeatedly drew off material from the magmatic reservoir, the roof of which gradually collapsed. Williams thinks that the Mono Basin, containing Lake Mono in California, may have been formed in the same way.

Tanakadate gives further Japanese examples of lakes in conche, namely Akanko in Hokkaido, which has a maximum depth of 36.3 m. and an area of 12.93 km.^{2} and Inawasiroko in Honsyu, maximum depth 94.6 m. and area 104.8 km.^{2}

Modification of caldera lakes by secondary activity. Subsidiary volcanic cones or domes often may develop, probably as the result of the subsidence (Tanakadate 1930) in the caldera, profoundly modifying the circular shape of the space in which a lake can collect. Such secondary structures are usually eccentric, being associated with the circular fault delimiting the caldera. Not infrequently they appear as islands, as Wizard Island in Crater Lake, Oregon. Occasionally they may remain as active volcanoes developing several craters, some of which may in turn become filled with crater lakes. Zuni Salt Lake partly fills the floor of a small caldera, in which at least one secondary cone also contains a salt lakelet (Darton 1905). Paoha Island in Lake Mono, which Williams believes to occupy a conca, has such crater lakelets now filled by seepage. The Taal Volcano in Taal or Bombon Lake (Fig. 14), Luzon, P.I., which has erupted on a number of occasions, also has usually born one or two crater lakes; a single such lake now replaces the Green and Yellow lakes that existed before the eruption of January 30, 1911 (Saderra Masó 1911). In the caldera lake of the still active island of Niuafoou in the Pacific, a row of marginal secondary cones has cut off a small lakelet from the margin of the main body of water. Not infrequently, as at Tengger, a secondary cone forms a large peninsula, which in this case bears a crater lake projecting into the main lake. Atitlan in Guatemala is modified in a comparable way by a very broad promontory. The Lago di Vico in the Roman Campagna, half encircling Monte Venere, a great dome-shaped secondary structure, provides another example. On Mayor Island or Tuhua in the Bay of Plenty, New Zealand, a large secondary dome again fills much of the caldera floor, leaving a narrow annular depression in the deepest part of which a small lake has collected. The great Aniakchak caldera in Alaska is largely filled with the ejecta of a secondary cone, the depression around which contains two widely separated lakes, the larger of the two, Surprise Lake, draining by a deep canyonlike valley through the caldera rim. The Newberry caldera in Central Oregon is likewise divided by pumice and cinders ejected from secondary vents, into two basins containing Pauline Lake and East Lake.

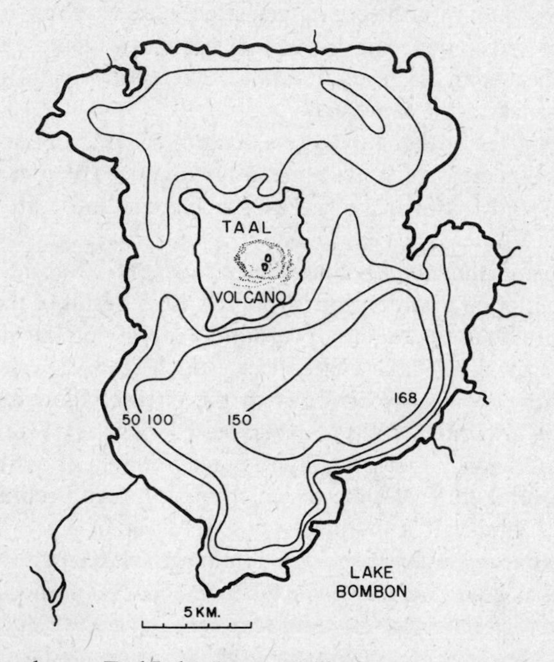

FIGURE 14. Bombon or Taal Lake, Luzon, Philippine Islands, prior to the eruption of January 30, 1911. The two small lakes in the crater of the Taal Volcano were replaced after the eruption by a single lake. (After Saderra Masó.)

At least one case (Type 15) is known in which secondary activity has filled the whole caldera, the outline of which is marked by an irregular ring of volcanic peaks which have coalesced, leaving a small basin in the middle. This case is Medicine Lake (Fig. 15) east of Mount Shasta, near the northern boundary of California, described by Anderson (1941). It is, moreover, possible that Thurston Lake near Clear Lake, California, provides another example of the same process, for Davis (1933) describes it as similar to but much smaller than Medicine Lake, and like the latter occupying a hollow accidentally enclosed by volcanic mounds.

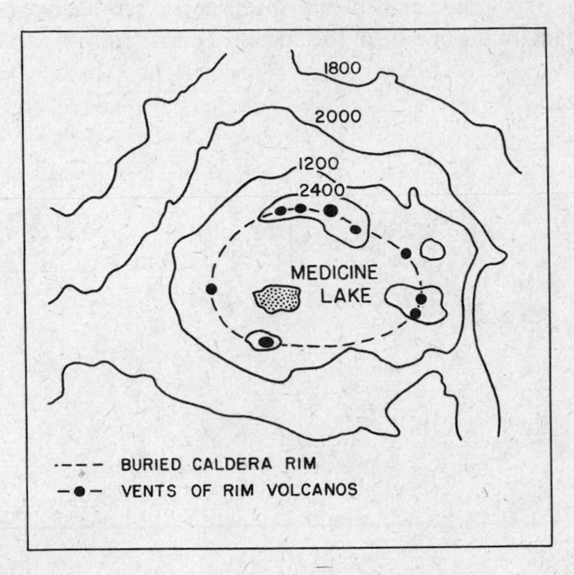

FIGURE 15. Map of the vicinity of Medicine Lake, California, showing position of the buried caldera rim and the vents of secondary volcanoes defining the drainage basin of the lake (after Anderson).

Volcano-tectonic basins (Type 16). The basin which contains Lake Toba in Sumatra has had a special and rather complex history (Fig. 16), which has resulted in the formation of the largest caldera in the world. This history has been succinctly described by Van Bemmelen (1930), whose paper is here followed. The original mountain landscape of the region now occupied by the lake was developed in the late Tertiary and was capped by a group of volcanoes. Paroxysmal eruptions of these volcanoes ejected an immense quantity of pumice and ash, which have formed beds of tuff covering an area of 20,000 km.^{2} and having a thickness of 600 m. in the vicinity of the lake. The emptying of the magma reservoirs implied by this huge quantity of ejected material must have led to a collapse which partly followed pre-existing fault lines, so producing what Williams calls a volcano-tectonic basin. This basin filled with water to produce a lake, the outlet of which cut a deep valley following the prevolcanic drainage and formed the existing effluent of Lake Toba, the Asahan River. This cutting process lowered the lake level. During the high-level stage the influents of the lake, laden with easily eroded material derived from the tuff beds, deposited deltas. The influent streams then cut small canyons in these deltas, which were left hanging as the lake level fell.

A new volcanic phase meanwhile intervened, producing a secondary peak as an island in the middle of the lake. This island volcano or series of

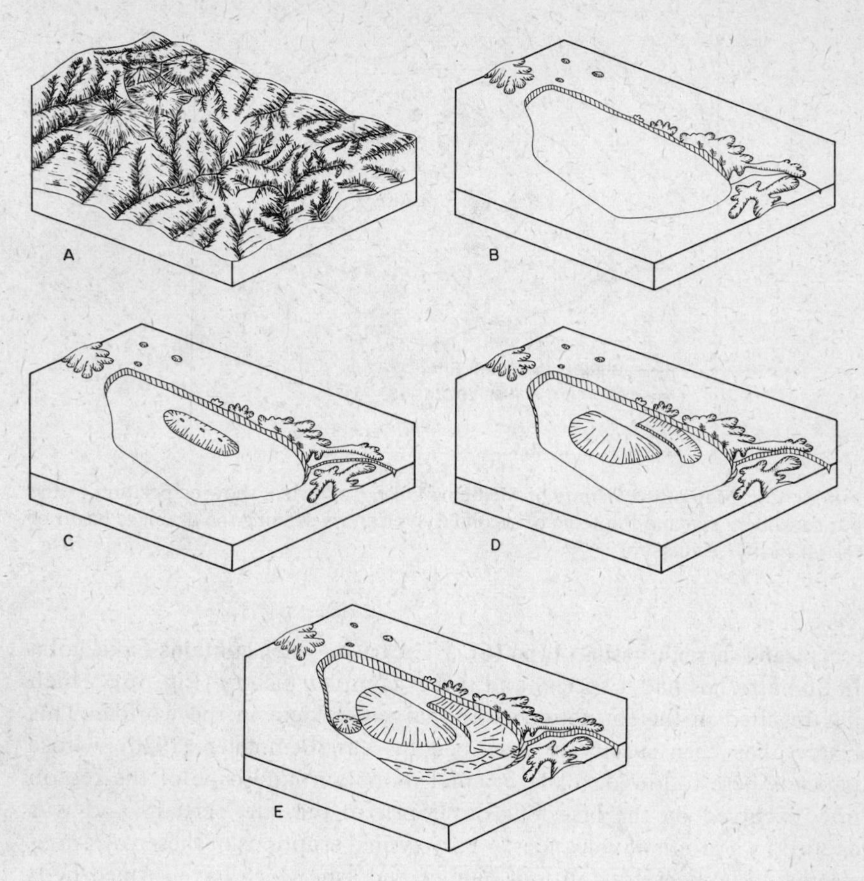

FIGURE 16. The history of Lake Toba, diagrammatic. *A*, original landscape, *B*, initial lake in caldera-like basin; *C*, formation of secondary volcanic peaks; *D*, subsidence of part of central island; *E*, modern lake modified by renewed minor volcanic activity. (Van Bemmelen, slightly modified.)

volcanoes erupted, and the subsequent subsidence cut the island in half. Later minor activity joined the eastern half of the island to the eastern shore of the lake along a broad stretch of shore. The western half, though described in the literature as the Island of Samosir, is joined by a very narrow isthmus to the western shore in both Van Bemmelen's and

Ruttner's (1931) maps.[6] The present lake is therefore an irregular and just incomplete ring. The area (Ruttner 1931) of the water surface is 1129.7 km.^{2}, but if the small island of Pardepur and the large "island" of Samosir are included to give the area of the caldera at the present water level of the lake, this figure is raised to 1776.5 km.^{2} The maximum depth of the lake is about 450 m.; it is reasonable to suppose that the original lake formed after the first phase of eruption was much deeper, probably exceeding Crater Lake, Oregon, in depth.

An immense area of the North Island of New Zealand, including Lake Taupo, Lake Rotorua, Lake Tarawera, Lake Rotoiti, and Lake Rotoehu, has undergone volcano-tectonic collapse. All of these lakes are to be regarded as having been produced by subsidence, in part gently downwarping, mainly in relation to pre-existing fault lines, after a series of Plio-Pleistocene eruptions. Lake Taupo has a well-developed secondary cone forming a promontory, so that its outline is somewhat reminiscent of Atitlan (Williams 1941, Cotton 1944).

Problematic caldera-like basins and cryptovolcanic structures. There are a few large isolated basins that have given rise to a good deal of controversy and of which the origin is still by no means clear. The three most important are Lake Bosumtwi in Ashanti, Gold Coast (Maclaren 1931, Rohleder 1936, and Junner 1937); Lonar Lake between Bombay and Nagpur in India (La Touche 1912); and the Pretoria Salt Pan in the Transvaal (Wagner 1922). The largest of these, Lake Bosumtwi, has a diameter of 8.4 km. and a maximum depth of about 73 m. It was attributed to the explosive impact of a meteorite by Maclaren, but Rohleder, on the basis of structural studies and of the finding of pumiceous agglomerate in the vicinity, thinks that a volcanic explanation is more probable. Junner, whose account is accepted by Williams (1941), believes that injection of a laccolith was followed by explosions which threw out a little volcanic material and great quantities of broken rock. Later subsidence occurred, producing the present caldera. The only anomalous feature of the structure is its isolation, though one somewhat similar though smaller depression, the Nebiewale caldera, exists in the Gold Coast. The other two examples are completely isolated examples of supposedly volcanic activity within their respective regions, and though most authorities appear to believe that they have had a history not unlike Lake Bosumtwi, Spencer (1933), Cotton (1944), and Baldwin (1949) have argued with some persuasiveness in favor of a meteoritic origin.

A large volcano-tectonic or caldera-like depression about 20 to 24 km. across, commonly spoken of as cryptovolcanic, existed during the Miocene

---

[6] Slight variations in water level may be involved (cf. Murray 1910).

in Germany and contained a fresh-water lake. This structure is now recognizable geologically as the Rieskessel; the not far distant but much smaller Steinheim Basin also contained a lake and doubtless resembled certain modern calderas. Still older examples have been described from North America (Bucher 1933, Baldwin 1949).

Lakes on collapsed or irregular lava flows (Type 17). Not infrequently the surface of a newly formed lava flow may cool and produce a crust while the lower layers are still fluid and are moving under the influence of gravity. Collapse of the crust may then occur to compensate for the lava that has flowed from beneath the solid layer, so producing a basin in which a lake can collect. Confluence of two lava streams or irregularity of flow consequent on pre-existing topography may also produce depressions on lava fields.

The most remarkable lake on a lava flow is probably Myvatn in Iceland, a body of water having an area of about 27 km.^{2} though the maximum depth is but 2.3 m. The shore line is very irregular and the lake contains about a hundred islands, some of which bear craters containing lakelets. The form of the basin was greatly modified during an eruption in 1729, when much lava ran into the lake and when some of the existing islands were formed. It is probable that the lake was sterilized at this time (Thoroddsen 1906, Ostenfeld and Wesenberg-Lund 1905). A few other less spectacular examples probably exist in the neighboring parts of Iceland.

In the Auvergne several lakes appear to have been formed in irregularities in lava flows, due either to collapse or to the confluence of two lava streams. Boule, Glangeaud, Rouchon, and Vernière (1901) mention the Lac de Bourdouze, the Lac de Chambedaze, the Lac des Esclauzes, and the Lac de la Godivelle d'en Bas as examples, none being over 5 m. deep. Glangeaud (1919) adds to this list the Lac d'Arcône. The contrast in depth between these lakes and the maars and crater lakes of the Auvergne is most striking. Yellowstone Lake, Yellowstone Park, provides a fine American example of a lake on an irregular lava flow.

Fenner (1918) and David and Browne (1950) indicate that many lakes and lakelets in the western part of Victoria, Australia, have been formed by the collapse of the surfaces of lava flows. Depressions in this region only contain water if they extend below the normal ground-water level. Some of the smaller examples figured by Skeats and James (1937, Pl. xiv) appear to be of this type, but all the larger lakes of the Colac and Stonyford districts which they consider seem to be calderas due to deep volcanic subsidence.

Lakes formed by volcanic damming. Volcanic activity may result in the building of a dam across a pre-existing valley, behind which dam a lake can collect. This can happen in two ways. Occasionally a volcanic peak or series of peaks may interrupt the pre-existing drainage (Type 18a); more usually a lava stream (Type 19a) or a mudflow (Type 19b) constitutes the dam. To these types of process may be added the case, already mentioned, of Niuafoou, where a small lakelet has been cut off from the margin of a caldera lake by a barrier formed by secondary peaks (Type 18b).

Damming by the building of new volcanoes has sometimes produced very important hydrographic changes. The formation of the valley of Mexico as a closed basin, containing the now largely drained Lake

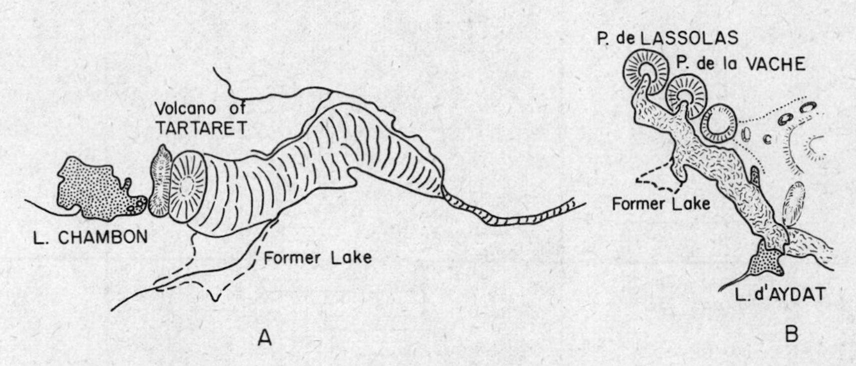

FIGURE 17. *A*, Lac de Chambon, dammed by the extinct volcano of Tartaret. *B*, lava flows damming Lac d'Aydat, Lac de la Cassière, and the extinct Lac de Randanne. (After Glangeaud.)

Texcoco, provides an example of the enclosure of a depression by volcanic mountain-building comparable to the formation of the Bonneville and Lahontan basins by tectonic activity. A neater but still vast example is the formation of Lake Kivu by the damming of the headwaters of the Ruchuru, one of the remoter tributaries of the Nile, when the Birunga Mountains, a range of snow-covered but still active volcanoes, were thrown up in the quite recent geological past.

A much less grandiose case of damming by the building of a volcano is seen in the Auvergne, where the Lac de Chambon (Fig. 17), a lake lying in the glacially remodeled valley of the river Couze, is held by the Pleistocene volcanic peak Le Tartaret. The lake has a maximum depth of 5.8 m. and appears to be doomed to a relatively rapid extinction by silting and by the backcutting of its effluent (Boule, Glangeaud, Rouchon, and Vernière 1901, Glangeaud 1913).

More often the damming is due to a lava flow filling or crossing a pre-existing valley, so ponding either a tributary or the main stream. A few examples of this occurring within historic time are known. In the Lassen National Park in California an area of old lake sediments, mainly diatomaceous earth, indicates the former position of a lake which existed to the east of the Cinder Cone, one of the newest volcanic peaks of the region. Subsequent to the senescence of the original lake a new lava flow produced

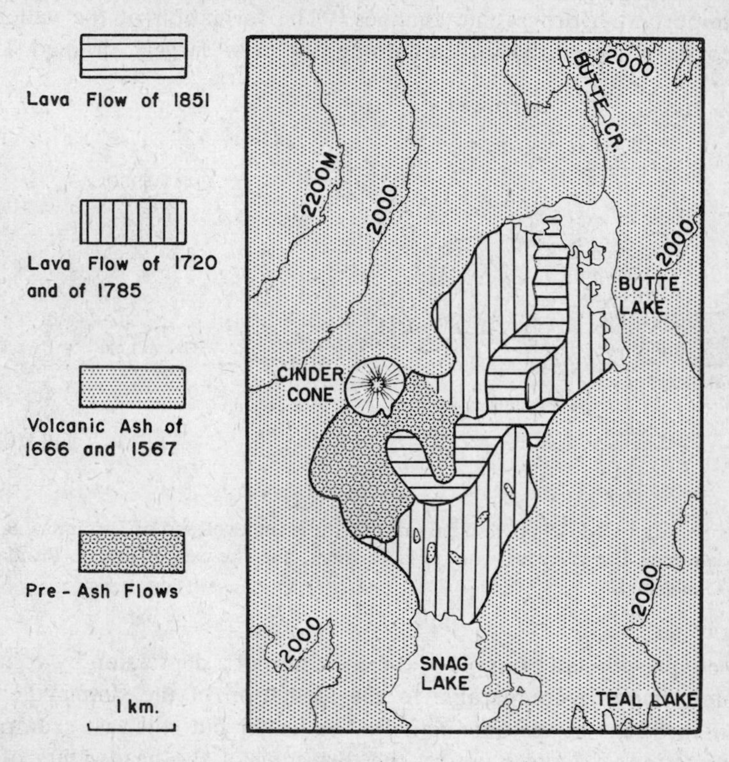

FIGURE 18. Snag Lake, Lassen National Park, California. Map showing the chronology of lava flows, obtained by tree-ring analysis, and including the eighteenth-century flows that dammed the lake (Finch 1937).

a dam behind which the water of Snag Lake collected, the name being derived from the numerous dead trees found by the first explorers of the region (Fig. 18). Finch (1937) has shown, by a study of the annual rings of trees that suffered inhibition of growth after various eruptions or which grew on lava flows of various ages, that the lake must have been formed subsequent to 1666, and probably about the year 1720.

Lakes dammed by lava flows are fairly common in volcanic regions in which much lava is produced. Admirable examples are known from the

Auvergne, where the Lac d'Aydat, maximum depth 14.5 m., and the smaller Lac de la Cassière, as well as several now extinct lakes, were formed when lateral valleys were dammed by an immense lava stream flowing from the Puy de la Vache and the Puy de Lassolas (Glangeaud 1913). Other lakes formed in this way are known from the region; in depth the existing examples appear to be intermediate between the crater lakes and maars on the one hand and the lakes on lava flows on the other, but many have become senescent. Deposits of diatomite are frequently found at the sites of such lakes.

A very fine example is provided by Lake Bunyoni in Uganda, formed by the damming of a steep valley in the Kigezi district to produce a lake about 28 km. long and 40 m. deep (Worthington 1932). The eruptions producing the lava are evidently of fairly recent date, and the lake has an irregular shore line with many islands, comparable to an artificial lake formed by damming a valley.

Volcanic dams have produced many of the lakes of Japan. Among the more remarkable of the examples given by Yoshimura (1938b), Penkeko and Pankeko, maximum depths 39.4 and 49.9 m. respectively, within the great conca of Akan and draining into Akanko, may be specially mentioned. The group of lakes formed after the eruption of Bandaisan in 1888 are also of interest as being formed in historic time and being held by mudflows produced by the explosive disruption of one side of the mountain. Such a mode of origin is in a way transitional to the landslide lakes (Type 20b) of the next section.

Cotton (1944) gives two good examples of New Zealand lakes, Lake Omapere in the upper Waitangi Valley, dammed by a basaltic flow, and Lake Rotoaira, dammed by lava from Tongariro. In the case of Lake Omapere, as more spectacularly in that of the Lake of Nicaragua in Central America, the formation of the dam has reversed the hydrographic pattern of the region around the lake. Lake Lanao in Mindanao in the Philippines is another good example, possibly the finest lake of this kind known. It was formed by lava damming a ravine that had cut into an upland plateau. The ravine above the dam and part of the adjacent plateau were flooded until a new effluent developed. The lake consists of a deep trough on the western side, which is supposedly[7] 300 m. deep at its southern end above the dam, and a wide expanse of shallow water, 4 to 10 m. deep stretching north and east of the old ravine, the whole area being 375 $km.^2$ (Herre 1933).

---

[7] An earlier survey quoted by Halbfass (1922) gives the maximum depth as 112 m. and the area as 900 $km.^2$ Herre's area may be more reliable, but his maximum depth is probably too great.

It must be borne in mind that in some valleys, such as those described in western Victoria by Skeats and James (1937), the presence of a basaltic lava barrier may merely cause a river to go underground and find its way along the old drainage beneath the lava. Not all the volcanic dams in Victoria are of this nature, however; good examples of lakes held by such dams, namely the Cockajemmy lakes and others near Mount Abrupt, are known (Fenner 1918). The possibility of the formation of lakes obviously depends on the porosity of the lava as well as on its presence. Moreover, lava-dammed lakes are often impermanent features, and in relatively old volcanic areas they may have disappeared before the maars and caldera lakes were breached or obliterated by silting. Cotton gives examples from Java and from Australia.

LAKES FORMED BY LANDSLIDES

Rockfalls, mudflows, or any of the intermediate kinds of landslides[8] may fill the floors of valleys and so dam streams (Type 20). The lake formed behind such a landslide dam is often transitory, for the unconsolidated material of the slide may be easily eroded by the effluent of the lake once the latter rises to spill over the top of the dam. The cutting back of the effluent may take place with extraordinary rapidity, discharging the water of the lake in a few hours and causing disastrous floods. The best chance of persistence of such a lake is when the dam is so high that water cannot discharge over it and develops an effluent that does not follow the original valley (*A* of Fig. 19).

At the end of 1840 or the beginning of 1841 the western side of the Lechar spur of Nanga Parbat in Kashmir collapsed into the Indus as the result of an earthquake. By the end of May 1841 a lake nearly 64 km. long and probably at least 300 m. deep had formed behind the obstruction. Late in May or early in June the dam gave way, presumably as the result of the re-establishment of flow over its lip, and the whole lake was discharged in twenty-four hours. "As a woman with a wet towel sweeps away a legion of ants, so the river blotted out the army of the Raja."[9] A later Indus flood in 1858 appears to have been due to the damming of the Hunza River by a landslide caused by snow and rain loosening a mountain side. The lake appears to have lasted for about six months (Mason 1929).

A comparable temporary lake was formed at Gohna in the headwaters of the Ganges as the result of a tremendous landslide in September 1893.

---

[8] For a classification of landslides and a general account of their incidence and nature, the reader is referred to Sharpe (1938).

[9] From a contemporary native account quoted by Mason (1929), who summarizes existing information on this disaster. A Sikh army was encamped by the river near Attock.

The dam is estimated to have been at least 250 m. high. When the effluent was re-established eight months later, it cut back so rapidly that the lake level fell about 100 m. in two hours. The resulting flood caused a rise of 15 m. at a distance over 100 km. downstream from the obstruction (Holland 1894, Strachey 1894, Lubbock 1894).

The available American examples of temporary landslide lakes are less spectacular than the North Indian cases just cited. The most impressive is the dam formed by a rockslide across the Gros Ventre River above Kelly, Wyoming. This slide occurred in 1925, and by 1927 a lake 8 km. long had collected behind the obstruction. Collapse of the dam in the latter year though causing serious floods was not complete, and the lowered lake persisted after the catastrophe (Alden 1928).

When the stream is small and the landslide large, a permanent dam may be produced, holding a lake that can run through the entire cycle of development, senescence, and extinction. Sufficiently large slides are most apt to occur in mountain valleys in which a stream is eroding a relatively soft rock overlaid by more resistant material, which thus may become undercut. Several lakes have been produced by rockslides originating in this way in the Warner Range of northeastern California, one of the few parts of the world where a lake district characterized by this type of basin exists. Russell (1927) describes Clear Lake,[10] Blue Lake, Pit Lake, and Lost Lake as formed in this way, and believes that the much larger Eagle Lake just outside the Lahontan Basin has a like origin. In the case of Clear Lake, the retaining barrier is formed by two landslides that have descended from opposite sides of the valley. Most of these lakes are filling rapidly. Lost Lake has been reduced to half its original area by the building of a delta, and several completely filled extinct lakes of comparable origin are known in the district. Other lakes in the west of North America are known, notably the two Kern Lakes west of Mount Whitney in the Sierra Nevada (Lawson 1904), Lakes Manzanilla and Reflection in Lassen Volcanic National Park (Williams 1932), and Janet Lake in Glacier National Park. Mudflow lakes (Type 20b) are doubtless always rarer than rockslide lakes (Type 20a). Lake San Cristobal on the Lake Fork of the Gunnison River provides an example from the west of North America.

Several good examples of rockslide dams are known from Europe. Ahlmann (1919, p. 125) has described a small lake in Norangsdal, Sunnmør, Sogn, in west Norway as dammed by a large rockfall in 1908.

[10] Not to be confused with Clear Lake in the coastal range of California, discussed below, nor with innumerable other Clear Lakes elsewhere in the relatively unpolluted parts of the English-speaking world.

Trueman (1938) mentions Mickleden Pond, Langsett, Derbyshire, as an English example of a lake held by a landslide dam. Delebecque (1898) cites several French examples, the deepest being the Lac de Sylans, maximum depth 22 m., between Bellegarde and Nantua. The effluent of this lake leaves from its original upper end, and the dam is therefore practically uneroded (*A*, Fig. 19). Two other small landslide lakes, the Lacs des Hôpitaux, exist in the same region. Penck (1894) notes several examples from the eastern Alps; among them the Lago di Alleghe in the Agordo Valley, Belluno, is noteworthy as having been formed relatively recently, and the much older, Lago di Molveno, Trentino, for its considerable depth of 118 m. The ancient landslide damming the Obersee above Nafels, Glarus, is described in some detail by Oberholzer (1900). The most interesting case in the Alps, however, is the Lac des Brenets or de Chaillexon (*B*, Fig. 19) in the Doubs Valley on the Franco-Swiss border. Collet (1925), who summarizes the literature on the origin of this lake, was with Buxtorf (1922) able to study the barrier holding it, under particularly favorable conditions when the water was very low. They conclude that two landslides occurred, one before and the other after the Würm glaciation. The effluent now flows over a rock lip between the two old channels, giving a deceptive appearance to the investigator studying the origin of the Lake.

Lake Sarez, the center of which lies about lat. 38°15′ N., long. 73° E., in the Murgab Valley, Pamir Mountains, is doubtless the finest example of a lake dammed by a landslide at present known. The dam, formed on February 18, 1911, has a volume of 2 km.^{3} and rises 750 m. above the floor of the valley. The lake is 75 km. long and has a maximum depth of 500 m. (Chuyenko in Berg [1938] 1950; Suslov 1947), thus ranking eleventh in maximum depth among the lakes of the world. The level rose slowly during the period from 1911 to 1934. The water discharges by seepage through the dam at a level 150 m. below the lake surface, and this effluent is now in equilibrium with the influent (Suslov 1947), so that a stable level is maintained.

Lake Busyû is a fine Japanese example, formed by a landslide in 1590 (Yoshimura 1932c).

Cotton (1941) cites Waikaremoana as a magnificent example of a lake held by a rockslide dam from New Zealand. The lake fills a branching valley system not unlike that of many artificially dammed lakes.

Though the foregoing examples indicate that landslide dams often form in glaciated mountains, the most interesting cases are the isolated examples that are occasionally met with in unglaciated regions, for the occurrence of landslides is the most usual way in which a small typical mountain-valley lake can be formed in unglaciated and nonvolcanic mountains. Mountain

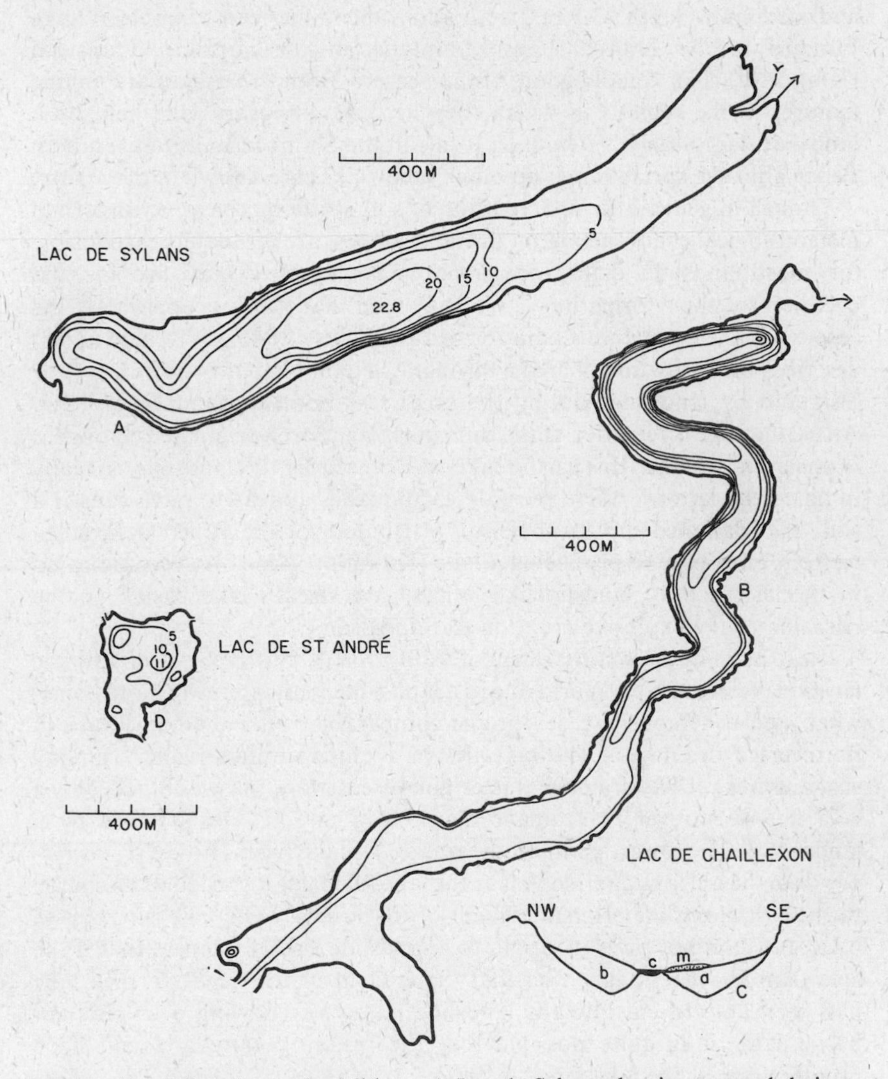

FIGURE 19. Lakes due to landslides. *A*, Lac de Sylans, showing reversed drainage with effluent discharging from bay enclosed by unilobate delta of influent (after Delebecque). *B*, Lac de Chaillexon (after Delebecque). *C*, diagram of a section across point of discharge of effluent: *a*, pre-Würmian landslide; *m*, moraine covering the same; *b*, post-Würmian landslide; *c*, effluent apparently flowing over a rock lip which formed an isolated projection in original valley (after Collet from Buxtorf). *D*, Lac de St. André, a small lake on the irregular surface of a landslide (after Delebecque).

Lake, Virginia (Stose in Hutchinson and Pickford 1932), area 0.47 km.^{2} and maximum depth 33.5 m., is an admirable American example. Lake Funduzi in the Zoutpansberg Mountains of the northern Transvaal (King 1942) is an equally good African case. Both these lakes are unique features in the regions in which they lie. A temporary lake held by a landslide dam nearly 250 m. high forms in the Shimbar Valley in southern Persia after the spring rains; no other valley lakes are known in the region.

Though in general the undercutting of soft strata by rivers, or abnormal meteorological events acting on unstable slopes, are ultimately responsible for most landslide dams, occasionally earthquakes may provide the occasion for their formation. We have seen that this was apparently the case when the great Indus dam formed in 1840 or 1841. Vivenzio (1788) describes the formation of a lake through the damming of two streams near Sitizzano by landslides during the great Calabrian earthquake of 1783. An earthquake is recorded at the time that Lake Sarez originated, though it is uncertain whether this earthquake was caused by the landslide or really initiated the latter. More recently earthquakes appear to have caused a slide that dammed the River Seisui, a tributary of the River Dakusui in western Formosa, so producing a lake (Kawada 1943). As was indicated in the last section, landslidelike lahars gave rise to lake basins on the occasion of the explosive eruption of Bandaisan.

Clear Lake in the coastal range of California (Davis 1933), probably the largest lake in the New World due to a landslide, is an interesting and somewhat special composite case, though comparable to the Lac de Sylans in its reversed drainage. The lake lies in an intermontane valley running east and west. The drainage was originally eastward but was blocked by a lava flow. A reversed drainage developed, but this was blocked by a landslide. The lowest point on the rim of the lake basin, however, corresponds to the old lava dam, so that as the lake filled up, it started to discharge to the east along the original effluent of the valley. The lake is thus held by a lava dam but owes its origin to a landslide. Davis thinks that if the lava dam had been higher, so that the effluent of the newly formed lake had been able to cut into the landslide instead of flowing over resistant basalt lava, it is quite possible that the resulting canyon would have rapidly drained the lake.

Ousley (1788) has described a case of a large part of a peat bog sliding across a shallow valley at Addergoole near Dunmore in County Galway, Ireland, producing a peatslide dam (Type 20c). The event occurred in March 1745 after abnormally heavy rain. The resulting lake had an area of at least 0.22 km.^{2} and possibly as much as 1.2 km.^{2} The dam was rapidly cut by the local inhabitants to drain the agricultural land that had been flooded.

The damming of valleys is not the only way in which landslides may produce basins. Small lakelets or ponds not infrequently are held between the mass of a slide and the valley wall from which it has been detached (Type 21). This is due to the fact that the inner part of a slide may fall faster and farther than the outer. Since there is little or no stream-cutting below such lakelets, they may persist longer than do the lakes dammed by the main mass of the slide. Russell (1893) implies the existence of such lakelets in the state of Washington, and Davis (1933) mentions that a series exists above the landslides that abound on the slopes of Mustang Ride northeast of Peachtree Valley near Salinas Valley in the southern coastal range of California. Delebecque (1898) mentions three small lakes, the deepest being the Lac de St. André (*D*, Fig. 19), 11.8 m. deep, formed in this way on the landslide that descended from Mont Granier in 1248. Delebecque notes that similar small lakes occur on the Goldau and Flims landslides in Switzerland.

In addition to the production of lakes by the catastrophic landslides discussed in the preceding paragraphs, a formally comparable but more or less continuous process must be recognized, namely the building of *scree dams* across valleys. A scree is an accumulation of angular detritus falling to the angle of repose from a steep slope. Occasionally the rate of formation of the scree is greater than the rate of its removal by the stream at the bottom of the slope, and a dam may be formed, holding a small lake (Type 22). According to Marr (1916), Goatswater below Coniston Old Man, and Hard Tarn below Nethermost Pike, Helvellyn, both in the English Lake District, are examples of small lakes formed in this way. The former drains by seepage through the dam, the latter over the scree or, when very high, over an adjacent rock lip.

GLACIAL LAKES

No lake-producing agency can compare in importance with the effects of the Pleistocene glaciations. During most of the earth's history, there must always have been some tectonic, volcanic, and solution basins, and some lakes due to the action of wind and to the building activity in mature river valleys. The immense number of small lakes produced by glacial activity which now exist is, however, a quite exceptional phenomenon, presenting to the limnologist many times the number of individual basins which would have been available for study during most of the Mesozoic or Tertiary eras.

The present glaciated land surfaces include the Antarctic continent and Greenland, which are covered by continental ice sheets; some of the arctic islands, which have similar but smaller ice sheets; and a very large number of small centers of glaciation in high mountains throughout the world.

The most important of these mountains are the more coastal ranges of western North America, the Andean chain of South America, the Scandinavian mountains, the Alps, the Caucasus, the Pamirs, the Hindu Kush, Karakoram, Himalayan, and some other high central Asiatic ranges, and the Southern Alps of the south island of New Zealand. Small but interesting glacial foci also occur in the Pyrenees; in three of the highest mountains of central Africa, namely Kenya, Ruwenzori, and Kilimanjaro; in the Carstensz Mountains of New Guinea; and on a single mountain, Ruapehu, in the north island of New Zealand.

During the Pleistocene glaciations, ice sheets comparable to those of Greenland and Antarctica developed in many parts of the northern hemisphere. In North America, most of the area westward from Long Island, north of the Ohio and Missouri Rivers, and in the west north of about lat. 48° N., was ice covered at some time during Pleistocene history.

In Europe, two major ice sheets, one in Scandinavia eastward into western Siberia and the other in the Alps, were developed at the height of glaciation, though in the last glacial the Scandinavian ice sheet did not reach Britain and a separate north British icecap was developed. An independent sheet was developed in eastern Siberia and another north of Lake Baikal. Most of the higher mountains south of these regions developed independent glacial systems. Seventy-one such independent centers in North America are enumerated by Flint (1947), and a similar abundance and complexity are recorded in the mountains of southern and central Italy, the Balkans, north Africa, and other parts of the old world.

From almost the beginning of the study of the former Pleistocene glaciation, it has been realized by many investigators that the process was multiple. It is now generally agreed that at least four major episodes were involved in the Northern Hemisphere. Many investigators consider that the glaciations which developed in the Southern Hemisphere, notably by the extension of the existing glaciers of South America and New Zealand and the development of glaciers in Tasmania and New South Wales, were also contemporaneous with those in the north. This synchrony is, however, more a pious hope than a well-established conclusion.[11]

Moreover, at various times in the Pleistocene, large lakes were developed in regions which now contain closed dry basins or basins with much smaller bodies of water than they once held. There is good reason to associate the high lake levels with glacial maxima in the Great Basin region already

[11] In view of the very rapid development of the study of Pleistocene chronology since the discovery of radiocarbon dating and O^{18} palaeothermometry, it has seemed wise to defer the presentation of a correlation table until Volume II, in which the subject is of importance in connection with the stratigraphy of lake sediments.

discussed, in the Ponto-Caspian Basin and perhaps in other parts of central Asia, notably Kashmir and western Tibet. It has also frequently and very reasonably been supposed that comparable high lake levels in equatorial Africa represent pluvial periods corresponding to the glacials. Recently (Solomon in O'Brien 1939) doubt has been cast on the African correlation, for in this continent the history of the lakes is complicated by Pleistocene earth movements which may have produced high levels in particular basins, such as those of Lake Kioga and Lake Victoria, where tilting can easily interfere with drainage patterns in the way already described. The present writer feels that the balance of the evidence is still in favor of the correlation of the pluvial high levels with the glacial periods (Nilsson 1940).

In addition to the pluvial periods in the tropical or extraglacial regions, the accumulation of ice lowered the level of the ocean, while the melting of ice during the interglacials raised the level of the ocean above its present position. The lowering of the base level during the later glacial stages is undoubtedly responsible for the presence of drowned valleys in every relatively stable coast throughout the world. Since such valleys often contain coastal lagoons, which in some cases may become lakes completely separated from the ocean, the variation in sea level during the Pleistocene is of some limnological importance. A supposedly postglacial high level has been widely recognized, at 1.5 m. or 5 ft. above present sea level. This is often attributed to a postglacial minimum in the ice sheets of the world, referred to as the climatic optimum or postglacial hypsithermal period (Flint 1947, Deevey and Flint 1957).

In considering the various ways in which glacial action can produce lake basins, it must be remembered that, while it is probable that in all cases glaciation began in mountain ranges, during glacial maxima the ice descended to form piedmont glaciers, which in many regions coalesced as immense ice sheets. The action of glaciers on regions of high relief may produce forms quite different from those due to the action of large ice sheets on regions of more mature and gentle relief. In either case lakes may be produced, but the way in which this happens may be quite different, and the resulting basins may have such diverse forms that throughout their entire histories their characters will reflect the difference in origin.

Strøm (1934, 1935a,b) has classified lakes of glacial origin into four categories. The first (*A*) includes lakes formed by the vigorous action of active ice in cirques and mountain valleys, producing simple or multiple cirque lakes and fjord lakes, lying in true rock basins. The second category (*B*) includes glacial rock basins formed by less active ice moving over peneplains or in shallow valleys, the position of the lake being primarily determined by variations in the lithology of the region or the existence of joints and shatter belts. The third category (*C*) includes lakes left after ice has melted, either in kettles, in subglacial channels or in irregularities on the ground moraine. The fourth category (*D*) includes all types of glacial dam, including ice dams.

In the following pages it has proved more convenient to use a somewhat different classification to conform to the regional outlook here adopted. The lakes held by existing ice dams have been treated separately, as geographically and historically they are obviously confined to regions and periods of actual glaciation. The glacial rock basins are all considered together, as too many of the largest examples combine features of Strøm's categories *A* and *B*. Strøm's third and fourth categories are accepted, after the removal of the ice lakes, but it is more convenient to treat the morainic dams before the kettles and other types of drift basins.

LAKES HELD BY ICE OR BY MORAINE IN CONTACT WITH ICE

It is convenient to begin with those cases in which the lake exists simultaneously with the ice that produces its basin and is often in contact with the ice itself. Such lakes are generally called *proglacial* lakes in the geological literature.

Lakes on or in ice (Type 23). Bodies of water are occasionally recorded as occupying basins consisting entirely of ice. Delebecque (1898) records such basins, up to 130 m. in diameter and filled with water, on the surface of the Gorner Glacier, Zermatt, Canton Valais. He considers that they are formed as moulins or glacial potholes, in which opinion Collet (1925), who has examined such depressions, concurs. Heim (1885) supposes that they are formed by subsidence. Lakelets of this sort are (Type 23a) of course temporary, and are interesting mainly as limnological curiosities.

Very rarely, bodies of water appear to collect under glaciers (Type 23b). Vallot and Delebecque (1892), who describe such a case in the small glacier of Tête-Rousse on Mont Blanc, which discharged on July 12, 1892, causing a catastrophic avalanche at Saint-Gervais (Haute Savoie), conclude that melt water collects between two converging arêtes below the glacier. The discharge of such subglacial lakes is of course the more dangerous for being entirely unexpected.

Crary, Cotell, and Sexton (1952) have described lakes (Type 23c) forming in depressions during the summer thaw on T3 or Fletcher's Ice Island, a floating mass of ice now drifting in the Arctic Ocean at about lat. 88° N. The lakes appear in early June and start to freeze in August.

Glacier-dammed lakes (Type 24). In glaciated mountains, streams may sometimes be dammed by the snouts of glaciers that descend well below the permanent snow line. This may happen when the glacier of a main valley extends far enough down to dam a tributary stream (Type 24a), or when a lateral glacier dams the main stream (Type 24b). In either case the dam is apt to be impermanent, being eroded when water spills over the top of the ice, or discharging when it makes its way through crevasses across the ice dam or when a tunnel develops under the glacier. Strøm (1938a) considers that the last-named event is due to the glacial dam floating as the lake increases in depth. When channels through crevasses or through a subglacial tunnel develop, catastrophic emptying may occur and the lower parts of the valley may suffer destructive floods.

The classical case of a lake at the base of a lateral valley dammed by the glacier in the main valley is the Märjelensee in the bottom of the Märjelen Valley, dammed by a branch of the great Aletsch Glacier in Canton Valais, Switzerland. The volume of the lake is variable, dependent on the form and history of the glacial dam at any given time. The maximum volume appears to have been prior to a catastrophic discharge in July 1878, of 10.7 million cubic meters of water. In July 1892 and in July 1913 the dam again failed, but only 7.5 and 3.1 million cubic meters of water were lost from the basin on these occasions. A very great amount of sediment is carried by these floods (Lütschg 1915, Collet 1925).

The finest example of a lake formed by a glacier in a lateral valley damming the main stream is doubtless the Gapshan Lake or Shyok ice lake (Fig. 20), which forms across the upper Shyok River, a tributary of the Indus draining the Muztagh-Karakorum Mountains in about lat. 3° N., long. 74° E. Three glaciers, the Chong Kumdan, Kichik Kumdan, and Ak-tash, approach the headwaters of the Shyok from the west, and it is probable that any of the three may constitute dams. Since the region has been under observation, it has usually been blocked by one of the glaciers; this does not necessarily mean that a lake is always present, for sometimes the river can escape below the ice. The ice barriers gave way, producing floods, about 1780, in 1835, 1839, 1842, 1903, 1926, 1929, and 1932. The last discharge, in 1932, caused much less serious floods than those on immediately preceding occasions. It is known that in 1929 a large crevasse opened, apparently along a line of weakness in the glacier, before the water reached the top of the dam; in 1932 a smaller channel developed under the ice, presumably initiated by the floating of the snout of the glacier. Subsequent to 1932 the secular retreat of the Chong Kumdan Glacier, which had begun to occur about that year, has ensured that no dam forms (Ludlow 1929, Mason 1929, Lyall-Grant and Mason 1940). Ludlow (1929) estimated the length of the Gapshan Lake to be 16 km. in 1928; Gunn found that just before the collapse in August 1929, the maximum depth in front of the glacial dam was 120 m. and the volume was 1.35 $km.^3$ (Gunn, Todd, and Mason 1930).

Mason (1929) has also considered the bursting of other less well-known ice dams in the Shingshal Valley, also in the Muztagh-Karakorum. He

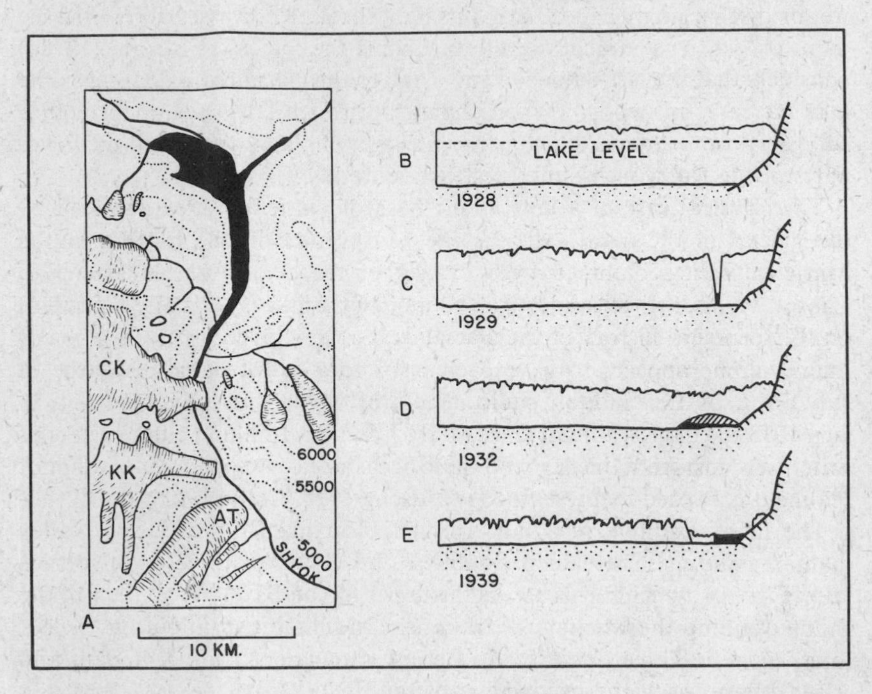

FIGURE 20. Gapshan or Shyok ice lake. *A*, condition of lake prior to rupture of dam: CK, Chong Kumdan Glacier; KK, Kichik Kumdan Glacier; AT, Ak-tash Glacier, any one of which may at times form a dam. *B*, diagram of dam in 1928 prior to its rupture; broken line indicates lake level as in *A*. *C*, broken dam in 1929. *D*, dam reformed in 1932, but floating at tip and eroded from below to form a tunnel. *E*, bridge over tunnel collapsed and stream unimpeded in 1939. (After Mason.)

considers that, in general, such dams develop when a lateral glacier not only crosses a narrow main valley but runs several hundred meters down the valley. Where a less complete block occurs, the lake formed in one year may drain in the spring of the next, by water percolating beneath the glacier before any discharge from the frozen headwaters of the stream can reach the lake to replenish it. Mason states that he has observed this happening in the Shaksgam Valley.

Strøm (1938a) lists four lakes held by ice-dams in Norway. The best known is Demmevatn in Hardanger. This lake may leak out through the ice dam, as happened in the summers of 1896 to 1898, or it may discharge cataclysmically, as in 1813, 1843, 1861, 1893, and 1937. Strøm indicates that sometimes, after discharge due to the floating of the glacier snout, the glacier may sink after only a part of the water has been lost. He also points out that if the glacier becomes reduced in volume, as has happened to many Norwegian glaciers since 1933, it may float at a much lower water level than would be needed to produce discharge before the wasting of the glacier. A tunnel supposed to keep the lake at a safe level, constructed in 1899, did not prevent the cataclysm of 1937, as the Rembesdalsskåki Glacier, which holds Demmevatn, had wasted so much in the interval.

A somewhat different series of cases is provided by the small lakes which may form between a glacier and its walls (Type 24c). Most cases seem to occur where two glaciers converge and fuse, the lake being held between the masses of ice and the spur that separates them. Very fine examples have been recorded as occurring between the Chaix Hills and the surrounding Malaspina Glacier in Alaska (Tarr and Martin 1914). Collet (1925) describes a similar case of a lake 70 m. long on the Rhone Glacier. Other examples have been known to form at the base of Mount Tacul, at the confluence of the Glacier du Géant and the Glacier de Léchaux (Delebecque 1894), and on Monte Rosa (Collet 1925).

A rather special type of ice-dammed lake is described by Collet (1925). This dam was formed by the Otemma Glacier in the valley of Bagnes, Valais, tributary to the Rhone, when its tributary glacier the Crête Sèche retreated, leaving a basin at the bottom of its lateral valley. Actually, the lake seems to have been held primarily by the old terminal moraine of the tributary, lying on top of the main glacier. The moraine, however, was thin and the barrier certainly only existed because of the underlying ice. Because the development of a large crevasse in the main glacier permitted the lake to discharge, a serious flood occurred in the valley, and in 1898 control measures were instituted to prevent the re-formation of the lake. The volume of water at its maximum height in 1898, before the flood, was about 1 million cubic meters, the maximum depth being then 30 m.

Another rather special case (Type 24d) of a lake held by a dam of frozen water is described by Lyell (1837). A lake about 200 m. long and 60 m. deep formed when the same valley of Bagnes was temporarily dammed by an avalanche in 1818. A dam of this sort being very unstable, efforts were made to drain the lake gradually, but after half of its water was drawn off, the center of the dam collapsed during the height of summer and the remaining part of the lake discharged in half an hour, causing a disastrous flood. Lyell notes that similar inundations, supposedly from the same cause, had been recorded earlier in this part of Switzerland, notably in 1595, when the town of Martigny was destroyed, and apparently also about 1545, when a still more destructive flood occurred. It is, of course, not certain that these older cases were not really the bursting of other kinds of glacial or landslide dams rather than avalanche dams.

In Scotland, as in Scandinavia and elsewhere, remarkable ancient examples of mountain valleys dammed at their lower ends by ice are known, producing lakes comparable to though longer lived than the Shyok ice lake of today. The Parallel Roads of Glen Roy and other terraces in valleys to the northeast of Ben Nevis were formed when the lower ends of these valleys were blocked by glaciers descending from the mountain. Initially four lakes were formed, the level of each determined by the col at the head of the valley. As the ice retreated and the lakes were put into communication one with another, the level progressively fell to that of the lowest col. In Glen Roy, which shows the finest terraces, the first lake stood at 351 m.; the second, when the lake communicated with the Glen Glaster Lake, at 328 m.; the third, when a single lake existed, drained at the top of Glen Spean at 274 m. above sea level. Glen Roy thus has three "roads," Glen Glaster two, Glen Spean one (Jamieson 1863).[12]

When an ice sheet is waning, the newly uncovered land will not immediately rebound to its preglacial elevation. If the original relief is very gentle, and even more if the preglacial drainage was directed towards the region still covered by ice, the newly uncovered valleys will form basins dammed by the retreating ice sheet (Type 24e). Small lakes of this sort are known today against the western part of the Greenland Icecap (Plate 1). A magnificent example of a large lake held in this way was provided by the Baltic ice lake already mentioned. In North America many examples are known. The reader interested in any particular region will be able to identify such lakes on the Glacial Map of North America (Geological Society of America 1945, 1948). Important examples are provided by certain developmental stages of the Laurentian Great Lakes (Fig. 27), and of the Finger Lakes (Fig. 30) in New York State. Magnificent ice lakes were also formed north and west of the Great Lakes during the deglaciation of North America. Two such lakes, later to become confluent, north of the Great Lakes, are named Lake Barlow and Lake Ojibway. Farther west, Lake Dakota formed in the James River Valley, Lake Souris in the Souris River Valley, and Lake Regina in the South Saskatchewan Valley. Lake Dakota drained as the ice front moved northeast, but the other two lakes became joined to an immense ice lake, Lake Agassiz, certainly the largest ice lake to form in the New World, in the valleys of the Red River of the North system. Two high levels of this lake are known, presumably corresponding to retreat and then some re-advance of the ice. It is possible that at some stage Lake Agassiz was connected with the Ojibway-Barlow Basin.

---

[12] Darwin (1839) made an attempt to prove that the roads were of marine origin, but later admitted Jamieson's explanation to be correct.

PLATE 1. Proglacial lake held by an existing ice sheet, south of Godthaab, west Greenland (*phot*. C. A. and A. M. Lindberg, courtesy of the National Geographic Society, Washington, D.C.).

Comparable, and probably even larger, glacial lakes appear to have developed at some stage against the southern margin of the Siberian ice sheet, between the Ural Mountains and the Yenisei River (Flint and Dorsey, 1945).

Smaller but nevertheless quite impressive lakes formed in England during the last glaciation, and have played a part in determining the modern drainage pattern of the country. The most notable cases of this effect are the course of the River Derwent south through Kirkham Abbey Gorge (Kendall 1902) instead of eastward, which was cut by the outflow of the glacial Lake Pickering in the Vale of Pickering, a large depression inland from Scarborough; and the course of the upper part of the Severn southward instead of northward through the Ironbridge Gorge, the outlet of the glacial Lake Lapworth. Both lakes had a complicated history and are associated with other smaller basins.

In some of these smaller basins in the North York moors, dammed by ice to the north, a complex drainage pattern joined lake to lake either along the ice front or over low points in the ridges separating the lakes. These lakes must have developed a rather characteristic morphology which is probably hardly known today, though paralleled on a much larger scale by certain stages of the ice lakes in the Finger Lakes region in New York State. Apart from the examples just mentioned, a large lake, Lake Humber, lay south of Lake Pickering; other ice lakes, not fully elucidated, may have existed in England.

Damming by moraines of existing glaciers (Type 25). In a few cases lakes are held not by the ice itself but by the moraine of an existing glacier. It may usually be supposed that in such cases the stability of the dam depends largely on the presence of the glacier. The lake of Mattmark in Switzerland provides an example of a lake held at least during part of its career in this way, though there was during the period of its highest level in the nineteenth century some damming by ice (Collet 1925). In this case the Allalin Glacier, the moraine of which held the lake, lies in a lateral valley, damming the lake in the main valley (Type 25a). But at least one magnificent case (*C* of Plate 2) of a lake in a lateral valley, held by the moraine of a glacier in the main valley, has been noted by Löffler in the Andes (Type 25b). Lakes of these types are perhaps really not distinguishable from those of Type 32.

GLACIAL ROCK BASINS

In discussing glacial rock basins it has proved convenient to begin with the small basins formed by ice moving over relatively flat surfaces of hard but jointed or fractured rock. A very characteristic type of small irregular lake may be developed on upland peneplains in this manner. After considering this type, the lakes produced as the result of cirque formation and by actively moving ice in glaciated mountain valleys are considered. It is then natural to consider the numerous and remarkable large lakes produced in piedmont regions, and by continental ice sheets when constrained by preglacial topography. Such lakes are generally not pure rock basins and lead easily to the category of basins formed by morainic dams.

Ice-scour lakes (Type 26). Ahlmann (1919) points out that in the mountains of Norway and south Sweden considerable areas of very ancient uplifted land surface have been scoured by ice, which, as well as removing all loose material, has scooped out small rock basins in zones of fracture and along shatter belts. These basins naturally hold small lakes on deglaciation. The view from the air (*A* of Plate 2) of hundreds of such small basins, contrasted with the enormous overdeepened valleys of the fjord lakes of the same region, constitutes one of the most impressive lakeland landscapes that the present writer has ever had the fortune to see. As Strøm (1935a) points out, such lakes, owing to the disposition of the jointing, frequently form a gridlike pattern. The disposition of the lakes of this type depends primarily on structure and not on any pre-existing drainage pattern.

In the Alps, the Saint Gotthard lakes and a few other small lakes situated among *roches moutonnées* on the Bernardin and Simplon cols are noted by Collet (1922, 1925) as having been formed by ice-scouring.

A large number of ice-scour lakes are known at a low level in western Scotland. Peach and Horne (1910) record many irregular basins formed in this way on the Lewisian gneiss of the western seaboard of Sutherland and Ross, and also in the Outer Hebrides. Soundings in these lakes indicate that the bottom is often very irregular, the weaknesses of the rock determining the position of the greatest depths. Loch Trealaval in the island of Lewis, an extraordinarily irregular body of water, maximum depth 10.7 m. and area 1.68 $km.^2$, with several smaller similar lakes only partially separated from it, provides a remarkable example of this type of lake. Numerous examples also occur in Ireland (Charlesworth 1953).

By far the most remarkable European examples of this kind of basin at a relatively low elevation are provided by the extraordinary lakeland landscape of Finland (Fig. 52). Some of the Finnish lake basins are primarily of tectonic or even volcanic origin, and a number are dammed by moraines. By far the greater majority, however, appear to lie in depressions formed by ice-scouring of fractures and shatter belts, which have been excavated when they lay parallel to the ice flow and filled with drift when transverse (Järnefelt 1938). Some of the larger lakes, such as Päijänne, which show

PLATE 2. *A*, glacial scour lakes on the upland peneplain north of Lysefjord, Norway (*phot*. Widerøes Flyveselskap og Polarfly A/S., through Professor K. Strøm). *B*, Laguna Llaca, Cordillera Blanca, Peru, a small lake held by terminal moraine of a glacier that has recently retreated (*phot*. H. Löffler). *C*, Lake Parron, Peru, a large lake, 9 km. long and more than 80 m. deep, held by the lateral moraine of a large existing glacier (*phot*. H. Löffler).

peculiar extensions along transverse faults are regarded by Järnefelt as in part of tectonic origin. A great deal of reversal of the initial postglacial drainage has occurred, owing to the greater rebound in the northwest than in the southeast of Finland, altering the positions of the effluents of the lakes.

In North America, ice-scour basins on upland surfaces are known in the glaciated parts of the High Sierras, notably in the Devils Basin, 11 km. southwest of Lake Tahoe, and in the Humphrey Basin, at the head of the south fork of the San Joaquin River, which basin contains Lake Desolation. Good examples are recorded from the upper Nugsuak Peninsula (lat. 74°12′ N. long. 56°40′ W.) in northern Greenland (Watson 1899). Tarr (1897) has recorded similar basins in the Cumberland Sound, Turnavik, and Hudson Straits areas; though he considered the formation of the basins to be pre-Pleistocene, it is reasonably certain that though the shattering was preglacial the actual excavation was glacial. A vast number of small lakes on the Archaean rocks of the Canadian shield, such as those in the vicinity of Great Slave Lake, where the drift is very thin and irregular, may reasonably be regarded as due to ice-scouring largely determined by structure.

Zumberge (1955) has pointed out that in the Rove area of Cook County, Minnesota, and on Isle Royale in Lake Superior, many elongate lake basins occur in rock basins excavated in the softer rocks and following the strike. In the former region the ice moved more or less across the axes of the lakes, in the latter region along the axes. The position of the lakes is thus determined by the underlying structure and lithology. He refers to the process of excavation as glacial quarrying, as distinct from abrasion by a debris-laden ice sheet. Zumberge's glacial quarrying would appear to be the same process as the ice-scouring of the present work.

It is possible that a few cases may be found where nivation, or the freezing and thawing of water round patches of snow, has produced small closed depressions in jointed or fractured rocks. The nivation cirques that have been described, for example by Russell (1933), do not, however, constitute basins. It has been supposed that lakes on the central plateau of Tasmania may have formed in this way, but the most modern opinion does not support such a view.

Cirque lakes (Type 27a). The heads of glaciated valleys are often modeled into amphitheaters by ice action, and similar amphitheater-like forms have frequently been produced by névé collecting in small initial irregularities not part of the preglacial drainage pattern. Such amphitheaters are called *cirques* in the French-speaking parts of the Alps, *Kars* in the German-speaking regions, *cwms* in Wales, and *corries* in Scotland.

All four terms have achieved some degree of international usage, but the first seems to have been the most widely employed.[13]

An enormous literature has grown up on the mechanisms of cirque formation. The development of knowledge of the subject, a development that has been marked by considerable controversy, is reviewed in several modern works on geomorphology; Cotton (1948) in particular gives an illuminating and balanced account. It is generally though not quite universally supposed that the process involved in the excavation of a cirque is fundamentally frost-riving due to freezing and thawing at the rock face of a névé-filled concavity on the mountain face, whether at the head of a valley or not.

Direct observation, notably by Willard Johnson (1899), who had himself lowered into the bergschrund or crevasse that nearly always develops between névé and the rock behind it, shows that at least by day in summer a continual shower of melt water is spraying onto the rock. When this freezes, it is obvious that continual detachment of small pieces of the back of the cirque must take place. The floor of the cirque is presumably largely excavated by glacial corrasion as a result of ice and the detached rock fragments being slowly pushed over this floor. Some investigators have claimed that since the depth of the bergschrund is limited and considerably less than is the depth of the vertical back wall of some cirques, a process of frost-riving must go on well below the level of the space between the névé and the wall; other investigators have tended to attribute almost the entire process to corrasion. From the standpoint of the present work these arguments, however interesting they may be in themselves, are of little direct limnological significance.

In many cirques the rock floor is excavated below the level of the entrance, so that when the ice disappears a rock basin is exposed. There is often, but not always, evidence of moraine at the entrance also. This may in part be due to the activity of the ice excavating the cirque, but also may represent debris undercut by ice, at the top of the cirque wall, that has slid over the ice surface and collected in front of the cirque entrance. When fully deglaciated, many cirques contain lakes that are held either by a rock lip or by moraine of the kinds just described. During the last decades of the nineteenth and the first decades of the twentieth centuries there was much controversy over the ability of ice to excavate a true rock basin. Few if any glaciologists would doubt this ability today, but the old warning given by Marr in his early work (1895), that the morainic dam can

---

[13] Some geomorphologists restrict the word corrie, and others the word cirque, to an amphitheater not part of a valley, and speak of the amphitheater-heads of valleys when the form occurs at a valley head. For the purpose of the present work, this distinction, which is often not made, is of no significance.

easily displace the effluent so that it runs over a rock lip, is still valid and should be borne in mind by anyone investigating such basins.

True cirque lakes are generally small and relatively shallow, though a few surprisingly deep examples are known. They occur in varying numbers in practically every glaciated mountain range. Innumerable examples are found in the Cordilleran chains of the Americas. Particularly fine examples are available in Glacier National Park (Dyson 1948a, b); Iceberg, Hidden, Avalanche, Gunsight and Ellen Wilson are among the best known in this remarkable lake district.

In the Scandinavian mountains many fine cirque lakes are known; in one limited region of this part of Europe, namely the island of Moskenesøy in the Lofoten chain, the entire landscape is dominated by cirque forms in a unique and fantastic way (Fig. 21). The geomorphology has been discussed by several writers, of whom the most recent, Strøm (1938b) has given an admirable account of the lakes. The whole island has been carved down from an unknown height during the process of cirque formation, so that it now consists of a series of basins separated by sharp walls. Except where the cirques open below sea level, forming bays, these basins all contain lakes which lie in complete rock basins with no morainic dam, the moraines of the cirque glaciers all lying beyond the rocky coast of the island. The cirque lakes may be of quite simple form, as in the case of Reinesvatn; or may represent the obvious fusion of several cirques, as in the case of Solbjørnvatn, in which there are three deep depressions; or may have been formed by such complete fusion of two cirques that no trace of multiple origin is found below water level, as in the case of Tennesvatn. The cirque lakes of Moskenesøy often occupy cryptodepressions, the bottom of the deepest, Solbjørnvatn, lying 101 m. below sea level. Practically all are extraordinary deep relative to their area. Tennesvatn, with an area of but 0.873 km.^{2}, is 168 m. deep; Reinesvatn, with an area of 0.170 km.^{2}, is 69 m. deep. In no other part of the world are cirque lakes developed on anything like this scale.

Numerous examples of small cirque lakes or tarns are known in the English Lake District and in the Snowdonia district of Wales. Most of these have morainic dams, and there has been some controversy as to whether they are fundamentally rock basins or not. Marr (1895, Marr and Adie 1898), who early was unconvinced that ice could excavate a rock basin, admitted later (1916) that some of the tarns of the Lake District were certainly glacial rock basins, Watendlath Tarn being a notable example. He believed, however, as has just been indicated, that many of the cirque lakes of the region, though their outlets ran over a rock lip, were really held by morainic dams, giving as examples Red Tarn, at least held at its present level by moraine, and Blind Tarn, Bran Pike, Coniston,

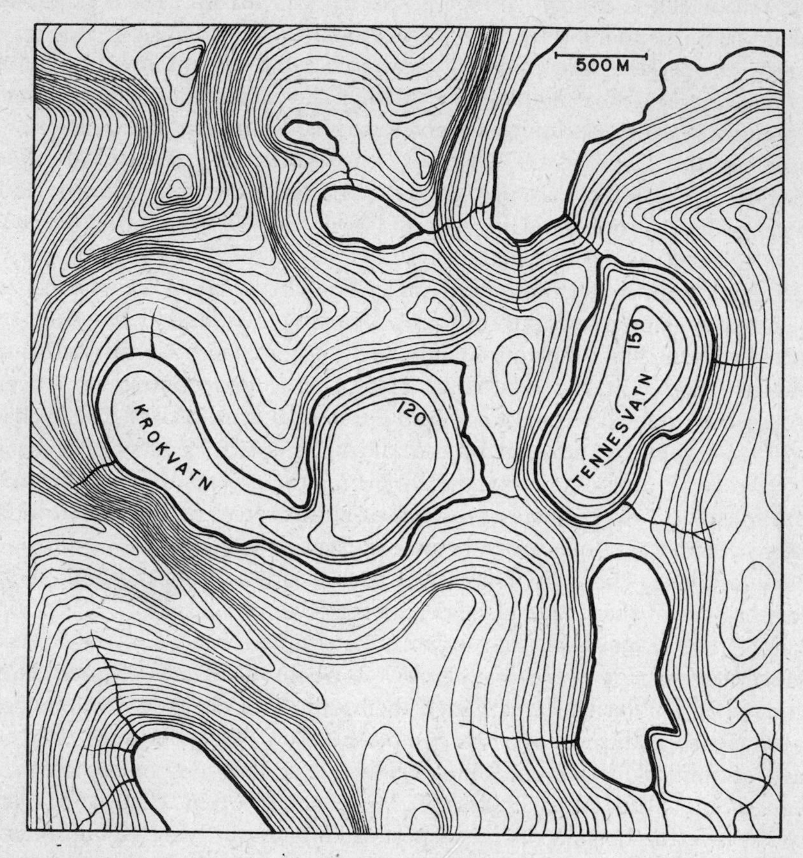

FIGURE 21. Moskenesøy, Lofoten Islands. A portion of the island around Tennesvatn to show excessive development of a cirque landscape. Contours every 30 meters, those of the lake basins measured from lake surface. (After Strøm.)

which is held by a beautiful crescent moraine through which surplus water seeps. Marr thought that if the effluent did not develop in a position where it could drain over a rock lip, as at Smallwater and Highhouse Tarn, the cutting of the moraine by the outflow would usually have destroyed the lake. It is reasonably certain that many of these small tarns lie in rock basins, though the water level is raised by morainic dams. In Scotland, Peach and Horne (1910) regard the cirque lake as an unimportant type, but mention about ten examples which they clearly regard as true rock basins, though in some cases the rock lip is said to be concealed by moraine. Fine examples in County Kerry, Ireland, were described long ago in a classical paper by Close (1871).

In the Central European Alps, it is often difficult to distinguish between pure cirque basins and basins formed wholly or in part by solution. Typical cirque lakes, however, occur in Canton Tessin, and in the eastern Alps where the Wildseelodersee (alt. 1900 m.), Kitzbühler Alpen, Austria, has been the subject of some limnological research. On the Italian side many beautiful examples in the Val d'Ossola have been studied in the course of the important investigations of Tonolli (1947a).

Many cirque lakes certainly exist in the glaciated mountains of Asia, but they have been little studied. A single example, Ororotse Tso (*A* of Plate 3), at an altitude of 5297 m. in Indian Tibet, may be mentioned as one of the highest lakes yet subjected to detailed limnological study (Hutchinson 1937b). Cirque lakes are known in the glaciated parts of Tasmania and the Kosciusko region of New South Wales. In Tasmania they are mainly due to the penultimate or Yolande glaciation; the latest or Margaret glaciation may, however, be responsible for some small tarns (David and Browne 1950). Numerous examples occur in the glaciated mountains of New Zealand. Cotton (1948) has figured a spectacular example, Lake Browne, hanging above the fjord wall of Doubtful Sound.

Multiple cirque lakes (*Type* 27*b*). Strøm (1934, 1935a,b) has called attention to cases in the Norwegian mountains of lake basins formed by the fusion of cirques. These he considers characteristic of areas which, like Scandinavia, had a relatively unfolded and gently modeled preglacial surface, for only in such regions will there be space enough for a group of cirque glaciers to excavate a compound basin. Strøm's best examples from the Norwegian mainland are Bessvatn, in which the cirque basins occur at the head of a long troughlike depression, presumably due to a valley glacier, and Flakevatn, composed of two cirque basins, one of which still contains ice.

While such forms are considered by Strøm not to occur in folded Alpine landscapes, the island of Moskenesøy, Lofoten Islands, already discussed briefly, developed a pseudoalpine landscape as a result of the confluence of cirques. Here Solbjørnvatn has formed as the result of the fusion of three cirque lakes, which remain separated, however, by subaqueous ridges; the deepest basin is 171 m. deep. Other compound cirque lakes on the island, such as Tennesvatn, consist of a pair of basins that have fused so completely as to be hardly distinguishable.

The cirque stairway. As a result of climatic oscillations, and notably of the general deglaciation of the last ten millennia, it is common to find cirques at more than one level. In still glaciated mountains, only the upper cirques will contain ice. The lower cirque, as at Quill Lake, Sutherland Falls, New Zealand (Cotton 1948), may contain a lake while the upper one is still ice bound. In Glacier National Park, Montana,

Grinnel Lake lies in the lower of two cirques, but recent recession has produced a small ice lake against the glacier in the upper cirque. In the same region Lake Ellen Wilson, depth 74 m. (Elrod 1912), in a fully deglaciated upper cirque, lies above Lake Lincoln in the lower cirque. Pairs of cirque lakes are also recorded from the Kosciusko region of New South Wales (David and Browne 1950). Pairs of lake basins of this sort differ from the lake basins of the ordinary glacial stairway, next to be described, in the extreme declivity of the back walls of both upper and lower members. A beautiful and biologically important three-step cirque stairway in the Vallone di Paione, Val d'Ossola, is described by Tonolli (1947a).

Valley rock-basins; paternoster lakes (Type 28). The formation of true cirques depends on the presence of névé and is a phenomenon occurring neither much above nor at all below the permanent snow line, which has lain, however, at different levels at different times. But valley glaciers may descend considerably lower, and by their corrasive powers can produce rock basins. As in the case of the formation of cirques, the mechanisms involved have long been debated; the relevant literature is quoted in the works already referred to. In the upper part of a glaciated valley a series of steps will often be found, constituting the so-called glacial stairway. Each step is bounded by a rock bar, riser, or *Riegel*. Behind the rock bars small basins may be excavated, so that a series of lakes (Type 28a) one below the other is developed. There may often be a true cirque lake at the head of the series, the last step separating this from the other basins being normally much the highest. The main argument about the production of such forms is whether the rock bar inevitably represents a stratum of more resistant material outcropping on the floor of the valley. Alternatively it has been supposed that small irregularities on the preglacial valley floor, or the incidence of tributary or distributary glaciers, or variations in the width of the valley may cause variations in the area and form of the cross section of the glacier and so in its thickness, velocity, and corrasive power at different points along the valley. It seems certain, for instance from Cotton's (1948) account of the glaciated mountains of New Zealand, that valleys can be greatly remodeled by ice action without producing the glacial stairway and its chain of lakes. It also seems probable that variation in the resistance of the rock floor along the valley is not the only factor determining the presence of such features when they occur. Von Engeln (1933) for instance thinks that any variation in declivity in the original preglacial stream may become the site of a step above which a basin might be excavated. It is evident that in mountains built of sedimentary rock the relation of the dip of the strata to the angle of inclination of the valley floor is important in determining the presence of rock bars, which consequently may be well marked on one face of a mountain range and absent on the other.

In Europe small lakes in a chain in a glaciated valley are often spoken of as *paternoster lakes*, from a fancied resemblance to the large beads of a rosary.

Cotton figures an impressive group of three lakes on a glacial stairway at the top of a troughlike valley on Baranoff Island in Alaska. Innumerable examples must occur in the cordilleran chains of North America. As particularly striking examples, the series of six or seven members found in Glacier National Park, Montana, may be mentioned (Fig. 22). Equally good examples could be found in the Bighorn Mountains of Wyoming, on the eastern side of the Continental Divide in Colorado (Hobbs 1912), or in the High Sierras (Von Engeln 1933). Even more dramatic series of lakes in glacial stairways formerly existed immediately after deglaciation, in certain valleys in which the basins have now been filled with silt, one of the finest of such series of extinct glacial stairway lakes being the old lake of the Yosemite Valley, with several lakes in the tributary valleys above it. European examples are abundant, particularly in the still glaciated Alps, and a few cases can be identified in the fully deglaciated mountains of Britain, notably in Wales.

The same pattern, naturally, is found in other glaciated mountains. In the region north of the Indus in Indian Tibet, Yaye Tso (*B* of Plate 3) behind a well defined rock bar, with an extinct lake above it represented by an alluvial plane, provides an example that has been studied limnologically in some detail (Hutchinson 1937b).

A curious and biologically interesting group of glacial rock basins is found on the central plateau of Tasmania, at altitudes around 1000 m. These lakes are all shallow, the largest of them, the Great Lake of Tasmania, had, prior to the construction of a dam, an area of 113.5 km.^{2} but a depth of only 6 m. Over the greater part of the plateau (Lewis 1933, Voisey 1949a,b, Fairbridge 1949), Jurassic dolerite is exposed, but in places this is covered by a much softer Tertiary basalt, now largely eroded away. Almost the whole of the lake district appears to have supported a thin icecap during the maximum Pleistocene glaciation. Lewis (1933) regards the lakes, which in general seem to occupy rock basins, as the sites where the ice remained longest during deglaciation. He supposes that they were hollowed by nivation, the last ice remnants behaving as cirque glaciers. He thinks, however, that the rock lips may have been raised by earth movements. The glacial part of this hypothesis is improbable, particularly during deglaciation.

More modern opinion (Voisey 1949a,b, Fairbridge 1949) inclines to the view that the lakes, except perhaps Lake Echo on the southern margin of

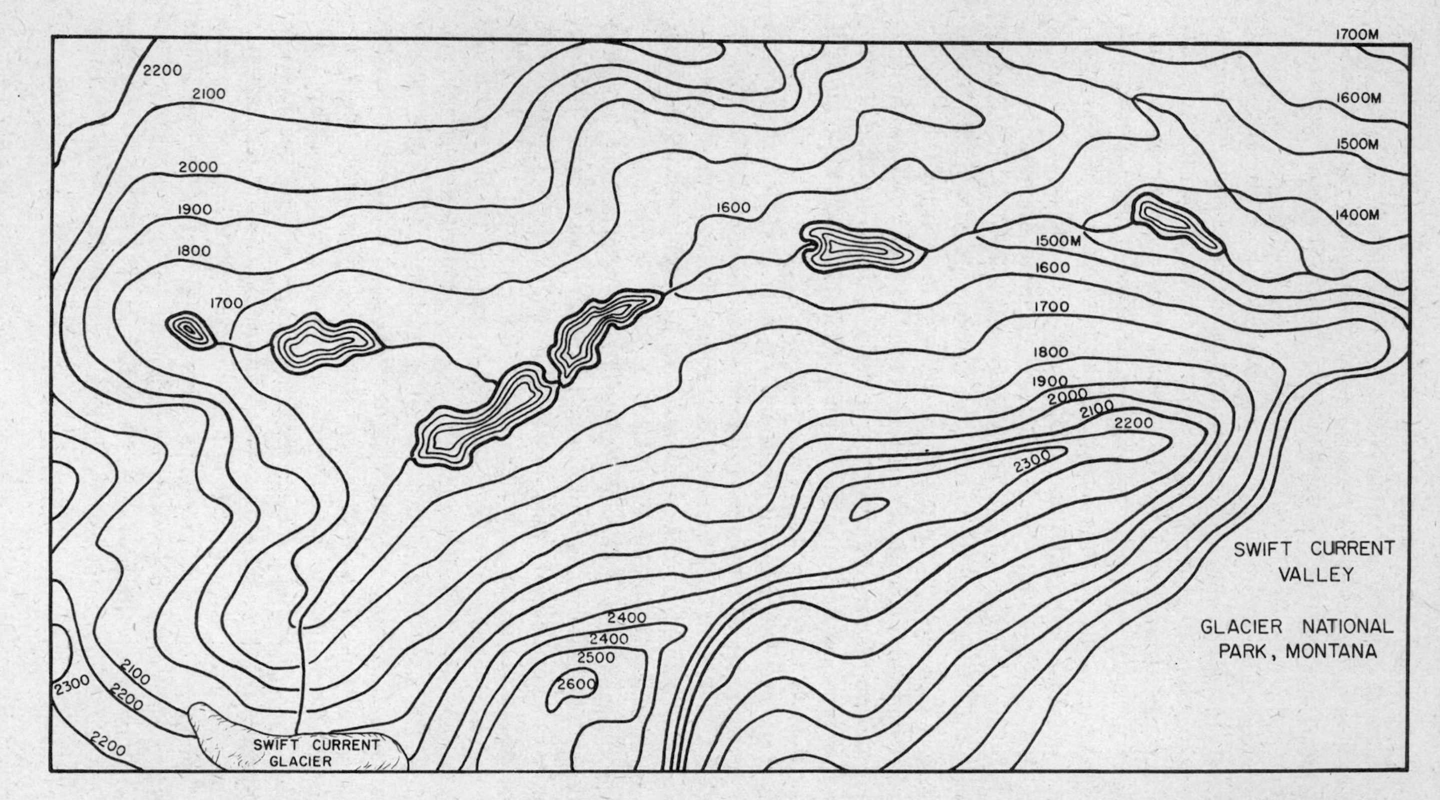

FIGURE 22. Swift Current Valley, Glacier National Park, Montana. Glacial stairway starting with a cirque lake, followed by a series of "paternoster lakes." (After Campbell.)

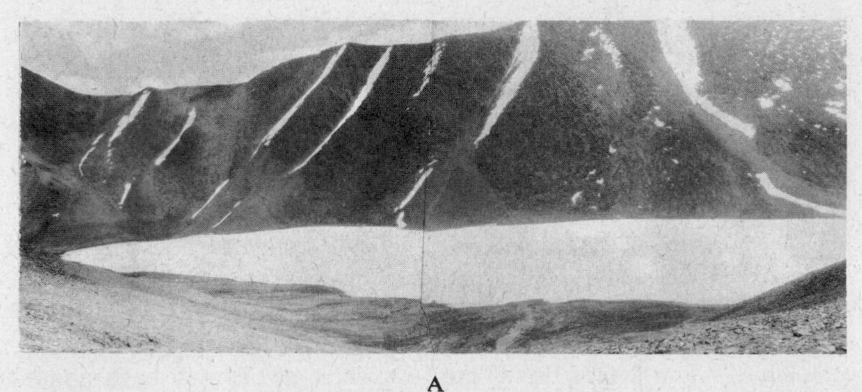

A

B

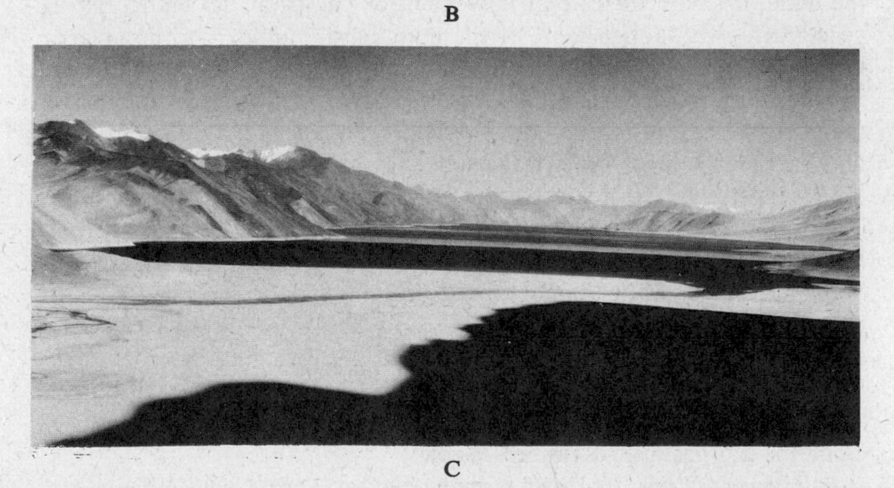

C

PLATE 3. *A*, Ororotse Tso, Indian Tibet, altitude 5297 m., July 11, 1932; a cirque lake, probably with a cold monomictic temperature regime. *B*, Yaye Tso, Indian Tibet, a small glacial valley lake. *C*, Pang-gong Tso, Indian Tibet, a large lake in a glacially remodeled graben. (*Phot*. G. E. H.)

the plateau, are primarily due to glacial corrasion by ice in shallow preglacial valleys. This process removed the softer Tertiary basalt. The Great Lake in its natural state was held by a dolerite ridge which had resisted ice action more successfully than the basalt and formed a rock lip over which the effluent discharged (Voisey 1949a). The majority of the Tasmanian lakes therefore appear to have been produced in the same way as the valley lakes of more rugged glaciated mountains. Lake Echo, however, may be of tectonic origin. The whole area obviously requires further intensive study, which should include investigation of the lake sediments. From a zoogeographic point of view it is very curious that the Great Lake, which is the only known locality for the archaic crustacean *Paranaspides*, and indeed that almost the whole of the plateau inhabited by the related but also endemic genus *Anaspides*, should have been ice bound and totally unsuitable for such forms during the Pleistocene glaciation.

Most of the lakes that have been discussed here tend to be small. Descending the glacial stairway, if such be present, they become larger and more elongate. A number of different types of large glacial lakes are known which are fundamentally rock basins but may have some moraine contributing to the damming of the lower end. It is convenient to discuss these in certain rather loose categories.

Glint lakes (Type 29).[14] Where, due to the preglacial topography, there is a tendency for ice to accumulate in large depressions or valleys of which the outlet is small, the rate of movement of the ice at the outlet may be very great, even though most of the mass hardly appears to be moving. Large rock basins are likely to develop in such places. The Scandinavian ice, moving westward from the Baltic over Swedish Lapland, encountered the mountain axis of Norway and forced its way through a series of passes, now marked by large lakes formed in this way. These lakes are sometimes termed *glint lakes* (*glint* being Norwegian for "boundary"). They are not necessarily pure rock basins; the largest, Torneträsk, is partly held by morainic dams. A number of very striking examples are known in Scotland, where local piling up of ice in certain wide valleys that served as reservoirs or "ice cauldrons" (Peach and Horne 1910) led to a radial escape of glaciers through pre-existing passes that were overdeepened. The best examples of lakes formed by such glaciers are Loch Rannoch, Loch Ericht, Loch Ossian, and Loch Treig, which radiate from the ice cauldron that occupied Rannoch Moor.

Fjord lakes (Type 28b) and piedmont lakes (Type 28c). Where large glaciers occupied long valleys at relatively low elevations, very impressive

[14] This type, being montane, is more conveniently considered prior to the piedmont lakes of Type 28c; it may perhaps not be worthy of special designation.

rock basins have been excavated. Often, as in the case of Loch Ness (maximum depth 230 m.) in a cryptodepression 214 m. deep, the position of the lake may be structurally determined, in this case by the Great Glen Fault. In other cases, lithology rather than faulting has determined the position of the basin. Lakes of this sort, often associated with extremely indented coastlines, are appropriately called *fjord lakes.* They are usually deep, and since they lie in U-shaped glacially modeled valleys, they are often steep-sided. When they lie near the coast, a small rise in sea level would convert them into arms of the sea. A small drop of sea level would often convert sea fjords, which frequently are bounded at their outer ends by rock bars, into long deep lakes. Such lakes are found in glaciated mountains in various parts of the world; fine examples are known in Scotland (Peach and Horne, 1910) and others in New Zealand (Keith Lucas 1904, Cotton 1948). Their most notable development, however, is in Norway, where the term fjord is applied to both salt-water and fresh-water basins of this sort. It will be appropriate to discuss in some detail certain very fine Norwegian examples, and then to consider certain less typical fjord lakes in North America.

The fjord lakes of Type 28b are, from one point of view, only a special subclass of the large class of *piedmont lakes* formed as a result of the activities of large glaciers that have descended to a low level from glaciated mountains. The more usual piedmont lakes (Type 28c) such as those of the Alps, usually appear to be held by dams of drift. However, in the progress of modern investigation such lakes appear more and more frequently as rock basins slightly raised in level by morainic dams.

The fjord lakes of Norway. The narrow fjordlike lakes of Norway provide the most magnificent examples of elongate, greatly overdeepened, glacially excavated lakes. A number of examples exist in the interior of the country, but from a purely limnological point of view none are more impressive than the Nordfjord lakes, admirably described by Strøm (1933a), that lie near the west coast of central Norway.

Nordfjord (Fig. 23) is a typical branching sea fjord, lying about lat. 61°40′ N. on the west coast of Norway. Three main and two subsidiary arms of this fjord run inland and at the heads of each of the main branches a lake is situated. The largest of these lakes, about 6.5 km. inland from the northern branch of Nordfjord, is Hornindalsvatn, the deepest lake (514 m.) in Europe and the ninth deepest lake of the world. The central branch of the fjord receives streams from three lakes entering near its head, namely Strynsvatn, Loenvatn, and Oldenvatn, while Breimsvatn drains into the southern branch.

The existing floors of all, at their deepest points, are below sea level. Hornindalsvatn, with a deepest point 461 m. below sea level, is the third

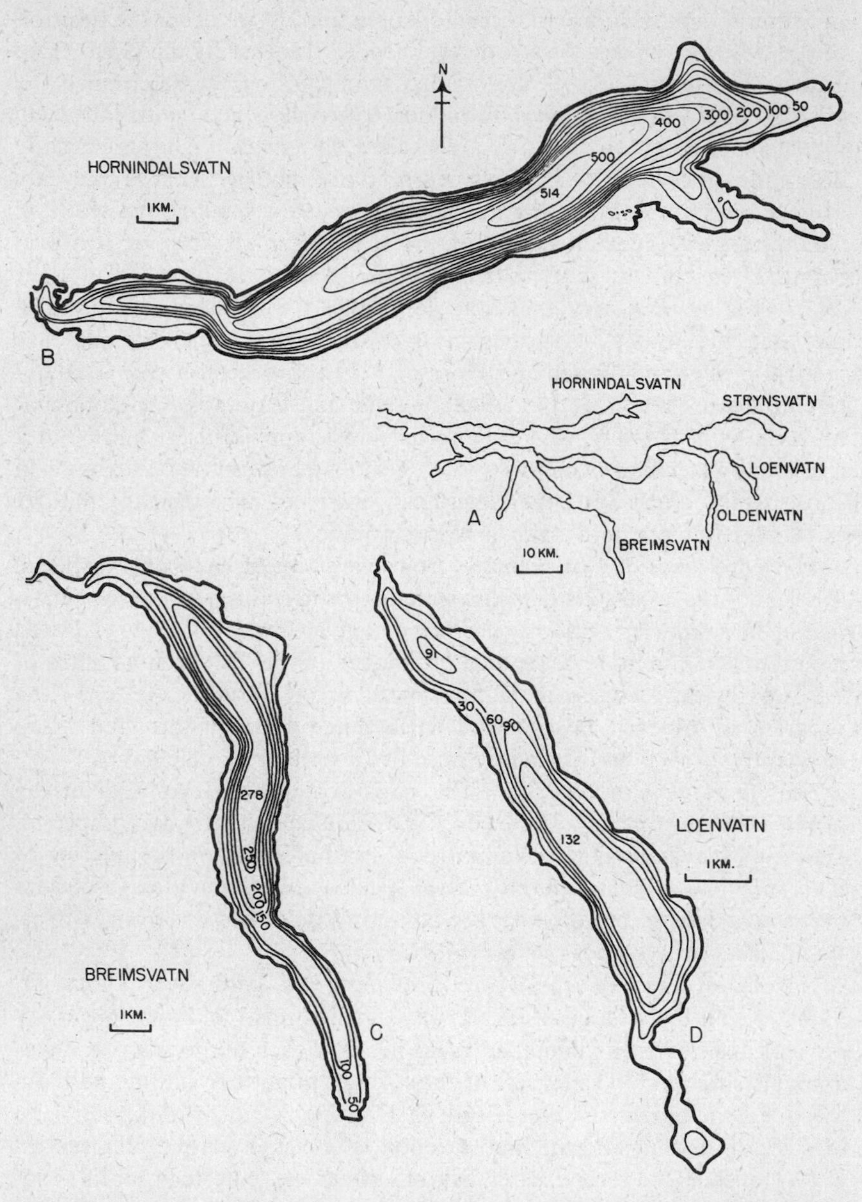

FIGURE 23. The Nordfjord lakes. *A*, general disposition of the lakes at the head of Nordfjord; *B*, Hornindalsvatn, the deepest lake in Europe; *C*, Breimsvatn; *D*, Loenvatn.

deepest cryptodepression known. The surfaces of the lakes lie at altitudes of 26.8 m. (Strynsvatn) to 61 m. (Breimsvatn). The five lakes differ characteristically in morphology. All are primarily rock basins, but only in the case of Loenvatn is there no trace of morainic damming. In all the other lakes the level is apparently retained a little above that of the rock lip by moraine. In the case of Hornindalsvatn, in which the channel of the effluent is deep, Strøm seems to imply that bedrock in that channel lies at 12 m. below lake surface, so that 97.7 per cent of the maximum depth is below the supposed rock lip. The general lie of the lake basins is presumably determined by preglacial drainage, though Strøm says that the ice has worked on the landscape so effectively that it is difficult to reconstruct any details of such preglacial hydrography.

In two cases, Loenvatn and Oldenvatn, the basins are divided by transverse ridges, presumably *Riegeln* or rock bars, analogous to the bars separating the lakes from the branches of Nordfjord itself. There are also several resistant rocky headlands or nesses projecting into the lakes; in such cases the contours indicate the movement of the ice around such resistant objects. Several of the lakes were excavated by confluent glaciers, just as Nordfjord was excavated by the confluence of the glaciers of the valleys now holding lakes. The existence and arrangement of such confluences has a considerable influence on the form of the basin. Three glaciers converged on the upper end of Hornindalsvatn and two on the upper end of Strynsvatn. In both cases the deepest area is well upstream from the mid-point of the lake. In the case of Breimsvatn, the main glacier was reinforced by a tributary about one third of the way downstream from the present head of the lake. Here the deep water is in the lower two thirds of the basin, and the floor of the upper third appears essentially as a hanging valley relative to the deep water.

The profiles of the lakes vary considerably. Strøm thinks that where the longitudinal profile as well as the transverse is very flat centrally, the basin has probably been filled with a considerable layer of sediment, while in the cases where the longitudinal profile descends regularly to the deepest point, as in Hornindalsvatn, the sedimentary filling is negligible.

All of the large Norwegian lakes, such as Mjøsa, Tyrifjord, and Eikeren, owe their existence mainly to the glacial overdeepening of valleys. Kolden and Strøm (1939) list sixteen Norwegian lakes known to be over 200 m. deep, all of which may be regarded as fjord lakes, and of which no less than twelve lie in cryptodepressions. While it is evident that most of the depth of such lakes is due to the corrasive action of glaciers, only a limited number, such as Loenvatn or Fyresvatn, are pure rock basins. In most, the water level is raised somewhat by morainic damming, or very rarely, as in the case of Eikesdalvatn (Strøm 1937), a lake 155 m. deep in a crypto-depression, by loose deposits, which, from Strøm's account, may be glacial outwash.

In the case of Tyrifjord, Strøm (1940) has shown how the form of the basin is determined largely by the relative strength of the Pre-Cambrian gneiss, Permian plutonics, and to a less extent Permian lavas, which surround the basin, and the relative weakness of various early Palaeozoic sediments into which the basin has been excavated.[15] Faulting plays a very minor part, if any, in determining the orientation of the basin.

Fjord lakes west of the Rocky Mountains. The most remarkable fjordlike lakes of North America are the group in British Columbia, of which the best known, in the southern part of the Province, are Kootenay, Upper and Lower Arrow, Okanagan, and Shuswap. Of these, only Okanagan has been adequately sounded. The history of the lake has been briefly considered by Schofield (1943). The valley of the Okanagan River is of early Tertiary origin. It has been greatly remodeled by glaciation and now contains several lakes, of which the largest is 108 km. long but averages only 3.4 km. wide. The lake surface lies at an altitude of 345 m., and though the basin is not a cryptodepression, the maximum depth is quite considerable, 232 m.

During the late Wisconsin, a large lake occupied even more of the valley than the present Lake Okanagan, and discharged northward into a river confluent with a similar lake in the Thompson Valley. These lakes may have been held by stagnant ice or may have been analogous to the Shyok ice lake. Schofield thinks that after the glacial lakes discharged, the valley was rejuvenated by uplift, and that the dam that holds the modern lake, postdating this uplift, must be quite recent. According to Schofield this dam was formed as the alluvial cone of a lateral stream. Flint (personal communication) indicates that such cones in this region are largely composed of glacial outwash, and that they often are pitted with kettles, indicating inclusion of ice. Such outwash cones probably collected over stagnant ice in the main valley. As well as tending to dam the valley, they deflected the effluent streams towards the valley wall so that they appear to be flowing over the lips of rock basins (cf. pages 44 and 61). It is, however, reasonably certain that outwash and fluviatile damming can account for only a small part of the depth of Lake Okanagan. Its considerable profundity, together with the fact that the deepest point in the lake is far to the north, little south of the region in which the two northern arms might have been expected to produce convergence and piling up of ice, make it practically certain that Okanagan is fundamentally a rock basin, not unlike the Norwegian fjord lakes. The other large British

[15] Cf. Brochu (1954) for Canadian examples of lake basins formed by differential corrasion; cf. also Zumberge (1955).

Columbian lakes doubtlessly have a like complex origin, in part dammed, in part excavated by ice. The history of Lake Kootenay has been further complicated by the formation of a lateral outlet. Originally the lake was dammed by moraine (Schofield 1946) at the southern end. As the ice retreated, a glacial lake collected between the moraine and the ice. When the present basin was about half free of ice, a lateral effluent became available. There are now two main influents, one at the south, which was formerly the effluent of the early stages of the ice lake, and another at the north end, representing the original drainage established in the Cretaceous and Eocene.

The English Lake District. The English Lake District, situated in a roughly circular area about 50 km. in diameter with its center in lat. 54°29′ N., long. 3°03′ W., in the counties of Cumberland, Westmorland, and Lancashire (Furness), constitutes a group of lakes of great beauty, which for their earlier literary and more recent scientific associations are second to none in any part of the world. It is of some interest to note that the remarkable radial arrangement of the hydrography of the region, "like spokes from the nave of a wheel," was described by Wordsworth, the most notable poet associated with the region, in his initially anonymous ([Wordsworth] 1810, Wordsworth 1820) introduction to the Rev. Joseph Wilkinson's *Views of the Lakes*.

The Lake District consists of an elevated dome of Ordovician and Silurian rocks on which Carboniferous and later Triassic rocks were deposited unconformably. The doming of the region is supposed to have taken place in the mid-Tertiary, when the drainage system noted by Wordsworth and all subsequent geomorphologists interested in the region was developed. Practically the whole of the Carboniferous and Triassic and all such later rocks as may have been deposited have now been eroded from the dome, and the present hydrography, due to the uplift, is superimposed on a landscape composed of ancient and much folded rocks (Marr 1916, Trueman 1938).

Nine principal valleys containing large lake basins radiate from the center of the district (Fig. 24), namely, from the north *cum sole*: (*A*), the Derwent Basin, containing Derwentwater and Bassenthwaite Lake, originally a single lake divided by the delta of the River Greta; (*B*), the Vale of St. John or Thirlmere Basin, containing the lake of that name; (*C*), the Eamont Basin, containing Ullswater; (*D*), the Hawes Water Basin, containing the lake of that name; (*E*) the Windermere Basin, containing that lake (*A* of Plate 4) and above it Rydal Water and Grasmere, and with the subsidiary Cunsey Beck or Esthwaite Water Basin draining into Windermere on the west; (*F*), the Coniston Basin, containing Coniston Water; (*G*) Wasdale, containing Wastwater; (*H*), Ennerdale, containing Ennerdale

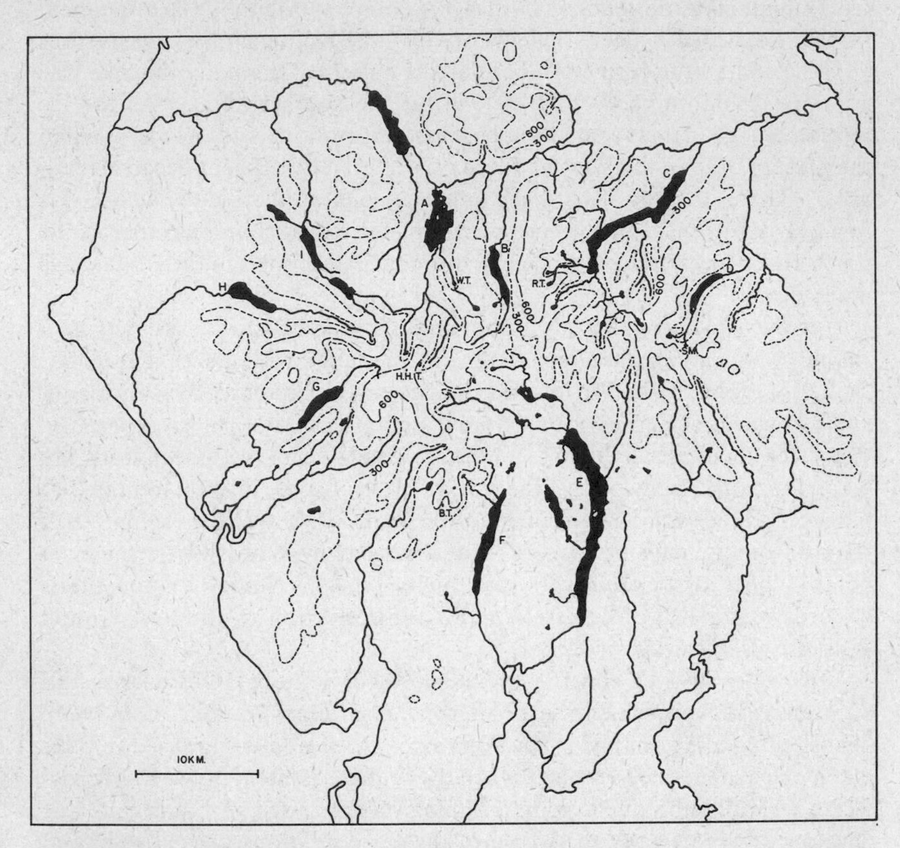

FIGURE 24. The English Lake District: A, Derwent Basin; B, Thirlmere Basin; C, Ullswater Basin; D, Hawes Water Basin; E. Windermere Basin, with Esthwaite Water to the west; F, Coniston Basin, G. Wasdale; H, Ennerdale, I, Lorton Vale, WT, Watendlath Tarn; RT, Red Tarn; HHT, Highhouse Tarn; BT, Blind Tarn; SM, Smallwater.

Water: and (*I*), Lorton Vale, containing Buttermere and Crummock Water, divided by a delta and receiving the drainage from Loweswater to the northwest. In addition to the lakes enumerated, some of which will play an important part in later chapters, there are a number of small lakes or tarns, including the beautiful examples of cirque lakes already mentioned.

Of the large lakes, Wastwater, Thirlmere, and Hawes Water are situated on Ordovician volcanic rocks (Borrowdale volcanics) and are steep-sided; Wastwater, with its magnificent screes and hanging valleys, can be regarded as a fjord lake. The others are mainly situated on the slightly older and

A

B

PLATE 4. *A*, Windermere, north basin, a glacially excavated lake of the subalpine type, with a warm monomictic temperature regime, photographed in a late winter (March) landscape (*Fox-Photos*, through the *Lady* magazine, London). *B*, Linsley Pond, Connecticut, a kettle in an outwash-filled valley (*phot*. G. E. H.).

less resistant Skiddaw Slates and occupy wider U-shaped valleys in which considerable quantities of morainic material and deltaic deposits have been laid down, giving areas of settled farm land which contrast picturesquely with the predominantly montane landscape. This difference between the two types is admirably indicated in the engravings reproduced in the frontispiece, from Harriet Martineau's *A Complete Guide to the English Lakes*, and will be found to be of fundamental importance in several later chapters of the present work, particularly in Volume II.

As in the case of other subalpine lakes, the basins of the English Lake District have given rise to controversy between those regarding them as due to morainic damming and those supposing that the main agency in their production has been glacial excavation. In the case of such a lake as Wastwater, which clearly occupies a rock basin, and a cryptodepression at that, there can be no doubt that action other than morainic damming has been involved. Marr, who was one of the strongest proponents of the morainic damming hypothesis, in his earlier work (1897) considered that in such cases tectonic agencies rather than glacial overdeepening were involved. Later (1916), in his definitive book on the region, he was prepared to admit the existence of glacially excavated rock basins. Recent investigators tend increasingly towards the view that glacial excavation has been of major importance in producing the principal lakes. In the case of Windermere, the south end seems to lie against a typical piedmont "apron" of drift, which appears to have diverted the drainage of the lake to the west instead of following the Cartmel Valley due south. Yet recent study by seismic methods (Coster and Gerrard 1947) has shown that the supposed dam of drift is actually very shallow, and that it covers a lip of rock which would retain the lake at its present level if the drift were removed. The anomalous drainage to the west can be explained by any one of three possible causes. The first is stream piracy; the second is the possible former existence of a dam of stagnant ice at the south end of the lake, which may have permitted a lateral overflow to cut the modern outlet before the basin was fully deglaciated; the third is the possibility that the glacier that filled the basin had a western distributary which cut the present outlet. According to the third hypothesis, the western end of Windermere suggests glacial diffluence, providing a small-scale model of the kind of process that occurred on a great scale in Lake Constance or the Lago di Como. The existing information does not permit a satisfactory choice between these possibilities, though by analogy with conditions in the Alps, the third is perhaps the most reasonable.

The lakes of the European Alps. The lakes of the European Alps form a group, more impressive though less regular, but in general comparable to those of the English Lakes. The Alpine massif, running east and west rather than forming a single round dome, divides the lakes into a rather definite northern and southern series, while the greater height and youth of the mountains, and their less resistant nature, has permitted the development of certain special morphological features in the lakes of the Alps which are hardly indicated in the English Lakes.

The great Swiss lakes have naturally been regarded as typical of piedmont or subalpine lakes, and as such their origin has been a matter of great interest and even more controversy. All the major lakes lie in rock basins, the direction of which is no doubt primarily of preglacial and structural origin. Two main theories as to their origin have been held.

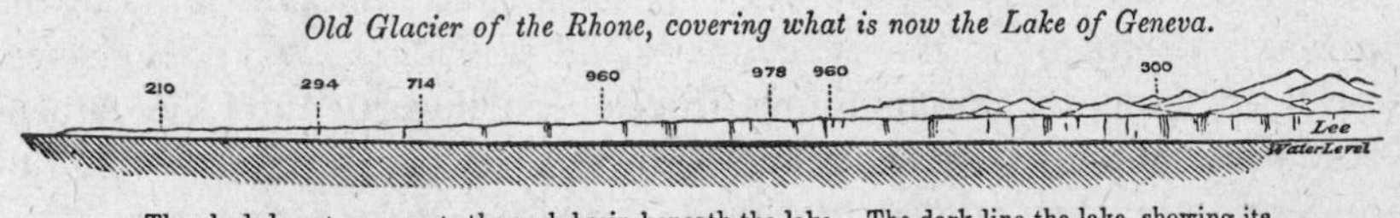

The shaded part represents the rock-basin beneath the lake. The dark line the lake, showing its depths on a true scale. The light part above represents the Old Glacier of the Rhone. Figures, depth of the lake in feet.

FIGURE 25. Profile of Lake Geneva, vertical and horizontal scales identical (after Ramsay 1862).

Lyell early suggested and Heim (1894), among Swiss geologists, strongly promulgated the idea that the basins were of tectonic origin, the warping of river valleys having depressed the inner ends and elevated the outer. The second view, that of glacial corrasion, was put forward by Ramsay in 1862 and was vigorously championed by Penck (1905, Penck and Brückner 1909). The main objection to Ramsay's idea seems ultimately to have been based on the general tendency of a ground-living organism such as man to exaggerate vertical distances, so that the great depths of the rock basins appeared to put an impossible demand on the excavating agency.[16] The modern reader is more likely to have experience of vertical distance, but it has seemed worth while to reproduce Ramsay's original figure (Fig. 25) reconstructing the excavation of Lake Geneva to indicate the very small depth, relative to the length of the lake, that has been excavated. Today, glacial excavation is almost universally accepted as the origin of the

---

[16] Dr. Paul Shepard tells me that he has found, from actual measurements of mid-nineteenth-century landscapes and of photographs of the scenes they depicted, that even in that period of rigorously representational painting when painters painted "what they saw," a more or less unconscious distortion of vertical to horizontal scales, in a ratio of 4:1, is by no means uncommon. An analysis of some of the optical phenomena involved is given by Cornish (1935).

basins. As Wallace (1893) pointed out, the hypothesis of warping implies the coincident warping of lateral valleys, which should provide a series of lakes with bays or subsidiary basins entering, particularly in the lower part of the lake, at a normal acute angle with the upper part of the lake axis above the bay. No such pattern is observed, and in so far as irregularities in the basin do occur, these are of an entirely different nature owing to glacial diffluence.

As well as glacial corrasion, all the main lakes have morainic dams and outwash at their lower ends, but where the thickness of this is known, it is inadequate to account for the great depths of the basin. The morainic thickness is greater and its area is of more limited extent on the southern than on the northern side of the Alps.

The most interesting feature of the basins of the subalpine lakes of the European continent is the effect of glacial diffluence. As the glaciers descended from the mountainous interior northward onto the Alpenvorland of southern Germany, or southward onto the Lombard Plain of northern Italy, they tended to diverge fanwise. Lake Constance in the Valley of the Rhine Glacier exhibits the effect of this diffluence at its northwestern lower end, which bifurcates to form the Überlingen Basin to the northeast and the more or less independent Untersee to the southwest. There is a clear hint of a similar effect of diffluence in the form of Lago di Garda, among the southern series while Lago di Como illustrates the phenomenon in its purest form on an extraordinary scale (Fig. 26). The complicated form of Lago di Lugano again in part is determined by glacial diffluence. Where a valley receives a number of glaciers from peaks from along the sides, it may become sufficiently filled with ice to spill over any pre-existing low passes at its head. When this happens, a considerable erosion of such passes can occur; if, at the same time, a lake basin is formed in the valley and is dammed with a considerable thickness of moraine, on deglaciation the lake may be left with reversed drainage, over the pass that lay at the head of the valley. This has happened in the case of Lago di Orta, where in common with the other Italian lakes the moraine is thick.

The large lakes of subarctic Canada. Certain large lakes, notably Lake Athabasca, Great Slave Lake, and Great Bear Lake, in northwestern Canada just south of or on the Arctic Circle, are apparently analogous to piedmont lakes excavated by ice, but differ from the examples just described in the lowness of the mountains around their upper ends. All were covered during their formation by a very large continental ice sheet, yet it seems probable that their existence as lakes is largely due to movement of parts of this ice sheet through pre-existing valleys which were greatly overdeepened (Type 28d).

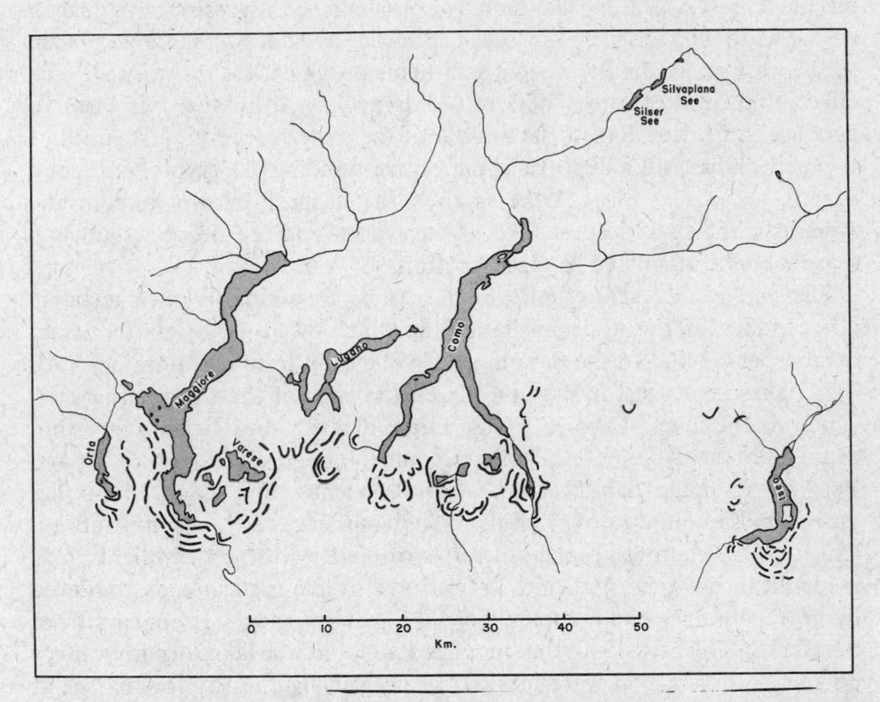

FIGURE 26. The Italian subalpine lakes, showing festoons of moraines at their southern ends and effects of glacial diffluence, notably in the Como Basin.

Great Slave Lake (Rawson 1950), which is the largest, deepest, and best known of the lakes, lies at about 62° N., 114° W., at the junction of the Mackenzie lowlands, a drift-covered area of Devonian, Silurian, and Ordovician sediments and the Archaean rocks of the Canadian shield. A pre-existing valley, which may have been determined structurally, undoubtedly directed the movement of ice in this region; further south the same is true of the Athabasca Valley. During deglaciation the ice margin formed lobes in these valleys, against which a well-defined series of ice lakes was developed. On complete deglaciation the Great Slave Lake Basin appeared as a complicated lake, which can be divided into two parts. The western wide, shallow part lies mainly on Palaeozoic sediments and, though reaching a maximum depth of 163 m., is for the most part less than 100 m. deep. The eastern part is elongate in a northeast-southwest direction and has numerous islands and peninsulas trending parallel to its shores. The largest of these peninsulas represent a huge diabase sill. It is reasonably certain that ice moving into the original valleys on the shield has immensely overdeepened them by removing the least resistant Archaean rocks, and then tended to fan out, producing a wider but shallower basin to the west where the Palaeozoic sediments provided less resistance. The effect of their excavation, determined largely by lithology, has been to produce in Christie Bay in the middle of the eastern section, just south of the main diabase sill, a depth of 614 m., corresponding to a cryptodepression of 452 m. Great Slave Lake is thus the deepest known lake in the Americas, the sixth deepest lake in the world, and the deepest that may reasonably be attributed to glacial action.

The large glacial lakes of Patagonia. There are a number of remarkable lakes situated in the Andes of Patagonia which owe their origin to glacial overdeepening and to the damming of valleys by terminal moraine. All these lakes lie at least in part on the eastern side of the Cordilleran axis. Three of the lakes, Lago la Plata, Lago Viedma, and Lagó Argentino, drain into the Atlantic; the others, namely Lago Buenos Aires, Lago Pueyrredón, Lago San Martin, Lago Sarmiento, and Lago Maravilla, drain into the complicated fjordlike system of the western coast of southern Chile. The relationship of both series of lakes with the Cordilleran axis is identical, however, and both series have simple eastern ends bordered by great moraines and complex fjordlike western ends. It appears from the work of Quensel (1910) that in all probability these lakes occupy valleys on the Cordillera that were marked in preglacial times by low passes at their upper ends. The valleys to the east of these passes have been overdeepened by glaciers moving eastward, which built immense moraines. The existing moraines, of which Caldenius (1932) identifies four, are presumably of late glacial origin. When deglaciation occurred, lakes were formed between the moraine and the retreating ice. Evidence of terraces and lacustrine sediments, according to Caldenius, indicates that these ice lakes drained into the original eastward-directed valleys of the Patagonian pampa. The effluents of the lakes that still drain to the Atlantic evidently cut down into the moraine as the ice receded, but in the case of the other lakes, deglaciation put the lake in communication with fjordlike valleys, cut at the heads of the glaciers down to depths below the moraine dam, so that when the ice had fully retreated, spillways to the west were established. These lakes thus repeat on a grand scale the type of phenomemon that produced the reversed drainage of Lago di Orta in Italy.

The Great Lakes of the St. Lawrence drainage. The Laurentian Great Lakes, taken together, constitute the largest continuous volume of liquid fresh water on the earth. The total volume, 24,620 km.^{3}, slightly exceeds that of Lake Baikal, though of course the Caspian Sea, which is moderately saline, is considerably larger. In area, Lake Superior, extending over

83,300 km.[2], itself exceeds all lakes save the Caspian. The origin of the immense Laurentian basins has naturally given rise to much discussion.

The basins of the Great Lakes lie in a region that in Tertiary times apparently drained towards the headwaters of the Mississippi. It has been supposed by some authors, notably Spencer (1891), that during part of the Tertiary the elevation of the North American continent was sufficiently great to permit cutting of the valleys to the present depths of the lake basins, which were closed by subsequent warping. Other authors have believed the lakes to occupy river-cut valleys of like origin, dammed by drift. Such hypotheses have been examined in the light of the bathymetry of the lakes by Shepard (1937) and by Thwaites (1947), both of whom believe that the only reasonable explanation of the basins in their present form is glacial erosion, so that the lakes may be referred to Type 28d. There is no evidence of extensive drift filling of the valleys toward the Mississippi drainage, and the rock floors of the valleys that drain southeast are well above the level of the bottoms of the lakes, which lie in true rock basins. There is, moreover, no clear independent evidence of the tectonic origin of such rock basins. Some subsidence due to solution of salt beds of the Salina formation may have occurred, but Thwaites considers that the main contribution of this process has been to brecciate, and so weaken, certain younger formations, and therefore to make them more easily excavated by ice.

Both Shepard and Thwaites emphasize the extremely irregular topography of the bottoms of Lake Michigan and Lake Superior, for which lakes good detailed bathymetric maps have been published. Thwaites shows, moreover, that it is possible to trace known escarpments, determined by the relative strengths of the various formations of the region, in the details of the sublacustrine topography of Lake Michigan. The fact that such a correlation appears possible strongly supports the view that ice action is largely responsible for the basin. There is evidently some drift deposited on the floors of the lake basins, and certain deep holes may be due to the incorporation of stagnant ice in this drift, such holes being in fact sublacustrine kettles.

During the glacial maxima the whole of the Great Lakes area was ice covered. The history of the modern lakes begins with the retreat of the Wisconsin ice sheet. Three major phenomena determined this history. The first was the development of ice lakes against the retreating ice fronts in the basins, which in general contained lobes of the main ice sheet. The second was the uncovering of pre-existing spillways at different levels, which controlled the levels of the glacial lakes and their direction of discharge. The third important phenomenon was the isostatic readjustment of the crust as the load of ice was removed. Since the ice was thickest to the north, the rebound was greater on the northern than on the southern sides of the basins, and northern spillways that initially lay at a low level could rise, when uncovered later, above the level of older more southern outlets. Late in the history of the lakes, levels were, of course, adjusted by the ordinary downcutting of the effluents. The most important work on the history of the basins is the magnificent monograph of Leverett and Taylor (1915). More recent work has been summarized by Flint (1947) and by Hough (1953); important papers by Bretz (1951), Hough (1955), and Zumberge and Potzger (1955) have considered specifically the events in the Lake Michigan Basin.

While the general outline of the history of the Great Lakes is now clear, there are still a number of small details that require elucidation. Radiocarbon dating is proving a very powerful adjunct to the more conventional methods of investigating these details, and a relatively complete history is likely to be available in the immediate future. For the present it has seemed best merely to present maps (Fig. 27), based primarily on Hough (1953), demonstrating the more important stages in development, deferring a detailed account of the history to Volume II of the present work, in which the subject becomes relevant to the problem of lacustrine endemism. In studying these maps it should be noted that Lake Erie has had an apparently continuous history from the time of the earliest phase of the proglacial Lake Maumee, late in the Cary substage of the Wisconsin glaciation, whereas in the Michigan Basin the earliest phase of Lake Chicago was completely obliterated by the advancing ice that produced the Tinley Moraine subsequent to the first stage of Fig. 27, and was again almost obliterated by an advance at the end of the second or Glenwood stage of the same figure. Marine transgression has affected only the Ontario Basin, perhaps during the Two Creeks interstadial, certainly at the time of Lake Stanley and Lake Chippewa in the fourth millennium B.C. The enormous Lake Nipissing in the third millennium B.C. provides an extraordinary case of a large lake with three effluents.

MORAINIC AND OUTWASH DAMS

The terminal or recessional moraines of valley glaciers may persist, in certain circumstances, in a sufficiently well-preserved state that they can dam the stream that replaces the glacier (Type 30).

Apart from certain rather obvious conditions, notably that a morainic dam is unlikely to be stable in a steep valley, it is probable that the accumulation of stagnant ice in the future lake basin, behind the moraine, is of great importance in preventing the basin from filling with outwash. Small but spectacular lakes (*B* of Plate 2) may form behind terminal

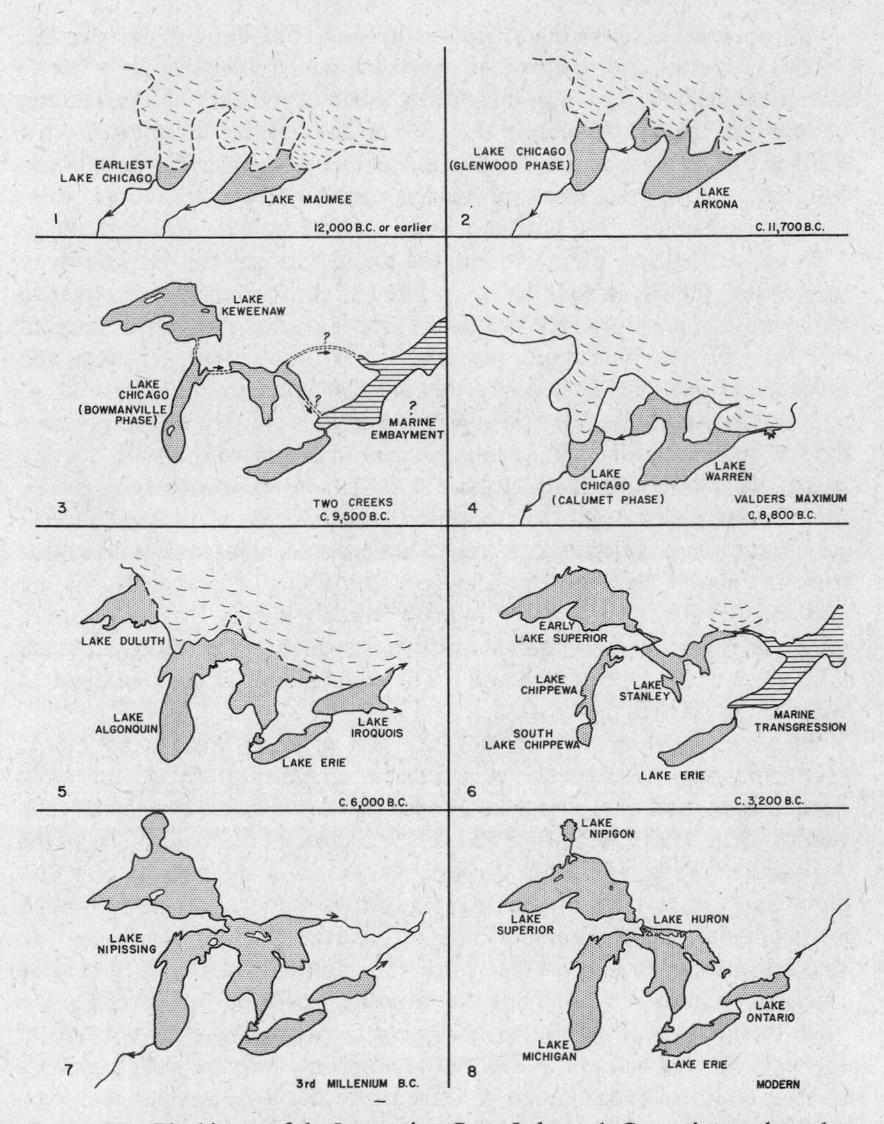

FIGURE 27. The history of the Laurentian Great Lakes. 1, Cary substage, later than Valparaiso moraine; 2, late Cary, about 11,700 B.C.; 3, Cary-Valders interstadial, about 9500 B.C., low water in eastern basins, marine transgression in Ontario Basin, limits of ice uncertain; 4, Valders maximum, about 8800 B.C.; 5, post-Mankato retreat, about 6000 B.C.; 6, postglacial thermal maximum, about 3200 B.C.; 7, Lake Nipissing, third millennium B.C., with triple drainage; 8, modern lakes. (Modified from Flint, Hough, and other sources, mainly by Jane VanZ. Brower.)

moraines solely as the result of recession, but they are usually short-lived.[17]

In some cases outwash deposited beyond and at the sides of dead ice in a valley, as the result of melting of either the main valley glacier or ice in lateral tributary valleys, may constitute a dam (Type 31). An interesting comparison is available among the long piedmont lakes draining into the Flathead River west of the Continental Divide in Glacier National Park, Montana. Here two series of basins can be recognized. The more elevated, exemplified by Bowman Lake at 1225 m. and Quartz Lake at 1338 m., are held by obvious morainic dams, though it is not known to what extent they lie in rock basins. The less elevated series somewhat to the west, comprising Kintla Lake at 1218 m. and McDonald Lake at 962 m., lack morainic dams but drain into outwash-filled valleys and presumably owe their existence to this outwash (Campbell 1914).

Strøm indicates that few Norwegian fjord lakes entirely lack a morainic dam, though it usually is quite unimportant in determining depth. A few small lakes, such as Liltvatnet (Strøm 1937) southeast of Eikesdalsvatn, owe their existence largely to morainic dams; Eikesdalsvatn itself, though seeming as typical a fjord lake as could be imagined, is apparently separated from the marine Eresfjord by a loose valley filling of unspecified nature and may belong to Type 31. Similarly Hurdalssjöen in Romerike, south of Mjøsa, is presumably dammed by outwash which, at the south end of the lake, covered dead ice in its basin. The region south of the lake is pitted with kettles (Holtedahl 1924).

Examination of the list of Scottish lakes given by Peach and Horne (1910) indicates the existence of a number of morainic dams, but, as in Scandinavia, most of the large lakes probably actually occupy glacial rock basins. The basins of the English lakes and those of the major lakes of the Alps and of Patagonia have all been attributed in the past to morainic damming. As has been indicated in previous sections, modern research has tended to reduce the importance of this factor, and it is now hard to find examples of large lakes held purely by morainic dams in well-studied glaciated mountains. The finest composite examples are certainly the great Italian lakes of the southern slope of the Alps, where the bottoms of the lakes are well below sea level but are certainly held by thick moraines at their southern ends. Lago di Garda provides a spectacular and well-studied (Penck and Brückner 1909) example. The moraine is at least 150 m. thick, but much of the 346 m. maximum depth of the lake must be due to glacial overdeepening of a rock basin.

Delebecque (1898) listed a number of lakes supposedly dammed by

[17] Several disastrous floods have been caused recently by the breaking of such morainic dams in the Andes (Löffler, verbal communication).

moraine in the Jura and Vosges Mountains, notably Lac de Chalain, Lac de Chambly, Lac du Val, and Lac de Nantua in the former range, and Gérardmer, Largemer, and Blanchemer in the latter. At least in the cases of the larger of these lakes, composite origin may be suspected. He also lists a few cases in which a lateral valley is dammed by either a terminal moraine in the main valley (Type 32a), as in the case of the Lac de Clairvaux in the Jura, or by the lateral moraine of a former glacier (Type 32b), as in the case of the Lac de Barterand in the same range. Such cases are presumably easily recognizable and valid.

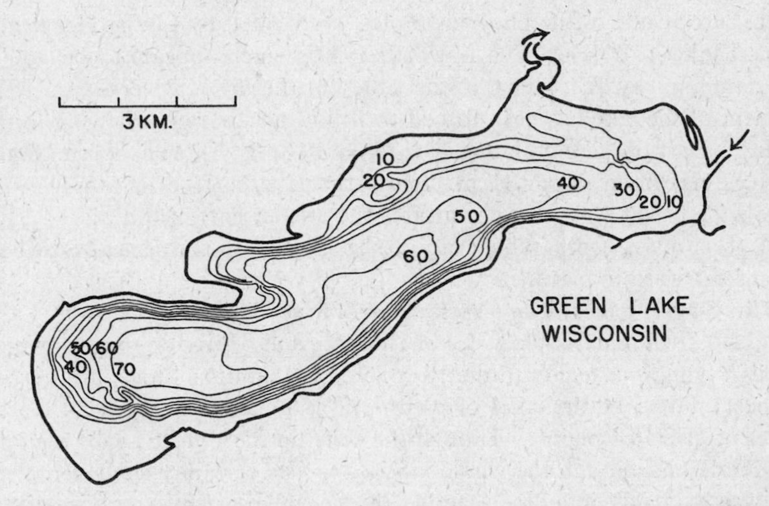

FIGURE 28. Green Lake, Wisconsin. The original valley was dammed by moraine at the western end, diverting its drainage to the north of the lake.

Genuine morainic dams are probably commoner in regions of moderate preglacial relief that has been covered by continental ice sheets or by large lobate glaciers. Here the damming is not necessarily due to well-defined terminal or recessional moraines. The best examples are perhaps the well-studied lakes of southeastern Wisconsin, namely Green Lake (Fig. 28), Lake Winnebago, and the four lakes (Mendota, Monona, Waubesa, and Kegonsa) of the Yahara drainage near Madison. All lie in preglacial valleys that have doubtless been somewhat overdeepened; all in their modern form owe their existence to dams of drift. Green Lake provides a good example of a morainic dam high enough to cause the lake to discharge by an effluent not following the original valley; such a drainage pattern obviously reduces the rate of erosion of the dam (cf. Lac de

Sylans, page 44). Usually a lake dammed by moraine is held only at one end (Type 30a), but a few striking examples of lakes dammed at each end (Type 30b) are known,

It is not possible to separate sharply the category of lakes held by morainic dams in landscapes of gentle relief, from that of lakes lying in irregularities of the ground moraine. Such irregularities mainly occur where the drift is thin and its surface largely determined by bedrock topography. The lakes of the Lake Simcoe District, Ontario (Deane 1950), are in general the result of morainic damming, but may be largely surrounded by ground moraine. Dalrymple Lake, in a basin of known bedrock topography, provides a good example (Fig. 29).

In Europe one of the finest examples is provided by Lough Neagh, the largest lake in Ireland (Charlesworth 1939), which appears to be due to the damming, by drift filling, of the valley of the lower River Bann. The effluent of the lake has been forced to follow a new course, owing to this filling. Part of the old course in the lake basin is visible as a remarkable ravine, maximum depth 31 m. below the lake surface or about 19 m. below the broad plain that constitutes the greater part of the lake.

A good many comparable examples must exist in the central European plains eastward into Russia.

The Finger Lakes, New York State (Type 30b). The Finger Lakes (Fig. 30) are a remarkable series of elongate and often deep basins, which, though lying in a region formerly covered by continental ice, are in most respects more like the lakes of piedmont and subalpine regions than the lakes of glaciated plains. Though the relief in the Finger Lakes region is moderate and though the whole landscape was certainly ice covered, the ice was evidently immensely active in the valleys, while on the tops of the intervening hills very little glacial erosion took place (Flint 1947). The Finger Lakes were thus formed essentially as if their valleys had been occupied by large independent glaciers.

The original drainage of the region was to the southward. During the Tertiary a westward flowing river, presumably tributary to the upper Mississippi as has already been indicated, occupied the valley of the St. Lawrence and the Great Lakes. The tributaries entering this Ontarian river (Fairchild 1926) gradually cut back into what is now the Finger Lakes region, forming a remarkable series of twenty-two longitudinal valleys, of which twelve contain lakes.

During the Pleistocene the whole area must have been subject to continual glaciation and deglaciation, and most modern authors reasonably attribute to glacial erosion the great deepening of the larger valleys, to well below sea level in the case of Lakes Seneca and Cayuga, though Fairchild (1926) has taken strong if unreasonable exception to this view.

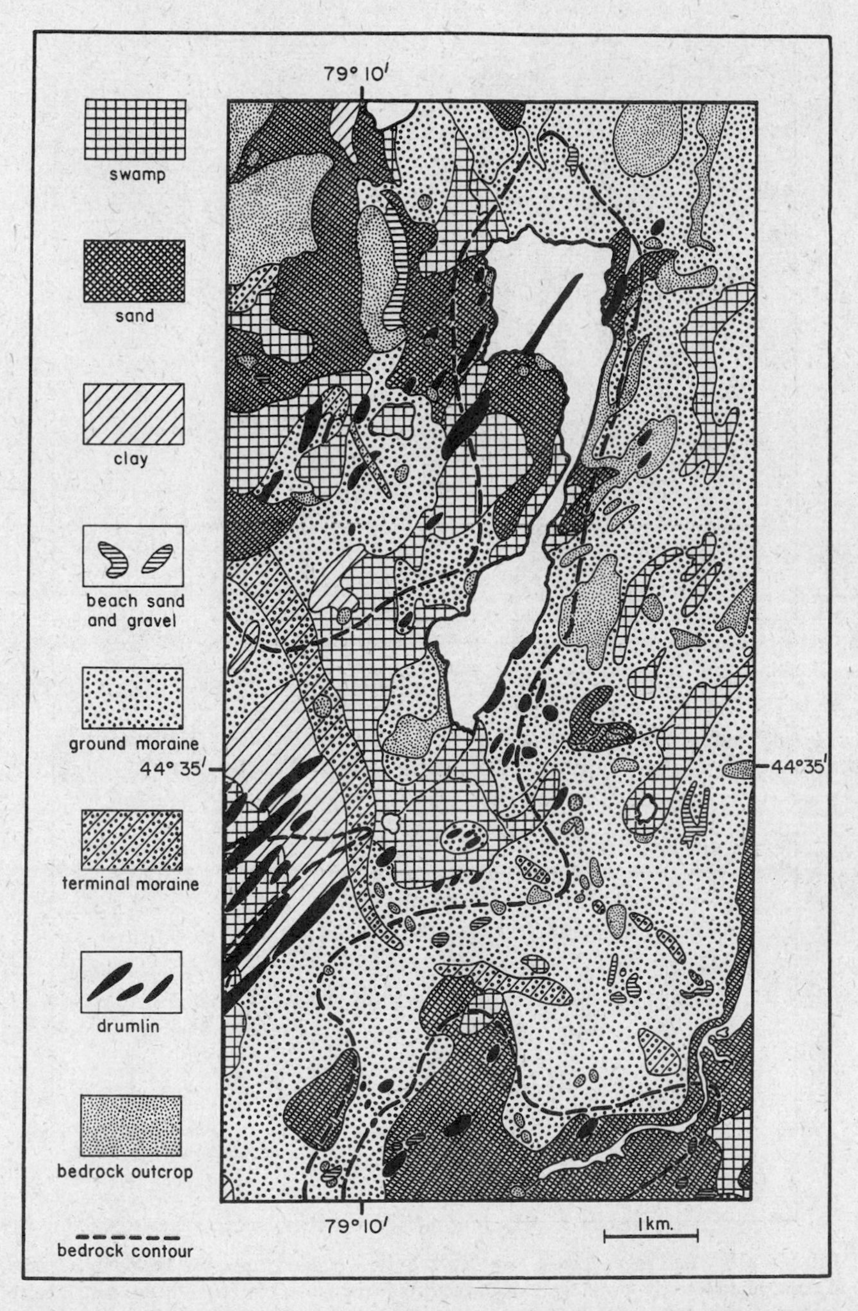

FIGURE 29. Dalrymple Lake, Lake Simcoe region, Canada. The lake is held largely by ground moraine; the bedrock topography, indicated by broken contours, shows the position of the preglacial valley so filled. Note the remarkable peninsula formed by a drumlin.

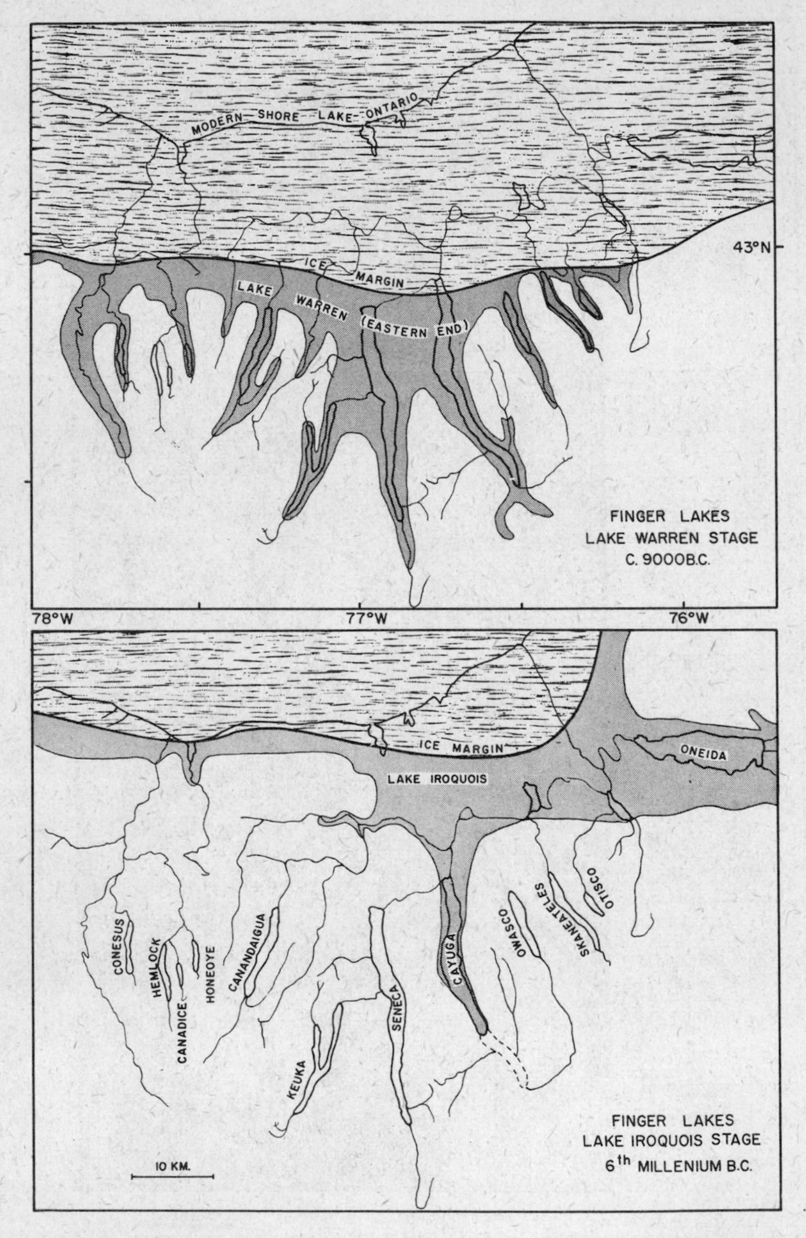

FIGURE 30. The Finger Lakes, New York State. *Upper panel*, the basins of all save Lake Oneida filled by prolongations of the proglacial Lake Warren. *Lower panel*, Lake Cayuga connected with Lake Iroquois, and Lake Oneida included in the latter; other basins all more or less as at present, but with lower lips at their northern ends (this feature is only indicated, by the dry south end, in the case of Lake Cayuga).

The southern ends of all the valleys are blocked by a festoon of moraines of the Cary stage of the Wisconsin glaciation, while the northern ends are dammed to a somewhat lesser height by more recent drift. The thickness of this drift at the northern end of Lake Seneca is certainly over 300 m. (Fairchild 1934, Flint 1947); it is not known if in any case there is a true rock basin below the drift fillings. The Finger Lakes appear, therefore, to provide some of the best known cases of lakes held exclusively by morainic dams.

During the early stages of deglaciation, proglacial lakes collected between the Cary Moraine and the receding ice. At first these lakes were separate, but retreat of the ice permitted them to coalesce along the receding ice margin. The first of the continuous lakes, Lake Newberry drained southward over the Cary Moraine south of Lake Seneca, into the Susquehanna River. Later recession led to the establishment of Lake Hall at a slightly lower level, which drained along the ice sheet westward towards the Great Lakes. Still further recession opened a lower eastern drainage. The composite lake of this stage, Lake Vanuxem I, fell gradually till all the modern lakes were isolated except Lake Oneida, the basin of which was still ice filled. The Valders readvance of the ice now blocked the eastern drainage, causing a refilling of Lake Vanuxem, and permitted the incorporation of Lake Vanuxem II into Lake Warren. A slight recession then reopened the eastern drainage, as Lake Warren became Lake Lundy. This event, as has been indicated in Fig. 27, occurred well after 9500 B.C. but before 7100 B.C. The Finger Lakes, excepting again the still glaciated Lake Oneida, then became reisolated as Lake Lundy fell, and the water in the Ontario Basin below the Niagara escarpment became Lake Iroquois. Lake Iroquois was, according to Fairchild, at first separated from Lake Cayuga (Cayuga II) as from the other lakes, but uplift at the Rome outlet of Lake Iroquois raised its level so that connection with Cayuga was reestablished, only to be broken again as the modern Lake Cayuga III was isolated. This must have happened prior to the third millennium B.C. Thus Cayuga has been a separate lake on three separate occasions, the other lakes except Oneida on two occasions, while Lake Oneida has had a relatively short uninterrupted history as a single lake. The differential uplift towards the north has raised the outlet of Lake Seneca, relative to its southern end, by about 20 m., and of Cayuga by about 15 m. The northern ends of the other, shorter lakes have suffered less relative uplift. In the case of Lake Seneca such differential uplift would have added about 8 m. to the maximum depth of the lake, now 188.4 m., while in the case of Cayuga the addition would be about 6 m. to bring the maximum depth to 132.6 m. In view of the considerable depths of these lakes, such additional depths due to postglacial rebound are unimportant (Birge and Juday 1914), but the process has undoubtedly extended all the lakes somewhat at their southern ends.

Lakes between terminal deposits (Type 33). Sometimes the retreat of the ice was very regular, producing a series of more or less parallel recessional moraines that can be traced for considerable distances and in certain cases may mark the position of the ice front during each winter of its retreat. In other cases there was much local production of dead ice, irregular wasting, and production of kames and kettles. Occasionally a somewhat intermediate type of retreat produced partially confluent ridges of terminal deposits between which a basin was formed, its long axis parallel to that of the ice border. Lakes produced in this way are very rare but are not quite unknown. Big Cedar Lake, Washington County, Wisconsin (Fig. 31), a long narrow basin of irregular shape and 31.9 m. deep, provides an example of a lake (Juday 1914) formed between moraines, though it probably was filled with stagnant ice and there is a good deal of outwash in the vicinity. Other small basins in the region were doubtless produced in the same way. Strøm (1935a,b; De Geer 1909) mentions Lilla Le, Dalsland, in Sweden as lying between two glaciofluviatile terraces which must represent outwash formed discontinuously during deglaciation. Ignatius (personal communication) points out that small lakes between annual recessional moraine deposited under water are known in the Replot Islands in the Gulf of Bothnia. In general such lakes are transitory unless, as in this case, the lake level is very close to base level. In the Chibaugamau area in the Mistassini district of Quebec, where comparable moraines laid down in Lake Ojibway-Barlow occur, a few intermorainic bogs still exist but the lakes in general have disappeared.

Ignatius also points out that occasionally there may be small lake basins between drumlins in the Mistassini district.

DRIFT BASINS

The majority of small lakes in areas covered by the ice of the fourth glaciation lie in basins in drift of various kinds. The exact nature of such basins depends largely on the details of the process of deglaciation, which varied considerably, producing an extraordinary number of kettle-hole lakes in North America, while in the region south of the Baltic a number of lakes were formed below the ice, essentially as plunge-pool basins.[18]

Irregularities in ground moraine (Type 34). A great many lakes have been supposed to have formed as irregularities in the ground moraine due

[18] Dr. W. Ohle informs me that current opinion in Germany is tending towards the view that most of the lakes of the area south of the Baltic represent basins occupied by stagnant ice and that the contrast between the two regions is less great than formerly believed.

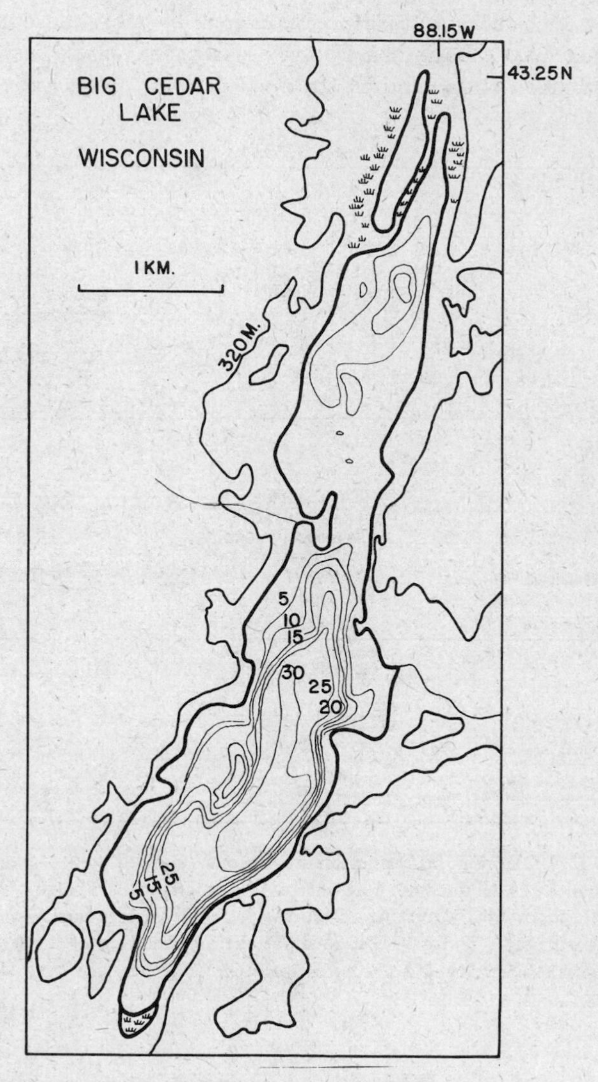

FIGURE 31. Big Cedar Lake, Washington County, Wisconsin, an elongate lake lying between subparallel terminal moraines (after Juday).

to uneven deposition (*A* and *B*, Fig. 32). Most of the cases given in the older literature are no doubt kettles. A few genuine cases doubtless exist, mostly in areas of thin drift where the irregularity is determined largely by preglacial relief. In discussing Dalrymple Lake in the Lake

Simcoe district, it has already been indicated that no formal distinction between such lakes and certain lakes held by morainic dams is possible. Pewaukee Lake, Waukesha County, Wisconsin, which lies in a preglacial valley blocked at one end by stratified drift but largely covered with thin

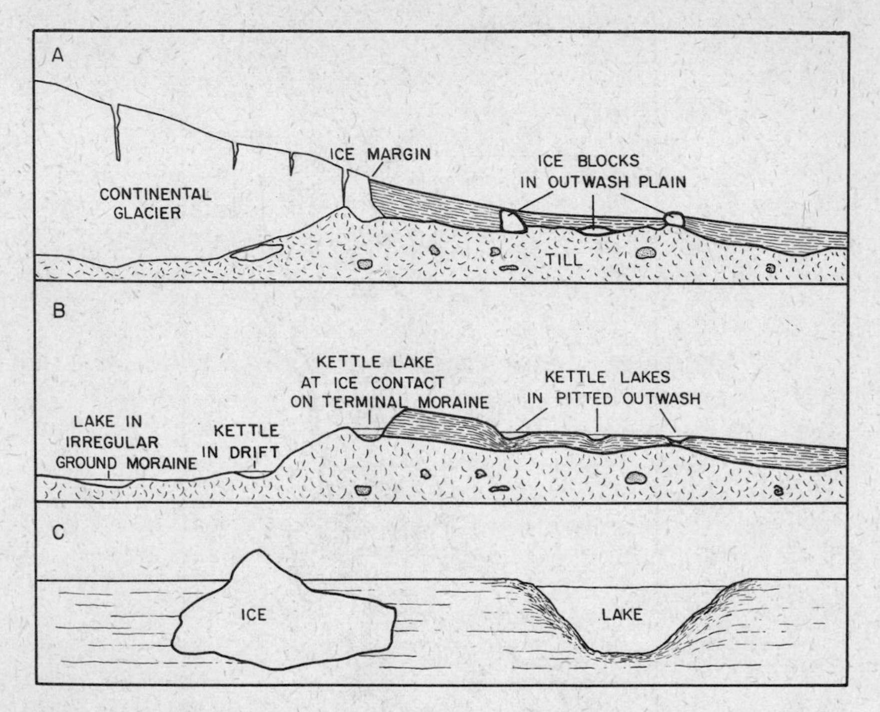

FIGURE 32. Diagram of formation of various types of kettle lakes. *A*, retreating continental ice, with outwash plain containing stagnant ice blocks. *B*, lakes formed in outwash and in till by melting of ice blocks, and as irregularity in ground moraine (after Zumberge). *C*, formation of slight shelf by melting of large, partly buried ice block, with much irregular sliding of outwash originally covering sides of block (cf. Fig. 190).

ground moraine, is apparently an example. It is, however, not entirely clear from Juday's (1914) account to what extent the outwash filling of the valley may have played a part in producing the basin. The shores of the lake lack the steep ice-contact slopes characterizing the numerous other lakes lying in kettles in this part of Wisconsin.

Scott (1916) concludes that Silver, Pike, and Little Eagle lakes in Kosciusko County, Indiana, are examples of lakes set in irregularities of the ground moraine. He also considers that Lingle Lake in the same region is of like origin, save for a deep hole, certainly a kettle, within the north margin. Zumberge (1952) gives Heron Lake, Jackson County, Minnesota, as an excellent example.

Strøm (1935b) gives no pure examples in his discussion of the genesis of the Norwegian lake basins, but indicates that Feforvatn owes its origin to excavation by feeble ice action combined with irregular deposition of moraine; he also thinks that irregular ground moraine has played a part in determining the existence of some basins otherwise in kettles or held by morainic or outwash dams.

Woldstedt (1926) likewise indicates that in spite of earlier opinion the type is rare in north Germany; he gives no examples. It may be questioned whether many of the numerous *Grundmoränenseen* listed by Halbfass (1922), particularly in the former province of East Prussia, are genuine examples of the type.

Kettles. The most characteristic type of lake in many areas of former continental glaciation has been formed by the incorporation of a mass of ice in the material that has washed out from a melting ice front or melting body of stagnant ice. Such glacial outwash was, of course, derived from the drift and moraine underlying or bordering the ice. When melting was sufficiently rapid, great quantities of water carried stones, gravel, sand, and silt some distance from the receding ice margin. Being now waterborne, this material became somewhat sorted. In it masses of ice, essentially icebergs detached from the receding ice, often became grounded. The spaces between two such masses might become filled up with gravel, and sometimes outwash was deposited over the tops of firmly anchored lumps of ice. As the masses of ice melted, basins were left, and if these basins penetrated below the water table, lakes occupied the site of the original ice masses.

Zumberge (1952) distinguishes four simple types, though it is often hard to assign, on the basis of the existing evidence, a given lake to one of the four. The simplest case is that in which outwash containing blocks of ice has discharged into a pre-existing valley. The ice on melting produces a series of lakes (*A* of Fig. 33) which lie along the stream following the valley (Type 35). When, however, the valley is filled with drift containing stagnant ice blocks an entirely different drainage pattern is apt to result (*B* of Fig. 33), which has no relation to the original hydrographic pattern (Type 36). Zumberge gives the maps, here reproduced, showing the two types as developed in Minnesota. The row of eighteen lakes of Type 36 is particularly impressive.

The other two types represent the formation of kettles in outwash or drift of continental ice uncontrolled by preglacial topography.

The pitted outwash plain, carrying a large number of small kettle-hole

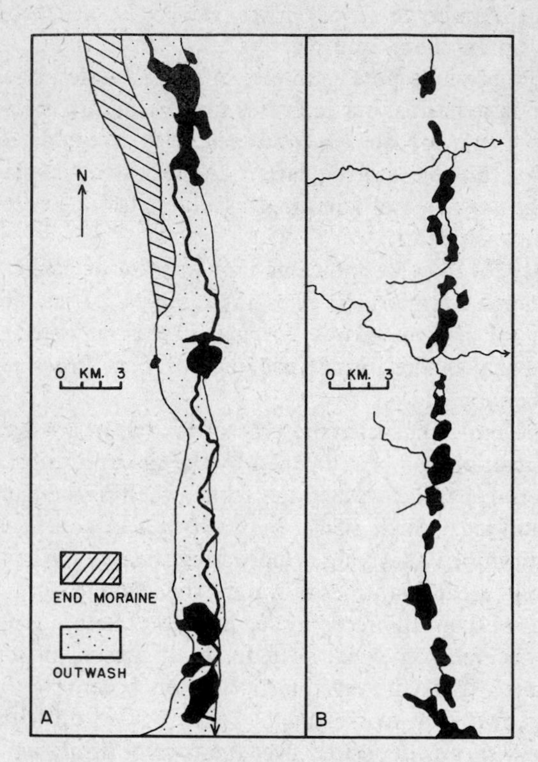

FIGURE 33. *A*, kettles in outwash filling of Pomme de Terre Valley, Minnesota. *B*, kettles in a valley running through Fairmont, Martin County, Minnesota, filled with till. Note the different relationships of the modern to the preglacial drainage. (After Zumberge.)

lakes (Type 37) derived from the melting of blocks of ice carried down the outwash, is characteristic of many regions that were covered by continental ice during the last Pleistocene glaciation. Zumberge mentions Round Lake and Lake Hubert in Crow Wing County, Minnesota; most of the lakes of Vilas County, Wisconsin, are of this type also.

Stagnant ice left in till likewise can give rise to an irregular landscape commonly termed knob-and-kettle topography. Lakes of this sort (Type 38) are hard to distinguish from those formed by irregularities in the ground moraine, but Zumberge thinks they are usually deeper. He lists a very large number, but with a warning that it is difficult to separate without close study lakes of Types 37 and 38.

Another kind of kettle, associated with eskers, which are gravel ridges deposited by subglacial streams, seems to be due (Norman 1938) to linear strips of ice from the lower parts of the side walls of such streams becoming embedded in the sides of the esker (Type 39). They occur in the Lake Chibougamau district of Quebec, and Zumberge has identified a few in Minnesota, notably Hill and Pine Lakes in Crow Wing County. The type has not long been recognized and requires further study.

In general the kettle lakes described prior to Zumberge's work, though usually belonging to Types 35, 38, and 39, have been inadequately differentiated. The following remarks apply to these common types more or less indiscriminately.

The shape and size of kettles are extremely variable. Some are only a few tens of meters across; others, such as the southern basin Trout Lake, Wisconsin, have a maximum diameter of 4.5 km. The smaller kettles are mainly fairly regular, but some of the larger basins may have a complex shore line and complicated underwater relief, generally no doubt due to more than one mass of ice having been involved in their formation. Flint (1947) has pointed out that in general the depth does not exceed 50 m., which represents the limiting depth of crevasse formation. This implies that the masses of ice involved are derived from glaciers or ice sheets that have thinned out until they consist solely of the upper fracture zone and lack a flow zone. They are, in fact, essentially stagnant.

Where a block of ice was completely buried, the resulting kettle is a gentle depression; where the block projected above the outwash, steep banks with indication of ice contact may have been preserved. It is not unusual for the lower part of the block to have extended a considerable way beyond the exposed upper part. This form, reminiscent of an iceberg, can lead to the production of sublacustrine terraces, as Fuller (1914) has pointed out (*C* of Fig. 32).

An enormous number of small lakes occupy kettles. In the glaciated parts of North America east of the Rocky Mountains, the number of such basins must be far greater than that of all other types of basins considered together. An extraordinary number occur in the regions of the Valders and Cary moraines of the middle western states, though in most cases the very fine details required for understanding the origin of a particular basin are lacking. In Vilas County, Wisconsin, the lakes of which have been studied intensively by Birge, Juday, and their associates, practically all these lakes lie in kettles, in outwash between moraines supposedly representing recession of the Cary stage. Other lakes are found on the Winegar Moraine, supposedly Valders, apparently representing the melting of ice incorporated in the terminal moraine of this, the last ice advance into the central United States. The exact relationship of the outwash to the Cary and Valders ice is not clear from the existing accounts (Thwaites 1929,

Fries 1938, Broughton 1941); it is certain, however, that under the outwash an older drift exists. A similar situation must prevail throughout the entire region, both to the west and the east, in which the Cary ice advanced over well-drained country of moderate relief.

An enormous number of glacial lakes exist farther north, throughout Canada. Many of the larger ones are doubtless dammed by moraine and outwash, a vast number of smaller ones must be kettles. The origin of any considerable number of Canadian lakes has, however, been examined only by Ignatius (personal communication), who has studied numerous kettles in the Mistassini district of Quebec. Some of these are somewhat modified in form by outwash deposition. In this region, kettles in eskers occur.

In New England the very striking pitted outwash plains are usually marked by small kettles containing tiny shallow ponds if any water remains perennially in them. There are, however, numerous isolated kettle lakes. Of these, perhaps the most famous is Walden Pond, Concord, Massachusetts (Thoreau 1854, Deevey 1942), a fine example of a double kettle. Many other kettle lakes exist in the vicinity. In northern New England these lakes are presumably of the same age as those in Wisconsin, but in southern Connecticut and Long Island the chronology of the late glacial is still doubtful. Fuller (1914) considered that kettle chains in old valleys filled with outwash are particularly characteristic of Long Island. In Connecticut, Upper Linsley Pond and Linsley Pond, North Branford (*B* of Plate 4), are a pair of such kettles in a valley, of greater importance than their size would suggest on account of the considerable amount of work that has been done on them.

In Europe, kettle lakes are known in most of the areas that experienced more than a limited mountain glaciation. They are perhaps more important in Ireland than elsewhere (Mitchell 1951). Dahl (personal communication) has called attention to the strange fact that neither kettles nor other kinds of glacial lake basin are now common along the drift borders in England. A few Scottish kettle lakes are listed by Peach and Horne (1910). In northern Europe, south of the Baltic, small kettles termed *sölle* are immensely abundant in some areas.

In the Scandinavian Peninsula good examples occur locally, notably in the plain south of Hurdalssjöen in Romerike, Norway, which has already been mentioned as a basin filled with stagnant ice in part covered by drift (Holtedahl 1924). Numerous examples occur in Sweden and Finland, including a number of lakes occupying Type 39 kettles in eskers (Ignatius, personal communication).

Collet (1925) indicates that a number of kettle lakes exist in Switzerland, particularly in the fluvioglacial deposits of Canton Zurich. He also cites Lac de Cauma as a particularly fine example in the Swiss Grisons.

Glacial tunnel lakes (Type 40). The whole of the glaciated plain south of the Baltic is characterized by lakes that appear to lie in channels excavated by melt water flowing under the ice. In general such lakes occur within the ice border of the Brandenburg substage of the last glaciation. They are usually elongate, often forming chains. Where their effluents discharged beyond the ice margin, characteristic sand plains were formed.

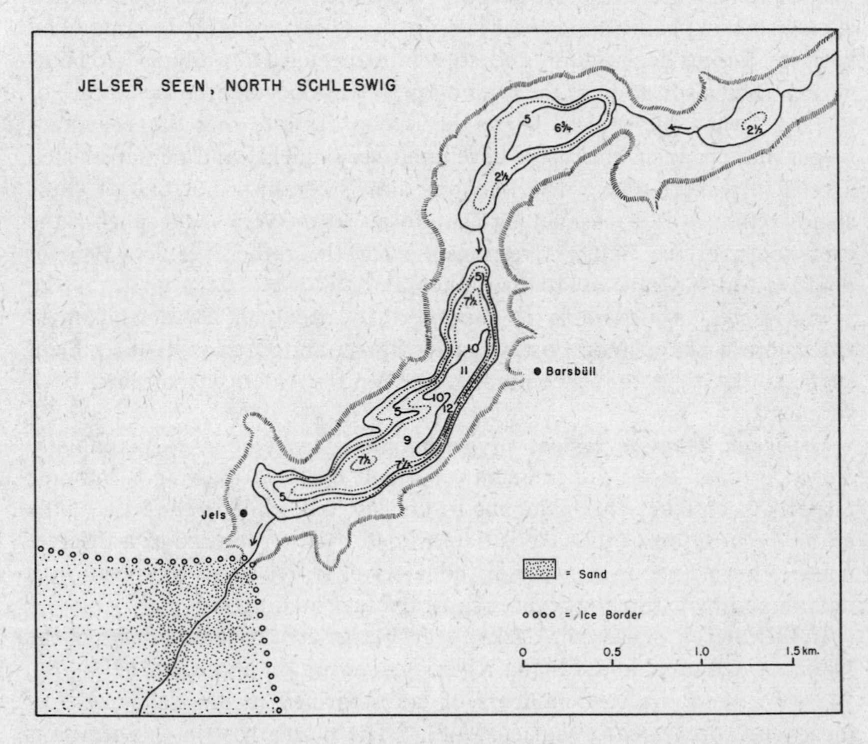

FIGURE 34. The Jelser Seen, in a subglacially eroded valley in which the lakes probably represent plunge pools. The sand plain to the southeast is formed of outwash from the subglacial stream in the course of which the lakes are formed. (After Woldstedt.)

The relationship between the sand plains and the lakes is admirably shown (Fig. 34) by the chain of lakes above Jels in North Schleswig (Woldstedt 1926). The system of glacial channel lakes extends from Denmark and northwestern Germany eastward into Poland, and apparently includes some of the classical localities of limnological research, notably the Grosser Plöner See and the Sakrower See.

The genesis of this type of lake has been considered by Woldstedt (1921,

1926), to name but one of the more recent investigators who have concerned themselves with the matter. It is probable that the channels were cut by water that ran down crevasses in the wasting ice sheet, and that at least the deeper parts of the lakes represent plunge pools (*Evorsionseen* of the German writers) or possibly the sites of large eddies in torrents rushing down from the melting surface of the ice. Quite definite potholes are sometimes observed; Ignatius (personal communication) has noted such in the Mistassini district of Quebec. Woldstedt thinks that where tunnel lakes occur in chains, the lowest lake in the series probably became filled with ice during deglaciation and so was protected from filling with outwash. The disposition of the sand plain, always beyond the ice margin of the time when the process began, seems to indicate that the system of basins and channels must have developed very quickly and perhaps lasted a relatively short time. The phenomenon is certainly not one of slow, steady retreat. It is reasonable to suppose that the very slight relief of the area south of the Baltic is responsible for the rather peculiar type of deglaciation that gave rise to the tunnel lakes.

Elsewhere, one example is supposed to occur in Sweden, namely Odensjön in Skane (Von Post, cited by Strøm and Østtveit 1948). Dahl suspects the type to be commoner in North America than has been supposed.

Cryogenic lakes in regions of permafrost. In regions of perennially frozen ground, lakes are frequently formed by local thawing. In some respects, such lakes are analogous to kettles; but while in the case of the latter the position of the lake is determined by the presence of a discrete mass of buried ice, in the case of the lakes of frozen areas the position is determined by extraneous events causing local melting.

In the interior of eastern Alaska, over a large area of the drainage of the Nabesna, Chisana, and Tanana Rivers, centering on about lat. $62^1/_2°$ N., 142° W., there are vast numbers of lakes formed by the subsidence of locally unfrozen areas (Wallace 1948). The permafrost in this region is unstable in the sense that if the soil is once thawed it will not reform a perennially frozen layer at the present time. Any accidental injury to the vegetation cover apparently may expose frozen ground sufficiently for melting to occur in summer; subsidence owing to decrease in volume occurs, and the melt water fills the hole. Once initiated, the embryo lake (Type 41) starts increasing in area. When a tree is growing at the edge of the invading lake, it will tip over somewhat as the ground on the lakeward side sinks (Fig. 35). Such trees remain growing for about ten years in the shallow water of the expanding lake, but the upper part, that has grown during these years, will be vertical. It is often possible to recognize a bend in the trunk and, if the tree is still growing, to estimate when the

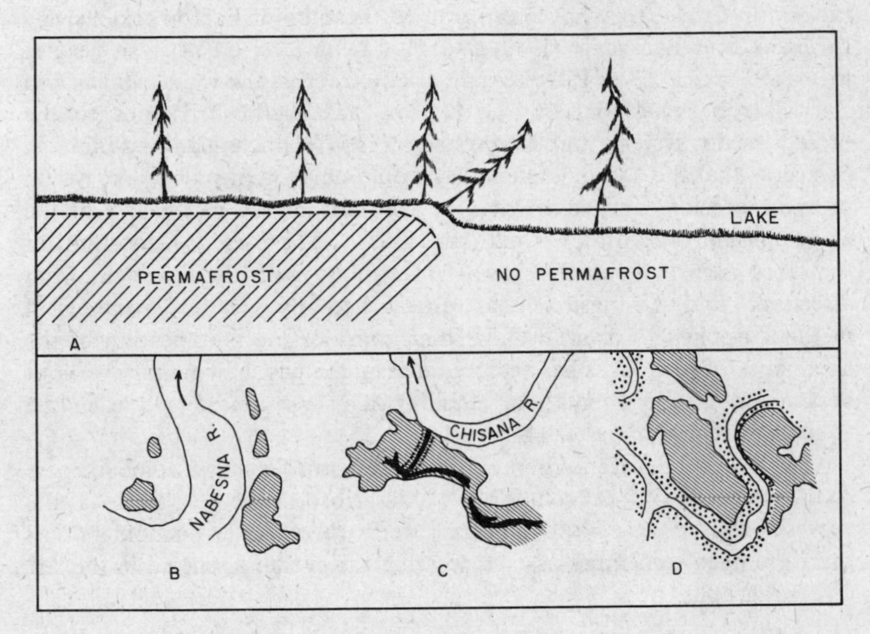

FIGURE 35. *A*, diagram showing the production of a double inflection in the trunk of a tree originally growing on soil underlaid by permafrost or when permafrost gives place to unfrozen soil beneath a shallow lake. *B*, thaw lakes in a fairly early stage of development; *C*, a later stage in which an integrated drainage system (black) with levees (dotted) is developing; *D*, late stage with lakes divided by levees. (After Wallace.)

margin of the lake had reached it. In this way Wallace concluded that the lakes increased in diameter at a rate of from 6 to 19 cm. per year, and that the largest lakes have taken from 500 to 1600 years to form.

When a few such more or less circular lakes have been initiated, they will finally fuse and the resulting complicated series of basins becomes part of an integrated drainage system. Finally levees may form along the water courses, and any lakes not integrated into the drainage system become isolated from the system by such barriers.

Farther northwest, in Alaska, around Imuruk, Hopkins (1949) has also described lakes formed by the melting of ice wedges in permanently frozen ground. These are believed to be initiated either by damage to vegetation or by small pools collecting over the polygonally arranged ice wedges. In some cases a stream running over an ice wedge can start the process. Hopkins believes such lakes to migrate down wind, as peat forms on the windward side, but his map shows little orientation.

The most extraordinary system of such lakes (Type 42) is found within the Arctic Circle; they have been studied, near Point Barrow, extensively by Black and Barksdale (1949) and by Livingstone (1954). In general they are elliptical (*B* of Plate 10) and oriented across the wind. Black and Barksdale believed that this orientation was established prior to the present wind system, but Livingstone (1954) concluded, undoubtedly correctly, that the current system that would develop in such a lake would promote erosion, and so melting, at the ends of the major axis of the ellipse, oriented across the wind (page 285). The present disposition of the lakes is in fact, on this basis, in harmony with the prevalent wind direction, and Livingstone has established by dendrochronological methods applied to dwarf willows that considerable modification of the shore lines of such lakes has occurred during the past half century without destruction of the orientation. Similar lakes presumably occur in the equivalent region of eastern Siberia.

The melting of frozen ground after local disturbance has some likeness to the processes of local solution to be described in the next sections; the expressive, if etymologically inelegant, term *thermokarst* phenomena has therefore been sometimes used to describe the events discussed in the last few paragraphs.

SOLUTION LAKES

Where any deposit of soluble rock exists, its solution by percolating water may produce cavities. In limestone districts, not only subterranean cavities but very characteristic depressions produced by local solution of superficial beds are often formed. Since the most dramatic development of such forms is in the Karst region just inland from the Dalmatian coast of the Adriatic, the production of solution basins and associated forms is frequently termed a karstic phenomenon.

Where rock salt or gypsum are in contact with moving ground water, they may be dissolved, leaving cavities which if large enough may collapse, and in limestone regions the collapse of caverns as well as solution by descending waters may produce depressions. Under certain circumstances any of the depressions formed in these ways may come to contain water.

Karstic phenomena: the various kinds of superficial basins produced in limestone areas. The action on limestone of superficial water containing carbon dioxide varies with the texture and purity of the stone and with the climatic regime of the region under consideration. In general the most striking karstic forms are produced in regions of markedly seasonal rainfall.

In some cases a *karrenfeld*, consisting of small pinnacles separated by crevices, may form. Often the crevices become completely filled with residual material, and there then may be little superficial indication of the existence of the structures. When jointing is well developed, water descends along the joints and may produce various kinds of sinks and swallow holes. These may sometimes be arranged linearly along the strike of the underlying rock or along fault lines. When such holes are deep circular pits, often leading into caves, they are of limnological interest in only the rare cases in which they now lie under water. Funnel-shaped depressions, commonly called in the European and in some American literature dolines (*dolina* in Carniola), usually drained by a number of small cracks, are frequent in many limestone regions. In some areas, as in the Pennines of northern England, they are small, usually only a few meters across, but in other places quite large examples are known. These may, if the drainage channels become blocked with residual material from the limestone or with exogenous water-borne sediment, become converted into ponds or small circular lakes.

Scott (1910), who has discussed the numerous small ponds occupying dolines on the Mitchell limestone (Mississippian) of southern Indiana and parts of Kentucky and Tennessee, suspects that the initial focus for production of such depressions was the decay of deep roots of the trees that formerly covered the area. Such a mode of origin may be widespread but can hardly be general. A variety of modes of initiation doubtless occurs. Zotov (1941), for instance, has noticed in New Zealand the formation of perfect cups up to 30 cm. deep in otherwise bare limestone, as the result of solution under colonies of the moss *Tortula phaea*. Scott found his ponds now to be filling rapidly with sediment; one that he studied, linear dimensions about 20 by 17 m., was about 1 m. deep at the time of his investigation but was known to have been about twice as deep twenty-four years previously.

Sometimes adjacent dolines may fuse to form a compound depression, for which the term *uvala* may be employed though its original Carniolan usage is perhaps somewhat looser. The occurrence of the original dolines along the strike of the rock or along a fault may produce an elongate uvala, but very irregular examples are also known.

In structurally determined valleys, which may lie in grabens, against fault scarps, or in synclines, karstic erosion may occur with particular intensity, producing tectono-karstic depressions, called *poljes* in Dalmatia. Comparable large karstic depressions, formed when a stream cutting through noncalcareous rock suddenly meets limestone in which sinks can develop, occur in Florida. As the sink forms, it drains not only the upper but also the lower reach of the stream, producing a closed basin.

Any of the larger types of karstic depressions, doline (Type 43a), uvala (Type 43b), or polje (Type 44), may under certain circumstances develop into a lake. Evidently, one important circumstance is variation in the depth of the water table, and this may be due to changes in climate or in base level. A number of solution lakes in Florida probably owe their existence to a lowering of the water table during a glacial eustatic fall in sea level, which permitted large dolines to form, followed by a rise in the water table in post glacial times, which permitted the basins to fill with water. It is evident that the same type of change has been important in the Karst of southeastern Europe. Here, moreover, many poljes show lake terraces on their flanks, so that there have been, presumably under pluvioglacial conditions, higher levels as well as lower since the depressions formed.

Solution lakes in the state of Florida. By far the most important groups of solution basins in the New World occur in the state of Florida. The whole of the Floridian Peninsula, though underlaid by ancient metamorphic rocks, is now covered with a considerable thickness of Tertiary sediments, most of which are limestones. The greater part of the noncalcareous rocks intercalated between or lying above these limestones are easily eroded and permeable sands. The area has been relatively stable for a long time and records in great detail, not always too well understood, the eustatic rises and falls in sea level during the Pleistocene. Practically the whole of the area is less than 100 m. above sea level at the present time. These circumstances combine to provide ideal conditions for the production of solution basins. A good many small lakes, often in very perfect round basins, occur in the northwestern part of the state (Cooke 1939), and a few in the southwestern part (Parker and Cooke 1944). Deep Lake, Collier County (Fig. 36), is a superb example of such a round lake in a doline. By far the greater number of the solution lakes of Florida occur, however, in the central highland region of low limestone hills covered with late Tertiary sands.

Most of the smaller lakes that exist in great numbers in Alachua, Putnam, Marion, Lake, Polk, and Osceola counties are either in simple or complex dolines. At least one case, namely Long Lake south of Chiefland (Vernon 1951), is known in which a series of sinks along a fault scarp has given rise to a tectono-karstic depression that would be termed a polje by European geomorphologists. In the Tallahassee Hills a rather different type of polje-like depression has been produced by the sudden development of a sink in a normal valley. Such a sink drains not merely the part of the valley above but also a section below it. The part of the valley draining by the sink is then eroded rather rapidly, forming a branching elongate closed basin. If the sink becomes blocked and the water table rises, a valley lake with a rather dendritic outline results.

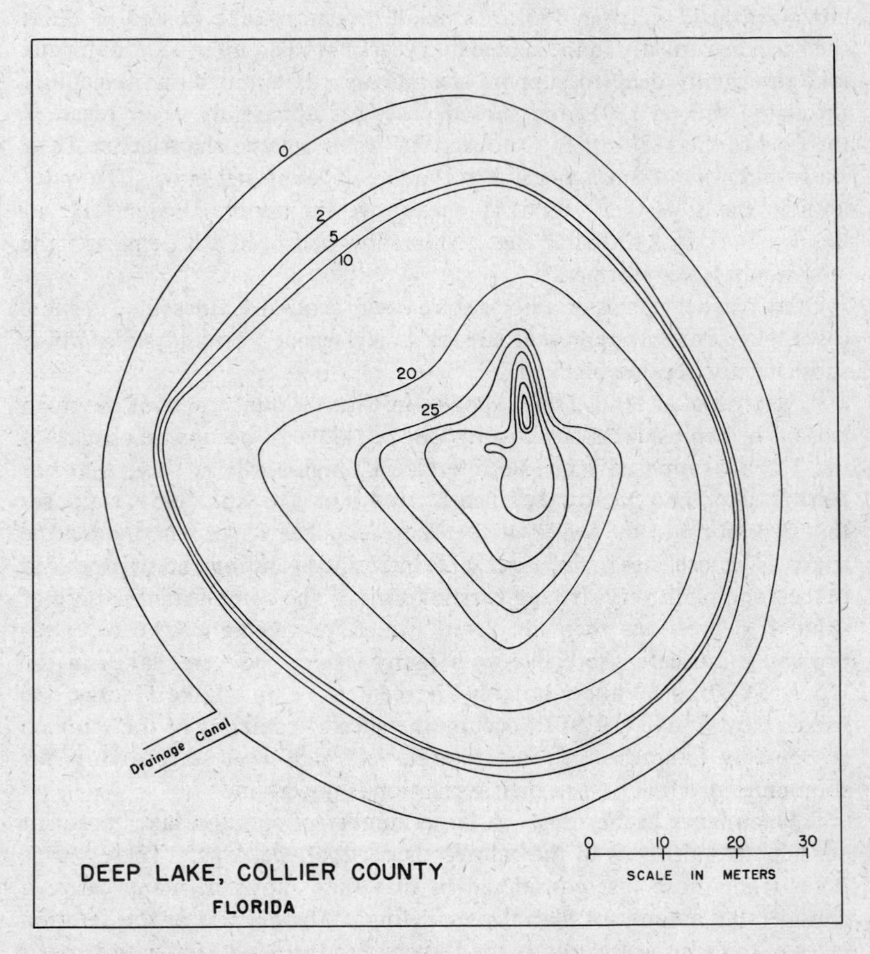

FIGURE 36. Deep Lake, Collier County, Florida, a simple doline lake; contours below 25 m. at 1 m. intervals (by courtesy of Florida State Geological Survey).

Iamonia Lake, Jackson Lake, Lafayette Lake, and Miccosukee Lake are the most important examples. They are all subject to great fluctuations in water level. In very dry seasons water may exist only in the sink, through which limited discharge still occurs.

Other polje-like depressions occur in central Florida. Alachua Lake, 13 km. long with an area of about 50 km.^{2}, is supposed to have been formed by flooding when a sink in an adjacent basin was blocked by visitors throwing logs into the swirling water as it disappeared underground

(Sellards 1914). The lake bed was certainly dry in 1776 and 1824, and contained little water in 1861. A small transitory lake existed in 1868; the basin became dry in the succeeding years but filled up in 1873 to form a lake sufficiently deep to support steamboats. It emptied spontaneously through a sink in 1891, and subsequently has apparently never regained the level of the 1880's. In October 1907 water entered the Alachua Lake Basin and ran out through a sink at the rate of 76 $m.^3$ per min. The water level in the sink then stood 61 cm. above the general water table at Gainesville. It is probable that if the inflow had been a little greater the lake would have reformed.

Other smaller basins of a like nature occur around Gainesville; in most cases they probably represent stream valleys eroded around sinks which subsequently became blocked.

A very peculiar lake, Tsala Apopka in Citrus County west of the main axis of the peninsula, is believed by Cooke (1939) to occupy an estuary of the Talbot (supposedly middle interglacial) high-sea-level stage that has become silted and greatly modified by solution. Vernon (1951) supposes that deposition in the valley of the Withlacoochee River has ponded the river. The local rise in the water table intensified solution and so produced in the flood plain very striking karstic forms. The complicated pattern of solution depressions then filled with water, producing a lake of extraordinarily intricate shore line with many islands. Several lakes in the Upper St. John's Valley, notably Crescent Lake and Lake George, are believed by Cooke (1939) to occupy the sites of estuaries of the Pamlico (supposedly Sangamon or last interglacial) high-level stage. How far solution is also involved in their formation is uncertain.

Solution lakes in the Alps. A large number of solution lakes occur at considerable altitudes in the calcareous parts of the Alps. These basins differ from those just considered in that they show, in many cases, a considerable amount of glacial remodeling. The deepest of the solution lakes of the Alps is apparently the Lünersee in the Rhätikon, with a maximum depth of 102 m. (Löwl 1888).

A detailed account of the solution basins in the Prealps between the Lake of Thun and the River Arve, in the High Calcareous Alps, and in Valais, Tessin, and the Grisons, has been given by Lugeon and Jérémine (1911). A large number of the smaller lakes that they discuss appear to lie in cirques but lack a superficially visible outlet. In some cases sublacustrine tunnels are known to carry the effluent waters, the lakes in fact occupying dolines in the cirque basins. It is, however, usually uncertain if the existence of the lake is really dependent on solution or whether it could not be due mainly to glacial excavation, as is so often the case in cirques in noncalcareous mountains. A very fine example is provided by three small lakes in the Fully Basin. Two of the lakes occupy very well-defined cirques at different levels; the lower Lac de Sorniot, at 1996 m., discharges by a fissure in Triassic limestone, but in other respects these lakes seem quite typical cirque lakes. Another beautiful example of a doline lake in a cirque is provided by the Seewlisee below the Grosse Windgälle, at 2045 m. Collet (1925) traced the discharge of the sub-lacustrine littoral outlets of this lake to springs in the valley of the Reuss, 1590 m. below the lake. South of the Wildhorn and the Geltenhorn there are several groups of lakes. The most eastern lake, the Lac des Audannes at 2460 m., is said by Lugeon and Jérémine to lie in a cirque in non-calcareous rock. The most western lakes, the Lacs des Grands Gouilles, at about 2456 m., consist of a group of dolines in a cirque. A third and intermediate group, the Lac de la Saourie with some smaller basins, lies in a region of many sinks and fissures and seems to be a typical doline lake. Many other typical doline lakes lie at lower altitudes; the Seewli of the Col der Brünig at 1160 m., is a good example among the many listed in Lugeon and Jérémine's memoir.

At least one magnificent multiple-doline or uvala lake is known in Switzerland, namely the Muttensee (Fig. 37), Glarus, at 2448 m., which consists of three main basins and some minor ones, all doubtless much remodeled by glacial action. The lake has an outlet, but it goes underground almost immediately.

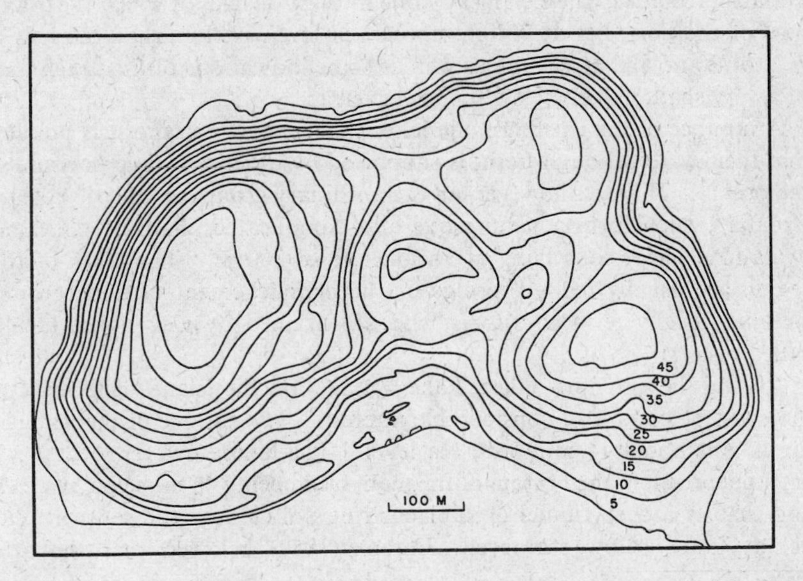

FIGURE 37. Muttensee, Glarus, Switzerland, a complex doline or uvala lake (after Collet).

In addition to the numerous small glaciokarstic lakes that we have discussed, there are in the Prealps and High Calcareous Alps of Switzerland a number of large closed basins in the bottom of which lakes are often found. These basins apparently represent the effect of chemical denudation in the floors of pre-existing valleys which are structurally determined and so are appropriately termed poljes. The karstic phenomena exhibited by these basins appear young; there is little alluvium deposited. The lake in such a depression may be a simple though very large doline, as in the case of the Seelisbergerseeli north of Seelisberg, which appears to be at the bottom of a synclinally determined polje (Collet 1925). In other cases the lake is larger and elongate, as in the case of the Dürrensee just to the west of the Seelisbergseeli, and of the Glattensee lying in the middle of a long basin west of the Ortstock, and in those of the Fählensee and Sambtisee in the Sentis chain, which lie at the eastern ends of their basins. Not infrequently such alpine poljes and their lakes have undergone a great deal of glacial action, the Daubensee lying in a polje in a quite typical glacial valley.

Solution lakes in the Balkan Peninsula. In the classical karst region of the northwestern part of the Balkan Peninsula, most of the poljes are now dry or only flooded intermittently. There are, however, often lacustrine terraces around the margins of such depressions, testifying to the existence of large permanent lakes during glaciopluvial times. Cvijić (1901), for example, records in the Livanjsko polje in West Bosnia, of which the lowest part of the floor lies at 701 m. above sea level, well-marked terraces at 712 to 720 m. and at 740 m., as well as some indications of the beach lines of a very shallow lake below the lower terrace.

A number of the existing polje lakes lie near the sea, and it is possible that their water, though fresh, is supported by the hydrostatic pressure of the ocean. There is, however, an extraordinary group, the Plitvic lakes in Croatia, which form a picturesque and complicated chain of polje and uvala-like depressions lying at an altitude of 700 m. The lakes of this region are usually fed by springs or the debouchement of underground streams, locally termed *estavels*, and are drained by sink holes, locally called *ponors*.

Among the perennial lakes, Lake Vrana[19] on the island of Cherso off the coast of Istria is of considerable interest. The surface of the water in the lake is about 14 m. above sea level. The lake, however, occupies a cryptodepression, the bottom of the main basin being 46 m. below sea level and that of a deep funnel or sublacustrine doline near the southern end being 70 m. below sea level. Lorenz (1859) believed that the area

19 Not to be confused with the Lake Vrana lying in a polje near the sea coast 35 km. southeast of Zara on the Dalmatian coast.

(5.59 km.^{2}) of Lake Vrana is too great to permit the lake to maintain a relatively constant level if the water is solely derived from rain falling on the island of Cherso, and concludes therefore that there must be continuous fresh water in subterranean channels between the island and the mountain of the mainland. This has been accepted by some subsequent authors and rejected by others (Gavazzi 1904), but seems reasonable.

Among the temporary lakes of the region, the Zirknitzer See, the *lacus lugeus* or *palus lugea* of the Romans, has excited attention on account of its irregular appearance and disappearance. It was, in fact, the subject of three of the first papers (Brown[20] 1669, 1674; Valvasor 1688) on a specifically limnological subject that ever appeared in a scientific journal. The lake (Fig. 38) is fed by a number of streams, springs, and estavels and is drained by a great number of ponors or sinks both along its margin and in its bed (Gavazzi 1904). The old accounts indicate that the water disappears from the sinks in a regular order whenever the level of the water table falls below that of the floor of the lake.

Farther south there is a belt stretching from Lake Scutari eastward into eastern Macedonia in which lie a magnificent series of tectonic basins. All of these exhibit some karstic features; this is notably true of Lake Scutari itself. It is the largest of the lakes of the Balkans, but over the greater part of its area has a depth of only 7 m. Along the southwestern shore of the lakes where the karst topography is best developed, there are not only many islands but also very deep holes, the deepest going to 44 m., or 38 m. below sea level. These holes are fundamentally sublacustrine dolines. Cvijić (1902) points out that the polje of Kupres in western Bosnia, in which there are several small lakes in dolines, provides a close analogue to Lake Scutari, differing only in the fact that the main basin is dry and the deep depressions alone are filled with water.

Lake Ostrovo has been elaborately considered by Hasluck (1936a,b, 1937), who showed that the lake has risen since ancient times and has exhibited irregular oscillations in level since the beginning of the nineteenth century. Though these variations can in part depend on rainfall, they are largely independent of it and may be attributed to the silting of the *katavothrai*[21] or sinks which drain the lake. Some choking of the outlet channels by fallen blocks of limestone, detached perhaps by earthquakes, also seems to take place. Discharge may occur when the channels are reopened by solution or by the bursting of sediment blocks weakened when their organic matter decays. The synchronous rise and fall of Lakes Ostrovo and Petersko are attributed, Hasluck believes correctly, to their discharge occurring through a common channel.

---

[20] Dr. Edward Brown, son of Sir Thomas Browne.

[21] *Καταβοθρα*, a hole going down.

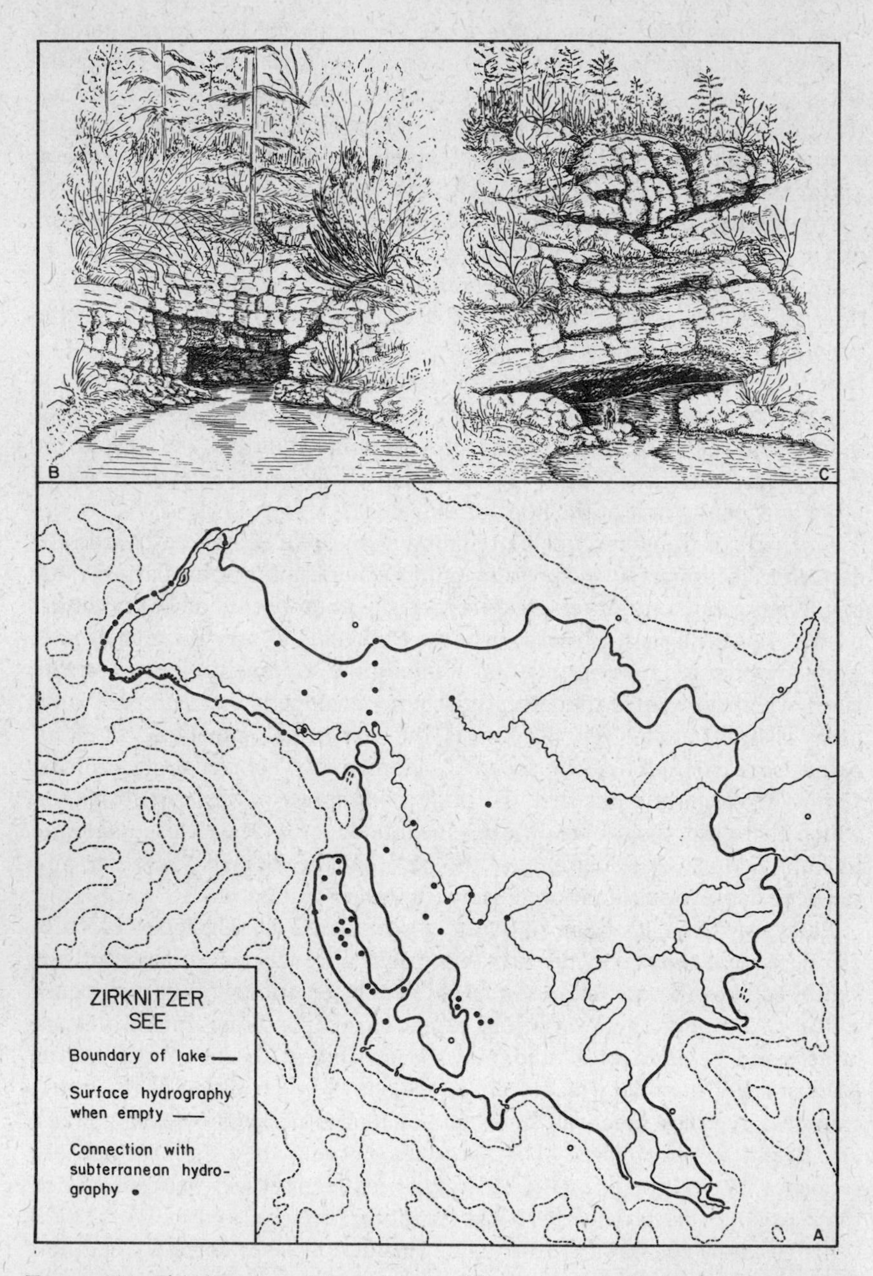

FIGURE 38. Zirknitzer See, a lake in a polje with many sinks in its floor and connections with a system of caverns along its margin. *A*, the ordinary upper limit of the lake when full, and the drainage pattern when empty; *B*, estavel Suha Dolica; *C*, marginal sink or ponor, Gr. Karlovica. (After Gavazzi.)

Farther south in Greece other such variable lakes exist, though many have been drained to provide agricultural land. In the Peloponesus there is an area of karstic interior drainage in which several large lakes exist from time to time, Lake Pheneos, Lake Stymphale, and Lake Karaklinou being the most important (Martel 1892). The first named discharged dramatically in 1892, apparently owing to the bursting of a block in one of its katavothrai (Frazer 1898). The term katavothra lake is sometimes (Philippson 1892) applied generally to lakes such as the Zirknitzer See or Lake Alachua, which vary greatly in level owing to variations in subaqueous drainage.

Solution lakes in other limestone areas. The distribution of karstic areas throughout the world has been summarized though not quite reliably and unfortunately without documentation, by Kosack (1952). Many such areas lack lakes, however, the ground water lying far below the present surface level. A few of the more important examples of solution lakes in limestone areas other than those previously discussed may be mentioned to give an idea of the distribution of this type of lake.

Delebecque (1898) records a certain number of karstic lakes near sea level along the Mediterranean coast of France. In Italy the Lago di Trasimeno is an important example of a very large but shallow solution lake (Penck 1894). In the Himalayas, Naini Tal in Kumaun, a beautiful and celebrated body of water, is regarded by Middlemiss (1890) as a solution lake, though faulting may also be involved in its formation. Lake Otjikoto and Lake Guina in southwest Africa are, according to Jaeger (1939), typical doline lakes. Lake Yojoa or Taulebé in Honduras is a solution lake (Sapper 1902); other examples probably occur in Central America. Numerous examples[22] are known in Australia, particularly in the Mallee district of Victoria (David and Browne 1950).

One or two cases of what appear to be solution lakes in raised atolls are mentioned on page 145.

Underground lakes in caverns. Underground lakes, while not really within the scope of the present work, may perhaps be recorded as a special type of lake (Type 45) formed in part by solution, in part by precipitation of calcareous sinter that can form barriers and delimit basins in caves.

Other kinds of solution lakes. The solution of calcium carbonate is not the only process by which solution basins can be formed. In some regions underlaid by extensive gypsum-bearing beds, depressions formed by the solution of this mineral may contain lakes (Type 46). The best known example is doubtless the Lac de la Girotte, southwest of Mont Blanc in the

---

[22] Lake George near Canberra is, in spite of Halbfass's (1923) tentative statement, apparently not a katavothra lake.

French Alps (Delebecque 1898), a lake only 0.57 km.[2] in area yet 99 m. deep. It is evidently meromictic, and the rich supply of hydrogen sulfide in its deep water must in part be derived from the reduction of $CaSO_4$. Other examples exist in the French Alps, notably the Lac de Tignes and at least in part the Lac du Mont Cenis (Delebecque 1898, Bourgin 1945).

Both gypsum and halite or rock salt may be present in sediments in which local solution is occurring. The solution of both these minerals appears to have been involved in the formation of the basin of the Mansfelder See (Ule 1893) in southwest Saxony, a slightly saline lake of some biological interest. In the vicinity of Besse, Department of Var, France, the solution of sodium chloride and calcium sulfate from sediments containing gypsum and halite has been responsible for the formation of several lakes, among which may be mentioned the Lacs du Grand et du Petit Lautien and the Lac de Besse. In the same district a pit 40 m. across was formed quite suddenly in January 1878, apparently as the result of solution by underground waters.

It is probable, according to Delebecque, that certain lakes near Biarritz and in the interior part of the plateau of Landes along the western coast of France are due to solution of sodium chloride from Tertiary saline sediments; modern cases of local subsidence are said to be known in the region.

The very large number of saucer-shaped lakes of the parklike forest steppe zone of western Siberia are attributed by Russian investigators (Suslov 1947) to solution of soluble salts.

The solution of ferric hydroxide and hydrous aluminum silicates in sandy sediments through the action of acid plant extracts apparently can occasionally form basins (Type 47). Smith (1931) has described a number of such basins in the coastal plains of South Carolina, and has shown by actual analysis of the sediments below the basin and in the vicinity of the basin rim that up to 23.66 per cent of the original mass may have been leached in the formation of the depression.

The process probably originated either in a very small irregularity in the coastal plain due to uneven sedimentation, or by a tree falling, leaving a small depression. Such depressions will retain moisture, produce richer vegetation, and promote more rapid leaching. The basins generally lie just beyond the heads of or at the sides of valleys, a situation that promotes rapid drainage initially. All seem to have contained water at some time and many still do; there is usually a dark organic layer rich in sponge spicules, even if water is now absent from the depression. The basins tend to be circular and are not oriented; they may be up to 800 m. across and 6.5 m. deep.

LAKES DUE TO FLUVIATILE ACTION

A number of different kinds of lake basin may be produced by the action of running water. In the less mature upper reaches of rivers, destructive excavation may produce rock basins which can persist as lakes if the course of the river is diverted. Such cases, to be considered first, are rare. In the more mature lower reaches of a large river, the flood plain may be rich in lakes produced by a variety of obstructive and destructive processes, while on deltas very remarkable constructive processes can produce large shallow basins.

Plunge pools and eversion lakes. A large waterfall has sufficient corrasive power to excavate a considerable basin below the fall. When such a pool has been formed in the course of a river that subsequently has become diverted, the rock basin left at the base of the fall may contain a lake (Type 48). The most impressive of such plunge-pool lakes are on that part of the former course of the Columbia River that now forms the Grand Coulee in the state of Washington. This gorge was formed at a time when the Columbia River was blocked by ice during some glacial antecedent to the last maximum in the region. During the last glaciation, renewed damming by ice caused the Grand Coulee to function again as the spillway of a glacial lake (Flint and Irwin 1939), but it is probable that the main excavation was done during the earlier phase of its life as an active watercourse. Bretz (1932) considers, and Flint and Irwin (1939) appear to agree, that the major excavation of the coulee was due to the water discharged at the end of the history of a large ice-dammed lake. Even for the work of such an unprecedented torrent, Bretz thinks the greatest torrent of which any record has survived, the resulting basins are very impressive. The most diagrammatic plunge pools, Falls Lake (Plate 5 and Fig. 39) and Castle Lake, lie at the bottom of dry scarps, formerly immense waterfalls, at the head of the lower coulee. Deep Lake, below Castle Lake, is believed by Bretz to represent a plunge pool of a steadily receding cataract. Lower down the coulee a series of lakes, of which Lenore Lake is the most impressive, are supposed by the same author to represent fluviatile plucking. The very large number of immense potholes in the lower coulee may indicate how a sufficiently violent and turbulent torrent of immense size could scoop out a basin by forming potholes that coalesced. The channel containing Lenore Lake lies in a zone of fractured basalt on a monoclinal slope and apparently cannot represent a preglacial drainage channel.

Fayetteville Green Lake (Miner in Eggleton 1956) provides another example, in eastern North America, near Syracuse, New York. It appears to have been formed by a cataract during deglaciation in late Wisconsin times.

PLATE 5. Falls Lake and associated basins, Grand Coulee, Washington (*phot.* U. S. Bureau of Reclamation, through Dr. W. T. Edmondson).

The erosion of hollows by cataracts or immense eddies in the channels of water courses is usually described as *eversion* in the continental literature. The eversion lakes of the northern central European plain have already been discussed as glacial tunnel lakes.

Fluviatile dams. Sometimes rivers may be dammed by sediment deposited across their courses by tributaries. Lakes formed in this way are not very common and are usually found only where a tributary, often with a fair gradient, enters a river flowing in a valley that is disharmoniously large. This may happen when the main river occupies a large structural depression, or flows in a valley that has been greatly modified by ice or one in which stream flow has been reduced owing to climatic changes or stream piracy. Lakes formed in this way are here said to be held by *fluviatile dams* (Type 49); the much larger class of lakes in which the lateral valley has been obstructed by the deposits of the main stream are referred to here as *lateral lakes* (Type 52).

Alluvial fans as dams. The conelike or fanlike structures built by lateral mountain streams descending into wide intermontane valleys, may sometimes act as dams. This is most likely to occur in valleys of tectonic origin in semiarid regions, for here the valley may be wide but the stream that it contains too feeble to cut away the lower edge of the fan. The finest example of a lake held in this way was probably Lake Tulare at the southern

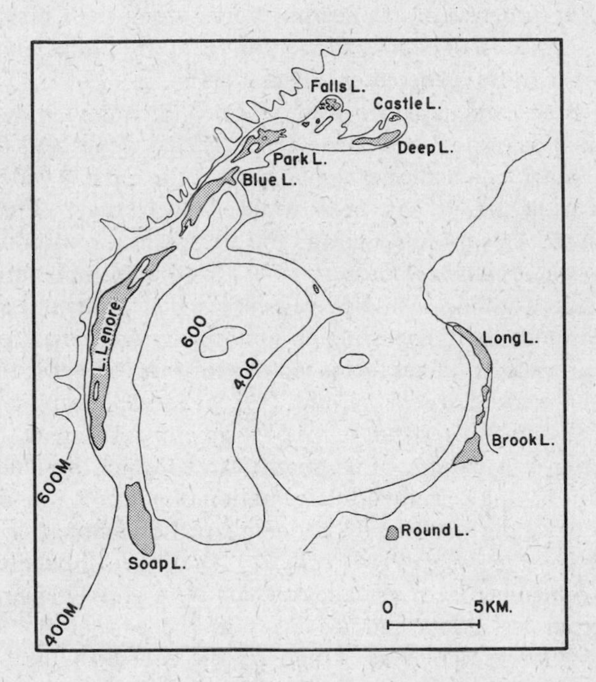

FIGURE 39. The lakes in the lower part of the Grand Coulee (after Bretz).

end of the Great Valley of California. During the middle of the last century, the lake occupied an area of about 1960 km.^{2}, dammed on the north side by the low alluvial fan of the Kings River. In recent years, diversion of the waters of its catchment basin into irrigation projects has reduced the lake to a swamp. Farther south two smaller and very variable lakes, Buena Vista Lake and Kern Lake, are still temporarily filled with water. Davis (1933), who describes these lakes, gives a few other small Californian examples.

An even more spectacular but, in its present condition, artificial case is provided by Salton Sea, which occupies a basin cut off from the northern end of the depression in which lies the Gulf of California. The Salton

Basin is separated from the gulf by delta deposits of the Colorado River. Although it is a cryptodepression, with its deepest point 84 m. below modern sea level, the basin appears never to have contained sea water. At one time a distributary of the Colorado fed the basin, forming a lake 93 m. deep which, at least in the later part of its history, became saline through evaporation, and the lake dried up completely before European exploration of the region. During 1905–1907 the basin was accidentally flooded in the progress of engineering works along the Colorado River; the lake so formed is, of course, evaporating and will ultimately disappear, just as did its natural predecessor (Davis 1933).

Alluvial cones and fans are often formed in humid mountains but seldom retain lakes. Peach and Horne (1910) describe Loch na Bi, Tyndrum, Argyll, as a Scotch example, lying at the top of a wide valley, the main stream of which has been reduced by piracy. Forel (1901a) mentions in the Alps the Silsersee and the Silvaplanasee, without details.

Catastrophic formation of fluviatile dams. A few examples are known of fluviatile dams forming in humid regions as the result of excessive but local floods suddenly transporting an abnormally great mass of sediment into the main valley. Lakes formed in this way are in a sense the fluviatile analogues of landslide lakes. Collet (1925) records the blocking of the Bavona in Tessin, Switzerland, in 1905 by an alluvial dam as the result of high waters in a tributary. Like many dams formed by landslides, the Bavona dam, having permitted the formation of a lake 2.5 m. deep, broke after a few days, the resulting flood doing a certain amount of damage.

Fluviatile dams in enlarged river valleys. The largest fluviatile dam lakes are those formed across river valleys which were greatly enlarged by the drainage from ice sheets and ice lakes at the end of the Pleistocene glaciation. Fine examples are found on the course of the old Warren River, which drained Lake Agassiz in late glacial times and entered the Mississippi by what is now the valley of the Minnesota River. Tributaries entering at a fair gradient suddenly slacken as they join the small modern occupant of the Warren Valley. These tributaries had no difficulty in blocking the Minnesota with their deposited sediment, so forming Traverse Lake and Big Stone Lake. Lake Pepin, between Minnesota and Wisconsin, on a reach of the Mississippi formerly fed by the Warren River and now reduced in volume, is similarly held by the deposits of the Chippewa River (Hobbs 1912, Zumberge 1952).

Fluviatile barriers in lakes. The discharge of a river into a lake gives excellent opportunity for sedimentation. A lateral valley may sometimes discharge at the middle of one side of a lake and may then build a delta of sufficient size to cut the basin in half. Two smaller lakes (Type 50) take the place of one large one, as the result of a process analogous to the formation of a lake by fluviatile damming. The most famous example is doubtless the separation of the Brienzersee from the Thunersee by the delta of the Lütschine, on which Interlaken stands. In the English lake district the separation of Derwentwater from Bassenthwaite Lake, and of Buttermere from Crummock Water, have already been mentioned.

A rather different type of small delta lake may often be seen on the *strath* or deltaic plain deposited by the influent at the head of a long narrow lake. As the stream entering the lake slackens, the sediment that is carried in may tend to be deposited as a pair of parallel ridges, really levees, continuing in the direction of the flow. Later, as the strath builds up, the space between such ridges and the lake shore may fill completely with sediment as a result of remodeling by wave action, but often a small lake may be included between part of such a ridge and the original lake shore. Such *strath lakes* (Type 51), as they may be called, are probably found in most regions where long narrow basins are fed by a principal influent at one end. Peach and Horne (1910) mention as good examples Loch Geal, cut off from the upper end of Loch Lomond by the advancing delta of the Fallock, and Loch Buidhe,[23] likewise separated (Fig. 40) from the upper end of Loch Lubnaig. The Italian subalpine lakes provide other cases, the most interesting being Lago di Mergozzo, separated from the northwestern arm of Lago Maggiore by the growth of the delta of the Toce. Baldi (1949) has, from a study of medieval records, obtained evidence that this lake was separated in the tenth century.

Lateral Lakes (*Type* 52). By far the most important type of fluviatile lake is formed when a large stream, by deposition of levees and of sediment elsewhere on its flood bed, aggrades its course faster than aggradation can occur in the lateral tributary valleys. This is very usual, since the smaller tributaries may not have had a chance to collect any sediment at all. The streams in the side valleys tend to become obstructed by the sediments deposited along the sides of the main valley. As aggradation increases, the lateral valleys become, in effect, drowned. Not infrequently, at high-level stages the main river actually flows into the side valleys, depositing at their lower ends reversed deltas pointing upstream and so increasing the obstruction.

As early as 1818 Darby suggested this mode of lake formation to explain the numerous lakes in the side valleys of the Red River. A different and more elaborate mechanism, involving the successive formation of log jams, was later proposed by Veatch (1906) but is now regarded as inacceptable (Vernon 1942). Davis (1882) developed the idea that lakes of this kind are indeed formed in the manner suggested by Darby, and

---

[23] Not to be confused with several other lochs of the same name.

pointed out that an extensive lake district of this sort resembles, when its hydrography is mapped, a drawing of a branch bearing leaves. Perhaps the finest examples are provided by the lakes on the tributaries of the Yang-tze-kiang in Hu-peh Province, China; the arrangement is most

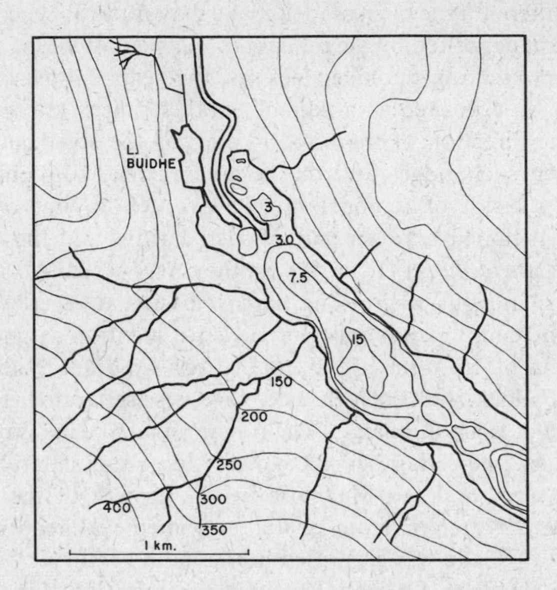

FIGURE 40. Strath lake at head of Loch Lubnaig (after Murray and Puller).

conspicuous in any map (Fig. 41) of the area, but otherwise little or nothing seems to be known about them. The largest, Lake Tung-ting, was subject to great seasonal variations in level; it is of considerable interest as the home of the peculiar dolphin *Lipotes*. The whole system has recently been greatly modified by flood-control works.

Magnificent examples of lateral lakes are also found along the north bank of the lower Danube in Bessarabia. Farther east, where the rivers parallel to these basins run into the Black Sea rather than the Danube, they end in limans or brackish lagoons in typical coastal drowned valleys.

A considerable number of lateral lakes occur along the course of the Darling River in western New South Wales (David and Browne 1950). The general appearance of several of these lakes, as indicated on maps, suggests some wind modeling; many appear to be rather regularly oval. Others are too far from the main river system to be pure lateral lakes and may be analogous to the end pans of the Kalahari (page 132).

A slightly different history from that of typical lateral lakes is recorded

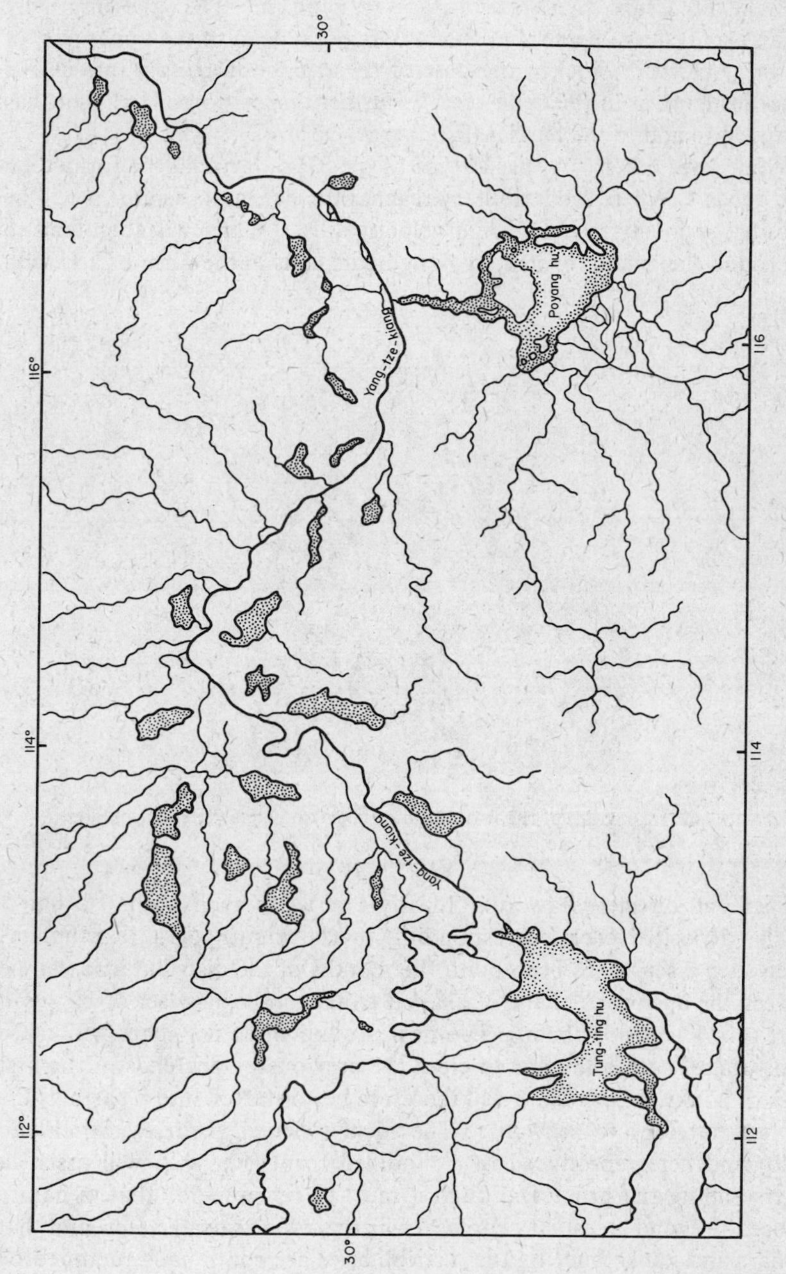

FIGURE 41. Lateral lakes on the Yang-tze-kiang, including Tung-ting Lake. (Re-drawn from various sources.)

for Lake St. Croix in Minnesota and Wisconsin. The lake originally was an arm of Lake Pepin, but the building of a delta in the upper part of that body of water blocked the base of the arm, converting it into a lake lateral to the river in the delta area. Further disposition of sediment has continued to add to the barrier (Zumberge 1952).

Deltaic levee lakes. A special type (Type 53) of levee lake is formed on large deltas. Where a distributary reaches the ocean, its sediment will be deposited suddenly as the current velocity falls. Water will then tend to flow round the bar so formed, prolonging its ends in the form of a U with

FIGURE 42. Diagram of the formation of lakes on deltas (after Strickland).

its open end directed seaward. Ideally a series of such U-shaped banks will be deposited each corresponding to the mouth of a distributary. When such a series has been built, the mouths of the distributaries will lie between the open ends of the U-shaped banks, and a new set of bars will start to form outside them. The flow around these new bars will cause deposition of such a kind as to close the previously open ends of the first series of banks. Each bank will therefore be converted into a basin as the new set develops to seaward. The ideal scheme, given by Strickland (1940) and here reproduced in a modified form (Fig. 42), will assist in understanding the process, though it must be remembered that in nature the ocean beyond an estuary can never be regarded as dead water, and that the bars and banks built by the distributaries are continually modified by tides and currents. When the river building the delta varies in volume, as is almost always the case, the banks deposited will be built up above mean sea level as levees, and lakes may collect in the central cavity or between a levee and the higher ground inland from the original coast. The continual subsidence of the alluvial deposits of the delta, owing to isostatic readjustment, maintains the basins even if there is a tendency for sediment to be washed over the levees. Since salt water can also easily be introduced, the deltaic levee lakes are frequently brackish or salt.

The major deltas of the world all bear lakes of this kind, but few detailed studies have been made of them. On the Mississippi Delta (Fig. 43), Lake Pontchartrain is a very fine example (Type 53a) of a lake held between the levee of an outgrown distributary, Bayou Sauvage, and the higher country north of the flood plain of the Mississippi. Other smaller lakes, such as Lake St. Catherine in the same region, are entirely enclosed (Type 53b) by levee walls of small distributaries (Steinmeyer 1939). The Camargue region of the Rhone Delta bears a large number of brackish lagoons, the largest of which is the Étange de Vaccarès, minimal area 64.8 km.^{2}, between the Rhône de Saint Ferréol and the Rhône d'Ulment. This and other examples in the same region have been well described by Russell (1942). Several conspicuous examples are present on the Nile Delta, the Mariotic Lake near Alexandria being the best known of those that are fully cut off from the sea. Strickland (1940) indicates the presence of a number on the Ganges Delta, though many of the *bhils* of this region are oxbows. Other examples can be found at the mouth of the Danube, of the Indus, and of the Yang-tze-kiang.

Meres and broads. Certain very shallow lakes of the eastern part of England, the meres of the Fenland, which were all drained during the eighteenth and nineteenth centuries, represent, according to the most recent researches (Jennings 1950), a type of levee lake (Type 54) peculiar in that the sediment forming the barrier was carried upstream by the tide rather than downstream by the flow of the river.

The northern part of the Fenland area was flooded by a slight marine transgression in Romano-British[24] times. The tidal water from this shallow arm of the sea, the southward continuation of the modern Wash, was evidently very turbid and deposited levees at the sides of the channels, known as *roddons*, along which it ran. Most of the meres collected as lakes inland from such levees. The section at the edge of the site of Red Mere (Fig. 44), formerly the second largest sheet of water in England and now the site of one of Messrs. Chivers' jam factories, indicates the roddon and levee that held the lake. Jennings finds that Willingham, Streatham, Ramsay, and Benwick Meres were formed in such a way. Soham Mere lay in a very shallow but quite definite stream valley and was therefore

[24] The archaeological evidence permits a fairly wide dating between 500 B.C. and 500 A.D.

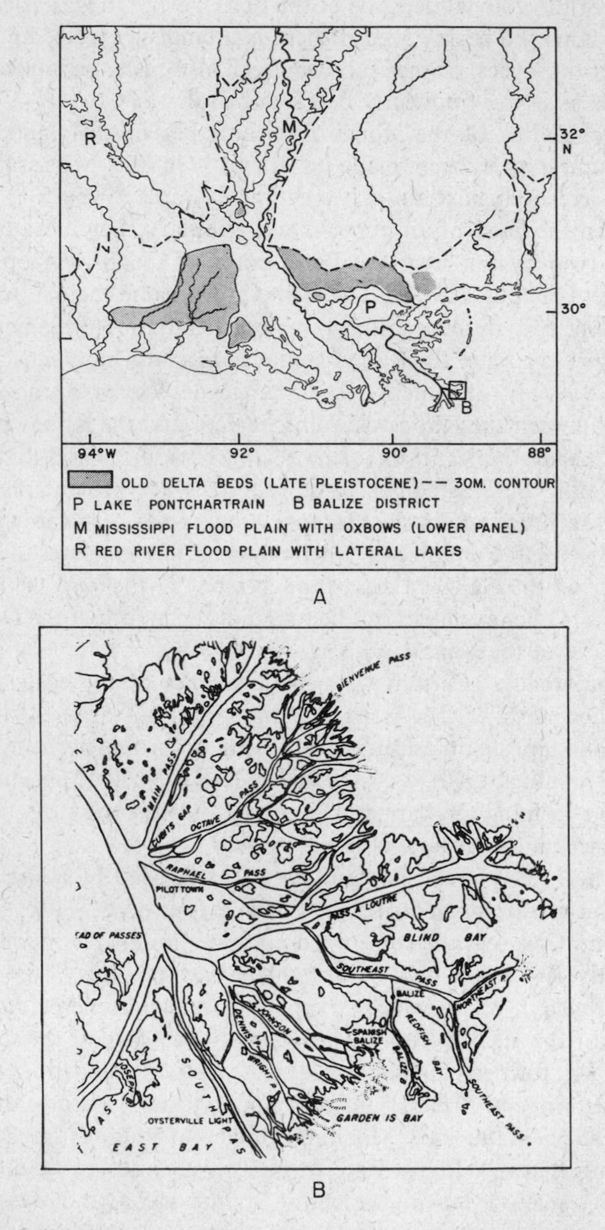

FIGURE 43. Mississippi delta to show: *A*, Lake Pontchartrain; *B*, numerous small lakes on the multilobate terminal part of the delta (after Russell and other sources).

analogous to the lateral lakes described earlier. The general relief of the country is so gentle that the distinction obviously was of little importance in determining the character of the lakes.

Some caution is perhaps needed in accepting this interpretation in its simplest form, because it was also believed by Jennings to apply to the Norfolk Broads, another series of bodies of water, happily still in existence,

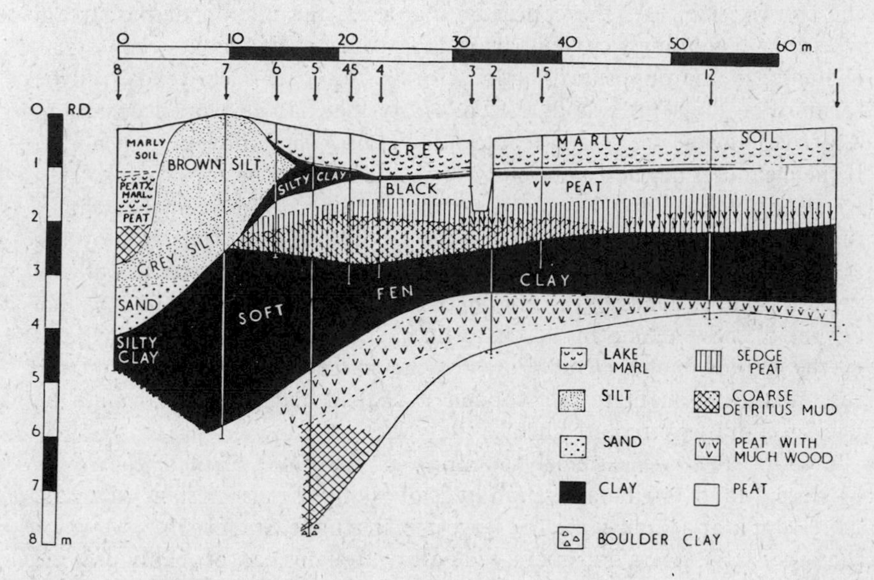

FIGURE 44. Section through the margin of the site of Red Mere (Jennings 1950).

that are separated from river channels by narrow alluvial barriers. As in the case of the Fenland meres, some of the broads, such as Hickling Broad, are lateral lakes in very shallow valleys, while others, such as Wroxham and Salhouse broads, lie in flood plains and are apparently held by levees. It is now known that in general the basins of the broads have vertical sides and therefore are apparently artificial (Lambert 1953, Jennings and Lambert 1953). It is believed that they were excavated in their present form by peat cutters during the Middle Ages. The peat which was removed does, however, appear to have formed in shallow depressions behind barriers of a marine clay. The medieval peat cutters may therefore have restored a system of lakes that originally formed in the manner suggested by Jennings.

Lakes of mature flood plains of rivers. In the lower parts of large river valleys, particularly those in which the main stream runs in a wide flood plain, several characteristic types of lakes tend to occur. These are associated partly with the kind of local erosion and partly with special types of deposition that occur in such flood plains.

Oxbows and other lakes in abandoned channels. Whenever a slight accidental variation in topography or structure produces the least sinuosity in the course of a river, greater turbulence is to be observed on the concave side of the stream and erosion begins. The concavity is therefore accentuated, while deposition starts in the slower, less turbulent water on the convex side. As the concavity increases, the section of the stream where the increasingly curved part returns to its original course now starts changing in the opposite direction, for in this part a convexity will be forming on the same side as the concavity that has developed upstream. Any river flowing in an easily eroded flood plain is therefore apt to meander. If the meanders develop sufficiently, whole loops may be cut off or isolated by silting. Since any loop has, during its formation, been the seat of maximum erosion in the middle of its concave side, this part of the abandoned channel will be slightly overdeepened, and even without any deposition at the ends of the loop may remain as a crescentic shallow lake (Type 55). Such lakes in old channels have long been known as oxbows in the United States (*B* of Plate 6), the name being derived from the resemblance in shape to the wooden U-shaped collar placed around the neck of a draft ox and attached to the yoke.

Oxbow lakes are extremely common in the flood plain of the lower Mississippi and its tributaries; hundreds of examples must exist in Arkansas, Mississippi, and Louisiana, the largest being Lake St. Joseph. Most of the other large rivers flowing in wide alluvial plains can probably provide examples. Numerous examples occur along the Darling and Murray rivers (David and Browne 1950) in Australia, where they are called *billabongs*; most of these are probably not permanently filled with water. Delebecque (1898) cites La Grande Mare north of Pont Audemor, very close to the mouth of the Seine, as a fine example in Europe, but most of the existing European examples are small ponds.

Occasionally rivers flowing over glacial outwash terraces far from their mouths may give rise to oxbows. Dahl (personal communication) cites Yersjoen in Ringerike, Norway, as an example. Ignatius (personal communication) has found beautiful examples in Quebec Province, Canada, in some cases showing evidence of two generations of oxbow formation.

It is evident that no very sharp distinction can be drawn between the category of lakes now being considered and the lateral levee lakes (Type 58) of the next section. Many oxbows, in fact probably all the most perfect ones of the Mississippi, must be dammed at their ends by levee deposits. Moreover, many lakes separated from river channels are simple slightly curved elongate basins occupying abandoned channels, not necessarily loops of meanders, formerly occupied by rivers and abandoned owing to breaking of levees or accidents of cutting and sedimentation during exceptional floods. Such basins, termed *lônes* in the Rhone Valley above Lyons, *bras morts* more generally in French, and *Altwasser* in German, are common in most large river valleys, though they seldom appear to be of great limnological importance.[25] Zumberge (1952) separates such lakes into those formed by uneven aggradation during floods (Type 56) and those formed by the abandonment of well-defined channels (Type 57). The former are usually very transitory; they are known along the bottom lands of the Mississippi River between Pike Island and southeastern Houston County, Minnesota. The latter, which when well developed are termed *meander scrolls* (*C* of Plate 6), are found along the Mississippi River in Benton County, Minnesota, and on the Minnesota River.

Lateral levee lakes. Large rivers carrying a considerable load of sediment tend to deposit levees or natural dykes delimiting their ordinary channels. This tendency is due to the fact that turbulence is minimal at the sides of a straight river and maximal in the middle. In the relatively quiet marginal waters, disturbance is frequently insufficient to keep a considerable part of the sediment in suspension. If deposition of such sediment occurs at high-level stages, the sediment may be left as a bank or levee when the water recedes to its normal level. Such levee formation is of course greatest where the river is, as a whole, aggrading or building up its valley. In the lower parts of most large rivers, such a condition is usual today, as the general base level determined by mean sea level has risen in the geologically recent past on account of deglaciation. Large flood plains are continually subsiding, moreover, owing to isostatic adjustment, and this permits aggradation to continue. It is also probable that owing to the large amount of easily eroded morainic material left on deglaciation, the majority of rivers draining glaciated regions still carry an abnormally great load of sediment in suspension. Many levees at the present time are of course largely or wholly artificial.

When a river is running close to the edge of its flood plain, which is

[25] The term *mortlake*, presumably derived from the place name, has been used in some English works as the equivalent of *bras mort* or *Altwasser*. Unfortunately, Mortlake in Surrey, though it is the site of the finish of the Oxford and Cambridge boat race, of certain curious and seemingly parapsychological phenomena of unknown nature demonstrated to Queen Elizabeth I by Dr. John Dee, and of the tentlike tomb of Sir Richard Burton, appears to offer nothing specifically of interest to the limnologist. The name, according to Ekwall (1947), means either young salmon stream or Morta's watermeadow, and is not derived, as was formerly fancied, from *mortuus lacus*. Since, moreover, it is the river rather than the lake that is "dead," the term, however convenient, is inappropriate.

FORT
SNELLING
PIKE ISLAND
MENDOTA BRIDGE
Lake
Minnesota River

A

N

B C

often delimited by a scarp, a lake (Type 58) may collect between the levee and the scarp. Catahoula Lake, on the course of the Little Red River, La Salle Parish, Louisiana, provides an example, recently described by Brown (1943). A scarp on the northwest delimits the flood plain, while a series of levees built, in part at least, by the French Fork, a distributary of the Little Red, delimits the lake on the southeast. The main stream of the Little Red River runs through the lake to the Red River, and so to the Mississippi. Like many fluviatile flood-plain lakes, Lake Catahoula is variable in area and depth. When the level is high in the Mississippi and Red Rivers, much water backs up into the lake, which then may be 12 m. deep and have an area of more than 100 $km.^2$ At times of low water the lake disappears completely. A large number of lakes of this sort occur in the valley of the Minnesota River, which, as the Warren River, was fed in late glacial times by the drainage of Lake Agassiz. The present valley is too large for the modern stream, and any sediment load deposited in the shallows of the modern river may easily isolate a basin between the modern stream and the old bank of the late glacial river. Grass Lake, Blue Lake, Fisher Lake, and Rice Lake, between Chaska and Fort Snelling, are examples given by Zumberge (1952), who also calls attention to a lake south of Pike Island formed in this way since 1894 (*A* of Plate 6).

Crescentic levee lakes (*Type* 59). In the lower reaches of a large river, where both marked meanders and levees are developed, the river may cut through its own levee at the concave side of a meander and form a new looplike channel enclosing the old. Lakes comparable to oxbows may be found in the abandoned channel, but there is also the possibility of crescentic lakes being formed between a successive series of levees formed in this way. Hobbs (1912) gives a map, here reproduced (Fig. 45), of Texas Lake enclosed by Walnut Bayou, the exact position of which has not been ascertained.

LAKE BASINS FORMED BY THE WIND

There are a number of basins, the origin of which appears to involve wind action, either through its effect on the distribution of sand or through the deflationary or erosive action of wind acting on broken rock. These basins occur mainly in dry regions, and when they contain perennial lakes it is usually on account of a relatively recent change in climate.

PLATE 6. *A*, lateral levee lake formed between the levee of the Minnesota River and an old bank south of Pike Island, Fort Snelling, Minnesota, since 1894. *B*, oxbow lakes along the Mississippi River, Aitkin County, Minnesota, on the floor of glacial Lake Aitkin. *C*, arcuate troughs or meander scrolls along the Mississippi River, Benton County, Minnesota. (*Phot.* U. S. Department of Agriculture, by courtesy of Dr. J. H. Zumberge.)

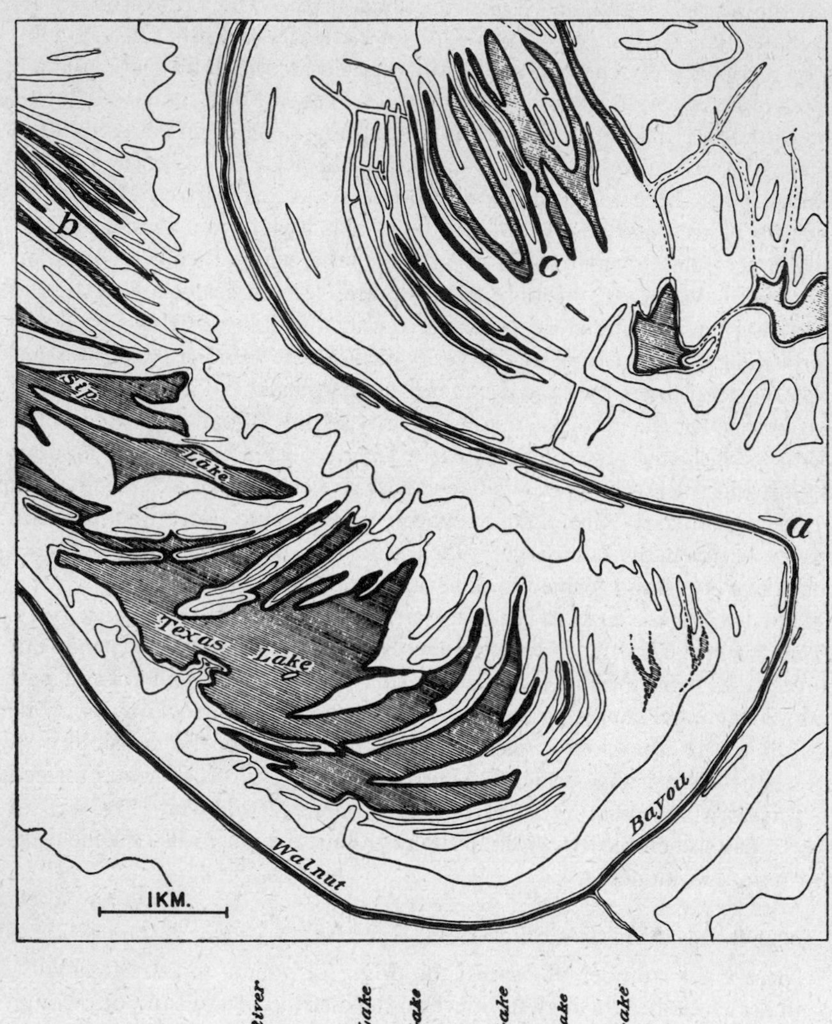

FIGURE 45. Texas Lake (Hobbs 1912).

Obstructive action of wind-borne sand. The best example of a lake in a valley dammed by wind-borne sand dunes (Type 60) appears to be Moses Lake in Crab Creek, Washington, briefly described by Russell (1893).

Several examples of normally dry basins formed by the piling up of dunes at the end of a river flowing into sandy desert are known in the Namib Desert of southwest Africa. Sossus Vley at the western end of the Tsauchab River is the best known of these (Jaeger 1939). When water collects in such basins, it seeps below the dunes rather than evaporating in the basin, which therefore does not become saline. The numerous basins of the interior of West Australia have been supposed by many investigators to have been formed by wind filling the channels of drying river systems, but it is probable that other forces have also been at work in these cases (page 135).

Uneven deposition of loess. According to Wilhelmy (1943), the "pods" of the south Russian steppes, shallow basins from 12 m. to 12 km. in diameter and from 10 cm. to 10 m. deep, have mainly been formed by uneven deposition of loess (Type 61), partly reflecting the subloess topography. The basins contain temporary lakes in the spring. Though they appear to resemble the pans of other semi-arid regions, Wilhelmy is skeptical of the significance of deflation and animal erosion. It is probable that at least some of these basins would be attributed by Russian investigators to solution (page 110).

Lakes formed between dunes. Rows of dunes may, if a suitable impermeable substrate exists, form parallel valleys between which lakes (Type 62) can collect. An example is provided by a series of long, narrow, shallow lakes (Fig. 46), up to 6 km. long, in Cherry County, Nebraska (Anderson and Walker 1920). The Pot Holes southwest of Moses Lake, Washington, occupy such depressions between dunes as reach below the prevalent water table (Harris 1954). Hedin (1904) has described very dramatic series of lakes of this kind in the Tarim Basin.

Deflation basins. The most important type of wind action in forming lake basins (Type 63) is deflation or wind erosion. The clearest evidence of this process is provided by those cases in which the deflated material is piled up as a curved mound of sand or *lunette* (Hills 1939) along the lee shore of the depression. The conditions for wind action to excavate a basin are not fully understood. It is certain that an arid climate is a necessary but by no means a sufficient condition. It is also likely that some initial action by an agency other than the wind is necessary to begin the erosion. The role played by animals visiting a small basin during wet seasons has been emphasized by some writers. Horizontal stratification of the bed rock has also been regarded as important. The evidence for these conditions will appear from the examples to be described.

Deflation basins in North America. The most characteristic deflation basins lie on the high plains that extend on the eastern side of the Rocky Mountains northward from Texas. Basins apparently due to wind action have been recorded from Texas, New Mexico, and Nebraska.

This is not the only part of the drier regions of the North American continent in which deflation basins occur. Gilbert (1895) noted many in Arkansas but gave no detailed description. Bryan (1925) records at Susuta, Sonora, just south of the Mexico-Arizona boundary, a lakelet about 470 m. across and about 1 m. deep in a round depression surrounded by sand hills. The sand had clearly been blown from an underlying

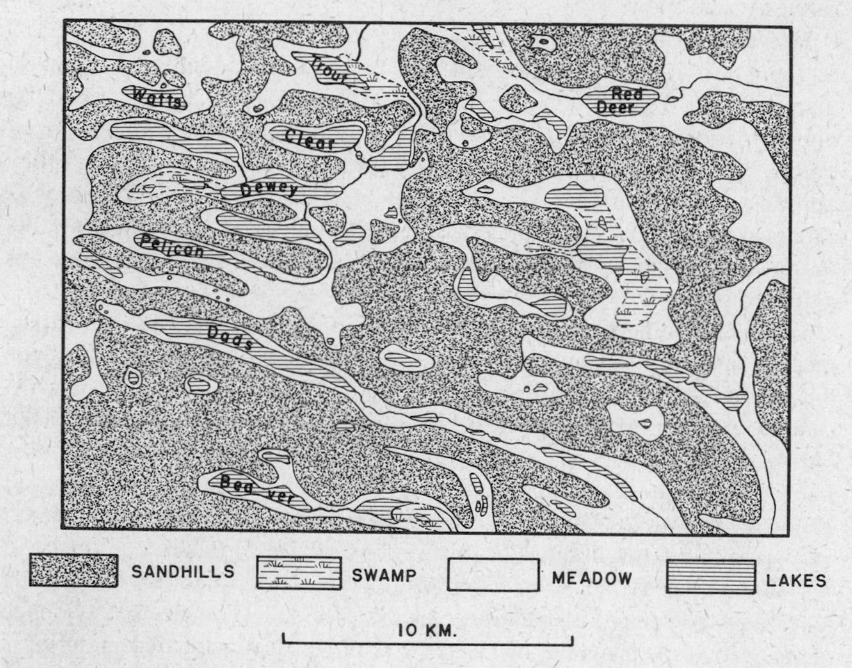

FIGURE 46. Lakes between sand hills, Cherry County, Nebraska (Anderson and Walker 1920).

sandy clay which held the lake. Farther north in Nevada, Big Washoe Lake, which looks very like one of the Transvaal pans, may in part be due to the uneven deposition of outwash, but there are sand hills on its eastern shore which presumably indicate remodeling by the wind. There must be other examples in the Great Basin.

The Texan examples, on the Staked Plains or Llano Estacado of northwestern Texas,[26] have been studied by Evans and Meade (1945). Previous

[26] The author has been struck with the similarity, at least from the air, of the basins of this region and those of the southern Transvaal.

workers had supposed them to be produced by solution or subsidence, but in most cases there is no evidence of such processes. In the larger, better developed depressions which tend to contain saline lakes, there is a lunette of sand on the eastern side. The depressions have a greater volume than do the lunettes, and much of the deflated material must have been carried away. Such depressions, if large, are ordinarily termed *playas*.

The lakes and dry depressions lie along the western sides of these dunes and are elongated transverse to the wind. Evans and Meade indicate that when water is present in them there is evidence of contemporary increase in area at the ends, though little deflation is now occurring. It is probable that these Texan basins combine the features of true deflation basins with those of the oriented Alaskan lakes described by Livingstone (1954) and on page 100. In the largest of the Texan basins a lacustrine clay fills a large part of the depression; this clay may contain bones of *Bison taylori*, *Equus* sp., *Camelops* sp., *Glyptodon* sp., and mammoth. It is certainly late Pleistocene and is regarded as of Wisconsin age by Evans and Meade. The initial deflation must antedate this late Pleistocene clay. Comparable clays of middle and early (Nebraskan) Pleistocene age are known in basins which became entirely filled and which Evans and Meade regard as representing lakes like the modern playas but formed at earlier stages in the complicated climatic cycle of the Pleistocene. There appear to have been two later cycles of deflation, separated by a minor wet phase, since the deposition of the clay. Both dry periods were somewhat more arid than is the present time.

The deflation basins of eastern New Mexico have been studied by Judson (1950), who examined not merely intact examples but certain basins that have been dissected by the backcutting of tributaries of the Canadian River. This river, a tributary of the Arkansas and so part of the Mississippi drainage, has cut into the high plains of eastern New Mexico. The localities examined by Judson lie on the southern side of its valley. One dissected basin, the San Jon depression, is an important archaeological site and has been carefully studied.

The basins examined by Judson, which lie in Quay and Curry counties, have a more irregular shape than most other deflation basins. They tend to be arranged in rows in long shallow valleys, apparently of Pliocene or early Pleistocene age, which now are broken up by the depressions and are mainly marked by the occurrence of a calcareous cap rock on the low ridges between the rows of basins. These basins usually contain temporary lakes, the largest, Hatfield Lake, being about 1 km. long. The eastern shores of the lakes are usually sandy and a well-marked sand-hill lunette may lie eastward of the lake. In addition to these large basins, small, shallow buffalo wallows occur throughout the region. They were probably initiated as local pools of water after rain, perhaps due to the grass retarding drainage. Once such a pool was visited by animals, some zoogenous erosion would occur, mud being carried away on feet and coats. Water would then more easily accumulate in the hollow after the next rain, and the process would continue at an accelerating rate.

The study of dissected depressions indicates that they were underlaid by a deposit of leached sand, continuous with the Ogallala sandstone, a Pliocene formation underlying much of the area. Judson supposes that during pluvial periods leaching, perhaps starting in the vicinity of buffalo wallows, locally removed the calcareous cap rock and calcareous cement from the Ogallala sandstone. During subsequent arid periods the loosened sand was easily removed by the wind.

The San Jon basin contains lake sediments in which an extinct species of bison is associated with Folsom-like artifacts. These must have been deposited before the basin was dissected from the north. A later sediment, apparently formed in pools in the partly drained basin, contains modern bison with Yuma-like artifacts. The main deflation periods must have been middle or early Pleistocene, but there seems to be no clear indication of one particular pluvial and subsequent arid period that were responsible.

Deevey (1953) notes an as yet unpublished account by Flint of oval deflation basins in Nebraska; the basins between sand hills in this region are likely to be due as much to deflation as to deposition.

The South African pans. An enormous number of shallow depressions, usually round, oval, or kidney-shaped in outline, occur scattered across a wide belt in South Africa, from the eastern Transvaal and northern part of the Orange Free State, westward across the Kalahari desert and northern part of the Karroo into southwest Africa. These basins are commonly called *pans*, though the term *vlei* is locally applied to many of them. Neither term can be used in its colloquial sense to designate any very satisfactorily defined class of basin. Etymologically, vlei refers to a part of a river system, though it is usually employed in a much wider context, while the term pan is used to include certain very large basins such as the Etosha and Makarikari basins, which are certainly of tectonic origin. The term pan will be restricted here to basins apparently of erosional rather than tectonic origin, without a downstream dam, the exact nature of the erosive forces being left for determination.

At the present time the pans of most of southern Africa are seasonally or perennially dry. The main exceptions are certain basins in the eastern (Lake Chrissie area) and southern parts of the Transvaal that hold perennial though not quite permanent water. At least in parts of the Transvaal (Plate 7) and in the case of a few examples in the Kalahari, the

PLATE 7. **Avenue Pan, Witwatersrand, Transvaal, a deflation basin (*phot.* South African Air Force).**

pans are true rock basins; it is reasonable that this will prove to be the case for many others in other regions. In size they vary from 100 m. or less up to a few tens of kilometers in diameter. Many are scattered about seemingly at random, but in the case of the Lake Chrissie pans in the eastern Transvaal (Wellington 1943) and many in the Kalahari Desert (Jaeger 1939), they are apparently parts of ill-defined ancient drainage systems.

Rogers (1922) points out that, in general, pans occur only where the bed rock is horizontal or has a very low dip. Three main agencies have probably been involved in pan formation. In parts of the Kalahari and in the Lake Chrissie area, ancient river erosion has doubtless been involved. In some cases in the Kalahari, when the pan contains water, a definite inflow and outflow are apparent (Jaeger 1939). It is probable that uneven intermittent fluviatile erosion and deposition have played a part in determining the sites of such pans, but it is almost equally certain that other agencies also have been involved. The two agencies that most investigators have invoked are deflation and animal erosion; probably both processes have operated in many cases in a complimentary manner. Whenever a small depression is formed in an arid region, any water entering it will tend to evaporate rapidly, so that the mud at the bottom, if it rests on impermeable rock, becomes saline or alkaline and cannot support vegetation. Wind action is obviously facilitated by the resultant bareness of the bottom.

The view that wind erosion is of primary importance has been expressed by most South African workers (Rogers 1907, 1908, 1922; du Toit 1907, 1908). Jaeger (1939) believes some of the pans of the Namib Desert to be pure deflation basins, and that in the Kalahari, wind erosion assisted by weathering processes in the beds of intermittent stream beds has played a part. Many of the Kalahari examples, termed end pans by Jaeger, receive drainage from a short valley terminating in the pan. Wyberg (1918) has observed tufa or calcareous sandstone on the south side of pans in the western Transvaal, which he attributes to sand blown from the pan floor during dry times, forming a lunette, which thus provides a mass of porous material up which lime-rich water moved in damper periods.

Passarge (1904, 1911) was the main proponent of animal erosion as a major factor in the production of pans, though the idea had been previously suggested by Alison (1899). Moreover, Dr. F. H. H. Guillemard, who visited the interior of South Africa in 1877, had come to the conclusion that the pans had been enlarged and deepened by animal erosion. Many years afterward he wrote (*in litt.*, see Hutchinson, Pickford, and Schuurman 1932) that at the time of his visit any naturalist would have come to such a conclusion as the "result of constantly seeing the blesbok and springbok in the pans of the district of the Western Free State and Kalahari border. . . . Yet the old Boers, even in those days, considered the game was finished. . . ." Jaeger (1939) is inclined to the view that Passarge overemphasized the role of zoogenous erosion, but it is quite certain that modern conditions give no hint of the significant possibilities of the process. Passarge (1911) was fully aware, however, of the part played by wind in removing material triturated by the feet of the vast herds of ungulates that would have visited any slight depression that might temporarily have harbored a little water. It is possible that in, for instance, the Chanse Veld of the Kalahari, where most pans are excavated in a relatively soft material, zoogenous erosion was more important than in the Transvaal.

In the greater part of the region where pans exist, it is probable that little deepening is now occurring. Passarge made a number of studies of the sediments in the pans of the Kalahari region and found that a layer of pan sandstone was often overlaid by one or two discrete calcareous layers, over which an alluvial deposit might be present. In some cases a smaller pan had formed in the calcareous deposits of a large pan in a rock basin. Passarge believed that the best developed sequences in Damaraland implied an initial dry period during which the original excavation occurred, a somewhat less dry phase in which the sandstone was formed, a drier period in which calcareous crusts were formed and siliceous cementing of the sandy sediments occurred, and then two wet phases producing the main calcareous deposits, each followed by a dry phase, the second dry phase being contemporary. No subsequent dry phase was as extreme as was the initial arid period. Similar though less complex stratification is indicated by Jaeger (1939).

Rogers (1922) has pointed out that the pans of the eastern Transvaal are quite out of harmony with the existing climate and must have been formed under conditions distinctly more arid than those obtaining today. The postulated period of aridity probably occurred at a time in the relatively remote past.

Smuts (1938, 1945) has published sections through the marginal deposits of several pans in the Witwatersrand region. The relationship of these sediments to the adjacent subsoil is not entirely clear; they appear to thin out toward the modern pan floor. In several pans, three or four periods of deposition of ferricrete or ferruginized grit can be recognized (Fig. 47). Such deposits are supposed to be the result of water-borne material, washed toward the pan in pluvial times, becoming cemented during the following interpluvial. All the layers contain pebble tools and the three uppermost ones flakes also. Some technological advance is apparent when implements from the upper layers are compared with those of the lower, and some of the latest are believed to have affinities with early Palaeolithic implements from other sites. The oldest implements are compared with the Kafuan of central Africa by Smuts, and appear to be more primitive than the pebble tools of Olduvai, which (Leakey 1951) are referrable to the Kamasian or second east African Pleistocene pluvial. Smuts thinks that his four beds record four Pliocene pluvials, a view that has the support of some South African authorities but is unlikely to find wide acceptance. A more conservative view would be to suppose that the pan deposits represent various episodes of one of the earlier pluvials, perhaps the Kamasian. The period of excavation might then be either the first or Kangeran-Kamasian interpluvial, or pre-Pleistocene. Further study of the localities is obviously necessary to elucidate their later history; one would expect that in some basins of this sort later pluvials would be recorded. The relationship of the obvious unconformities in the sections, and of the thinning of the layers toward the modern pan floor, to interpluvial periods of deflation also requires investigation.

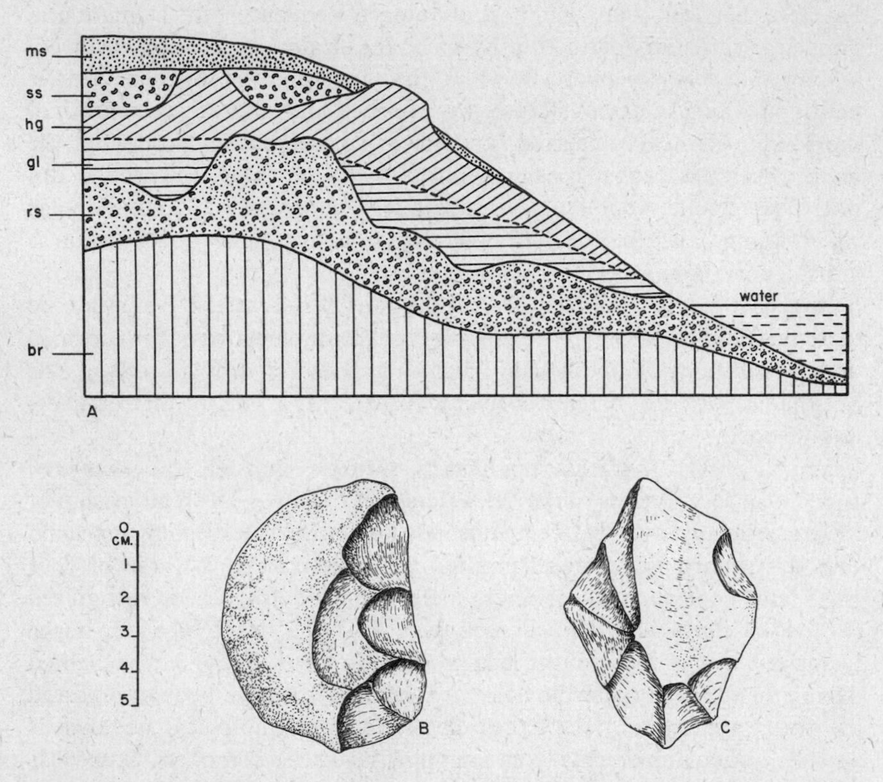

FIGURE 47. *A*, rough section through marginal deposits of a pan at Brentwood Park, Transvaal. The stratified deposits are regarded by Smuts as indicating a period of decreasing aridity, passing from loose red soil (*rs* lying on quartzitic bedrock *br*), into gritty loam *gl* and hard gravel *hg*, which is taken to indicate the first pluvial. The subsoil *ss*, lying unconformably on the hard gravel and capped by modern soil *ms*, is believed to represent a later second pluvial. *B*, *C*, artifacts from a second pluvial bed, equivalent to *ss*, in the less perfectly developed section at Benoni Pan; comparable implements were found at Brentwood Park but were not figured; the scale indicates the size of these implements. (After Smuts.)

Deflation basins in Australia. In the central division of Western Australia there are a very large number of irregular basins which after rain are filled with transitory saline lakes. The area (between about lat. 24°30′ S. and 33° S. and long. 117° W. and 123° W.) in which they occur is termed Salinaland by Jutson (1934) and the Salt Lake Division by David and Browne (1950). Jutson's map indicates at least twenty-seven such basins over 14 km. long, and there must be an enormous number of smaller ones. Many others are scattered more sparsely to the north and east of this area. They are often elongate and may lie in rows as if they were part of a drainage system. They have extremely flat, smooth floors when dry. The floors are covered with fine silt, which is generally very thin and in places may expose bedrock. There is usually only a small deposit of halite and gypsum left when the lake evaporates. There are frequently cliff-like shores on the western sides of these basins and sand hills on the eastern sides. In some cases deep depressions filled with smoothed-over detritus occur in the lake floor.

There is evidence that one or two of the lakes, notably Lake Disappointment and Lake Cowan, are of tectonic origin. It has usually been supposed that most of the more elongate basins are remains of river valleys choked with drifting sand; this view appears probable to David and Browne (1950). Jutson (1934) believes, however, that it is far from established, and that the elongate form of the basins may be due to a migration westward, the west ends being deflated while sand tends to drift into the eastern end. He is of the opinion that rapid erosion after rare rainstorms is of considerable importance in the area even today, and that the ancient drainage pattern postulated to explain the distribution of the basins is largely imaginary, consisting of the basins themselves and of quite recent channels running into or connecting them.

Though the large basins of South Australia are tectonic in origin and many of the larger lakes of western New South Wales are lateral to the Darling River, smaller deflation basins filled intermittently with water are common in the last named region (Collins 1923) and are evidently very like those of southern Africa. In the Barrier Region in the extreme west of New South Wales, the pans are described as being due to the deflation of superficial sand exposing a hard clay floor. Evidence of fusion of two adjacent pans was noted by Collins, who suspected that the largest pans were formed in this way. In such large pans, saline water may remain for many months without entirely evaporating. Lunettes or irregular crescentic mounds of deflated material on the eastern sides of deflation basins have been noted also in northern Victoria, on Kangaroo Island, in the Peterborough district of South Australia, and in southwestern West Australia (David and Browne 1950). The maturity of the soil profiles on some of these structures suggests deflation during Pleistocene interpluvials.

Pans in other regions of inland drainage. It is probable that some of the numerous depressions in other endorheic regions, notably in central Asia and in parts of South America, are deflation basins.

Large desert depressions. In the deserts of north Africa, south Africa, and Australia there are a number of immense depressions, most of which now rarely if ever contain water, but which have contained important lakes during the Pleistocene pluvials. In north Africa there is a series of inland drainage basins or shotts between the Greater and Lesser Atlas Mountains. Other basins occur farther east, the Fayum depression in Egypt being the best known. The large irregular depressions of southern Africa, notably the Etosha Pan in southwest Africa and some of the very large pans in the Makarikari Basin of the Kalahari, are certainly tectonic (Jaeger 1939). In South Australia, Lake Eyre and its associated basins are further remarkable examples, dry remnants of a Pleistocene lake, Lake Dieri (Mawson 1950). The basins north of Pyramid Lake in the Lahontan Basin of western North America are comparable. These large basins presumably are nearly always in part structurally determined, but it is probable that, in some cases at least, deflation has been involved in the production of their present form. In the very similar shotts of north Africa, a small amount of very saline water may be present. The position of the water is variable, as it is blown about over the flat surface of the shott by the wind (Gauthier 1928). The same phenomenon has been noted in the much less extreme environment of the Carson Sink, Nevada (Hutchinson 1937c). Caton-Thompson and Gardner (1929), Gardner (1929), and Huzayyin (1941), following Beadnell (1905), believe the Fayum depression to be primarily due to late Tertiary deflation. Though Sanford and Arkell (1928) have strongly urged that it was cut by water, the most recent evidence seems to be against this hypothesis (Huzayyin 1941). The depression contained a well-developed fresh-water lake during part of the paleolithic and again in neolithic times, and today its bottom is occupied by the brackish Lake Quarun. If it is really a deflation basin, it is much the most scientifically important of any of this class of basins still containing a lake.

LAKES ASSOCIATED WITH COAST LINES

The ordinary way in which a lake associated with a coast line is formed is by the building of a bar across some irregularity or indentation of the coast line. The mechanics of bar formation are dealt with in relation to lake coast lines in a later section of the present work. For the present it is only necessary to note that if any longshore current is flowing along a strait coast indented by a bay, the inertia of the current will tend to carry it across the mouth of the bay. If the current is carrying material disturbed by wave action, this material will be deposited as a spit, following the coast line. Provided other erosive action does not take place, the spit can finally cut off the bay as a coastal lake or lagoon. Such action is possible along either a sea coast or a lake coast.

Maritime coastal lakes. In many cases both the river discharge and tidal currents prevent the complete separation of a bay as a lagoon or coastal lake (Type 64). When the building is sufficiently vigorous, any river discharge into the bay may be insufficient to keep open a channel across the bar. Either a small stream may remain flowing over a bar or on low-lying coasts a new drainage system behind the bar and parallel to the coast may develop, ultimately discharging into some bay or estuary which is so placed that tidal currents or fluviatile outflow is great enough to keep open a channel to the sea. In this way a series of lakes, corresponding in most cases, though not invariably, to old bays and estuaries, may be built up behind a coastal ridge. The main conditions favoring such a development appear to be, first, a history of rising sea level drowning old estuaries, followed by a slight lowering of sea level which adds to the width and so the stability of the bar; and, second, the development of extensive sand hills, which tend to blow over and fill any drainage channels that may develop. The first condition is met by any stable coast line outside the areas glaciated in the Pleistocene, because there has ordinarily been a considerable rise in sea level subsequent to the culmination of the last glaciation, followed by a very small fall, probably since hypsithermal postglacial times. Many glaciated coasts also fulfill the condition because initially the water liberated by the melting ice rose faster than the land, but later, after the volume of the ocean has become nearly constant, a slow plastic recoil of the continental blocks released from ice pressure has caused a slight rise in land level.

Outside the glaciated areas the finest series of examples may be found (1) along the western coast of France, in the Landes area, where the bar is covered with very recent sand hills, and where the lakes, though they retain a triangular estuarine form (Fig. 48), drain into streams parallel with the coast; (2) along much of the Mediterranean coast, notably that of the Peloponesus (Philippson 1892), where they exhibit karstic features; (3) along much of the southeastern and eastern coasts of Australia; and (4) along parts of the coasts of southern Africa. In some of the Australian cases, and probably some of the African also, there is evidence of two or more generations of bars holding lakes, the inner ones being now largely filled up and dried (Mawson 1929). In North America some salt lakes of this sort are present along the coasts of the Gulf of Mexico.

Beautiful examples of isolated lagoons formed by the building of bars across recently drowned valleys are found on Cape Cod and on the shores of Nantucket Island. The irregularity of the coast line, which is apparently being drowned at the moment, is here largely of glacial origin; many of the depressions separated from the sea by bars doubtless started as kettles. Other examples combining glacial history with recent changes

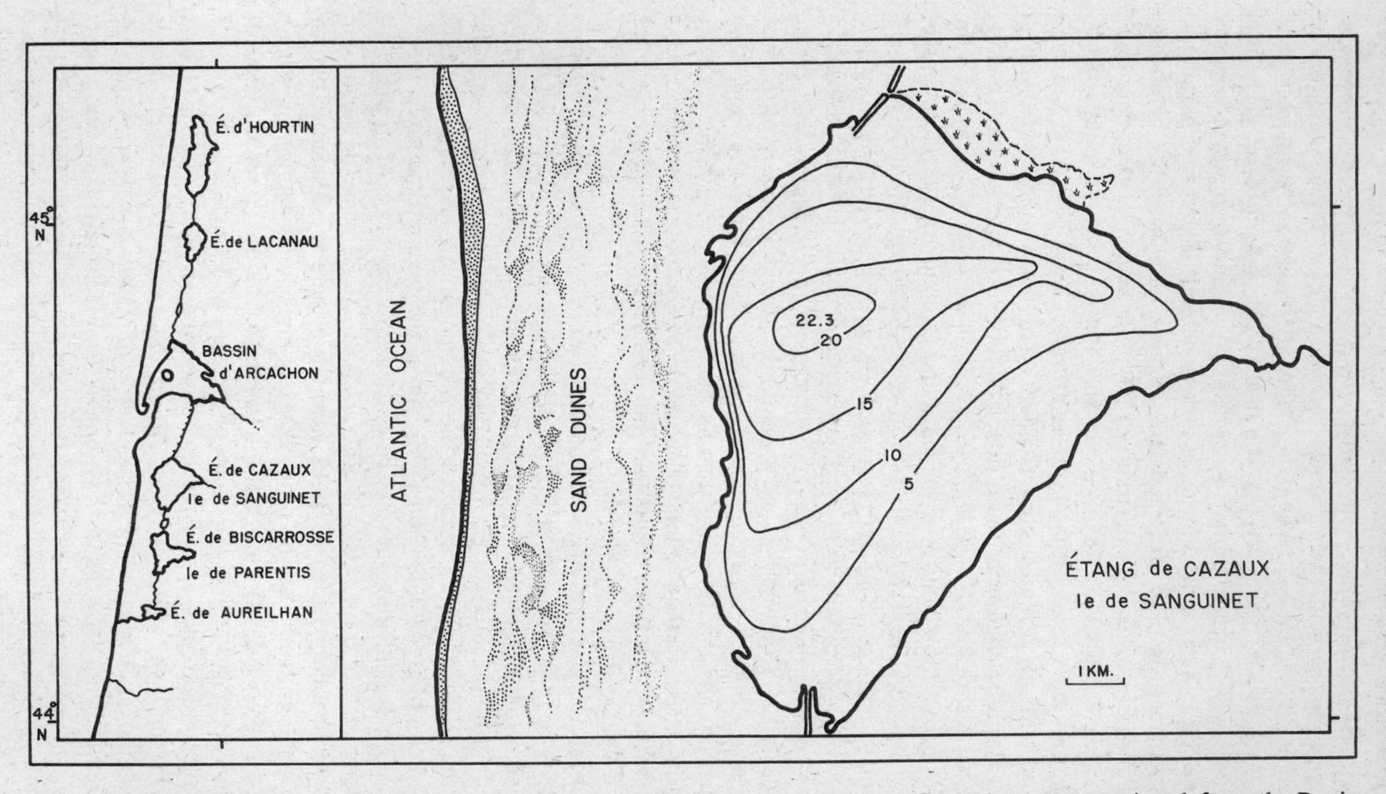

FIGURE 48. Coast of the Landes district of western France, showing the secondary parallel drainage pattern (canal from the Bassin d'Arcachon southward is artificial) and the bathymetry of the Étang de Cazaux et de Sanguinet (after Delebecque).

in the relative levels of land and sea are found along the Baltic coast of Eastern Germany.

The formation of a single bar across an inlet is not the only way in which a coastal lake can be cut off from the sea. Not infrequently, bars form between the coast and adjacent islands. Such a bar is ordinarily called a *tombolo.* Often, if the island is of any size, it may be united to the coast by two tombolos, which then enclose a lagoon or lake (Type 65). Salt lakes of this sort are well known along the coast of Italy; Johnson (1919) gives as illustrations of this form the Stagno di Orbetello, enclosed between two tombolos which unite Monte Argentario to the coast of Italy (Fig. 49), and a small lagoon in a compound Y-shaped tombolo at Morro del Puerto Santo, Venezuela.

A spit growing from an angle on a coast line frequently may turn at the end and rejoin the coast. Such *cuspate spits* enclose small depressions, but they are ordinarily of little limnological significance.

Lakes produced by lacustrine shore processes. The same kinds of processes that have isolated many coastal lakes from the ocean have also acted to separate small lakes from large, or to divide lakes into two or more parts.

Zumberge (1952) has considered three possible cases. In the first (Type 66), a bar is built across a bay, which becomes isolated as a small lake. He believes several hundred examples might be found within the State of Minnesota alone, the most dramatic perhaps being Buck Lake, cut off from Cass Lake in Beltrami County (*B* of Plate 8). Many examples can be found along the shores of the Laurentian Great Lakes (Gilbert 1885). In some cases, such as that of Lake Nabugabo, cut off from the northeastern coast of Lake Victoria in central Africa, the existence of the small secondary lake is of considerable biological interest. In the development of bars along a lake coast, though tidal influences are negligible, Krecker (1931) has pointed out that seiche currents may often play a similar role in preventing the complete isolation of small coastal lakes; it is probable that the action of such currents has limited the number of completely isolated lakelets along the margins of many large lakes.

The second case (Type 67) considered by Zumberge is that of a lake being bisected by two spits approaching each other from opposite shores. Though such a form is by no means uncommon (Raisz 1934) it is rare for the spits to meet. Zumberge figures one case of almost complete bisection of a lake in this way, namely Marion Lake, Otter Tail County, Minnesota (*A* of Plate 8). Here the barrier consists of a very long spit developed from the south shore and a far less well-developed one from the north shore. The lake is separated into two basins at times of very low water. This

A

Plate 8. Lakes formed from pre-existing lakes by littoral processes *A*, division of Marion Lake, Otter Tail County, Minnesota, into two almost separate basins through the growth of spits. *B*, Buck Lake, a bay cut off from Cass Lake, Beltrami County, Minnesota. *C*, Gould Lake, behind the spits of a double tombolo in Leech Lake, Cass County, Minnesota. (*Phot*. U. S. Department of Agriculture, by courtesy of Dr. J. H. Zumberge.)

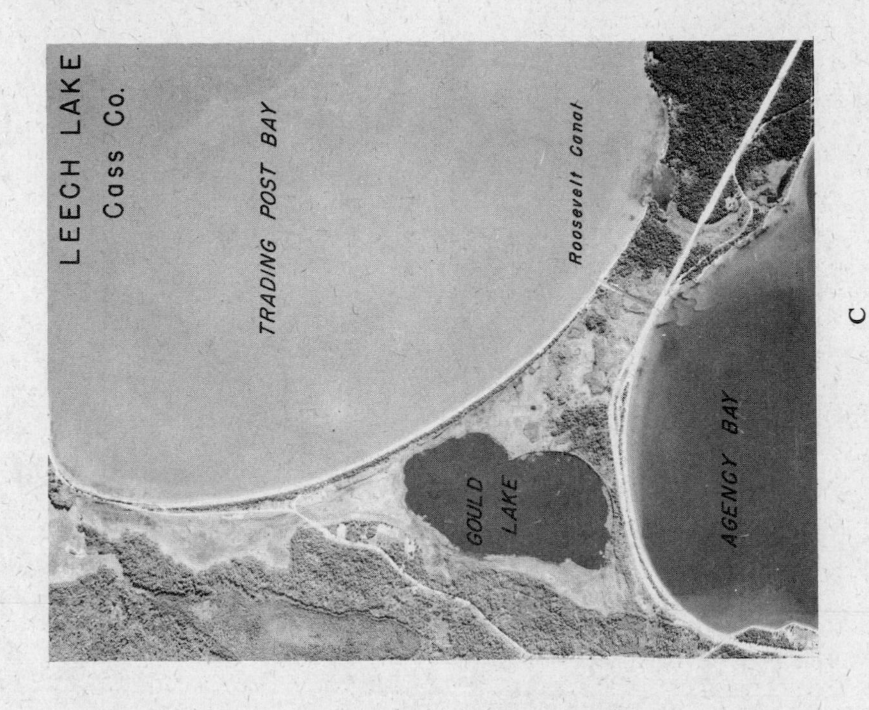
LEECH LAKE
Cass Co.
TRADING POST BAY
Roosevelt Canal
GOULD LAKE
AGENCY BAY

C

CASS LAKE
BUCK LAKE

B

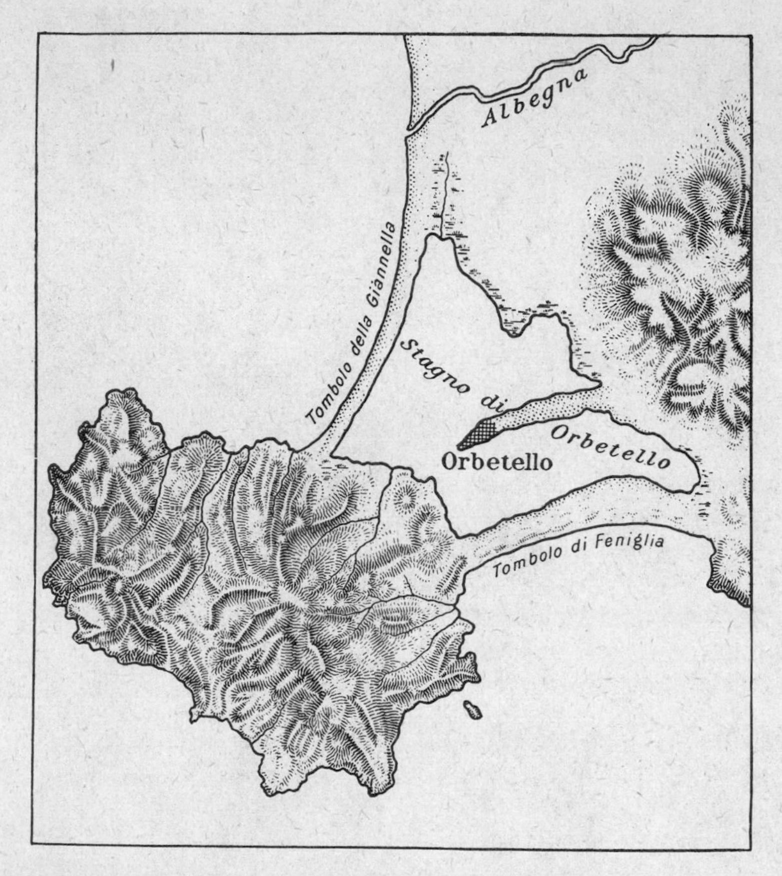

FIGURE 49. Stagno di Orbetello on the western coast of central Italy, a lake held by a pair of tombolos (Johnson 1919).

type of separation must be clearly distinguished from the bisection of a lake by a delta or pair of deltas.

The third case (Type 68) distinguished by Zumberge is the formation of a lake behind a cuspate spit. Such cuspate spits are known as small-scale features along the shores of many lakes (Gilbert 1885, Fenneman 1902, Meinertzhagen 1927). Usually the depression enclosed is very small, but sometimes, probably only when the bars form a double tombolo, a quite definite lake may occupy the depression. Zumberge gives as an example Gould Lake (*C* of Plate 8), cut off from Leech Lake, Cass County, Minnesota, between the two bars of an obvious double tombolo.

The processes that have produced marginal lakes along the shores of large existing lakes naturally operated also in the past. Gilbert (1885) has described the shore formations of Lake Bonneville in detail, and notes one remarkable case in which a small and now somewhat impermanent lake, Rush Lake, still exists behind a large littoral bar at Stockton, Utah. A similar case of a littoral lake behind a bar formed along an extinct lake shore was formerly provided by Humboldt Lake in Nevada, on the margin of Lake Lahontan. Artificial changes in the basin appear to have caused this lake to become perennially dry.

LAKES FORMED BY ORGANIC ACCUMULATION

A small and generally unimportant category of lakes has been formed by the accumulation of the less easily decomposed parts of organisms, deposited either as a dam or as a complete rim round a basin. The category contains two quite different series of lakes, those in which the obstruction or construction is caused by plant remains, and those in which construction by corals has produced a basin.

Phytogenic dams (Type 69). It has already been pointed out that blocking of the drainage of Lake Okeechobee in Florida by vegetation may be responsible for the persistence of that very large but shallow body of water. In many low-lying humid tropical areas, similar action may be expected to have played a part in preserving if not in forming lake basins. Murray (1910) mentions the flooding of the valley of the Upper Nile when the course is obstructed by "sudd" as a less permanent example of the blocking of a river by plant debris.

Livingstone (unpublished) believes that the same kind of effect may occasionally take place in north temperate or subarctic latitudes. According to Livingstone, Silver Lake, Halifax County, Nova Scotia, is due to a dam formed by a mat of *Sphagnum* and *Andromeda* that has developed so extensively that the lake has spilled over one side to form a new outlet.

Closed phytogenic basins (Type 70). Murray (1910) believed that the vast number of basins of the tundra regions of the far north were formed by local uneven accumulation of snow, whereby some parts of the tundra vegetation started to grow earlier than other patches uncovered a little later in the spring. He supposed that the patches which started growing late would form peat less rapidly than occurred round about, so that closed depressions surrounded by rather high peat deposits would be formed. There seems little doubt that most of the basins of the tundra are actually due to local thawing of buried ice wedges or of frozen soils.

It is possible that sometimes ponds on bog surfaces are formed by unequal rates of growth of peat, as for instance on the bog at Rivière du Loup, Quebec (Auer 1930). The formation of peat may control the creep of bogs on slopes, and sometimes fissures at right angles to the slope may develop and form small ponds. The process is probably an indication of slow growth relative to the rate of creep. Such a pond is really a very special case of a lakelet on a landslide surface. The presence of such ponds can lead to striking solifluction if the pressure of the water on the down-slope wall is great enough, but such a process does not build a lake. Livingstone notes this process as occurring on the east shore of Yale Lake in the Brooks Range, Alaska.

Lakes in coral atolls (Type 71). A few coral atolls are known in which a complete ring of coral rock exists, or in which the apertures in the original incomplete ring have become so choked with sand that the lagoon is isolated from the ocean. Apparently isolated lagoons may retain underground connections with the ocean. The lagoon of Malden Island, which apparently is salter than the sea, rises and falls tidally but lacks a visible connection with the ocean (Dixon 1878). The same is true of a saline lake or pond apparently in a doline on Nauru (Power 1905). Where the lagoon is entirely isolated, it may become very salt as on Laysan, on which island there also was once a fresh-water pond. Other examples of shallow, closed, and probably saline lagoons are known on the islands of the Phoenix group (see Hutchinson 1950 for a summary of available information).

By far the most remarkable fresh-water lake in an atoll is that of Washington Island, an island 6.1 km. long and 2.3 km. wide, in the central Pacific at lat. 4°43′ N., long. 160°25′ W. The rim is complete and encloses a lake (Fig. 50) said to be 9 m. deep (Wentworth 1931), with a

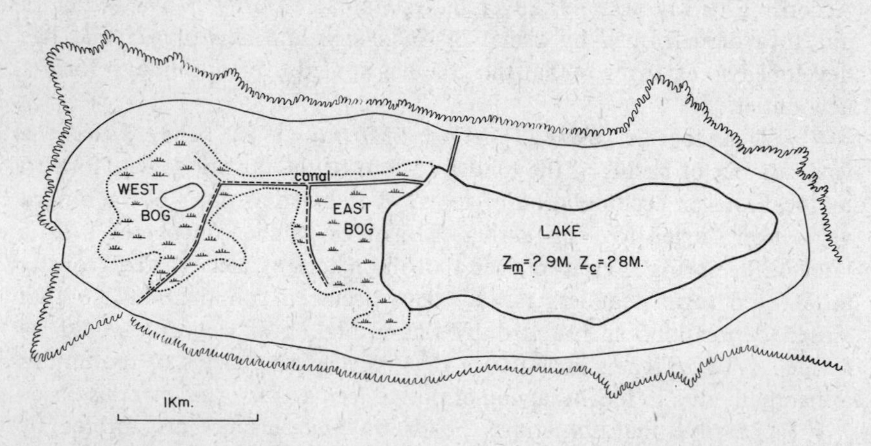

Figure 50. Washington Island, central Pacific, a lake with marginal peat bog in an atoll (after Christophersen).

considerable development of peat bog formed from *Scirpus riparius* Presl. on its western side. The maximum height of the land rim is about 6 m. above sea level, so that the lake occupies a cryptodepression. Elschner (1915, 1922) believed that the lake is sealed by a calcium phosphate mud derived from guano deposited by sea birds. In 1906 when a channel was cut in the peat, a canoe was allegedly found at a depth about 1.4 m. below the bog surface. This canoe, believed by Emory (1934) to be perhaps of medieval Tongan manufacture, appears to be not more than 250 years old (Preston, Person, and Deevey 1955). There is evidence, from the pre-European introduction of the parrot *Vini kuhlii* (Vigors), that the island had been visited from the Austral Islands, while others of the Line Islands seem to have been occupied for a time from the Tuamotu and the Hawaiian Islands. If the find of the canoe is genuine, the peat clearly must have formed at a quite recent date. Christophersen (1927) concludes that it cannot advance further, owing to the depth of the lake. Though the lake as such must postdate the emergence of the land rim, presumably late in the Pleistocene, it was the home of an endemic duck, *Anas streperus couesi*, a dwarf subspecies of the circumboreal gadwall; unhappily, this duck is known from only two immature birds collected in 1874, and it is now evidently extinct. The lake of Washington Island clearly requires modern investigation.

An atoll lake that is comparable to that of Washington Island is apparently now forming in the lagoon of Clipperton Island off the western coast of Mexico, in lat. 10°17′ N., long. 109°13′ W. The island consists of a pan-shaped land rim, on which there is apparently no raised coral *in situ*. Part of the volcanic substrate of the island is visible as a large trachyte rock on the southeastern side of the rim. In 1840 there were two boat channels through the rim into the lagoon, but prior to 1898 these became choked with coral sand (Wharton 1898), producing an entirely closed lake at least 37 m. deep. This has gradually freshened by dilution with rain water; Snodgrass and Heller (1902) call it brackish, and Taylor (1939) has described a remarkable algal flora consisting of species of *Lyngbya*, *Calothrix*, *Closterium*, *Cosmarium*, *Oöcystis*, and *Oedogonium*, an assemblage strongly suggesting relatively dilute water at least in the upper layers of the lake.

A few old coral islands possess small fresh-water lakes; that on Nauru has been mentioned. Two lakes in the central depression of Kita Daito Jima, namely Akaike and Oike, have been described by Yoshimura (1938a). It is probable that in these cases the lake basins are due at least as much to solution as to the original structure of the island. The same is likely to be true of a small fresh-water basin on Odtia in the Marshall Islands, described by Chamisso (1821).

LAKE BASINS PRODUCED BY THE COMPLEX BEHAVIOR OF HIGHER ANIMALS

Two mammals, namely the American beaver and man, can construct dams across valleys flooding previously dry areas and so producing lakes. Man, moreover, by making excavations undertaken to exploit mineral resources, may produce large holes in the ground that can fill with water. The detailed consideration of artificial methods of producing lakes is beyond the scope of this work, but since the events in artificial lakes are of almost as great interest to the limnologist as those in natural basins, it is desirable to call attention to certain peculiarities of man-made lakes.

Beaver dams (Type 72). A large number of ponds and very small lakes in North America have been formed wholly or partly by the activities of the American beaver (*Castor canadensis*). The dam-building activities of the European beaver (*C. fiber*), though they may on occasions produce ponds, seem to be on a much smaller scale; in the Swedish colonies studied by Fries (1943), the largest dams are about 15 m. long.

Although in many regions in which beavers still exist the area of the water ponded behind their dams is of very limited extent, when a dam is built in a relatively wide valley a fairly large area may be flooded. The best examples are apparently found in the region from south of the Laurentian Great Lakes northward to Hudson Bay, but Morgan (1868), whose classical work on the American beaver provides some of the best information on beaver dams in the virgin landscapes that existed in the middle of the last century, indicates equally fine examples existed in more montane areas farther west, for example in the headwaters of the Missouri, in the Big Horn and Laramie Mountains, and in the Black Hills. Good examples are still to be seen in the Yellowstone region and in northern New England.

The dam is built of tree trunks, sticks, and mud. The largest examples are evidently the result of the work of many generations of beavers. Mills (1913) records one example of evident antiquity on the Jefferson River near Three Forks, Montana, that was 650 m. long. Warren (1927) records that the beaver dam holding Beaver Lake, Yellowstone, is 321 m. long, and quotes Shiras as finding the beaver dam holding Echo Lake, Grand Island, Lake Superior, to be 457 m. long. Such great lengths are clearly exceptional, but dams 30 to 50 m. long must have been quite common under natural conditions.

Morgan has given a map of part of the region south of Lake Superior; a section of this map, including a reach of the Carp River, is reproduced in Fig. 51. The lakelet behind Morgan's dam No. 19 had an area of about 0.1 km.^{2}, the dam itself being about 150 m. long and the water behind it about 2.8 m. deep. An even larger lakelet, area 0.24 km.^{2}, called Grass

Lake, existed to the east of the area shown in the figure. The dam here was shorter, about 80 m. long, and the lake not over 2 m. deep. It is probable that all the ponds smaller than these, that existed along every stream studied by Morgan, are due to the work of beavers; the levels of some of the larger lakes, doubtless in kettle holes, may have been raised or at least maintained by the building of beaver dams.

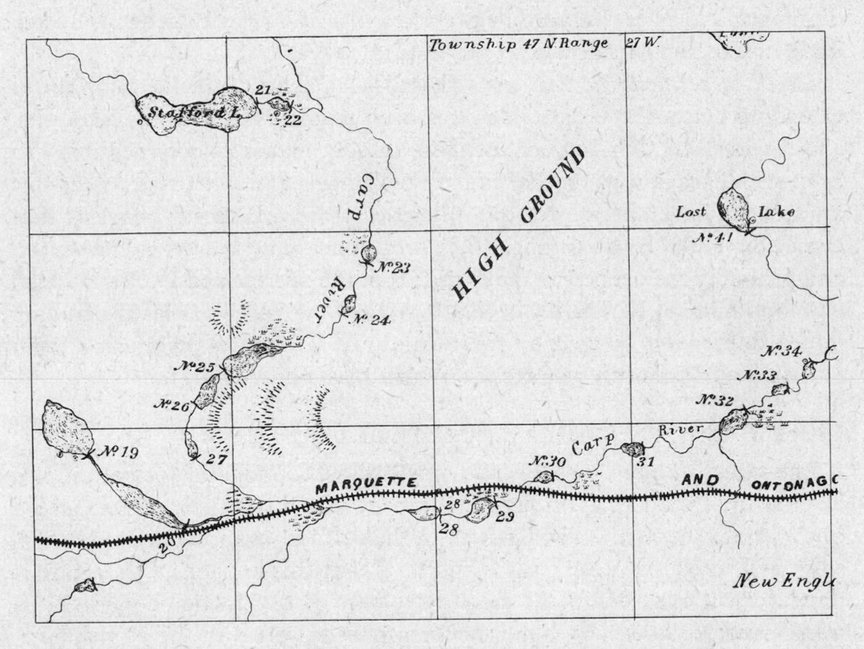

FIGURE 51. Beaver dams on the Carp River (Morgan 1868).

Morgan considered that ordinarily the dam is convex upstream; this form is used in the construction of dams by man on account of its stability. Warren, however, dismisses this conclusion. In view of the great number of dams studied by Morgan, it is difficult to believe that he was entirely in error on this matter.

Human construction of dams (Type 73). The damming of streams to produce artificial lakes was probably practiced by the Egyptians prior to 2000 B.C. The Lake of Homs, area 50 km.^{2}, in the Orontes Valley of Syria, has a dam that may date from *c.* 1300 B.C. Ancient dams are also known in Mesopotamia, where the convex upstream pattern just mentioned was adopted (Drower 1954).

Artificial lakes have been constructed to serve as water supplies, to provide power, to aid in navigation, to increase the supply of fish and other aquatic products, for defense, and for purely decorative reasons. As in nearly all types of engineering, enormous strides have been taken in the art of construction of large dams during the past century. All the really large artificial lakes of the world are very young. This means that their shore lines are virtually unmodified and follow the contours of the slopes of the valleys that they occupy. If they are built in areas of very dendritic drainage pattern, an extraordinary complexity of outline may be produced, quite unlike that of the majority of natural lakes.

Lakes in artificial depressions (*Type* 74). Where artificial excavations extending below the water table have been made, as in quarries, clay pits, and the like, on abandonment of the pumping that is usually necessary to keep such excavations dry, lakes may develop. Small depressions of this kind occur in temperate regions wherever there are useful minerals that can be exploited by quarrying. The largest example known to the writer, and probably the deepest in the world, is in the abandoned De Beers Deep diamond mine at Kimberley in South Africa. A case of a lake subsiding into a depression formed by the burning of a layer of lignite has been recorded at Chomutov in Czechoslovakia by Kuchař (1947).

LAKE BASINS FORMED BY METEORITIC IMPACT

The rarest, though certainly the most dramatic mode of formation of a lake basin (Type 75), is through the impact of a meteorite. When such a body strikes the earth, it will penetrate the surface a short distance but will soon be brought to rest. At the moment that this happens a tremendous pressure will exist below the meteorite; part of its kinetic energy will be transferred to shock waves in the earth, part to waves in the atmosphere, and part will be dissipated locally as heat. The result of the intense heating is to expand water vapor and other gases in the rocks, producing an explosion. The resulting explosion crater will be far greater in diameter than the meteorite producing it. Since the pressure will be exerted in all directions, the crater will be roughly circular whatever the angle of impact of the meteorite. The floor of the crater may be elevated centrally, and around the crater wall concentric elevations and depressions may occur. The floor of the crater will consist of a considerable thickness of crushed and, in places, fused rock. Most of the meteorite itself is likely to be blown out of the crater by the explosion. If the maximum depth of artificial explosion craters is plotted against the diameter on double logarithmic paper, a smooth curve is obtained, and this curve appears to be continuous with that describing the same relationship for the craters of the moon. The region between the artificial and lunar craters is neatly filled by the few terrestrial meteorite craters, the original depth of which can be ascertained; one of these, the Chubb Crater, Ungava, which contains a large lake,[26] lies well in the lunar part of the curve.

Modern meteoritic crater lakes. Most of the best known meteorite craters are dry, probably because the preservation of such earth forms is better in arid than in humid regions. The Great Arizona Crater or Coon Butte contained a lake during some part of the Pleistocene, its floor being formed of about 30 m. of lake sediments (Barringer 1905).

Three reasonably well-established meteorite-crater lakes exist. (*a*) Kaalijärv, on the Island of Ösel or Saaremaa in the Baltic, is a tiny lakelet in the largest of six pits from one of which meteoritic material has been recovered (Reinwaldt and Luha 1928; Kraus[27] Meyer, and Wegener 1928; Spencer 1933; Fisher 1936; Reinwaldt 1938). (*b*) Laguna Negra, Campo del Ciele, about lat. 27°28′ S., 61°30′ W., in the Gran Chaco of Argentina, occupies one of a series of small flat craters; most are circular, but Laguna Negra is elliptical, about 320 m. long and 215 m. wide. The crater rim rises only 1.2 m. above the surrounding plain, and the depth of water appears to be little over 1 m. (Nagéra 1926). (*c*) Ungava or Chubb Lake, Ungava, Quebec (Plate 9), occupies the largest well-established terrestrial meteorite crater. The diameter of the top of the crater wall is about 3350 m. and the over-all depth about 410 m. The lake in the crater has a maximum depth of 251 m. The crater apparently is of late Pleistocene origin (Meen 1950 and 1952, Harrison 1954, Martin 1955).

In addition to these cases, a "buffalo wallow" near Brenham, Kansas, was found (Nininger and Figgins 1933) to contain meteoritic material, and so to be a small meteorite crater. The shower of meteorites that fell on June 30, 1908, in the area about lat. 61° N., long. 103° E., between the Yenisei and Lena Rivers in Siberia, produced swampy depressions but apparently none that can be termed lakes (Kulik in Merrill 1928).

Several lakes other than those listed above have been supposed by some investigators to be of meteoritic origin, though there is no concensus of opinion definitely in favor of the hypothesis. Lake Bosumtwi in the Gold Coast has already been mentioned (page 37). The Pretoria salt pan, an isolated craterlike structure in the Transvaal containing a saline deposit and a little water, is perhaps more likely to be meteoritic in origin than is Lake Bosumtwi (Rohleder 1933, Cotton 1944, Spencer 1933).

---

[26] This was first pointed out by Dr. Matt Walton in a manuscript note on the margin of the copy of Baldwin's *The Face of the Moon* in the Yale University Library.

[27] Kraus was not able to subscribe to the meteoritic hypothesis and believed solution of subsurface salt had produced the pits. This was before Reinwaldt had discovered meteoritic material in the vicinity.

PLATE 9. Ungava or Chubb meteorite crater lake, a basin like a lunar crater covered with ice, contrasting strikingly with the surrounding irregular lakes of a recently glaciated landscape (*phot*. Canadian Royal Air Force).

Fossil lakes in cryptovolcanic structures. Several ancient depressions now filled with sediments, but otherwise comparable to the craters just described, are known. They are usually designated as *cryptovolcanic* because they superficially resemble large calderas, but independent evidence of vulcanism is lacking. It is probable that at least some of them are meteorite craters. Baldwin (1949) lists twelve such cryptovolcanic basins, ranging in age from the late Cambrian or early Ordovician to the Miocene. The two most recent, both of Miocene age, namely the Steinheim Basin (Branca and Fraas 1905, Rohleder 1933) and the Reiskessel (Kranz 1934, Williams 1941) in Southern Germany, have already been mentioned (page 38). Both contained lakes during part of the Miocene. The sediments of these lakes have yielded fossil mollusks of considerable biogeographic interest.

The Carolina bays. It is necessary to consider separately one of the most extraordinary groups of lake basins known anywhere on the earth's surface, namely, the vast series of shallow elliptical depressions (Type 76) found scattered along the coastal plain of the southeastern United States from New Jersey to Florida. The greatest number of these basins occurs in the Carolinas and in the contiguous parts of Georgia. Following a rather peculiar local usage, these depressions are commonly called bays. Their origin has been the source of a great deal of discussion; no consensus of opinion has yet been reached by students of the matter. The two most convincing hypotheses postulate that they are due either to deflation or to meteoritic impact; the problem is therefore conveniently discussed after the better authenticated cases of meteorite craters have been considered.

The Carolina bays constituted a series of oriented elliptical depressions most of which are now filled with deposits of a shallow lacustrine or paludine character. Many have been drained and cultivated, but a sufficiently large number still contain enough open water to be regarded as lakes. It is possible that all of them started as lake basins. The term bay refers to the dominant vegetation of those parts of the surface lacking open water.

Depressions of this sort are known from the extreme northeastern part of Florida, northeastward into the Chesapeake Bay region of Maryland; a few doubtful examples have been noted in New Jersey. Prouty (1952) estimates that at least 140,000 large and medium-sized depressions exist, and that if small examples overlooked by topographers are added, there may be as many as a quarter of a million.

The bays are limited to the coastal plain and appear to have been formed prior to the retreat of the sea from the Suffolk Scarp or Pamlico Terrace, the youngest terrace on this coast. Frey (1950) points out, however, that they are much more numerous well inland from this scarp. They are all elliptical and are oriented approximately northwest-southeast. There is a slight tendency for the long axis to undergo a clockwise rotation as one proceeds from northeast to southwest. The ellipticity[27] varies, but appears to have a modal value between 0.3 and 0.4, provided single bays alone are considered. There are innumerable cases of superposition of bays, implying that the formation of adjacent depressions was not strictly contemporaneous, but the general appearance of the whole series suggests that there are no regional differences in age; in particular the more coastal bays are not manifestly younger than the most inland. In view of their distribution, predominantly on the less recent Sunderland and Coharie Terraces, Frey is unconvinced that they are really of the same approximate age. All are underlaid by sandy sediments, but apart from this there seems to be no geological determination of their occurrence. They normally have a rim of sand around them, best developed along the eastern part of the southeastern end of the depression, but the steepness of the rim may be greatest on the northwest side. Where there is intersection of two superimposed bays, the rim is highest at the intersection point (Prouty 1952).

The water-filled bays generally have a very shallow bench at the southeast end, due apparently to wave action (Frey 1949). The depressions are all very shallow; there is an evident correlation of maximum depth with length, but the deepest basin so far explored, that of Singletary Lake, is only 6.6 m. deep. The deepest point may be central or at the southeastern end. When a lake is still present in a bay, its long axis is usually inclined more north and south than is that of the bay (*A* of Plate 10). This may well be attributed to the action of westerly winds, though Prouty believed that it is due to the rotation of the earth in the formation of the bays. Frey (1955a) has obtained evidence that the original basin of Singletary Lake, which appears today to be regularly elliptical, was actually formed by the fusion and remodeling of three far less regular depressions.

Greater details of the stratigraphy of the lake sediments will be given when the problem of lake development is discussed in Volume II. Wells and Boyce (1953), who consider that pollen-analytical studies in the basins are unreliable owing to the production of unconformities by burning, believe that the lakes themselves are depressions left after fires in peat-filled bays. However, Frey's (1953) profiles from several of the bays, including Singletary Lake, show a remarkably consistent postglacial sequence within the range of radiocarbon dating, but below this there are two bands of organic sediments that are older than 40,000 years (Rubin and Suess 1955). If the Pamlico Terrace is the equivalent of the Sangamon interglacial, there is nothing inconsistent in these dates. The only stratigraphic difficulty raised by the bays that have been studied is the extraordinary thinness of their sediments.

---

[27] (Long axis – short axis)/long axis. This will be zero for a circle.

Omitting theories which appear to be too improbable to consider seriously, such as that of Grant (1945) that they are modeled by shoals of ovipositing fish, which cannot explain the sand rim, implies an unbelievable population, and is otherwise inconsistent with much that is known of the bays (Prouty 1952), the theories of origin can be grouped as meteoritic or extraterrestrial on the one hand and as terrestrial on the other.

The first group of theories, due originally to Melton and Schriever (1933), postulates a meteoritic origin for the bays. In its original form the theory requires modification, because the ordinary shape of a meteorite explosion crater is circular. The modified theory (Prouty 1935, 1952; MacCarthy 1937) supposes that the bays were formed, not as the result of an explosion when the meteorite had entered the ground, but by the compression wave in the air accompanying the meteorite. Experiments in which rifle bullets are fired at low angles into clay covered with powdered plaster of paris, can give very good models of elliptical basins generated by such shock waves. Prouty supposes that the numerous superimposed bays represent the successive impact of a number of meteorites in a shower of some size, and that though they differ in age, the difference may be of the order of a few minutes. He considers that the process would occur only in regions of loose and relatively unconsolidated surficial deposits, and that the limitation to the region of the coastal plain is explained by the distribution of such material. The final site of rest of the meteorite is believed to be represented by a magnetic high found to the southeast of the bay. The evidence of such spot highs is adequate, but Johnson (1942) believes that there are a sufficient number of such highs not associated with bays to cast much doubt on the significance of the association. Proponents of the nonmeteoritic theories, moreover, have supposed highs to be due to precipitated iron in the seepage downslope from the bay. Prouty has advanced a rather speculative theory to account for heart-shaped bays supposedly formed by a pair of meteorites in tandem, the air-shock cone of one sucking in that of the second under the influence of the rotation of the earth.

The most cogent objection to the meteoritic hypothesis would appear to be cases in which the bays are clearly younger than other topographic features but seem to be arranged with respect to such features. Johnson emphasizes this situation. Prouty calls attention to it in relation to Cooke's ideas that the bays resulted from the development of elliptical lakes in pre-existing depressions, and figures a case in which six bays are alligned in a row at the side of a river. He points out that if Cooke were correct one would expect the bays to replace the river, but does not indicate why six meteorites hit in a row. It is possible, though Johnson doubts it, that such arrangements are merely special samples of a very large random

A

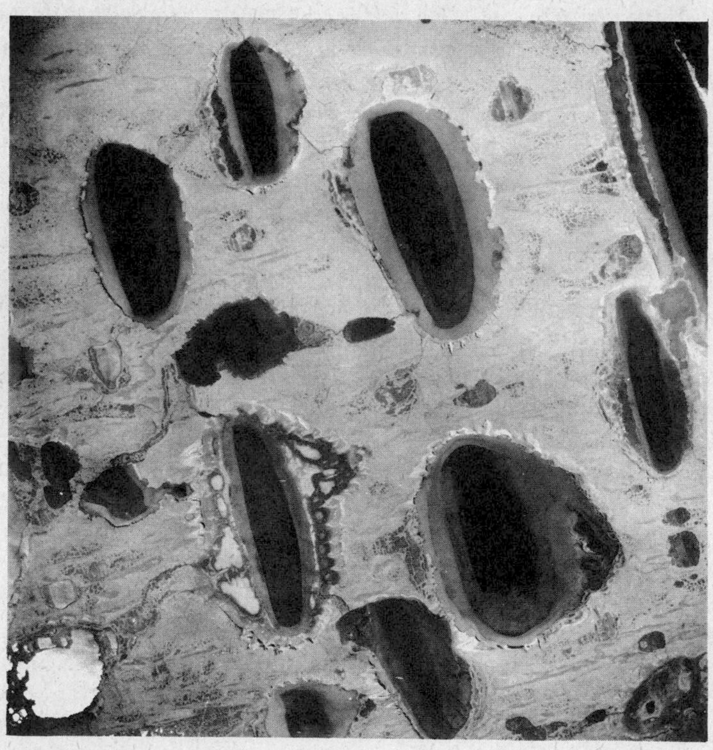

B

array. Johnson's other main argument against the meteoritic hypothesis, namely that it implies formation of circular basins, takes no cognizance of the modified theory of Prouty.

The theories of purely terrestrial forces are varied, but all of them consider the wind as an important agent in bay formation. At the present time it seems certain that the bays are not growing, most of them having, in fact, become filled with peat, and that if they owe their origin to wind action, it was at some definite time in the past. When, as is usually the case if a lake is present, there is peat between the lake shore and the bay rim, there is evidence in some cases of a recent increase in lake area (Frey 1954). Frey thinks increases and decreases of this sort have alternated over a long period of time. The simplest theory, and one which to the present writer has a certain appearance of probability, is that the bays are deflation basins excavated by the wind. They have, indeed, been compared by Shand (1946) to the pans of the High Veld in the Transvaal.

Odum (1952) has reconstructed a weather map for the winter during a glacial maximum, and concludes that the prevalent wind direction fits the prevalent long axis of the bays. Odum considered that the wind effect was primarily one of modeling the basins when they contained water. If so, it is reasonably certain that the basins would be oriented at right angles to the prevalent wind. The present form of some bay lakes, which seem to be expanding and which have a slightly different axis from that of the depression that contains them, might involve modeling across a southwest wind. If, as Odum supposes, the prevalent winds were northwestern at the time of the modeling of the bays, and if the latter contained water, they should have a southwest to northeast orientation. If his northwestern wind is responsible for the bays, then they must have been dry when formed. This is admittedly quite possible, though it is worth noting that during a large part of the Cary they were certainly filled with water; to adopt the deflation hypothesis it would be necessary to suppose that during some earlier glacial stage the region was much drier than during the second half of the Wisconsin glaciation. If the general orientation of the long axes was developed by the wind when the bays contained water, it would seem that Odum's weather map is irrelevant to the process.

The main proponent of modeling by water currents has been Cooke (1933, 1934, 1940), who in an ingenious series of papers finally concluded that the bays originated as irregular depressions between slight ridges and

---

PLATE 10. *A*, Salter's Lake, Bladen County, North Carolina, a typical bay lake; note difference in orientation of the bay and the lake (from W. F. Prouty, by courtesy of Dr. C. E. Prouty and the Geological Society of America). *B*, elliptical oriented cryogenic basins in lat. 70°26′ N., long. 153°48′ W., near Point Barrow, Alaska (by courtesy of U. S. Navy and U. S. Geological Survey).

that the present form is due to the interaction of wind stress from any direction with the rotation of the earth. It is, however, most unlikely that the processes that he postulates could occur in shallow basins filled with water of ordinary eddy viscosity. Schriever (1955) concludes that Cooke's final theory is hydro-dynamically impossible. Where lake basins are being remodeled by winds of variable direction elsewhere, the net effect is a circular rather than an elliptical form (Raisz 1934).

The most elaborate theory of terrestrial origin has been developed by Douglas Johnson (1936, 1937, 1942, 1944). Johnson's ideas developed considerably as his study progressed; in their final form, the essential elements in his theory were that the bays originated as artesian springs, which tended to migrate inland up-dip. The seepage of the springs increased the size of the basins by solution of any soluble material in the surficial sediments. Wind action played an important part in modeling the elongate solution lakes so produced and in producing the sand rims. Prouty, in particular, has examined the hypothesis critically and finds it deficient for a variety of reasons. He specifically points out that Johnson's hypothesis would suggest that the more coastal bays should be younger and probably larger, as the hydrostatic head would be greater, than the more inland bays; that the direction of backward migration that defines the major axis is hardly likely to be constant; and that some correlation both with bedrock geology and with topography would be expected. None of these expectations appear to be fulfilled in nature.

At the present time it is obviously not possible to reach any adequate solution of the problem; Frey's recent discovery of the irregular multiple structure of the original Singletary Lake depression may indeed render untenable all the ideas so far put forward.

SUMMARY

The existence of lakes depends on a variety of processes which produce depressions surrounded on all sides by a rim higher than the deepest enclosed point. Since in any region in which water is available to fill such a depression there is usually also the possibility of erosion, the rim holding the lake starts to be destroyed as soon as it is formed. This will happen most rapidly if there is an outlet or effluent. Meanwhile the influent is carrying sediment into the lake and filling its basin, and organic production within the lake does the same. It is evident that the lifetime of a lake is limited. So many lakes exist at present mainly because we live in a period that has been geologically somewhat catastrophic.

The actual processes involved in lake formation have been classified as *constructive* when the rim is actively built, *destructive* when the lake is excavated, and *obstructive* when a pre-existing valley is dammed.

It is more convenient, however, to classify according to the general nature of the processes responsible for building, excavation, and damming. Since these processes have acted locally, the resulting classification tends to be regional, certain types of process occurring in certain areas of the earth's surface. It may best be summarized in the following formal presentation.

TECTONIC BASINS

Type 1, relict lakes cut off by gentle or epeirogenetic uplift of the sea bottom, the basin having an original structural identity retained from its marine period, e.g., Caspian Sea, Sea of Aral.

Type 2, newland lakes formed after epeirogenetic uplift of marine surfaces on which there were irregularities due to uneven sedimentation, e.g., Lake Okeechobee, Florida.

Type 3, lakes formed by movement reversing drainage patterns, e.g., Lake Kioga in central Africa.

Type 4, lakes formed by upwarping all around a basin, e.g., Lake Victoria in central Africa.

Type 5, lakes in areas of local subsidence due to earthquakes, a somewhat inadequately analyzed category, e.g., Reelfoot Lake, Tennessee.

Type 6, lakes in basins in tectonically dammed synclines. A valley formed by folding may be dammed by an anticline thrust across its lower end, e.g., Fählensee in Switzerland.

Type 7, lakes on old peneplain surfaces in intermontane basins. Occasionally a surface may remain undisturbed or be uplifted with little tilting, between much folded areas which then enclose the old peneplain remnant, e.g., Lake Poso in Celebes, though this in part is transitional to Type 4.

Type 8, basins on tilted fault blocks, e.g., Abert Lake, Oregon.

Type 9, basins in grabens between faults, e.g., Lake Baikal, Lake Tanganyika, and Pyramid Lake in Nevada. The most important type of tectonic basin. The distinction between types 8 and 9 is often obscure.

LAKES ASSOCIATED WITH VOLCANIC ACTIVITY

Type 10, lakes in relatively unmodified craters in cinder cones, e.g., lake in Crater Butte, Mount Lassen National Park, California.

Type 11, maars. Lakes in basins formed by single explosive eruptions, e.g., the maars of the Eifel, Lake Avernus, and Lac d'Issarlès in the Auvergne.

Type 12, crater lakes in stratovolcanoes modified during the terminal phases of eruption.

Type 12*a*, single, the ordinary kind of crater lake, e.g., Big Soda Lake, Nevada; Lac de la Godivelle d'en Haut, the Auvergne.

Type 12*b*, multiple, e.g., Lake Rotomahana, North Island, New Zealand.

Type 13, caldera lakes, in large basins due to collapse of the central part of a volcano after the ejection of magma, e.g., Crater Lake, Oregon; Tôyako, Japan.

Type 14, lakes in conche, large basins formed by slower collapse with step faulting over an emptied magma chamber, e.g., Lago de Bolsena, Italy; perhaps Mono Lake, California.

Type 15, lakes between secondary peaks filling a caldera, e.g., Medicine Lake, California.

Type 16, lakes in volcano-tectonic basins comparable to calderas or conche but with the plan of the lake determined by pre-existing faults, e.g., Lake Toba, Sumatra.

Type 17, lakes on collapsed lava flows. Lake basins may form as the lava solidifies superficially while the lower layer tends to flow under the influence of gravity. E.g., Myvatn, Iceland; Yellowstone Lake in Yellowstone National Park; a number of shallow lakes in the Auvergne.

Type 18, lakes formed by a barrier constituted by a volcano or group of volcanoes.

Type 18*a*, across a valley, e.g., Lake Kivu in central Africa.

Type 18*b*, along margin of a caldera lake, e.g., Niuafoou in the Pacific Ocean.

Type 19, lakes formed by damming of a valley.

Type 19*a*, by lava flows, e.g., Snag Lake, Mount Lassen National Park; Lake Bunyoni, Uganda, central Africa.

Type 19*b*, by volcanic mud flows or lahars, e.g., lakes around Bandaisan, Japan (these may be regarded as transitional to Type 20b).

LAKES FORMED BY LANDSLIDES

Type 20, lakes held by landslide dams.

Type 20*a*, by rockslides, e.g., Lake Sarez in the Pamir Mountains; Lac des Chaillexon, Doubs, on the Franco-Swiss boundary.

Type 20*b*, by mudflows, e.g., Lake San Cristobal on Lake Fork of Gunnison River.

Type 20*c*, by peatslides, as near Dunmore, County Galway, Ireland.

Type 21, lakes on the irregular surface of landslides, e.g., Lac de St. André, Mount Granier.

Type 22, lakes held by scree dams formed by prolonged rockfall, e.g., Goatswater and Hard Tarn in the English Lake District.

LAKES FORMED BY GLACIAL ACTIVITY

Lakes held by ice or by moraine in contact with existing ice

Type 23, lakes on or in ice. Transitory but have occurred on or in various alpine glaciers, and on T3 or Fletcher's Ice Island.

Type 23*a*, on the surface of glaciers.

Type 23*b*, within glaciers.

Type 23*c*, on ice sheets.

Type 24, lakes dammed by ice.

Type 24*a*, in a lateral valley dammed by ice in main valley, e.g., Märjelensee, Switzerland.

Type 24*b*, in main valley dammed by ice from lateral valley, e.g., Gapshan or Shyok ice lake in south central Asia.

Type 24*c*, between glacier and valley wall, e.g., lakes formerly associated with the Malaspina Glacier in Alaska.

Type 24*d*, held by avalanches; transitory but occasionally formed in the Swiss Alps.

Type 24*e*, against the margins of continental ice sheets, mainly extinct, but a few exist in west Greenland.

Type 25, lakes held by the moraine of an existing glacier (transitional to Type 30, but in some cases their stability doubtless depends on the presence of ice).

Type 25*a*, in a lateral valley with the lake in the main valley, as in the lake of Mattmark in Switzerland.

Type 25*b*, in the main valley with the lake in the lateral valley, as in at least one example in the Andes.

Glacial rock basins

Type 26, ice-scour lakes on shattered and jointed mature surfaces, e.g., Loch Trealaval and numerous other examples in the Hebrides and western Scotland; many in Norway, Finland, and northern Canada.

Type 27, cirque lakes, formed at about the snow line in glaciated valleys.

Type 27*a*, simple, e.g., Iceberg Lake, Glacier National Park; Watendlath Tarn, English Lake District; Wildseelodersee, Austrian Alps; Ororotse Tso, Indian Tibet; and innumerable other examples in glaciated mountains.

Type 27*b*, multiple, formed by fusion of two or more cirques, e.g., Tennesvatn, Moskenesøy, Lofoten Islands.

Type 28, valley rock basins formed below the snow line by glacial corrasion.

Type 28*a*, small montane valley lakes, often in chains (paternoster lakes), as in Glacier National Park.

Type 28*b*, fjord lakes in very greatly overdeepened narrow valleys, e.g., the Nordfjord lakes in western Norway.

Type 28*c*, piedmont lakes marginal to mountain ranges that were heavily glaciated, e.g., the larger lakes of the English Lake District and the Alps.

Type 28*d*, large rock basins produced by continental ice, e.g., Great Slave Lake and the Laurentian Lakes. These lakes show some of the features of glacial scouring on an enormous scale.

Type 29, glint lakes or ice-cauldron lakes produced by glacial corrasion in specific sites where the ice flow is impeded by the pre-existing topography, e.g., Törneträsk in Swedish Lapland, at least in part; Loch Rannoch in Scotland.

Morainic and outwash dams

Type 30, lakes dammed by terminal or recessional moraines.

Type 30*a*, a single dam blocking a valley, e.g., Quartz and Bowman lakes in Glacier National Park, at least in part; Green Lake and the Madison lakes, Wisconsin; Lough Neagh in northern Ireland.

Type 30*b*, a dam at each end, e.g., the Finger Lakes of New York State.

Type 31, lakes dammed by outwash filling a valley below the former glacier, e.g., Kintla and McDonald lakes in Glacier National Park.

Type 32, lakes formed in lateral valleys by:

Type 32*a*, terminal moraine in main valley, e.g., Lac de Clairvaux, Jura Mountains.

Type 32*b*, lateral moraine in main valley, e.g., Lac de Barterand, Jura Mountains.

Type 33, lakes between parallel recessional moraines, usually with a greater or less amount of outwash which contained stagnant ice, e.g., Big Cedar Lake, Wisconsin; Lilla Le, Dalsland, Sweden.

Drift basins

Type 34, lakes in irregularities in ground moraine, e.g., Heron Lake, Jackson County, Minnesota. The type is probably much rarer than often supposed.

Type 35, lakes in kettles or cavities left by melting of ice blocks in outwash discharged into a pre-existing valley, e.g., Barrett Lake and others in the Pomme de Terre Valley, Minnesota; probably Linsley Pond and Upper Linsley Pond, Connecticut. When several lakes are present, they form a chain with drainage following the valley.

Type 36, lakes in kettles in drift-filled valleys, the drainage having no relationship to the pre-existing hydrography, e.g., numerous lakes in Martin County, Minnesota.

Type 37, lakes in kettles in pitted outwash plains, e.g., most of the lakes in Vilas County, Wisconsin. Probably simple (37a) and compound (37b) types may be distinguished.

Type 38, lakes in kettles in till of continental ice sheets, e.g., numerous but frequently not well-characterized examples in Minnesota.

Type 39, lakes in kettles in eskers, e.g., Hill and Pine lakes, Minnesota. (Types 35 to 39 have been inadequately distinguished in the literature; together they represent the commonest types of small glacial lakes on drift-covered plains.)

Type 40, glacial tunnel lakes formed as plunge pools in ground moraine where water descends a crevasse to continue as a stream below the ice, e.g., Jelser Seen and other examples in northern Germany.

Type 41, thaw or thermokarst lakes in regions of permafrost, unoriented, initiated by local melting by water, or by destruction of plant cover, e.g., many lakes in eastern Alaska.

Type 42, elliptical oriented thaw or thermokarst lakes in regions of permafrost, the form being primarily due to current pattern, e.g., many lakes in northern Alaska around Point Barrow.

SOLUTION LAKES

Type 43, lakes formed in depressions due to solution of limestone by water, by water draining along joints, etc.

- *Type* 43*a*, doline lakes in simple circular depressions, e.g., Deep Lake, Collier County, Florida.
- *Type* 43*b*, uvala lakes, formed by fusion of several dolines, e.g., Muttensee, Glarus, Switzerland.

Type 44, polje or tectono-karstic lakes, formed mainly by solution in large tectonically determined basins, e.g., Lake Scutari.

Type 45, lakes forming in caves by solution and deposition of calcareous sinter.

Type 46, lakes formed by subsidence after solution of underground soluble salts (mainly NaCl and $CaSO_4$), e.g., Mansfelder See in southwest Saxony, Germany.

Type 47, lakes formed by action of acid water on sediments containing ferric and aluminum hydroxide, apparently formerly and to some extent now occurring in the coastal plains of South Carolina.

LAKES DUE TO FLUVIATILE ACTION

Plunge-pool lakes

Type 48, plunge-pool lakes in basins excavated below water falls now dry, e.g., Falls Lake and Castle Lake in the Grand Coulee.

Fluviatile dams

Type 49, fluviatile dams holding lakes, due to deposition by a lateral tributary, either temporarily or perennially, of more sediment than the main stream can remove, e.g., Lake Pepin, Minnesota-Wisconsin.

Type 50, fluviatile deposits of deltas dividing an original lake into two, e.g., Brienzersee and Thunersee in Switzerland, Derwentwater and Bassenthwaite in the English Lake District.

Type 51, strath lakes, formed between a growing delta and the adjacent margin of the lake (perhaps not formally separable from Type 50, but giving a very uneven division), e.g., Loch Geal, cut off from the upper end of Loch Lomond.

Type 52, lateral lakes, formed when the sediments of the main stream, deposited as levees, back water up a tributary stream, e.g., Lake Tung-ting and other lakes on the Yang-tze-kiang; numerous examples on the lower Danube and on the Red River in the Mississippi drainage.

Type 53, deltaic levee lakes.

Type 53*a*, between a levee and higher ground beyond the delta, e.g., Lake Pontchartrain.

Type 53*b*, entirely surrounded by levees, e.g., Lake St. Catherine on the Mississippi Delta, and Étang de Vaccarès on the Rhone Delta.

Type 54, meres formed behind levees (roddons) built by sediment carried upstream by the tide, e.g., the former Red Mere in the English fens, and probably the original Norfolk Broads.

Lakes of mature flood plains

Type 55, oxbows or isolated loops of meanders, e.g., numerous examples on the Mississippi.

Type 56, lakes in depressions formed on flood plains by uneven aggradation during floods; examples are described from the flood plain of the Mississippi in Minnesota.

Type 57, lakes in abandoned channels, common on most large flood plains but seldom important.

Type 58, lateral levee lakes lying between a levee and the scarp defining the flood plain, e.g., Catahoula Lake, Louisiana.

Type 59, crescentic levee lakes, supposedly formed between levees along meander loops which have occupied a series of different positions, a somewhat doubtful category.

LAKE BASINS FORMED BY WIND

Type 60, basins dammed by wind-blown sand, e.g., Moses Lake, Washington.

Type 61, basins in uneven aeolian deposits; somewhat doubtful, with possible examples in the south Russian steppes.

Type 62, lakes between well-oriented sand dunes; numerous examples in Cherry County, Nebraska; well developed in the Tarim Basin of central Asia.

Type 63, deflation basins formed by wind action under previously arid conditions, with or without some degree of erosion by visiting ungulates, e.g., numerous basins in northern Texas and New Mexico, South Africa, and parts of Australia.

LAKES ASSOCIATED WITH SHORELINES

Type 64, maritime coastal lakes, ordinarily in drowned estuaries, e.g., the lakes of the Landes area on the west coast of France.

Type 65, lakes inclosed by two tombolos or spits joining an island to the mainland, e.g., the Stagno di Orbetello.

Type 66, lakes cut off from larger lakes by a bar built across a bay, e.g., Buck Lake, Beltrami County, Minnesota; Lake Nabugabo on the north-east coast of Lake Victoria in central Africa.

Type 67, lakes divided by the meeting of two spits, e.g., Marion Lake, Otter Tail County, Minnesota.

Type 68, lakes formed behind cuspate spits or double tombolos, the lacustrine equivalent of Type 65, e.g., Gould Lake, Cass County, Minnesota.

LAKES FORMED BY ORGANIC ACCUMULATION

Type 69, phytogenic dams, formed by dense growth of plants, e.g., Silver Lake, Halifax County, Nova Scotia.

Type 70, closed phytogenic basins, a somewhat doubtful category, probably of less importance than has been supposed, but perhaps exhibited by ponds on Canadian peat bogs.

Type 71, lakes in slightly raised, completely closed coral atolls, e.g., the lake on Washington Island in the central Pacific.

LAKES PRODUCED BY THE COMPLEX BEHAVIOR OF HIGHER ORGANISMS

Type 72, beaver dams, e.g., Beaver Lake, Yellowstone; Echo Lake, Grand Island, Lake Superior.

Type 73, dams built by man, e.g., Lake Mead.

Type 74, excavations made by man, as the abandoned diamond mines at Kimberley, South Africa.

LAKES PRODUCED BY METEORITE IMPACT

Type 75, meteorite craters due to explosion on impact, e.g., Ungava or Chubb Lake, Ungava, Quebec.

Type 76, the bay lakes of southeastern North America, perhaps due to compression waves in the air around a meteorite as it approaches sandy sediments, e.g., Singletary Lake, North Carolina. The origin of the bays is, however, still quite problematical.

CHAPTER 2

The Morphometry and Morphology of Lakes

The form of a lake basin and of the lake that occupies it will depend partly on the forces that produced the basin, which have been considered in detail in the previous chapter, and partly on the events that occur in the lake and its drainage basin after it has been formed. Before proceeding to a discussion of these events it is convenient to give an account of various ways of expressing aspects of the form of a lake quantitatively so that comparison may be made between different lakes.

MORPHOMETRY

The bathymetric map and the morphometric parameters. The primary method of recording lake morphology is the bathymetric map. In the construction of such a map it is first necessary to have a good topographic map of the lake basin or at least an accurate outline map of the lake. The ordinary methods of surveying can be used in constructing such maps; the application of such methods to limnology is treated excellently by Welch (1948). Most cartographers tend to be careless in the drawing of lake shores except on very detailed large-scale maps; this must be remembered when maps not prepared for limnological purposes are used. The most useful single datum on the outline of a lake is a well-centered aerial photograph of known scale.

The construction of a bathymetric map from an outline map can be done in three ways: (1) the fixing of the position at which soundings are taken on or from the shore; (2) the use of ordinary land-survey methods on ice; and (3) the counting of oar strokes or other method of dead reckoning on a course across the lake. These methods are discussed by Welch, as are the various precautions that must be taken in the making of soundings, the care and treatment of sounding lines, and other very useful information which because it may seem so elementary to the novice may be overlooked in a disastrous manner.

From the soundings and the bathymetric map, suitably contoured, a number of parameters may be calculated. These parameters and the symbols used in denoting them in the present work are as follows.

Maximum depth (z_m). The maximum depth will vary slightly with variations in water level, and ideally it should be referred to some independent datum level. In engineering practice and in some purely limnological work, depths have sometimes been given as altitudes above sea level. This practice, while theoretically defensible, is very inconvenient. In much American work, feet are used as the unit of depth; it is not difficult even for someone educated in a purely Anglo-Saxon tradition to learn to think in the metric system, and the convenience later experienced in multiplication far outweighs the initial effort. The use of the metric system, moreover, is in line with the international point of view, on which all progress in science depends.

Length (l) is defined as the shortest distance, through the water or on the water surface, between the most distant points on the lake shore.

Breadth (b_x) is defined as the length of a line from shore to shore cutting the line defining length at right angles at any point x. The mean breadth is given by

$$\bar{b} = \frac{A}{l} \tag{1}$$

Mean depth ($\bar{z}$). The mean depth is obtained by dividing the volume of the lake by its area.

Depth of cryptodepression (z_c). When the bottom of the lake lies below mean sea level, it is said to occupy a cryptodepression, the depth of which is the depth below sea level of the deepest point of the basin.

Area (A) is determined by planimetry from the outline of the map. In computing volumes and for certain other purposes, it is necessary to measure the area of each contour on the map. The area of the contour of depth z is then designated as A_z.

Volume (V). The volume of a lake is given by

$$V = \int_{z=0}^{z=z_m} A_z \cdot dz \tag{2}$$

Two standard procedures are available for the evaluation of the integral. In both, the areas of contours, as closely spaced as the soundings permit, are measured. These may be plotted against z, and the area of the curve so obtained may then be measured planimetrically. Alternatively, the volume of a series of elements is computed and summed. The volume of a given element between $z = m$ and $z = n$ is taken as

$$V_{n-m} = \tfrac{1}{3}(A_m + A_n + \sqrt{A_m A_n})(n - m) \quad (3)$$

The volume between the surface and a horizontal plane of depth z is conveniently designated V_z.

Shore line (L). The shore line may be measured on the map by means of a rotometer. It must be remembered that its length depends on the fineness of detail of the map, and for this reason the values obtained on maps of different scale or by different cartographers may be somewhat different. The length of any contour of depth z may be conveniently designated as L_z.

Development of shore line (D_L) is the ratio of the length of the shore line to the length of the circumference of a circle of area equal to that of the lake.

$$D_L = \frac{L}{2\sqrt{\pi A}} \quad (4)$$

The development of the shore line obviously cannot be less than unity. The quantity can be regarded as a measure of the potential effect of littoral processes on the lake, area being constant. It suffers from the same sources of uncertainty as measurements of L.

A somewhat similar concept, the *articulation* of a lake, or ratio of the areas of inlets and bays to the total area, is now little used.

Development of volume (D_V). Several related parameters have been used as expressions of the form of the basin. The development of volume is defined as the ratio of the volume of the lake to that of a cone of basal area A and height z_m. Since the volume of a cone is one third the product of the height and basal area, we may write

$$\begin{aligned} D_V &= A\bar{z}(\tfrac{1}{3}z_m A)^{-1} \\ &= 3\bar{z}/z_m \end{aligned} \quad (5)$$

The ratio of mean to maximum depth therefore provides as good a measure of the quantity under consideration, and leaves out all reference to an arbitrary ideal conical form.

Cvijić (1903) used as a measure of the basin form $(3\bar{z} - z_m)/z_m$, which reduces to $(D_V - 1)$. In the tables of morphometry the ratio $\bar{z}:z_m$ is given. The other formulations can easily be obtained by anyone desiring them.

Relative depth (z_r). Delebecque (1898) used as a measure of relative depth the ratio of the maximum depth in meters to the square root of the area in hectares. This is a somewhat arbitrary procedure. In the present work the relative depth is defined on a percentage basis as

$$z_r = 50 z_m \sqrt{\pi} (\sqrt{A})^{-1} \tag{6}$$

or the maximum depth as a percentage of the mean diameter.

Mean slope (ϵ) approximated as

$$\tan \epsilon = \frac{(\frac{1}{2}L_0 + L_1 + L_2 \ldots L_{n-2} + \frac{1}{2}L_{n-1})z_n}{nA} \tag{7}$$

Rate of change of area with respect to volume may be given for any depth z by

$$\frac{dA_z}{dV_z} = \frac{dA_z}{dz} \cdot \frac{dz}{dV_z} = \frac{1}{A_z} \cdot \frac{dA_z}{dz} \tag{8}$$

The value of $\frac{dA_z}{dz}$ can be obtained from the curve used in the planimetric determination of volume, by mechanical differentiation[1] rather than integration. The resulting quantity has proved most useful in the study of the chemical effect of mud on the water in stably stratified lakes.

It is sometimes desirable to construct a *generalized profile* by obtaining for any depth z, r_z the radius of the circle of area A_z and plotting r_z against z.

The **curvature of the earth and the form of a lake surface.** The surface of a lake when undisturbed is not a plane but is nearly an arc of a sphere in equilibrium in the earth's gravitational field. Cvijić (1903, Peucker 1903) has computed that for Lake Scutari the highest point on the water surface lies 30.5 m. above the shortest line through the water joining the most distant points on the lake shore, and so defining the length of the lake. If we imagine a circular lake of the same area A as a real lake, this highest point will lie at the same height above all possible diameters or lengths. We may then consider the volume of the arc of the sphere above the plane, and the ratio of the whole volume V to the volume of this arc will give a new measure of relative depth. The ratio was found by Cvijić to vary from 1.06 for Lake Scutari to 43.10 for Lake Ohrid. The rather abstract argument used in obtaining this measure has probably prevented its adoption, but it may well prove useful.

Morphometric tables for lakes of particular interest. In Table 1 the morphometric data for the lakes known to be over 400 m. deep are given; in Table 2 the same quantities are set out for lakes having an area in excess of 15,000 km.[2] that do not appear in Table 1. A good deal of morpho-

[1] The Richards-Roope tangentimeter is ideal if it can be obtained.

TABLE 1. *Morphometry of lakes over 400 m. deep*

Lake	Type of Basin	A, km.²	A_i, km.²	z_m, m.	$\bar{z}$, m.	z_c, m.	$\bar{z}:z_m$	z_r, %	V, km.³	L, km.	D_L
Baikal	Tectonic (graben), Type 9	31,500	800	1741	730	1279	0.43	0.85	23,000	2200	3.4
Tanganyika	Tectonic (graben), Type 9	34,000	...	1470	572	647	0.39	0.70	18,940	1900	3.1
Caspian	Tectonic (epeirogenetic), Type 1	436,400	2340	946	182	972	0.19	0.13	79,319	6000	2.55
Nyasa	Tectonic (graben), Type 9	30,800	...	706	273	242	0.49	0.36	8,400	1500	2.7
Issyk Kul	Tectonic (graben), Type 9	6,200	...	702	320	...	0.46	0.79	1,732	760	2.8
Great Slave	Glacial corrasion, Type 28d	30,000	...	614	...	464	...	0.31	...	2200	3.6
Crater	Caldera, Type 13	55	5.2	608	364	...	0.60	7.29	20	35	1.33
Matana	Tectonic (graben), Type 9	164	...	590	240	208	0.41	4.08	39	80	1.76
Hornindalsvatn	Fjord lake, Type 28b	508	...	514	237	461	0.49	6.38	12	65	2.6
Tahoe	Tectonic (graben), Type 9	499	...	501*	249	...	0.50	1.99	124	125	1.58
Sarez	Landslide dam, Type 20a	...	...	500	...	...	...	...	...	...	...
Chelan	Glacial corrasion and damming, Type 28b,c, 30a	150	...	458	...	129	...	3.32	...	180	3.6
Toba	Volcano-tectonic, Type 16	...	...	450	...	...	...	...	...	...	...
Mjøsa	Fjord lake, Type 28b	362	...	449	187	328	0.42	2.09	...	290	4.3
Manapouri	Fjord lake, Type 28b	145	6	445	100	262	0.23	3.33	145	130	3.1
Salsvatn	Fjord lake, Type 28b	45	...	445	...	432	...	5.87	...	130	5.5
Tinn	Fjord lake, Type 28b	54	...	438	...	255	...	5.35	...	80	3.34
Tazawa	Caldera, Type 13	25.65	...	425	267	175	0.63	7.42	6.8	14.6	1+
Como	Glacial corrasion and damming, Type 28c, 30a	146	...	410	185	212	0.45	2.99	27	175	4.07

* Kemmerer, Bovard, and Boorman (1923), confirming older observations of Le Conte (1883). Map 5001 of the U. S. Coast and Geodetic Survey, on which Fig. 7 is based, gives no depth over 486 m.

TABLE 2. *Morphometry of lakes over 15,000 km.² in area not included in Table 1*

Lake	Type of Basin	A, km.^{2}	A_i, km.^{2}	z_m, m.	$\bar{z}$, m.	z_c, m.	$\bar{z}:z_m$	z_r, %	V, km.^{3}	L, km.	D_L
Superior	Glacial corrasion, Type 28d	83,300	1710	307	145	124	0.47	0.095	12,000	3000	2.93
Victoria	Tectonic (epeirogenetic), Type 4	68,800	2550	79	40	...	0.51	0.027	2,700	3440	3.7
Aral	Tectonic (epeirogenetic), Type 1	62,000	1200	68	15.6	...	0.23	0.024	970	2300	2.6
Huron	Glacial corrasion, Type 28d	59,510	4400	223	76	46	0.34	0.081	4,600	2700	3.1
Michigan	Glacial corrasion, Type 28d	57,850	400	265	99	88	0.27	0.099	5,760	2210	2.6
Great Bear	Glacial corrasion, Type 28d	29,500	...	>137	...	>18	...	>0.071	...	2100	3.3
Erie	Glacial corrasion, Type 28d	25,820	60	64	21	...	0.33	0.035	540	1200	2.1
Winnipeg	Glacial corrasion, Type 28d	24,530	750	19	13	...	0.69	0.011	3,110	1900	3.4
Ontario	Glacial corrasion, Type 28d	18,760	320	225	91	151	0.40	0.146	1,720	1380	2.8
Ladoga	?Glaciotectonic (epeirogenetic), ?Type 1, 28d	18,734	343	250	52	245	0.21	0.162	920	930	1.95
Balkhash	Tectonic (graben), Type 9	17,575	...	26.5	6.13	...	0.23	0.018	112	2384	5.08
Chad	?Tectonic, Type ?	16,500±	...	12±	1.5±	...	0.13	0.008	24	700	1.56

metric data is also presented in Table 53, in which a number of lakes of particular limnological interest are included. These tables give a good idea of the way that the various parameters vary in lakes of different sizes and origins. A very large amount of additional data can be obtained from the summaries prepared by Halbfass (1922, 1937[2]), which have been extensively used in drawing up the tables given in the present chapter. Horie's (1956) important summary for 312 Japanese lakes appeared too late to use.

While Table 2 is presumably complete for natural lakes, it is possible that a few additions may be made to Table 1. Apart from innumerable legendary bottomless lakes,[3] two doubtful examples of lakes over 400 m. deep omitted from Table 1 merit brief consideration.

Nahuel Huapi in Patagonia is said by Marchetti (1949) to be at least 1000 m. deep; no details are given. It may be the third deepest lake in the world, but the available information does not yet justify such a claim.[4]

Kasenoi-Am or Retlow-Am, a very small lake in Transcaucasia, is stated by Büdel (1926), on the authority of Radde, to be 420 m. deep. The area of the lake is about 2 km.^{2}, so that the maximal depth implies a relative depth of 25 per cent, much in excess of that of any other known lake. The original publication has not been found and seems to have been unknown to Halbfass, who was skeptical of the validity of the record. Suslov (1947) clearly does not include the lake in the four deepest of the U.S.S.R.[5] It is evident that Kasenoi-Am is a deep lake, but most improbable that it is over 400 m. deep.

MORPHOLOGY

Form of shore line in relation to origin and history. A number of fairly characteristic outlines can be recognized, though there are naturally complete intergradations among the different types.

---

[2] There are a number of misprints or errors of transcription in the second, supplementary paper, for which the reader must be on his guard.

[3] In 1932 the writer was assured that a hermit once lived on the shore of Lake Manasbal in Kashmir (z_m = 12.7 m.), devoting his life to cording a rope with which to sound the lake. When the rope finally appeared long enough, he proceeded to his investigation, but feeling no bottom when he had payed out the entire length, he threw himself after the rope and was never seen again. The reader may draw any desired psychological implications from this or from similar stories.

[4] The number of lakes of maximum depth of over 400 m. will, however, be greatly increased when the large lakes of southern South America become known morphometrically (Löffler, verbal communication).

[5] He states that Lake Teleskoye, with a maximum depth of 325 m., is the fourth deepest. Since Baikal, Issyk-kul, and Sarez are all deeper and are all mentioned at length in his book, it is clear that Kasenoi-Am was not known to him as a very deep lake.

(*a*) *Circular*, exemplified by most crater (Types 10 to 12) and caldera (Type 13) lakes; by the more perfect doline lakes (Type 43a); by the smaller deflation basins (Type 63), at least in some regions; and by one or two meteorite craters (Type 75), notably Chubb Lake. The development of shore line approaches unity in such lakes, being 1.04 to 1.15 in the maars of the Eifel and 1.02 to 1.04 in those of the Auvergne. In other respects such lakes are highly diversified, the volcanic examples and some doline lakes being deep, the deflation lakes shallow.

(*b*) *Subcircular*, a less perfect approach to circular form, exhibited by many cirque lakes (Type 27a) and by kettle-hole lakes (Types 35 to 38) and other small lakes in relatively unconsolidated material easily remodeled by shore processes. In some cases of remodeling, the effect of the prevalence of winds in particular directions may lead to more deposition on one side of the lake than on the other, producing a kidney-shaped basin. Great Pond west of Wellfleet, Cape Cod (Raisz, 1934), provides a beautiful example of an originally irregular kettle hole modeled by the building of bars across the irregularities, with a slight tendency to deposition on one side.

(*c*) *Elliptical*, exemplified by the oriented lakes of the Arctic (Type 42) and by the Carolina bays (Type 76). The development of shore line is only a little greater than in circular lakes.

(*d*) *Subrectangular elongate*, exemplified by most lakes in grabens (Type 9) and by the lakes of glacially overdeepened valleys (Type 28), whether or not a small morainic dam is present. Most of the larger graben lakes and fjord lakes have $D_L = 2.5$ to 5.0.

(*e*) *Dendritic*, representing the flooding of a valley that is not overdeepened, when its lower end is blocked by damming or tilting. Lake Waikaremoana in New Zealand, held by a landslide dam, is strikingly dendritic. Lake Kioga provides a spectacular specimen due to tilting (Type 3). Lake Mälar, a complicated basin between Stockholm and Uppsala, which has been transformed from an arm of the Baltic to a lake by uplift at the coast within historic time, is another remarkable example. Glacial diffluence, notably exhibited by some of the Italian subalpine lakes, may produce a similar pattern with high values of D_L.

(*f*) *Lunate*, exemplified by oxbow lakes (Type 55), and by a few calderas where a secondary cone set eccentrically occupies a large part of the basin to one side of which it is joined.

(*g*) *Triangular*, exemplified by the lakes in drowned valleys behind bars (Type 64), as in the Landes district of western France, and by most lateral lakes (Type 52) in mature valleys.

(*h*) *Irregular*, in extreme cases mainly in areas where the fusion of basins has occurred, or in regions of glacial scouring of shattered rock (Type 26).

Lake Lahontan provided an extraordinary case of a complex irregular tectonic basin due to faulting, now partly occupied by smaller lakes in separate fault troughs.

The very irregular lakes in Wisconsin and other middle western states usually occupy compound kettle holes. In most of such lakes the development of the shore line lies between 1.5 and 2.5. This may seem low until it is remembered that a lake formed of four circular basins just

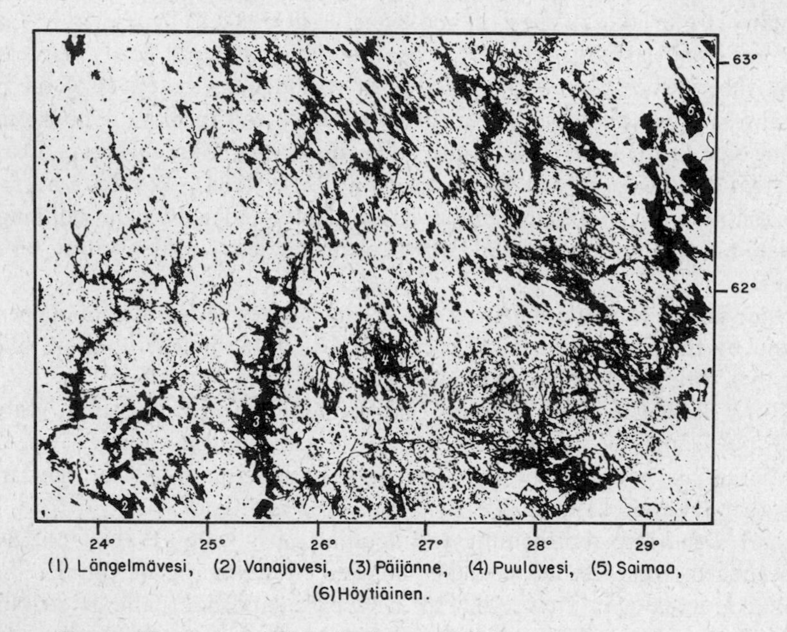

FIGURE 52. Pattern of lake landscape of Finland (after Järnefelt).

in contact would have a development of only 2.0. Elongation is generally much more important than mere sinuosity of shore line in giving a high value of D_L. The highest value in Wisconsin is for Beulah Lake ($D_L = 3.18$), which represents the partly artificial fusion of five kettle holes in a row.

The irregularity of many of the large lakes of northwestern Canada is probably due to glacial scouring acting on a complicated system of ancient rocks; this is certainly true of the eastern part of Great Slave Lake. The small glacially scoured lakes of the Outer Hebrides and western Scottish mainland certainly owe their outline to differences in resistance in shattered and jointed rock. In the extreme cases (Fig. 52) known in Finland, such causes are partly no doubt involved, but there is here a further source of irregularity in differential postglacial elevation, which tends to produce dendritic patterns, while parallel end moraines add to the linearity of the basins. Two small Finnish lakes, Iso Tarjänno and Iso Kangasola, were found by Halbfass to have $D_L = 21$!

The very complicated patterns developed in the lakes of the Erne River valley in Ireland (Fig. 53) are due mainly to the disposition of drumlins.

Insulosity. The insulosity of a lake is the percentage of the area within the shore line that is occupied by islands (A_i). Halbfass (1922) has tabulated all values known to him. Insulosity over 30 per cent is very rare. Upper Lough Erne in Ireland, an extraordinary plexus of lacustrine channels between drumlins (Charlesworth 1953) in a broad valley, has an insulosity of 40 per cent; Lake Mälar, 36 per cent; Lake Lojo in Finland, 32 per cent; and Lake Saimaa in the same country, 30 per cent. All are very irregular lakes with a high development of shore line.

Islands may be formed in lakes by a number of processes. Where tectonic events have been involved in forming a basin the islands may be defined by subsidiary faults scarps or other structural features. The large Olkhan Island and the four small Ushkani Islands in Lake Baikal represent the axis of the Academic Mountains cutting across the complex graben in which the lake lies. A magnificent *structural island* in the center of the ancient Lake Lahontan is now represented by the Trinity Mountains.

Volcanic islands are common, particularly as secondary cones in calderas, though they are not confined to this type of basin. They not infrequently bear craters containing lakes. Wizard Island in Crater Lake, Oregon, is a good example in a typical caldera.

Most islands in glacial lakes are probably *residual*, around which corrasion or, in the case of dammed valleys, prelacustrine erosion has taken place, leaving a boss that was more resistant than the surrounding rock or was favorably placed for survival. Irregular lakes formed by glacial scouring are rich in such islands. Drumlins occasionally provide another type of residual island inherited from the prelacustrine landscape.

Coastal islands formed by cutting behind promontories, so producing isolated stacks, occur along the margins of large lakes. Suslov (1947) indicates that apart from the Ushkani Islands, all the small islands of Lake Baikal are of this type; they are being rapidly cut away by wave action.

Depositional islands are formed when spits become broken late in their development, as seems to have happened to Long Point, Lake Erie (Wilson 1908). Such islands have, of course, some of the features of coastal islands. Local conditions may occasionally permit the building of hardly submerged banks away from the shore. Wilson (1938) has

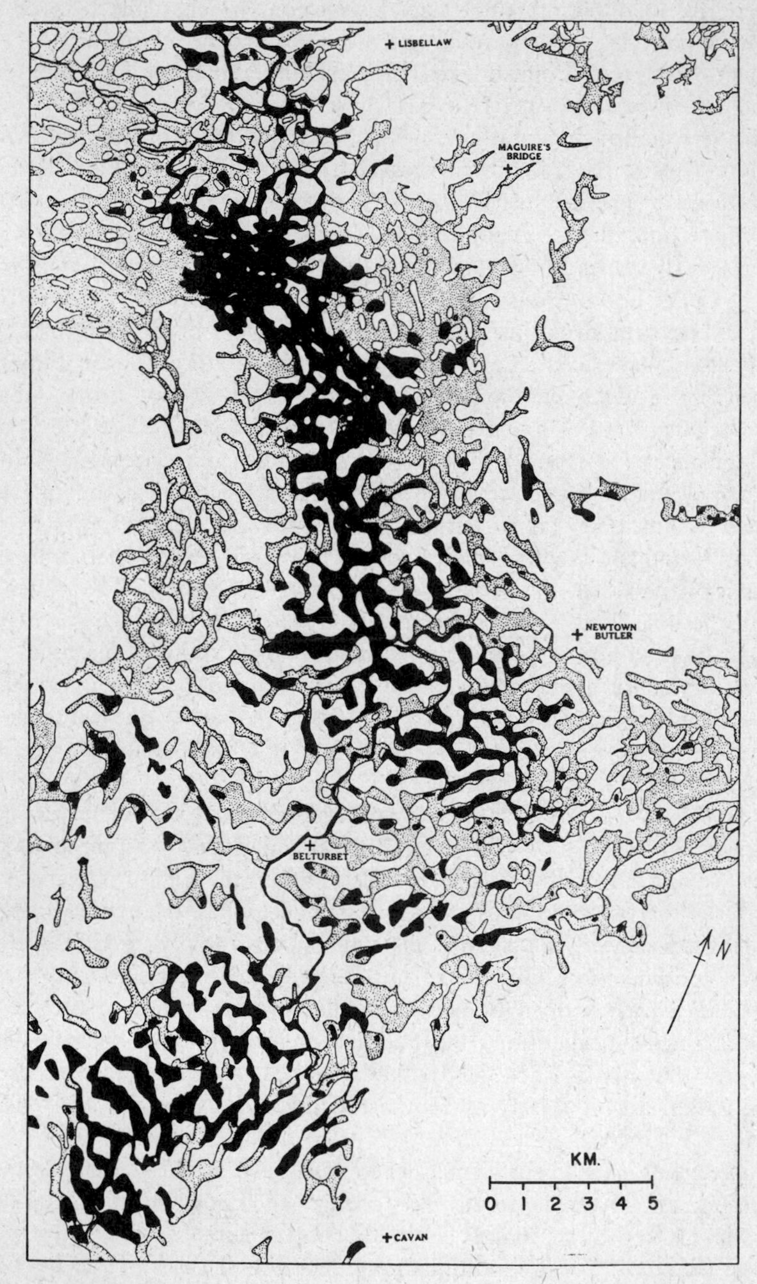

FIGURE 53. Lough Erne, Ireland, showing elaborate pattern of drumlins (after Charlesworth).

described as marl islands certain banks in Tippecanoe Lake, Indiana. Since they are founded on irregularities on the old drift floor of the lake, they are in a sense residual. He believes that these banks were always shallow enough to bear some vegetation, which acted as a sediment trap so that the banks grew in height faster than the surrounding areas. They also could acquire marl precipitated photosynthetically. When the banks built up to near the surface, wave action prevented further growth, but if the effluent were to cut back rapidly it would be possible for such a pure depositional island to emerge.

Potonié (1913) has described a very curious case in the Ögel-see near Beeskow in Brandenburg. Here an island 60 m. long and 30 m. wide suddenly emerged from water about 3 m. deep on October 23, 1910. The island was apparently continuous with the bottom and not floating; its appearance involved the formation of a trough 10 m. deep on its northern side. The bottom consisted of organic sediment under sand. Potonié thinks an accumulation of gas under the sand first occurred, and the soft material of the bottom was squeezed by the pressure of the surrounding water and sediment into the compressible gas cavity. Other cases where this sort of buckling seems to have occurred are discussed by him and by Von Koenen (1913).

The floating island of Derwentwater in the English Lake District, which appears at irregular intervals, has been described by Symons (?1889). It appears in the same place in the southeast corner of the lake during the summer; Symons records eleven appearances between 1773 and 1884. In 1798 it had an area of about 7800 $m.^2$ but is usually smaller. It often has a cleft across the middle, so that two islands really exist. It appears to be due to the peeling off of part of a layer of organic sediment, the free edge of which is buoyed up by the gaseous products of anaerobic decomposition (Fig. 54). Gas collected from the material of the island was analyzed early in the last century by no less a chemist than Dalton (Otley 1819, 1831), who found approximately equal volumes of CH_4 and N_2 with 5 to 10 per cent CO_2. Later analyses given by Symons are similar. An apparently comparable case in the Zahren-see near Dabelow in Mecklenberg is figured by Potonié (1913), who notes one or two other examples.

True mobile *floating islands* are produced when fragments of a marginal floating mat break off. They are not uncommon in bog lakes. Potonié (1913) describes several German examples, and figures a very persistent one on the Hautsee near Eisenach, though in an apparently stranded condition. Halbfass (1923) figures (Fig. 55) a quite spectacular example, bearing a solitary tree, on Lake Pakkasch in Estonia. Symons indicates that a well-authenticated example existed during part of the last century on Esthwaite Water, and that there is a legend of such a floating island on

Loch Lomond in medieval times. Zumberge (1952) implies that floating islands occur on the bog lakes of Minnesota. An example about 100 m. long, covered with *Typha*, was present in Kingston Harbor on Lake Ontario about 1951 (Seward Brown, personal communication). Gay (1833) observed several floating islands in Taguaagua in Chile. A number of statements collected by Symons from Latin literature appear to suggest that floating islands were well known in classical antiquity. Seneca (*Quaest. Nat.*, III, 25), the elder Pliny (*Hist. Nat.*, II, 94), and the younger

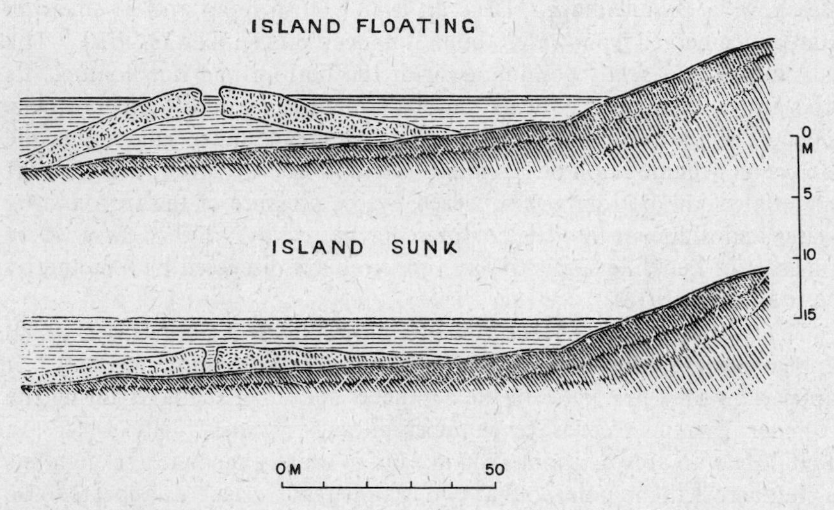

FIGURE 54. Formation of stationary floating island in Derwentwater (Symons [1889]).

Pliny (*Litt.*, VIII, 20) all appear to have observed the phenomenon on Lake Cutilias, now Lago de Contigliano near Riezi, and on Lake Vadimon, later Lago di Bassariello but now no longer a lake. It is unreasonable to dismiss such observations, but the nature of these ancient floating islands is obscure.

Shore processes. The modifications that occur in the form of a lake basin that has become filled with water are due in part to processes occurring within the lake itself and in part to events of external origin. The work of waves and currents on the shore belongs to the first category, the delivery of sediment load to form a delta to the second.

Processes of the first kind in general tend towards an equilibrium at which the basin may be regarded as adjusted to the events within it. The adjustment is never perfect, for there is never time to permit equilibrium to be established and the internal events may change owing to external events altering the nature of the lake. The tendency towards an equilibrium form is, however, certainly real and entirely expected. The processes occurring in the lake encounter resistance, their action tends to reduce the resistance in an irreversible way so that later there is less resistance, and so on asymptotically. Meanwhile external events, mainly deposition of

FIGURE 55. Mobile floating island, Lake Pakkasch (after Halbfass).

sediment and cutting at the effluent, tend to reduce the life span of the lake and may provide further occasions of resistance to the internal factors.

Following the concepts of Johnson (1919), it is customary to regard a shore line as one of submergence if its characters are primarily determined by a rise of water relative to the land, of emergence if the opposite is true; neutral shore lines dominated by processes such as the building of deltas, not related to water level, also occur and shore lines of compound character are known.

Any lake at its time of origin may be regarded as possessing a shore line of submergence; most open lakes may possess late in their history, when cutting has occurred at the effluent, shore lines of emergence; neutral processes are certain to be of importance. Compound shore lines involving submergent and emergent features are likely only in lakes in closed basins, but almost all lakes of any age will combine submergent or, more rarely, emergent with neutral features.

A newly formed lake will in general occupy a basin and possess a shore line of submergence that has no relation to the hydromechanical processes that occur in its water. This situation immediately begins to change once the lake is formed.

Normally the slope of the basin, both above and below water level, will be greater than the equilibrium slope, defined as that slope that is unaltered by shore processes. The result will be that at times of considerable wind, when high waves are formed, these will break against a sloping land surface, which they will tend to erode. The extent of this erosion depends on the original slope, on the depth of the water, on the nature of the material forming the land, and on the fetch and so the height of the waves.

A vertical cliff face rising from deep water, as may be found on the shores of some tectonic and glacially excavated basins, will tend to reflect waves virtually unchanged and will suffer practically no wave erosion, particularly if it is unjointed.

As material is removed by cutting, the larger fragments are of course transported a negligible distance and the finest clay may be suspended in the water for a long period; the material usually moved in this way is gravel, sand, and silt. The result is a terrace formed by both cutting peripherally and building centrally. The part of the terrace above water will be termed the *beach*, the part below water the *littoral shelf*. The beach of a lake is dry during quiet weather. When waves break upon it, the swash carries particles upward from the quiet-weather waterline, the smallest particles being carried furthest. A beach may therefore consist of progressively smaller particles as one ascends from the waterline, while the shelf bears progressively smaller particles as one descends towards the deep water. This effect is best seen subsequent to the initial stages of beach formation during which much material is falling from the cliff face.

The formation of a wave-cut cliff, beach, and sublittoral shelf is possible on any scale. The cliff may be less than a meter high in lakes in regions of very gentle relief, or hundreds of meters high along the coasts of some of the Norwegian fjord lakes. The shelf likewise can be of any width up to about 20 km. as in one part of Lake Michigan (Andrews 1870), but Johnson (1919) estimates that only a part of this width is due to cutting and at least half to building.[6]

[6] These estimates may well be quite excessive, as they do not take into account changes in water level (cf. Hough 1955).

The rate at which cutting takes place likewise varies enormously. On Geneva Lake, Wisconsin (Fenneman 1902), a slight artificial raising of lake level resulted in rather widespread cutting at the rate of about 5 m. in 67 years or 7.5 cm. per year. The circumstances of this case are interesting as an example of the complexity of what may seem at first to be rather obvious physical processes. It is unlikely that the observed rate represents a natural rate of erosion, even in the unresistant drift that forms the shores of the basin. Initially the water lay at the modern level, but cutting at the effluent lowered the lake a meter or so during postglacial times. During this period the shore line would be one of emergence, so that the waterline retreated from the original cliffs, and wave action would merely remove material from the littoral shelf. A rise of water level when a dam was built in 1835 brought the waterline back to its original position, but waves reaching the shore would now travel over deeper water than initially and would therefore have more energy at breaking.

The processes just discussed are dependent on wave action and can occur if the waves follow a direction normal to the coast. They will tend to act particularly strongly on isolated headlands and promontories, owing to wave refraction. The net result is to increase the area of the lake, to reduce the relative length or development of the shore line, and to reduce the mean depth. Sedimentation will tend to counteract the first but intensify the second and third of these effects; cutting at the effluent will tend to reduce both area and mean depth. Other events increase the effect of shore processes. Whenever there is a longshore current, which will develop if the wind drift meets the coast at an angle, or if the type of circulation described by Livingstone (1954) prevails, with rip currents determined by the morphology of the shore line, movement of debris eroded by waves occurs. The movement apparently mainly concerns particles disturbed by wave action that are carried by the current during this brief sojourn in the water; in this way waves and current collaborate in the transport. Moreover, waves reaching a shore line at an angle can by themselves have the same effect, because the direction of particles in the swash depends on both the direction of their initial velocity and the component of gravity down the beach. Such particles will therefore loop along the shore in parabolic steps.

The rate at which any shore process dependent on waves occurs is controlled as much or more by the size of the waves as by the over-all period of action of waves of any given size. Wave action may therefore occur much more rapidly along a coast exposed to a very long fetch over which strong winds occasionally blow than on a coast perennially exposed to the action of waves generated over a short fetch. Johnson (1919) points out that along the coast of Lake Michigan shore drift is predominantly to the south on both coasts of the southern part of the basin, owing to the long fetches of northeast and northwest winds, and towards the north in the northern part of the basin, owing to the long fetches of southeast and southwest winds, in spite of the fact that the prevalent wind is westward.

Spits and bars. It is reasonably certain that in any lake over which winds of variable direction blow, there will at some time be a wind drift arriving at an angle to the shore line. This will inevitably produce longshore currents, so that material stirred by waves may be expected to travel along the shore of the lake. At any angle in the coast line the momentum of the current will tend to make the latter continue in its path rather than bend round the coast, and for some distance particulate matter will be sedimented along this path, forming a spit. The same effect will of course occur if the angle marks the opening of a bay. If the bay becomes completely enclosed, the spit is referred to as a bar. In the formation of a spit the outer end growing into deeper water will form progressively more slowly. This is generally believed to make it progressively more susceptible to forces other than those producing it. Since the fetch will be greatest on the lake side, such forces will tend to turn the spit inward, producing a recurved spit. Since the coast line along which the current is running may recede and the main axis of the spit itself will follow each successive position of the current, rather complex forms may be produced at the apex, various successive recurved points being incorporated in a compound recurved spit. A number of spits of varying degrees of complexity formed on lake shores are illustrated in the maps of Figs. 56 and 57. Presque Isle and Toronto harbor are two of the finest known compound recurved spits on any coast lines, marine or lacustrine.

A rather curious structure known as a cuspate bar is found, usually as a small-scale feature of some lake shores. It usually consists of a looped bar built out a short way into the lake, then taking a sharp re-entrant turn joining the shore again. A small lagoon may be inclosed by the bar. Cuspate bars can certainly be formed by wave action alone; Comstock (1900) noted a small but perfect example formed at the side of an arcuate delta in Lake George as the result of two trains of waves set up by a steamer that followed an angulate course off the site of the building of the bar. A fine example in Lake Balaton, opposite the peninsula that separates the western and eastern ends of the lake, is explained by Johnson as due to alternate beach drifting provoked by waves from the west in the western basin and from the east in the eastern basin. The peninsula is believed to interrupt the fetch too much to permit the opposite destructive processes. Occasionally the two sides of a bar formed in this way are inseparable, as in Crowbar Point in Lake Cayuga (Tarr 1898). Cuspate bars occur

A

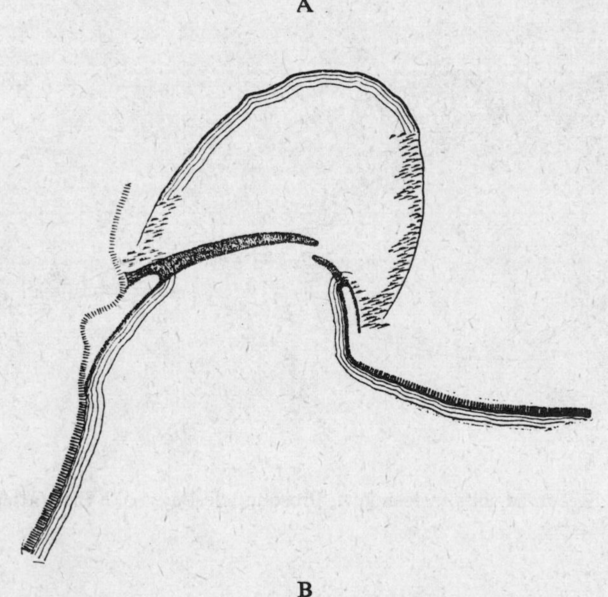

B

FIGURE 56. Bays partly separated by bars. *A*, near Duluth on Lake Superior (after Johnson); *B*, Dartford Bay, Green Lake, Wisconsin (after Fenneman).

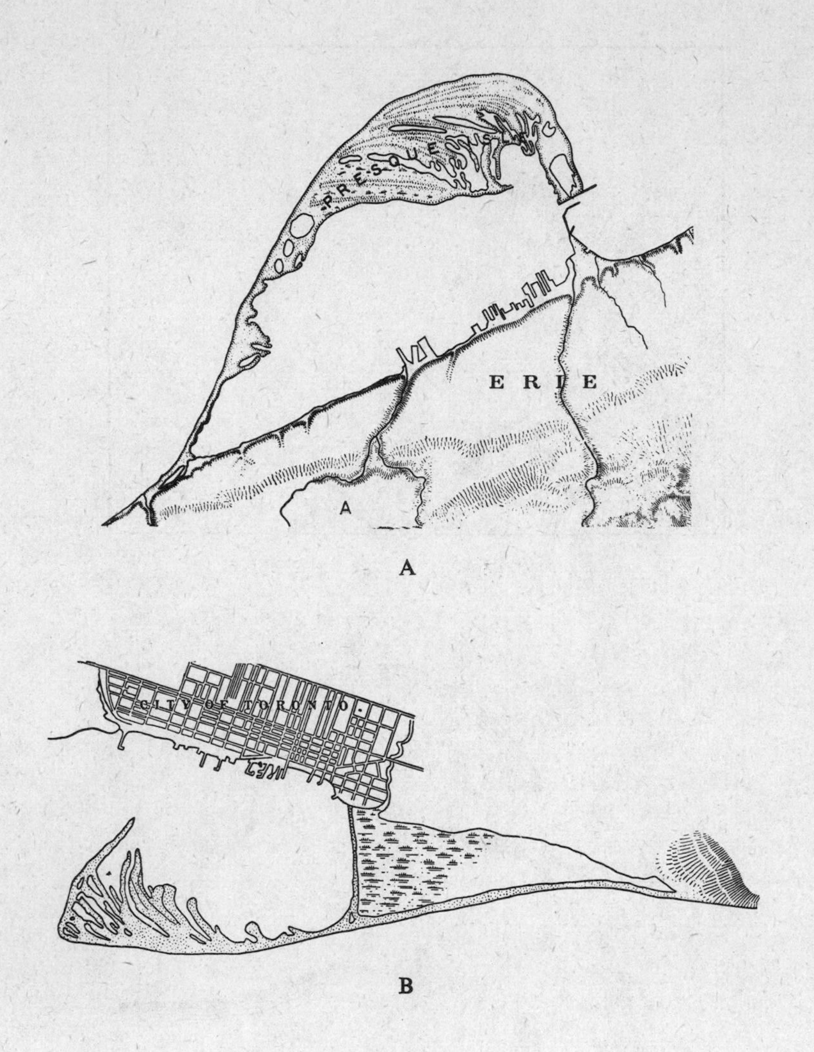

FIGURE 57. Recurved spits enclosing, *A*, Presque Isle Bay, Lake Erie (after Johnson); *B*, Toronto Harbor.

mainly on submerged shore lines. Meinertzhagen (1927) noted fine examples on Pang-gong Tso, which had formed after the rise in level in the earlier decades of this century (De Terra and Hutchinson 1934).

Deltas. The building up of a platform of detrital material to form a delta at the mouth of an influent stream is extremely common in lakes, and may perhaps be regarded as one of the most characteristic processes of the lake littoral. Deltas occur at the mouths of rivers entering the ocean; indeed, the type of all such structures, the Nile Delta, lies on a sea coast. Salting out of charged silt particles by water of compensation currents is doubtless important in such cases, as it is said also to be in the Sea of Aral (Suslov 1947). Small rivers seldom produce deltaic structures where they enter the sea, but in lakes, where tidal action is absent and wave action usually less fierce, quite small streams build recognizable deltas.

The fundamental process in the formation of a delta involves the passage of a river with a definite velocity and concomitant turbulence into a large volume of water of lower velocity and lower turbulence. The load of sediment suspended in the river tends to settle as the velocity and turbulence are reduced, the larger particles falling out rapidly. When the river mouth opens on a fairly steep lake shore, the rapidly sedimenting larger particles will pile up on the bottom of the lake at the angle of respose, forming what are termed the foreset beds of the delta. As more and more foreset beds are laid down, the mouth of the river is carried out into the lake. The river now flows over a flat area in which its flow is somewhat reduced. Sedimentation, largely of the coarser particles, takes place on this flat area, producing topset beds at the surface of the delta. In section, therefore, a delta formed on a relatively steep slope consists of a lower part of foreset beds dipping at an angle up to 20 degrees toward the lake, with a veneer of almost horizontally stratified material often of coarser texture. If, through cutting at the outlet or other causes, the lake level falls, the river will cut a deep valley through the delta, exposing this structure. The diagram here presented (Fig. 58) is based on Cotton's (1941) photograph of such a trenched delta built into Lake Wakatipu in New Zealand. Deltas of this simple sort occur mainly in lakes in regions of fairly high relief in which the streams carry a considerable load of coarse sediment. In relatively young landscapes the rate of growth of the delta may be quite rapid. Stumpf (1923) found in the Alps that the Rhine added annually about 2,790,000 $m.^3$ of sediment to its delta in Lake Constance. This corresponds to 456 $m.^3$ per $km.^2$ of drainage basin.

If the river is of moderate velocity or if the water into which it is flowing is moderately disturbed by wind and wave action, the sediment will usually be deposited over a front wider than that of the river mouth. The margins of the delta, if no restrictions are imposed by the coast line, will form a convex projection into the lake. Such a delta is termed an *arcuate delta* (*A*, Fig. 59) by Johnson (1919). When wave action is more intense, the mouth of the river tends to be flanked by triangular projections into the lake which have concave shores, forming a *cuspate delta* according to Johnson's terminology; such deltas are probably commoner on sea

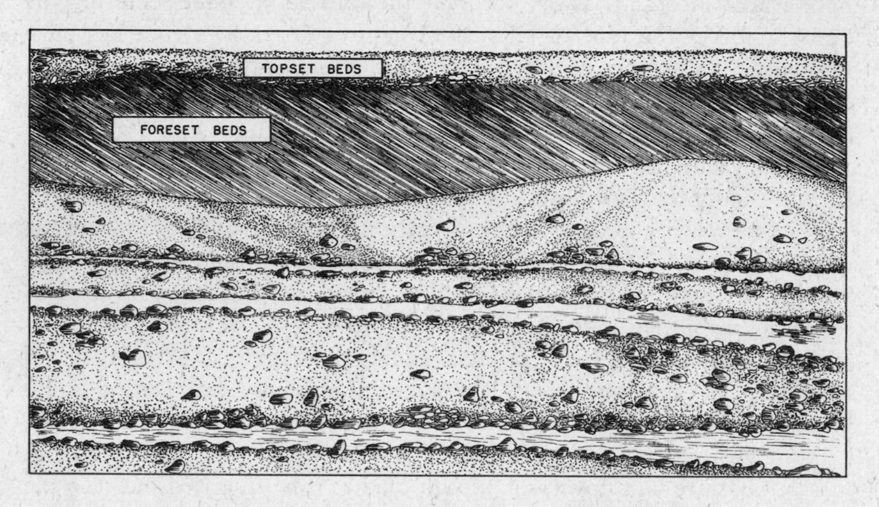

FIGURE 58. Foreset and topset deltaic beds (after Cotton).

coasts than on the shores of lakes. When wave action is greatly restricted, as on some small lakes, the momentum of the incoming stream carries it well out into the lake; deposition occurs mainly at the sides of this stream, where it is losing momentum by turbulent transfer to the free water, forming a pair of bars projecting into the lake. Johnson points out that on large deltas, rivers carrying much fine sediment usually break up into numerous distributaries, each one of which may be carried into the sea as such a pair of bars. He terms such a delta lobate. The process in lakes, as for instance in the Lago di Vannino or in Bogstadvatn northwest of Oslo (*D*, Fig. 59), usually leads to the production of a single channel, forming what may be called a *unilobate delta*. Lacustrine *multilobate* deltas are rarer and usually not well formed (*C*, Fig. 59). The formation of lakes on a multilobate delta such as that of the Mississippi has already been described.

In long narrow lakes the influents usually enter at the upper ends. Here there may not be sufficient room for the development of the full width of an arcuate delta; the delta deposits fill the whole width of the lake,

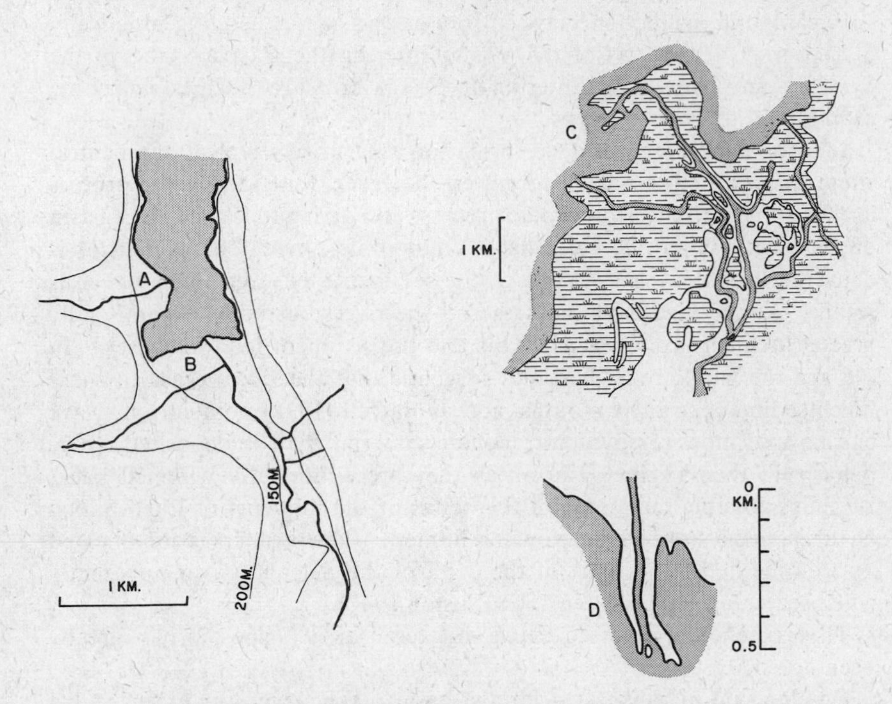

FIGURE 59. *A*, arcuate delta of Glenridding Beck; *B*, bay delta at head of Ullswater, English Lake District; *C*, irregular multilobate delta of River Jhelum in Wular Lake, Kashmir; *D*, unilobate delta of Sörkedalselva in Bogstadvatn, Oslo, Norway. (After Mill and various official surveys.)

producing what Johnson terms a *bay delta*, or the landscape form known as a strath at the head of many lakes (*B*, Fig. 59) Where a unilobate delta develops under such circumstances, one of the bars may fuse with the neighboring shore, enclosing a strath lake (page 115).

Arcuate deltas built on the sides of long narrow lakes may occasionally grow right across the lake, dividing it in two. The division of Buttermere and Crummock Water by the delta of Mill Beck provides a good example. Occasionally such a division is achieved by the fusion of deltas from either side of the lake, as in the division of the Brienzersee and Thunersee by the deltas of the Lütschine and Lombach.

The deltas built by rapid intermittent streams on the shores of deep lakes are little more than steep-sided gravel cones. At the opposite extreme are the immense deltas built by large rivers in mature landscapes, rivers that carry much fine sediment and break into numerous distributaries as they enter the basins of relatively shallow and morphologically old lakes. In such cases the distinction between foreset and topset bedding practically disappears. The deltas of the Volga, entering the Caspian, and of the Syr-daria and Amu-daria, entering the Sea of Aral, are the finest lacustrine examples of this extreme type.

Johnson (1919) regards delta formation as a process typical of a neutral shore line. It must not be forgotten, however, that ideally the process itself is subject to the geomorphic processes occurring in the drainage basin of the lake. When the basin has a young relief, much coarse detritus is delivered, but the total volume of the sediment delivered in unit time is small. As the landscape matures and the rivers cut back, a continually greater area for erosion is available and the sediment load increases. In old age the whole relief becomes so gentle that the rivers cease to carry much sediment to their mouths, and as Barrell (1912) pointed out, wave cutting may now remove more deltaic material than is deposited. The deltas will then begin to shrink as they are continually trimmed back. It is just possible that some of the deltas of the influents of Lake Baikal have reached a stage where growth is limited. The problem is complicated by tectonic activity; a part of the delta of the Selenga River apparently collapsed catastrophically in 1861 (Suslov 1947).

The existence of reversed deltas in lateral lakes (Type 52) has already been noted.

Development of volume and form of basin. Whenever any extensive list of lakes is examined, by far the greater number have a ratio of $\bar{z}:z_m > 0.33$, or a development of volume greater than unity. A basin of which the volume is less than that of a cone of basal area A and height z_m is relatively rare. The ratio $\bar{z}:z_m$ is greatest in shallow lakes with flat bottoms, such as the Carolina bay lakes. Among deep lakes it is maximal in calderas and other crater lakes, but is also high, over 0.5, in many rock basins, whether flat-bottomed graben lakes, glacially overdeepened valleys and fjord lakes, or cirque lakes. Any extensive action of shore processes is apt to reduce the ratio; most lakes in relatively easily eroded terrain have a ratio between 0.33 and 0.50. Extremely small values of the ratio are found in only a few lakes in which highly localized deep holes, representing in some cases (as Lake Scutari) ponors or sinks, in other cases (as Douglas Lake, Michigan) sublacustrine kettle holes, occur in the floor of otherwise shallow lakes.

Although the relative depths and the ratio $\bar{z}:z_m$ provide useful indices, they indicate nothing about the distribution of the deep areas in a lake. In the simplest basins there is usually a more or less central deep area. In glacially overdeepened valleys its exact position is likely to be determined by the convergence of tributary glaciers, as has been pointed out for the Nordfjord lakes, and in some cases, as Lake Okanagan, it is far above the center of the mid-line of the lake. In any kind of dammed lake one may expect the deep water somewhat below the center and towards the dam, as in Green Lake, Wisconsin (Fig. 28). There are, however, many exceptions to this, particularly in relatively wide valleys in areas of low relief; the deep of Lake Mendota, in such an area, is almost central.

The presence of more than one deep area, with a submerged bar or bars separating the deeps, is common and may be due to many causes. The two deepest graben lakes known, Baikal and Tanganyika, both have three markedly deep areas, presumably of tectonic origin. Some lakes in glacially overdeepened valleys have sublacustrine bars, presumably representing specially resistant material set across their axes, as in Oldenvatn in the Nordfjord group (Strøm 1933a). Glacially scoured basins on shattered and jointed peneplains, such as Loch Trealaval, often have more than one deep. Many kettles appear to be composite, with highly irregular outlines and several deep holes. The multiple karstic lakes also provide extraordinary examples of a number of depressions in a single basin.

Welch (1935), working on Douglas Lake, Michigan, in which there are six basins, has emphasized the fact that during thermal stratification the deep layers of different depressions may develop different thermal and chemical properties, and therefore he speaks of *submerged depression individuality*.

Sublacustrine shelves. The most usual cause of a shelf in the profile of a lake is the cutting and building process along the shore, already discussed. Many lakes, however, exhibit deeper shelves which are doubtless usually present from the time of formation of the basin. An example of such a shelf forming a hanging valley in Breimsvatn in the Nordfjord group has already been given (*C*, Fig. 23). In the formation of kettle holes the collapse of drift covering over the peripheral parts of the original ice block may produce a shelf (*C*, Fig. 32). A shelf in Linsley Pond, probably of this nature, is of importance in the interpretation of the chemistry of lakes. This shelf is described in terms of the variation of dV_z/dA_z in Fig. 190.

Variation of morphometric parameters with time. Wilson (1936, 1938, 1945) has made numerous borings in the sediments of three complex kettle-hole lakes in Indiana and Michigan, which permit mapping of the original basins before sedimentation began. In the two simpler cases, Winona Lake and Tippecanoe Lake in Indiana, the parameters in Table 3 can be derived from his[7] measurements.

[7] There are slight discrepancies in the data for the same lake in different papers. In Table 3 the computations given in Wilson and Opdyke (1941) are used for Tippecanoe Lake, and in Wilson (1938) for Winona Lake.

TABLE 3. *Variation of morphometric parameters with time in Tippecanoe and Winona lakes*

	A_z, km.^{2}	z_m, m.	$\bar{z}$, m.	$\bar{z}:z_m$	z_r, %	V, m.$^3 10^6$
Tippecanoe						
Original basin	3.63	54.9	13.3	0.242	2.41	48.3
Modern basin	2.87	37.2	11.2	0.301	1.84	32.9
Winona						
Original basin	3.26	39.1	10.4	0.266	1.81	33.9
Modern basin	2.04	24.4	9.4	0.385	1.43	19.1

The decrease in maximum depth is due in part to sedimentation, and in part to cutting at the outlet which has apparently lowered the water level two or three meters; a very considerable part of the original littoral, now filled with sediment, has been exposed round the edge. The increase in development of volume implies that such shallow-water sedimentation has predominated as the areas of the lakes contracted, so that the mean depths suffered little reduction in spite of the filling of deep holes. The original basin of Winona Lake had five individual depressions, though the modern has but one. In the case of Tippecanoe Lake, ten original depressions are now reduced to three.

Douglas Lake, Michigan (Wilson 1945), presents a more complicated problem, because lacustrine sedimentation began when the site of the lake lay on the floor of Lake Algonquin. When the margin of the latter retreated, the original Douglas Lake had a level about 3 m. higher than at present. The comparisons possible are very arbitrary owing to the difficulty of deciding what shore lines may be taken for the Algonquin and post-Algonquin high-level stages. In Table 4 all quantities are taken as referring to present mean lake level; the Algonquin high-level transition is used as the only easily identifiable intermediate stage.

TABLE 4. *Variation of mean and maximum depth of Douglas Lake during postglacial time*

	z_m	$\bar{z}$	$\bar{z}:z_m$
Original	61.3	14.6	0.234
Algonquin high-level transition	58.8	8.8	0.148
Modern	27.2	5.5	0.204

The very low volume development, due to the presence of a number of blocks of ice set in drift that was subsequently covered by a large lake, became even lower as the Lake Algonquin sediments accumulated in the shallow water. The modern deposition of largely autochthonous organic sediments has filled these holes totally or in part, reducing their number from ten to seven. The very peculiar type of sedimentation during the Algonquin stage is interpreted by Wilson to mean that ice persisted in the deepest depressions until some time after the site of the lake became flooded.

Sublacustrine channels. In a few lakes the bottom topography is characterized by remarkable channels or sublacustrine valleys. These appear to be of two sorts. One kind has formed under subaerial conditions and later has been flooded, the other kind represents subaqueous erosion by density or turbidity currents.

The channel of the old Mackinaw River, which ran from Lake Michigan to Lake Huron during the Lake Chippewa phase, provides a magnificent example of the first type. Elsewhere, in northern Ireland the valley of the Bann River can be traced on the floor of Lough Neagh, which is held by the drift filling of the lower part of the old river valley (Charlesworth 1939).

Very spectacular examples have been described in Lake Tanganyika by Capart (1949). They all continue the course of existing river valleys (*B*, Fig. 9), but not all modern rivers lead to submerged channels. In the case of the Ruzizi River there are two very unequal distributaries; the sublacustrine channel runs from the mouth of the smaller of these, which, however, is known to be older than the larger distributary. There is no evidence of special subsurface water movements in the channels today. None descend below a depth of 500 m. Capart concludes that these channels were cut when the lake stood at a much lower level than at present; it is, indeed, not impossible that the two really deep basins of the lake were once almost or entirely separate, a possibility that will appear later to be of considerable zoogeographic interest.

The second kind of sublacustrine channel occurs in a few of the large subalpine lakes of Europe, on the floors of which river channels can be traced far beyond the entry of the main influent. The most remarkable example is provided by the sublacustrine channel of the Rhone at the eastern end of Lake Geneva (Fig. 60). The basin of the lake has been filled in by fluviatile sediments forming a strath far to the east of the present shore of the lake. Moreover, it is evident from the bathymetric map or the profile of the eastern end that much of the present lake basin has also been filled with fluviatile sediments. Across this broad shelf of sediments a well-defined valley has been cut by the very stream that has deposited the sedimentary filling of the basin. The main ravine can be traced at least 8 km. beyond the mouth of the main distributary, and smaller, shorter, less well-defined ravines can be made out beyond former mouths of the Rhone. The modern Rhone, as it enters the lake, can be seen to persist as a turbid stream for 150 to 200 m., sinking below the surface of the lake. At some seasons it may persist at middle depths and appear far down the lake. There can be no doubt that the main sublacustrine ravine is genetically related to the presence of the modern river. Forel (1892b) explains its existence essentially as the work of a turbidity current which is slowly mixed with the quieter waters of the lake. Laterally, where mixing and a turbulent transfer of momentum occur,

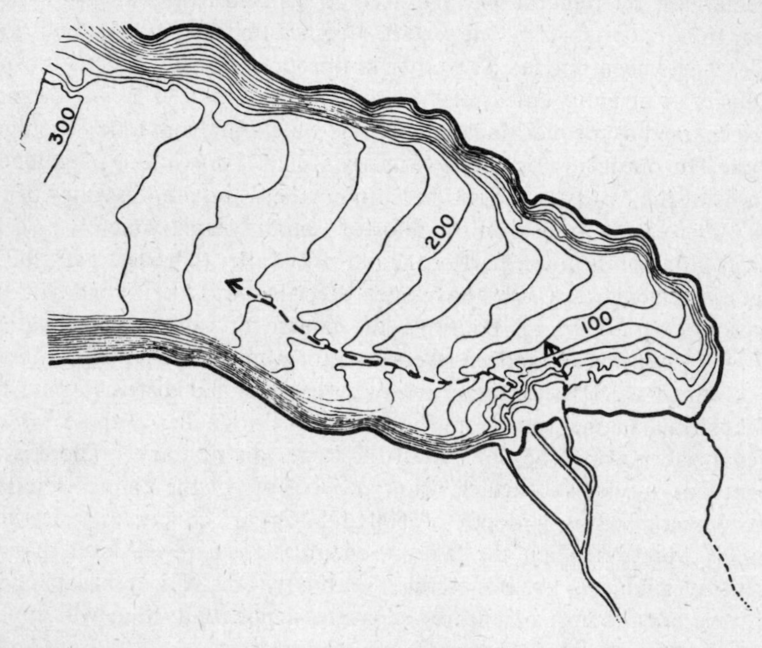

FIGURE 60. Eastern end of Lake Geneva, showing main sublacustrine channels of the Rhone, marked with broken arrows (after Delebecque).

sediment is deposited, but the main stream for a long distance is sufficient to keep a channel relatively clear, so producing the ravine. Delebecque (1898) accepts this explanation, which is in line with modern hydrogeological concepts. A comparable sublacustrine ravine has been formed by the Rhine in Lake Constance, but such forms are otherwise not developed in the subalpine lakes of Europe. Delebecque believed that the phenomenon was determined not only by the presence of an appropriate load of fine sediment, but also by the chemical composition of the water. If the content of dissolved salts is too low, he considered that the sedimentation of more or less colloidally dispersed clay proceeds so slowly that the sublacustrine delta bounding the ravine is not deposited.

Cryptodepressions. A number of lakes lie in basins whose deepest point lies below modern mean sea level. Such a depression is termed a *cryptodepression*, and the depth (z_c) of the cryptodepression is the depth of the bottom of the lake below sea level. Most of the world's deepest lakes occupy cryptodepressions. The most striking cases are in tectonic lakes, such as Baikal (z_c = 1279 m.), the Dead Sea (z_c = 793 m.) the Sea of Galilee (z_c = *c.* 260 m.), and Matana (z_c = 208 m.), all lying in grabens, and the Caspian (z_c = 972 m.) in an epeirogenetic basin.

There are also very remarkable examples in fjord lakes, where glacial overdeepening has excavated valleys far below sea level, as is true also of marine fjords. The deepest Norwegian examples are Hornindalsvatn (z_c = 461 m.), Salsvatn (z_c = 432 m.), and Mjøsa (z_c = 325 m.). Loch Morar (z_c = 301 m.) provides almost as spectacular a case, and a number of shallower cryptodepressions of the same kind are known in Scotland. In New Zealand there are comparable cases, Manapouri (z_c = 262 m.) being the deepest. Most of the large Chilean lakes probably occupy cryptodepressions; this is certainly true of Fagnano (z_c = 309 m.) and Elena (z_c = 122 m.). Of the subalpine basins of central Europe, only the large Italian lakes have depths below sea level (Garda, z_c = 281 m.; Como, z_c = 212 m.; Maggiore, z_c = 176 m.; Iseo, z_c = 65 m.).

The North American examples of this kind are less impressive. Chelan (z_c = 128.8 m.) is the finest North American cryptodepression in an overdeepened valley. The four deep Laurentian lakes occupy cryptodepressions (Superior, z_c = 124 m.; Michigan, z_c = 88 m.; Huron, z_c = 46 m.; Ontario, z_c = 151 m.), as do Seneca (z_c = 54.4 m.), and Cayuga (z_c = 17.4 m.) among the Finger Lakes. Among glacial corrasion lakes not in fjordlike valleys, Great Slave Lake (z_c = 464 m.) is unique.

A few maars and calderas occupy basins whose deepest point is below sea level, notably Tazawako (z_c = 175 m.) and Tôyako (z_c = 96.3 m.) in Japan (Yoshimura 1938b), and Apoyo (z_c = 110 m.) in Nicaragua (Sapper 1913). A good many of the smaller lakes of Holstein lie in shallow cryptodepressions; in the case of the Grosser Plöner See, z_c = 39 m. Any coastal plain with small glacial lakes is in fact likely to provide some examples either of modern cryptodepressions or of basins that would be so but for the sediments they have acquired. A few cases of fairly deep ponors or sink holes going below sea level are known; in the case of Scutari, z_c = 38 m. The surfaces of most coastal lakes lie little above sea level, and such lakes thus usually occupy shallow cryptodepressions. The total number of lakes in cryptodepressions is probably at least of the order of 1000.

A few lakes lie entirely below sea level. The most extraordinary cases are given in Table 5. Of these, all except the Sea of Galilee are closed and therefore subject to marked change in level.

TABLE 5. *Surface levels of lakes entirely below sea level*

	Meters
Dead Sea	– 980
Sea of Galilee (Tiberias)	– 210
Lage Enriquillo, Haiti	– 48
Lake Quarun, Faiyum, Egypt	– 44
Caspian Sea	– 25

SUMMARY

The form of a lake can be most perfectly expressed by a bathymetric map. From this map or from the data used in its construction, certain quantities that may be termed morphometric parameters can be obtained. The most important of these are the area (A), volume (V) the maximum depth (z_m) and the mean depth ($\bar{z} = V/A$), the length of the shore line (L), the development of shore line ($D_L = L/2\sqrt{\pi A}$) and the ratio $\bar{z}:z_m$, which is a measure of departure of the shape of the lake basin from that of a cone.

Only two lakes, Baikal and Tanganyika, are known to have maximum depths over 1000 m. and mean depths over 500 m. At least nineteen lakes have maximum depths in excess of 400 m. The Caspian Sea has the greatest area (436,400 km.^{2}) and volume (79,319 km.^{3}) of any body of water separated from the ocean. Lake Superior (A = 83,300 km.^{2}) has the greatest area of any purely fresh-water lake, and Lake Baikal (V = 23,000 km.^{3}) the greatest volume.

The form of a lake as seen from above may be approximately (*a*) *circular*, as in the case of most crater (Types 10 to 12) and caldera (Type 13) lakes, doline lakes (Type 43a), the smaller deflation basins (Type 63), and one or two of the supposed meteorite craters (Type 75); (*b*) *subcircular*, as in the case of cirque lakes (Type 27a) and of kettles (Types 35 to 38), particularly when remodeled by shore processes; (*c*) *elliptical*, as in the case of the oriented lakes of the Arctic (Type 42) and the Carolina bays (Type 76); (*d*) *subrectangular elongate*, as in the case of most lakes in grabens (Type 9) and overdeepened valleys (Type 28); (*e*) *dendritic*, when a dam is formed of sufficient height to flood tributary valleys (particularly Type 3); (*f*) *lunate*, as in the case of oxbow lakes (Type 55); (*g*) *triangular*, as in the case of drowned valleys behind bars (Type 64); (*h*) *irregular*, as in regions of glacial scouring and where several basins have been fused to form a lake.

Islands may be formed by a number of processes. *Structural* islands, due to faulting, etc., occur mainly in tectonic basins. *Volcanic* islands, formed as secondary volcanic cones, are frequent in caldera lakes. Most islands in glacially produced lakes are *residual* in the sense of being formed prior to the flooding of the basin. *Coastal* islands, formed by cutting behind promontories and other types of shore processes, are frequent. *Depositional* islands may be formed when spits are cut late in their development, and more rarely by sedimentation away from the coast.

Temporary anchored floating islands, formed when gas produced in the sediments buoys up a layer of the latter, are recorded in a few lakes, and quite large floating islands detached from marginal floating mats are frequently observed.

When a lake is first filled with water, its coast line will be out of harmony with the water movements in the lake. Shore processes will tend to cut a cliff with a beach and littoral shelf below it, and particularly to destroy any projecting promontories. Whenever wind drift follows a coast, it will carry fine particles of sediment from the shallow bottom along with it. If any indentation occurs in the coast, the momentum of the wind-driven current will carry it into deeper water across the opening of the indentation; sedimentation will therefore occur, and a spit or bar will develop. Such spits are often recurved at their free ends.

Deltas are formed where rivers enter lakes; the velocity and turbulence of the river being suddenly reduced, sedimentation takes place rapidly. The deposits so formed pile up on the slope of the lake, at the angle of repose, until they reach the surface, forming the foreset beds of the delta. Further deposition in the meanders and distributaries on such a delta will lead to horizontal or topset beds. The form of the delta depends on the amount of wave action that can influence it. Occasionally, under quiet conditions, the river builds a pair of levees out into the lake, forming a *unilobate* delta. With some wave action, a convex *arcuate* delta is formed; with still more, a biconcave *cuspate* delta. When there is little space for delta formation, as in a narrow lake with an influent at one end, the whole breadth of the lake may be filled in gradually, forming a *strath*.

The majority of lake basins have a ratio $\bar{z}:z_m > 0.33$, the value that would be given by a conical depression. In many caldera lakes, graben lakes, and fjord lakes, the ratio exceeds 0.5. Most lakes in areas of easily eroded rock have ratios between 0.33 and 0.5. Very low values of the ratio are found only in lakes with deep holes, either sinks due to solution or sublacustrine kettles due to the former presence of buried stagnant ice blocks. Where a number of deeps of differing area and profundity occur within a given lake, the chemical events in the water at the bottom of the different depressions may differ somewhat at times of stratification, giving rise to *submerged depression individuality*. During the history of any lake, the number of the sublacustrine depressions usually decreases with time.

Sublacustrine channels on the floors of lakes may be old river channels, formed prior to the lake or when the water of the lake stood at a lower level than at present, or may be due to the entering river retaining its individuality, owing to its lower temperature or greater silt load, and so greater density, than the lake water. In the latter case, deposition occurs laterally to the main stream, producing a kind of unilobate delta under water, the channel of which is still kept clear by the river.

A large number of lakes occupy basins of which the deepest point is below sea level. Such basins are said to be cryptodepressions. The deepest, that of Lake Baikal, has its bottom 1279 m. below sea level. The majority of very deep cryptodepressions are graben lakes of Type 9, or fjord lakes of Type 28b. A few lakes lie entirely below sea level, the Dead Sea lying lowest with its surface at – 980 m.

CHAPTER 3

The Properties of Water

It is appropriate, having considered the formation and shape of lake basins, to turn briefly to the physical properties of water, on which the whole economy of a lake depends. In this relation, it will also be convenient to discuss those chemical aspects of water that make the simple characterization of the substance as H_2O so inadequate to the modern investigator.

STRUCTURE

The curious physical properties of water were a favorite subject for speculation among the natural theologians of the early nineteenth century. Whewell (1833) and Prout (1833), in their Bridgewater treatises, both emphasized such matters. Prout, indeed, considered the anomalous expansion of water below 4° C. and its consequences "as presenting the most remarkable instances of design in the whole order of nature—an instance of something done expressly, and almost (could we indeed conceive such a thing of the Deity) at second thought, to accomplish a particular object." L. J. Henderson (1913), in his *Fitness of the Environment* revived many of the ideas of the natural theologians, but in a form largely devoid of explicit metaphysical connotations; the literary style of his celebrated work, however, echoes that of his theological predecessors. If the same facts are to be set forth again, it is mainly because they can be seen in perspective, in relation to modern ideas about the structure of liquids and to the much more detailed information now available about the behavior of other substances.

The water molecule. The water molecule,[1] which exists as unassociated H_2O in water vapor, consists of an oxygen atom to one side of which two hydrogen atoms are attached. The distance from the center of the oxygen atom to the center of either hydrogen atom is 0.96 A., and the angle between the lines joining the centers of the atoms is 105 degrees. This angle is somewhat greater than the 90 degrees suggested theoretically from quantum mechanical considerations, the difference being attributable to electrostatic repulsion of the hydrogens, which are to some extent ionized. The molecule resonates between the following structures

```
  H         H+         H          H+
  ..        .. –       .. –       .. =
 :O:H      :O:H       :O:H+      :O:H+
  ..        ..         ..         ..
```

The first structure, completely covalent, contributes about 56 per cent to the over-all structure of the molecule.[2] In several other aspects of the discussion, it will be legitimate to regard the hydrogens as ionized, that is to say, as protons.

A molecule of this kind has a considerable electrical dipole moment. It will act as if the oxygen had two weak unsatisfied valencies, indicated by the unshared pairs of electrons in the structural formulas just given; the hydrogens, owing to their capacity to form hydrogen bonds or hydrogen bridges between sufficiently small electronegative atoms such as oxygen atoms, will act as if they had weak unsatisfied single valencies. This permits the association of the H_2O molecules in liquid water, a phenomenon which underlies practically all the anomalous properties of that remarkable substance.

The structure of liquid water. The idea that the anomalous properties of water could be explained by variation in degree of association of H_2O molecules was first put forward in detail by Röntgen (1892), who supposed that ordinary water consisted of a solution of ice in a liquid of simpler molecular constitution. A number of hypotheses as to the nature of the more polymerized component were soon put forward, the most popular being that of Sutherland (1900), who supposed liquid water to consist of dihydrol H_4O_2 with trihydrol H_6O_3 in solution in increasing quantities as the freezing point is approached. Other workers believed that hydrol H_2O was also present, at least near the boiling point, or that the variation

[1] Most of the general discussion on the structure of water is derived, when no other reference is given, from Pauling's (1939) already classical work *The Nature of the Chemical Bond.*

[2] A single molecule may be thought of as spending 56 per cent of its time in the covalent condition.

in properties was exclusively due to variations in the proportions of H_4O_2 and H_2O rather than of H_6O_3 and H_4O_2.

These older concepts have been summarized in detail by Barnes and Jahn (1934) and by Dorsey (1940). They are, however, not now generally accepted, having been replaced by theories based on the fact that a beam of X rays passed through liquid water is diffracted and that the diffraction pattern, which varies with the temperature, gives evidence of a quasi-crystalline structure.

Bernal and Fowler (1933) first put forward a comprehensive theory of the structure of liquid water based on X-ray diffraction crystallography, and though this theory has not found acceptance in every detail, it has been the foundation of all subsequent discussion. In the simplest liquids, such as molten sodium or a liquefied rare gas, the atoms are packed together almost as closely as is possible, Ideally, in the arrangement of closest packing in a collection of spherical atoms, each atom will have twelve closest neighbors. In the case of metallic sodium, experimental studies indicate that the average number is ten, which thus closely approximates the ideal. Near the critical point, 150° C., under appropriate pressure, Bernal and Fowler believe water to have such a structure. This form of water, with the closest packing structure, they designated water III. Ordinary ice has a crystalline structure of the kind typically exhibited by the SiO_2 mineral tridymite. In this structure as exhibited by ice (Fig. 61), the oxygen atoms are arranged in such a way that any atom is associated with four other atoms set at the apices of a tetrahedron, in the center of which lies the atom taken as a point of reference. The interatomic distances from the center of the central atom to that of any of the four tetrahedrally coordinated with it is 2.76 A. Farther away from the central atom, at a distance of 4.51 A., there will be twelve atoms, three coordinated with each member of the tetrahedrally placed group of nearest neighbors. In this way a tetrahedrally coordinated open lattice is built up.

Between any two oxygen atoms lies a hydrogen atom, forming a hydrogen bond or bridge. This hydrogen atom is not midway between the two oxygens, at 1.38 A. from the center of either, but is 0.99 A. from one, almost as in steam, and 1.77 A. from the other. Pauling (1939) points out that, for any four-coordinated oxygen atom, if the position of each bond could be marked, there would be sixteen possible arrangements. Thus all the four hydrogen atoms could be at 0.99 A., corresponding to H_4O^{++}; there can obviously be only one such arrangement. Three of the hydrogen atoms could be at 0.99 A. and one at 1.77 A., corresponding to the hydroxonium ion H_3O^+; there are four possible arrangements in this category. Two hydrogens can be at 0.99 A. and two at 1.77 A., corresponding to the neutral molecule H_2O; this is possible in six ways. There will then be four possibilities corresponding to the hydroxyl ion OH^- and one to the doubly charged ion $O^=$, around which all the bonding hydrogens can be regarded as at 1.77 A. It is reasonable to suppose that the hydroxonium and hydroxyl ions are at least as rare in ice as in water at 0° C.; the two doubly charged ions must be much rarer still. Obviously,

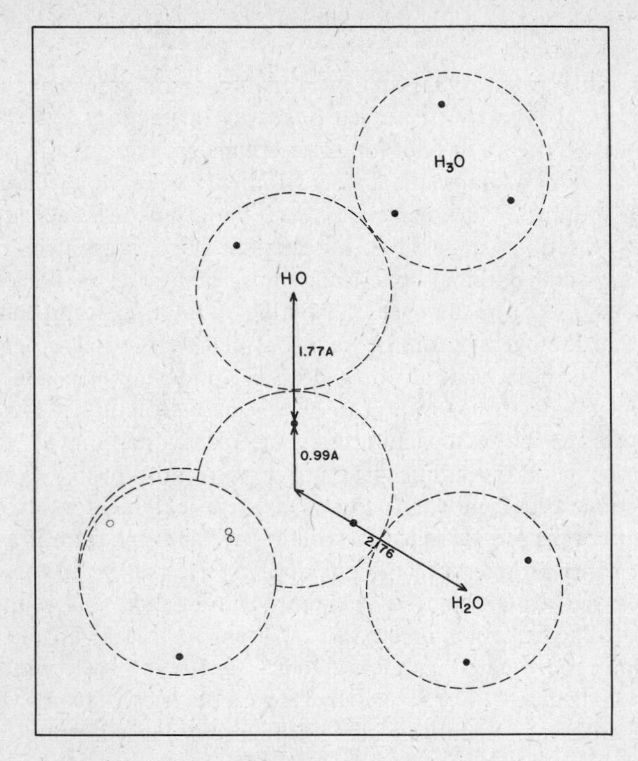

FIGURE 61. The tridymite structure of ice.

nearly all the oxygen atoms are associated with two close and two remote hydrogens. As there is no preferred arrangement determining which two of the four hydrogens should be close and which remote, a certain amount of randomness exists in the ice lattice, even at low temperatures, and is responsible for the residual entropy of the substance.

Bernal and Fowler supposed that in liquid water at about 0° C., some of the material designated water I retained this icelike structure, though the lattice bonds were believed to be continually breaking and reforming in the liquid state. Most of the liquid water termed water II was believed to exhibit an arrangement relative to ice of the same kind as quartz bears to tridymite, which implies a slightly closer packing than is found in ice or in the hypothetical water I. This point of view has been widely disseminated in the literature, but is now superseded by a somewhat simpler scheme, based on a new study of the X-ray diffraction pattern of water at various temperatures by Morgan and Warren (1938).

Water can be considered an almost monatomic substance, the hydrogen atoms being very small relative to those of oxygen. The X-ray diffraction pattern of such a substance can be translated into a curve showing the variation with distance r from an arbitrary reference atom, of a function which indicates the rate of change of the number of atoms within the sphere of radius r. If a substance with the ideal structure of ice were being considered, starting with any atom and increasing r, the value of the function which outside the reference atom is at first zero would rise to a maximum at $r = 2.76$ A., and then rapidly fall again to zero. The, integral of the function over the range of r corresponding to this peak will give the number of the atoms present in the closest neighbor position, in this case four. Proceeding to higher values of r, at 4.15 A. there will be a second maximum corresponding to the second closest neighbors, and the area under this peak will, when it is suitably isolated, be 12, and so on, with the more complex pattern developed as r increases.

The X-ray diffraction patterns and values of the distribution function just discussed are given for water at 1.5° C. and at 83° C. in Figs. 62 and 63, from the work of Morgan and Warren. The first maximum, which corresponds to an interatomic distance of 2.90 A. at 1.5° C., increasing to 3.05 A. at 83° C., is not sharply defined on the right, indicating that although few if any atoms are likely to be present within the interatomic distance 2.90 to 3.05 A., beyond this distance atoms may be found nearer than are the next closest neighbors in ice. If the left-hand sides of the first peaks are reflected about the maximum, areas corresponding to the number of closest neighbors can be obtained, perhaps somewhat arbitrarily, and the number of such neighbors is found to be 4.4 at 1.5° C., 4.6 at 30° C., and 4.9 at 85° C. Each oxygen, therefore, tends to be associated with rather more than the four tetrahedrally arranged neighboring oxygens present in ice. With increasing values of r, the pattern tends rapidly to become more diffuse, particularly at high temperatures, indicating that as one progresses from any oxygen atom outwards, there is a considerable probability of reaching a place fairly rapidly where the structure has broken down. At 1.5° C. there is still a fairly high probability of encountering a high density of atoms at 4.5 A.; at the higher temperatures this becomes much less marked. Bernal and Fowler's structure for water II would involve a displacement of the second maximum from 4.5 to 4.2 A.; the experimental evidence clearly excludes this possibility.

Morgan and Warren interpret their results as follows. There is good evidence from other work that the coordination is below rather than above 4, as might have been supposed from the diffraction pattern. Thus Cross, Burnham and Leighton (1937), from a study of the Raman

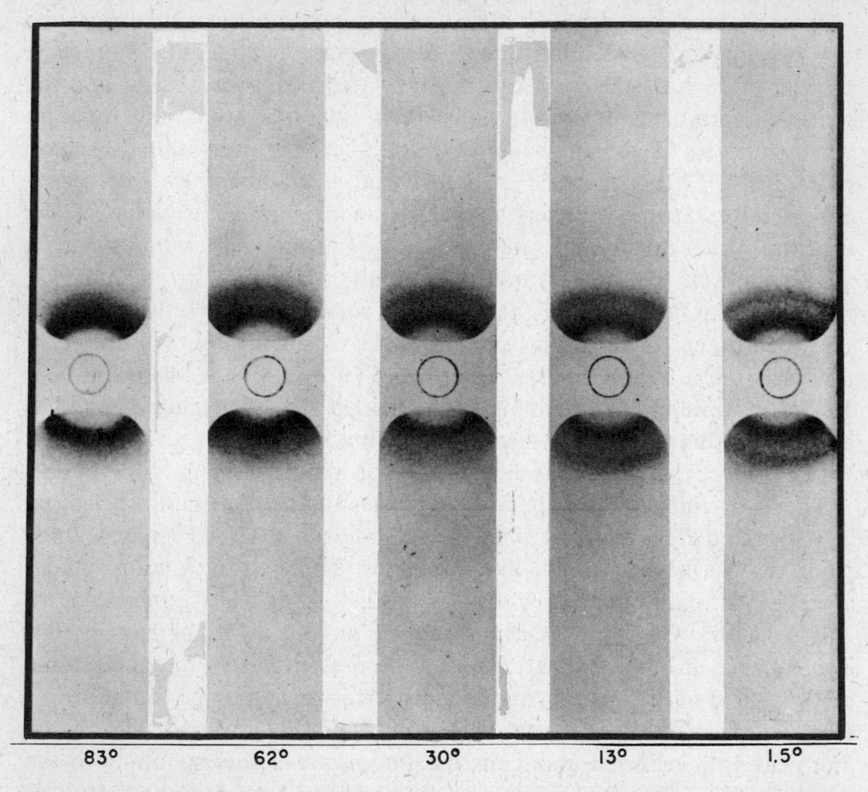

FIGURE 62. X-ray diffraction pattern of water at different temperatures. Note the striking loss of structure at high temperatures, indicated by the disappearance of the outer band. (Morgan and Warren 1938.)

spectrum, conclude that the coordination is slightly greater than two; from a study of viscosity, Ewell and Eyring (1937) obtain a value of 2.5 at 0° C., falling to 1.5 at 50° C. This implies a breaking down of the ice lattice so that many oxygens become less than four coordinated. At the same time, the somewhat excessive number of closest neighbors indicates that the breakdown permits increased packing together of the uncoordinated oxygen atoms.

Increasing interatomic distance on melting of ice and heating of water will lead to a lowering of the density, while increased bond rupture and disorganization of the lattice will permit the filling in of spaces open in the complete ice lattice, increasing the density. From 0° to 4° C. the latter effect predominates, and above 4° the former effect predominates, producing the maximum density at 4° C. At least at low temperatures, a

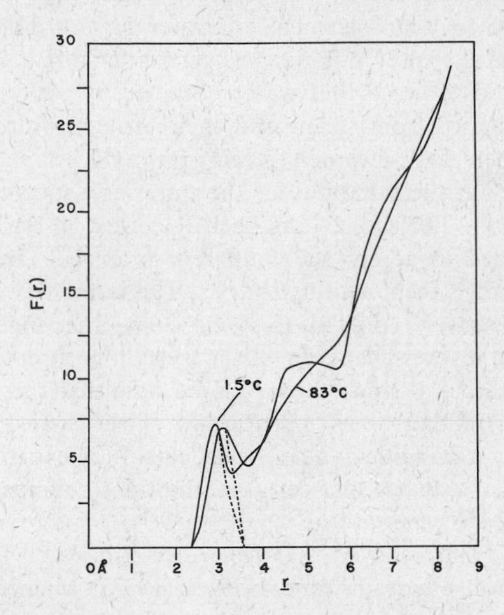

FIGURE 63. Distribution function F(*r*) plotted against *r*, at 1.5° C. and at 83° C. (After Morgan and Warren.)

fairly definite if continually changing lattice must exist, bonds breaking and reforming with great rapidity. At 85° C., though any given oxygen is still likely to be surrounded by over four other oxygens, the pattern is so indefinite that very little of the more peripheral arrangement can be distinguished. The most recent work (Finbak and Viervoll 1943) suggests that, in so far as a lattice exists in liquid water, it consists of branched chains of tetrahedra united corner to corner in such a way as to permit the individual members to rotate freely. These chains form an intermeshed net and run in any direction in the free liquid.

In visualizing any structure deduced from X-ray diffraction studies of a liquid, it is important to bear in mind, as Dorsey (1940) points out, that the diffraction patterns imply a condition lasting a time of the order of the period, or reciprocal of the frequency, of the X rays employed, i.e., about 10^{-18} sec. The pattern observed may therefore refer to a structure which is changing with great rapidity, though maintaining throughout such changes certain average relationships.

Establishment of associative equilibrium. In the past, especially when it seemed likely that individual molecular species such as H_4O_2 and H_6O_3 existed in liquid water, there was much discussion of the possibility that the equilibrium appropriate to any temperature might take an appreciable time to establish, so that specimens of water having different thermal histories might have slightly different properties for a few hours or days. In particular, it was claimed that water obtained by melting ice differed from water condensed from steam and then brought unfrozen to 0° C. The supposed differences concerned freezing rate and certain other physical properties, as well as the suitability of the water as a medium for certain aquatic organisms. The subject has been reviewed by Barnes and Jahn (1934), who tended to accept such differences as real, and by Dorsey (1940), who regarded them as illusionary. The latter view now appears to be all but universal; it is difficult to see how any other conclusion can be reconciled with the ever-changing quasi-crystalline structure of liquid water just described. It must, however, be admitted that some of the observations on freezing times, particularly those carried out in large tanks, though far from critical, have never been fully explained, and that it is still just possible that under certain conditions colloidal ice particles might be dispersed in water at 0° C.

Barnes (1932, 1933) believed that ice water stimulated the growth of certain of the lower organisms, and Harvey (1933) obtained partial confirmation of this with the diatom *Nitzschia closterium*. Much of the biological work may be vitiated by the fact that in many cases the ice and steam waters used were not obtained from the same source and may well, in spite of their supposed purity, have contained different trace elements. Later investigators seem to have thought the whole subject too unprofitable to merit reinvestigation.

PHYSICAL PROPERTIES OF WATER

Liquid nature of water. The existence of liquids in nature is rather exceptional. Apart from water, the only inorganic liquids known at the surface of the earth are elementary mercury, sometimes found in cinnabar, and liquid CO_2, which is supposed to be present under pressure in cavities in quartz crystals (Hawes 1881, Wright 1881). The only other non-aqueous natural liquids are the complex mixtures of hydrocarbons and other organic compounds known as petroleums. These are now universally acknowledged to be of organic origin. The anomaly of the liquid nature of water, at the temperatures and pressures normally obtaining at the earth's surface, can easily be seen from a comparison of the melting and boiling points of the dihydrides of oxygen, sulfur, selenium, and tellurium.

	H_2O	H_2S	H_2Se	H_2Te
Melting point, °C.	0	− 82.9	− 64	− 48
Boiling point, °C.	+ 100	− 59.6	− 42	− 2

Extrapolation from the other three compounds would suggest that water should boil at about − 80° C.

Similarly, proceeding horizontally along the periodic table,

	H_4C	H_3N	H_2O	HF
Melting point, °C.	− 182.6	− 77.7	0	− 83
Boiling point, °C.	− 161.4	− 33.4	+ 100	+ 19.4

Both ammonia and hydrogen fluoride, which associate through hydrogen bond formation between the small and electronegative N and F atoms respectively, show some raising of the melting and boiling points, but much less than in the case of water. In the case of ammonia, the effect is less because the degree of association is less; in the case of hydrogen fluoride, the associated molecules behave as units and pass into the vapor phase.

The latent heat of sublimation of ice is 679 cal. per g. or 12,200 cal. per mole. Of the latter figure, about 3000 cal. are reasonably attributed, by comparison with other substances, to ordinary Van der Waals forces retaining the molecules in the solid, leaving some 9000 cal. per mole due to the rupture of hydrogen bonds.

The latent heat of fusion of ice is, however, 79.7 cal. per g. or 1440 cal. per mole, which is about 15 per cent of the heat of rupture of the hydrogen bonds in vaporization. Though there is some evidence that some such bonds may be broken in ice at 0° C. before melting begins, it is evident that

TABLE 6. *Thermal properties of water and of certain comparable compounds*

Compound	Boiling Point, °C.	Latent Heat of Evaporization, cal. g.$^{-1}$	Melting Point, °C.	Latent Heat of Fusion, cal. g.$^{-1}$	Specific Heat (c_p) of Liquid cal. g.$^{-1}$ °C.$^{-1}$	at °C.
H_2O	100	539.55	0	79.67	1.01	0
H_2O_2	...	...	− 1.7	74	0.58	− 1·7
CH_4	− 159	138	− 182.6	14.0	0.86	− 177
NH_3	− 33.4	327.1	− 77.6	83.9	1.05	− 73
CH_3OH	64.7	262.8	− 97.8	16.4	0.601	20
C_2H_5OH	78.3	204	− 114.4	24.9	0.579	23

TABLE 7. *Density of water as a function of temperature from 0° C. to 35° C. at a pressure of* 1 *atm.*

°C.	0.0	0.1	0.2	0.3	0.4
0	0.9998679	0.9998746	0.9998811	0.9998874	0.9998935
1	0.9999267	0.9999315	0.9999363	0.9999408	0.9999452
2	0.9999679	0.9999711	0.9999741	0.9999769	0.9999796
3	0.9999922	0.9999937	0.9999951	0.9999962	0.9999973
4	1.0000000	0.9999999	0.9999996	0.9999992	0.9999986
5	0.9999919	0.9999902	0.9999883	0.9999864	0.9999842
6	0.9999681	0.9999649	0.9999616	0.9999581	0.9999544
7	0.9999295	0.9999248	0.9999200	0.9999150	0.9999099
8	0.9998762	0.9998701	0.9998638	0.9998574	0.9998509
9	0.9998088	0.9998013	0.9997936	0.9997859	0.9997780
10	0.9997277	0.9997189	0.9997099	0.9997008	0.9996915
11	0.9996328	0.9996225	0.9996121	0.9996017	0.9995911
12	0.9995247	0.9995132	0.9995016	0.9994898	0.9994780
13	0.9994040	0.9993913	0.9993784	0.9993655	0.9993524
14	0.9992712	0.9992572	0.9992432	0.9992290	0.9992147
15	0.9991265	0.9991113	0.9990961	0.9990808	0.9990653
16	0.9989701	0.9989538	0.9989374	0.9989209	0.9989043
17	0.9988022	0.9987848	0.9987673	0.9987497	0.9987319
18	0.9986232	0.9986046	0.9985861	0.9985673	0.9985485
19	0.9984331	0.9984136	0.9983938	0.9983740	0.9983541
20	0.9982323	0.9982117	0.9981909	0.9981701	0.9981490
21	0.9980210	0.9979993	0.9979775	0.9979556	0.9979335
22	0.9977993	0.9977765	0.9977537	0.9977308	0.9977077
23	0.9975674	0.9975437	0.9975198	0.9974959	0.9974718
24	0.9973256	0.9973009	0.9972760	0.9972511	0.9972261
25	0.9970739	0.9970482	0.9970225	0.9969966	0.9969706
26	0.9968128	0.9967861	0.9967594	0.9967326	0.9967057
27	0.9965421	0.9965146	0.9964869	0.9964591	0.9964313
28	0.9962623	0.9962338	0.9962052	0.9961766	0.9961478
29	0.9959735	0.9959440	0.9959146	0.9958850	0.9958554
30	0.9956756	0.9956454	0.9956151	0.9955846	0.9955541
31	0.9953692	0.9953380	0.9953068	0.9952755	0.9952442
32	0.9950542	0.9950222	0.9949901	0.9949580	0.9949258
33	0.9947308	0.9946980	0.9946651	0.9946321	0.9945991
34	0.9943991	0.9943655	0.9943319	0.9942981	0.9942643
35	0.9940594	0.9940251	0.9939906	0.9939560	0.9939214

TABLE 7 (*Continued*)

°C.	0.5	0.6	0.7	0.8	0.9
0	0.9998995	0.9999053	0.9999109	0.9999163	0.9999216
1	0.9999494	0.9999534	0.9999573	0.9999610	0.9999645
2	0.9999821	0.9999844	0.9999866	0.9999887	0.9999905
3	0.9999981	0.9999988	0.9999994	0.9999998	1.0000000
4	0.9999979	0.9999970	0.9999960	0.9999947	0.9999934
5	0.9999819	0.9999795	0.9999769	0.9999741	0.9999712
6	0.9999506	0.9999467	0.9999426	0.9999384	0.9999340
7	0.9999046	0.9998992	0.9998936	0.9998879	0.9998821
8	0.9998442	0.9998374	0.9998305	0.9998234	0.9998162
9	0.9997699	0.9997617	0.9997534	0.9997450	0.9997364
10	0.9996820	0.9996724	0.9996627	0.9996529	0.9996428
11	0.9995803	0.9995694	0.9995585	0.9995473	0.9995361
12	0.9994660	0.9994538	0.9994415	0.9994291	0.9994166
13	0.9993391	0.9993258	0.9993123	0.9992987	0.9992850
14	0.9992003	0.9991858	0.9991711	0.9991564	0.9991415
15	0.9990497	0.9990340	0.9990182	0.9990023	0.9989862
16	0.9988876	0.9988707	0.9988538	0.9988367	0.9988195
17	0.9987141	0.9986961	0.9986781	0.9986599	0.9986416
18	0.9985295	0.9985105	0.9984913	0.9984720	0.9984526
19	0.9983341	0.9983140	0.9982937	0.9982733	0.9982529
20	0.9981280	0.9981068	0.9980855	0.9980641	0.9980426
21	0.9979114	0.9978892	0.9978669	0.9978444	0.9978219
22	0.9976846	0.9976613	0.9976380	0.9976145	0.9975910
23	0.9974477	0.9974235	0.9973991	0.9973747	0.9973502
24	0.9972010	0.9971758	0.9971505	0.9971250	0.9970995
25	0.9969445	0.9969184	0.9968921	0.9968657	0.9968393
26	0.9966786	0.9966515	0.9966243	0.9965970	0.9965696
27	0.9964033	0.9963753	0.9963472	0.9963190	0.9962907
28	0.9961190	0.9960901	0.9960610	0.9960319	0.9960027
29	0.9958257	0.9957958	0.9957659	0.9957359	0.9957059
30	0.9955235	0.9954928	0.9954620	0.9954312	0.9954002
31	0.9952127	0.9951812	0.9951495	0.9951178	0.9950861
32	0.9948935	0.9948612	0.9948286	0.9947961	0.9947635
33	0.9945660	0.9945328	0.9944995	0.9944661	0.9944327
34	0.9942303	0.9941963	0.9941622	0.9941280	0.9940938
35	0.9938867	0.9938518	0.9938170	0.9937820	0.9937470

of the large number remaining, only about 15 per cent are broken in the actual transition from solid to liquid. The breaking of the remainder of the hydrogen bonds in liquid water contributes to the high specific heat of the substance and to the high latent heat of evaporation. The very considerable thermal inertia of the hydrosphere and the great effect of water masses on the temperature of adjacent air and land are largely due to the very great amounts of heat required to effect phase transitions and to change the temperature of the liquid, and are therefore indirectly due to the association by means of hydrogen bonds characteristic of the liquid. The exceptional nature of these values for the latent heats and heat capacities can be seen from Table 6, in which values for other inorganic and simple organic compounds of relatively low molecular weight are set out. With regard to the heat capacity, it has long been known that this quantity exhibits a minimum at 30° C., a phenomenon which led Rowland (1880) to suspect that water in the lower part of its temperature range had not fully recovered from freezing. In more modern terms, a greater but decreasing amount of the heat capacity is involved in rupture of hydrogen bonds, a lesser but increasing amount of heat in increasing the kinetic energy of the molecules, as in an unassociated liquid.

Density relationships. By far the most important anomalous properties of water from the standpoint of the present work are those involving changes in density. As has already been indicated, these receive a very convincing explanation from the structure of ice and water. It may, however, be convenient to present the actual data on this matter, for they will be used on many later occasions.

The density of ordinary ice at 0° C. is 0.9168, whereas that of water at the same temperature is 0.9999. The contraction on melting is not unique, being exhibited by a few metals, such as gallium and bismuth, and by some organic liquids.

As water is warmed from 0° C., it increases in density until a maximum of 1.000 is reached at 3.94° C.; above this temperature, expansion occurs at an increasing rate as the temperature is raised. The full data, which are of the greatest importance in the study of the hydrosphere, are presented in Table 7, from Stott and Bigg (1928). These data relate to a pressure of 1 atm., or 1.032 kg. per cm.^{2}, corresponding to the surface of a lake at sea level.

Water is by no means incompressible; the density of pure water varies with pressure (Smith and Keyes 1934), over the range of limnological interest, in a practically linear manner at any given temperature. The values given in Table 8 are computed to correspond to convenient depths, and are tabulated with respect to depth rather than to pressure.

TABLE 8. *Density of water as a function of depth*

Temperature, °C.	Density* at Depths of 0 m.	250 m.	500 m.	1000 m.	1500 m.	2000 m.
0	0.9998	1.0009	1.0020	1.0042	1.0063	1.0084
10	0.9995	1.0006	1.0017	1.0037	1.0058	1.0080
20	0.9980	0.9991	1.0002	1.0023	1.0043	1.0064
30	0.9955	0.9966	0.9977	0.9998	1.0018	1.0040

* It will be observed that the data given by Smith and Keyes, from which these values are computed, differ slightly from the standard values (Table 7) of densities at 1 atm. pressure.

The change in volume when water is moved from a high pressure or a great depth to a lesser pressure or depth, produces adiabatic cooling. In oceanography, it is customary to describe as the potential temperature of water at any depth, the temperature which the water would reach if allowed to expand adiabatically when brought to a pressure of 1 atm. The amount of change in temperature per unit change of pressure is known as the Joule-Thomson coefficient. The coefficient will be negative below the temperature of maximum density, but, as the latter is a function of pressure, a rather complicated situation exists at low temperatures. The coefficient has been determined directly by a number of workers, and has been computed from the specific heat and coefficient of thermal expansion, but the results are not concordant. The computed values, based on Bridgman's data (Dorsey 1940), appear to be the only values within the limnologically interesting range which are likely to be satisfactory. They indicate a practically linear increase in the coefficient with temperature at any given pressure, as indicated in Table 9.

TABLE 9. *Joule-Thomson coefficient of water*

Pressure or Depth	Coefficients, °C. per 1000 m. at 0° C.	20° C.	40° C.
1 kg. cm.$^{-2}$ or 0 m.	− 0.016	+ 0.137	+ 0.287
500 kg. cm.$^{-2}$ or 4999 m.	+ 0.068*	+ 0.175	+ 0.300

* If, as seems probable, water at 0° C. and 500 kg. per cm.^{2} is below the temperature of maximum density, it is apparent that this value is too high.

It will be observed that in Lake Baikal, in which all the deep water is near the temperature of maximum density, the effect is likely to be very small, but in very accurate work on the thermal properties of tropical lakes of great depth, adiabatic heating might have to be considered. In Tanganyika, where the greater part of the water is at about 23° C., the potential temperature at 1000 m. is evidently about 0.16° C. lower than the observed temperature.

The variation of the temperature of maximum density with pressure is of considerable interest in relation to lake temperatures. Strøm (1945) has reviewed all the available laboratory determinations and concludes that the only reliable values are 3.94° C. at a pressure of 1 atm. or roughly 1 kg. per cm.^{2} and 0° C. at a pressure of 600 kg. corresponding to a depth of 5999 m. By examining the temperature curves (Fig. 140) of very deep lakes, which curves should never cross the line relating temperature of maximum density to depth as a measure of pressure, Strøm concludes that, in default of laboratory data, the best provisional values are those given in Table 10.

TABLE 10. *Temperature of maximum density as a function of pressure of depth*

Pressure		Temperature,	°C. per
Kg. cm.$^{-2}$	Meters	°C.	100 m.
1	0	3.94	
			0.12
11	100	3.82	
			0.11
21	200	3.71	
			0.11
31	300	3.60	
			0.11
41	400	3.49	
			0.10
51	500	3.39	
			0.10
61	600	3.29	
			0.09
101	1000	2.91	
			0.08
171	1700	2.32	

Values down to 600 m. are evidently reasonably satisfactory, but the last two entries are based on extrapolating the rather inadequately defined curve toward the known value of 0° C. at 600 kg. per cm.^{2} and are obviously not very trustworthy. It will be observed that the temperature of maximum density falls with increasing pressure more rapidly in the range of moderate pressures of wide limnological interest, than at the great pressures which are present only at the bottoms of the very deepest lakes.

Dielectric constant and solubility. The capacity of water to form ionic solutions is dependent largely on the high dielectric constant of the liquid. In general, the dielectric constant increases in a regular way with the value of the dipole moment, but in liquids associated by means of hydrogen bonds, very much higher values are observed than would be expected. The observed value of 80 cgs units at 20° C. is exceeded only by H_2O_2, 87 units, and HCN, 116 units. The high value of the dielectric constant also permits the formation of *colloidal sols* in which large charged particles are dispersed through the liquid phase. The third kind of solution, namely the true *solution of nonelectrolytes*, mainly involves molecules which, like water itself, can form hydrogen bridges with other water molecules. The molecular structure of the liquid is thus seen to underlie the exceptionally great range of materials which can be dispersed through water as solutions of one kind or another.[3]

Other properties. Of the other properties of water, those most relevant to the present work are the surface tension and viscosity. The surface tension at the liquid-air interface, 73.5 dynes at 15° C., is higher in the case of water than in those of any other liquid save mercury. The viscosity of water, 0.0114 poise at 15° C., is also very high for a liquid of low molecular weight. Though the theory of the relationships is fairly complex, it takes little intuitive insight to realize that both these quantities are likely to be higher in an associated than in an unassociated liquid. It has already been pointed out that viscosity data permit an estimate of the coordination of oxygen atoms in water.

Naturally occurring forms of ice. The problems of the formation and disappearance of ice on lakes will be discussed in Chapter 7. It is appropriate, however, while discussing water as a substance, to add a few paragraphs on its solid phases.

As the result of long series of research by Tammann and by Bridgman, five forms of ice, commonly called ice II, ice III, ice V, ice VI, and ice VII, are known to be stable over various ranges of pressure in excess of 2000 kg. per $cm.^2$ At least one unstable form, ice IV, which cannot exist in the presence of the others, has been described; other unstable phases may exist.[4] All these forms of ice are denser than the water with which they are in equilibrium. Although it is probable that some of them play an important part in the formation of the outer layers of the major planets, they are not known to occur naturally on earth.

Two claims (Cox 1904, Shaw 1924) that ice denser than water can form

---

[3] The dissolved material, particularly the electrolytes, will in turn effect the properties of the liquid phase. The oceanographer, physiologist, or limnologist concerned with saline lakes may profitably consult Robinson and Stokes (1955) and Chapters 7 and 8 of Cohn and Edsell (1943).

[4] The considerable literature of the high-pressure ices is reviewed by Dorsey (1940). The belief (Seljakov 1936, 1937) that ice I occurs in two modifications in nature has proved to be erroneous (Owston and Lonsdale 1948).

under certain undetermined conditions, at a pressure not greater than 1 atm., have been made. Both claims rest on observations in the cryophorous apparatus during classroom demonstration. Dorsey suspects that, in the confined space of the apparatus, capillary phenomena depressed the ice and deceived the observers. Shaw's account is very convincing, however, and suggests that the possibility of the formation of an unstable heavy ice, forming at ordinary pressures under unknown circumstances, must be taken quite seriously.

Another possible form of solid water is the vitreous, which is easily produced in an impure condition if thin films of various solutions are supercooled rapidly. Pure vitreous water is very difficult to prepare, and the substance is not likely to be formed in nature.

Dissociation of pure water. It has already been indicated that, although most oxygen atoms in ice or water are associated with two close and two remote hydrogen atoms or protons, such protons may easily jump from the remote to the close position or vice versa, forming the hydroxonium, oxonium, or hydrated hydrogen ion H_3O^+ and the hydroxyl ion OH^-, and much more rarely the doubly charged ions H_4O^{++} and $O^=$. Both the hydroxonium and the hydroxyl ions are usually regarded as more or less hydrated, which is only another way of saying that they take part in the formation of the transient lattice structure of water. Electrolysis of water involves not the actual migration of permanent H_3O^+ and OH^- ions, but the continual formation of these ions by proton jumps, with a tendency for the protons to move from oxygen atom to oxygen atom towards the cathode while the holes left by the protons move in the opposite direction. Bernal and Fowler (1933) have shown that the abnormally high ionic mobilities of these ions, when migrating in water but not in other substances such as HF, can be explained on quantum mechanical principles, according to such a scheme.

According to ordinary principles of physical chemistry, if $[H_3O^+]$ is the concentration of hydroxonium ions in moles per 1000 g. water, and if $[OH^-]$ is the concentration of hydroxyl ions in like units, then in dilute solution

$$(1) \qquad [H_3O^+]\ [OH^-] = K_w$$

Since in pure water the concentrations of the hydroxyl and hydroxonium ions will be identical,

$$(2) \qquad [H_3O^+] = [OH^-] = \sqrt{K_w}$$

It is usual to employ the symbol pH to denote the common logarithm of the reciprocal of the concentration of the hydroxonium or "hydrogen" ion expressed as moles per liter.

$$(3) \qquad p\text{H} = \log (\rho\sqrt{K_w}^{-1})$$

where ρ is the density and log K_w^{-1} is often written P_w. The concentration of the hydroxonium and hydroxyl ions increases with the temperature. The best values, those of Harned and his collaborators (Dorsey 1940) are given in Table 11.

TABLE 11. *Dissociation constant and* pH *of neutrality of pure water as a function of temperature*

Temperature, °C.	K_w	pH
0	0.113×10^{-14}	7.47
5	0.185	7.37
10	0.292	7.27
15	0.450	7.17
20	0.681	7.08
25	1.008	7.00
30	1.468	6.92
35	2.089	6.84
40	2.917	6.77

It will be remembered that in this case, in contrast to what happens when an acid is added to a solution, the concentration of hydroxyl ions increases at the same time as does that of the hydroxonium or hydrated hydrogen ions.

THE ISOTOPES OF HYDROGEN AND OXYGEN: LIGHT AND HEAVY WATER.

Hydrogen has two stable isotopes, H^1, sometimes called protium, and H^2 or D, usually called deuterium. There is also a radioactive isotope H^3 or T, usually termed tritium, a β-emitter with a half life of about 12.5 years. Owing to the great relative differences between their atomic masses, the isotopes of hydrogen are chemically at least as different as are some of the rare earth elements. The use of special symbols and names is therefore convenient, if not quite logically consistent. It is always much easier to determine the relative difference in the deuterium content of two waters than to determine the absolute concentration in either one of them. The best work is that of Kirshenbaum, Graff, and Forstat (in Kirshenbaum 1951), who calibrated their mass spectrograph by means of water samples of known deuterium content prepared synthetically. The results of this investigation indicate that most fresh waters in temperate latitudes have a concentration of deuterium near 0.0148 mole per cent of the total hydrogen present. Lake Michigan water is believed by Friedman (1953) to contain such a quantity of deuterium, and in his work is used as a standard with which the other waters that he studied were compared. Such a concentration is, moreover, in accord with the earlier results of Gabbard and Dole (1937), Swartout and Dole (1939) and Greene and Voskuyl (1939).

Tritium content of natural waters. The tritium present in minute amounts in the hydrosphere is formed by nuclear reactions occurring in the atmosphere. Neutrons produced by cosmic radiation can react with nitrogen atoms in several ways (Libby 1946, Fireman 1953). The reaction

$$N^{14} + n \rightarrow C^{14} + p$$

is responsible for the very important fact that radiocarbon exists in the atmosphere. The less probable reaction

$$N^{14} + n \rightarrow C^{12} + H^{3}$$

is presumably an important source of tritium, but some part of the latter may be produced by reactions in which a number of particles are produced which, when they are recorded in a photographic emulsion, are recognized as cosmic ray "stars."

Tritium decays to produce He^3, which is lost from the atmosphere to space. The half life of tritium being 12.5 years, very little accumulates in the hydrosphere. Kaufman and Libby (1954) believe the production rate to be of the order of 0.14 atom cm.$^{-2}$ sec.$^{-1}$, and the total natural terrestrial reserve about 1800 ± 600 g. most of which is present in the ocean. The number of atoms relative to hydrogen atoms in various waters of interest are given in Table 12, based on Kaufman and Libby (1954; see also Grosse, Johnston, Wolfgang, and Libby 1951). Maritime rain

TABLE 12. *Tritium content of natural water (atomic proportions per atom protium)*

Source	Tritium per Atom Protium
Surface sea water	$0.54 \pm 0.02 \times 10^{-18}$
Maritime rain (estimated)	$0.54 \pm 0.07 \times 10^{-18}$
Mean precipitation at Chicago	5.5×10^{-18}
Mississippi River	$5.2 \pm 0.2 \times 10^{-18}$
Lake Michigan	$1.7 \pm 0.06 \times 10^{-18}$

evidently contains less tritium than does precipitation falling on the continents. Such continental rain is condensed from water that has had appreciable time to equilibrate with tritium in the atmosphere. There is presumably a latitudinal effect in tritium production, dependent on the variation of cosmic-ray flux with geomagnetic latitude. The tritium content of a lake may therefore be expected to depend on the history of the rain falling on its drainage basin, on the rate of replacement of the water of the lake, and on the geomagnetic latitude.

Distribution of stable isotopes of oxygen and hydrogen. Oxygen obtained by the total electrolysis of water has been studied by various investigators. Murphey (1941) found that commercial tank oxygen contained 0.20 mole per cent O^{18} and 0.041 mole per cent O^{17}. More recent studies by Voskuyl, Ingraham, and Rustad (Kirshenbaum 1951) give, when averaged with the earlier determinations, 0.198 ± 0.003 mole per cent O^{18} and 0.042 ± 0.003 mole per cent O^{17}. Craig and Boato (1955) are apparently somewhat skeptical of such absolute determinations. They are useful, however, in permitting one to gain a rough idea of the relative abundance of the stable molecular species in pure water, as indicated in Table 13.

TABLE 13. *Approximate distribution of various kinds of molecules in natural water*

Molecule	Mole Per Cent
H_2O^{16}	99.745
H_2O^{18}	0.198
H_2O^{17}	0.042
HDO^{16}	0.015
HDO^{18}	0.000,029
HDO^{17}	0.000,006
D_2O^{16}	0.000,002
D_2O^{18}	0.000,000,004
D_2O^{17}	0.000,000,000,8

It will be observed that the content of HDO^{16}, which differs from H_2O in its chemical properties sufficiently to be of biochemical interest, is less than the contents of the two protium compounds of the heavier oxygen isotopes. The amount of HDO^{16} is, however, quite considerable relative to the concentrations of some of the constituents of greatest biochemical importance in inland waters. Slight variations observed when various natural waters are examined are discussed in the succeeding paragraphs.

After the discovery of deuterium (Urey, Brickwedde, and Murphy 1932) it was natural to attribute to variations in the deuterium content the slight variations in density observed when specimens of water purified by total distillation were compared. More recent work has shown that variations in O^{18} are in most cases more significant in determining such density differences, though at the surface of the earth both deuterium and O^{18} tend to vary together.

A great many determinations of density anomaly have been published; the conclusions from such determinations have in general later been confirmed by more accurate studies of the individual isotopes, but there are enough cases in which the newer work is not concordant with the old to lead Craig and Boato (1955) to suggest that practically all the determinations of density anomaly should be treated with suspicion.

It was early discovered that most ordinary fresh waters have about the same density. Goto and Okabe (1940), examining tap waters from Asia, Africa, and America, recorded a range of only 0.7 γ cm.$^{-3}$ difference; in a more limited collection, Greene and Voskuyl (1939) found a range of 0.15 γ cm.$^{-3}$

Water formed by condensation of atmospheric moisture was often found to be lighter than local surface waters, particularly when the precipitation formed as snow (Riesenfeld and Chang 1936, Oana 1948). Ocean water (Dole 1936) shows a systematic greater density, of the order of 1.5 γ cm.$^{-3}$ Many of the earlier investigators (Eméleus, James, King, Pearson, Purcell, and Briscoe 1934; Washburn and Smith 1934) believed that salt lakes formed by evaporation, rather than by the flooding of dry salt deposits, contained water of higher heavy isotope content than normal. This was recorded of Pang-gong Tso, the Dead Sea, and Great Salt Lake. The most modern work on the latter locality (Friedman 1953, Epstein and Mayeda 1953) does not support this belief.

Vereščagin, Gorbov, and Medelejev (1934) believed they had found a great concentration of heavy isotopes in the deep water of Lake Baikal, but the later work of Teis (1939) entirely failed to confirm this.

Some Russian work by Teis and his associates has been based on the possibility of separating the effects of deuterium and O^{18} by simultaneous determinations of density and refractive index. Using this technique, Teis (1946) examined various natural snow crusts and firn samples, and found anomalies due to HDO ranging from $-$ 6.4 to 3.7 γ cm.$^{-3}$, and due to H_2O^{18} from $-$ 8.1 to 6.3 γ cm.$^{-3}$ Teis and Florenskii (1940) found that, in general, water running from melting glaciers is higher in H_2O^{18} and lower in HDO than ordinary river water, but that the normal composition is soon established downstream. The over-all anomaly in such glacier waters was generally positive. Thus, in the upper part of the Ingur River, an over-all positive anomaly of 3.4 γ cm.$^{-3}$ corresponded to 9.6 γ cm.$^{-3}$ due to H_2O^{18} and $-$ 6.2 γ cm.$^{-3}$ due to HDO. Farther down, after tributaries had entered the river, the small negative anomaly of $-$ 1.3 γ cm.$^{-3}$ was due solely to a deficiency of HDO. Any consistent and complete explanation of these observations is difficult to give. It would obviously be interesting to make further studies of the isotopic composition of water in regions of continued melting and regelation.

The more modern mass spectrographic determinations,[5] notably those of Friedman (1953), Epstein and Mayeda (1953), and Craig and Boato (1955; Craig, Boato, and White 1954), indicate very clearly that unless some peculiar processes such as exchange of oxygen between water and silicate have occurred, the D and O^{18} contents vary concomitantly (Fig. 64). Ocean surface water from subtropical latitudes contains about

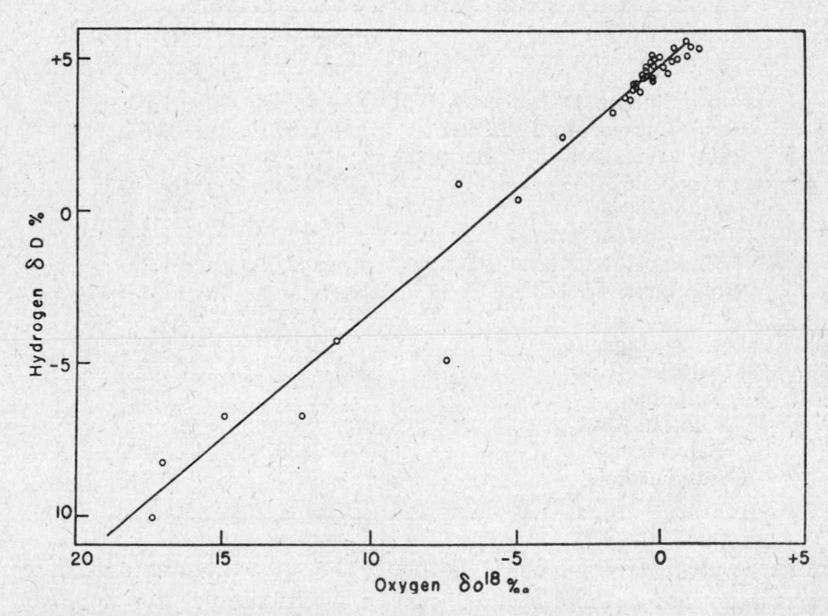

Figure 64. Divergence of O^{18} content plotted against divergence of D content from arbitrary standards for numerous surface waters both fresh and salt (Epstein and Mayeda 1953).

5 per cent more deuterium than does Lake Michigan water, but at high latitudes the excess may be only half this amount. The O^{18} content of subtropical water differs little from that of the arbitrary standard employed by Epstein and Mayeda, but at high latitudes there is a deficiency due to dilution by melting ice, which may be as great as 0.335 per cent.

The divergencies of certain fresh waters for which both deuterium and

[5] The reader must bear in mind that the distribution of the isotopes of hydrogen and oxygen is under intense study. Any account is certain to be out of date before it appears. The recent issues of relevant journals, particularly *Geochimica et Cosmochimica Acta,* must be consulted for the most modern work.

O^{18} analyses exist are given in Table 14, with reference to Lake Michigan as the standard for deuterium and Epstein and Mayeda's standard for O^{18}.

The data, some as yet unpublished, considered by Craig and Boato (1955) indicate a range of variation of about 15 per cent in deuterium content and up to 3 per cent in O^{18} content. Since the quantity of D in

TABLE 14. *Percentage deviation of deuterium and* O^{18} *from arbitrary standards in waters of various origins*

	Deuterium, per cent	O^{18}, per cent
Snow, Chicago, December 14, 1951	− 8.2	− 1.70
Rain, Chicago, April 12, 1952	+ 0.89	− 0.70
Rain (thunderstorm), Bermuda, March 26, 1953	+ 0.53	− 0.655
Juneau Glacier		
47 m. below surface	− 6.75	− 1.490
72 m. below surface	− 6.75	− 1.275
Violin Lake, Trail, B. C.	− 6.75 to − 8.83	...
Lake Michigan	0.00	− 0.613
Mississippi River		
St. Louis	...	− 0.889
Clinton, Iowa	− 1.63	...
Baton Rouge	+ 0.39	− 0.490
Great Salt Lake	− 4.92	− 0.743

ordinary water is less than one tenth that of O^{18}, the variations in the latter are as important as those in the former in determining density anomaly. Very few natural waters appear richer in heavier isotopes than sea water, but thermal and volcanic waters may exhibit such enrichment. In one such case from New Zealand, the deuterium content was found by Craig and Boato to be 4 per cent greater than in the ocean. This is apparently the greatest certain enrichment recorded in natural terrestrial water.

The variations in isotopic composition of ordinary surface waters can in general be attributed to the lower vapor pressures of molecular species containing the heavier isotopes, so that fresh waters, derived ultimately from the evaporation of the ocean, are depleted in D and O^{18} relative to the latter. The variation apparently involves a multiple-stage distillation process, and tends to increasing depletion in moving from the tropics to high latitudes, evaporation being excessive in the former and precipitation in the latter (Craig and Boato 1955). The considerable difference between Chicago snow and rain is attributed to the fact that the former has come from the north Pacific; the latter, moving by a shorter route more directly from an Atlantic air mass, has undergone fewer steps in the depletion process.

The data for Great Salt Lake depart considerably from the linear plot relating D to O^{18} enrichment or depletion, the deuterium content being considerably depleted, as in Rocky Mountain waters and Chicago snow, while the O^{18} content is comparable to that of Chicago rain. It is evident that nonequilibrium conditions in evaporation must be involved, but even so the low heavy-isotope content of this water is curious.

All the different isotopes present in ordinary water will, of course, be involved in the production of hydroxonium and hydroxyl ions. The dissociation constant for the production of D_3O^+ and OD^- in D_2O is 0.16×10^{-14} at 25° C., a value considerably lower than that given above for H_2O. In view of the greater stability of the deuterium bond, indicated by the lower dissociation constant, there will be present a disproportionate amount of H_2DO^+ relative to the amount to be expected from random recombinations. It is, however, not very likely that this and similar effects have any significance in nature; the curious may consult Gross, Steiner, and Suess (1936) for a discussion of the physical chemistry of the matter.

Supposed biological significance of deuterium. When the temperature coefficient is positive, reactions involving deuterium proceed more slowly than do the equivalent reactions with ordinary hydrogen compounds. Any marked concentration of deuterium in the aqueous medium is therefore likely to reduce enzymatic activity. High concentrations of deuterium in the water supplied to organisms cause death (Lewis 1933; Taylor, Swingle, Eyring, and Frost 1933). It is entirely consistent with the known properties of HDO to find that, under conditions of starvation, the life of such lower organisms as are capable of negative growth may be greatly prolonged by the presence of additional deuterium in the medium. This has been dramatically demonstrated by Barnes and Larson (1934) on a flatworm presumably referrable to *Phagocata woodworthi* Hyman.[6] When these animals were kept without food in ordinary water, they underwent an immense decrescence; but when starved for weeks in water in which the hydrogen:deuterium ratio was 2000:1, and which therefore contained about 3.5 times the normal content of the isotope, little reduction in size took place (Fig. 65).[7]

---

[6] The species was given as *P. gracilis* by Barnes and Larson, but it has been shown by Hyman (1937) that the true *P. gracilis* does not occur in New England, while *P. woodworthi* is common around New Haven, where Barnes and Larson's specimens presumably were collected.

[7] Since these experiments have apparently been received with a certain amount of skepticism, the writer would like to place on record the fact that he followed the course of the work with interest, and in so far as mere visual observation permitted a judgment, was satisfied as to its correctness.

Certain other experiments, particularly a series on *Spirogyra* maintained under unfavorable conditions, are supposed to have given comparable results, but this work would seem to have been less critical than that on *Phagocata*. A number of claims have also been made of a stimulating effect of very dilute HDO on lower organisms; good reviews are given by Barnes and Jahn (1934) and by Barnes (1937). The best case is probably

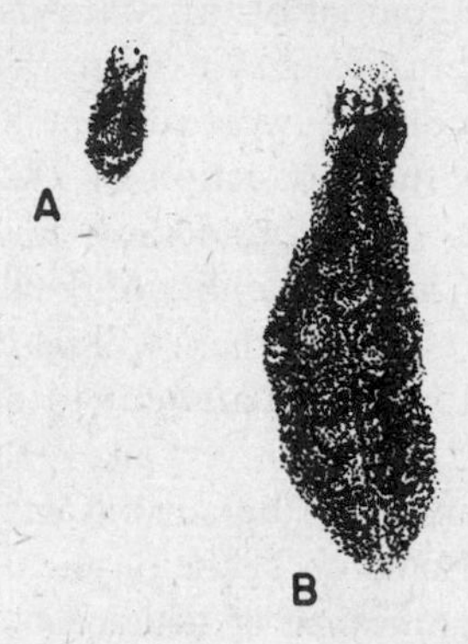

FIGURE 65. *Phagocata woodworthi*: *A*, starved in ordinary water; *B*, in water enriched in deuterium by a factor of about 3.5 (after Barnes and Larsen).

an experiment by Richards (1934), who found a slight increase in the yield of yeast in a medium containing a quantity of deuterium identical to that used by Barnes and Larson in the experiments on *Phagocata*. It is not unlikely, however, that all the work which was done when only dilute deuterium oxide preparations were available is vitiated by the fact that the water of the experimental and control series inevitably came from different sources and may well have contained different impurities. None of this work, even if it should be confirmed, is likely to have much significance in explaining events in nature. But since the quantities of deuterium supposed to be effective are only a few times greater than those found in the hydrosphere, the categorical denial of some slight ecological significance on rare occasions would be inappropriate.

SUMMARY

The properties of water are remarkable, and relative to the substances with which chemical comparison is justifiable, rather unexpected. They can be explained by the tendency of H_2O molecules to associate by means of hydrogen bridges, so that liquid water consists of continually changing branched chains of imperfect oxygen tetrahedra, linked by hydrogen atoms. Most oxygen atoms have two hydrogen atoms 0.99 A. from their centers, as in gaseous H_2O, and two 1.77 A. from their centers. In ice an almost perfect lattice is built up in this way. At low temperatures, near freezing, the tetrahedral arrangement in the liquid is marked; as the temperature is increased, there is less structural association, so that the possibility of close random packing increases at the same time that increasing thermal agitation makes for a looser arrangement. The interaction of the two processes results in contraction on melting, so that ice is less dense than water, and a further slight contraction up to 3.94° C., the temperature of maximum density at 1 atm. pressure. The existence of this maximum is of fundamental importance in limnology. With increasing pressure the temperature of maximum density is depressed, an effect very clear in the temperature relations of deep lakes. The rate of increase in density with increasing temperature increases with the latter, being very small near the temperature of maximum density; this is also a matter of great limnological importance. The compressibility of water is small but not negligible; adiabatic heating on compression is just significant in the thermal history of the deepest lakes.

The association of water by means of hydrogen bridges is responsible for its high dielectric constant; on this depends the possibility of ionic solution and of formation of charged colloidal sols. The solution of nonelectrolytes depends primarily on their capacity to form hydrogen bridges with water molecules. The structure of water therefore underlies the great variety of aqueous solutions that can exist.

Water is one of the very few natural substances that can exist as a liquid at the surface of the earth. Under ordinary pressures the only other liquid not of biological origin is mercury. Much water is of course present as ice, and although several varieties of solid water are known in the laboratory, ordinary ice I is the only kind known to exist in nature on earth.

Although most of the oxygen atoms in water are associated with two close and two remote hydrogen atoms, formally denoted by H_2O, a few are associated with three (H_3O^+) or one (HO^-) close hydrogen atoms, producing the hydroxonium and hydroxyl ions. The hydroxonium ion is often thought of as $H^+ \cdot H_2O$, or a hydrated hydrogen ion. In pure water the concentration of H_3O^+ and HO^- will be equal, increasing with temperature. At 25° C. the concentration is almost exactly 10^{-7} moles per liter, corresponding to *p*H 7. Very minute amounts of oxygen associated with no close hydrogens ($O^=$) or with four such hydrogens (H_4O^{++}) exist.

In addition to ordinary H^1 and O^{16} there are present in water small quantities of deuterium H^2 or D, minute quantities of tritium H^3 or T, appreciable amounts of O^{18}, and a small amount of O^{17}. Tritium is formed by nuclear reactions in the atmosphere generated by cosmic rays; it is β-radioactive with a half life of 12.5 years. The concentration is greatest in continental rain (5.5×10^{-18} atom per atom H) and least in the ocean (0.54×10^{-18} atom per atom H). Lake Michigan water contains $1.7 \pm 0.06 \times 10^{-18}$ atom per atom H. The other isotopes are all much more abundant and less variable. Most fresh waters contain about 0.0148 atoms D per 100 atoms H, but there is a slight variation. The quantities of the heavier oxygen atoms are greater, O^{18} being about 0.20 per cent of the oxygen atoms present. Ordinary water thus contains about 0.20 per cent H_2O^{18} molecules, 0.04 per cent H_2O^{17}, and 0.015 per cent HDO^{16}. Atmospheric precipitation, particularly snow, is usually somewhat impoverished in the heavier isotopes; sea water is definitely enriched relative to fresh water, the over-all range of variation in all surface waters studied being of the order of 15 per cent for deuterium and 3 per cent for O^{18}.

In the case of deuterium, water from Violin Lake, British Columbia, contained up to 8.83 per cent less than Lake Michigan waters, and sea water 5 per cent more. In general the O^{18} content varies with the deuterium. Saline lakes, contrary to earlier results, seem not to show marked enrichment. Great Salt Lake is much poorer in D and slightly poorer in O^{18} than Lake Michigan. Although there is a little evidence that moderate enrichment in deuterium in the medium of fresh-water organisms may have striking physiological effects, the variations in nature do not seem great enough for the differences in deuterium content to be ecologically significant.

CHAPTER 4

The Hydrological Cycle and the Water Balance of Lakes

In the present chapter the distribution of water at the surface of the earth is considered in so far as it affects the distribution of lakes and their variation in volume.

THE HYDROLOGICAL CYCLE

The water content of the earth. Of the four inner terrestrial planets, Mercury has no known atmosphere, Venus has a dry and dusty atmosphere in which CO_2 is the only identifiable constituent, while Mars has a little water vapor in equilibrium with ice rather than the liquid, and so cannot be said to have a hydrosphere. Only the earth is favored with an abundance of liquid water. The major planets undoubtedly contain large quantities of ice, but their whole chemistry is so different from that of the terrestrial planets that they need not be considered in the present work.

That abundant water should exist at the surface of the earth is cosmochemically anomalous. Neon, of atomic and molecular mass 20, has practically disappeared from our planet, which at some stage in its history must have consisted of material either too hot or fragmented into too small pieces to hold gravitationally this cosmically abundant gas (Brown 1949b). The most reasonable explanation of the presence of water in the earth is that the planets were put together from relatively cold pieces of matter, the water initially being incorporated as ice. Urey (1952), whose recent book provides by far the richest and most convincing account of the chemistry of the formation of the planets, believes that they have formed from the aggregation of small particles. Mercury was probably too close to the sun to receive any water; Venus and the earth perhaps received water in the form of ice particles which tended to melt and so become sticky. This would provide rather favorable conditions for aggregation.

In the case of Mars and the asteroids, still further from the sun, the ice may well have been too cold for such stickiness to develop, so that aggregation was less easy. The major planets lie far enough from the sun for ammonia to have taken the place of water as a binding agent. As the protoplanets grew in mass and shrank in diameter, the release of gravitational energy must have raised their temperatures. The surface of the earth probably reached a temperature of about 1200° C., at which time most of the water would be present in the interior as water of hydration in equilibrium with vapor in the atmosphere. Mars may have lost more water than the earth by photochemical decomposition and by the subsequent escape of hydrogen, and of oxygen atoms which probably can leave the planet under some conditions. The disappearance of water on Venus must have been due to chemical events, the nature of which is not entirely clear, but which are presumably to be associated with the higher surface temperature of the planet relative to the earth. The complete absence of water in the present atmosphere of Venus, in contrast with its abundance on the surface of the earth, serves to show how critical were the conditions permitting the evolution of a planet capable of supporting highly developed organisms.[1]

The water content of the major part of the lithosphere, the great mantle of ultrabasic rock which composes most of the earth, is unknown. Kulp (1951) supposes that a value can be derived from the hydrogen content of silicate meteorites, estimated at about 0.06 per cent by Brown (1949a). This amount of hydrogen, if entirely oxidized, corresponds to about 0.54 per cent H_2O. Such an estimate is quite probably too low for the earth's mantle. Assuming its correctness, and taking the mass of the earth without the metallic core as 4.0×10^7 Gg., Kulp concludes that the water content of the earth as a whole is about 2.5×10^5 Gg. For sedimentary rocks, various estimates have been given; Rubey's (1951) figure of 2100 Gg. is doubtless the most satisfactory. The water contents of the ocean and of the atmosphere are well known, and plausible figures can be derived for the icecaps and other glaciated regions (Hess 1933).

The total quantity of inland waters, with which this book primarily

[1] Impressive evidence in favor of a flora of simple type on Mars has been recently discussed by Kuiper (1952).

deals, is relatively small. According to Penck (1894), whose estimate is accepted by Halbfass (1933, 1934), the total area of the lakes of the world is 2,500,000 $km.^2$ Halbfass (1933) also quotes another estimate, from H. Wagner, of 1,700,000 $km.^2$ Of the total area, 438,000 $km.^2$ are contributed by the Caspian and 62,000 $km.^2$ by the Sea of Aral. The endorheic (page 226) basins thus must contribute about 600,000 $km.^2$, as Halbfass points out. A like area of about 590,000 $km.^2$ is contributed by the lakes of the glaciated regions of Europe (160,000 $km.^2$) and North America (430,000 $km.^2$), according to G. Wagner (1922). It seems probable that Penck's total estimate is a little large. Halbfass regards 10 m. as a reasonable mean depth, giving the value entered in Table 15 for inland waters. In obtaining this value no allowance was made for the volume of rivers, but in view of the suspicion that the value for the area of the lakes is too great, the estimate is not unreasonable. The value for circulating ground water given by Kalle (1945) and used in Table 15 is a plausible estimate due to Meinhardus (1928). Kalle appears to assume 100 m. as the mean depth of the lakes and rivers, a value that is obviously much too great.

TABLE 15. *Water content of the various parts of the earth*

	Water*	
	kg. $cm.^{-2}$ e.	Gg.
Primary lithosphere	4900	250,000
Sedimentary rocks	41	2,100
Ocean	269	13,800
Polar caps and other ice	3.24	167
Inland waters	0.0049	0.25
Circulating ground water	0.049	2.5
Atmospheric water vapor	0.0026	0.13

* In this and all subsequent presentations, the symbol $cm.^{-2}$ e. means per square centimeter of earth's surface; Gg. means geogram or 10^{20} g.

No agreement has been reached as to the constancy of the quantity of water, of the order of 16,000 Gg., at the surface of the earth during geological time. A small loss must occur owing to photochemical decomposition and subsequent loss of hydrogen from the upper atmosphere; Kuiper (1952) gives a full discussion of this phenomenon. Volcano gases consist largely of water, but it is uncertain how much is truly juvenile and how much is derived from the hydrosphere. Rubey (1951), who gives the most complete recent discussion, estimates the total output of hot springs as 66×10^{15} g. per year, or, assuming constant delivery during 3000 million years, 2×10^{6} Gg. during the history of the earth. If

0.8 per cent of this water was juvenile, the whole of the ocean and other waters at the earth's surface could have been produced from thermal springs alone.

The smallest fraction in Table 15, namely the water vapor of the atmosphere, is, owing to its great mobility, in many ways the most important, since from it is derived the whole of the atmospheric precipitation. The distribution of lakes depends no less on the distribution of this precipitation than on the presence of lake basins. The geography of rain must therefore be briefly considered.

Atmospheric precipitation and its distribution. Rain and snow require for their production a supersaturation with water vapor in the air from which they originate. This is not the only condition which must be satisfied, but it is a necessary one and also one which determines the distribution of rainfall and so of rivers and lakes on the surface of the earth. In general, supersaturation can occur whenever a mass of air containing water vapor is uplifted, undergoing adiabatic expansion and cooling sufficiently to reduce the temperature below that corresponding to saturation with the concentration of vapor present in the air. If the air is practically saturated at ground level, very little elevation will be needed to produce the required cooling; if the air, say, were half saturated at 20° C., it would have to be cooled below 9.2° C. to produce supersaturation, which usually would correspond to an uplift of the order of 2000 m.

Ordinary uplift and cooling may occur in three major ways. If the ground below calm air is heated by solar radiation, convection will be started as the air in contact with the hot ground becomes unstable. A series of cells in which updraughts occur will be produced, and the tops of these convection cells may be high enough to permit condensation. This is easily observed in the great cumulonimbus cloud systems of summer thunderstorms, each of which cloud systems constitutes such an updraught cell. Many of the warmer steppe regions of the world receive their entire rainfall in this way. Such rain is often called *convective.*

When high mountains stand in the path of a prevalent wet wind, the upturning of the wind by the mountains produces supersaturation, resulting in *orographic rain.*

In the large circulation systems, the formation of cyclonic eddies and the production of discontinuities where major elements of the systems or large eddies meet, may give rise to complex patterns of uplift. In both the equatorial rain belt and in the humid maritime parts of the temperate regions, *frontal-cyclonic rain* of this sort is of paramount importance.

The distribution of rainfall is thus determined primarily by the circulation pattern of the atmosphere. Ideally this pattern consists of certain well-defined major elements which might be expected to develop in the atmosphere of any planet revolving round a source of radiation and rotating about an axis set at a large angle to the plane of its orbit. The most constant elements are the trade-wind systems, which are air currents moving toward the equatorial belt of high temperatures and, since the earth is rotating, also moving toward the west. Thus, at ground level, a northeast wind is constantly blowing in subtropical and northern tropical latitudes of the Northern Hemisphere, and a southeast wind in the equivalent latitudes of the Southern Hemisphere. Between the trade winds there is an equatorial region of calm, known to mariners as the doldrums.

North and south of the trade winds lie the regions of the westerlies. Here the circulation pattern is the reverse of that in the trade-wind belts, southwestern winds blowing in the Northern Hemisphere, northwestern in the Southern Hemisphere. In the polar regions, a third system of easterlies is developed.

Though much of the conventional theory which has been developed to explain the observed pattern has been found deficient during the past decade, the general scheme just outlined is sufficient for the purposes of the present work. The reader desiring a fully analytical summary, treating particularly the jet-stream mechanisms believed to maintain circulation in the region of the westerlies, should consult Rossby (1952).

The trade winds, moving from cooler to warmer latitudes, tend to take up moisture whenever they move over water or humid land, but rarely deliver this moisture as rain until they reach the equatorial region, where frontal convergences and a general upward movement produce adiabatic cooling and cloud formation.

The westerlies, blowing from warm to cool latitudes, may develop supersaturation very easily. The great contrast between the rainfall regimes in the trade-wind and westerly belts can be recognized throughout large parts of the continents, but is nowhere so apparent as on their western margins, which are very dry in the trade-wind belts and very wet at higher latitudes.

Superimposed on the simple planetary system are a number of features due to the distribution of land and water as it happens to be developed at the surface of the earth. The most striking of these secondary patterns are the monsoon systems, which are due to the development of high pressures in winter and low pressures in summer in the centers of continents where the cooling and heating of the air masses is very great. Though recognizable elsewhere, the most striking monsoons are those controlled by the thermal regime of the interior of the Eurasiatic continent, which determine the climate of most of southern and southeastern Asia. The development of a wet monsoon blowing across the Indian Ocean during the summer, in what would otherwise be a part of the trade-wind belt, is of the greatest importance in producing summer rain in areas which would be desert if the monsoon did not blow.

Examining the land masses of the earth, De Martonne and Aufrère (1928) have recognized three types of hydrological region; these types may be termed, by a modification[2] of their usage: *exorheic regions*, from which rivers reach the sea; *endorheic regions*, within which rivers arise but from which these rivers never reach the sea, losing themselves in dry water courses or entering closed lake basins; *arheic regions*, within which no rivers arise.

It is important to note that a large river may flow through an arheic region; the lower part of the Nile provides an excellent example of this.

The distribution of these kinds of area shows a characteristic pattern, dependent on the distribution of rainfall just outlined. Two large desert zones tend to develop in the latitudes of the trade winds. These constitute the main arheic regions of the world. Between them lies the zone of equatorial rains; north and south, the zones of temperate humid climates, which pass into the arctic without interruption. The endorheic regions tend to lie between the two arheic and the three exorheic regions. In the Northern Hemisphere there is a clearly marked zonal distribution, endorheic belts fringing both northern and southern margins of the desert zones. In the Southern Hemisphere the attenuation of the land masses southward makes the pattern less clear, but if the area of the endorheic regions as a percentage of total land surface is considered, there is an indication of a bimodal distribution of the same kind as is found in the Northern Hemisphere.

The Caspian Basin, into which the Volga drains, adds greatly to the area of the endorheic regions of the north temperate zone. Actually, most of the Volga Basin is, in its hydrological relations, typical of the adjacent exorheic areas, and from a climatic point of view it may be regarded as a geographical accident that the river runs south into a closed basin. The dotted line of Fig. 66 has therefore been drawn to indicate the relative distribution of endorheic regions, omitting the Volga Basin.

The exorheic regions contain the main lake districts of the world. Most of the lakes with which this work is concerned lie in the exorheic part of the north temperate zone, and a good minority in the equatorial and monsoon belts of the tropics.

---

[2] This arrangement is simplified terminologically when compared with previous expositions. De Martonne and Aufrère themselves first divided the earth's surface into exorheic and endorheic regions, and then the latter into the endorheic regions *proprement dites* and the arheic. Hutchinson (1937c) used euendorheic for the endorheic *proprement dites* of the French geographers. Subsequent consideration indicates no reason for performing the classification in two stages.

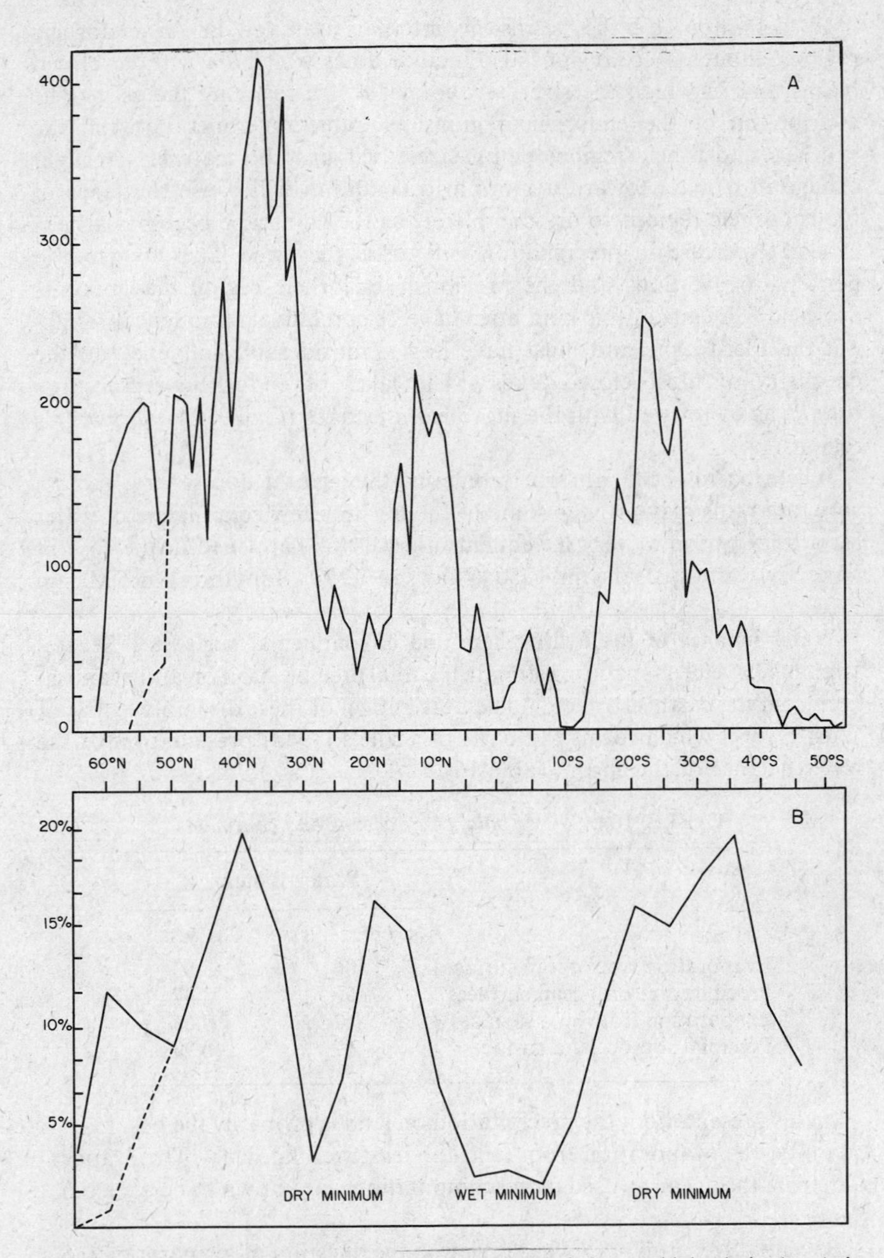

FIGURE 66. Distribution of the endorheic regions of the world by latitude, *A*, in absolute area; *B*, as smoothed percentages of land surface. The dotted line indicates areas excluding the Volga Basin. (Hutchinson 1937c, from data of De Martonne and Aufrère.)

By definition, all the basins of internal drainage lie in endorheic regions, though these regions also include areas which are without closed basins and in which the rivers evaporate before reaching the sea. The distribution of the endorheic regions as transition zones between the exorheic and arheic regions emphasizes their unstable nature. A slight change in climate toward a more arid condition will cause the lakes of the endorheic regions to dry completely as the landscape becomes arheic. A slight increase in precipitation will cause the same lakes to rise and perhaps to overflow, and the previously endorheic region may become exorheic. Events of this kind must have been relatively frequent throughout the Pleistocene and must have had a considerable influence on the chemistry of most closed lakes. The lakes of endorheic regions are *astatic*, as contrasted with the more *eustatic* lakes (Gajl 1924) of exorheic regions.

Where basins occur in arheic regions, they are of course dry, though very rare rains may provide some of them with a few centimeters of water for a brief period at very infrequent intervals, as happened in the case of Lake Eyre in Australia in 1950 (Mawson 1950, Bonython and Mason 1953).

Water balance of the hydrosphere and of continental surfaces. Several attempts have been made to estimate the total precipitation on and evaporation from the earth's surface. The best known of these attempts is that of Wüst (1936), which forms the basis of Kalle's (1945) presentation of the water balance of the earth (Table 16).

TABLE 16. *Water balance of oceans and continents*

	Water Balance	
	g. cm.$^{-2}$ yr.$^{-1}$	Gg. yr.$^{-1}$
Evaporation from ocean surfaces	106	3.83
Precipitation on ocean surfaces	96	3.47
Evaporation from land surfaces	42	0.63
Precipitation on land surfaces	67	0.99

In this presentation, the precipitation on land is probably the best known quantity, the evaporation from land the least well known. The evaporation from the ocean as used in the computation was known to be a maximal figure.

Recently Sverdrup (1952) has reviewed the question of evaporation from the ocean, and indicates that by independent methods values of 96 ± 12 and 99 ± 10 g. cm.$^{-2}$ yr.$^{-1}$ can be obtained. The best estimate would therefore appear to be 97.5 g. cm.$^{-2}$ yr.$^{-1}$, though for the purposes of the present computation the error in such an estimate is evidently inconveniently large. Jacobs (1951) has also reconsidered both evaporation and precipitation, but his treatment of evaporation is limited to the Northern Hemisphere and cannot be used in the present discussion. His estimate of precipitation on the ocean, which is doubtless the best so far available, is 81.2 g. cm.$^{-2}$ yr.$^{-1}$ Hence, the most recent data suggest a runoff from land to sea of 0.59 Gg. yr.$^{-1}$, in contrast to the 0.36 Gg. yr.$^{-7}$ implied by Wüst and Kalle. Since the estimates of evaporation might be 10 per cent too high, though this is unlikely on account of their concordance, the runoff on the basis of the present data might conceivably be as low as 0.22 Gg. yr.$^{-1}$, a figure referred to herein as the minimum computed value.

A number of independent estimates of total runoff have been made by attempting to sum the rates of discharge of the rivers of the world. The best of these attempts is that of Henkel (1912), brought up to date by Kalle (1945), which gives a total of 5.1×10^6 m.^{3} sec.$^{-1}$ or 0.16 Gg. yr.$^{-1}$. This value must be too small, as it neglects innumerable small streams and also the direct movement of ground water into the ocean. Neither of these errors is likely to be very serious. The catchment area of the small streams directly discharging into the ocean is obviously very small compared to that of the great rivers. Most of the ground water entering the ocean must be moving as hidden rivers in the alluvial filling of river valleys. The discussion of such movement given by Halbfass (1934) indicates that it is not likely to be in excess of 10 per cent of the visible flow. It is probably valid to raise the observed runoff to 0.18 Gg. to correct for these omissions. It is difficult to see how the runoff can be greater than this figure, but also difficult to see how such a figure can be reconciled with present knowledge of evaporation and precipitation on the ocean.

As a provisional value, it has seemed best to take the mean of the corrected observed figure of 0.18 Gg. yr.$^{-1}$ and the minimum computed value of 0.22 Gg. yr.$^{-1}$, giving 0.20 Gg. yr.$^{-1}$ This is somewhat lower than the value of 0.28 Gg. yr.$^{-1}$ given by Murray (1888) and used extensively in geochemical computations by Clarke (1924), Conway (1942, 1943), and others. The provisional value suggests that the inland waters of the exorheic regions are replaced rather over once in the course of a year, which is not unreasonable.

The details of the movement of water from the ocean to the land surfaces are not well known. It has sometimes been supposed that much of the rain falling at any given locality represents previous evaporation at or near that locality. An investigation of this hypothesis underlies the most intensive study hitherto made of the water balance of any particular region, that of Benton, Blackburn, and Snead (1950) on the Mississippi watershed.

This area, though it cannot be regarded as a perfect sample of the exorheic parts of the continents, certainly is reasonably representative, so that the results of its study have far more than local interest. The whole balance sheet, converted into metric terms, is given in Table 17.

TABLE 17. *Water balance of the Mississippi catchment area*

	Values, cm. yr.$^{-1}$	Mean Total Precipitation, %
Vapor entering in maritime air	254.0	335
Vapor entering in continental air	116.9	153
Total vapor entering air above basin	370.9	488
Evaporation into maritime air	26.4	35
Evaporation into continental air	32.2	43
Total evaporation	58.6	78
Precipitation from maritime air	67.8	90
Precipitation from continental air	7.6	10
Total precipitation	75.4	100
Precipitation from maritime air evaporating in continental air	7.6	10
Runoff	16.8	22

It will be observed that in this case the estimate of the proportion of the rainfall running into the sea, 22 per cent, is close to that of the mean estimate for the earth, 0.20 Gg. from 0.99 Gg. of precipitation. Very roughly, it would seem that about one fifth of the rain which falls on land surfaces may be expected to appear in lakes and rivers. It will also be observed that about 90 per cent of the rainfall in the Mississippi comes from maritime air; Benton, Blackburn, and Snead conclude that very little of this rain can represent water which has evaporated from the surface of the basin and has then been reprecipitated therein. The fraction which undergoes such a secondary cycle might be a little greater in the case of continental air, but the whole of the precipitation from such air is so small that this slightly greater proportion is not very significant in the whole precipitation. Their final estimate is that 90 per cent of the rain falling within the basin is of external origin, and 86 to 88 per cent is directly derived from the ocean. Most of the lake water which will be studied here thus seems to have come directly from the sea by a single process of distillation. It is possible that further studies of isotopic fractionation along the lines already discussed on p. 216 may modify this conclusion somewhat.

THE WATER BALANCE OF LAKES AND VARIATIONS IN LAKE LEVEL

Introductory qualitative considerations. The water balance of a lake is obviously expressed by an equation indicating that the rate of change of volume of the lake is equal to the rate of inflow from all sources, less the rate of water loss.

Before proceeding to a more formal statement of the details of such a relationship, certain preliminary aspects of the hydrology of lakes must be made clear.

The sources of income are (1) precipitation falling on the lake surface, (2) water in surface influents, (3) ground water seeping through the floor of the lake, and (4) ground water entering by discrete springs.

It is probable that lakes exist in which nearly all the water enters in one of these ways. Halbfass, using rather old data, concluded that 76 per cent of the water entering Lake Victoria is precipitation on the lake surface; in the case of the Dead Sea the proportion would be practically zero, and in most of the large lakes of central Europe only a few per cent.

Birge and Juday (1934) emphasized the distinction between *drainage* lakes with an outlet, and *seepage* lakes into which ground water enters and from which water leaves by seeping through the wall of the lake basin. The terms are convenient but must be used with discretion. Broughton thinks some of the bog lakes of Wisconsin receive almost no ground water, and that in all cases the seepage is probably of very superficial ground water. The main significance of the distinction relates to the influent, though the original definition refers to the outlet only. Many lakes in semi-arid regions lie in basins without any kind of effluent, losing water only by evaporation. Such lakes may be termed *closed*, in contradistinction to *open* lakes having an effluent. All seepage lakes are in this sense almost certainly open.

Many lakes in karstic landscapes fill and empty mainly by sublacustrine channels, and at time of high water have received nearly all their contents from sublacustrine springs.

It is commonly held that many other lakes are spring fed, but most of the evidence is purely suppositious. Most of these lakes are probably seepage lakes. A few crenogenic meromictic lakes (page 486) show adequate chemical evidence of sublacustrine springs of sufficient magnitude to be important water sources. In a few cases the presence of marginal springs or less definite but localized areas of influent ground water may be made known by the slightly higher temperature of their water preventing the formation of ice. Halbfass (1923, Fig. 89) figures such weak ice and open water near the margin of the Lunzer Mittersee, but Forel (1898b) concluded that most supposed cases in Switzerland were due to disturbance by water fowl delaying freezing.

In most small lakes not in rock basins, the water is seperated from the ground water by a seal which represents the initial deposition of clay and very fine silt that has settled out of the lake during its early stages of development (Broughton 1941). This seal not only permits the lake to retain water regardless of seasonal variations in ground water level, but also allows considerable chemical differences between the lake water and the ground water. Broughton found in northeastern Wisconsin marked differences in the water in test pits and wells drilled around some of the lakes, both among themselves and when compared with the lakes as indicated in Table 18.

TABLE 18. *Calcium and magnesium contents of lake water and of adjacent ground water*

	Ca, mg. per liter	Mg, mg. per liter
Bog No. 1	0.87	0.2
Pits near Bog No. 1	1.0–1.8	0.3–0.9
Witcher Lake (seepage)	1.35	0.2
Pits near Witcher Lake	1.1–1.9	trace–0.1
Scaffold Lake (seepage)	3.29	0.5
Pits near Scaffold Lake	5.1–13.4	0.2–4.5
Erickson Lake (drainage)	6.02	2.5
Wells and pits near Erickson Lake	3.5–6.4	1.0–2.7
Arbor Vitae Lake (drainage)	14.27	5.0
Wells near Arbor Vitae Lake	10.2–14.8	3.0–6.2

While it is evident that some correlation between the ground water and lake water exists, determined no doubt by the irregular distribution of calcareous drift, the water of Scaffold Lake for instance is obviously deficient in calcium and must be effectively separated from the adjacent ground water. Broughton thinks that in typical bogs the water is all superficial runoff and local rain, while true seepage lakes receive very superficial ground water from above the seal.

The modes of loss of water are discharge at the effluent and evaporation. Discharge is normally from a single effluent. If a lake initially were to have two effluents, it is most unlikely that cutting at both lips would proceed at the same rate; one effluent would therefore finally collect all the drainage. The process can be documented from geological data on the Laurentian Great Lakes. A few very young lakes in the Canadian Arctic and Labrador are known, from which discharge occurs by two (Cabot 1946) or in one case (Watson 1897) even five channels. In some cases the drainage is so uncertain that when the lakes fill up with melt water they may form a ring rather than a linear drainage system. Very occasionally the same superficial channel may act as an influent in some seasons and an effluent in others. This is likely to occur only in regions of very irregular rainfall, such as the Lake Ngami Basin of Bechuanaland. Sublacustrine effluents or katavothrai (page 107) which run down directly to the ground-water level can of course easily behave in this way. The normal effluents of some lateral lakes can serve temporarily as influents when the main stream is high.

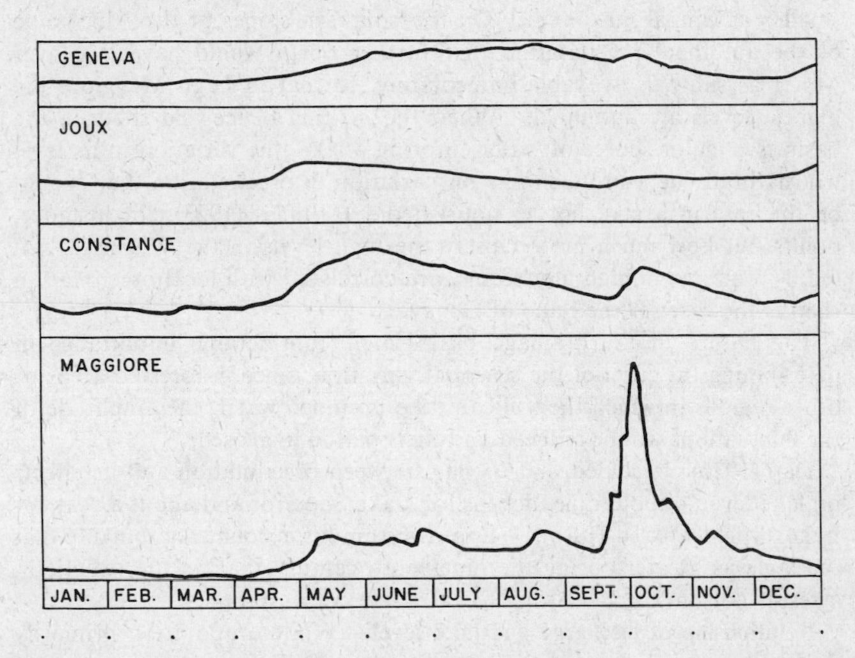

FIGURE 67. Examples of annual variation in level of certain large subalpine lakes (after Collet).

Seasonal variation. Both the rate of entry of water into a lake and the rate of loss vary seasonally in practically all lake basins, producing an annual variation in level. It will be apparent later than in temperate regions evaporation is generally greatest in the summer. On this seasonal evaporative cycle, a seasonally determined period of melting of ice and snow in the spring is often superimposed, and there is almost always a seasonal variation in rainfall. In Switzerland, Collet distinguishes several types of annual variation (Fig. 67).

When the main influents are fed primarily from melting ice and snow, the minimum level is in February or March and the maximum in the late summer. The natural level of Lake Geneva is determined in this way.

In the lakes of the Jura, such as the Lake of Joux, there is a vernal high level determined by the melting of snow in the vicinity of the lake, and an autumnal high level due to rains during that part of the year.

In a number of lakes both the presence of ice and snow fields and the seasonal incidence of rain determine the variation, which may vary from year to year. In Lake Constance, rain falling in the summer may greatly augment the melt water and produce floods in June or July, followed by a smaller autumnal maximum. On the southern slopes of the Alps some of the autumnal precipitation that farther north would have lain over winter as snow, is available immediately, so that in Lago Maggiore the highest levels are autumnal. When the melting of ice and snow in the basin is a major source of water entering a lake, the variation in its level will naturally depend much less on variation in precipitation than would be the case in a lake not so nourished. Halbfass (1923), for instance, points out how much more closely the high levels in the Würmsee, not fed by water from high mountains, are correlated with local precipitation than is the case for the Lake of Geneva.

The nature of the drainage basin is of considerable importance in determining the form of the seasonal variation, since in forested areas or other regions in which the soil can take up much water, the amplitude of the fluctuations will be reduced and their period increased.

Day (1926) concluded that the lag between precipitation and its effects on lake level is about nine months for Lake Superior and about a year for Lakes Michigan and Huron; it appears from his account not unlikely that the lag was greater during the nineteenth century than in the twentieth, a result not unexpected owing to extensive deforestation.

Relationship of discharge and lake level. While evaporation primarily depends on events outside the lake, the rate of discharge is a function of the lake level.[3] We may write the fundamental equation for the water balance as

$$\mathrm{A}I_n = \mathrm{A}E_f + \mathrm{A}\frac{\mathrm{d}z}{\mathrm{d}t} \tag{1}$$

where I_n = rate of entry of water from all sources per unit area of lake surface.

E_f = rate of discharge per unit area of lake surface.

$\mathrm{A}E_f$ is a function of z. $\mathrm{dA}E_f/\mathrm{d}z$ can be determined empirically, and depends on the dimensions of the effluent as well as on the area. For

[3] The present mode of approach is ultimately derived from that of Lombardini (1845), who did not proceed, however, in a formal mathematical manner.

several Italian lakes the values of this derivative are known to be fairly constant, as shown in Table 19.

TABLE 19. dAE_f/dz *for four subalpine Italian lakes* (*De Marchi* 1949)

	Flood Stage, m.^{3} sec.$^{-1}$ m.$^{-1}$	Normal Level, m.^{3} sec.$^{-1}$ m.$^{-1}$
Maggiore	600 (2 m. above mean level)	330
Como	260 (1 m. above mean level)	220
Iseo	90 (1 m. above mean level)	90
Garda	58 (0.6 m. above mean level)	50

If the lake were to undergo a sudden rise, the rate of fall would be given by

$$e^{-t\,dE_f/dz}$$

assuming A to be essentially constant. For the four lakes in question, after a rise of 1 m., the excess level in centimeters will be given by the figures in Table 20.

TABLE 20. *Excess level at various times after a rapid rise of* 1 *m. in four large Italian lakes*

		Excess Level (cm.) after		
	dE_f/dz	2 Days	5 Days	10 Days
Maggiore	1.5×10^{-6} sec.$^{-1}$	77	52	25
Como	1.59×10^{-6} sec.$^{-1}$	77	52	27
Iseo	1.47×10^{-6} sec.$^{-1}$	78	53	28
Garda	1.84×10^{-7} sec.$^{-1}$	97	92	85

The slowness of discharge from Lago di Garda is made evident by this hypothetical example. De Marchi (1949) has given a somewhat more elaborate mathematical analysis based on an earlier study of Fantoli (1897) that, however, assumes a sinusoidal variation in level over the year, which the data appear not to warrant.

Secular changes in lake level; general and theoretical considerations. The simplest case, and the only one for which a theoretical model can be constructed at all easily, is one in which the water falling as rain in the drainage basin reaches the lake within a time short compared to the time intervals used in the graduation of the data. An ideal case would be provided by the consideration of mean annual water levels in a lake in a region of short, well-defined rainy seasons. It so happens that a number of the most interesting cases approximately satisfy this condition.

Let

P_r = the mean precipitation over the lake surface.

P_r' = the mean precipitation over the rest of the drainage basin.

E_v = the mean evaporation from the lake surface.

E_v' = the mean evaporation from the rest of the drainage basin.

E_f = the mean discharge by any channel per unit area of lake.

$\dot{z}$ = the mean rate of increase of z.

A′ = the area of the drainage basin including the lake.

A = the area of the lake, as before.

Ā = the mean area of the lake over a range of values of the depth z.

We may then write in all ordinary cases in which A is small relative to A′ and the lake is fed by precipitation falling within the drainage basin (cf. Jentzsch 1912)

$$\text{(2)} \qquad \mathrm{A}\dot{z} = \mathrm{A}(P_r - E_v) + (\mathrm{A}' - \mathrm{A})(P_r' - E_v') - \mathrm{A}E_f$$

It is to be noted that A and E_f are monotonic functions of z, both increasing as z increases.

It has often been supposed that any high lake level reflects a period of high precipitation or low evaporation, any low level the reverse. It is desirable, however, to inquire whether two lakes of different form in adjacent drainage basins undergoing identical climatic variation will, if they conform to the theoretical model, necessarily have their maxima and minima at the same time. The condition for this is that in the two lakes $\dot{z}$ is always positive under one set of meteorological conditions, always zero or negative under all other meteorological conditions.

The condition for $\dot{z}$ to be positive is that

$$\text{(3)} \qquad (P_r - E_v) + \frac{\mathrm{A}' - \mathrm{A}}{\mathrm{A}}(P_r' - E_v') - E_f > 0$$

It would appear therefore that in an open lake, since increasing z increases both the second term and the third term, but according to quite unrelated laws, one lake might continue to rise while the other had started to fall, even if A represented the same function of z in both lakes, simply on account of differences in the variation in the discharge with varying hydraulic head at the outlet, along the lines that have already been indicated. The great difference in behavior of Lakes Vener and Storsjön (Fig. 68) recorded by Bergsten (1949) may well provide an example, though Bergsten attributes the difference to the small ration A/A′ in the case of Storsjön. Other factors being equal, however, according to theory this should operate in the opposite way from what he suggests. Perhaps delayed runoff is involved in this case.

In a closed lake, in which $E_f = 0$, the required inequality becomes

$$(P_r - E_v) + \frac{A' - A}{A}(P_r' - E_v') > 0 \tag{4}$$

In such a lake $(P_r - E_v)$ is obviously negative, while if any water reaches the lake $(P_r' - E_v')$ must be positive. It is therefore always possible, by choosing suitable values of $(A' - A)/A$, to obtain a satisfaction of the inequality (4) in one basin, in which the water is rising, and satisfaction of the converse inequality in the other basin, in which the water is falling.

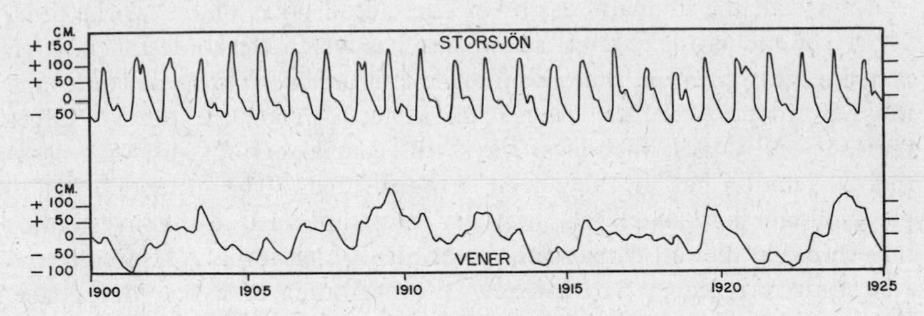

FIGURE 68. Variations in level of Lakes Vener and Storsjön in Sweden over a period of years (after Bergsten).

This can be done by increasing A′, which will increase the second term, or reducing the dependence of A on z by making the lake more steep-sided. Small steep-sided lakes in large drainage basins will thus tend to go on rising after saucerlike lakes with a gentle slope have started to contract. While a more complete theoretical treatment runs into many difficulties, it is most important to realize the diversity of response that is possible. In general, the presence of an effluent, particularly one which is highly sensitive to increases in z, a high ratio of $\bar{A}/A'$, and a gentle slope will make the lake more sensitive to short-period changes in meteorological variables, whereas the smallness or absence of an effluent, a small ratio $\bar{A}/A'$, and a steep-sloping lake basin will make the lake less sensitive to minor variations and perhaps permit its variation in level to pick out selectively long-period variations in climate. Where a large group of lakes in a region behave in the same way, one is probably justified in drawing some climatic conclusions, but single lakes, unless the full water balance can be computed, will give little climatic information of value.

It is very important to note, moreover, that the whole of the above theoretical treatment assumes a rapid discharge of water from the drainage basin into the lake. Delayed drainage due to forest cover, which might vary from basin to basin in a partly cleared area, would tend to invalidate the theory; this may be involved in the case of Lakes Vener and Storsjön. The theory is, in fact, likely to be primarily applicable to basins in semi-arid regions, which are climatologically of greatest interest. A special case of great importance, in which, however, it does not hold, is where the drainage basin contains permanent ice and snow fields. In such a case the type of climatic change, involving a lowered precipitation, higher evaporation, and increased temperature, that may otherwise be deduced from a falling level, may by increased melting of ice and snow actually cause an increase in lake level.

Almost all the drainage basins of the closed lakes of the world bear, above the modern lake level, raised beaches which clearly testify to high lake levels at a previous time; Bonneville and Lahontan are only two of the more dramatic examples. It is natural to correlate the pluvial stages represented by these high-level lakes with glacial periods; in some cases this correlation has, in fact, been demonstrated. The general belief in glaciopluvial high lake levels, however, must not mislead the investigator into thinking that all climatically determined high levels reflect cold and wet climatic phases. At least near the beginning of a hot, dry phase, while much ice still exists in the basin, it is obvious that rising lake levels might be observed and would mean just the opposite of a glaciopluvial climatic oscillation.

The Caspian Basin and central Asia. The variations of the level of the Caspian Sea have been the subject of continuous discussion since the early work of Brückner (1890) and of Huntington (1907). This discussion has largely been influenced by a widespread but probably erroneous belief in a continual process of desiccation in central Asia since protohistoric times.

Huntington believed that geomorphological, historical, and archeological evidence existed for a relatively high level, about 40 m. above that prevalent in the early twentieth century, around 600 B.C. This high level is supposed to have sunk to a minimum, in the sixth century A.D., to somewhat below the modern datum level (25.8 m. below sea level). The later medieval history of the Caspian is complicated by evidence that at times the Oxus has discharged into the Caspian, at other times into the Sea of Aral. Huntington, however, believed that climatically determined high levels about 900 A.D. and 1300 A.D., above any recorded since 1700 A.D., were implied by the data.

Huntington's work has been severely criticized by a number of later investigators.[4] Berg's (1934, [1938] 1950) argument that subfossil shells of *Cardium edule*, which has inhabited the lake since early postglacial times, are never found in undisturbed Caspian sediments more than 5 m. above the twentieth-century level (or at −21 m. relative to sea level) is a particularly impressive one against Huntington's high levels in protohistoric and medieval times. The distribution of this mollusk is somewhat mysterious, however, for it inhabits the Sea of Aral at an altitude of 52 m. and is found subfossil in the Aral Basin at 55 m. above sea level. Berg supposes that the alleged archeological and geomorphological evidence of change in the level of the Caspian is actually evidence of earth movements.

The history of the Caspian since 1700 is fairly well known (Fig. 69). Brückner (1890) obtained qualitative evidence of high eighteenth-century levels, and of a decline early in the nineteenth century. Quantitative data are available since 1851. There was a rapid rise from the low level of the middle nineteenth century during 1864–1869; thereafter, the relatively high level sank irregularly and slowly until 1911, when there was a sudden drop. More recently the surface has fallen considerably. Berg (1950) speaks of the 1925 level as lower than any other indicated by historic records, and Shorygin (1943) indicates a lowering of about 2 m. between 1929 and 1940. The recent changes may in part be due to utilization and regulation of the Volga. It is clear that at least the 1911 fall reflected a deficiency of rainfall in the catchment area of the Volga (Schokalsky 1914).

The levels of the Sea of Aral (Suslov 1947) do not seem to vary concurrently with those of the Caspian. The evaporation is very great, of the order of 1000 mm. $yr.^{-1}$ or ten times as great as the local rainfall. The lake receives most of its waters from the east by the Amu-Darya or Oxus River, reaching its highest levels in summer. The mean level rose more than 2 m. from 1900 to 1915, during which years the Caspian was falling slightly; a fall of 1.13 m. occurred between 1915 and 1920.

Lake Balkhash, to the east of the Sea of Aral, varies in level, but there is less old data for this lake than for the Caspian or a few lakes farther to the east. The lake level is known (Domratschev 1933) to have fallen between about 1875 and 1890, to have risen from about 1890 or somewhat later to about 1910, and then to have fallen until 1930, after which date a new rise began. It is evident that the rises and falls are irregular. The over-all fall between 1910 and 1930 was approximately 2.8 m., but there were probably occasional years of rise during this period. The meager data suggests a behavior different from that of the Caspian but possibly like that

[4] Gregory (1914) gives a good summary with an extensive bibliography; see also Herbette (1914).

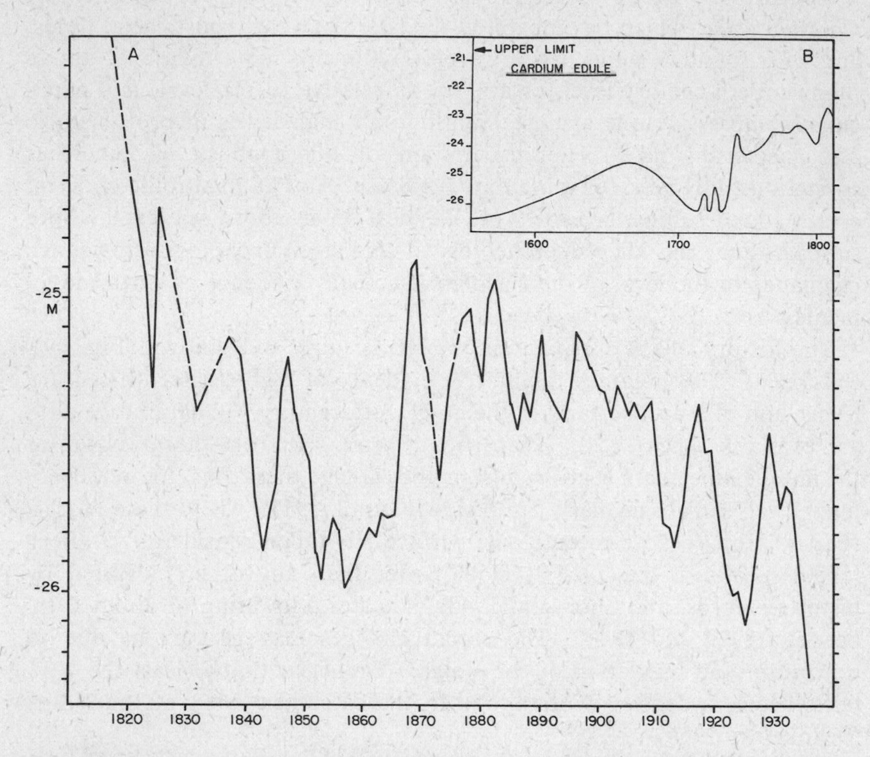

FIGURE 69. *A*, variation in level of the Caspian during the nineteenth and twentieth centuries (from data of Berg and of Brujewicz.) *B*, Berg's interpretation of the historic data prior to the high levels of the early nineteenth century; there were probably high levels in the early fifteenth century, but prior to the middle of the sixteen century all that can be certainly said is that the water did not transgress, in historic or protohistoric times, the upper limit of subfossil *Cardium edule*.

of the Sea of Aral. Panov (1932), studying rainfall records during part of the period of decline in level, from 1912–1918, concluded that 80 per cent of the variation in lake level could be attributed to variations in the discharge of the River Ili, the main influent, and in the rain falling on the lake surface itself.

In south central Asia, the longest record that can be in any way reconstructed relates to Lake Manasarovar, and to Rakas Tal into which the former lake flows and which can itself discharge into the Sutlej (Hedin 1917). The observations of these lakes are, however, at intervals far too long to provide any important information about a system that is likely to be very sensitive to short-period change. The data strongly suggest higher levels prior to 1804 than at any subsequent time. Some information has been assembled by De Terra and Hutchinson (1934) on Pang-gong Tso (Fig. 70) which makes a fairly consistent story of a high level prior to 1821, a fall in level after 1841 to a minimum in 1863 and a rise from 1869 to 1901. There may have been a slight minor oscillation during 1901 to

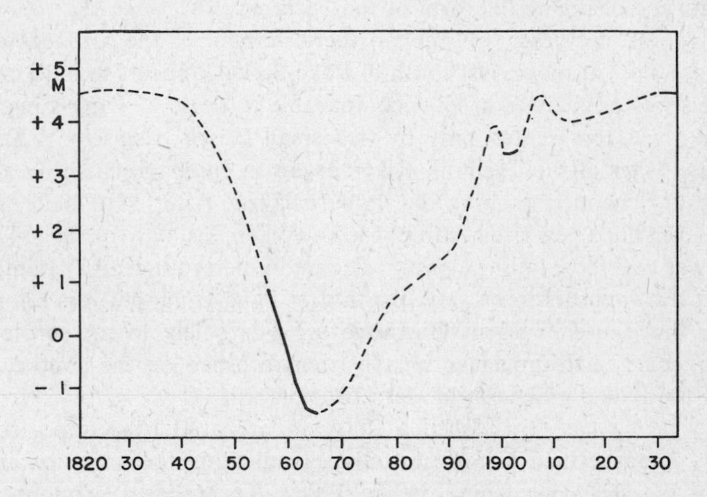

FIGURE 70. Variation in level of Pang-gong Tso. Solid line, direct observations; broken line, inferred from qualitative data. Datum level is the top of a submerged rock, which at low levels appears as an island, at the western end of the lake. (After De Terra and Hutchinson.)

1903, but the high level achieved about 1900 was evidently maintained relatively unchanged at least till 1932. Other closed lakes in Indian Tibet all gave evidence of relatively recent increases in level in 1932.

The evidence from the extreme southern part of central Asia thus suggests falling levels in the first half of the nineteenth century and, at least in Indian Tibet, a rise in the last third of that century which apparently led to a high modern level. There is nothing in the history of Pang-gong to suggest a low level at the time of the very low Caspian level of 1925.

De Terra and Hutchinson concluded that this rise of Pang-gong Tso could not be attributed to the melting of ice or névé in the drainage basin, as there is no evidence of recent retreat glaciers, and a relatively thin patch of old snow known in 1848 north of the lake still existed in 1932. They attribute the rise to an increase in precipitation, mainly at the end of the nineteenth century, recorded at the meteorological station at Leh, and point out that Brooks (1919) concluded that there was a general rise in rainfall during the period from 1880 to 1910 in the temperate areas of the Eurasiatic continent.

Very striking changes in level and position of Lop Nor have been recorded, but Hedin (1904), whose experience of the region was unrivaled, believed them to be due primarily to the drift of sand, blocking or opening influents and changing the form of the basin.

Far to the northeast, on the northern border of the arid center of Eurasia, Lake Gusinoye, just south of Lake Baikal, appears to have undergone a series of changes in level comparable to those of Pang-gong Tso. The lake was represented only by two small closed basins in 1720. In 1730 the water of the Tyemnik River began to enter the basin by a previously dry channel, forming the modern lake. After 1810 there was a fall in level, but a new rise began in 1865. When Suslov wrote in 1947, the water was said to be falling again. The mid-nineteenth-century minimum level at least appears to be very like that of Pang-gong, but it is not quite certain that the changes in Gusinoye are due solely to meteorological factors, since earth movements have been recorded on the southeastern side of the Baikal graben in modern times.

Enge (1931) has reproduced a diagram prepared by Koppe from a variety of unspecified historical sources, indicating the variation in the level of the Dead Sea from 1650 to 1906. There was a pronounced if slightly irregular rise from a minimum early in the last century to a very high level in 1900. Enge shows that at least the later part of this rise is due to increase in precipitation. Low levels are indicated about 1760, 1730, and 1670, moderately high levels in 1780 and 1740.

Central Africa. Good information is available for Lake Albert and Lake Victoria (Brooks 1923, Walker 1936, Hurst and Phillips 1931), Lake Nyasa (Dixey 1924, Kanthack 1941), and Lake Tanganyika (Tison 1949).

Both Lake Victoria and Lake Albert have oscillated in an irregularly periodic manner (Fig. 71) throughout the past sixty years. In general the major rises and falls are synchronous in the two lakes. Brooks, with only two and a half cycles available, believed the variation in lake level to be closely correlated with the sun-spot cycle. Since the available rainfall records showed no such periodic oscillation, he concluded that the lake levels were controlled primarily by variation in evaporation. Subsequent observations have shown that the correlation with solar activity must have been accidental, but the argument that the variation is primarily one of evaporation seems valid.

The history of Lake Tanganyika is complicated by the fact that the lake rose from the time of its discovery in 1854 until 1878, when it established a

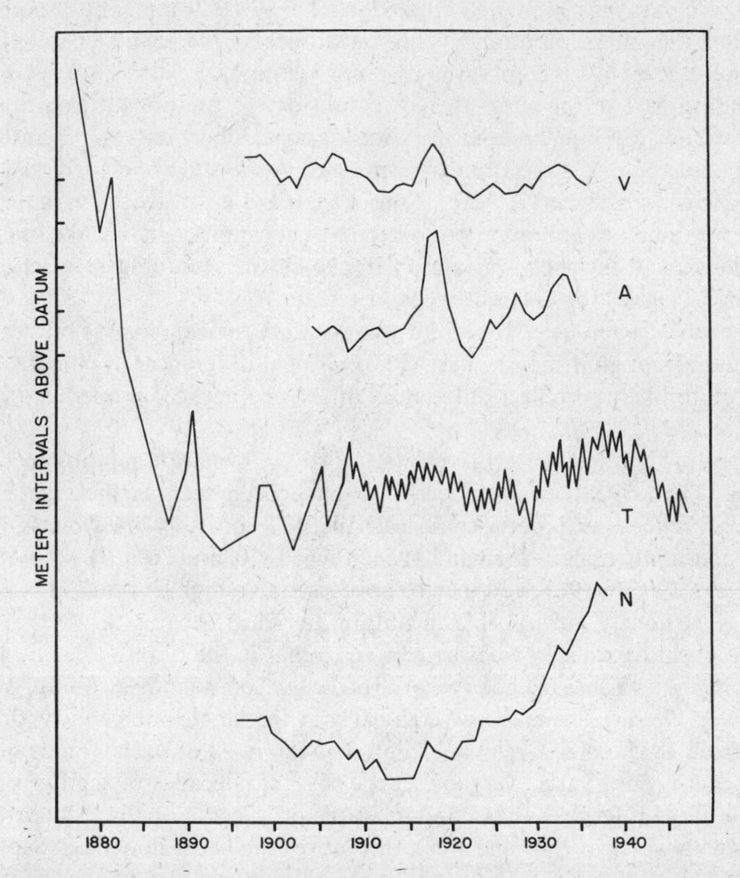

FIGURE 71. Variations in the levels of the lakes of central Africa: V, Victoria; A, Albert; T, Tanganyika; N. Nyasa. The great fall in the level of Tanganyika in the earlier years of the record is due to the cutting of the dam blocking the Lukuga River. The later part of the Tanganyika record reflects annual change due to wet and dry seasons. Note that in spite of considerable differences in behavior, all lakes indicate maxima in level about 1917 and rising levels after 1929, suggesting the superposition of a regional effect, probably in evaporation, on local meteorological and hydrographic influences.

discharge over a dam of silt that had blocked the effluent, the Lukuga River, at its exit from the lake. As soon as the discharge was established, the obstruction was very rapidly cut away, leading to a dramatic fall in level, at least mainly not determined climatically. The modern and well-established variations in level are less striking than in Victoria and Albert, but possibly exhibit the effects of the same climatic variation that operate on these two lakes.

Lake Nyasa has been well studied by Dixey (1924) and by Kanthack (1941). The main variation has been a slow fall since 1895 or before, to a minimum level in 1915, followed by a rise lasting at least to 1940. On the ascending part of the curve there are subsidiary variations that probably correspond to the maxima in the Victoria and Albert curves. Kanthack computed the discharge from the drainage basin and concluded that the increase in level since 1915 can be more than explained by the variation in precipitation. Evaporation was taken to be constant from the lake surface, but to depend inversely on rainfall in the basin, according to a series of empirical curves for different types of terrain.

The variation in behavior of the great central African lakes is interesting because although it may be partly due to local differences in climate, it is also quite likely to reflect differences in the geometrical properties of the lake basins.

There is, in addition to these data, a little information relating to Lake Chad (Tilho 1928), which appears to have been high in the eighteenth and early nineteenth centuries, at least till 1823; possibly low from 1840 to 1850; certainly high in 1853 and 1854, 1866, 1870, and from 1892 to 1898; low from 1905 to 1915; and then to have risen to a high level in 1921.

It is obviously not possible to obtain any clear idea as to whether the variation in Africa is in any way related to that in the Eurasiatic continent.

Australia. There is a good record for Lake George in New South Wales (Walker 1936). The lake was high early in the nineteenth century, dry or low from 1835 to 1863, relatively high through most of the later part of the nineteenth century, and very low in the first two decades of the twentieth.

The large interior basins contain water only occasionally. Lake Eyre filled with water in 1949 and 1950, the entire lake being flooded in September of the latter year. The evaporation rate appears to have been about 200 to 250 cm. per year, and the lake had practically disappeared early in 1952 (Fig. 72). There is a possibility that the lake contained water also in 1890 and 1891, when an abnormally high rainfall comparable to that in 1949 and 1950 is recorded (Bonython and Mason 1953).

Western North America. Such information as is available from the closed basins of western North America indicates clearly very low levels in the middle of the nineteenth century. Bowman (1935) points out that the wagon tracks of the forty-niners on the floor of the then dry Goose Lake on the Oregon-California boundary were disclosed when the lake level started to fall rapidly in 1920.[5] It is apparent from tree-ring records that in this region the two driest periods in recent centuries were from 1809 to 1854 and from 1917 to 1934.

[5] This may have been in part caused by irrigation (cf. Harding 1949).

Harding (1949), the most recent writer to review the matter, concludes that Pyramid Lake, which was not flowing into the dry basin of Winnemucca in 1844, cannot have done so for at least twenty-five years, or there would have been a little water in the Winnemucca Basin. He assumes, therefore, a dry period from at least 1809 till the 1860's, when

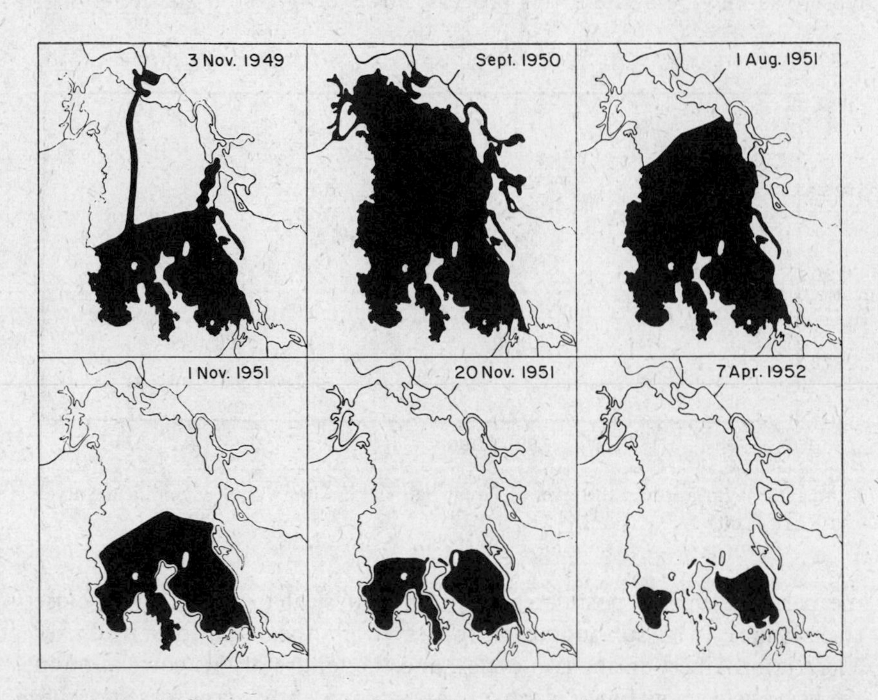

FIGURE 72. Incidence of water in Lake Eyre, Australia (Bonython and Mason 1953).

Winnemucca began to fill. He also points out that below the modern level of Lake Tahoe, tree stumps exist in which 100 annual rings can be counted. They cannot be very old, as sand and gravel moving in the littoral zone destroy such wood by abrasion. Harding therefore supposes that the low-level period of their growth culminated in the 1850's, and that the whole period 1750 to 1850 was dry in and around the Lahontan Basin.

High levels are recorded in Mono and Eagle Lake in 1917, and in many cases part of the subsequent low-water level is due to diversion of water for irrigation. Harding considers that Great Salt Lake (Fig. 73), for instance, would have exhibited as high levels in the present century as at the time of its known maximum level in the 1870's if the water that would naturally have flowed into the lake had not been diverted.

General considerations. The various attempts, from that of Brückner onward, to fit the oscillations of lake levels to some periodic function that is physically significant have so far failed. It is always possible, by the superposition of sine curves, to approximate any irregular curve as closely as may be desired; this process, however, gives no guarantee that

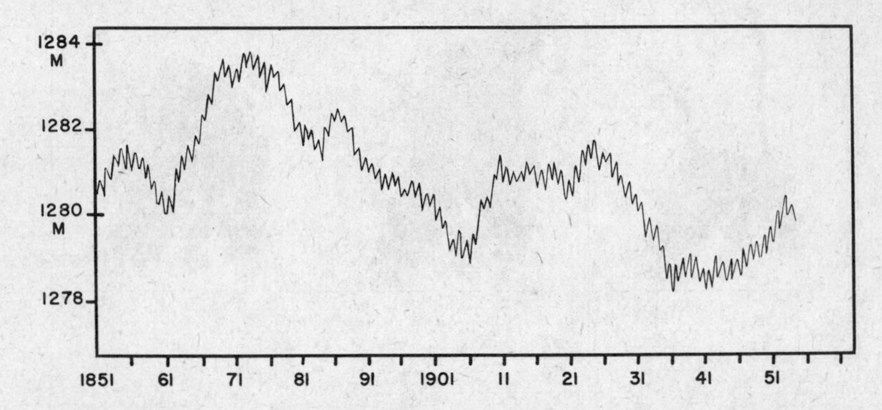

FIGURE 73. Variation in the level of Great Salt Lake, Utah (U. S. Geological Survey and R. F. Flint).

the period elements postulated have any physical meaning. The close resemblance to a sun-spot cycle observed in the variations in Lakes Victoria or Albert during two cycles, and the failure of this correlation in the next cycle, show how deceptive even an apparently simple case may be. As an example of the analysis of lake-level records into cyclical components, the paper of Shuman (1931) may be mentioned; the procedure was for a time fashionable but has now been largely abandoned as too unreliable. In default of a satisfactory confirmation of a prediction based on any hypothesis that may be advanced—and in the study of lake levels, hypotheses have in general been falsified rather than verified—the most satisfactory simple method of procedure is by autocorrelation. The series of numbers constituting the data are paired off successively, as

a b c d...	a b c d...	a b c d...	
. . . .	. . .	. .	
a b c d...	a b c...	a b...	etc.

and the correlation coefficient is determined for the paired numbers for each successive pairing. The first correlation coefficient will obviously be unity. If there is any truly periodic element of period length Π in the series, when the numbers are displaced by $\Pi/2$, $3\Pi/2$ and so on, there should be a significant negative correlation; when by Π, 2Π, etc., a significant positive correlation. Bruno (1938) has used this method on the levels of Lake Vetter, in which Bergsten (1928) believed several periods to exist, but signally failed to demonstrate any nonrandom variation.

The failure to obtain a simple set of numerical relationships from lake-level data does not imply, however, that such data lack general interest. The problem of whether they exhibit evidence of a tendency to synchronic behavior throughout the world is independent of the problem of solar or other types of cyclical control, but is in its own right an important problem. The theoretical expectation that different basins under the same variable climatic regime will behave differently complicates the problem of synchrony very greatly. All one can say at present is that the hypothesis of some degree of synchronic control of lake levels appears to have sufficient probability to justify further study. The culmination of the middle nineteenth-century dry period, around 1845 to 1855, which appears in the Caspian, Pang-gong Tso, Gusinoye, western North America, and apparently in Lake George, New South Wales, is at least very suggestive of synchrony, even though the time of the onset of aridity may have varied from place to place. It is unfortunate that the African data are inadequate to demonstrate or refute the occurrence of an arid period at this time. The decline of water levels during part of the present century, culminating in the 1920's, appears also to have been rather general, but it is noteworthy that Pang-gong Tso, which was very low in the 1850's, was very high in 1932 and seems to have varied little for some time prior to the latter date.

SUMMARY

The earth is peculiar among the terrestrial planets in its large supply of water. Most of this water is probably in the primary lithosphere. In the accessible parts of the earth, the ocean has a mass of about $13{,}800 \times 10^{20}$ g.; the ice of polar caps and other ice fields, about 167×10^{20} g.; and the inland waters with which this books deals, only 0.25×10^{20} g. The quantity of water vapor in the atmosphere is of the order of 0.13×10^{20} g., but on account of its mobility, is far more important than its small mass would indicate.

The distribution of lakes depends partly on the distribution of basins and partly on that of water. We may divide the earth into *exorheic* regions, in which rivers originate and from which they reach the sea; *endorheic* regions, in which rivers arise but do not reach the sea; and *arheic* regions, in which no rivers originate. Lakes are obviously more likely to occur in exorheic regions than in the others. Very roughly the arheic regions occupy the subtropical trade-wind belts, and the endorheic regions lie between these subtropical deserts and the tropical and temperate humid regions. Almost all basins of internal drainage lie in endorheic regions. By virtue of the transitional nature of the latter, small oscillations in climate may convert endorheic areas into exorheic or arheic, and the lakes of such areas may sometimes acquire outlets and at other times dry up completely.

The total rain falling on land has been estimated as about 0.99×10^{20} g. yr.$^{-1}$, of which about 20 per cent runs to the sea and the rest evaporates. In a single large basin such as that of the Mississippi, the proportion appears the same as for the whole world.

Lakes may receive water as precipitation on their surfaces, from surface influents, from seepage of ground water, and from ground water entering as discrete springs. Very large lakes may receive most of their water as precipitation; lakes in endorheic regions receive all or almost all from their influents. Lake Victoria and the Dead Sea provide extreme examples of these two situations. The rate and time of changes in level may be greatly modified by the nature of the vegetation cover in the drainage basin. The rate of discharge after a given rise in lake level varies greatly from lake to lake, owing primarily to the variations in area of the lake surface relative to the dimensions of the effluent channel.

It is possible to show, by an elementary mathematical analysis, that in two lakes existing under the same varying climatic regime the level of one may be rising, that of the other falling. This is of particular interest in the case of closed basins in endorheic regions, which are very sensitive to climatic change. Small steep-sided lakes in large drainage basins will tend to continue rising after large saucer-shaped lakes have started to contract. Only when a considerable group of lakes exhibits more or less parallel fluctuation are we justified in drawing conclusions as to the time and nature of the climatic change causing the fluctuations.

Lakes from which water leaves only by evaporation are termed *closed*; those with any kind of effluent, *open*. Among open lakes there are often cases with no visible outlets; where the water passes out of such a lake as superficial ground water, the lake is said to be a *seepage* as opposed to a *drainage* lake. Most seepage lakes also receive their water by seepage. Most small lakes not in rock basins are separated from the ground water by a clay seal, formed as an early lake sediment.

Discharge of drainage lakes is ordinarily by a single effluent; cases of more than one effluent are ordinarily either temporary or occur in tundra regions of very uncertain drainage.

The rate of delivery of water into a lake and its loss by evaporation ordinarily varies with the seasons. The influents are usually high in rainy seasons or when ice is melting; the level of the lake tends to rise at such times.

There is a good deal of evidence to suggest world-wide low levels in the middle nineteenth century, and rather widespread low levels also at the end of the first quarter of the present century. There is no unequivocal evidence that variations in lake levels exhibit any objectively significant regular periodicities. The only safe way of investigating such periodicities is by autocorrelation; the single case so far analyzed in this way failed to demonstrate periodicities supposed to exist in the data.

CHAPTER 5

The Hydromechanics of Lakes

In the present chapter the water of a lake is considered as a fluid body, wholly or partly in motion. It is convenient to begin such a study with an examination of turbulent movement, to pass on to the more or less unidirectional currents generated largely by the wind, and then to consider periodic movements, notably waves and seiches. The phenomenon of the internal seiche, discussed in the later sections of the chapter, appears from the recent work of Mortimer (1952a, 1953a) to constitute a very important type of movement in most lakes. The biological limnologist, who may find much of the earlier part of the chapter wearisome, should at least consider this part of the argument.

The general results of deductive studies have been used throughout, but no attempt to derive these results has been made. Such a restriction will probably be welcomed by most limnologists. The more mathematically minded student of lake physics will probably already have learned most of what he needs from Proudman's (1953) excellent work on dynamic oceanography.

TURBULENCE

Classical hydrodynamics developed as a deductive science during the eighteenth and early nineteenth centuries, without extensive experimental confirmation of its theorems. The early mathematical work assumed, for example, that the velocity at any point in a fluid could be expressed as a continuous monotonic or regularly oscillating function of time. Anyone who has looked down from a boat into the water of a lake containing macroscopically visible plankton, or who, for that matter, has watched the movement of tobacco smoke in a supposedly still room, will realize that this is not the case. At any point, the velocities and accelerations keep changing, and the only simple way of observing regular streamline or laminar flow is adjacent to some smooth surface over which a slow current is passing. The irregular complexity of the turbulent movements in large masses of fluids in nature at first made investigation of real fluids very difficult. Some empirical work was done in the eighteenth century, largely by engineers, but it seldom conformed to the expectations of the deductive dynamicists.

Actually, a more realistic approach to the problem is less difficult than might at first appear. The beautifully regular movements observed when water flows slowly down a narrow tube or air moves slowly over a solid, smooth surface are regular only because of our inability directly to appreciate the random movements of water or air molecules. The laws of classical hydrodynamics, in so far as they are applicable to actual laminar movement, must be statistical laws, and the suggestion therefore is reasonable that comparable statistical laws applicable on a large enough scale might hold for turbulent flow. Most of the study of real movements in water in the past eighty years has been based on this idea. In view of the important part played by turbulent movement in the hydromechanics of all actual bodies of water, it is convenient to begin the study of water movements with a consideration of turbulence.

Reynolds number as a criterion of turbulent flow. The first important theoretical study, that of Boussinesq (1877), assumed turbulence to be generated in moving fluids by the roughness of the boundaries, and to be transmitted from these boundaries through the moving liquid. This approach led to the introduction of the coefficient of eddy viscosity as a measure of the transmission of shearing stress through a moving liquid. Somewhat later Reynolds (1883, 1894) investigated the question as a matter of the conditions under which either laminar flow or turbulent flow is stable. Reynolds developed a general criterion, namely, that if the quantity

$$R_e = \frac{\rho v l}{\mu} \tag{1}$$

is less than some critical value, flow is laminar; if it is greater than the critical value, flow is turbulent. Here ρ is the density of the liquid, v its velocity, μ its viscosity, and l a length, which in the case of a pipe or of a falling particle is the diameter but which is less easily defined in certain other cases. The number R_e is termed Reynolds number and will also be of considerable importance when we are considering the hydromechanics of plankton in Volume II.

Jeffreys (1925a) found that in a shallow channel much wider than deep, turbulent flow begins when $R_e = 310$, l here being the depth of the water. If this result may be applied to a shallow lake, say 100 cm. deep, the critical velocity will be about 0.03 cm. sec.$^{-1}$. We clearly need not concern ourselves with laminar flow in limnology except perhaps actually on the bottom or, as Jeffreys suggests, in weed beds.

In recent years there has been an enormous amount of work done on turbulent movement, particularly in air in relation to aeronautics and meteorology. The modern tendency has been to emphasize the differences between molecular and turbulent movement, or perhaps more properly to regard molecular movement as a special limiting case of a whole spectrum of turbulence. This approach is certainly realistic but it is extremely difficult; most of the useful if approximate results have been obtained by employing more or less proper analogies between molecular and turbulent movement.

Turbulent velocity, intensity, and kinetic energy of turbulence. If we consider a fixed point in a fluid and measure the velocity along, say, the x axis, this velocity may be found to be either stationary or varying systematically, or it may be found to vary in a random, irregular manner.

Let $u(t)$ be any instantaneous measure of velocity.[1] Then we can, over any time interval t^*, define a mean velocity as $\frac{1}{t^*}\int_0^{t^*} u(t)\,\mathrm{d}t$. If there is no systematic variation in the mean velocity we can, following the usual statistical convention, define a basic velocity $\bar{u}$ by

$$\bar{u} = \lim_{t^*\to\infty} \frac{1}{t^*}\int_0^{t^*} u(t)\,\mathrm{d}t \tag{2}$$

The instantaneous velocity may then be written

$$u(t) = \bar{u} + \mathbf{u}(t) \tag{3}$$

where $\mathbf{u}(t)$, which can be positive, negative, or zero, is called the turbulent velocity at time t. It is evident that

$$\lim_{t^*\to\infty} \frac{1}{t^*}\int_0^{t^*} \mathbf{u}(t)\,\mathrm{d}t = 0 \tag{4}$$

The time mean square turbulent velocity will not vanish, and we may write

$$\bar{\mathbf{u}}^2 = \lim_{t^*\to\infty} \frac{1}{t^*}\int_0^{t^*} \mathbf{u}^2(t)\,\mathrm{d}t \tag{5}$$

[1] The notation and the general introductory exposition follow Stommel (1949) save for a few details.

where $\bar{\mathbf{u}}^2$ may be regarded as a measure of the intensity of turbulence along the x axis. We may similarly write for the y and z axes respectively

$$v(t) = \bar{v} + \mathbf{v}(t) \tag{6}$$

and

$$w(t) = \bar{w} + \mathbf{w}(t) \tag{7}$$

and obtain $\bar{\mathbf{v}}^2$ and $\bar{\mathbf{w}}^2$.

The kinetic energy of turbulence may then be defined as

$${}^1/{}_2\rho(\bar{\mathbf{u}}^2 + \bar{\mathbf{v}}^2 + \bar{\mathbf{w}}^2) \tag{8}$$

From considerations of continuity it is obvious that if one of the intensities is nonzero, at least one of the other two must be nonzero.

Eddy diffusivity. If some property $s(t)$ other than velocity varies along the x axis, it is clear that in a series of observations at any point along the axis the observed value of $s(t)$ will vary irregularly if the flow is turbulent. There is, moreover, no reason to suppose that when a particular value of $u(t)$ is observed it will inevitably imply a particular value of $s(t)$. In general, if

$$s(t) = \bar{s} + \mathbf{s}(t) \tag{9}$$

$$\overline{\mathbf{us}} = \lim_{t^* \to \infty} \frac{1}{t^*} \int_0^{t^*} \mathbf{u}(t)\mathbf{s}(t)\, dt \neq 0 \tag{10}$$

If s be defined in terms of the concentration of some substance per unit mass of water

$$\overline{\mathbf{us}} = -A_{sx} \frac{ds}{dx} \tag{11}$$

where A_{sx} is defined as the *coefficient of eddy diffusivity* in the direction of the x axis. Comparable coefficients A_{sy} and A_{sz} may be likewise defined for the y and z axes.

If the concentration of the dissolved substance be given per unit volume of water rather than per unit mass, we may write for the rate of passage of the substance[2] across unit area

$$\frac{\partial S}{\partial t} = -\frac{1}{\rho} A_{sx} \frac{\partial s}{\partial x} \tag{12}$$

The coefficient of eddy diffusivity as defined by equation 12 has the dimensions $[ML^{-1}T^{-1}]$ and is ordinarily given as g. cm.$^{-1}$ sec.$^{-1}$ If the reciprocal of the density is assimilated into this ordinary or dynamic

[2] The convention will be adopted throughout of using capital letters for quantities transported across areas, and lower case letters for intensities or concentrations. Thus Θ will be a quantity of heat, $d\Theta/dt$ the rate of transport of heat across unit area, and θ a temperature.

coefficient, we obtain a kinematic[3] coefficient, having the dimensions $[L^2T^{-1}]$. The density ordinarily being practically unity, many investigators give the eddy diffusivity in $cm.^2\,sec.^{-1}$ Equations of the form of equations 11 and 12 are comparable to that expressing passage of a diffusing solute in an undisturbed liquid according to Fick's law; such equations are therefore commonly called Fickian. An analogous expression for the movement of heat by turbulent processes may be written as

$$\frac{\partial \Theta}{\partial t} = -\frac{1}{\rho} c_p A_{\theta x} \frac{\partial \theta}{\partial x} \tag{13}$$

where c_p is the specific heat at constant pressure, and $A_{\theta x}$ the *coefficient of eddy conductivity.* Comparable coefficients $A_{\theta y}$ and $A_{\theta z}$ may be likewise defined. In general, the coefficients of eddy conductivity and diffusivity are practically identical, and when there is no advantage in their distinction, the symbols A_x, A_y, and A_z, or in some cases merely A, will be used,[4] and such coefficients will be spoken of as coefficients of turbulent transport.

At least in limnology, the values of c_p and ρ for water may be taken as unity in all computations. The resulting values of the coefficients of turbulent transport must not be thought of as constants; it is often an advantage, however, in the construction of theoretical models, to assume their constancy. Actual values of A_z, obtained mainly from heat transport in the deep waters of lakes, will be found in Tables 48 to 51 of Chapter 7.

Eddy viscosity. Not only heat and dissolved materials but momentum also can be transferred by means of the eddy system. In laminar flow we have the molecular viscosity defined by

$$\tau = -\mu \frac{du}{dx} \tag{14}$$

where τ is the shearing stress exerted on a plane normal to the x axis. Similarly in turbulent water

$$\tau = -A_{vx} \frac{d\bar{u}}{dx} \tag{15}$$

where the stress τ is now often called the Reynolds stress, and where A_{vx}, the coefficient of eddy viscosity, plays the same role in turbulent water as the molecular viscosity does in still water. It is easy to realize qualitatively that virtual or eddy viscosity due to turbulence plays a great

---

[3] Analogy to a kinematic coefficient of viscosity will be apparent from the next section. In the study of heat transfer, the dynamic molecular coefficient is usually called the coefficient of conductivity, the kinematic coefficient that of thermal diffusivity. In limnology the difference is only of significance in dealing with sediments (page 505).

[4] The German term *Austausch* has been frequently employed, in writing in English, to express turbulent exchange; whence the use of the symbol A.

role in nature, retarding the settling of sediments or suspended micro-organisms, even though they be denser than the medium. Much of the difference between the results of classical hydrodynamics and of observations and experiments in nature and the laboratory is in fact explicable by the great values of virtual viscosity that are encountered in large masses of fluid. It is, for instance, historically interesting that before the significance of eddy viscosity was realized, it was supposed that vast periods of geological time would be needed for the establishment of the wind-driven circulation of the ocean.

In water of indifferent stability, lacking density gradients, A_v and A are of the same magnitude, A_0. When we consider the vertical coefficients in stably stratified water, however, it must be remembered that any mass of water of low density, displaced downwards, will tend to return to its original level. During its temporary sojourn below that level, it may have given up all its momentum but have suffered no appreciable loss of heat, oxygen, or other substance before returning to its mean horizontal position. If $\mathbf{l}$ be defined as the average distance traveled by small masses of water before they attain the momentum of their surroundings, Prandtl (1925) shows that

$$A_v = \rho \mathbf{l}^2 \left| \frac{d\bar{v}}{dz} \right|$$

The distance $\mathbf{l}$ is called the *mixing length.*

Taylor (1931b) points out that the rate at which potential energy is supplied at any level in a vertical density gradient is

$$gA_z \left(\frac{\partial \rho}{\partial z} \right)^2$$

while the rate at which energy is lost from the main current of mean velocity $\bar{v}$ is

$$\rho A_{vz} \left(\frac{\partial \bar{v}}{\partial z} \right)^2$$

Since some of the turbulent energy will be dissipated as heat, we may write

$$gA_z \frac{\partial \rho}{\partial z} \leqslant \rho A_{vz} \left(\frac{\partial \bar{v}}{\partial z} \right)^2 \tag{16}$$

or writing E for the stability $\dfrac{1}{\rho} \dfrac{\partial \rho}{\partial z}$

$$\frac{A_{vz}}{A_z} \geqslant \frac{gE}{\left(\dfrac{\partial \bar{v}}{\partial z} \right)^2} \tag{17}$$

The expression on the right in equation 17 is termed Richardson's number R_i, having been introduced as a criterion of the stability of vertical turbulence in meteorology by Richardson (1926).

A semi-empirical treatment (Munk and Anderson 1948) suggests that we may write

(18) $$A_{vz} = A_0(1 + 10\,R_i)^{-1/2}$$

(19) $$A_z = A_0(1 + 3.33\,R_i)^{-3/2}$$

or

(20) $$\frac{A_{vz}}{A_z} = \sqrt{\frac{(1 + 3.33\,R_i)^3}{(1 + 10\,R_i)}}$$

the last expression being unity when $R_i = 0$, and greater than unity for all positive[5] values of R_i. These expressions indicate that both A_{vz} and A_z decrease as the density gradient $\frac{\partial \rho}{\partial z}$ increases and as the velocity gradient decreases, the change in A_{vz} being less great than in A_z. Such conclusions, which are certainly qualitatively true irrespective of the exact validity of equations 18 and 19, are of great importance in limnology. It is, however, necessary to realize that we are not dealing with a single independent variable, and that we cannot assume that as stability increases, the coefficients of turbulent transport must decrease. If the velocity gradient were proportional to $E^{1/2}$, the stability could vary over any range for which such a proportionality held, without altering the coefficient of vertical turbulent transport. This has not always been appreciated.

In water under the influence of the wind, the eddy viscosity may be regarded as a function of the wind stress. Since the stress τ_a that the wind exerts on the water is equal to that exerted by the water on the air, we may write

(21a) $$\tau_{ax} = -A_v \frac{du}{dz}$$

(21b) $$\tau_{ay} = -A_v \frac{dv}{dz}$$

Since the stress due to the wind depends on its velocity W, attempts have been made to relate A_v to W. According to Ekman (1905), for a wind of over 600 cm. sec.$^{-1}$

(22) $$A_v = 4.3 \times 10^{-4}\,W^2$$

Thorade (1914) obtained results for less strong winds, leading to

(23) $$A_v = 1.02 \times 10^{-6}\,W^3$$

[5] Negative values can occur in meteorology.

These relationships give the following results:

W cm. sec.$^{-1}$	200	400	600	800	1000	1500	2000
A_v g. cm.$^{-1}$ sec.$^{-1}$	8	65	(218)	375	430	970	1720

But Ekman's relationship depends on the stress being proportional to W^2, whereas much modern work to be discussed later suggests that the exponent is rather less than 2 or that the stresses cannot be expressed as monodic functions of W. These values are therefore not to be taken as better than rough approximations.

If we may assume in a current system a vertically constant value of A_v, we may define a very important quantity D, the *depth of frictional resistance*, as

$$D = \pi \sqrt{\frac{A_v}{\Omega \sin \phi \cdot \rho}} \tag{24}$$

where Ω is the angular velocity of rotation of the earth, and ϕ the geographical latitude. Evaluation of equation 24 for wind velocities between 200 and 2000 cm. sec.$^{-1}$ and latitudes between 30° and 70° gives a range of D between 11 and 216 m. This range is, however, certainly too great for average conditions; for most lakes in temperate latitudes, D may be regarded as lying between 20 and 100 m.[6] The full significance of this will appear in later sections.

The turbulence spectrum. A fundamental objection can be raised against the whole of the approach to turbulence already presented. The Fickian type of equation refers to the statistical effects of independent rectilinear movements of molecules. In turbulent transport the movements are not rectilinear and are increasingly less independent as the distances considered become smaller (Richardson 1926, Stommel 1949). When two floats are placed very close together, their movements are obviously more likely to be correlated than when they are placed farther apart. In the close position, their tendency to behave independently will be due only to the smaller eddies; they will be moved more or less in parallel by most of the larger eddies. In the distant position, the larger eddies will contribute to their independent random movement. This effect can occur at any distance of separation, the major current systems of the oceans providing the largest eddies possible in the terrestrial hydrosphere.[7]

---

[6] Grote (1934) takes D as the depth of the epilimnion, ordinarily less than 20 m. The theory, however, properly applies only to unstratified water.

[7] Big swirls have little swirls
That prey on their velocity,
And little swirls have lesser swirls
And so on . . . to viscosity.
(Attributed to L. F. Richardson.)

Consideration of the distribution of turbulent velocity, energy, and viscosity among eddies of different sizes, or in more formal terms the energy, velocity, and viscosity spectra of turbulence, has become important not only in considering the terrestrial atmosphere and hydrosphere, but also in cosmogenic speculation.

It is now generally believed that the solar system owes many of its peculiarities to a pre-existing pattern of turbulence in a cloud of gas and dust from which the system formed. The breakdown of a single large eddy into smaller and smaller eddies has been studied by a number of investigators, notably Kolmogoroff, Onsager, Von Weizsäcker, and Heisenberg, whose mathematical results, largely obtained independently, have been summarized, in so far as they may apply to the hydrosphere, by Stommel (1949). If we consider a square of side l_0, divided up into smaller squares of side l_1, l_2, etc., in such a way that

$$l_n = c l_{n+1} \tag{25}$$

where c is a constant, it can be shown that averaging over a square of side l_n, the turbulent energy varies as $l_n^{2/3}$, the eddy viscosity as $l_n^{4/3}$, and the characteristic average velocity[8] as $l_n^{1/3}$. These relationships are true only when we consider all the energy originally to have been present in the largest eddy in the system. Stommel points out that in the ocean there is likely to be a good deal of horizontal turbulent energy added in the region of the spectrum where l_n is small, perhaps of the order of 10 m. to 100 m. This will cause the spectral distributions to vary less strikingly with eddy size than if all the energy was derived from a single major eddy breaking up.

These theoretical results are of considerable interest in that Richardson (1926) long ago concluded that it is possible to obtain a quantity $F(l)$, called neighbor diffusivity, which can be substituted for eddy diffusivity in Fickian theory and which empirically is proportional to the $^4/_3$ power of the distance between floats or other marked points in turbulent water. Stommel (1949) has obtained evidence that such a relationship actually describes the separation of pieces of wet paper floating on the surface of the sea. This relationship is presumably dependent on the variation of eddy viscosity with $l_n^{4/3}$ in the results of Von Weizsäcker's theory summarized above. Stommel (1949; see also Richardson and Stommel 1948) finds the law to hold not only in the open ocean but in more or less enclosed arms of the sea. The tendency toward a flattening of the spectral distribution of energy, velocity, and eddy viscosity due to small scale disturbance

[8] The average of the observed velocities within the square l_n less the average over each succeeding larger square.

is likely to play an even greater role in lakes than in the sea; but in considering horizontal turbulent diffusion in large lakes, variation of diffusivity or viscosity with the $^4/_3$ power of the distance may well be expected.

As will be indicated later, there is evidence (Brooks 1947) that some aquatic organisms respond specifically to the turbulence of their environment. In such cases, eddies having a radius of curvature of a few centimeters are probably of considerable importance. Unfortunately, at the present time nothing is known empirically about the distribution of turbulence in this or any other part of the spectrum in inland waters.

CURRENTS

In the preceding paragraphs it has been assumed that the water of lakes is in motion, and that as well as the continually varying turbulent velocities, residual average velocities also exist. Nothing, however, was said as to their nature and origin.

The larger average water movements that must now be considered constitute the current systems of lakes. In general such current systems are of two kinds. The first kind is nonperiodic; such currents are generated by external forces, namely, the influent-effluent system of the lake, unequal heating, the entry of dissolved material from the sediments, and variations in atmospheric pressure and winds. The second kind is periodic and is generally due to the disturbance of the lake by wind or atmospheric pressure changes and a subsequent oscillation of the lake, or a part of it, as frictional forces slowly degrade the energy delivered earlier by the external forces. The word *current* is frequently considered to apply primarily to those movements maintained by the action of external forces, while the periodic movements which tend to develop when the external forces cease to act are usually termed *seiches*. In practice, however, it is extremely difficult to distinguish aperiodic deepwater currents from internal seiches of long period and large amplitude, and although the distinction is convenient enough to maintain, it is imprecise and often impractical.

In discussing the theory of water movements it is important to bear in mind that most lakes lie between two extreme ideal types of water mass, the unlimited mass of infinite depth, and the long channel of finite depth and inappreciable width. Since the properties of the currents in large lakes generally show some features of the types of circulation found in large unbounded oceanic masses of water, it is convenient to start by considering the relevant aspects of the elementary theory of the movement of such a water mass.

General equations of motion: cyclonic and anticyclonic swirls. The equations of motion of an element of water in an unbounded ocean on a rotating planet must express the effects of gravitational forces, pressure gradients which may be the result of external forces, the deflecting forces of the earth's rotation, and frictional forces.[9]

Using a coordinate system with the z axis directed downward, the Euler-Navier equations of motion of classical hydrodynamics, when set up to describe these effects, become

$$\frac{du}{dt} = 2\Omega \sin \phi v - \alpha_s \frac{\partial p}{\partial x} + \alpha_s R_x \tag{26a}$$

$$\frac{dv}{dt} = -\, 2\Omega \sin \phi u - \alpha_s \frac{\partial p}{\partial y} + \alpha_s R_y \tag{26b}$$

$$\frac{dw}{dt} = -\, 2\Omega \sin \phi v_E - \alpha_s \frac{\partial p}{\partial z} + \alpha_s R_z \tag{26c}$$

where u, v, w = the components of velocity along the x, y, and z axes.
v_E = the eastward horizontal component of velocity.
Ω = the angular velocity of rotation of the earth = 0.729×10^{-4} radians $sec.^{-1}$
ϕ = the geographic latitude.
α_s = the specific volume of water.
R_x, R_y, R_z = components of frictional force per unit volume.
p = the hydrostatic pressure at any point x, y, z.

The horizontal components of the geostrophic acceleration[10] represented by the first term on the right in equation 26a and 26b are commonly called Coriolis forces, though the horizontal geostrophic component in tidal theory was first considered by Laplace in 1775.

If we imagine a current of velocity u, generated by the transitory action of some external force and then continuing uninfluenced by friction, pressure gradients, or any force other than gravity, hydrostatic pressure, and the rotation of the earth, the last two terms on the right of equation 26 will become zero, the current will be rotated to the right in the Northern Hemisphere, to the left in the Southern Hemisphere, describing a circle of radius r_i given by

$$r_i = \frac{u}{2\Omega \sin \phi} \tag{27}$$

---

[9] The mathematically minded beginner can find no better work on the basic theory of these effects than Proudman (1953).

[10] Consider a particle at the equator, moving north. On the equator it will have a certain momentum, directed eastward, due to the earth's rotation. This eastward momentum is conserved as the particle moves to higher latitudes. The mass of the particle, moreover, is assumed to be constant. The velocity of the particle relative to the earth's surface, which has a constant angular velocity and so an apparent surface velocity decreasing with latitude, will increase. The particle relative to the earth's surface will therefore be accelerated to the right.

The period of rotation of such a current T_p is given by

$$T_p = \frac{\pi}{\Omega \sin \phi} \tag{28}$$

and is half the period of rotation of a Foucault pendulum. The quantity T_p is consequently often called half a pendulum day. It is obviously π/Ω or half a sidereal day, at the pole; and at lat. 30°, approximately 24 hours. As the equator is approached, the period becomes rapidly longer and the theory rapidly unapplicable. *Inertia currents* of this sort cannot exist in ideal purity in nature, but good approaches to such currents have occasionally been recorded, notably in the Baltic (Gustafson and Kullenberg 1936). The radius of the inertia circle for a current of 100 cm. sec.$^{-1}$, which is very rapid, would be 10 km. in lat. 45°, and less at higher latitudes and lower velocities. As Ekman (1905) pointed out, this radius gives a convenient measure to assess the effects of the radius of curvature of shores and other barriers in modifying the effect of the earth's rotation.

Returning to the more general case of equations 26a, b, and c, when we can legitimately consider the motion as steady, the terms on the left become zero. In an unlimited mass of water, the vertical velocities are negligible. We accordingly have

$$2\Omega \sin \phi v - \alpha_s \frac{\partial p}{\partial x} + \alpha_s R_x = 0 \tag{29a}$$

$$-2\Omega \sin \phi u - \alpha_s \frac{\partial p}{\partial y} + \alpha_s R_y = 0 \tag{29b}$$

Where the frictional forces are also negligible, we have

$$v = \frac{\alpha_s}{2\Omega \sin \phi} \cdot \frac{\partial p}{\partial x} \tag{30a}$$

$$u = -\frac{\alpha_s}{2\Omega \sin \phi} \cdot \frac{\partial p}{\partial y} \tag{30b}$$

If we now measure x along the pressure gradient, so that $\partial p/\partial y = 0$, we are left with (equation 30a)

$$v = \frac{\alpha_s}{2\Omega \sin \phi} \cdot \frac{\partial p}{\partial x}$$

indicating that the movement is at right angles to the pressure gradient, or along the isobars. As the observer looks from a point of high to a point of low pressure, the movement is to the right in the Northern Hemisphere and to the left in the Southern Hemisphere. This means that when we have any area of low pressure in a sufficiently unrestrained system, we shall have circulation around the area, counterclockwise in the Northern, clockwise in the Southern Hemisphere, or if we prefer, *contra solem*.

Such circulation is termed *cyclonic.* If we have an area of high pressure, we shall have circulation in the opposite direction, *cum sole*; this is termed *anticyclonic.* It is of importance to remember that equations 30a and 30b merely describe the relationship that must exist, if the velocity is steady, between that velocity and the pressure gradient and latitude. They tell us nothing about the causes of the velocity or of the pressure gradient. They do tell us that whenever we have a cyclonic swirl at the surface, we may expect a depression of the lake's surface in its center, and whenever we have an anticyclonic swirl we may contrariwise expect an elevation of the lake surface in its center. Provided we are really dealing with a reasonable approximation to steady motion, it will obviously be possible to calculate velocities from pressure gradients or vice versa. It is also not difficult to see qualitatively that if the angular velocity of the cyclonic swirl decreases with depth, as is usually the case, the deeper water, since it is now moving anticyclonically relative to the surface water, will be pulled up into the middle of the swirl, while in an anticyclonic swirl the deep water will be depressed and surface water accumulated in the middle. With marked temperature gradients there will be a rising mound of cold water below the center of a cyclonic swirl, and a pool of warm water in the middle of an anticyclonic swirl. More generally, in the Northern Hemisphere, whenever a current is flowing in water in which there is a density gradient, the less dense water will tend to be on the right of the current, the denser water on the left. In the Southern Hemisphere the reverse conditions hold.

Gradient or slope currents. In a long narrow lake in which transverse movement is negligible owing to the restraint imposed by the sides, the simplest type of current is a gradient or slope current due solely to the hydrographic slope of the lake as part of a river system. For steady flow, equation 29a becomes merely

$$\frac{\partial p}{\partial x} = \mathrm{R}_x \tag{31}$$

The frictional forces and the force due to the pressure gradient must balance. In unstratified water, the velocity of such a gradient current is independent of depth except in the vicinity of the bottom and sides, which exert a frictional force reducing the rate of flow. The velocity will be mainly dependent on the turbulence, or specifically the eddy viscosity. The velocity of flow under such conditions was empirically found by Chézy (1775) to be proportional to the square root of the gradient. In lakes there is an approximate relationship of the form

$$u = \mathrm{c}\sqrt{\bar{z}\, \mathrm{d}z/\mathrm{d}x} \tag{32}$$

where c is a constant, which is dependent, however, on the mean depth.

According to Hellström (1941), c may be expected to have the following values:

$\bar{z}$	=	1 m.	10 m.	100 m.
c	=	25	37	54

when lengths are measured in meters and times in seconds.

It is evident that the simple hydrographic slope current is usually unlikely to be of great importance. If it were, the locality would probably be spoken of as a widening of a river rather than as a lake. Consider a hypothetical example of a lake 2 km. wide and of 50 m. mean depth, fed by a river 100 m. wide and 10 m. deep in which the velocity is 1 m. $sec.^{-1}$ If the lake is unstratified and of the same density as its influent, the resulting mean velocity will be 10^{-2} m. $sec.^{-1}$ Taking for such a lake c = 50, the gradient is 1:1,250,000,000 or 1 mm. in 1,250 km., which is negligible compared with the gradients often imposed by external forces such as the wind.

In any actual case the width of the lake will always be appreciable, and if the depth is of the order of *D*, the depth of frictional resistance (page 257) or greater, the slope current will veer significantly to the right. This will cause piling up of water on the right bank, looking downstream, the transverse slope being in extreme cases much greater than the original hydrographic slope. A bottom current to the left will then compensate the movement at the surface to the right (Ekman 1905).

In view of the ease with which wind and pressure changes can produce denivellation far in excess of the hydrographic slope, it is probable that in most cases the contribution of the latter to the current system of the lake is not great. It must not, however, be forgotten that the incoming water may retain its individuality as an inertia current, particularly if the lake is stratified and the influent is at a lower temperature than the surface water or is heavily laden with silt. Cases of this sort will be discussed in a later paragraph.

Gradient currents may be set up by any process of denivellation; the most striking denivellations are those produced by changes in atmospheric pressure and wind. The variation in atmospheric pressure from place to place on a lake may easily (page 326) depress or elevate the surface by several millimeters. When such a denivellation is established, no current flows, but changes in the pattern of atmospheric pressure must lead to a series of temporary slope currents. These are chiefly known by the periodic oscillations that they generate. The very important gradient currents set up when water is piled up downwind are best considered in connection with the wind drift.

Internal pressure currents. When isosteric surfaces or surfaces of equal density or specific volume do not correspond to the equigeopotential or level surfaces, a current must be flowing; and when a current is flowing in water with a density gradient, there is always a disturbance of the isosteric surfaces. If we consider two adjacent stations **A** and **B** and assume that, at a certain depth corresponding to a pressure p_m no current is flowing between the stations, then the relative current between **A** and **B**, v_{AB} is given by Helland-Hansen's formula

$$v_{AB} = \frac{1}{2\Omega \sin \phi . \mathbf{AB}} \int_{p_z}^{p_m} (\alpha_{zB} - \alpha_{zA}) \, dp \tag{33}$$

In computing the integral it is sufficient to assume p_z to be proportional to depth and, if measured in decibars, to be essentially the depth in meters. This formula is valid only so far as local accelerations due to wind stress, friction on the bottom, and the like, are not imposed on the system. It has been used in oceanography extensively, but so far but little in limnology. Its actual application to Lake Constance by Elster (1939) is discussed later, in connection with the current system of that lake. In fresh-water lakes the density differences are caused primarily by temperature changes; in the sea, changes due both to temperature and salinity are involved. Ayers (1956), in a paper published too late for detailed discussion, has introduced certain refinements appropriate to fresh water and has applied his method to Lake Huron.

In any extensive ocean there is obviously an opportunity for marked density gradients to develop along a level surface as the result of climatological variation. In most lakes this effect will be negligible. It is possible that in some lakes heating by reflection from the bottom may produce a horizontal temperature gradient from the littoral to the central part of the lake, and that this gradient might be responsible for a cyclonic circulation in summer. The effect would be expected only in a large lake the bottom of which sloped very gently to a considerable depth. Even under such circumstances, complications due to friction at the bottom might entirely obscure the effect, and disturbances by wind drift would very likely make this type of current inappreciable in practice. The anticyclonic current to be expected in the winter, with a mean temperature well below 4° C., a light, free-floating ice cover, and marginal heating leading to an increase in density peripherally, might perhaps be more easily detected in the field.

The general absence of conditions that could lead to a primary horizontal density gradient in lakes of ordinary size and the extreme importance of local accelerations at the surface in generating lacustrine currents, suggest that some caution is desirable in the use of Helland-Hansen's relationship in limnology. Density current are almost certainly generated by exchange of heat and electrolytes at the bottom water interface in the deeper parts of a stratified lake. These are discussed in particular in connection with their thermal effects in a later chapter. Though their existence must involve horizontal density gradients, it may be suspected that in most cases frictional forces would be far more important than Coriolis forces in determining their behavior, and that they will nearly always flow downward very close to the bottom.[11]

The effect of the wind in generating currents. By far the most important type of current known in lakes is that caused by the wind. Whenever a wind blows over a lake, it exerts a shearing stress at the air-water interface, in consequence of which the water is accelerated. When a steady wind has been blowing for some time on a lake, this movement of water will result in the piling up of water in one part of the basin and a lowering of surface level in another part. The slope so developed will cause a gradient current to start flowing, and the current system will come to equilibrium when the over-all transport due to the wind drift is balanced by the return transport due to the gradient current. The difference between the level or equigeopotential surface and the observed surface is termed by Hellström the *wind effect*; the term wind denivellation is perhaps less vague. It is positive in some parts of the lake and negative in others, the line separating the two areas being the *nodal line*. In engineering practice the difference between the extreme observed levels, negative and positive, is spoken of as *set-up*.

Since the geostrophic terms in the equations of motion are responsible for a very remarkable type of current system in sufficiently large bodies of water, it is convenient to begin our discussion of the wind drift with a consideration of an unlimited body of water. The probable relevance of conclusions based on such a theory for very large lakes will be apparent. In most other cases a simplified treatment without geostrophic terms will be sufficient.

Wind drift and the Ekman spiral. At the surface the stress that the wind exerts on the water must be equal to the stress the water exerts on the wind. We may write the latter as (equation 21)

$$\tau_{ax} = -\mathrm{A}_v\left(\frac{du}{dz}\right) \qquad \tau_{ay} = -\mathrm{A}_v\left(\frac{dv}{dz}\right)$$

If we may assume an unlimited surface, no piling up of water can take place, the water surface remains level, and no pressure term occurs in the equations of motion. With a steady wind and current

[11] Certain large lakes, such as Great Slave Lake, and particularly some of the larger closed bodies of water, such as Lake Balkhash, exhibit considerable horizontal variations in chemical composition; the hydrodynamic consequences of this type of variation appear to be uninvestigated.

(34a) $$2\Omega \sin \phi v + \frac{d}{dz}\left(A_v \frac{du}{dz}\right) = 0$$

(34b) $$-2\Omega \sin \phi u + \frac{d}{dz}\left(A_v \frac{dv}{dz}\right) = 0$$

If we assume that τ_a is measured along the y axis, that A_v is invariant with respect to z, and that the water is so deep that the effect of the bottom can be neglected, we can obtain by integration and appropriate consideration of the boundary conditions[12]

(35a) $$u = v_0 e^{-\pi z/D} \cos (45^\circ - \pi z/D)$$

(35b) $$v = v_0 e^{-\pi z/D} \sin (45^\circ - \pi z/D)$$

(36) $$v_0 = \frac{\tau_a}{\sqrt{A_v 2\Omega \sin \phi . \rho}} = \frac{\pi \tau_a}{D\Omega \sin \phi . \rho \sqrt{2}}$$

where D has already been defined (equation 24) as

$$D = \pi \sqrt{\frac{A_v}{\Omega \sin \phi . \rho}}$$

Equations 35a and 35b mean that at the surface the current sets 45 degrees to the right of the wind. Its velocity decreases logarithmically with depth, and at $z = D$ it is flowing in the direction opposite to the direction at the surface. The depth D is commonly called the "depth of frictional resistance" (page 257), since practically all the flow occurs above this depth.

The theoretical results outlined above were derived by Ekman (1905) to explain the observations of the drift of polar ice to the right of the wind recorded by Nansen on the Fram during 1893 to 1896.

Assuming that the stress of the wind is proportional to the square of the wind velocity

(37) $$\frac{v_0}{W} = 3.2 \times 10^{-6} \frac{\pi W}{D\rho\Omega \sin \phi \sqrt{2}}$$

Ekman, however, concluded empirically that

(38) $$\frac{v_0}{W} = \frac{0.0127}{\sqrt{\sin \phi}}$$

which implies a current velocity of 1.8 per cent of the wind speed at 30° lat. and 1.4 per cent at 60° lat., and thence

(39) $$D = \frac{7.6\ W}{\sqrt{\sin \phi}}$$

---

[12] The mathematically minded novice should consult Ekman's (1905) paper, which gives a masterly presentation; Sverdrup, Johnson, and Fleming (1942) give a good summary.

For winds of less than about 600 cm. sec.$^{-1}$, a better empirical result (Thorade 1914) is

$$D = \frac{3.67\sqrt{W^2}}{\sqrt{\sin \phi}} \tag{40}$$

It is obvious that these relationships cannot hold very close to the equator. As has already been indicated, in most lakes D is between 20 and 100 m.

The initial assumption that the eddy viscosity is constant vertically, though not impossible in homogeneous water, is probably an oversimplification. Fjeldstad (1929) has examined the case of water in which the eddy viscosity falls to zero in the bottom layer. For such a case he concluded that A_v varied very nearly as the $^3/_4$ power of the distance from the bottom. An eddy viscosity gradient of this sort produces a marked increase in the angle of deviation between current and wind, so that very marked spirals with deviations at the surface in excess of 45 degrees can be produced in quite shallow water. Fjeldstad's results at least suggest that quite well-marked modified Ekman spirals may occur in lakes of sufficient area even if the depth is not great.

Rossby and Montgomery (1935) have examined a model in which the eddy viscosity is low at the immediate surface, rises rapidly to a maximum, and then falls off proportional to the depth. This model, which they believe to be more realistic than the original model of Ekman, gives an increasing angle of deviation with increasing wind velocity and increasing latitude. At latitude 5° and a wind of 500 cm. sec.$^{-1}$, the angle is 35 degrees; at latitude 60° and a wind of 2000 cm. sec.$^{-1}$, the angle is 52.7 degrees. The wind factor or ratio v_0:W decreases with latitude and wind velocity.

If the surface water consists of a layer of depth z_1, lying over a denser layer of indefinite depth, the Ekman spiral develops, and the angle of deviation depends on the relation of z_1 to D. If $z_1 \gg D$, the angle approaches 45 degrees; if $z_1 = D$, the angle is 34 degrees; for $z_1 \ll D$, the current and wind directions are practically identical (cf. Proudman 1953, p. 181).

Ekman has also examined a number of special cases, of which the one of greatest limnological interest is that of an enclosed water mass of finite depth. In such a case the surface does not remain level, and a slope or gradient current, as well as the wind drift, must be running at equilibrium. The effect of the gradient is to produce a uniform current at all depths, but as this runs over the bottom, the shear produced by the latter exerts a force on the bottom of the lake qualitatively comparable to the wind at the surface. A second Ekman spiral therefore develops near the bottom. If $z_m > 2\,D$, we can distinguish, between the two spirals, a region of homogeneous gradient current flowing to the left of the wind. As the ratio of z_m to D decreases, the angle of inclination of the surface current to the wind decreases, the spiral at the bottom is likewise largely suppressed and we have a wind drift at the surface almost along the direction of the wind, and a region near the bottom of current flowing against the wind, the so-called return current. The velocity of this current falls to zero at the bottom. The current patterns calculated by Ekman for three values of $z_m : D$ are given in Fig. 74. Ekman obtained the very remarkable result that although in a deep lake the current may set at a considerable angle to the wind, the direction of inclination of the sloping water surface hardly deviates from the wind direction. When $z_m : D = 0.5$, the slope is hardly reduced (98 per cent) below its value in the absence of geostrophic forces; when $z_m : D = 1.25$, it is 77 per cent; and as the ratio approaches infinity, the slope approaches 66.7 per cent of its value in the absence of the effect of the earth's rotation. In most cases in limnology, therefore, only the effect of Coriolis forces on current direction is likely to be of any importance.

Empirically it is interesting that Witting (1909) found that the mean deviation of current from wind direction in a series of 109 observations from a lightship anchored in only 9 m. of water in Lake Ladoga, was 33 degrees to the right. Witting does not give a mean wind velocity, but indicates that his data suggest a proportionality between wind and current velocities of the form

$$v = 0.48\sqrt{W} \tag{41}$$

From the mean current velocity at the station in Ladoga, 8.7 cm. sec.$^{-1}$, we may obtain a mean wind speed of 310 cm. sec.$^{-1}$ From equation 40, which relates D to W at low wind speeds, and from the latitude of the station, 60°36′ N., we obtain for D a value of 21.5 m. The observed deflection is therefore excessive; no deflection would in fact be expected from Ekman's theory, since $z_m < 0.5\ D$. Observations in deeper water in the Baltic generally gave a smaller deflection than the determinations in Lake Ladoga. It is possible that Witting's observations are explained by Fjeldstad's theoretical studies of the effect of a gradient in eddy viscosity in increasing the angle of divergence between wind and current. It may be noted that Witting found in general, considering all his stations, that the angle between the wind and current decreased with wind and current velocity. This finds a natural explanation in the decrease in the ratio $z_m : D$ as D, which depends on these velocities, increases. Olson (1950, 1952) found no significant deviation on Lake Erie; Verber, Bryson, and Suomi (1953) recorded a small but definite mean deviation on Lake Mendota.

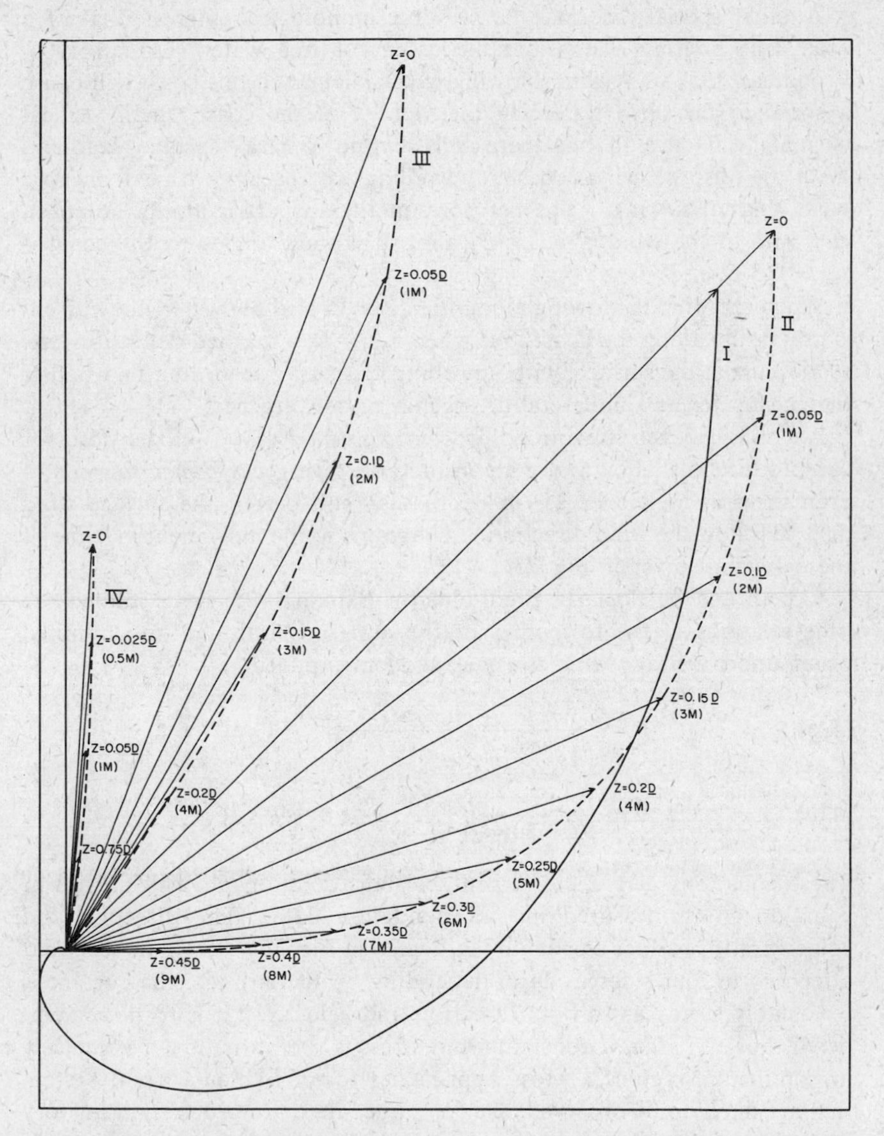

FIGURE 74. The Ekman spiral, or envelope of the ends of the lines representing current vectors, projected onto the x,y plane. I, the unmodified spiral in an ocean of indefinite area and depth; II, for an ocean of indefinite area, depth $0.5D$; III, the same for depth $0.25D$; IV, the same for depth $0.1D$. Arrows indicate the directions and relative velocities of the water movements in terms of D. If D is taken as 20 m., a reasonable minimum for lakes, the depths are those given in parentheses (meters). Wind direction is upward, parallel to the ordinate (Northern Hemisphere). Models of this sort should be regarded as limiting cases in limnology, though possibly more applicable than those of Fig. 75 in lakes of great area, such as Lake Ladoga, in which more deflection is observed than would be expected in a basin of limited area.

Another special case may be of some limnological interest, that of a coast line limiting a considerable extent of free water (Ekman 1905, Proudman 1953). A wind blowing perpendicular to this coast will cause a spiral in the open water; in the vicinity of the coast, small vertical currents will develop but there will be no general surface gradient. With an offshore wind, coastal upwelling will occur; with an onshore wind, coastal sinking. The net horizontal transport is along the coast, *cum sole* to the wind if $z_m > D$, almost at right angles to the coast if $z_m = 0.5\, D$.

When the wind is blowing along the coast in shallow water, the current is practically along the coast, but when $z_m > D$, a marked deflection *cum sole* occurs at the surface, with upwelling or sinking according to whether the current forms a unilateral divergence or convergence.

An obliquely set coast trending *contra solem* away from the observer can produce in shallow water a considerable divergence *contra solem*, and even in water well over $2D$ in depth may still permit the surface wind drift to follow the wind direction. Diagrams of the movement in some of these cases are given in Fig. 75.

Ekman and particularly Fredholm (in Ekman 1905) have studied the time relations of the formation of the current spiral. In an unlimited ocean under a steady wind stress τ_a, suddenly applied,

$$u = \frac{\pi\tau_a}{\rho D\Omega \sin\phi} \int_0^{t_p} \frac{\sin 2\pi\zeta}{\sqrt{\zeta}} e^{-\pi z^2/4D^2\zeta}\, d\zeta \tag{42a}$$

$$v = \frac{\pi\tau_a}{\rho D\Omega \sin\phi} \int_0^{t_p} \frac{\cos 2\pi\zeta}{\sqrt{\zeta}} e^{-\pi z^2/4D^2\zeta}\, d\zeta \tag{42b}$$

These equations give a convergent periodic solution, in which t_p as in equation 28 is measured in half pendulum days. The direction and velocity of the surface current are indicated in Fig. 76. It will be seen that after two to four sidereal days, depending on the latitude, the current is reasonably steady as to both direction and velocity. It is to be remembered, however, that a constant wind stress is rare in nature, particularly in continental regions. Most approaches to an Ekman current system that are likely to be observed, even in large lakes, are probably transitory phases in the development of the steady theoretical pattern. Ekman's investigations indicate that in a closed body of water the development of the wind denivellation or set-up is likely to be very rapid.

Wind drift in a long lake of moderate depth: wind stress and current velocity. If the depth is less than $0.5D$ and the wind is blowing along the axis of the lake, which will be taken as the x axis, we may neglect the geostrophic components or indeed any components along the y axis. The

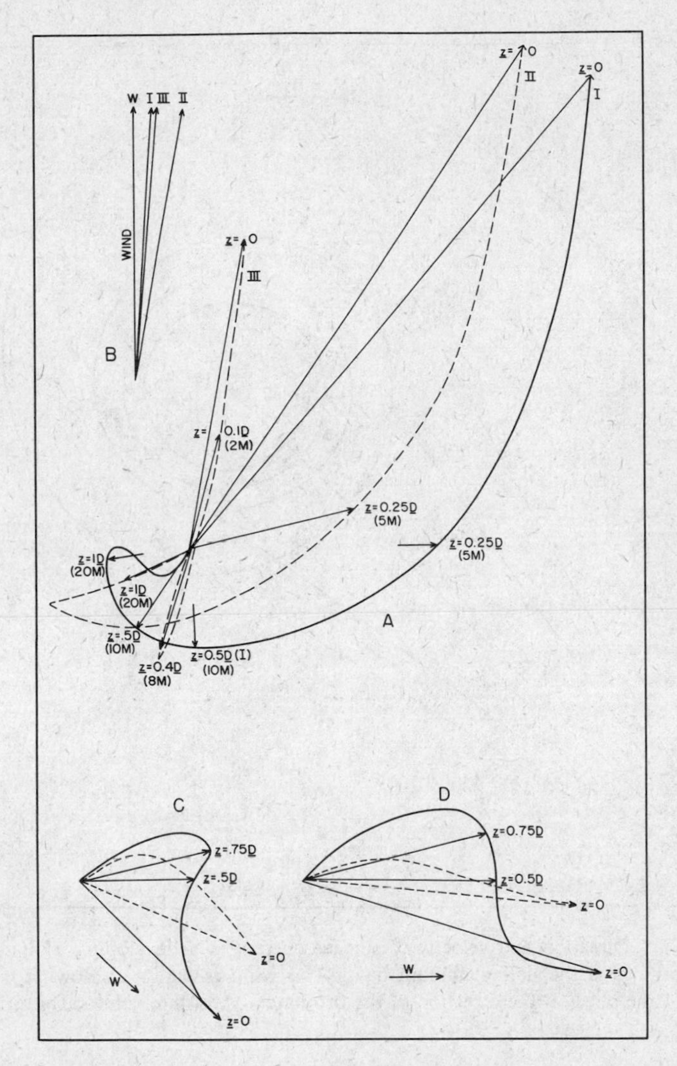

FIGURE 75. *A*, the Ekman spiral in a completely bounded lake, permitting the development of wind drift denivellation or set-up. I, the otherwise unmodified spiral for depths $>2.5D$; II, the same for depth $1.25D$; III, the same for depth $0.5D$. Note that at the last named depth the geostrophic effect is almost undetectable. Conventions as in Fig. 74. *B*, wind direction, *W*. I, II, III, directions of slope of wind denivellations for cases I, II and III, the amount of denivellation being respectively 67, 77, and 98 per cent of that which would develop in the absence of the geostrophic effect. *C*, Ekman spirals for maximum depths $0.5D$ (broken) and $1.25D$ (solid) off a coast trending 45 degrees *contra solem* from wind direction; note in shallow water the current direction is moved *contra solem* from that of wind, while in deep water the current flows with the wind. *D*, longshore wind with current in deep water rotated slightly *cum sole*, producing a unilateral convergence.

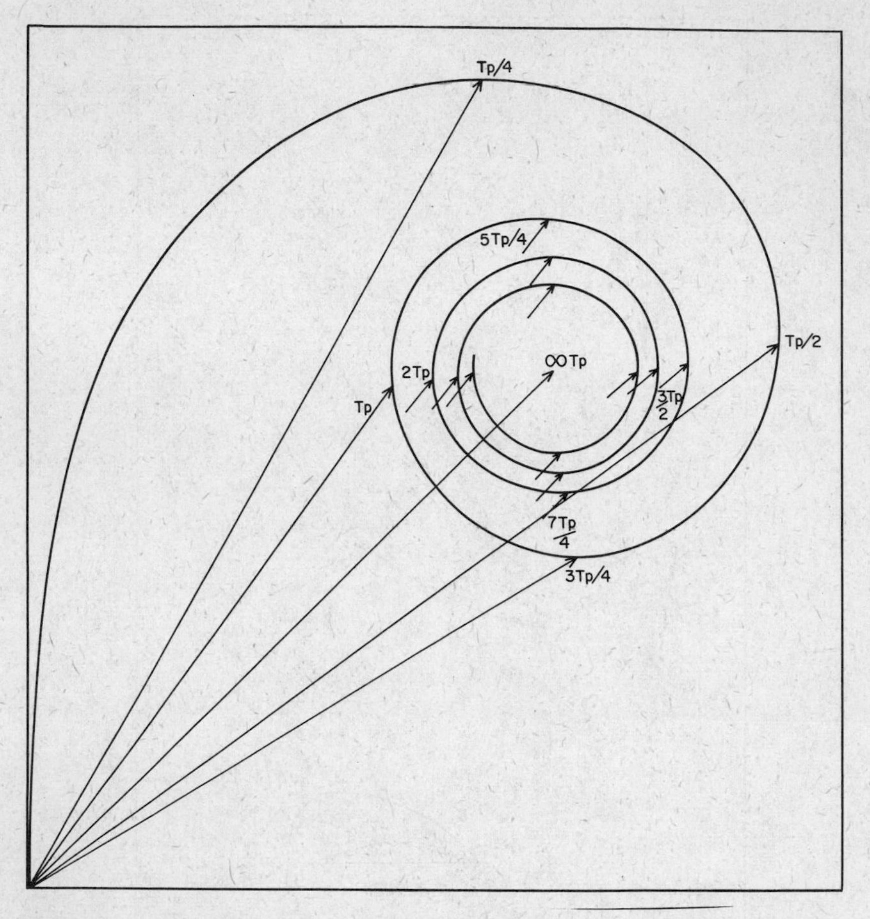

FIGURE 76. Direction and velocity of surface current in a developing Ekman spiral; unit of time t_p is the half pendulum day. The wind is taken as blowing from the bottom of the page, in the direction of the ordinate. Absolute values arbitrary.

equation of horizontal motion now takes the form (cf. Sverdrup 1942, p. 121)

$$g\rho i_x + \frac{d}{dz}\left(A_v \frac{du}{dz}\right) = 0 \tag{43}$$

where i_x is the inclination of the water surface along the x axis.

Taking A_v as constant, we have, analogous to equation 21a,

$$\tau_b = A_v\left(\frac{du}{dz}\right)_{z_{b\alpha}} \tag{44}$$

where τ_b is the stress on the bottom. The total stress is the integral of the second term of equation 43, so that

$$\tau_a + \tau_b = -\int_0^{z_b} g\rho i_x \, dz \tag{45}$$

which for homogeneous water becomes

$$\tau_a + \tau_b = -g\rho i_x z_{bx} \tag{46}$$

or writing $c_s = 1 + \frac{\tau_b}{\tau_a}$

$$\tau_a = \frac{-g\rho i_x z_{bx}}{c_s} \tag{47}$$

In this expression, though the eddy viscosity has conveniently disappeared as an explicit quantity, it has become necessary to know in its place the ratio of bottom to surface stress. It is important to note that the total depth z_{bx} is now a function of x even if the bottom is level, and consists of the depth z'_{bx} at x when the surface is level, plus or minus the wind denivellation h_x. In practice it is usually not necessary to consider this variation in depth, which ordinarily is very small.

If the flow were laminar, c_s would be 1.5, and for turbulent flow in general $1 < c_s < 1.5$ (Hellström 1941). Ekman deduced that for uniform eddy viscosity $c_s = 1.46$; Hellström made his calculations with $c_s = 1.5$; and Keulegan (1951) with $c_s = 1.25$. The only attempt at an empirical determination in nature is by Van Dorn (1953), who in a rectangular pond 240 m. long, mean depth 1.84 m., estimated the stress directly on a glass plate covered with sand placed on the bottom of the pond. He concluded that $1.0 < c_s < 1.1$, and that in general the bottom stress was of little or no importance. From an experiment in a tank, Francis (1951) reached a like conclusion.

Whenever a series of measurements of wind denivellation and of wind velocity under really steady conditions are available, it should be possible to evaluate τ_a from equation 44, at least within the limits of uncertainty of c_s. In this way the relationship between wind stress and wind velocity should be determinable.

Most investigators have concluded that the stress varies as some power of the wind velocity or, more accurately, of the wind velocity relative to the water, as

$$\tau_a = \gamma^2 \rho_a (W - v_0)^n \tag{48}$$

Where γ^2 is defined as a coefficient of resistance. In general v_0 is small compared to W and may be neglected. Taylor (1916) concluded that for turbulent air $n = 2$ and

$$\tau_a = \gamma^2 \rho_a W^2 \tag{49}$$

Ekman obtained empirically, from a study of data assembled by Coldring (1876) relating to denivellation of the Baltic during a severe storm,

(50) $$\tau_a = 3.2 \times 10^{-6} W^2$$

Palmén and Laurila (1938) found practically the same relationship from more recent observations, and Rossby has concluded from aerodynamic theory and measurements, without considering the water movements at all, that this relationship with the coefficient 3.2×10^{-6} holds for moderate or strong winds.[13] Variations in the constant recorded by various investigators may be due to variations in anemometer height. Not all empirical studies, however, give results conforming to equation 49. Both Hellström (1941) and Francis (1951) found in laboratory tanks proportionality to W^3 rather than W^2. Hayford's (1922) elaborate series of observations on Lake Erie seemed to imply that the stress is proportional to $W^{2.4}$. Hellström, who also had a great range of field observations at his disposal, concluded that for these the stress varied as $W^{1.8}$. His expressions, if it is assumed[14] that $c_s = 1$, becomes for a 6-m. anemometer level

(51) $$\tau_a = 9.3 \times 10^{-6} W^{1.8}$$

Bryson, Suomi, and Stearns (1952) have used an ingenious method on Lake Mendota, which differs from those of other investigators. If three floats on drags are placed in the surface water, of such a size that their movement records the mean velocity $\bar{u}$ of the top z cm., and the area of the triangle defined by these floats is measured at two time intervals, if $\bar{A}$ is the mean area and ΔA is the increment in area of the triangle in time Δt, and w_z is the velocity at z, it can be shown that

(52) $$w_z = -\frac{1}{\bar{A}}\frac{\Delta A}{\Delta t}z$$

and that

(53) $$\tau_a = 2\rho\bar{u}w_z$$

In the observations on Lake Mendota, the current drags used measured the mean velocity over the top 10 cm., so that w_z is the vertical velocity at that depth. Most of the determinations using this method were made on smooth water. A method based on the rate of damping of seiches was better adapted to rough water. The values for rough water follow

[13] Aerodynamic theory leads to a more complicated form for a smooth water surface (cf. Sverdrup, Johnson, and Fleming 1942).

[14] Hellström used $c_s = 1.5$; the closeness of the curves for equations 50 and 51 may reflect agreement of empirical data with Rossby's theory, but this, as it stands, is accidental, since different values of c_s are involved.

essentially a law of the form deduced by many previous workers, the stress varying with W^2, but on smooth water Bryson, Suomi, and Stearns concluded that

(54) $$\tau_a = 1.5 \times 10^{-4} W^5$$

W being given in cm. sec.$^{-1}$ and measured 5 m. above the water surface.

Munk and Anderson (1948) have presented in graphic form, for low and moderate wind velocities, the investigations of T. Saur, which are said to consider the roughness of the water surface and gustiness of the wind. Except at low wind velocities, their curve gives slightly higher results than does equation 50 but is very close to that of equation 51. Van Dorn (1953), however, in the latest investigation of the matter, following in part Keulegan (1951), concludes that no monodic function of the form of equation 49 can be used at moderate or high wind velocities. Van Dorn believes the best estimate of stress can be obtained from an equation of the form

(55) $$\tau_a = \gamma_1{}^2 \rho_a W^2 + \gamma_2{}^2 \rho_a (W - W_c)^2$$

in which the second term is disregarded when $(W - W_c)$ is negative. All the constants vary with the anemometer level; for an anemometer at 10 m., the equation becomes

(56) $$\tau_a = 1.21 \times 10^{-6} W^2 + 2.25 \times 10^{-6} (W - 5.6)^2$$

This gives considerably smaller results than does equation 55. If W_c is very small but the second term is still retained, equation 56 becomes practically the same as Ekman's expression (equation 50). Keulegan's work suggests that this is what happens on large bodies of water, where he supposes that waves form at much lower wind velocities than on small ponds. It is therefore best to obtain the approximate wind stress, when required, from one of the upper curves of Fig. 77, except when dealing with very small bodies of water.

Langmuir (1938) has made two computations of the rate of transfer of momentum from wind to water in Lake George, by measuring the current velocity at different depths at the beginning and end of a period of relatively constant wind. Simply by integration from the surface to the depths at which no change in velocity occurs, in the present case about 6 m., the change in momentum can be estimated. Langmuir found that on July 27, 1929, a wind of approximate velocity 500 cm. sec.$^{-1}$ delivered 7000 g.-cm. sec.$^{-1}$ momentum to each cm.^{2} of the lake surface in three hours, corresponding to a force of 0.65 dynes cm.$^{-2}$ This is in rough agreement with Munk and Anderson's curve. On another occasion, however, when a shift in wind direction caused a marked change in current direction, it appears that 8300 g.-cm. sec.$^{-1}$ cm.$^{-2}$ were delivered by a wind of velocity

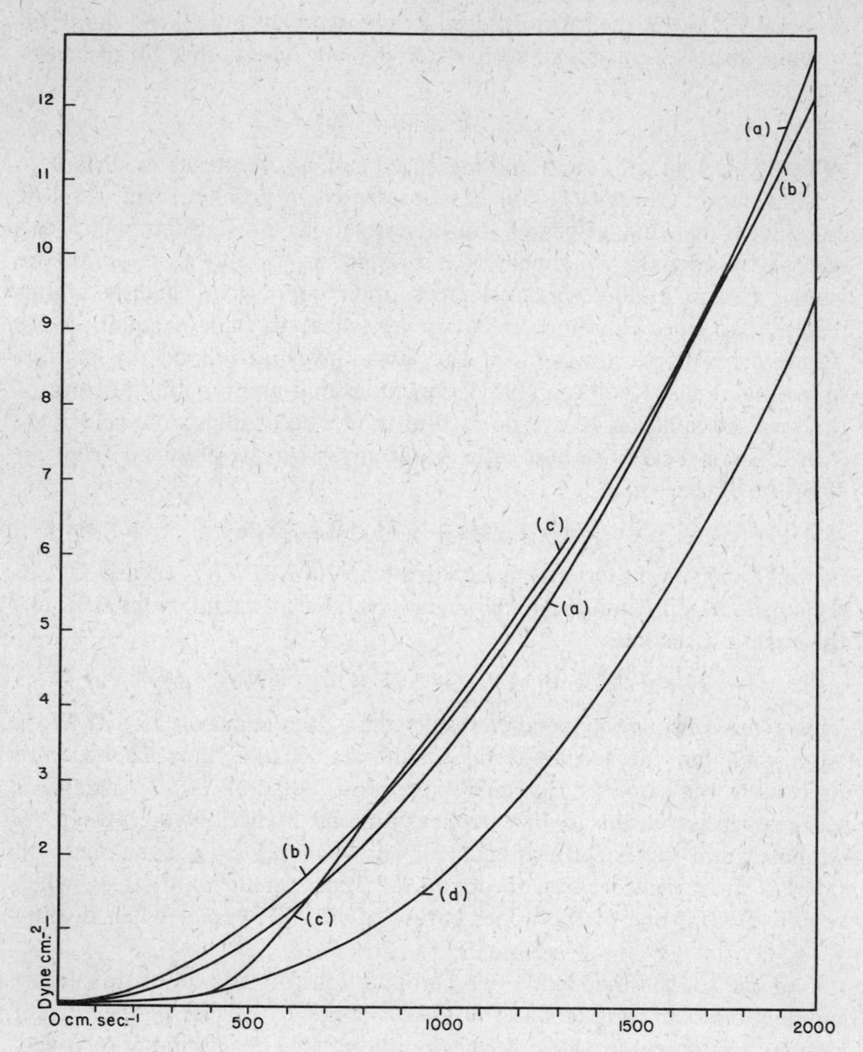

FIGURE 77. Relation of wind stress τ_a to wind velocity W, based on (a) equation 50, (b) equation 51, (c) Saur's computions after Munk and Anderson, (d) equation 56.

falling from 400 to 330 cm. sec.$^{-1}$ in 22 minutes, corresponding to a force of 6.3 dynes cm.$^{-2}$ Such an estimate is quite discordant with any others. Langmuir pointed out that the rate of transfer of momentum falls rapidly with a steady wind, and supposed that once wave trains are set up they carry energy faster than the current and dissipate it against the banks of the lake. This is not in accord with the observations of Keulegan (1951) and

Van Dorn (1953) that the surface velocity is the same for a given steady wind in the presence, or artificial absence, of waves.

Velocity of the wind drift. If the flow in a lake were laminar, it can be shown that

$$u_{x,z} = u_{x,0}\left[3\left(\frac{z_u - z}{z_u}\right)^2 - 2\left(\frac{z_u - z}{z_u}\right)\right] \tag{57}$$

This implies that in immediate contact with the bottom there will be no current; at a depth two thirds of the maximum, the return upwind current will be flowing with its maximum velocity ($-\,{}^1/_3 u_{x,0}$); and that at a depth one third of the maximum, there will be no current. The plane of this depth is often called the *plane of shear.* Considerable departure from this scheme may be expected in turbulent water, though some trace of the laminar current pattern will be retained.

Hellström (1941), working on a somewhat insecure theoretical foundation, has derived a complicated expression for the velocity distribution in an unstratified turbulent lake, which leads to the typical results presented in Table 21.

TABLE 21. *Computed velocity distribution in an unstratified lake as a function of wind stress and of depth*

τ_a, dynes cm.$^{-2}$		2			10	
W,* cm. sec.$^{-1}$		750–900			1870–2250	
z_u, m.	1	10	100	1	10	100
Surface current, cm. sec.$^{-1}$	10	13	17	23	30	37
Maximum return current, cm. sec.$^{-1}$	−4	−5	−7	−8	−11	−15
Bottom current, cm. sec.$^{-1}$	−3	−4	−6	−6	−10	−14
Depth of plane of shear, m.	0.27	3.92	40	0.27	3.92	40

* The lower figure gives the value from Munk and Anderson's curve, the higher from Hellström's calculation.

The plane of shear is lower than in the laminar case, clearly approaching 40 per cent of the maximum depth. The return current extends but little diminished even to the bottom, and all current velocities increase with increasing values of z_u.

Hellström concludes that the figures for the surface current are in accord with observations, particularly those made in the Baltic. Most investigators of the open ocean, moreover, have concluded that the current velocity lies between 1 and 2 per cent of that of the wind, some of the variation being due to the latitude. Theoretical studies by Rossby and Montgomery suggest values about 2.3 to 3.2 per cent, dependent on latitude and wind velocity.

In Lake George, Langmuir found that after a windless night a current of 2 to 3 cm. sec.$^{-1}$ could be detected; after a wind of 400 to 800 cm. sec.$^{-1}$ measured at a height of 2 m. had been blowing, a current of velocity 10 to 20 cm. sec.$^{-1}$ was generally flowing downwind. This implies, if the wind drift is corrected for inertia current or whatever type flows in dead calm, that the wind drift has a velocity of about 2 per cent of the wind. On some occasions much greater velocities than are implied by this relationship occur; the current velocity may even be 10 per cent of that of the wind. In normal circumstances, however, it is clear that the wind factor or ratio v_0:W for Lake George is not very different from that observed in the ocean. On Lake Erie, Olson (1952) likewise obtained a mean wind factor of 2 per cent; the much higher values, with a mean of 10 per cent, obtained by Diénert and Guillerd (1949) on Lake Geneva may be due to the wind influencing their drift bottles directly.

Recently Keulegan (1951) in his experimental tank and Van Dorn (1953) in his artificial pond have made a number of observations which are likely to be of fundamental importance in limnology.

Keulegan concluded that if the ratio v_0:W be plotted against Reynolds number R_e, in which the depth of the tank, z_u, supplies the linear dimension, up to $R_e \simeq 1000$

$$\frac{v_0}{W} = 7.6 \times 10^{-4} \left(\frac{v_0 \rho z_u}{\mu}\right) \tag{58}$$

or

$$v_0 = 57.8 \times 10^{-8} \frac{\rho z_u W^2}{\mu} \tag{59}$$

At values of $R_e \geqslant 2000$ the ratio becomes a constant, so that v_0 is given by 0.033W, the velocity being independent of the depth of the water in the tank.

Since in water of viscosity 0.01 and depth 1 m., $R_e = 2000$ implies a water velocity of only 0.2 cm. sec.$^{-1}$, and so a wind velocity of only 6 cm. sec.$^{-1}$, it is evident that in limnology we usually may take $R_e > 2000$. Van Dorn, in his shallow pond, has obtained very striking confirmation of Keulegan's conclusions. It is to be noted that the relationship obtained for $R_e > 2000$ implies, for any given wind, a somewhat greater velocity than is postulated by Hellström or found in Lake Erie or Lake George under normal conditions. Though Hellström may be correct that in passing, say, from water of depth 10 m. to that of depth 100 m. the velocities increase at all levels, it is also conceivable that in passing from 1 m. to perhaps

10 m. they decrease. The possible meaning of this will be apparent later. It is a very remarkable empirical fact, established by both Keulegan and Van Dorn, that though the production of waves can be inhibited by detergents, such treatment has no significant effect upon the surface current produced by a given wind. Van Dorn, however, concluded that when waves are not present, there is a very rapid drop in velocity below the surface, whereas in the presence of waves the top 30 cm. or so appeared to have a fairly uniform mean velocity. The waves presumably transfer momentum to the layers immediately below the surface. Their apparent lack of influence on velocity may be the result of a greater transport of momentum across a wavy than a smooth surface, and at the same time a transport of this momentum into a greater depth of water. The balance between these two factors might in fact be fortuitous.

Helical structure of the wind drift. Langmuir (1938) has pointed out that in Lake George, as in the ocean and probably in all but the smallest, shallowest bodies of water, the wind drift is not a uniform movement on which random turbulent velocities are superimposed, but has a definite structure. The water flows in a series of parallel helices, their long axes along the wind, the helices being alternately clockwise and counterclockwise. The surface of the lake is thus marked by linear divergences, with a counterclockwise helix on the left, a clockwise on the right looking downstream, alternating with convergences. Leaves, oil films, and the like collect in the convergences, which appear as, and are often called, *streaks.* In November, when the lake is freely circulating to a considerable depth, the convergences are about 25 m. apart; in June, when the thermocline is becoming established and the isothermal epilimnion is thin, they are but 5 to 10 m. apart. With a horizontal wind velocity of 600 cm. sec.$^{-1}$ and a horizontal surface current of 15 m. sec.$^{-1}$, the downward descent in a convergence was found to be 1.6 cm. sec.$^{-1}$ In other cases, when vertical movement was measured in both convergences and divergences at the same time, the downward movement in the former had a velocity of 2 to 3 cm. sec.$^{-1}$, or just about three times the upward velocity in the divergences.

Observations made by placing a cord with floats across the current system showed that the horizontal velocity of the wind drift is greater in convergences than in divergences. There is evidently a transfer of momentum downward at the convergences; the helices probably become less and less definite with depth, but they serve to transfer momentum from the surface all through the freely circulating part of the lake. It is therefore possible that in very shallow water less momentum can be transferred downward, giving a v_0:W ratio or wind factor somewhat greater than in deep water, as Van Dorn found on his shallow pond.

In the open ocean there is evidence (Woodcock 1944) of a geostrophic effect on the helices, the clockwise series in the Northern Hemisphere and the counterclockwise in the Southern Hemisphere being larger than the opposing helices, so that the convergences do not lie in the center between divergences. The very extraordinary adaptation of *Physalia*, the Portuguese man-of-war, which is asymmetrical and forms antimeric populations north and south of the Equator, is related to this phenomenon. As Woodcock, who discovered the antimerism, has shown, the two forms drifting downwind are adapted in their respective hemispheres to stay in recently upwelled water rich in plankton. Although it is very unlikely that any biological consequence of the helical movement of the wind drift as spectacular as this will be discovered in lakes, the general effect of this kind of movement is likely to be of importance in the life of the plankton. Stommel's study of the theory of such an interaction is discussed in Volume II.

Form of the wind-denivellated water surface. Returning to equation 47, we may write

$$\frac{dz_{bx}}{dx} = \frac{c_s \tau_a}{g \rho z_{bx}} \tag{60}$$

writing z'_{bx} as the total depth from the level water surface to the bottom at x, the wind denivellation h_x will be

$$h_x = z_{bx} - z'_{bx} \tag{61}$$

and the set-up S_h will be, in a lake of length l,

$$S_h = |h_0| + |h_l| \tag{62}$$

If z'_b is constant and great with respect to S_h, the surface of the lake is an inclined plane, and

$$h_x = \frac{c_s \tau_a}{g \rho z'_b}\left(x - \frac{l}{2}\right) \tag{63}$$

Or if $c_s = 1$ and $\rho = 1$, and τ_a be given by equation 50,

$$S_h = \frac{3.2 \times 10^{-6}}{g z'_b} W^2 l \tag{64}$$

Volker (1949) finds good agreement on Lake Ijssel, the artificial lake on the site of the Zuider Zee, if the coefficient is slightly higher, namely 3.5×10^{-6}.

If, however, we are dealing with a shallow lake, in which by equation 60 the gradient—and so, for any x, the value of h_x—will be greater as the total depth is less, we cannot neglect the effect of wind denivellation on the right-hand side of the equation. Any longitudinal section of the lake surface, which is now convex, can be expressed as a parabola. The full mathematical treatment is given by Hellström (1941), who has also treated some types of basin in which the bottom is not at a uniform depth. In cases in which the bottom slopes upward downwind, concave surfaces can be obtained with sufficient wind stress. Hellström has confirmed his

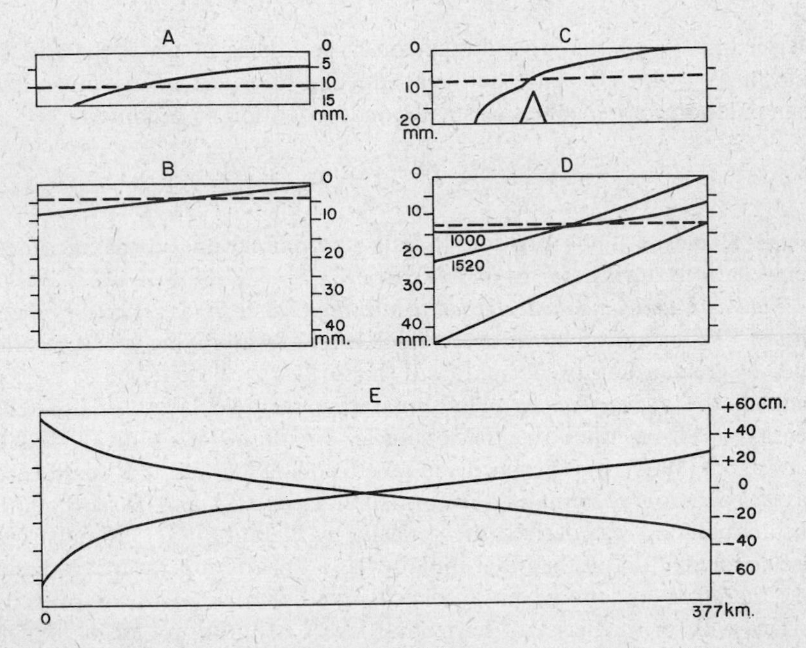

FIGURE 78. Form of water surfaces under wind stress. *A*, parabolic surface in a very shallow (5 mm.) experimental tank, W = 860 cm. sec.$^{-1}$. *B*, practically plane tilted surface in a deep vessel, W = 850 cm. sec.$^{-1}$. *C*, distortion produced by a barrier. *D*, concave surface produced by a sloping bottom with W = 1000 and 1520 cm. sec.$^{-1}$. *E*, computed surfaces for Lake Erie with east or west winds; note greater denivellation at shallower western end. (After Hellström.)

theoretical conclusions with models in the laboratory, some of his experimental results being presented in Fig. 78.

In a shallow lake the denivellation of the water surface may lead to a considerable area of the bottom being exposed during storms, a phenomenon well shown in Lake Okeechobee (Fig. 98). Hellström also elaborated a method of computing the denivellation in a lake of fairly irregular bottom profile by dividing the lake into a number of sections. His computation for the denivellation of Lake Erie with eastern or western winds, based on the data of Hayford,[15] is given in *E* of Fig. 78. Keulegan (1951) has criticized Hellström's exhaustive work on the ground that the relationship between the gradient of the surface and the wind stress cannot, except at very low stresses, be expressed as a monodic equation of the form of equation 60. Keulegan used an expression of the form

$$\text{(65)} \qquad \frac{dz_b}{dx} = 3.30 \times 10^{-6}\frac{W^2}{gz_b} + 2.08 \times 10^{-4}\frac{(W - W_c)^2}{gz_b}\left(\frac{z_b}{l}\right)^{1/2}$$

Neglecting W_c on the grounds that on large bodies of water the critical velocity W_c, which depends on the production of waves, is very much less than in laboratory channels or small ponds, equation 65 becomes

$$\text{(66)} \qquad \frac{dz_b}{dx} = 3.30 \times 10^{-6}\left(1 + 63\sqrt{\frac{z_b}{l}}\right)\frac{W^2}{gz_b}$$

which Keulegan finds provides a fair approximation to the recorded denivellations during storms on Lake Erie.

Wind drift and wind denivellation in stratified water in long, narrow model basins. When a wind blows over a thick layer of water lying over a second layer of greater density, not only will the surface level be raised at the lee end but the *pycnocline*, or plane separating the two layers of different density, will be tilted in the opposite direction (cf. Murray 1888). Sandström (1908), in experiments in tanks, concluded that it is possible to have circulation patterns of the form indicated in *A* and *B* of Fig. 79. Such circulation was derived theoretically by Defant (1932). Sandström also concluded that if the water initially has a linear temperature gradient, and therefore an almost linear density gradient, wind blown over its surface will form a series of horizontal layers of uniform density, which constitute circulation cells of the kind shown in the figure. There seems, however, to be considerable doubt about these observations. According to Hellström (1941), Hans Pettersson was unable to repeat them, and considered that the viscosity, either molecular or turbulent, at the pycnoclines is too low to permit any appreciable vertical transfer of momentum from one layer to the next. The numerical example given by Defant (1932) involves quite excessive values of A_v, namely, 100 cm.^{2} sec.$^{-1}$ in the upper and 20 cm.^{2} sec.$^{-1}$ (!) in the lower layer, producing a deep-water current of 1 cm. sec.$^{-1}$ Alsterberg (1927) attempted to apply a scheme of the same kind as that of Sandström, but with an indefinitely large number of elementary circulating layers, to explain the hypolimnetic water movements of lakes. It is reasonably certain that Pettersson's objection,

[15] Whose theoretical approach Hellström shows to be in part erroneous.

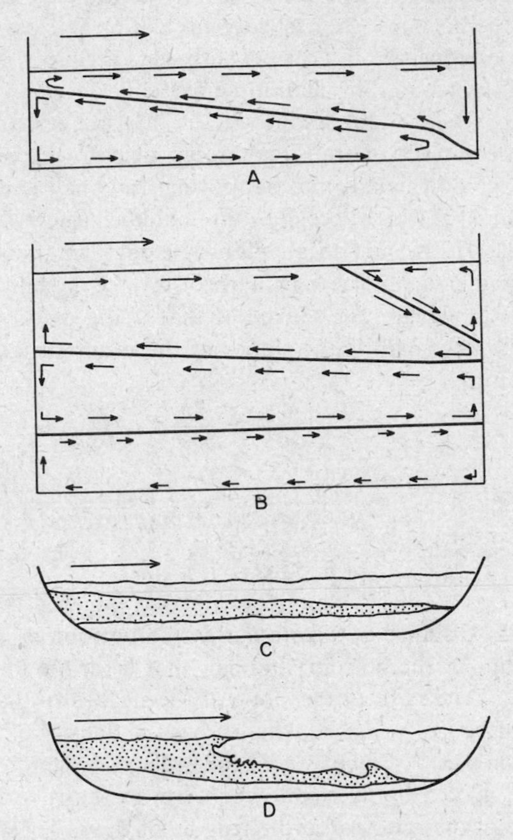

FIGURE 79. *A*, *B*, circulation pattern in stratified model, according to Sandström. *C*, *D*, steady-state positions of pycnoclines in stratified models, according to Mortimer.

reiterated by Johnsson (1949), to any such model would be particularly relevant to Alsterberg's scheme.[16]

Johnsson (1946, 1949), in a model in which a fresh-water layer over salt water formed a two-layered model with a fairly sharp density gradient between the layers, observed that as the upper layer started to pile up downwind, there was flow in the upper part of the lower layer in a direction

---

[16] The present writer, in experiments in which strong brine, dyed at one end, in a small rectangular tank, was covered with fresh water over which air was blown, has occasionally observed interdigitating streamers flowing from the dyed and undyed ends of the strong brine. The effect evidently occurs only under conditions that are hard to standardize, and more often than not cannot be observed. The matter clearly requires further study.

against the wind, and a little later, as this continued, some movement downwind at the bottom. As a steady circulation was set up, movement at the bottom disappeared. These experiments certainly give no support to the type of deep-water circulation supposedly observed by Sandström and deduced by Defant; Johnsson believes that the earlier workers were misled, by observations of transitory adjustment to the piling-up of the least dense layer downwind, into supposing that they had set up a true steady-state model in which accelerations could be neglected.

Hellström (1941) has made an elaborate mathematical study of the equilibrium gradients observed in a three-layered system over which a steady wind is blowing. He concludes that if the eddy viscosity in the middle layer is very small,[17] the slopes of the water surfaces and pycnoclines are given by

(67a)
$$\frac{dz_x}{dx} \simeq \frac{2c_s\tau_a}{3g\rho_1 z_1}$$

(67b)
$$\frac{d(z_x - z_{1,x})}{dx} \simeq -\frac{2c_s\tau_a}{3g(\rho_2 - \rho_1)z_1}$$

(67c)
$$\frac{d(z_x - z_{1,x} - z_{2,x})}{dx} \simeq 0$$

where $z_{0,x}$ is the distance at x from surface to bottom, $z_{1,x}$ that from the upper pycnocline to the bottom, and $z_{2,x}$ that from the lower pycnocline to the bottom. The ratio of the slope of the surface to that of the upper pycnocline will be $(\rho_1 - \rho_2):\rho_1$ where ρ_1 is the density of the upper and ρ_2 of the middle layer. Since ρ_1 is of the order of magnitude of unity and the difference in density is small when we are considering water, the absolute value of the ratio will ordinarily be of the order of magnitude of 100 to 1000. A denivellation measured in centimeters at the surface will correspond to one of meters or tens of meters at the upper pycnocline. It is important to note that so long as the middle layer is completely covered by the upper, the lower pycnocline remains horizontal. In view of the dependence of τ_a on the wind velocity, the tilt of the upper pycnocline will

[17] If the viscosity approaches zero in both the middle and the lower layers, three coefficients in Hellström's equations approach indeterminacy, as all the terms in both their numerators and denominators approach zero. Hellström believes that in a stratified lake the thermocline represents a region of minimal eddy viscosity. Most other authors who have considered the matter are inclined to agree. There is, however, no clear empirical evidence that this is the case, and some observations (page 475) suggest the contrary. We have already seen (page 256) that such a minimum does not inevitably follow from theory. Hellström's results appear to be confirmed experimentally, at least qualitatively, by Mortimer, though in his three-layered model there is no evidence of greater turbulence in the lower than in the middle layer, and probably a somewhat greater molecular viscosity exists in the latter than in the former.

vary as about $W^{1.8}$ or W^2, so that a small increase in wind velocity can produce a considerable increase in the tilt.

Mortimer (1951, 1952b) has performed a series of very elegant experiments (*C* and *D* of Fig. 79) in which the qualitative results of Hellström's theory are confirmed. His experimental system consisted of a model lake basin with glass sides, flat bottom, and sloping ends, over which blowers were placed which permitted winds of varying velocity to play over the liquid surface. The model was filled with either two or three liquids of different density. In the more complex three-layered experiments the best liquids were: at the bottom, 25 per cent glycerine ($\rho = 1.06$); in the middle, commercial cresol or phenol ($\rho = 1.04$ or 1.02 respectively); and at the top, pure water ($\rho = 1.00$). The two-layered system showed the expected piling up of water downwind. The tilted pycnocline developed complex irregular waves when a strong wind was blowing. In the three-layered system the same phenomena occurred, but with a moderate wind the lower surface of the middle layer showed no appreciable tilting. With very strong winds the upper layer was carried downwind so much that the middle layer was exposed at the windward end. When this happened, the lower surface of the middle layer started to tilt. The upper and middle layers being miscible, the exposure of the middle layer caused it to be mixed, as wind drift, into the upper layer. This result is of great significance in relation to the heating of lakes, as will be shown later (page 449). For the present, the movement of the lower pycnocline when the middle layer is exposed at the surface, and its horizontal position when no such exposure occurs, are the most important results of the experiments.

As will be shown in a later section, it is possible to treat the metalimnion of Windermere as a middle layer of the kind used in the experiments, thereby explaining some of the properties of the internal seiches observed in that lake. It is to be noted that Mortimer appears to have found that a deformation of the lower pycnocline could be produced by a very sudden strong gust of wind before the middle layer is exposed. This is presumably not a steady-state condition, but it implies that the movements in the middle layer could not take place fast enough during the rapid rise in the wind to produce a steady state without distorting the lower pycnocline. At all times during the passage of wind, more or less complex traveling waves were produced at the upper pycnocline, and to a less extent at the lower when the middle layer was exposed to windward. These waves disappear quickly when the wind falls.

Horizontal wind drift in a shallow circular lake. Livingstone (1954) has considered the special case of a circular lake of small depth over which a wind is blowing. It is supposed that the motion is all horizontal and that a constant wind stress imposes a constant acceleration $\dot{v}$ on the water in the direction of the wind. If the radius of the lake is r and x is measured from the diameter normal to the wind direction, it can be shown that the acceleration producing a longshore gradient current at any point x, $(r^2 - x^2)^{1/2}$, on the lee shore is given by [18]

$$\dot{v}' = g\left[\dot{v}\frac{(r^2 - x^2)}{x^3}\right]^{1/2} - \dot{v}\left(\frac{r^2 - x^2}{r^2}\right)^{1/2} \tag{68}$$

It is evident that at the midpoint of the windward shore where $x = r$, $\dot{v}' = 0$; and at the points where the x axis cuts the shore and $x = 0$, $\dot{v}' = \infty$. Other forces, notably deflection by the lee shore, are certain to be involved, but since only horizontal movement is permitted, we may expect a longshore current to develop which increases in velocity towards the points at which the x axis cuts the shore. For all negative values of x, moreover, $\dot{v}'$ will be negative, and therefore windward of these points the velocity also falls and water must leave the shore. The expected circulation consists, in fact, of a pair of antimeric gyrals. Strong indirect evidence of this type of circulation is provided by the form of the elliptical lakes set across the wind in arctic Alaska. Von Arx (1948) has described a somewhat similar pattern of circulation in the lagoons of Bikini and Rongelap atolls.[19]

The two accelerating longshore return currents are analogous to the rip currents that develop in certain places on marine coasts on which waves, involving some mass transport of water, are breaking.

Vertical distribution of current velocities in moderate sized lakes. At the time of the bathymetric survey of the Scottish lochs, Wedderburn made a number of studies of currents in Loch Ness (Wedderburn and Watson 1909) and Loch Garry (Wedderburn 1908, 1910). The observations were made at a number of depths with an Ekman current meter, with which both velocity and direction could be estimated. The results obtained in Loch Ness are certainly complicated by the presence of temperature seiches of considerable amplitude, but in Loch Garry most of the movement appears to be due to the wind drift.

One series of observations in the latter lake, made in March when the water was unstratified and a strong wind was blowing from the west, appear to show a decreasing current with progressive rotation to the right (Table 22). When the lake was thermally stratified, considerable

---

[18] The first term of this expression strictly should be multiplied by a dimension constant, of value in the c.g.s. system of 1 sec. This is omitted to avoid introduction of unnecessary symbols.

[19] Ormerod (1859) described minute elliptical basins on granite surfaces on Dartmoor in England, which just possibly reflect the type of erosion produced by Livingstone's current system.

TABLE 22. *Current velocities and directions in Loch Garry when unstratified*

Depth, m.	Set	Velocity, cm. sec.$^{-1}$
0	E.	28.0
0.3	E. 22° S.	25.3
7.6	E. 50° S.	3.9
30.5	W. 50° S.	3.7
61.0	W. 10° S.	3.1

irregularity was observed; but after the wind had been blowing steadily from either the east or the west, the current in the top 5 m. or so tended to set with the wind, and a marked return current was registered in the lower epilimnion.

A wind drift with a definite return current in the lower epilimnion also probably underlies the observations recorded in Loch Ness, and was noted by Langmuir in Lake George also. Wedderburn and Watson found evidence of a deeper wind drift immediately after a wind had started than existed later at equilibrium, when the gradient or return current was well established. They also found that at the windward end of the lake the return current might be detected at the surface, owing to tilting of the isotherms. The incidence of a calm after a strong wind may produce a gradient current at the surface in the direction opposite to that of the wind that has been blowing. This is in accord with the ideas that are generally held about the origin of seiches (page 299).

When a hypolimnetic current was recorded, it was generally in the direction of the wind but was very variable. Wedderburn concluded that in Loch Garry the strong but variable deep-water currents were of the nature of eddies. It is practically certain that whenever such currents are well developed they are due to internal seiches, as Mortimer (1953a) points out.

A number of other investigators have made observations of the depth distribution of currents, though in most cases during only a few days of a season. It is probable that in all cases in stratified lakes the deep water currents do not represent steady-state circulation systems maintained by a more or less constant wind stress, but rather continually changing re-adjustments to the changing positions of the isosteres.

Möller (1928, 1933), working on the Sakrower See near Potsdam, found in July a very marked oscillatory current in the hypolimnion, certainly due to a temperature seiche. On two occasions in August, measurements showed a wind drift at the surface more or less in the direction of the wind, with a return current at the bottom of the epilimnion. In the best studied case, August 23, 1919 (Table 23), there was a great variety of movements in the thermocline. Demoll (1922) gives similar observations on the Walchensee.

TABLE 23. *Current velocities and set in Sakrower See. The thermocline lay between* 5 *and* 10 *m., over which depth range the temperature drops about* 10°*C.*

	August 9, 1919, Wind SSW.		August 23, 1919, Wind WSW.	
Depth, m.	Velocity, cm. sec.$^{-1}$	Set	Velocity, cm. sec.$^{-1}$	Set
0	6.5	N. 70° E.	5	N. 46° E.
1	6.7	N. 38° E.	...	...
1.5	7.2	N. 63° E.	...	...
3	1.5	N. 168° E.	...	...
5	3.7	N. 230° E.	7.6	N. 229° E.
6.5	2.8	N. 230° E.	...	...
7	1.8	N. 298° E.	6.2	N. 209° E.
8	1.7	N. 66° E.	3.7	N. 242° E.
8.5	...	...	4.2	N. 144° E.
9	...	...	6.4	N. 64° E.
			4.0	N. 332° E.
13	...	...	1.3	N. 152° E.
20	...	...	3.3	N. 100° E.
30	1.5	N. 262° E.	7.0	N. 60° E.

Cross currents comparable to those observed by Möller were noted by Wedderburn and Watson in Loch Ness and by Langmuir in Lake George.

Bryson and Suomi (1952) have considered the vertical distribution of currents in Lake Mendota. Unfortunately their treatment is over-condensed, and the few data given for the actual lake are presented in rather obscure diagrams. They believe that their results can be best explained in terms of a wind drift that tends a little to the right (Verber, Bryson, and Suomi 1953) of the wind, inducing by the disturbance of the isosteres a type of circulation superficially comparable to that described by Sandström. The upwind current below the pycnocline is, however, to be regarded as determined by the displacement of the isosteres rather than by frictional forces. Owing to the variation of the wind direction, steady-state conditions can hardly be expected. Bryson and Suomi point out that if a current system of this sort is present, there is divergence upwind at the top of the upper layer and convergence upwind near the pycnocline. By using a method employed in meteorology it is possible to compute the divergence[20] from three velocity determinations in any plane at the apices of a triangle (page 274). A series of such determinations is given in Fig. 80 for the windward end of Lake Mendota. It is by no means clear to what extent these results really indicate a vertical circulation involving mixing in the deep layer as a whole, rather than displacements that will lead to a periodic reversal of movement as an internal seiche.

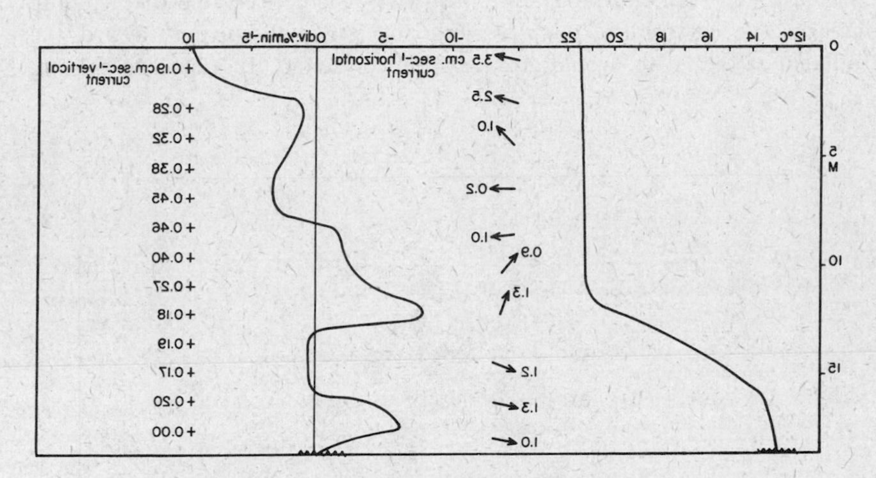

FIGURE 80. Vertical distribution of temperature, horizontal and vertical currents, and divergence in Picnic Point area of Lake Mendota, August 31, 1950 (after Bryson and Suomi).

Surface currents in large lakes. In very large lakes there is usually a tendency for the currents to form large swirls or gyrals. In some cases, as in Lake Constance or the Caspian Sea, this is facilitated by differences in meteorological regime at either end of the lake, but it is probable in other cases, such as Lake Michigan, that patterns such as those described by Livingstone under uniform wind disturbance may play a part.

An extensive study of the surface currents of the Great Lakes was made in the summer months of 1892 by the U. S. Weather Bureau (Harrington 1894). The winds over the lakes are irregular, but there is nearly always a strong westerly component. The method involved drift bottles that were just not submerged. Except in bays in which surf forms, it seems unlikely that the part of the bottle projecting above the water surface was

[20] Considering any velocity that can be resolved into components u, v, and w along the x, y, and z axis, the divergence is defined as $\partial u/\partial x + \partial v/\partial y + \partial w/\partial z$. For a steady flow in any rectilinear direction, this will be zero.

moved by the wind in a direction significantly contrary to the current. The over-all results are presented in *A* of Fig. 81. In general the current sets eastward, following the direction of the hydrographic gradient and of the predominant wind component. Except perhaps along the western coast of Lake Huron and in Georgian Bay, there is no clear evidence of any large-scale geostrophic effect. Nearly all the swirls that develop are markedly cyclonic. In the case of Lake Michigan, in the southern part of which there is an enormous cyclonic swirl, it is conceivable that a piling up on the east shore, coupled with the gradient current toward the effluent at the north, would give a cyclonic system. It is hard to believe,

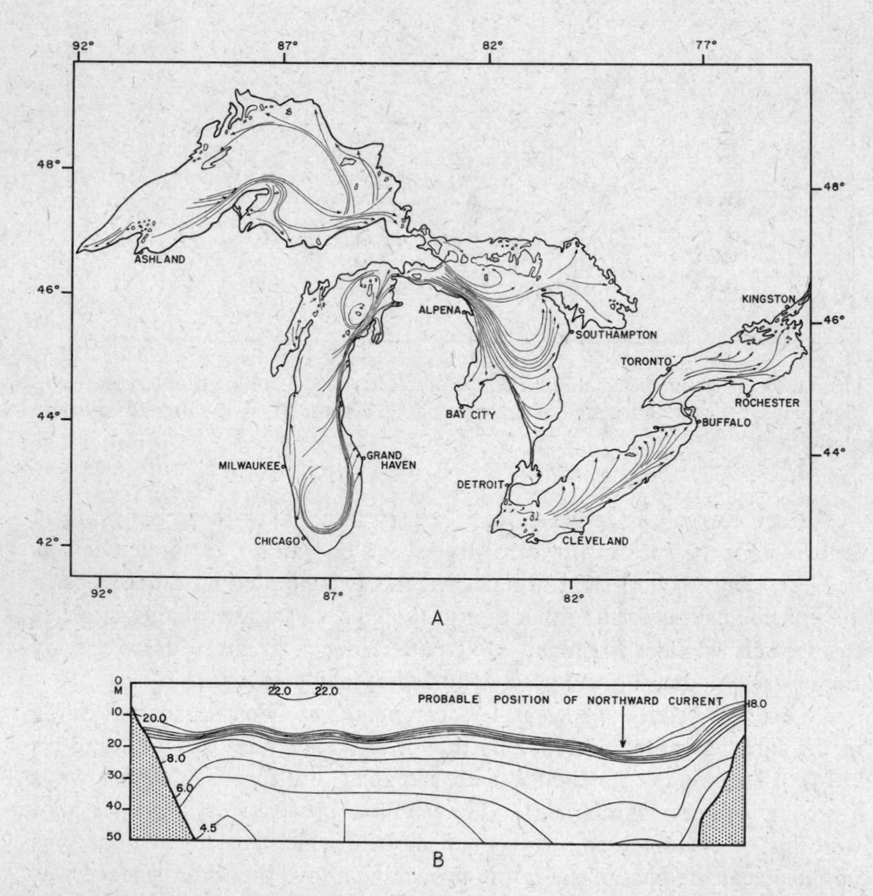

FIGURE 81. *A*, the general circulation pattern of the Laurentian Great Lakes (after Harrington). *B*, temperature profile across the main gyral of lake Michigan, August 15, 1942, showing slight downward bend of isotherms at the top of the metalimnion in the northward setting limb of the cyclonic swirl (after Church).

however, that the hydrographic gradient would be sufficient to produce so marked an effect. Scale-model experiments might be extremely interesting. It is just possible that the hypothetical cyclonic circulation set up by more intense warming of surface waters in the littoral region may play some part in producing the circulation. The large swirls appear as longshore currents along considerable stretches of the coast lines of the lakes.

A very marked tendency for drift bottles to collect at the heads of bays was explained by the bottles being controlled in their movements in shallow water by a tendency of surf to "seek the shores." Part of this tendency may be due to the heads of bays providing the best type of shore on which a bottle can be grounded and discovered. Only 5 to 10 per cent of the bottles were actually recovered, so that the statistical aspects of the circumstances that led to discovery may be very significant.

Olson (1950, 1952) has made a thorough study of the western end of Lake Erie, using drift cards floating in plastic envelopes. Here the lake is not merely under the influence of the wind but also of a complicated but only partly understood seiche pattern, and of the discharge of the Detroit River from Lake St. Claire and so from Lake Huron. His general conclusions are indicated in Fig. 82. The water entering the lake by the Detroit River tends in general to move eastward north of Pelee Island. This drift, however, can be separated into three regions. Just to the east of the Detroit River there is a clear tendency for water to converge on the coast at Colchester. Farther south there is a tendency to move somewhat north of east into Pigeon Bay; still farther south the current goes directly westward through the Pelee Passage. A unilateral convergence also occurs along the western coast of the lake. Most of the area of the south-western corner of the lake, termed by Olson the Maumee Flowage, shows no definite directional drift, save that water from the Maumee River moves into the southern half of this area. There is clear evidence of a very marked clockwise gyral around Pelee Island and also of divergence near Kelly Island. These movements are believed to be due mainly to seiches (Verber 1955).

The current velocity was positively and significantly correlated with the wind velocity, the mean value of v being about 2 per cent W. No significant geostrophic effects could be detected. The current system described fits in general with that of Harrington if allowance is made for inaccuracies in the latter. It is also in accord with the distribution of turbidity and phytoplankton in the various water masses in the lake. In particular, the water of the northwestern part of the Maumee Flowage appears to be relatively stagnant and can develop a greater phytoplankton crop than the water that is passing eastward more rapidly and which is often turbid from the outflow of streams.

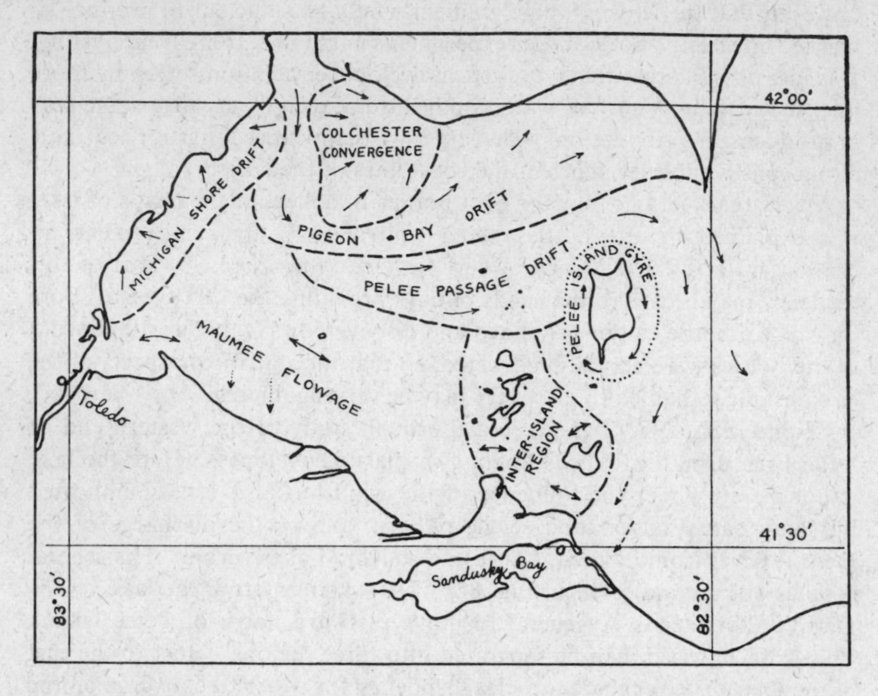

FIGURE 82. Circulation, largely due to seiche currents, in western Lake Erie (after Olson).

Wasmund (1927, 1928) has made an elaborate study of Lake Constance, using records of drifting of fishing nets. Though this method of study is subject to various errors, the general results are confirmed by later studies, particularly of Auerbach and Ritzi (1937) and of Nümann (1938), who studied the movement of the water of the Rhine through the lake by conductivity measurements. The ordinary type of circulation is indicated in Wasmund's schematic figure (Fig. 83). The inflow from the Rhine tends to move northward before following the northern coast westward. A subsidiary branch of the Rhine water moves to the right towards Bregenz, and may form either a cyclonic or an anticyclonic swirl at the extreme eastern end of the lake. There is usually an immense counter-clockwise or cyclonic swirl in the wide eastern central part of the lake, and a smaller clockwise swirl in the west central part. One or two subsidiary swirls form in the northwestern arm or Überlinger Basin. The main swirl is explicable in terms of southern winds down the Rhine Valley and predominantly northeastern winds from the north coast in the vicinity of Friedrichshafen, together with a geostrophic tendency of the Rhine water

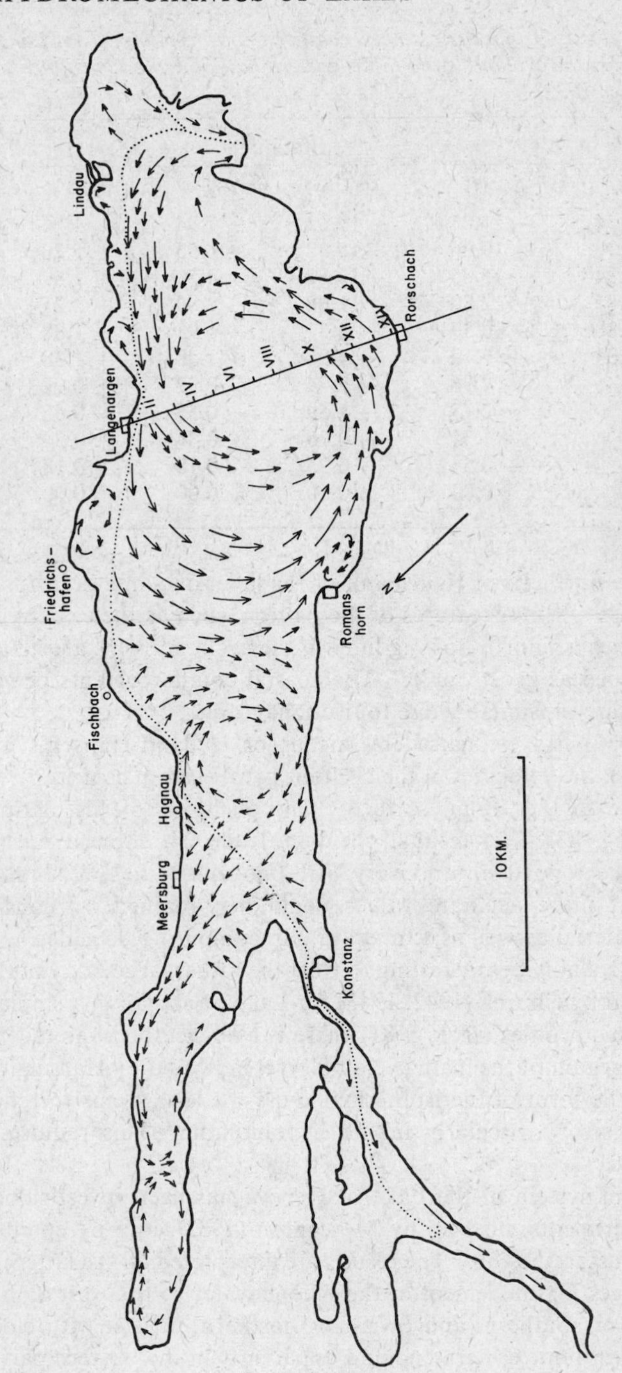

FIGURE 83. Lake Constance, ordinary disposition of surface currents (after Wasmund and Nümann), with position of stations between Langenargen and Rorschach studied by Elster. Dotted line, typical course of Rhine water, according to Auerbach and Ritzi.

TABLE 24. *Velocity profile (cm. sec.$^{-1}$ relative to 50-m. plane) from Langenargen to Rorschach across the cyclonic swirl in the eastern half of Lake Constance, on October 28, 1937 (Elster 1939)*

	Station					
Depth	I–II	II–III	III–IV	IV–V	V–VI	VI–VII
0	− 15.34	− 21.03	− 17.03	− 7.94	+ 4.46	+ 13.99
5	− 14.98	− 19.06	− 15.32	− 6.55	+ 5.26	+ 13.21
10	− 14.34	− 15.85	− 13.82	− 4.59	+ 3.89	+ 11.91
15	− 10.81	− 12.80	− 11.68	− 2.86	+ 2.25	+ 8.63
20	− 4.88	− 10.59	− 8.31	− 1.97	+ 1.88	+ 5.07
25	− 1.03	− 7.79	− 4.98	− 1.38	+ 1.38	+ 3.09
30	...	− 4.96	− 2.84	− 0.92	+ 0.92	+ 1.84
35	...	− 2.75	− 1.59	− 0.55	+ 0.55	+ 0.84
40	...	− 1.34	− 0.75	− 0.34	+ 0.34	+ 0.32
45	...	− 0.5	− 0.25	− 0.18	+ 0.18	+ 0.11
50	...	0.00	0.00	0.00	0.00	0.00

to follow the northern or right bank of the lake, on some occasions for a great distance. Minor features of the pattern, such as the eastern current leaving the eastern north-flowing limb of the cyclonic swirl, are also tentatively explained as geostrophic. The coastal countercurrents detected in some places are presumably due to frictional forces.

Elster (1939) has estimated, by means of Helland-Hansen's formula (equation 33), the velocities in the cyclonic gyral eastern central part of the lake, relative to the 50 m. surface. The positions of his stations are shown in Fig. 83. His results, given in Table 24, show a fairly swift current in the top 10 m. and very little movement below 35 m. It is possible that more accurate values could be obtained by considering dissolved material as well as temperature in computing densities.

It would probably be interesting to treat the rather large body of thermal data collected by Church (1942, 1945) for Lake Michigan in a similar way. In some of his profiles (*B*, Fig. 81) there is evidence of a mound of cold water in the region of the main cyclonic gyral indicated by Harrington, but more often the form of the isotherms across the lake seems to reflect the bottom contour, particularly in the hypolimnion. This requires more dynamic study.

The current system of the Lake of Geneva has been investigated both by drift experiments, notably by Mercanton (1932), and by conductivity studies by Dussart (1948). The results are interpreted by the latter investigator as suggesting movement of the Rhone water to the west along both the French, or southern, and Swiss, or northern, coasts, with clockwise gyrals at the extreme eastern end and apparently in the western part of the

TABLE 24 (*Continued*)

Depth	Station VII–VIII	VIII–IX	IX–X	X–XI	XI–XII
0	+ 22.57	+ 27.00	+ 22.56	+ 10.66	− 0.09
5	+ 21.41	+ 25.66	+ 21.81	+ 11.09	− 1.30
10	+ 18.13	+ 24.98	+ 21.17	+ 11.09	− 1.55
15	+ 12.05	+ 23.45	+ 19.57	+ 10.82	− 1.55
20	+ 6.03	+ 18.82	+ 18.34	+ 10.11	− 1.37
25	+ 2.47	+ 11.19	+ 15.79	+ 9.20	− 1.14
30	+ 1.33	+ 5.28	+ 9.86	+ 6.67	− 0.89
35	+ 1.22	+ 2.32	+ 3.73	+ 3.30	− 0.46
40	+ 0.86	+ 0.91	+ 1.09	+ 1.43	...
45	+ 0.39	+ 0.36	+ 0.36	+ 0.52	...
50	0.00	0.00	0.00	0.00	...

lake. The published data do not appear to exclude a circulation system essentially like that of Lake Constance.

Density currents and the influx of rivers. When a river enters a lake, particularly if the lake is statified, the incoming water, owing to its temperature, dissolved material, and suspended load, will have a density that may be different from that of any layer in the lake, or may be identical with that of some restricted layer. Any of the properties of these rivers may vary seasonally. When the density of the river is less than that of the water at any level in the lake, the river tends to flow over the surface but becomes rapidly mixed by wind-generated turbulent movement. When the river is denser than any layer, it will descend over the floor of the lake, sometimes excavating a subaqueous channel, as is the case at the mouths of both the Rhine and the Rhone in Lakes Constance and Geneva. If the river water corresponds to an intermediate density, it may remain as a discrete layer in the middle depths of the lake; in large lakes in the Northern Hemisphere such a layer will tend to the right, owing to geostrophic forces. On some occasions the Rhine can be detected as a band of water of different conductivity and temperature, though doubtless at almost the same density as the rest of the water at the layer in which it is found, up against the right bank of Lake Constance.

In winter the rivers flowing into the Swiss lakes are cold and concentrated. They are usually denser than the lake waters into which they flow, but at times of high water, dilution coupled with a temperature well below 4° C. may make them less dense than the lake for a time. The usual winter pattern, which is interrupted at high water, is flow along the bottom

PLATE 11. "*La Bataillère*," the entrance of the Rhone into the Lake of Geneva (by courtesy of Mons. B. Dussart).

until mixing in the freely circulating lake destroys the identity of the incoming water.

In summer, when an enormous dilution by glacial melt water occurs, the turbid waters flow in the thermocline or lower epilimnion, at about 10 to 20 m., though their temperatures are equivalent to that of the lake at 25 to 30 m. The identity of the Rhine as a definite current on the right of Lake Constance seems clearer than the Rhone in the Lake of Geneva. The Rhone, however, descending below the lake surface as a discrete stream locally called *La Bataillère* (Plate 11), often presents a remarkable appearance, and its water may be recognizable far down the lake on the windward shore as a slightly turbid and therefore green area from which the clearer blue water has been blown by the wind (Forel 1895, Dussart 1948).

In the large artificial reservoirs of the southeastern United States, notably Norris Reservoir, Tennessee, Wiebe (1939a,b, 1940, 1941) has supposed that very striking oxygen minima in the thermocline region are determined by inflowing water rich in organic matter, moving downstream at the appropriate level as the deep water is drawn off from the base of the dam. In contrast to Wiebe's observations, Bryson and Suomi

(1951) have noted an increase in the oxygen content of the hypolimnion of Lake Mendota during summer stagnation, which they attribute to turbidity currents from influents after exceptionally heavy summer rain.

By far the most remarkable system of density currents due to inflow is found (Anderson and Pritchard 1951) in the large artificial Lake Mead in southeastern Nevada and northwestern Arizona, held by the Hoover Dam on the Colorado River. The lake is an irregular branched body of water, very long compared with its maximum width. The mean total volume at any time is approximately equal to the inflow over a period of 1.64 years. This rate is great enough for the current due to the influents to have an effect on events in the lake, but small enough to permit great differences in composition between lake and river water at certain seasons.

The water in the lake during the winter is almost homogeneous, with a temperature around 10° C. and a total electrolyte content of 700 mg. per liter. The water entering the lake in spring becomes warmer and more dilute, so that at its least concentrated it has a temperature of about 19.5° C. and an electrolyte content of 275 mg. per liter. Further warming to the height of summer coincides with an increase in concentration, so that typical summer water has a temperature of 24.5° C. and an electrolyte content of 800 mg. per liter. Cooling in the fall is at first accompanied by an increase in concentration, the maximum of 900 mg. per liter being reached at a temperature of about 19° C. It is possible, by analogy with the temperature-salinity characterization of water masses in oceanography, to describe any water in the lake in terms of mixtures of these four arbitrary primary types. In winter the incoming water is slightly cooler and more concentrated than that of the lake; it flows down the slope of the old river channel, producing a single circulation cell with a return current upstream at the surface. In spring the dilute and relatively cool water spreads over the surface of the lake; Anderson and Pritchard suppose that below this there is probably a circulation cell with an upstream current near the bottom. During summer and autumn a progressively more concentrated inflow leads to a movement of the incoming water below the surface, with some return circulation above and below this stream. During summer a density current determined by the silt load may also flow as a *turbidity current* at a lower level than the main influent. The general circulation patterns deduced by Anderson and Pritchard are shown in Fig. 84.

As will be pointed out later (page 476), there is some evidence for the rather general occurrence of density currents running down the slope of the bottom of a lake, due to the diffusion of dissolved material from the mud into the water immediately in contact with the sediments. Such currents are probably best developed when reductive processes liberate a considerable supply of ionic material during summer stagnation.

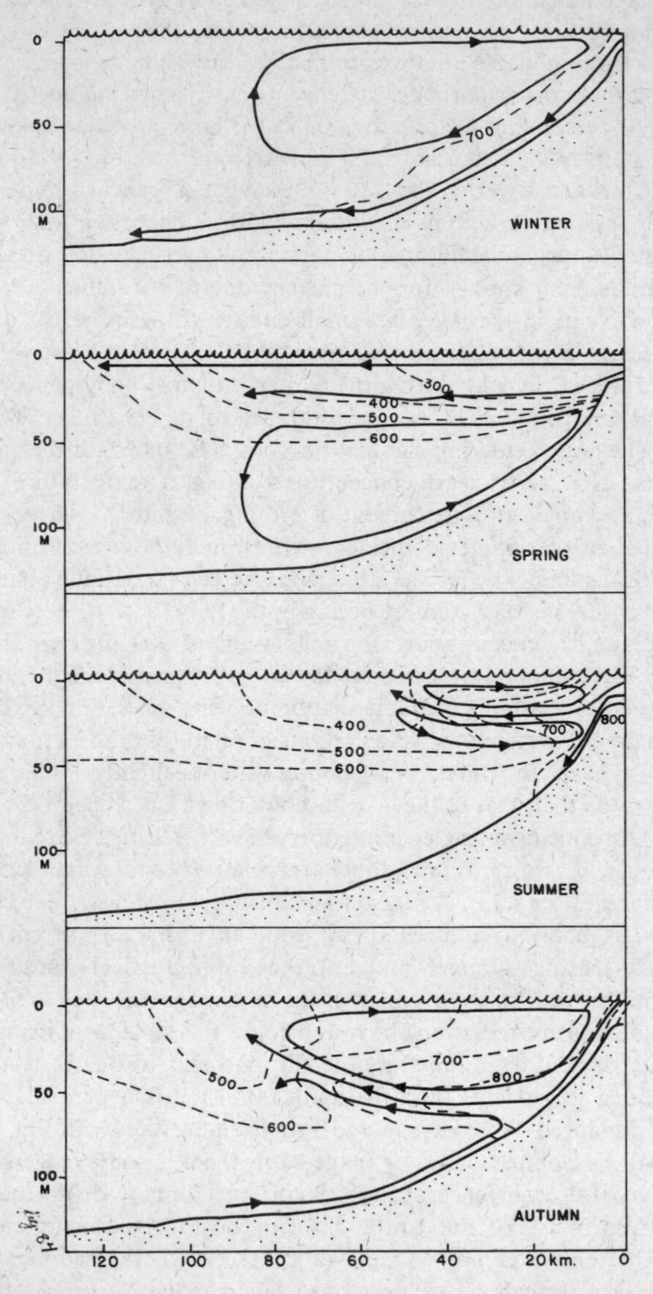

FIGURE 84. The concentration of electrolytes, milligrams per liter (broken lines), and the characteristic types of vertical circulation in Lake Mead during the four seasons (after Anderson and Prichard).

SEICHES

The word *seiche* is used to express a stationary oscillation of a lake or a large independent part of a lake. If a wind that has piled up water at one end of a lake suddenly dies down, a current will flow from leeward to windward, momentarily restoring the lake to its former level. The current, however, will not have lost its energy when the surface is level, but will continue, now piling water at the former windward end. This will cause a new current in the direction of the original wind drift. Being gradient currents, these currents are independent of depth except near the bottom, where the stress on the basin will gradually slow the movement. The current system thus constitutes an oscillation about a nodal line, determined by the shape of the basin.

As with other oscillating systems, there is a tendency for the formation of harmonics. The simple case described above is a uninodal seiche, but binodal, trinodal, etc., seiches are well known; in several cases the higher harmonics, certainly up to the octinodal, have been recorded. In the case of Loch Earn, periods have been identified which may belong to a nodality as high as the sixteenth.

While the periods and the positions of the nodes depend on the form of the lake basin, the amplitude of the seiche depends also on the source of energy that generates it, and is therefore very variable. The phenomenon has long been recognized locally in the Lake of Geneva, and the term seiche was recorded by Fatio de Duillier (1730) as applied to the oscillation in that lake. The Italian *sesse*, clearly the same word as seiche, has been explained as derived from *siccus*, from the exposure of the littoral zone at the downswing. Of the Lago de Bolsena it is said that *il lago trenfia*. Around Lake Constance the word *Rühss* has been employed, another local German word being *Laufe*. In Norway the term *floing* is used (Halbfass 1923); in Sweden, around Lake Vetter, *lunken*, which more generally means jogtrot.

The earlier records, largely summarized by Forel (1895), refer to specific events at a few lakes which are particularly favorable to demonstrate them. In these cases peculiar combinations of external events have produced oscillations of considerable amplitude.[21] Deevey (1957) has pointed out

---

[21] Some of the most conspicuous disturbances may be due, however to resonant exchange of energy between ordinary surface waves and atmospheric disturbances, such as pressure jumps traveling at the same velocity as the waves. It is known that under such conditions very large waves can be produced, which when arriving at the shore might simulate a seiche. Ewing, Press, and Donn (1954) have suggested such a mechanism to explain the abrupt rise of Lake Michigan at Chicago on June 26, 1954. The spectacular example noticed by the Aztecs on Lake Texcoco may have been comparable.

that such oscillations on Lake Texcoco had apparently been observed by the Aztecs prior to 1519. The earliest European record (Christoph Schultheiss in Forel 1893) refers to a seiche observed on February 23, 1549, at Constance, which had a range of about 2 ft. A very dramatic series of cases was observed in November 1755 in northwestern Europe, caused by the great earthquake that devastated Lisbon on that day. The seiche of Loch Lomond had a period of about ten minutes and an amplitude of $2^1/_2$ ft. on this occasion; several other Scottish lochs and numerous small bodies of water in England, Germany, and the Low Countries were also made to oscillate on the same day (anon. 1755, Gemsege 1755). These seiches are of peculiar interest, as they still provide the best examples of lakes being set into oscillatory motion by an earthquake.[22]

A remarkable seiche was later recorded on Loch Tay in 1784, having a period of seven minutes and a maximum range of 5 ft. Recognition of unexplained movements on some of the Swedish lakes is also evident in the writings of Hearne (1705) and Tiselius (1722).

The seiche of Lake Geneva has attracted more attention than those of other lakes because, owing to the narrowing of the western end of the lake towards Geneva, it is often very conspicuous at that city. Several explanations were put forward, in terms of melt water ([Addison] 1705, Jallabert 1745), wind denivellation (Fatio de Duillier 1730), electrostatic attraction of the water surface to thunderclouds (Bertrand unpublished, see De Saussure 1779), and local variations of atmospheric pressure (De Saussure 1779). The first author to consider the general problem of seiches was Vaucher (1833), who pointed out that they are not confined to the Lake of Geneva but can commonly be recognized in other lakes. Like De Saussure, he believed them to be due to disturbances of atmospheric origin. Some other eighteenth-century and early nineteenth-century observations and theories of no scientific interest are discussed by Forel (1895). No real progress in the scientific study of seiches was made until 1869, when Forel began observations on the Lake of Geneva. From that date for nearly fifty years no other branch of limnology received so much attention. If there are fewer recent contributions, it is largely because the subject has been so well studied by the earlier investigators. In common with certain other branches of science assiduously cultivated at the end of the last century, what has been put into textbooks is well known, but many quite fascinating details have been forgotten.

Theory. The fundamental differential equations for the water movements in a seiche in a long narrow lake, in which frictional forces, geo-

---

[22] Sea seiches, due to local resonance to waves coming shoreward from disturbances at sea, are often set up by seismic events and may be very destructive.

strophic forces, and vertical accelerations can be neglected, are

$$\frac{\partial(Hu)}{\partial x} + b\frac{\partial h}{\partial t} = 0 \tag{69a}$$

$$\frac{\partial u}{\partial t} = -g\frac{\partial h}{\partial x} \tag{69b}$$

where at any point x along the long or x axis of the lake the water is displaced above or below the level equilibrium surface by h, the breadth along the y axis is b, and the area of cross section of the lake is H. Both b and H are functions of x, generally of a complicated nature.

The simplest case is that of a rectangular lake of uniform depth z_m, in which H and b are constant. Dividing equation 69a by b, we obtain

$$z_m\frac{\partial u}{\partial x} + \frac{\partial h}{\partial t} = 0 \tag{70}$$

Since the velocity is due to a gradient current, it is independent of depth. From equations 70 and 69b we can obtain

$$u = u_m \sin\frac{n\pi x}{l}\sin\frac{2\pi t}{T_n} \tag{71}$$

$$h = \frac{T_n z_m u_m}{2l}\cos\frac{n\pi x}{l}\cos\frac{2\pi t}{T_n} \tag{72}$$

$$= h_m \cos\frac{n\pi x}{l}\cos\frac{2\pi t}{T_n}$$

$$T_n = \frac{1}{n}\frac{2l}{(gz_m)^{1/2}} \tag{73}$$

It is apparent u_m is the maximum velocity across the nodal lines at the first and third quarters of an oscillation, and h_m the *amplitude*[23] or the maximum displacement from the level at the ends of the lake or at the *ventral points* between the nodal lines of a multinodal seiche. These results follow, for the special case of $l \gg z_m$, from a more general treatment of the oscillation of a liquid in a rectangular tank, developed by Merian (1828). The theory appears to have been applied to seiches for the first time by Forel (1876).

In the majority of cases the contour of the bottom cannot be taken as

[23] The *amplitude* is the maximum displacement from the equilibrium position; the *range* is the distance between positive and negative maximum displacements and is therefore ordinarily twice the amplitude.

that of a rectangular basin of uniform depth, and it is necessary to consider the variation of b and H with x.

Du Boys (1891) gave a more general though theoretically inadequate approximation

$$\text{(74)} \qquad T_n = \frac{2l}{n}\int_0^l \frac{dx}{\sqrt{gz_x}} = \frac{2l}{n\sqrt{g}}\int_0^l \frac{dx}{\sqrt{z_x}}$$

The integral can be evaluated planimetrically. The formula gives almost perfect results for the higher nodalities, but for the uninodal and binodal seiches it predicts periods that are too great if the lake is concave, too small if convex.

Terada (1906) gives the sounder but more complicated expression

$$\text{(75)} \qquad T_n = \frac{2l}{n\sqrt{g\bar{z}}}\left[1 + \frac{1}{2l}\int_0^l \left(\frac{b_x - \bar{b}}{\bar{b}_x} + \frac{H_x - \bar{H}}{\bar{H}_x}\right)\cos\frac{2\pi n x}{l}\,dx\right]$$

The most complex and precise theoretical treatment is due in the first place to Chrystal (1904, 1905a,b, 1908; Chrystal and Wedderburn 1905).[24] A subsequent development of Chrystal's theory was published by Proudman (1915).

The approach adopted by Chrystal is to consider the motion in long stationary waves in basins of analytically known shape, and then to find methods of converting the forms of actual lake basins into equivalent forms which can be treated more conveniently than can the actual lake basin. Chrystal's investigations gave periods and nodal lines for parallel-sided basins of analytically known longitudinal profile, most of which are summarized here. Except in the case of the uniformly deep basin, where the origin is taken in the middle, and of the convex parabolic basin, where it is taken over the shallowest central point, the origin of the curves expressing the form of the basin is in the surface of the undisturbed lake over the deepest point (Fig. 85).

(*a*) Rectangular basin of uniform depth, defined by $z_x = z_m$.

$$\text{(76)} \qquad T_n = \frac{1}{n}\frac{2l}{(gz_c)^{1/2}}$$

or

$$T_1:T_2:T_3:T_4:T_5 = 100:50:33.3:25:20$$

Nodes at $x = 0$ (uninodal); $x = \pm 0.167l$ (binodal); $x = 0$, $\pm 0.25l$ (trinodal); $x = \pm 0.1l$, $\pm 0.3l$ (quadrinodal); $x = 0$, $\pm 0.167l$, $\pm 0.333l$ (quinquenodal).

---

[24] Good general expositions are given in two reviews (Chrystal 1909, 1910) adapted to the nonmathematical reader.

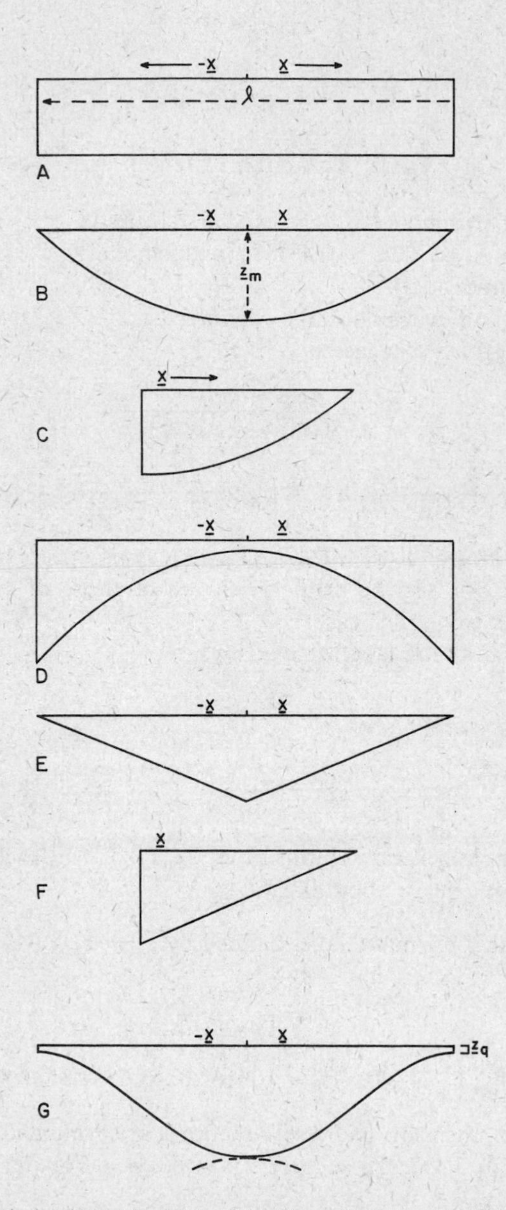

FIGURE 85. Sections of ideal basins used in Chrystal's hydrodynamic theory of seiches. *A*, rectangular basin of uniform depth; *B*, symmetrical concave parabolic basin; *C*, concave half parabolic basin; *D*, symmetrical convex parabolic basin; *E*, symmetrical prismatic basin; *F*, half prismatic basin; *G*, symmetrical truncated quartic basin.

(*b*) Symmetrical concave parabolic basin, defined by

$$z_x = z_m[1 - (4x^2/l^2)].$$

(77) $$T_n = \frac{\pi l}{(n[n + 1]gz_m)^{1/2}}$$

or

$$T_1 : T_2 : T_3 : T_4 : T_5 = 100 : 57.7 : 40.8 : 31.6 : 25.8$$

nodes at $x = 0$ (uninodal); $x = \pm 0.2887l$ (binodal); $x = 0, \pm 0.3873l$ (trinodal); $x = \pm 0.1700l, \pm 0.4311l$ (quadrinodal); $x = 0, \pm 0.2692l, \pm 0.4531l$ (quinquenodal).

(*c*) Concave half parabolic lake, defined by $z_x = z_m[1 - (4x^2/l^2)]$ for x positive, $z_x = 0$ for x negative.

(78) $$T_n = \frac{2\pi l}{(2n[2n + 1]gz_m)^{1/2}}$$

or

$$T_1 : T_2 : T_3 : T_4 : T_5 = 100 : 54.8 : 37.8 : 28.9 : 23.4$$

These are the ratios of $T_2 : T_4 : T_6 : T_8 : T_{10}$ in a symmetrical parabolic lake; the nodal lines will also be in the positions of those of seiches of even nodality in the symmetrical case.

(*d*) Convex parabolic lake, defined by

$$z_x = z_m\left(1 + \frac{4x^2}{l^2}\right) \quad \text{for} \quad |x| < \frac{l}{2}$$

$$z_x = 0 \quad \text{for} \quad |x| \geqslant \frac{l}{2}$$

There is no simple general solution, but $T_1 : T_2 : T_3 : T_4 = 100 : 47.2 : 31.2 : 23.4$. The binode lies at about $0.236l$.

(*e*) Symmetrical prismatic lake, defined by $z_x = z_m\left(1 - \left|\frac{2x}{l}\right|\right)$.

(79) $$T_n = \frac{2\pi l}{j_n(gz_m)^{1/2}}$$

$$T_1 : T_2 : T_3 : T_4 : T_5 = 100 : 62.8 : 43.6 : 34.3 : 27.8$$

where j_1, j_2, etc., the roots of Bessel functions, are evaluated as $j_1 = 2.405$, $j_2 = 3.832$, $j_3 = 5.520$, $j_4 = 7.016$, $j_5 = 8.654$. For larger values of n, $j_n \simeq (2n + 1)\pi/4$.

The nodes are at $x = 0$ (uninodal); $x = \pm 0.3029l$ (binodal); $x = 0, \pm 0.4051l$ (trinodal); $x = \pm 0.1905l, \pm 0.4413l$ (quadrinodal); $x = 0, \pm 0.2965l, \pm 0.4614l$ (quinquenodal).

(*f*) Half prismatic lake, defined by $z_x = z_m[1 - (2x/l)]$ for x positive, $z_x = 0$ for x negative.

$$\text{(80)} \qquad T_n = \frac{4\pi l}{j_{2n}(gz_m)^{1/2}}$$

$$T_1 : T_2 : T_3 : T_4 : T_5 = 100:54.6:37.7:28.8:23.3$$

As in the case of the half parabolic, these are the ratios of the seiches of even nodality in the symmetrical case.

The nodes are likewise at the positions of even nodality, remembering that the lake is half as long: $0.6057l$ (uninodal); $0.3809l$ and $0.8825l$ (binodal); $0.376l$, $0.7086l$, and $0.9441l$ (trinodal).

(*g*) Truncated symmetrical concave quartic lake, defined by

$$z_x = z_m\left(\frac{l_q^2}{4} - x^2\right)^2 \qquad \text{for} \quad x \leqslant \frac{l}{2}$$

$$z_x = 0 \qquad \text{for} \quad x > \frac{l}{2}$$

Let the depth at the ends for $x = l/2$ be z_q.

$$\text{(81)} \qquad T_n = \frac{2\pi l}{\gamma_q \sqrt{gz_m(4n^2\pi^2/k_q^2 + 1)}}$$

where

$$\gamma_q = 2\sqrt{(1 - \sqrt{z_q/z_m})}$$

$$k_q = l_n\left(\frac{1 + \sqrt{1 - (z_q/z_m)}}{1 - \sqrt{1 - (z_q/z_m)}}\right)^2$$

Chrystal gives a more elaborate treatment for asymmetrical truncation. This case is extremely interesting theoretically. It corresponds to a lake with very gently shelving ends and a deep concave central depression. Such a shape approximates to many actual lakes. As l is increased and approaches l_q, and the amount of truncation decreases, it can be shown (Chrystal 1905a) that for *seiches of every nodality* the period approaches $\pi l(gz_m)^{-1/2}$. Chrystal terms this the period of the anomalous seiche. While a close approximation to this condition cannot be expected, the quartic form explains why it is sometimes possible for T_2 to be a very appreciable fraction, well over one half (or even over the 0.628 of the symmetrical prismatic lake) of T_1.

In dealing with an actual basin, Chrystal replaced it with an ideal model of uniform breadth and vertical sides, having as its longitudinal profile what he called the normal curve of the lake. This curve is constructed by imagining a line perpendicular to the x axis, moving along that axis from the origin. The area of the lake ($A_{0,x}$) between the line at x and its original position at the origin is used as the abscissa, and the ordinate is defined by the product $b_x H_x$. A lake of uniform breadth and rectangular cross section having this longitudinal profile can be shown to have longitudinal seiches identical with those of the actual lake.

Chrystal then divided the normal curve into sections, each of which could be approximated to an analytically known curve; these sections are linked together by considerations of continuity across their boundaries. Proudman's development of the method dispenses with this approximation which is unsatisfactory in certain cases. The full treatment involves a good deal of computation and has been applied to very few lakes. In very irregular basins any practical application of Chrystal's original method may lead to less valid results than some of the approximate methods, as Nakamura and Honda (1911) found in the case of Lake Hakone. The cases in which a full analysis, and adequate observations to check the analysis, have been made are presented in Table 25. These results indicate that, at least in fairly simple elongate basins, the water movements in a seiche are well understood. It may be doubted whether, in general, the detailed application of the hydrodynamic analysis of seiches will indicate anything further of particular interest except in cases in which there is known empirically to be some very peculiar feature.

In many complicated cases the use of laboratory models (Endrös 1903, White and Watson in Chrystal 1910, and particularly Nakamura and Honda 1911) is likely to be instructive.[25] In simpler cases it may often be possible (Endrös 1908) to approximate the actual lake profile, rather than the normal curve, to one of the simple analytic forms discussed above, or to approximate the normal curve by a single expression.

The great value of Chrystal's investigations is that they permit certain broad general principles to be deduced. It is clear that a shallow area attracts the nodal lines, a deep area repels them. A very marked shallow near a node may render a seiche unstable, however, or prevent its occurrence entirely. Concave profiles tend to reduce the difference between the periods of seiches of different nodalities, convex profiles tend to increase them. The particular combination of concavity and convexities discussed under (*g*), in which the periods approach each other, gives of course the least difference. Since convex basins are rarer than concave, and since many lakes are likely to have littoral shelves at least at their upper ends, roughly simulating the quartic form of (*g*), it is not surprising that in

[25] Jeffreys (1919) suggests that since the oscillation of a lake is formally equivalent to vibration of air in an organ pipe, a model might be made by constructing a pipe with three plane sides and a fourth that could be varied according to an appropriate function of the lake basin. The pitch of the pipe thus could be related by simple proportionality to the seiche period.

TABLE 25. *Periods of longitudinal seiches, calculated from hydrodynamic theory and observed*

		Calculated	Observed
Loch Treig	T_1	9.14	9.18
(Chrystal 1910, Chrystal and Maclagan-Wedderburn 1905)	T_2	5.10	5.15
	T_3	3.59	
A period of 1.18 minutes also recorded			
Loch Earn	T_1	14.50	14.52
(Chrystal 1908, Chrystal and Maclagan-Wedderburn 1905)	T_2	8.14	8.09
	T_3	5.74	6.01
Periods of 1.70 (n = 10),	T_4	4.28	3.99
1.54 (n = 12), 1.36 (n = 13),	T_5	3.62	3.54
1.31 (n = 14), 1.15 (n = 15),	T_6	2.90	2.88
1.09 (n = 16) also recorded			
Lago di Garda	T_1	42.92	42.83
(Defant 1908)	T_2	28.58	28.00
A period of 3.06 minutes also known,	T_3	21.79	20.10
which may represent a transverse	T_4	14.96	14.83
seiche	T_5	12.07	11.9
	T_6	9.87	10.1
	T_7	8.80	8.75
	T_8	7.33	7.6
Lake of Geneva	T_1	74.45	73.5
(Doodson, Carey, and Baldwin 1920)	T_2	35.1	35.5
Transverse seiche at Morges, 10.3 minutes; others probably present	T_3	28	
Lake Vetter	T_1	177.94	178.99
(Bergsten 1926)	T_2	95.95	97.52
A second transverse seiche of 60-	T_3	79.17	80.74
minutes period recorded at Motala,	T_4	59.83	57.89
where the lake is wider	T_5	49.77	48.10
	T_6	42.66	42.59
	Transverse (Hästholmen)	31.00	30.8

Table 26, $T_2 \geqslant {}^1/_2T_1$ in the majority of lakes. It is worth noting that the earlier students of seiches often recorded a seiche of shorter period than that of the uninodal, but of longer period than half that of the latter. Supposing that in all cases $T_2 = {}^1/_2T_1$, Forel considered such seiches to be analogous to the fifth in musical theory and termed them *seiches à la quinte*. It is reasonably certain, however, that such seiches are binodal, the supposed binodal seiches accompanying them being really trinodal (Chrystal 1905b).

Subsequent work (Jeffreys 1923, 1928) has mainly been concerned with the plan of the lake surface. It appears that the most important variations concern the form of the ends of an elongate lake. An elliptical lake of length l and breadth $0.6l$ has a uninodal longitudinal period only 2 per cent less than that of a circular lake of diameter l. In a narrow elliptical lake, however, the period tends to be shortened when comparison is made with a narrow rectangular lake. This may be regarded as due to a reflection of energy all the way along the shores of the lake. A sharp angular end and still more a cuspate end give even greater reductions of period. The effect of the form of the ends is in fact equivalent to the reduction of l in the Merian formula; under certain conditions it is possible to have practically no movement at the ends, even though the main body of water is in uninodal oscillation. This has been realized by some investigators of particular lakes, notably Hayford (1922) and Bergsten (1926).

The seiches discussed have all been longitudinal, with the nodal lines at right angles to the long axis of the lake. Transverse seiches are also known, though they are often confined to restricted stretches of the lake which are defined, at least along one shore, by a bay. The theory of the transverse seiche differs in no essential way from that of the longitudinal. In a lake in which the breadth is not very different from the length, a rather complicated pattern may be set up. Proudman deduces for a broad rectangular lake of length l and breadth b that the combination of the m-nodal longitudinal and the n-nodal transverse oscillation gives a compound seiche of period

$$T = \frac{2}{(gz_m)^{1/2}}\left(\frac{m^2}{l^2} + \frac{n^2}{b^2}\right)^{-1/2} \tag{82}$$

For the simple case of a square rectangular lake this reduces to Merian's formula. In cases of this sort the points of high and low water travel around the lake, the direction being ordinarily determined by geostrophic forces. In Lake Mendota, in which the seiche periods along the north-south and east-west axes are nearly identical, Bryson and Kuhn (1952) found evidence of a clockwise rotation of the seiche. In some cases, when the periods are of the same order of magnitude but are not quite identical, interference producing beat phenomena has been described (page 320).

Very elaborate oscillations can develop in circular basins. Here the nodal lines may be concentric circles or diameters or both, and a great variety of periods is possible. In a model it is possible to generate a number of such oscillations by attaching the circular basin to a channel down which a train of waves of appropriate period can be transmitted. While such a case is of great interest to harbor engineers, it is probably of very restricted limnological significance. The reader desirous of more information can consult the work of McNown (1952).

When a lake is so large that the period of the uninodal transverse seiche is much greater than half a pendulum day, it can be shown (Proudman 1953) that any ordinary uninodal oscillation of the lake as a whole degenerates into a simple geostrophically controlled inertia current. The condition

(83) $$\frac{1}{T_p^2} \gg \frac{1}{T_s^2}$$

depends on the square of the period, so that the difference of the period and one half pendulum day does not in effect have to be very great. No basin completely isolated from the ocean appears to exist at present for which the condition is satisfied for the ordinary seiche.

Energy and damping of seiches. The total energy (E_s) of a seiche is at any moment partly potential (E_p) and partly kinetic (E_k), except of course that the kinetic energy is zero at the moment of maximum displacement and the potential energy is zero at the moment of zero displacement from the level. In the general case

(84) $$E_k = {}^1/_2\rho \int_0^l Hu^2\, dx$$

(85) $$E_p = {}^1/_2 g\rho \int_0^l bh^2\, dx$$

For a rectangular basin of uniform depth z_m, for the uninodal seiche

(86) $$E_k = {}^1/_4\rho b z_m l u_m^2 \sin^2 \frac{2\pi t}{T_1}$$

(87) $$E_p = {}^1/_4 g\rho b l h_m^2 \cos^2 \frac{2\pi t}{T_1}$$

For the uninodal seiche of Loch Earn, if the maximal displacement (h_m) at the ends is 10 cm., the total energy is about 2.4×10^{15} ergs or 2.4×10^5 ergs cm.$^{-2}$

Since the potential energy at the time ($t = T_1/4, 3T_1/4, \cdots$) of maximum displacement must equal the kinetic energy at the time of zero displacement ($t = 0$, $T_1/2$, $T_1, \ldots$)

(88) $$u_m = h_m\sqrt{g/z_m}$$

Assuming that the rate of loss of momentum per unit area from the system is proportional to the bottom stress τ_b, Bryson and Kuhn (1952) write

(89) $$\tau_b = \rho\sqrt{g/z_m}\,\frac{dh_m}{dt}$$

whence the bottom stress at the moment when the current is maximal can be determined from the rate of decline of h_m with time. Values for Lake Mendota vary from 0.10 to 0.49 g.-cm. sec.^{2} It is probable that they are a little high, as some damping due to eddy viscosity must occur. Proudman (1953) has examined the damping of the uninodal seiche of Loch Earn, using an approach which involves an empirical relationship for the bottom stress of somewhat doubtful applicability, and concludes that the total energy would be reduced by a factor $1/e$ in 250 days, a time that seems rather long. The longest continuous record that exists appears to be one for the Lake of Geneva, of the uninodal seiche continuing with a few small irregularities for seven days, 17 hours, or through 150 complete

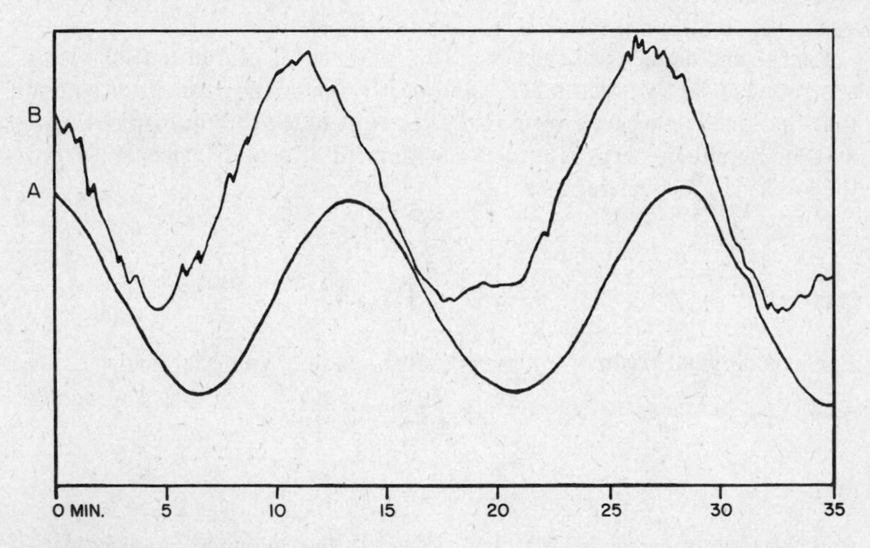

FIGURE 86. Loch Earn. *A*, isolation of a very pure uninodal seiche by means of a limnograph with a narrow access tube; *B*, the oscillations recorded with a wide tube, disturbed by much secondary embroidery. (After Chrystal.)

periods (Forel 1895). Since all the energy is potential at maximum displacement, the rate of loss of energy over a time long compared with T_1 can be obtained by differentiation of equation 87 for $t = T_1/4$. Bryson and Kuhn (1952) concluded that during July and August 1951, seiches on Lake Mendota lost energy at a rate of about 0.2 erg cm.$^{-2}$ sec.$^{-1}$ In very shallow lakes of large size and great turbulence, the energy loss due to eddy viscosity may be so rapid that the period is affected (page 324).

Empirical observations. The fundamental data are obtained by means of various instruments in which a float, protected from casual disturbance, is attached to a suitably counterpoised pointer or stylus. Such an instrument is termed a limnometer if read directly, or a limnograph if a stylus writes a continuous record. The details of various types of instruments may be found in Forel (1895) and Chrystal (1910), and in the papers cited in these works. From a theoretical point of view it is important to realize that the system used in measurement may have considerable inertia, and that different instruments may therefore give different types of records. In most cases the float is protected by a shield. If this shield consists of a vessel with an access tube,[26] it is possible, by making the tube small

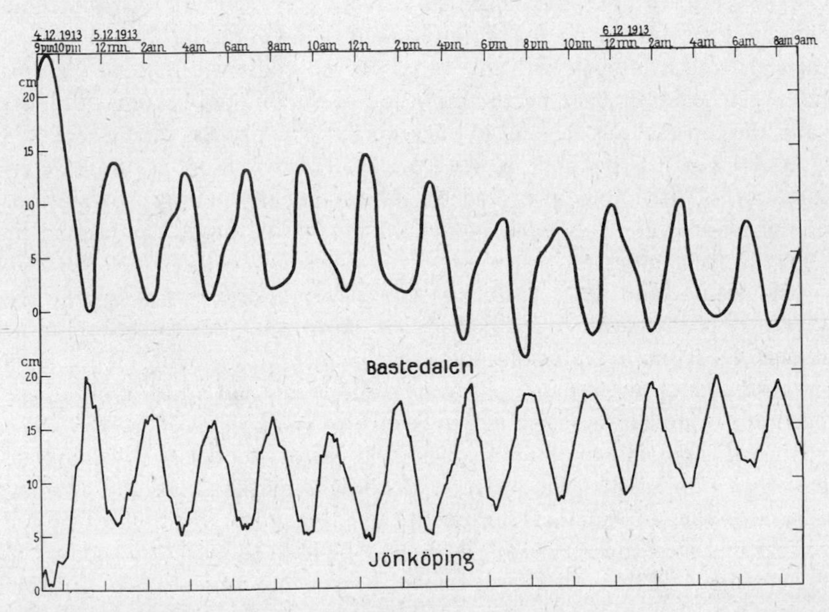

FIGURE 87. Seiche records of great regularity taken simultaneously at either end of Lake Vetter, at Bastedalen and at Jönköping (Bergsten 1926).

enough, to damp out the shorter period oscillations. Such an instrument can indicate only part of what is happening, but the part isolated may be particularly clear and important. (Fig. 86.)

In some lakes, suitably placed limnographs may give extremely regular records approaching simple harmonic motion (Fig. 87), but in the majority of cases the seiche recorded is a complex of several oscillations. The commonest type of record, termed by Forel the *dicrote*, is a superposition of a uninodal and a binodal seiche. The analysis of such records, in which initially the amplitude, wavelength and phase are unknown, presents all

[26] Errors due to the direction of the access tube in certain portable limnographs are discussed by Endrös (1908).

the difficulties familiar to the student of natural oscillations. Careful placing of the limnographs will often permit stations to be found near the binode that give fairly pure uninodal seiches. The method of damping out all short oscillations is also sometimes helpful, and if periods can be calculated, their appearance in the limnogram is often recognizable. There may be times when a virtually pure seiche occurs. Methods of semi-empirical mathematical analysis, once an approximate uninodal period is available, have been described, notably by Defant (1908), Chrystal (1910), and Nakamura and Honda (1911). In essence these consist of moving the record initially about half a uninodal wave length to the right and averaging with the initial curve. This will remove the uninodal seiche, so that the mean amplitude of the binodal becomes in most cases the conspicuous element in the record. The process can be repeated (Fig. 88) and it is possible to work backwards to a more accurate determination of the uninodal period by subtracting the first approximations to the binodal and other plurinodal seiches. This process is termed by Chrystal *residuation.*

The periods and the positions of the nodal lines depend purely on the geometry of the basin; as the periods are more easily determined than the nodal lines, more information exists concerning them. The most important data on periods are given in Table 26, and an example of the positions of nodal lines is shown in *A* of Fig. 89.

Since it is possible to obtain seiches in laboratory models, or even in cups of coffee[27] in a cafeteria, there is nothing remarkable in the smallest natural bodies of water exhibiting the phenomenon. Such short period seiches in ponds and very small lakes may be hard to detect on account of the smallness of their amplitude, since the extreme range of any pressure or wind denivellation or set-up will decrease as the length of the lake is decreased. Seiches with periods of 76.8 to 93.6 sec. have been detected by Endrös (1904) in a small pond near Traunstein, 112 m. long; the period varied with the depth of from 80 to 50 cm. Van Dorn (1953) observed a period of 112 sec. in a pond 1.83 m. deep and 240 m. long.

Seiches in very large lakes. At the opposite end of the scale we have the very long period oscillations of the large but shallow Lake Erie and the Sea of Aral.[28]

---

[27] For a coffee cup of radius 5 cm. and depth 6.5 cm., Merian's approximate formula gives four uninodal oscillations per second, while the symmetrical concave parabolic expression of Chrystal gives 3.6 per second. The actual oscillation appears to be rather slower, most of the movement occurring in the upper part of the liquid in the cup, presumably owing to friction at the bottom.

[28] The still longer period of 42 hours (2520 min.) recorded by Fülleborn (1906) for Lake Nyasa (see Halbfass 1923) must be erroneous.

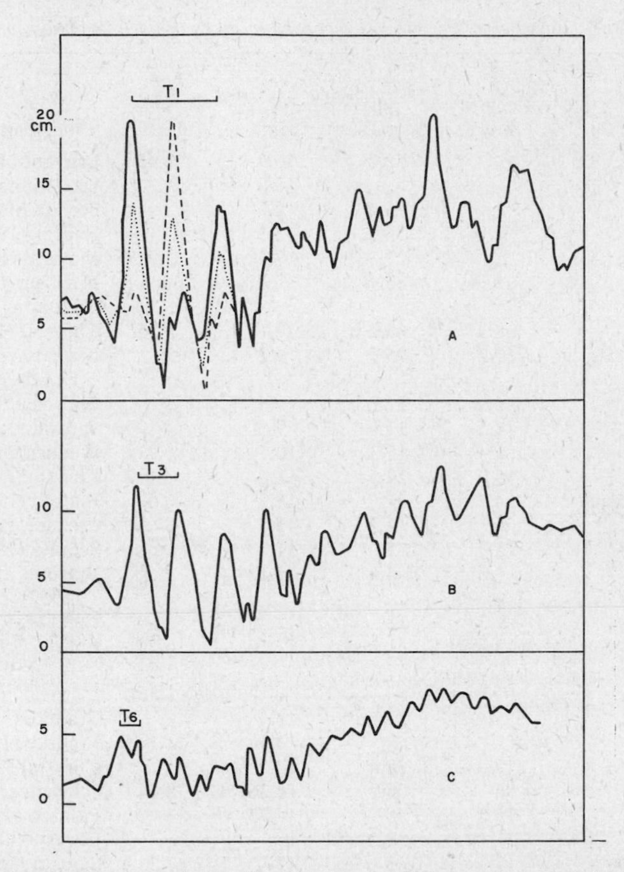

FIGURE 88. Residuation of a highly complicated limnogram recorded at Jönköping, Lake Vetter, September 4–5, 1914. *A*, the uninodal period is determined by inspection (T_1), and the curve is moved half a period to the right (broken line); the means of the original and displaced (broken) lines give a limnogram (dotted line) essentially free of the uninodal period. *B*, the full record, free of the uninodal period, shows a trinodal period (T_3), determined by inspection. The curve is moved half this period to the right and averaged as before, giving *C*, in which an almost pure sexinodal seiche of period T_6 is superimposed on a nonperiodic denivellation. (After Bergsten.)

The seiches of a still larger basin, namely the Caspian, have been discussed by Knipowitsch (1922), who concluded that at Petrovsk on the west coast, north of the middle of the basin, periods of 5.5, 2.75, and 1.5 hours could be detected. The limnogram that he published, purporting to show the 5.5-hour period, is most unconvincing, but the other two periods seem to be real. By far the most striking oscillations on the limnograms from Petrovsk are minor disturbances of considerable

TABLE 26. *Periods of seiches recorded in lakes of various dimensions**

Lake	Length, km.	z_m, m.	T_1, min.	Other Periods, min.	Author
Pavin	...	52.2	0.9	0.45	Bruyant (1903)
Bret	...	...	3.0	...	See Halbfass (1923)
Silser	...	35.4	4.7	...	See Halbfass (1922)
Altausseer	2.8	34.6	5.3	...	Endrös (1906c)
Santa Croce	...	23.3	7.4	5.0	Magrini (1903)
Chuzenji	5.9	85.4	7.77	...	Nakamura and Honda (1911)
Treig	8.3	63.2	9.18	5.15	Chrystal (1910)
Toya	10.2	95.8	9.29	...	Nakamura and Honda (1911)
Grundl	5.9	32.2	9.5	...	Endrös (1906c)
Brienzer	14.5	17.6	9.8	...	Sarasin (1895)
Garry	6	15.3	10.5	...	Wedderburn (1908)
Chroisg	5	22.5	11.2	...	Chrystal (1910)
Zeller (Pinzgau)	4.7	37	11.2	6.5	Endrös (1906c)
Fada	6	3.2	11.5	6	Chrystal (1910)
Gmundener	12.9	...	11.7	...	Schultz (1899)
Kegonsa	3.6	4.6	12.1	...	Bryson and Kuhn (1952)
Joux	9.3	18	12.4	...	Forel (1897)
Tachinger	3.6	...	12.56	6.25 3.5 1.56†	Endrös (1905, 1906a)
Tschetsch	...	22.5	13.3	...	Bradtke (1910)
Lagower	...	...	13.9	...	Bradtke (1910)
Morar	19	38.4	14	...	Chrystal (1910)
Walen	16	103	14.26	8.04	Schweitzer (1909)
Bolsena	13.8	78	14.75	7	Palazzo (1905)
Maree	21	38.2	15	...	Chrystal (1910)
Thun	17	135	15	7.5	Sarasin (1895)
Sims	6	13.4	15	7.6 5.7 4.5 3.9	Endrös (1913)
Mond	11.0	36	15.4	9.6 7.5 3.3	Endrös (1906c)
Hakone	6.6	32	15.38	6.76 4.63 3.90 3.11	Nakamura and Honda (1911)
Yamanaka	5	8.2	15.61	10.57 5.46	Nakamura and Honda (1911)
Hallstätter	8.8	64.9	16.4	6.4	Endrös (1906c)
Waginger	7.0	15.6	16.80	11.78 7.5 6.0 4.67 3.87‡ 3.0†	Endrös (1905)

TABLE 26 (*Continued*)

Lake	Length, km.	z_m, m.	T_1, min.	Other Periods, min.	Author
Atter	20.7	84.2	22.4	11.8	Endrös (1906c)
				7.4	
Kawaguchi	5.9	9.63	22.98	11.50	Nakamura and
				10.66	Honda (1911)
				8.58	
				7.82	
				6.36	
Arkaig	19	46.6	24	...	Chrystal (1910)
Lubnaig	6	13	24.4	...	Chrystal (1910)
Würm (Starnberger)	20.2	54	25	15.8	Ebert (1900)
Biwa					
Main basin	46	46.2	72.6	30.5†	Nakamura and
				25.2	Honda (1911)
				22.7	
				16.7	
				12.0	
				10.6	
				9.4	
				8.5	
South basin	14	...	230	...	
Mendota					
NS. axis	6.8	12	25.60	...	Bryson and Kuhn
E. axis	9.1	12.8	25.80	...	(1952)
Laggan	11	20.7	26.6	...	Chrystal (1910)
Tay	24	60.7	28.4	16.4	Chrystal (1910)
Ness	38.6	132	31.5	15.3	Chrystal (1910)
				8.8	
St. Wolfgang	11.2	47.1	32	6.24	
				4.6	Endrös (1906c)
Hemmelsdorfer	...	5.5	34.5	4.9	Griesel (1920)
				1.7	
Madü	16.6	18.7	35.5	20.3	Halbfass (1902, 1903)
				15.3	
				13.4	
				8.4	
				5.3	
Chiemsee	14.7	24.5	43.21	(See Fig. 93)	Endrös (1903, 1906b)
Lucerne (Lucerne-Flüelen axis)	38.6	104	44.25	24.25	Sarasin (1897–1900)
Lucerne (Küssnacht-Stansstad axis)	...	...	18.5	9.3	Sarasin (1897–1900)
Zurich	29	44	45.6	23.8	Sarasin (1886)
Neuchâtel	37.5	64	49.6	24.64	Sarasin and Pasquier
			39.4	20.6	(1895)
				15	
				8.3	
Constance	66	90	55.8	28.15	Forel (1893)
			39 (36)		

(*Continued*)

TABLE 26 (*Continued*)

Lake	Length, km.	z_m, m.	T_1, min.	Other Periods, min.	Author
Michigan	...	99	112†	...	Perkins (1893)
George, NSW.	30	5.5	131	...	See Chrystal (1910)
Baikal	665	680	231.2	31.9	See Halbfass (1923)
			72.6		
Huron	310	76	289	...	Denison (1898)
Balaton (whole)	77.2	...	600–720	117	Cholnoky (1897)
			43†		
Upper basin	...	...	143		
Lower basin	...	...	60		
Erie	400	21	786	222	Hayford (1922)
			156†		
Aral	430	15.6	1365	516	Berg (1908)

* Norwegian data given by Halbfass has been omitted as in part probably erroneous (Endrös 1908).
† Transverse seiche.
‡ Wiessee, southern basin of lake.

regularity having periods from 11.5 to 15.8 min., the great majority being within the range 12.5 to 14.1 min. with a mean about 12.9 min. These are presumably vibrations due to local resonance, dependent on the form of the profile of the bottom near the coast. Comparable phenomena are known in other lakes and in the sea.

At Krasnovodsk, south of the middle of the basin on the east shore, none of these periods were observed. At this station all oscillations were feeble and irregular, the long periods usually being 1 to 1.33 hours, the short from 1.5 to 8.5 min. but most commonly about 3.25 min. In a basin about 1000 km. long and with a mean depth of about 200 m., Merian's approximation would give 12.5 hours. This is well below half a pendulum day at 42° N., and is in fact less than the longitudinal seiche of the Sea of Aral. Nevertheless it is reasonably clear that Knipowitsch found no evidence of a uninodal longitudinal seiche. It is also obvious that both stations cannot be on the uninodal line. At Petrovsk the transverse uninodal seiche would probably have a period between 3.7 to 4.7 hours, the binodal between 2.14 to 2.35 hours. None of these roughly computed periods seem to bear significantly on the data. As far as can be ascertained, no recent work on the matter exists, though in view of the fact that the Caspian is the largest body of water cut off from the ocean, the subject is not without interest.

The seiches of Lake Erie have been investigated more completely than those of any other lake in the New World or, indeed, of any other really large lake. They are, however, still imperfectly understood. This is partly because no proper theoretical study of the basin has been published.

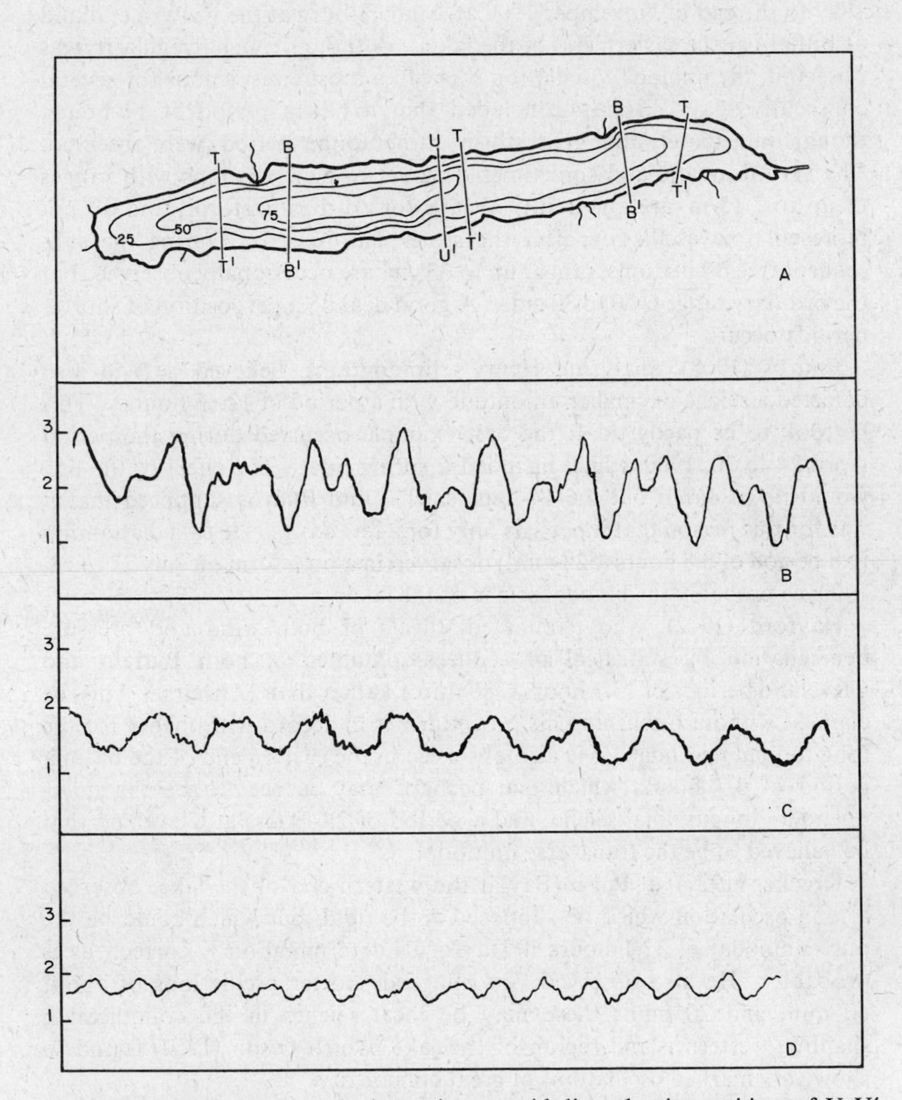

FIGURE 89. Loch Earn. *A*, bathymetric map with lines showing positions of U–U′ uninode, B–B′ binodes, T–T′ trinodes; *B*, limnogram of a dicrote seiche at the eastern end of the lake; *C*, limnogram of the pure uninodal component, recorded at the binode; *D*, limnogram of the pure binodal component recorded at the uninode. All limnograms were taken simultaneously. (After Chrystal, and Chrystal and Wedderburn.)

Henry (1902) gave a long series of limnograms made from December 1899 to the end of November 1900 at Amherstburg at the western end and at Buffalo at the eastern end of the lake. Although much irregularity was observed, the uninodal oscillation was often most conspicuous for several consecutive days. Henry concluded that it had a period of 14 hours, though immediately after great storms rather longer periods were observed. The very dramatic and sometimes destructive denivellations with ranges of up to 2.15 m. are apparently always forced during storms and do not represent free oscillation after the storm has passed. During the subsequent free oscillations, ranges up to 63 cm. are occasionally observed, but the ordinary range is 20 to 30 cm. A good deal of superposition of shorter periods occurs.

Endrös (1908), analyzing Henry's limnograms, believed he had also detected a seiche of smaller amplitude with a period of 12.65 hours. This he took to be partly tidal; the best example occurred during the period April 24 to 30, 1900, which included a spring tide. The effect of the tide would be to damp out the 14-hour seiche, and Endrös supposed that it was for this reason that it persists only for a few days. He called attention to a period of 8.8 hours (528 min.) observed in a pure form on July 27 to 28, 1900, as probably the binodal seiche of the lake.

Hayford (1922), who eliminated effects of both wind and pressure denivellation by statistical procedures, obtained at both Buffalo and Cleveland periods of 13.1 hours (786 min.), rather than 14 hours. This, he claimed without publishing his procedure, is in accord with theory for the longitudinal uninodal. He also observed in the eastern end of the basin a period of 3.7 hours, which one perhaps may suspect to be the quinquenodal longitudinal seiche, and a period of 2.6 hours at Cleveland that he believed to be the transverse uninodal.

Krecker (1928), at Put-in-Bay at the western end of the lake, observed a long oscillation which he supposed to be tidal, but which could be the pure uninodal of 13.1 hours if Hayford's determination is correct, as is probable. He also observed very marked shorter oscillations of about 60 min. and 20 min.; these may be local seiches in the complicated, shallow, western island region of the lake, which Olson (1952) found to show very marked oscillations of great complexity.

The other Laurentian lakes are less well known. Perkins (1893) recorded a seiche of 112 min., ranging up to 1.5 m., in Lake Michigan; Endrös takes this to be transverse. Krecker (1931) has published a complicated limnogram showing short oscillations. Denison (1897, 1898) found a seiche period of 289 min. in Lake Huron and another, supposedly transverse, of 45 min. Endrös (1908) computed from the Du Boys and symmetrical parabolic formulae that the uninodal longitudinal for Michigan should have a period between 7 and 11 hours, for Huron between 6 and 10 hours. It is possible that the seiche period of 7 hours observed in the Straits of Mackinac by Hayford (1922) is really one of these longitudinal uninodals. Hayford himself, on the basis of unpublished theory, considered it to represent the oscillation of Huron and Michigan as a whole, for which Endrös had computed from the convex parabolic formula a period of 39.2 hours.

In view of Young's (1929) observation of a strongly marked oscillation having a period of about 12.4 hours, which he took to be tidal, in the north channel of Lake Huron, a detailed restudy of the Michigan-Huron system is very desirable.

Ratio of uninodal to binodal period. The ratio of $T_1:T_2$ in natural bodies of water varies from 100:77 on one of the axes of the Chiemsee, studied by Endrös (1903, 1906b, 1908), to 100:19.5 for the St. Wolfgangsee, studied by the same investigator. Several cases, including Lake Constance supposed by Forel to have a seiche *à la quinte*, have ratios of about 100:70. That such high values are not infrequently to be expected is evident from Chrystal's treatment of the truncated concave quartic lake form. The extremely short binodal seiche of the St. Wolfgangsee has been investigated by Endrös. The lake is divided into two basins by deltaic deposits. The normal curve can be practically represented by a pair of concave parabolas. For the parabolic basins so defined, the theoretical uninodal seiche periods are 6.10 and 6.17 min. The observed binodal seiche for the lake as a whole with a period of 6.24 min. is therefore in good agreement with theory. For the uninodal seiche of the whole lake, we are dealing with a very convex basin, and though its period has not been computed, it would be expected to be greatly in excess of $2T_2$.

Transverse seiches. A certain number of transverse seiches have been recorded. They can be regarded as properly established only when evidence of alternating changes in level on opposite sides of the lake is forthcoming, as in the case of the Lake of Lucerne and of the Tachingersee (Fig. 90) (Endrös 1908). Presumptive evidence of a transverse seiche is sometimes obtainable by computing periods and then identifying them in the limnogram, as in the case of Lake Michigan. Well-established transverse seiches often occur across sections of lakes defined by strongly marked bays, and are not characteristic of the basin as a whole. The periods at different stations along a long lake may therefore be somewhat different, as is conspicuously exhibited in Lake Vetter (Fig. 91). Here the well defined transverse seiche across the lake at the level of Motala, which lies at the head of a wide, deep bay, has twice the period of a transverse seiche recognized in the narrower middle section of the lake.

Seiches in complex basins. A number of striking complications due to

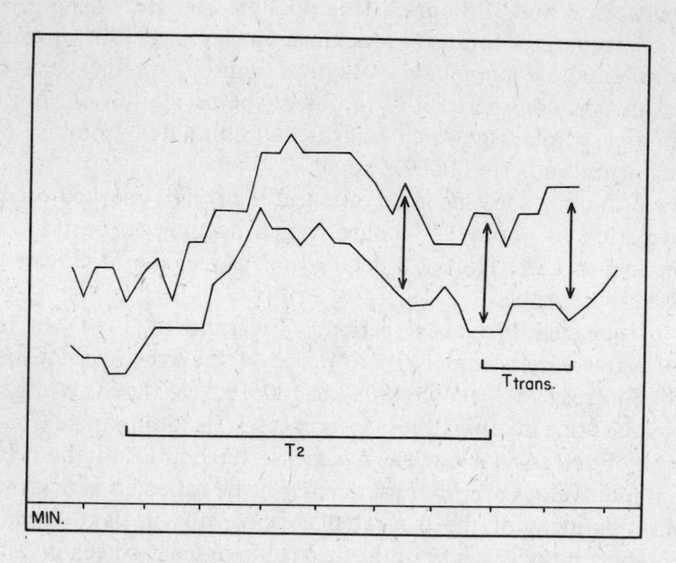

FIGURE 90. Tachingersee. Limnograms recorded simultaneously on opposite shores, showing a transverse seiche superimposed on part of the record of a binodal longitudinal seiche. The instruments used were not self-recording, observations being made every twenty-five seconds. (Endrös).

the form of the basin have been studied. A peculiarly beautiful case is provided by the annular caldera lake Toyako in Japan, which was investigated by Nakamura and Honda. Experiments with models suggest the possibility of two seiches at right angles, with almost identical periods of 10.3 and 11.2 min. An observed seiche with a period of 9.29 min. showed remarkable beat phenomena (Fig. 92), as if interference between two independent oscillations of similar but not identical period were occurring. Since the limnograph was set rather close to one nodal line, as indicated in the model, it is possible that the axes in the model are not quite the same as in the real lake, for one would have expected less interference so close to a supposed nodal line.

The most extraordinary case is no doubt that of the Chiemsee in Bavaria, in which Endrös (1903, 1906b) has recorded seventeen different seiches, excluding those with periods of less than 3 min.

For the main uninodal, which has a period of 41 min. the nodal line is displaced towards the shallow western end of the irregular basin, and the maximum amplitude occurs in a small bay at the north west corner of the lake (Fig. 93). Two other uninodals with periods of 54 min. and 36 min.,

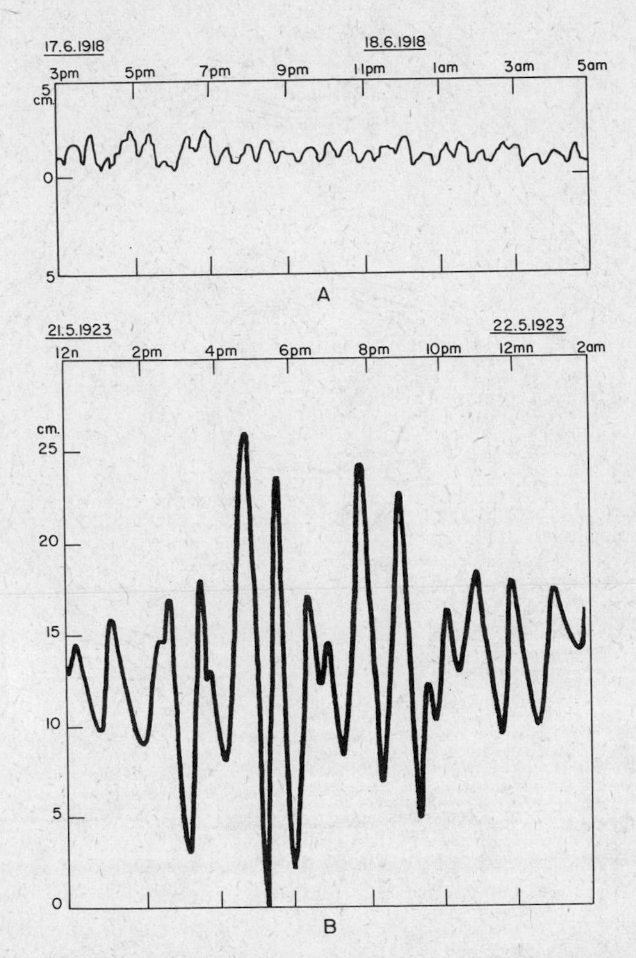

FIGURE 91. Transverse seiches in Lake Vetter: *A*, at Hastholmen; *B*, at Motala (after Bergsten.)

respectively, apparently represent separate oscillations in the shallower water north of and the deeper water south of the latitude of the large Herreninsel.

Three main binodal seiches occur, one of 28.5 min., which is longitudinal, the other two of 18 and 15.5 min. period, more or less transverse. The shorter period seiches give rise to extremely complicated and only partially analyzed patterns, owing to the presence of islands across the nodal lines, as is indicated in the case of the 8-min. quadrinodal. The

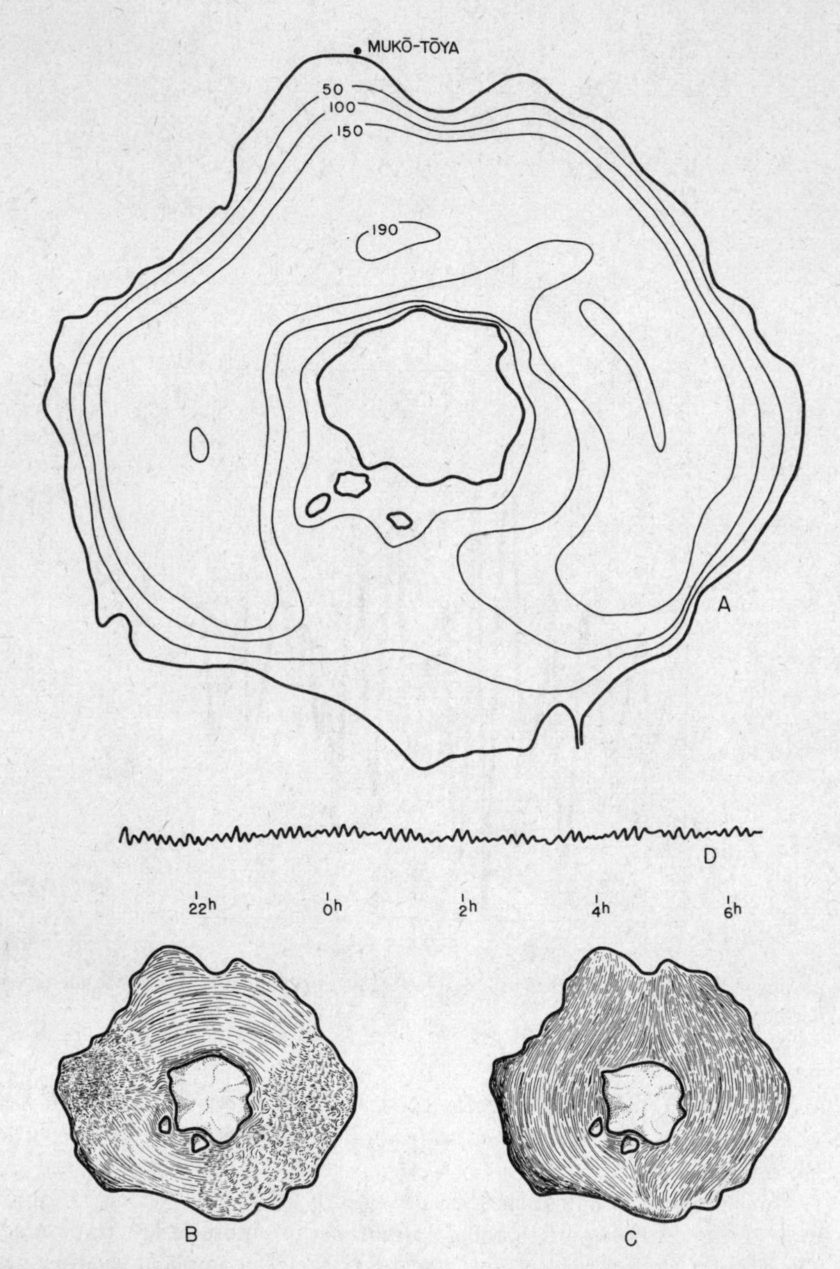

FIGURE 92. Toyako. *A*, bathymetric map of lake; *B*, *C*, vibration patterns of aluminum powder on the surface of water in a model of the basin; *D*, limnogram taken at Muko-Toya, showing apparent interference of two seiches of approximately equal period as developed in the model experiments. (After Nakamura and Honda.)

FIGURE 93. Chiemsee. *A*, position and relative percentage amplitude of positive and negative simultaneous denivellations in the uninodal seiche of 41 minutes period; *B*, in the binodal seiche of 28.5 minutes; *C*, in the binodal transverse seiche of 15.5 minutes; *D*, in the binodal transverse seiche of 18 minutes; *E*, in the quadrinodal seiche of 8 minutes. *F*, bathymetric map of the lake. (After Endrös.)

periods of some of these seiches are very sensitive to changes in water level. The main 41-min. uninodal can vary from 39.7 min. at low water to 43.9 min. at high water levels; this is in the opposite direction from that expected from elementary theory, as given by Merian or Du Boys, and clearly expresses changes in shape as littoral shelves are exposed.

The Wagingersee and Tachingersee in Bavaria provide a remarkable example of two practically independent lakes oscillating at times as a single unit (Endrös 1905). The two lakes are elongate basins, communicating by a passage of 10 m. long, and 20 m. wide, with a maximum depth of 5 m. Each lake has its own seiches; in the larger Wagingersee, uni-, bi-tri-, quadri-, quinque-, and perhaps septinodal seiches can be recognized, as indicated in Table 26. There is evidence that the uninodal of one lake can force oscillations in the other. There is also a very long uninodal of 62 min. period, which represents the uninodal oscillation of the two lakes as a whole. This never persists for more than ten successive oscillations, being evidently damped out by the frictional resistance provided by the channel. As has been indicated, Hayford believed that he had detected a comparable seiche with a period of 420 min. involving Lakes Michigan and Huron, but this needs reinvestigation.

A further interesting case is provided by Lake Biwa in Japan, a southern very shallow arm of which has its own long-period seiche of 230 min. The main uninodal seiche of the lake, period 72.6 min., is much more conspicuous in this basin though it is hard to observe in the lake itself. Nakamura and Honda (1911), in experiments with models, concluded that the 72.6-min. seiche is a free oscillation in the main basin but is forced with a considerably greater amplitude in the shallow southern arm, which is not, however, part of the resonating system actually producing the seiche.

Effect of turbulence on period. Since the damping effect of turbulence will start acting as soon as the seiche begins, if such damping is sufficiently great it will not merely cause the oscillation to die out but will have an appreciable effect in increasing the period. Defant (1932) has examined the rather complicated theory of this process for a simple rectangular lake of uniform depth z_m. If we define

$$\beta_s = \frac{A_v^2 l^2}{g z_m^5 \pi^2} \qquad (90)$$

it is possible to compute, as a function of β_s, the ratio of the period T_t, allowing for turbulence, to the seiche period T_m derived from the Merian formula (equation 73), using Table 27 prepared by Defant.

For Lake Balaton (Cholnoky 1897) the observed uninodal seiche has a period for mean water level of 690 minutes; Merian's formula for a lake

4 m. deep and 77.2 km. long would give T_m 411 min. The ratio is therefore 1.68, and β_s has approximately the value of 0.3, corresponding to an eddy viscosity coefficient of 45 cm.^{2} sec.$^{-1}$, a value of the same order as those known in the open ocean.[29] If the lake were twice as deep, β_s

TABLE 27. *Effect of turbulent damping on seiche period*

β_s	$T_t:T_m$
0.5370	∞
0.4732	3.217
0.3376	1.817
0.2132	1.424
0.1284	1.269
0.0615	1.174
0.0244	1.128
0.0094	1.106
0.0027	1.083

would be only 0.01 and the increase in the seiche period would be only about 10 per cent over that computed by the Merian formula. It is obvious that since the value of β_s involves the square of the length and the reciprocal of the fifth power of the depth, this effect of turbulent viscosity will be marked only in very large but extremely shallow lakes. Moreover, only in such lakes is the value of A_v likely to be sufficiently high to be significant. Lake Okeechobee in Florida should show the effect; this lake and Lake Balaton may well be the only bodies of water in which the phenomenon can occur on a considerable scale.

Origin of Seiches. Though a number of possible processes clearly may start a lake oscillating, it is reasonably certain that pressure denivellations and wind denivellations are the commonest. The theoretical adequacy of these causes has been demonstrated by Chrystal (1908) and by Bergsten (1926). The relative importance of the two depends greatly on the geographical position and surroundings of the lake. Forel (1895) believed barometric pressure changes to be most effective on Lake Geneva. Chrystal concluded that small traveling changes in pressure were more important than wind denivellation on Loch Earn, which like Lake Geneva lies on a west-east axis; Bergsten found the reverse to be true on Lake Vetter, which lies on a north-south axis. Bergsten points out that a sudden reversal of wind direction is theoretically very effective. There is often some difficulty in distinguishing wind effects from pressure effects, because a large wind denivellation is likely to be accompanied by disturbances in pressure distribution. Hayford's record for August 6 and 7,

[29] Part of the effect may be due to damping in the Straits of Tihany.

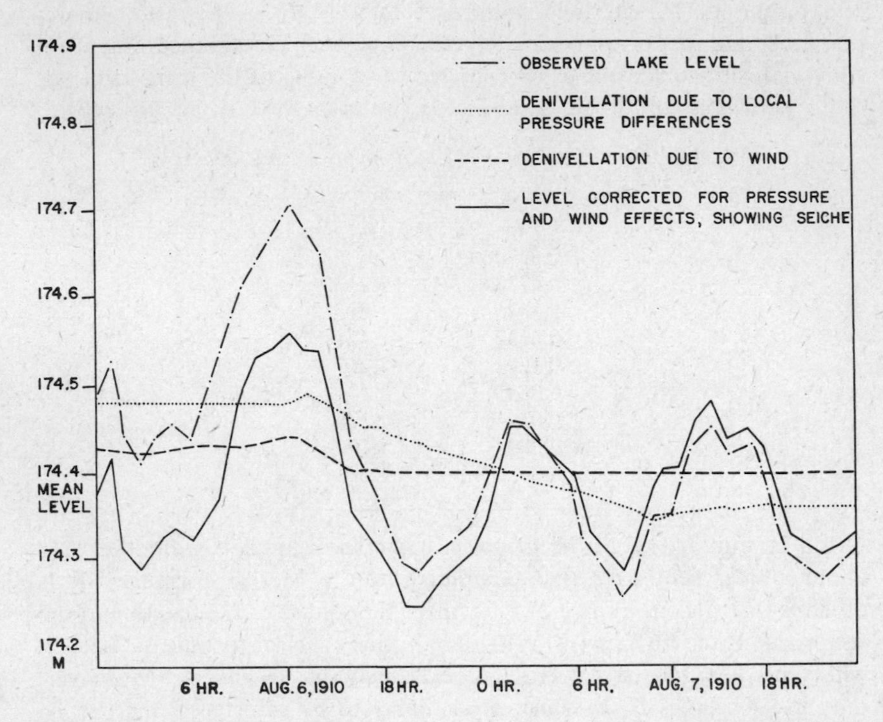

FIGURE 94. Denivellation of Lake Erie, August 6 and 7, 1910, analyzed statistically into a component due to wind denivellation and a component due to pressure (after Hayford).

1910, on Lake Erie (Fig. 94) shows a very striking denivellation in which both wind and pressure components are considerable, but in this case the wind component fell before the pressure component, and this fall started a very regular seiche. Bergsten, though he found wind denivellation to be the more frequent cause of seiches on Lake Vetter, gives good records of various kinds of pressure change initiating or amplifying a seiche (Fig. 95). Very small variations may have a quite remarkable effect if they happen to come in an irregular quasi-periodic series, because when the pressure changes happen to occur at intervals corresponding to the period of the seiche, they can force a persistent oscillation. It seems likely that this is the major method of genesis of the seiches of Loch Earn.

Endrös (1903) and Chrystal (1910) have given examples from the Chiemsee and Loch Earn, respectively, of seiches apparently generated by local showers of rain. It can be shown that the impact of such a shower is, in certain cases, quite adequate to account for the effect. Local

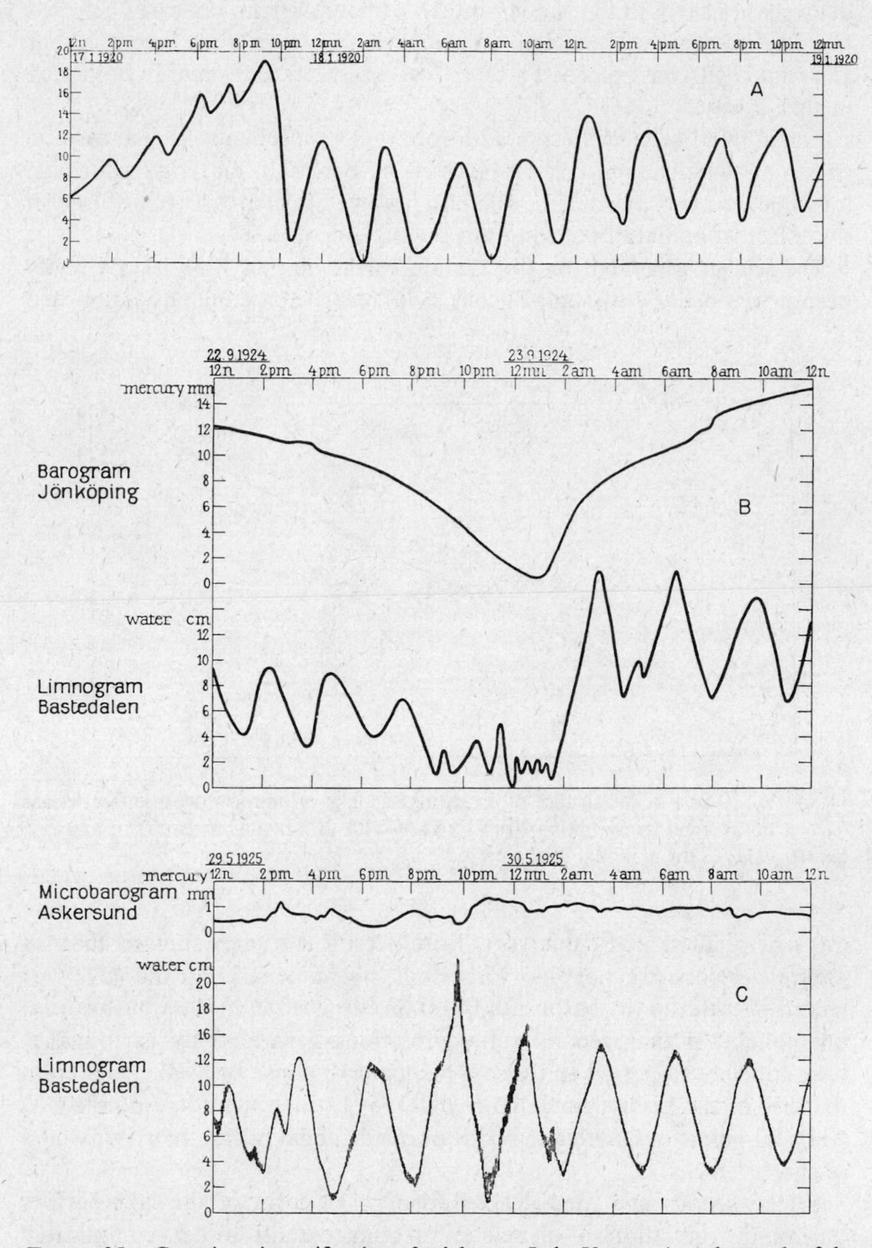

FIGURE 95. Genesis or intensification of seiches on Lake Vetter: *A*, as the result of the release of a strong wind denivellation; *B*, as the result of a single barometric low traveling across the end of lake opposite to that of the limnograph station; *C*, as the result of continuous small quasi-periodic pressure changes (Bergsten 1926).

denivellations due to the sudden inflow of flood water were early believed to cause the seiches of the Lake of Geneva; this is certainly incorrect, but Chrystal (1910) has recorded a case from Loch Earn reasonably explained in such a way.

Forel (1895) believed that in addition to these mechanisms, electrostatic effects of thunderclouds, first suggested by Bertrand, may be responsible for some seiches. It would be difficult, however, in such cases to distinguish this effect from that of concomitant pressure changes.

The seiches generated by the Lisbon earthquake of 1755 have already been mentioned. Although Fuchs (1876) recorded a number of supposed

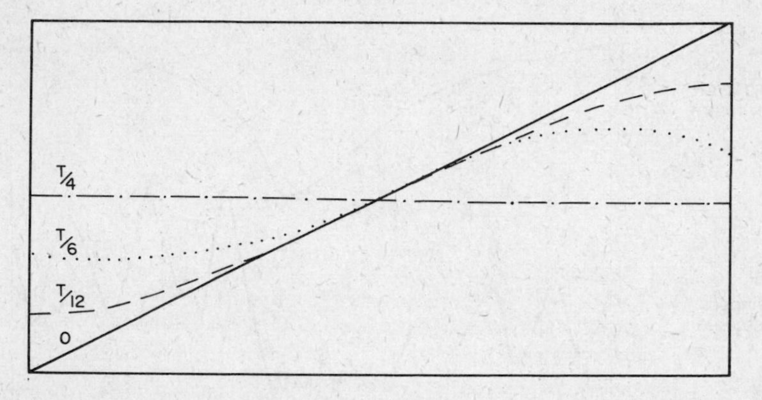

FIGURE 96. Form of the surface of a rectangular lake of uniform depth, after release from a linear wind denivellation (0), at one twelfth (T/12), one sixth (T/6), and one quarter (T/4) of the uninodal seiche period. (After Haurwitz.)

cases of oscillations of this sort, Forel's critical studies showed that, in general, seiches are not due to seismic phenomena. He did give one record of water in the basin of a fountain being set in motion in this way, but concluded that generally the vibrations generated by earthquakes have too short a period to cause a resonant response by a lake. Even in the case of the Lisbon earthquake of 1755, Loch Lomond responded, as Chrystal points out, with a trinodal or quadrinodal rather than a uninodal seiche.

Seiche periods and wind denivellation. The form of the lake surface during the initiation of a seiche presents certain rather complicated problems. These have been examined by Haurwitz (1951) for a rectangular lake of uniform depth. When the wind produces a linear denivellation and stops suddenly, the water surface moves towards the level most rapidly at the ends. Figure 96 shows the successive positions in a square lake of the area of Lake Okeechobee, with a uniform depth of 2.15 m. The seiche period is taken as 5.2 hours.[30]

When the wind changes in direction or in force, the rate of the change is of great importance. A rapid fall and rise produces little effect, but when the wind veers 180 degrees in a period of the order of $T_1/2$, rather large and superficially unexpected denivellations may occur. A computed example of the angle between wind and slope is given in Fig. 97, and a series of maps of actual denivellations on Lake Okeechobee during a hurricane are shown in Fig. 98.

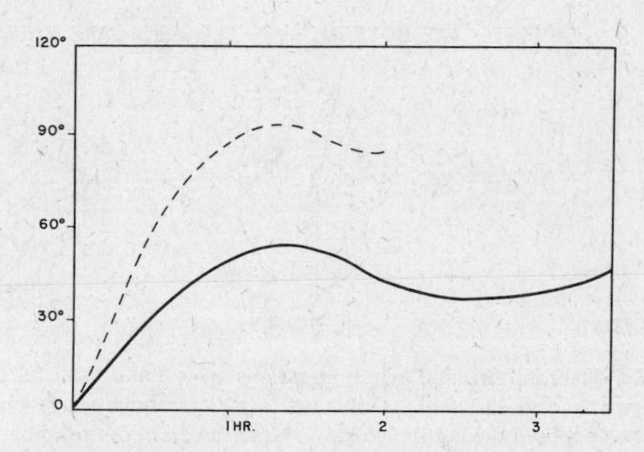

FIGURE 97. Computed angle between wind stress and the slope of the center of a water surface in a lake with a uninodal seiche period of 5.2 hours, when (broken line) the wind reverses its direction in 2 hours and (solid line) in 3.5 hours. (After Haurwitz.)

Effect of seiches on the form of lakes. Krecker (1931) has pointed out that the reversing currents produced by the 60-min. seiche observed at the western end of Lake Erie are of some value in keeping channels leading from the lake to marginal lagoons open and unsilted. This is worth noting, as these lagoons are important nurseries for young fish. It is quite probable that whenever a marked constriction occurs in a lake basin, seiche currents may tend to maintain the passage. The very strong seiche current (Cholnoky 1897) observed in the narrow Strait of Tihany in Lake Balaton may well provide an example.

Vibrations. A very rapid if transitory oscillation on the limnograph, usually of small amplitude, is often observed. Such a disturbance of the lake surface was termed by Forel a *vibration*. On Lake Geneva at

---

[30] It may well be longer in Lake Okeechobee owing to turbulent damping.

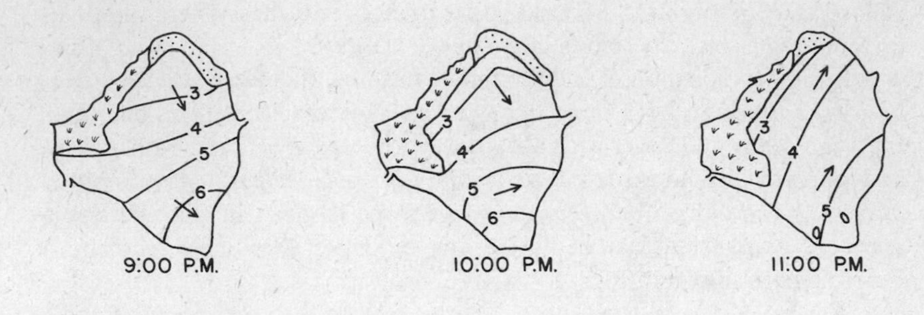

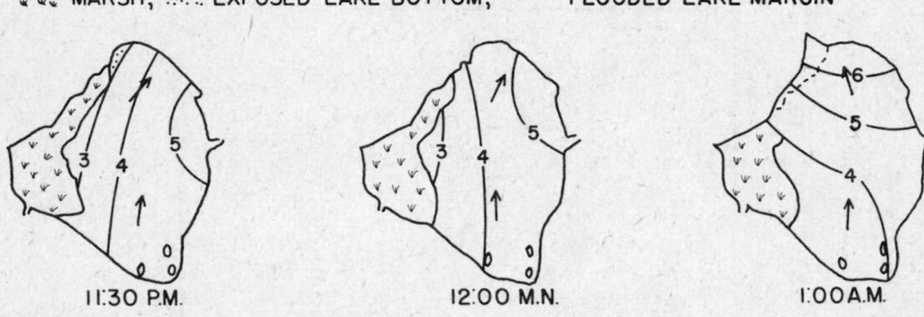

FIGURE 98. Surface contours and wind directions over Lake Okeechobee during a hurricane, August 26 and 27, 1949. Contours in meters above sea level; dotted area indicates lake bed exposed during the storm. Note the contours running parallel with the wind arrows from 11 P.M. to midnight. (After Haurwitz.)

Morges, Forel found that such vibrations may be started by the wind or by steamers passing at some distance from the coast. He supposed that they were analogous to the so-called sea seiches that have frequently been observed in marine coastal waters. Harris (1909) and Honda, Terada, Yoshida, and Isitani (1909) treated sea seiches as resonance phenomena due to the configuration of bays on the coast. It has been shown by Proudman (1925) that sea seiches can occur on rectilinear coasts, provided the depth suddenly decreases going from the open water to the coast line. If we imagine a series of any kind of propagated disturbance originating in the open water, it can be shown that if the wave length of any such disturbance is l_w and it travels toward the coast in water of depth z_d, suddenly reaching a littoral zone of depth z_s and breadth b_s, the condition for the excitation of a maximal disturbance at the coast is

$$l_w = \frac{4b_s}{2n + 1}\sqrt{\frac{z_d}{z_s}} \quad (91)$$

Thus for a lake 100 m. deep with a shelf 25 m. broad and 10 m. deep, the wave lengths to which resonance would occur would be 316 m., 105 m., 63 m., 45 m., 35 m., 29 m., 24 m., etc. The smaller wave lengths would be the commoner, but their resonant vibrations would die out more quickly. Embayed coast lines can, as has been indicated, have the same effect; for the more complicated theory the interested reader may consult Proudman (1925). It is evident that in actual cases the conditions are never likely to be simple; the theory is mainly of interest in showing that the effect can, under reasonable conditions, occur.

The short periods in the limnograms of the Caspian are, as has already been indicated (page 316), doubtless vibrations of this sort; other examples are given by Forel (1895), Chrystal (1910) and Endrös (1908). Some of these are reproduced in Fig. 99.

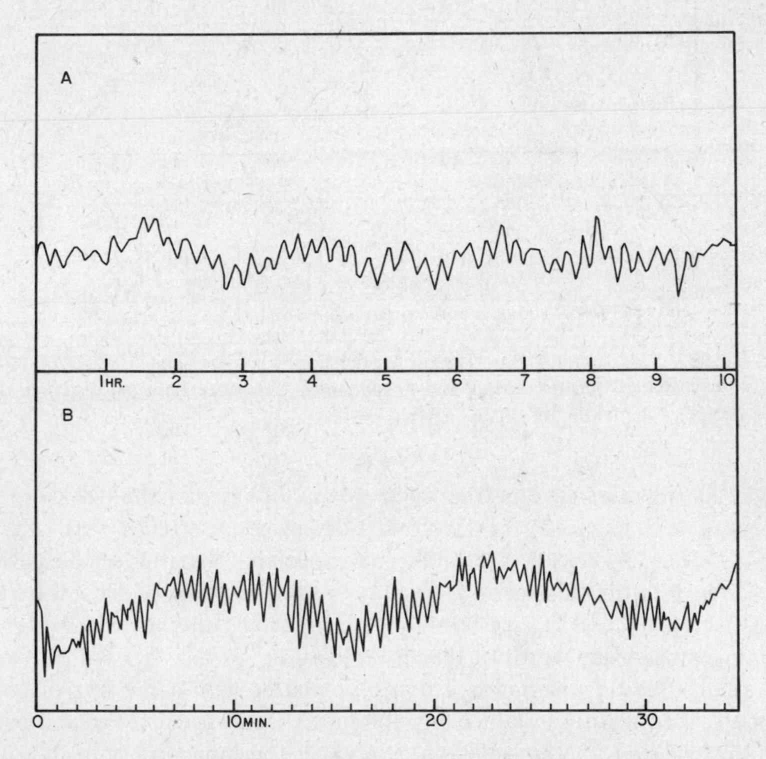

FIGURE 99. *A*, short-period oscillations, probably analogous to sea seiches, superimposed on the 2.75-hour seiche at Petrovsk, Caspian Sea (after Knipowitsch). *B*, very short-period vibrations of uncertain origin, perhaps due to trains of surface waves, superimposed on the uninodal seiche of Loch Earn (after Chrystal).

TIDES

The question of the occurrence of true tides due to the gravitational attraction of the moon and sun in lakes has not been adequately investigated.

It can be shown (cf. Proudman 1953) that if the longest surface-seiche period is short compared to the tidal period of about 12.5 hours, the gravitational effects will be negligible. In general the existence of true tides that can be detected would be expected only in large lakes with a fairly long seiche period. It has been possible, however, to demonstrate

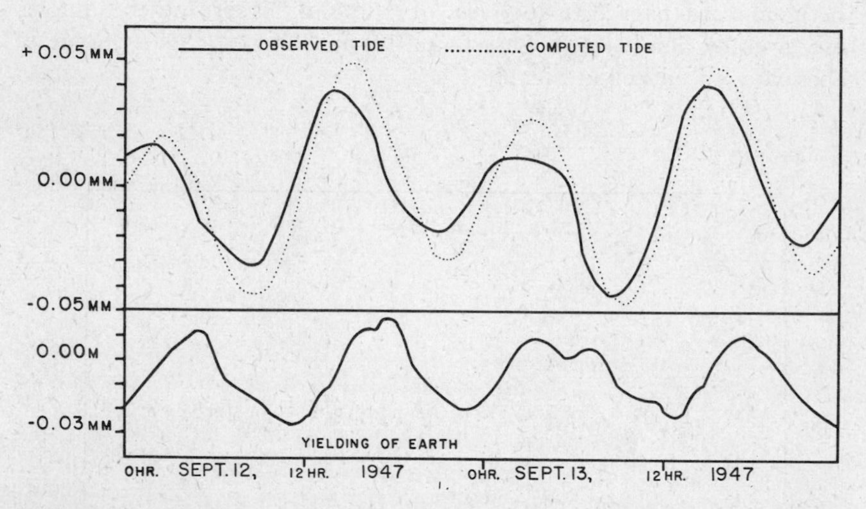

FIGURE 100. Tide in the David Taylor Model Basin. *Upper panel*, observed (solid line) and computed (dotted line) tides; *lower panel*, difference between observed and computed, due to yielding of earth. (After Zerbe.)

(Fig. 100) the presence of a true lunar tide in an experimental tank 813 m. long and 6.71 m. deep; the maximal tidal range was 0.084 mm. (Zerbe 1947, 1952). A special case might be expected when the seiche period coincides with the tidal period, so that the seiche could be forced by the tide; no such case has certainly been reported, though Lake Huron seemingly would be worth further investigation.

Actually the tidal period is a complex matter with a number of components, the most important being the lunar semidiurnal (M_2), and solar semidiurnal (S_2), the lunisolar diurnal (K_1), the lunar diurnal (O_1), and the solar diurnal (P_1).

The Great Lakes. Attention has naturally been turned to the Laurentian Great Lakes, which, lying east and west and being very large, may be expected to exhibit small tides. The early history of tidal investigations on these bodies of water is reviewed by Harris (1909).

Many of the nineteenth-century observers failed to distinguish seiches from true tides. Harris submitted several series of measurements of level to harmonic analysis and obtained, for the most important components the amplitudes given in Table 28.

TABLE 28. *Amplitudes of the more important tidal components at Milwaukee, Marquette, and Duluth*

	Lake Michigan		Lake Superior		
	Milwaukee, Sept.–Oct. 1852, mm.	Milwaukee, Oct.–Nov. 1852, mm.	Marquette, 1904, mm.	Duluth, 1901–2, mm.	Duluth, 1902–3, mm.
M_2 lunar semi-diurnal	7.5	7.0	3.9	19.2	20.1
S_2 solar semi-diurnal	3.1	6.3	2.6	10.2	10.3
K_1 lunisolar diurnal	...	...	2.0	9.1	10.5
P_1 solar diurnal	...	...	0.8	4.7	5.1
O_1 lunar diurnal	...	...	1.2	7.6	6.5

The lunitidal interval at Milwaukee appears to be about 30 min. The two Duluth series of observations each extend over a year and overlap over a period of six months. In spite of this overlap it seems reasonable to regard the closeness of the values obtained as confirming the reality of these five tidal components. Harris concluded that the Lake Superior tides accord with theory and imply little elastic yielding of the earth.

Young (1929) reported larger tides in the north channel of Lake Huron; although the mean interval in the strongly marked oscillation that he observed was 12.4 hours, the lunitidal interval recorded was somewhat irregular. The range, up to about 10 cm., seems very large. The oscillation may be a seiche, possibly forced by the tide. Young rejected the idea of a seiche, as the period seemed too long. No proper study of the seiches of Lake Huron has been made, but the matter is obviously of interest.

Hayford (1922) found no evidence of a tide on Lake Erie. Endrös computed a lunar tide of 35 mm. and believed the total range might be 50 mm. He believed part of Henry's (1902) record to show such a tide, but that in general, interference with the main uninodal seiche obscured the tide and damped the seiche. Olson (1950) examined the matter by means of periodogram analysis and found no evidence of a period near 12 hours but shorter than the seiche.

Lake Baikal. The only other well-analyzed independent lacustrine tide is that of Lake Baikal. Here Jekimov and Krawitz (1926) found a semidiurnal lunar component of 5 mm. in the Pestschannaja Bay. Sterneck (1928), analyzing the water-level data provided by Schostakowitsch, obtained an empirical analysis of the components, and has also predicted the amplitudes from theory (Table 29).

TABLE 29. *Amplitudes of various components of the tide in Lake Baikal*

Tides	From Observation, mm.	From Theory, mm.
M_2 lunar semidiurnal	4.5	8.6
S_2 solar semidiurnal	2.9	4.0
N_2 lunar elliptic semidiurnal	0.8	1.7
K_2 lunisolar semidiurnal	0.7	1.1
K_1 lunisolar diurnal	3.5	4.8
P_1 solar diurnal	0.8	1.6
O_1 lunar diurnal	2.2	3.4

The total displacement evidently never exceeds 1.5 cm., and is markedly less than the theoretical. Sterneck suggests that frictional damping explains the discrepancy; Proudman believes that elastic yielding of the earth may be involved. Zerbe (1952), in his studies of the tide in the David Taylor Model Basin, showed that the observed tide was less in range than the predicted, and concluded that tidal yielding of the earth was involved. The interested reader may consult Sterneck (1928) and Proudman (1953) for details, which though theoretically interesting to students of tides, are not likely to be of importance to limnologists interested in any other aspect of very large lakes.

INTERNAL SEICHES

Not only can a lake oscillate as a whole, but if it is stratified, the various layers of different density can oscillate relative to one another (*A* of Fig. 101; *A* of Fig. 103). The existence of such oscillations in thermally stratified lakes was first fully established for Loch Ness by Watson (1903, 1904a), though Thoulet (1894) had earlier hinted at their existence and explanation. A very detailed series of studies was conducted by the limnologists of the bathymetric survey of the Scottish lochs (Wedderburn 1907, Wedderburn and Young 1915), with theoretical (Wedderburn 1911, 1912) and model investigations (Wedderburn and Williams 1911).

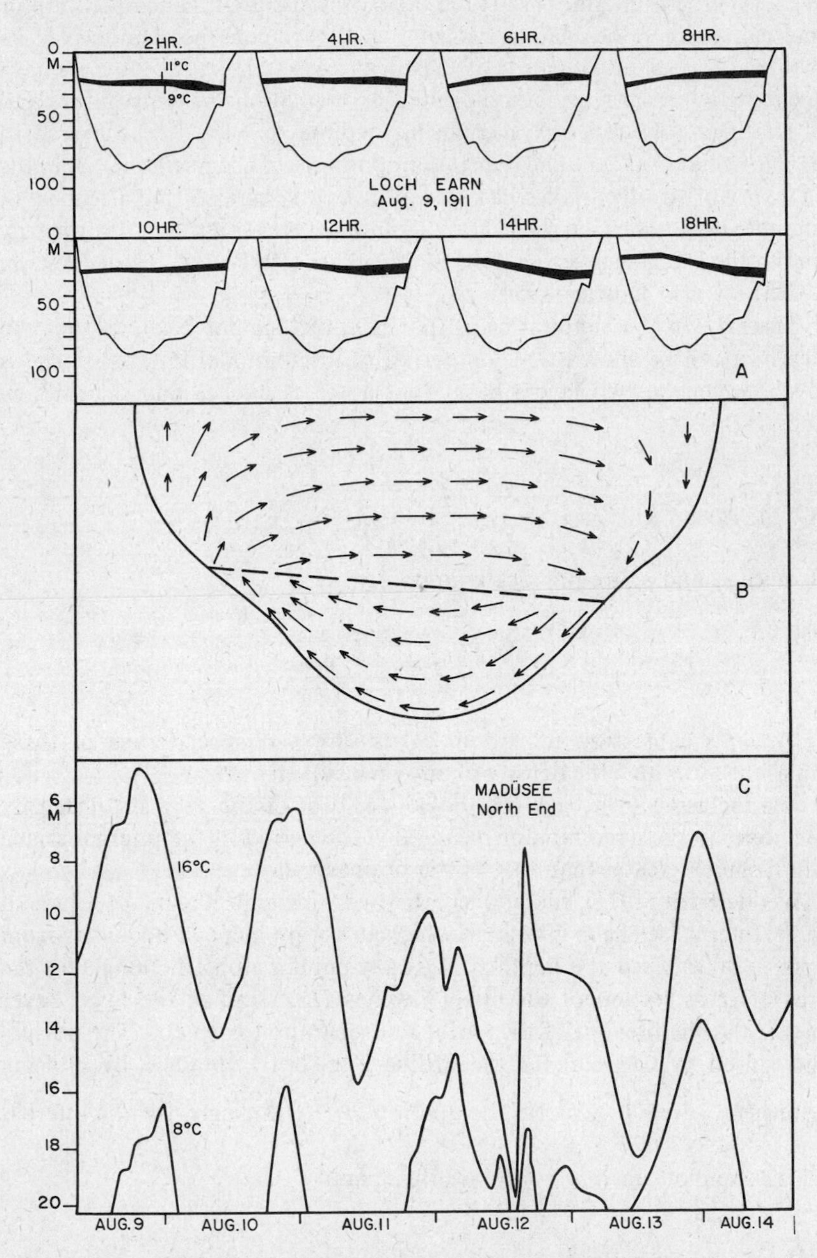

FIGURE 101. *A*, variations in the position of the layer bounded by the 11° and 9° C. isotherms in Loch Earn. *B*, generalized pattern of movements underlying such an internal seiche. *C*, variation in position of 16° and 8° C. isotherms at a single station at the north end of the Madüsee. (After Wedderburn.)

Wedderburn (1911, Halbfass 1910b) also demonstrated the temperature or internal seiche in the Madüsee (*C* of Fig. 101) to convince Halbfass of its reality. Several investigators of European lakes in more recent years have encountered internal seiches, indicated by oscillations of temperature and of stratified plankton. A magnificent example in Lake Erie was studied by Parmenter (1929). The immense importance of the phenomenon in the economy of stratified lakes has, however, largely escaped the attention of limnologists until recently. The recognition of this importance is primarily due to the brilliant investigations of Mortimer (1941–1942, 1951, 1952a,b, 1953a; see also Dussart 1954).

Theory. In the simplest case, that of a rectangular basin of uniform depth, it can be shown that the period of the uninodal internal seiche is given, when the two layers have thicknesses z_e and z_h and densities ρ_e and ρ_h, by

$$\text{(92)} \qquad \mathrm{T_i} = \frac{2l}{\sqrt{\dfrac{g(\rho_h - \rho_e)}{\rho_h/z_h + \rho_e/z_e}}}$$

or since ρ_h and ρ_e are practically unity, by

$$\text{(93)} \qquad \mathrm{T_i} = \frac{2l}{\sqrt{\dfrac{g(\rho_h - \rho_e)}{1/z_h + 1/z_e}}}$$

Merian's expression (equation 73) is simply a special case of these equations, in which the density of air can be taken as zero.

The inclusion of a term for the square root of the very small density difference in the denominator means that the period of the internal seiche will be much greater than that of the ordinary seiche.

Wedderburn (1911) has applied the hydrodynamic theory of Chrystal to the internal seiche in a basin in which an abrupt increase in density from ρ_e to ρ_h takes place at a depth z_e. At any point x along the long axis, the area of cross section of the upper layer is $H_e(x)$ and of the lower layer $H_h(x)$, and the breadth of the surface of separation is $b_e(x)$. The normal curve used by Chrystal for the ordinary seiche is obtained by plotting against $\mathrm{A}_{0,x} \left(= \int_0^x b_e(x)\,dx \right)$ the quantity $b_x H_x$; similarly, for the internal seiche one plots against $\int_0^x b_e(x)\,dx$ the quantity

$$\text{(94)} \qquad b_e(x)\left(\frac{\rho_e}{H_e(x)} + \frac{\rho_h}{H_h(x)}\right)^{-1} \simeq b_e(x)\left(\frac{1}{H_e(x)} + \frac{1}{H_h(x)}\right)^{-1}$$

The resulting curve is approximated to one or more analytically known curves. If the normal curves for the ordinary seiche and the temperature seiche are of the same form, i.e., parabolic, and the ratio of the maximum value of the ordinate of the former to that of the latter is R (which is always greater than unity), one can obtain an approximate relationship between the periods of the ordinary and internal seiches of the form

$$\text{(95)} \qquad \mathrm{T_i} = \mathrm{T_1}\frac{\mathrm{A_e}}{\mathrm{A_0}}\sqrt{\frac{R}{\rho_h + \rho_e}}$$

In a later paper Wedderburn examined the case of a continuous density gradient, which is broken up into a series of steps $\rho_1, \rho_2, \rho_3 \cdots \rho_m$ for layers of breadth $b_i(x)$ above some plane of oscillation, and a series $\rho_1' \ \rho_2' \cdots \rho_n'$ for layers of breadth $b_j'(x)$ below the plane of oscillation. The ordinate for the normal curve, corresponding to the right-hand part of equation 94, now becomes

$$\text{(96)} \qquad \sum_{j=1}^{n} \rho_j' b_j'(x) + \sum_{i=1}^{m} \rho_i b_i(x) \left(\frac{1}{H_e(x)} + \frac{1}{H_h(x)} \right)^{-1}$$

Applying Chrystal's method, using equation 94 Wedderburn obtained for the uninodal internal seiche of the Madüsee a period of 24.8 hours; using the approximation (equation 95) with an ordinary seiche of 35.5 min., a period of 24 hours. The observed period is 25 hours.

In Loch Earn, taking $z_e = 16$ m. and using expression 96, the uninodal period is 14.99 hours; with $z_e = 17$ m., 14.7 hours; and with $z_e = 15$ m., 15.3 hours. The observed period was 15.2 hours. The binodal period, with $z_e = 16$ m., was calculated as 8.44 hours and observed to be 8.2 hours.

The pattern of movement in a more or less parabolic basin, as deduced by Wedderburn for two layers, is given in *B* of Fig. 101. There is probably some horizontal vortical or turbulent motion at the surface of separation when the density change is not abrupt. Mortimer (1953a) has concluded, from his study of the internal seiches in Windermere, that a two-layer theory is usually adequate and that for most purposes the seiche nodality can be identified by computation from a formula of the form of equation 93, in which the period is inversely proportional to the nodality.

Empirical studies. The observational data have recently been summarized and discussed by Mortimer (1953a). It is reasonably certain that when other lakes are studied internal seiches will prove to be a universal concomitant of stratification. As has already been indicated, both periods and ranges are far greater than is the case of the surface seiche, and though the internal seiche is invisible at the surface, it is of far greater general importance. The relevant observations are recorded in Table 30, the calculated value being based on the mean depth of the hypolimnion along the longitudinal profile of greatest depth.

TABLE 30. *Periods of internal seiches in various lakes*

Lake	Year Observed	Observer	Period Observed, hours	Period by Equation 93, hours	Remarks
Lunzer Untersee	1927	Exner (1928)	3.7	3.9	Trinodal perhaps present
Windermere					
North basin	1947	Mortimer (1953a)	15	14.8	
North basin	1951	Mortimer (1953a)	12–14	13.8	
South basin	1950	Mortimer (1953a)	23–24	19.2	
South basin	1951	Mortimer (1953a)	23–25	22.2	
Loch Earn	1911	Wedderburn (1912)	15–16	16.4	Range up to about 8 m. Wedderburn's theory gives 15 hours
St. Wolfgangsee	1907	(Exner 1908a,b)	25	20.6	Binodal and other oscillations observed. Range up to about 10 m.
Madüsee	1910	Halbfass (1910b), Wedderburn (1911)	25	25.8	Range 10 to 15 m.
Würmsee (Starnbergersee)	1894	Ule (1901), Mortimer (1953a)	30	29.2	Range over 4 m.
Loch Ness	1903	Wedderburn (1907), Mortimer (1953a)	57.6–60	60	Range up to 60 m.
Lake of Geneva	1941–44	Mortimer (1953a), Dussart (1954)	72–108	96	
Lake Baikal	1914	Schostokowitsch (1926), Mortimer (1953a)	912	1812	Range up to 150 m. Presumably binodal

The abnormally long period in the St. Wolfgangsee is in accord with the form of the basin, already discussed. The data for Windermere, discussed in detail by Mortimer, show in general separate internal seiches in the north and south basins. Mortimer applied an instrumental frequency analysis to his records. In the south basin in 1950, the uninodal seiche showed as a strong period of about 23.5 hours (*G* of Fig. 104). The slight excess of this period over that calculated from elementary theory is attributed to the narrowness of the deep water at the north end of the lake. There is also a subsidiary peak at 13.5 hours, presumably the binodal, and perhaps another harmonic at about 11 hours. There is another subsidiary peak at 43 hours and perhaps at 19 hours. It is probable that these last two periods are uni- and binodal oscillations that originate in the movement of a third deeper layer. In the summer of the next year the picture is much less regular, and after the beginning of August the binodal seiche appears to be dominant. Mortimer supposes that as the thermocline is depressed, bottom relief tends progressively to determine the nature of the oscillation. The periodogram of the north basin exhibited not only the calculated uninodal between 12 and 15 hours but also a number of minor peaks between 16 and 40 hours, of uncertain significance (*H* of Fig. 104).

Geostrophic effects on the internal seiche. In addition to these observations, Mortimer (1955a) has also studied the geostrophic effect observable in the internal seiche of Loch Ness. The range of the seiche may exceed 50 m., and the period is rather over 48 hours, in accordance with theory. Both of the currents of the seiche will be deflected to the right, but as the water in the upper layer moves in one direction, that of the lower layer moves in the opposite direction. The result (Fig. 102) is a rotatory movement tilting the thermocline in both a long and a transverse axis. With a density difference of 0.0004, a two-layer model suggests a maximum relative velocity of 9 cm. sec.$^{-1}$ between the layers. This is found to correspond to a transverse range between 4 and 5 m., in agreement with observations.

Internal seiches in very large lakes. Apart from these detailed studies, two very dramatic cases must be mentioned. Parmenter (1929) found that the temperature seiche of Lake Erie, in which at rest the hypolimnion is confined to a relatively small deep area of the lake, could cause the margin of the cold deep layer to move horizontally up to 64 km.

One of the most recent reports ([Beauchamp] 1954) on internal seiches is a note on a very remarkable example with a period of about 30 days in Lake Victoria. The currents of this seiche force water into channels leading to somewhat isolated peripheral parts of the lake and tend to cause turbulent exchange with water over the bottom, bringing nutrients into the free water of these littoral regions.

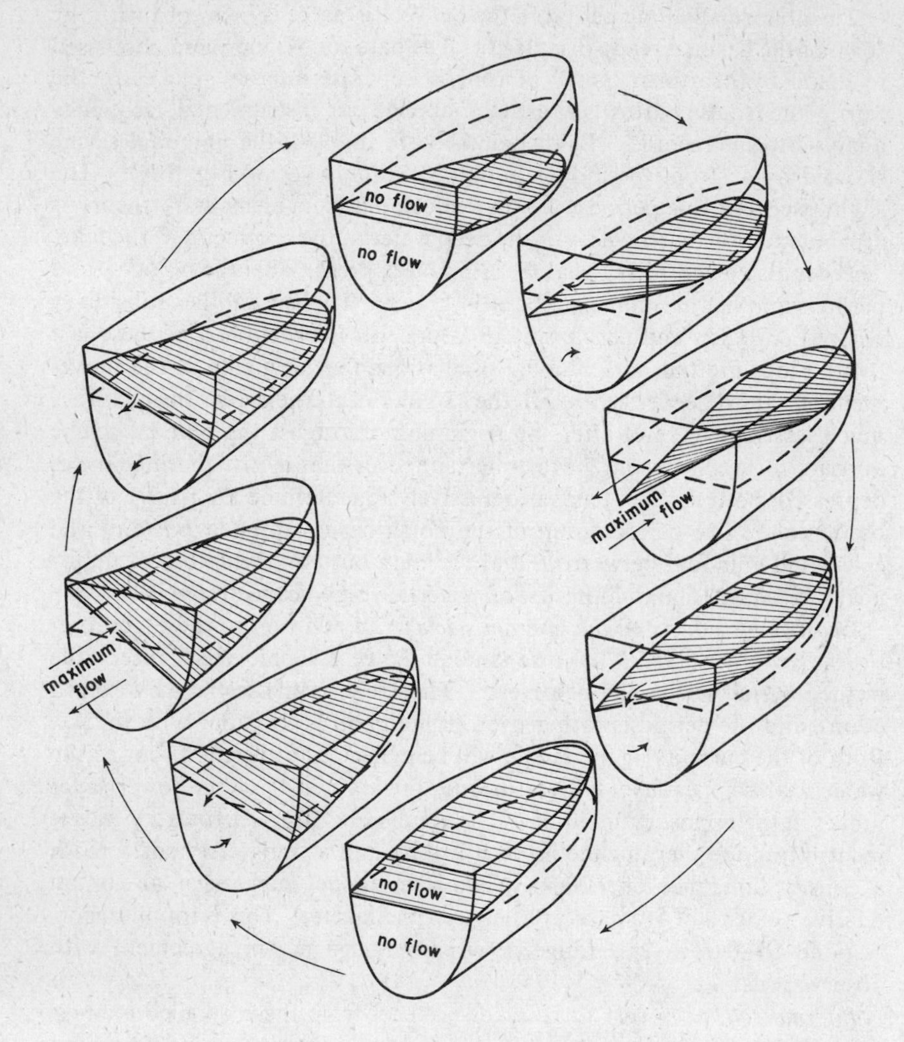

FIGURE 102. Geostrophic effect on the internal seiche in Loch Ness (Mortimer).

The so-called plankton seiche. Demoll (1922) has observed movements of isotherms in the Walchensee, certainly due to a temperature seiche as has already been indicated, which are accompanied by differences in the quantity of zooplankton that can be collected at a particular station in the lake. Demoll supposes that the layer inhabited by *Daphnia* in the thermocline is at any position periodically thickened or attenuated by the water movements, concentrating or rarefying the zooplankton available in a vertical column at the station investigated. The data presented are very inadequate.

Significance of internal seiches in causing horizontal water movements. It is reasonably certain that the turbulence of the hypolimnia of all lakes carefully studied, as well as the horizontal movements deduced by Hutchinson (1941) in his study of the stratified water of Linsley Pond and directly demonstrated by McCarter, Hayes, Jodney, and Cameron (1952), are due to the internal seiche current system, which in nature will not exhibit laminar flow and will therefore not lead, as the seiche dies down, to an exact return of every water particle to the place at which it lay before the seiche started. The very surprising currents sometimes recorded, as by Verber, Bryson, and Suomi (1953, Bryson and Bunge 1956) in Lake Mendota, in isolated studies of the deep water of stratified lakes, are also almost certainly due to internal seiches. Demoll's (1922) observations, made in water clearly exhibiting a temperature seiche, suggest that even in a relatively small lake the movements are not simply up and down the long axis, but may involve spiral current structure.

The theory of the three-layered system. For certain purposes it is desirable to consider the lake as a three-layered system; we have already seen that long periods suggesting a third layer occur in Windermere. Longuet-Higgins (in Mortimer 1952a), following in part Makkaweev (1936), has given a theory for such a system in a rectangular parallel-sided basin of depth $z_1 + z_2 + z_3 = z_u$. If we consider the seiches as stationary waves of period T and wave length l_s, we may write

$$T = \frac{2\pi}{\phi_s}, \qquad l_s = \frac{2\pi}{\kappa_s} \qquad \kappa_0 = \frac{\pi}{l_s}$$

We now define a quantity H as

$$H = -\frac{\phi_s^2}{g\kappa_s^2}$$

It is possible to show that H is given in terms of the densities and depths of the layers by the cubic

$$\text{(97)}\quad H^3 + H^2(z + z_2 + z_3) - H\left[z_2 z_3\left(\frac{\rho_2}{\rho_3} - 1\right) + z_1 z_3\left(\frac{\rho_1}{\rho_3} - 1\right) + z_1 z_2\left(\frac{\rho_1}{\rho_2} - 1\right)\right] + z_1 z_2 z_3\left(\frac{\rho_1}{\rho_2} - 1\right)\left(\frac{\rho_2}{\rho_3} - 1\right) = 0$$

One solution, $H^{(1)}$, is given approximately by

$$\text{(98)}\qquad H^{(1)} = -(z_1 + z_2 + z_3)$$

which corresponds to the ordinary surface seiche and is of no further interest.

Since the other two roots of equation 98 are very small quantities, the term in H^3 is negligible when they are sought, and we obtain

$$(99)\quad H^2(z_1 + z_2 + z_3) - H\left[z_2z_3\left(\frac{\rho_2}{\rho_3} - 1\right) + z_1z_3\left(\frac{\rho_1}{\rho_3} - 1\right) + z_1z_2\left(\frac{\rho_1}{\rho_2} - 1\right)\right] + z_1z_2z_3\left(\frac{\rho_1}{\rho_2} - 1\right)\left(\frac{\rho_2}{\rho_3} - 1\right) = 0$$

The two roots of this equation, which can be obtained by ordinary elementary algebra, are designated $H^{(2)}$ and $H^{(3)}$; they refer to the seiches set up at the top and bottom of the middle layer.

We now define the displacements at the surface, the upper pycnocline, and the lower pycnocline, due to any of the three possible seiches, as $h_1^{(i)}$, $h_2^{(i)}$, $h_3^{(i)}$, where i = 1, 2, or 3 according to whether the seiche considered is generated by displacement of the surface, by the upper pycnocline, or by the lower pycnocline. It can be shown that

$$(100)\qquad h_1^{(i)}:h_2^{(i)}:h_3^{(i)} = H^{(i)}:z_i:\frac{H^{(i)}z_1z_3}{z_2z_3\left(\frac{\rho_3}{\rho_2} - 1\right) + H^{(i)}(z_2 + z_3)}$$

The ratios of the displacements can therefore be computed solely from data on the depths and the densities of the layers.

We may write

$$(101a)\qquad \beta^{(2)} = \frac{h_2^{(2)}}{h_3^{(2)}} = z_2H^{(2)}\left(\frac{\rho_3}{\rho_2} - 1\right) + \left(\frac{z_2}{z_3} + 1\right)$$

$$(101b)\qquad \beta^{(3)} = \frac{h_2^{(3)}}{h_3^{(3)}} = z_2H^{(3)}\left(\frac{\rho_3}{\rho_2} - 1\right) + \left(\frac{z_2}{z_3} + 1\right)$$

Moreover, if the surface seiche is negligible compared to the two internal seiches, we can obtain for the displacements h_2 and h_3 at any value of x and t, as Fourier expansions,

$$(102a)\quad h_1 = 0$$

$$(102b)\quad h_2 = \sum_{n=1}^{\infty} \cos n\kappa_0 x(\beta^{(2)}A_n^{(2)} \cos n\phi_0^{(2)}t + \beta^{(2)}B_n^{(2)} \sin n\phi_0^{(2)}t) + \sum_{n=1}^{\infty} \cos n\kappa_0 x(\beta^{(3)}A_n^{(3)} \cos n\phi_0^{(3)}(t) + \beta^{(3)}B_n^{(3)} \sin n\phi_0^{(3)}t)$$

$$(102c)\quad h_3 = \sum_{n=1}^{\infty} \cos n\kappa_0 x(A_n^{(2)} \cos n\phi_0^{(2)}t + B_n^{(2)} \sin n\phi_0^{(2)}t) + \sum_{n=1}^{\infty} \cos n\kappa_0 x(A_n^{(3)} \cos n\phi_0^{(3)}t + B_n^{(3)} \sin n\phi_0^{(3)}t)$$

When the motion starts from rest, the terms in $\sin n\phi_0^{(1)}t$ are zero, and $A_n^{(2)}$ and $A_n^{(3)}$ are given by

$$\text{(103a)} \qquad A_n^{(2)} = \frac{2}{l}\int_0^l \frac{h_2'(x) - \beta^{(3)}h_3'(x)}{\beta^{(2)} - \beta^{(3)}} \cos n\kappa_0 x \, dx$$

$$\text{(103b)} \qquad A_n^{(3)} = \frac{2}{l}\int_0^l \frac{h_2'(x) - \beta^{(2)}h_3'(x)}{\beta^{(3)} - \beta^{(2)}} \cos n\kappa_0 x \, dx$$

where $\kappa_0 = \frac{\pi}{l}$, $\phi_0' = \frac{\pi}{l}(-gH^{(i)})^{1/2}$, and $h_2'(x)$ and $h_3'(x)$ are the initial displacements of the upper and lower pycnoclines at any point x.

Finally, when the motion starts at rest, the velocities u_1, u_2, and u_3 in a horizontal plane in the middle of the lake are given by

$$\text{(104a)} \qquad z_1u_1 = \sum_{n=1}^{\infty} \frac{\phi_0^{(2)}}{\kappa_0} \sin \tfrac{1}{2}n\pi\beta^{(2)}A_n^{(2)} \sin n\phi_0^{(2)}t + \sum_{n=1}^{\infty} \frac{\phi_0^{(3)}}{\kappa_0} \sin \tfrac{1}{2}n\pi\beta^{(3)}A_n^{(3)} \sin n\phi_0^{(3)}t$$

$$\text{(104b)} \qquad -z_3u_3 = \sum_{n=1}^{\infty} \frac{\phi_0^{(3)}}{\kappa_0} \sin \tfrac{1}{2}n\pi A_n^{(2)} \sin n\phi_0^{(2)}t + \sum_{n=1}^{\infty} \frac{\phi_0^{(3)}}{\kappa_0} \sin \tfrac{1}{2}n\pi A_n^{(3)} \sin n\phi_0^{(3)}t$$

$$\text{(104c)} \qquad z_2u_2 = -(z_1u_1 + z_3u_3)$$

Application to Windermere. Mortimer has considered the case of oscillations caused by a gale on June 9, 1947, which exposed, at its maximum, water at a temperature of 6.7° C. at the surface, though the ordinary surface temperature at the time was 14.1°. The initial condition is taken rather later when the gale was moderating, as there is no complete profile during the morning of the day of the gale. The arrangement of isotherms in Fig. 103 is taken as the zero time position from which the motion starts. The top and bottom of the middle layer are taken as defined by the 11° C. and 6.7° C. isotherms. The positions of these isotherms can be expressed by Fourier analysis as cosine series, for the 11° C. isotherm as

$$8.5 - (2.08 \cos \pi x/l + 0.28 \cos 2\pi x/l + 0.46 \cos 3\pi x/l)$$

and for the 6.7° isotherm as

$$20.5 - (6.40 \cos \pi x/l + 2.62 \cos 2\pi x/l + 1.33 \cos 3\pi x/l)$$

Three terms are found to be sufficient. The mean position of the 11° C. isotherm, namely 8.5 m., gives z_1; the difference in mean positions $(20.5 - 8.5 = 12.0)$, z_2; and z_3 is taken as 21.5 m. (*sic*).

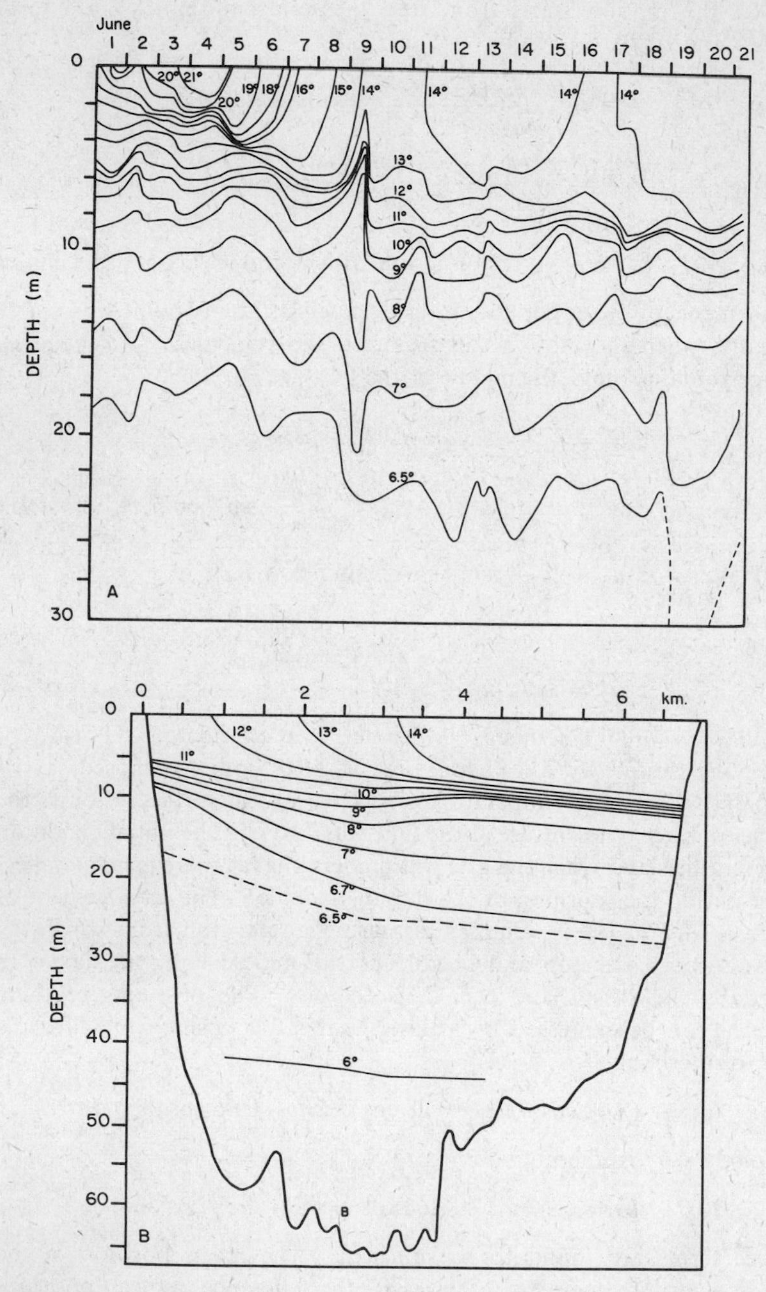

FIGURE 103. *A*, isotherms at station B in the northern basin of Windermere, June 1 to 21, 1947. *B*, isotherms throughout the basin, June 9, 1947, representing the initial condition for the analysis of an internal seiche. Position of station B is indicated. (After Mortimer.)

$H^{(i)}$ can be found from equation 100, knowing z_1, z_2, z_3, and ρ_1, ρ_2, and ρ_3, which can be obtained from the mean temperatures of the layers. Both from the elementary theory of the temperature seiche and from observation, the periods may be taken as $T^{(2)} = 19$ hours and $T^{(3)} = 41$ hours.

All the required quantities for evaluating equations 102a, b, and c, for any value of x can thus be obtained. It is assumed that three terms in t are sufficient as they are in x, i.e., the summation is done over $n = 1, 2$, and 3.

The resulting curves for Mortimer's station at the northern end of the basin are shown in *A*, *B*, and *C* of Fig. 104. The times of maximum elevation of both isotherms are clearly predicted. The position of the 11° C. isotherm is, however, progressively nearer the mean than the curve would suggest. If the theoretical curve is reduced continuously by 23 per cent per cycle, the fit is excellent. This reduction presumably measures damping, largely by turbulent viscosity. Both the damped and

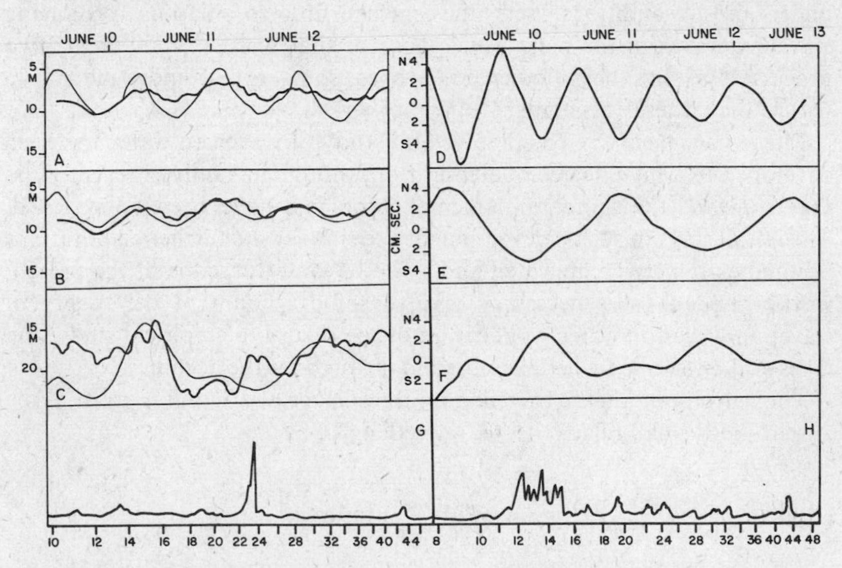

FIGURE 104. Temperature seiche in Windermere. Variation in position (heavy line) of the 11° C. isotherm as observed, and (thin line) as calculated for *A*, undamped theoretical variation, and *B*, an arbitrary damping of the theoretical variation. *C*, the same for the 6.7° isotherm, the calculated line damped as in *B*. *D*, horizontal velocity at uninode, computed for damped oscillation, in upper layer; *E*, the same in middle layer; *F*, in lower layer. *G*, frequency periodogram of temperature change at 13 m., south basin, July 1 to August 19, 1950; *H*, the same, north basin, 14 m., August 14 to Oct. 17, 1951. (After Mortimer.)

the observed curves show some influence, indicated by the dotted line, of the long-period seiche of 40 hours on the upper isotherm. A comparable rate of damping applied to the 6.7° C. isotherm leads to a curve following the observed curve in its broad outlines but without the detailed embroidery that was actually present.

From these damped theoretical values of the displacements, the theoretical horizontal current velocities due to the seiche in the top, middle, and bottom layers at the uninodal near the middle of the lake can be computed. These are shown in *D*, *E*, and *F* of Fig. 104. Calculations based on theory due to Taylor (1931a,b and to Goldstein (1931) suggest that these velocities are stable against general disintegration of the pattern, but are nevertheless certain to be accompanied by small-scale turbulent mixing. If, as Mortimer believes, this will be reduced in the thermocline region,[31] damping should be less rapid when the thermocline is very sharp.

SURFACE WAVES

Wind blowing over water not merely produces a wind drift but also, under most conditions, sets the surface into oscillation, producing traveling waves in the most usual sense of that word. Though the two processes, as has been indicated, are to some extent independent, in nature they usually accompany one another.

Elementary theory. The theorectical study of waves on water has been developed to a high degree of elaboration, though it is only in recent years that some of the more fundamental problems have been approached. The initial steps in such a development are simple; the further elaborations become extremely complicated and quite beyond the scope of the present work. Though it is necessary to give some account of the theory of traveling waves on water, application of the theory to empirical studies on lakes will be stressed wherever possible throughout the account.

The velocity of a wave traveling on the surface of a liquid is given, if the amplitude is small relative to the wave length, by

$$(105) \qquad v_w = \sqrt{\frac{g}{\kappa} + \frac{T}{\rho}\kappa \tanh \kappa z}$$

where κ is defined by

$$(106) \qquad l_w = \frac{2\pi}{\kappa}$$

l_w being the wave length, and T the surface tension of water.

---

[31] Cf., however, pages 256, 284 fn. and 475.

For wave lengths in excess of $2\pi = 6.28$ cm., $\kappa < 1$ and $\frac{T}{\rho}\kappa$ is small relative to $\frac{g}{\kappa}$. Moreover, if $2z > l_w$, $\kappa z > \pi$ and tanh κz is practically unity, so that for ordinary waves in open water we may write

(107) $$v_w = \sqrt{\frac{g}{\kappa}} = \sqrt{\frac{gl_w}{2\pi}} = 12.5\sqrt{l_w} \text{ cm. sec.}^{-1}$$

Similarly, the period will be $0.08\sqrt{l_w}$ sec.$^{-1}$ Such waves, in which the velocity of propagation is proportional to the square root of the wave length, are termed *deepwater gravity waves.*

If the depth of water is a small fraction less than one quarter of the wave length and the latter is still in excess of about 2π, we may approximate tanh κz by κz and obtain

(108) $$v_w = \sqrt{gz}$$

Such shallow-water waves are commonly called *long*, since they are long compared to the depth, but the term must be used carefully to avoid confusion with the epithet *long-crested*, which has a very different significance. If $\pi > \kappa z > \pi/2$, the full formula must be used. It will be apparent that as the wave length becomes shorter this full expression for v_w passes a minimum, which for ordinary water corresponds to a wave length of 1.8 cm. All waves of length greater than the minimum are called *gravity waves*, since g/κ is progressively more important with increasing wave length. All waves of length less than the minimum are called *ripples* or capillary waves, T becoming progressively more important with decreasing wave length. It will be apparent that velocity increases with decreasing wave length for ripples, and with increasing wave length for gravity waves. Empirically it is found to be difficult to produce wind-generated gravity waves shorter than 4 cm. on ordinary water.

The small deepwater waves of negligible amplitude, for which equation 107 is valid, produce virtually no mass transport of water, the water particles oscillating in circles and the section of the waves at the surface forming a sine curve. The diameter of the circles described by the water particles falls very rapidly with depth. For such a nontranslatory motion, the ratio of the height (h_w) of the wave from crest to trough to the wave length must not exceed 1:100. Such a wave is said to be of *small height.*

As the amplitude increases and the ratio $h_w:l_w$ comes to lie above 1:100 but below 1:25, the waves approximate in section to a trochoid, the curve described by a point on a disk rolled along a plane. Such a wave is said to be of *moderate height.*

As the height increases still more, the crest becomes sharpened, forming obtuse angles. Such waves are said to be of *great height.* Classical theory shows that the angle cannot be reduced below 120 degrees before the wave becomes unstable, breaking at the top to form a *whitecap*. The critical angle corresponds to a ratio $h_w:l_w = 1:7$. It is improbable that this ratio is actually approached; most whitecaps are high waves the tops of which are blown off by the wind. Sverdrup and Munk (1947) give a theoretical limit, based on the relative rates of growth in height and length, of about 1:10, and it is possible that even this is too high.

Waves of finite amplitude involve translatory movement of the water, at a velocity ($\overset{*}{v}$) given by

$$\overset{*}{v} = v_w\left(\pi\frac{h_w}{l_w}\right)^2 \tag{109}$$

Irrotational[32] waves of finite amplitude are generally termed Stokes waves, and travel in deep water with a velocity given by

$$v_w = \frac{gl_w}{2\pi}\left[1 + \pi^2\left(\frac{h_w}{l_w}\right)^2 + \frac{5}{4}\pi^4\left(\frac{h_w}{l_w}\right)^4 + \cdots\right]^{1/2} \tag{110}$$

Since $h_w:l_w$ so rarely exceeds 0.1, the expression in brackets seldom exceeds 1.1, and for most ordinary cases no great error will be introduced by using the expression already given (equation 107) for the actual cases encountered in nature. Longuet-Higgins (1953) has reinvestigated the problem of mass transport in a wave train in a channel, using modern concepts of turbulence. He finds that there is a slight forward movement of water at the surface but most of the forward transport occurs at the bottom. There is a slight motion against the direction of the wave train at intermediate depths.

In view of the fact that instability depends on the ratio $h_w:l_w$ rather than on the absolute value of the height, long waves can grow, without breaking, to much greater heights than can small waves. If in a disturbed area a whole spectrum of wave lengths is formed, the long waves, traveling with greater velocity, can obviously become much higher before breaking as whitecaps than can the small. The result is that at some distance from the site of the disturbance, only trains of long waves will be observed. When such trains have passed out of the region of the generative disturbance and are slowly loosing energy, they are collectively termed *swell*.

The energy of a wave is at any time partly kinetic and partly potential. The potential energy travels with the wave, the kinetic energy of a deep-water long-crested wave stays behind to be added to the energy of the next wave. The total energy of any given wave will therefore be reduced by half for each wave length traveled.

---

[32] Waves not involving rotation of water particles.

If we imagine a *train* of uniform waves set up by some disturbance such as the regular movement of a plunger at the end of a very long trough, the initial energy of any wave produced will be E. After traveling one wave length, the first wave has an energy $^1/_2E$; meanwhile a second wave will be formed, and as the second wave has traveled a wave length, a third wave will also. At this moment the first wave will have an energy of $^1/_4E$, the second of $^3/_4E$. When the nth wave has moved a complete wave length from the plunger, its energy will be $E(2^n - 1)/2^n$, which for large values of n is nearly equal to E, while the original first wave of the train has an energy $E/2^n$. If the waves have not traveled too far their height will be proportional to their energy.

We may now inquire more generally what will be the energy of the mth wave of a train of n waves. It is shown by Sverdrup and Munk that this is given with fair accuracy by

$$(111) \qquad {}^nE_m = E\left(\frac{1}{2} - \frac{1}{2\sqrt{\pi}}\int_0^\zeta e^{-\zeta^2/2}\,d\zeta\right)$$

where $\zeta = 2m - n - 1/\sqrt{n}$.

The second term in the parentheses is of course the well known error function.

When equation 111 is evaluated, it is found that the train consists approximately of a younger half of waves of almost the original energy and height and an older half of low waves of very small energy, the transition zone being short (for a train of 900 waves it is less than 90 wave lengths). This means that when any disturbance sets up any kind of wave train, the first recognizable waves to arrive at a distant spot have appeared to travel to that spot at half the wave velocity.

If a group of waves is formed by the interference of two trains of waves of similar but not identical wave length, these groups will travel with half the wave velocity in deep water. New waves are continually being formed at the back of the group and dying at the front. The energy of the group remains constant. The rate of travel of the group, half of v_w in deep water, is called the *group velocity*.

In any actual case of wind disturbance a very complex spectral distribution may be expected, and rather irregular groups due to interference must be of common occurrence.

The waves so far considered are supposed to have indefinitely extended crests. When the turbulent horizontal velocities in the air over the water surface are sufficiently great, there may be breaking of such *long-crested waves* into segments or *short-crested waves*. The latter can also be produced as an interference pattern when two wave trains traveling in slightly different directions meet. Conversely, short-crested waves set up by turbulence may resolve themselves into separate trains of long-crested waves traveling in somewhat different directions. The properties of short-crested waves differ in several respects from those of long-crested; notably, the energy travels with the same velocity as the wave. Much of the sea surface bears short-crested waves, and it is probable that they occur on very large lakes in stormy weather, though empirical studies of the matter are lacking. They are stable only in fairly deep water, and the waves ordinarily encountered on smaller lakes are typically long crested, the length of the crest apparently increasing with the fetch, as is indicated by Johnson's (1948) studies of aerial photographs of Clear Lake, California.

The origins of waves. The above discussion provides a classification of waves and introduces the important concepts of the wave group and the group velocity. It tells us, however, nothing about the origin of waves. It is possible, under the assumptions of classical hydrodynamics, to show that when one fluid moves over another with sufficient speed, the slightest irregularity may involve the formation of waves. Unfortunately, the forces required are much too great to afford an explanation of the waves observed in nature.

The fundamental investigations of the theory of the origin of waves are those of Jeffreys (1925b, 1926), and although they are certainly incomplete and apparently quantitatively incorrect, they have been the basis of all subsequent studies.

If we imagine a sufficient but small disturbance of the water producing a local rise in one place and fall of the surface in another, any initially laminar wind will tend to flow over the disturbance, but produce an eddy on the lee side. The effect of this eddy will be to deflect the main wind momentarily upward. Beyond the disturbance, the wind will move back to its original level with a downward component of momentum. The result is to lower the atmospheric pressure on the sheltered side of the disturbance and to raise it on the unsheltered side of the next wave. As a result, the water will be rising where the air pressure is falling and falling where the air pressure is rising. The wave, in consequence, will grow in amplitude, as long as the wave velocity is less than that of the wind. The condition for wave growth depends on the rate of supply of energy by this mechanism exceeding the rate of turbulent dissipation of energy. Jeffreys shows that, on his assumptions, this condition is equivalent to

$$\text{s}\rho_{\text{a}} v_{\text{w}}(\text{W} - v_{\text{w}})^2 > 4\mu g \tag{112}$$

where s is termed the *sheltering coefficient*. Since the term on the left is at a minimum when

$$3v_{\text{w}} = \text{W} \tag{113}$$

no waves will form at all unless

$$W^3 \geqslant \frac{27\mu g}{s\rho_a} \tag{114}$$

Jeffreys concluded from observations made on the River Cam at Cambridge and in a flooded clay pit at Barnwell nearby, that the minimum wind velocity producing waves is 110 cm. sec.$^{-1}$, which by equation 113 would correspond to $v_w = 37$ cm. sec.$^{-1}$, and so by equation 107 to $l_w = 8.8$ cm., and by equation 114 to $s = 0.27$.

It is very doubtful, however, that Jeffreys' theory can be correct. Cornish believed he had observed waves of length 2.5 cm. set up by wind. Keulegan (1951), in a very careful study of wave formation in an experimental tank, obtained a minimum wave length of 4 cm., corresponding to $v_w = 25$ cm. sec.$^{-1}$ By equation 113 this would correspond to $W = 75$ cm. sec.$^{-1}$ Keulegan, however, obtained an empirical expression of the same form as equation 114 but with different numerical coefficients, implying a much smaller value of s. Evaluating his expression, the minimum wind to raise waves on water should be 400 cm. sec.$^{-1}$

Other workers who have attempted to evaluate the sheltering coefficient have also run into difficulties, or have at least obtained inconsistent results. From the standpoint of the present work, the important empirical finding is that a less strong wind was needed in Jeffreys' cases, to obtain waves on water where the fetch[33] may have been of the order of 10^2 m., than on Keulegan's experimental tank 20 m. long. Whether the required wind velocity would decrease as the fetch became still greater, of the order of 10^5 m., is not known.

In his theoretical investigations Jeffreys neglected the tangential stress of the wind on the water. Sverdrup and Munk (1947) have pointed out that although his interpretation of the genesis of waves may be correct, his theory cannot account for their growth, as it requires a far greater transfer of energy than actually takes place. Sverdrup and Munk conclude that at least over rough water the tangential stress of the wind, not considered in Jeffreys' theory, must be taken into account in dealing with any actual wave of finite amplitude. Inclusion of this factor leads to a condition

$$\pm s\rho_a v_w (W - v_w)^2 + 2\gamma^2 \rho_a W^2 v_w \geqslant 4\mu g \tag{115}$$

which permits v_w to increase[34] above W, apparently in accordance with oceanographic experience. The expression is valid only when W exceeds 500 cm. sec.$^{-1}$, and tells us nothing about the conditions of origin of waves.

[33] Both the localities studied by Jeffreys are known in a general way to the writer, and a fetch of more than 100 m. is doubtless possible on the Cam and probable at Barnwell.

[34] Cf., however, Schaaf and Sauer (1950), who have reinvestigated the theory, including second order terms, and have reached the contrary conclusion.

Francis (1951) suggests that increasing aerodynamic roughness as rippling develops on the surface of large waves at high wind speeds increases the tangential stress.

It is quite possible that a more careful experimental study of exactly what happens when waves form will throw more light on the matter than will theoretical studies not supported by adequate empirical definition of the problem. Keulegan has found in his experimental tank that as the wind velocity increases, the water surface is thrown into a tremor and the initial waves then appear at the lee end of the tank. For any steady wind velocity there is a section of fetch free of true waves at the windward end of the tank, and the length of this initial fetch varies inversely as $\sqrt{W}$. At high wind velocities the initial fetch may be covered with ripples of wave length about 2 cm., which are actually just at the dividing line between capillary and gravity waves, but the true gravity waves of wave length 4 cm. emerge from this initial fetch quite suddenly at its leeward end. It is evident that the nature of the tremor or rippling of the initial fetch deserves a much more careful experimental study.

Relationship of velocity and form to age and fetch. The relationship of the velocity and form of waves to the velocity of the wind and to the fetch or distance over which the wind has blown has been the subject of considerable speculation, some mathematical study, and a little empirical work since the time of Leonardo da Vinci, who noted the greater velocity of large waves. It is only quite recently, as the result of the investigations of Sverdrup and Munk (1947) and their successors, that any approach to a reasonable theory has been possible.

In considering this problem, most attention has been given to the highest waves observed at a given time. It has been estimated that in making observations of such waves the experienced observer concentrates on a subclass of waves of which the mean height and wave length are equal to the mean height and wave length of the top third of an array of all waves present if they are arranged in order of height. The waves of this subclass are termed by Sverdrup and Munk *significant waves*. Their treatment refers primarily to such significant waves, for which they have considered certain properties that can be expressed as dimensionless numbers and can also be calculated from their theory of the generation and growth of waves, taking into account both air pressure and tangential stress. The actual functions deduced are quite complicated and are by no means certainly valid. The interested reader who is prepared to follow through the mathematics is referred to their original paper. While subsequent research suggests that considerable modification of their theory will be necessary, there can be no doubt that they have directed research along the right lines.

It is assumed by Sverdrup and Munk that, since initially there is no wave and therefore no wave velocity and later the wave is traveling almost as fast as if not faster than the wind, the ratio v_w/W is a dimensionless measure of the age of the wave. The empirical data supporting this inherently reasonable assumption are limited, but if the few cases of waves of known age are considered, time being represented as gt/W, and v_w/W is plotted against this dimensionless measure of time, the distribution of the points is clearly not discordant with the hypothesis (*A* of Fig. 105).

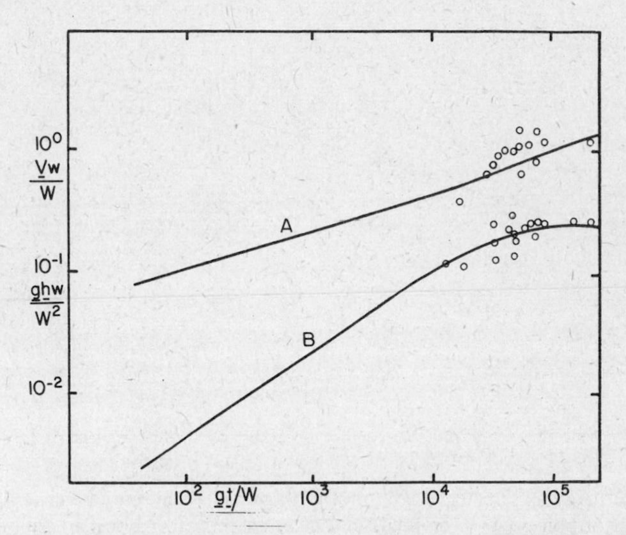

FIGURE 105. *A*, age parameter v_w/W, and *B*, height parameter gh_w/W^2, plotted against time parameter gt/W. All observations are marine. (After Sverdrup and Munk.)

We may now plot v_w/W against gx/W^2 as a dimensionless measure of fetch. For such a procedure there is a very large body of data from the open ocean, large lakes, a small brackish lake Abbott's Lagoon on Point Reyes Peninsula, 80 km. north of San Francisco, California (Johnson 1950), and from a number of studies in experimental tanks (Johnson and Rice 1952). The results are shown in Fig. 106. The broken line shows Sverdrup and Munk's functional relationship between the two numbers plotted; the solid line is Bretschneider's (1952) estimate of the best fitting line. It seems certain that the points lie slightly above Sverdrup and Munk's line, but later authors have attempted no improved theory. Since the wave period for waves of small amplitude is proportional to the wave velocity, and even for waves of finite amplitude such a relationship will hold roughly, a plot of a dimensionless estimate of the period on logarithmic paper will be roughly parallel to the line for v_w/W, and the same is true for a suitable estimate of l_w^2.

We may also plot gh_w/W^2 as a dimensionless measure of the height, against gx/W^2 as a measure of fetch (Fig. 107), or gt/W as a measure of time (*B* of Fig. 105). Again we find a close but by no means accurate correspondence with Sverdrup and Munk's theoretical curve.

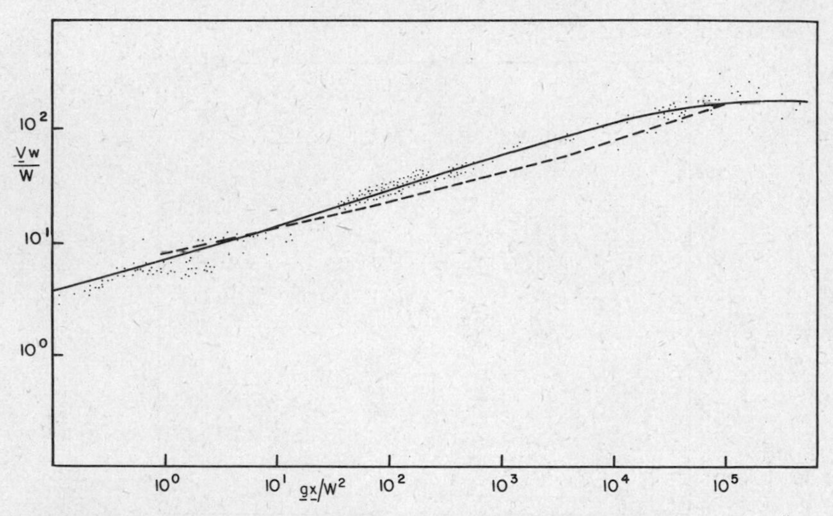

FIGURE 106. Age parameter v_w/W plotted against fetch parameter gx/W^2. The large number of points (some omitted) in the center refer to Abbott's Lagoon; most of the data for longer fetches is from the ocean, most from shorter fetches from experimental tanks. (Johnson and Rice.)

If we now consider h_w/l_w as a measure of steepness, we find that according to Sverdrup and Munk's theory there should be a maximum steepness of about 1:10 in the region of wave age 0.4. The divergence of the Abbott's Lagoon data from Sverdrup and Munk's theoretical values, together with the rather large scatter, greatly obscures this maximum. It would seem probable that some sort of maximum does exist, but possible that it can be expressed only when the fetch is very long. It seems certain that lake waves are never likely to be as steep as the steepest observed on the ocean.

A somewhat different approach has been adopted by Neumann (1952), who has considered not significant waves but definite spectral classes of waves isolated in frequency periodograms. Neumann concludes that such a periodogram is normally trimodal. The main mode consists of the waves that build up to a length and height as a characteristic degree of roughness is developed. This remains the main type of wave, but later a certain number of waves of characteristic higher wave lengths develop. The longest of these waves may have velocities greater than the wind.

Neumann developed a theory based on the energetics of wave formation that takes into account dissipation of energy by eddy viscosity. This

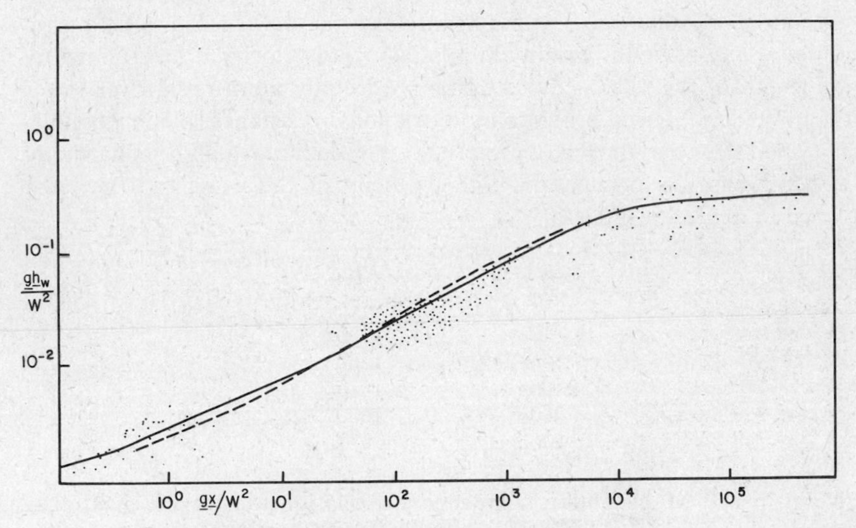

FIGURE 107. Height parameter gh_w/W^2 plotted against fetch parameter, from the same bodies of data used in Fig. 106.

theory indicates the maximum age or v_u/W possible for a given fetch. The values obtained are lower than would be given by Sverdrup and Munk. Neumann supposes, for example, that the waves observed on Lake Michigan and reported by Sverdrup and Munk from data of the U. S. Army Engineers have reached a maximum age for the prevailing wind velocity, and that any increase in fetch would not increase their velocity relative to the wind, unless waves of the higher spectral classes developed. As the wind velocity increases, the fetch required to produce a maximum "age" increases rapidly. For the strongest winds, no lake is large enough to give maximum ages.

Finally mention must be made of a single purely empirical relationship that has often been used. Stevenson (1852), largely as the result of studies on the Firth of Forth, the Moray Firth, and an unspecified small freshwater loch, concluded that during the strongest winds the height of the highest waves is proportional to the square root of the fetch. In centimeters his relationship becomes

$$h_w = 0.105\sqrt{x} \tag{116}$$

x being measured from the margin of the lake downwind to the point of observation. For Lake Superior, with a fetch of 482 km., the formula gives 7.3 m. as the height of the highest waves, in good agreement with the 6.9 m. reliably recorded (Cornish 1934). Sverdrup and Munk point out, however, that there is no theoretical reason for the relationship unless at any place the maximum possible wind velocity is determined by the length of the fetch. Since winds are usually more rapid over water than over land, which provides more frictional resistance, this is possible. They point out that the relationships that they deduce would be in harmony with Stevenson's formulation if the velocity of the wind and the fetch varied in the following way

Fetch (x), km.	Max. Wind (W), cm. sec.$^{-1}$
10	1230
50	1260
100	1370
250	1600
500	1870

Sverdrup and Munk obtain from theory a relationship for the maximum possible height h_{wm} with a given wind, irrespective of fetch or duration,

$$h_{wm} = \frac{0.26W^2}{g} \tag{117}$$

This is in fair accord with observations at sea; other formulations have been proposed but are unsatisfactory.

Waves in shallow water, breakers. Decay of waves due to eddy viscosity is ordinarily too slow to be of importance in a lake. The usual fate of a lake wave is to enter shallow water and to break near the shore. Waves in very shallow water are often called *long waves* because the wave length is much greater than the depth. If the amplitude is sufficiently small and the wave length is over a few centimeters, these waves travel with a velocity approximately proportional to the square root of the depth (equation 108). As a wave approaches the shore and enters very shallow water, the velocity at any instant tends to be what it would be if the wave had been traveling over water of the depth encountered at that instant. Simultaneously the wave length is reduced. The height of the wave at first decreases slightly, then increases[35] as the wave approaches breaking. During the final phase of the rise in height, the sinusoidal or trochoidal form is lost and a strong asymmetry develops. This appears to be due to the advancing part of the wave entering a progressively shallower region which cannot provide enough water to form a complete wave.

During its passage in shallow water, the water particles of the wave

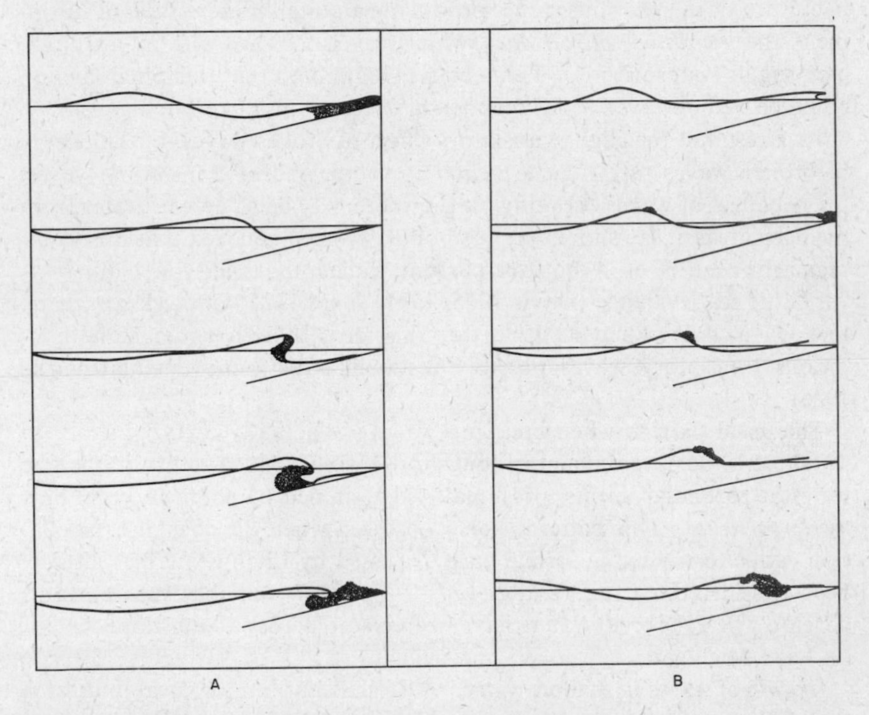

FIGURE 108. The breaking of waves: *A*, plunging breaker; *B*, spilling breaker.

travel in ellipses whose major axes are set horizontal, and whose minor axes are proportionately shorter as the bottom is approached. At the bottom, a purely horizontal movement back and forth occurs.

Breakers may be of two extreme or limiting types (Fig. 108), though there is a complete series of transitional forms. In the so-called *spilling breaker*, the top of the wave spills over its forward face; in the *plunging breaker*, the anterior face of the wave becomes convex as the crest curls over. There appears to be a considerable range of angles on either side

[35] The elementary theory, which does not accord too well with the empirical data, is summarized by Iversen (1952).

of the theoretical 120 degrees realized by waves just before they break. The wave surface, moreover, becomes irregular, and small quite vertical regions seem, from high-speed cinematographic studies, to initiate the breaking (Mason 1952). No very definite information as to breaking of lake waves appears to exist.

When a breaker is large and breaks in relatively deep water, it may suddenly add a considerable quantity of water to the relatively smooth backwash of the previous wave, producing a quite different type of movement, the *translatory* or *roll wave*, which travels forward without a trough, carrying its water with it. Fenneman (1902) noted that this phenomenon could be well observed at Jackson Park, Chicago, on Lake Michigan.

As a result of the slight translatory effect of Stokes waves, of the energy of broken waves carrying the *swash* up shore, and of translatory waves, the net effect of waves breaking on the shore may be to deliver water from the deep areas at the shore. It has popularly been believed that this water regularly returns as a bottom current, called the *undertow*. There is very little real evidence (Davis 1925, 1931; Jones 1925; Craig, Brant, Hite, and Davis 1925; Shepard 1936) for this idea, though under some conditions translatory waves possibly result in such a movement (Quirke 1925).

The usual pattern when breaking waves are actually carrying water to the shore is the development of longshore currents which return to the free water at particular points, presumably determined by shore morphology, as *rip currents*. This pattern is also observed when the coastal denivellation is due to wind; the formal case discussed by Livingstone has already been considered. Craig (Craig *et al.* 1925) points out that such currents develop wherever breakwaters have been placed across a longshore current on Lake Michigan.

Growth of waves in shallow water. Of particular limnological interest is the problem of the growth of waves in quite shallow water, which has been studied empirically by Thijsse (1952) in numerous shallow lakes in Holland. For any given wind velocity a certain fetch is required in order to develop waves of maximum size, and this fetch depends on the depth. For a given wind velocity W and a given depth z, the relative depth is computed as gz/W^2, and this quantity can be used as a dimensionless measure of depth against which other dimensionless parameters can be plotted.

Thijsse considers the case of a series of lakes 1.5 m. deep compared with a similar series 7.5 m. deep, each series being disturbed by a wind of 1700 cm. sec.$^{-1}$ The relative depths of the two series are 0.05 and 0.25 respectively. With different values of fetch, the wave lengths and heights of the significant waves are given in Table 31.

TABLE 31. *Effect of increase in fetch on wave length and wave height in very shallow and in moderately deep lakes*

	Shallow Series (1.5 m.), m.			Deep Series (7.5 m.), m.		
x	150	600	7500	150	600	7500
h_w	0.20	0.33	0.48	0.20	0.39	1.41
l_w	2.8	4.7	7.3	2.8	5.5	16.4

The small deep and small shallow lakes behave in the same way, but the large deep lakes produce much longer and higher waves than do lakes of the same fetches but smaller depths. This work when reported was in its initial stages. It is purely empirical, representing actual measurements on a number of unspecified bodies of water. It is of some practical importance; in parts of Holland where sand is scarce, lake sand may be dredged commercially in quantities great enough to increase the waviness of a large lake.

Miscellaneous empirical observations. The data in Table 32, in part including values incorporated in Fig. 106, give some of the properties of the highest waves observed on a variety of lakes. The data are derived from Cornish (1934), except for the three observations for Lake Michigan by the U. S. Army Engineers, given by Sverdrup and Munk. Many other data from this source are given by the same authors, but all are incomplete, lacking wave lengths or velocities which are interconvertible by equation 107. The single incomplete entry refers to the highest waves recorded at this station.

TABLE 32. *Properties of waves on lakes*

Lake	Fetch (x), km.	W, cm. sec.$^{-1}$	l_w, m.	h_w, m.	$h_w:l_w$	v_w, cm. sec.$^{-1}$	v_w/W
Superior	480	~2600	93 (86–106)	6.9	0.07	1200	0.46
Michigan	362	820	28.7	2.4	0.08	670	0.82
Milwaukee Light	152	1130	28.7	2.7	0.09	670	0.59
Milwaukee Light	120	1600	...	4.3	...	...	...
Geneva	49–72	...	39	2.5	0.06	780	...
Coniston	8.8	...	20 ± 2	~1.5	0.07	...	...
Round Pond, Kensington Gardens, London	~0.2	~2200	1.55	0.12	0.08	1930	0.88

PLATE 12. Lake of Geneva from Thonon, showing slicks or *taches d'huile* (*phot.* C. H. Mortimer).

It may also be noted that Gaillard (1904) found in the large waves of Lake Superior, about 3 m. high, dynamic pressures due to the water movement of 0.232 to 0.487 kg. cm.$^{-2}$, which can be considerably greater than the static pressure due to the weight of water. The latter was determined, 275 cm. below the crest of a wave, to be 0.225 kg. cm.$^{-2}$ Owing to the centrifugal force on the water particles undergoing orbital circulation in the wave, this value may be expected to be less than that due to the weight of the water between the surface and the instrument.

Slicks (*taches d'huile*). The surface of almost any lake may, in moderately calm weather, exhibit patches of water which present a smoother texture (Plate 12) than the rest of the lake, though ordinary gravity waves are present throughout the lake surface. Close examination indicates that the capillary ripples ordinarily present on water surfaces are absent in such areas, which may also perhaps have slightly different optical properties than the rest of the surface. There can be little doubt that these patches, commonly known in English as slicks and termed by Forel (1873) *taches d'huile*, are actually due to the presence of an oil film on the water, which damps out the ripples. They are commonly observed in shallow seas, and though they can be produced by artificial contamination, probably in nature are mainly derived from the liberation of fatty material from decomposing diatoms (Dietz and LaFond 1950).

Dietz and LaFond concluded that all natural coastal waters have at least a monomolecular oil film at their surfaces; in the slicks this film becomes piled up to a considerably greater thickness.

The general tendency of slicks to form a linear pattern has been examined by Dietz and LaFond (1950) and by Ewing (1950). Ewing concludes that in a thermally stratified sea over which the wind has a velocity of less than about 350 cm. sec.$^{-1}$, the linear arrangement results from the slicks forming over the troughs of internal waves traveling along the thermocline; under greater wind velocities, both Dietz and LaFond, and Ewing believe the linear pattern to indicate the convergence between the spiral circulation cells of the wind drift. The same types of explanation doubtless hold for lakes.

LAKE GUNS OR SEESCHIESSEN

A final section may be added to the present chapter to notice a phenomenon which has apparently never been explained and which, in the absence of a convincing causal analysis, cannot be appropriately discussed in any other chapter.

The phenomenon in question is the apparent production of sounds like distant thunder or gunfire that have been recorded as sporadically produced in some lakes. These sounds have been noted from the shores of a number of the subalpine lakes of Europe, covering a range in size and depth from Lake Constance (area 538 km.^{2}, z_m 252 m.) to Lago di Tovel, Trentino (area 0.52 km.^{2}, z_m 35m.). The same type of sound is well known on Seneca Lake, New York, where the term *lake guns* is used to designate the phenomenon. There appears, in most cases, to be a seasonal variation in the incidence of the sounds. They are reported to be rarer during winter than at other seasons, and to be less frequent at night than during the day. They have been supposed in some localities to occur chiefly in the early morning and evening. On Lake Seneca the phenomenon is believed to occur mainly by day in the autumn. The supposed diurnal variation in incidence suggests a very superficial origin for the sounds.

There seems to be little doubt that the sounds emitted or heard over lakes constitute a special case of a more widely distributed category of sounds usually termed *brontides*. This category includes the *mistpoeffer* of the coasts of Holland, similar sounds known on the Baltic coasts, the Barisal Guns of the Ganges delta, and various comparable sounds known on land far from any bodies of water. Brontides in general have often been explained as due to microseisms, and the lake guns as produced by escaping bubbles of gas, in some cases perhaps liberated by seismic vibrations. Fairchild (1934) supposes that in the case of Lake Seneca the gas is methane derived from a natural gas reservoir in the Oriskany sandstone, which is cut by the preglacial valley in which the lake lies. He supposes that the actual noise is produced by an explosion when the gas reaches the surface, but there is no evidence that a spontaneous explosion would occur in such a case. Ingalls (1934) tends to accept this explanation in so far as the origin of the gas is concerned, but supposes that the noise is due to the gas bursting through the sediment-water interface.

From the existing information it is difficult to judge whether the phenomenon involves the existence of lakes in any essential way. It is quite possible that in some cases liberation of gas may be involved, though the lakes which produce the sounds do not appear to be those in which a large endogenous production of gas is probable. It is also possible that a general low noise level over a quiet water surface, combined perhaps with other favorable acoustic conditions provided by some lake basins, may enhance the audibility of brontides not actually originating in the lake. The interested reader may consult Halbfass (1923) and Pesta (1929) for the older European observations, and Davison (1921) for a brief account of brontides in general.

SUMMARY

The movement of water in lakes is for the most part turbulent, that is to say, at any point, in addition to an average steady flow in a definite direction, variable velocities in any direction may be observed. Turbulence permits the transfer of material and heat in any direction in the lake according to laws that in first approximation are of the form of the laws of diffusion of dissoved substances in undisturbed liquids and the conduction of heat in solids. But the coefficients of diffusivity and conductivity involved, which are in general interchangeable, are variables, dependent on the state of the water.

Not only heat and material exchanges occur but also exchange of momentum; the effect of the turbulence is to increase the apparent viscosity of the liquid. A coefficient of eddy viscosity can be defined which, in water of indifferent stability, is identical with the coefficients of eddy diffusivity and conductivity, but in water with a density gradient is greater than the latter coefficients. The concept of eddy viscosity is of great importance in providing an explanation of wind-driven currents, which in a nonturbulent medium would be far shallower and would take far longer to reach observed velocities than is actually the case. The irregular swirls implied by turbulence can be of all sizes; the upper limiting case in a lake may be the major current system of the lake, the lower limit is irregular molecular movement.

The currents of a lake are of various kinds, but in general they reduce to the usually unimportant hydrographic slope current from influents to effluent, and the dominant wind-driven circulation. However the surface currents may be setting, if they tend to rotate *cum sole*, they produce an anticyclone with high pressure or a slightly raised water surface in the middle of the swirl; if *contra solem*, they produce a cyclonic swirl depressed in the middle. There is an adjustment so that in the Northern Hemisphere denser water collects on the left, less dense on the right of the current. This permits, under some circumstances, determination of current velocities from temperature profiles, and suggests that littoral warming may produce cyclonic circulation.

The effect of the wind on the lake surface is to move water downwind and to the right (in the Northern Hemisphere). The first effect is due to the stress of the wind on the water, the second to the rotation of the earth or to geostrophic forces. In an unrestricted water surface of great depth, the resulting wind drift always sets 45 degrees to the wind. If the eddy viscosity is constant with depth, reducing the depth of the water reduces this angle, so that in lakes 10 to 20 m. deep it is likely to be inappreciable. If, however, certain particular vertical patterns of eddy viscosity exist, a considerable divergence of current from wind may occur in quite shallow water. Empirically, geostrophic effects are shown by the surface currents of Lake Ladoga and probably those of Lake Mendota also; there is no recognizable geostrophic effect on the wind drift of western Lake Erie. In general the velocity of the wind drift in lakes is usually about 2 per cent of the wind velocity. The stress producing the drift varies as $W^{1.5}$ to $W^{2.0}$, the most acceptable empirical value being $W^{1.8}$. The surface velocity appears to be markedly independent of the waviness.

In shallow wide lakes the effect of the wind denivellation of the water on the lee side is to produce return currents that ideally curve around the sides of the lake and converge in the windward half of the water surface. In deep unstratified lakes there is a marked return current upwind. This is a gradient current dependent on the extra pressure exerted at the lee end by the piled-up water. The observed wind drift has a velocity that is due to the wind less the gradient current. Since the wind effect decreases rapidly with depth while the velocity of the gradient current is independent of depth save near the bottom, there is a plane at which they exactly balance, and below this, a return current upwind. If geostrophic forces operate, the current system forms a spiral. In actual lakes there is sometimes evidence of a simple return current and sometimes of spiralling. When the water is stratified, the piling of the water at the lee end produces a great depression of the uppermost surface of discontinuity of density. Provided no water of high density is exposed at the surface, no lower discontinuity layers are tilted. It is probable that the circulation in the top layer of such a lake involves a simple return gradient current; the nature of the deepwater currents in stratified lakes is not adequately elucidated. When the lake is well stratified, the depression downwind of the uppermost surface of discontinuity will cause water movements upwind, but whether there is a vertical rotary movement about a horizontal axis in deep water, as has been postulated, is uncertain.

In addition to the wind drift, currents due to effluents can be important if the density of the incoming water differs from that of the bulk of the lake. In summer the Rhine and Rhone enter Lakes Constance and Geneva at densities greater than the surface water and move downstream at a depth appropriate to that density, the Rhine particularly hugging the right side of Lake Constance on account of the geostrophic effect. Such density currents are an important feature of large artificial lakes, especially when the entering water is turbid.

After water has been piled up downwind and the wind has dropped, it flows back as an unopposed gradient current. Since in the process the momentum is not dissipated, a denivellation is produced at the former windward end, and a new flow starts from the former windward end to the former leeward end. In this way a periodic rocking movement or seiche is produced. Any other kind of local denivellation, notably that due to differences in atmospheric pressure over parts of the lake, can likewise start a seiche. The recurrence of disturbances at intervals near the period of oscillation of the lake is peculiarly effective. The periods of seiches are determined entirely by the shape of the lake, and for basins of the same form are directly proportional to the axis of oscillation and inversely proportional to the square root of the depth. A general tendency to the production of harmonics exists. The periods of the harmonics relative to the fundamental depend on the form of the lake; in cases that are known empirically, the period of the first harmonic or binodal seiche varies from 19.5 to 77 per cent. Theoretically it is possible to conceive lakes in which all the harmonics have periods approaching that of the fundamental as nearly as is desired. Seiches play some part in maintaining channels and openings into lagoons.

True tides can be detected on the very largest lakes, but have been little studied. Even in Lake Baikal and Lake Superior they have a maximum range of about 3 cm. and can hardly play any part in the economy of the lakes. Larger ranges may occur when a seiche period corresponds to the tidal period of 12.5 hours. This is possibly the case on Lake Huron. A somewhat larger seiche may just possibly be damped out in Lake Erie by tidal interference.

The very great displacement of the uppermost surface of discontinuity of density, normally at the top of the thermocline in a stratified lake, when a wind is blowing, leads to a very large internal seiche when the wind dies down. The period and amplitude of such an internal seiche are much greater than those of the ordinary seiche. The currents rhythmically changing in direction due to the internal seiche are believed to be the most important type of deepwater movement in lakes. Such currents certainly are turbulent and lead to both vertical and horizontal transport of heat and dissolved substances.

Surface waves arise when the wind over the water reaches a certain critical speed, which seems to be of the order of $n10^2$ cm. sec.$^{-1}$ They continue to grow in height and length, the process involving pressure on the part of the wave that is sinking and tangential stress on the surface. Aerodynamic roughening by small ripples probably increases the action of the stress. The ratio of height to length seldom exceeds 9:100 on lakes. It is supposed to reach a maximum at a particular point in the history of a wave, but the empirical evidence for this is confused. Ordinary surface waves in deep water travel with a velocity in centimeters per second equal to 12.5 times the square root of the wave length in centimeters. Waves of great wave length therefore leave the generating area most rapidly and become swell.

In small lakes the size of the waves is virtually independent of depth, but in lakes of great area it increases with the depth. The height of the highest waves observed on a lake appears to be proportional to the square root of the fetch of the wind generating the waves. No good theoretical explanation for this relationship has been advanced. Certain more complex theoretical relationships between wind and wave properties have been deduced. Near the coast the breaking of an ordinary wave may produce a translatory wave without a trough. Some translation is also effected by ordinary waves of moderate or great height, so that a slight piling up of water along a coast on which waves are breaking can be expected in the absence of wind. This leads to longshore and return rip currents. The existence of a return undertow is very doutbful.

The peculiar acoustic phenomenon of *lake guns*, booming noises recorded as apparently emitted by certain lakes, is without an adequate explanation. It has been attributed to sudden evolution of methane by the mud of the lake, but is not confined to localities where this is likely. The phenomenon may be a special case of the sounds known as brontides, which are not necessarily associated with water and are probably due to microseisms. Typically, however, lake guns are heard mainly by day and rarely in the winter; this suggests some relatively superficial cause operating in the lake.

CHAPTER 6

The Optical Properties of Lakes

Almost everything considered in the succeeding chapters of this book depends on the fate of the light of the sun and sky that falls on lake surfaces. The currents generated by wind and by incoming and outgoing streams are, in their inception, not related to the radiation impinging on the lake, though in their pattern at most seasons they reflect the thermal structure and so the previous optical history of the lake. With this partial exception, probably all the events within a lake are directly or indirectly determined by the radiation which that lake receives.

Cosmic radiation, which is the only other form of radiation received at the earth's surface that is of any significance in the present work, consists of a mixture of several components, most of which do not belong to the electromagnetic spectrum. A few studies of the penetration of cosmic rays into lakes have been made, for the purpose of learning about the nature of the radiation more than about the properties of lakes. A short section on these studies is added as a tailpiece to the present chapter.

SUN AND SKY AS SOURCES OF INCOMING RADIATION

For almost all purposes, it may be assumed that the radiation to be considered is solar. The radiation from the full moon is from one thirty-thousandth to one fifty-thousandth of that from the sun. Moonlight is, of course, reflected sunlight, so even if the supposed minute effects of the moon on the temporal variation of the phytoplankton, and possibly on certain other biological phenomena in inland waters, are due to moonlight, these slight effects are in their turn regulated by the timetable of the main nocturnal reflector of the light of the sun. The intensity of starlight is so very much less than that of moonlight that it may be safely neglected, as may those other forms of natural nocturnal illumination, the aurora and the imperceptible glow of the night sky.

Nature and intensity of solar radiation. The ordinary energy distribution of the solar spectrum at the earth's surface is indicated in Fig. 109, where the relative sensitivity of the human eye is also indicated. In general, the maximum energy in the solar spectrum as received at the earth's surface is in the blue-green. Toward the short-wave end of the spectrum, the energy curve declines very rapidly owing to the selective absorption of ultraviolet by the oxygen of the atmosphere. Toward the long-wave end, a very gradual decline toward the extreme infrared occurs. In

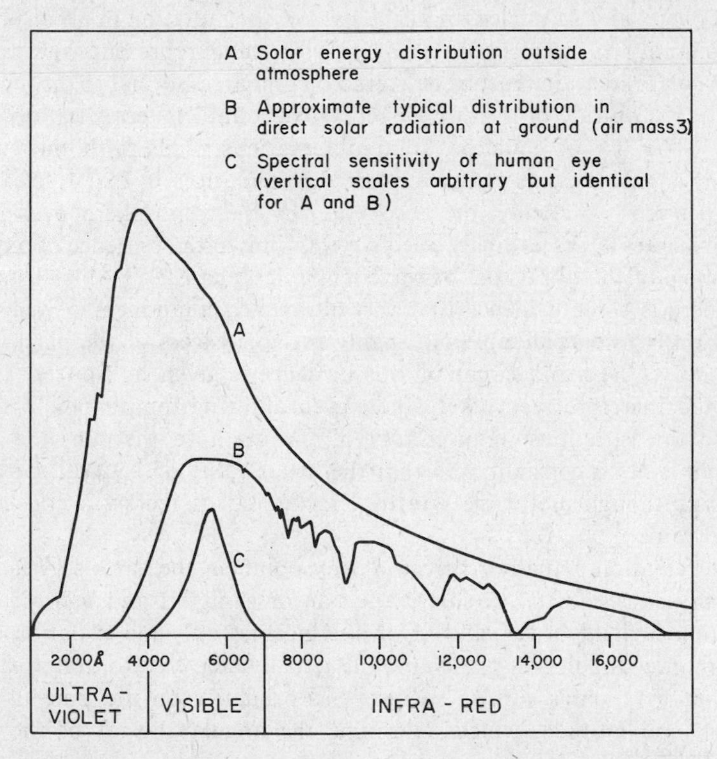

FIGURE 109. Spectral distribution of radiant energy outside the atmosphere and at the earth's surface (air mass 3), with the spectral sensitivity of the human eye.

general, the infrared region accounts for just under half the radiation received at the earth's surface. In the very long wave length region, it is reinforced by thermal radiation emitted by the atmosphere.

There are slight seasonal variations in spectral composition, and very marked variations dependent on the elevation of the observer. These involve primarily the ultraviolet component, and depend on the length of the path of the radiation through the atmosphere. Atmospheric turbidity may also cause some variation (Ross and Utterback 1939). As will appear in Volume II, it is not impossible that the variation in ultraviolet with elevation is of some biological importance.

For most purposes, the radiation of the sun may be regarded as that of an incandescent black body radiating at a temperature of about 6000° C. The radiation received at the surface of measuring instruments set on the earth's surface obeys the ordinary law of transmission

$$I = I_c e^{-\eta_a M} \tag{1}$$

where η_a is the extinction coefficient of the air referred to one standard atmosphere, and M the length of the path of the radiation in atmospheres. The quantity I_c is called the *solar constant*, and represents the rate at which solar radiation must be delivered on unit area normal to the incident rays, at the outside of the atmosphere, to account for terrestrial observations. The use of equation 1 on observations made with the sun at varying angular heights permits a mean determination[1] of I_c as 1.92 cal. per cm.^{2} per min. Actually, the absorption by the atmosphere, even by its tenuous outer layers, is highly selective. Owing to the presence of oxygen, the whole of the ultraviolet of wave length less than 2950 A. is absorbed, and there is some evidence that this ultraviolet component of the solar spectrum is in considerable but highly variable excess of the black-body radiation. A good account of this evidence is given by Spitzer (1949). It would therefore seem likely that, even allowing for periodic changes due to the variations in proximity of the earth to the sun, the *solar constant* is not a constant, and that the mean value of 1.92 cal. per cm.^{2} per min., though useful in terrestrial reckoning, is merely a convenient virtual value.

The radiation actually received at any point on the earth's surface by day may consist of *direct* solar radiation or sunlight, and *indirect* solar radiation or light of the sky. Considering first only the radiation from the sun in a cloudless sky, the amount actually delivered to unit area of a plane at the earth's surface will depend primarily on the time of day, season, and latitude, which determine the angular height of the sun;

[1] Fritz (1951) gives $1.90 < I_c < 1.94$ but indicates that direct measurements from rockets may imply a value as high as 2.0.

on the elevation of the observer, which together with the angular height determines the amount of the atmosphere through which the radiation has passed; and on the transparency of the atmosphere. The angular height ψ_s' is involved (Table 33) in two ways. First, the intensity on a plane horizontal surface will be sin ψ_s' times the intensity on a plane normal to the ray. Second, the path traversed by the ray through the atmosphere will be greater for low angles than for high angles. The altitude is involved in reducing the thickness of the absorbing layer (Table 34).

TABLE 33. *Length of path of solar radiation for various solar heights*

Angular height ψ_s'	90°	80°	70°	60°	50°	40°	35°	30°	25°	20°	15°	10°	5°
Path, atm. (*M*)	1.00	1.02	1.06	1.15	1.30	1.55	1.74	2.00	2.36	2.90	3.82	5.40	10.40

TABLE 34. *Relative thickness of atmosphere above a given altitude from earth's surface*

Elevation, km.	0	0.9	1.9	2.9	4.0	5.4	7.0	9.0	11.6	16.0
Relative thickness of atmosphere	1.0	0.9	0.8	0.7	0.6	0.5	0.4	0.3	0.2	0.1

The transparency of the atmosphere is quite variable and of some importance in regulating solar radiation. In clean dry air, the primary cause of opacity to radiation of the wave lengths normally present at the earth's surface is molecular scattering. In damp and dusty air, there is considerably more scattering and absorption. The turbidity of the atmosphere is generally lower at great altitudes than at sea level, and lower in winter than in summer. The presence of a large water surface and consequent local high water-vapor contents may reduce the intensity of the total solar radiation. This is particularly likely to happen over large lakes, where much of the lower atmosphere over the water can have been influenced by the presence of the lake. Very great turbidities are often found in the vicinity of industrial towns. Various methods of expressing turbidity have been proposed (Fritz 1951); none are entirely satisfactory.

The radiation from the sky is due to the scattering of sunlight by the atmosphere. The scattering by very small particles of molecular dimensions is proportional to the fourth power of the frequency of the radiation (Rayleigh's law), and therefore affects the shorter wave lengths more than the longer. This results in the blue color of the sky. The presence of water vapor and of dust decreases the saturation of the hue. Ordinarily, the most intense blue is in the vertical plane of the sun, 90 degrees from that body. Remote from this point, particularly near the horizon, the blue is less intense, tending to become whitish.

The proportion of indirect radiation from the sky to the total radiation is variable. It naturally depends on the angle of the sun, for with a long path opportunity for scattering is great and the intensity of the direct radiation reduced. The proportion also depends on the elevation, for at high altitudes there is less atmosphere to produce scattering. The data in Table 35 indicate the mean percentage of indirect radiation from a

TABLE 35. *Percentage of indirect radiation as a function of the sun's altitude*

		Elevation (approx.),		Indirect radiation, %, at Sun Altitudes of			
Place	Season	m.	Observer	10°	20°	30°	40°
Washington, D. C.	Summer	50	Kimball (1924)	40.1	29.6	24.2	21.3
	Winter	50	Kimball (1924)	37.1	21.8	16.7	13.0
Zugspitze, Bavaria	Winter	2962	Lipp (1929)	19.3	11.5	10.1	8.3

cloudless sky in the total radiation per unit area, under different geographical and seasonal conditions and with the sun at different altitudes.

The presence of cloud has a great effect on the radiation of the sky. A partial covering of white cloud can increase the radiation considerably, owing to diffuse reflection from the clouds if the sun is not covered by them. In extreme cases, with a low sun a moderate degree of cloudiness may increase the total radiation falling on the earth's surface by as much as 40 per cent.

Total energy delivered. For many limnological purposes, particularly in connection with thermal problems, the most interesting method of presenting the radiation flux from sun and sky, is as the total energy delivered over a whole day or over a long period of days. Unfortunately, the data available for this purpose are by no means extensive. For cloudless days, the direct radiation from the sun depends largely on the angular height of the sun and on the length of day, that is to say primarily on geometrical properties. It is therefore possible to estimate the variation with elevation and latitude with some accuracy, and to apply the data collected at a limited number of stations to a general discussion of variation over the earth's surface. Comparison with observations enables the application of mean corrections for variation in atmospheric turbidity. Perl (1935) estimated the mean values given in Table 36 in cal. cm.$^{-2}$ day^{-1} for the fifteenth day of each month for sea-level stations at various latitudes.

TABLE 36. *Estimates of total direct radiation on the 15th day of each month at various latitudes*

Lat.	Jan.	Feb.	Mar.	Apr.	May	June	July	Aug.	Sept.	Oct.	Nov.	Dec.
0°	480	512	530	518	500	485	490	520	550	547	507	475
15°	425	480	545	575	575	567	570	567	540	485	427	400
30°	297	380	482	565	617	625	610	567	490	400	312	270
45°	152	245	367	495	582	625	605	537	415	285	180	130
50°	112	175	310	460	550	585	570	490	350	215	117	55
60°	77	155	240	420	560	627	602	477	295	140	47	23
75°	0	7	85	280	515	637	595	397	155	32	0	0
90°	0	0	0	155	482	667	610	303	27	0	0	0

It will be observed that the interaction of the effects of the angular height of the sun and the day length produces a quite complicated distribution in daily radiation received at midsummer.

Steinhauser (1939) gives for varying altitudes in lat. 47° N. the mean monthly estimates in cal. cm.$^{-2}$ day^{-1} set out in Table 37, the last column of which represents the sum of the period from April 1 to July 31.

TABLE 37. *Estimates of mean daily direct radiation in each month as a function of altitude in Europe at lat. 47° N.*

Altitude, m.	Jan.	Feb.	Mar.	Apr.	May	June	July	Aug.	Sept.	Oct.	Nov.	Dec.	Heating Period
200	115	201	322	431	536	588	558	484	382	240	148	102	64,484
500	128	219	343	465	564	618	586	512	404	262	160	110	68,140
1000	148	247	373	509	606	659	623	550	434	290	178	124	73,139
1500	159	264	403	547	645	696	660	585	460	314	194	136	77,745
2000	167	278	422	575	684	729	700	614	484	332	206	144	83,408
3000	175	293	445	599	717	773	752	656	516	350	217	152	86,699

It has already been indicated that the direct solar radiation is augmented by the scattered light of the sky; this very variable quantity is usually of the order of 20 per cent of the total radiation. When, however, the sun itself is obscured by clouds, much of the radiation is neither transmitted, absorbed, nor scattered, but is reflected back into space. In consequence of such overall loss due to cloudiness, the values just given, though they neglect sky radiation, are evidently usually excessive, as is demonstrated by the figures given by Sauberer and Ruttner (1941) for certain European stations (Table 38).

It is evident that in middle latitudes in Europe the total radiation received during the heating period is unlikely to exceed 60,000 cal. cm.$^{-2}$ except under very favorable circumstances at high altitudes. The large

TABLE 38. *Mean monthly radiation (cal. cm.$^{-2}$ day^{-1}) and total radiation received between April 1 and July 31 at two lowland and two alpine localities in middle latitudes*

	Latitude N.	Jan.	Feb.	Mar.	Apr.	May.	June	July	Aug.	Sept.	Oct.	Nov.	Dec.	Heating Season
Karlsruhe	49°00′	68	148	240	342	502	518	525	370	302	202	69	52	57,637
Boulogne-sur-Seine	46°50′	84	154	246	381	517	529	471	418	303	214	101	48	57,928
Zugspitze, Bavaria	47°27′	130	234	325	424	473	562	433	509	368	277	160	107	57,666
Davos	46°44′	154	307	470	528	538	621	640	511	468	358	191	141	70,988

body of data published by Aurén (1939) for Scandinavia, which permits a map (Fig. 110) to be drawn of the distribution of radiation during the four months at which most heating occurs at low altitudes, shows the steady reduction northward in the total radiation received during spring and early summer. These Scandinavian data may instructively be compared with the heat budgets of the lakes of the region.

The American data, presented graphically from Hand (1941), indicate the same general phenomenon (Fig. 111). In the cases of certain North American lakes to be discussed later (page 525), for which numerous spring and summer temperatures exist, the heating period can be defined more accurately than the arbitrary four months used in Tables 37 and 38. In these cases, however, the total radiation received is almost the same, about 60,000 cal. cm.$^{-2}$

In tropical regions the seasonal variation is greatly reduced and becomes bimodal. It is important to note that owing to the length of the period of potential sunshine at high latitudes, the maximum radiation that can be received at midsummer at the pole is greater than that delivered at any other latitude on any clear day.

REFLECTION, BACK-RADIATION, AND NET RADIATION SURPLUS

Reflection. Part of the radiation falling on the surface of a lake will be reflected and will therefore play no further part in events of limnological interest.

The full theoretical treatment of reflection is complicated by the fact that light from both the sun and the sky must be considered. As far as the sun is concerned, it appears that Fresnel's law, namely that the reflectivity, as a fraction of the incident light, is given by

$$R_f = \tfrac{1}{2}\left[\frac{\sin^2(\psi_i - \psi_r)}{\sin^2(\psi_i + \psi_r)} + \frac{\tan^2(\psi_i - \psi_r)}{\tan^2(\psi_i + \psi_r)}\right] \qquad (2)$$

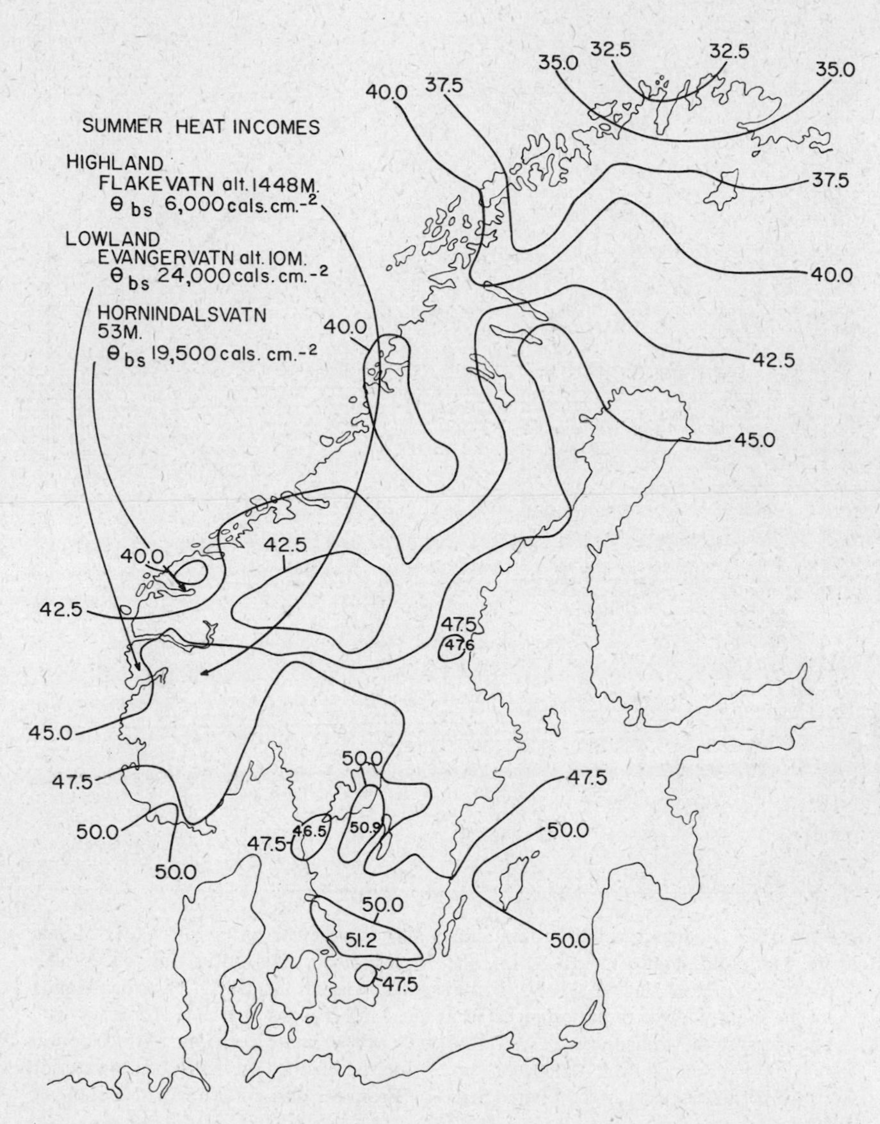

FIGURE 110. Isophots in 1000 cal. cm.$^{-2}$ received during April, May, June, and July on the Scandinavian Peninsula, also typical summer heat incomes of two lakes near sea level, taking up about half the available heat, and of one at a high level.

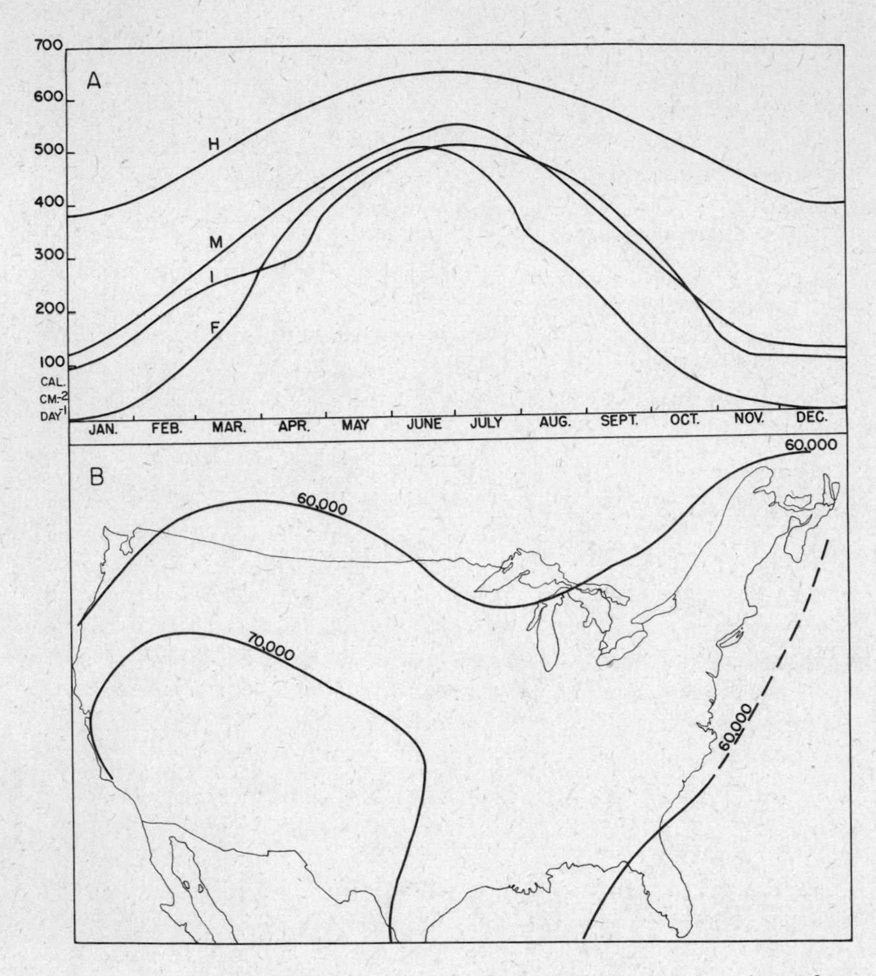

FIGURE 111. *A*, mean radiation (cal. cm.$^{-2}$ day^{-1}) received at various times of year at H, Honolulu, Hawaii, lat. 21°18′ N.; M, Madison, Wisconsin, lat. 43°05′ N.; I, Ithaca, New York, lat. 42°27′ N.; F, Fairbanks, Alaska, lat. 64°52′ N. (after Hand). Note the slightly lower radiation received at nearly all seasons at Ithaca (Lake Cayuga) as compared with Madison (Lake Mendota). *B*, approximate positions of the isophots for 60,000 cal. cm.$^{-2}$ and 70,000 cal. cm.$^{-2}$ for the entire temperate heating period from April 1 to August 1 in the United States. (From the data of Fritz and MacDonald 1949.)

where ψ_i is the angle of incidence and ψ_r the angle of refraction, holds for reflection from an undisturbed water surface (Lunelund 1924, Ångstrom 1925). Earlier observations on the Lake of Geneva by Dufour (1873), which suggested considerable divergence from theory, are presumably due to instrumental imperfections (Ångstrom 1925). Ångstrom found rather greater reflection from a naturally disturbed water surface, probably the Baltic near Stockholm, than equation 2 predicts. The effect was of the order of 20 per cent at low solar angles ($\psi_s' = 5°$), falling to 10 per cent at high solar angles. Since the reflectivity falls very rapidly with increase in ψ_s', the effect of the state of the surface is negligible in estimating the total radiation entering the water for $\psi_s' > 15°$. Moreover, since little radiation is ordinarily delivered during those hours when $\psi_s' < 15°$, the over-all effect on the radiation budget of moderate disturbance of the surface is very small. Ångstrom concluded experimentally that with very strong disturbance a slight decrease in reflectivity occurred, presumably owing to reflected light entering the elevated waves.

The effect of the light of the sky is rather difficult to treat theoretically (Anderson 1952). Various empirical determinations (Poole and Atkins 1926, Powell and Clarke 1936, Whitney 1938b) indicate that 6 to 8 per cent of the light from an overcast sky returns from the surface. A mean value of 7 per cent is not unreasonable. Sauberer and Eckel (1938) obtained lower values, between 3 and 5.5 per cent in small Austrian lakes, but point out that in such lakes the surrounding mountains cut off part of the low-angle radiation of the sky which contributes disproportionately to the reflectivity.

Several attempts have been made to define the reflectivity from sun and sky together, in terms of ψ_s'. Anderson (1952) gives a complicated expression in which terms are introduced both for Fresnel reflection and the reflection of the diffuse light of the sky, both corrected for variation in atmospheric turbidity. In practice, however, he employed an empirical approach, concluding that for clear days at Lake Hefner

$$R_t = 1.18\,\psi_s'^{-0.77} \tag{3}$$

where R_t is the total reflectivity of light, apparently including upward-scattered light, from sun and sky together. No detectable effect of wind velocity was observed, and determinations made at times when a Canadian polar air mass lay over the lake did not differ from those made when a tropical air mass was present. In spite of the theoretical expectation that high atmospheric turbidity should increase the reflectivity for high values of ψ_s' and decrease it for low, the variations at Lake Hefner due to differences in turbidity are probably negligible, and the curves (Fig. 112) given by Anderson would probably apply approximately on clear days in any region of moderate altitude and temperate latitude. The presence of a partial cloud cover raises complications, as the light from the sky may be greatly increased. The empirical curves of Anderson (1952) probably permit satisfactory estimates of reflectivity, including upward scattering under cloud.

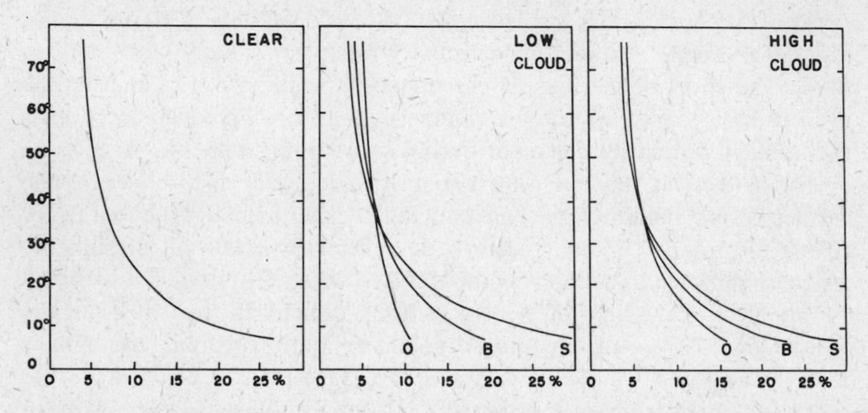

FIGURE 112. Reflectivity and upward scattering as percentage of total sun and sky radiation for various solar altitudes. Clear sky corresponds to $R_t = 1.18\psi_s'^{-0.77}$. Low clouds: S, clouds scattered (1/10 to 5/10), $R_t = 2.17\psi_s'^{-0.96}$; B, cloud cover broken (6/10 to 9/10), $R_t = 0.78\psi_s'^{-0.68}$; O, overcast (10/10), $R_t = 0.20\psi_s'^{-0.30}$. High clouds: S, $R_t = 2.20\psi_s'^{-0.98}$; B, $R_t = 1.14\psi_s'^{-0.68}$; O, $R_t = 0.51\psi_s'^{-0.68}$. (Anderson.)

Empirical determinations, as for instance those of Davis (1941), indicate that between 5 and 6 per cent of the total incident radiation is reflected or scattered during a summer day. Davis concluded that about half this loss was due to scattering from the water. Johnsson (1946) used an arbitrary value of 6 per cent for the average reflection and scattering in summer and 10 per cent in winter, which cannot be far from the truth.

The spectral composition of the total reflected light when the sun is high is little different from that of the incident, but at low solar angles the proportion of red rises, apparently because the lower angular zones of the sky are deficient in red. Sauberer and Ruttner (1941) and Davis (1941) have detected this effect empirically on lakes.

The presence of a water surface can, through reflection, increase the radiation falling on a vertical surface such as a vertically set leaf or parts of the human body when the sun is low; the effect is small, however, not exceeding 0.1 cal. cm.$^{-2}$ min.$^{-1}$ (Ångstrom 1925).

Thermal radiation from water and from atmosphere. The radiation from a black body is, by the Stefan-Boltzmann law, proportional to the fourth power of the absolute temperature, or

$$Q = 8.26 \times 10^{-9}(\theta_K)^4 \quad (4)$$

the numerical coefficient being the Boltzmann constant in cal. cm.$^{-2}$ min.$^{-1}$ deg.$^{-4}$ Water departs slightly from the behavior of an ideal black body, owing to reflection at the smooth water-air interface. Sauberer and Ruttner (1941) consider that the best empirical data on this matter are those of Falkenberg (1928), who concluded that the radiation emitted was 96 per cent of the black-body radiation for the observed temperature. Anderson (1952) gives 97 per cent as the result of renewed investigations.[2]

The atmosphere does not radiate as a black body; there is very little absorption or emission in the region of wave lengths 7.5 to 12.5μ. The maximum energy for pure black-body radiation at ordinary temperatures would be at wave lengths 9 to 11μ, but since this is in the region where little emission occurs, two marked bands occur, separated by the "window" from 7.5 to 12.5μ. An empirical expression of which the form is due to Ångstrom, but with coefficients determined by Falckenberg and Bolz (Bolz and Fritz 1950), gives the radiation of the atmosphere with a cloudless sky as

$$Q_A = 8.26 \times 10^{-9}\theta_{KA}{}^4(0.820 - 0.025 \times 10^{-0.126p_w}) \quad (5)$$

where θ_{KA} is the absolute temperature and p_w the vapor pressure of water.

Several modifications have been suggested as the result of empirical studies on atmospheric radiation with a cloudy sky. Bolz and Fritz (1950) use

$$Q_A = 8.26 \times 10^{-9}\theta_{KA}{}^4(1 + k_c B^{2.5})(0.820 - 0.250 \times 10^{-0.126p_w}) \quad (6)$$

when B is the fraction of the sky, in tenths, covered by cloud.

In general the water and air will not be at the same temperature; the effective back-radiation will be

$$Q_W - Q_A = 8.26 \times 10^{-9}[0.97\theta_{KW}{}^4 - \theta_{KA}{}^4(1 + k_c B^{2.5})(0.820 - 0.250 \times 10^{-0.126p_w})] \quad (7)$$

where k_c depends on the nature of the cloud cover, being for

Cirrus	0.04
Cirro-stratus	0.08
Altocirrus	0.17
Altostratus	0.20
Cumulus	0.20
Stratus	0.24

[2] Anderson thinks that occasionally oil derived from plankton might lower this factor a little.

Only Neumann (1953), among students of lake optics, has considered these refinements worthwhile. Johnsson (1946), in his study of Lake Klämmingen, was content to assume black-body radiation of both air and water, writing

$$Q_W - Q_A = 8.26 \times 10^{-9}(\theta_{KW}{}^4 - \theta_{KA}{}^4) \simeq 11(\theta_{KW} - \theta_{KA}) \tag{8}$$

The net radiation surplus. The surface of any lake will receive direct solar radiation (Q_S), indirect or scattered and reflected solar radiation (Q_H) from the sky and clouds, and long-wave thermal radiation (Q_A) from the atmosphere, and in some cases from mountains around the lake (Q_M). A part (Q_R) of the radiation impinging on the lake will be reflected, and a part (Q_U) will be scattered upward from below. The water will also emit long-wave radiation (Q_W). Regarding all the quantities as positive when not zero, regardless of direction, the radiation surplus Q_B is defined as

$$Q_B = Q_S + Q_H + Q_A + Q_M - Q_R - Q_U - Q_W \tag{9}$$

By night, since Q_S, Q_H, Q_M, Q_R, and Q_U are negligible,

$$Q_B = Q_A - Q_W \tag{10}$$

or accepting Johnsson's approximation

$$Q_B = -11(\theta_W - \theta_A) \text{ cal. cm.}^{-2} \text{ day}^{-1} \tag{11}$$

It is important to note that, even though the mean value of Q_B may be positive over a period of days, the lake is not necessarily gaining heat over this period, as the positive radiation surplus may be more than compensated by losses of heat through evaporation and the convective heating of the air. All the quantities which are involved in the estimate of the net radiation surplus are, however, also involved in determining the heat budget.

The net radiation surplus can be determined directly (Albrecht 1933, Sauberer 1937c). Some investigators, notably Franssila (1940), have therefore found the empirical estimate of the radiation balance to be a convenient method of determining $Q_W - Q_A$, when the total net radiation flux, the rate of heat accumulation or loss, and the rate of evaporation can also be ascertained. Franssila indicates for Lake Puujärvi in the parish of Karjalokja (lat. 60.2° N., long. 23.7° E.), Finland, that at the height of summer the balance is positive from 5 A.M. to 8 P.M., reaching a maximum value of 0.91 cal. cm.$^{-2}$ min.$^{-1}$ at midday, and negative during the night hours from 8 P.M. to 5 A.M., the minimum value being $-$ 0.15 cal. cm.$^{-2}$ min.$^{-1}$ between 11 P.M. and 1 A.M.

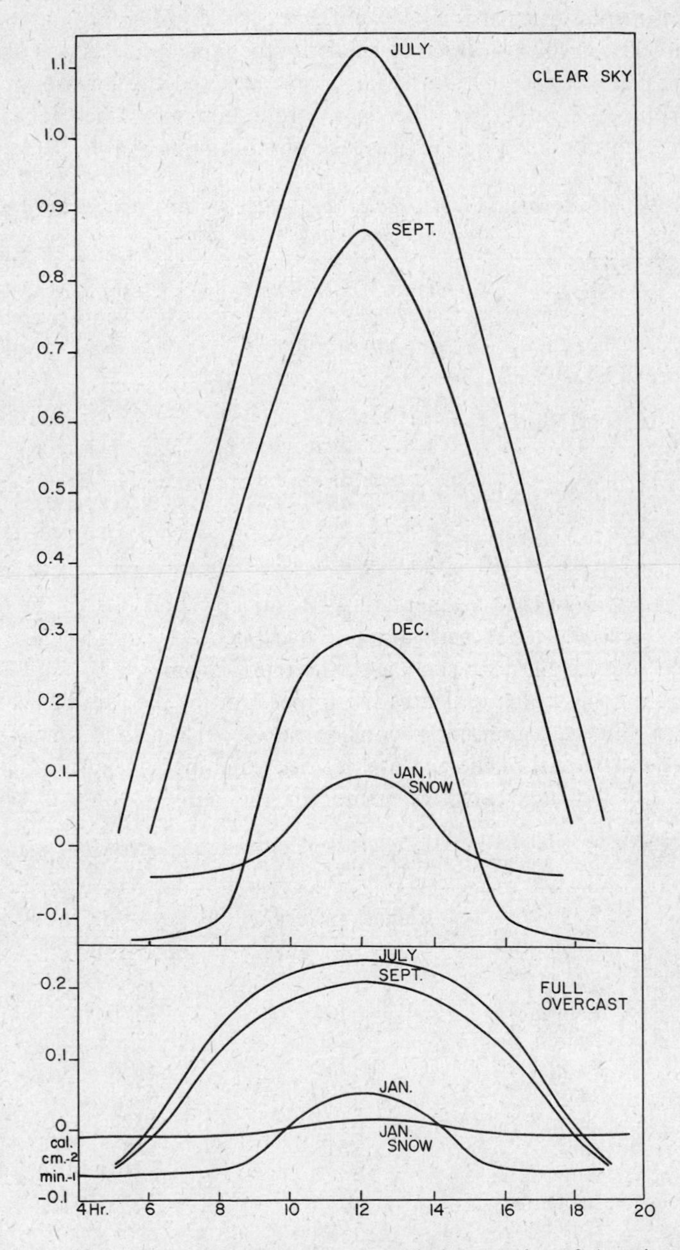

FIGURE 113. Mean diurnal variation in net radiation surplus of Austrian lakes at altitudes of 300 to 700 m., under clear sky and under complete overcast at different seasons of the year (after Sauberer).

Sauberer (1937a) has published some more generalized curves for the radiation surplus of Austrian lakes under a clear sky (Fig. 113), at altitudes between 300 and 700 m., which cover about the same range as was observed by Franssila. Sauberer (1937b) has also estimated the maximum radiation surplus integrated over the twenty-four hours for such lakes if the horizon is not modified by the presence of mountains (Table 39).

TABLE 39. *Maximum daily net radiation surplus for Austrian lakes at different seasons on clear and cloudy days*

	Cloud 3/10 or Less, Bright Sun, cal. cm.$^{-2}$ day^{-1}	Cloud 8/10 to 10/10, Sun Covered, cal. cm.$^{-2}$ day^{-1}
January–February (open water)	− 40	− 31
January (ice with snow cover)	− 53	− 4
April	+ 336	+ 71
July	+ 485	+ 93
September	+ 293	+ 68

It will be evident that the mean daily positive surplus during the heating period is unlikely to exceed about 400 cal. cm.$^{-2}$ day^{-1}. Since non-radiative sources of heat such as conduction from the air are usually unimportant, this value indicates the upper limit of the rate of heating to be expected in the region under consideration. It is also noteworthy that during the early part of the cooling period, which begins before the end of August, a fairly high positive value for the daily surplus is possible.

TABLE 40. *Mean net daily radiation surplus in the lakes of the Jordan Valley*

	Radiation Surplus, cal. cm.$^{-2}$ day^{-1}	
Month	Sea of Galilee	Lake Hula
January	61	110
February	188	157
March	250	239
April	353	363
May	400	439
June	523	537
July	513	540
August	422	445
September	367	363
October	206	258
November	87	103
December	39	55

Johnsson's (1946) analysis of the thermal cycle in Lake Klämmingen in south Sweden indicates a daily positive surplus from the time of the breaking of the ice until October, even though the lake starts losing heat late in July. In Lake Mead (Anderson and Pritchard 1951) the balance is positive until November, though the lake starts cooling early in August. The two fresh-water lakes of the Jordan Rift Valley appear from the beautiful work of Neumann (1953) always to exhibit mean daily positive net surpluses (Table 40), whether they are warming or cooling.

LABORATORY STUDIES ON TRANSMISSION AND ABSORPTION OF LIGHT BY WATER

The intensity of a beam of monochromatic light entering water at right angles or normal to the surface of the water is given by

$$I_z = I_0 e^{-\eta z} \tag{12}$$

or

$$\ln I_0 - \ln I_z = \eta z \tag{13}$$

where I_0 is the intensity of the beam as it crosses the water surface, I_z the intensity at depth z and η a constant for any given wave length, termed the *extinction coefficient*. In nature the sun is rarely vertical, and its rays therefore seldom are normal to a water surface. Moreover, of the total radiation coming from the sky, there is always a component which is never normal to the water surface. Sunlight also is not monochromatic. The conditions required for the verification of equation 12 are therefore somewhat imperfectly realized in nature, though easily obtained in the laboratory. It is accordingly convenient to start a discussion of extinction coefficients and transmission by a consideration of laboratory data on pure water and on various lake waters.

The absorption and transmission of light by pure water. The data for chemically pure water have been assembled from a number of sources, some unpublished, by James and Birge (1938), who have added a new series of determinations for wave lengths from 3650 to 8000 A. Such determinations (Table 41) are usually expressed in terms of the extinction coefficient η, or as percentile transmission

$$\frac{100 I_z}{I_0} = 100 e^{-\eta}$$

or percentile absorption

$$\frac{100(I_0 - I_z)}{I_0} = 100(1 - e^{-\eta})$$

the path z being taken as 1 m.

TABLE 41. *Optical properties of water* (*room temperature*)

Wave Length (Color), A.	η	100 $(1 - e^{-\eta})$	Refractive Index
8200 (infrared)	2.42	91.1	
8000	2.24	89.4	
7800	2.31	90.1	
7600	2.45	91.4	1.329
7400	2.16	88.5	
7200	1.04	64.5	
7000	0.598	45.0	
6800 (red)	0.455	36.6	
6600	0.370	31.0	1.331
6400	0.310	26.6	
6200 (orange)	0.273	23.5	
6000	0.210	19.0	
5800 (yellow)	0.078	7.0	1.333
5600	0.040	3.9	
5400	0.030	3.0	
5200 (green)	0.016	1.6	
5000	0.0075	0.77	
4800	0.0050	0.52	1.338
4600 (blue)	0.0054	0.52	
4400	0.0078	0.70	
4200	0.0088	0.92	
4000 (violet)	0.0134	1.63	1.343
3800 (ultraviolet)	0.0255	2.10	

There is quite good agreement as to the *general* form of the curve relating any of these quantities to wave length. The transmission is low and the extinction coefficient high in the infrared; there is a rapid increase in transmission between 7400 A. and 7000 A. and again about 6000 A. to 5800 A. in the orange. The transmission is maximal in a region about 4600 A. in the blue, and falls very slowly in the violet and more rapidly in the long ultraviolet. Within this general pattern, however, a considerable variation is shown when individual sets of determinations are compared. Considering only the region within which James and Birge worked, 3650 to 8000 A., marked differences in the blue and violet are observable between their determinations and those of Sawyer (1931) and Ewan (1894). At 3650 A., Sawyer finds a reduction of 21.6 per cent per meter, James and Birge only 3.6 per cent. It is practically certain that these differences are due to minute quantities of impurities in the less transmissive waters. Paraffin-lined absorption tubes may increase the absorption in this region; James and Birge evidently got low values because they used silver tubes, which contained purer water than those of the other experimenters.

At the red end of the visible spectrum, there is less glaring disagreement than at the violet end, but comparison of the extremely careful work of Collins (1925), and of Ganz, and of Baldock, whose previously unpublished data are given in full by James and Birge, indicates that at 7000 A. values between 36.6 and 48.3 per cent transmission per meter have been reported. This variation is attributed tentatively by James and Birge to unexplained differences in the aggregation or molecular structure of the water samples purified and observed under varying conditions. Such a conclusion, however, can by no means be regarded as established.

Absorption by natural waters in the laboratory. The absorption spectra of lake waters and some other kinds of natural waters have been studied in the laboratory by Von Aufsess (1903), Witting (1914), Pietenpol (1918), Erikson (1933), James and Birge (1938), and Åberg and Rodhe (1942). Though almost all these studies are of significance, that of James and Birge is so much more extensive than the other investigations that it provides a natural basis for all further discussion. In considering the effect of dissolved and suspended matter in lake water on the transmission of light, it is necessary to remember that if the extinction coefficients are additive, the effect of each class of substance on the percentile transmission will be multiplicative.

If

$$I_z = I_0 e^{-\eta_t z} \tag{14}$$

and

$$\eta_t = \eta_w + \eta_p + \eta_c \tag{15}$$

where η_w is due to water, η_p to suspended particles, and η_c to dissolved material, then

$$I_z = I_0 e^{-\eta_w z} e^{-\eta_p z} e^{-\eta_c z} \tag{16}$$

If $z = 1$, then η_w, η_p, and η_c can be considered three factors which reduce I_0 to I_z owing to the passage of light through *water* containing *particles* and *dissolved color*. In an experiment with a mixture of humus solution and mastic suspension, Åberg and Rodhe found the extinction coefficients to be additive, justifying the type of treatment just outlined.

James and Birge used only waters that had been allowed to stand, to permit the settling out of the larger suspended particles. Unfortunately, the time during which settling was allowed to take place was extremely variable, varying from one week to one year. There is no adequate evidence that during the settling process the extinction coefficient for any given wave length really approaches asymptotically a value significantly above the value for pure water. It is quite possible that the coloring matter in solution may be dispersed stably enough not to aggregate slowly during settling. In the only case studied (Fig. 114), a practically uncolored water from Devil's Lake, Wisconsin, which was examined at intervals throughout a year, there was a great difference in absorption between water that had settled for 14 weeks and water stored for 52 weeks. Comparison of this water with others stored for some time does, however, show that it is possible for a water to be at least as transparent in the middle wave lengths as well-settled Devil's Lake water, but at the same time to exhibit much greater absorption in the short-wave region. It would be in accord with the entire work if subsequent studies showed a very stable material in colloidal or true solution and not removable by settling, absorbing most strongly in the blue and violet.

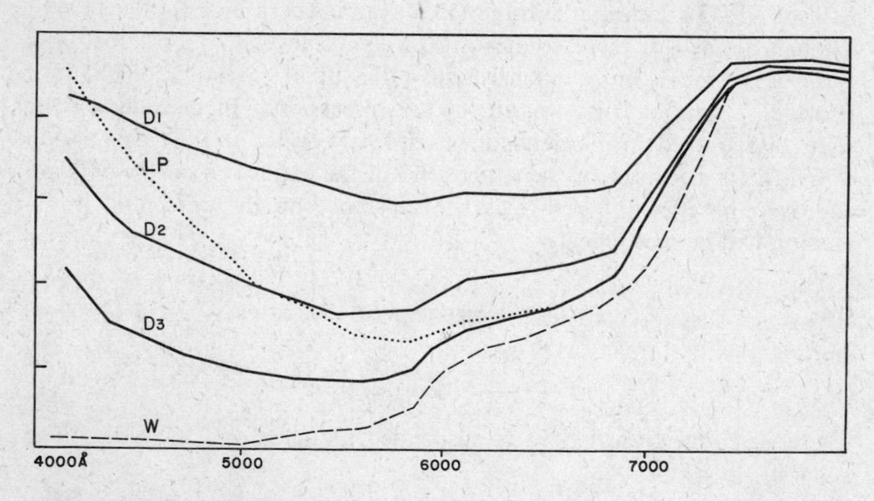

FIGURE 114. Mean percentile absorption at different wave lengths. D1, Devil's Lake, Wisconsin, water recently collected; D2, the same after 14 weeks settling; D3, the same after 1 year settling. LP (dotted line), Little Papoose Lake after 10 months settling. W (broken line), pure water. (After James and Birge.)

James and Birge studied the absorption of their settled waters at twenty-two wave lengths between 3650 A. and 8000 A. They then filtered the water through a Berkefeld's No. V filter, which removes particles down to a size corresponding to the removal of 75 per cent of the bacteria, and after filtration they again determined the absorption spectrum. The color on the platinum-cobalt scale (see p. 413) was determined visually for both settled and filtered water. This mode of analysis permits the presentation of the data in terms of a three-component analysis for each wave length, namely $e^{-\eta_w}$, $e^{-\eta_p}$, and $e^{-\eta_c}$, or η_w, η_p, and η_c. The effect due to pure water is known from the initial experiments, $e^{-\eta_c}$ is determined by dividing the transmission for pure water into that for the filtrate, and $e^{-\eta_p}$ by dividing that for the filtrate into that for settled water. Since the transmissions in the red are very low owing to the opacity of pure water to the longer visible wave lengths, considerable irregularities of little or no significance may be expected in this region, an unfortunate fact, for it is just here, at wave lengths in excess of 7000 A., that the results of greatest chemical interest may be expected.

Apart from the three-component analyses just outlined, James and Birge use several other modes of presentation, but they do not materially assist in the interpretation of their data. They present all their three component analyses in terms of $100e^{-\eta_p}$ and $100e^{-\eta_c}$ plotted downwards on coordinate paper, so that if the curves are read upward in the ordinary way, giving $100\ (1 - e^{-\eta_p})$ and $100\ (1 - e^{-\eta_c})$, they give the percentage effect of suspended or retained and dissolved or filter-passing material on the absorption. Åberg and Rodhe prefer to give their data in terms of the extinction coefficients.

The three-component analysis of the waters of the Wisconsin lakes indicates that although filtration almost always reduces the color of the water, the suspensoids remaining in settled waters are relatively unselective in their optical effects. Moreover, for low colors (Pt units < 20), the variation in absorption due to suspensoids is apparently largely independent of color. This is well shown in Fig. 115, in which mean values are given for the percentage effect due to suspensoids, for lake waters of colors 4, 6, 10, and 20, and including at least five lake waters in each class. The curves are not regularly ordered according to the color classes, and show a rather feeble decrease in absorption toward the red end.[3] The curves for $100\ (1 - e^{-\eta_c})$, the percentage effect on absorption due to dissolved color in the same groups of lakes, are strikingly different. It is evident that here the effect is markedly selective, and that in the violet and blue the curves are ordered according to the colors. This ordering, however, is not apparent at the red end. It is quite evident that the material producing the main absorption in the filtrate at the violet end varies systematically in concentration with the visually determined color, but it is extremely doubtful that the same is true of the material producing the absorption at the red end. It would therefore seem likely that the absorption of long wave lengths is produced by substances quite different from and varying independently of the dissolved color responsible for the principal absorption of the shorter wave lengths. Examination of the curves for the effect

[3] Some degree of selectivity is to be expected. For a very recent treatment for ocean water, see Burt (1955).

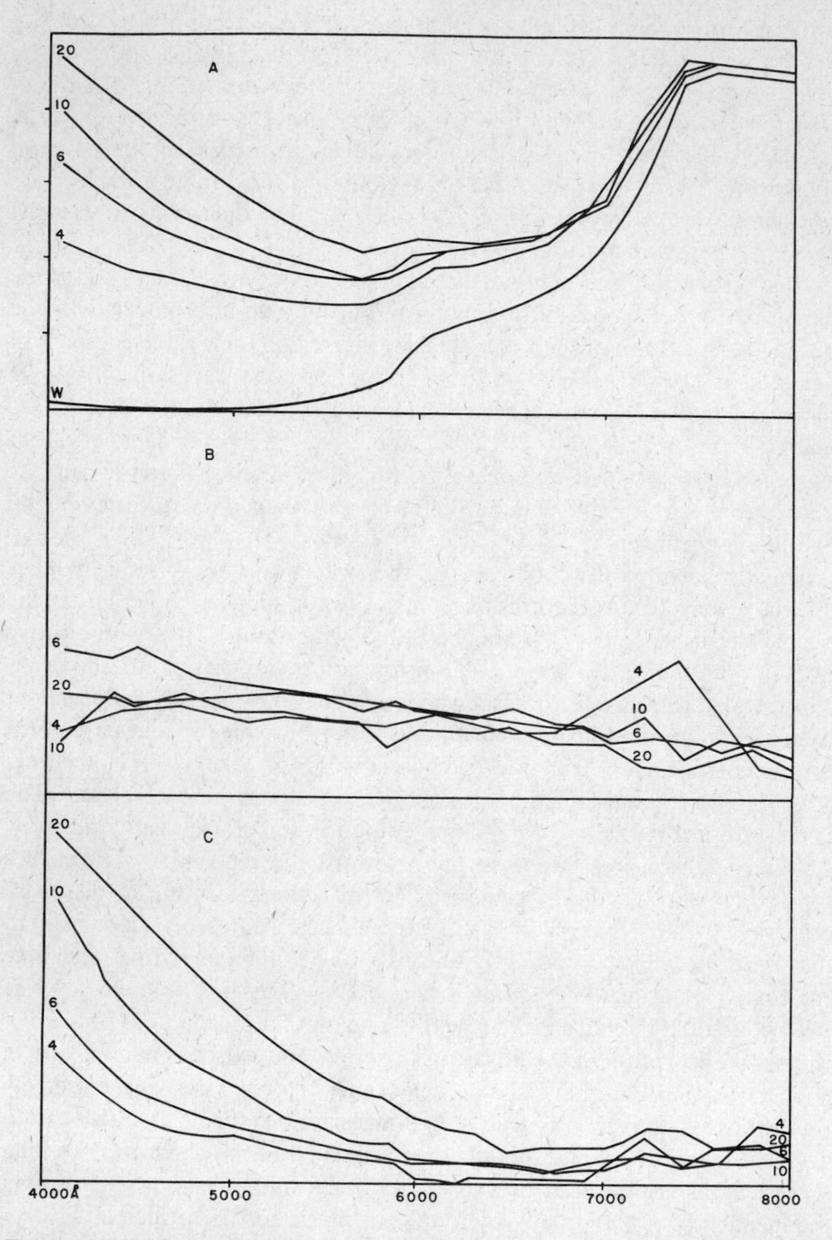

FIGURE 115. Mean percentile absorption at different wave lengths. *A*, for settled unfiltered waters of groups of lakes of color 4, 6, 10, and 20 Pt units; and W, pure water. *B*, contribution of suspended matter to total absorption. *C*, contribution of dissolved matter. Note low, irregular selectivity of absorption by suspended matter, and the regular increase in absorption of violet, blue, and green light with increasing dissolved color. (After James and Birge.)

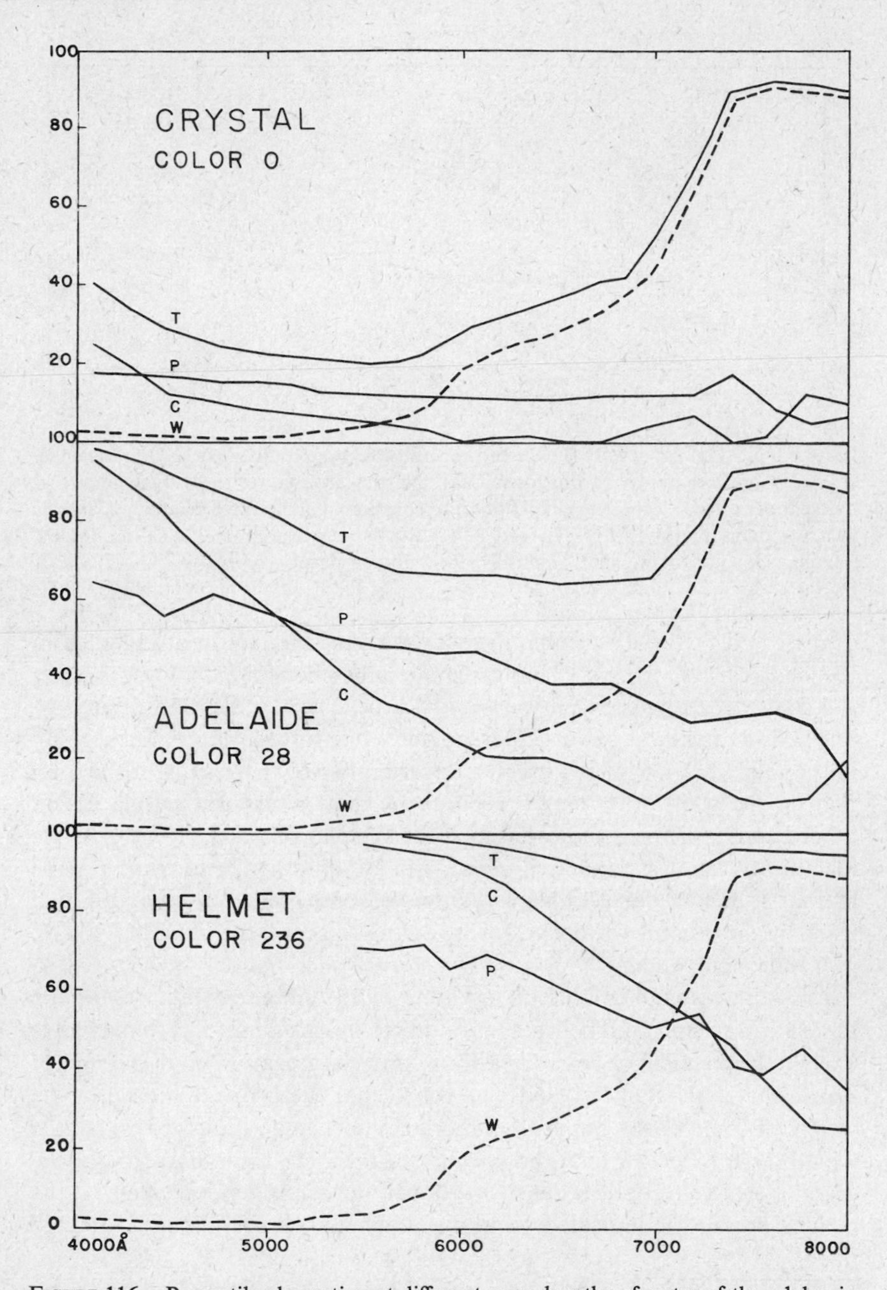

FIGURE 116. Percentile absorption at different wave lengths of water of three lakes in northeastern Wisconsin, ranging from visually uncolored to very dark (236 Pt units). T, total absorption; P, due to suspended particulate material; C, due to dissolved color; W (broken line), pure water. (After James and Birge.)

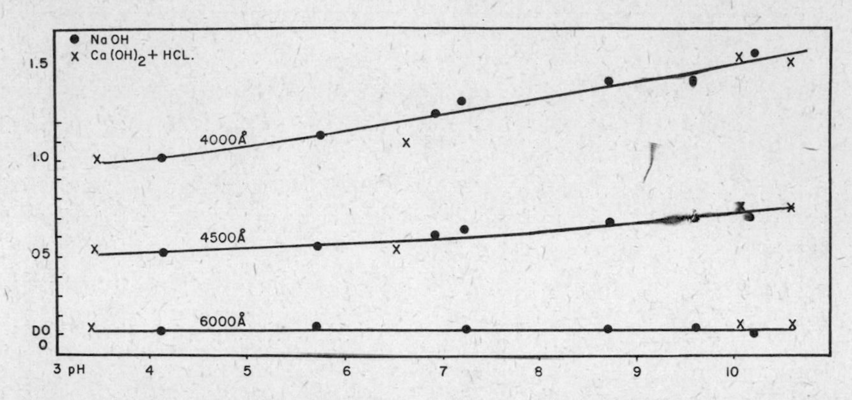

FIGURE 117. Optical density (DO arbitrary units, relative to distilled H_2O) of strongly colored, initially acid water from Jones Lake, North Carolina, for three wave lengths at different *p*H values. The intensification of absorption at short wave lengths apparently increases continuously with *p*H, independent of the nature of the cation added. (Hutchinson, Vallentyne, and Townsley, unpublished observations.)

of dissolved material on some individual lakes (Fig. 116) not included in the mean curves of Fig. 115 indicates the same phenomenon.

James and Birge believed there to be a persistent indication of at least one definite band, at about 7200 A., due to dissolved material absorbing in the red. In view of the greater uncertainties of the results in the long than in the short wave-length region, no great stress can be placed on other bands which are indicated in some samples of water. The possible identification of the band at 7200 A., and of any other bands that may be present, is of considerable chemical interest; some preliminary unpublished work by the present writer, however, did not disclose the band at 7200 A. in Connecticut waters.

According to James and Birge, none of the observed features of the absorption spectra of lake waters are due to inorganic solutes, but there is likely to be an effect of *p*H and cations on the aggregation or dispersion of brown material. Ohle (1934a) has found that the brown material of the waters of some of the north German lakes acts as an indicator, changing from brown to yellow over the *p*H range 4.0 to 5.0; a comparable if more gradual change of optical density with *p*H, apparently independent of the cations present, is indicated by some preliminary experiments presented in Fig. 117.[4]

[4] Shapiro (verbal communication), in an extension of this work, finds that the behavior indicated in Fig. 117 is characteristic of some of the yellow water-soluble fluorescent materials that he has isolated from lake water.

THE TRANSMISSION OF LIGHT BY LAKE WATER IN NATURE

The problems of transmission of radiation and of illumination under water can be considered in several different ways. First, it is possible to transfer some of the essential elements of a laboratory experiment to the field, as is done with the Petersson transmission meter, in which a light source and a photocell fixed at either end of a horizontal frame are lowered, preferably at night, into the lake. It might seem logical to begin by considering such experiments on transmission *in situ*, but inasmuch as they have so far been used to obtain purely relative values of the transmission or extinction and derive their great interest from other than purely optical considerations, Sauberer's and Whitney's remarkable studies by this method will be summarized at the end rather than at the beginning of this section.

Going one step farther from the laboratory, we find a great volume of work in which the sun and sky are the light source. Such work can be considered from two points of view. It is possible to present the results, as before, in terms of the optical properties of the water, using the sun and sky as if they were no more than an artificial light source. In view of the well-defined spectral composition of sunlight and of the varying spectral extinction coefficients of different waters, great variations in the energy of different wave lengths actually delivered from sun and sky into the depths of lakes will be found. It is also possible to proceed still further from the laboratory, not merely replacing the monochromatic artificial light source by sunlight, but also replacing the recording instrument by a photosynthesizing organism. This is indeed the most interesting optical situation that a lake can afford, but as it involves so many purely biological considerations, it is best deferred until Volume II.

Illumination and extinction coefficients, basic concepts and terminology. It is first necessary to distinguish between three possible meanings that might be given to the words *illumination at a given depth*. These three meanings are to some extent arbitrary, but may all be useful. It is therefore best to distinguish them by definite terms, following Poole and Atkins (1926).

Vertical illumination, I_v, at depth z is defined as the illumination on a horizontal plane placed at depth z.

Maximum illumination, I_m, at depth z is defined as the illumination on a recording surface set at such an angle that the surface receives the maximum possible illumination. If the illumination consisted of a single beam, the angle would be such that the recording surface would be normal to the beam. In nature, since both sun and sky are involved, on a clear day the surface must be set slightly more horizontal than the position normal to the refracted rays of the sun. The concept of maximum illumination is not of great importance.

Total illumination, I_t, at depth z is the illumination received by a point recorder from all directions. If the illumination due to an elementary pencil of rays of solid angle $d\Xi$ be $i\, d\Xi$, then the total illumination will be given by

$$I_t = \int_0^{2\pi} i\, d\Xi \tag{17}$$

This is the illumination incident on a small phytoplanktonic organism suspended in the water and performing photosynthesis there.

Except where otherwise stated, the illumination to be discussed here will always be vertical illumination at depth z, for which the symbol I_z will be used. In nature, the light from the main source of illumination, the sun, will seldom fall normally on the water surface or on any recording instrument set horizontally below the surface. The height of the sun is, of course, variable. If the angular distance of the sun from the zenith, or angle of incidence, be ψ_i and the angle of refraction be ψ_r, then

$$\frac{\sin \psi_i}{\sin \psi_r} = \mu_R \tag{18}$$

where μ_R is the refractive index of water, which has, if variation with wave length (Table 41) is not considered, an approximate value of 1.33. The distance traveled through the water by light from the sun to reach a depth z will therefore be bz, where

$$b = \frac{1}{\cos \psi_r} \tag{19}$$

The vertical illumination as defined above, in so far as it comes only from the sun and in so far as the variation of the extinction coefficient with wave length can be neglected, may therefore be expected to decline according to

$$I_z = I_0 e^{-\eta b z} \tag{20}$$

Actually, some of the light incident on the lake, even on a clear day, does not come from the sun but from the sky, and when only the sun is covered with clouds this proportion is much increased. Although such light from the sky is incident on the lake from all angles, the area of the sky corresponding to the difference between any two angular heights ψ' and $\psi' + \Delta\, \psi'$ decreases progressively, and is in fact proportional to $\cos \psi'\, \Delta\, \psi'$. The amount of light from unit sky area falling on a horizontal photometer of given area is, however, proportional to $\sin \psi'$. The total radiation from a sky of uniform brightness in which unit area at the zenith gives an illumination of i per unit area of photometer will therefore be given by

$$I_H = i\int_0^{\pi/2} \sin \psi' \cos \psi' \, d\psi' \tag{21}$$

Under water, ψ' must be replaced by $\psi_r'(= \pi/2 - \psi_r')$, where $\mu_R \cos \psi_r' = \cos \psi'$. The lower limit of the integration is moreover not now zero, but is $\psi_r'_{\text{min.}}$, given by

$$\cos \psi_r'_{\text{min.}} = \frac{\cos 0°}{\mu_R} = \frac{1}{\mu_R} \tag{22}$$

Taking $\mu_R = 1.33$ and introducing a term R′ to allow for loss by reflection the light of the sky is given by

$$I_H = i\int_{\psi_r'_{\text{min.}}}^{\pi/2} \sin \psi_r' \cos \psi_r' (1 - R') \, d\psi_r' \tag{23}$$

This detailed theory of the light of the sky is due to Lauscher (1941). Somewhat less formal treatments have been given by Poole and Atkins (1926) and by Whitney (1938b). Poole and Atkins computed b, the mean optical path per meter, as 1.19; Whitney, as 1.185; Lauscher, as 1.22. The corresponding angular height of the sun ψ_s' is about 55 degrees, the zenith angular distance ψ_s being about 35 degrees. A slight error is introduced into all these treatments by assuming the sky to be uniformly bright. This assumption leads to a slight overestimate of the optical path so that, of the various computations, the lower values of about 1.19 are to be preferred. It will be observed that the mean angle of incidence of the light of the sky is not widely different from the angles of incidence of sunlight at midday during the summer in temperate latitudes.

Birge and Juday, in a series of papers (1929, 1930, 1931, 1932) on penetration of radiation into the Wisconsin lakes, have tended to treat all radiation as if it came from the sun. They plotted the observed vertical illumination at depth z against bz. The resultant curves in effect give the illumination that would be received at depth z if all the light of sun and sky were derived from a point source at the zenith. The slope of such a curve gives the value of the extinction coefficient η' in

$$I_z = I_0 e^{-\eta' z} \tag{24}$$

This value will be very close to the laboratory value of η, but a separate symbol is convenient to indicate values derived from field observations by methods involving approximation relating to the light of the sky. Such values will be called *zenith extinction coefficients*, and the equivalent percentile transmissions and absorptions will be zenith transmissions and absorptions per meter.

In contrast to this method of treatment, some other investigators, notably Poole and Atkins (1926,), Sauberer (1939, 1945), and also Åberg and Rodhe (1942), have concluded that it is better to neglect supposed variations in vertical illumination due to the height of the sun. The optical path is taken to be more or less constant, variations in the angle of incidence being regarded as of little importance. The extinction coefficient η'' is then termed the *vertical extinction coefficient* by Poole and Atkins. The argument in favor of adopting this point of view is in part empirical; it is claimed by Poole and Atkins that the observed values of the slope of the line obtained by plotting intensity logarithmically against depth on semilogarithmic paper do not in fact vary with the angular height of the sun.

The theoretical interpretation of this empirical conclusion is that when the sun is low a disproportionate amount of radiation comes from the sky, which light has, as has been indicated, a mean optical path of about $1.19z$. As the sun gets higher in the sky, more light from the sun and proportionately less from the sky is involved, but at the same time, the optical path of the direct rays of the sun decreases, so that in the temperate summertime the midday height is not very far from that corresponding to an optical path of $1.19z$, as in the case of the light of the sky. Moreover, the more oblique rays due mainly to scattering, will, unless they were scattered only a short distance above the photometer, be markedly absorbed before reaching the instrument, so that little very oblique light is likely to be recorded. Poole and Atkins indeed suggest that the mean optical path per meter is more dependent on the ratio of scattering to absorption than on the angular height of the sun, from which the major part of the illumination comes. The *vertical extinction coefficient* η'' is measured simply as the slope of the line obtained by plotting photometer readings on the logarithmic axis against depth on semilogarithmic paper. Corresponding to this coefficient are the percentile vertical transmissions or absorptions along the mean optical path per meter of depth. Assuming the correctness of the argument, this is obviously the most convenient extinction coefficient to use.

Transmission of white light from sun and sky. A number of investigators have made studies of the penetration of natural white light. Strictly speaking, only monochromatic light can undergo constant absorption by homogeneous water, but in many cases, after the rapidly absorbed ultraviolet and infrared have been removed by the top meter, the variation of η' or η'' with wave length is sufficiently small for a mean value for white light to be useful (Fig. 118). The long series of researches by Birge and Juday (1929, 1930, 1931, 1932) provide the most convenient body of data. In nearly all of these studies, the transmission by the top meter was much

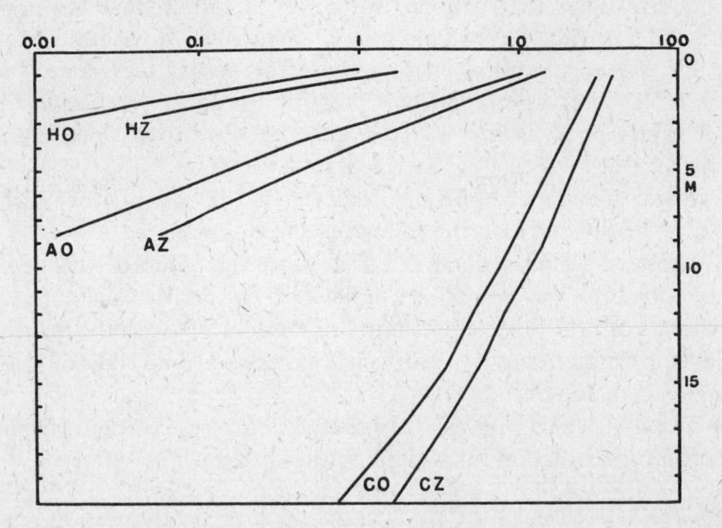

FIGURE 118. Intensity as a percentage of the radiation in air at various depths, as observed (O series), corresponding to vertical extinction, and (Z series) corrected to zenith extinction. C, Crystal Lake, color 0, Secchi disk transparency 13 m.; A, Adelaide Lake, color 36, transparency 4.8 m.; H, Helmet Lake, color 260, transparency 0.8 m. Note in Helmet and Adelaide a slight increase in transmission as short-wave light is progressively removed in the upper layers; in Crystal a slight decrease in the hypolimnion, presumably due to suspended material or seston. (From data of Birge and Juday.)

less than that below 1 m., owing, as has just been indicated, to the opacity of water outside the visible part of the spectrum. It is possible that in some cases small bubbles at a disturbed water surface may also reduce transmission in the top meter; Powell and Clarke (1936) believe this occurs in the sea. In view of the differences that probably exist between sea and fresh waters in their capacity to foam, it is not safe to carry over such results from the ocean to lakes.

Below a depth of 1 m., a characteristic transmission or extinction is often achieved. In the Wisconsin lakes this varies greatly; the following extreme values are recorded.

Crystal Lake	$\eta' = 0.192$ or zenith transmission 82 per cent per meter
Little Star Lake	$\eta' = 3.900$ or zenith transmission 2 per cent per meter

Crystal Lake is certainly not the most transparent lake known, for Utterback, Phifer, and Robinson (1942) find that in Crater Lake, Oregon, the extinction coefficient varies from 0.311 in the red to 0.031 in the blue. The value for the red is essentially as in pure water; in the blue, the value is about five times that of pure water. It is certain that the apparent mean value for white light would be well below that of Crystal Lake. As it is, the Wisconsin values overlap those for neritic sea water, and the Crater Lake observations indicate that for the lakes of the world as a whole, values of η' as low as those encountered in the open ocean are possible.

Birge and Juday's observations indicate that in general the Wisconsin lakes may be divided into the following three groups.

(1) Those in which beyond the first meter the value of the percentile transmission or of the extinction coefficient is essentially constant. As far as the limits of depth or the possibility of observation with the instruments employed permitted study, the very clear lakes and some lakes of moderate transparency belong here.

(2) Those in which there is a considerable increase in the transmission or decrease in extinction for white light in the deeper water.

	1–2	2–3	3–4	4–5	5–6	meters
Little Long (color 96)	16	30	26	38	38	per cent transmission per meter

In such cases the water is always highly colored (68 to 268 Pt units) and the increase in transmission is, as far as is known, apparent only and due to the change, with increasing depth, of the spectral composition of the light penetrating the lake. All but the longer wave lengths are removed by the color in the top few meters, and the values for transmission or extinction therefore become, as the depth increases, more and more characteristic of the narrow range of frequencies in the red for which such water is most transparent. This is merely an accentuation of the effect normally exhibited in the top meter.

(3) Those in which there is a considerable decrease in transmission or increase in the extinction coefficient in the deeper water, a characteristic transmission being attained only throughout part of the epilimnion.

	1–2	2–5	5–7	7–9	9–11	11–13	13–14	meters
Day Lake	70	69	72	67	53	39	30	per cent transmission per meter
White Sand Lake	67	65	64	38	51	46		

In these cases there is clearly a real increase in the opacity of the water in the hypolimnion.

Little systematic work has been done on the seasonal variation of transmission. It is likely, however, that the results obtained by Pearsall and Ullyot (1934) in Lake Windermere, which indicate an inverse variation of transmission with the density of the phytoplankton, are generally applicable to reasonably productive lakes not subject to great seasonal variations in turbidity of nonbiological origin. In the Wisconsin lakes the transmission showed no systematic changes from year to year, at least during the period before 1932; but in Windermere, Pearsall and Hewitt (1933) describe a systematic fall in the proportion of incident radiation received at 4.5 m., from 2 to 5 per cent in 1922 to 0.11 to 0.18 per cent in 1932, due to an upward trend in phytoplankton productivity.

Spectral composition of light reaching deep water. In the very clearest lakes, in which, as in pure water, absorption is least at the short-wave end of the visible spectrum, the light that penetrates to the greatest depths will be blue, as in the ocean. This is obviously the case in Crater Lake (Fig. 119). In the most deeply colored lakes, light penetrates only a short distance, and the last light to remain will be practically all of long wave length. Thus in Lake Mary, Wisconsin (Birge and Juday 1931), with a color of 123 Pt units, even at a depth of 1 m. there is no detectable light of wave length less than 4000 A., and 78 per cent of the radiation present has a wave length in excess of 6000 A. in the orange and red.

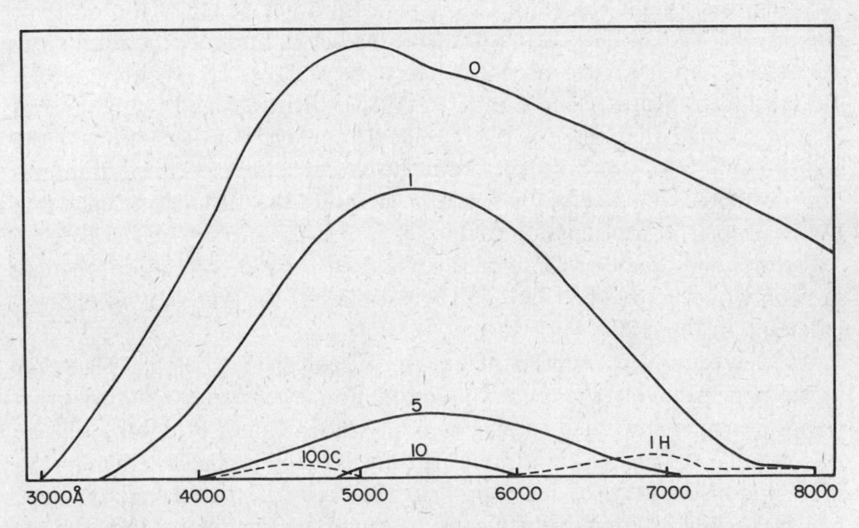

FIGURE 119. Variation in spectral composition of light with depth. The outer curve, 0, gives an approximate standard spectral energy distribution for sunlight; solid lines (1, 5, and 10), energy distribution at 1, 5, and 10 m. in the Lunzer Untersee, Austria, calculated from 0 and the data of Sauberer; 1H (broken), the same at 1 m. in Helmet Lake, Wisconsin; 100C (broken), the same at 100 m. in Crater Lake, Oregon. All somewhat approximate.

At 2 m. practically all the light present falls in the latter range. Helmet Lake, with a color of 236 Pt units, provides a still more extreme case (Fig. 119).

Even in lakes in which the visual color is zero, there is usually a considerable divergence in the spectral composition of the light actually present in the deeper layers from the distribution that would be implied by light of the spectral composition of sunlight penetrating pure water. In most small and clear lakes the light reaching to greatest depths is in the green part of the spectrum, as in Crystal Lake; in those of low but detectable visual color and moderate transparency, in the yellow. This is well shown (Fig. 119) in the case of the Lunzer Untersee, in which practically the whole of the small amount of radiation present at 10 m. is in the range between 5000 and 6000 A. The variation in wave length, from short to long, of the most penetrating part of the illumination in lakes, as the color increases and the depth to which a given proportion of total radiation reaches decreases, is of considerable biological importance and will be discussed again in Volume II.

Extinction as measured in a horizontal beam. If an apparatus consisting of a rigid rod with a light at one end and a photometer at the other is suspended horizontally and lowered into the lake carefully at night, it is possible to obtain a record of the transmission per meter of the water *in situ* in the lake. The method was introduced into oceanography by Petersson, and has been used in the study of lakes by Whitney (1937, 1938a) and by Ruttner and Sauberer (1938). Ruttner and Sauberer and Whitney found that there is a great deal more vertical variation in the transmission than would appear from the curves relating vertical illumination to depth, which curves essentially integrate the variations detectable by the Petersson transmission meter.

Ruttner and Sauberer found a very marked inverse correlation of tranmission with the total plankton in their studies of the Austrian lakes, as is indicated in Fig. 120.

Whitney examined a number of lakes in Wisconsin (Fig. 121). Wherever a well-marked, freely circulating epilimnion occurred, the horizontal transparency in this region was essentially independent of depth. There was usually one marked minimum in transparency just below the epilimnion, and also a very sudden drop near the bottom. In very many lakes, however, elaborate microstratification occurred throughout the hypolimnion. Two sets of readings on White Sand Lake, at closely adjacent stations about 15 m. apart, indicate the reality of even the fine embroidery on the transparency curve. It is quite evident from this and some of the other cases given by Whitney that horizontal strata of characteristic transparency, not more than 30 cm. thick, extend across the hypolimnion

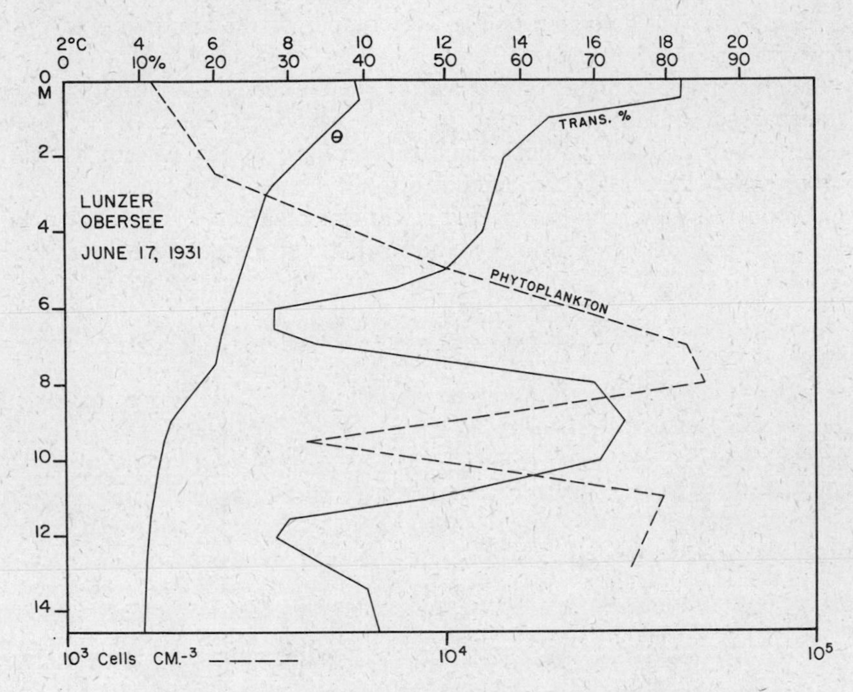

FIGURE 120. Relative horizontal transmission and number of plankton organisms at various depths, Lunzer Obersee (after Ruttner and Sauberer).

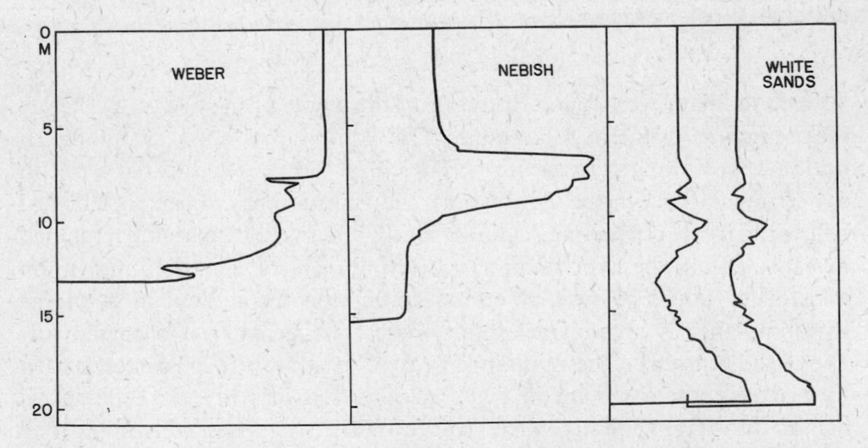

FIGURE 121. Relative horizontal transmission in Weber, Nebish, and White Sands lakes, in the last case at stations 15 m. apart (Whitney).

of many lakes. Ruttner and Sauberer and Whitney, however, find evidence that when remote points are compared, the continuity of the finer details is less impressive.

Whitney made a number of observations on the changes in the horizontal transparency during the autumn in Lake Mendota (Fig. 122). These observations naturally indicated a disintegration of the pattern as the progressively deeper layers entered into the epilimnetic circulation. Early in October, however, an unexpected result was obtained, for the very low values of transparency previously characteristic of the water just over the

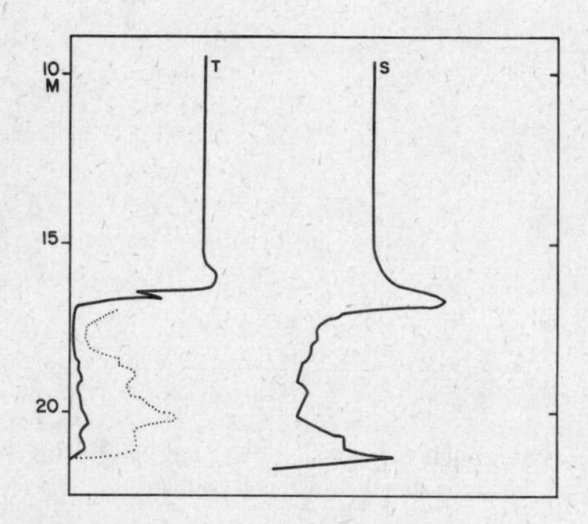

FIGURE 122. Horizontal transmission, T, and scattering, S, at various depths in Lake Mendota, October 3, 1938. Arbitrary scales, that of dotted part of T being magnified to indicate detail. (Whitney.)

bottom suddenly were found throughout the whole of the five or six meters that remained of the hypolimnion. On October 3, a study of the distribution of scattering material in the lake (S of Fig. 122) indicated a rise in scattering material in the layer where the transparency began to fall, just below the freely circulating region. But while the transparency remained low throughout the hypolimnion, scattering reached a marked maximum and fell to values below that of the epilimnion throughout most of the hypolimnion. A second maximum in scattering material occurred just above the bottom. The minimum in scattering at 19.3 m. corresponded to a secondary maximum in transparency, but this inverse behavior is confined to the hypolimnion, the water at 20.2 m. having only about one twenty-fifth of the transparency of the epilimnion and considerably greater scattering.

Whitney made some determinations of the organic seston and suspected that the minimum in transparency below the epilimnion was usually accompanied by an increase in particulate organic matter. In many cases this layer was observed to contain large *Daphnia* populations, which were absent in the succeeding layer of greater transparency. In one case, however, namely Nebish Lake, a large population of *D. pulex* appears to have been associated with a transparency maximum. Here it is reasonable to suppose that these *Daphnia* have kept the water transparent by their grazing activity. It is to be noted that a given amount of matter distributed as "dust" is more effective in optical extinction than the same quantity in "lumps." Some evidence of the dependence of the hypolimnetic opacity on the bacterial population was also obtained. These studies therefore seem in general to indicate that in the hypolimnia of stratified lakes, an optical microstratification usually develops as the result of a complex stratification in phytoplankton, bacteria, and perhaps dead seston and colloidal or dissolved coloring matter of various kinds. The existence of this complex stratification is of considerable interest in connection with the chemical and biological events in lakes.

Sauberer and Ruttner (1941) have followed the descent of suspended mineral matter through the Lunzer Untersee after turbid flood water had entered the lake during thermal stratification. The suspended material fell rapidly if irregularly through the epilimnion during the course of a day, but passed the thermocline slowly, influencing the hypolimnion to a comparatively small extent even after seven days. Their diagrams (Fig. 123) suggest a striking but not entirely obvious effect of the epilimnetic current system, which requires investigation. The same authors detected hypolimnetic changes in transparency due to precipitation of ferric hydroxide, a phenomenon which may underlie some of Whitney's observations in Lake Mendota.

Secchi disk transparency. Though the study of the transmission of light by means of suitable photosensitive instruments let down into the water has added enormously to our knowledge of limnological optics, the very simple procedure of determining transparency, in a restricted sense, with the Secchi disk still retains its value. However, it depends on more complicated theoretical considerations than do the instrumentally more elaborate studies of transmission, and is not free of subjective factors. The procedure is simply to observe the depth at which a white disk let down from the surface just disappears from view. The observations must be made through a shaded area of water surface. It is usual to determine the point of disappearance as the disk is lowered, allow it to drop a little farther, and then determine the point of reappearance as the disk is raised. The mean of the two readings is taken as the *Secchi disk transparency*.

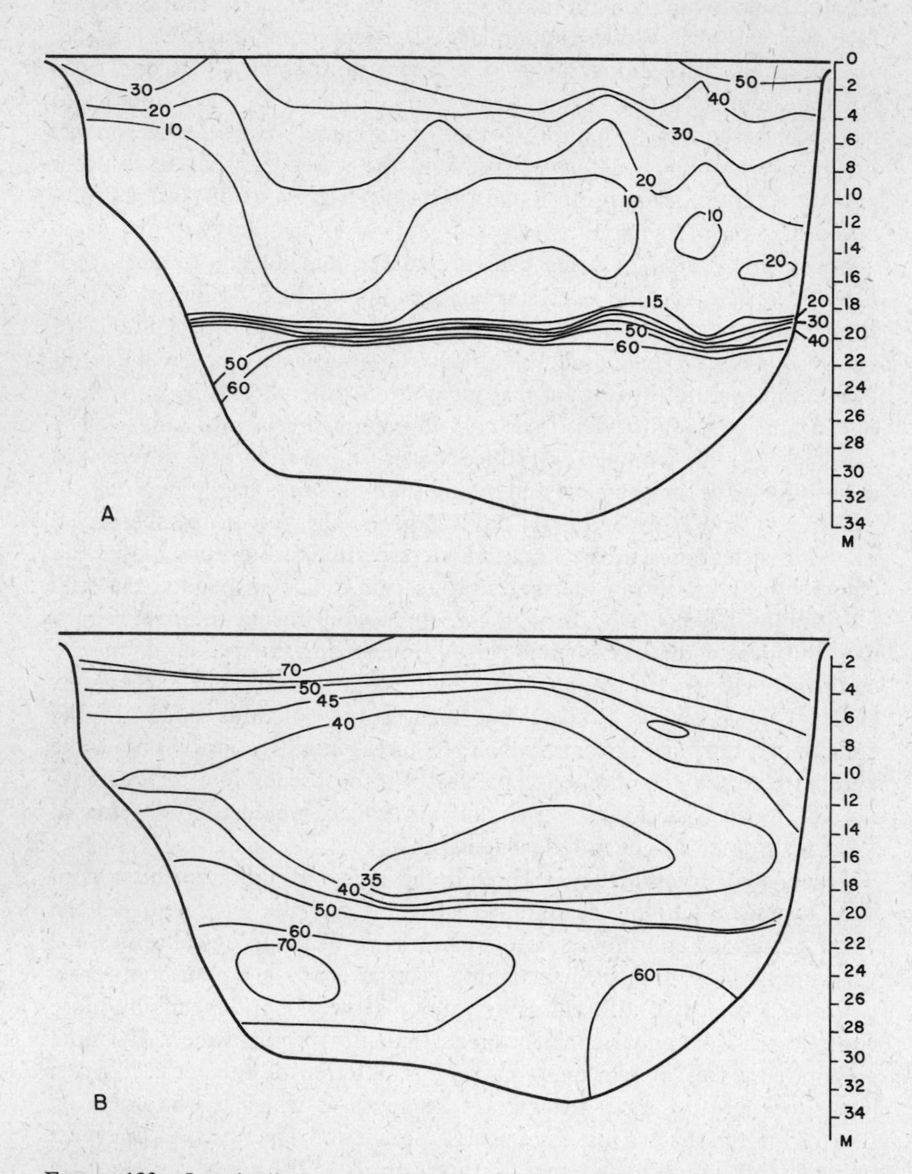

FIGURE 123. Longitudinal sections of the Lunzer Untersee, showing percentage of horizontal transmission after the inflow of turbid flood water. *A*, August 25, 1937, one day after the flood; *B*, August 31, 1937, seven days after the flood. The contours indicate percentage. (After Sauberer.)

The observations should not be made early in the morning or late in the afternoon, though both theory and observation (Fig. 124) show that the result is largely independent of the illumination.

Determinations of this sort were first[5] suggested by Horner (1821). The modern circular disk was introduced by Secchi in 1865, on a cruise of the pontifical steam corvette *Immacolata Concezione* in the Mediterranean

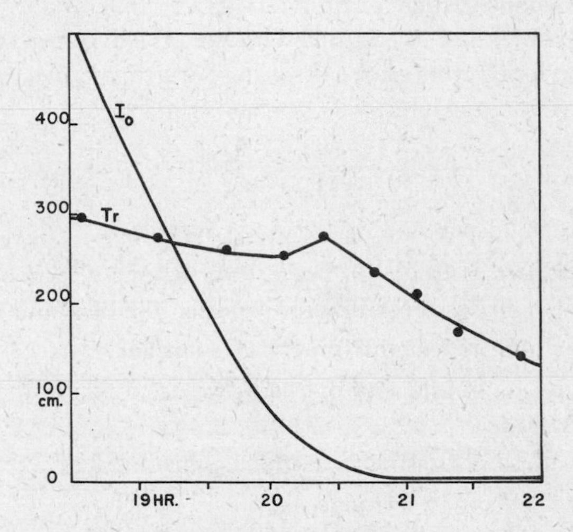

FIGURE 124. Secchi disk transparency, Tr, at various times during the late afternoon and evening, Lake Stråken, Sweden, July 27, 1936. The light intensity, I_0, at the surface is plotted on an arbitrary scale, but at 21.31 hours is 11×10^{-5} of the value at 18.45 hours. (After Åberg and Rodhe.)

(Cialdi 1866). The size of the disk is usually considered to make some difference to the value obtained. Since scattering of light from around the disk is of importance in determining its visibility, this is indeed to be expected. Von Aufsess (1903) found that in the Walchensee a disk 1 m. in diameter disappeared at 14 m., and one 30 cm. in diameter at 10 m. Juday and Birge (1933), however, found no difference in transparency in comparing disks 25 cm. and 75 cm. in diameter, and only a 5 per cent reduction when a disk 10 cm. in diameter was employed. Most workers seem to prefer the 20-cm. disk; Juday and Birge employed for routine work a 10-cm. disk. Disks painted with black and white quadrants have

[5] Casual observations were certainly made earlier. Thus Hearne (1705) writes of Lake Vetter in Sweden, "Ipse Magister *Ericus Simonius* album vel denarium 60 immersum cubitis aere sereniore se fatetur observasse," implying a transparency of the order of 30 m.

also been recommended, as have lamps that could be submerged until they became invisible. None of these elaborations have found favor with recent workers, although the black and white disk is said to give a sharper end point but generally at a smaller depth (Åberg and Rodhe 1942).

It is important to remember that in making an observation, the investigator is really comparing the brightness of the disk with the brightness of the water around it. Light reflected from the bottom in shallow water may introduce a considerable error.

The apparent difference in brightness between the disk and its surrounding water when viewed from above the water surface, relative to the brightness of the surrounding water, will be (Sauberer 1939)

$$\frac{I_0 d_1 d_2 r_d - I_u}{I_u + I_u' + I_R} \tag{25}$$

where I_0 = the light penetrating the water surface.

I_u = the light scattered upward from below the level of the disk.

I_u' = the light scattered upward between the disk and the surface.

I_R = the light reflected from the lake surface.

d_1 = the loss in intensity of the light passing from the surface to the disk.

d_2 = the loss in passing from the disk to the eye.

r_d = the reflectivity of the disk.

The human eye in general can just appreciate a difference in intensity if the quantity defined by (25) is not less[6] than 1/133.

If the transmission of the water were decreased solely by an increase in absorption, as by increasing soluble color, only $d_1 d_2$ would be decreased. If the same reduction in transmission occurred by increase in scattering, I_u and I_u' would both be increased. In the first case the ratio would be decreased much less than in the second. Accordingly, two lakes can exist, differing in scattering, in which the transmission is greater in one, the Secchi disk transparency in the other. Sauberer confirmed this expectation in the case of the more transmissive, turbid, less colored, and less transparent Leopoldsteiner See and the less transmissive, less turbid, more colored, and more transparent Lunzer Obersee.

In view of these conclusions, we should not expect to find more than a rough correspondence between the transmission and Secchi disk transparency of a series of lakes. Nevertheless, when a relatively homogeneous group of lakes is compared, there is a high correlation between the

[6] The value given is that used by Sauberer, after Helmholtz, but perhaps implies greater discrimination than is likely to be possible under ordinary conditions of observation.

Secchi disk transparency and transmission. Yoshimura (1938b) using the data of Birge and Juday, concluded that the Secchi disk disappears at about the level of penetration of 5 per cent solar radiation. Kikuchi (1937) had previously obtained values of 12 to 15 per cent; the discrepancy is presumably due to the greater spectral sensitivity of the Wisconsin workers' pyrlimnometer compared with that of Kikuchi's photocell.

The range of Secchi disk transparencies recorded is from a few centimeters in very turbid lakes to 41.6 m. in the Japanese caldera lake Masyuko. Crater Lake, Oregon, likewise has a transparency up to 40 m. (Hasler 1938). Values of over 30 m. are very rare, however, though Martin (1955) found a value of 35 m. in Chubb or Ungava Crater Lake and Le Conte (1883) a value of 33 m. in Lake Tahoe. Several Japanese caldera lakes fall in the 20 to 30 m. range. The open water of Tanganyika has a transparency of 22 m. (Capart 1952). The large subalpine lakes of Europe, for which Halbfass (1923) has summarized a great deal of data, ordinarily have transparencies between 10 and 20 m., Lago Maggiore and the Lake of Geneva occasionally reaching 21 m. in the winter, but all may give values somewhat under 10 m. in the summer, when melt water and plankton presumably increase scattering. The comparable lakes of North America fall in the same range. Among the small lakes of Wisconsin the modal transparency for drainage lakes is in the class 1.0 to 1.4 m.; for seepage lakes, 3.0 to 3.4 m. All the most transparent lakes are seepage lakes. This is no doubt largely due to the absence of silt-laden influents. The transparency is inversely correlated with both water color and seston content; it is not possible from the mean data presented by Juday and Birge (1933) to separate the effects of dissolved and particulate matter. Most European observers have found transparency to be higher in winter than in summer. Riley (1939) found in Linsley Pond and Tressler, Wagner, and Bere (1940) in Lake Chautauqua that variations in seston accounted for most of the seasonal variation in Secchi disk transparency. Within a single lake this is to be expected, both because such variations are likely to be greater than are variations in color and because, owing to the role played by scattering in determining the Secchi disk transparency, they are likely to be more effective.

SCATTERING OF LIGHT IN LAKE WATER

When a lake is examined on a bright day by looking down into it with a water telescope or in the shadow of a boat, it will generally appear, unless the water is very turbid, to be flooded with light. This light may be blue, green, yellow, or reddish brown in color. Since light is obviously reaching the eye from the lake, there must be a definite quantity of radiation directed upward from the water. Such illumination, sometimes referred to as the *light of the water*, is due to the scattering, both by the water and by material suspended in it, of the radiation which enters the lake from the sun and the sky. The major part of the scattering in most lakes is evidently due to materials other than water, and the capacity to scatter varies greatly from lake to lake.

Theory. The elementary theory of scattering has been set out by Whitney (1938c) in a form suitable for application to natural bodies of water, in the following way.

The light incident on a horizontal plane at depth z is, as in equation 20,

$$I_z = I_0 e^{-\eta_t' bz}$$

where η_t' is the mean extinction coefficient and bz the mean optical depth, or average path through which light must travel to reach depth z under the conditions of the observation. Differentiating,

$$dI_z = -\eta_t' bI_0 e^{-\eta_t' bz}\, dz \tag{26}$$

or, considering unit horizontal area at a depth z,

$$dI_z = -\eta_t' bI_0 e^{-\eta_t' bz}\, dV \tag{27}$$

where dV is an element of volume removing energy according to equation 20. Part of this energy is removed by absorption by the water, part by absorption by suspended matter or seston, and part by scattering. Let the energy scattered in all directions from a beam of unit energy traveling unit distance be S_s. The energy scattered in all direction by any element of volume dV will be

$$S_s bI_z\, dV = S_s bI_0 e^{-\eta_t' bz}\, dV \tag{28}$$

If one considers a recording instrument of unit area, outside the element of volume dV and subtending a solid angle Ξ at any point of the volume, the light scattered toward the instrument will be

$$\frac{\Xi}{4\pi} S_s bI_0 e^{-\eta_t' bz}\, dV \tag{29}$$

Let the recording instrument be a horizontal inverted photometer at depth 0, just below the surface film of the lake. This photometer will be receiving light from points at all depths and distances throughout the lake, so that the elements of volume to be considered can be of any form, which will, when integrated with respect to the three coordinate axes, constitute the entire lake. The most convenient form is found to be an annulus or flat ring, of internal radius r, external radius r + dr, and thickness dz. It will appear from Fig. 125 that

$$dr = \frac{z}{\cos^2 \psi_w}\, d\psi_w \tag{30}$$

The volume of the ring is given by

$$dV = \frac{2\pi z^2 \sin \psi_w}{\cos^3 \psi_w} d\psi_w\, dz \tag{31}$$

The solid angle subtended by the receiver at $z = 0$ at any point in the ring is given by

$$\Xi = \frac{\cos^3 \psi_w}{z^2} \tag{32}$$

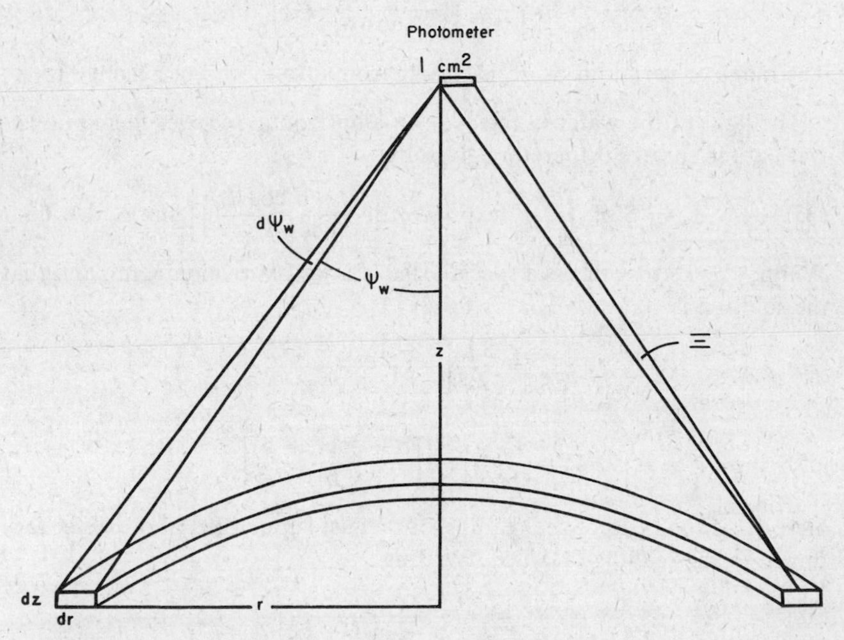

FIGURE 125. Geometry of Whitney's scattering theory. See text.

The total energy scattered in the ring is, from equation 28,

$$S_s b I_0 e^{-\eta_t' bz}\, dV$$

and the total energy scattered towards the receiver is, as before,

$$\frac{\Xi}{4\pi} S_s b I_0 e^{-\eta_t' bz}\, dV$$

From equations 31 and 32

$$\frac{\Xi}{4\pi} S_s b I_0 e^{-\eta_t' bz}\, dV = \frac{S_s b I_0}{4\pi} \cdot \frac{2\pi z^2 \sin \psi_w\, d\psi_w\, dz}{\cos^3 \psi_w} \cdot \frac{\cos^3 \psi_w}{z^2} e^{-\eta_t' bz} \tag{33}$$

$$= \frac{S_s b}{2} I_0\, e^{-\eta_t' bz} \sin \psi_w\, d\psi_w\, dz$$

However, this energy is being absorbed and scattered on its way from the ring to the receiver. In traveling to the receiver, it traverses a path of length $\frac{z}{\cos \psi_w}$; the element of energy dI_w received from any element of volume dV is therefore given by

$$(34) \qquad dI_w = \left(\frac{S_s b I_0}{2} e^{-\eta_t' bz} \sin \psi_w \, d\psi_w \, dz\right) \exp\left(\frac{-\eta_t' z}{\cos \psi_w}\right)$$

$$= \frac{S_s b I_0}{2} \exp\left[-\eta_t' z\left(\frac{1 + b \cos \psi_w}{\cos \psi_w}\right)\right] \sin \psi_w \, d\psi_w \, dz$$

The range of variation of ψ_w is clearly from 0 to $\frac{\pi}{2}$, and of z from 0 to ∞.

The light of the water I_w incident on a horizontal receiver facing downward at the surface is therefore given by

$$(35) \qquad I_w = S_s b I_0 \int_0^{\pi/2} \int_0^{\infty} \exp\left[-\eta_t' z\left(\frac{1 + b \cos \psi_w}{\cos \psi_w}\right)\right] \sin \psi_w \, d\psi_w \, dz$$

Whitney notes that this is a special case of a well-known integral, and that the solution is

$$(36) \qquad I_w = \frac{S_s I_0}{2\eta'}\left[\frac{b - \ln(1 + b)}{b}\right]$$

or

$$(37) \qquad R_p = \frac{I_w}{I_0} = \frac{S_s}{2\eta'}\left[\frac{b - \ln(1 + b)}{b}\right]$$

Here R_p is the ratio of the reading of the just submerged photometer face down, to the reading face up. Writing

$$(38) \qquad F_s(b) = \frac{2b}{b - \ln(1 + b)}$$

$$(39) \qquad S_s = F_s(b)\eta_t' R_p$$

it is found that $F_s(b)$ varies relatively slowly with b.

b (mean optical path), meters	1.0	1.1	1.2	1.3	1.4	1.5
$F_s(b)$	6.52	6.14	5.83	5.57	5.35	5.15

For most purposes, a mean value of 5.9 for $F_s(b)$ is sufficiently accurate.

$$(40) \qquad S_s = 5.9\eta_t' R_p$$

Since only light scattered upward can have any effect on the inverted receiver, and only half the scattered energy has an upward component in its path, the total extinction coefficient may be regarded as composed of

$$(41) \qquad \eta_t' = \eta_w' + \eta_{pc}' + \frac{S_s}{2}$$

where η_w' is the value of absorption by pure water, η_{pc}' by particulate and dissolved matter in the water, and $S_s/2$ the upward scattering.

The percentage scattering P_s may therefore be defined as

$$P_s = 100\frac{S_s}{\eta_t' + S_s/2 - \eta_w'}, \tag{42}$$

where, if spectral composition is not being considered, η_w' has a mean value of 0.03. The percentage absorption may be defined as

$$P_a = 100 - P_s \tag{43}$$

Observational data. Whitney determined the values of P_s and S_s for fifteen lakes in northern Wisconsin. He made measurements with instruments facing upward and downward, and in general found the light from below to be proportional to that from above or, for any lake, that R_p is more or less constant, in accord with equation 40.[7]

For the most part, his detailed discussion relates to the middle of the epilimnion, in which region the extinction coefficient is practically independent of depth. A few determinations at different depths were also made. It is found that there are very considerable differences in the value of S_s, which measures the amount of scattering per unit volume, and of P_s, which measures the qualitative nature of the material in the water with respect to scattering or absorption. The figures in Table 42 indicate this variation.

TABLE 42. *Scattering in Wisconsin Lakes*

	At Depth, m.	η'	S_s	P_s
Crystal	5	0.192	0.022	12.7
Muskellunge	5	0.300	0.070	22.2
Trout	5	0.375	0.045	12.2
	10	0.375	0.056	15.0
	16	0.375	0.035	10.9
Allequash	3	0.714	0.177	22.9
Ruth	3	0.810	0.025	3.2
Mud	3	0.944	0.031	3.3
Little John	3	0.994	0.258	23.6
Little Star	3	3.900	1.060	24.0

The values for P_s are a measure of the optical properties of the scattering particles, independent of their number; S_s varies with both number and kind. Although in a very rough way it is evident that where total extinction or absorption is very low, as in Crystal Lake, or very high, as in Little

[7] This does not hold if very turbid lakes are considered, such as glacial-fed lakes in high mountains (Löffler unpublished).

Star Lake, the scattering coefficients are also very low or very high; no close correlation over the intermediate values exists. This lack of correlation is clearly due to the great differences in the optical properties of the scattering material in different lakes. In Muskellunge and Allequash lakes, the values of P_s, which depend on the optical properties of the scattering material but not on its concentration, are practically identical. The values of S_s indicate about 2.5 times as much scattering material in Allequash as in Muskellunge. Little Star, with an extremely high value for both the extinction and scattering coefficients, also has a high P_s, comparable to these two lakes, and must contain about 14 times as much scattering material as does Muskellunge.

It is unfortunate that no attempt has apparently been made to interpret the very considerable variations in the values of P_s. Such an interpretation might yield quite significant information about the colloidal and suspended matters in lakes. It may be noted that Utterback, Phifer, and Robinson (1942) concluded that the rather large amount of minute fragments of volcanic glass suspended in the waters of Crater Lake, Oregon, is of little optical significance. In the ocean there is evidence of forward scattering by refraction by large transparent particles (Jerlov 1951).

Åberg and Rodhe (1942) published a series of determinations in Stråken in July 1936, and a single determination in Lammen made about the same time. These determinations fall within the Wisconsin range; for Stråken, S_s varies from 0.050 to 0.071, P_s from 5.2 to 6.3; for Lammen, at a depth of 1 m., $S_s = 0.117$ and $P_s = 6.8$. As in the case of Whitney's work, the Stråken observations also indicate very clearly that R_p is independent of depth, as is required by equation 40.

Spectral composition of scattered light. The spectral composition of the light scattered by Crater Lake water in the laboratory has been studied by Pettit (1936). Sauberer (1939, 1945) and Sauberer and Ruttner (1941) have examined the composition of the light scattered from the water *in situ* on a

TABLE 43. *Percentage light scattered at various wave lengths*

	3770 A.	4350 A.	5250 A.	5900 A.	6300 A.	6500 A.	7000 A.	7300 A.
Achensee	0.5	1.5	2.1	1.0	0.4	0.3	0.3	0.1
Lunzer Obersee	...	0.3	0.9	1.0	0.7	0.5	0.1	...
Lunzer Untersee	0.4	0.9	1.4	1.0	0.6	0.5	0.2	...
Leopoldsteiner See	1.2	3.8	7.2	6.5	4.8	3.0	2.0	1.1
Krottensee	...	0.6	0.9	1.0	0.6	0.3	0.2	...
Mondsee	...	1.2	2.0	1.3	0.8	0.5	0.4	...
Irrsee	0.4	1.0	2.0	1.4	0.7	0.7	0.3	...

number of Austrian lakes (Table 43). Pettit's laboratory results, are expressed relative to the ratio of scattered to direct light at 6000 A., which ratio is taken as unity. Sauberer's results are expressed as the percentage of incident light scattered upward at each wave length studied. In order to permit easy comparison, three of Sauberer's cases have been recomputed to approximately the same form as Pettit's (Table 44). Of

TABLE 44. *Relative spectral composition of scattered light* (Ratio $i_0:I_0$ at 6000 A. = 1)

A.	Pure H_2O	Crater Lake Water	Achensee	Krottensee	Leopoldsteiner See
6000	1.00	1.00	1.00	1.00	1.00
5500	1.35	1.20	2.30	0.99	1.18
5000	1.85	1.85	2.54	0.88	1.10
4500	2.80	2.80	2.11	0.72	0.80
4000	4.50	4.10	1.18	< 0.63	0.36
3500	6.00	6.10	< 0.63	...	0.23

these three cases, the Achensee exhibits the greatest relative quantity of shorter wave radiation, the Krottensee of longer wave radiation, the Leopoldsteiner See exhibits the greatest total scattering, and the Krottensee probably the least.

It will be observed from the foregoing determinations that while Crater Lake water scatters light almost in the same manner as does optically pure distilled water, the light scattered upward from the other lakes has its energy maximum in the green (Achensee), green-yellow (Leopoldsteiner See, Lunzer Untersee, Mondsee, Irrsee), or yellow-orange (Lunzer Obersee and Krottensee).

The nature of the spectral distribution depends not only on the variation of the scattering coefficient for different wave lengths, but also on the variation in absorption with wave length. In the almost optically pure water of Crater Lake, in which the transmission is greatest in the blue, and at the same time the scattering is mainly of short-wave light, the resulting light from a considerable mean depth will be conspicuously blue. In lakes with larger and more numerous scattering bodies and with a maximum transmission in the green or yellow, not only will longer wave lengths be scattered more easily than in pure water, but there will be proportionately less short-wave radiation to scatter and more will be absorbed on the upward path when it is scattered.

Effects of reflection and of angle of incidence. Not quite all the scattered light that reaches the surface emerges from the lake, part will be reflected back into the water. Sauberer (Sauberer and Ruttner 1941), by inverting his photometer above the lake with suitable shielding to exclude reflection of incoming radiation, concluded that in the Lunzer Untersee 82 per cent of the blue component, 90 per cent of the green, and 95 per cent of the red component of the scattered light or light of the water emerged from the lake. This is in accord with the greater refractivity of blue light, which will lead to total reflection of blue rays reaching the surface at angles which would permit transmission of red rays to the air.

Davis (1941) finds that the proportion of light scattered upward varies almost linearly with the angle of incidence, rising from 2 per cent when $\psi_i = 36$ degrees to 2.9 when $\psi_i = 75$ degrees. At higher angles of incidence, the light scattered upward declined greatly. The effect was less marked for red than for the entire spectrum. Since both the reflected and scattered components increase as percentages of the incident illumination with increasing ψ_i while the total illumination decreases, there is a long period, apparently from 7 A.M. to 5 P.M. in summer, during which the upward visible radiation changes little in absolute amount. During windy weather this period is longer and more constant than during calm (Fig. 126).

POLARIZATION

Diffuse reflected or scattered light is in general partially polarized, and the light of the sky exhibits a complicated pattern of polarization.

If an observer is well below the water surface and looks upward, the whole of the sky will be visible as a circle, the circumference of which makes an angle, the critical angle of about 48 degrees, with the vertical at the point of observation. In considering the problem of underwater polarization, one may consider how far the pattern of the polarization of sky light is present within the critical angle, and how far scattering by the water produces polarization outside the critical angle. The available data are all due to Waterman (1954, 1955), who has worked both visually and instrumentally in the ocean. The original papers may be consulted for details of the polarimeters employed.

Within the circle defining the critical angle (Fig. 127), Waterman found clear evidence of the sky pattern, at least in clear water to a depth of 6 m. Beyond the critical angle, polarization in the scattered light was easily observed, and persisted to depths in turbid water beyond those at which the sky pattern could be seen. The most recent measurements have indicated considerable polarization at depths of 200 m.

With an almost vertical sun, polarization of the scattered light of the water was maximal in the horizontal plane and declined toward the antisun near the nadir. The plane of polarization, as defined by the electric vector, was horizontal at all azimuths. With the sun near the horizon, the maximal polarization was found along an arc running through all points outside the critical angle that lay at right angles to the bearing of the

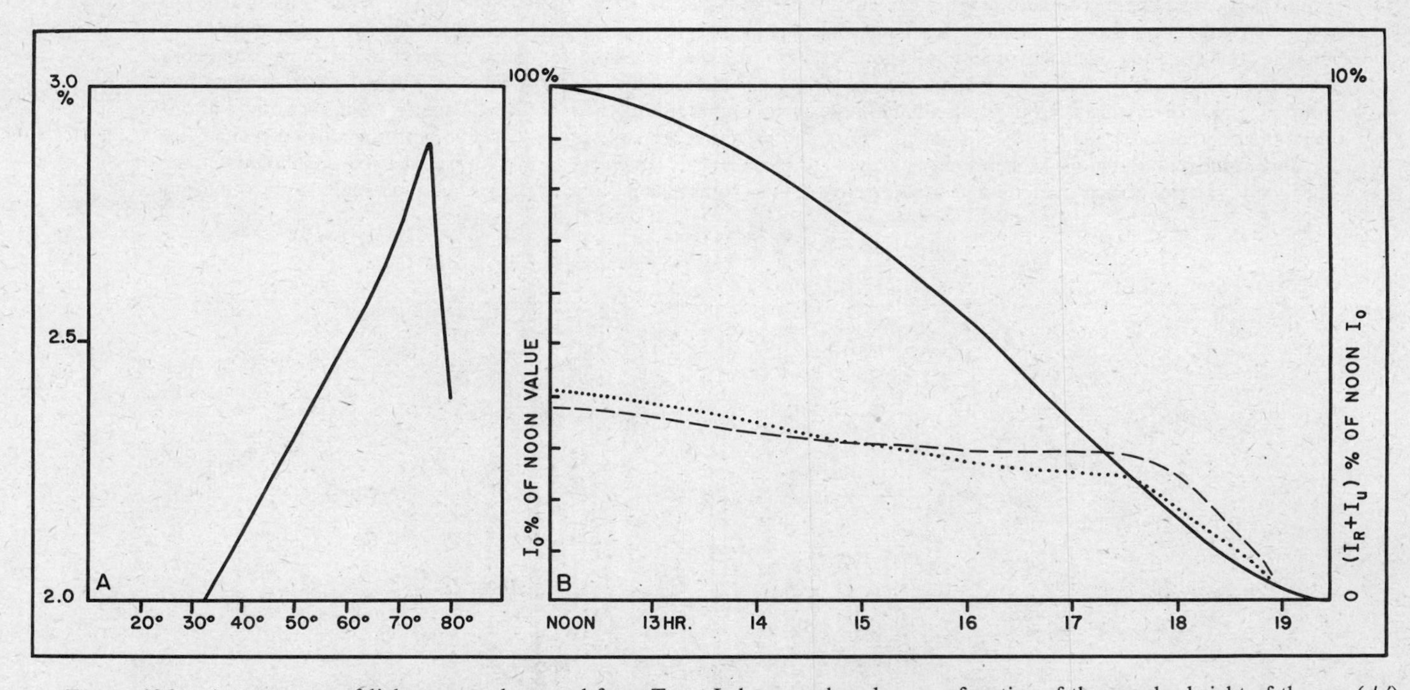

FIGURE 126. *A*, percentage of light scattered upward from Trout Lake on a clear day, as a function of the angular height of the sun (ψ_s'). *B*, relative intensity as percentage of incident light at noon: solid line, incident light falling on the lake throughout afternoon; broken line, light lost from a calm surface; dotted line, light lost from a surface with 20 cm. waves. (After Davis.)

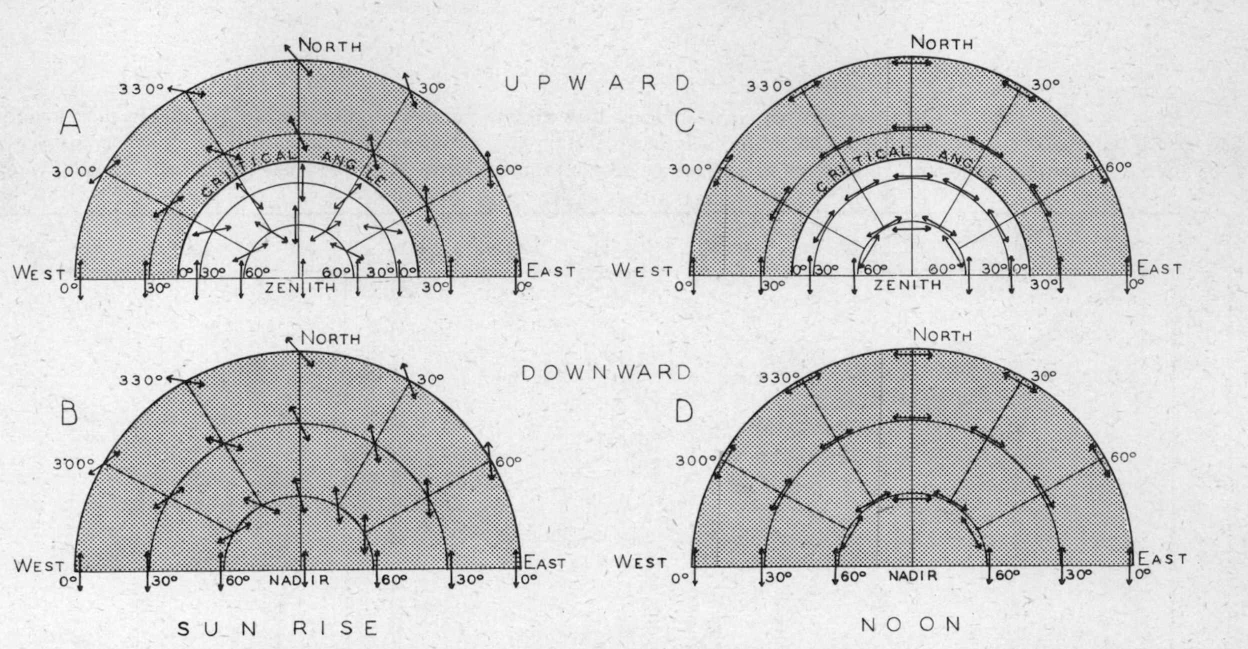

FIGURE 127. Underwater polarization patterns diagrammed as they would appear at a depth of several meters. Each semicircle represents in polar plot a quarter of the whole sphere of vision. *A* and *C* are quarters looking up from the horizon (zero elevation shown in the outermost circles) to the zenith; *B* and *D* are quarters looking down from the horizon (again zero elevation is shown by the outermost circles) to the nadir. Bearings are represented in the diagrams by radii; elevations and depressions of the line of sight from the horizon are given by distances along these radii. The shaded portions of each figure indicate directions in which the polarization arises primarily in the water by scattering. The clear parts of *A* and *C* are areas within the critical angle where the polarization of the sky light itself, distorted by refraction, is seen as part of the over-all pattern. The plane of polarization for representative points is indicated for two characteristic times of day: sunrise (*A* and *B*) and noon (*C* and *D*) on a clear, cloudless day. At the former time, the sun is assumed to be on the horizon and due east; for the latter, the sun is assumed to be in the zenith. The vectors are drawn for each point plotted at an angle made by the *e*-vector of the polarization with a plane determined by the observer, the vertical, and the point observed. (Waterman.)

sun. The plane of polarization was horizontal in the sun's azimuth, and tilted toward the sun at about 45 degrees when observed 90 degrees from the sun. In such a case, the magnetic vector of the polarization coincides with the direction of the sun's rays in the water. No clear evidence of vertical polarization due to reflection and refraction at the air-water interface was observed.

Waterman's studies are still developing; in view of the fact that the compound eyes of arthropods are efficient polarization analyzers, polarization is likely to have great biological as well as physical interest.

COLOR OF LAKE WATERS AND OF LAKES

Color of water. The color of lake waters viewed by transmitted light may be matched against a series of dilutions of an arbitrary standard of appropriate tint. In the United States the most usual standard is prepared by dissolving 2.492 g. K_2PtCl_6 and 2 g. $CoCl_2 \cdot 6H_2O$ in 200 ml. concentrated HCL, and making up to 1 liter (Hazen 1892). Such a standard has a color of 1000 units, commonly spoken of as platinum units. Color is in practice usually determined by comparison of a column of lake water with distilled water viewed through a set of filters. Ohle (1934a) introduced a standard based on alkaline methyl orange, which owing to its cheapness and availability has found much favor in Europe. According to Åberg and Rodhe, one Ohle unit, corresponding to 0.01 mg. methyl orange per liter, is equal to 2.8 Pt units.

Occasionally, with very dark waters it may appear convenient to dilute the lake water rather than the standard. James (1941) found that in diluting lake waters, Beer's law held until solutions containing 95 per cent distilled water were reached. Irregularities that then appeared were attributed to changes in *p*H or ionic concentration altering the state of colloidal coloring matter. It is sometimes useful to distinguish between *true color* after filtration and *apparent color* before filtration, but this has seldom been done in limnological work.

In general, very clear waters give a color of zero; the darkest bog waters, colors of up to 340 Pt units.

Both Spring (1897) and Whipple (1899) obtained evidence that the brown material coloring lakes can be decomposed photochemically. In Whipple's experiments, up to 65 per cent of the color would be removed by exposure of a bottle of water set about 15 cm. below the surface of a lake for a month. The effect (Fig. 128) was just detectable at 3 m., but is clearly due to easily absorbed wave lengths, in part, no doubt, ultraviolet. Åberg and Rodhe (1942) found that sterilized and unsterilized water in the light lost color at about the same rate, but the unsterilized water in the dark showed a slight increase rather than a decrease in color.

There is frequently a marked increase in color with depth in a stratified lake, and sometimes a maximum in color in the metalimnion or upper hypolimnion (Fig. 129). The increase in color near the bottom may in part be due to organic materials derived from the sediments; in some cases, however, it is likely that hypolimnetic color may be due to oxidation of ferrous iron to ferric hydroxide as the former substance diffuses upward from the mud.

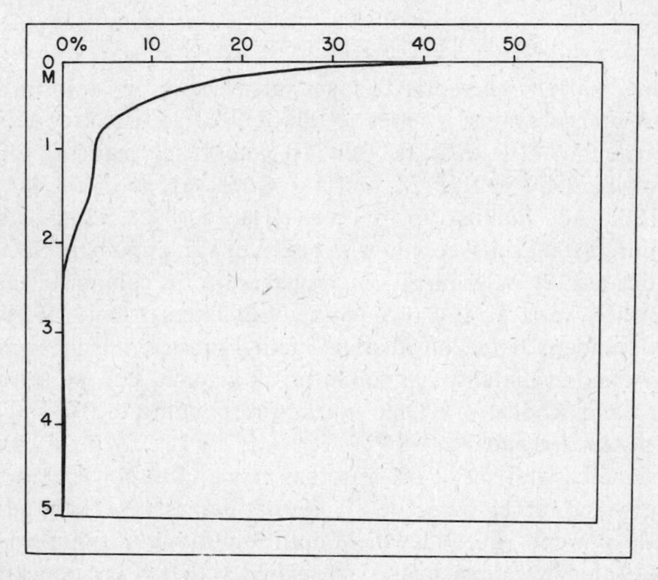

FIGURE 128. Percentage reduction of color in water initially of color 170 Pt units, exposed for a month at different depths in a reservoir of color 44 Pt units (from data of Whipple).

As has been indicated (page 385), there is probably more than one class of organic compounds producing color in lake waters. Brown materials extracted either from the lake littoral or from allochthonous soil and peat are certainly very important. The organic matter produced by plankton is apparently less colored per gram of carbon than are soil and peat extractives, and may well be heterogeneous. The fluorescent compounds found by Shapiro (page 889) in Connecticut lakes are yellow and are evidently involved in determining water color.[8]

[8] The most recent results suggest that the substances are not the same as the *Gelbstoff* recorded in the sea.

The colors of lakes. The *color of a lake* refers to the color of the light of the water as it emerges from the lake surface. This varies from the clear blue of pure water, through greenish blue, bluish green, to pure green, and then through yellowish green and greenish yellow and yellow to yellow brown or occasionally a clear brown.

Some empirical scales for recording such colors have been suggested. The only one that has come into wide use is due originally to Forel (1889), modified and extended by Ule (1892). It consists of a series of

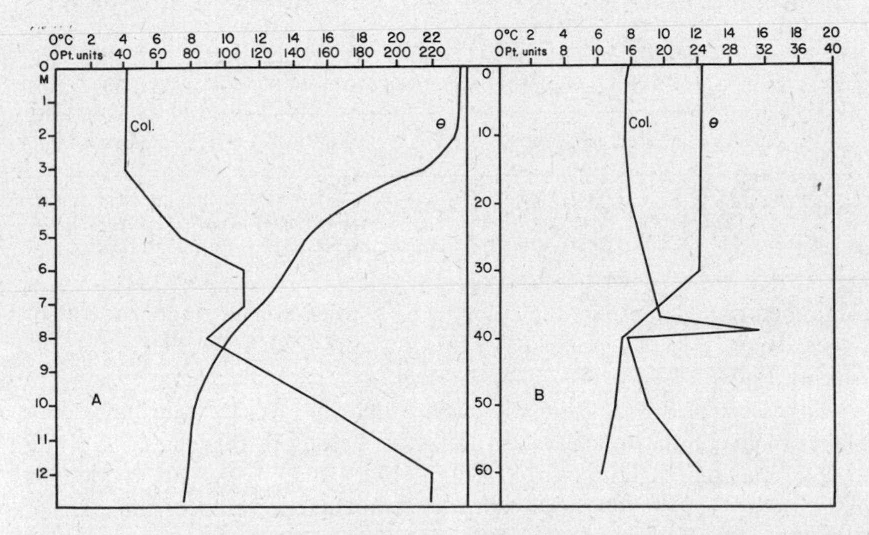

FIGURE 129. Vertical distribution of color. *A*, Skärshultsjön, Sweden, late in summer (August 15, 1938), stratification showing strong increase in color in hypolimnion with a small secondary maximum in metalimnion (after Åberg and Rodhe). *B*, the same in the Grosser Plöner See at the beginning of autumnal circulation (October 15, 1931), with a very strong localized metalimnetic color maximum (after Ohle and Åberg and Rodhe).

tubes containing solutions varying from blue through green to brown, prepared by mixing cuprammonium sulfate, potassium chromate, and cobalt ammonium sulfate, in varying proportions, as given in Table 45. The color is supposedly best observed by looking into the water through a water telescope, but it may also be compared with the color of the Secchi disk before it disappears, the standards being on a white background. In spite of considerable subjectivity in observation, the results appear to be more useful than Juday and Birge (1933) believe.

Kalle (1938), in his studies of the color of the sea, has introduced a

TABLE 45. *The Forel-Ule color scale*

Solution*	I	II	III	IV	V	VI	VII	VIII	IX	X
1	100	98	95	91	86	80	73	65	56	46
2	0	2	5	9	14	20	27	35	44	54
3	0	0	0	0	0	0	0	0	0	0
Color	Blue		Greenish blue		Bluish green			Green		

Solution*	XI	XII	XIII	XIV	XV	XVI	XVII	XVIII	XIX	XX	XXI	XXII
1	35	35	35	35	35	35	35	35	35	35	35	35
2	65	60	55	50	45	40	35	30	25	20	15	10
3	0	5	10	15	20	25	30	35	40	45	50	55
Color	Greenish yellow					Yellow				Brown		

* Solution 1 = 0.5 g. $CuSO_4 \cdot 5H_2O$ + 5 ml. strong NH_4OH + to 100 ml. H_2O
2 = 0.5 g. $K_2CrO_4 \cdot 5H_2O$ + 5 ml. strong NH_4OH + to 100 ml. H_2O
3 = 0.5 g. $CoSO_4 \cdot 7H_2O$ + 5 ml. strong NH_4OH + to 100 ml. H_2O

tintometer which permits the color to be expressed as a single equivalent wave length. In blue ocean water this is about 4750 A. (Kalle 1938, 1939; Jerlov 1951).

Lakes of color I are evidently extraordinarily rare, though the caldera lakes of maximum transparency doubtless belong in this class. Garbini (1897) claimed that Lago di Garda is bluer than I, but his account is confused. Tanganyika has a color of II in the open part of the lake, of III–IV in rocky bays. The rather inadequate data (Delebecque 1898, Halbfass 1923, Pesta 1929, Hacker 1933) suggests that most small alpine lakes in the mountains of Europe are blue green (II–VII), while the large subalpine lakes vary over the range from III to IV (Annecy, Geneva, Orta) to VII (Constance, Zurich). The highly productive lakes of cultivated plains generally are distinctly yellowish, the Grosser Plöner See having a color of XIV, Linsley Pond usually XVII–XVIII (occasionally XV or XX). The very brown extreme colors XXI-XXII are likely only in lakes containing peat extractives.

It is reasonably certain that the blue color of pure water is due to molecular scattering, which, being dependent on the fourth power of the frequency, is very much greater for short than for long wave lengths. Scattering from larger particles is less selective; it is possible that in some cases very finely divided suspended matter may produce a green color, such as that observed in the very hard waters, which well may contain colloidal $CaCO_3$, of limestone quarries and chalk pits.

In general, however, it is reasonable to suppose, following ideas originally developed by Wittstein (1860) and Forel (1895), that the main determinant of the color is dissolved organic matter. Spring (1897) showed that a very small amount of humic material could give a green color to pure water in a long tube, which without the addition would have appeared blue by transmitted light. The color ordinarily observed is, of course, that of the light scattered upward, but this has passed through the water and so has lost part of those wave lengths most easily absorbed.

The fact that many lakes look clear green is thus dependent on their containing matter of biological origin, but not on their chlorophyll content. The pure organic matter producing color would in fact look brown or yellow. When a lake is highly productive, it may be in fact disappointingly yellow.

Seston color. These considerations do not apply to colors which are due to the reflection spectrum from suspended particles of microscopic or submacroscopic size, rather than to the effects of particles of ultramicroscopic size. *Seston color* of this sort is often observed in highly productive lakes, and may also give a characteristic color to lakes containing large quantities of suspended inorganic matter. Naumann (1922) has summarized a good deal of information on seston color, and Jermakoff (1926) has given a classification of dispersed color in waters, which seems somewhat pedantic.

The shallow lakes of semi-arid regions frequently have a gray color, owing to sediments that are grayish yellow and do not match any of the Forel-Ule colors. More rarely suspended inorganic sediment can give a red color, as in the case of the red claylike material of Triassic origin suspended in the Rotsee, 10 km. south of Witzenhausen, Kassel, Germany (Von Bülow and Otto 1931). Crater lakes containing water of volcanic origin may be red, owing to suspended ferric hydroxide, or yellowish green, owing to elementary sulfur particles.

In most cases, seston color is due to large concentrations of phytoplanktonic organisms. The densest accumulations in ordinary productive lakes are due to blue-green algae, but the over-all appearance is generally a dull green with little suggestion of blue. Yellowish-brown colors are frequently produced by very large populations of diatoms. Very striking cases of red or purple coloration due to the mixing of hypolimnetic water containing *Oscillatoria rubescens* occur during circulation periods in some Swiss lakes, where the phenomenon is termed the appearance of *Burgunderblut*. Both *Euglena sanguinea* and *Haematococcus pluvialis* may produce "blood lakes" in the mountains of central Europe (Klausener 1908, Huber-Pestalozzi 1936). Purple sulfur bacteria may color some salt lakes in which much H_2S is produced in the bottom sediments by reduction of sulfate, as in the extraordinary case of Son-sakesar-kahar (page 773). Naumann (1922) indicates that occasionally reddish zooplankton, such as *Daphnia*, or diaptomid copepods, in cold water, may be abundant enough to give a red tint; in small bodies of water, milky aggregates of *Spirostomum* and other ciliates and black aggregates of *Stentor* have been recorded.

In the cases of *Euglena sanguinea*, *Chromulina*, and some blue-green algal blooms, the organisms may be associated with the surface film and the resultant coloration should then be termed *neuston color* rather than seston color.

PHENOMENA OF REFLECTION AND REFRACTION AT AND ABOVE THE LAKE SURFACE

A few phenomena due to the optical properties of the lake surface or of the air immediately above it are sufficiently striking and interesting to merit a brief account in the present chapter.

The sun path. When the sun or moon is shining on a disturbed water surface, the reflection is drawn out into a band of light, running from near the observer toward a point on the horizon below the reflected celestial body. This appearance is termed the sun path[9] or moon path.

In the plane of the sun and the eye, normal to the water surface, light will be reflected along the water by any element having an inclination to the mean surface of $^1/_2\psi_s'$. On either side of this plane, increasingly steep elements will reflect the sun along the water at increasing angular distances from the line joining sun and eye, so forming the path. The angular width of the path is thus qualitatively easily seen to depend on the height of the sun and the maximum slope of the available reflecting surfaces. The complete quantitative treatment results in a very lengthy expression for the dependence of angular breadth on the two variables; the curious reader may consult Griesseier (1953) for the spherical trigonometry involved, or an earlier paper of Hulburt (1934), who gives the results in the form of a series of graphs. Observing actual sun paths, Hulburt concludes that over the sea a wind of 450 cm. sec.$^{-1}$ can produce reflecting surfaces inclined up to 20 degrees to the horizontal; a wind of 820 cm. sec.$^{-1}$, surfaces inclined up to 38 degrees to the horizontal. The distribution of intensity in the path is clearly related to the distribution of slopes on the disturbed water surface. Hulburt points out that theoretically the form of wave-covered water could be studied in this way, but that it would be a

[9] The German terms *Die goldene Brücke* and *Der sonniger Weg ins Glück* are more poetic; they find their English equivalent, as Dr. E. S. Deevey has pointed out to the writer, in the "path of gold" of Robert Browning's "Parting at Morning."

hard way of going about the problem. Recently, the distribution of *glitter*, or flashes of light reflected from transitorily inclined surfaces, as observed from the air, has provided valuable information about the form of the disturbed surface of the sea (Cox and Munk 1954).

Iris or the horizontal rainbow. The horizontal rainbow, or iris, is a relatively rare phenomenon, though it has been observed on a number of lakes in many parts of the world. It is produced by the reflection and refraction of light in minute droplets of water derived from mist adhering to the surface of the lake, especially when the surface is covered with slicks or *taches d'huile*, or by macroscopically visible planktonic scum. Ponds contaminated with oil or coal dust can also easily exhibit an iris under appropriate conditions, and it can also sometimes be observed over ice surfaces or over wide expanses of dew-covered grass; Minnaert (1954) gives a good elementary summary. It is possible that sometimes a thin layer of mist just above the water surface can also give rise to the phenomenon (Fujiwhara 1914), though it seems certain that in many cases this is absent.

The theory of the horizontal rainbow (Humphreys 1929; cf. also Buffle, Jung, and Rossier 1938) is in general the same as that of the more generally known celestial rainbow. The primary bow appears where the surface of a cone, whose axis runs from the sun through the eye of the observer and whose angular spread (axis to element of surface) is 42 degrees, cuts the water surface. The secondary bow is similarly formed where the cone of angular spread 51 degrees cuts the water surface. If the sun is below 42 degrees in the sky, as is usually the case when other conditions for an iris are fulfilled, the bows appear as parabolas or fragments of parabolas; if the sun is high in the sky, as ellipses or parts of ellipses. Supernumerary bows due to diffraction may be observed. Juday (1916) records on Mendota a primary bow with two supernumeraries and a secondary bow. There is also the possibility of seeing the bows reflected in the water surface, the reflected primary lying on the cone of angular spread $2\psi_s' - 42$ degrees. Light entering from the water may contribute to the reflected bows, or after reflection, to the direct bows. A fine reflected bow with primary and secondary direct bows was apparently observed on Lake Monroe, Florida, by E. D. Ball on October 23, 1914 (Humphreys 1929).

The conditions for the observation of the horizontal rainbow are obviously, as Juday points out, the coincidence of scum or an oily surface, fog followed by bright sun, and dead calm. These conditions are most frequently encountered in the autumn, but the phenomenon has been observed on Lake Mendota in May; they are also most likely to be present in the early morning, but one case on the same lake has been noted at 2 P.M. The horizontal rainbow has been recorded on Lake Mendota in about half the years during which close observation has been kept. It has been noted on the Lakes of Geneva, Constance, and various other Swiss and German lakes, on Loch Lomond and on various smaller bodies of water in Britain, on several inland waters in Japan, and on the American lakes mentioned. Forel (1895), Juday (1916), and Halbfass (1923) have summarized the published records of the phenomenon.

Mirages and the fata morgana or castles in the air. Various optical phenomena may be produced by thermal gradients set up in the air over the surface of a lake. If the air is initially colder than the lake, its lowest layer will be warmed and an inverse thermal gradient established. This is frequently the case in the early morning on clear days in spring or early summer, or during the whole day in autumn; in the case of monomictic lakes, in winter also. A ray of light passing through the stratified layer in which the optical density has decreased downward, will be refracted along a path convex from above (*A* of Fig. 130). As a result of such refraction over warm water, the plane of the apparent horizon is depressed, the water surface appears convex, and distant objects may appear above an inverted image or *inferior mirage*, over the apparent horizon, as in desert mirages. As the thermal gradient decreases, the symmetry of such an image is lost, its apparent height decreasing.

If the air is warmer than the water, as may be the case on summer afternoons, various refractive phenomena are observed owing to the formation of a direct thermal gradient in the air above the water. In such a gradient the path of a refracted ray is convex from above (*B* of Fig. 130). Objects even beyond the true horizon can appear lifted up, but reduced in height, the reduction most affecting the lower part of the object. The water surface appears concave, and the apparent horizon lies above the true horizon. It is possible, if a sufficiently strong thermal gradient exists well over the water surface, as sometimes happens at sea, for an inverted image or *superior mirage* to appear above a distant object as the result of the opposite process to that producing the typical mirage over a warm surface.

During the process of transition between inverse and direct thermal gradients, it is possible for the rays from a single point to be refracted along both concave and convex paths, so that they appear to reach the eye from a vertical line (*C* of Fig. 130). The result is that the waves at the horizon, and distant objects in general, are distorted into rectangular blocks of varying color and intensity, which may look like buildings bearing turrets, the so called *castles in the air*. This appearance is characteristically observed in the vicinity of the Straits of Messina, and is there spoken of as the *fata morgana*, a name derived from the Arthurian Morgan le Fay, whose somewhat improbable entry into Sicilian folklore presumably

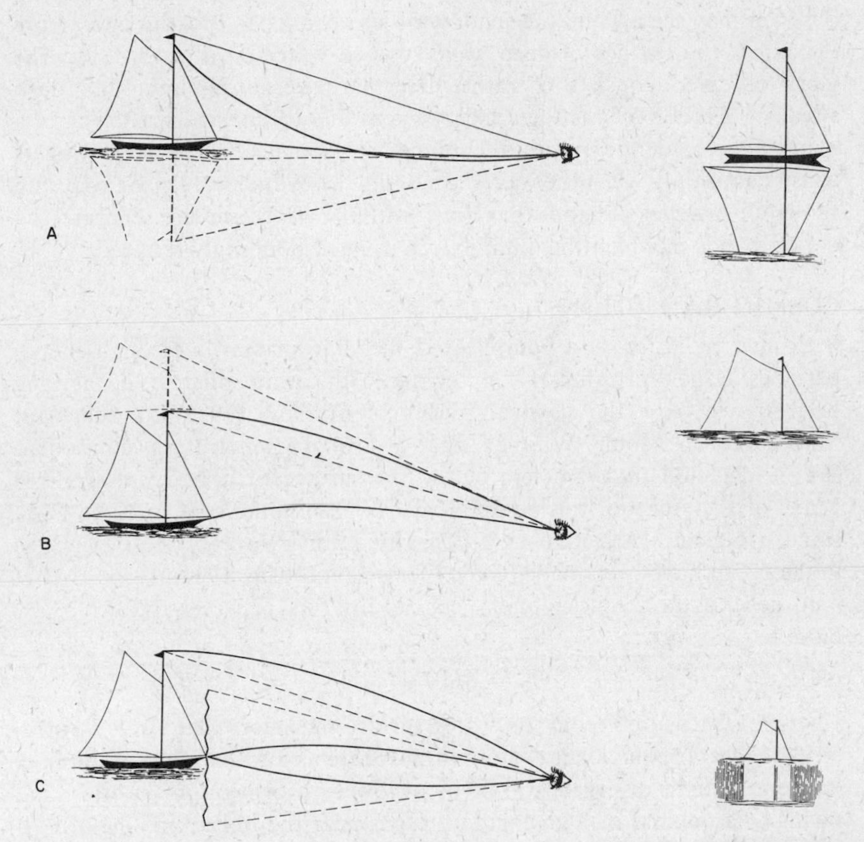

FIGURE 130. *A*, formation of an inverted mirage in an inverse thermal gradient over warm water. *B*, refraction in a direct thermal gradient. *C*, fata morgana formed during transition between inverse and direct gradients.

dates from the Norman kings of the Island. The fata morgana has been extensively studied on the Lake of Geneva by Forel (1895, 1897, 1912), where it normally appears in the afternoon of warm spring days. A less compact and more nebulous form of the phenomenon was termed by Forel the *fata brumosa*. These appearances can be expected on any large lake over which sufficient thermal variation in the air is possible. For a more detailed discussion the reader is referred to R. W. Wood's (1911) elegant account of the experimental production of the inferior mirage, to Perntner and Exner's (1910) standard treatise, which provides an admirable guide to the older literature, and to Minnaert's (1954) delightful book on the optics of nature.

Glories. Forel (1874) has described a type of radiance which an observer may see around the shadow of his head when looking away from the sun at the surface of deep wave-covered water on a clear day. The glory observed consists of rather irregular alternating light and dark streaks. Forel explains the appearance as due to upward scattering in a plane almost identical with the plane of incidence. The scattering is conspicuous only when the waves are acting as cylindrical lenses, focusing sunlight in a series of bands that travel with the waves and are separated by darker bands in which the illumination is correspondingly reduced.

COSMIC RADIATION

Cosmic radiation is a complicated flux due primarily to high-energy particles, largely protons, that enter the earth's atmosphere. It has long been known from the classical studies of Millikan (1926) conducted at Muir Lake on Mount Whitney, that a strongly penetrating component can be detected many meters below the surface of a body of water.[10] Much of the study of this component has been conducted in Lake Constance (Regener 1932, Ehmert 1937), though more recently observations in the ocean have been made (Clay, Gemert, and Clay 1939).

In general the intensity of the cosmic ray flux in water follows a law close to

$$I_z = I_0(10 + z)^{-1.9} \quad (44)$$

where z is measured from the water surface in meters and $10 + z$ represents the water equivalent of both atmosphere and water layer of depth z. There is a slight divergence in favor of lower absorption in the top 3 m., owing to an excess of high-energy mesons near the bottom of the atmosphere. The primary component of the flux in water is believed to be due to mesons, but it has been suggested that the residual flux below 400 m. (or its equivalent of rock in a mine) may be due to some unrecognized component (Wilson 1938).

SUMMARY

The total radiation falling on a lake surface during any day depends on geographical position, elevation, season, and the state of the atmosphere. At the equator, the direct radiation received from the sun may be expected

[10] During the bathymetric survey of the Scottish lochs, Watson (1904b) submerged an electroscope in Loch Ness. He concluded that the rate of discharge implied that the ionization of the air in the electroscope chamber was not reduced to less than 75 per cent of the surface value when the instrument was covered by 120 ft. of water. He does not seem to have regarded the observed reduction as particularly significant, though it is now apparent that it must have been due to absorption of the cosmic ray flux by the water.

to vary at sea level from about 480 cal. cm.$^{-2}$ day^{-1} in December and June to about 540 cal. cm.$^{-2}$ day^{-1} in March and September during clear weather. At the poles, the great day length at midsummer permits a daily flux of about 670 cal. cm.$^{-2}$ day^{-1} for a short midsummer season greater than at any other latitude, but during half the year no direct solar radiation is received. At high altitudes the radiation received will be greater than at sea level. Not all the radiation comes from the sun; the indirect solar radiation from the sky accounts for an appreciable though very variable fraction of the total flux, roughly of the order of 20 per cent. In north temperate regions, at low or moderate altitudes the period from April 1 to July 31 is generally the period when heating of lakes takes place; in these regions about 60,000 cal. cm.$^{-2}$ is, in most seasons, delivered to the lake surface in such a four-month period. Part of the radiation delivered is reflected back from the lake surface and part scattered back from below the surface, into the atmosphere. The over-all reflection and back scattering appear to vary in temperate latitudes from about 10 per cent in winter to 7 per cent in summer. Very roughly, the reflected and scattered fractions are of the same order of magnitude. The rest of the radiation is absorbed by the water, the solutes, and the suspended material, and so directly or indirectly heats the lake. The water is always emitting long-wave radiation in the far infrared, as is the atmosphere above it. The difference between all the radiation entering a lake and all that leaving it is called the *net radiation surplus.* The mean daily radiation surplus may be positive at all seasons, as in warm temperate latitudes, or may be negative during the winter. The detailed variation of the net radiation surplus is probably of importance in the freezing of lakes. Empirically it seems that the balance is unlikely to exceed a mean value of 400 cal. cm.$^{-2}$ day^{-1} over a period of a week or so in temperate latitudes, a figure which probably indicates the upper rate of heating possible in a lake in ordinary temperate regions.

Solar radiation entering pure water is absorbed selectively, the minimum absorption being at about 4700 A. in the blue. Long-wave ultraviolet is absorbed a little more rapidly, but the main absorption by water is at the red end of the spectrum. At wave lengths $>$ 7500 A., about 90 per cent of the radiation is absorbed by a meter of pure water. The large number of lake waters that have been studied in the laboratory all have higher absorptions than pure water, due to a relatively unselective effect of suspended particles and a very highly selective effect of dissolved coloring matter. The most important dissolved substances absorb strongly in the violet and blue, moderately in the middle wave lengths, and much less strongly at greater wave lengths. When such substances are present in small quantities, the water will be most transmissive in the green; when they are present in large amounts, the transmission will be greatest in the orange and red, though much less great, of course, than for the same wave lengths in pure water. As the transparency decreases, the last light to be instrumentally detectable will be present at smaller and smaller depths and will consist of visible radiation of progressively longer wave length.

In general, when I_0 is the light incident at the surface and I_z is the light present at depth z, we may write

$$I_z = I_0 e^{-\eta' z}$$

where η' is the extinction coefficient. The percentage transmission per meter is defined as $100e^{-\eta'}$.

The radiation actually received on a horizontal surface at any depth z in a lake has passed through a mean path longer than z, even if the sun is at zenith, because some light will have come from all parts of the sky. In moderate temperate latitudes, at all times of day the mean equivalent path is about $1.19z$. If measurements are considered without correction for the path length, the implied transmission is called *vertical transmission* ($100e^{-\eta''}$); if the path is corrected to imply all light from the zenith, the transmission is termed *zenith transmission* ($100e^{-\eta'}$). Both systems are used. However computed, the radiation, except just at the surface, where there is a great absorption of infrared, falls off more or less logarithmically with depth. When transmission is determined in a horizontal beam generated artificially at one end of an apparatus which has a photosensitive element at the other, a quite elaborate variation in opacity is often recorded, particularly in the stable layers of stratified lakes.

The simplest optical measurement that can be made on a lake is the determination of *Secchi disk transparency*, or the depth at which a white disk is just visible. This depends on the ratio of the intensity of light scattered from around the disk to light reflected off the disk and scattered upward above the disk. It is therefore possible for one of two lakes to have a greater transmission and the other a greater disk reading. Very roughly, the light intensity at the depth of disappearance of the disk is about 5 per cent of that at the surface.

Some of the light lost from a lake is scattered upward rather than absorbed. This is due both to highly selective molecular scattering and to nonselective scattering by larger bodies. The scattering S_s in a beam of unit energy traveling unit distance is proportional to the ratio of upward scattered light to total incident light, and to the total extinction coefficient. The scattering so estimated may be expressed as a percentage of $\eta_t' + S_s/2 - \eta_w'$, where η_t' is the total extinction, $S_s/2$ the light scattered forward, so reaching the photometer, and η_w' the mean extinction for pure water. The results vary from 3.2 to 22.9 per cent, indicating little scattering power in some cases, much in others. The nature of the scattering particles is obviously very variable. The scattered light of natural waters has been shown to be partially polarized, to a depth of at least 200 m. in the ocean.

The majority of lake waters show a greater or less visual color when viewed in transmitted light. This is usually measured by an arbitrary scale based on dilutions of a mixture of potassium chloroplatinate and cobaltous chloride. The clearest lake waters have a color of zero; the darkest waters in bog lakes, over 300 Pt units. The color actually observed looking into a lake from above is that of the upward scattered light. This will depend partly on selective scattering, which gives pure water its blue color, and partly on absorption of the upward scattered light by the coloring matter of the water. A small amount of the latter produces a green lake; a larger amount, yellow or brown lakes. Occasionally *seston color* produced by reflection from large quantities of organisms or other suspended material is of significance.

Certain optical phenomena of a striking nature may be observed at the surface of lakes. The *sun or moon path*, which is the reflection of the sun or moon drawn out into a band of light on disturbed water, depends on the distribution of reflecting surfaces on the waves, and is related to the maximum slopes of the latter in a complicated manner.

Very occasionally *horizontal rainbows* can be seen on lake surfaces. They are apparently formed in the same way as are ordinary rainbows, by refraction in droplets of mist stuck to the oily surfaces of lakes that have produced much plankton.

Mirages have often been noted over lakes. Usually they are due to cold air lying over warmer water, which produces a more or less perfect inverted mirage of distant objects. When warm air lies over cold water, objects otherwise beyond the horizon may become visible. When there is a transition from one type of thermal gradient to the other, rays from a distant object can be received at the eye by both concave and convex paths, so that they appear to come from a vertical line rather than a point, giving rise to the illusion known as *castles in the air* or the *fata morgana*.

As well as ordinary electromagnetic radiation, some components of cosmic radiation penetrate considerable depths into water. This penetrating cosmic ray flux consists mostly of mesons.

CHAPTER 7

The Thermal Properties of Lakes

It has undoubtedly been known from prehistoric times that small shallow lakes tend to feel warmer to a swimmer than do large deep lakes, and in some cases the phenomenon of strong surface heating must have been similarly experienced. No attempt at a scientific exploration of lake temperatures was made until the late eighteenth century, when De Saussure (1779, 1796), by means of an alcohol thermometer in a wood casing of great thermal capacity exposed to the water for many hours, demonstrated the low temperatures prevalent at the bottoms of the Swiss lakes. The first series of vertical temperature determinations were made on some of the Scottish lakes by Jardine during 1812–1814. These were not published at the time; they were discussed by Leslie (1838) but not printed in full until much later (Buchan 1871). Leslie understood the small part played by radiation and the importance of mass transport of water in the heating of the depths of a lake.

The first published contributions demonstrating what now would be called thermal stratification were those of De la Beche (1819, 1820), an English geologist, who studied the lakes of the Swiss Alps using a recording thermograph lowered to the bottom at a series of depths. De Fischer-Foster and Brunner (1849) published the first temperature curve, for the Thunersee.

Series of vertical temperatures were also made by Simony (1850) in

Austria, who for the first time appreciated the importance of the shape of the basin in regulating lake temperatures. Deevey (1942) has pointed out that Thoreau (1906) knew of thermal stratification in 1860, when he made the first temperature measurements in the deep water of an American lake; he did not, however, take enough readings to elucidate the shape of the curve.

During the last quarter of the nineteenth century several important studies, notably those of Buchanan (1886), Richter (1892), and Fitzgerald (1895) definitely established the generality of the kind of temperature stratification observed by the earlier workers on temperate lakes of sufficient depth.

The development of knowledge of lake temperatures has grown concomitantly with the development of suitable thermometers. For an extended account of the various techniques that have been employed, the reader is referred to Welch (1948) and particularly to Mortimer (1953b). The very useful new thermistor resistance thermometers, which are likely to provide the most adaptable instruments for much future research, have been discussed at length by Mortimer and Moore (1953).

THERMAL STRATIFICATION

Qualitative statement of the heating of a lake in temperate latitudes, in spring and early summer. It is convenient to begin by considering a lake in a temperate region, in which the entire body of water is at some time in the spring at a uniform low temperature, at or slightly above 4° C., the temperature of maximum density. If such water were of uniform transparency and were quite undisturbed, radiation entering the water surface and being absorbed exponentially would heat the water at a rate falling exponentially from the surface, and so would produce an exponential temperature curve. Two principal factors prevent such a process from taking place or indeed ever being approached. First, evaporation will always cool the surface layer, setting up convection currents; these will be enhanced by back-radiation and loss of sensible heat, particularly at night. Second, the surface of the lake will never be undisturbed by the wind, and the currents set up by the wind stress on the water will generate turbulent motion, leading to mixing and the downward transport of momentum and heat.

The form of the resulting temperature distribution is exceedingly characteristic. In all lakes of sufficient depth, heating in the spring from a low temperature, the water tends to become divided into an upper region of more or less uniformly warm, circulating, and fairly turbulent water termed the *epilimnion*, and a deep, cold, and relatively undisturbed region termed the *hypolimnion* (Birge 1910b). The region of rapid decrease in temperature separating the epilimnion from the hypolimnion was termed the *Sprungschicht* by Richter (1892), the discontinuity layer by Wedderburn (1907), and the thermocline by Birge (1897). In introducing the last term, Birge defined it as that layer in which the fall in temperature exceeds 1° C. per meter. Such a definition is quite arbitrary. Since the decrease in density per degree Centigrade, and so the increase in thermal resistance or stability, increases progressively with increasing temperature above 4° C., the region of decrease in temperature in two lakes, in one of which all temperatures are relatively high and in the other relatively low, may be a thermocline in Birge's sense in the cool lake but not in the warm lake, without implying differences in stability. Since the original definition is unsatisfactory in its arbitrariness, and since the word is too well known and too widely used to be abandoned, it must be redefined to give it a maximum of precision and utility. This was in effect done by Brönsted and Wesenberg-Lund (1911), whose usage is here adopted.

Following these authors, the *thermocline* is defined as the plane of maximum rate of decrease in temperature, or in more formal terms the plane defined by

$$\theta'' = \frac{d^2\theta}{dz^2} = 0 \qquad (1)$$

It will appear in the subsequent discussion that the region below the thermocline, though its temperature may be falling very rapidly with depth, is in certain respects more comparable with the upper part of the hypolimnion than with the narrow zone including and just above the thermocline as here defined. It is included in that part of the hypolimnion later to be defined as the *clinolimnion*, in which the rate of heating falls approximately exponentially with depth.

Since in the top of the clinolimnion the temperature gradient is often very great, and since an organism migrating through a temperature gradient is not likely to be particular about the physics of its production, it is convenient to define the widely used term *metalimnion*[1] to designate the whole of the region in which the temperature gradient is steep, from the upper plane of maximum curvature, termed by Munk and Anderson

[1] Jenkin (1942) uses the older term discontinuity layer in approximately the same sense, and Yoshimura (1936b) the term inflection layer. Metalimnion is more convenient, if only on account of the adjectival form *metalimnetic*. Birge and Juday (1914) protested against the redefinition of the word thermocline by Brönsted and Wesenberg-Lund; Dr. Wilhelm Rodhe has also expressed to the writer some unhappiness over the proposed distinction between thermocline and metalimnion. In the course of preparing the present work several possible terminologies were tried; the one adopted appeared to be the most practical.

the *knee* of the thermocline, to the lower plane of maximum (inverse) curvature.

Though it is probable that a variety of different processes are involved in the production of any given summer temperature curve, it is convenient to consider first the effects of wind-generated currents and the resultant turbulent exchanges of heat and momentum in the water, for such turbulent exchanges constitute the essential mechanism in the production of thermal stratification.

When the ice melts in the spring and the whole body of water is below 4° C., heating by incident radiation will cause an increase in density at the surface and an unstable stratification, which will break down, water mixing by convection until the whole lake is at the temperature below 4° C., as is indicated in a later section. For the present we shall assume a homoiothermal condition at 4°. Any heating at the surface now produces an expansion of the heated water, and the lake becomes stably stratified, with the least dense water at the surface. Work, other than that to be done against viscosity, is now needed to mix the newly heated water into the depths of the lake. It is obvious, moreover, that more work is needed to mix newly warmed surface water throughout the whole depth of a deep lake than of a shallow lake. Indeed, in a very shallow lake we might expect the wind to keep the entire water mass mixed and more or less homoiothermal during the whole period of heating, while in a very deep lake we would not be surprised to find a negligible increase in temperature in the deepest water throughout the entire summer. As Birge (1916) himself put it, ". . . the wind creates surface currents which, guided by the shores and bottom, would establish and maintain a complete circulation. The sun warms the surface stratum and so tends to confine the wind currents to the surface. From the interaction of these two forces result the phenomena of the actual warming of the lake."

In lakes of moderate or great depth, the actual type of distribution observed, with a freely circulating epilimnion, a thermocline, and a hypolimnion, satisfies intuitively such expectations, and a very shallow lake can be thought of as one in which the whole water mass is epilimnion. It is, however, very much less easy to see intuitively in full detail why such an arrangement is inevitably the result of heating by solar radiation from above and heat transfer by wind-generated turbulence throughout the water. While convective streaming due to evaporation and nocturnal cooling certainly assists in sharpening the stratification in most cases, and while the fact that the increase in stability per degree Centigrade is greater at high temperatures than at low temperatures is also of importance in accentuating thermal stratification, particularly in warm lakes in which small temperature differences produce quite stable stratifications, neither of these factors appears to be essential. The fundamental process involved in producing an epilimnion and thermocline would rather appear to be a twofold effect of the establishment of a stable stratification in density of the kind that would result simply from exponential heating.

If initially the water movements are confined to wind-generated currents, the velocity of such currents, in so far as the boundaries of the lake are not involved, will also decrease exponentially. Initially, no density gradient is supposed to be present, and the coefficient of turbulent viscosity can therefore be independent of depth, as appears from equation 5 below. As soon as the water begins to heat from above to a temperature in excess of 4° C. and so acquires a stable density gradient, the turbulence begins to fall in the stabilized upper layers, reducing the rate of transfer of heat and momentum to the deeper layers. One might therefore expect a distribution in which θ' fell throughout with depth. However, the presence of a density gradient, or of elementary layers of different densities, greatly facilitates horizontal movement, one elementary layer slipping over another; this increase in horizontal movement generates turbulence in just those layers in which the increase in stability tends to decrease turbulence. It is apparently the interplay of these two opposing processes, one increasing and the other decreasing vertical turbulent transport of momentum and heat, that is ultimately responsible for the production of a freely circulating epilimnion of limited thickness, lying over a thermocline region in which a very marked temperature gradient is developed in any lake of sufficient depth and with a pronounced seasonal cycle of temperature.

Quantitative theory of the thermocline. At the surface of an unbounded water mass of infinite depth on a rotating planet, it has been pointed out (page 265) that according to the Ekman current law, the current sets at 45 degrees to the wind direction, while below the surface the azimuth of the current varies linearly with depth and its velocity falls off exponentially. This is, however, only to be expected in an unstratified body of water of neutral stability, in which the eddy viscosity does not vary with depth. As soon as heat is delivered to the surface of the water and a thermal gradient develops, this condition no longer holds. As the stability increases in any layer, A_v will tend to decrease, but, as has just been indicated, the stable layers will tend to slip over each other. Munk and Anderson (1948) have examined the steady-state condition set up by the tendency for the increase in stability to decrease eddy viscosity directly and the tendency for the increase in horizontal movement to increase A_v indirectly. The results of their study, though it involves some quite unnatural restrictions, indicate that a thermocline can be generated by the processes under consideration.

Let the initial eddy viscosity coefficient, characterizing indifferent stability, be A_0, and suppose that

$$A_v = A_0 \mathrm{f}(g, \mathrm{E}, v') \tag{2}$$

where g is the acceleration due to gravity, which may be taken as constant, and

$$\mathrm{E} = \frac{1}{\rho}\frac{\mathrm{d}\rho}{\mathrm{d}z}$$

or stability, increase in which decreases turbulence, and

$$v' = \frac{\mathrm{d}v}{\mathrm{d}z}$$

or shear, increase in which increases turbulence.

It can be shown by dimensional analysis that the only way in which these quantities can enter into an expression for A_v/A_0 is as Richardson's number $\mathrm{R_i}$, defined as

$$\mathrm{R_i} = \frac{g\mathrm{E}}{(v')^2} \tag{3}$$

In fresh water, over a moderate range of temperatures we may replace E by $-\,\mathrm{k}\theta'$, obtaining

$$\mathrm{R_i} = g(-\,\mathrm{k})\theta'(v')^{-2} \tag{4}$$

As A_v approaches A_0, $\mathrm{R_i}$ approaches zero, since the stability is then zero whatever the shear may be; and as A_v approaches zero, $\mathrm{R_i}$ approaches infinity, since any eddy viscosity would disappear if the stability were infinite. A suitable expression satisfying these conditions is

$$\frac{A_v}{A_0} = (1 + \beta_v \mathrm{R_i})^m \tag{5}$$

and an analogous equation can be written for heat transfer as

$$\frac{A}{A_0} = (1 + \beta_\theta \mathrm{R_i})^n \tag{6}$$

As indicated before (page 256), Munk and Anderson conclude that we may specifically write, as in Chapter 5, equations 18 and 19,

$$\frac{A_{vz}}{A_0} = (1 + 10\mathrm{R_i})^{-0.5} \tag{7}$$

$$\frac{A_z}{A_0} = (1 + 3.3\mathrm{R_i})^{-1.5} \tag{8}$$

These may be combined with

$$\tau = A_{vz}v' \tag{9}$$

and

$$\frac{\partial \Theta}{\partial t} = - c_p A_z \theta' \tag{10}$$

which give the stress and heat flux at depth z.

For a steady-state condition

$$\tau_x' = \frac{d}{dz}(A_{vz}u') = - 2\Omega \sin \phi v \tag{11a}$$

$$\tau_y' = \frac{d}{dz}(A_{vz}v') = - 2\Omega \sin \phi u \tag{11b}$$

When $z = 0$, we have by equation 36 of Chapter 5,

$$u = v = \frac{\tau_a}{\sqrt{4\rho A_v \Omega \sin \phi}} \tag{12}$$

It is possible to define the upper boundary condition by equation 12 and to obtain solutions of the system of simultaneous differential equations 3, 5, 6, 9, 10, and 11a,b for stresses, velocities, temperature, A_z and A_{vz} as functions of z, by means of a differential analyzer. Without such aid the problem is obviously hopelessly intractable.

From equations 4, 7, and 8, we obtain

$$- \frac{gk\Theta_z A_0}{c_p \tau_z^2} = R_i(1 + \beta_v R_i)(1 + \beta_\theta R_i)^{-3/2} \tag{13}$$

Since the terms on the left-hand side of the equation are either constants or functions of z, while the right-hand side is a function of R_i, we may write

$$F_1(z) = F_2(R_i) \tag{14}$$

If it be now supposed that the heat flux Θ_z is constant at all depths, which corresponds to a limiting steady-state condition of obvious artificiality but one which is not likely to increase the probability of thermocline formation,

$$\frac{A_0}{A_z} = \frac{\theta'}{\theta_0'} = (1 + \beta_\theta R_i)^{3/2} \tag{15}$$

The temperature gradient, given a constant heat flux, will thus be a monotonic function of R_i, and there will be an inflection point, corresponding to the thermocline as defined in the present chapter, at the point where R_i is greatest or the stress τ_z least.

The shear, or rate of change of velocity with depth, can be shown to be minimal at a depth somewhat above the thermocline, and defined by

$$v_{\text{min.}} = \sqrt{\frac{3\sqrt{3}}{2} \cdot g(-\,k)\theta'\beta_\theta} \tag{16}$$

corresponding to $R_i = 0.6$.

The top of the metalimnion, or upper point of maximum curvature where $\theta''' = 0$, corresponds to the solution of

$$R_i'' + \tfrac{1}{2}\beta_\theta(1 + \beta_\theta R_i)^{-1}(R_i')^2 = 0 \tag{17}$$

It can be shown that except for very large values of R_i the corresponding depth is given by the point of zero curvature in the curve relating $F_1(z)$ to z. Thus, if the function $F_1(z)$ can be computed from the constants on the left-hand side of equation 13 and from the solution for the stress, a calculated depth for the base of the epilimnion and of the thermocline under steady-state conditions can be obtained. Model computations for various values of wind velocity, which imply certain empirically determined values of wind stress (page 275) and of A_0 taken with various values of the geographical latitude and of heat flux, indicate that wind stress and therefore wind velocity are by far the most important variables determining the depth of the thermocline. Latitude would appear to be second only to wind stress in determining the form of the temperature curve. In a model calculation with a high initial value of $A_0 = 155$ g. cm.$^{-1}$ sec.$^{-1}$, changing the latitude from 30° to 60° decreased the depth of the thermocline from 22 m. to 14 m. The magnitude of the heat flux proved relatively unimportant in determining the shape of the temperature curve.

Munk and Anderson have analyzed observations on Sweetwater Lake near San Diego, California, at lat. 32° N. The relevant data and results of computations are given in Table 46. It is evident that while there is no

TABLE 46. *Empirical test of Munk and Anderson's theory of the thermocline on Sweetwater Lake*

W(cm. sec.$^{-1}$)	10^3 dΘ/dt (cal. cm.$^{-2}$ sec.$^{-1}$)	10^4k (°C.$^{-1}$)	v_0 calc. (cm. sec.$^{-1}$)	τ_a calc.	A_0 calc. (g. cm.$^{-1}$ sec.$^{-1}$)	Depth of min. Shear, calc. (m.)	Depth of Top of Metalimnion (m.) Calc.	Obs.
350	1.18	1.89	2.58	0.15	43.7	3.5	7.2	6.5
330	2.28	2.07	2.63	0.14	36.6	...	4.5	6.5
365	2.00	1.33	2.59	0.16	49.6	2.5	7.5	5.5
325	2.20	2.23	2.50	0.13	35.0	...	3.0	4.0
300	1.18	1.83	2.39	0.11	27.5	2.3	4.5	4.0

close correlation between variation of the depth of the top of the metalimnion as computed and as observed, the general order of magnitude of the depths in the last two columns is the same. Since W had to be determined at San Diego rather than over the lake, and since the Ekman theory implied by equation 12 can hardly be applicable to the lake even at times of indifferent stability, it is perhaps not surprising that the agreement is not better. It may be pointed out that analyses of observations made at sea clearly indicate the existence of the latitudinal effect.

Ertel's theory of the thermocline. A mathematically simpler if physically less satisfying treatment of the thermocline has recently been given by Ertel (1954).

Considering the temperature at any depth z, averaged over the twenty-four hour period about time t, we may write (equation 30, page 468), if temperature change is due entirely to uniform vertical turbulence,

$$\frac{\partial \theta_z}{\partial t} = A_\theta \frac{\partial^2 \theta_z}{\partial z^2} \tag{18}$$

Actually, during the heating period other sources of heat, notably solar radiation and horizontal advection on the positive side, evaporative cooling, back-radiation, and convection on the negative side, are significant at least in the epilimnion. The mean rate of change of temperature can therefore be regarded as due on the one hand to these sources of heat, positive and negative, and on the other to redistribution of heat due to turbulent mixing. The actual rate of change will be

$$\frac{\partial \theta_z}{\partial t} = A_\theta \frac{\partial^2 \theta_z}{\partial z^2} + F \tag{19}$$

where F may be called the mean source function.

Ertel supposes that the change in F with time may be expressed by means of a coefficient analogous to A_θ, here designated A_F. Writing equation 19 in operational form we obtain

$$\left(\frac{\partial}{\partial t} - A_\theta \frac{\partial^2}{\partial z^2}\right)\theta_z = F \tag{20}$$

Writing the equation analogous to equation 18 for the rate of change of F, we obtain

$$\left(\frac{\partial}{\partial t} - A_F \frac{\partial^2}{\partial z^2}\right)F = 0 \tag{21}$$

whence

$$\left(\frac{\partial}{\partial t} - A_F \frac{\partial^2}{\partial z^2}\right)\left(\frac{\partial}{\partial t} - A_\theta \frac{\partial^2}{\partial z^2}\right)\theta_z = 0 \tag{22}$$

Measuring θ_z from the temperature of vernal circulation, ordinarily 4° C. in dimictic lakes, we take as boundary conditions

$$\text{(22a)} \qquad z \to 0 \qquad -\frac{\partial \theta_z}{\partial z} \to 0$$

$$\text{(22b)} \qquad z \to \infty \qquad \theta_z \to 0$$

$$\text{(22c)} \qquad z \to \infty \qquad -\frac{\partial \theta_z}{\partial z} \to 0$$

$$\text{(22d)} \qquad \theta_0 = \alpha'(\sqrt{t})^n, \qquad \text{where } n = 1, 2, 3 \cdots$$

The solution of equation 22 when $A_\theta \neq A_F$ is now known to be

$$\text{(23)} \quad \theta_{tz} = \frac{\alpha' t}{(\sqrt{A_\theta} - \sqrt{A_F})\mathrm{J}^{(n)}\,\mathrm{erfc}(0)}\left[\sqrt{A_\theta}\,\mathrm{J}^{(n)}\,\mathrm{erfc}\left(\frac{z}{2\sqrt{A_\theta t}}\right) - \sqrt{A_F}\,\mathrm{J}^{(n)}\,\mathrm{erfc}\left(\frac{z}{2\sqrt{A_F t}}\right)\right]$$

where

$$\mathrm{J}^{(n)}\,\mathrm{erfc}(z) = \int_z^\infty \mathrm{J}^{(n-1)}\,\mathrm{erfc}(\xi)\,\mathrm{d}\xi \qquad (n = 1, 2, 3 \cdots)$$

and[2]

$$\mathrm{J}^{(0)}\,\mathrm{erfc}(z) = \mathrm{erfc}(z) = \frac{z}{\sqrt{\pi}}\int_z^\infty e^{-\xi^2}\,\mathrm{d}\xi$$

This rather formidable expression enables one to show that z_e, the depth of the epilimnion to the thermocline as defined above, is given by

$$\text{(24)} \qquad z_e = 2\omega_n\sqrt{A_\theta t}$$

where ω_n is defined by the transcendental equation

$$\mathrm{J}^{(n-2)}\,\mathrm{erfc}(\omega_n) - \sqrt{\frac{A_\theta}{A_F}}\,\mathrm{J}^{(n-2)}\,\mathrm{erfc}\left(\omega_n\sqrt{\frac{A_\theta}{A_F}}\right) = 0$$

Assuming constant values for the coefficient of eddy conductivity A_θ and for the mixing coefficient A_F of the mean source function F, the depth of the thermocline during the heating period should be proportional to the square root of the time since vernal circulation, whatever the law defined by boundary condition 22d may happen to be.

Figure 131 presents an attempt to test this aspect of Ertel's theory on the thermal data for Linsley Pond. The lake in 1936 was homoiothermal at 4° C. on or about March 29. Curve II gives the distribution of mean temperature with depth for a period June 1 to June 15, and probably

---

[2] erfc(z) is known as the complementary error function of z, ξ being an arbitrary variable.

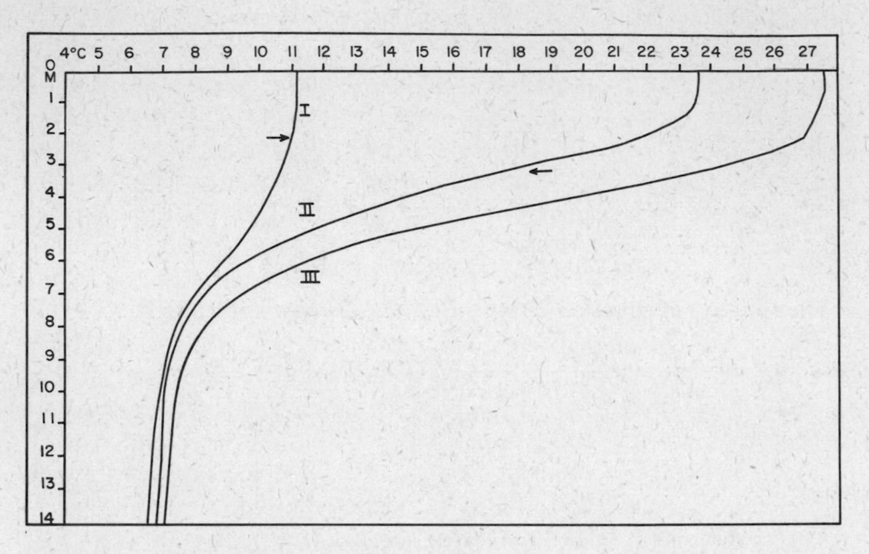

FIGURE 131. Temperature curve (I) for April 30, and mean temperature curves for (II) June 1–15 and (III) August 3–17, 1936, Linsley Pond, Connecticut. The arrow to the left indicates the observed position of the thermocline in the June series; that to the right, the position of the thermocline, calculated from Ertel's theory, for the April series, which is obviously not in accord with observation.

constitutes the best estimate of the temperature curve for June 8. The inflection point defining the thermocline is just below 3 m.; a depth of 3.1 m. may be regarded as a satisfactory estimate. Curve III gives the mean temperature distribution for the period August 3 to August 17, when the surface temperature was maximal. According to Ertel's theory, the thermocline should be at

$$3.1 \times \sqrt{\frac{134}{71}} = 4.26 \text{ m.}$$

which is obviously not very far from the observed inflection point. A detailed study of the descent during late June and July shows considerable irregularity, but t is too large to give really satisfactory variation in $\sqrt{t}$. Earlier in the year, the less adequate data exemplified by curve I for April 30 appear quite discordant with the theory, the thermocline apparently lying lower than in the summer. It is possible that Ertel's theory describes relatively small changes at the height of stratification fairly well; the physical complexity of the processes determining F and A_F perhaps justify a priori scepticism of its applicability early in the season, when large changes in radiation, windiness, and temperature distribution are occurring.[3]

Terminology of Stratification. Forel (1892a, 1895) pointed out that in any ordinary lake in the temperate regions under continental climatic conditions, with a cold winter and a warm summer, full circulation at or near the temperature of maximum density would occur twice a year—in the spring, after the ice melted, and in the autumn or, actually often by terrestrial standards, the early winter, before the lake froze. He called any lake in which the water passed through the temperature of maximum density twice a year a *temperate* lake. Forel designated as *tropical* all lakes in which there is a single circulation period, in the winter time, the water never being cooled below 4° C., and as *polar* all lakes in which there is a single circulation period, in the summer, the water never rising above 4° C. Forel realized that occasionally a lake could be temperate in shallow bays, tropical in the open water.

The terminology is very unfortunate; not only are many of the best examples of tropical lakes to be found in western Scotland, but in recent years limnological studies in countries more genuinely tropical than Scotland have indicated that many lakes in such regions are very irregular in their circulation.

An improved series of categories was given by Yoshimura (1936a), who recognized:

Tropical lakes, with a high surface temperature of 20° to 30° C., a small annual amplitude of variation, and a small thermal gradient at any depth, though the density gradient may be sufficient to impart considerable stability. Circulation is irregular, but usually occurs only in the coldest time of year.

Subtropical lakes, with surface temperature never below 4° C., annual variation large, thermal gradient large, one circulation period in winter. This is essentially Forel's tropical category.

Temperate lakes, with surface temperature above 4° C. in summer and below 4° C. in winter, thermal gradient large, seasonal variation large, two circulation periods in spring and late autumn.

Subpolar lakes, with surface temperature above 4° C. only for a short time in summer, thermal gradient small; thermocline, if present, poorly developed and generally near surface; two circulation periods, ideally in early summer and early autumn, but temporary cooling throughout the summer generally permits fairly frequent mixing.

---

[3] If there is any significance in the apparent agreement between observation and theory in July or August, it is obvious that the origin of t must be a virtual date of vernal circulation which happens to coincide with the actual date.

Polar lakes, with surface temperatures always below 4° C., ice-free period very short, and circulation only at the height of summer.

The temperature curves of Fig. 132, most of which were used by Yoshimura to illustrate this classification, indicate the utility of his scheme. It is frequently convenient, however, to have terms to indicate temperate, tropical, and polar in Forel's sense, but without the objectionable geographical connotation of his original terms. It is therefore proposed to use the term *dimictic* for any lake circulating twice a year; *warm monomictic* will be employed as a substitute for tropical, implying winter circulation above 4° C.; and *cold monomictic* as a substitute for

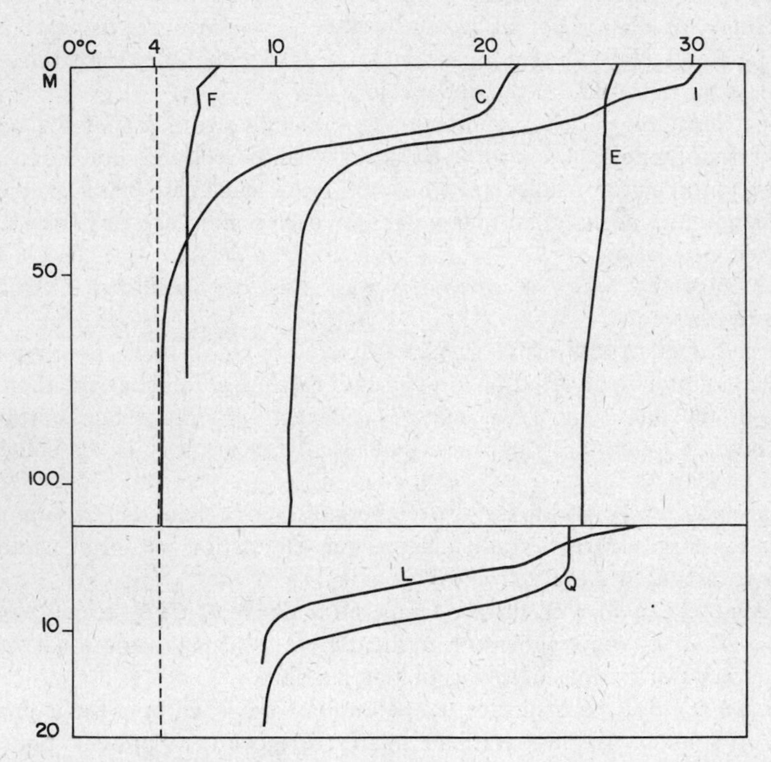

FIGURE 132. Temperature curves at height of summer stratification. *Upper panel:* F, Flakevatn, Norway, a subpolar dimictic lake; C, Lake Cayuga, New York, a first-class temperate dimictic lake; I, Ikedako, Japan, a warm monomictic (subtropical) lake; E, Lake Edward, Uganda, an oligomictic or more or less meromictic lake. *Lower panel:* L, Linsley Pond, and Q, Lake Quassapaug, shallower dimictic lakes of the second class, at about the same latitude as Lake Cayuga; note the identical slope in the metalimnion in spite of the greater area, and the more definite epilimnion in Lake Quassapaug in contrast to Linsley Pond.

polar, implying summer circulation below 4° C. When a lake is spoken of as monomictic, without qualification, it will be a warm monomictic lake or tropical in Forel's sense. Additional categories are required (p. 462) for certain lakes at very high altitudes and very low latitudes.

Whipple (1898) introduced three other terms, which have come into wide use to describe the thermal regime of lakes but have been much misunderstood:

Lake of the first order, one in which the bottom temperature remains at 4° C. and circulation periods are often absent.

Lake of the second order, one in which marked thermal stratification occurs, the bottom water being somewhat or greatly above 4° C. in summer, and with one or two full circulation periods per year.

Lake of the third order, one in which thermal stratification never develops, the whole lake remaining in circulation even during the summer.

The definition of the lake of the first order is evidently based on a misconception. Any lake in which the temperature of the surface water passes from below 4° C. to above 4° C. in the course of a year will have a circulation period when this transition occurs, unless the deep water is stabilized by a chemically determined increase in density. If the typical lake of the first order does not circulate, it owes this property to nonthermal characters which have no place in a classification ostensibly set up to arrange lakes in thermal categories. The classification must obviously be first set up for *holomictic* lakes, which can go through freely circulating periods, before it is extended to *meromictic* lakes, in which part of the deep water, the *monimolimnion* of Findenegg (1935), is stabilized by dissolved substances.

The fundamental idea behind Whipple's orders is valuable, however, and his categories can be suitably redefined and used. The orders express the fact that while some lakes are so shallow that they never stratify thermally, others are so deep that they take up all the heat from their environment that is possible under the existing climatic conditions. Lakes of intermediate depth are thermally stratified, but if made deeper they would take up more heat. That increasing the depth or the area or both should not increase the amount of heat taken up per unit area is essentially the requirement of a lake of the *first class* as defined by Birge (1915), but Birge never went on to define any other thermal classes, though he presumably had Whipple's orders in mind. For the purposes of the present work, it is convenient to combine the concepts of these pioneer limnologists and to define for dimictic lakes:

(1) *First-class lake*, bottom temperature so near 4° C. that extrapolation of the summer temperature curve would involve a negligible increase in the heat entering the lake.

(2) *Second-class lake*, thermally stratified but with the bottom temperature sufficiently above 4° C. for the extrapolation of the summer temperature curve to involve a significant increase in the heat entering the lake.

(3) *Third-class lake*, not thermally stratified.

If the minimum winter temperature is known and is substituted for 4° C. in the above definitions, they can be also used for monomictic lakes, at least of the subtropical type.

Geographical limits of dimictic temperature regimes. In most general terms, the temperature cycle in a lake will be determined by latitude, altitude, and the continental or oceanic nature of the climate, which appears on the continents as a somewhat complicated function of longitude.

Provided a lake is sufficiently deep to exhibit thermal stratification, the required depth being dependent on the area, it may be expected to be dimictic anywhere in continental Europe from Lapland south into the Alps. In Scotland, the English Lake District, and presumably in Ireland, warm monomictic lakes are well known, on account of the oceanic climatic regime. Some of the large subalpine lakes of the Swiss Alps, notably the Lake of Lucerne (lat. 47° N.) with a minimum temperature of 4.3° C., and the Lago di Lugano (lat. 45°57′ N.) with a minimum temperature of 5.2° C. at an altitude of 657 m., are clearly monomictic; others, such as Lake Constance (lat. 47°39′ N.) and the Lake of Geneva (lat. 46°27′ N.) with minimum temperatures of 3.7° and 3.9° C. respectively (Halbfass 1923), are just dimictic. All the large subalpine lakes on the Italian side of the Alps, lying south of latitude 46° N., appear to be monomictic. Farther south, dimictic lakes can be expected only at considerable altitudes.

In western North America, all first-class and second-class holomictic lakes from north of about 40° N. to well into the Arctic Circle are likely to be dimictic, except when they lie at the snow line or close to the Pacific Coast. Lake Tahoe, lat. 39° N., altitude 1890 m., is apparently just monomictic (Kemmerer, Bovard, and Boorman 1923); Pyramid Lake, which spans the 40° parallel but lies lower than Tahoe, at 1153 m., is definitely monomictic (Jones 1925). Farther south, dimictic lakes would be expected only at considerable altitudes. Making due allowance for elevation, the latitudinal limits appear to be comparable in continental Europe and western continental North America. Near the coast the warm monomictic type appears as far north as British Columbia (Ricker 1937).

In Japan, the very extensive data of Yoshimura indicate no monomictic lakes north of 36° N.; a few dimictic lakes occur as far south as lat. 32° N.

(Yoshimura 1936b,c). Yôdako on Horomusiro Island, at lat. 50°29′ N., altitude only 130 m., exhibited a subpolar stratification in the middle of July, the surface being at 7.6° C., the bottom (29 m.) at 6.6° C. This case emphasizes the southern displacement of lacustrine climatic zonation on the eastern margin of the Eurasiatic continent.

In eastern North America there are not enough deep lakes south of the Pleistocene ice border to permit definite conclusions, but the shallow Tom Wallace Lake, Kentucky, lat. 38°05′ N., altitude just under 200 m., is monomictic in some years, dimictic in others (Cole 1954). The limit for dimictic lakes thus seems well south of the European and western North American limiting latitudes, though perhaps not as far south as in Japan.

The effect of altitude. So far as the temperature regime is concerned, an increase in altitude, except in equatorial latitudes, is very roughly equivalent to an increase in latitude. Very small differences in altitude may be reflected in detectable temperature differences. Halbfass (1923) calls attention to Loch Hope and Loch Laoghal, two neighboring lakes of comparable area and depth on the west coast of Scotland, the former at an altitude of 3.8 m., the latter at 113 m. The small altitudinal difference apparently accounts for the fact that at all depths Loch Hope is 0.8° to 2.8° C. warmer than Loch Laoghal.

In central Norway, at lat. 60°49′ N., Strøm (1934) found Flakevatn, a lake at 1448 m., to be just dimictic except perhaps in abnormally cool summers. As observed by Strøm, the lake clearly belongs to Yoshimura's subpolar type, with very little thermal stratification. Strøm points out that a few small lakes in Scandinavia, evidently at comparable elevations, may belong to the cold monomictic type.

In the Central European Alps, between 44° N. and 48° N., it appears from the observations of Pesta (1929), Hacker (1933), and Steinböck (1938), to name but three important investigators, that a few small lakes between 2100 m. and 2900 m. have a cold monomictic regime. These lakes are usually associated with glacier margins or the edges of snow fields, or may be fed by short streams of melt water. Other lakes in the same altitudinal range, however, have maximum surface temperatures up to 16° C. or, in the cases of a few shallow examples under 5 m. deep, even higher.

In Bulgaria, at about 42°11′ N., Leutelt-Kipke (1935) observed some indefinite stratification in the highest lake, 2780 m., that she investigated.

At the western end of the Tibetan plateau, between 33° N. and 34°30′ N., Hutchinson (1937b) observed a thick ice cover on Ororotse Tso, at 5297 m., in the middle of July. The rich chironomid fauna of the bottom of the lake suggests that it cannot be permanently frozen, and it is presumably at least partly free of ice during a few weeks in August. Mitpal Tso, at 4875 m., is clearly dimictic. At about the same latitude, Lake Manasbal in Kashmir, altitude 1584 m., was thermally stratified on April 29, 1932, with a temperature at 11.5 m. of 7.8° C. This lake is presumably a rather warm subtropical monomictic lake. The boundary between monomictic and dimictic lakes in this region must be somewhere between 2000 and 3000 m. Considering the available data, in the more continental parts of the Eurasiatic continent the limits for dimictic circulation appear to be:

At Latitude	Lower Limit, m.	Upper Limit, m.
60° N.	...	c. 1500
46° N.	500–1000	2100–2900
33° N.	2000–3000	c. 5300[4]

The upper limit will be little below the snow line. Many local climatic and geographical factors are likely to cause considerable divergence from such figures, which are at best only rough approximations.

In the western part of North America, Rawson (1942, 1953) found either typical or subpolar dimictic circulation at altitudes of 1900 to 2000 m. in the Jasper National Park region of the Canadian Rocky Mountains, about lat. 53° N. Pennak (1944) records at about lat. 40° N. a subpolar temperature curve in Summit Lake, altitude 3884 m., while Grand Lake, which is 62 m. deep at 2550 m., had a bottom temperature of about 7.5° C. on September 6, 1941. The former lake is certainly dimictic, the latter probably warm monomictic. These figures imply somewhat higher limits than in the Old World.

No lakes are available at high altitudes in eastern North America. In Japan, Yoshimura found no cold monomictic lakes; his highest lake, Sannoike, on Mount Ontake, at lat. 35°54′ N., altitude 2720 m., is only 12.5 m. deep. It should probably be regarded as a third-class lake, having a temperature on September 17 of 9.7° C. in the top 2 m. and 9.5° C. at all greater depths. Ôike, on Mount Sirouma, at 2379 m. in lat. 36°47′ N., is shallower, but exhibited strong thermal stratification. The warm but subpolar nature of Sannoike is certainly due to persistent low stability in the upper layers of the lake.

In the mountains of central Europe, Steinböck believes that what he

[4] In the Elburz Mountains, Iran, about lat. 36° N., the upper limit is certainly far above 3000 m. (H. Löffler, verbal communication). In Ceylon, lat. 7° N., bodies of water at altitudes between 1900 and 2500 m. may freeze occasionally, and so must be dimictic (S. Dillon Ripley, verbal communication; Holsinger 1955).

calls the warm alpine lakes, which would here be designated dimictic, either lack definite temperature stratification if they are shallow, or develop a relatively shallow thermocline, often with a transitory rapid fall in temperature at the surface during the day. In Indian Tibet, Mitpal Tso, the highest dimictic lake so far described, had a very thick epilimnion with a surface temperature of 12.49° C., and a thermocline at about 11 m. It is possible that this difference between the European and south central Asiatic alpine lakes is, if real, due to differences in humidity and cloud cover, and conceivably to the direct effect of latitude (Munk and Anderson 1948). It is perhaps not accidental that Rawson found a temperature curve not unlike that of the smaller Tibetan lakes in Amethyst Lake in the Jasper National Park region, though the other lakes in the region were barely stratified. Here the rainfall is markedly less than in central Europe, though much higher than in Indian Tibet. Too much stress should not be put on such comparisons, however, for the number of extra-European lakes involved is small, while the range of variation in Europe is very great.

It should finally be noted that in considering the thermal regime of alpine lakes, the shading of the lake surface by high mountains may often reduce the daily period of insolation sufficiently to reduce the heat income.

Continental and oceanic thermal regimes. In any given latitude, as the data just presented on the distribution of monomictic and dimictic lakes indicate, the distinction between *oceanic*[5] and *continental* climatic regimes is of great importance in determining not only circulation periods but also the form of the summer temperature curve.

In a temperate continental climate with cold winters and hot summers, the lake generally circulates for some time in the early winter below 4° C., freezes when much of the water is below that temperature, and breaks up to go into vernal circulation relatively late. The heating of the lake begins during relatively warm and calm spring weather. In an oceanic climate the lake may never freeze, or if it does, soon becomes ice-free. Heating starts early, when the weather is often still stormy, and though the heating period is prolonged, the daily rate of heating may be small because of the many cool, wet summer days; the low summer air temperatures, moreover, may permit greater loss of sensible heat than in a continental climate. The general low temperature of the epilimnion developing under

[5] Strøm (1931) and Jenkin (1942), who of previous authors have most strongly emphasized the distinction, use the terms *Atlantic* and *coastal* respectively. The former would be inappropriate in New Zealand or Patagonia, while the latter word, though inherently no more misleading than *oceanic*, lacks the sanction as a term in descriptive climatology that *oceanic* appears to possess.

such circumstances reduces the stability in the metalimnion and facilitates turbulent transport of heat into the depths of the lake. In extreme cases, even in quite high latitudes, winter cooling is insufficient to bring the water below 4° C. and the lake has a warm monomictic cycle.

Though many examples can be chosen, the effect is perhaps demonstrated most dramatically in the comparison made by Strøm (1931) between two Norwegian lakes, Evangervatn in western Norway (long. 6°05′ E.) and Holsfjord in eastern Norway (long. 10°10′ E.) the former under an oceanic, the latter under a continental climate (*D*, Fig. 133). Both take up about the same quantity of heat per unit area during the spring and summer, but the distribution is extraordinarily different. In Evangervatn the epilimnion is quite indefinite and not properly separated from the metalimnion, while at the same time of year Holsfjord exhibits as sharp a division into epilimnion, metalimnion, and hypolimnion as could be desired.

Many less dramatic examples could be found in the literature. Most lakes in the eastern and central United States and Canada and in the interior of Europe exhibit the sharp continental type of stratification, but most of the deeper lakes of Scotland, such as Loch Ness (Wedderburn 1907) and Loch Awe (Jenkin 1930), and to a lesser extent those in the English Lake District (Jenkin 1942), exhibit less sharp stratifications. The prevalence of the oceanic type of stratification in the Scottish lochs, in fact, misled British workers into believing the classical threefold division of a stratified lake to be a phenomenon of autumnal cooling. Many of the larger subalpine lakes are, as has been indicated, warm monomictic, but when they are compared with the Scottish lochs there is also clear evidence of the differences due to oceanic as opposed to continental climatic regime.

Effect of mean depth. The depth and volume development of a lake may have a considerable effect on its temperature curve, but this appears to vary with the nature of the climatic regime of the region in which the basin is situated.

In regions of cold winters, in which the spring circulation occurs late and heating then takes place rapidly, the greater thermal capacity of the hypolimnion of a deep lake may ensure colder deep water than in a lake of less great depth or volume development. This is well shown in the case of Lake Canandaigua as compared with Lake Cayuga. The latter is a much larger lake than the former, but in spite of the greater area it remains colder below 50 m.

If, however, the winter is mild and the lake is of the subtropical monomictic type or approaches that type, a great mean depth and therefore a great thermal capacity may ensure very little fall in temperature during

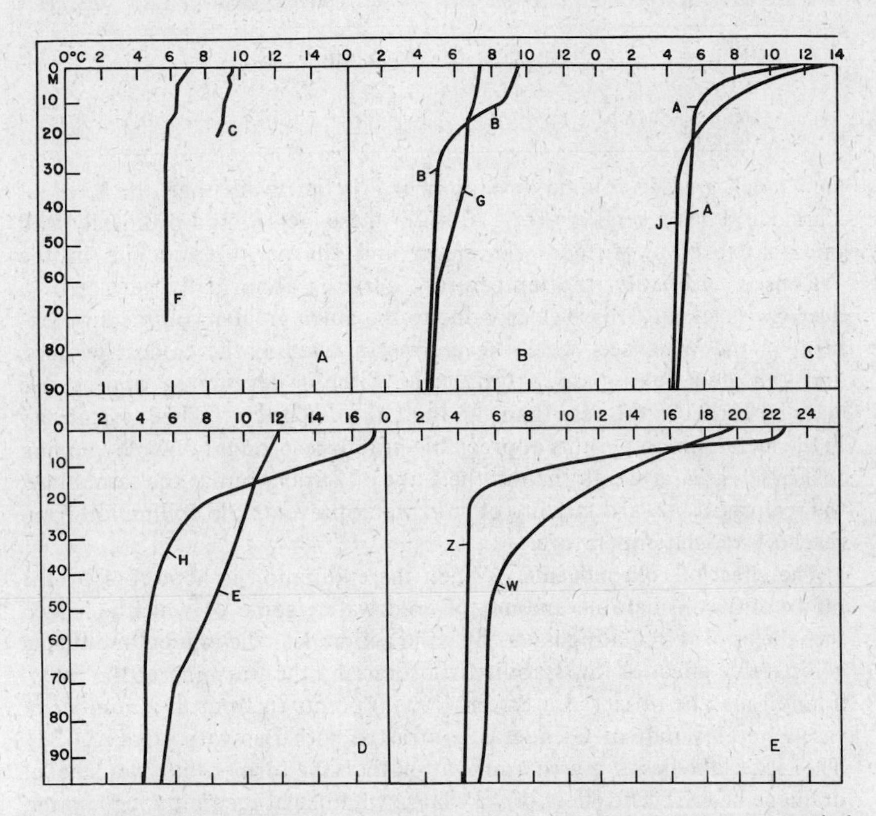

FIGURE 133. Effects of various factors on summer temperature curves. *A*, typical subpolar curves at high latitudes and altitudes: C, Chandler Lake, Alaska (Livingstone); F, Flakevatn, Norway (Strøm). *B*, effect of melt water on adjacent lakes: B, Bessvatn, G, Gjende, Norway. *C*, effect of melt water during a single season, Oldenvatn, Norway: J, June 17, 1931; A, August 18, 1931. *D*, effect of continentality in Norway: H, Holsfjord, relatively continental; E, Evangervatn, markedly oceanic (Strøm). *E*, effect of volume development: *Z*, Lake of Zurich; W, Walensee. (Data in Halbfass.)

the few really cold days of the winter. Heating starts early, at relatively low air temperatures, and proceeds slowly. The great heat capacity of the circulating water may for a long time prevent a stable epilimnion from developing, and the final result is a relatively warm hypolimnion, a cool epilimnion, and a gentle thermal gradient at the thermocline. Following Halbfass (1923), the Walensee may probably be compared with the Lake of Zurich in this respect (*E*, Fig. 133). They have comparable maximal depths (Table 47), but the Walensee has a much greater mean depth and

TABLE 47. *Morphometric comparison of the Walensee and the Lake of Zurich*

	Altitude, m.	Area, km.^{2}	z_m	$\bar{z}$	$\bar{z}:z_m$
Walensee	42.3	24.2	151	103	0.68
Lake of Zurich	409	88.7	143	44	0.31

so a much greater volume development. It lies lower than the Lake of Zurich and has a smaller area. Both of these facts would ordinarily lead one to expect a warmer epilimnion and sharper thermocline in the Walensee. Actually, the temperature curve is abnormally gentle for a deep Swiss lake. This is clearly due to the much greater volume development in the Walensee, which never freezes, even in the coldest winters, and can start taking up a little heat without developing appreciable stability earlier in the year than can the Lake of Zurich. The low stability in the metalimnion permits appreciable heat to be brought down by mixing to a much greater depth than in the Lake of Zurich, but at the same time, the reciprocal upward mixture of cold water prevents the epilimnion from reaching a high temperature.

The effect of cold influents. When the epilimnion is kept cool by the influx of a considerable amount of cold water, some of which is mixed into the upper circulating layers by wind action near the influent mouths, a comparable effect of low stability in promoting the warming of the hypolimnion may be observed. Strøm (1935a) points this out in a number of cases, notably that of Gjende as contrasted with Bessvatn (*B*, Fig. 133). The latter lake lies at a greater elevation than the former, but in a smaller drainage basin. The effect of very large volumes of incoming melt water, not only in lowering the surface temperature of Gjende but also in raising the hypolimnetic temperatures, is apparent when the temperature curves of the two lakes are compared.

In the case of two temperature curves for Oldenvatn (Strøm 1933a), the decrease in epilimnetic temperatures and the increase of those of the hypolimnion can be observed occurring during the course of a summer as glacial melt water is added to the lake (*C*, Fig. 133).

Windiness and area. In theory the degree of exposure or shelteredness should be expected to play a part in determining the thickness of the epilimnion. This in effect appears to be true. It is difficult, however, to separate the effect of windiness as such from the effect of the area over which the wind is blowing. Yoshimura (1936b), after excluding certain lakes with abnormal inflowing water, has divided his data on summer temperatures into four groups, exposed to mean winds (*a*) under 200 cm. sec.$^{-1}$, (*b*) between 200 and 300 cm. sec.$^{-1}$, (*c*) between 300 and 500 cm. sec.$^{-1}$, and (*d*) over 500 cm. sec.$^{-1}$. In each group a strong correlation of the temperature at 15 m. with area was found, the regression equations being

(25a) $$\theta_{15} = 1.6(\log A - 1.5)$$

(25b) $$\theta_{15} = 1.86(\log A - 1.89)$$

(25c) $$\theta_{15} = 2.7(\log A - 2.52)$$

(25d) $$\theta_{15} = 4.36(\log A - 3.31)$$

These equations generalize the fact that in very small lakes the wind velocity has but little effect, but in larger lakes, of area 1 to 20 $km.^2$ the thickness of the warm circulating water is much greater, all lakes of the class (*d*) having temperatures over 13° C. at 15 m., while in class (*b*) none had temperatures in excess of 12° C. In the former case, the 15-m. depth was evidently ordinarily epilimnetic, in the latter case metalimnetic. A more refined study might have indicated some effect of oceanic climate in the wind-swept lakes of the Kurile Islands, and the mode of approach is perhaps not the best to bring out the expected relationship. In general, however, Yoshimura's work appears to provide confirmation, on a large scale, of theoretical expectation.

The effect of the area presumably depends on the greater velocity at the water surface when the wind has a long fetch over water, as has already been pointed out in connection with wave formation (page 356).

Innumerable examples of neighboring lakes differing in area and degree of exposure, exhibiting thermoclines at different depths, are known. Grote (1934) compares the Grosser Plöner See, with a thermocline at 15 m., with the small and sheltered Kleiner Uklei See, in which the thermocline lies at 3 m. In Fig. 132 the seasonably comparable temperature curves for Linsley Pond and Lake Quassapaug indicate that the metalimnetic part of the curve of the latter lake is virtually identical with that for Linsley Pond, pushed down 4 m. into the water.

Grote supposes that one can take the depth of the epilimnion as approximately equal to *D*, the depth of frictional resistance (Chapter 5, equation 24), in which case in any two lakes at the same latitudes the ratio of the coefficients of eddy viscosity will be roughly proportional to the ratios of the squares of the thicknesses of the epilimnia. The ratio or *modulus of turbulence* would thus be 25 in the case of the Grosser Plöner See against the Kleiner Uklei See, and 4 in the case of Lake Quassapaug against Linsley Pond. The more detailed treatment of the theory of the thermocline given here makes this approach of doubtful utility. It is certain, however, that A_v will be greater in the wind-swept epilimnion of a large exposed lake than in the relatively undisturbed epilimnion of a small sheltered lake. It is therefore of considerable interest to find that not only is the slope of the temperature curve in the metalimnion of Lake Quassapaug almost the same as that of the comparable part of the curve in Linsley Pond, but also that the coefficients of eddy conductivity, computed by the method to be presented later (p. 472) for the clinolimnia of the two lakes, do not differ significantly. The effect of the greater exposure would seem to be solely to deepen the epilimnion and, within certain limits, not to have a direct effect on what happens below the thermocline.

Secondary and multiple thermoclines. The most frequent divergence from the classical temperature curve is due to the formation of secondary thermoclines.

At the surface there is often sufficient heating by radiation during a warm day to produce a marked temperature gradient; in hot weather such a gradient may persist for several days. Epilimnetic thermoclines are often recorded in alpine lakes, into which the radiation can be very intense when the sky is cloudless. When the epilimnion is thin, the effect of intense heating on hot days may, at least during the daylight hours, result in the metalimnetic temperature curve being produced at almost its maximum gradient, right to the surface.

Recent work with the bathythermograph, particularly by Rawson (personal communication) has indicated that the formation of what are apparently true thermoclines at several levels is not very unusual. Presumably this happens when a lake becomes thermally stratified just before a spell of cool weather. A deep circulating epilimnion is produced. A subsequent spell of warm and relatively calm weather may now set up a shallow thermocline in the epilimnion, which behaves as if it were a complete homoiothermal lake. In Great Slave Lake, Rawson found two very marked thermoclines and indications of a third on August 7, 1947 (Fig. 134). He has also observed in Lac la Ronge (Fig. 135) the development at the end of August 1948, from an austerely classical type of stratification, of a temperature curve with at least three thermoclines, due apparently to changes in weather leading to periods of cooling that reduced the temperature and increased the thickness of the epilimnion, alternating with warm periods that re-established thermal stratification in the shallower layers. In yet another case, that of Lake Minnewanka, Banff, in which four thermoclines were recorded, the complex thermal stratification is probably due to the fact that the lake is dammed and that water is intermittently drawn out from a depth about 10 m. below the surface. It is quite likely, moreover, that as more continuous records of the variation of temperature with depth are made with thermistors or bathythermographs, steplike discontinuities will be found to be far more common in temperature curves for any season at which stratification occurs than has pre-

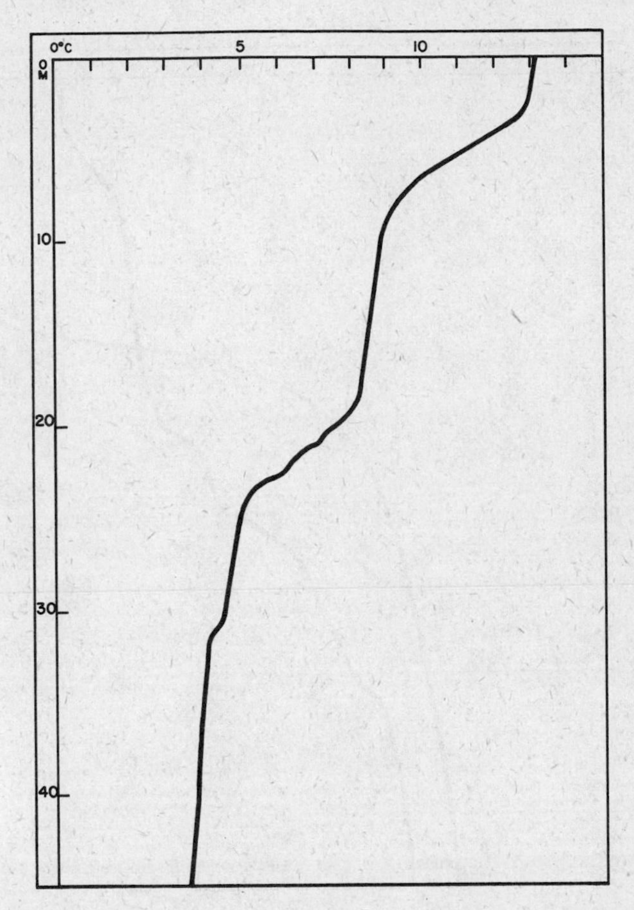

FIGURE 134. Great Slave Lake, August 7, 1947, showing strong development of two thermoclines (Rawson, personal communication).

viously been believed. Such discontinuities will raise interesting hydrodynamic problems.

The effect of exposure of cold water upwind. Mortimer (1952a) has shown, in his studies of Windermere, that exposure of deeper cooler layers at the windward end of a lake during periods of high wind may play an important part in determining the form of the summer temperature curve. The effect of the piling up of the epilimnion downwind is to force the metalimnetic layers upwind, as has already been indicated. When this process continues until there is an appreciable exposure of cool metalimnetic water at the windward end of the lake, some of this water will be

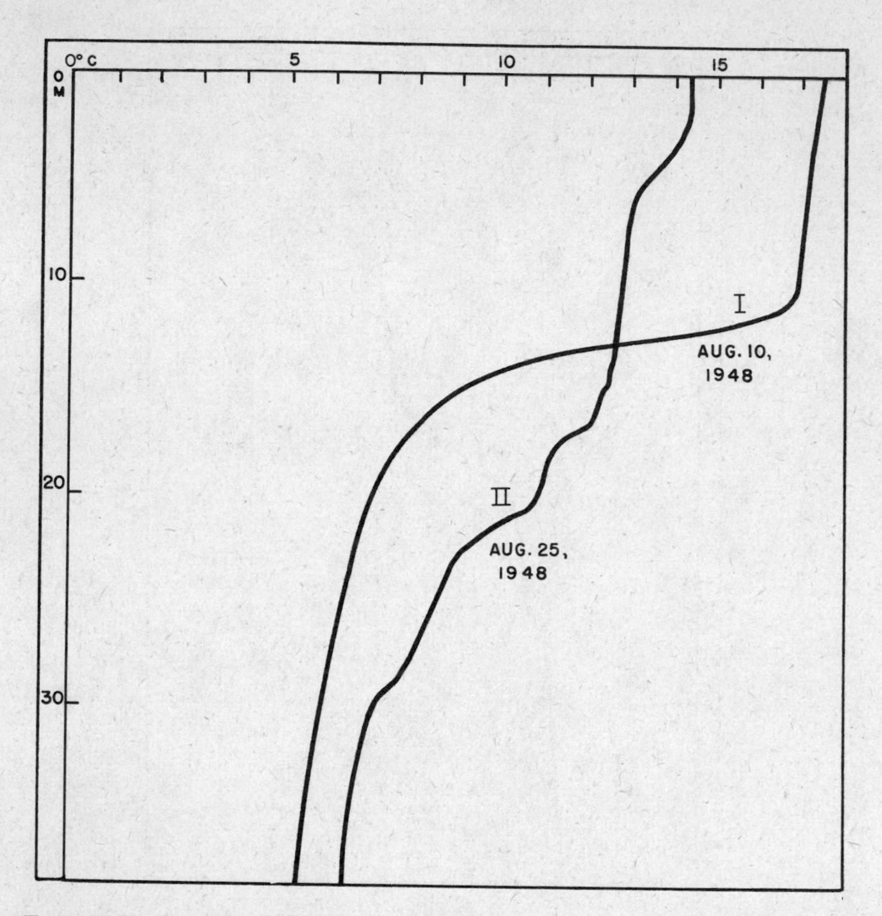

FIGURE 135. Lac la Ronge, August 1948. I, Typical summer stratification, August 10; II, complex stratification, August 25. (Rawson, personal communication.)

driven downwind and mixed into the epilimnion. The result is that, at the same time, the epilimnion is cooled and thickened but the thermocline sharpened, results that are comparable to those that could be obtained merely by convectional mixing after cooling from the lake surface, but which in this case are achieved without any net loss of heat. The resulting lower surface temperature in fact facilitates gain in heat after the wind has dropped. It is not unlikely that the possibility of operation of this mechanism is largely responsible for the magnitude of the heat budget of a large, deep, wind-swept first-class lake.

It is evident from Suslov's (1947) account that this process must occur to a very remarkable degree in Lake Baikal, where prevalent northwest winds may drive the epilimnion onto the eastern shore, causing a drop in surface temperature from 13° to 4° C.

Autumnal cooling and circulation. In dimictic lakes there is generally a period in the late summer, during August in the Northern Hemisphere, during which there is little loss or gain of heat and after which the lake begins to cool. The cooling process is very characteristic, consisting primarily in a steady increase in thickness of the epilimnion as its temperature slowly falls. This is evidently due largely to heat loss at the surface. Such simple cooling would, by reducing the surface temperature, set up convection currents that produced an epilimnion of indifferent stability, its thickness being determined by the depth on the summer temperature curve of the new low autumnal surface temperature. This process, at least initially, may accentuate the thermocline. There is, however, another process occurring, for not only is the epilimnion cooled, but the metalimnion is warmed, evidently on account of greater turbulent movement at the bottom of the completely circulating epilimnion. Initially the deeper layers of the hypolimnion may be hardly disturbed. An extreme case is provided by the temperature curves of Fig. 136, which refer to dates before and after the hurricane of September 21, 1938. It is evident that although winds of up to 100 km. $hr.^{-1}$ had blown over the surface of Linsley Pond, no significant disturbance of the water below 10 m. had taken place.

As the cooling proceeds, the epilimnion, now at a quite low temperature, finally includes the entire lake, which is then in full autumnal circulation. The temperature at the beginning of circulation is very variable, being much higher in small shallow lakes of the second class than in large deep lakes of the first class. In the former group, as exemplified by Lake Mendota (Fig. 137), full autumnal circulation may start in October at a temperature as great as 14° C. The bottom water will therefore be warmer at this time than at any other time of year. In a large, deep first-class lake, such as Lake Constance, the full autumnal circulation may not be achieved until midwinter, at temperatures little above 4° C. (Auerbach, Maerker, and Schmalz 1924), but even here the mean temperatures below 100 m. are very slightly higher in January than at any other season.

Winter temperatures: inverse stratification and winter heating. Monomictic lakes of the subtropical type, such as the large lakes of the southern slopes of the Alps, circulate throughout the winter at a low temperature little above 4° C., and some lakes that do not freeze fail to establish any definite inverse stratification in winter even if the temperatures are below

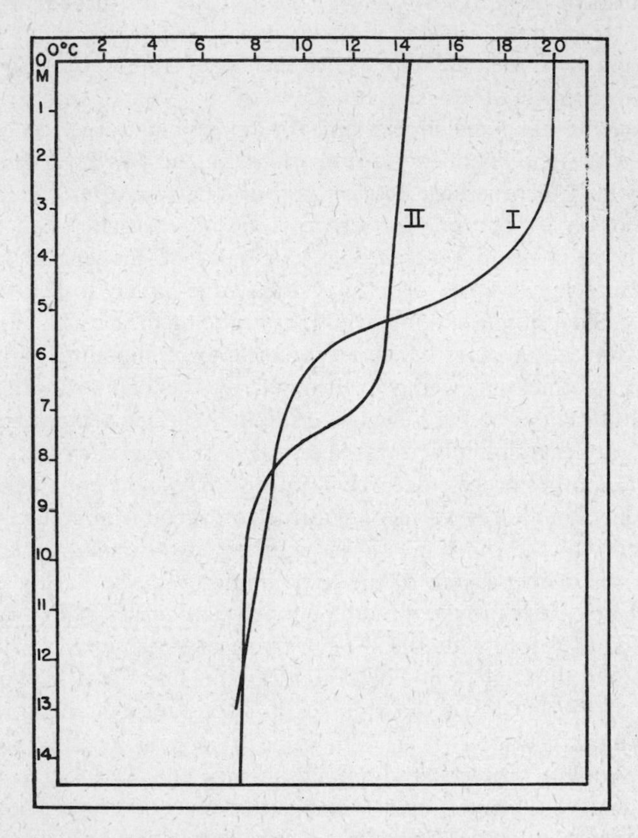

FIGURE 136. Linsley Pond, temperature curves before and after the hurricane of September 21, 1938, to show extreme effective stability below 10 m. I, September 16, 1938; II, October 13, 1938. (Hutchinson.)

that of maximum density, the stability imparted by a given change of temperature below 4° C. being much smaller than at a higher temperature. In any case, one may expect a relatively long period of autumnal circulation until the lake is well below 4° C. before freezing and winter stratification are established. In Furesø, Brönsted and Wesenberg-Lund (1911) found circulation on January 19, 1909, at 0.9° C.; the lake froze four days later. In Lake George, Langmuir (1938) observed the circulation to continue till the lake is at about 1.2° C. In Lake Vetter, circulation at about 0.3° to 0.4° C. seems possible (Liljequist 1941). Lake Mendota appears to remain homoiothermal till its temperature is about 2° C. Lake Kawagutiko (Yoshimura 1936c), a lake 12 m. deep, circulated till the water was at

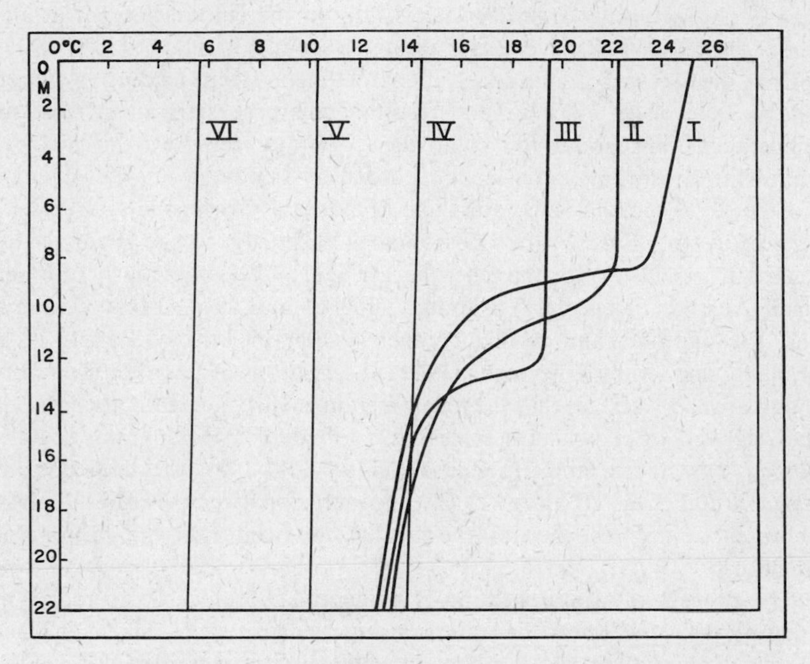

FIGURE 137. The cooling of Lake Mendota, temperature curves: I, July 30, 1906; II, August 30, 1906; III, September 27, 1906; IV, October 11, 1906; V, November 2, 1906; VI, November 24, 1906. Note prolonged autumnal circulation at temperatures falling from 14° C. (After Birge and Juday.)

about 2.5° C. In the Obernberger See at 1590 m. in the Austrian Alps, Leutelt-Kipke (1934) observed complete circulation at 3.4° C. in November. It is thus usual to find that in any dimictic lake that freezes, the main mass of water covered by ice will lie well below the temperature of maximum density.

The process of freezing generally depends on a sudden loss of heat at the surface during a still night; a stratification of definite but low stability is set up in the top few meters. When an ice cover forms, this inverse stratification is preserved, the water in contact with the ice being at 0° C., while that below a depth of 1 or 2 m. exhibits the temperature of free circulation just before freezing (Fitzgerald 1895, Yoshimura 1936c). The time at which this happens will depend partly on the temperature of the circulating water; a deep lake will take longer to lose heat than will a shallow one. It will also depend on the probability of the surface being undisturbed during a cold night, which will be greater in a lake of small than of large area. In some regions, small and shallow lakes and ponds may freeze regularly, large deep lakes only in the coldest years if at all. The characteristics of the ice cover are discussed in detail later.

Once the lake is frozen, a series of characteristic events leading to winter heating take place. The general result of this process is an increase in temperature throughout the main mass of water remote from the ice. Such winter warming under ice is recorded by Langmuir (1938) for Lake George; by Birge and Juday for Lake Mendota; by Brönsted and Wesenberg-Lund for Furesø; by Rossolimo (1932a) for Lake Beloye; by Leutelt-Kipke (1934) for the Obernberger See, and less obviously in other small Austrian lakes; by Yoshimura (1936c) in Lakes Osima Oonuma and Kawagutiko; and by Hutchinson (1941) in Linsley Pond. It is probably implied by Liljequist's (1941) observations in Lake Vetter. It is thus reasonably certain that the phenomenon is fairly general. In some of the relatively shallow lakes studied by Fitzgerald, Rossolimo, Leutelt-Kipke, Yoshimura, and Hutchinson (Figs. 138, 178), the final bottom temperatures are well above 4° C., and occasional temperatures slightly above 4° C. are recorded at the end of the winter at the bottom of Mendota.[6]

The process of winter heating is evidently twofold. Solar radiation entering the ice will heat the water, raising its density. If this takes place marginally in shallow water, a body of water slightly denser than that at the same level in the central parts of the lake will be produced, and will flow down the slope of the basin toward the depths as a density current (Birge, Juday, and March 1928; Alsterberg 1930b, 1931). Since the main body of water is practically homoiothermal when the process starts, any heating will raise the density above that of the water at any depth, and the movement down the slope will continue to the deepest point in the basin. At the same time, heat stored in the mud will tend to be conducted into the water. If it enters pure water at 4° C., the density will fall causing instability and convectional mixing; but since the process is occurring at the mud-water interface, there may be considerable opportunity for simultaneous diffusion of solutes out of the mud, raising the density. A chemical density current can then be set up which, if enough heat can be conducted from the mud, will raise the temperature of the deep water well above 4° C. without causing instability. As these currents run, they will displace water upward, warming the whole water mass and in some cases raising its concentration of solutes. It is probable that the movement of dissolved matter from the mud will take place mainly from the deeper

[6] Church (1942) even records 4.25° C. in the deepest water of Lake Michigan on one occasion when most of the surface was at about 4° and the marginal water cooler and bearing ice. This may represent a transitory instability.

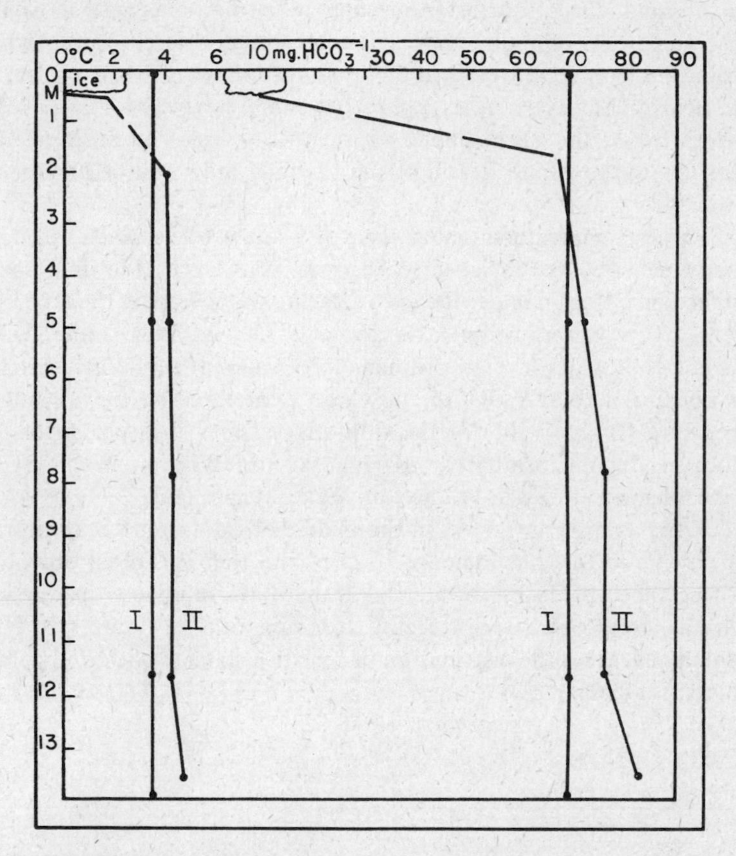

FIGURE 138. Linsley Pond, temperature and bicarbonate (HCO_3 mg. per liter): I, during late autumnal circulation December 14, 1935; II, at the end of winter stagnation under ice, March 7, 1936. The dilution of the surface water in II is due to dilute melt water having run in under ice. Note increase in temperature above 4° C. at bottom, and concomitant rise in bicarbonate. (Hutchinson.)

parts of the slope, where the oxygen is being depleted and the mud has developed a reduced surface. In the case of the Obernberger See, studied by Leutelt-Kipke, there is evidence of a small increment in alkalinity at the bottom at a temperature of 4.3° C., with an actual O_2 deficit of 42.3 per cent. In the other known cases of bottom water markedly over 4.0° C. in holomictic lakes in winter, the oxygen deficit is probably always considerably greater.

As is indicated on page 507, Birge, Juday, and March concluded that in the winter heating of Lake Mendota, about one fourth of the heat comes from the mud, three fourths from solar radiation. Because most lakes have a cooler hypolimnion than Lake Mendota, it is probable that the contribution from heat stored in the mud in summer is less. The mud of Lake Beloye, however, takes up rather more heat than that of Lake Mendota, while the mean depth of the lake is less. In such cases the proportion of the winter heating from the mud may well be greater than one fourth.

Anomalous temperatures under ice. If a thaw takes place during the period when the lake is ice-covered, melt water very low in dissolved substances may run in under the ice, replacing water leaving the lake by the effluent. This water, being little above 0° C. and very dilute, will be considerably less dense than ordinary lake water of moderate dissolved-solid content. Solar radiation now can penetrate the ice, raising the temperature (Birge 1910c) of the dilute layer until it is well over 4° C. without producing instability. Koźmiński and Wisniewski (1935) have recorded temperatures in studies on Lake Wigry (Fig. 139) and other Polish lakes as high as 7.5° C. in the middle of the top meter of unfrozen water, owing to this phenomenon. They find that it is most obvious, as would be expected, in hard-water lakes that have relatively opaque water which has circulated before freezing at temperatures below 3° C. In all cases it appears that the distribution of densities, in spite of the temperature anomaly, is stable.

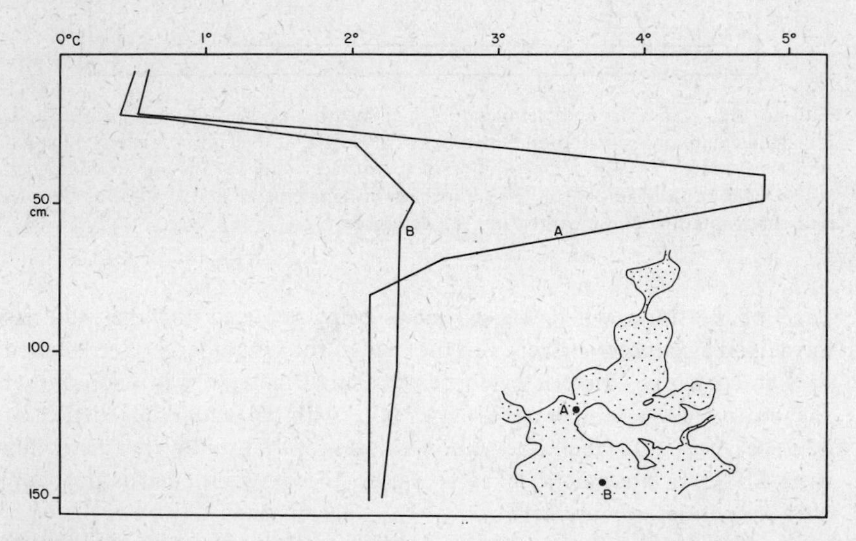

FIGURE 139. Anomalous temperature curves immediately below ice in Lake Wigry, March 26, 1933: *A*, marginal; *B*, center of lake. (After Koźmiński and Wisniewski.)

Temperatures at the water-air interface. Woodcock (1941) has measured the temperatures of a number of fresh, brackish, and salt waters on Cape Cod with a special thermometer permitting readings at depths of a few millimeters. So long as the air temperature was less than that of the water, as was usually the case in September and October, the temperature at 3 mm. was from 0.1° to 1.0° C. less than at 20 mm. This inverse gradient produced instability, so that when air moved over the water a series of longitudinal streaks could be produced by scattering lycopodium powder on the water surface. The powder was carried into the streaks with great rapidity, so fast that its motion could hardly be followed with the eye. The streaks, normally separated by a few centimeters, were inconstant, changing or disappearing in a few seconds. They certainly represent cold water converging and sinking, and form a small-scale model of the streaks observed by Langmuir in the wind drift. On the few occasions when the air was warmer than the water, the inverse thermal gradient was not observed.

Though the observations are consistent with transport of sensible heat from water to air, it is reasonable to suppose evaporation also to be involved.

In the quietest bodies of water there is evidence that evaporative cooling at the surface can produce enough instability and consequent downward movement of water to promote exchange of gases across the air-water interface (Adeney, Leonard, and Richardson 1922; see also page 589). McEwen (1929) has attempted to incorporate such evaporative cooling into a general theory of heat transfer across a lake surface, but as is pointed out on page 466, his mode of procedure appears not to be valid.

The deepwater temperatures of very deep dimictic lakes. The temperature of maximum density of water is not a constant, but varies with the pressure. The available direct and indirect determinations of the temperature at high pressures in the laboratory are somewhat discordant. Strøm (1945), who has reviewed the whole of the data, considers that the work of Pushin and Grebenshchikov (1923) alone gives any satisfactory information. This information, however, only refers to the pressure, 600 atm., at which the temperature of maximum density is 0° C. No certain experimental data exist to indicate whether the fall in the temperature of maximum density from 3.94° C. at 1 atm. to 0° C. at 600 atm. can be legitimately represented by linear interpolation.

It has long been known that in many very deep lakes the hypolimnetic temperatures may lie well below 4° C. The matter was investigated as long ago as 1883 in Mjøsa (Schiøtz 1887), and casual observations have been made by other workers. All these results have been summarized by Strøm,[7] who has added many accurate data of his own. The observed lowering of the deepwater temperatures in very deep lakes is to be expected since, if a lake is homoiothermal in the spring at any temperature below 3.94° C., the deep water is nearer the temperature of maximum density appropriate to its pressure than is the surface water. Any slight acquisition of heat at the surface will therefore tend to produce convectional mixing only down to a certain level, dependent on the amount of heating. A stable stratification can be set up in this way, even though the temperatures are all below 3.94° C. and are falling with depth.

The nearness of the approach to the temperatures of maximum density in the deep water will depend on several factors, the most important being those that control the temperature of the lake in winter and spring, for all heat loss must occur from the surface of the lake and often the mixing of a rather limited amount of cold water near 0° C. from the surface with the deeper water of the lake will be inadequate to cool this deep water to the temperature of maximum density appropriate to the pressures prevalent in the lower parts of the lake. Moreover, the stability conferred by any stratification in the neighborhood of 4° C. is not great, and vertical mixing will easily obscure the effects of pressure on the temperature curves of very deep but highly disturbed lakes.

Nevertheless, some effect may nearly always be observed in dimictic lakes over 100 m. deep, and Strøm has shown that when the available temperature curves for a number of very deep lakes are drawn on the same paper (Fig. 140), an envelope line can be drawn cutting 3.94° C. at 0 m., 3.82° C. at 100 m., and 3.50° C. at 400 m., so that all points on all summer temperature curves lie to the right of this line, though several temperature curves, notably that for Mjøsa, approach the envelope very closely. The envelope is, moreover, similarly approached from the cold side by a vernal temperature curve for Tyrifjord, representing a time when the density of the shallow water of the lake was still increasing with increasing heat intake. It is evident that the envelope line drawn by Strøm indicates the relationship of the temperature of maximum density to pressure, and gives in effect the lowest temperature that is possible in a deep lake during summer stratification, or alternatively the highest possible during winter stagnation or the early stages of spring circulation. The envelope indicates a far more rapid rate of change in the lower part of the pressure range than is exhibited by the straight line joining the two well-established laboratory points, which are 3.94° C. at 1 atm. and 0° C. at 600 atm. Strøm's envelope is evidently part of a nonlinear curve,

[7] Yoshimura (1936b) has given special attention to the bottom temperatures of very deep Japanese lakes, but they are mostly monomictic.

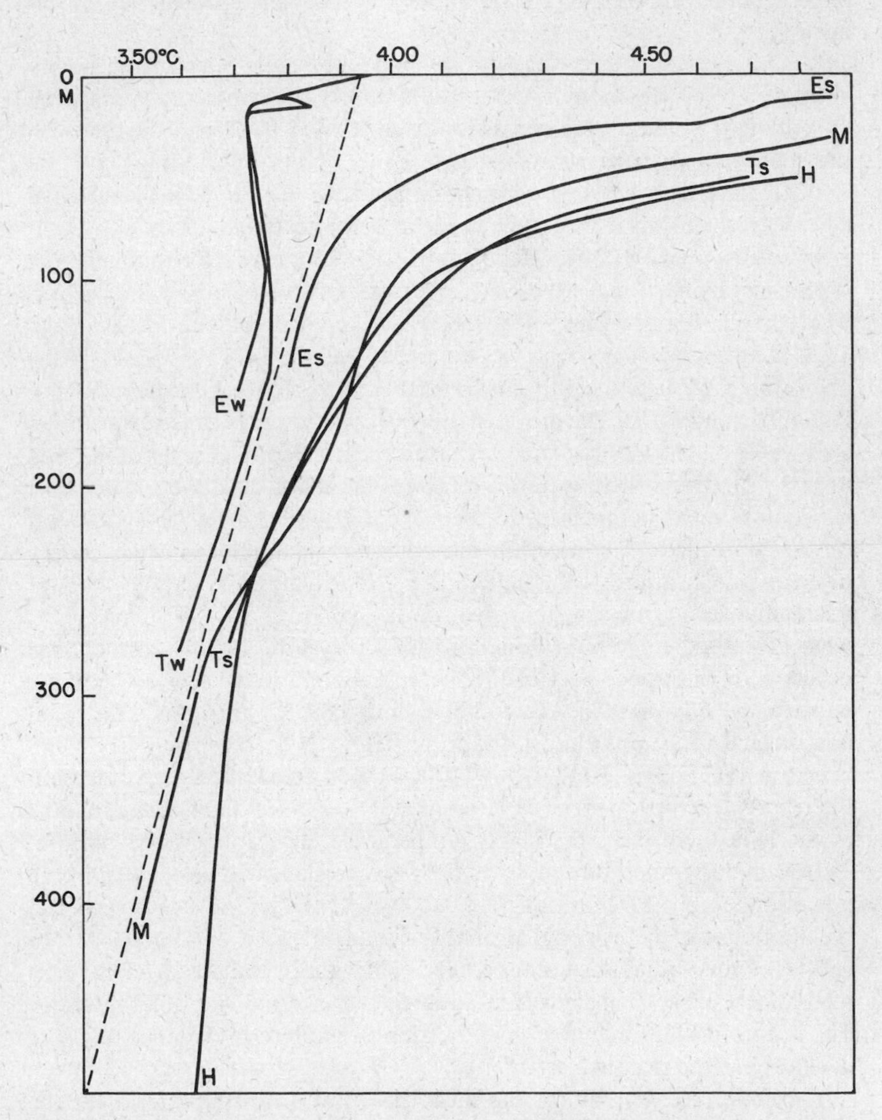

FIGURE 140. Temperature curves for very deep dimictic lakes. Tw, Tyrifjord, May 25, 1933, just before vernal circulation; Ew, Eikeren, May 24, 1935, just before vernal circulation; Ts, Tyrifjord, August 27, 1930, in summer stratification; Es, Eikeren, June 12, 1935, in summer stratification; M, Mjøsa, July 9, 1936; H, Hornindalsvatn, August 25, 1931. Envelope line (broken), not cut by these or any other recorded European temperature curves, gives the best estimate of the temperature of maximum density as a function of pressure. (After Strøm.)

but over the greater part of the range between 0 and 600 atm. no data either from nature or from the laboratory exist for the construction of this curve.

It is interesting to note that the envelope drawn by Strøm from a consideration of all the modern data hardly differs from the curve computed by Schiøtz from data obtained as long ago as 1851 by Grassi for the compressibility of water between 0° and 4° C.

The vertical distribution of temperatures in Lake Baikal. In Lake Baikal (Vereščagin 1937) the situation is more complicated than in any other lake, on account of the great depths involved. In the top 100 m. there is an ordinary seasonal cycle of the kind to be expected in any large dimictic lake. The surface temperature is variable, of course, reaching occasionally 19.1° C. in the Selenga Delta; it can rise at least to 14.5° C. locally in the open water, though the July mean temperature is 9.0° C. (Suslov 1947). As has been indicated, the prevalent northwest winds may cause exposure of cold water on the western shore, derived from a depth of several hundred meters. Mixing of this water into the epilimnion is no doubt responsible for its low mean temperature. Between 200 and 300 m. there is what Vereščagin called the zone of the mesothermic maximum, in which during inverse stratification the temperature lies at approximately that of maximum density or a little higher, about 3.6° to 3.8° C. Below this is what he calls the perennial zone, in which very little temperature change occurs. In the upper part of this perennial zone there is a very minute seasonal cycle, about 0.20° C. at 300 m. and 0.09° C. at 600 m. Seasonal and vertical variations almost disappear below this. The range at 800 m. is between 3.38° and 3.41° C.; at 1200 m., 3.34° to 3.38° C. At 1600 m. the recorded temperatures are 3.16° and 3.19° C. In the extreme depths of the lake there is a very slight temperature inversion; at the bottom, in 1698 m., a temperature of 3.23° C. was recorded. In passing from the minimum temperature of 3.18° C. at 1600 m. to the temperatures recorded on the bottom, there would probably be an increase of about 0.01° to 0.02° C. due to adiabatic compression. The rest of the observed increase of 0.03° to 0.04° C. may well be due to the internal heat of the earth. The resulting small inversion may be stable, as there is evidence of a rise in silicate in the deep water.

The most recent value for the mean heat flux from the interior of the earth is 1.23×10^{-6} cal. cm.$^{-2}$ sec.$^{-1}$ or 38.7 cal. cm.$^{-2}$ yr.$^{-1}$ (Bullard 1954). If the mean temperature of a water column 200 m. above the deepest point in the lake were 0.02° C. above the minimum temperature, 400 cal. cm.$^{-2}$ would be present in excess of the heat storage implied by that minimum. Given the mean heat flux of 38.7 cal. cm.$^{-2}$ yr.$^{-1}$, this heat would accumulate in about ten years. It is possible that the observed gradient represents a steady state between a decrease in density as heat is delivered from the earth and an increase in density as dissolved material is added from the bottom and from mineralized falling plankton. If the deepest part of the lake is really otherwise undisturbed, it is evident, in view of the very small difference in temperature between the bottom and the zone of minimum temperature, that a slow convective circulation due to the internal heat of the earth must be occurring. The high oxygen content of 9.12 mg. per liter at 1698 m., however, perhaps suggests more mixing with superficial strata than the temperatures appear to indicate.

Cold monomictic lakes. There are apparently no really good data on the thermal cycles of cold monomictic lakes. A certain number of such lakes certainly occur at elevations over 2100 m. in the mountains of central Europe. They are apparently always fed by newly melted glacier water, and in many cases the waters of the lake are perennially in contact with ice. In north Sweden, Ekman (1904) noted a few lakes on the Truollo Plateau in the Sarek Mountains, at about lat. 62°30′ N. and an altitude of 1300 m., which were said only to become ice-free in the warmest summers; the same may be true of Ororotse Tso and a few smaller lakes examined by Hutchinson in Indian Tibet at just under 5300 m. When such lakes are ice-free, they may always circulate under 4° C., but there is no evidence that this is the case. Far within the Arctic Circle, Greeley (1888) noted lakes on Grinnell Land at 82° N. which were apparently ice-free for a month or two; the ice on Lake Hazen appears to have started to break up by May 22, 1915 (Ekblaw 1918), which is very extraordinary. There are, however, no temperature records. Anneks Sø at an altitude of 40 m., lat. 77° N., in northeast Greenland is supposed by Trolle (1913) to circulate at 2° C. in the summer, the bottom temperature under ice being 1.98° C., but no temperatures from the ice-free period are known. Some of these lakes may be monomictic, but it is more likely that at least the shallower ones take up enough heat in the surface waters to reach temperatures above 4° C. and are therefore formally dimictic, though of the subpolar type with irregular summer circulation periods.

The general impression left by the very fragmentary data is that cold monomictic lakes are rare even at very high latitudes, and that they usually owe their thermal character to immediately adjacent ice rather than to the limitation of the circulation period to a few weeks in summer. It must be remembered that over the whole twenty-four-hour period more radiation may be delivered per unit area at midsummer in the Arctic than in temperate latitudes, while at very high altitudes radiation is intense unless the cloud cover is heavy. The conditions during the supposed single circulation period are thus quite unlike those during the circulation periods of ordinary temperate dimictic lakes.

Perennially frozen or amictic lakes. It is known that in the Antarctic, notably in the vicinity of Cape Royds on Ross Island (lat. 77°31′ S.), there are well-defined lakes that seem to be permanently frozen (Murray 1909, Shackleton 1909). Blue Lake was found to be about 8 m. deep, but the upper 6.4 m. was occupied by ice. Lakes in this region less than 1.5 m. deep may become open during summer, but under present conditions it is inconceivable that the ice on Blue Lake would ever melt. In one locality in this region, ice lay over frozen peat in a lake basin (David and Priestley 1909). It is most remarkable that even the permanently frozen lakes had a benthic flora of undetermined microscopic species, and a fauna of bdelloid rotifers. A full study of such localities, including radiocarbon dating of peat deposits, would obviously be of extraordinary palaeoclimatological significance.[8]

Circulation in lakes in equatorial regions; polymictic and oligomictic lakes. Within the tropics there are two annual maxima in solar radiation, but in general the variation in the radiation flux is small, and factors other than solar radiation are likely to be of major importance. Two extreme situations apparently can develop.

In lakes of great area, moderate or little depth, in windy regions of low humidity, or at great altitudes (Löffler, unpublished) no persistent thermal stratification develops. A considerable amount of superficial, or at great altitudes deep, heating may occur by day, followed by loss of heat and complete mixing at night. This type, in its extreme high-altitude expression, is termed *polymictic* by Hutchinson and Löffler (1956).

In lakes of small or moderate area or of very great depth, or in regions of high humidity, a very small temperature difference between surface and bottom suffices to maintain stable stratification, and the lake may circulate only at very rare irregular intervals when abnormal cold spells occur. This type is termed *oligomictic* by Hutchinson and Löffler.

Various intermediate conditions can of course occur; in these, annual circulation may often occur but its incidence is likely to be determined by seasonal variations in windiness and perhaps humidity, as much as by solar radiation or air temperatures. There is some evidence from Lake Tanganyika that a sudden fall of cool rain may reduce the stability sufficiently to induce local mixing well into the otherwise stagnant water below the thermocline (Kufferath, in Capart 1952).

All possible conditions seem to be found in equatorial Africa. In Lake Naivasha (lat. 0.45′ S., long. 36°24′ E.), a shallow lake about 20 m. deep, at an altitude of 1900 m., there seems (Beadle 1932) to have been a hint of chemical stratification in November, but in February, during the

---

[8] Löffler (unpublished) has found one amictic lake in the Andes.

hot, dry, windy season, the lake must have circulated in the early hours of every morning, as the result of the instability induced by nocturnal cooling. Lake Rudolf (4° N., 36° E.), a deeper and much more extensive lake, seems to have circulated irregularly at night during April. Worthington (1930) believed Lake Albert also to circulate completely. Lake Victoria, according to Worthington (1930), circulates completely to 20 m. every night, and there is probably some mixing of surface waters into the deeper layers going on continually, at least in the open water. These three lakes appear to be more or less polymictic.

Lake Mohasi in Ruanda, a lake only 13.8 m. deep, at an altitude of about 1450 m., was sharply stratified (Damas 1955) in the rainy season, beginning in May, with the top 5 m. at 24.15° C., falling to 22.45° C. at 11 m. The hypolimnion was anaerobic. During the dry season in January, irregular circulation took place, and the lake may be regarded as warm monomictic.

The other large deep lakes that have been studied all appear to possess a perennially more or less stagnant deep layer, often stabilized by a slight increase in dissolved matter. They are in fact oligomictic, with a strong tendency to meromixis. The small deep lakes of Ruanda, namely Luhondo ($z_m = 68$ m.) and Bulera ($z_m = 173$ m.) studied by Damas (1955), perhaps fall in the same category as Edward (Beadle 1932), Nyasa (Beauchamp 1953a), Tanganyika (Beauchamp 1939, Capart 1952), and Kivu (Damas 1937), in having deep and stable stagnant layers. But in the case of Bulera there is evidence of continual replacement of the deep water by water from a cold influent. Kivu, which is highly mineralized, presumably is the most meromictic of all these lakes, and owes its characteristics as much to its chemistry as to its tropical latitude.

Beauchamp thinks that Lake Nyasa may occasionally circulate. It is reasonably certain that irregular mixing of considerable amounts of stagnant water into the freely circulating layers must occur. In 1939 the temperature below 200 m. was about 22.1° C. During August 1946, surface temperatures as low as 20.75° C. were recorded at the south end of the lake. A month later an immense bloom of *Anabaena* developed, thicker than had been observed within living memory. This event strongly suggests some unusually great circulation and consequent fertilization of the surface waters. Against the view apparently held by Beauchamp, though in a tentative manner, that such circulation was total, may be set the lack of observations of extensive fish deaths, which would be expected on an unprecedented scale if the enormous mass of H_2S present in the deep water had really been mixed freely through the lake.

The immensely rich faunas of both Nyasa and Tanganyika, in fact, strongly suggest that these lakes have never fully circulated in the ordinary way, since they developed large monimolimnia. The more or less permanently stagnant water begins at a somewhat higher level in Tanganyika than in Nyasa, and there is some biological evidence (Brooks 1950) suggesting that there has been an encroachment of this water on the freely circulating zone since the discovery of the lake. The great depth of the lake raises certain special problems, which will be discussed in a later section.

In central and northern South America several lakes have been studied. Gessner (1955) finds in Lake Valencia in Venezuela (lat. 10°10′ N.) practically no thermal stratification, with evidence of nocturnal circulation affecting the water at 20 m. At a greater altitude (3600 m.) the Lago Mucubaji had a bottom temperature in February of 10° C. and rather feebly developed stratification in the top 10 m. Deevey (1957) found Lake Atitlan (lat. 14°40′ N.) to have in August a temperature of 24° C. at the surface and 20° at the bottom. There is evidence of full circulation in February, during the dry season. The total heat delivered in the period between February and August is comparable to that gained by a temperate lake, as will be shown. These tropical American lakes are clearly comparable to the less permanently stratified lakes of Africa in being either polymictic or warm monomictic.

The largest body of thermal data for any equatorial region, that of Ruttner (1931) for Indonesia, provides almost exclusively examples of strikingly stratified oligomictic lakes, possibly on account of the more humid climate, with a mean relative humidity usually between 80 and 90 per cent. The surface temperatures of these lakes vary from 21.3° to 31.4° C., the bottom from 20.1° to 27.0° C. The bottom temperatures are almost linearly dependent on the altitude from 10 m. to 1531 m. above sea level, and presumably represent the coldest temperatures at the surface during the recent histories of the lakes. One lake, Danau Bratan in Bali, at 1231 m., was clearly not stratified in June 1929; some degree of thermal stratification was found in all the others. In general, the smaller lakes, area less than 2 km.^{2}, showed very striking metalimnia at depths of between 3 and 10 m., involving temperature falls of 2° to 3° C. Larger lakes, area of the order of 100 km.^{2}, had less striking metalimnia below 10 m., involving falls in temperature of 1° to 2° C. The largest lake, Lake Toba, in its main basin exhibited a well-defined epilimnion, a feebly marked metalimnion from 25 m. to 50 m., and a very gentle fall in temperature in the deep water, the whole range being from 26.4° to 23.9° C. It is probable that circulation is quite irregular, depending solely on the incidence of short spells of cool weather, which may occur in some years but not in others. In the less stably stratified lakes of large area there is doubtless continual turbulent mixing of surface water into the depths, but the hypolimnetic oxygen content of the small lakes suggest that in them such a process is negligible.

The temperatures of Lake Tanganyika. Lake Tanganyika offers certain special problems owing to its enormous depth. Some of the earlier investigators (Stappers 1915, Marquardsen 1916) made temperature observations, but it was not until the investigations of Beauchamp (1939) that it became clear that the whole volume of the lake below about 200 m. is anaerobic and does not circulate. This fact, which is in line with what is now known about other deep equatorial lakes, is essential for an understanding of the thermal regime of the lake.

The most important series of temperature observations are those of Capart (1952), at this time published only in part. The surface water temperatures are very variable; values as low as 23.3° and as high as 29.5° C. have been recorded. In shallow water, as at Capart's station 36 near Moba, in 40 m., the diurnal variation may be from 26.30° to 29.50° C.; in deeper water the corresponding variation is from 26.30° to 26.90° C. There is evidence of some nocturnal mixing at least to 80 m. Seasonal variation is confined to the top 200 to 250 m. Below 200 m., where the mean temperature is 23.45° C., there is a fall to a minimum between 500 and 800 m. of 23.27° C. At depths below 1000 m. the temperature is found to have risen very slightly, to 23.32° to 23.35° C., implying an increase of 0.06° C. in about 600 m. The entire body of water below about 200 m. is definitely richer in dissolved salts than is the surface layer, and is entirely devoid of oxygen; little oxygen is present below 150 m. at any season.

Though the vast mass of water below 200 m. appears to be stagnant, it is believed by Kufferath (1952) actually to undergo a very slow exchange with the surface layers. The first mechanism involved in this exchange is the discharge of warm but saline water from Lake Kivu by the Ruzizi River at the north end of the lake. This water is denser than the surface water and moves along the gentle slope of the shallow north end of the lake, bringing heat, mineral matter, and oxygen into the monimolimnion. The second mechanism is the falling of heavy showers of cool rain, which (Kufferath, see Capart 1952) may cause some mixing down to 400 m. The temperature inversion below 800 m. is presumably due to adiabatic heating, which would be expected to cause a rise of about 0.09° C. in a water column 600 m. deep and of uniform potential temperature. The flux of heat from the interior of the earth seems therefore to produce no recognizable effect on the temperature curve and must be distributed upwards by feeble movements, providing a further argument that the water of the monimolimnion is not entirely stationary.

DEEP WATER TURBULENCE AND THE MECHANISMS OF HYPOLIMNETIC HEATING

In a water column in an infinitely large lake, if Θ_z be the total heat that has passed through unit area of any plane at depth z and θ_z be the temperature at z, then

$$\frac{d\Theta_z}{dt} = - c_p \rho A_{\theta_z} \frac{d\theta_z}{dz} \tag{26}$$

A similar expression can be written for the transport of any other conservative property. Within the limits of error of observation, the density ρ and the specific heat c_p may be taken as unity and the coefficients of eddy conductivity and diffusivity as identical.[9] We may therefore write

$$\frac{d\Theta_z}{dt} = - A_z \frac{d\theta_z}{dz} \tag{27}$$

Hypolimnetic values of A_z obtained from equation 27. In the epilimnion the effects of direct heating by solar radiation and of cooling by evaporation and back-radiation cannot be neglected. Moreover, the diameter of some of the eddies involved in the transport of heat in the epilimnion appears to be of the order of magnitude of the epilimnetic thickness. It is therefore not surprising that, in spite of an attempt by McEwen (1929), no satisfactory comprehensive theory of the heating of the epilimnion has been developed.

In the thermocline and in the hypolimnion it would at first sight appear reasonable to apply equation 27 to data on temperature and heat transport, so obtaining values of A. Several investigators, notably Schmidt (1925) and Mortimer (1941–1942), have in fact proceeded in this way, simply computing the rate of increase in the heat content of a column of water of unit area below a certain depth over a designated period and dividing by the value of the mean thermal gradient at the depth under consideration. The values for certain lakes as obtained by this simple method by Mortimer are given in Table 48.

As Mortimer points out, these values of A are clearly correlated with the areas and depths of the lakes considered. Since the areas and depths are themselves highly correlated, it is reasonable to suppose that the primary relationship is between the coefficient and area, or rather the fetch of the wind which supplies the energy for turbulent transport. Mortimer regards his values of A as rough approximations. As will be indicated, they are almost certainly too great, though the excess may be significant mainly in

---

[9] Some authors (McEwen 1929, Hutchinson 1941) have used μ^2 to indicate coefficients of turbulence expressed as $m.^2$ $month^{-1}$. Such a practice has little in its favor and may be abandoned.

TABLE 48. *Coefficient of eddy conductivity computed from heat transfer in the hypolimnia of various lakes*

	Area, km.^{2}	Maximum Depth, m.	Depth at Station, m.	Depth Considered, m.	Period Considered	A_z, g. cm.$^{-1}$ sec.$^{-1}$ ($\times 10^{-2}$)
Holsfjord, Norway	121	295	295	100	26.vi–26.ix	310
Geneva, Switzerland (main basin)	503	303	285	100	21.vi–23.x	190
Loch Lomond, Scotland	71	195	185	56	22.ix–14.xi	53
Windermere, England (North basin)	8.2	67	55	30	14.vi–13.ix	39
(South basin)	6.7	44	30	15	6.vi–20.ix	9
Mendota, Wisconsin	39	25.6	23.5	12	15.vi–15.viii	7
Maxinkuckee, Indiana	7.5	27	26	18.5	17.vii–28.ix	7
Kizakiko, Japan	1.4	29	25	15	vi–viii	5
Lunzer Untersee, Austria	0.68	34	32	20	29.iv–29.vi	5
Esthwaite Water, England	1.0	16	14	12	vii–viii	3
Schleinsee, Bavaria	0.15	11.6	11.6	11	iv–viii	2

the cases of the smallest lakes. An arbitrary element would appear to enter into the table in the selection of the depths to which the calculations refer. In spite of these reservations, the immense variations in the value of A_z recorded in the table do emphasize the difference in the degree of disturbance to be expected when the hypolimnion of a large deep lake is compared with that of a small shallow lake.

In the simple method of calculation used by Schmidt and by Mortimer there is an implicit assumption that no heat has entered the lower part of the water column under consideration by any mechanism other than vertical turbulent transport. The values of Θ used in the procedure are in fact heat budgets calculated for unit area of planes at various depths by the method employed by Forel. As is indicated in a later section (page 493), this method of calculation gives impossibly high results when applied to the summer heat income of Lake Ladoga and certain other bodies of water, and therefore appears to be invalid. The correct procedure, applicable to any plane above the depths at which lateral heating is significant, would be to determine the rate of increase for the entire body of water and mud below the plane in question and then express this per unit area of the plane. In practice, the heating of the mud must usually be neglected and the heat budget to be used is that which is designated here as Θ_{bz}, or the Birgean heat budget per unit area at depth z. The correct expression for A_z if lateral heat transport into the bottom of the column is suspected, is therefore

$$A_z = -\frac{\partial \Theta_{bz}}{\partial t} \cdot \frac{\partial z}{\partial \theta_z} \tag{28}$$

and this is probably valid only at depths above those at which lateral transport of heat becomes significant. Unfortunately there is no way of telling, from any of the quantities involved, over what range of depths the use of this expression is permissible.

The eddy conductivity coefficient derived from temperature changes. If, instead of the transport of heat across a plane, the temperature changes at a series of depths be considered, differentiation of equation 27 with respect to depth gives

$$\frac{\partial \theta_z}{\partial t} = \frac{\partial A_z}{\partial z} \cdot \frac{\partial \theta_z}{\partial z} + A_z \frac{\partial^2 \theta_z}{\partial z^2} \tag{29}$$

This equation is obviously of little value if A_z varies with depth; but if, over a given range of depths, A_z is a constant A, then

$$A = \frac{\partial \theta_z}{\partial t} \Big/ \frac{\partial^2 \theta_z}{\partial z^2} \tag{30}$$

McEwen has pointed out that in general the temperature curve in the thermocline and the upper part of the hypolimnion of a thermally stratified lake can be fitted to an expression of the form

$$(\theta_z - C) = C_1 e^{-az} \tag{31}$$

$$\ln(\theta_z - C) = \ln C_1 - az \tag{32}$$

where C_1, C, and a are constants to be determined empirically. C_1 expresses the virtual surface temperature ($z = 0$) which would be observed if the temperature curve of the hypolimnion were extrapolated to the surface; C represents the virtual temperature of isothermal circulation prior to the development of stratification in the early summer; a determines the temperature gradient. Whenever the temperature curve can be legitimately expressed by this equation and when A is also constant

$$\frac{\partial \theta}{\partial t} = Aa^2 C_1 e^{-az} \tag{33}$$

$$= Aa^2(\theta_z - C)$$

If C is determined from the distribution of temperatures by the method of least squares[10] and $(\theta_z - C)$ is plotted against depth on the logarithmic axis of semilogarithmic paper, the observed values will fall on a straight line over the range of depths for which equation 31 legitimately expresses the temperature curve. If the observed values of the rate of change of temperature with time $\partial\theta_z/\partial t$ for each depth are now plotted against depth on the same sheet of semilogarithmic paper, the points representing $\partial\theta_z/\partial t$ will be found to fall on a parallel straight line if A is constant and equation 33 is valid. The straightness and parallelism of the two lines giving log $(\theta_z - C)$ against z and log $\partial\theta_z/\theta t$ against z thus together constitute a *criterion of the validity of the assumptions that the fall in temperature is exponential and the coefficient of turbulence constant* within the depth zone under consideration.

In Fig. 141 semilogarithmic plots of $(\theta_z - C)$ and $\partial\theta_z/\partial t$ against z are given for Lake Mendota, using mean values of θ over a number of years, for the heating period from the third week in June to the second week in August. The plot of $(\theta_z - C)$ shows that over practically the whole of the region below the thermocline, equation 31 may be legitimately used. The plot of $\partial\theta_z/\partial t$ can be used to divide the lake into three depth zones. The zone from 0 to 9 m. corresponds to the epilimnion and metalimnion above the thermocline. Here equation 33 does not hold, and the rate of heating varies somewhat irregularly with depth. The second zone, from 9 to 16 m. corresponds to the lower part of the metalimnion and upper part of the hypolimnion. Equation 31 clearly expresses the fall of temperature with depth with considerable accuracy; the rate of heating expressed by equation 33, moreover, falls exponentially and parallel to $(\theta_z - C)$. In this second depth zone the assumption of a constant value of A over a range of depths is therefore valid. The third depth zone extends from 16 m. to the bottom. The temperature curve is expressed fairly well, except near the bottom, by equation 31, but the rate of change of temperature with time now approaches a constant value as the depth increases, and the criterion of the validity of assumptions that A is constant is no longer satisfied.

Since it is evident that the physics of heating are different in the second and third zones, it has proved convenient to use the term *clinolimnion* for the region of exponential fall in $\partial\theta_z/\partial t$, and the term *bathylimnion* for the deep third zone, where $\partial\theta_z/\partial t$ tends to be greater and to assume a more constant value than extrapolation from the clinolimnion would suggest.

The general phenomena implied by the results of this treatment of the mean summer temperatures of Lake Mendota are exhibited by at least

[10] Full directions can be obtained in McEwen (1929) and Hutchinson (1941). The whole procedure is not recommended unless weekly series of temperatures at 1-m. intervals are available over at least one whole summer heating period.

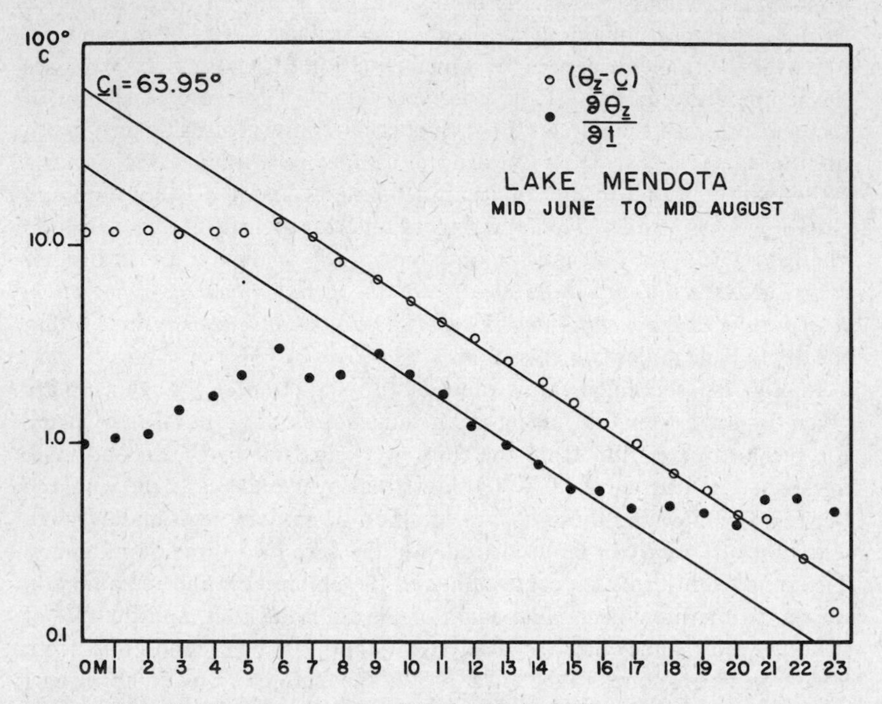

FIGURE 141. Semilogarithmic plots of $(\theta_z - C)$ and of $\partial\theta_z/\partial t$ against z, for the mean temperatures and temperature changes over a series of years in Lake Mendota, during the period from the third week of June to the second week of August. $C = 11.91°$ C. (After Hutchinson.)

some other thermally stratified lakes. The diagram given in Fig. 142 was constructed from the mean temperatures and mean rates of increase in temperature in Linsley Pond for the heating period of June to August 1937. This diagram exhibits the essential features of that constructed for Lake Mendota, save that the divergence of the temperatures from the curve of equation 32 is greater in the bathylimnion of Linsley Pond than in that of Lake Mendota. Much less satisfactory data from Lake Quassapaug seem to imply the same relationships as are exhibited by the other two lakes. Ricker's (1937) data for Cultus Lake, during the period from the beginning of May to the end of August, for the layer from 20 to 30 m. have also been analyzed. Above 20 m., heating by solar radiation is certainly significant in this very transparent lake.

Newcombe and Dwyer (1949) are so far the only authors other than the present writer who have used the method of analysis just outlined. For the layer from 2 to 5 m. in the mixolimnion of the meromictic Sodon Lake,

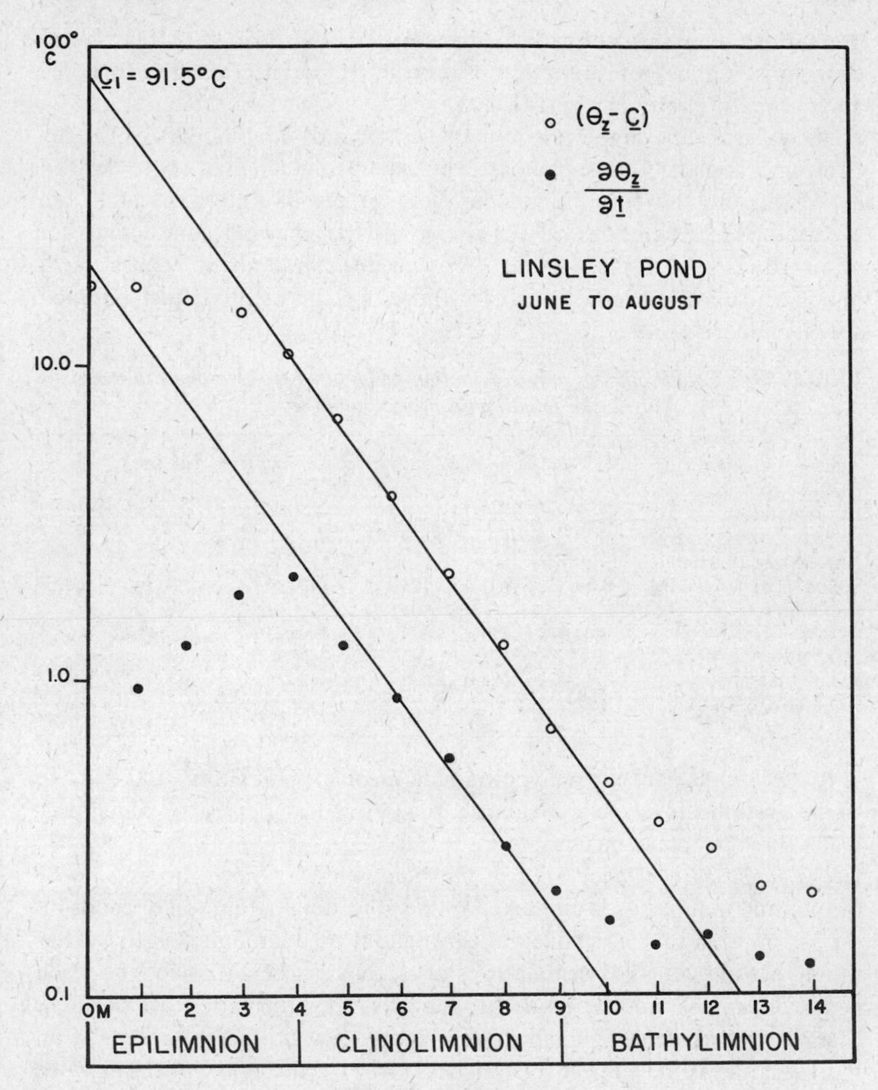

FIGURE 142. Semilogarithmic plots of $(\partial_z - C)$ and of $\partial\theta_z/\partial t$ against z, for the mean temperatures and temperature changes in Linsley Pond, June to August 1937. $C = 6.82°$ C. (After Hutchinson.)

Michigan, they obtain a good parallelism between semilogarithmic plots of $(\theta_z - C)$ and the rate of increase of temperature. Their resulting coefficient of turbulence may be a little high owing to direct heating from solar radiation. It must be remembered that the isothermal surfaces in any lake are not fixed horizontal planes but are continually undergoing oscillations and distortions by internal seiches. For this reason an enormous number of observations appear to be necessary before the diagrammatic clarity of Fig. 141 is achieved.

The values of the clinolimnetic coefficient of eddy conductivity of Linsley Pond and of Lake Quassapaug are very much smaller than those derived by Mortimer, and for this reason these values, if not those for Lake Mendota, must be corrected for the molecular thermal conductivity of water (0.12×10^{-2} c.g.s units). The mean clinolimnetic values of A obtained from the five lakes so far studied by the methods just outlined are given in Table 49.

TABLE 49. *Coefficients of eddy conductivity computed from temperature change in the clinolimnia of various lakes*

Cultus Lake (v–viii), 20–30 m.	$85.7 \times 10^{-2} - 0.12 \times 10^{-2} =$	85.6×10^{-2}	g. cm.$^{-1}$ sec.$^{-1}$
Lake Mendota (3rd week vi–2nd week viii), 9–16 m.	$2.52 \times 10^{-2} - 0.12 \times 10^{-2} =$	2.40×10^{-2}	g. cm.$^{-1}$ sec.$^{-1}$
Sodon Lake (iv–vii), 2–5 m.	$0.69 \times 10^{-2} - 0.12 \times 10^{-2} =$	0.57×10^{-2}	g. cm.$^{-1}$ sec.$^{-1}$
Linsley Pond (vi–viii), 4–9 m.	$0.33 \times 10^{-2} - 0.12 \times 10^{-2} =$	0.21×10^{-2}	g. cm.$^{-1}$ sec.$^{-1}$
Lake Quassapaug (vi–viii), 8–14 m.	$0.4 \times 10^{-2} - 0.12 \times 10^{-2} =$ (approx.)	0.3×10^{-2} (approx.)	g. cm.$^{-1}$ sec.$^{-1}$

Ricker thinks part of the hypolimnetic turbulence of Cultus Lake is due to the movements of sockeye salmon, the population density of which may reach three per cubic meter.

The supposed constancy of A in the clinolimnion. Since the validity of the method that has been employed in the preceding paragraph depends on the assumption that A is constant throughout the clinolimnion, the variation in the individual determinations of A_z that can be obtained from each depth cannot be considered as providing valid information about slight changes in turbulence. Each determination must rather be taken as an individual measure, subject to statistical error, of the constant mean value within the clinolimnion. It is of interest, however, to compare the vertical variation of A_z throughout the clinolimnion when calculations are made by different methods. For Lake Mendota, values (uncorrected for molecular conduction) obtained from individual values of $\partial\theta_z/\partial t$ by the use of the second part of equation 33 are given in the second column of Table 50; the values in the third column are obtained from the uncorrected first-order equation 27, as used by Mortimer. In the last column the mean stability at each clinolimnetic depth is entered.

It will be observed that the determinations from equation 33 do justify

TABLE 50. *Coefficients of eddy conductivity and stability in the clinolimnion of Lake Mendota*

Depth, m.	A_z, Equation 33, g. cm.$^{-1}$ sec.$^{-1}$ ($\times 10^{-2}$)	A_z, Equation 27, g. cm.$^{-1}$ sec.$^{-1}$ ($\times 10^{-2}$)	E_z stability, g. cm.$^{-3}$ cm.$^{-1}$ ($\times 10^{-6}$)
9	2.42	2.93	2.6
10	2.52	3.03	2.2
11	2.68	3.19	1.8
12	2.43	3.36	1.2
13	2.47	3.62	0.85
14	2.47	3.95	0.73
15	2.45	4.33	0.52
16	2.75	4.82	0.37

the assumption of a constant value of A throughout the clinolimnion of Lake Mendota, as is of course also implied in Fig. 141. The determinations by the much simpler method (equation 27) employed by Mortimer indicate, however, a steady rise in the value of A_z with increasing depth.

The determinations in Table 51 indicate the vertical variation in the values of A_z (uncorrected for molecular conductivity or diffusivity) when calculated for Linsley Pond by four different methods. The second column contains values obtained from the rates of change of temperature using equation 33. The third column contains values derived from the rates of increase of heat budgets Θ_b computed on a Birgean basis. It is possible that loss of heat in the upper levels and some gain in heat at 9 m. by the nonturbulent transport of heat downward and centripetally, by the mechanism postulated in a later paragraph, introduces errors into the

TABLE 51. *Coefficients of eddy conductivity, apparent coefficient of eddy diffusivity, and stability in the clinolimnion of Linsley Pond*

Depth, m.	$A_{\theta z}$, Equation 33, g. cm.$^{-1}$ sec.$^{-1}$ ($\times 10^{-2}$)	$A_{\theta z}$, Equation 28, g. cm.$^{-1}$ sec.$^{-1}$ ($\times 10^{-2}$)	$A_{\theta z}$, Equation 27, g. cm.$^{-1}$ sec.$^{-1}$ ($\times 10^{-2}$)	A_{sz}, Equation 27, Applied to HCO_3, g. cm.$^{-1}$ sec.$^{-1}$ ($\times 10^{-2}$)	Stability, E_z, g. cm.$^{-3}$ cm.$^{-1}$ ($\times 10^{-6}$)
4	0.27	0.30	0.35	...	7.7
5	0.29	0.21	0.30	1.2	3.4
6	0.32	0.27	0.46	2.3	2.2
7	0.37	0.24	0.47	4.9	0.84
8	0.32	0.27	0.50	10.2	0.32
9	0.42	0.59	1.20	11.9	0.24
Mean	0.33	0.31	0.55	...	...
Corrected for molecular conductivity	0.21	0.19	0.43	...	...

individual estimates of this column, but the mean value agrees extremely well with that derived from the second column.[11] The use of the uncorrected first-order equation 27 applied to the heat budget again leads to a result higher than that given by the use of the second-order equation, but the increase in values of $A_{\theta z}$ with depth is less regular than in the case of Lake Mendota, where the crude data are certainly superior. The fourth set of estimates is derived from the alkalinity data, using the analogue of the first-order equation 27 for eddy diffusivity. The values of A_{sz} have no relation to those derived from the thermal data and clearly indicate other mechanisms than vertical turbulent diffusion.

In spite of the general satisfaction of the criterion of validity in the clinolimnion of Lake Mendota, and to a less degree in that of Linsley Pond also, some doubt evidently remains in the minds of limnologists as to the propriety of the method for determining A derived from McEwen's treatment.

The use of the second-order equation 33 certainly is more laborious than is that of equation 27. A very great amount of field data is needed, and the subsequent mathematical treatment of these data is tedious. This, however, is no valid objection if the method is correct and the application of the first-order equation is incorrect. In many cases good approximations to values of A at depths at which lateral heat transport is insignificant can probably be obtained fairly easily from the rate of increase of Θ_b, which can be computed if the morphometry of the lake is known. The most serious objection to the procedures here recommended is not that they are laborious, but that they lead to what is, to most investigators, an entirely unexpected result.

In general, given a constant distribution of horizontal velocities, the value of A_z should be inversely proportional to the stability. Since in the thermocline the stability is maximal and since it rapidly decreases to very small values in the lower hypolimnion, it would be reasonable to expect a minimum value of A_z in the thermocline and a steady increase in the value as the stability decreased with depth in the hypolimnion. Grote (1934) has emphasized this theoretical scheme, and Yoshimura (1936b), whose results on the turbulence of Japanese lakes do not seem to be published, has indicated that he has found minimal values of A in the thermocline. If the values of A_z for Lake Mendota based on equation 27 are considered acceptable, then the expected relation undoubtedly appears to hold; for when so computed, A_z appears to increase with depth throughout the entire hypolimnion, and since the epilimnion is certainly the most turbulent part of the lake, the coefficient must have a metalimnetic minimum in the

[11] Slightly different results are obtained according to whether differention of the temperature curve is performed mechanically or by differentiating equation 31.

region of greatest stability. The application of the simple first-order equation 27 thus leads to the expected conclusion. The writer therefore viewed the results of the application of equation 33 with great misgivings when they were first encountered, but confronted with the regularity of Fig. 141 for the mean Mendota data, it became increasingly difficult to disregard the phenomena as presented by nature merely because they did not agree with a theoretical preconception. In choosing between the two possible modes of approach, it is necessary to bear in mind that there is an alternative explanation available for the high increasing values of A_z obtained by the use of the first-order equation 27, which is that lateral transport of heat occurs in the bathylimnion. If, however, equation 27 is assumed to give correct results, it seems impossible to account for the internally consistent results of the application of equation 33. For this reason it appeared better to accept the relatively invariant values of A_z in the clinolimnion, in spite of the great variation in stability with depth.

It is now apparent from Munk and Anderson's treatment that some distribution like that observed is actually what one might expect. In the most stable layers the horizontal velocity will be greatest, and it will fall off with depth throughout the clinolimnion. If we may suppose that equation 33 is satisfied, the observational data are explained if the rate of decline of velocity with depth in the clinolimnion is proportional to the square root of stability, as has been pointed out (page 256). For the present at least, this seems to be the most reasonable interpretation of the observations. Below the clinolimnion the criterion of validity, namely that the graphs of equations 31 and 33 on semilogarithmic paper are parallel straight lines, is by definition unsatisfied. The second-order equation therefore cannot be used to determine A_z. There is no more reason to suppose that the first-order equation 27 is valid in the bathylimnion than in the clinolimnion. Moreover, if the existence of lateral transport of heat, below the clinolimnion, is suspected as an explanation of the discrepancies between the use of equation 27 and equation 33, the corrected first-order equation 28 cannot be applied in those layers in which such lateral heating occurs. There appears, therefore, to be no way of ascertaining what the bathylimnetic values of A_z really are.

The most important problem raised by the phenomena that have been considered in the previous paragraphs is, however, not that of determining the bathylimnetic values of A_z but that of providing a reasonable mechanism of heating in the bathylimnion in addition to vertical turbulent transport. Before considering such a mechanism it is desirable to discuss certain aspects of the temporal variation of A throughout the annual cycle.

Seasonal variation of A_z. Though the middle layers of a stratified lake appear to constitute a self-adjusting system in which A_z is independent of the vertical variation in stability, there is clearly an over-all decrease in the clinolimnetic values of A as the stability of the entire lake increases during the heating period. In the Lunzer Untersee, Schmidt, using the approximate method of equation 27, found rather uniform values of of 200×10^{-2} g. cm.$^{-1}$ sec.$^{-1}$ between 5 m. and 15 m. in April, while in July these values had sunk to 18×10^{-2} at 5 m. and as low as 1×10^{-2} g. cm.$^{-1}$ sec.$^{-1}$ at 20 m. In Lake Mendota, from Birge and Juday's mean temperatures, calculation by equation 33 indicates that the value of A falls from 21×10^{-2} g. cm.$^{-1}$ sec.$^{-1}$ in the clinolimnion (9 to 15 m.) in the first week in June, to 1.9×10^{-2} g. cm.$^{-1}$ sec.$^{-1}$ (9 to 16 m.) in the last week in July.

Mortimer has made some interesting approximate computations of the values of A under an ice cover.

Esthwaite	13 m.	c. 0.5×10^{-2} g. cm.$^{-1}$ sec.$^{-1}$
Blelham Tarn	12 m.	1.0×10^{-2} g. cm.$^{-1}$ sec.$^{-1}$
Schleinsee	11 m.	$1.0\text{–}1.8 \times 10^{-2}$ g. cm.$^{-1}$ sec.$^{-1}$

This turbulence is probably an expression of the convection currents set up by the transfer of heat from the mud to the water, though in the case of Blelham Tarn disturbance of the water by methane bubbles is also likely to be involved. In Esthwaite Water, the value of A under ice was about one sixth of that derived from the summer data, but no such reduction is apparent in the Schleinsee when winter values are compared with those obtained from summer observations. In view of the probable existence of profile-bound density currents in small ice-covered lakes, Mortimer's quantitative comparisons, though suggestive, are not entirely convincing, as he now agrees (personal communication).

Mechanisms of bathylimnetic heating. If the argument that has been developed in the preceding paragraphs is valid, it is obvious that some heating mechanism in addition to vertical turbulent transport is operating in the bathylimnia of the few lakes that have been submitted to detailed thermal analysis. The most reasonable explanation of the observed phenomena is that they are due to density currents running down the slope of the lake basin towards its deepest point.

It has already been pointed out that during the winter, when the free water of the lake is at temperatures below 4° C., the warming of water in shallow regions, often by solar radiation through ice, and the conduction of heat from the mud to the water in contact with the bottom, will raise the density of this water, so that it will tend to flow down the sides of the basin. The existence of this type of density current is well established in lakes varying in size from the small Swedish examples studied by Alsterberg (1930b, 1931) to Lake George, studied by Langmuir (1938).

When, as is occasionally the case, the water at the bottom of a lake at the end of winter has a temperature in excess of 4° C., it is obvious that the increase in density above unity implied by such a temperature must be due to dissolved substances that have passed from the mud into the water. During the summer, the water in contact with the mud at intermediate depths will tend to lose heat rather than gain heat from the sediments. The result, however, will still be to increase the density, the water being above the temperature of maximum density. The movement of dissolved materials will still occur, and will contribute to this increase in density. Such water will tend to flow down the slope of the basin just as in winter, but if its increase in density is due in part to solutes from the mud as well as to loss of heat to the mud, it should take up a level at which the free water is slightly cooler than the water brought down by the density current. In this way additional heat will be brought into the deepest layers of the lake.

Alsterberg has obtained evidence of a slight upward distortion of the isotherms at the water-mud interfaces in the hypolimnia of small lakes in Sweden, which he attributes to cooling of the contact water by the mud; he points out that such cooling should set up profile-bound density currents. In addition, an increase in density due to dissolved substances would increase the tendency for such a current system to develop, and once this mechanism is in operation some downward nonturbulent heat transfer is inevitable. It is possible, however, that no chemical density current could start descending while the mud retained an oxidized surface. This aspect of the matter deserves further study, which may perhaps reveal severe limitations to the interpretation here advanced.[12]

Further objections to this explanation of bathylimnetic heating can be raised. Mortimer suspects that the increase in density would usually not be sufficient to overcome friction at the gently sloping mud-water interface. This objection perhaps has some force, but it should be remembered that the rate of heat delivery by such a current gives no information about the density differences causing the current. Thus, if in a small lake such as Linsley Pond, water were cooled by the mud at 8 m. from 8° to 7° C., this water would tend to flow to the level of the isothermal surface for the latter temperature, which in the example in question would lie at about 12 m. No heating would result. If, however, at the same time, 2.8 mg. per liter HCO_3 diffused from the mud into the water, then the descending current could pass to an isotherm of about 6.9° C. and would therefore be delivering heat to the bathylimnion.

---

[12] A chemical density current would not be expected, for instance, in Cultus Lake, which appears to have a bathylimnion.

In practice there will also be a chemical as well as a thermal gradient in the free water, so that more bicarbonate would be needed to produce the effect than is indicated in this example. It will be apparent, however, that the density differences involved could be much greater than would be suspected from the very small contribution of heat made by nonturbulent mechanisms, which in Linsley Pond corresponds to a rate of change of temperature of the order of a tenth of a degree per month. Inasmuch as the depth at which heat can be delivered to the bathylimnion appears to depend directly on both the heat lost to the mud and the dissolved material gained by the water, while the quantity of heat that can be delivered depends on the difference between the heat loss and the solute gain when all quantities are expressed in terms of densities, it is difficult to predict how the process would vary seasonally. As the temperature differences between mud and water are usually likely to be greatest early in the stratification period, the rather large contribution to the nonturbulent heating of the bathylimnion of Lake Mendota made early in the summer is perhaps not as surprising as was believed when the theory was first advanced.

It must also be pointed out that Mortimer has questioned the theory of bathylimnetic heating here presented, on the grounds that it should lead to higher apparent values of A_θ, when these are calculated by equation 27 from thermal data, than of A_s obtained by the same equation from chemical data. He finds that the chemical data actually give somewhat higher results. It has already been pointed out that in Linsley Pond the apparent value of A_s calculated from the bicarbonate transport is of the order of ten times that of A_θ calculated from the thermal data, confirming or rather surpassing Mortimer's observations. But this is not a valid objection to the theory as here developed, for there is every reason to believe that the large discrepancy observed in Linsley Pond and the smaller discrepancies observed by Mortimer are actually due to the transport of dissolved substances from the mud into the free water at the observation stations by horizontal currents, the existence of which is generally admitted and which are in fact required to produce the turbulence on which Mortimer relies. That the effect should be so much greater in Linsley Pond than in the lakes studied by Mortimer is certainly due to the extraordinarily low clinolimnetic turbulence of Linsley Pond, in the almost undisturbed hypolimnion of which certain consequences of the horizontal current system are expressed in a more striking manner than in any other lake yet investigated.

It must be pointed out that unsuspected modes of heat transfer may be discovered. Ricker (1937), for instance, has indicated that in Cultus Lake migrating zooplankton might, at the height of summer, transport 0.07 cal. cm.$^{-2}$ day^{-1} into the hypolimnion.

ANOMALOUS TEMPERATURE CURVES AND MEROMIXIS

In all the cases that have hitherto been considered, the thermal stratification involves a regular monotonic change in temperature, the temperature decreasing from below upward when the water is below the temperature of maximum density, from above downward when the water is above the temperature of maximum density, except for the very slight inversions observable at the surface and due to sensible heat loss or evaporation. However, when sufficient vertical variations in the chemical composition of the water occur, it may be possible to observe stable stratifications in which maxima or minima or both occur in the temperature curve.

Yoshimura (1936b, 1937) has introduced the terms *dichothermy* to indicate the condition when a minimum occurs in a temperature curve, and *mesothermy* to indicate the condition when there is a maximum. The very rare condition when one or more maxima coexist with one or more minima, he describes as *poikilothermy*. Anomalies, particularly *dichothermy*, are most often associated with meromixis or the presence of permanently stagnating water in a lake, but they are also known as the result of transitory conditions. Mesothermy may develop in the late summer or early autumn, when one of the influents of a lake contains more material in solution than does the lake itself. The influent may become as warm or warmer than the epilimnetic water of the lake but, owing to its greater mineral content, may have the same density as a much cooler layer in the lake. The entering influent will flow as a density current at the depth of appropriate density, the temperature of which will be raised, producing a mesothermal condition. If later further heating occurs as the result of a spell of hot weather in the autumn, a poikilothermic condition of a minimum between two maxima, one at the surface and one in the intermediate water, may be produced. Examples of mesothermy and poikilothermy of this sort are provided by Yoshimura's temperature curves for Bentennuma (August 16, 1935) and Yarokunuma (August 18, 1935), two lakes on Mount Bandai, Japan. The former shows an almost pure mesothermal condition, the latter is definitely poikilothermal (Fig. 143).

Very extraordinary mesothermy can sometimes be produced in saline lakes after rain. If a shallow zone of fresh water comes to lie on the top of the salt lake, it will form a greenhouse roof for the saline water, which receives heat by radiation but loses none by evaporation and little in any other way. Temperatures as high as 56° C. can be produced in this way at the top of the saline layer in Lake Medvetó near Szováta in Hungary (Kaleczinsky 1901).[13] On the Norwegian coast this phenomenon is

[13] The various investigations on this and nearby lakes are summarized by Halbfass (1923).

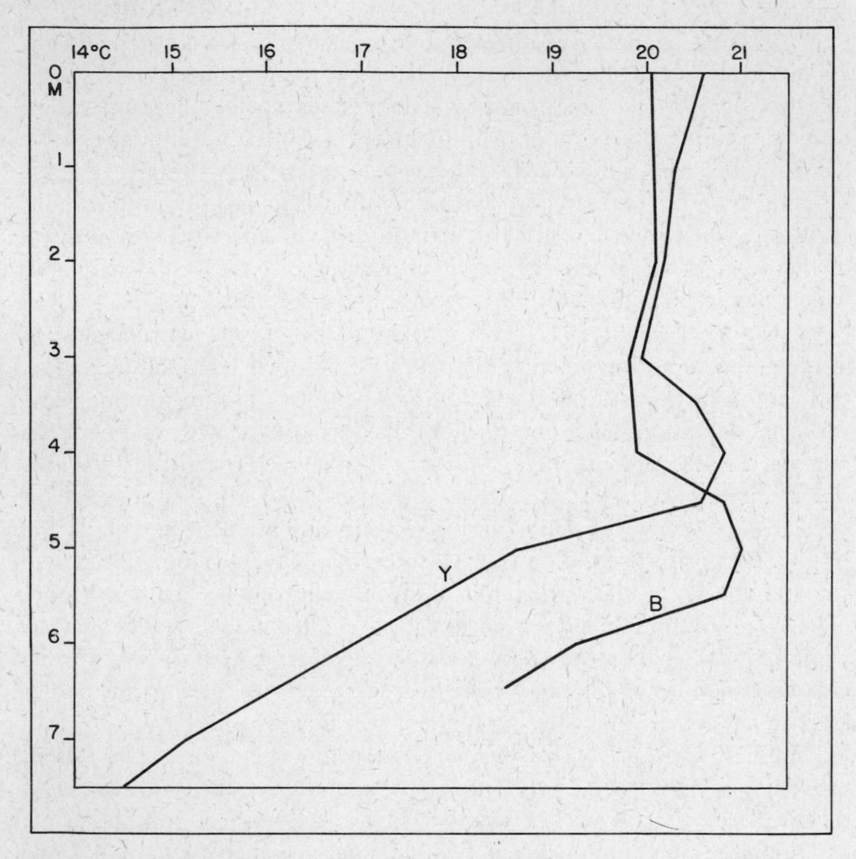

FIGURE 143. Temperature curves for Bentennuma (B) and Yarokunuma (Y), Mount Bandai, Japan, August 16, 1935 (after Yoshimura).

exploited in the culture of oysters, which are grown in ponds of salt water protected by a layer of fresh at the surface.

Terminology of meromixis. A lake in which some water remains partly or wholly unmixed with the main water mass at the circulation periods is said to be *meromictic*; the more ordinary type of lake, in which the whole of the water is mixed at such periods, is termed *holomictic*. Findenegg (1935), who introduced these terms, designated the perennially stagnant deep layer of a meromictic lake the *monimolimnion*, and Hutchinson (1937c) introduced the term *mixolimnion* for the remaining part in which free circulation periodically can occur. The boundary between the mixolimnion and the monimolimnion is known as the *chemocline*.

Findenegg (1937), to whom so much of our knowledge of this kind of lake is due, classified the processes leading to meromixis as *static* if they are dependent on the geologically determined presence of saline water from which a stable monimolimnion can be formed, or *dynamic* if the ordinary processes of decomposition and solution in the lake sediments lead, in special cases, to increasing permanent enrichment of the lower layers of the lake in solutes such as bicarbonate and silicate, which under more ordinary circumstances would be freely distributed throughout the lake at circulation.

Yoshimura (1937), who gave a list of the known meromictic lakes of the world, believed that the static, or as he called it, the nonbiochemical type of stratification was always characterized by enrichment in sodium, chloride, and sulfate ions, while the dynamic or biochemical type of stratification exhibited no enrichment in these ions. Such a generalization is probably essentially correct so far as sodium and chloride are concerned, though it is conceivable that cases may be discovered where mineral springs very poor in these elements, but nevertheless of high density, are responsible for the formation of the monimolimnion of a meromictic lake. In the case of sulfate, it seems certain that quite marked biochemical stratifications can occur even in holomictic lakes, in the absence of mineral springs.

Hutchinson (1937c), who defined meromictic stability as the amount of work needed to mix a meromictic lake to a uniform concentration if it were closed, or to mix the monimolimnion into the perennially freshened mixolimnion if open, assuming an initial isothermal condition, preferred the following threefold division.

(1) *Ectogenic meromixis*, due to some external catastrophe bringing either salt water into a fresh-water lake or fresh water into a saline lake, so establishing a deeper, saline, denser layer covered by a less saline, less dense, superficial layer. Although very little mixing may occur at the chemocline, such as does occur inevitably reduces the meromictic stability of the lake.

(2) *Crenogenic meromixis*, due to saline springs delivering dense water into the depths of a lake. This water will displace the fresh water of the mixolimnion, forcing it out of the effluent; but before this happens, mixing is likely to take place near the surface of the lake, so that a chemocline is formed at a depth determined by the rate of discharge from the springs, the density difference between monimolimnion and mixolimnion, and the amount of work available for mixing. The meromictic stability implied by such a steady-state condition will be constant or may perhaps vary seasonally about a mean value.

(3) *Biogenic meromixis*, due to the accumulation in the monimolimnion

of salts liberated from the sediments, primarily as the result of chemical changes of biochemical origin. Once the monimolimnion has been initiated and provided abnormal meteorological events do not intervene, the meromictic stability will tend to increase with time. It is, however, possible that a limiting value may be reached. It is very probable, from the recent studies of Frey (1955b) on the Längsee, that many cases of dynamic or biogenic meromixis were initiated ectogenically, by prehistoric clearing of forested land around the lake causing turbidity currents which descended to the bottom and formed an initial stable layer. This remarkable work will be discussed in greater detail when lake history and development is taken up.

It is not unlikely that the definition of crenogenic meromixis may have to be widened to include cases where solid salts in the bottom deposits are going into solution, or where saline rivers or dense industrial wastes enter the lake at the surface and flow into the monimolimnion as density currents. It is also obvious, as Findenegg pointed out, that the dynamic processes leading to biogenic meromixis may start to occur in the monimolimnia of lakes of the ectogenic and crenogenic types. The general utility of the threefold scheme will, however, be apparent from the detailed account of certain examples presented in the following sections.

Ectogenic meromixis. The most spectacular case is provided by the history of the Hemmelsdorfersee near Lübeck. This lake occupies a deep cryptodepression very near the flat coast of the Baltic. The greater part of the lake is shallow, but a single funnel-shaped hole reaches a depth of 45 m. Marine floods which have occurred at irregular intervals, averaging one hundred and twenty years apart, have from time to time since the Middle Ages covered the lake with Baltic water. When the marine flood subsides, the basin is left filled with saline water which has displaced the less dense fresh water of the basin, but a layer of fresh water soon runs in over the top of the Baltic water, and the lake becomes divided into two sharply defined layers. The last of these floods to be recorded took place in 1872. The conditions during the late stages of freshening of the basin after this flood were studied in great detail by Griesel (1920, 1935).

In 1914, when Griesel began his investigation, a sharp chemocline delimiting the boundary between the fresh mixolimnion and the salt monimolimnion lay at 32 m.; by 1934 the monimolimnion had practically disappeared. The mean rate of descent of the chemocline was as follows:

1872–1914	76.2 cm. per year
1914–1919	67.0 cm. per year
1919–1934	55.8 cm. per year

The amount of work needed to mix the monimolimnion into the mixolimnion of the Hemmelsdorfersee can be computed from the density of the saline layer and the morphometry of the basin, and was found to have been 2690 g.-cm. cm.$^{-2}$, or a mean amount of 44 g.-cm. cm.$^{-2}$ per year between 1872 and 1934.[14] Early in the process, although more work was needed to lower the chemocline through unit depth than later, the rate of lowering was actually greater, as the figures just presented clearly show, and the amount of work available per annum must therefore have been greater also. This is not unexpected, as the momentum transmitted from the surface to the chemocline must fall progressively as the depth of the latter increases. Moreover, during the late stages of freshening, the monimolimnion was confined to the deep hole and this may also have made mixing less easy. Griesel obtained evidence that the whole of the mixing took place during a comparatively short time during autumnal circulation, when the mixolimnion was unstratified. An irregular rhythmical variation in the rate at which the chemocline descended between 1919 and 1934 is attributed by Griesel to variations in the meteorological conditions during this limited period in the autumn. The lake exhibited a slightly dichothermic temperature curve in 1914, the deep water lying between 4.5° and 5.4° C. Very large quantities of H_2S, formed presumably by reduction of sulfate in the sea water, were present in the monimolimnion, and a purple bacterial layer was present in the chemocline region.

Lakes showing a similar but less spectacular type of ectogenic meromixis are known along the Japanese coast, but as they receive sea water far more frequently than does the Hemmelsdorfersee and as they may be artificially modified, no very regular phenomena can be recognized in their histories. Harutoriko, near Kusiro, is the best example, but being only 8 m. deep it does not develop such stable meromixis as is possible in the Hemmelsdorfersee. In 1932 the chemocline in Harutoriko lay at 4 m. Certain other coastal Japanese meromictic lakes (Yoshimura 1938b) are discussed (page 777) in the chapter on the sulfate cycle.

It is not impossible that certain crater lakes close to the sea in the Philippine Islands are comparable with the Japanese examples. Woltereck, Tressler, and Buñag (1941) describe briefly, under the name of Seitsee, a lake on the north coast of the Isthmus of Jols. This lake is 47 m. deep, but its surface is only 8 m. above sea level. The surface temperature was 30.8° C., the bottom 28.9°. There was no oxygen below 8 m. The rather meager chemical data indicate a thirteenfold increase in alkalinity from surface to bottom. The bottom water, however, contained only 13 mg.

---

[14] Dr. W. T. Edmondson most kindly pointed out an error in the estimate of the density difference in the calculation given in Hutchinson (1937c).

per liter chloride. Another crater lake in a cryptodepression, Lake Singuyan on the south coast of Cagayan, Sulu, which has a depth of 69 m. of which 59 m. are below sea level, also shows a very marked increase in alkalinity. These may be lakes of a typical equatorial oligomictic kind, but there may alternatively be a little sea water that has been mixed in and has given stability to the deepest layers.

A more certain case is provided by Saelsø in northeast Greenland, presumably a cold monomictic lake, in lat. 77°03′ N., long. 20°35′ W., in a cryptodepression 111.75 m. deep, the lake surface lying 4 m. above sea level. There is a definite chemocline at 58 m., below which the basin contains dilute sea water of chlorinity 15.2 to 15.7 per cent. Trolle (1913) concluded that sea water had entered the lake during a period of high seas, which he supposed, from a consideration of the rate of diffusion of chloride, to have occurred some fifty years before the time of his writing. The argument is not fully convincing, but the general explanation probable.

An entirely different kind of ectogenic meromixis is responsible for the condition described by Hutchinson (1937c) in Soda Lake, Fallon, Nevada. This lake, occupying a crater, has undergone a great rise in level during this century, owing to an upward movement of the local water table consequent on irrigation. Originally its basin was occupied by a very saline lake (Cl 45.06 g. per liter, CO_3 1856 g. per liter). The rise in water level occurred gradually between 1905 and 1925. The new water seems to have moved in over the surface of the old, but some salt was mixed into it. Meanwhile there was evidently a slow flow through the entire basin, so that the bottom layer was diluted and about 80×10^9 g. Cl and about 18×10^9 g. CO_3 were lost between the years 1882 (Russell 1885) and 1933 (Hutchinson 1937c).

In 1933 (*A* of Fig. 144) the monimolimnion had a temperature of 12.6° C. between 40 m. and 60 m., with a chloride content of 27.3 g. per liter and a carbonate content of 12.2 g. per liter. The temperature curve above the monimolimnion was strongly dichothermic, with a minimum of 8.9° C. at 20 m., while the mixolimnion above this depth, containing 8.2 g. Cl per liter and 1.83 g. CO_3 per liter, exhibited ordinary thermal stratification.

The total stability of the lake, practically entirely due to the chemical stratification, was 60,000 g.-cm. $cm.^{-2}$ This implies that to mix the water of Big Soda Lake completely would require about twenty-three times as much work as was needed to remove the monimolimnion of the Hemmelsdorfersee, a process that took sixty-two years. In the case of Big Soda Lake, since the mixolimnion is closed, the density gradient across the chemocline must fall as mixing proceeds. Moreover, ground water is washing salt out of both mixolimnion and monimolimnion, a

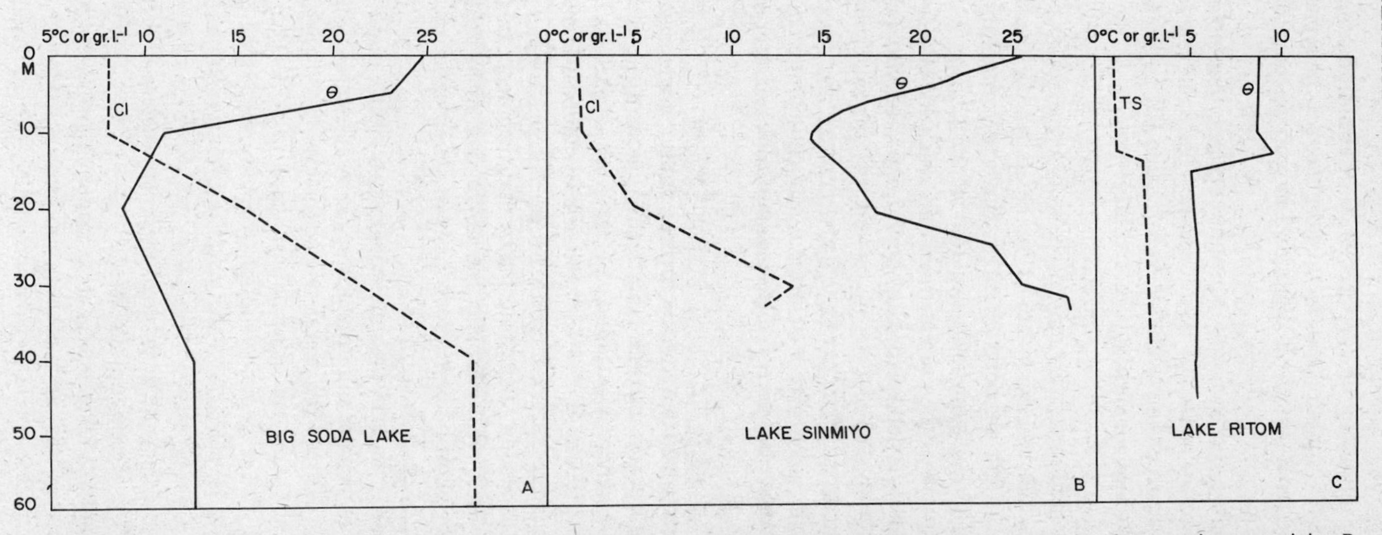

FIGURE 144. Temperature and chloride (broken line); *A*, Big Soda Lake, Nevada, July 23, 1933, an example of ectogenic meromixis; *B*, Lake Sinmiyo, Miyake Island, Japan, June 5, 1932, probably an example of coastal ectogenic meromixis superimposed on crenogenic meromixis due to a hot spring. *C*, Lake Ritom, Switzerland, October 8, 1913, full circulation of mixolimnion in a lake exhibiting typical crenogenic meromixis; broken line, total solids. (After Hutchinson, Yoshimura, Collet.)

process which may ultimately be of importance in abolishing the stratification. It is, however, not unreasonable to suppose that the separation of the monimolimnion from the rest of Big Soda Lake may persist for several centuries.

Strøm (1955) has recently found a saline monimolimnion in Lake Tokke, south Norway, which must have been derived from sea water left as the lake and its surroundings were elevated during the past 6000 years. This extraordinary case emphasizes the stability possible in meromictic lakes.

Another type of meromixis that may perhaps be regarded as ectogenic has recently been studied by Edmondson (1955) in Soap Lake, Washington. In this lake, which is 25.6 m. deep, there is a layer of mirabilite crystals on the bottom. There is some evidence that crystallization of the salt takes place on occasions at the present time. If, when the lake is concentrated, such crystals are deposited, dilution of the lake will produce a mixolimnion lying on the top of water in contact with the salt, which water will become saturated.

In the case of Soap Lake there is at present a fairly definite chemocline at about 20.5 m. Above this there is a transitional zone. The transitional zone began at 18.43 m. in July 1955, but by December 1955 its top had been depressed about 80 cm. and the salinity gradient sharpened. The formation of the transitional zone requires further study; some evidence of movement of ground water into the lake in the region of the chemocline has been obtained. There is evidence of internal seiches rocking the monimolimnion in the autumn. Radiorubidium, added locally at the 22-m. level, was found to spread laterally at the rate of 2 to 3 m. day^{-1}, in a rather irregular manner, suggesting horizontal turbulence of the kind which will later be invoked to explain the movement of bicarbonate in Linsley Pond and of phosphate on those lakes in which experiments with radiophosphorus have been performed.

Another locality, Hot Lake in the same region (Anderson, unpublished), occupying an old magnesium sulfate working, shows features reminiscent of Lake Medvetó. The bottom water may be at 30° C., when the fresh mixolimnion freezes. This type of lake, in which a large mass of soluble material is present at the bottom, though formally exhibiting artificial ectogenic meromixis is perhaps to be regarded as transitional to the crenogenic type.

Further work on the meromictic lakes of the northwestern United States should provide very interesting information.

Crenogenic meromixis. The classical example is provided by Lake Ritom in Switzerland, at least in its natural condition prior to the building of a hydroelectric plant in 1918 (Bourcart 1906; Collet, Mellet, and

Ghezzi 1918; Collet 1925). The lake was 45 m. deep, and below 20 m. the temperature was always about 6.3° C. In summer there was a well-marked dichothermic temperature curve with a minimum temperature of about 5° C. at 13 m. (*C*, Fig. 144). Chemical studies indicated a very sharp chemocline at about 13 m., which remained remarkably constant in depth between 1904 and 1914. Since the visible outflow of the lake is greater than the visible inflow, it is evident that the balance between water entering and leaving the lake is made up by subaquatic springs; some such springs, discharging dense mineralized water, were exposed during the lowering of the level of the lake after 1918.

Assuming that the discharge of the mineral springs into the monimolimnion represents the balance between inlets and outlet, and knowing the density of the highly mineralized monimolimnion below the chemocline, it is possible to compute the amount of work that must have been done in unit time by the wind to maintain the chemocline at a steady-state level. This is found to be 479 g.-cm. $cm.^{-2}$ per year (Hutchinson 1937c). Such an amount of work is much more than was done in a year in depressing the chemocline in the Hemmelsdorfersee during the late stages of the freshening of that lake. It must be remembered, however, that Griesel's observations were all made when the chemocline lay at a much greater depth than the steady-state level in Lake Ritom; moreover, that the monimolimnion of the Hemmelsdorfersee was, when the lake came under observation, confined to the bottom of the funnel-shaped hole which constitutes the deepest part of the basin. That the amount of work required to maintain the old condition in Lake Ritom is not unreasonable is seen when it is compared with the considerably greater quantity of 1209 g.-cm. $cm.^{-2}$ needed to distribute heat throughout Lake Mendota during the summer heating period.

A number of other cases of crenogenic meromixis are known in Europe. The Lac de la Girotte (Delebecque 1898) and the Ulmener Maar (Thienemann 1913b, 1915) may be specifically mentioned; the latter has a large monimolimnion at 7° C. A detailed list of other cases is given by Yoshimura (1937), who adds some Japanese cases. Of these, the most extraordinary is Lake Sinmiyo, Miyake, Riu-Kiu Islands (*B*, Fig. 144), perhaps the most striking example known anywhere. Here the monimolimnion is undoubtedly fed by a warm saline spring. The bottom temperature of the lake, recorded by Yoshimura and Miyadi (1936) as 26.8° C., may actually be higher than the surface temperature; and intermediate minimum of 14.1° C. occurred at 11 m.

Biogenic meromixis. It was early noted by Richter (1892) that slight temperature inversions of the order of 0.1° to 0.2° C. often occurred in the temperature curves at the bottom of two Austrian lakes, the Wörthersee and the Millstättersee. Renewed investigation of the subalpine lakes of Austria has more recently demonstrated that very many of them are meromictic and exhibit slight temperature inversions.

In Carinthia, Findenegg (1932, 1933, 1934a,b,c, 1935, 1936a,b, 1937, 1938) finds the Längsee, Klopeinersee, Weissensee, Wörthersee, and Millstättersee to be meromictic. The same is true of the Zellersee in Tyrol (Einsele 1944) and of the Hallstättersee and the Krottensee in the Salzkammergut, though the Traunsee in the same region apparently owes its meromictic condition to the discharge of a soda factory creating a sort of artificial crenogenic meromixis (Ruttner 1937); in the case of the Toplitzsee, chloride leached from the basin may stabilize the deep water. It seems reasonably certain that the Lago di Lugano (Baldi, Pirocchi, and Tonolli 1949) belongs with the same series of meromictic lakes as does the Türlersee (Thomas 1948), though in this case there is no temperature inversion.

In North America, Lake Mary, Wisconsin (Juday and Birge 1932; Juday, Birge, and Meloche 1938), Canyon Lake, Huron Mountains, Michigan (Smith 1941), Sodon Lake, Michigan (Newcombe and Slater 1949, 1950), and Fayetteville Green Lake, New York (Eggleton 1931, 1956), are doubtless all examples of the biogenic or dynamic type of meromixis.

In the tropics it is probable that all very deep lakes are somewhat meromictic, as has already been indicated. In most cases such meromixis is doubtless biogenic, but in the very striking case of Lake Kivu external events in the history of the basin may be involved (Damas 1937).

In the temperate zones the conditions which tend to produce biogenic meromixis are proportionately great depth, a basin sheltered from the wind, and a markedly continental climate. During a prolonged winter under an ice cover, the deep water of a lake may acquire sufficient electrolytes, liberated by predominantly biochemical processes in the mud, to increase appreciably in density. If spring comes late but is warm, and if the lake is deep and sheltered, the ice may melt and the lake begin to warm at the surface before the chemically stabilized water has been fully mixed with the main body of water in the lake. The lake thus enters the period of summer stagnation with some excess electrolyte already present in the deep water. During summer stagnation, there will be a considerable further accession of dissolved material by the deeper part of the hypolimnion. The partial suppression of vernal circulation is probably fairly common in small, deep, sheltered lakes in regions of hard winters. It has been recorded by Karcher (1939) in some of the lakes of the Masurian plain, by Findenegg for the small alpine Turachersee on the boundary of Carinthia and Styria, and in Linsley Pond after the prolonged winter of 1935–1936 by Hutchinson (1941).

In such lakes the autumnal circulation, which we have already seen to be the effective period of mixing in the case of the Hemmelsdorfersee, is always complete; but it is easy to see that if the lakes were somewhat deeper and better protected, the increased stability acquired during summer stagnation might permit the lake to pass through the critical autumnal period incompletely mixed, carrying over some excess electrolyte into the succeeding winter stagnation period.

That a process of this sort is actually involved in the maintenance of biogenic meromixis in the subalpine lakes of Austria is beautifully shown by Findenegg's data for the temperatures and oxygen concentrations of the Wörthersee between 1933 and 1935. During the early part of the period in question the lake showed a temperature inversion. In January 1935 the top 30 m. were freely circulating at 4.2° C., while the deeper part of the lake lay between 4.4° and 4.6° C. In other years an ice cover formed at about this time, but in the winter months at the beginning of 1935 the lake did not freeze, and at the end of winter the whole lake was at or below 4.2° C. and no inversion developed until after the succeeding winter provided an ice cover (Fig. 145). The oxygen data (Findenegg 1936b) are even more instructive (Fig. 184). In 1934, after a prolonged period with an ice cover, the oxygen at 45 m. remained below 3.5 mg. per liter throughout the entire spring and summer, and at 70 m. only 0.1 to 0.2 mg. per liter were present. After the winter of 1933, with a relatively short period when the lake was ice-covered, the oxygen content at the time of the spring circulation rose to 5.5 mg. per liter at 45 m., and a slight ventilation is evident at 70 m. In 1935, after an ice free winter, the oxygen rose to 6.6 mg. per liter at 45 m., and a marked ventilation at 70 m. brought the oxygen content of that level up to 2.4 mg. per liter. Even under these conditions of enhanced circulation when the dichothermy disappeared, it is evident that the lake is still markedly meromictic.

The very remarkable work of Frey (1955b) on the Längsee has, however, demonstrated that in this lake, and therefore possibly in some other Austrian lakes, the initiation of meromixis was due to turbidity currents incident on the clearing of land around the lake when agriculture was first practiced in the first few centuries B.C.

The most interesting problems presented by these lakes are the source of the excess heat in the monimolimnion, which produces the frequent dichothermic condition, and the nature of the materials stabilizing the monimolimnion. In Lake Mary, Wisconsin, a small sheltered lake 22 m. deep, the temperature was found by Juday and Birge (1932) to rise from 4.0° C. at 10 m. to 4.7° C. at 22 m. Considering the water at the bottom of the lake, at least 700 cal. per liter must have been added since the water was freely circulating. If it is assumed that originally the water was

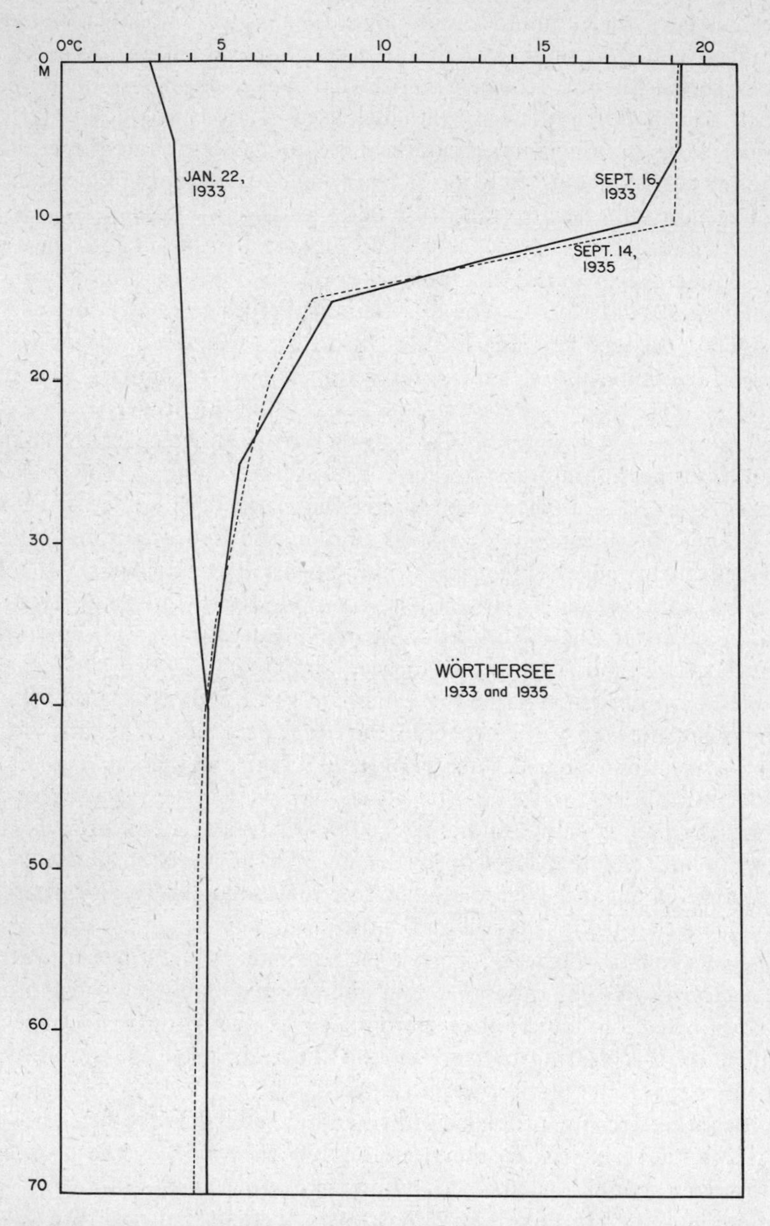

FIGURE 145. Temperature distributions in Wörthersee, 1933 and 1935. Solid lines, January 22 and September 16, 1933, circulation not occurring below 40 m., with an inversion of 0.05° C. below this level; dotted line, September 14, 1935, after partial circulation during the mild winter of 1934–1935, which permitted the mixing of some cold water into the monimolimnion. (After Findenegg.)

saturated with oxygen, it is evident that about 13 mg. of oxygen have been lost from the water, since it is apparently free of the gas. The loss of 13 mg. O_2 in oxidation of organic matter may be expected to produce about 46 cal. and to add 18.2 mg. CO_2 to the water. Actually the amount of CO_2 that has been added, either as free gas or in converting carbonate in sediments to bicarbonate, would appear to be 47.2 mg. per liter. Assuming that the additional CO_2 has been produced by anaerobic processes comparable to alcoholic fermentation, which probably gives an upper limit of heat production, we find that the anaerobic CO_2 corresponds to about 9 cal. per liter. If the greater part of the carbonate as well as bicarbonate CO_2 is actually derived from metabolic processes and not from pre-existing carbonate, this figure would have to be raised, but under no circumstances is it likely that more than about 64 cal. or less than 10 per cent of the excess heat could be of biochemical origin. A similar calculation for Canyon Lake, based on Smith's (1941) data, indicates that in this case 400 cal. must have been added to the bottom water and that 150 to 235 cal. might have been produced metabolically. The discrepancy is less great than in the case of Lake Mary but is probably great enough to indicate that in Canyon Lake, as in Lake Mary, there is a major source of heat producing the inversion, other than oxidation and anaerobic fermentation.

Findenegg (1934c) has described the condition present in September 1934 in the Wörthersee. Three stations were considered: A, the deepest point in the eastern basin of the lake; B, a slight ridge 65 m. deep; and C, a very slight depression north of the ridge. The water temperatures at 64 m. were 4.5° C. at all stations. The temperatures and silicate contents at the deeper levels are given in Table 52. It is evident that the silicate

TABLE 52. *Temperatures and silicate concentrations in the deep water of the Wörthersee*

	A		B		C	
Depth, m.	Temp., °C.	SiO_2, mg. per liter	Temp., °C.	SiO_2, mg. per liter	Temp. °C.	SiO_2, mg. per liter
60	4.5	3.3	4.5	3.5	4.5	3.5
65	4.5	3.8	4.55*	4.5	4.55	4.8
66.5	4.5	4.5	...	...	4.7*	8.5
73	4.6	9.0	...	...	...	...

* Bottom.

increases rapidly close to the bottom, but that this effect is much less on the top of the low ridge than at the bottoms of the depressions. It is therefore reasonable to assume that the water in the depressions has collected there from higher up the slope, and that in some circumstances water which is sufficiently dense to move downward is also sufficiently warm to convey heat into the deepest parts of the basin.

In all the larger glacial rock basins a contribution from the internal heat of the earth may be expected. This would be, as indicated earlier (page 460), about 38.7 cal. cm.$^{-2}$ yr.$^{-1}$ Wherever there is reason, as in the case of the Längsee, to believe that the meromixis is of ancient origin, it is clear that the internal heat of the earth must be perennially causing some instability and convective circulation. In the case of a lake like Lake Mary, set in a layer of drift in which ground water is circulating, it is exceedingly improbable that this source of heating is possible. Since biochemical heating appears to be inadequate, the best explanation of the inversion is that the monimolimnion is essentially a perennial bathylimnion, and has formed by the descent of density currents down the slope, carrying dissolved material and, when an inversion is present, heat from a higher level in the hypolimnion.[15]

HEAT BUDGETS

Comparisons of the conditions exhibited by a lake during the different stages of the heating cycle may be made, as in the previous sections of this chapter, in terms of intensity factors alone, by comparing tables of temperatures or temperature curves. It is, however, also useful to introduce more explicitly a capacity factor, namely, the amount of water in the lake to be heated, not merely in the hypolimnion, as in the last section, but in the lake as a whole. This was first done by Forel (1880, 1895, 1901b), who considered the total heat income of a column of water of unit area of cross section in the deepest part of the lake. As Wojeikoff (1902) and Birge (1915) have pointed out, this method is not satisfactory.[16] Later Halbfass (1905, 1910a) considered the total heat content of the lake above 0° C., obtained by summing the product of temperature and volume for a series of layers. Division by the volume of the lake then gives a mean temperature. This mode of computation is theoretically above reproach, but permits comparisons of significance only when mean depths are available.

Forel's method is apparently theoretically unsound. He assumed that the whole of the heat that has entered a column of the deep water passed into the column at its top. All heating was therefore considered to be vertical. Using this mode of computation he obtained quite impossible

[15] Löffler (unpublished) has recorded an anomalous type of temperature curve in some Andean lakes into which glaciers flow. The extreme bottom water is well below 4° C., and must be melt water stabilized by suspended material.

[16] The earlier work, prior to Birge's study, is mainly of historic interest; it is well summarized by Halbfass (1923).

values for certain lakes. Thus he concluded that Lake Ladoga could gain heat at the rate of 1010 g.-cal. $cm.^{-2}$ day^{-1} during the heating period, though as has been made clear in Chapter 6, it is very improbable that half this amount could ever be delivered by radiation from the sun and sky.

If, however, Forel's method and Halbfass's method of estimation are combined by computing the heat income of the entire lake layer by layer and then expressing this per unit area, the result is the quite reasonable average rate of heating of 160 g.-cal. $cm.^{-2}$ day^{-1}, as computed by Birge for this lake. It is evident that in the shallows, heat is delivered to the bottom by radiation and part is then lost by conduction and convection to the freely moving epilimnetic water in contact with the bottom. In this way an appreciable amount of the heat ultimately computed by Forel as delivered to the deepest possible vertical column has actually entered the surface of the lake, not at the top of the column, but lateral to it, in the peripheral shallow parts.

It has been pointed out in the previous section that there is probably also a lateral contribution to the heating of the deep bottom water of many lakes. As in the earlier sections dealing with hypolimnetic heating, on the rare occasions when it is necessary to discuss the total amount of heat that has entered a deep column of water by any path whatever, the term Forelian heat budget, Θ, will be used. For the more correctly computed budget of a mean column of water, obtained by dividing the total heat that has entered the lake by the total area, the term Birgean heat budget, Θ_b, will be employed.

Terminology. Since the mechanics of heating a large body of fresh water below and above the temperature of maximum density are quite different, for a dimictic lake it is convenient to distinguish the following quantities.

Annual heat budget Θ_{ba}, the total amount of heat that enters the lake between the time of its lowest and its highest heat content.

Summer heat income Θ_{bs}, the amount of heat needed to raise the lake from an isothermal condition at 4° C. up to the highest observed summer heat content. Most of this heat is delivered by the mixing of less dense, warm, superficial water with more dense, cool, deep water. The summer income of heat therefore implies work done by the wind against gravity, as well as heating of the surface layers by direct radiation. The term wind-distributed heat is sometimes used in place of summer heat income, but the latter term is preferable because radiation does play a part in heating the top few meters.

Winter heat income Θ_{bw}, the amount of heat needed to raise the water from the temperatures characterizing the time of minimum heat content up to 4° C. Since the winter income can be distributed throughout the vertical extent of the lake by convectional streaming and other types of flow not involving direct work by the wind, it is usually considered that the winter heat income is not wind-distributed. Actually, though this is strictly correct, the winter heat income does imply a previous wind-distributed heat expenditure. During the time between isothermal circulation at 4° C. in the autumn and the time of minimum heat content, the loss of heat at the surface lowers the density of the water, and for any significant cooling of the deeper layers to take place, cold surface water must be mixed, against gravity, by the wind. Since as much heat will be needed to bring the water back to isothermal circulation at the temperature of maximum density as was lost between the time of autumnal circulation at 4° C. and the time of minimum heat content, the winter heat income actually reflects wind action, but action that has previously cooled the lake and is not operating during the time of heating.

It is sometimes of interest to consider the total heat content in a mean column, irrespective of the seasonal variation. This may be done in several ways. The only way to exclude any arbitrary element from such an estimate would be to consider the heat above absolute zero, which would be impractical and of little meaning. It is usual to take 0° C. as an arbitrary base line for such computations and call the result the *gross heat budget*. It would be equally reasonable though equally arbitrary to take 4° C. The quantity is of interest only in relation to monomictic lakes, where it gives a measure of equatorial as against temperate nature. For this purpose the heat content above 4° C. is probably the best figure to use. The minimum or winter value of the total heat content above 4° C. may be regarded as a sort of negative winter heat income. In the monomictic lakes of Table 53 such a quantity has been entered with a negative sign, and within brackets, in the column for Θ_{bw}.

High values, well over the annual budget of dimictic lakes, distinguish the deep lakes of the equatorial region. Thus for some of the deeper lakes of Sumatra, Strøm (1944a) indicates that this quantity must be of the order of 300,000 cal. cm.$^{-2}$, and for Tanganyika of the order of 1,500,000 cal. cm.$^{-2}$ The interest of such figures is exemplified by the remarkable case of Atitlan, which Deevey (1957) has shown has a winter heat content above 4° C. of the same order as those of the deep Sumatran lakes, but at the same time Atitlan exhibits an annual heat budget of the order of those of temperate dimictic lakes.

Methods of calculating the heat budget. The easiest way of determining the heat budget is to plot against depth z the product $A_z \cdot (\theta_{sz} - \theta_{wz})$, where θ_{sz} and θ_{wz} are the summer and winter temperatures at depth z, and then integrate by measuring the area of the resulting curve by means of a planimeter, and divide by the area A_0. When summer or winter heat incomes are desired, 4° C. is substituted for θ_{wz}, the winter temperature at depth z, or for θ_{sz}, the summer temperature, respectively. In warm monomictic lakes, the entire heat budget is, of course, summer heat income; in cold monomictic lakes, the entire heat budget is winter heat income.

Strictly, correction for the latent heat of melting of ice should be made to the winter heat income. If this is done, it should be remembered that the resulting addition does not represent compensation of the wind-distributed heat expenditure in the early winter. When the ice cover is 14.3 cm. thick, the correction will amount to about 1000 cal. cm.$^{-2}$ Moreover, heat used in heating the ice from its lowest temperature to its melting point should theoretically be considered also. Strøm (1944a), however, alone appears to have employed these corrections.

Birge (1916) has shown that in Lake Mendota an extra 2035 cal. cm.$^{-2}$ must be added to the summer heat income to allow for the heating of the lake sediments, but as 650 cal. cm.$^{-2}$ of this heat is lost to the water during the period of winter heating rather than during the period of autumnal cooling, and so is included in the winter heat income, the correction to the gross heat budget is only 1350 cal. cm.$^{-2}$ It is probable that the analogous correction for most other large deep lakes would be less, as the high hypolimnetic temperatures encountered in Mendota must lead to an unusually great heating of the sediments of the deeper parts of the lake. In shallow lakes, however, the heating of the mud may be proportionately more important than in Lake Mendota. In Lake Beloye ($\bar{z} = 4.15$ m.) the total heat taken up by the sediments is 2500 cal. cm.$^{-2}$, the annual heat budget of the water being 8000 cal. cm.$^{-2}$ (Rossolimo 1932a). In Lake Hula ($\bar{z} = 1.7$ m.) the heat taken up by the sediments is calculated from assumptions as to their heat conductivity as 1400 cal. cm.$^{-2}$ (Neumann 1953), the annual budget for the water being 2240 cal. cm.$^{-2}$

Heat budgets of the lakes of the world. The mean heat budgets for the various lakes of the world for which adequate information exists are set out in Table 53, the data being derived from Trolle (1913), Birge (1915), Scott (1916), Halbfass (1923), Rossolimo (1932a), Scheffer and Robinson (1939), Strøm (1944a), Ricker (1937), Church (1945), Rawson (1942, 1950, 1953), Hutchinson (1937b,c), Vereščagin (1937), Johnsson (1946), Neumann (1953), Livingstone (1957), and Deevey (1957).[17]

The largest annual and summer incomes are those for Lakes Baikal and Michigan. It is possible that the summer income for Lake Baikal

[17] The material in Table 53 will be discussed in a forthcoming paper on the upper limits for heat budgets by Dr. J. Neumann.

TABLE 53. *Heat budgets of the adequately known lakes of the world*

Lake	Lat.	Alt.	A_0	z_m	$\bar{z}$	Θ_{bs}	Θ_{bw}	Θ_{ba}
ARCTIC								
Anneks Sø	77°15′ N.	40	~40	90	45	...	2,000	2,000
Chandler	67°40′	886	15	21	13.5	5,760	...	...
SCANDINAVIAN PENINSULA AND FINLAND								
Tennesvatn	67°53′ N.	230	1	168	93	5,000	...	...
Eikesdalsvatn	62°30′	16	23	155	83	21,000	...	...
Hornindalsvatn	61°56′	53	50.8	514	237.2	19,500	...	...
Strynsvatn	61°55′	27	22.3	209	130	19,800	...	...
Loenvatn	61°47′	48	10.1	132	68.9	13,700	...	...
Oldenvatn	61°45′	36	7.4	92	42.6	8,600	...	...
Breimsvatn	61°42′	61	23.0	278	130.4	18,600	...	...
Rondvatn	62°00′	1163	...	...	...	8,000	...	...
Lemonsjön	62°00′	858	4	45	~15	12,000	...	...
Bessvatn	61°31′	1374	4	102	30	8,000	...	...
Feforvatn	61°30′	878	3	54	19	10,000	...	...
Flakevatn	60°49′	1448	3	75	30	6,000	...	...
Mjøsa	60°40′	124	362	449	187	23,000	14,000	37,000
Evangervatn	60°37′	10	4	107	46	24,000	...	...
Vangsvatn	60°36′	50	5	60	37	20,000	...	...
Ladoga	60°36′	5	18,150	223	56	18,000	15,300	33,300
Holsfjord	60°02′	64	121	295	114	22,000	...	...
Eikeren	59°40′	16	26	154	94	23,000	18,000	42,000
Vettern	58°10′	88	1,898	119	39	16,200	15,800	32,000
Klämmingen*								
Klövsta basin	59°06′	9.4	5.38	37	16.3	18,860	1,840	20,700
Central basin	59°07′	9.4	3.08	11	5.0	10,900	900	11,800
Laxne basin	59°09′	9.4	2.08	29	16.2	18,460	1,840	20,300

TABLE 53 (*Continued*)

Lake	Lat.	Alt.	A_0	z_m	$\bar{z}$	Θ_{bs}	Θ_{bw}	Θ_{ba}
NORTHWESTERN CANADA AND ALASKA								
Great Slave	61°31′ N.	150	27,200	614	62	15,600	8,700+	24,300+
Karluk	57°24′	< 100	39.5	126	48.6	18,900	4,600	33,500
BRITAIN								
Ness	57°20′ N.	16	56.7	238	133	37,200	[− 14,600]	37,200
Garry	57°04′	78.4	4.5	65	24	19,000	2,600	21,600
Lochy	56°59′	28	15.3	162	70	31,500	[− 9,000]	31,500
Morar	56°57′	9	26.7	310	87	29,400	[− 13,000]	29,400
Katrine	56°15′	111	12.4	151	61	25,800	[− 2,700]	25,800
Windermere	54°20′	39.3						
North basin			8.16	67	26.0	17,500	[− 2,900]	17,500
South basin			10.5	44	17.7	15,680	[− 2,450]	15,680
NORTHERN CONTINENTAL EUROPE								
Beloye	57°00′ N.	142	0.27	13.5	4.15	6,800	1,200	8,000
Furesø	55°47′	19	9.9	36	12.3	14,400	2,700	17,100
Marien	54°12′	188	1.5	11	4.9	17,000	6,000	23,000
Hemmelsdorfer	53°57′	− 0.2	5.0	44.5	5.5	6,700	900	7,600
Dratzig	53°36′	128	18.6	83	20	17,200	0	17,200
Müritz	53°26′	62	115.3	33	6.3	10,600	2,200	12,800
Madü	53°16′	14	36	42	18.7	15,400	0	15,400
Arend	52°54′	21	5.4	49.5	29.7	21,600	5,800	27,400
Pulvermaar	50°08′	414	0.35	74	37.6	16,200	2,600	18,800
Schalkenmehrener-maar	49°44′	420.5	0.22	21	11.4	12,600	1,800	14,400
SIBERIA								
Baikal	53°00′ N.	453	31,500	1741	730	42,300	22,800	65,500

TABLE 53 (*Continued*)

Lake	Lat.	Alt.	A_0	z_m	$\bar{z}$	Θ_{bs}	Θ_{bw}	Θ_{ba}
WESTERN CANADA								
Waskeseu	53°56′ N.	530	70	24	11.1	15,900	...	...
Kingsmere	54°00′	536	47	47	21.2	20,500	...	...
Amethyst (South basin)	52°42′	1965	1.74	21	9.7	5,650	...	...
Maligne	52°40′	1663	21.8	96	40.5	8,900	...	...
Bow	51°45′	1990	3.6	48	17.6	7,500	...	...
Minnewanka	51°16′	1454	13.0	80	38.1	25,900	...	...
Paul	50°45′	777	3.9	56	34.2	18,500	...	...
Okanagan	49°55′	345	370	235	69.5	32,700	1,300	34,000
Cultus	49°04′	41	6.26	42	32.2	24,000	...	...
CENTRAL EUROPEAN ALPS								
Weisser	48°10′ N.	1054	0.28	58.7	22.9	11,100	3,400	14,500
Gérardmer	48°04′	660	1.15	36.2	16.9	12,300	1,800	14,100
Lunzer Untersee	47°57′	607	0.68	33.7	19.8	12,300	1,400	13,700
Gmundner	47°53′	422	25.7	197	89.7	30,400	3,000	33,400
Atter	47°52′	469	46.7	170.6	84.2	24,900	2,700	27,600
Tegern	47°45′	725	9.12	71	40	22,200	3,200	25,400
Schlier	47°44′	778	2.19	37	24.9	14,100	600	14,700
Staffel	47°42′	648	7.65	40	10.7	13,100	2,100	15,200
Constance	47°39′	395	538.5	252	90	29,000	0	29,000
Kochel	47°38′	600	5.95	65	28.5	17,100	3,800	20,900
Walchen	47°35′	802	16.4	196	79.3	22,400	0	22,400
Hallstätter	47°35′	508	8.58	125.2	64.9	26,600	0	26,600
Greifen	47°23′	439	8.56	34	17	14,000	2,000	16,000
Zürich	47°15′	409	88.66	143	44	20,500	1,200	21,700
Walen	47°09′	42.3	24.2	151	103	36,000	[– 2,000]	36,000

TABLE 53 (*Continued*)

Lake	Lat.	Alt.	A_0	z_m	$\bar{z}$	Θ_{bs}	Θ_{bw}	Θ_{ba}
CENTRAL EUROPEAN ALPS (*Continued*)								
Zuger	47°04′ N.	416.6	38.2	198	84	29,500	0	29,500
Lucerne	46°58′	437	113.8	214	104	24,500	[− 14,500]	24,500
Thun	46°41′	560	47.8	217.2	135	24,200	[− 9,200]	24,200
Wörther	46°37′	548	28.2	84.6	43.2	21,600	2,700	24,300
Oeschinen	46°30′	1581	1.16	56.6	34.6	10,400	700	11,100
Geneva	46°27′	375	581.5	310	154.4	36,600	[− 23,200]	36,600
Nantua	46°09′	457	1.41	43	28.4	16,900	300	17,200
Como	46°00′	198	146	410	185	32,800	[− 50,000]	32,800
Lugano	45°57′	274	48.9	288	130	40,000	[− 17,000]	40,000
Annecy	45°54′	447	27	80.6	41.5	23,200	3,300	26,500
Orta	45°50′	290	18.15	143	71.3	25,400	[− 6,400]	24,500
Bourget	45°45′	231	44.6	145	81	32,000	[− 2,800]	32,000
Aiguebelette	45°33′	314	5.45	71.1	30.6	19,300	1,200	20,500
Bolsena	42°35′	305	114.5	146	78	31,600	[− 29,000]	31,600
EASTERN AND CENTRAL NORTH AMERICA								
Michigan	44°00′ N.	177	57,850	282	77†	40,800	11,600	52,400
Green	43°48′	278	29.7	72.3	33.1	26,200	7,800	34,000
Mendota	43°07′	259	39.2	25.6	12.1	18,240	c. 5,260	c. 23,500
Geneva	42°34′	263	22.1	49.3	19.7	20,849	...	...
Skaneateles	42°56′	264	35.9	90.5	43.5	28,200	9,800	38,000
Owasco	42°53′	217	26.7	54.0	29.3	27,050	8,400	35,450
Canandaigua	42°45′	209	42.3	83.5	38.8	26,980	...	...
Cayuga	42°45′	116	172.1	132.6	54.5	29,480	9,600	39,080
Seneca	42°45′	135	175.4	188.4	88.6	34,020	5,400	39,420
Keuka	42°30′	316	47.0	55.8	30.5	23,850	...	...
Plew	c. 41°19′	c. 250	0.42	18.6	7.3	10,172	...	...

TABLE 53 (*Continued*)

Lake	Lat.	Alt.	A_0	z_m	$\bar{z}$	Θ_{bs}	Θ_{bw}	Θ_{ba}
EASTERN AND CENTRAL NORTH AMERICA (*Continued*)								
Little Eagle	41°15′ N.	c. 250	...	...	...	10,304	...	...
Yellow Creek	41°18′	c. 250	0.58	22	10.6	11,458	...	...
Big Barbee	41°18′	c. 250	1.06	15	6.7	10,563	...	...
Silver	c. 41°08′	c. 250	0.38	10.4	4.1	9,438	...	...
Manitou	41°08′	235	3.27	14.8	3.0	5,361	...	...
WESTERN NORTH AMERICA								
Washington	47°40′ N.	6	128	65	18	43,000	[− 2,600]	43,000
Pyramid	40°10′	1173	532	104	57	33,600	[− 7,000]	33,600
Tahoe	39°09′	1890	499	501	249	34,800	0	34,800
Mead	36°12′	365	472	58.6	137	46,200	[− 22,000]	46,200
JORDAN VALLEY								
Hula	33°04′ N.	67	14	4	1.7	2,240	[− 1,600]	2,240
Galilee (Tiberias)	32°50′	− 210	167	50	24	33,500	[− 26,400]	33,500
HIGH SOUTH CENTRAL ASIA								
Pang-gong Tso	33°45′ N.	4241	279.2	50	26.1	22,000	...	...
Manasarovar	30°40′	4602	558	81.8	49.5	26,000	...	...
CENTRAL AMERICA								
Atitlan	14°40′ N.	1555	136.9	341	183	22,110	[− 288,300]	22,110
Amatitlan	14°25′	1189	8.23	33.6	18.8	8,510	[− 29,670]	8,510
Güija	14°13′	426	44.25	26.0	16.5	5,410	[− 32,090]	5,410
INDONESIA								
Ranu Klindungan	7°55′ S	10	2	134	90	c. 3,410	[− c. 189,000]	c. 3,410

* Winter income computed and subtracted from Johnsson's total budget (May–July) to obtain summer income.
† Church (1945), who gives an even higher budget of 62,000 cal. for 1943. The data in the table are for 1942.

derived from Vereščagin's temperatures for August 15, 1934, in deep water off Listvenicnoye, is a little too high;[18] some purely local warming may have raised the surface temperature to 14.54° C., and such a temperature therefore might not have been typical of the whole of the lake. However, if the mean July temperature of 9° C. given by Suslov (1947) is substituted for the surface temperature, and the water is supposed to be homoiothermal at 9° C. down to 30 m., as seems reasonable from Vereščagin's curve, the annual budget is 57, 300 cal. cm.$^{-2}$, and the summer heat income is 34,500 cal. cm.$^{-2}$; the former value is still greater than that of any other lake. Lake Baikal appears to have been circulating in 1934 in its upper, thermally variable layers, at the end of April.[19] The observed summer heat income therefore represents about 117 days heating or an uptake of up to 362 cal. cm.$^{-2}$ day^{-1}. This quantity, though high, is by no means impossible. The mixing of deep water exposed upwind into the epilimnion, in the manner described by Mortimer for Windermere, must occur on a grand scale on Lake Baikal, and is doubtless responsible for the high heat budget. The situation in Lake Michigan is clearly similar. In both lakes a considerable amount of heat may be due to condensation of fog in early summer, which possibly may be of special significance in very large lakes.

Since the ice on Lake Baikal has a mean thickness of 92 cm. (Suslov 1947), an additional 6720 cal. cm.$^{-2}$ are needed to supply the latent heat of melting, making an over-all budget, if Vereščagin's temperatures are accepted, of 72,200 cal. cm.$^{-2}$

Birge considered that in temperate North America, any lake not at an excessive altitude or otherwise abnormally situated, with dimensions of at least 10 × 2 km. and a mean depth of 30 m., would take up the climatically maximal amount of heat. Such a lake he called a first-class lake. In view of the information now available about Lake Baikal, it is doubtful if he was quite correct in this view. But it is obvious from the data in the table that smaller shallower lakes all have low heat budgets.

The lakes that Birge would have considered first class have mean annual budgets between 20,000 and 40,000 cal. cm.$^{-2}$ As is indicated in Table 54, a few lakes may provide individual values in excess of 40,000 cal. cm.$^{-2}$ during particular years, but it seems unlikely that apart from Lake Baikal and Lake Michigan any well-authenticated[20] budget is in excess of 50,000 cal. cm.$^{-2}$

[18] Halbfass gives a still larger and clearly erroneous estimate.

[19] Suslov's figures suggest that this is unusual.

[20] A value of 60,600 for Mjøsa is certainly erroneous (Birge 1915). A few other cases of individual budgets in excess of 40,000 cal. cm.$^{-2}$ represent autumnal heat loss rather than vernal and aestival gain.

TABLE 54. *Maximal individual heat budgets*

Lake	Year	Θ_{ba}, cal. cm.$^{-2}$	Θ_{bs}, cal. cm.$^{-2}$
Gmundner (Traun)	1895	49,500	38,000
Geneva	1907	43,700	Monomictic
Skaneateles	1911	41,900	29,300

Geographical and climatic factors affecting the heat budget. There seems to be no consistent latitudinal variation within the temperate zone. The largest mean budgets in Europe are, on the one hand, for two Norwegian lakes, Mjøsa and Eikeren; on the other, for two monomictic Swiss lakes, Geneva and Lugano.

Outside the temperate regions, at low altitudes the budgets fall, though it is evidently possible to have in subequatorial regions a considerable annual budget in spite of very high winter temperatures. This is demonstrated by the case of Lake Atitlan.

In the polar regions very low budgets may be expected, but if the latent heat of melting of the 250 cm. of ice on Anneks Sjø, which is 18,200 cal. cm.$^{-2}$, is included in its annual income, the resulting value of 20,200 cal. cm.$^{-2}$ is comparable to that of some temperate lakes. In the subpolar Chandler Lake (Livingstone 1957) in Alaska, the warming and melting of the ice sheet are believed to involve about 17,500 cal., giving a total budget, if the lake circulated at about 0° C., of the order of 27,000 cal. cm.$^{-2}$ This is probably a little too great, but it indicates how in such a lake the latent heat of melting is the most important term in the annual budget.

The effect of altitude is very marked; in fact, Strøm defines a high mountain lake as one with a heat budget markedly reduced below what a comparable lake would have at sea level. Dealing only with summer heat incomes, it appears that in Norway, at lat. 61° to 62° N., at altitudes in excess of 800 to 1000 m. the summer heat income is not more than half what it would be at sea level. In the Rocky Mountains, in lat. 50° N., a comparable reduction in summer heat income is apparently achieved at about 1660 m. It is probable that if annual budgets could be compared and the latent heat of the melting of ice were included, much of the altitudinal effect would disappear.

At low latitudes the effect of great altitude is much less striking, as a comparison of the Sea of Galilee with Pang-gong Tso or Lake Manasarovar indicates.

The two lakes of the Nordfjord group in Norway with very low budgets are kept cool by melt water. This emphasizes the critical nature of the heat budget, for in other cases a general low temperature promotes heat uptake by facilitating mixing, and leads to a higher budget. The effect of an oceanic climate is for similar reasons very poorly indicated in Table 53. The slow heating and low summer temperatures permit mixing of surface water deep into the lake, while the high winter temperatures naturally reduce the annual income. The difference between the distribution of heat in a more oceanic and more continental lake of almost identical budgets is well shown by Strøm's (1944a) figures for Eikesdalsvatn and Eikeren (Table 55).

TABLE 55. *Distribution of heat at height of summer in Norwegian lakes under oceanic (Eikesdalsvatn) and continental (Eikeren) climatic regimes*

Depth, m.	Eikesdalsvatn	Eikeren
0–10	8,427	11,803
10–20	5,488	4,255
20–50	5,583	2,438
50	2,815	930
Total	22,313	19,426

In this case, cool water from cold glacial streams, as well as the more oceanic climate, probably keeps the epilimnion of Eikesdalsvatn at a low temperature, promoting the transfer of heat into the depths.

Rate of heating. Some observations are available as to the rate at which heat enters and leaves lakes at different times of year. In the cases of Lake Mead, the Sea of Galilee, Lake Hula, and Lake Klämmingen (page 514), it is possible to give a relative complete analysis of the energy changes in the lake. In all these cases the rate of heating declines much more rapidly than the rate of delivery of solar radiation falls after the summer solstice. In Lake Mead and the Sea of Galilee, a loss of heat is in fact already occurring before the solstice is reached. This is largely due to the greater evaporation from a warm than a cold water surface. It is also evident that as thermal stratification develops, it is progressively harder to mix warm surface water into the depths. In the Lunzer Untersee, Schmidt (1928) found in 1912 that 85.7 per cent and in 1913 that 75.7 per cent of the total budget had been taken up by the end of June. In Green Lake, Wisconsin, Birge and Juday (1911) found 82 per cent of the total budget had entered by July 5. In Windermere, Jenkin (1942) found heating at a rate of 200 cal. cm.$^{-2}$ day^{-1} from April 21 to May 21, but 118 cal. cm.$^{-2}$ day^{-1} from May 21 to July 21. In the large, deep Lake Eikeren at a somewhat higher latitude, Strøm's smoothed curve shows a more uniform rate of heating, but even here 61 per cent of the heat had entered by July 1. In contrast to these cases, Hutchinson (1937b) concluded that over 50 per cent of the heating of Pang-gong Tso occurred after July 1.

Variations from year to year. The observations of the preceding paragraph clearly have a bearing on the constancy of the heat budget. In a strongly continental climate, in which the lake freezes every year and in which the additions made to the budget after July 1 are small however hot the weather may be, one would expect a much greater constancy than in an oceanic climate or one in which the lake is just monomictic. From Birge's (1915) data it is evident that Green Lake has a far more constant budget than most European lakes for which adequate data exist. This is shown by the standard deviations for the data for Green, Como, and Geneva as given in Birge's tables[21] (Table 56).

TABLE 56. *Variation in heat budgets of one American and two European lakes*

Lake	Θ_{ba}, cal. cm.$^{-2}$	σ	Range, cal. cm.$^{-2}$
Green (5 observations)	34,200	1,500	32,300–36,400
Geneva (4 observations)	32,300	7,900	22,000–40,200
Como (5 observations)	33,000	3,800	29,400–37,700

There is little systematic data available on the pattern of annual variation. Stromsten (1927), however, noted in Lake Okoboji, Iowa, a low summer income of about 17,300 cal. cm.$^{-2}$ in 1915, high values around 20,000 to 21,000 in 1916, 1919, and 1922, and then a steady decline in each subsequent year to a minimum of about 18,000 cal. cm.$^{-2}$ in 1926; there was a feeble rise in 1927. The lake is not a first-class lake in Birge's sense (z = 12.3 m.) and is probably more sensitive to meteorological variations than a deeper lake would be.

The heat budget of lake sediments. In shallow water, particularly in lakes of the second and third class in which most of the bottom is in contact with warm water in summer, an appreciable amount of heat may be expected to enter the mud from the water at the time of maximum temperature, and to pass from the mud to the water when the latter is cold. At any point on the bottom of a lake, unless it is of the extreme equatorial type,[22] the annual cycle of temperature may be regarded as having a wave-

[21] Only cases involving uptake in a given year, with minimum heat content recorded *earlier* than the maximum heat, are admitted. Birge gives many budgets which are really integrals of heat loss rather than gain.

[22] Points at the monimolimnion-sediment interface in meromictic lakes are likewise excluded.

like form that can be approximated by harmonic analysis to the sum of a number of sine curves or Fourier waves. The equation of the annual march of temperature at the mud surface thus becomes

$$(34) \qquad \overset{*}{\theta}_0 = \sum_{n=1}^{m} \mathrm{a}_n \sin(\overset{*}{\mathrm{A}}_n + n\mathrm{w}t)$$

where a_n is termed the harmonic coefficient and $\overset{*}{\mathrm{A}}_n$ the phase angle of the nth Fourier wave, and w is in this case $2\pi/3.16 \times 10^7$, or 2π divided by the number of seconds in the wave period, namely one year.

In the cases that have been considered (March in Birge, Juday, and March 1928; Neumann 1953), for relatively shallow water in which the temperature variation is little complicated by prolonged thermal stratification, it appears that a good approximation can be obtained by summing three or four Fourier waves ($m = 3$ or 4).

From the surface of the mud, heat is conducted downward into the sediments. At any depth $\overset{*}{z}$ in the sediment, it follows from well-known theory that

$$(35) \qquad \overset{*}{\theta}_z = \overset{*}{\theta}_m + \sum_{n=1}^{m} \mathrm{a}_n e^{-\sqrt{n\mathrm{w}/2\overset{*}{\mathrm{k}}}\cdot\overset{*}{z}} \sin(\overset{*}{\mathrm{A}}_n + n\mathrm{w}t - \sqrt{n\mathrm{w}/2\overset{*}{\mathrm{k}}}\cdot\overset{*}{z})$$

where $\overset{*}{\theta}_m$ is the mean annual temperature, taken as constant throughout the sediment below any point, and $\overset{*}{\mathrm{k}}$ is the thermal diffusivity[23] of the sediment. It will be observed that as $\overset{*}{z}$ increases, $e^{-\sqrt{n\mathrm{w}/2\overset{*}{\mathrm{k}}}\cdot\overset{*}{z}}$ decreases toward zero, so that the amplitude of the temperature variation is continually reduced with depth and the phase is continually displaced by variation in $\sqrt{n\mathrm{w}/2\overset{*}{\mathrm{k}}}\cdot\overset{*}{z}$ in the parentheses.

By fitting the progression of temperatures observed at 0, 1, and 2 m. below 8 m. of water in Lake Mendota to equations 34 and 35, March (Birge, Juday, and March 1928) concluded that for Lake Mendota sediment $\overset{*}{\mathrm{k}} = 0.00325$ cm.^{2} sec.$^{-1}$, though laboratory studies, not reported in detail, made by L. R. Ingersoll for Birge, Juday, and March, gave values of the thermal conductivity lower than would be expected from this value. Neumann (1953) believes that in a shallow lake such as Lake Hula, in which some sand is present in the sediments, a value of $\overset{*}{\mathrm{k}} = 0.004$ cm.^{2} sec.$^{-1}$ is appropriate.

Birge, Juday, and March, using an electrical resistance thermometer that could be forced into the mud, made a number of studies of the

---

[23] This quantity is the thermal conductivity divided by the product of the specific heat and density. The thermal diffusivity bears much the same sort of relationship to the thermal conductivity as does the kinematic to the dynamic coefficient of viscosity (page 254).

temperature of the sediments of Lake Mendota. The longest series of data, and the only one analyzed theoretically, was obtained below 8 m. of water. Here an annual variation of 21.5° C. is apparent at the mud-water interface, and 1.6° C. at 5 m. below the interface. The curves are essentially in opposite phase, the warmest time at 5 m. below the mud surface being in January, the coldest in July. The variation at the different depths studied is shown in A of Fig. 146; the curves fitted by harmonic analysis are given in B of the same figure.

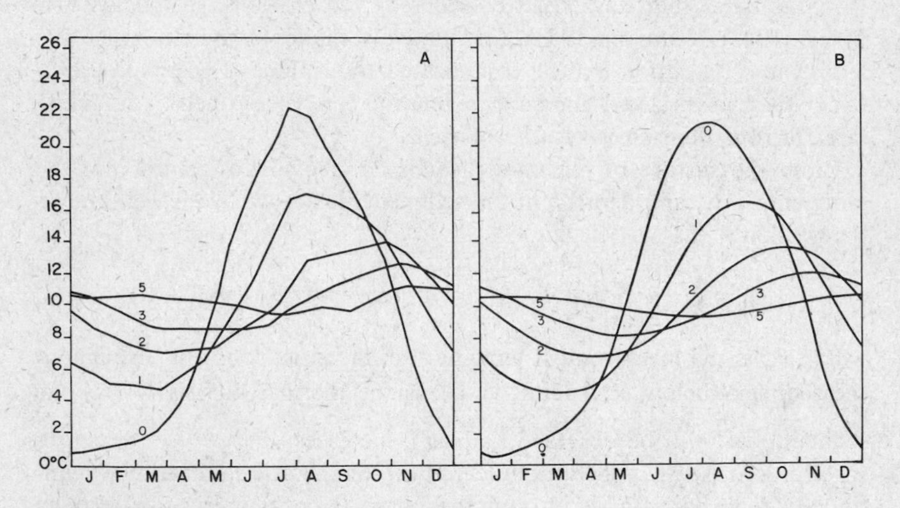

FIGURE 146. Annual variation in temperature at different depths in the sediments below 8 m. of water in Lake Mendota: *A*, observed; *B*, smooth curves fitted by Fourier analysis. (After Birge, Juday, and March.)

Below greater depths of water, the interface temperature varies less smoothly, owing to thermal stratification. The amplitude is of course less, and the annual variation in mud 5 m. below the interface at 23.5 m. is only 0.7° C. (Fig. 147). Because the highest bottom temperature occurs in the lake at the autumnal overturn, the highest temperature at 5 m. below the interface at 23.5 m. is in the spring rather than winter. Birge, Juday, and March concluded that in the zone from 0 to 8 m., about 3000 cal. cm.$^{-2}$ entered and left the mud annually, whereas in the deep water below 18 m. the annual budget was about 1100 cal. cm.$^{-2}$ The mean budget was taken to be 2000 cal. cm.$^{-2}$, of which 650 cal. cm.$^{-2}$ left the mud to enter the water during the winter period from December 15 to April 1. Since the ice-covered lake gained during this period 2400 cal. cm.$^{-2}$, only a small amount of which was likely to be advective gain, the contribution of the mud to winter heating seems to be only just over one third of the contribution of solar radiation through the ice.

Rossolimo (1932a) has made studies of the annual change of mud temperature in Lake Beloye. His results refer to only the top meter of the sediments. He concludes that in this lake the annual budget for the sediments is about 2500 cal. cm.$^{-2}$ Neumann (1953), using the annual variation of temperature at the mud-water interface of Lake Hula, fitted to equation 34, found from an integral form of equation 35 a budget of

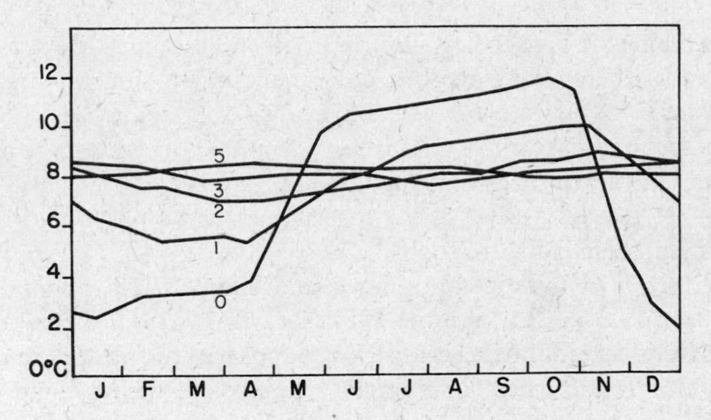

FIGURE 147. Annual variation in temperature at different depths in the sediments below 23.5 m. of water in Lake Mendota (after Birge, Juday, and March).

1400 cal. cm.$^{-2}$ The great relative importance of the sediment budget Θ_m in a shallow lake is indicated when the foregoing estimates are compared with the total annual budget for the water (Table 57).

TABLE 57. *Heat budgets of lakes and lake sediments*

	$\bar{z}$	Θ_{ba}	Θ_m
Mendota	12.1	23,500	2,000
Beloye	4.15	8,000	2,500
Hula	1.7	2,290	1,400

A rather large vertical increment in temperature, up to about 5° C., has been recorded (Anderson and Pritchard 1951) in the bottom sediments of Lake Mead. ZoBell, Sisler, and Oppenheimer (1953) conclude that bacterial metabolism in these muds could produce in a gram of mud 2.6 cal. yr.$^{-1}$ This source of heating is likely to be of some importance in warm monomictic lakes with rather high bottom temperatures, in the case of Lake Mead around 12° C.

The work of the wind. Birge (1916) introduced the useful method of studying the thermal history of a lake by computing the amount of work per unit area (B) needed to distribute the summer heat income. Assuming that heating by radiation is negligible except at the surface, an assumption which introduces a small but usually unimportant error, the work per unit area is given for a dimictic lake circulating fully in spring at 4° C. by

$$B = \frac{g}{A_0}\int_0^{z_m} \bar{z}A_z(1 - \rho_z)\,dz \tag{36}$$

For a monomictic lake the expression in parentheses must be replaced by the density difference at depth z corresponding to summer and winter temperatures. The integral can be evaluated planimetrically. If the result is required in g.-cm. cm.$^{-2}$, the g is omitted, including the acceleration due to gravity, giving the results in ergs. The curve giving $\bar{z}A_z(1 - \rho_z)$ plotted against z, the direct-work curve of Birge, is in itself sometimes instructive.

Relatively few investigators (Birge and Juday 1921; Schmidt 1928; Hutchinson 1937b,c; Livingstone 1957; Deevey 1957) have computed the work of the wind, doubtless because it has appeared too abstruse a quantity to interest the biological limnologist and too elementary and crudely empirical a parameter to concern the physical limnologist. The results set out in Table 58 are, however, clearly of some interest. It is evident that

TABLE 58. *Summer heat incomes and work required to distribute heat*

Lake	A, km.^{2}	Θ_{bs}, cal. cm.$^{-2}$	B, g.-cm. cm.$^{-2}$	B/Θ_{bs}, g.-cm. cal.$^{-1}$
Baikal	31,500	42,300	2330	0.055
Tahoe	499	34,800	3100	0.089
Seneca	175.4	34,000	2876	0.085
Pyramid, 1914	560	33,750	4055	0.120
Pyramid, 1933	503	32,100	3700	0.111
Cayuga	172.1	29,480	2446	0.083
Green	29.7	27,316	2023	0.074
Canandaigua	42.3	26,980	1930	0.072
Geneva, Wisc.	22.1	26,695	2337	0.087
Atitlan	136.9	22,110	3741	0.169
Pang-gong	279	22,000	1320	0.083
Okoboji	15.35	20,849	1600	0.077
Mendota	39.4	18,370	1209	0.066
Lunzer Untersee	0.68	13,700	535	0.039
Amatitlan	8.23	8,510	752	0.088
Chandler	15.00	5,800	175	0.030
Güija	44.25	5,410	958	0.177

the quantity B/Θ_{bs} in the last column of the table will be vanishingly small as heating begins at 4° C., and will increase with temperature increase in the lake.

In subpolar dimictic lakes, in which there is virtually no stratification, all heating occurs at a low temperature and very easily. This is graphically demonstrated when Chandler Lake, a subpolar lake in Alaska, is compared with Güija in El Salvador. In very large, deep, temperate lakes the great thermal capacity, coupled with the opportunities for exposure of cold water to windward when a strong wind is blowing, likewise promote efficient heating.

In small shallow lakes heating is certainly inefficient, as the low budgets indicate. The low value of B/Θ_{bs} is doubtless due to the loss, by evaporation and back-radiation at high surface temperatures, of a greater proportion of the heat that has been gained than is the case in an ordinary first-class lake.

In the tropics, heating becomes more and more difficult as the density difference for 1° C. increases (cf. Birge 1910a). When heating is achieved, it is an uneconomical process. In Atitlan the uptake of so much heat above 19° C. involves the formation of a very thick epilimnion, and this in turn may reflect the theoretical inverse relationship between the depth of the top of the metalimnion and geographical latitude discovered by Munk and Anderson (page 433).

The figures given for Green Lake, which has a summer heating period of about 122 days, indicate that the mean rate of work in heating the lake is about 16.5 g.-cm. cm.$^{-2}$ day^{-1}, or 1600 ergs cm.$^{-2}$ day^{-1}. This is equivalent to about 0.02 erg cm.$^{-2}$ sec.$^{-1}$, or a force of 0.02 dyne cm.$^{-2}$ Langmuir (page 275) found empirically that winds of 300 to 700 cm. sec.$^{-1}$ exerted forces of 0.65 to 6.3 dynes cm.$^{-2}$ on Lake George. His lower values are in rough accordance with marine experience and with laboratory investigation. It seems reasonable, therefore, to suppose that the work done by the wind in distributing the summer heat income represents only a small fraction, of the order of a few per cent, of the work done by the wind on the water surface.

The direct-work curves (Fig. 148) for Pyramid Lake are of some interest, as they indicate conditions in two different years with practically identical heat budgets, but with different areas and different values for the work of the wind. Although it is by no means certain that the 1933 curve is representative of the entire lake in that year, as it is based on a single series of observations, the differences between the direct-work curves do at least suggest that the study of successive years in a desert lake with a fluctuating water level, and so a fluctuating area, might give a good deal of empirical information about the mechanics of heating. Accepting the curves as they stand, it appears that a reduction in area of 10.3 per cent is accompanied by a decrease in work of commensurable amount, or 8.75 per cent. This implies a thinner epilimnion in the later year after the area was reduced.

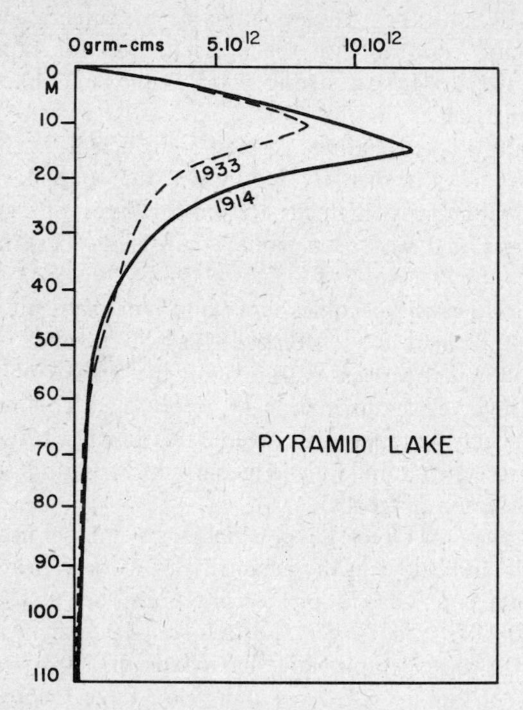

FIGURE 148. Direct-work curves, showing quantity of work needed to distribute the heat present at different depths in Pyramid Lake, Nevada, in 1914 and 1933, referred to the entire lake surface (Hutchinson).

Stability of a lake. Schmidt (1915, 1928) has defined the stability[24] of a lake as the amount of work needed to mix the entire body of water to uniform temperature without addition or subtraction of heat. Measuring the work in ergs, we obtain for S, the stability per unit area,[25]

$$S = \frac{g}{A_0}\int_0^{z_m}(z - z_g)A_z(1 - \rho_z)\,dz \tag{37}$$

[24] Not to be confused with the stability of a layer used in Chapter 5.

[25] Schmidt gives B, S, and G in terms of the whole lake. When, as is usual in limnological writing, the work is given in g.-cm., the g is omitted.

where z_g is the depth of the center of gravity of the lake, given by

$$z_g = \frac{1}{V}\int_0^{z_m} zA_z\,dz \tag{39}$$

Both integrals can easily be evaluated planimetrically.

The quantity S is a measure of the work that would have had to be done by the wind to prevent a lake from developing thermal stratification. It obviously increases as the lake becomes more and more stratified, but since during heating some mixing of water from the surface into the deep water occurs, the process of heating does not imply the production of maximum stability. In the late summer, as the thermocline begins to descend, there may be a small accession of heat with a decrease in stability. It is probable, therefore, that the maximum stability is achieved before the heat content is maximal. Schmidt's data for the Lunzer Untersee suggest this, but owing to the irregularity of the heating of this lake, which is not within Birge's first class, the progress of both B and S through the season is very ragged. The data given in Table 59 are the means for the years

TABLE 59. *Work required to distribute heat (B), to render lake homoiothermal (S), and to maintain hypothetical homoiothermal conditions throughout heating season (G) for lakes of large and moderate size*

Lake	B, g.-cm. cm.$^{-2}$	S, g.-cm. cm.$^{-2}$	G, g.-cm. cm.$^{-2}$
Atitlan	3741	21,500	25,241
Pyramid (1914)	4055	10,255	14,310
Pyramid (1933)	3700	7,640	11,340
Mendota	1209	514	1,723
Amatitlan	752	415	1,167
Lunzer Untersee	535	390	925
Güija	958	175	1,133

1907, 1910, 1912, and 1913, for which complete information exists, at the time of maximum heat content. Schmidt showed that in this lake variations in stability were clearly inversely correlated with prevalent wind velocities.

The sum of B and S, given as G in Table 59 (Schmidt 1928, Birge 1915, Hutchinson 1937c, Deevey 1957), represents the total work that the wind would have to do to heat a lake from 4° C. (or from the minimal winter temperature if the lake is monomictic) in order to give the observed mean temperature without thermal stratification. It will be observed that in the two large lakes the amount of work required to achieve homoiothermy is vastly greater than any of the observed values of B in Table 58. In the four smaller lakes, the values of G are well within the possible range of values of B. This emphasizes the extreme importance of size and form in determining how much work is actually done, and so the summer heat income of a lake.

Meromictic stability. It is of interest to compare the amount of work than can be delivered by the wind with the amount needed to mix permanently stratified or meromictic lakes. Two cases must be considered. If the mixolimnion is open, its density will remain constant throughout the mixing; this is the usual condition. If the mixolimnion is closed, its density will increase as mixing occurs; this condition is approached by Big Soda Lake, Nevada.

Applying essentially the methods developed by Schmidt for thermal stratification, Hutchinson (1937c, see also p. 483) calculated that to mix the monimolimnion of the Hemmelsdorfersee into the mixolimnion required 2690 g.-cm. cm.$^{-2}$ The process would be easy at first, when the distance that an element of monimolimnion would have to be lifted was small. It would become more difficult as the chemocline descended and came to lie below an ordinarily tranquil hypolimnion, but this would ultimately be compensated by the decreasing area of the chemocline. The process took about sixty-two years, so that an average of 144 g.-cm. cm.$^{-2}$ yr.$^{-1}$ of work was needed. Towards the end of the process, when it was under observation, mixing took place only during the short period of autumnal circulation.

In the case of the more or less closed Big Soda Lake, it would require 60,000 g.-cm. cm.$^{-2}$ to mix the lake completely. Reference to Table 58 will indicate that this would take many years.

In the case of crenogenic meromixis, the interesting quantity is the amount of work needed to keep the chemocline at a constant level. In Lake Ritom, in its natural condition, with a chemocline at about 13 m., this amounts to 480 g.-cm. cm.$^{-2}$ yr.$^{-1}$, a reasonable figure but a quite considerable fraction of the values of B given in Table 58.

Analytical energy budgets. Though it is possible to learn something of the thermal history of a lake from the simple considerations presented in the last two sections, a much more detailed treatment is available for a few lakes, which is far more illuminating. In this treatment the rates of transfer of various forms of radiant and thermal energy into and out of the lake are considered throughout the annual cycle.

The sources of energy are the radiation from the sun (Q_S) and sky (Q_H), the heat advected by influents, long-wave radiation from the atmosphere (Q_A) and from mountains (Q_M) in the basin, and sensible heat transfer from the atmosphere. The losses of energy are due to reflection ($-$ Q_R)

from the water surface, and scattering ($-Q_U$) long-wave back radiation ($-Q_W$) from the water, evaporation ($-Q_E$), transfer of heat in evaporated water vapor ($-Q_e$), sensible heat transfer from the water to the atmosphere, and heat advected from the lake by the effluent. The difference between the losses and gains gives the rate of storage (Q_θ) of heat in the lake and its sediments, so that if Q_s is the net sensible heat transfer and Q_i the net advective transfer, all quantities being referred to unit area,

$$Q_\theta = (Q_S + Q_H - Q_R - Q_U) + (Q_A + Q_M - Q_W) + Q_s + Q_i - Q_E - Q_e \tag{39}$$

It will also be realized that

$$\Theta_{ba} = \int_{Q_{\theta\,min.}}^{Q_{\theta\,max.}} Q_\theta\,dt \tag{40}$$

The sum of the first four terms in parentheses in equation 39 is referred to as the net radiation flux (Q_S'); the sum of the two sets of terms in parentheses has already been defined as the net radiation surplus (Q_B).

Ordinarily Q_M is negligible, and in most cases, since the limnologist attempts to study lakes rather than rivers, he will hope that Q_i is small also. The terms ordinarily to be considered are therefore Q_S' the net radiation flux; ($Q_A - Q_W$) the net back radiation; Q_s the transfer, positive or negative, of sensible heat; Q_E the loss of heat in evaporation; and Q_e the loss of heat into evaporating water vapor. The last term, which obviously involves knowing the temperature of the incoming water as well as of the vapor leaving the lake, has only been considered by Anderson (1952) and Neumann (1953). It is ordinarily quite small; the greatest value recorded by the latter investigator, for Lake Hula in July, was -9.2 cal. cm.$^{-2}$ day^{-1}, at a time when the radiation surplus Q_B was 540 cal. cm.$^{-2}$ day^{-1}, the net back radiation -173 cal. cm.$^{-2}$ day^{-1}, the evaporative loss -533 cal. cm.$^{-2}$ day^{-1}, and the sensible heat transfer $+16.0$ cal. cm.$^{-2}$ day^{-1}. Ordinarily no great error will be introduced by neglect of this term, which is difficult to estimate properly.

The data required for evaluation of equation 39 will obviously include estimates of total radiation flux and reflection, and a sufficiently detailed series of lake temperatures. If flow rates are significant, they will be needed for estimation of the advective terms. Back-radiation can be obtained from theory, as explained on page 377, or the net surplus radiation (Q_B) can be measured directly and the back-radiation determined by difference.

The most difficult quantities to evaluate are the evaporative and sensible heat losses. If all other terms are known, the Bowen ratio R_b, which is the ratio of sensible to evaporative heat loss, can be used. This is defined as

$$R_b = \frac{Q_s}{Q_E} = 61 \times 10^{-5} P_h \frac{\theta_W - \theta_A}{p_w' - p_w} \tag{41}$$

where θ_W is the surface temperature of the lake, θ_A is the temperature of the air at some standard height, p_w' is the saturation pressure of water vapor at θ_W, p_w the observed pressure of water vapor in the air at the same standard height, and P_h the pressure of the atmosphere, measured in the same units as p_w and p_w' at the altitude h of the lake surface.

There has been considerable discussion of this quantity (Anderson 1952), which is not entirely satisfactory. Anderson has shown in his studies of Lake Hefner that if evaporative heat loss is estimated from equation 39 and the Bowen ratio, the results averaged over a ten-day period are essentially the same as those from estimates made by direct water-balance studies. The latter mode of measurement of evaporation is applicable only to reservoirs, for which there are very accurate independent estimates of advection in and out of the lake.

A third method of procedure is the estimation of both heat and water-vapor transfer in the air above the lake, from the wind profile. Johnsson (1946) employs semi-empirical expressions of a widely used form

$$Q_E = 0.18 L_e (p_w' - p_w) W^{0.8} \tag{42}$$

and

$$Q_s = 4.4(\theta_W - \theta_A) W^{0.8} \tag{43}$$

where p_w and θ_A are measured 2 m. above the lake surface, and the wind velocity W is measured 6 m. above the lake surface. L_e is the latent heat of evaporation at θ_W.

More elaborate expressions have been used in oceanography and have been examined by Marciano and Harbeck (1952). At the present time the use of the Bowen ratio seems empirically a better method of procedure than the mass transfer methods, but as the latter develop they are likely to displace the methods based on the use of R_b.

Analytical energy budgets of monomictic lakes. The most complete studies have hitherto been made on warm monomictic lakes, in which the whole process of heating and cooling can be studied without the complication of a winter ice cover. Three cases have been studied: Lake Mead, above Hoover Dam in Arizona and Nevada (Anderson and Pritchard 1951), and the two fresh-water lakes of the Jordan Valley (Neumann 1953), In all cases, estimates of back-radiation were made from air and water temperatures, and the sensible heat and evaporative heat losses were distributed according to Bowen's ratio. The annual contributions of various processes to heating and cooling are set out in Table 60, and in Figs. 149 and 150 the seasonal changes in rate of gain and loss of energy through these processes can be followed.

TABLE 60. *Analytical heat budgets for warm monomictic lakes*

	Sea of Galilee	Lake Hula	Lake Mead
Altitude, m.	− 210	67	366
Latitude	32°50′ N.	33°04′ N.	36°10′ N.
z_m, m.	50±	4±	137±
$\bar{z}$, m.	24	1.7	58.6
$Q_S + Q_H$	177,000	184,300	180,891
Q_R	− 10,700	− 11,300	− 11,683
$(Q_A - Q_W)$	− 63,400	− 63,000	− 93,106
Q_B	102,900	110,000	76,102
Q_E	− 94,586	− 97,666	− 94,834
Q_e	− 1,546	− 1,673	...
Q_s	− 6,758	− 6,161	− 6,238
Q_i	...	− 4,500	11,854
Θ_b (total)	33,500	3,640*	46,200
Θ_b (nonadvective)	...	...	22,100

* Including sediments.

In the case of Lake Mead the advective contribution of heat due to warming of the influents before they reach the lake is very great, so that just over half the annual heat budget is due to this process. Some advective loss occurs in the winter. Neither the uncorrected budget nor the nonadvective budget is likely to give a fair idea of the heat income of a lake with a relatively small rate of replacement in the same region. Pyramid Lake, a little north of Lake Mead, has a heat budget of 34,000 cal. cm.$^{-2}$ Lake Mead, being slightly farther south, would probably have a slightly smaller budget, owing to higher winter temperatures, if large amounts of heat were not brought in by the influents.

In spite of the difference in thermal regime there is an extraordinary resemblance in the evaporative and sensible heat losses from the three lakes. Back-radiation, however, is markedly greater in Lake Mead than in the two lakes of the Jordan Valley. The great similarity in the evaporative loss, back-radiation, and sensible heat loss from the Sea of Galilee and Lake Hula is of particular interest in view of the fact that the former is a deep lake, thermally stratified in summer, the latter a shallow lake, unstratified and with a very small annual heat budget.

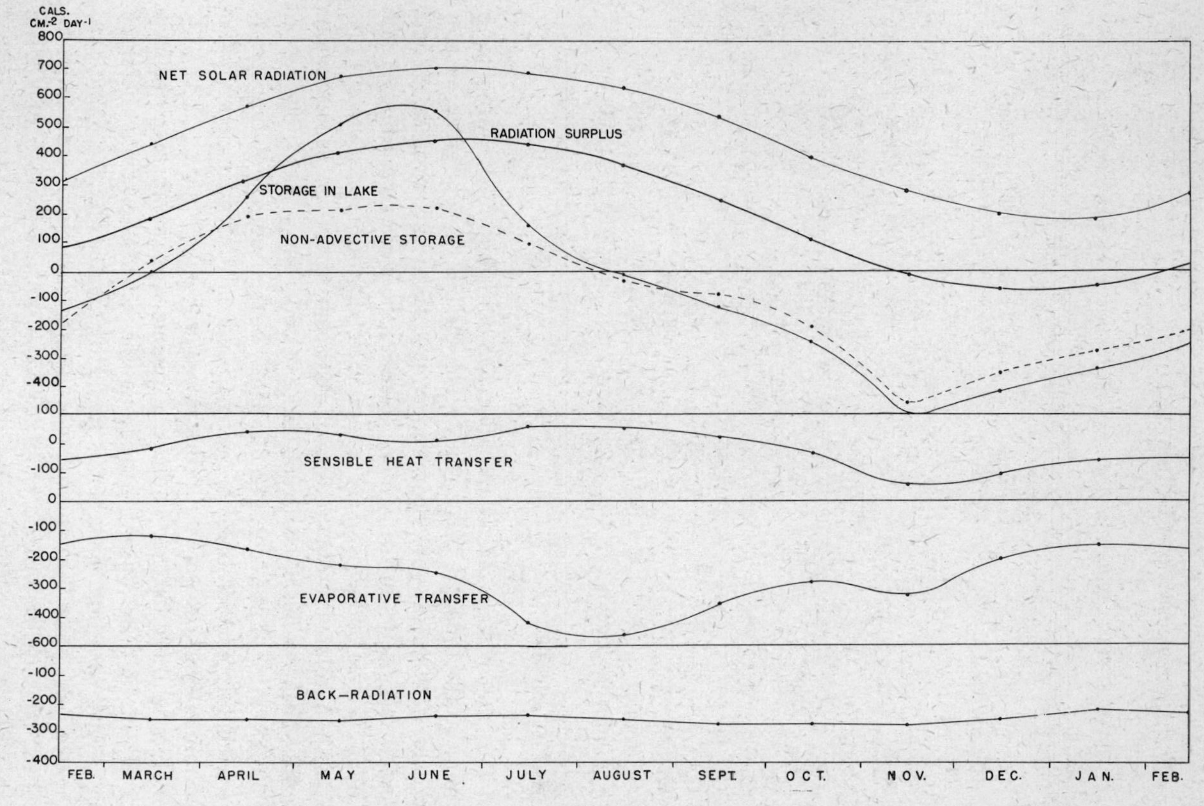

FIGURE 149. Annual variations in energy transfer in Lake Mead (after Anderson and Pritchard).

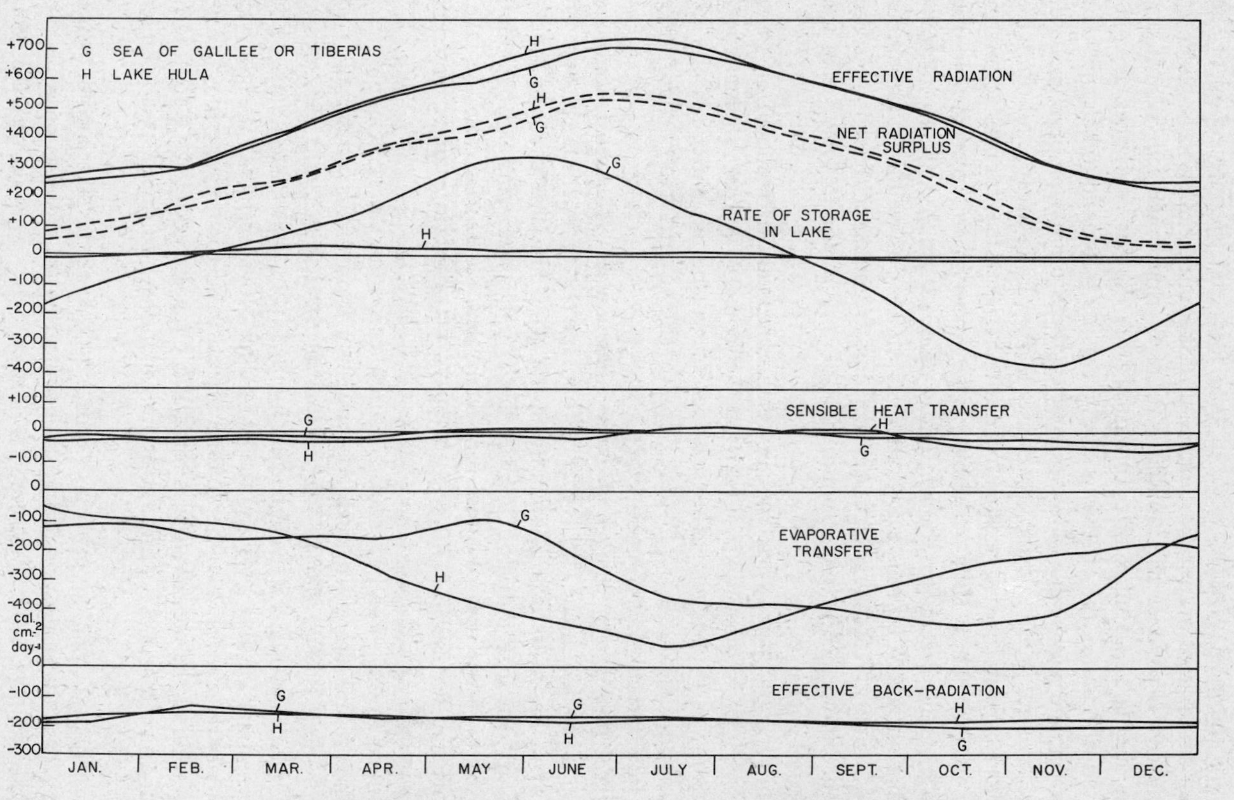

FIGURE 150. Annual variation in energy transfer in the Sea of Galilee or Tiberias (after Neumann).

The seasonal variation differs somewhat in the three lakes. Evaporative loss is the most variable mode of heat loss, and the interaction of the seasonal cycle of evaporation and that of incoming radiation is evidently sufficient to produce the familiar type of thermal regime in such lakes. The very late maximum of evaporation from the Sea of Galilee as compared with that of the other two lakes is said by Neumann to be somewhat suspect, and may be due to inadequacies in the autumnal temperature data. The back-radiation is extremely constant from Lake Mead, and can have no part in determining seasonal variation; it exceeds evaporative loss in the winter. Lake Hula resembles Lake Mead in this respect, but the Sea of Galilee seems to exhibit slightly more seasonal variation in back-radiation, which exceeds evaporative loss in May as well as in December and January. Although sensible heat loss from Lake Mead is of some slight importance in the autumn, it is of very minor importance in the over-all energy budget.

In their published form the very elaborate studies conducted on Lake Hefner, Oklahoma, do not permit the construction of an analytical annual energy budget. The seasonal trend and absolute quantity of evaporative heat loss are, however, comparable to those recorded for Lake Mead (Harbeck and Kennon 1952).

Analytical energy budget of Lake Klämmingen. The most complete analysis for a dimictic lake is that of Johnsson (1946) for the deep southern Klövsta Basin of Lake Klämmingen, about 50 km. southwest of Stockholm. The total radiation flux was based on measurements at Stockholm, raised by 5 per cent to allow for slightly less cloudiness over the lake. Reflection was taken as 10 per cent of $Q_S + Q_H$ in winter and 7 per cent in summer. Evaporative loss and sensible heat transfer were computed from equations 42 and 43, back-radiation from equation 8 of Chapter 6.

The results are given by Johnsson in the form of histograms for two week periods over the ice-free seasons of 1941, 1942, and 1943. The mean rates of Fig. 151 have been computed from these histograms. For the first half of May 1941 and the last half of November 1942 the lake was not ice-free, so only two sets of data are available for these periods. The variation of the mean values is remarkably regular and gives a good idea of the mechanism of heating and cooling in the lake during the period under consideration.

Early in the heating process evaporation is low; the lake is colder than the air, so that a net conductive gain of heat by the water is possible and back-radiation is the main mode of loss of heat. During May the total radiation entering the lake is just over 400 cal. $cm.^{-2}$ day^{-1} and almost three fourths of this amount is retained by the lake. As the early summer advances there is a slight increase in the radiation delivered, but the

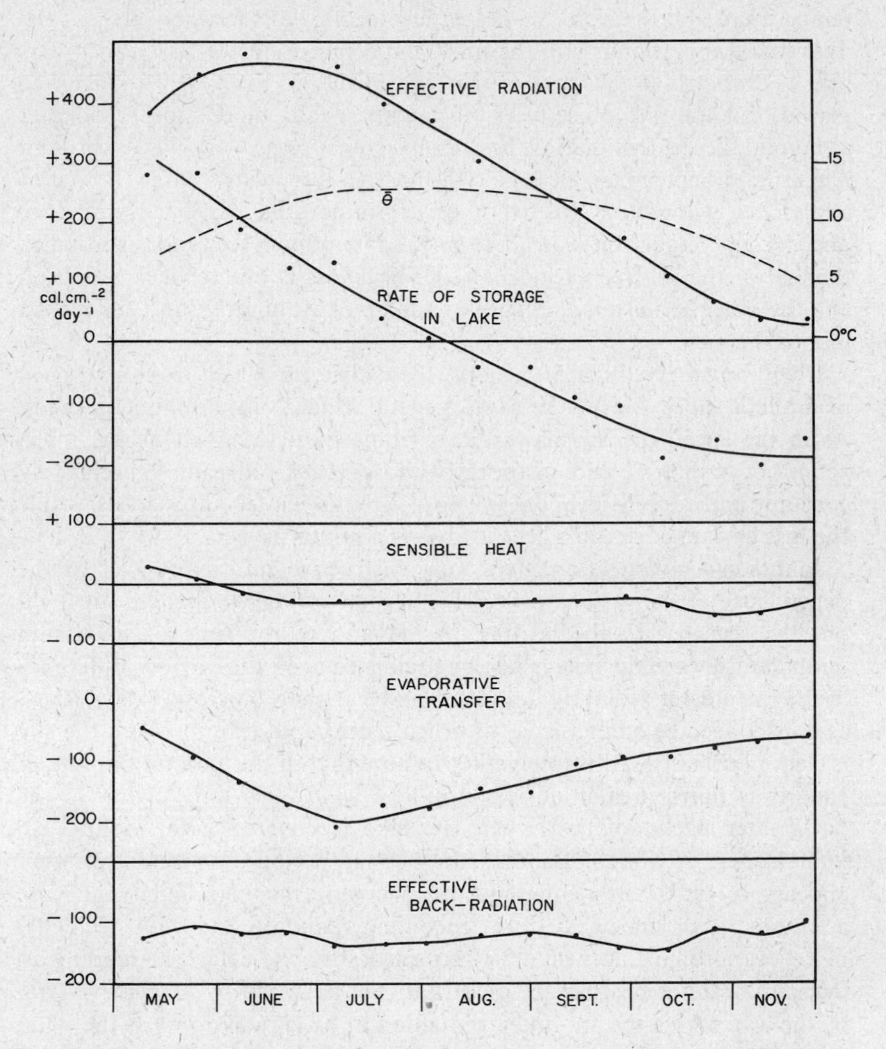

FIGURE 151. Variation in energy transfer during open season in Lake Klämmingen (after Johnsson).

progressively warmer lake loses more and more heat by evaporation and a small convective loss to the air is also evident after the beginning of June. The rate of heating of the lake therefore falls throughout June. Evaporation continues to rise until the beginning of July, by which time the rate of delivery of radiation to the lake surface has begun to decline. At the beginning of August when the lake has reached its maximum mean temperature and the curve of net rate of heating of the water crosses the abscissa, evaporation is still the most important mode of heat loss. The rate of evaporation falls progressively from July to the end of the ice-free period, but the rate of delivery of radiation falls more rapidly, so that although the back-radiation and convective losses vary little through August and September, the lake continues to lose heat. After the end of September evaporative loss becomes less important than back-radiation, and during October there is a rise in the rate of convective loss so that at the end of the ice-free period half the heat loss is due to back-radiation and the other half is about equally made up of evaporative and convective loss to the air.

It will be observed that in general these changes, which must be typical of dimictic lakes, are of the same general kind as in monomictic lakes. As in the latter, the thermal cycle is primarily determined by the interaction of radiation and evaporative loss; back-radiation is relatively constant and exceeds evaporative loss only as winter approaches, while the loss or gain of sensible heat is always a minor process.

Johnsson's investigation gives very instructive information as to the dependence of the various processes on meteorological changes throughout the season. In the heating period and at the time of maximum temperature when the lake is just beginning to cool, the obvious difference in the far greater radiation delivered on clear than on cloudy days leads, as hardly need be emphasized, to much more rapid heating when the sky is clear (Table 61). It is interesting to note that in the autumn the rate of cooling is much greater during cloudless periods, mainly owing to the far greater back-radiation when the lake is covered with cooling air, unblanketed by cloud. Early in the heating period the amount of wind appears to make little difference to the heat income; high winds are associated with somewhat lower incoming radiation and with decidedly more evaporation, as would be expected, but these factors tending to lower the rate of heating are quantitatively balanced by convective gain by the water from the air, and a reduction in back-radiation[26] as the wind mixes the warming water at the surface into the deeper layers of the lake. Later in the year, increase in winds leads to an immense increase in evaporation; this is particularly striking in July, when the evaporation is maximal. Wind direction is extremely important in the spring, when

---

[26] Although both back-radiation and convective cooling depend on the difference between air and water temperatures and must at any moment be of the same sign, in the expression for convective loss or gain the presence of the term for wind velocity, which may vary greatly over short periods, makes it possible for the mean values over a period of days to be of opposite sign.

TABLE 61. *Energy exchange, Lake Klämmingen, under different meteorological conditions (cal. cm.$^{-2}$ day^{-1})*

Conditions	Q_S' Net Radiation Income	$Q_A - Q_W$ Net Back-Radiation	Q_s Convection	Q_E Evaporation	Q_θ Rate of Accumulation*
May–June mean	434	− 111	− 3	− 110	210
May–June clear	588	− 185	10	− 138	275
May–June cloudy	210	− 59	− 25	− 94	33
July–August mean	356	− 126	− 33	− 166	31
July–August clear	491	− 182	− 17	− 188	104
July–August cloudy	157	− 83	− 49	− 143	− 118
September–November mean	105	− 117	− 42	− 77	− 131
September–November clear	179	− 234	− 55	− 93	− 203
September–November cloudy	48	− 70	− 33	− 58	− 113
May wind					
1.2 m. sec.$^{-1}$	400	− 124	− 2	− 41	233
7.5 m. sec.$^{-1}$	365	− 81	55	− 105	234
July wind					
2.0 m. sec.$^{-1}$	458	− 121	− 12	− 118	207
7.2 m. sec.$^{-1}$	481	− 143	− 26	− 354	− 42
October wind					
0.8 m. sec.$^{-1}$	121	− 132	− 16	− 28	− 55
7.2 m. sec.$^{-1}$	105	− 100	− 55	− 116	− 166
May					
SW. 3.9 m. sec.$^{-1}$	471	− 102	33	− 70	332
N. 4.2 m. sec.$^{-1}$	398	− 117	− 7	− 120	154
July					
SW. 4.4 m. sec.$^{-1}$	458	− 129	− 23	− 216	90
N. 3.7 m. sec.$^{-1}$	429	− 138	− 33	− 218	40
October					
SW. 4.2 m. sec.$^{-1}$	81	− 88	− 24	− 59	− 90
N.3.0 m. sec.$^{-1}$	86	− 136	− 57	− 94	− 201

* Calculated.

more than twice as much heat may be taken up with a southwest wind than when a north wind of the same velocity is blowing. This is largely due to the greater evaporation occurring with a north wind. During July, wind direction makes little difference to the rate of heating, but in October all forms of heat loss are greater with a north than with a southwest wind.

Miscellaneous data relative to the relation of the heat budget to evaporation and radiation. It is possible to gain a little further information about the relative importance of back-radiation, evaporation, and convectional cooling in a lake under relatively dry tropical conditions from Morandini's (1940) data relating to Lake Tana in Ethiopia. On the night of March 2–3, 1937, at Morandini's station II the lake lost from a water column of 8 m. about 416 cal. cm.$^{-2}$ between 6 P.M. and 6 A.M. On the night of March 4–5, at station IV the loss during the same period was about 100 cal. cm.$^{-2}$ from a column 10 m. deep. The mean value of 258 cal. cm.$^{-2}$ for the nocturnal twelve hours is probably a fair estimate for the lake as a whole at this season, the mean depth being about 9 m. The mean difference between the air and water temperatures during these nights was 2.9° C. Using the approximation of equation 8 of Chapter 6, this corresponds to 16 cal. cm.$^{-2}$ net back-radiation from the lake into the atmosphere during the nocturnal twelve hours. Evaporation studies indicate that at this season 5.0 mm. per day are lost from the lake. This would correspond to 293 cal. cm.$^{-2}$ day^{-1} latent heat of evaporation. It is reasonably certain that not more than half of this loss took place during the nocturnal twelve hours. The cooling of the lake from 6 P.M. to 6 A.M. may therefore be analyzed roughly as:

	Heat Loss, cal. cm.$^{-2}$
Observed cooling	258
Cooling due to evaporation	147
Back-radiation	16
Unaccounted for	95

It is reasonable to suppose that the unaccounted for fraction is sensible heat loss.

If the lake were neither gaining nor losing heat in a twenty-four hour period, as seems to have been the case at the time, the net radiation surplus required would be at least equal to the nocturnal cooling of 258 cal. cm.$^{-2}$ together with the evaporative loss by day of 147 cal. cm.$^{-2}$, or 405 cal. cm.$^{-2}$ This assumes no net sensible heat loss. It is a high but by no means impossible value. As in previous cases, evaporation is the main cause of heat loss, but the figures suggest that nocturnal sensible heat loss is greater at the season of observation than is back-radiation.

Miscellaneous evaporation studies. There is some information available on evaporation from lakes without other thermal data; much of this information is based on very dubious attempts to transfer determinations of evaporation in pans of various sizes and shapes to the estimation of lake evaporation. Such estimates are in general excessive, though the best pan data may be only about 10 per cent too great (Kohler 1952). More acceptable water-balance studies have been performed in a few cases.

For Lake Hjälmaren in Sweden, annual variation is comparable to that of Lake Klämmingen (Wallén 1914). Similarly Volker (1949a), studying Lake Ijssel, the artificial lake now replacing a large part of the Zuider Zee, found the maximum evaporation to take place in June and July. His results are set out in Table 62; the values of the heat losses are approximate, computed for a latent heat of 590 cal. g.$^{-1}$

TABLE 62. *Evaporation from Lake Ijssel*

Month	Evaporation, mm. month^{-1}	Heat Loss, cal. cm.$^{-2}$ day^{-1}
January	10	19
February	20	42
March	30	57
April	60	118
May	100	190
June	110	216
July	110	210
August	90	171
September	70	137
October	30	57
November	20	39
December	10	19

As in the other cases discussed, and in marked contrast to the condition in the ocean, the maximum evaporation is at the time of maximum temperature. In very large lakes there is probably a tendency for windiness to influence the period of maximum evaporation more than in the cases so far considered, but the freezing of temperate lakes under continental climates will always prevent such lakes from showing the winter evaporative maximum characteristic of the ocean. The Laurentian Great Lakes show a trace of this effect; Lake Constance, which does not freeze, seems to exhibit it more clearly.

An elaborate study of the evaporation from the surface of the Great Lakes was initiated by Hayford and completed by Folse (1929). The method employed began with an estimation of the effect, either aperiodic or periodic, of the winds and of changes in barometric pressure on the levels observed at a number of stations. Correcting for these and for the very slight tide, it was possible to estimate changes in level due to advection and evaporation. Knowing the rates of advection into and out of the lakes, evaporation could be estimated as a simple problem in water balance. The whole procedure involved a very elaborate series of least square fits as the effect of one variable after another was removed. The final conclusion was that when the wind velocity was below 480 cm. sec.$^{-1}$

the rate of evaporation was proportional to $(p_w' - p_w)$, but that when the wind velocity exceeded 480 cm. sec.$^{-1}$ a second term proportional to (W − 480) must be added. The constants in the resulting equations depend on the nature of the basin, and the equations themselves would now be held to have little theoretical justification. Nevertheless, the estimates which they produced for evaporation from the surfaces of Lakes Huron and Michigan, considered as a unit, are doubtless reasonably accurate. For the entire ice-free period, the mean monthly rates are given in Table 63. In default of exact temperature data, the corresponding heat losses can be estimated only approximately; a latent heat of 590 cal. g.$^{-1}$ has been employed, corresponding to 10° C.

TABLE 63. *Evaporation from Lakes Huron and Michigan*

Month	Rate of Evaporation, cm. day^{-1}	Heat Loss, cal. cm.$^{-2}$ day^{-1}
May	0.168	99
June	0.168	99
July	0.216	127
August	0.155	92
September	0.116	68
October	0.161	95

At the height of summer the vapor pressure potential $(p_w' - p_w)$ is maximal. Both this quantity and the windiness decline in August and September. In October, however, a further fall in $(p_w' - p_w)$ is more than compensated by increasing winds. If the lakes remained open, one might suppose a second maximum would develop in winter. The values are, on the whole, lower than those recorded for Lake Klämmingen. The total heat loss during the ice-free months as the result of evaporation is about 17,800 cal. cm.$^{-2}$

Halbfass (1923) has calculated from the difference in inflow and outflow the rate of evaporation from Lake Constance. The data are not presented in absolute terms, but relative to the inflow. It appears, however, that in this large lake in which the water remains unfrozen, the maximum evaporation is in November, the minimum in the spring from March to May. Maurer (1913) found in the Ägerisee and Zuger See maximal evaporation of rather over 100 mm. in July, and minimal values of 25 to 35 mm. in November and December. There was a hint of a secondary minimum in March or April.

In Pyramid and Walker lakes, Nevada, a little north of Lake Mead, the evaporative maximum is recorded in September, the minimum in February

(Langbein 1951). Neumann (1954) found evidence of an evaporative minimum in spring for the Dead Sea.

Suslov (1947) gives some data for Lake Baikal, stating that the mean discharge at the effluent is 1716 m.^{3} sec.$^{-1}$, which corresponds to 54.11 km.^{3} yr.$^{-1}$ The loss by evaporation is said to be 6 per cent of the total water loss, so that it should correspond to 3.45 km.^{3} yr.$^{-1}$ Since the area of the lake is 31,500 km.^{2}, the evaporation would be 109 mm. yr.$^{-1}$ This quantity, corresponding to about 6000 cal. cm.$^{-2}$, seems extremely low, but if it is correct[27] it is of considerable interest in suggesting that the high heat budget may in part be due to low evaporation during the heating period. There is indeed said to be more condensation than evaporation in June, July, and early August; the evaporative maximum is in September and October (Forsh 1944).

Summer heat income in relation to radiation flux, miscellaneous data. The summer heat income for Green Lake, Wisconsin, taken up apparently during the 122 days between about May 1 and August 20, is 26,200 cal. cm.$^{-2}$, whereas the mean annual radiation received at Madison during such a period is, by planimetric integration of the relevant part of the curve for Madison given by Hand (1949), 62,300 cal. cm.$^{-2}$ In the cases of Lake Cayuga, with a summer heat income of 29,480 cal. cm.$^{-2}$, and Lake Seneca, with an income of 34,020 cal. cm.$^{-2}$, comparison may be made with the mean value of 58,000 cal. cm.$^{-2}$ received during the period May 1 to August 20 at Ithaca, New York. The supposedly first-class lakes of eastern North America must take up an amount of heat during summer warming that corresponds very roughly to 40 to 60 per cent of that which is delivered as radiation to their surfaces. In the case of Lake Michigan a figure of just under 70 per cent is probable. Comparison of Strøm's heat budgets for the lowland Norwegian lakes with the values for the radiation received between the beginning of April and the end of July, as shown in Fig. 110, reveals much the same sort of relationship as in Wisconsin or the Finger lakes. In May, as has already been indicated, Klämmingen retains three fourths of the radiation that it receives, but at the time of maximal heat content in July, the retention is obviously zero. The American figures therefore appear to be reasonable. The high-altitude lakes of Norway, however, obviously retain a much smaller proportion than lowland first-class lakes. It is probable that in the solitary case of Lake Baikal at least 70 per cent of the radiation delivered is used in heating.

---

[27] The annual precipitation is given as 316.8 mm. over the lake; there is, however, an obvious error in converting this figure into the total volume added by rain and snow to the lake.

Effect on surrounding climate. The slow uptake of heat during the spring and summer and its loss during the autumn and early winter lead to a reduction in the amplitude of annual temperature variation in the vicinity of the lake. Since the total heat budget is a very respectable fraction of the radiant energy falling on the lake during the heating period of a first-class lake, the effect will obviously be of climatological importance in the vicinity of a large lake. Indeed, it can often be observed locally, as in parts of northern New York State where the incidence of small lakes may increase the growing season for crops (Geren 1941). When very large lakes are considered, the effect can be quite spectacular.

The most extraordinary case is certainly provided by Lake Baikal, as can be seen from the disposition of isotherms for December and for July given in Figs. 152 and 153.

Comparable but less regular effects are observed on the area around the Great Lakes in North America (Gruissmayr 1931), but owing to their

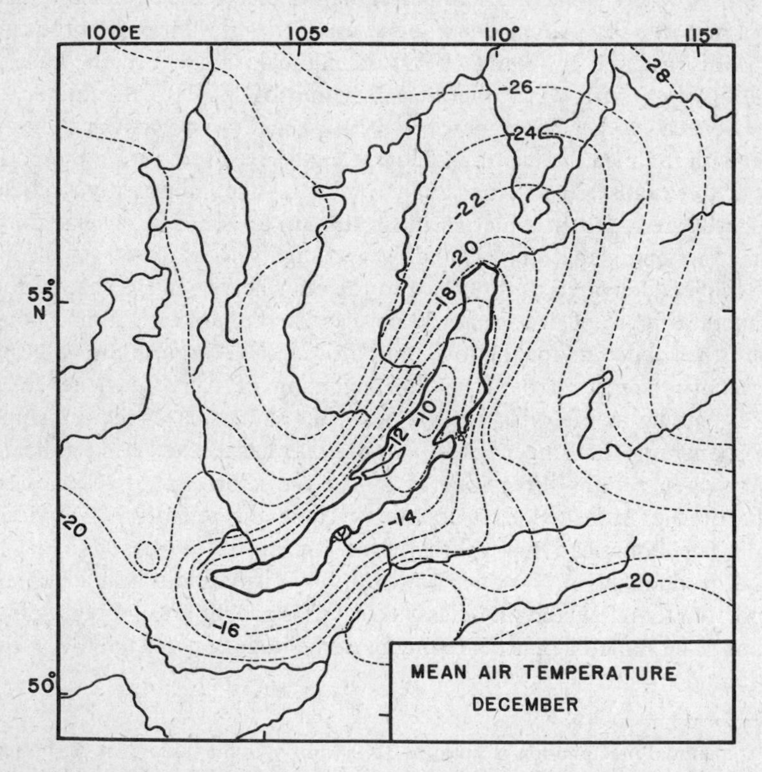

FIGURE 152. Isotherms for mean ground-level temperatures in December in the region around Lake Baikal (after Suslov).

irregular form and their relation to prevalent winds, their influence results in a rather less neat pattern than in the case of Lake Baikal. The seasons of minimum winter and maximum summer temperatures are a week or two later near the lakes than at the same latitudes far from the water (Leighly 1942). Cold air from the northwest moving southeast over Lake Superior and Lake Michigan is often warmed considerably, increases in temperature up to 12.5° C. being recorded. Because of this the eastern shore of Lake Michigan has a less severe winter climate than does the western Wisconsin shore (Wills 1941).

The effect of large masses of water in providing maritime climates around lakes is not of course limited to temperature changes. Humidity, cloudiness, and rainfall may all be increased by the presence of the lake, and the increase in water-vapor tension and cloud cover may affect the radiation received. In the Great Lakes region this effect will be greater to the southeast than to the west, owing to the prevalent westerly winds,

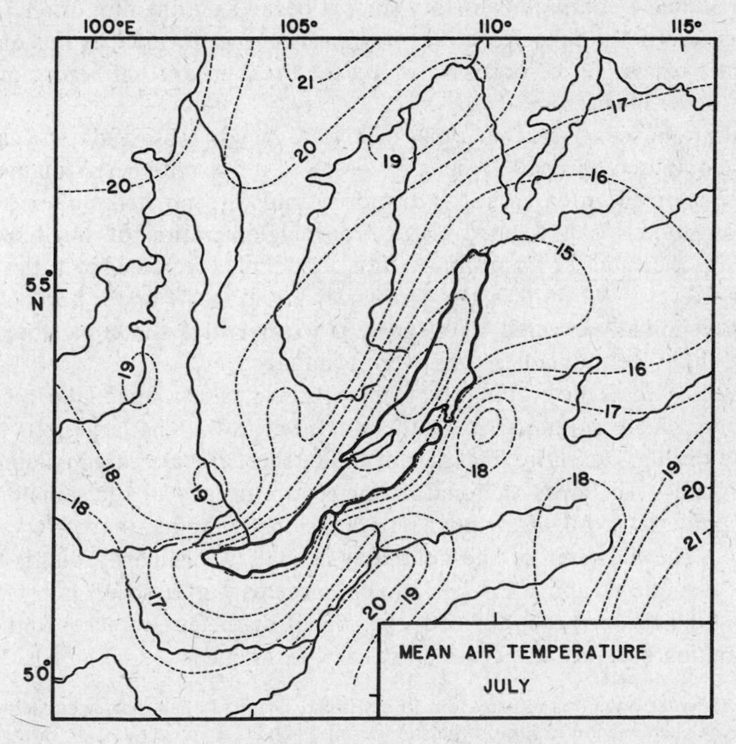

FIGURE 153. Isotherms of mean ground-level temperatures in July in the region around Lake Baikal (after Suslov).

and presumably underlies the greater radiation received (Fig. 111) at Madison, in spite of the Yahara or Madison Lakes, than at Ithaca, New York, southeast of the immense water surfaces of the Great Lakes (Hand 1949).

The climatic effect of a lake on its surroundings may occasionally have a striking biogeographic effect. The northward extension of mistletoe (*Viscum album* L.) along the shore of Lake Mälar in Sweden constitutes a striking example.[28]

LAKE ICE

Most lakes in cold temperate regions freeze after a longer or shorter period of autumnal circulation. The resulting ice cover has interesting properties of some geological and engineering significance.[29]

Classification of lake ice. Wilson, Zumberge, and Marshall (1954) classify lake ice according to the following scheme.

Two main gross types of ice cover can be recognized: (1) *sheet ice*, typically a smooth-surfaced homogeneous covering that has had, except for cracking, a continuous history since it began as a fine film or *ice skim* formed when the lake froze; (2) *agglomeritic ice*, formed by fusion of separate masses of ice or snow, or by the breakup and refreezing of an initial cover of sheet ice.

Within the ice cover, four types of texture may be observed. *Granular ice* is composed of small crystals of ice up to a few millimeters diameter, poorly oriented, often entirely at random, and with no striking tendency to elongation. It is formed either from agglomeration of small freely floating ice crystals or from snow. It is commonly formed when turbulent water freezes slowly. It is also frequently found in the upper parts of an ice sheet initially covered with snow, if water enters the snow through cracks in the ice and then freezes over the initial cover.

Columnar ice is the ordinary type of ice formed when a small lake freezes. The crystals are columnar, with the *c* axis vertical. The largest crystals extend through the entire thickness of the ice sheet and increase in diameter downward. The cross-sectional diameters are usually of the same order of magnitude in all directions, but occasionally wedge-shaped crystals occur. The diameter of the columnar crystals is ordinarily distributed about a mode of about 2.5 cm. in the lower part of a sheet, but as the crystals fill the spaces so formed, the modal diameter decreases and the distribution becomes less symmetrical and less regular.

---

[28] The writer owes this information to his friend Dr. G. Lohammar, who demonstrated the phenomenon to him in the summer of 1949.

[29] It is probable that there is a Russian literature on this subject which it has not been possible to consult.

Porphyritic ice, observed once on Lake Huron, is probably a variety of columnar ice in which large isolated crystals up to 25 cm. in diameter are included in a generally columnar matrix. Its mode of genesis is unknown.

Tabular ice differs from columnar in being formed entirely of large crystals, much wider than deep and of very irregular outline, though with vertical sides. It is formed when quiet water undergoes very rapid freezing. When this type of ice is breaking, it forms large, discrete, flat floating plates, and is therefore referred to as *flagstone ice* by Wilson, Zumberge, and Marshall.

These four types characterize sheet ice. The less well-defined agglomeritic ice is formed, as has been indicated, either from the breakup of sheet ice or from snow, in the form of slush balls. In the breakup of ice floes, large irregular pieces may be formed or the fragments may become rounded by continual collisions, producing pan ice. Either type may be reincorporated into an ice sheet by refreezing, giving a dynamic *abrasional* ice sheet. Slush balls are most perfectly formed during snow squalls in shallow water where waves are breaking. They may become connected by thin ice, giving a dynamic *accretional* ice sheet. The genesis of the various textural types is neatly summarized by Wilson, Zumberge, and Marshall in the diagram of Fig. 154.

When a sheet of ice is covered by a considerable amount of snow, it may be depressed by the snow sufficiently to allow water to run in under the snow but above the original clear ice. Further cooling may produce a layer of névé-like "frozen" snow above the water. Occasionally more than one watery layer may form; Götzinger (1909) and Morandini (1949) give examples from the Austrian and Italian Alps. Along the cracks, through which water rises, a series of depressions in the snow often develop.

Duration of the ice cover. The length of time a lake is ice-covered depends on the severity of the winter, on the depth and situation of the lake, and on the exact incidence of suitable freezing weather and thaw weather.

Freezing weather is in general determined by low temperatures, clear skies, and low wind velocities. Johnsson (1946) gives the following mean conditions for the night of freezing of the central and Klövsta basins of Lake Klämmingen:

	Temperature, °C.	Cloud Cover, in tenths	Wind Force (Beaufort)
Central	– 10.9	1.2	1.3
Klövsta	– 16.8	3.2	1.2

The central basin, being shallower than the Klövsta, freezes earlier at a high temperature, but may thaw and refreeze before the whole lake is

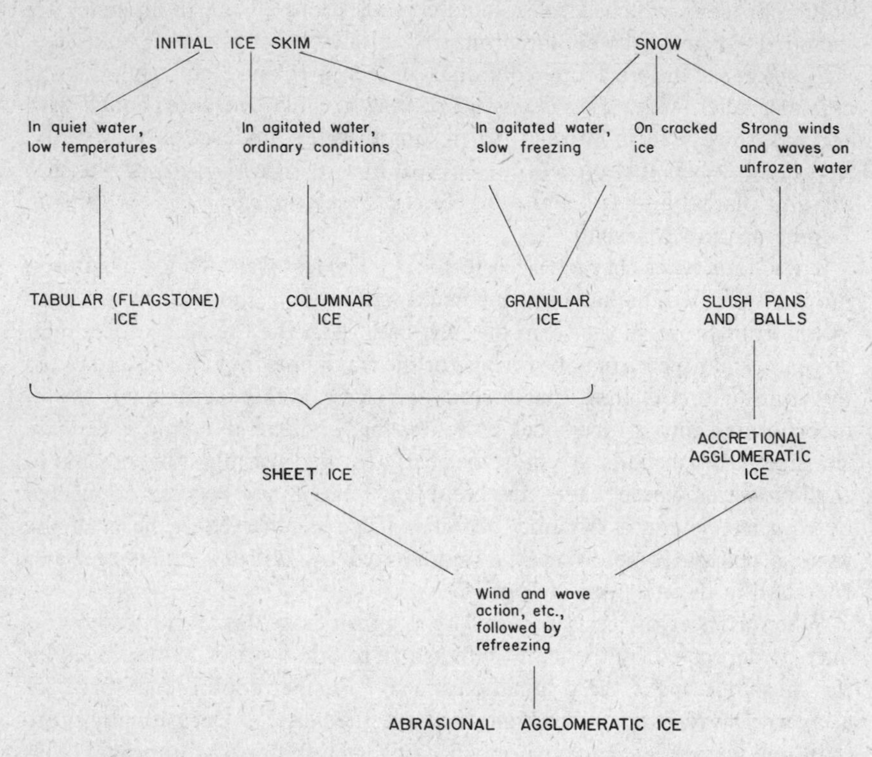

FIGURE 154. Genetic relationships of the various types of lake ice (Wilson, Zumberge, and Marshall).

ice-covered. The days prior to thaw in the spring naturally have temperatures well above 0° C., with a greater cloud cover and stronger winds.

The initial stages of thawing generally involve the "rotting" of the ice, which becomes very weak while preserving a considerable thickness. Humphreys (1935) suggests that the rotting process involves depression of the freezing point of interstitial water by inorganic matter, precipitated out during freezing, that has accumulated between the ice crystals which constitute the main volume of the sheet. The final breakup of the rotting ice may be very rapid, giving rise to the popular belief that the ice sheet sinks overnight.

Birge (1910c) believed that absorption of radiant heat by the ice sheet is of primary importance in melting. The usual meteorological conditions accompanying a thaw suggest that an increase in the net radiation surplus owing to lowering of effective back-radiation is critical. The thaw usually starts in shallow water, in which radiation is absorbed by or reflected from the bottom. If the open marginal water is warmed above 4° C., it may flow laterally under the ice, carrying a considerable amount of heat to the lower surface of the sheet. This possibly may be the explanation of the very irregular eroded nature of the lower surface of some old ice sheets, as for instance that of Ororotse Tso (Hutchinson 1937b), where water 1 cm. from the ice at the margin had a temperature of 8° C. at 11 A.M. in July. Heating of a layer of dilute melt water below the ice, producing a warm layer which would be stable relative to the water beneath it but unstable relative to the transition at 4° between the ice and the warm layer, may play a part (Birge 1910c) in melting from below in any lake which can develop such a layer. One would expect, both from Humphrey's idea about ice rotting and from Birge's suggestion just noted, that the ice on a hard-water lake would break up faster than that on a soft-water lake.

Remarkable patterns, apparently indicating the existence of thermal convective cells over the centers of which melting is initiated, have been described by Woodcock and Riley (1947, Brunt 1946) in small lakes on Long Island, New York.

The duration of the ice cover on a number of lakes has been recorded for many years. There is, as is to be expected, great irregularity, but a long series of records (Wing 1943) for the number of days Lakes Mendota and Monona were ice-covered between 1850 and 1941 shows a clear tendency towards a short period of ice cover during the twentieth-century as contrasted with the nineteenth century.

Maurer (1924) has been able to determine, from an analysis of a variety of historic records, years of abnormally protracted ice cover on the lakes of southern Germany and Switzerland since about 1400 A.D. He concludes that there were eight such years in both the fifteenth and sixteenth centuries, six in both the seventeenth and eighteenth centuries, and only four in the nineteenth century. There is thus evidence of a general amelioration in winter climate over the period considered, but Maurer points out that this is, in part, only of local significance.

A third and the most remarkable of all studies of the freezing dates of lakes relates to Lake Suwa in central Japan (Arakawa 1954), for which a large body of data exists, going back to 1443. To 1682 it is particularly complete; the eighteenth- and nineteenth-century information is less detailed and often unsatisfactory. The record as a whole, however, indicates that the number of completely open winters, in which the lake did not freeze, was 27 during the period from 1700 to 1954, as against 13 during the period from 1443 to 1700. Nearly all the open winters in the sixteenth century were between 1505 and 1515, when the lake froze only in 1506–1507. During the entire period between 1557 and 1700, the lake froze every year for which records are available (no information exists for 1594–1595 and 1682–1683), except in 1670–1671. In contrast to this, from 1800 to 1954 there were at least 17 open years. Subsidiary information on *Omiwatari* or pressure ridges, which normally form a few days after freezing, provides corroborative evidence of the accuracy of the freezing dates. The modal date for the formation of such ridges lay in the first ten days of January prior to 1700, and during the second ten days subsequent to 1700. It is quite evident that the winters of the late fifteenth, the late sixteenth, and the whole of the seventeenth centuries were usually more severe in central Japan than during the past two and a half centuries. There is, however, no clear evidence from the record of mildness developing progressively during the past century, and the only consecutive run of open winters since 1515 was from 1865 to 1868.[30]

Effect of temperature changes on the ice cover. Very characteristic effects of temperature changes may be observed on an ice-covered lake. After the ice has covered a lake, it is likely to be cooled as winter temperatures fall. A slow fall in temperature apparently permits the ice to flow plastically over the lake surface without cracking, but if the fall is rapid, cracks develop as the cooling ice contracts. According to Zumberge and Wilson (1953) there is a tendency for the cracks to form two systems, one radial and the other at right angles. These tension cracks expose water, which rapidly freezes. The optimum condition to produce this effect is a very large and sudden drop in temperature after a relatively mild period has permitted all the ice to approach 0° C. Zumberge and Wilson figure intersecting cracks formed on Wamplers Lake, Michigan, after the air temperature had fallen from about − 1° C. to about − 13°, at a mean rate of 1.06° C. per hour. The cracks varied in width from a few millimeters to nearly 4 cm. across.

When a lake has frozen, cracked, and refrozen, heating to the original freezing temperature of 0° C., if sufficiently rapid, will produce an ice sheet of greater area than before cooling and crack formation occurred. This sheet may then exert pressure on the shore, forcing gravel and stones landward and building an *ice push* or *ice rampart* (Buckley 1900, Hobbs 1911, Scott 1926). These earlier investigators all realized that sudden expansion was essential to produce an ice push; Zumberge and Wilson (1953) found that a rise in air temperature of 0.9° C. per hour for 12 to 14 hours under an overcast sky was sufficient to produce an ice push on Wamplers Lake. The ice on this lake, when 20 cm. thick, may push almost half a meter beyond its original position on the shore, sliding over and

---

[30] Important information on the ice cover of Finnish lakes is given by Simojoki (1940) in a work that became available to the writer too late to use in this book.

tending to move landward such boulders and gravel as it may encounter. On larger lakes a greater effect may be expected. The direction of the expansion is evidently partly determined by the form of the basin; Zumberge and Wilson, who followed the position of markers set in ice, found that in Wamplers Lake such markers moved up to 80 cm. between January 23 and March 1, 1952. The direction was about midway between a line normal to the coast and the direction of the long axis of the lake.

If the ice is not over about 30 cm. thick, it may buckle as it expands, producing pressure ridges in which cracks are formed. These are at right angles to the directional movement of the expanding ice. When ice having a surface temperature well below 0° C. is warmed suddenly by solar radiation, the resulting compression is released by fracturing, particularly if a slight additional force, such as that due to someone walking on the ice, is applied. The effect does not occur if there is a light snow cover. The cracking is accompanied by a rumbling sound, which has doubtless been noted by everyone who has ventured onto lake ice at the end of winter. The loudness of the noise of cracking on Lake Baikal, noted by Suslov (1947), must be very dramatic and terrifying to anyone previously unfamiliar with the lake.

Wilson, Zumberge, and Marshall (1954) have examined the theoretical aspects of the production of an ice push. The ice is considered initially as a flat plate with a temperature gradient; the lower side of the plate will always be at 0° C., the upper side will have a varying temperature. If the surface of a flat ice sheet is heated from a temperature below 0°, the ice will take the form of the surface of a segment of a sphere. The central part will be unsupported by the buoyancy of ice in water, and the strain set up will tend to cause cracking. By making certain reasonable though oversimplified assumptions, an elementary theory is developed which indicates that the rise in temperature at the surface needed to cause cracking is independent of thickness, though the size of the segments resulting from cracking will increase with the thickness of the ice. Thus a sheet 10 cm. thick, raised in temperature 3.6° C. at the surface, will break into free-floating pieces which will be stable when their radius is about 13 m.; a sheet 300 cm. thick will break, for the same temperature rise, into pieces having a radius of about 155 m. It is possible also to compute the linear extent of the ice push produced by the collapse of a sheet for a given rise in temperature, but the computed results are apparently too great by a factor of 4 or 5.

All the movements of ice caused by contractions and expansion are likely to be influenced by local topography. Suslov (1947) states that on Lake Baikal large cracks appear in sufficiently constant position from winter to winter to have received names.

SUMMARY

The thermal history of any lake depends primarily on its geographical position and depth. The most complete annual cycle is exhibited by relatively deep lakes in temperate latitudes, in which the water can all lie at or below 4° C. (the temperature of maximum density) in the winter, and all at or above 4° C. in the summer. In such a lake, in the spring there will normally be a period of vernal circulation at which the water is of the same density throughout, and in which vertical circulation requires only minimal forces sufficient to overcome viscosity. As the season progresses, heat is delivered to the lake at its surface, mainly as solar radiation which is rapidly absorbed by the superficial water. In a completely undisturbed lake of vertically uniform transparency, this would lead to a temperature curve falling exponentially with depth.

In any actual lake the observed distribution of temperatures as the summer heating proceeds is not, in the upper part of the lake, of this form. What is observed is an upper layer or *epilimnion* of relatively freely circulating water with a small and variable temperature gradient, lying over a deep and more or less undisturbed cold layer or *hypolimnion* in which the gentle vertical fall in temperature is roughly exponential. The plane in which the temperature falls most rapidly is called the *thermocline*, and the layers on either side of the thermocline, loosely embracing the whole region of rapid fall in temperature with depth, are called collectively the *metalimnion*. (This usage is not strictly that of the originators of the terms but proves to be convenient.)

When a lake has developed this type of temperature distribution it is said to exhibit direct or summer *thermal stratification*. The exact method by which thermal stratification comes into being has proved very hard to elucidate. It is clear, however, that it depends on the fact that water heated at the surface is less dense than deeper, colder water, and that work must be done to mix the warm upper with the cold lower water. The most reasonable explanation of the observed type of temperature curve also involves the idea of variation in turbulence with depth. As the temperature gradient develops, the consequent gradient in density will reduce vertical turbulence, but at the same time will tend to cause horizontal movement when any momentum is transferred from the upper to the lower layers by such turbulent exchange as is possible. The horizontal movement, however, promotes vertical turbulence. There is reason to believe that the interaction of these processes results in the observed form of temperature distribution, though the simplified equations involved require mechanical aid for their solution and are still too complicated to permit any really clear physical insight into the details of the process.

If the variations of geographical latitude, altitude, and depth of actual basins are considered, we can regard lakes as belonging thermally to the following types:

Amictic, sealed off by ice from most of the changes at the earth's surface; a very rare type known in the Antarctic and occasionally in high mountains.

Cold monomictic, with water temperature never in excess of 4° C. at any depth, freely circulating in summer at or below 4° C., ice-covered with an inverse stratification in winter; the polar lakes of earlier investigators, probably only clearly exemplified by lakes which, though open in the summer, always have ice in contact with the water.

Dimictic, freely circulating twice a year in spring and autumn, inversely stratified in winter, directly in summer; the temperate lakes of older investigators, characteristic of most of the cooler parts of the temperate zones and some high mountains in subtropical latitudes.

Warm monomictic, with the water never below 4° C. at any level, freely circulating in winter at or above 4° C., directly stratified in summer; the tropical lakes of older investigators, characterizing, however, the warmer and more oceanic parts of the temperate zones and some mountains in subtropical latitudes.

Oligomictic, in which the water is always well above 4° C., with rare circulation periods at irregular intervals; the characteristic type of the humid tropics at low altitudes.

Polymictic, continually circulating at low temperatures little above 4° C. whatever the depth; the characteristic type of high mountains in equatorial latitudes, in which there is little seasonal change in air temperature but always enough heat loss to prevent the development of stable stratification.

The foregoing classification refers to lakes which are of sufficient depth to develop a hypolimnion. It is also convenient to have a series of categories, applicable at least to dimictic and warm monomictic lakes, that express the fact that the completeness of thermal stratification depends on the mean depth of the lake:

First-class lake, bottom temperature during summer stratification so near 4° C. (or the minimum temperature during winter circulation of a warm monomictic lake) that extrapolation of the temperature curve would involve a negligible increase in heat entering lake.

Second-class lake, thermally stratified but with sufficient seasonal variation in bottom temperature that extrapolation towards a limiting value of 4° C. (or the minimum winter circulation temperature) would involve an appreciable increase in heat entering lake.

Third-class lake, not thermally stratified.

It is to be noted that any mechanism which tends to keep the surface waters cool while heating is taking place will decrease the over-all temperature gradient but may increase the amount of heat that can enter the lake. The most important of such mechanisms are oceanic climatic regimes, the presence of ice in the drainage basin, very great area exposed to wind, and a large volume development. Of even greater importance in large lakes, particularly elongate lakes in overdeepened valleys or grabens, is the effect, during strong wind, of exposure of the metalimnion at the surface upwind and consequent horizontal mixture of warm and cold water. This probably underlies the enormous heat uptake by Lake Baikal.

The ordinary type of thermal stratification is subject to many modifications. If cooling starts in the late summer, a thick epilimnion may be formed, which can develop one or even more secondary thermoclines if subsequent warm spells intervene. It is, in fact, quite likely that the conventional smooth temperature curves drawn by most limnologists would be replaced in many cases by curves consisting of a series of small steps if more data were available from the records of continuously recording instruments.

In the autumn, cooling at the surface causes a thickening of the epilimnion by convective mixture as its temperature falls; at first the hypolimnion may be very little affected. Cooling at the surface in autumn can sometimes produce unstable thermal gradients of up to 0.5° C. per centimeter in the top few centimeters of a calm lake. If the lake freezes, it will generally be after it has circulated for some time below 4° C. The water at the ice contact will be at zero, and a slight *inverse stratification* will be set up immediately below the ice. Usually this is accentuated, as the winter continues, by marginal heating through the ice, causing water near 4° C. to flow down the slope to the bottom of the lake, and by heating from the mud. In small lakes these processes may raise water above 4° C., but at the same time sufficient dissolved material from the mud may enter to make such water dense enough to flow to the bottom, raising the bottom temperature several tenths of a degree above 4° C. When melt water flows under clear ice, particularly in lakes with hard water, the difference in density between the lake water and very dilute melt water may permit temperatures up to 7.5° C. under the ice without disturbing the stable density gradient.

Since the temperature of maximum density decreases with increasing pressure, any lake in which cold water can be mixed at circulation to a very great depth will have, during summer stratification, bottom temperatures somehwat below 4° C. This effect can usually be observed in dimictic lakes of maximum depth over 100 m.

In Lake Baikal there is a very thick layer, from 300 m. to 1600 m., in which the temperatures show practically no seasonal variation but fall very slowly with depth to a minimum of 3.16° to 3.19° C. In the extreme bottom layer there is a slight temperature inversion, perhaps stabilized by dissolved material from the sediments; at 1698 m. the temperature is 3.23° C., or about 0.04° to 0.07° higher than at 1600 m. This is probably due to heating from the earth. Adiabatic heating should be insignificant so close to the temperature of maximum density.

Most lakes at low altitudes in the humid tropics are oligomictic. The polymictic lakes are so far best known from the Andes. In tropical central Africa a great variety of intermediate types occurs. It is probable that all very deep tropical lakes are *meromictic*, having a chemically stabilized zone of dead water at the bottom. In the case of Lake Tanganyika there is a minute temperature inversion below 1000 m., which is probably due to adiabatic heating. The additional heating from the earth's interior must cause very slow circulation.

Though the temperature changes in the hypolimnion throughout the summer may be very small they are of great interest in throwing light on other events in a lake. They have been best studied by consideration of the coefficients of eddy conductivity involved. Simple assumption of downward turbulent conduction of heat leads to values of the coefficient that are minimal in the metalimnion. It also appears to lead to inconsistencies. If the study of turbulence begins with a consideration of temperature change at specific levels, the coefficient usually seems to be relatively constant in spite of stability decreasing with depth through the lower metalimnion and upper hypolimnion, a region which may be termed the *clinolimnion*. This point of view leads to difficulties with respect to the deepest part of the lake, the *bathylimnion*, which difficulties may perhaps be resolved in some cases by postulating heating by chemically determined density currents. As far as the results of such studies go, large deep lakes have more turbulent hypolimnia than do small shallow lakes. The only reasonable general interpretation of hypolimnetic turbulence is that it is due to nonlaminar flow in internal seiches. In some special cases other mechanisms, including the movements of fish or the rising of methane bubbles, may be involved.

Most of the lakes discussed herein are *holomictic*; that is to say, when they circulate, the circulation is complete to the bottom. A large class of *meromictic* lakes exists; in such lakes a greater or lesser volume at the bottom is stabilized by dissolved or even, for a time, suspended matter, so that there is a permanent increase in density across a *chemocline*. The deep stable layer is called the *monimolimion*; the rest of the lake, which can stratify as a holomictic lake does, is called the *mixolimnion*. In general three kinds of meromixis may be recognized:

Ectogenic meromixis, in which an external event fills the basin with saline water over which fresher water flows, or initiates a flow of fresher water over original saline water. The meromictic stability slowly decreases as the saline water is mixed into the fresher water (e.g., Hemmelsdorfersee near Lübeck).

Crenogenic meromixis, where a submerged source of saline water exists. As saline water enters the lake, the chemocline is forced up till an equilibrium is established between the rate of entry of denser water and the rate of its loss by mixing at the top of the chemocline. This is possible, of course, only where there are influents and an effluent. The meromictic stability will be more or less constant (e.g., Lake Ritom, Switzerland).

Biogenic meromixis, where the results of bacterial mineralization lead to a concentrated layer of water at the bottom of the hypolimnion. In a year having low wind velocity in autumn, this layer persists into the next year's cycle. The effect is then cumulative, and the monimolimnion may survive permanently (e.g., many cases in Austria; Lake Mary, Wisconsin).

It is possible that biogenic meromixis is artificial, facilitated by clearing of forest and the subsequent delivery of slowly settling silt to the bottom of the lake, which initiates the condition.

Some artificial lakes in salt pits and similar places are perhaps intermediate between the ectogenic and crenogenic types, as for example, Hot Lake, Washington.

There is often, though not always, a stable temperature inversion in the monimolimnion which may sometimes be very dramatic.

Temperate first-class lakes over a wide range of latitudes usually take up 30,000 to 40,000 cal. cm.$^{-2}$ yr.$^{-1}$ during their complete annual cycle. Of this, 20,000 to 30,000 cal. cm.$^{-2}$ yr.$^{-1}$ are taken up after vernal circulation and require work to distribute them into the deeper water of the lake. This summer heat income usually appears to correspond to about half the heat delivered to the lake surface in the warming period. Occasionally larger heat budgets are recorded: Lake Baikal appears to have an annual budget of nearly 60,000 cal. cm.$^{-2}$ yr.$^{-1}$ and a summer income of at least 34.500 cal. cm.$^{-2}$ yr.$^{-1}$ The budgets of lakes within the Arctic Circle are probably little below those of comparable temperate lakes if due allowance is made for the melting of ice. The annual income and loss of the oligomictic lakes of the tropics is of course very small, and that of the tropical polymictic lakes not estimable on an annual basis. In general, the rate of entry of heat is greatest in the late spring and early summer, but at high altitudes in central Asia more than half the heating occurs after July 1, and the same must be true in the Arctic.

In shallow lakes the heat budget of the sediments is often of importance. In general, the temperature in the sediments follows a wave which has a decreasing amplitude, with its maximum displaced later and later in the year, with increasing depth. In Lake Mendota, with an annual budget of 23,500 cal. cm.$^{-2}$ yr.$^{-1}$, about 2000 cal. cm.$^{-2}$ yr.$^{-1}$ enter and leave the sediments; in the much smaller Lake Hula, with an annual budget of 2240 cal. cm.$^{-2}$ yr.$^{-1}$, 1400 cal. cm.$^{-2}$ yr.$^{-1}$ enter and leave the sediments.

The work needed to produce the observed distribution of temperatures must be due primarily to the wind. In most first-class lakes 0.08 to 0.12 g.-cm. of work are needed to mix 1 cal. of heat. In Lake Atitlan in Guatemala a larger figure of 0.16 g.-cm. cal.$^{-1}$ has been recorded; this may prove characteristic of warm monomictic lakes at fairly low latitudes but high enough altitudes to have fairly large annual budgets. In the case of Green Lake, Wisconsin, the work required to distribute the summer heat income corresponds to an average force of about 0.02 dyne cm.$^{-2}$, and similar figures would be obtained from other dimictic first-class lakes. This probably represents only a small fraction, of the order of a few per cent, of the average wind stress.

The stability of a lake may be defined as the amount of work needed to mix it completely, so that its temperature becomes uniform without gain or loss of heat. The stability will thus be zero for a third-class lake. Adding the stability to the work of the wind will give a figure indicating how much work would have to be done to keep the lake circulating, given its heat budget. For large deep lakes such a quantity of work may be very great, several times the actual observed maximal values of the work of the wind on any lake, but in the case of second-class lakes the work needed to keep them circulating is often well within the limits possible. Lake Mendota always stratifies very stably in summer, but to maintain it in full circulation would take less work than is actually done to produce the observed stratification in the deeper but smaller Green Lake in the same general region.

In the few cases in which it is possible to follow the heating of a lake analytically, it appears that the annual cycle of radiation, with a maximum in June in the northern temperate zone, coupled with the annual cycle of evaporation, which generally implies a maximum evaporative heat loss in the summer or autumn, is sufficient to account for the general form of the curve relating heat content to time of year. Back-radiation is almost as important as evaporation as a major mechanism of heat loss, but it appears to exhibit very much less seasonal variation. The conductive heating of air by water is ordinarily unimportant. Lakes can start losing heat by evaporation in summer, well before their net radiation surplus is negative; the Sea of Galilee, which has a well-developed warm monomictic temperature cycle, always has a net positive surplus whether it is gaining or losing heat.

The presence of large lakes in continental areas has effects on the local climate comparable to the effects of the ocean. Summer temperatures are lowered, winter temperatures are raised, and the times of mean maximum or minimum temperature are delayed.

Ice forms on many dimictic and on all cold monomictic lakes during the winter. The conditions for ice formation are, in general, air temperatures well below 0° C., water temperatures well below 4° C., clear skies, and low wind velocities. Two main kinds of ice may be recognized: *sheet ice*, with a continuous history since it formed as an *ice skim*; and *agglomeritic ice*, formed by the aggregation of masses of broken sheet ice or of snow. Four types of texture may result. *Granular ice* is composed of poorly oriented small crystals; it may be formed directly in turbulent water or from snow. *Columnar ice* with crystals set vertically in the sheet along their *c* axes, is ordinarily formed when a small lake freezes in the absence of snow. *Porphyritic ice* is a very rare type, differing from columnar ice in the inclusion of very large crystals of unknown origin. *Tabular ice*, composed of very wide crystals with vertical sides, is formed when quiet water freezes very rapidly.

The conditions for the breakup of an ice cover are roughly the reverse of those for its formation. The development of a positive net radiation surplus is no doubt important. The ice usually becomes very friable just before it melts. This is supposed to be due to the presence of trapped interstitial water containing dissolved mineral matter; such water, if it has frozen, will melt below 0° C.

The duration of the ice cover is a rough measure of the severity of the winter. Records exist that indicate longer ice covers in Wisconsin in the last century than in the present century, a progressive decline in the severity of freezing in south Germany and the Alps since the fifteenth century, and very marked periods of intense freezing in Japan in the fifteenth, late sixteenth and seventeenth centuries.

The thermal contraction of ice, if fast enough to preclude plastic flow, causes cracks to develop. These tend to form two systems, one radial, one circular. The cracks fill with water and refreeze. When the temperature rises again, if the rise is rapid, the ice cover exerts a force on the shore, producing an ice rampart. If the ice is less than about 30 cm. thick, the expansion may produce buckling or pressure ridges. Rising temperatures may produce compression, which is relieved by fracturing when a small additional force is applied, as by someone walking on the ice. The fracturing is usually accompanied by a loud sound. All the results of contraction and expansion are influenced by local topography. On Lake Baikal the cracks are constant enough from year to year to have acquired geographical names.

CHAPTER 8

The Inorganic Ions of Rain, Lakes, and Rivers

This chapter presents information on the main inorganic salts dissolved in the part of the hydrosphere with which the limnologist is concerned. It begins with a consideration of rain, which is much more like certain dilute lake waters than might be supposed, then develops the idea of Rodhe and of Gorham, of a standard river water to which existing river and lake waters approach more or less closely, and finally considers what happens when such waters evaporate in closed basins.

THE COMPOSITION OF ATMOSPHERIC PRECIPITATION

Nuclei of water droplets. Owing to surface tension, the equilibrium vapor pressure in the air in contact with a small drop is greater than in air in contact with a plane liquid surface. In consequence of this, the minute droplets of water which constitute clouds cannot form unless nuclei are present in the supersaturated air from which cloud formation may be expected. The necessity of such nuclei has been recognized for more than half a century, but no agreement as to their nature has yet been reached. The most effective nuclei are presumably hygroscopic electrolytes, over which the vapor pressure of water will be lowered. It is well known, however, that charged aggregates of molecules, forming gaseous ions, can act as such condensation nuclei, and there is experimental evidence that at least in some circumstances nonhygroscopic dusts are also effective.

The work prior to 1941 has been the subject of a masterly review by Simpson (1941a).

The majority of workers in the past have supposed the most important kind of nucleus involved in ordinary cloud formation to be very finely divided sea salt particles derived from spray. Woodcock's studies (1950, 1953) strongly suggest that cloud droplets form around such nuclei over the ocean. Owing to the presence of magnesium chloride, salt particles are hygroscopic and go into solution at a relative humidity of about 70 per cent. Chloride is almost invariably present in rain, and Köhler (1937) has been able, by the optical analysis of cloud and the chemical analysis of rime formed from cloud at the Hallde Observatory in northern Norway, to estimate the number of droplets per unit volume of cloud and the chloride concentration within them. If chloride particles are assumed to be the nuclei of the droplets of the cloud, they must have a mean mass of 1.85×10^{-14} g., and a mean diameter of 17.7 μ. The rime frozen from such clouds contains, when melted, 3.6 mg. Cl per liter, which, though a little high, is a reasonable figure for coastal rain. Wright (1940), by means of a most ingenious study of the opacity and humidity of the air at Valentia in Ireland, a locality far from industrial contamination, has provided clear evidence that the nuclei involved in the formation of mist at that locality are hygroscopic. He assumed them to be formed of sea salt, crystallizing as solid particles at a relative humidity below 70 per cent. Simpson (1941b), reconsidering Wright's data, has shown very clearly that the nuclei involved cannot be sea salt, for Wright observed practically no reduction in opacity as the relative humidity fell below 70 per cent; if crystallization occurred at that point, there should be a sudden reduction in opacity to about half the value which was observed just above 70 per cent relative humidity.

Since the products of industrial combustion, which Findeisen (1937) has regarded as the most important source of nuclei, are likely to be present in minimal amounts at Valentia, and since there is no reason to suppose that the existence of rain is dependent on man's invention of fire, Simpson believes that nitrous acid probably forms most of the nuclei in natural air. This idea has indeed been held by several earlier investigators (Pringal 1908; Coste and Wright 1935) who have studied cloud and nucleus formation in the laboratory. In at least one case, cloud has been noted to form over a lake along the path of a lightning flash; this is reported by Pringal to have been observed by J. Kiessling over the Lake of Lucerne. Sulfuric acid as well as nitrous acid is probably quantitatively significant in industrially polluted air. Ammonium chloride may be involved also, but the occurrence of this compound in uncontaminated air is uncertain.

In view of the improbability of either sulfuric acid or sea salt being of general significance as cloud nuclei, it is evident that the presence of cloud, and so of rain, is dependent ultimately on nitrogen compounds, either ammonia or HNO_2. It is becoming apparent (Hutchinson 1954) that oxidation in thunderstorms is not the only, nor indeed the most important, source of oxides of nitrogen in the atmosphere and in rain. Photochemical production of nitrite from ammonia must be taking place continuously in the atmosphere, probably in association with dust particles. It is reasonable to suppose that either ammonia brought into the atmosphere as chloride, or HNO_2 and perhaps also HNO_3 formed from ammonia by photochemical oxidation, are the common natural sources of cloud nuclei in unpolluted air. (See, however, Junge 1956.)

Formation of raindrops. There has been a great deal of discussion as to the mechanisms by which the small droplets of cloud are aggregated into true raindrops. Although there must be a complete set of transitions between the smallest cloud particles and the largest raindrops, there is an important functional difference, for Findeisen (1939) has shown that drops smaller than 0.1 mm. in diameter can evaporate after falling a few hundred meters, whereas drops of the order of 1 mm. in diameter must fall many tens of kilometers before they evaporate completely in an unsaturated atmosphere. All drops with a diameter less than 0.1 mm. are very unlikely to reach the earth's surface, and are therefore to be regarded as cloud; the larger, more rapidly falling drops are to be regarded as rain. The major argument has been whether true rain can be formed without the prior production of ice crystals at the top of the cloud. Simpson's review may be consulted for a clear discussion of the matter. It would seem that, in general, ice crystals, which will have a lower vapor pressure than supercooled cloud droplets and so will collect water, do provide one good mechanism for rain production, but it is most unlikely that it is the only one. Langmuir (1948) has produced a characteristically ingenious theory in which the growth of a single water drop, falling and sweeping into itself enough cloud droplets to grow, may produce instability when the drop reaches a diameter of about 3 mm. If the drop fractionates, small daughter drops may be carried to the top of the cloud in an updraft, to repeat the process until a very large part of the cloud is aggregated and can fall as rain.

The chemistry of cloud and rain water. From the standpoint of the present work, the exact mode of formation of a raindrop is of less importance than the nature of the materials which it carries to the ground. The available analyses of cloud waters are limited to Köhler's studies (1937) of the composition of frozen cloud forming rime at Hallde in north Norway, and a few analyses by Houghton published by Rankama and

Sahama (1950). The determinations in which both sulfate and chloride are given are presented in Table 64.

TABLE 64. *Composition of cloud and fog*
(All determinations in mg. per liter)

	Hallde, Norway	Fog, Kent Is., N.S.	Fog, SE. coast Mass.	Cloud, Mt. Washington, N.H.
Na	1.98	...	...	...
Mg	0.28	...	...	...
Ca	0.09	...	...	...
Cl	3.56	7.5	35	0.1
SO_4	0.57	13	18	7
*p*H	...	7.0	...	4.5
$Cl:SO_4$	1:0.16	1:1.73	1:0.51	1:70

The cations are present in the Hallde rime in practically the same proportions as in sea water. The ratio of sulfate to chloride is in every case higher than the oceanic value of 0.1396, though the Hallde value is doubtless not significantly different from that given by sea water.

No really complete analyses of single samples of rain water exist. Some substances, notably nitrate and nitrite, have been determined hundreds or perhaps thousands of times; others, no less interesting geochemically, have been estimated on a single occasion, as is the case for strontium. It is therefore impossible to consider single samples as units; the available data must be summarized by constituents. But wherever possible, the more significant matter of ratios will be considered in such detail as the data permit. Eriksson (1952a,b) gives an excellent review.

Chloride. Chloride is known as a practically constant constituent of rain and snow, the amount recorded varying from zero in a very few samples of clean continental rain to 3920 mg. per liter in a sample collected at the Butt of Lewis, Outer Hebrides, by Miller (1914). The mean content has been estimated by various authors to lie between 0.3 mg. per liter (Collins and Williams 1933) and 3.0 mg. per liter (Riffenburg 1925). The lower value is probably a better mean for uncontaminated continental rain than is the higher. The value for any given station certainly depends greatly on distance from the ocean and prevalent direction of rain-bearing winds.

Drischel (1940) has considered the available data for the variation of chloride with distance from the coast, and concludes that the following concentrations may be expected.

Distance, km.	Chloride Variation, mg. per liter
< 0.1	70–700
1–2	15–30
5–10	6–13
50	4–9
100	3–5
500	1–2
1000	0.5–1.5
2000	< 0.3

Conway (1942) has examined the isochlor maps for surface waters prepared for the northeastern United States by Jackson (1905). Assuming that these maps indicate the variation of chloride in rain with distance (x) from the coast in kilometers, he concludes that the concentration [Cl] may be expressed[1] in mg. per liter, by

$$[\mathrm{Cl}] = 5.7e^{-0.037x} + 0.55e^{-0.0024x} \tag{1}$$

This equation was derived by assuming two different diffusion processes to be operating with virtual diffusion coefficients, the first process being predominant near the coast and the second predominant far inland. The resulting equation fits the North American data on which it is based with considerable accuracy, but Drischel's estimates all lie well above the curve obtained by plotting equation 1. It is reasonable to suppose that this is because Drischel considered a number of European localities in making his estimates, and that in Europe a greater proportion of rain is derived from air which has moved in perpendicular to the isochlors than is likely to be the case in North America. Drischel's figures, if plotted on semilogarithmic paper do, however, indicate a very rapid fall from the coast in the first few kilometers or tens of kilometers, and then a more gradual fall as the rains of the interior are considered. Eriksson (1952a,b) has obtained an equation for the chloride in rain falling over Holland, again taking the form of the sum of two exponential terms, but in this case there is also a constant implying at least 3.0 mg. Cl per liter, however far from the sea the rain is collected. Conway's suggestion of a multiple process is certainly reasonable. It is tempting to suppose that the coastal process represents the transport inland of relatively large droplets of spray, while the process which continues far inland may involve minute charged droplets which can be considered as the dispersed phase of an aerosol.

Leeflang's data from Holland, as treated by Eriksson (1952a,b), suggest

[1] Originally given with the coefficients of x, measured in miles, as -0.059 and -0.0039.

a similar relationship for the alkali metals, as would be expected. A third type of aerial transport of chloride is known in the few cases in which the wind picks up dry salt from a wide expanse of arid coastal land on which sea water has evaporated. The most remarkable case is provided by the Rann of Cutch, in the northwest corner of peninsular India. Holland and Christie (1909) compute that a quantity of salt of the order of 100,000 metric tons per year is transported from this region into Rajputana. In such a process, there appears to be little fractionation of sulfate from chloride, but to judge from the composition of the Sambhar Lake, which is believed to have received its salt in this manner, there is a loss of bromine, iodine, and magnesium. Magnesium chloride, bromide, and iodide being the last salts to crystallize from evaporating sea water and being very hygroscopic, it is reasonable to suppose that they are most easily retained at the coast and may seep back into the ocean.

At least in some localities, the mean chloride content and the chloride delivered per unit area per month, show a marked seasonal trend. At Rothamsted, the maxima for both variables occur in the winter and the minima in the summer (Russell and Richards 1919). The less extensive data for the Hebrides and Iceland (Miller 1914) also show summer minima. In all these cases, the tendency for low chloride to occur in summer is doubtless due to the greater disturbance of the surface of the eastern North Atlantic during the winter months. Drischel's rather broken set of data for Reinerz shows somewhat irregular variation, with the winter months tending to exhibit lower values than do the early summer, the maximum chloride concentrations being recorded in May. He reasonably supposes that a marked seasonal cycle is characteristic of maritime regions. Variations in the Na:Cl ratio are probably due to oxidation of Cl^- to Cl_2 by ozone, and diffusion of the chlorine into the gaseous phase (cf. Rossby 1955).

Bromine and iodine. Bromine and iodine, in addition to chlorine, are known to occur in rain water. The data for iodine, which are far more numerous than those for bromine, have recently been summarized from the very extensive literature by Eriksson (1952a,b) and Low and Hutchinson (*in press*). The better determinations for the iodine concentration in uncontaminated rain appear to vary from zero to about 10 micrograms per liter. Owing to the fact that the iodine content of air is very sensitive to industrial operations, even at a great distance from the site under investigation, it is difficult to give a satisfactory mean value for natural rain; 1 or 2 micrograms per liter would seem reasonable. Selivanov (1946), who alone has studied the bromine content of rain and snow, finds in atmospheric precipitation from near Moscow the quantities given in Table 65.

TABLE 65. *The halogens in rain falling near Moscow*

	Chlorine, mg. per liter	Bromine, mg. per liter	Iodine, mg. per liter
From maritime air	0.5	0.03	0.001
Ratio	100	6	0.5
From continental air	0.04	0.002	0.0002
Ratio	100	5	0.5
Sea water, ratio	100	0.34	0.00021

The chlorine contents are rather low, but even if they were tenfold too small, the ratios would still indicate a slight enrichment of bromine and an immense enrichment of iodine over the relative quantities in sea water. This is in marked contrast to the condition just noted for dry wind-borne salt. There is good reason to believe that most of the excess iodine in rain is derived from soil and that there is a relatively rapid movement, involving diffusion of iodine from soil into the air and so into the rain, which in turn returns the iodine to the soil. The bromine contents of rain given by Selivanov, and the iodine contents recorded by a number of investigators are of the same order of magnitude as the concentrations of these elements known to occur in lakes and rivers (page 562).

Sulfate. Sulfate has been determined less frequently than chloride, but is apparently always present in rain water. In continental precipitation not contaminated by sea spray, the ratio of sulfate as SO_4 to chloride as Cl is apparently normally greater than unity, and always much greater than the ratio characteristic of sea water. Thus, Collins and Williams (1933), who give analyses of both anions from the same samples of uncontaminated North American rain, found the values given in Table 66.

Smith (1872) gives for inland rural localities in Britain a mean chloride content of 3.99 mg. per liter and a mean SO_4 content of 6.3 mg. per liter,

TABLE 66. *Sulfate and chloride in North American rain*

	SO_4, mg. per liter	Cl, mg. per liter	Ratio
Alfred, Maine	2	0.26	7.7
Blacksburg, Virginia	1.4	0.30	4.7
Flagstaff, Arizona	2	1.0	2.0
Grand Canyon, Arizona	2	0.37	5.4
Kearn, Nebraska	2	0.14	14.3
Mean (each station equally weighted)	1.9	0.41	4.5

the ratio thus being 1.58. Drischel (1940) records a number of determinations of both anions in summer rain at Reinerz in Silesia, the mean values being 1.97 mg. Cl per liter, and 4.15 mg. SO_4 per liter. He also gives means for winter rain at Oberschreiberhau in the same region, but in these means far more analyses of chloride than of sulfate were included; the results, 1.32 mg. Cl per liter and 0.966 mg. SO_4 per liter are exceptionally low in sulfate for continental rain. The ratio for the whole region, weighting the two stations equally, is 1.55, much as in Britain.

Sugawara (1948; Sugawara, Oana, and Koyama 1949; Koyama and Sugawara 1953b) has obtained similar results in Japan, where he finds the SO_4:Cl ratio to be greater inland than near the coast and greater during heavy thundershowers than in light rain. He appears to have compared samples of suspended salts from air with those of rain, finding the following proportions.

	Mg:Cl	Ca:Cl	SO_4:Cl
Atomic proportion			
Air	0.30	1.36	0.68
Rain	0.33	0.34	0.38
Proportion by weight			
Air	0.20	1.53	1.78
Rain	0.22	0.38	1.03

It is evident that an SO_4:Cl ratio greater than that exhibited by sea water is the most salient feature of the chemistry of atmospheric water droplets. In a later chapter it will be shown that this same characteristic is also exhibited, as might be expected, by the dilute waters of soft-water lakes.

Three explanations of the presence of excess sulfate in rain water have been advanced. Some of the older workers, as MacIntire and Young (1923), implied that the excess sulfate recorded in their analyses had entered the atmosphere with the products of the combustion of fuel containing sulfide. This is certainly the explanation of the very high values often recorded in industrial regions. It must, however, be pointed out that while there can be no doubt that in urban districts sulfate is formed from the products of combustion, Coste and Wright (1935) indicate that it is difficult to find SO_4 in London air containing appreciable amounts of SO_2. The sulfate detected corresponded to about 1 per cent of the SO_2 present. The oxidation of the latter gas to SO_3 or H_2SO_4 can hardly be as rapid as has often been supposed. Aitken (1911) has given evidence that in nature photochemical processes are involved.

Conway (1942) has shown that, as a source of excess sulfate in uncontaminated country rain, the combustion of sulfide-containing fuels is quantitatively inadequate though Gorham's most recent results indicate a close correlation between SO_4 and soot. Smith (1872), many years before, had suggested that the major source of such sulfate is the oxidation in the atmosphere of H_2S derived from organic decomposition. Smith believed such decomposition to occur mainly on land. Conway suggests that blue mud on the continental shelves is the main source of the H_2S which he supposes to pass into the air, there to become oxidized and so to constitute the source of the sulfate in natural rain.

A third possibility arises from the studies of Sugawara (1948; Sugawara, Oana, and Koyama 1949), which indicate that the salts crystallizing out when sea spray evaporates tend to produce two kinds of solid particles dispersed in the atmosphere. As a droplet of sea water evaporates, calcium sulfate will first be precipitated as gypsum. Later, sodium and magnesium will crystallize out. Sugawara's observations lead him to believe that sulfate and chloride may become separated in the atmosphere. The calcium sulfate particles, not being hygroscopic, will form a stable aerosol, while the chlorides will act as condensation nuclei. The figures already given for the SO_4:Cl ratio in air and rain show the process at work. Over a sea surface, more of the chloride than sulfate will be rapidly returned to the ocean, and the air circulating over the continents will in consequence be enriched in sulfate relative to the proportion of the two ions in sea water. This enrichment will ultimately be exhibited in the rain that washes out the suspended particles in the air. It should be possible to discriminate between Conway's hypothesis and that of Sugawara, by a study of the isotopic constitution of the sulfur in the sulfate of rain.

Other anions. Of the other anions, it is convenient to treat nitrite and nitrate with other nitrogen compounds after considering the cations. There are a few records of phosphate in rain, ranging from a trace, recorded by MacIntire and Young, to the very improbable value of 0.15 mg. per liter $P \cdot PO_4$ said by Ingham (1950) to represent the mean value in rain falling near the coast of Natal. Gorham (1955b) found minimal value usually less than 1 mg. $m.^{-3}$ Conway points out that the recorded *p*H values of rain, which can lie between 3.5 and 7.7 (Riffenburg 1925, Potter 1930, Drischel 1940, Atkins 1947) but generally appear to be between 4.0 and 5.0, indicate a very low bicarbonate content. Gorham (1955b) finds 0.4 mg. per liter in average and 1.2 mg. per liter in clean rain in the English Lake District. Odum and Parrish (1954) found 0.009 to 0.015 mg. per liter of boron in rain at Gainesville, Florida. The high B:Cl ratio, of the order of 40×10^{-4}, suggests a source of boron other than the borate of the sea.

Cations. Sodium, potassium, calcium, and magnesium, as well as sulfate and chloride, were determined a century ago by Pierre, quoted by

Clarke (1924). These results are not likely to be particularly reliable, and are of interest mainly in view of the paucity of modern analyses. Relative to sodium, the quantities recorded, as compared to sea water, are:

	Rain Water	Sea Water
Na	100	100
K	4.5	3.6
Ca	13.8	3.8
Mg	10.3	12.0
Cl	169.5	180.0
SO_4	110.0	25.0

These analyses indicate the main salt to be sodium chloride, but also that there is some enrichment of potassium and calcium, as well as of sulfate. A marked enrichment in both potassium and calcium is recorded by MacIntire and Young, who observed in rain collected about 2 km. from Knoxville, Tennessee, 1.5 mg. per liter of sodium, and 2.6 mg. per liter of potassium. It is reasonable to attribute such results to industrial pollution, or possibly to dissemination of commercial fertilizers containing potash. Bertrand (1945) found but 0.02 to 0.04 mg. K per liter in rain water collected in Paris.

The enrichment of rain waters in calcium appears to be usual and has been studied by Leeflang (1938), by Bertrand (1943a,b, 1944), and by Sugawara, Oana, and Koyama (1949). It is probable that most of the calcium is derived from calcareous dust. Both Ca and Mg increase with distance from the sea, but the calcium concentration rises more than that of magnesium. The ratio of magnesium to calcium is ordinarily greater in rain than in terrestrial calcareous materials, and when dust and rain water do contain about the same proportions of the two elements, the terrigenous materials from which the calcium is apparently derived lose calcium to distilled water much more readily than they lose magnesium. It would therefore seem reasonable to attribute much of the magnesium content of rain to sea spray.

Odum (1950) has determined the Sr:Ca ratio in a sample of snow falling in New Haven, Connecticut, and finds a value comparable to that in terrestrial calcareous rocks rather than to the much higher value given by sea water. This finding is in line with those of Pierre and of Bertrand, suggesting that most of the calcium in atmospheric precipitation is of terrigenous origin.

Nitrogen compounds. The nitrogen compounds of rain have been far more carefully investigated than any of the other dissolved substances which may be present. Most of the older work will be found reviewed in Miller (1914), and Dhar and Ram have summarized interesting observations in their consideration of the photochemical origin of the oxidized nitrogen in rain. Hutchinson (1944b, 1954) has discussed certain of the more general aspects of the atmospheric nitrogen cycle. Four forms of nitrogen, namely nitrate, nitrite, ammonia, and organic or albuminoid nitrogen, are usually considered in the more complete analyses. Of these, the organic nitrogen, which may constitute one fourth of the total, must obviously be derived from the ground. It is probable that most of the ammonia is of like origin. In temperate regions, the nitrate and nitrite content of rain in general is markedly less than the ammonia content, but this is not true of tropical regions. In either group of localities, there is a marked tendency for the ammonia and the nitrate to be correlated over a wide range of values, estimated either as concentrations or as masses of nitrogen delivered per unit area per year. This suggests some genetic connection between ammonia and nitrate. While some of the latter substance is probably formed in lightning flashes, as has been believed for over a century, it is difficult to reconcile even the data on the enhanced precipitation of nitrate in thunderstorms with the hypothesis that a considerable part of the nitrate in rain is of such origin. The most probable theory of the origin of oxidized nitrogen in atmospheric precipitation is certainly that of the photochemical oxidation of ammonia, championed particularly by Dhar and his associates.

TABLE 67. *Composition of rain and of dilute lake water**

	Rain Water (Provisional Estimate), mg. per liter	Rain, Mean, Lake District, mg. per liter	Rain, Uncontaminated, Lake District, mg. per liter	Min. Concentration, Wisconsin Lakes, mg. per liter
Cl	0.5	3.3	5.1	0.1
Br	0.03	...	...	...
I	0.001	...	...	...
SO_4	2.0	3.2	1.7	0.75
B	0.01	...	...	...
Na	0.4 or more	1.9	3.1	0.13
K	0.03 or more	0.2	0.2	0.25
Mg	0.1 or more	0.3	0.3	< 0.5
Ca	0.1–10	0.2	0.2	0.13
$N \cdot NH_3$	0.5	...	...	< 0.01
$N \cdot NO_3$	0.2	...	...	< 0.004

* The two columns for the Lake District are based on Gorham's (1955b) most recent paper, received after this chapter was written. The criterion of contamination is the presence of soot. It seems likely that in the English Lake District clean rain reflects the proportions of chloride and sulfate in sea water more closely than would be expected from the material discussed in the earlier papers mentioned in this chapter.

The mean ammonia-nitrogen content of rain falling in temperate regions, omitting large industrial towns, is, from Drischel's tabulation, 0.64 mg. per liter; the mean nitrate-nitrogen content of the same samples is 0.196 mg. per liter. Tropical rains contain more nitrate and less ammonia the mean values being 0.308 mg. $N \cdot NH_3$ per liter and 0.267 mg. $N \cdot NO_3$. Such nitrate figures normally include nitrite, which may vary from 1 or 2 per cent up to about 40 per cent of the total. It will be observed that these quantities of nitrogen compounds are of the same order of magnitude as those found in many lake waters.

Estimated mean values. While it is obviously impossible to give any satisfactory mean values for the concentration of substances dissolved in the atmospheric precipitation of the earth as a whole, the quantities given in Table 67 evidently represent the orders of magnitude involved, which may be compared with the minimum values found in the surface waters of the lakes of northeastern Wisconsin (Juday, Birge, and Meloche 1938; Lohuis, Meloche, and Juday 1938). Though the comparison is very rough, it is evident that a basin filled solely with rain water could exhibit many of the chemical, and so the biological, properties of the more dilute lakes.

THE COMPOSITION OF LAKES AND RIVERS

Normal fresh waters are dilute solutions of alkali and alkaline earth bicarbonate and carbonate, sulfate, and chloride, with a variable quantity of largely undissociated silicic acid (page 789) which is often present in excess of sulfate and chloride. There are also a number of minor constituents in true solution, some of them being of great biological interest, and a variety of colloidal materials both inorganic and organic. It will be necessary to discuss a number of these components in detail in later chapters. The present section is mainly devoted to a consideration of the relative abundance of the various major ionic constituents in so far as they determine the over-all chemical characteristics of a water.

In general the inorganic composition of closed lakes is very different from that of lakes with effluents through which water is passing. In the latter type of lake, the primary determinant of the chemical composition is the nature of the influent water; in the former type, the composition is greatly modified by precipitation of the less soluble salts as evaporation proceeds. It is therefore necessary to consider the two types separately.

The composition of the influents may be supposed to depend on three factors. Firstly, atmospheric precipitation is not pure water. We have seen that, at least in its anionic composition, it may approximate to the most dilute lake waters. Secondly, the solution of materials in the drainage basin determines to a large extent the nature of the influent solution.

Thirdly, the equilibrium set up between ions in the water and the exchange systems in soils and lake sediments through or over which much of the water may be percolating on its way to the lake, modifies the resulting solution.

In spite of the vast mass of material available, satisfactory attempts to reduce it to a rational order are of comparatively recent date.

Terminology; the salinity of inland waters. The total inorganic material dissolved in a natural water is usually estimated by evaporating the water to dryness, so obtaining the *total solids* per unit mass or volume, and then igniting the residue to obtain the *nonvolatile solids* per unit mass or volume. The difference, called the loss on ignition, is mainly organic matter and CO_2 lost from alkaline earth carbonates and in some cases from alkali bicarbonates. There may be some slight loss of alkali and chloride also. The total solids represents a useful pragmatic quantity, but if it is to be considered further, the CO_2 that is lost in evaporation and ignition must be known.

The salinity of an inland water may be regarded as the concentration of all the ionic constituents present. Clarke (1924) includes SiO_2, Al_2O_3 and Fe_2O_3 in his statements of salinity. Since most of the silicate is probably present as undissociated silicic acid and the iron and aluminum oxides are rarely in an ionic condition, it would be more satisfactory to define salinity as the concentration of the Na, K, Mg, Ca,[2] $CO_3^=$, $SO_4^=$, and halide present, all bicarbonate being converted to carbonate. The salinity can be expressed in milligrams or milliequivalents per liter or per kilogram. In ordinary fresh water the difference between reference to mass and to volume is negligible. Clarke (1924), Conway (1942), and many other authors have used essentially this definition but with the inclusion of silica and sesquioxides, but for the most part these make little difference and no attempt has been made to correct their estimates.

The mean salinity of all the rivers for which published analyses were collected by Clarke is 146 mg. per liter, but a weighted mean of 100 mg. per liter, allowing for the great discharge of dilute tropical rivers such as the Amazon, is considered preferable by Conway. The individual values are enormously variable, the lowest being from regions of bare igneous rock fed mainly by melting snow or freshly fallen rain, the highest from regions from which soluble salts can be leached. Conway, considering all the data published by Clarke, concludes that waters up to a salinity of 50 mg. per liter have a mean composition representing igneous drainage.

---

[2] A slight error will often be introduced by the fact that many calcareous waters contain some colloidal $CaCO_3$ (page 670).

At higher salinities, calcium, magnesium, carbonate, and sulfate increase relative to the other ions until a salinity of 200 mg. per liter is reached. In still more concentrated waters in open lakes and rivers, a relative increase in sodium and in chloride is attributable to either human contamination or the leaching of salt beds.

It is customary to speak of waters containing small concentrations of alkaline earths as *soft*, those containing large concentrations as *hard*. The soft waters are mainly derived from the drainage of acidic igneous rocks, the very hard from the drainage of calcareous sediments.

The major cationic constituents. The quantities of the major solutes, ordinarily determined in water analysis, in a number of representative waters are presented in Table 68. The waters chosen do not include a number from which analyses might be desired. In some cases, notably those from the English Lake District studied by Pearsall (1921), sodium and potassium are not determined separately; in other cases, notably the Wisconsin Lakes studied by Birge and Juday (1911), the great discrepancy between the total cations and anions, when expressed as equivalents, strongly suggests gross technical error. These less perfect analyses will be considered later in the discussion.

In Table 68 the percentage equivalent proportions are computed separately for the four cations and three anions[3] which comprise the salinity of the water. These proportions have been calculated for the mean water derived from igneous rocks, as computed by Conway; for the mean of four lakes (Crystal, Muskellunge, Weber and Nebish) containing dilute soft water in northeastern Wisconsin (Juday, Birge, and Meloche 1938; Lohuis, Meloche, and Juday 1938); for the group of lakes numbered 1 to 21 in Uppland, Sweden (Lohammar 1938), considered by Rodhe (1949); for the larger lakes of Bavaria, Austria, and the Swiss and Italian Alps, analyses of which are assembled by Clarke (1924); for the mean fresh water of the world's river systems, as given by Clarke; and for soft north German lakes, given by Ohle (1955b). The ratio of the same ions in the mean terrestrial sediment, also according to Clarke,[4] is likewise given; in this case the chloride content is negligible. The ratios of the cations for the primary igneous source material as derived by Kuenen (1941) are also included in the table.

---

[3] There is often a slight discrepancy between the equivalent content of anions and cations, usually no doubt owing to a part of the silicic acid being ionized at high *p*H values, and to cations being held by organic matter.

[4] This is based on a mixture of 80 per cent shale, 15 per cent sandstone, and 5 per cent limestone. Other mixtures have been regarded as more realistic (Kuenen 1941), but Conway (1942) has shown that Clarke's estimate is the most satisfactory for the present purpose.

TABLE 68. *Mean equivalent proportions of cations and anions in certain important natural waters*

	Mean Igneous Source Material*	Mean Sedimentary Source Material†	Water from Igneous Rock‡	Wisconsin Soft Waters§	Uppland‖	Central Europe¶	Mean River**	N. German Soft Waters††
Na^+	20.1	4.8	30.6	10.9	13.6	4.5	15.7	43.0
K^+	9.8	8.0	6.9	4.8	2.2	1.9	3.4	6.7
Mg^{++}	33.1	34.0	14.2	37.7	16.9	25.4	17.4	14.3
Ca^{++}	37.1	53.2	48.3	46.9	67.3	68.2	63.5	36.0
$CO_3^=$	...	93.8	73.3	69.6	74.3	85.4	73.9	42.4
$SO_4^=$	...	6.2	14.1	20.5	16.2	10.8	16.0	14.1
Cl^-	...	0 (?)	12.6	9.9	9.5	3.9	10.1	43.5

* Kuenen 1941.
† Clarke 1924.
‡ Conway 1942.
§ Juday, Birge, and Meloche 1938; Lohuis, Meloche, and Juday 1938.
‖ Lohammar 1938, Rodhe 1949.
¶ Computed from Clarke 1924.
** Clarke 1924.
†† Ohle 1955b.

It is immediately evident, in passing from the source material to the water, that calcium and in most cases sodium tend to be enriched in the solution, while magnesium and potassium tend to be depauperated. It is possible, however, that some highly productive lakes are somewhat enriched in potassium (Höll 1951).

In very soft waters directly derived from igneous areas, the alkalies may be more abundant than the alkaline earths; and in some very dilute seepage lakes, potassium may be several times as abundant as sodium, but in all the more concentrated waters of open river systems there is a tendency for Ca > Mg > Na > K. It is reasonable to suppose that all fresh waters moving through rivers and lakes are gradually tending to such a composition. As Rodhe (1949) and in greater detail Gorham (1955a) have pointed out, this general tendency to an over-all uniform composition, in spite of differences in the lithology of the drainage basin, must reflect changes that occur after the initial solution of the cations from their parent rock. Such changes are most naturally referable to base-exchange reactions with clay minerals in soils and fresh-water sediments. The over-all composition of soil solutions is in fact very similar to that of average river water.

In individual basins some idiosyncracies may often be found. The low sodium and high magnesium and calcium of the central European alpine lakes is doubtless due to the fact that in so far as these waters drain from sedimentary rocks, such rocks are likely to be limestone, which may be dolomitized. This is usually true of any montane area of late Mesozoic or Tertiary age.

If adequate data were available for southeastern Wisconsin, it is reasonably certain from the defective analyses that have been published that magnesium would be at least as abundant as calcium, and possibly more so. This is explicable by the considerable amount of ancient magnesian limestone outcropping in the drainage basins or incorporated in the drift in which the Wisconsin lakes lie. The relatively high magnesium contents of the northeastern Wisconsin lakes are doubtless to be accounted for in the same way. A few individual cases, such as that of the Hallstätter See, in which high sodium accompanies high chloride, are presumably due to saline springs or the leaching of saline sediments in the basins. The existence of these exceptional lakes only serves to emphasize the remarkable similarities of the average fresh waters when Uppland and central Europe are compared with each other or with the average for the rivers of the world as a whole. It is evident that in the mean river water we have a standard fresh water which would be approached by the water of any river system, at least in temperate latitudes, if sufficient time were available.

In the large number of small lakes of northeastern Wisconsin, it appears from the work of Juday, Birge, and Meloche (1938) that the ratio Mg:Ca falls regularly with increasing calcium content and increasing total concentration, from 0.77 by weight or 1.26 by equivalents in the large group of lakes with a mean Ca content of 0.65 mg. per liter.

Similarly, in the data given for the sodium and potassium contents of the same lakes, the ratio of potassium to sodium falls with increasing sodium contents. In the eight lakes containing less than 0.5 mg. Na per liter, the ratio is 4.1 by weight or 2.4 by equivalents; in the eleven lakes containing more than 1.5 mg. Na per liter, the ratio is 0.69 by weight or 0.4 by equivalents. The change is quite smooth through the various intermediate categories. Sodium apparently increases slightly with the other major inorganic ions, but except in the most dilute waters potassium has a mean concentration of about 1 mg. per liter.

In the northeastern Wisconsin lake district neither potassium nor magnesium is depauperated relative to calcium and sodium as much as in mean river water, but the changes in passing from very dilute waters to somewhat concentrated are in the direction of the production of standard river water. In this case it is probable that a number of the most dilute waters have had most of their ions removed by adsorption on organic peaty material.

Whatever process is involved in the production of these very dilute waters seems to favor retention of potassium and magnesium in solution more than sodium and calcium, whereas in the leaching of sedimentary rock giving the relatively concentrated water of the mean river, the exchange with clay minerals leads to a solution containing more sodium than potassium.[5]

In view of the general similarity of bicarbonate waters wherever they are found, Rodhe (1949) has determined the relationship of total salinity to conductivity in such waters, as the conductivity is often the most easily determined measure of concentration. The results of his study are given in Table 69.

Graphical presentation. A convenient method of indicating the ionic composition of waters graphically has been introduced by Maucha (1932). A sixteen-sided regular polygon is constructed, having a standard area, say 200 $mm.^2$ This is divided vertically to give an anion field on the left, a cation field on the right. Each field is divided into four equal sectors.

---

[5] A very important paper by Ohle (1955b) has appeared, in which the effect of organic soft water sediments in removing Ca^{++} and HCO_3^- ions from waters is considered. Ohle gives a standard composition for such waters which it has been possible to add to Table 68. His experiments with the mud of the Kleiner Uklei See indicate that magnesium is not adsorbed, in accord with the field observations from northeastern Wisconsin.

TABLE 69. *Composition of standard bicarbonate waters of varying specific conductivities*

(All determinations given in mg. per liter)

Specific Conductivity (20° C.) × 10^6	Salinity	Na	K	Mg	Ca	CO_3	SO_4	Cl
20	10.5	0.7	0.3	0.4	2.5	4.4	1.5	0.7
40	21.5	1.5	0.5	0.9	5.1	9.0	3.0	1.5
60	32.9	2.2	0.8	1.3	7.9	13.8	4.7	2.2
80	44.7	3.0	1.1	1.8	10.8	18.7	6.3	3.0
100	56.5	3.8	1.4	2.3	13.5	23.7	8.0	3.8
120	68.3	4.6	1.7	2.7	16.3	28.7	9.7	4.6
140	80.3	5.4	2.0	3.2	19.2	33.7	11.4	5.4
160	92.1	6.2	2.3	3.7	22.0	38.6	13.1	6.2
180	104.7	7.1	2.6	4.2	25.0	43.8	14.9	7.1
200	117.3	7.9	2.9	4.7	28.1	49.2	16.6	7.9
220	129.7	8.7	3.2	5.2	31.1	54.4	18.4	8.7
240	142.4	9.6	3.5	5.7	34.1	59.7	20.2	9.6
260	155.1	10.4	3.8	6.2	37.1	65.2	22.0	10.4
280	167.5	11.3	4.1	6.7	40.1	70.2	23.8	11.3
300	180.8	12.2	4.4	7.3	43.3	75.8	25.6	12.2
320	193.9	13.1	4.7	7.8	46.4	81.3	27.5	13.1
340	206.4	13.9	5.0	8.3	49.4	86.6	29.3	13.9
360	219.7	14.8	5.4	8.8	52.6	92.1	31.2	14.8
380	233.1	15.7	5.7	9.4	55.8	97.7	33.1	15.7
400	246.2	16.6	6.0	9.9	59.0	103.2	34.9	16.6

The four sections on the left are used for representing $CO_3^{=}$, HCO_3^-, Cl^-, and $SO_4^{=}$; those on the right, for K^+, Na^+, Ca^{++}, and Mg^{++}. In any diagram the line passing through the center of the polygon and bisecting a sector is measured off from the center of the polygon to a point given by the concentration of the ion as equivalent per cent of the total ionic concentration, divided by 8.082 sin 22.5 degrees. Lines are drawn from this point to the outer angles of the section, and the quadrilateral defined by these two angles, the intersection of the lines, and the center of the polygon is blacked in. A little elementary trigonometry will show such a quadrilateral to have an area in square millimeters equal to the concentration.

In Fig. 155 the ionic composition of sea water, of a typical calcium bicarbonate water (Lunzer Untersee), of a sodium magnesium bicarbonate lake (Lake Velence), and of a sulfatocarbonate lake (Lake Balaton) are given from Maucha (1932).

Where very large numbers of analyses of surface values exist, contoured maps indicating the distribution of total solids or of particular elements may be prepared, as has been done by Möller (1955) for parts of Germany.

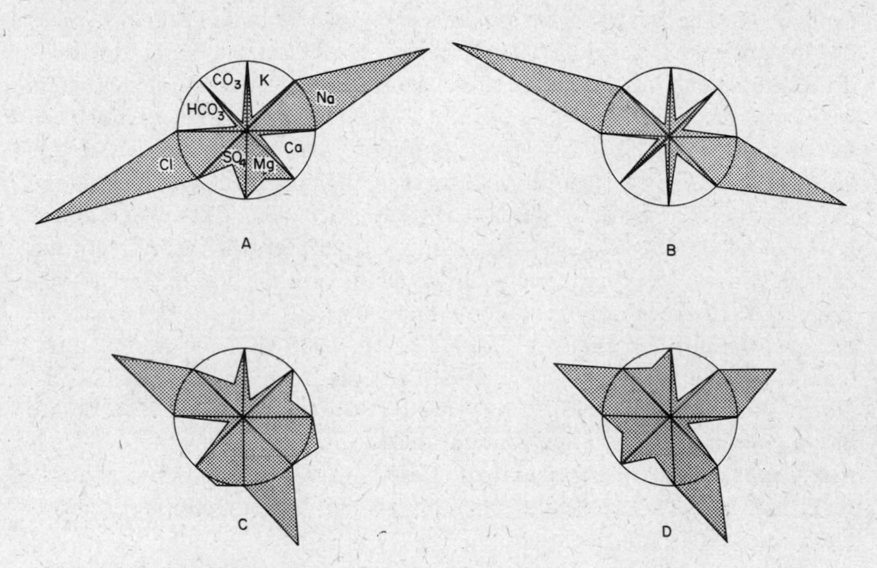

FIGURE 155. Ionic diagrams in which the areas of the segments are proportional to the equivalent concentration of the major anions and cations: *A*, sea water; *B*, Lunzer Untersee, a normal calcareous water; *C*, Lake Balaton; *D*, Lake Velence. The two last examples, from Hungary, show striking enrichment of magnesium, sodium, and sulfate. (After Maucha.)

Rarer alkalies and alkaline earths. The rarer alkalies are very little known. Lohuis, Meloche, and Juday (1938) indicate that enough lithium occurs in some of the lake waters of northeastern Wisconsin to introduce an error into their published data for sodium. Since the ratio Na: Li has a value of about 1000:2.3 in primary rock, any significant error in the very low sodium values encountered must imply a considerable enrichment in lithium. No quantitative data were published, and the matter might repay further study. Lithium is a well-known constituent of mineral springs.

Rubidium has occasionally been detected, notably by Schmidt (1882), who gave the rubidium content of the water of Lake Peipus as 0.055 mg. per liter, or about 1 per cent of the potassium. This Rb: K ratio is by no means geochemically impossible.

Of the rarer alkaline earths, strontium alone has been adequately studied. Lohammar (1938) detected the element in the waters of all but two of the thirty-seven lakes in which it was sought. The maximum quantity recorded was 0.34 mg. per liter in Glisstjärn, a shallow richly vegetated lake in Dalarna. No other lake water contained more than 0.099 mg. per liter, and the anomalous high value for Glisstjärn is unexplained. The mean value for twenty-two lowland lakes in the provinces of Uppland and Dalarna is 0.038 mg. per liter, Glisstjärn being excluded; if the anomalous value for the latter lake is included, the regional mean is 0.050 mg. per liter. In the generally less mineralized lakes of northern Sweden the mean is 0.014 mg. Omitting the exceptional value for Glisstjärn, the highest strontium contents are in the more calcareous waters, but the ratio Sr:Ca falls systematically with increasing Ca. Thus, in the group of lakes containing less than 2 mg. Ca per liter, the mean strontium content was 0.0086 mg. per liter and the atomic ratio Sr:Ca was 2.45:1000; in the group containing more than 50 mg. Ca per liter, the mean strontium content was 0.0707 mg. per liter and the atomic ratio 0.52:1000. Odum (1950) indicates that the fresh waters of eastern North America from New England to North Carolina give absolute Sr contents and Sr:Ca ratios, comparable to those of Lohammar. This is partly explicable by the fact that the ratio is very much higher in primary rock than in calcareous sediments, though the former material dissolves much less easily than the latter.

Odum (1951a) reports higher values for Florida than those obtained from most other localities that have been studied, Lake Kanapaha containing 0.257 mg. per liter with an atomic ratio of 1.8:1000. He believes that ground waters are richer in the element than surface waters, and that the high values reported in Florida are due to the large amount of ground water appearing in lakes and rivers. The majority of the springs studied in Florida contained even more Sr than did the lakes. Even highly calcareous springs are recorded with up to 0.78 mg. Sr per liter. It is possible that the Sr:Ca ratio is modified as soon as ground water leaves a spring and starts percolating over sediments containing clay minerals.

Barium is very little known; Bowen (1948) found an amount in the water of Linsley Pond of the order of 0.01 mg. per liter, of which only one third appeared to be in solution. The quantity in the mud appeared considerably less than that in the residue after ignition from the water. In certain hard waters from Connecticut, the element was undetectable.

Anion contents. In the mean analyses previously given, the carbonate present in the original sample, mainly as HCO_3, exceeds sulfate and the latter exceeds chloride. This appears to be the normal situation, and the standard fresh water may be characterized as a *bicarbonate* water. Less commonly, sulfate or chloride may be the most abundant anion. Both types may arise from the influence of volcanic waters and from the solution of sulfate or chloride in sediments of the drainage basin; sulfate may be formed from organically precipitated FeS_2 in peat deposits, and chloride may be added from sewage in polluted waters.

As will be further discussed, most extremely acid lake waters probably owe their acidity to sulfuric acid, and in such cases the $SO_4^{=}$ ion is likely to be the prevalent anion. Triassic calcium sulfate deposits may cause high sulfate contents in a number of the smaller lakes of the Swiss Alps. Lake Ritom forms a striking example, for in this case the mineral springs producing meromixis deliver a water very rich in $CaSO_4$ into the basin.

Halide content. Chloride is very rarely the dominant anion in an open drainage system. Conway supposes that all the chloride in uncontaminated water flowing from either igneous rock or from normal sediments is of atmospheric, and so ultimately of oceanic, origin.

In uncontaminated inland waters not flowing from saline sediments, the quantity present is ordinarily very low. In regions near the coast, however, cyclical salt carried in the air and precipitated in the basin may raise the chloride as well as the sodium content, so that these two ions are the dominant inorganic constituents. Within relatively uniform lake districts the effect is regarded by Yoshimura (1936d) as linearly dependent on distance up to 2 to 3 km. from the coast, though his relationship may be part of an exponential rather than a linear fall. The quantity of sea spray involved may be considerable; Mackereth and Heron (1954) find that Ennerdale Water, 13 km. from the coast, may receive during a two-day westerly gale the salt of 3000 tons of sea water, though the chloride concentration of the lake is raised by only 0.7 mg. per liter in the process. Juday, Birge, and Meloche (1938) detected the element in all of the 474 lakes of northeastern Wisconsin that they studied, the range of concentration being 0.1 to 4.5 mg. per liter; 68 per cent of the lakes, however, contained less than 1 mg. per liter. Slight seasonal and vertical variations were recorded but are possibly not outside the error of the method of determination. The vertical variation involved a decrease with depth in some lakes and an increase in others. Ruttner found a range from 0 to 3.3 mg. per liter in the lakes of the Austrian Alps, but in some of the meromictic lakes very great increases in chloride content, due to mineral springs, occurred in the monimolimnion. In the Toplitzsee the deep water below the temperature inversion contained about 40 mg. per liter, the surface waters only 2.1 mg.

In striking contrast to these figures, the lakes of Holstein studied by Ohle (1934a) provide an example of a group of otherwise normal lakes receiving chloride-rich ground water and having a chloride content of from 16.2 to 155.0 mg. per liter (Table 70).

TABLE 70. *Anions in north German lakes (milliequivalents per liter)*

	CO_3	SO_4	Cl
Hard water			
Grosser Plöner See	2.37	0.105	0.92
Herta	2.19	0.294	0.304
Soft water			
Schwarzsee	0.01	0.290	0.316
Grundloser Kolke	0.01	0.129	0.124
Saline			
Trammer See	3.12	0.147	4.95

Atomic proportions in the order $CO_3 > Cl \geqslant SO_4$ are usual in the hard-water lakes, $Cl \geqslant SO_4 > CO_3$ in the very soft waters, and $Cl > CO_3 > SO_4$ in the single lake which receives greatest chloride.

Fluoride has been studied rather extensively in surface and ground waters, as its concentration is related to the incidence of dental caries and mottled enamel. High incidence of dental caries in children is usually found in areas where fluoride in potable water is well below 1 mg. per liter and mottling is characteristic of areas in which the waters contain several milligrams per liter. The optimal quantity from the standpoint of dental hygiene is about 1 mg. per liter. There are a few determinations (Juday, Birge, and Meloche 1938) for the lakes of northeastern Wisconsin ranging from 0.08 to 0.51 mg. per liter. Mackereth and Heron (1954) found a comparable amount of the order of 0.15 mg. per liter in the English Lake District. It is reasonably certain that in inland fresh waters the ratio F:Cl is of the order of 0.01–0.1:1, much higher than the 0.00005:1 characterizing sea water. Kobayashi (1954), who has attempted fractionation of the fluorine in Japanese waters, finds 0.4 mg. per liter of dissolved and about 0.2 mg. per liter of suspended fluorine in Kizakiko.

Bromine has been determined in a number of Russian lakes and rivers by Selivanov (1939a,b, 1944, 1946), who found a range of from 0.5 to 140 mg. $m.^{-3}$ in river water and from 2 to 10.1 mg. $m.^{-3}$ in a more restricted sample of lake waters. The means were 21 and 4.5 mg. $m.^{-3}$ respectively, and the ratios of the mean Cl to mean Br contents were 574 for the rivers and 941 for the lakes. These figures indicate less bromine relative to chlorine than in the ocean, in which the ratio is 294:1. This difference may be due to the fact that some of the Russian rivers studied receive their chloride from rock salt, which is very low in bromine.

The iodine content of lake and other fresh waters has been extensively studied in goiter investigations. It appears to vary from about 0.01 mg. $m.^{-3}$ in some Finnish lakes (Adlercreutz 1928) to supposedly over

1 mg. m.$^{-3}$ in some samples from Lake Superior. Closed basins in arid regions naturally contain more (McClendon and Hathaway 1924; Jarvis, Clough, and Clarke 1926). The mean value would probably be about 0.2 mg. m.$^{-3}$ Relative to bromine or chlorine, fresh waters are probably enriched in the element, while sea water is greatly depauperated. Both bromine and iodine are slightly enriched in humic lake sediments, relative to the concentration in the water (Selivanov 1946).[6] Sugawara (in press) finds concentration of iodine in lake muds and believes the element is sedimented adsorbed to ferric hydroxide or alumina.

Other anions. Silicate, phosphate and the oxyanions of nitrogen are of such great importance that they require extended treatment in special chapters. Boron as borate is likely to be of some biochemical interest, but has been very little studied except in salt lakes. Odum and Parrish (1954) find that in Florida the boron content varies with chlorinity in those waters that receive chloride from residual salt water trapped in sediments. In such waters the ratio B:Cl is of the order of 3×10^{-4}, as in the ocean. In waters not influenced by this source of chloride, the boron content is lower but the ratio is an order of magnitude greater. In five fresh-water lakes the boron content varied from 12 to 27 mg. m.$^{-3}$, and the B:Cl ratio from 17×10^{-4} to 68×10^{-4}. Odum and Parrish point out that this high ratio suggests that sources of boron other than cyclical salt are involved in such cases.

Selenate is of both geochemical and biochemical interest, but has hardly been studied in lakes. Some seepage and well waters in seleniferous regions are known to contain selenium, but Beath, Eppson, and Gilbert (1935), working in Wyoming, have shown that such waters draining into lakes tend to lose their selenium, probably by precipitation as a ferric selenite, near $Fe_2(OH)_4SeO_3$, which is apparently the ordinary insoluble selenium compound in soils (Trelease and Beath 1949).

Arsenic, which is known to be present in sea water, supposedly largely as arsenite, in quantities not greatly less than the amount of phosphorus, seems to have been neglected in the study of fresh waters. According to Von Bülow and Otto (1931), the normal arsenic content of fresh waters in Germany is 2 to 3 mg. m.$^{-3}$, a quantity which, if general, is sufficient to have influenced the accuracy of some total phosphorus determinations. In the Rot See, a peculiar quarry pond 10 km. south of Witzenhausen, Kassel, Germany, the same authors found very large amounts of arsenic associated with a dispersed red ferric aluminum silicate which colors the water of the pond. Algae from this locality contained 1.7 mg. As kg.$^{-1}$

---

[6] A more detailed summary is presented by Low and Hutchinson (in press).

COMPOSITION OF THE WATERS OF CLOSED BASINS

The waters of closed basins may be expected to increase in concentration as the water running into the basins evaporates and the saline material that it brings in accumulates.

Total concentration. It has been pointed out, however, that in general, closed basins containing water occupy a relatively restricted area of endorheic drainage, between the exorheic areas from which rivers reach the ocean and the arheic areas in which there are no surface waters. Any climatic change is likely to change the disposition of this intermediate zone. Increasing humidity will cause a dilution and finally convert a closed basin into an open one; increasing aridity will dry the basin up entirely. Either process may cause a loss of salts from the basin, for when a lake bed is quite dry, it can lose solid material by deflation. There is some evidence that both processes have occurred in the Lahontan Basin and that the low salinity of Pyramid Lake is due to overflow, that of Walker Lake to desiccation (Hutchinson 1937b).

The result of these processes is to produce relatively few highly concentrated lakes in which saturation with the very soluble salts of sodium and magnesium is approached. Most lakes in closed basins are saturated only with respect to alkaline earth carbonates, and in a few cases to $CaSO_4 \cdot 2H_2O$ or gypsum. Hutchinson (1937b) pointed out that in a series of lakes investigated by limnologists the modal logarithmic class contained 1 to 10 g. per liter of total dissolved solids. More recently Rawson and Moore (1944) have studied a number of closed lakes[7] in Saskatchewan; these may be added to the data assembled by Hutchinson. In the same paper the latter pointed out that all the closed lakes considered by Clarke give a rather higher modal concentration. The choice of lakes studied by limnologists may be biased by accessibility to centers of learning, while that of Clarke may be biased by interest in the commercial exploitation of salts. Since there are sixty-six lakes in both groups, the combined percentages in Table 71 give equal weighting to each group.

The solubilities at 10° C. of six pure salts well known in saline lakes and occurring as stable minerals[8] in contact with water are given in Table 72.

---

[7] The closed fresh-water lakes are included, but three apparently open lakes are omitted. It is possible that the eight lakes so included are seepage lakes, but they make little difference to the total.

[8] Nahcolite and bischofite are quite rare, the ordinary sodium carbonate mineral being *trona*, $Na_3H(CO_3)_2 \cdot 2H_2O$, and the magnesium chloride mineral being *carnallite*, $KMgCl_3 \cdot 6H_2O$. The rather complicated conditions determining crystallization from a solution of several salts prevents easy exposition; for the purposes of the present comparison, the simple situation implied by Table 72 is preferable.

TABLE 71. *Distribution of closed lakes in logarithmic classes of concentration*

Concentration, mg. kg.$^{-1}$	Limnological Studies	Clarke	Per Cent of Total
10–100	0	0	0
100–1000	31	3	25.8
1000–10,000	21	13	25.8
10,000–100,000	13	29	31.8
>100,000	1	21	16.6

TABLE 72. *Solubilities of important salts in salt lakes*

		Solubility g. kg.$^{-1}$
Halite	$NaCl$	357
Mirabilite	$Na_2SO_4 \cdot 10H_2O$	88.7
Nacholite	$NaHCO_3$	81.5
Bischofite	$MgCl_2 \cdot 6H_2O$	536
Epsomite	$MgSO_4 \cdot 7H_2O$	305
Gypsum	$CaSO_4 \cdot 2H_2O$	1.93

It will immediately be apparent how far from saturated with the common soluble salts are the majority of the lakes of closed basins.

Types of saline waters. It has become customary to divide the saline waters of inland basins into three extreme types based on the prevalent anion, and therefore termed carbonate, sulfate, and chloride waters. All possible intermediate mixtures exist.

In a very large number of closed basins the sequence expected on evaporation of a normal fresh water is complicated by the presence of old salt beds, which may greatly modify the composition of the influents. For two regions, the Lahontan Basin and certain adjacent basins in western North America[9] (Clarke 1924) and the mountains bordering the western edge of the Tibetan plateau (Hutchinson 1937b), there are a number of analyses of both fresh and mineralized waters, and a fair presumption that salt beds and saline springs have not been an important complicating factor. The relative proportions of the three major anions in the waters of these two regions are set out in Figs. 156 and 157, on the type of triangular diagram used by petrologists.[10]

[9] Modern conditions in the Lahontan and adjacent basins are so complicated by irrigation that only the older analyses are of value for the present purpose.

[10] The percentage of any one of the three anions is proportional to the length of the perpendicular from any point on to the side opposite the angle marked by the symbol of the anion. This angle will thus correspond to 100 per cent of the anion in question.

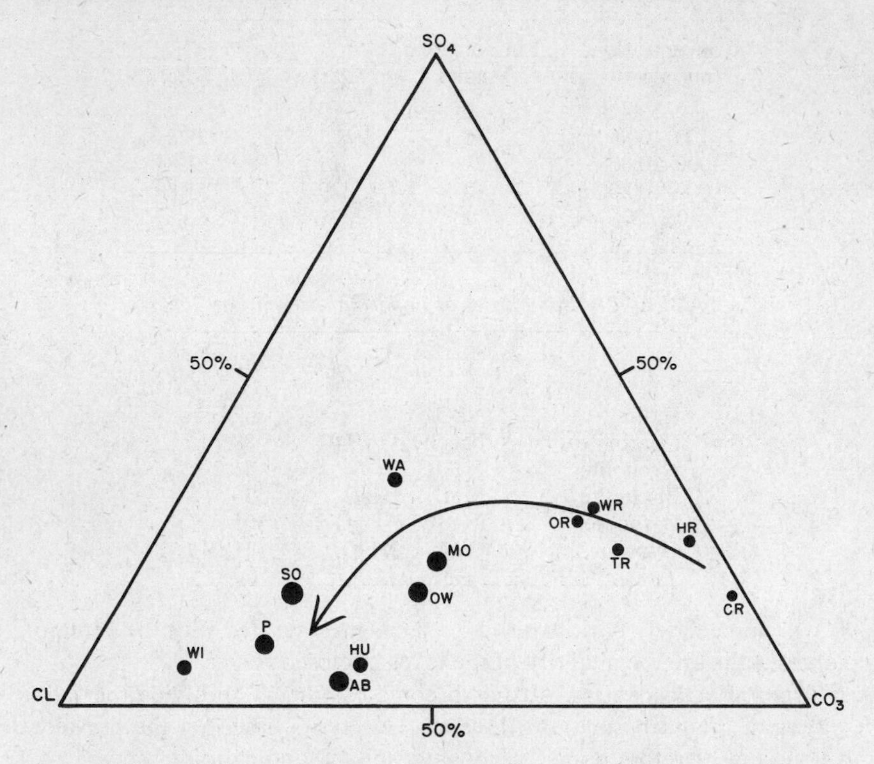

FIGURE 156. Anionic compositions of rivers and lakes in the Lahontan Basin and in closed basins to the north and west. CR, Chewaucan River, influent of Abert Lake; HR, Humboldt River, influent of Humboldt Lake; TR, Truckee River, influent of Pyramid Lake; WR, Walker River, influent of Walker Lake; OR, Owens River, influent of Owens Lake; OW, Owens Lake; WA, Walker Lake; MO, Mono Lake; SO, Big Soda Lake; HU, Humboldt Lake; AB, Abert Lake; P, Pyramid Lake; WI Winnemucca Lake. (All nineteenth and early twentieth-century analyses, collected by Clarke.)

It has long been realized (Clarke 1924) that starting with any water of the kind that may be expected in an ordinary river system, concentration will first result in the precipitation of calcium carbonate. This will lead to a relative enrichment of chloride and sulfate in the remaining liquid phase. As sulfate accumulates, the solubility product of calcium sulfate will finally be exceeded and gypsum, $CaSO_4 \cdot 2H_2O$, will begin to precipitate.[11] The solution will therefore be enriched in chloride relative to the other

[11] In ordinary solutions *anhydrite*, $CaSO_4$ does not form until a temperature of 42° C. is reached, but in very concentrated brines this mineral rather than gypsum is precipitated at much lower temperatures.

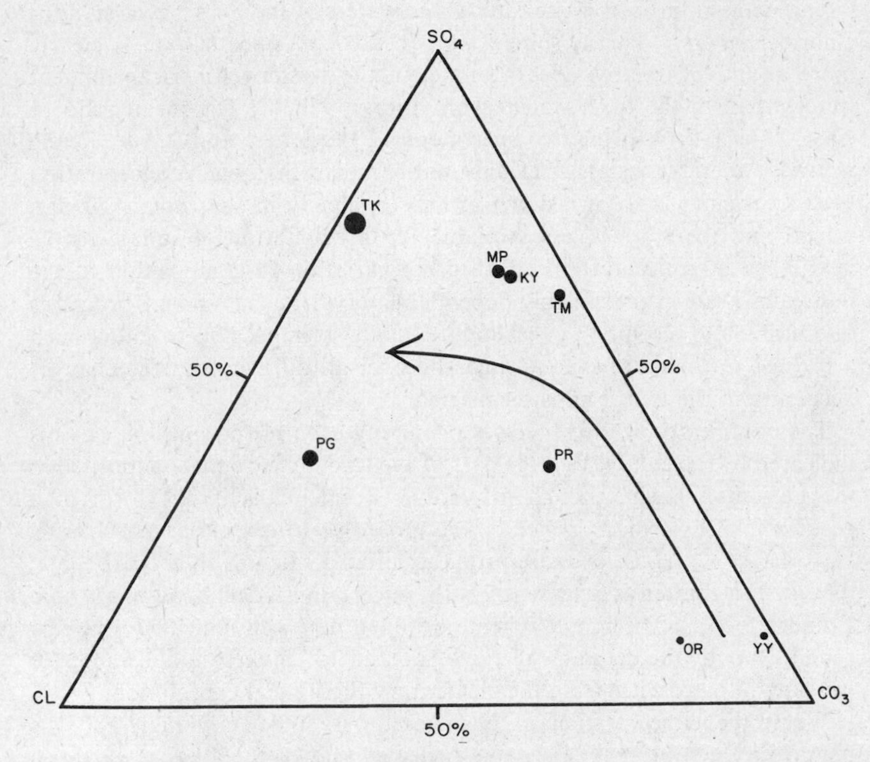

FIGURE 157. Anionic composition of open and closed lakes in Indian Tibet (Ladakh and Rupshu). YY, Yaye Tso; OR, Ororotse Tso; TM, Tso Moriri; PR, Pangur Tso; KY, Khyagar Tso; MP, Mitpal Tso; PG, Pang-gong Tso; TK, Tso Kar. (Data from Hutchinson.)

anions, so that the normal sequence appears to be from carbonate to sulfatochloride to chloride waters. The exact path taken by any given water will depend on the amount of calcium present at the time when most of the carbonate has precipitated.

In Figs. 156 and 157 the diameter of the circle indicating any analysis is proportional to the logarithm of the salinity of the water. It must be remembered, however, that dilution of any water, occurring relatively rapidly, will have very little effect on the anionic ratios, so that there is no guarantee that the sizes of the circles increase uniformly with the chemical history of the water.

It will be apparent that in each diagram a history of change in the anionic proportions, roughly following the direction of the arrows, would provide a reasonable interpretation. The Lahontan waters appear to end up as chlorocarbonate waters, the waters of Indian Tibet as sulfato-chloride waters. This is somewhat paradoxical, since sulfate is clearly more abundant, relative to carbonate, in the western American influent waters than in the fresh waters from Indian Tibet. The mean ratio of $Ca:CO_3$ is 1:2.14 in the two specimens of the latter, and 1:1.88 in the western American waters. This probably means that when concentration occurs, calcium is thrown down more completely in the Indian Tibetan waters and therefore is less available for precipitation of sulfate in the next stage of concentration. The great variation in composition of the individual waters, presumably dependent on various secondary processes of dilution and resolution of carbonate, makes it impossible to follow such a process in the data, even though the over-all difference in the changes occurring in the two regions seems clear.

The complexity of the processes occurring even in the simplest cases is indicated by the fact that a large part of the calcium carbonate deposited in the Lahontan Basin, though present as a calcite, is in the form of a pseudomorph termed *thinolite*, supposedly after either gaylussite, $Na_2Ca(CO_3)_2 \cdot 5H_2O$, or more probably after a calcium chlorocarbonate, though the matter is apparently still uncertain (Palache, Berman, and Frondel 1951). The simple scheme outlined here and indicated in a very general way by the diagrams of Figs. 156 and 157 makes no allowance for this sort of divergence from the elementary theoretical expectation.

When the concentration of bicarbonate is sufficiently high and the quantity of alkalies is of the same order as that of the alkaline earths, a precipitation of practically all the calcium as $CaCO_3$ and presumably most of the sulfate as $CaSO_4 \cdot 2H_2O$ can leave a water that contains sodium bicarbonate and carbonate as its major constituent, as in some of the lakes of the northwestern United States. Moses Lake and Soap Lake, Washington, are important examples.

It will be apparent from this discussion that relatively small variations in the initial ionic proportions can have striking effects in regulating the final composition of the waters of closed basins. Further complications are introduced when the influent receives drainage from pre-existing salt beds. Most of the almost pure chloride waters, unless they represent the evaporation of sea water,[12] appear to have had such an origin.

The marked differences in composition in the water of Great Salt Lake, Utah (Table 73), compared with the more concentrated of the Lahontan waters is due to additional chloride in the influents. This is well indicated

---

[12] The first change in evaporating sea water is a reduction in sulfate content as calcium sulfate precipitates.

TABLE 73. *Composition of mineral salts of chloride, sulfate, and carbonate lakes and their influents*

	Na	K	Mg	Ca	CO_3	SO_4	Cl	SiO_2 A	$(AlFe)_2O_3$	Salinity
Chloride waters										
Bear River										
Upper, Wyoming	4.49		6.86	23.69	52.68	5.76	2.68	3.84	...	185
Lower, Utah	20.54		4.76	10.12	21.53	8.16	32.36	...	2.53	637
Great Salt Lake, Utah	33.17	1.66	2.76	0.17	0.09	6.68	55.48	...	...	203,490
Jordan at Jericho	18.11	1.14	4.88	10.67	13.11	7.22	41.47	1.95	1.45	7,700
Dead Sea	11.14	2.42	13.62	4.37	Trace	0.28	66.37*	Trace	...	226,000
Sulfate waters										
Montreal Lake, Saskatchewan	4.9	2.3	10.8	16.8	56.5	1.8	2.5	3.9	0.5	150.5
Redberry Lake, Saskatchewan	12.0	0.85	12.3	0.56	2.58	70.5	1.1	0.03	0.07	12,898
Little Manitou Lake	16.8	1.0	10.9	0.48	0.47	48.4	21.8	0.019	0.21	106,851
Carbonate waters										
Silvies River, Oregon	10.42	2.45	3.13	12.88	34.76	7.35	2.88	25.13	0.08	163
Malheur Lake, Oregon	24 17	5.58	4.13	5.58	44.63	7.64	4.55	2.89	Trace	484
Pelican Lake, Oregon	29.25	3.58	2.62	2.27	30.87	22.09	7.97	1.21	0.02	1,983
Bluejoint Lake, Oregon	37.70	2.62	0.63	0.53	38.68	5.67	13.85	0.55	0.02	3,640
Moses Lake, Washington	19.86		7.25	8.41	51.56	2.87	3.88	5.06	1.11	2.966

* And 1.78 per cent Br.

by the composition of Bear River in its upper reaches in Wyoming, where carbonate is the dominant anion and calcium the dominant cation, and in its lower part, where chloride exceeds carbonate and the alkalies exceed calcium.

The same situation prevails in the Dead Sea; the lower Jordan is clearly enriched in chloride. It is supposed (Rankama and Sahama 1950) that the high bromide content of the lake is due to bromine-rich springs flowing into the Sea of Galilee.[13] In most other cases it is probable that the Br:Cl ratio would be low in such chloride waters, as bromine is excluded from halite. This, however, is only known to be the case in the Caspian, where the low ratio reflects the bromine-deficient salt deposits of the Urals. This fact is of considerable interest biogeographically, as it confirms the view that the salts of the Caspian have accumulated from the drainage basin after the last high level and presumably fresh stage during the Pleistocene glaciation and therefore do not represent Tertiary sea salt.

Some sulfate waters undoubtedly owe their composition to the leaching of soils and sediments; this is in fact the only reasonable explanation of the low relative chloride in lakes such as Redberry in Saskatchewan, studied by Rawson and Moore (1944). This lake, though less concentrated than some of its neighbors, seems to be the purest sulfate water for which a complete analysis exists. Little Manitou Lake, though richer in chloride, is approximately saturated with sodium sulfate and, in winter, deposits *mirabilite*, $Na_2SO_4 \cdot 10H_2O$, on its floor. The very large amount of sodium and magnesium sulfate in these waters is directly due to an accumulation of these salts in the prairie soils of the region.

Nelson and Thompson (1954) have pointed out that when sea water freezes, the first salt to be deposited is mirabilite. They suggest that some sulfate deposits have been formed by freezing; later solution might give very pure sulfate waters. There is evidence from the saline lakes of Australia that loss of sulfate by bacterial reduction to H_2S in the mud at times of low water can produce chlorocarbonate waters from concentration of wind-borne salt of marine origin.

Boron in saline lakes. Borate is the fourth in abundance of the anions of sea water. Borax, $Na_2B_4O_7 \cdot 10H_2O$, is known to crystallize from saline lakes in several parts of the world. According to Palache, Berman, and Frondel (1951), the mineral has been obtained commercially from early times from saline lakes in Tibet. The supposed occurrences in Ladakh (Indian Tibet) appear doubtful to the present writer; borax is

[13] The suggestion that this bromine was concentrated biologically as dibromindigo in gastropod mollusks appears to the writer to be wildly improbable.

apparently brought into that region from Rudok to the east and is also known in central Tibet. Very large amounts have been found in California lakes, notably Borax Lake and Searles Lake, and in comparable lakes in South America. A number of other boron minerals are known in playa deposits of western North America. At least in Great Salt Lake, the ratio of B:Cl is very close to that in the ocean (Odum and Parrish 1954), but it is hard to avoid the conclusion that such a resemblance is accidental.

Possibility of occurrence of chlorine-36 *in saline lakes.* The radioactive isotope Cl^{36} is to be expected to form in minute amounts in localities in which neutrons in cosmic ray cascades can be captured by Cl^{35} atoms. The conditions for the effect being detectable in a lake are primarily great age, high sodium chloride content, and small mean depth. Davis and Schaeffer (1955) were unable to detect the isotope in Great Salt Lake. They computed that at radioactive equilibrium the chlorine of the lake would contain enough Cl^{36} to give 0.028 disintegration per gram per minute, which would be just detectable. They suspect that the salt has not remained long enough in the basin to show the effect.

Cations of closed basins. It will be evident from this discussion and from the analyses of Table 73 that ordinarily sodium and magnesium are the dominant cations in the late stages of concentration. Late in the history of a concentrated salt lake, magnesium salts, being more soluble than sodium, may be enriched, as in the Dead Sea. Very occasionally, quite extraordinary quantities of potassium undoubtedly occur. Some of the saline lakes of Nebraska provide the best examples. In Floyd Lake and Jesse Lake no less than 23 per cent of the dissolved inorganic content is potassium (Clarke 1924).

The great variations in ionic ratios in all waters is probably of biological significance, though the matter has been little studied. The most likely effect would be in lakes rich in potassium. Boone and Baas-Becking (1931) found that the hatching of *Artemia* eggs is inhibited by excess potassium, and Byrstein and Beliaev (1946) claim that the relatively high potassium content of Lake Balkhash water is responsible for the depauperated fauna of the lake. Bond (1935) refers to the inland salt lakes rich in anions other than chloride, and often in cations other than sodium, as *athalassohaline*, in contrast to the waters of marine origin or simulating concentrated sea water, as in the Great Salt Lake, which are *thalassohaline*. Too little is yet known of the ecological importance of the ionic ratios to estimate how useful such terms may be, but they do emphasize the fact that the waters of most saline lakes are very unlike sea water.

SUMMARY

Rain water is not pure water but contains in solution atmospheric gases or their reaction products, and various salts derived from dust and from sea spray. It is probable that natural raindrops condense around nuclei provided either by marine salt or by nitrous acid, the latter probably being formed by the photochemical oxidation of ammonia adsorbed on dust particles. In industrial regions sulfuric acid nuclei are also important.

Continental rain usually contains more sulfate than chloride. The chloride is probably of marine origin. The source of the sulfate in excess of the expected marine proportion is uncertain. Part may be of industrial origin even in supposedly clean rain, part may be due to oxidation of biochemically produced H_2S; there is also the possibility that $CaSO_4$ may be separated from more soluble salts in evaporating sea water and then blown as an aerosol for great distances. Under some circumstances oxidation of Cl^- to Cl_2 by ozone may lower the chloride content of rain relative to other constituents.

The main cation is sodium. Magnesium and calcium increase relative to sodium as the distance from the ocean increases. Nitrogen compounds are always present. The ammonia is almost certainly of terrestrial origin, and is apparently associated with dust. Nitrite and nitrate, ordinarily regarded as formed from N_2 and O_2 by electric discharges in the atmosphere, are usually more likely to have been produced by the photochemical oxidation of ammonia. A basin filled solely with rain water would have a chemical composition not very different from that actually recorded in the most dilute soft-water lakes.

The total inorganic matter in solution in a lake water may be expressed in various ways. The *total solids* per unit mass or volume refers to the total residue on evaporation, the *nonvolatile solids* to the residue after the total solids are ignited at dull-red heat. The *salinity* of a water is best defined as the concentration of all cations, significantly Na, K, Mg, and Ca, and of the anions $CO_3^{=}$ and $SO_4^{=}$ and halide, all HCO_3 being converted to $CO_3^{=}$. Silica and sesquioxides are usually also included, but this usage has little to recommend it in theory; the error involved is usually very small. The mean salinity of the rivers of the world appears to be about 100 mg. per liter, but individual variations are enormous. Water having a salinity of less than 50 mg. per liter represents, in general, drainage from igneous rocks. The average equivalent percentage composition of the cations of such water is Na 30.6, K 6.9, Mg 14.2, Ca 48.3. As the salinity rises, the proportion of alkalies falls and that of magnesium, and particularly of calcium, rises. In the average river water the percentage proportions are Na 15.7, K 3.4, Mg 17.4, Ca 63.5. In most lake districts the mean cationic composition of the waters of open lakes tends to approximate to the average river. This is partly due to the present distribution of source materials in igneous and sedimentary rocks, and partly due apparently to base-exchange phenomena. When base-exchange reactions take place with organic peaty sediments, very soft waters enriched in alkalies and much depleted in calcium may be produced. Usually bicarbonate is the dominant anion in inland waters, but soft waters that have lost Ca^{++} to organic sediments also appear to have lost HCO_3^-.

Lithium is known to occur in lake water, but few data on its occurrence have been published. Rubidium has once been recorded in a concentration about 1 per cent of the potassium, which is geochemically reasonable. Strontium has been detected whenever it has been sought. In Sweden the ratio of Sr to Ca by atomic proportions falls from 2.45:1000 in soft waters to about 0.5:1000 in calcareous waters. This is to be expected, as limestones are deficient in the element. The general pattern of distribution observed in Sweden probably will be found elsewhere. Barium has been very little studied and appears to occur in minimal quantities.

The main variations in anions are found in acid waters, in which sulfate is often dominant. This is also true in a limited number of lakes receiving water from gypsum-bearing rocks. Chloride is ordinarily very low in uncontaminated waters, but it may be locally enriched by the influx of ground water containing chloride. Near the ocean the influence of wind-borne sea spray is significant.

Such lake waters as have been studied usually contain 0.1 to 0.5 mg. per liter of fluorine. The ratio F:Cl is clearly of the order of 0.01–0.1:1, much higher than in the ocean. Bromine has been determined mainly in Russian fresh waters, where the mean amount for a series of lakes was 4.5 mg. $m.^{-3}$, implying a ratio Cl:Br of 941:1. This figure implies less Br relative to Cl than in the ocean, but may be atypical owing to drainage from strata bearing rock salt, which is very low in bromine. Iodine varies from 0.01 to about 1 mg. $m.^{-3}$ The mean value would appear to be about 0.2 mg. $m.^{-3}$ This implies considerable enrichment relative to Cl and Br over what is found in the ocean. Both Br and I may be slightly concentrated in organic lake sediments.

The silicate and phosphate concentrations and those of the oxyacid anions of nitrogen are discussed in later chapters. Boron occurs in lakes as borate, the amount present in fresh waters being about 20 mg. $m.^{-3}$ Of the other possible anions, it seems likely that selenite, which might be brought into some lakes, would be precipitated as ferric selenite. Arsenic has been supposed to occur in quantities of the order of 2 mg. $m.^{-3}$ but requires investigation.

In closed basins the inorganic composition is dependent not only on the sources of material dissolved in the influents, but also on their fate on evaporation. Since most closed lakes occupy regions of rather unstable climate, they easily dry up, losing salts by deflation. They may also at times overflow and lose salts through the newly formed effluent. For these reasons lakes saturated with the more soluble salts are relatively rare. Over half the basins for which analyses can be assembled seem to have salinities under 10 g. $kg.^{-1}$, while saturation for sodium chloride, bicarbonate, or sulfate, or for magnesium chloride or sulfate, would imply salinities over 80 g. $kg.^{-1}$

The mineralized lakes of closed basins are best classified by their anion content, the extreme types being chloride, sulfate, and carbonate (or bicarbonate) waters. All possible intermediates exist. When an ordinary fresh water evaporates, calcium carbonate is first deposited. If excess calcium remains in solution, further evaporation leads to a precipitation of calcium sulfate. Ideally, a water containing chloride as its main anion would result by this process. In practice, the initial quantities of bicarbonate, sulfate, and calcium will be rather variable, and the final results will usually be the formation of various types of sulfatochloride water. Variations in the initial calcium content are apparently as important as variations in the ratio of SO_4 to Cl in determining the final proportion of anions. With sufficient alkali in the initial water, practically all the calcium can be removed as $CaCO_3$ and $CaSO_4 \cdot 2H_2O$ before the relative bicarbonate concentration is greatly reduced, giving lakes of the carbonate or carbonatochloride type. In a great many cases the sequence is modified by the inflow of water from old salt beds, which may greatly raise the chloride, as in the Great Salt Lake and the Dead Sea.

Many sulfate lakes owe their dominant anion to leaching of sulfate deposits, which may initially have been formed by the freezing of sea water. It is not unlikely that carbonatochloride waters may be produced by loss of sulfate through bacterial reduction in shallow basins with richly organic mud. Wind-borne salt is certainly of significance in the production of salt lakes. It is possible that in some cases when salt has been picked up by wind from a wide expanse of coastal land, the most soluble components, the magnesium halides, may be deficient; this provides a further mechanism in differentiation of saline waters, operating, for instance, in the Sambhar Lake in India. Borate may be considerably enriched in inland saline lakes. In Great Salt Lake the ratio B:Cl is very close to that in the ocean, but in view of the history of the salt in the lake, this resemblance is probably accidental.

CHAPTER 9

Oxygen in Lake Waters

The concentration of no substance present in lake water, except perhaps in a rough way the hydroxonium ion, has been studied as much as that of oxygen. A skillful limnologist can probably learn more about the nature of a lake from a series of oxygen determinations than from any other kind of chemical data. If these oxygen determinations are accompanied by observations on Secchi disk transparency, lake color, and some morphometric data, a very great deal is known about the lake.

THE SOLUBILITY OF OXYGEN IN WATER

Oxygen is moderately soluble in water, the solubility being directly proportional to the partial pressure in the gas phase, by Henry's law, and decreasing in a nonlinear manner with increasing temperature. These relationships to pressure and temperature raise certain problems of limnological importance which must be discussed before proceeding to a consideration of the passage of oxygen from the atmosphere into the lake, and the origin and fate of such oxygen as may be dissolved in the lake at any given time.

The effects of atmospheric pressure. The proportion of oxygen in dry air is essentially constant and may be taken as 20.95 per cent by volume. Water in nature, however, cannot be regarded as tending to come into equilibrium with dry air; it would in fact be more reasonable to suppose that the air in contact with any exposed water surface is saturated with water vapor. It is possible that in some cases this condition is not actually realized, but in default of empirical information, it is better to assume that equilibrium will tend to be established with air saturated with water vapor, so that the partial pressure of the oxygen will be not $0.2095P_h$, but rather $0.2095(P_h - p_w)$, when at the locality under consideration at altitude h meters, the total atmospheric pressure is P_h, and p_w is the partial pressure of water vapor at the temperature of the observation.

Atmospheric pressure varies irregularly according to the meteorological conditions, and these irregular variations show a seasonal trend. Since the variations are in general small and the various values succeed one another rapidly, they may generally be disregarded. A departure from the mean atmospheric pressure at lake level at any time during summer stratification is certainly quite irrelevant to the conditions under which oxygen entered the deep water during vernal circulation or by slow turbulent exchange during many subsequent weeks.

For very accurate work, it might be desirable to use observed pressures over a period of a few weeks in the spring in calculations relating to the hypolimnion, and over a like period immediately preceding the time of observations for the epilimnion, but Minder (1941) and Tonolli (1947b) alone appear to have regarded a refinement of this sort as worthwhile. The value of the atmospheric pressure ordinarily employed in limnological calculations is the mean pressure at the altitude of the lake surface. The pressure P_h at any altitude h is given by

$$\log P_h = \log P_0 - \frac{273h}{18421\theta_K} \tag{1}$$

where P_0 is the pressure at sea level, namely 760 mm. Hg or 1 atm. and θ_K is the mean absolute temperature of the air column between sea level and the level h. The major difficulty involved in the use of this equation is the correct evaluation of the term θ_K, for when dealing with a lake or other locality not in the free atmosphere, θ_K is essentially the temperature of an imaginary column of air. Alsterberg (1930a) and Yoshimura (1938b) took $\theta_K = 273°$ K., which simplifies the equation but does not improve the resulting value of P_h. Juday and Birge (1932) took $\theta_K = (273 + \theta_0)$, where θ_0 is the surface temperature of the lake. For small values of h this is doubtless the best procedure. Ricker (1934) points out that, since meteorological experience indicates that an increase of 200 m. in elevation is usually accompanied by a decrease in mean temperature of 1° C., a good approximation to the pressure at the lake surface will be given by

$$\log P_h = \log P_0 - \frac{273h}{18421(\theta_0 + 273 + 0.0025h)} \tag{2}$$

Ideally, if very accurate results are required, this equation should be evaluated for each temperature in a vertical series of observations, for the surface temperature, which is doubtless a good approximation to the mean air temperature at the time of observation, is irrelevant to the conditions under which the deeper layers of the lake were aerated. Minder adopts a workable compromise by computing the pressure for the epilimnion and hypolimnion separately. In many cases it seems probable that the mean temperature $\bar{\theta}$ of the lake would give a suitable approximation applicable to all levels.

When a meteorological station exists at altitude h_1 near the lake, Minder (1941) computes the mean pressure at the lake surface for any season from the observed mean pressure P_1 recorded at the meteorological station, the mean absolute temperatures at the lake and at the station for the season in question being θ_{Kh} and θ_{Ks} from

$$\text{(3)} \qquad \log P_h = \log P_1 - \frac{273(h - h_1)}{18421(\theta_{Kh} + \theta_{Ks})0.5}$$

In general, if P_h is not known from direct observation, θ_{Kh} will also not be recorded, so that approximations of the form $\theta_{Kh} = 273 + \theta_z$ will usually have to be employed.

If stations exist at altitudes h_1 and h_2, below and above the level of the lake h respectively, and if the highest of these stations lies below 2500 m., Minder (1941) suggests using the approximation

$$\text{(4)} \qquad P_h = P_1 - \frac{P_1 - P_2}{h_1 - h_2}(h - h_1)$$

As an example of the use of these last two expressions, Minder computes that during the month of July the mean pressure at the surface of the Davosersee, which lies at an altitude of 1562 m., between stations at 910 m. and 1811 m., is by equation 3, 634.9 mm. Hg, and by the approximate equation 4, which assumes a linear fall in pressure with increasing altitude, 635.3 mm. Hg. The difference is negligible for limnological purposes. If no stations in the vicinity had been available and equation 2 had been used, taking $\theta_0 = 14.9°$ C., the surface temperature at the beginning of August, a value of 628.0 mm. Hg would have been obtained, and for most purposes such an estimate, which is about 1 per cent too low, would be quite adequate.

For lakes at very high altitudes in inaccessible regions, it will usually be necessary to employ either equation 2 or equation 3 relative to some quite distant station at a much lower altitude than the lake. In such cases, much less good agreement between the different modes of computation is likely to be encountered. Thus for Yaye Tso in Indian Tibet, at an altitude of 4686 m. and with a surface temperature on August 19, 1932, of 14.19° C. and a mean temperature on the same day of 11.68° C. (Hutchinson 1937b), the evaluations of equations 1 and 2 give

$$\theta_K = 273^\circ \text{ K.} \qquad P_h = 423 \text{ mm. Hg}$$
$$\theta_K = (273 + \theta_0)^\circ \text{ K.} \qquad P_h = 436 \text{ mm. Hg}$$
$$\theta_K = (273 + \theta_0 + 0.0025h)^\circ \text{ K.} \qquad P_h = 445 \text{ mm. Hg}$$
$$\theta_K = (273 + \bar{\theta})^\circ \text{ K.} \qquad P_h = 433 \text{ mm. Hg}$$
$$\theta_K = (273 + \bar{\theta} + 0.0025h)^\circ \text{ K.} \qquad P_h = 442 \text{ mm. Hg}$$

These values may be compared with those derived from the meteorological data for the station at Leh, about 125 km. to the northwest of the lake and at an altitude of 3510 m. Here the mean August temperature is 17.5°C. and the mean pressure for the same month is 500.5 mm. Hg; for the whole period from May to August the mean temperature is 14.9° C. and the mean pressure is 499.3 mm. Hg. From these data and from the surface and mean temperatures of the lake, the pressures at the surface of Yaye Tso were computed by equation 3.

August $P_h = 435$ mm. Hg
May–August $P_h = 434$ mm. Hg

It is difficult to know in this case which values are preferable, though it would not be unreasonable to take the values derived by reduction of the observations at Leh to the altitude of the lake; the value of 434 mm. has accordingly been used on page 632.

Tables of solubilities. The solubility of oxygen in water has been determined by a number of investigators, and the results have been expressed in a variety of ways. The most usual mode of presentation in the physical literature is in terms of the Bunsen absorption coefficient or volume of gas, reduced to 0° C. and 760 mm. Hg, taken up by unit volume of solvent when the solvent comes into equilibrium with a constant partial pressure of 1 atm. of the gas. For oxygen, this figure when multiplied by 209.5 will give the number of milliliters of the gas that are in equilibrium in 1 liter of water, with 1 atm. of dry air. For some purposes, the Henry's law constant is found to be a convenient mode of expressing the solubility. For all practical purposes, the Bunsen coefficient for gases dissolving in water (molecular weight 18 and density 1) may be found by dividing 946,200 by the value of the Henry's law constant. Since the latter quantity is used in the tabulation of the solubility of gases in the International Critical Tables, this relationship may be of use to limnologists dealing with the less abundant gases such as methane or hydrogen, occasionally of importance in the hypolimnion.

If tables of absorption coefficients or values of saturation relative to dry air at 760 mm. Hg are used, the saturation relative to wet air at any pressure P_h will be given by multiplication by $(P_h - p_w)/760$; if tables of solubility or saturation relative to wet air are used, the equivalent factor is $(P_h - p_w)/(760 - p_w)$. Some workers may prefer the former slightly simpler factor, but the difference in labor in the two operations is negligible. Moreover, for altitudes below 500 m. the error introduced in applying the first factor to a solubility table relative to wet air is so small as to be of no importance except for the most accurate work in the upper part of the limnological temperature range. It is therefore believed that the presentation of saturations relative to wet air is likely to be preferred by most workers (Table 74).

The various solubility tables which have been employed by limnologists have been critically discussed by Ricker (1934), Minder (1941), Ohle (1952), and Mortimer (1955b). The laboratory data on which these tables are based are those of:

(1) Roscoe and Lunt (1889), referring to wet air between 5° C. and 30° C. This series of observations was extrapolated to 0° C. by Whipple and Parker (1902). The original data and the extrapolation form the basis of the table given by Birge and Juday (1911). This table has been widely used, even as recently as 1928 by Thienemann. The extrapolated values are certainly too low.

(2) Winkler (1889), presented for dry oxygen and for wet air, for the temperature range 0° to 30° C.

(3) Jacobsen (1905), given for dry air over the range 0° to 25° C.

(4) Fox (1907, 1909), given originally for dry air and so used by Birge and Juday (1914) and, following them, by a number of other investigators. The data were obtained from experiments in which pure oxygen was used, and were computed on the basis of 20.90 per cent O_2 by volume. Minder has recomputed the table on the basis of the more acceptable 20.95 per cent by volume;[1] Ohle (1952) has used a value of 20.93 per cent. Whipple and Whipple calculated a table for wet air from Fox's data, and this has been given in the many editions of the excellent *Standard Methods of Water Analysis* (American Public Health Association, 1917 *et seq.*). Fox's determinations appear to have been confirmed by Whipple and Whipple in some preliminary experiments, and practically identical values were also given by Bohr and Bock (1891). The nomograms of Ricker (1934) and of Rawson (1944) for wet air and that of Schmassmann (1948) for dry air are based on these values.

(5) Krogh and Lange (1932) who determined the oxygen concentration of water derived from Furesø when it has supposedly come into equilibrium with air.

(6) Truesdale, Downing, and Lowden (1955; Truesdale and Downing 1954), who determined the oxygen concentration in contact with air after slow equilibration.

---

[1] The best value seems to be 20.946 per cent.

TABLE 74. *Solubility of oxygen, from a wet atmosphere at a pressure of 760 mm. Hg, in mg. per liter, at temperatures from* 0° *to* 35° *C.**

Temp.	0.0	0.1	0.2	0.3	0.4	0.5	0.6	0.7	0.8	0.9
0	14.16	14.12	14.08	14.04	14.00	13.97	13.93	13.89	13.85	13.81
1	13.77	13.74	13.70	13.66	13.63	13.59	13.55	13.51	13.48	13.44
2	13.40	13.37	13.33	13.30	13.26	13.22	13.19	13.45	13.12	13.08
3	13.05	13.01	12.98	12.94	12.91	12.87	12.84	12.81	12.77	12.74
4	12.70	12.67	12.64	12.60	12.57	12.54	12.51	12.47	12.44	12.41
5	12.37	12.34	12.31	12.28	12.25	12.22	12.18	12.15	12.12	12.09
6	12.06	12.03	12.00	11.97	11.94	11.91	11.88	11.85	11.82	11.79
7	11.76	11.73	11.70	11.67	11.64	11.61	11.58	11.55	11.52	11.50
8	11.47	11.44	11.41	11.38	11.36	11.33	11.30	11.27	11.25	11.22
9	11.19	11.16	11.14	11.11	11.08	11.06	11.03	11.00	10.98	10.95
10	10.92	10.90	10.87	10.85	10.82	10.80	10.77	10.75	10.72	10.70
11	10.67	10.65	10.62	10.60	10.57	10.55	10.53	10.50	10.48	10.45
12	10.43	10.40	10.38	10.36	10.34	10.31	10.29	10.27	10.24	10.22
13	10.20	10.17	10.15	10.13	10.11	10.09	10.06	10.04	10.02	10.00
14	9.98	9.95	9.93	9.91	9.89	9.87	9.85	9.83	9.81	9.78
15	9.76	9.74	9.72	9.70	9.68	9.66	9.64	9.62	9.60	9.58
16	9.56	9.54	9.52	9.50	9.48	9.46	9.45	9.43	9.41	9.39
17	9.37	9.35	9.33	9.31	9.30	9.28	9.26	9.24	9.22	9.20
18	9.18	9.17	9.15	9.13	9.12	9.10	9.08	9.06	9.04	9.03
19	9.01	8.99	8.98	8.96	8.94	8.93	8.91	8.89	8.88	8.86
20	8.84	8.83	8.81	8.79	8.78	8.76	8.75	8.73	8.71	8.70
21	8.68	8.67	8.65	8.64	8.62	8.61	8.59	8.58	8.56	8.55
22	8.53	8.52	8.50	8.49	8.47	8.46	8.44	8.43	8.41	8.40
23	8.38	8.37	8.36	8.34	8.33	8.32	8.30	8.29	8.27	8.26
24	8.25	8.23	8.22	8.21	8.19	8.18	8.17	8.15	8.14	8.13
25	8.11	8.10	8.09	8.07	8.06	8.05	8.04	8.02	8.01	8.00
26	7.99	7.97	7.96	7.95	7.94	7.92	7.91	7.90	7.89	7.88
27	7.86	7.85	7.84	7.83	7.82	7.81	7.79	7.78	7.77	7.76
28	7.75	7.74	7.72	7.71	7.70	7.69	7.68	7.67	7.66	7.65
29	7.64	7.62	7.61	7.60	7.59	7.58	7.57	7.56	7.55	7.54
30	7.53	7.52	7.51	7.50	7.48	7.47	7.46	7.45	7.44	7.43
31	7.42	7.41	7.40	7.39	7.38	7.37	7.36	7.35	7.34	7.33
32	7.32	7.31	7.30	7.29	7.28	7.27	7.26	7.25	7.24	7.23
33	7.22	7.21	7.20	7.20	7.19	7.18	7.17	7.16	7.15	7.14
34	7.13	7.12	7.11	7.10	7.09	7.08	7.07	7.06	7.05	7.05
35	7.04	7.03	7.02	7.01	7.00	6.99	6.98	6.97	6.96	6.95

* From Truesdale, Downing, and Lowden (1955).

Even among the more modern data there is considerable variation in the values obtained by different investigators. Fox's values, apparently confirmed by Whipple and Whipple and in accord with the earlier series of Bohr and Bock, are definitely higher than the others. Mortimer

(1955b) has considered the problems involved and concludes that the most recent data, those of Truesdale, Downing, and Lowden, confirmed at a temperature of 25° C. by Allen (1955), represent the conditions most closely comparable to those found in nature. The values of Krogh and Lange on Furesø water are in very good agreement with those of Truesdale, Downing, and Lowden. Most of the illustrations for the present chapter had been drawn before the appearance of Truesdale and Downing's (1954) initial contribution. In all calculations for such illustrations, Fox's values, referred to wet air containing 20.95 per cent by volume O_2, following Minder, were used. These saturation values are evidently of the order of 3 per cent too great. As far as was possible, the text has been rewritten using the new values, but it is not always possible to correct derived data given by other workers. Through the kindness of Dr. Mortimer and the Council of the International Association for Pure and Applied Limnology, Mortimer's revised nomogram for determination of saturation is reproduced in Fig. 158.

Absolute saturation. The term absolute saturation was introduced by Ricker (1934),[2] whose general mode of presentation is here followed, to express the saturation corresponding to the actual pressure P_z, both atmospheric and hydrostatic at any depth z. In general,

$$P_z = P_0 + 0.0967z \tag{5}$$

when the pressures are measured in atmospheres. The quantity of gas in solution at equilibrium at such a pressure is of interest, because if the quantity is increased beyond the equilibrium concentration, there is a possibility of bubble formation. Owing to the fact that several gases of varying solubility are present in lake water, it is not adequate to compute the pressure at which a given concentration of pure oxygen or other gas will just permit the formation of a bubble; all the gases present in the water must, in theory, be considered, for they will all be present in the gas phase of the bubble also. In practice, certain gases may be neglected, and argon and nitrogen may be treated together as atmospheric nitrogen.

Let unit volume of water at depth z m., temperature θ_z° C., and pressure P_z, contain concentrations of several gases

$$[G_1]_z, \quad [G_2]_z, \quad [G_3]_z, \quad [G_4]_z, \quad \text{etc.}$$

and let

$$[G_1]_S, \quad [G_2]_S, \quad [G_3]_S, \quad [G_4]_S, \quad \text{etc.}$$

be the concentrations of the pure gases in water in equilibrium with 1 atm. of the pure gas at θ_z° C. Let the concentration $[G_1]_z$ of one of the gases

[2] Miyake (1944) has used a similar concept independently.

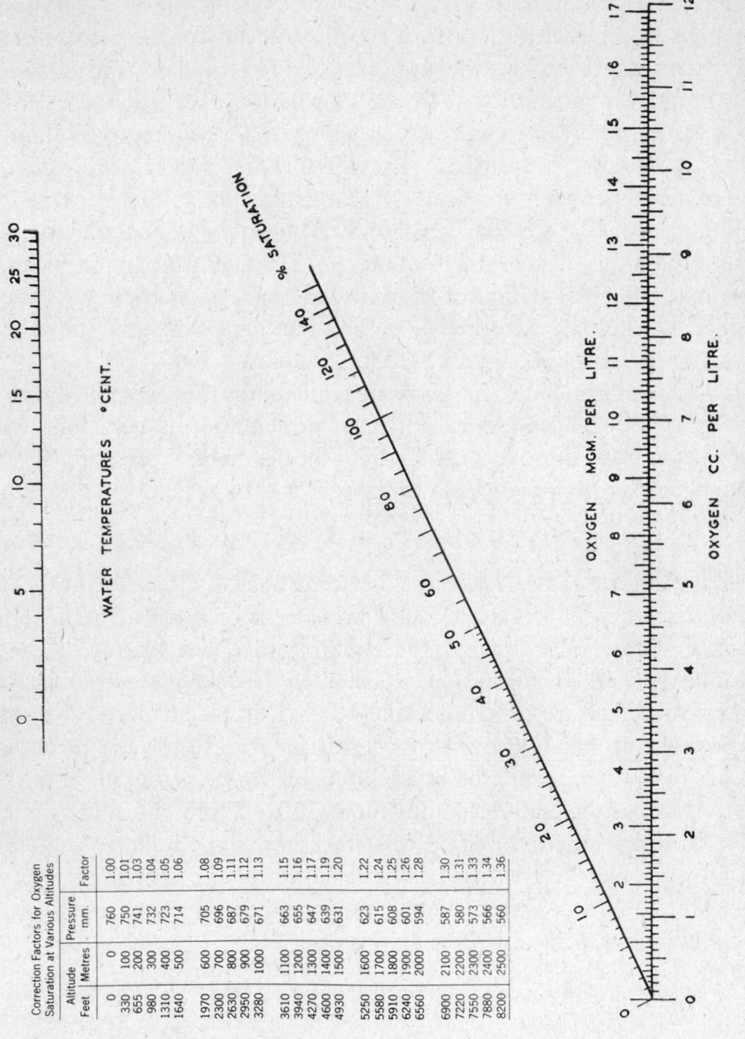

Correction Factors for Oxygen Saturation at Various Altitudes

Altitude Feet	Altitude Metres	Pressure mm.	Factor
0	0	760	1.00
330	100	750	1.01
655	200	741	1.03
980	300	732	1.04
1310	400	723	1.05
1640	500	714	1.06
1970	600	705	1.08
2300	700	696	1.09
2630	800	687	1.11
2950	900	679	1.12
3280	1000	671	1.13
3610	1100	663	1.15
3940	1200	655	1.16
4270	1300	647	1.17
4600	1400	639	1.19
4930	1500	631	1.20
5250	1600	623	1.22
5580	1700	615	1.24
5910	1800	608	1.25
6240	1900	601	1.26
6560	2000	594	1.28
6900	2100	587	1.30
7220	2200	580	1.31
7550	2300	573	1.33
7880	2400	566	1.34
8200	2500	560	1.36

FIGURE 158. Nomogram for determining O_2 saturation at different temperatures and altitudes (Mortimer).

be increased until a critical value $[G_1]_c$ is reached at which, under appropriate conditions, bubbles can just form. Let the proportions of the gases in these bubbles be

$$g_1 + g_2 + g_3 + g_4 \cdots = 1 - \frac{p_w}{P_z} \tag{6}$$

where p_w is the pressure of water vapor at θ_z° C.

The partial pressures of the gases will be

$$g_1P_z, \quad g_2P_z, \quad g_3P_z, \quad g_4P_z, \quad \text{etc.}$$

At the moment of the formation of bubbles

$$\begin{aligned} [G_1]_c &= g_1P_z[G_1]_S \\ [G_2]_z &= g_2P_z[G_2]_S \\ [G_3]_z &= g_3P_z[G_3]_S \\ [G_4]_z &= g_4P_z[G_4]_S \end{aligned}$$

whence, from equation 6,

$$[G_1]_c = [G_1]_S\left(P_z - p_w - \frac{[G_2]_z}{[G_2]_S} - \frac{[G_3]_z}{[G_3]_S} - \frac{[G_4]_z}{[G_4]_S} \cdots\right) \tag{7}$$

In many cases p_w will be very small compared with P_z and may then be neglected.[3]

In general, the two gases which are most likely to be responsible for the formation of bubbles are oxygen, produced by photosynthesis in the trophogenic layers of the lake, and methane, produced by anaerobic decomposition in the tropholytic layers.[4] In the case of oxygen, only

---

[3] Ricker refers $[G_1]_S$, $[G_2]_S$, etc., to 1 atm. of pure gas saturated with water vapor, and omits all reference to the water vapor in the bubble. This treatment implies that p_w/P_z is constant, which is of course incorrect, though the error introduced will be small.

[4] The formation of bubbles under conditions of physiological interest has been admirably discussed in a review by Harvey, Barnes, McElroy, Whiteley, Pease, and Cooper (1944). In a free volume of water containing no minute masses of gas that can act as nuclei, an enormous excess tension is needed for bubble formation. Experiments indicate that water must be saturated under at least 250 atm. of oxygen if bubbles are to form in the free liquid on return of the pressure to 1 atm. In most laboratory experiments, solid surfaces that have been exposed to the air carry sufficient gas to provide nuclei. In lakes, one may perhaps suppose that such nuclei are carried down on sedimenting particles from the turbulent water near the surface. It is, moreover, statistically possible for nuclei to form at the bottoms of cracks or pits in hydrofuge solids, initially quite clean of gas, by random Brownian movement of water and dissolved gas molecules. This process conceivably could produce nuclei in wet sediments that have never been exposed directly to the air.

The composition and rate of re-solution of bubbles produced by cavitation or local reduction of pressure, at different depths, from water initially saturated with air at 1 atm. has been studied by Wyman, Scholander, Edwards, and Irving (1952). Their paper is not concerned with the gas mixtures likely to be of greatest interest to limnologists, but would provide an invaluable point of departure for anyone seriously interested in the chemical physics of bubbles in nature.

oxygen itself and atmospheric nitrogen need be considered, because under the conditions which permit great photosynthetic production of oxygen, methane will certainly be absent, and the very great value of $[CO_2]_S$ makes the term for this gas of negligible value for any concentration $[CO_2]_z$ likely in the illuminated regions of the lake. If the water under consideration is assumed to be approximately saturated with nitrogen at the pressure of the lake surface P_0 and the observed temperature θ_0, for a lake at sea level

$$\frac{[N_2]_z}{[N_2]_S} = 0.791$$

whence

$$[O_2]_c = [O_2]_S(0.209 + 0.0967z - p_w) \tag{8}$$

or, more conveniently, writing $[O_2]_s$ for the value of oxygen saturation with respect to 1 atm. of dry air of normal composition,

$$[O_2]_c = [O_2]_s(1 - 4.78p_w + 0.462z) \tag{9}$$

At 25° C., $p_w = 0.0312$ atm.; the production of a saturation of 131 per cent of the value at the surface of the lake, necessary for bubble formation at a depth of 1 m. at this temperature, is not usual; the required 270 per cent saturation in terms of the surface value, necessary to produce bubbles at 4 m. depth, is very exceptional indeed. It is reasonably certain that the production of bubbles by photosynthesis, if it occurs at all, is likely to be limited to shallow and warm water.

In the case of methane, produced by anaerobic decomposition of sediments at low redox potentials, oxygen need not be considered. Disregarding small amounts of hydrogen, equation 7 becomes

$$[CH_4]_c = [CH_4]_S\left(P_z - p_w - \frac{[N_2]_z}{[N_2]_S} - \frac{[CO_2]_z}{[CO_2]_S}\right) \tag{10}$$

Allgeier, Peterson, Juday, and Birge (1932) found that almost as much carbon dioxide as methane can be produced during the low-temperature fermentation of lake mud, but even this large quantity is insufficient to make the term for carbon dioxide of great significance. In approximate work, at the low temperatures of the deep hypolimnion of a stratified lake, p_w may also be neglected. Allgeier, Peterson, Juday, and Birge found that a small production of molecular nitrogen occurred in their experiments. In their shorter run of 101 days, the quantity produced amounted to about 18 per cent by volume of the methane. Assuming a reasonable upper limit for nitrogen production of 20 per cent by volume of the methane, the critical value or absolute saturation for methane is likely to lie between the critical value assuming no nitrogen to be formed, given by

$$[CH_4]_c = [CH_4]_S(0.209 + 0.0967z) \tag{11}$$

and the critical value assuming the nitrogen produced is 20 per cent by volume of the methane, given by

$$[CH_4]_c = \frac{[N_2]_S[CH_4]_S}{[N_2]_S + 0.2[CH_4]_S}(0.209 + 0.0967z) \qquad (12)$$

If $z = 10$ m. and $\theta_z = 7°$ C., $[N_2]_S = 19.9$ cc. per liter and $[CH_4]_S = 45.4$ cc. per liter, whence the critical concentration of methane may be expected to lie between 36.7 and 53.5 cc. or 26.1 and 38.2 mg. per liter.

If $z = 20$ m. and $\theta_2 = 4°$ C., $[N_2]_S = 21.3$ cc. per liter and $[CH_4]_S = 49.5$ cc. per liter, whence the critical concentration is likely to lie between 72.6 and 106.1 cc. or 51.8 and 75.9 mg. per liter. Since Allgeier, Peterson, Juday, and Birge found that 1 liter of Lake Mendota mud covered with water, in the absence of air, could produce 28 to 29 cc. of methane in 100 days at 7° C., and since the volume of the water present to dissolve this gas must be less than the total volume of the mud, the production of methane bubbles at a depth up to 10 m. during prolonged summer stagnation is reasonable. The process is unlikely to take place at depths greater than 20 m. except in very polluted waters over sediments generating much more of the gas than was produced in the experiments of Allgeier, Peterson, Juday, and Birge.[5]

THE PASSAGE OF OXYGEN ACROSS THE LAKE SURFACE, AND THE OXYGEN CONTENT AT FULL CIRCULATION

The passage of oxygen across the air-water interface of a lake can be discussed, in the light of theory and of laboratory experiments, for three possible cases: (1) the water may be supposed to be quite undisturbed save for molecular motion; (2) the water may be supposed to be so well and so continually mixed that no oxygen gradient ever develops; or (3) the water may be supposed to behave in the same way that quiescent columns of water exposed at the top to the atmosphere are known to behave in the laboratory. An examination of the available field data will suggest that the conditions normally encountered in nature usually fall between alternatives (2) and (3).

[5] The remarkable work of Ohle on the effect of cultural eutrophication in causing massive methane production is as yet unpublished but will be considered in relation to successional changes in lakes in Volume II. A very important paper by Koyama (1955) has recently been received, in which it is shown that methane production by lake mud in the laboratory is probably due to bacterial reduction of CO_2 by hydrogen or by hydrogen donators. The production of methane appears somewhat pressure-dependent in a complicated way, with a minimum at about 50 atm. and another at about 300 atm. CO_2 production appears to rise at high pressures. Nitrogen may be fixed during the fermentation process. The amount of methane produced by the mud of Kizakiko seems to be greater than that produced by the mud of Lake Mendota.

Exchange through an undisturbed water surface: the insignificance of molecular diffusion. In a perfectly undisturbed lake in which only molecular processes are operating, an infinitely thin layer at the surface may be supposed to become saturated with gas relative to the pressure of the atmosphere. The part played by molecular diffusion from this saturated layer into the rest of the lake may now be examined. Grote (1934) has presented the full treatment, which is summarized here, though, in accordance with the conventions adopted in the present work, some of the symbols employed are different from those that Grote uses.

The oxygen flux or rate of transport of oxygen $\partial O/\partial t$ by molecular diffusion across any plane will be given by the Fick diffusion equation

$$\frac{\partial O}{\partial t} = -\,\mathrm{Ka}\,\frac{\partial[O_2]_z}{\partial z} \tag{13}$$

where a is the area of the part of the plane under consideration, ordinarily taken as unity, $\partial[O_2]_z/\partial z$ is the oxygen gradient across the plane, and K is the coefficient of molecular diffusion, which for oxygen is 1.98×10^{-5} $\mathrm{cm.}^2\ \mathrm{sec.}^{-1}$ The rate of change of concentration at any depth z will be given by

$$\frac{\partial[O_2]_z}{\partial t} = \mathrm{K}\,\frac{\partial^2[O_2]_z}{\partial z^2} \tag{14}$$

Let the concentration of oxygen initially in the lake be $[O_2]_0$ and let $[O_2]_s$ be the saturation concentration corresponding to the temperature of the surface layer. Now writing

$$[\Delta O_2] = [O_2]_s - [O_2]_0$$

equation 14 may be integrated to give for the concentration $[O_2]_{tz}$ at time t and depth z

$$[O_2]_{tz} = [O_2]_0 + [\Delta O_2]\{1 - \Phi(\mathrm{U})\} \tag{15}$$

where $\Phi(\mathrm{U})$ is the so-called error function

$$\Phi(\mathrm{U}) = \frac{2}{\sqrt{\pi}}\int_0^{\mathrm{U}} e^{-\zeta^2}\, d\zeta$$

In this case

$$\mathrm{U} = \frac{z}{2\sqrt{\mathrm{K}t}}$$

and ζ is an arbitrary variable. Whence

$$[O_2]_{tz} = [O_2]_0 + [\Delta O_2]\left\{1 - \frac{2}{\sqrt{\pi}}\int_0^{\frac{z}{2\sqrt{\mathrm{K}t}}} e^{-\zeta^2}\, d\zeta\right\} \tag{16}$$

Grote has computed curves showing the depth at which the oxygen concentration will have changed by $0.25[\Delta O_2]$, $0.5[\Delta O_2]$, and $0.75[\Delta O_2]$ at various times t after the surface concentration has been changed by $[\Delta O_2]$. As an example, suppose that a lake containing uniformly 11 mg.

per liter of oxygen comes into equilibrium with the atmosphere, so that its surface layer now contains 12.6 mg. per liter. Examination of Grote's curves indicates that after diffusion had proceeded for one month, the concentration would be 12.2 mg. per liter at 3.1 cm. depth, 11.8 mg. at 6.7 cm. depth, and 11.4 mg. at 11.4 cm. depth. Moreover, to raise the oxygen concentration to 11.4 mg. per liter at a depth of 10 m. would require 638 years. The example chosen approximates the conditions observed by Birge and Juday (1911) in Lake Mendota in March and April, 1907, when the oxygen concentration of the lake actually changed during spring circulation from just under 11 mg. per liter to just over 12.5 mg. per liter at all depths during a period of three weeks. It is obvious that the very slow changes due to molecular diffusion play no important part in the observed oxygenation. The whole of the above presentation is indeed little more than a limnological *jeu d'esprit*, because under the most carefully controlled laboratory conditions even quite narrow water columns, maintained at a uniform temperature, never remain undisturbed if their upper surfaces are exposed to the atmosphere. Grote's treatment is here reproduced mainly because the writer has noticed that many students, particularly those inclined towards physiological investigation, require to be convinced of the small role played by molecular diffusion in the natural distribution of atmospheric gases through the hydrosphere.

Exchange through a moderately disturbed water surface into water of uniform oxygen concentration: entrance and exit coefficients. If water is in contact with air and is sufficiently disturbed[6] so that the concentrations of

---

[6] According to the Lewis and Whitman two-film theory, the gas forms a nonturbulent film on the gaseous side of the interface, the water a nonturbulent film *essentially always saturated with gas* on the liquid side of the interface. In dealing with a very soluble minor constituent such as CO_2, the rate of entry of gas molecules into the gas film has to be considered, but in the case of oxygen in air this aspect of the system may be neglected. The exit coefficient is found to be given by the ratio of the diffusivity of the gas in water to the thickness of the film. When the film is renewed too fast, the rate of invasion is determined primarily by the rate at which more or less saturated fragments of film are mixed into the free water. Adeney and Becker's treatment, discussed hereafter, assumes a static saturated film. The fact that the coefficients that they obtained are of the same order of magnitude as those found by Redfield for the open Atlantic suggests that the theory, as here developed, will be useful at least temporarily in limnology, for no lake surface is likely to be more disturbed than is that of the ocean. The limnological appropriateness of this kind of treatment seems, moreover, to be supported by the experience of Truesdale and Downing (1954).

The theory does not apply to small ascending bubbles (cf. Krogh 1910) nor to water surfaces violently disturbed artificially; in such cases the transfer of gas across the interface is much faster than Adeney and Becker's results would indicate. The admirable contribution of Holroyd and Parker (1949) may be consulted for a discussion of the limitations of the theory as here set out and as an introduction to the rather large literature that has grown out of studies of chemical engineers interested in sewage purification. See also Odum (1956), Downing and Truesdale (1956).

the dissolved gases remain uniform throughout the water mass at any time during the course of the experiment, the rate of movement of a gas across the interface is given by Bohr's equation

$$\frac{dO}{dt} = a\alpha(P - p_t) \tag{17}$$

where a is the area of the interface under consideration, ordinarily taken as unity, P is the partial pressure of the gas in the atmosphere, p_t is the pressure at which the concentration of gas at time t in the water would be in equilibrium, and α is a coefficient which may be termed the *entrance coefficient*. Since

$$p_t = \frac{P[O_2]_t}{[O_2]s}$$

equation 16 may be written

$$\frac{dO}{dt} = a\left(\alpha P - \alpha P\frac{[O_2]_t}{[O_2]s}\right) \tag{18}$$

or writing β for $\dfrac{\alpha P}{[O_2]s}$

$$\frac{dO}{dt} = a(\alpha - \beta[O_2]_t) \tag{19}$$

In this form α may be regarded as a virtual coefficient indicating the tendency of oxygen to enter the water from the air, and β a virtual coefficient indicating the tendency of oxygen in solution to leave the water. Following Dorsey (1940), α may be termed the *entrance coefficient* and β the *exit coefficient*. These two *exchange coefficients* are of course interdependent, and the choice of which one to use depends primarily on mathematical convenience. Adeney and Becker (1919), from whose work the principal laboratory data are derived, use only the exit coefficient, which they designate as *f*.

For limnological purposes, the most convenient integral form of equations 16 to 19 is

$$[O_2]_t = [O_2]_0 + ([O_2]_s - [O_2]_0)(1 - e^{-t\beta/\bar{z}}) \tag{20}$$

where $\bar{z}$ is the mean depth of the lake. Or writing

$$[\Delta O_2]_\infty = ([O_2]_s - [O_2]_0)$$

and

$$[\Delta O_2]_t = ([O_2]_t - [O_2]_0)$$

$$t = -\frac{\bar{z}}{\beta}\ln\left(1 - \frac{[\Delta O_2]_t}{[\Delta O_2]_0}\right) \tag{21}$$

Adeney and Becker examined the rate of solution from a large bubble into water kept continually mixed by the passage of the bubble up the containing tube, which was inverted after each ascent.[7] The following values of the coefficients for oxygen are derived from their expression for the variation of the exit coefficient with temperature, which is

$$\beta = 0.0096(\theta + 36) \tag{22}$$

where β is in[8] cm. min.$^{-1}$ and θ is the temperature in °C.

Temp.,	β		α		
°C.	cm. min.$^{-1}$	m. day^{-1}	mg. cm.$^{-2}$ min.$^{-1}$ atm.$^{-1}$ ($\times 10^{-3}$)	mg. m.$^{-2}$ day^{-1} atm.$^{-1}$ ($\times 10^{5}$)	cm.^{3} m.$^{-2}$ month^{-1} atm.$^{-1}$ ($\times 10^{6}$)
0	0.346	4.98	24.3	3.50	7.4
4	0.384	5.53	23.8	3.43	7.2
10	0.442	6.36	24.2	3.50	7.3
20	0.538	7.75	24.6	3.53	7.4
30	0.634	9.14	24.6	3.53	7.4

The values of α given in the last column conform to the system of units employed by Redfield (1949); many limnologists will prefer units based on mass rather than volume, and the use of 1 day and 1 m. as units of time and length also proves convenient. It will be observed that as the exit coefficient rises with temperature, the saturation value falls, so that the entrance coefficient remains almost constant.

Laboratory studies of quiescent columns: oxygenation by convective streaming. As has already been indicated, evaporation tends to cool any exposed water surface, even if the water and air are at the same temperature The exposed surface of a lake will therefore tend to be cooled a little below the temperature obtaining at a depth of a few centimeters, as Woodcock (1941) found often to be the case in nature. The result of this cooling, and the concomitant increase in salinity in saline waters, is to raise the density of the water at the surface, setting up convection streams of cellular form, cold water streaming downward and warm water rising to take its place. The process results in an exchange of gases, which has been investigated by Adeney and his co-workers (Adeney 1905, 1926; Adeney and Becker 1920; Adeney, Leonard, and Richardson 1922; Becker and Pearson 1923). Using a cylindrical column of water about 27 cm. high and 4 cm. in diameter, Adeney and Becker found that the rate of entrance of air followed the Bohr equation (equation 17), but that the virtual coefficients α_1 and β_1 were very much smaller than the molecular coefficients α and β. At 15° C. they obtained for β_1

[7] These experiments appear to satisfy the condition that the film is renewed slowly.
[8] In Adeney (1926) erroneously given as cm. sec.$^{-1}$

	Values, cm. min.$^{-1}$
Experiment 1	
Tap water, laboratory air	0.0065
Sea water, laboratory air	0.0085
Experiment 2	
Water, dry air	0.0100
Water, wet air (almost saturated)	0.0038

It is evident that these values of what may be termed convective exit coefficients are of the order of $^1/_{100}$ of the molecular exit coefficient, which for air at 15° C. is 0.485 cm. min.$^{-1}$ In consequence, the values of the convective entrance coefficient α_1 also will be of the order of $^1/_{100}$ of the value of the molecular entrance coefficient α. The variation of the convective coefficient with salinity and with the degree of saturation of the air follows the expected course, at least qualitatively. It is reasonable to suppose that both the convective exit and entrance coefficients will be considerably lower at low temperatures than at high temperatures, since they so clearly depend on factors influencing evaporation.

The exchange across water surfaces in nature and in large-scale experiments. The results of Redfield (1949) for the ocean may first be examined. In the Gulf of Maine, Redfield compared the observed rates of change of oxygen content in a column of water with the calculated rates of change due to known physical and biological processes in the column. The differences were taken as measures of rates of exchange with the atmosphere. Considering intervals of a few weeks' duration, separating series of determinations, values of α were computed from these rates of exchange and from the mean of the surface oxygen concentrations at the beginning and end of the period. In winter, the values of the entrance coefficient were uniformly high, the mean being 6.1×10^5 mg. m.$^{-2}$ day^{-1} atm.$^{-1}$ or 13.0×10^6 cm.^{3} m.$^{-2}$ month^{-1} atm.$^{-1}$ In summer much lower values were obtained, the mean being 1.3×10^5 mg. m.$^{-2}$ day^{-1} atm.$^{-1}$ or 2.8×10^6 cm.^{3} m.$^{-2}$ month^{-1} atm.$^{-1}$

Adeney and Becker had found that the coefficients for oxygen entering sea water were about 5 per cent lower than those they gave for distilled water. The very high winter values for the Gulf of Maine, which are about twice the value to be expected from Adeney and Becker's experiments, therefore require some explanation other than the nature of the liquid phase. Redfield supposes that during the winter the conditions for penetration into water without an oxygen gradient at the surface are indeed realized, but that the area of the sea through which movement of oxygen takes place is actually about twice that of the plane projection of the ocean surface on a map, owing to the formation of waves, spray, and bubbles. During the summer the low virtual coefficients are probably to be explained by the presence of a gradient in oxygen at the surface, even the top meter being sufficiently stable for the recorded surface oxygen concentrations, based on water actually taken at a depth of 50 cm., not to give a true value for the water of the interface.

The only specifically limnological study that has been made is by Vinberg (1940), who investigated the changes in oxygen concentration in lake water enclosed in a tank open to the atmosphere, moored at the surface of Lake Beloye, Kossino, U.S.S.R. By this method Vinberg excluded the influence of both vertical and lateral mixing, though it is doubtful whether the water surface in his tank gave a true picture of the changes occurring at the surface of the lake. Vinberg worked only at night, when the water, supersaturated by photosynthesis during the day, was losing oxygen. By making a large number of successive determinations of the oxygen concentration at the surface of the tank, he was able to fit the rate of change of oxygen to an expression of the form

$$\frac{dO}{dt} = a\alpha(P - p_t) - azq \tag{23}$$

where z is the depth of the rank and q the rate of biochemical consumption of oxygen per unit volume of water, which rate is assumed to be constant. The results obtained by Vinberg indicate that under his experimental conditions α had a value of 0.38×10^5 mg. m.$^{-2}$ day^{-1} atm.$^{-1}$ or 0.8×10^6 cm.^{3} m.$^{-2}$ month^{-1} atm.$^{-1}$ This value is about $^1/_{10}$ of the value given in the preceding section for entrance through a constantly disturbed water surface, and is considerably less than any of the values for the surface of the Gulf of Maine during the summer. It can hardly be doubted that the surface in Vinberg's tank was reformed far less rapidly than were any of the sea surfaces studied by Redfield. It is quite likely that in the unenclosed surface of Lake Beloye around the tank much more movement occurred than is indicated in Vinberg's experiment, for the epilimnion of the lake is sure to have been somewhat turbulent. In general, it is probable that convective streaming played a part in the observed oxygen exchange; at night in nature, such convection is likely to be due both to evaporation and to other kinds of heat loss. It is not unreasonable to suppose from Vinberg's experiment and from Redfield's observations in the sea, that virtual coefficients of the order of $^1/_{10}$ the laboratory values of α and β as determined by Adeney and Becker represent the lower limit of possible values of the two exchange coefficients for extensive water surfaces in nature, at least in temperate regions.

The difficulties of this kind of investigation are well illustrated by the following circumstance. Dr. G. A. Riley kindly put at the writer's disposal his data on the oxygen concentration in Linsley Pond, and his estimates of the net rate of production of oxygen based on experiments with suspended bottles of lake water. Ideally, it should be possible to estimate virtual exchange coefficients from these data. Unfortunately, it appears that the surface of the lake is significantly undersaturated throughout the spring months, even at times when the net oxygen production of the water of the previous week was sufficient to raise the concentration to full saturation within the experimental period. Moreover, later in the year, during June, the opposite phenomenon can occur, the water being supersaturated and yet still able to gain oxygen more rapidly than the estimated net rate of production allows for. There can be little doubt that these observations must be explained by lateral movement of water from the margins towards the observation station. Before the marginal vegetation has reached its maximum development, there is presumably a net consumption in the littoral, owing to decomposition of organic detritus; later, after the weed beds have developed, there is evidently a net production of oxygen in the littoral. In order to determine the virtual exchange coefficients, it would be necessary to isolate the water from these effects, as indeed Vinberg did. Such isolation, however, can clearly be criticized on the grounds that by excluding wind-generated currents, and so the turbulent movement due to these currents, the renewal of the air-water interface in the experimental tank is likely to be much reduced. A solution of the problem in the field would therefore be extremely difficult.

OXYGEN CONTENT OF EPILIMNETIC AND UNSTRATIFIED WATERS

It is commonly assumed that a dimictic lake at the time of vernal circulation will come to contain a quantity of oxygen corresponding to saturation at 4° C. at the altitude of the lake. Since in many lakes under ice, particularly if the ice is covered with snow, consumption of oxygen is in excess of photosynthetic production, the water at the beginning of the vernal circulation period is likely often to be undersaturated. In view of the limited number of observations usually available in the early spring, it has often been the practice to use the saturation value at 4° C. as the primary constant in computing so-called absolute oxygen deficits, in the way to be explained later, or to assume that the so-called actual deficit may be obtained by subtracting the observed concentration from the saturation concentration of water at the temperature recorded *in situ.* If, however, the water never became saturated during vernal circulation it is evident that a considerable error might be introduced into such calculations.

Theoretical aspects of oxygenation during vernal circulation. It has become apparent from the discussion of the preceding sections that, while virtual entrance and exit coefficients for oxygen have never been adequately determined for any lake, these values at the time of vernal circulation are likely to lie between the limits set by Redfield's determinations for the sea in winter and by Vinberg's determination for a tank moored in Lake Beloye in summer. The upper limit set in this way is almost certainly too high for any except the largest lakes, and it is reasonable to think that the lower limit is too low for vernal circulation except perhaps in very protected small lakes. Within the extreme limits, it would be reasonable to expect most values for spring circulation to lie between Adeney and Becker's determinations on continuously reformed films and some large fraction of such determinations, possibly of the order of one-third, as in the sea in summer. If it is supposed that for most lakes Adeney and Becker's value of β for 4° C., namely 5.53 m. day^{-1}, is seldom if ever likely to be exceeded, we can estimate the probable minimum time taken to reoxygenate to any degree of saturation a lake at sea level that has been partially deoxygenated during winter stagnation. In most cases it is unlikely that the water at the beginning of vernal circulation, will be less than 80 per cent saturated. Assuming such a minimum percentage saturation, from equations 20 and 21 Table 75 has been constructed.

TABLE 75. *Oxygenation of ideal lakes at vernal circulation*

Percentage saturation at time t	90.0	95.0	99.0	99.9
$e^{-\beta t/\bar{z}}$	0.5	0.25	0.05	0.005
$\beta t/\bar{z}$	0.69	1.39	3.00	5.30
$t/\bar{z}$ when $\beta = 5.53$ m. day^{-1}	0.124	0.250	0.542	0.958
t when $\bar{z} = 12.5$ m., in days	1.55	3.12	6.78	11.98
t when $\bar{z} = 100$ m., in days	12.4	25.0	54.2	95.8

These figures indicate that a comparatively shallow lake such as Mendota ($\bar{z} = 12.14$ m.) may be expected to become practically reoxygenated by purely physical means within a reasonable time, provided it is maintained in full circulation and sufficiently disturbed to prevent the establishment of an oxygen gradient at the surface. The full reoxygenation of a very deep lake, however, presents real difficulties, even if the virtual exchange coefficients are of the order of those found by Redfield for the ocean surface in winter, that is to say, nearly twice the value used in preparing Table 75. It will readily be apparent that, even with such a favorable condition for reoxygenation as a highly disturbed lake surface might sometimes afford, the exchange coefficients being twice the value employed in constructing the table, it would take a lake of mean depth 100 m. 14.6 days to pass from 95.0 to 99.0 per cent saturation, and during this period it is reasonably certain that some thermal stratification would have been established, retarding the rate at which oxygen could be delivered by turbulent exchange to the depths.

Observed oxygen concentration at vernal circulation. These ideas can be applied critically only to lakes in which biological consumption and production are known from experimental studies, or in which such processes play so small a part in determining the oxygen content that they may be neglected. Linsley Pond, the only lake for which there is any experimental data relating to the period of vernal circulation, proves unsuitable for this kind of analysis, owing to the great effect of the littoral zone. The large, deep unproductive lakes of Norway, however, provide cases in which there is so little biological activity that the physical factors involved are clearly apparent.

The effect of circulation time on large unproductive lakes. In the Nordfjord lakes (Strøm 1933a), situated at the head of Nordfjord on the western coast of Norway, the oxygen concentration at the time of vernal overturn must hardly depart from saturation. In the upper parts of the hypolimnia of the three larger lakes, the concentrations in August 1931 were:

	Concentration, mg. per liter		Saturation, per cent	
	At 50 m.	At 100 m.	At 50 m.	At 100 m.
Hornindalsvatn	12.9	12.7	103.6	100.5
Breimsvatn	12.9	13.1	104.9	104.4
Strynsvatn	13.0	13.0	105.2	102.2

The temperatures in the layers under consideration lay between 4.10° and 4.92° C., much too low to permit the operation of either of the effects that may give slight apparent increases in saturation without changes in concentration (page 601). It is quite certain that at the time of vernal circulation these lakes must have been saturated with oxygen.[9]

In marked contrast to the conditions found in the Nordfjord lakes are those of certain equally unproductive lakes in southeastern Norway. In Hurdalsø, studied by Gran and Ruud (1927), the deep-water oxygen content in June lay between 10.9 and 11.0 mg. per liter. Since at all the hypolimnetic depths studied the variations throughout the summer were small and irregular, leading at most to a deficit of 0.83 mg. per liter in

[9] The values given above are recalculations based on Truesdale, Downing, and Lowden's saturation values. The Fox-Whipple values would give a range of 97 to 101.4 per cent at circulation.

three and a half months, it is quite certain that the departure of the June figures from the saturation value at 175 m., the altitude of the lake, namely 12.47 mg. per liter at 4° C., cannot have been due to recent consumption. At the period of vernal circulation, which evidently lasted during 1926, the year of the investigation, for a relatively short period in the middle of May, the concentration cannot have exceeded 11.3 mg. per liter, or 91 per cent saturation. Similarly in Holsfjord, the main basin of Tyrifjord, at an altitude of 64 m., a lake with a maximum depth of 295 m. and a mean depth of 114 m., Strøm (1932a) found 11.4 to 11.7 mg. per liter in solution in the deep water on June 26, corresponding to 92 to 94 per cent saturation. Very little change in the oxygen concentration occurred during the summer, except certain irregular oscillations due to water movements. It is quite unreasonable to suppose that the lake could have been more than 94 per cent saturated at the time of vernal circulation, a few weeks prior to Strøm's first visit.

The difference between the maximum vernal oxygen contents of the western and southeastern Norwegian lakes is due primarily, as Strøm has most clearly indicated, to the more oceanic climate enjoyed by the western lakes, the large examples of which freeze completely only in exceptional winters, whereas even the larger lakes of southeastern Norway, under a more continental climatic regime, normally remain frozen until April or May, and on melting warm up so rapidly that the full circulation period is relatively short.

Modification by biochemical events. From other lake districts in which the waters are so unproductive biologically that purely physical phenomena can be easily recognized, little or no information of the kind obtained by Strøm exists. In lake districts in which biochemical events modify the physically determined concentrations, both supersaturations and undersaturations are likely to be recorded at the time of vernal circulation. Thienemann (1928), for instance, records in the Grosser Plöner See on May 5, 1919, 13.21 mg. per liter at 41 m. corresponding to a supersaturation of 107 per cent, and 13.94 mg. at 20 m., corresponding to a supersaturation of 116 per cent. It is possible that the 41-m. sample had already lost some oxygen, owing to biochemical oxidations, and it seems likely that during April, when the lake must have been circulating at a temperature of about 5° C., the whole body of water was about 116 per cent saturated. Such supersaturations are, of course, reasonable when it is remembered that the vernal circulation period is likely to correspond with the beginning of the spring bloom of diatoms. In other lakes, biochemical oxidations, either in the free water or at the mud surface against which the turbulent water is continually being brought, are likely to reduce the oxygen content even when the lake is freely circulating. Minder's (1926) data for the lower Lake Zurich indicate that although this lake may circulate intermittently throughout the winter, it is undersaturated throughout most of the period from January to March and only begins to show supersaturations in the top 10 m. in April. In 1917, when some inverse thermal stratification was apparent in February, the lake remained not more than 91 per cent saturated in April; and even in 1919, when there seems to have been prolonged homoiothermy at temperatures from 5.2° to 4.0° C. during January, February, and March, the lake was under 91 per cent saturated on March 11.[10]

Although Minder is certainly correct in believing that the main rise in oxygen which he records during the winter is due to physical rather than biological events, the slowness of the rise, at least in some years, would seem to indicate that a biochemical removal of oxygen is occurring and in part annulling the effects of purely physical aeration. This is confirmed by the fact that loss of oxygen occurred during the winter homoiothermal period in 1919, even though the lake was always undersaturated. The lower Lake Zurich is probably the most productive of the large Swiss lakes, owing to the cultural influences of the city of Zurich. In the less productive Lake Constance, somewhat higher values for the period of winter homoiothermal circulation are given by Auerbach, Maerker, and Schmalz (1924), who found an average saturation of about 93 per cent for January and February during the period 1921–1923.[11]

Even where there is likely to be a good deal of consumption and production and where the mean depth of the lake is relatively small so that aeration should proceed rapidly, there still may be recognizable differences from year to year in the oxygen concentration at the vernal circulation, which differences appear to reflect the varying meteorological character of the season. Thus, in Lake Mendota in 1906, when the lake became homoiothermal in the first week of April and was above 4° C. at all depths by April 20, the oxygen concentration during the circulation period, at about 5.0° C., seems to have been about 11.4 mg. per liter, or 97 per cent saturated; but in 1907, when the lake became homoiothermal earlier, about the end of the third week in March, and, heating more slowly, was still circulating at about 4° C. in the middle of April, the oxygen content at the time of this circulation was 12.5 mg. per liter, or 104 per cent saturated.

---

[10] Minder gives 86.9 per cent saturated as the maximum on this date, at 10 m. This is said to be corrected for barometric pressure and is referred to Birge and Juday's (1914) table, which gives values about 4 per cent too low at the temperatures in question.

[11] The figures are said to be corrected for variations in atmospheric pressure and to be referred to the table given by Birge and Juday (1911); this table gives values in the neighborhood of 4° C. very close to those of Truesdale, Downing, and Lowden, though the agreement is fortuitous.

It is evident from the foregoing discussion that although the oxygen concentration of the waters of a lake frequently approach saturation at the altitude of the lake surface and the temperature of homoiothermal circulation in the spring, the approach is often not very exact. On purely physical grounds, divergences are to be expected in large deep lakes whenever any winter oxygen deficit occurs, and such divergences will be greater where the lake heats rapidly and late in the spring. Experience indicates that this expectation is in fact realized. Where biochemical oxygen uptake or production occurs, no general rule as to the sign of the divergence from saturation will be possible, but even in a relatively shallow lake such as Mendota, in which biological processes are likely to be of great importance, there is some evidence that differences in the physical events determined by variations in season still may be recognizable. Such observations emphasize that whenever it is desired to estimate the rates of change of oxygen in a given layer of the lake, due to the operation of all causes, physical and biological, for accurate work it is essential to have determinations made at the beginning of the season to provide a suitable base line from which all subsequent changes can be measured. The assumption of saturation at 4° C. at the period of vernal circulation is a relatively crude approximation, to be used only when no other and more accurate information is available.

Oxygen content of surface waters. Though the surface of a lake is in contact with an unlimited supply of oxygen, samples collected from the top few centimeters are frequently not found to be exactly saturated relative to the prevailing temperatures and pressures.

Surface concentration in unproductive lakes. In extremely unproductive lakes of very low color, values close to saturation may be expected and are in fact generally observed. Thus, in the surface waters of six lakes on the island of Moskenesøy in the Lofoten group, Strøm (1938b) found from 10.44 to 11.88 mg. O_2 per liter, which, when allowance is made for slight variations in altitude and in the vapor pressure of water at different surface temperatures,[12] correspond to a range in saturations from 99 to 104 per cent, the mean being 102 per cent. When a marked departure from saturation is found in an uncolored surface water in a biologically sterile lake, it is presumably due to change in temperature having occurred too quickly to permit concomitant equilibration with the atmosphere. In the Nordfjord lakes, which are extremely unproductive, Strøm (1932a) found up to 116 per cent saturation. In Oldenvatn, 12.55 mg. O_2 per liter was observed at both 0 m. and 10 m., but this absolute quantity

[12] Strøm's own estimates of saturation admittedly are not corrected for the effects of these variables.

corresponded to 112 per cent saturation at the surface and to 105 per cent saturation at 10 m., owing to the slightly greater surface temperature. Presumably, the rise in surface temperature had occurred shortly before the observations were made.

Deviations from saturation due to biochemical oxidation and photosynthesis. In regions in which marked biochemical effects may be expected, considerable deviations from saturation often occur. In the most extensive investigation yet made, that of Juday and Birge (1932) on the lakes of northeastern Wisconsin, 1056 samples were analyzed.[13] The range in percentage saturations that they observed is very great, from 38 per cent in Cardinal Bog to 146 per cent in Little St. Germaine Lake. The modal class is about 97.0 to 98.0 per cent saturation, the mean is about 93 per cent saturation, and there appear to be 384 clearly supersaturated[14] samples, or 36.4 per cent of the total number.

Ohle (1934a), examining 21 samples from the region around Plön, found an irregularly distributed range of values from 70 to 120 per cent saturation, the mean value being 102 per cent; ten of his lakes, or 48 per cent, were supersaturated.

Lönnerblad (1931), analyzing 41 summer surface samples from the open water of lakes in the Aneboda region in south Sweden, found a range of from 78 to 111 per cent saturation, with a mean of 94.2 per cent and a mode between 95 and 105 per cent. Only three of his samples, or 7.3 per cent, were supersaturated.

Yoshimura (1938b) examined 208 samples from Japanese lakes. In his complete series the mode lies at about 100 per cent saturated,[15] and more than 93, or 44.7 per cent, are supersaturated. His data are, however, more instructively considered according to the categories into which he has arranged them (Table 76).

The first category contains very unproductive lakes. The brackish lakes are always fairly productive, often very productive; they may be considered along with the mesotrophic and eutrophic lakes as forming a more productive group. It is evident that very great supersaturation occurs only in such more productive lakes, but by no means all are super-

[13] Juday and Birge give their percentage saturations in terms of dry air at sea level. In recomputing their statistical data, an altitude of 500 m. at 20° C. and saturation with water vapor at 20° C. have been assumed. Using the results of Truesdale, Downing, and Lowden (1955), it is found that 88.6 per cent saturation as given by Birge and Juday corresponds to 100 per cent saturation as here given.

[14] All those at least 90 per cent saturated with respect to dry air at sea level, as given by Juday and Birge's Figure 2.

[15] It is not possible to correct Yoshimura's summary table for the revised values of saturation concentration, but its striking features, as given in Table 76, are dependent on the comparison between classes and not on absolute concentrations.

TABLE 76. *Percentage of saturation in surface waters of Japanese lakes**

	Oligo-trophic (47 lakes)	Meso-trophic (64 lakes)	Eutrophic (37 lakes)	Brackish (29 lakes)	All More Productive (130 lakes)	Dystrophic (13 lakes)	Mineral Acid (20 lakes)
Range	86–120	68–126	87–166	77–194	68–194	36–112	36–127
Mode	95–100	100–105	90–95	100–105	100–105	Indefinite	95–100
Super-saturation	36.2	54.6	54.1	51.8	53.9	23.1	$\leqslant$ 15

* Saturations based on Whipple's tables.

saturated; some, in fact, are markedly deficient in oxygen. High organic production seems to increase the range of possible surface oxygen content, as might be expected.

Wherever really marked supersaturation is encountered, it is presumably attributable to photosynthesis. The causes of the subsaturation observable in rather more than half the samples analyzed by the investigators whose work has just been considered are presumably various. In Lake Katanuma, one of the Japanese lakes containing mineral acid, Yoshimura found only 2.9 mg. O_2 per liter at 24.6° C., corresponding to 35 per cent saturation. The water contained suspended sulfur of volcanic origin; oxidation of this material, or of suboxidized sulfur compounds or ferrous compounds in solution, presumably leads to a low concentration of oxygen at the surface of this and a few comparable lakes.

Catastrophic oxygen deficiency in summer. In some highly productive lakes, low surface values may be found at the ends of periods of development of massive algal blooms. Olson (1932) has reported cases in which the nocturnal respiration of algae can produce a very marked oxygen deficit at the surface. In Clearwater Lake, Minnesota, a dense population of *Hydrodictyon* raised the surface oxygen at one station to 18.7 mg. per liter, or 248 per cent saturation, at 4 P.M., and then lowered the concentration to 2.2 mg., or 27 per cent saturation, at 5 A.M. Similarly, a bloom of *Oscillatoria ågardhii* lowered the oxygen sufficiently in Lake Albert Lea, Minnesota, to produce a mass mortality of fish during quiet weather. When the bloom is decomposing during hot, dull, windless weather, the effect can be catastrophic (Tomlinson 1935, Sears 1936, Hutchinson 1936, Moore 1942). Less dramatic variations in the intensity of algal and bacterial respiration must be continually occurring, and in quiet weather must often produce slight but detectable variations in the oxygen content of the surface waters of productive lakes.

Effect of coloring matter. In view of the low values of the surface oxygen concentrations in Yoshimura's dystrophic, or brown, humic lakes, as well as in Lönnerblad's series derived for the most part from highly colored lakes, it seems probable that humic coloring matter undergoes sufficient oxidation to lower the oxygen content at the surface of many lakes. The data, admittedly not comprehensive, published by Juday and Birge show traces of the same sort of relationship. Of the 72 surface samples recorded by Juday and Birge as having a color less than 50 on the U.S.G.S. scale, 22, or 31 per cent, were supersaturated; of the 25 waters with a color greater than 50, only 1, or 4 per cent, was supersaturated.[16] In view of Lönnerblad's observations that *dy* or humic sediment can take up oxygen after sterilization, it is possible that purely chemical oxidation may be involved in lowering the surface oxygen content of humic lakes. The possibility that the bleaching reaction (page 413), presumably due to ultraviolet radiation, may be a photo-oxidation may also be kept in mind. In all cases, meteorological factors—particularly wind, the incidence of bright, hot weather, causing superficial heating, or of sudden cold spells, causing convection—as well as the indirect effects of such factors on photosynthesis, must play so great a part in determining the oxygen concentration immediately below an exposed water surface, that it is perhaps remarkable to find any trace of systematic effects due to productivity and color.

THE DISTRIBUTION OF OXYGEN IN A STRATIFIED LAKE DURING THE SUMMER

The first oxygen determinations in the deep waters of lakes, those made by Walter for Forel (1885) in the Lake of Geneva, showed little difference between surface and bottom values. Hoppe-Seyler (1896), however, demonstrated by comparing the oxygen and nitrogen contents of the deep water of Lake Constance that although a considerable amount of oxygen is present, evidence of oxygen uptake can be obtained. His suggestion that this uptake is a reasonable measure of the biological activity in the water initiated an important aspect of modern limnology. The first detailed tables of oxygen content at different depths in two lakes, demonstrating marked differences in the distribution of the gas, were published by Delebecque (1898).

The modern study of dissolved gases in lake waters was initiated by Birge and Juday (1911; see also Birge 1906, 1907, 1908) in a classic monograph which has become the basis of all further studies and will be referred to repeatedly in the present work. Further very important contributions were made by Thienemann (1913a, 1913b, 1915, 1920, 1928), whose

[16] That is, more than 89 per cent saturated relative to dry air at sea level, which is equivalent to more than 100 per cent saturated at 500 m. at 20° C. and in contact with wet air.

treatment of the theoretical aspects of the problem has played a fundamental part in the development of limnology. Some other work was published early in the present century; it is in general technically unsatisfactory. The curious may consult Birge and Juday (1911) for references.

The physical effects of summer stratification. If a lake were to heat by radiation alone and no water movements occurred in it, the temperature would drop exponentially from the surface (Fig. 159). If such a lake is

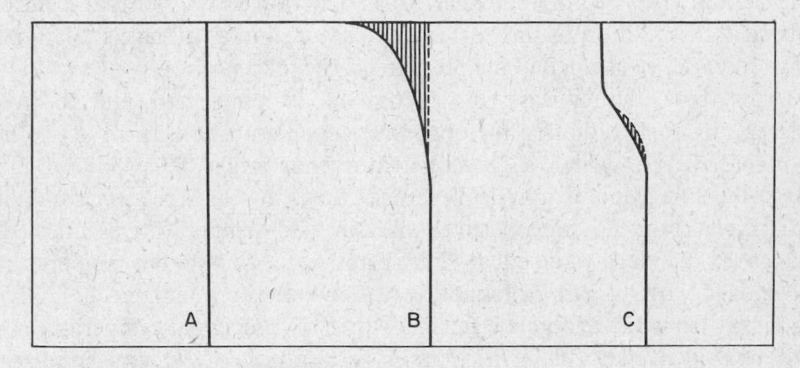

FIGURE 159. *A*, initial uniform saturation with oxygen, in the absence of a temperature gradient; *B*, saturation curve determined hypothetically by nonturbulent heating, the shaded portion representing supersaturation; *C*, turbulent distribution of heat at surface, with a very thin deeper zone supposedly heated slightly by radiation and exhibiting supersaturation. (After Yoshimura.)

imagined to have been initially saturated with oxygen at 4° C., it is evident that in view of the slowness of molecular diffusion, the superficial layers of the lake would soon appear supersaturated with oxygen relative to their increased temperatures and to the pressure at the lake surface, though no gain or loss of oxygen would have taken place. Such a situation is of course impossible, as lakes do not heat in this way. Yoshimura (1938b), however, believes that the slight supersaturations sometimes observed in stable layers just below the epilimnion in very unproductive transparent lakes are due to heating *in situ* by solar radiation, lowering the saturation value relative to the pressure of the lake surface without any change in oxygen concentration taking place. Strøm (1931) as has already been indicated, has observed similar but presumably quite transitory effects at the surface of very unproductive lakes by day.

In a lake heating by wind-generated turbulence in the ordinary way, the main direct physical effect of the establishment of thermal stratification will be a loss of oxygen from the freely circulating epilimnion, the water of which will tend to come into equilibrium, at its new, enhanced temperature, with the air at the surface of the lake. The turbulent transport of heat into the deeper water will also involve the exchange of warmer water containing less oxygen for colder water containing more oxygen per unit volume. The final results of such a process will be the production of a vertical distribution which has an inverse relationship to the temperature curve, the warm epilimnion containing less oxygen per unit volume than the cold hypolimnion, and the concentration therefore rising with increasing depth across the thermocline. It is generally supposed that a purely physical distribution of this sort implies saturation at all levels relative to the temperature *in situ* and the pressure at the lake surface.

Apart from the few cases in which a rise in temperature due to direct heating by solar radiation may produce supersaturation, there is another possible way in which slight supersaturation might be produced in a stratified lake by purely physical events. Since the curve relating solubility of oxygen from air, at any given pressure, to temperature is markedly concave, and since when equal volumes of water at different temperatures are mixed without gain or loss of heat they come to a temperature hardly different from the arithmetic mean of their two original temperatures, it will be evident that if the two volumes were saturated with oxygen relative to their original temperatures, they will now be supersaturated relative to their new intermediate temperature. The magnitude of this effect can be estimated with sufficient precision by drawing a chord on the saturation-temperature curve between the two points representing the two original volumes of water. Drawing such a chord on Fig. 160 will indicate that a supersaturation of the order of 0.5 mg. per liter, or 105 per cent, can be produced by mixing equal volumes of water at 4° and 24° C. Normally, the various components of the mixed water of the metalimnetic layers of a lake will all have come into equilibrium at temperatures very much closer together than the two of this extreme example, but it is just possible that in some cases in which strong winds have mixed recently warmed water into the lake in the early stages of the establishment of summer stratification, supersaturations of the order of 102 or 103 per cent might be produced in this way.

Types of vertical distribution and their nomenclature. As is usual in the formative period of any science, the terminology employed at first was somewhat unsatisfactory. This has finally been rectified by Åberg and Rodhe (1942), whose usage will be followed in all essential points. The purely physical events described in the preceding section produce a distribution which, though not uniform in the epilimnion, is very nearly so in the hypolimnion. For this type of uniform distribution, dependent solely

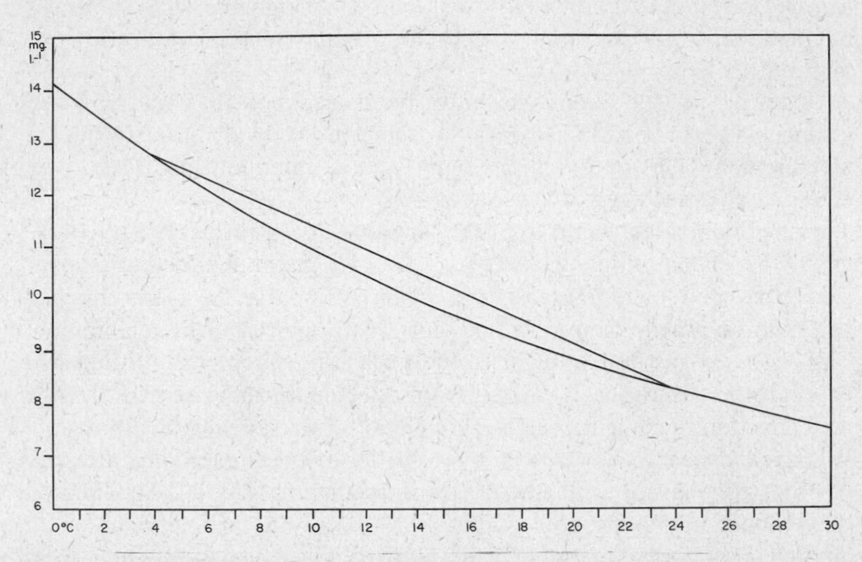

FIGURE 160. Variation of oxygen dissolved from a wet atmosphere at a pressure of 760 mm., with temperature. The chord indicates possible compositions produced by mixing different proportions of water saturated at 4° and 24° C. (After Mortimer.)

on conditions at circulation and on subsequent physical events, Åberg and Rodhe's term *orthograde* is convenient. The use of the term can be generalized to cover other cases of nearly uniform distribution without any appreciable decrease in concentration in the deep water. Such a distribution was term *oligotrophic* by Thienemann, but modern investigation has indicated that this usage is unfortunate.[17]

In the great majority of the lakes which have been studied, oxidative processes occurring in the hypolimnion remove some of the oxygen. Usually such a loss is most noticeable at the bottom of the lake. As the temperature curve falls to some limiting value near 4° C., so the oxygen curve in extreme cases falls off to zero. Such an extreme distribution is

---

[17] Åberg and Rodhe distinguish between the extreme type of orthograde distribution, which they term α-orthograde, and the less extreme type with some decline in concentration in the deep water, which they call β-orthograde. Since alteration of the vertical scale can make any β-orthograde curve look more or less clinograde, and since there is a complete series of intermediates between the orthograde and clinograde types of distribution, the term orthograde will be used exclusively in an extreme sense unless qualified, and no formal division of the type will be attempted.

termed by Åberg and Rodhe *clinograde*;[18] this term is the equivalent of *eutrophic* as applied to an oxygen curve in Thienemann's sense, but does not necessarily involve eutrophy in the original sense of well-provided with nutrients.

Under some conditions later to be considered in detail, oxygen disappears from a restricted layer in the middle depths, usually in the metalimnion. Distributions involving such a minimum are termed by Åberg and Rodhe *negative heterograde*.

In the illuminated part of the lake, oxygen concentrations are of course greatly dependent on photosynthesis. Usually the illuminated or trophogenic zone corresponds closely to the epilimnion, so that any water enriched with photosynthetic oxygen will be mixed with the rest of the epilimnion. The dissolved gas thus brought to the surface by turbulent diffusion will pass into the atmosphere. Apart from a feeble diurnal rise and fall, the final effect on the lake is usually, though not always, negligible, though in swamps and weed beds very marked variation between night and day may be observed (Fig. 161). If a lake is transparent enough and the epilimnion thin enough for some photosynthesis to occur in the metalimnion, a marked oxygen maximum may develop in this zone, producing what Åberg and Rodhe call a *positive heterograde* distribution.

Examples of vertical distributions of oxygen. In Figs. 162 to 165, 172, and 174 oxygen curves obtained at the height of summer from a variety of lakes are presented, covering most known types of distribution. It is evident, when the whole body of available data is considered, that the extreme orthograde type of distribution is confined to deep lakes.[19] The two examples given in Fig. 162 are derived from the deepest lake in Europe and in Japan. Hornindalsvatn, as has been indicated previously, was presumably saturated at the time of the spring circulation; there is evidence that this was not so in the case of Tazawako. In the first case, the metalimnion is approximately saturated with oxygen; in the second, it is clearly supersaturated. Yoshimura (1938b), as has already been indicated, gives a reasonable physical explanation of the supersaturation.

In all cases, however orthograde the curve may appear to the eye, there is some evidence of an oxygen deficit or loss of oxygen from the water of the hypolimnion by the end of summer stagnation. In the less extreme orthograde and transitional oxygen curves, there is evidence that two

[18] It will be convenient when considering the accumulation of ammonia, ferrous iron, or other substances whose presence in the deep water is the result of the same group of reactions as are responsible for the removal of oxygen, to speak of an *inverse clinograde distribution*.

[19] Unless, of course, the uniform distribution of oxygen in unstratified lakes of the third class is considered orthograde.

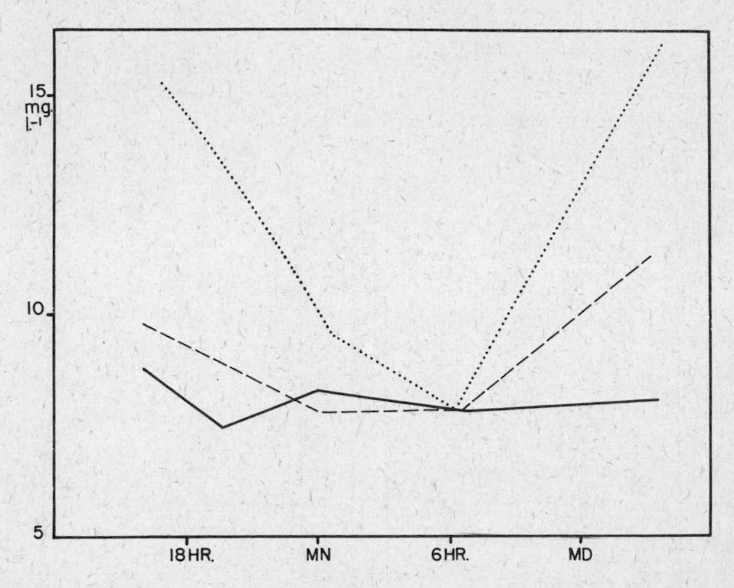

FIGURE 161. Diurnal variation in oxygen concentration in (solid line) surface water of Lokut Dal Lake, Kashmir, a very shallow lake with vegetation over the whole bottom; (broken line) the same at 1 m., in vicinity of rooted vegetation; (dotted line) the same in a weedy pond at Gangribal, Kashmir, separated from the lake by an embankment. April 5–6, 1932. The immense changes in the small pond are in striking contrast to the very small irregular variations in the water away from weeds in the lake. (Hutchinson, unpublished.)

main types of divergence from the strict orthograde distribution may occur. One type, exemplified by Tyûzenziko (Fig. 163) and a few other lakes of great depth and small area in Japan, shows a striking reduction of oxygen concentration near the bottom. The resulting oxygen curve, here spoken of as convex, is described by Yoshimura as microzonal, though this use of the word is wider than that originally introduced by Alsterberg (1927). A few examples of such convex curves are known from other lakes, notably Breiter Lucin (Fig. 164), studied by Thienemann (1925b), but in general the type is rare.[20] More usually, the lower part of the curve is concave at the bottom, as in Seneca Lake, New York (Fig. 163), Crystal Lake, Wisconsin (Fig. 165), and Green Lake, Wisconsin (Fig. 164), with an inflection point in the hypolimnion. Such lakes connect in an unbroken series with those having fully developed clinograde curves.

---

[20] It is possible that in some cases, squeezing out of intermediate layers by internal seiches may produce temporary convex oxygen curves (cf. Birge and Juday 1911, Lake Mendota, May 25, 1906).

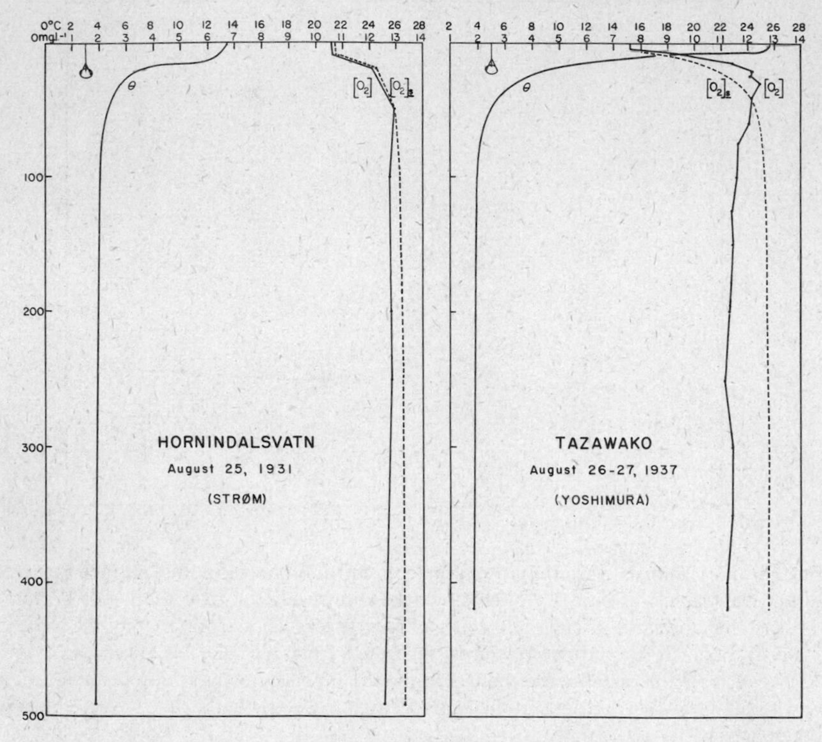

FIGURE 162. Temperature and oxygen distributions in Hornindalsvatn and in Tazawako at end of summer stratification; temperature θ, $[O_2]$ oxygen observed, and $[O_2]_s$ saturation after Fox. The depth of the Secchi disk transparency is also indicated. The two curves indicate extreme orthograde distributions, without (Hornindalsvatn) and with (Tazawako) a very slight metalimnetic maximum leading to supersaturation, probably largely of physical origin. (After Strøm, Yoshimura.)

Typical clinograde curves are shown in Fig. 165. So long as holomictic lakes alone are considered, thus excluding the deep meromictic lakes of the tropics, there can be no doubt that clinograde curves are more characteristic of shallower lakes than are orthograde. Other factors are also involved, however, so that a variety of types of oxygen curve is possible within a restricted depth range.

Explanation of the clinograde oxygen curve. There can be no doubt that in practically all lakes[21] the hypolimnetic oxygen deficit is due to oxidation of organic matter that may enter the lake in suspension or solution, may be

[21] A few cases may occur in which inorganic reducing substances enter the hypolimnion from subaqueous springs.

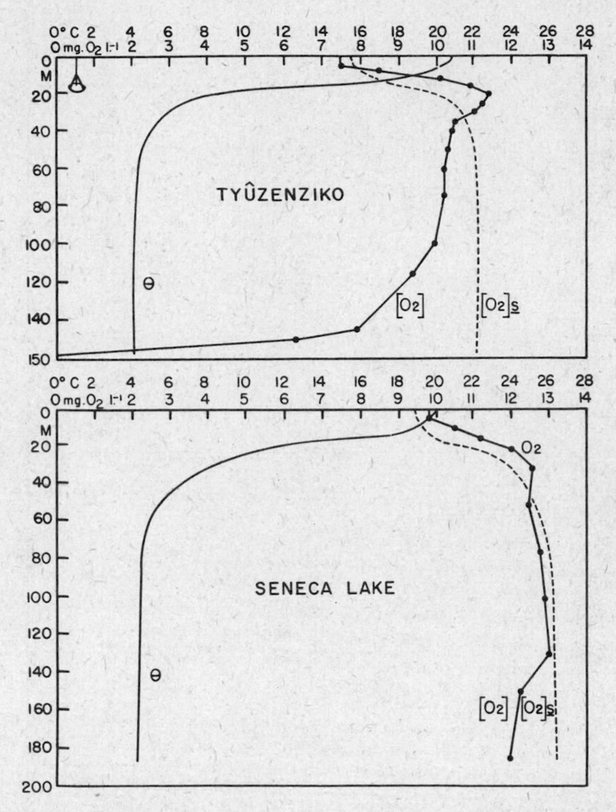

FIGURE 163. Temperature and oxygen distribution in Tyûzenziko, Japan, and Seneca Lake, New York. Orthograde curves, but with uptake of oxygen near bottom. (After Yoshimura and Birge and Juday.)

eroded from the shore, or, as is usually the case, may be produced photosynthetically in the lake itself. The exact site of the oxidation has been a matter of much discussion.

Oxygen can be lost from the hypolimnion either by uptake at the mud-water interface or by uptake in the free water. Following in part the scheme of Thienemann (1928), the agents of deoxygenation may be classified as: (1) animal respiration; (2) plant respiration at night, or when the respiring organism has sedimented below the compensation point and is using reserves or is heterotrophic; (3) bacterial respiration in the decomposition of sedimentary organic matter; and (4) purely chemical oxidation of organic matter in solution, either produced in or brought into the lake.

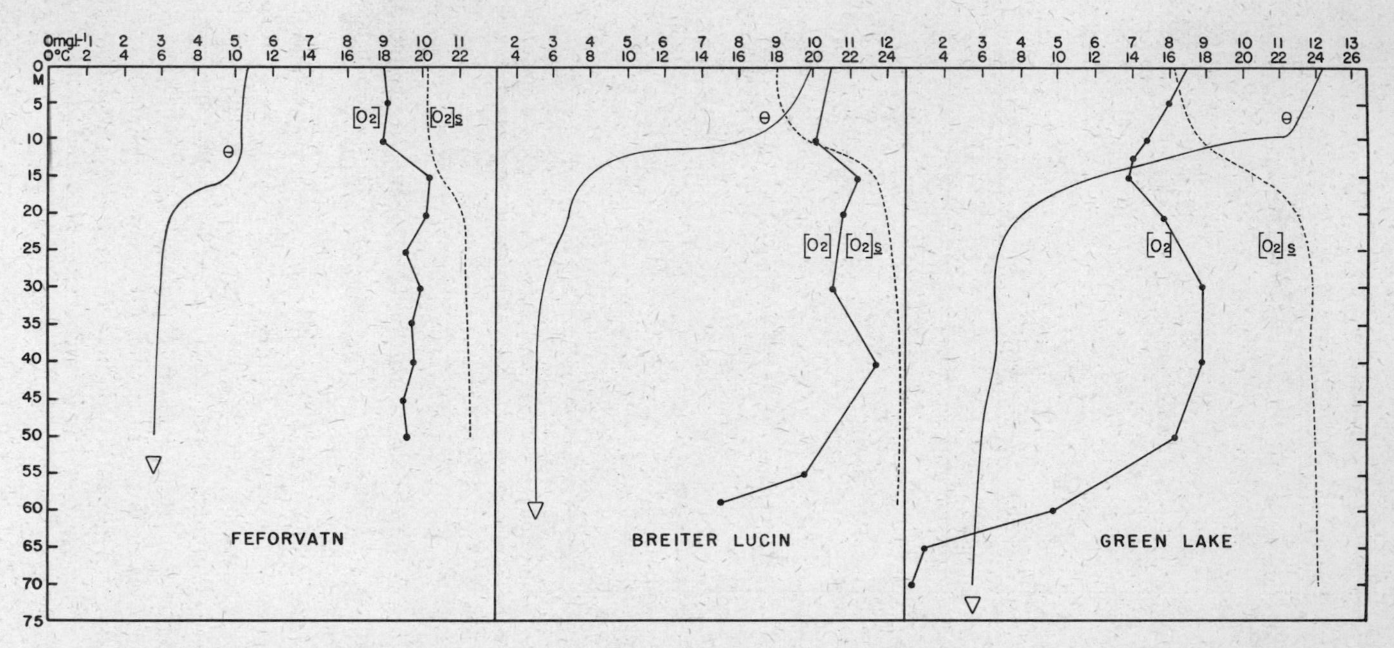

FIGURE 164. Temperature and oxygen distributions at the height of summer stratification in dimictic lakes 50 to 75 m. deep: Feforvatn, a Norwegian mountain lake; Breiter Lucin, northern Germany; and Green Lake, Wisconsin. The three lakes form part of a series ranging from the biologically sterile to the highly productive. (After Strøm, Thienemann, and Birge and Juday.)

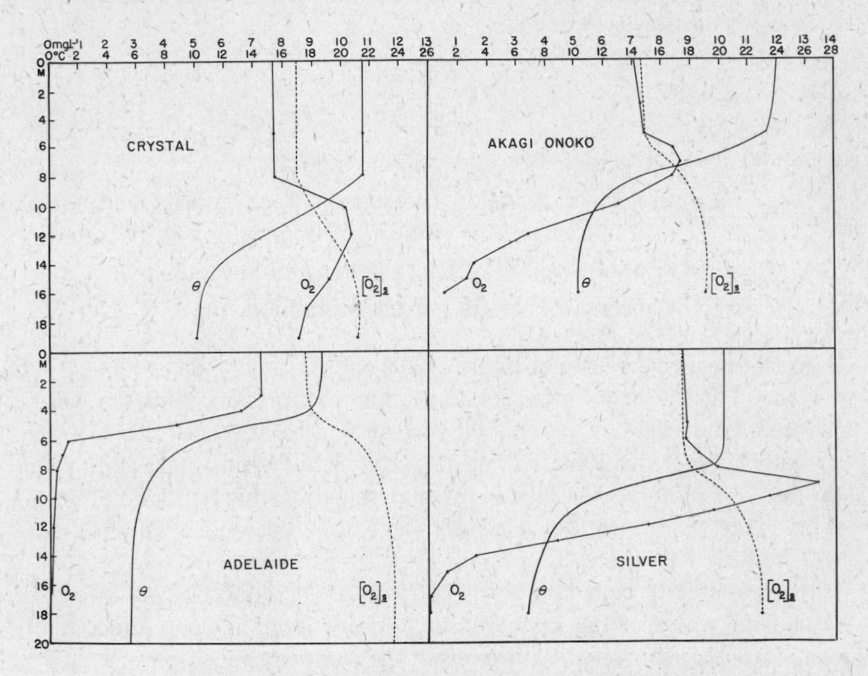

FIGURE 165. Temperature and oxygen distributions at the height of summer stratification in dimictic lakes 15 to 20 m. deep. Crystal Lake, Wisconsin, is unproductive and transparent, some photosynthesis occurring on the bottom. The other three lakes are typical of small moderately or highly productive lakes with clinograde oxygen curves, with a metalimnetic maximum moderately developed in Anagi-Onoko and quite strongly developed in Silver Lake, Wisconsin. (After Juday and Birge, Yoshimura.)

Oxygen uptake at the mud-water interface. It is convenient to consider the possible events at the mud-water interface before proceeding to the free water. Alsterberg (1922) has been the main proponent of the idea that the major oxygen loss occurs at the mud surface. By covering lake mud in the laboratory with oxygenated water, he was able to show that a microzone of oxygen-poor water was rapidly produced over the mud. Lönnerblad (1930), using the mud (*gyttja*) of productive but relatively uncolored lakes, found that this uptake could be abolished by poisoning the bacteria in the mud; and in experiments with the peaty sediments (*dy*) of brown humic lakes, he determined that much purely chemical oxygen uptake occurs.

The formation of a microzone deficient in oxygen at the mud-water interface by bacterial respiration in the mud has been considered by Grote

(1934). With a linear gradient in the mud, the depth h_0 at which oxygen disappears entirely will be given by

$$h_0 = \sqrt{\frac{\hat{k}[O_2]_b}{r_b v_b}} \quad (24)$$

where $[O_2]_b$ = oxygen concentration at the mud surface.
$\hat{k}$ = molecular diffusion coefficient for oxygen in mud, which will not exceed the value already given for K, in water.
r_b = respiratory rate of bacteria per unit volume.
v_b = volume of bacteria per unit volume of mud.

Assuming a very reasonable rate of 400 cm.^{3} O_2 cm.$^{-3}$ day^{-1} for r_b, and a value of 1:125 or 0.8 per cent for v_b, the oxygen concentration will be reduced from 11.5 mg. at the mud surface to zero at 2 cm. To maintain a steady state of this kind, 9.2 mg. O_2 cm.$^{-2}$ day^{-1} must pass from water to mud. So long as the lake is in full circulation this is possible, but as soon as the water becomes relatively stagnant, a deoxygenated microzone will form.

It is possible to calculate how this microzonal reduction will affect the rest of the water of the hypolimnion. If the latter is deep and initially contains water saturated with oxygen, the same approach as was used in considering molecular diffusion into the lake can be used in treating the upward spread of an oxygen-deficient microzone. Only turbulent diffusivity need be considered, since the molecular coefficient of diffusivity K is about 1.98×10^{-5} cm.^{2} sec.$^{-1}$, while the turbulent coefficient A is rarely less than 2×10^{-3} cm.^{2} sec.$^{-1}$ The mud-water interface is to be taken as perennially free of oxygen.

At any depth z and time t, the oxygen concentration will be given by

$$[O_2]_{tz} = [O_2]_s \Phi(u) \quad (25)$$

where $[O_2]_s$ = the initial concentration.

$$u = \frac{z_m - z}{2\sqrt{At}}$$

Evaluation of $\Phi(u)$ for different values of A or t gives the curves of Fig. 166, which curves are, for a typical hypolimnion with little variation in temperature, essentially percentage saturation curves. It was pointed out by Alsterberg (1929) that molecular diffusion into the deoxygenated water of the microzone is totally inadequate to explain the production of any significant macrostratification. Examination of Fig. 166 indicates that even simple vertical turbulent diffusion requires quite high values of A, in excess of 10^{-2} cm.^{2} sec.$^{-1}$, to produce a significant deficit 5 m. from the bottom. A deficit at this distance could be produced in Lake Mendota, where $A \simeq 2 \times 10^{-2}$ cm.^{2} sec.$^{-1}$, but not in Linsley Pond, where $A \simeq 0.25 \times 10^{-2}$ cm.^{2} sec.$^{-1}$. It will also be apparent that the very convex theoretical curves do not accord in shape even with the most favorable convex curves, for instance for Tyûzenziko or for Breiter

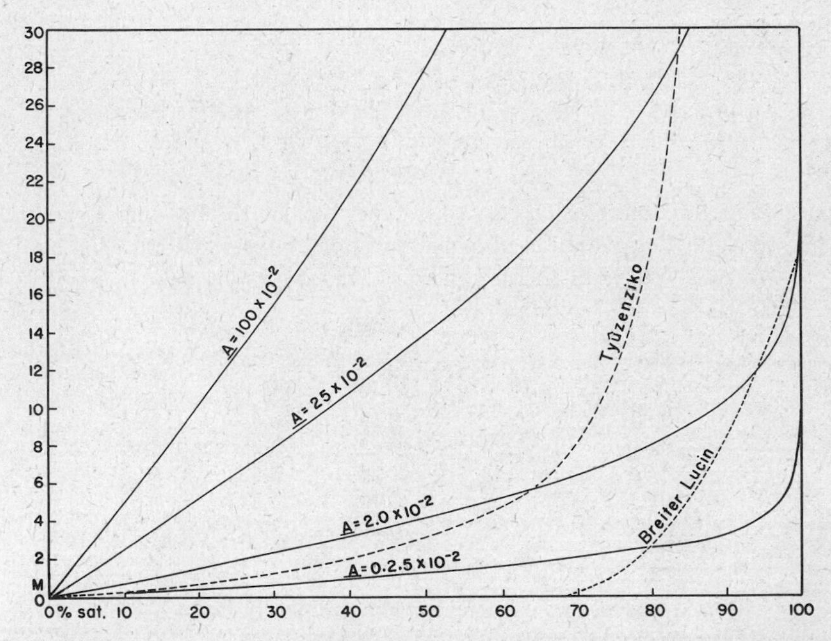

FIGURE 166. Oxygen saturations computed for various values of A after 100 days stagnation of an initially saturated hypolimnion over completely oxygen-free reducing mud. The broken lines indicate observed oxygen curves, computed as percentages of the concentration at the top of the hypolimnion, for two lakes having strongly concave distributions of oxygen near the bottom.

Lucin, observed in nature. Something more than uniform vertical turbulent exchange is clearly needed to explain the typical clinograde oxygen curve.

We can also estimate the rate of uptake of oxygen by the mud in a hypolimnion of finite thickness z_h. Assume that at the top of the hypolimnion, at depth $z_m - z_h$, the oxygen concentration is maintained at $[O_2]_s$ by photosynthesis, diffusion from a more turbulent epilimnion, or any other process. Then the amount of oxygen O passing in time t

into unit area of a perennially oxygen-free sediment at the bottom will be given (Mellor 1922, pp. 488-489, Ex. III and IV), by

$$(26)\quad O = \frac{8z_h[O_2]_s}{\pi^2}\left[(1 - e^{-W}) - \frac{1}{3(3 \times 2 - 1)}(1 - e^{-3^2W}) + \frac{1}{5(5 \times 2 - 1)}(1 - e^{-5^2W}) - \cdots\right]$$

where $W = \left(\frac{\pi}{2z_h}\right)^2 At$

or writing

$$X = \frac{8}{\pi^2}\left[(1 - e^{-W}) - \frac{1}{15}(1 - e^{-9W}) + \frac{1}{45}(1 - e^{-25W}) - \cdots\right]$$

$$(27)\quad O = z_h[O_2]_s X$$

Taking, for example, $[O_2]_s$ as 10 mg. per liter or 10^{-5} g. cm.$^{-3}$, we find that the following quantities of oxygen would be consumed, after 864×10^4 seconds or 100 days of stagnation, by 1 cm.^{2} of the mud.

A,	O_2 Consumed, g. $\times 10^{-3}$ cm.$^{-2}$, at z_h			
cm.^{2} sec.$^{-1}$	10 m.	20 m.	50 m.	100 m.
0.002	0.25	0.12	0.05	0.03
0.01	1.24	0.62	0.25	0.13
0.02	2.48	1.24	0.50	0.25
0.05	...	3.10	1.24	0.62
0.1	...	6.20	2.49	1.24
0.5	...	...	12.4	6.2

The vacant spaces indicate regions in which the values of the hypolimnetic coefficient of turbulence and of the depth of the hypolimnion seem mutually inconsistent for a stratified lake. Since X varies almost in direct proportion to W when the latter is small, it is evident that in lakes with moderately deep hypolimnia in which the coefficient of turbulence does not exceed 0.5 cm.^{2} sec.$^{-1}$, the development of an oxygen deficit over anaerobic mud would be directly proportional to time and to A, and inversely proportional to the depth of the hypolimnion. Actually, in Lake Mendota, in which the bottom deposits are highly reducing, direct proportionality to time (Fig. 167) appears to be confirmed, but further examination indicates that in this case, if we may legitimately treat the whole hypolimnion below 5 m. as a unit of mean depth 10.7 m. and with a mean value of A of 2.3×10^{-2} cm.^{2} sec.$^{-1}$, the uptake should be 2.7 mg. per liter per cm.^{2} in 100 days, but the observed uptake for 1906 was 10.9 mg. per liter per cm.^{2} in 100 days. It is evident, therefore, from both types of quantitative treatment, that the hypothesis of vertical

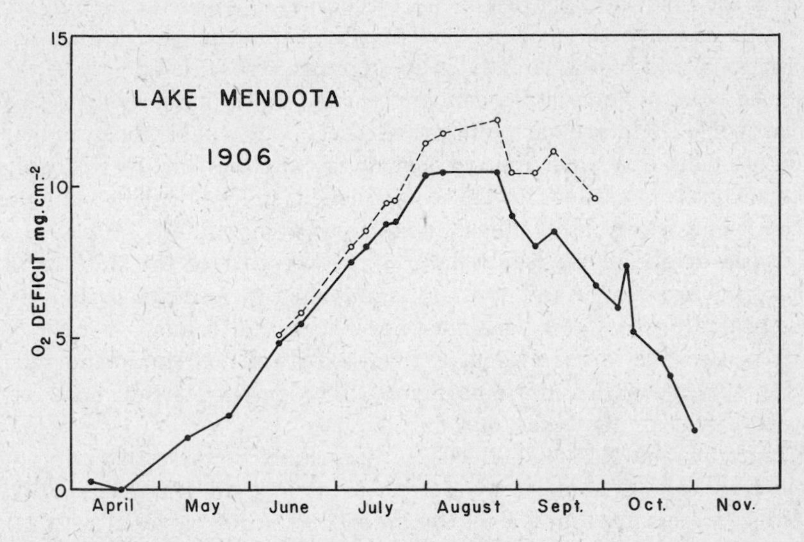

FIGURE 167. The development of the relative deficit per unit area of hypolimnion surface during summer stratification in Lake Mendota. The broken line is the deficit corrected for methane production. (Hutchinson.)

turbulent exchange is inadequate to account for the observed phenomena in those lakes for which a fairly detailed analysis is possible.

Since any stratification dependent solely on vertical diffusion, either molecular or turbulent, should follow the profile of the basin, and since, in general, macrostratification is horizontal unless it is distorted by wind-generated depression of the isotherms and by subsequent temperature seiches, Alsterberg has supposed that horizontal water movements play a considerable part in mixing the oxygen-deficient water of the profile-bound microzone with the free water of the hypolimnion, slowly lowering the oxygen concentration of the latter and producing the observed clinograde curve. Although Alsterberg's scheme is hydrodynamically very unlikely, there is a great deal of independent evidence of horizontal translation in the hypolimnion of even the most stably stratified lakes, and it is reasonable to suppose that in many lakes such movements do play a part in the development of the clinograde oxygen curve.

Alsterberg (1930b, 1931) has also called attention to the fact that density currents set up at the water-mud interface by diffusion of material from the mud should cause a slow descent of the water of the microzone into the deepest part of the lake. Such density currents were later evoked by Hutchinson (1941) to explain certain thermal phenomena, as has already been indicated (page 476). Considering together the three types of postulated hypolimnetic movement—vertical turbulence, horizontal translations, and profile-bound density currents—the development of the observed types of horizontal secondary macrostratification is by no means unreasonable. Unfortunately, in the next section it will be apparent that changes in the free water independent of the mud surface are probably often adequate to account for the oxygen deficit, so that the latter can, on the basis of existing observations, be explained in more than one way.

Oxygen uptake in the free water. Since any part of the water of a holomictic lake may be and normally is inhabited by respiring organisms, some respiratory loss of oxygen must normally occur from any volume of water in the lake. It is evident that in lakes with extreme orthograde oxygen curves such losses are negligible. The question arises, however, to what extent are they ever important?

Kusnetzow and Karsinkin (1931), in their study of the distribution of bacteria in Lake Glubokoye by a direct-count method after evaporation at reduced pressure, found a maximum of 2.85×10^6 bacterial cells per $cm.^3$ at 8 m., and a mean of about 2.3×10^6 cells per $cm.^{-3}$ between 5 and 8 m., and concluded that the fairly marked metalimnetic minimum observed in the lake is due to this metalimnetic maximum in bacteria. In general, the rate of uptake of oxygen in the deeper water of the lake is about 0.2 mg. per liter per day. From a consideration of known bacterial rates of metabolism, they conclude that 2×10^9 bacterial cells, taken as a supposedly conservative estimate for the population of each liter of the metalimnion, would consume at least 0.24 mg. O_2 per liter per day. The bacterial population, unless it has, owing to lack of food, an abnormally low rate of oxygen consumption, seems adequate to account for the observed oxygen deficit. It must be borne in mind, however, that in other lakes such large bacterial populations appear to be seldom found. There also seems to be no very close negative correlation between the details of the bacterial and oxygen distributions (Fig. 168).

ZoBell and Stadler (1940) found that in natural populations of bacteria in Lake Mendota water free of coarse particulate matter, the initial rate of O_2 uptake was about 0.28 mg. per liter per day at 25° C., whatever the oxygen concentration. This rate fell off rapidly, and clearly depended on depletion of respirable organic matter. In special experiments, the O_2 uptake proved independent of O_2 concentration from 0.3 to 36.5 mg. per liter. It is evident that even at one fourth of this rate at lower temperatures, such bacterial populations could reduce the oxygen of the hypolimnion to very low levels in three to four months' stagnation, provided a rapid renewal of respirable organic matter took place. At any one time, the

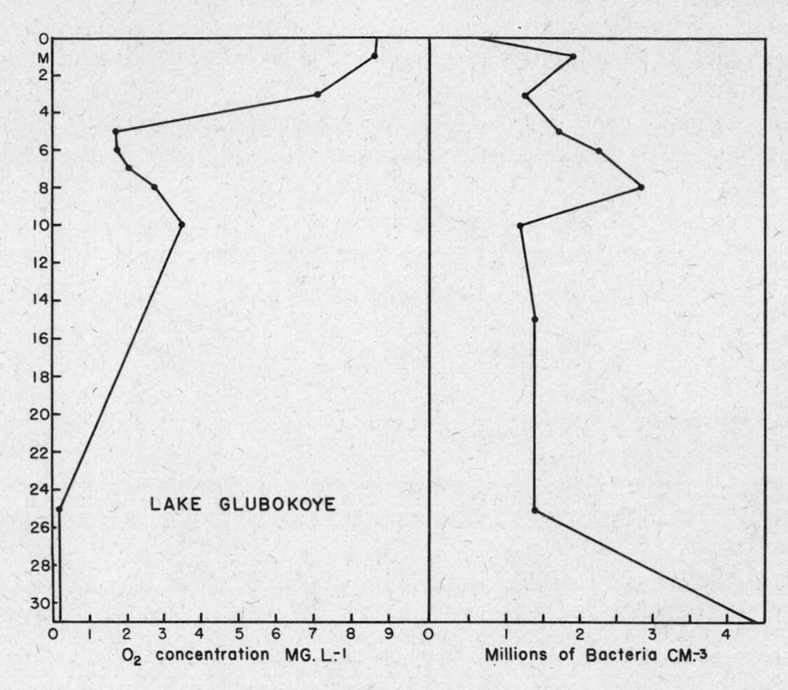

FIGURE 168. Lake Glubokoye, oxygen curve and bacterial population at height of summer stratification (after Kusnetzow and Karsinkin).

water appeared to contain respirable organic matter equivalent to 2 to 5 mg. O_2 per liter. As Krogh and Lange (1932) earlier pointed out, not all the organic matter can be used, and water stored in darkness for long periods may retain some oxygen and some very stable organic material also. The presence of adequate organic matter that can be metabolized is therefore obviously of the greatest importance in causing oxygen deficits. Hutchinson and Riley (unpublished) found that black bottles filled in the spring at about the time of circulation retained large quantities of oxygen later in the summer. From April 15 to June 7, the O_2 content declined from 10.18 mg. per liter to 8.49 mg., or 1.69 mg. per liter, whereas in the lake at this level the concentration had fallen to 4.21 mg. per liter.

For Linsley Pond there are a number of analyses of water exposed in dark bottles at the depths from which the water was obtained. In general these determinations show practically no vertical variation in oxygen uptake over such a period. As the temperatures at the surface and bottom are very different in summer, such determinations presumably measure oxidizable organic matter rather than initial bacterial respiration. The deficit developed in a week in a bottle is, however, of the same order of magnitude as the utilization of oxygen in the lake during the same length of time.

In the upper panel of Fig. 169 a series of determinations (Riley, ms.) is given for two dates during the development of a clinograde curve in Linsley

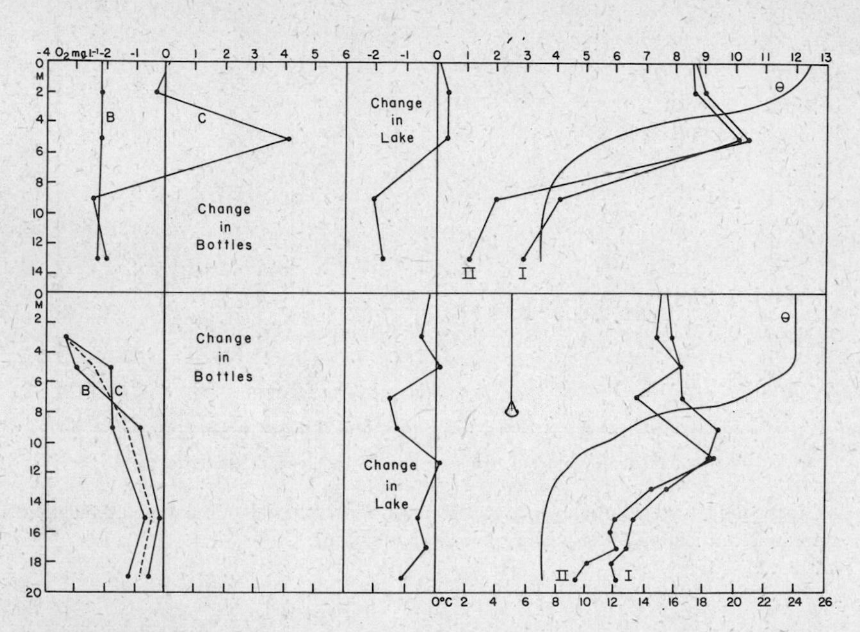

FIGURE 169. Oxygen balance experiments. *Upper panel*, Linsley Pond, Connecticut, July 12–19, 1938; *lower panel*, Lake Quassapaug, Connecticut, July 15–22, 1938. θ, temperature (mean of beginning and end); I and II, initial and final concentrations of O_2; B and C, changes in dark and clear bottles, respectively; broken line, mean of these differences. The Secchi disk transparency is indicated for Lake Quassapaug. (Riley and Hutchinson, unpublished.)

Pond. The changes in black and clear bottles are also indicated. In the black bottles (B) the oxygen uptake is, as has just been indicated, very uniform at all depths. In the top 2 m. production approximately balances consumption and there is almost no change in the clear bottles (C). In the lake, this region is freely circulating and any excessive production or consumption would be largely equalized by exchange with the air. The higher value in the metalimnion of the lake is known to be largely an expression of such physical loss in the epilimnion, on the one hand, and of biochemical uptake in the hypolimnion, on the other. The 5-m. bottle showed a considerable excess of production over consumption. This did not occur in the lake;[22] if it had occurred in the free water, it would have produced a very marked positive heterograde curve. So far as the hypolimnion is concerned, comparison with the black bottles indicates that the change between the two dates is reasonably accounted for solely by the consumption in the water, without any direct effect of the mud.

A similar experiment in Lake Quassapaug, recorded in the lower panel of Fig. 169, a much more transparent (Secchi disk transparency 3.4 to 10.5 m. as compared with 0.8 to 3.8 m. in Linsley Pond) and greener lake (Forel-Ule color V to X as compared with XV to XX), is complicated by the fact that oxygen uptake in the clear bottles is actually higher than in the black at the metalimnetic and hypolimnetic levels studied. This is presumably due to a photokinetic effect on the zooplankton in the bottles. The clear bottles doubtless give a better idea of events independent of turbulent exchange in the lake than do the black. It is to be noted that at the three deeper levels studied, the clear-bottle consumption is of the same magnitude as the consumption in the free water. In both cases there is a minimum consumption in the middle part of the hypolimnion, though the more numerous determinations in the lake place the minimum at 11 m. rather than the 15 m. of the bottle series. Again, no direct effect of the mud is needed to explain the oxygen distribution, but the increase in demand near the bottom, both in the two series of bottles and in the free water. may perhaps be explained in terms of metabolizable material derived from the mud being present in the deep water.

Most workers, notably ZoBell and Stadler, and Kusnetzow and his associates, have regarded all the uptake in the free water as bacterial. It remains, however, to consider whether spontaneously oxidizable material may not exist in lake water in solution. Lönnerblad finds that brown peaty sediments (*dy*) lose far less of their oxygen-absorbing capacity after poisoning than do black organic sediments (*gyttja*). There is some evidence that low surface oxygen is correlated with high color. If this is so, it is perhaps reasonable to suspect that brown extractives behave like *dy* sediments, and that at least part of the uptake is due in such cases to purely chemical oxidation or conceivably to photochemical oxidation. Lönnerblad (1931) clearly believed that some such chemical uptake occurred in brown waters, though his experiments on the matter are not reported in detail. It has been supposed, largely in consequence of his work, that in brown or humic lakes a considerably hypolimnetic deficit can develop as a result of such oxygen uptake. Strøm (1931) thought that in a few cases

---

[22] It was observed in other experiments that *Anabaena* filaments might multiply greatly in bottles but not in the ambient water.

of very unproductive lakes the observed hypolimnetic deficit could be best explained on this hypothesis. He draws attention to the case of four lakes, Lønavatn, Øvre and Nedre Vangsvatn, and Evangervatn, near Voss, which lie one below the other in the same river system. The first named appears, from its color and from the nature of its bottom deposits, to receive most allochthonous humic organic matter, the last named least. When the mean oxygen uptake per unit volume of hypolimnion is calculated at the height of summer stagnation, it is found to vary from 2.10 mg. per liter to 0.57 mg., proceeding from the first or most humic to the last or least humic. Expressing the deficit on an areal basis, as explained in a later section, discloses no such regularity, so that in all probability the humic material here is responsible for the uptake.

In small lakes, humic material on the bottom might add to the deficit. In Dead Pike Lake, one of the northeastern Wisconsin lakes, with a color of 70, 4 to 5 mg. O_2 per liter remain in the hypolimnion in August. In other brown lakes of the same region, such as Adelaide, extreme clinograde curves develop. Although most of such extreme clinograde curves in northeastern Wisconsin appear to occur in colored lakes, Juday and Birge regard them as originating in the same way as in small lakes with water much less colored. It is reasonable to suppose that in very sterile lakes chemical oxidation may produce such deficit as is observed, whereas in more productive lakes such effects may be masked by the oxidation of endogenous organic matter. The subject clearly requires further investigation.

The development of the clinograde curve in an individual lake. In the case of Lake Mendota, for which many observations exist, it is possible to follow the uptake of oxygen throughout the early summer of 1907 with reasonable accuracy (Fig. 170). The best data are the mean monthly values for May, June, and July; individual series may in some cases exhibit distortion owing to temperature seiches. The differences between the monthly means give a rough measure of rates of change and are plotted in Fig. 171. It is evident that while in the early part of summer there is a marked increase in the rate of uptake as depth increases, this increase is not uniform, being greater in the metalimnion between 8 and 10 m. and in the middle hypolimnion between 15 and 18 m. than at intermediate or at greater depths. The falling off of the increase of the rate with depth at the bottom of the lake may be determined by falling O_2 concentration. The decline between 10 and 15 m. cannot be due to this and does not seem to be related to the morphometry of the lake. The kind of argument used by Alsterberg (1927) relative to oxygen and by Hutchinson (1941) relative to bicarbonate, discussed in a later section, would lead one to expect a maximum increase of the rate with depth

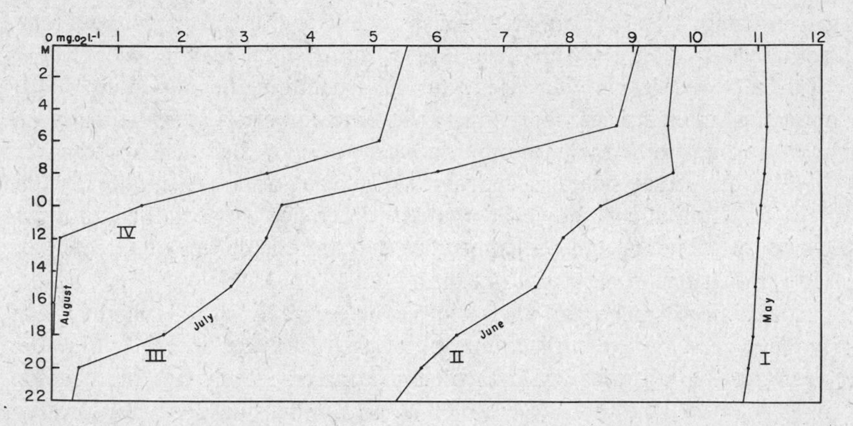

FIGURE 170. Lake Mendota, mean oxygen curves for May, June, and July 1907 and August 1906.

between 10 and 15 m., where there is evidence of a slight shelf, rather than a minimum as is observed. The slope of curve *B* in Fig. 171 between 8 and 15 m. is therefore more likely to be determined by those factors, whatever they may be, acting in a vertical column of water, that produce the higher oxygen uptake in Lake Quassapaug water from the metalimnion and lead to a negative heterograde curve in many lakes.

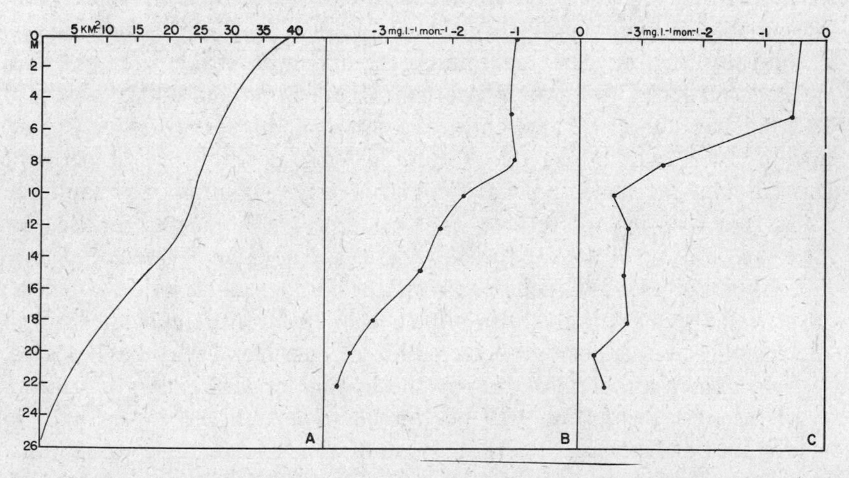

FIGURE 171. Rate of uptake of oxygen in Lake Mendota during summer stratification in 1907. *A*, area (A_z) of contours of depth *z* plotted against *z*; *B*, difference of mean O_2 content between May and June; *C*, difference, June and July.

In the middle of summer the same factors that operated earlier are accentuated. The oxygen at 20 m. is very low and its rate of disappearance actually less than at 18 m., while at 10 m. a small maximum in rate of uptake is observable. In Mendota this metalimnetic maximum in the uptake is small, appears late, and at this time involves a region already low in O_2; a metalimnetic minimum in concentration is therefore not apparent in Birge and Juday's data. The form of the curve of rate of uptake between May and June shows, however, that the difference between the clinograde curves of Mendota and the somewhat heterograde curves of Green Lake (Fig. 164) is one of degree rather than kind.

Metalimnetic maxima: the positive heterograde curve. Slight metalimnetic maxima are very common. Loss of oxygen in the epilimnion, owing to the fall in saturation concentration, together with the development of a clinograde curve, gives a maximum but does not produce supersaturation. A slight degree of supersaturation, which in theory need not imply the existence of a maximum in the concentration curve, can be produced by purely physical causes. Most of the supersaturation in Tazawako is doubtless of such an origin.

In most lakes, however, supersaturations are doubtless due to photosynthesis in stabilized layers. Maxima are pronounced, of course, only in lakes containing a fair amount of phytoplankton, though a very slight effect may be responsible for part of the supersaturation involving a true maximum in the concentration curve in a lake as sterile as Tazawako. In highly productive lakes the effect may be very striking; in Beasley Lake, Birge and Juday indicate that a marked uptake of silica and loss of calcium, presumably by precipitation at a high *p*H produced by photosynthesis, accompany the oxygen maximum. Even more striking cases were observed in Knight's Lake, Wisconsin, where Birge and Juday observed 36.5 mg. per liter or 353 per cent saturation at 4.5 m. at the end of August, and in Otter Lake, where they found 35.7 mg. per liter or 381 per cent saturation at the same depth at the end of July. A number of Japanese cases have also been described by Yoshimura. The curves for Beasley Lake and for one of the Japanese examples are given in Fig. 172.

Yoshimura (1938b) concludes that if the compensation point, which he takes roughly as a depth 1.2 times the Secchi disk transparency, lies within the metalimnion, a positive heterograde oxygen curve may be expected. If the compensation point lies in the epilimnion, an excess of photosynthesis over respiration will be limited to unstablized layers and no maximum can develop. There seems to be a marked tendency to regional incidence of these conditions (Fig. 173). The Baltic lakes of north Germany usually have clinograde curves; lakes of similar depths in Japan usually have a well-marked maximum. Yoshimura attributes the fre-

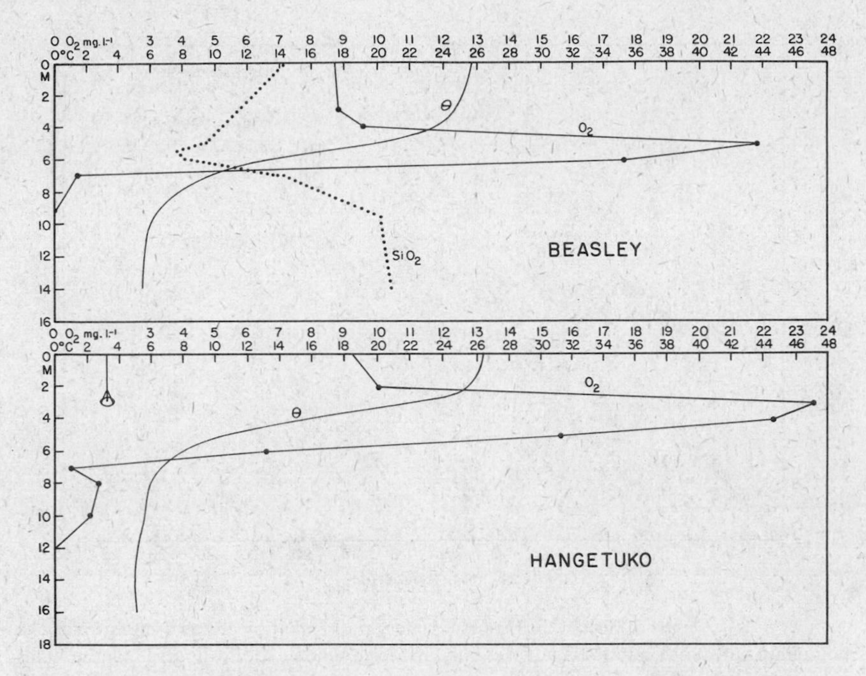

FIGURE 172. Extreme positive heterograde curves with metalimnetic maxima, in Beasley Lake, Wisconsin, and in Hangetuko, Japan (after Birge and Juday, Yoshimura).

quency of positive heterograde oxygen curves in Japan to clear water combined with weak winds and often rather high summer temperatures. The phenomenon does not occur in brown water lakes, in which the compensation point is at a relatively small depth, the vegetation is sparse, and abstraction of oxygen by brown organic material occurs throughout the water column.

Metalimnetic minima: the negative heterograde curve. In a great many lakes there is a marked tendency for the oxygen uptake to be greater in the metalimnion than in the upper hypolimnion, thus giving rise to a negative heterograde distribution. A complete series exists between very slight indications of the negative heterograde condition, indicated by the rates of change during the late developments in Mendota already discussed, and the extreme type (Fig. 174) in the Sakrower See (Antonescu 1931). Intermediate stages are well exemplified by the curves presented in Fig. 175 for Green Lake, in Fig. 164 for Breiter Lucin, and Fig. 168 for Lake Glubokoye. No good examples occur in northeastern Wisconsin and the type is rare in Japan; in both these regions, metalimnetic maxima are

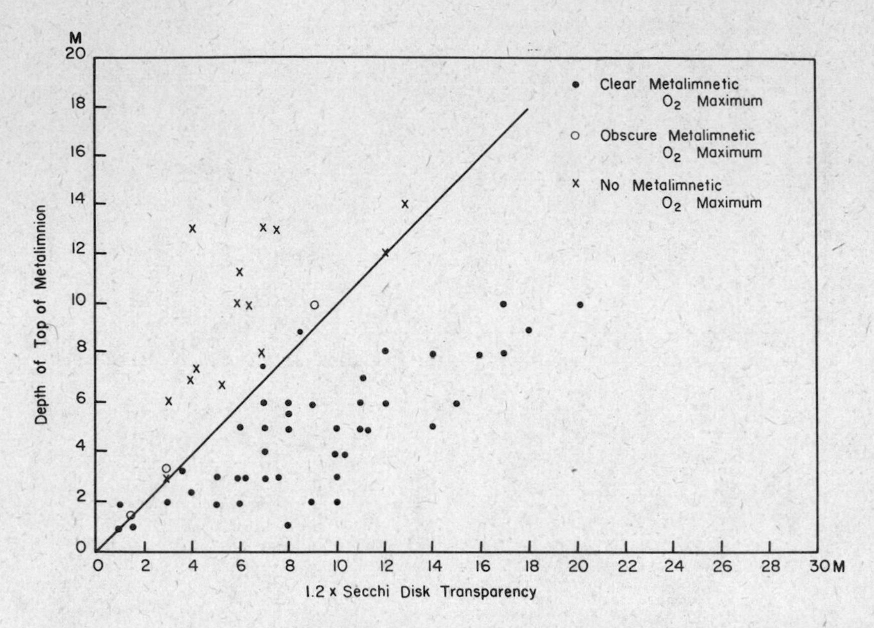

FIGURE 173. Depth of Secchi disk transparency, multiplied by 1.2 as a measure of the compensation point, plotted against depth of thermocline. Solid dots, Japanese lakes with a clear metalimnetic maximum, or positive heterograde oxygen curve; open circles, Japanese lakes with an obscure metalimnetic maximum; crosses Baltic lakes with no metalimnetic maximum. (Yoshimura.)

common. A fair number are known in the Baltic region (Riikoja 1929, Järnefeld 1932); in some of these there is little fall in the lower hypolimnion, so that a strictly dichotomic oxygen curve results. Cases are known in which a minimum succeeds a metalimnetic maximum, as in Lake Wigry (Litynski 1926), or in which two maxima are separated by a minimum, as in Unagiike (Yoshimura 1938b). In the latter case, the lower maximum is due to photosynthesis early in the summer, the upper maximum to photosynthesis after the epilimnion had been made turbid by silt carried in as the result of unusually great precipitation.

Two types of theory have been put forward to explain the more ordinary type of metalimnetic minimum. The first is based on events in a vertical water column, independent of the form of the basin and of the horizontal movements of water in thermally stratified layers. The second is based on morphology of the basin in relation to such horizontal movements. There is little doubt that both types are to a large extent correct, but the applicability of one or the other varies from basin to basin.

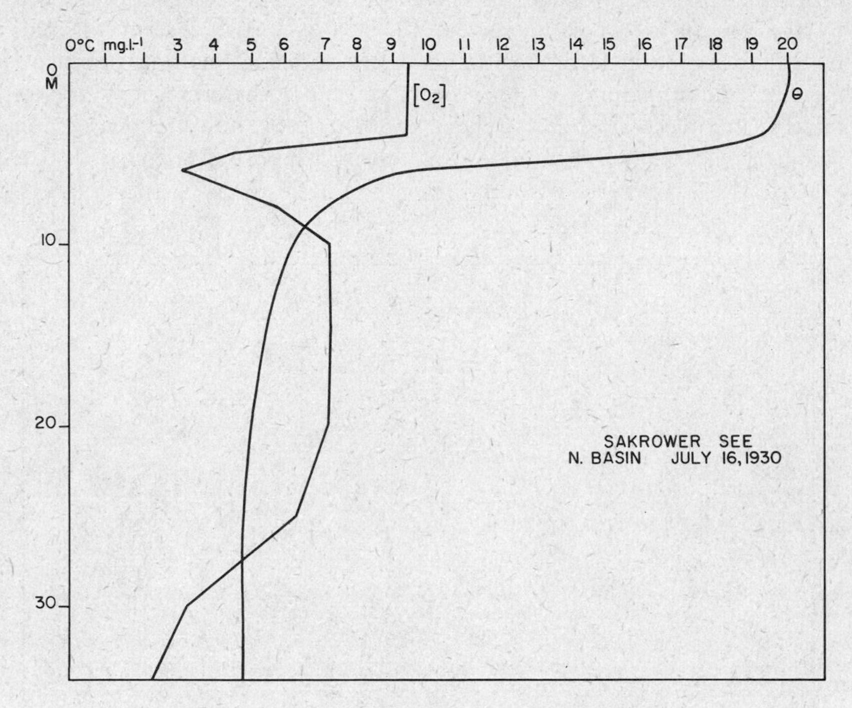

FIGURE 174. Sakrower See, oxygen curve showing extreme metalimnetic minimum or negative heterograde type (after Antonescu).

Falling seston as a cause of the minimum. The most widely known theory of the metalimnetic minimum is that it represents a region into which oxidizable material, such as dead plankton, faeces, and perhaps living plankton metabolizing its reserves or living saprobiotically, has fallen from the epilimnion. Birge and Juday (1911) suppose that the fall of such material would be checked on entering the cold hypolimnion, owing to the greater viscosity of the region. The descent of falling particles being suddenly slowed, the easily oxidizable material would all be destroyed in the upper layers of the hypolimnion, removing oxygen and leaving only rather resistant debris to fall through the hypolimnion. Actually, it is obvious that in the turbulent epilimnion the sinking speed will be determined solely by the turbulence; throughout the entire freely circulating layers, any tendency to abstract oxygen will be counterbalanced by turbulent diffusion from the surface, if not by photosynthesis. Most small particles that reach the bottom of the epilimnion will be returned by turbulent movements to the upper layers. Only when a particle descends below the epilimnion—when, as Thienemann (1928) puts it, it is captured by the less turbulent stable layers—is its downward descent assured. Only then can initial decomposition of easily oxidizable material produce a metalimnetic minimum. The process depends, therefore, not on an increase in molecular viscosity checking the descent, but on a decrease in turbulent viscosity permitting descent. It is of course qui[illegible] possible that

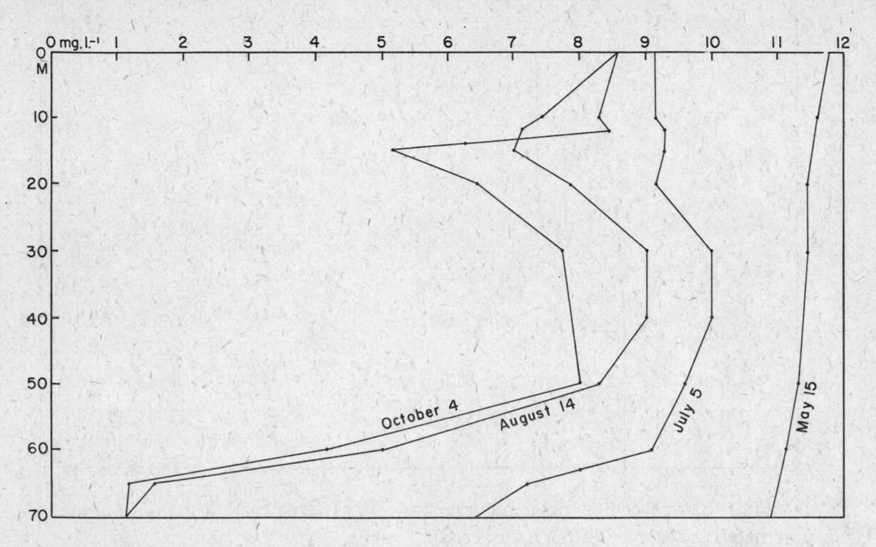

FIGURE 175. Green Lake, Wisconsin, development of a moderate negative heterograde curve (from the data of Birge and Juday).

where the coefficient of turbulent viscosity, which may ordinarily be taken as approximated by A, is not much greater than the coefficient of kinematic molecular viscosity, some slowing of particles may take place as the temperature falls in the lower metalimnion. Since the kinematic viscosity of water is 1.57×10^{-2} cm.^{2} sec.$^{-1}$ at 4° C. and 1.31×10^{-2} cm.^{2} sec.$^{-1}$ at 10° C., an effect due to molecular viscosity is not unlikely in well-stratified lakes.[23]

It must also be borne in mind that not only is a particle inherently likely to have a higher rate of oxygen uptake early in its descent while the most unstable materials are still present, but in the metalimnion the temperature is somewhat higher than in the lower layers, so that oxidation

[23] The contention of Schmidt (1925), Thienemann (1928), and other workers that there should be a turbulence minimum in the metalimnion is, as has been indicated (page 474), empirically doubtful.

will proceed faster, as Auerbach, Maerker, and Schmalz (1926) point out.

Morphometry as a cause of the minimum. The second type of theory of the cause of metalimnetic oxygen minima is proposed by Alsterberg (1927). He supposes that below the epilimnion there is a system of horizontal currents, each horizontal element of the lake being supposed to exhibit a circulation pattern of its own, comparable to that of a shallow isothermal lake under wind stress. While Alsterberg's scheme is, as has been pointed out, most improbable hydrodynamically, there is abundant evidence of the reality of horizontal water movements in the metalimnia and hypolimnia of stratified lakes. Such movements will be evoked to explain the vertical distribution of bicarbonate and phosphate in later chapters. They are probably best developed in the metalimnion, where vertical stability is greatest, and are presumably due to internal seiches. Irregular damping of such seiches in the presence of slight turbulence well might produce horizontal mixing.

Alsterberg points out that, considering any elementary layer of the lake, the ratio of element of mud surface to element of water volume decreases as the steepness of the slope increases. If movement is predominantly horizontal, there is clearly a greater tendency for the mud to influence the water at those levels where the slope is gentlest. This concept is used later

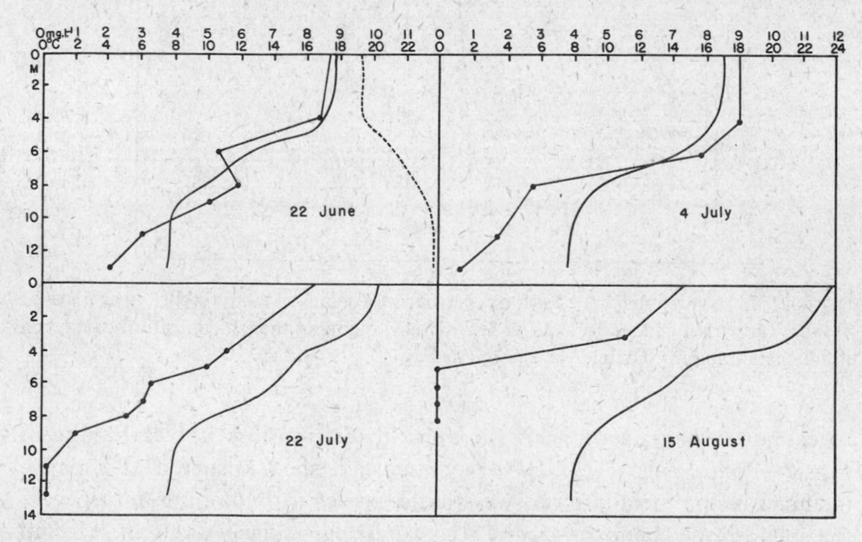

FIGURE 176. Skärshultsjön, Sweden. Development of clinograde oxygen curve, June 22, 1938, slight heterograde condition probably due to shelf at 3 to 6 m.; July 22, development of clinograde curve with two concavities, the upper one probably corresponding to the minimum of July 22; August 15, extreme clinograde curve. (After Åberg and Rodhe.)

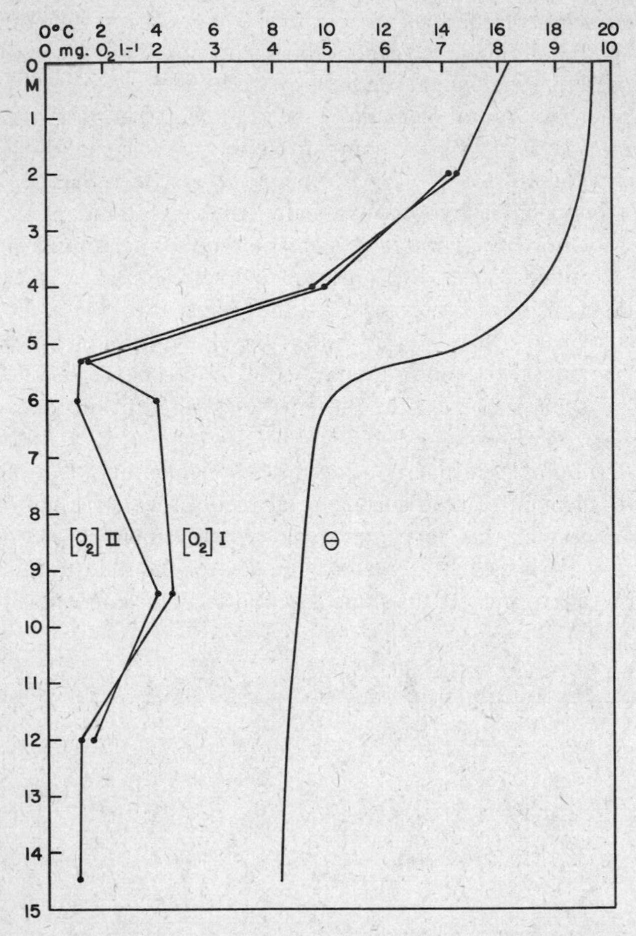

FIGURE 177. Skärshultsjön: $[O_2]$ I, oxygen curve in centeı of lake; $[O_2]$ II, near margin. To show apparent effect of a shelf at 3 to 6 m. in reducing the concentration of oxygen in the metalimnion. (After Alsterberg.)

to explain certain features of the vertical distribution of bicarbonate in Linsley Pond (page 675). If a very marked shelf occurred at a depth included in the metalimnion, we should expect to find some influence of the shelf on the chemistry of the free water, particularly as the horizontal streaming is likely to be most pronounced where the vertical stability is greatest. Alsterberg concluded, therefore, that the presence of such shelves bearing reductive organic sediments is an important cause of metalimnetic minimum. He put forward this explanation to account for the oxygen distribution in Skärshultsjön (Figs. 176 and 177). In this lake there is a strong shelf between 3 and 5 m., in the middle of the metalimnion. This shelf appears to influence the oxygen curve, strongly near the margin ($[O_2]$ II of Fig. 177), less strongly in the middle of the lake ($[O_2]$ I of Fig. 177). While the effects of falling plankton at present appear to provide a more general explanation of the minimum, other cases like that of Skärshultsjön are likely to be discovered, and Alsterberg's theory will probably be found to apply at least in a limited number of lakes.

THE VERTICAL DISTRIBUTION OF OXYGEN DURING WINTER STAGNATION

When a definite winter stratification is established in a dimictic lake, and particularly when a persistent ice cover forms during the winter, the mixing of freshly oxygenated water from the surface of the lake into the depths becomes restricted or impossible. In a lake that has retained an orthograde distribution of oxygen throughout the summer, one may expect a winter distribution of the same sort. When the summer curve is markedly clinograde, a considerable loss of oxygen from the deep water is likely to take place during the winter also. But the resulting winter oxygen curves are usually of a different shape from those encountered in summer, being in most cases markedly less concave and often strikingly convex.

Form of the winter oxygen curve. Several good Japanese examples published by Yoshimura (1938b), which permit comparisons between the winter and summer oxygen curves, are given in Fig. 178. A typical winter curve for Lake Mendota is given in Fig. 179. In such a curve there is nothing that could not receive a formal explanation in terms of rather feeble vertical turbulence, due perhaps to the flow of water through the lake.

In the case of the Mendota curve, it is reasonable to suppose that before winter stratification and freezing occurred, the water had reached a fairly uniform concentration of about 13.28 mg. O_2 per liter. If a completely oxygen-free layer is assumed just below the mud surface, at say 22.5 m. in the station under consideration, it is possible to plot the oxygen concentrations observed as percentages of 13.28. This now gives a curve that is comparable to those of Fig. 166. Comparison of the two figures indicates that the observed Mendota curve falls neatly between the curves of Fig. 166, computed for 100 days' stagnation, for $A = 0.25 \times 10^{-2}$ cm.^{2} sec.$^{-1}$ and $A = 2 \times 10^{-2}$ cm.^{2} sec.$^{-1}$ The time taken to achieve the distribution of Fig. 179 is not known; it must have been more than 60 but less than 100 days. In default of more accurate information, the conclusion is justified that vertical turbulence, with $A \simeq 10^{-2}$ cm.^{2} sec.$^{-1}$,

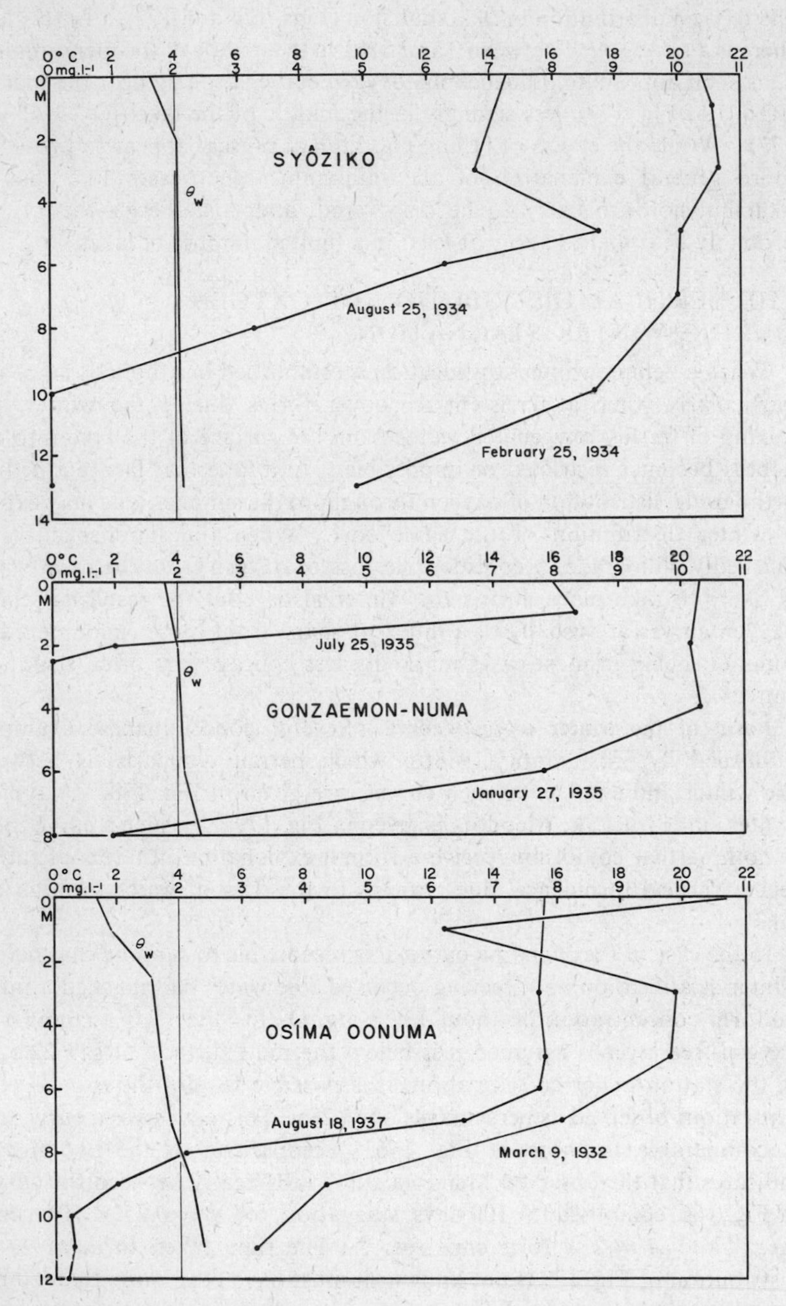

FIGURE 178. Winter and summer oxygen curves for three small Japanese lakes: θ_W, winter temperature; $[O_2]$ curves designated by dates. (After Yoshimura.)

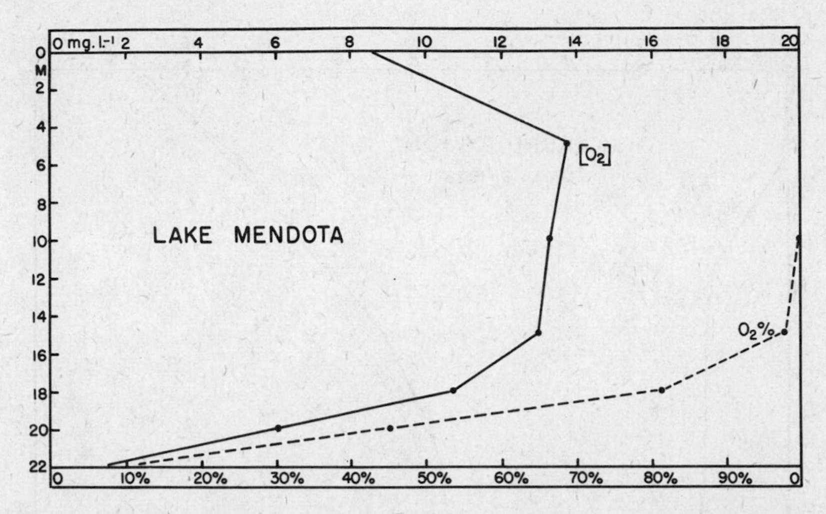

FIGURE 179. Lake Mendota, winter oxygen curve under ice, March 8, 1906. The reduced value at the surface is certainly due to melt water, which lowered the alkalinity below the ice and may have carried oxidizable soil extractives. The broken line gives deep-water values in percentages of the concentration at 10 m., for comparison with Fig. 166. (From the data of Birge and Juday.)

could have produced the observed distribution. It is, however, not very likely that it did.

Significance of density currents. There is often evidence of a profile-bound density current running when the whole of the lake is below 4° C. and conveying heat from the shallow to the deeper parts of the basin. In most lakes this heat is derived from solar radiation penetrating the ice and reaching the bottom in shallow water; in some cases, heat from the mud stored during the summer is also significant. The effect of such a density current on the distribution of oxygen in the lake is likely to depend on the speed of the current and on the reductive capacity of the mud over which it flows. A relatively rapid current, particularly if it descends towards largely deoxygenated water, is likely to raise the oxygen content near the mud-water interface; a relatively slow current, if passing over highly reductive mud at a time when the greater part of the free water is still rich in oxygen, will lower the oxygen content at least in the deepest part of the lake. The former situation is apparently demonstrated by the data for Skärshultsjön (Fig. 180) given by Lönnerblad (1931) and discussed by Alsterberg (1930b). The effect here is most marked in the middle depths. Both effects are apparently indicated by the oxygen profiles given

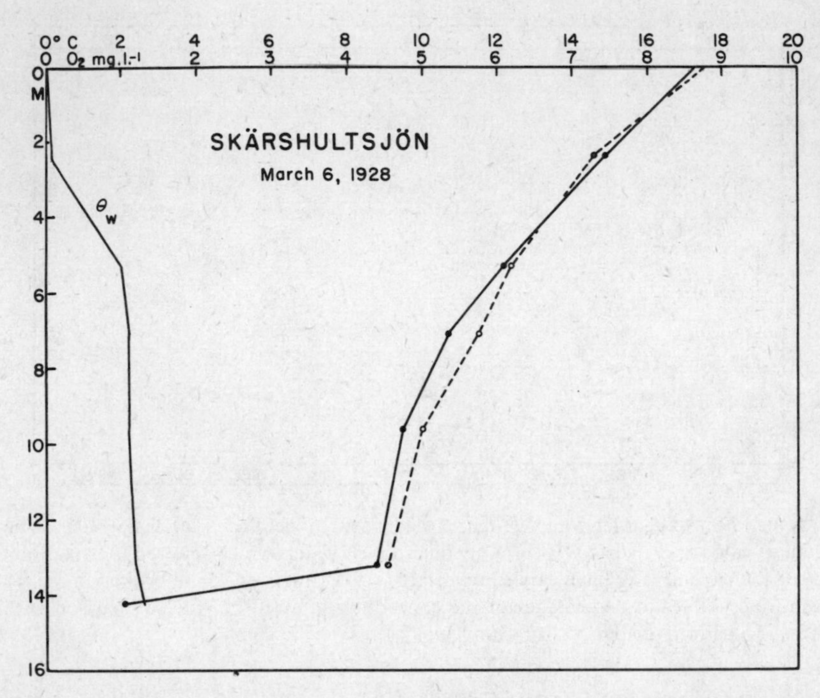

FIGURE 180. Skärshultsjön, Sweden, March 6, 1928. Vertical distribution of temperature and oxygen in center of lake (solid line), and of oxygen at a more peripheral station (broken line), to show supposed transfer of oxygen downward by density currents due to marginal heating. (After Alsterberg.)

by Greenbank (1945) for Green Lake, Waterloo area, southern Michigan, a shallow dam about 3.3 m. deep. In this lake, as in Skärshultsjön, it is probable that a good deal of oxygen is lost under the ice by respiratory decomposition in the free water. Early, on February 2, 1941, in Greenbank's series of profiles, however, clear evidence of a profile-bound density current raising the temperature and lowering the oxygen concentration appears to exist. On February 5, 1943, the mean lay at 0.6 mg. per liter, and such evidence of the effect of a density current as exists suggests a raising of the oxygen concentration along the bottom profile. Unfortunately, it is not certain that this effect can be separated from an effect due to the influent. Three days later a still more striking deflection of the oxygen isopleths downward at the upper end of the basin was noted.

By far the most remarkable series of observations bearing on these matters are those of Rossolimo (1935) on Lake Beloye and some other lakes in the vicinity of Kossino. These lakes appear to be frozen from sometime in November to sometime in April. The oxygen determinations, made at several stations, permit some idea of the form of the isopleth surfaces under an ice cover. In Lake Pereslavskoye, one of the two deeper lakes studied, in which the oxygen remains high, the isopleths appear to dip slightly over the deepest point. The density current, if present, presumably here loses oxygen fast enough and flows slowly enough to lower the oxygen concentration peripherally. The same situation occurs most strikingly in Lake Glubokoye (*A* of Fig. 181).

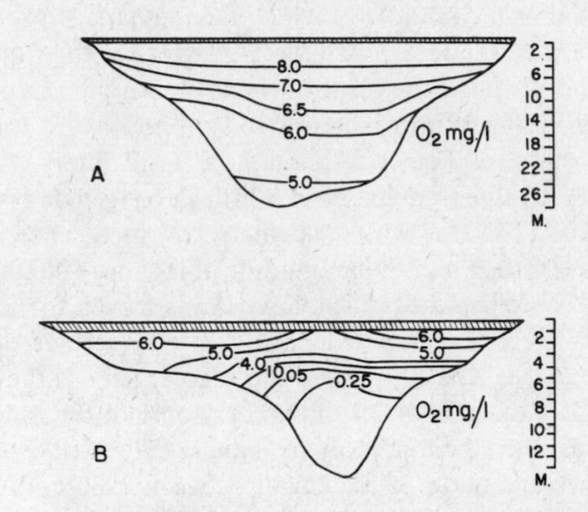

FIGURE 181. Distribution of oxygen at the end of winter stagnation under ice: *A*, Lake Glubokoye, March 5, 1931; *B*, Lake Beloye, March 15, 1928. Methane bubbles collect in the ice formed over the deep depression of Lake Beloye. (After Rossolimo.)

In all the other Kossino lakes there is a convexity over the deepest point; this is very marked in some lakes, such as Lake Beloye (*B* of Fig. 181) where there is a localized area somewhat deeper than most of the rest of the basin. The explanation of this convexity is apparently twofold. In the first place, warm concentrated water, deoxygenated by flowing over the mud, collects in the deepest part of the basin and slowly forces the water above it upward and outward. This, however, is clearly not the whole explanation in extreme cases such as Lake Beloye. In this lake we find a marked and continuous production of methane, hydrogen, and other gases by the mud in the deeper parts of the basin. This production occurs most rapidly in the summer, when the water at the bottom of the lake is at 10° to 12° C., but it is considerable at any season. The area producing the methane is clearly indicated by the presence of bubbles in the ice cover, and lies over that part of the lake more than about 9 m. deep. It is greatest above two holes over 10 m. deep. The bubbles, when formed, contain 74.1 to 83.5 per cent CH_4 and 5.1 to 18.4 per cent H_2. During their ascent they lose their hydrogen, and their methane content decreases to 20 to 24 per cent. There is a great gain in nitrogen and a small gain in oxygen. Most of the CH_4 and H_2 must be oxidized bacterially in the water, reducing the oxygen concentration and producing the extreme patterns of Fig. 181. It is probable that the presence of the density current, which must maintain more or less anaerobic conditions in the deepest part of the lake during a large part of the winter, is responsible for low redox potential of the mud from which the methane is produced.

In the case of Lake Mendota, since three-quarters of the winter heating is due to insolation through the ice, followed by movement downward of the warmed water, it is not unreasonable to suppose that in this lake also the form of the curve of Fig. 179, though it could imply pure vertical turbulence, is actually due to delivery of a little deoxygenated water to the depths of the lake. The true winter coefficient of turbulence is therefore quite likely much less than half the magnitude of the mean summer value.[24]

Winter fish kills. When a very shallow lake is frozen for a long time, particularly if the ice is covered with snow, a very serious depletion of oxygen may take place (Knauthe 1899) and a great part of the fauna may be destroyed. Greenbank (1945) summarizes some of the older observations and records the phenomenon as having been reported at least sixty-three times in the State of Michigan between 1930 and 1941, more than half the instances relating to the winter 1935–1936.

SPECIAL TYPES OF OXYGEN REGIME

The effect of great altitudes. It is worth noting that if a lake is saturated with respect to atmospheric pressure at great altitudes, the development of a moderate clinograde oxygen curve will reduce the absolute concentration at the bottom of the lake to a value that at low altitudes would only occur when a very strong clinograde curve develops. Thus in Yaye Tso, at an altitude of 4686 m. and an atmospheric pressure of about 434 mm. (page 578), the saturation value at the temperature of 8.78° C. at the bottom of the lake would be 6.41 mg. per liter. The observed concentration was 2.0 mg. per liter, implying a deficit of 4.4 mg. If the lake had lain at sea level with a similar thermal regime, the oxygen concentration implied by this deficit would have been 6.9 mg. per liter. It is reasonably certain that with such a concentration the nature of the bottom fauna would be quite different from that observed.

[24] Mortimer (verbal communication) now accepts this argument relative to values of A in winter.

Monomictic subtropical lakes. In deep lakes situated in subtropical latitudes in which a single circulation period may be expected, the existence of such a circulation period will to some extent be determined by the severity of the winter. In a cold winter the surface waters may lose heat sufficiently to produce instability fairly early in the season, and the whole lake may circulate for long enough to become saturated. In a very mild season, the surface layers may never be cooled enough to produce instability relative to the bottom, so that circulation is partial and the deep water is not fully aerated. In a winter of normal severity, though full circulation may occur, it may last for such a short time that the lake is not saturated. A very considerable variety of distributions is therefore possible.

Yoshimura has studied a number of lakes in southern Kyushu which tend to be monomictic in some years, meromictic or perhaps oligomictic in others. The early months of 1936 were abnormally cold in southern Japan, and the lakes under observation showed evidence of some mixing right to the bottom, though in the cases of several lakes, notably Sumiyosiike (Fig. 182), the circulation period was too short to permit complete oxygenation in the deeper water. The succeeding winter was much milder,

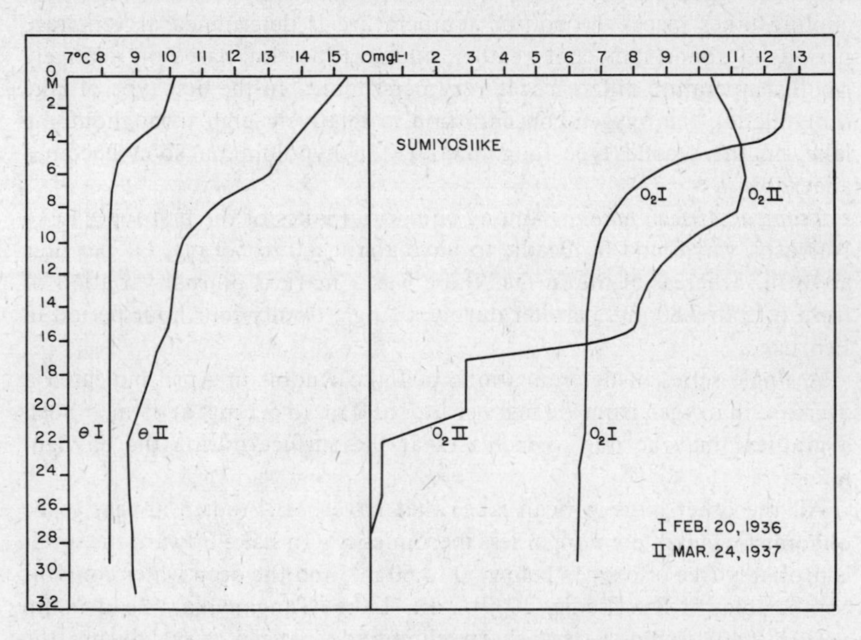

FIGURE 182. Sumiyosiike, southern Kuyshu, Japan. Vertical distribution of oxygen and of temperature: I, after a cold winter when the lake was more or less monomictic; II, after a mild winter when the lake was meromictic, with a slight temperature inversion below 25 m. and a complete lack of oxygen on the bottom. (After Yoshimura.)

and in nearly all the lakes studied there was evidence of incomplete circulation. In Sumiyosiike evidently no mixing occurred below 25 m., and a small temperature inversion that must have developed during the latter part of 1936 was preserved. A very little mixing may have occurred in the 22 to 25-m. zone, and rather more in the 17 to 20-m. zone; only the top 8 m. circulated freely. A remarkable clinograde curve with several steps thus resulted. In the same region, Ikedako, a deep lake (z_m = 230 m.), behaved in a like manner. These two lakes are evidently on the border line between holomixis and meromixis; Ikedako, in fact, has had a monimolimnion below 40 to 75 m. in every year studied (1917, 1919, 1929, 1930, 1937) except 1936, in which year it was homoiothermal at 10.7° C. right to the bottom.

Oxygen distribution in lakes of equatorial regions. Under the relatively invariant climatic conditions of the equatorial regions, two extreme types of temperature stratification can develop. In lakes of great area and moderate depth, particularly in regions of great elevation and high evaporation, in the absence of marked thermal differences between the seasons, nocturnal cooling ensures circulation throughout the year. In deep lakes, particularly in the humid tropics where insufficient nocturnal cooling takes place, the bottom temperature is determined at very rare periods of abnormally cool weather, and stratification based on relatively small temperature differences is very persistent. In the first type of lake (polymictic), the oxygen concentration is relatively high throughout the lake; in the second type (oligomictic), the hypolimnion soon becomes anaerobic.

Tropical African lakes. Among equatorial lakes of the first type, Lake Naivasha was found by Beadle to have about 5.0 to 5.3 mg. O_2 per liter at 18 m., whereas at the surface there was a marked diurnal variation of from 6.15 to 7.80 mg. per liter during a single twenty-four hour period in February.

A single series of determinations in Lake Rudolf in April indicated a decrease in oxygen from 7.6 mg. per liter at 0 m. to 5.1 mg. at 45 m. Such a gradient may be due to increases at the surface during the daylight hours.

All the other large African lakes that have been studied appear to be oligomictic and often more or less meromictic. In Lake Edward the water is probably free of oxygen below 50 to 60 m., and the deep water contains appreciable H_2S (Beadle 1932). In Lake Tanganyika, Beauchamp (1939, 1940) found a very sharp clinograde oxygen curve during the warm season, when the top 50 m. of the lake lay at about 25.7° C. and the whole of the water below 100 m. a little below 24° C. The oxygen content of the circulating epilimnion lay at about 7.2 mg. per liter, falling to zero at 20 m. During the cool season, the surface temperature fell to 25.0° C. and there was a considerable amount of mixing at least to 100 m. The oxygen curve then fell from about 7.5 mg. per liter at the surface, to about 0.3 mg. at 90 m. and to zero at 110 m. The monimolimnion appeared from Beauchamp's account to contain H_2S. There is some evidence, from the distribution of the mollusk *Tiphobia horei*, that the mixolimnion extended to 300 m. at the end of the nineteenth century (Brooks 1950).

In Lake Nyasa (Beauchamp 1940, 1953a) there is a considerable quantity of oxygen at all seasons in a fairly thick zone below the 55-m. level, which level marks the uppermost position of the thermocline during the warm season. The upper water of the lake is continually disturbed during the cool season, April to September, by the trade winds, and by sudden changeable storms during the warm season. In view of the immense distortion of the isotherms produced by the resulting temperature seiches, it is best to consider the variation of the oxygen content at levels defined by particular isotherms. During the earlier part of the warm period, from October to January, the oxygen content of the water at the 23° isotherm, which lies well below the developing thermocline is practically constant at 4.9 to 5.5 mg. per liter. Later, very considerable fluctuations appear at all levels. Beauchamp believes that initially a small steady renewal is taking place by turbulent diffusion from the epilimnion; later in the warm season much greater irregularity in the turbulence occurs. In general, there is a little oxygen still present at 250 m. and traces sometimes at 275 m. In the still deeper anaerobic water, H_2S appears to exist. There is a suggestion, as indicated earlier, that Lake Nyasa may have circulated completely in 1946.

Indonesian lakes. In Indonesia, Ruttner (1931) found that practically every lake exhibited a more or less clinograde oxygen curve. Danau di Atas, at 1531 m. in central Sumatra, and Danau Bratan, at 1231 m. in Bali, alone provide analogies with the freely circulating African lakes such as Naivasha.

The other lakes may be divided into three groups (Fig. 183). In the first are a number of small though often deep lakes not exceeding 2 $km.^2$ in area, in which there was virtually no oxygen below 10 m. Occasionally, secondary maxima are recorded in their oxygen curves, such as the transitory metalimnetic maximum below a slight minimum in Ranu Lamongan in east Java. A small but rather definite maximum of 0.77 to 0.87 mg. per liter at 30 to 35 m. was observed in the upper hypolimnion of Ranu Pakis in the same area; the minimum above this, 0.0 to 0.2 mg., lay at 20 m., well below the rather feeble thermocline observed during October. None of these divergences from an extreme type of clinograde curve are particularly impressive.

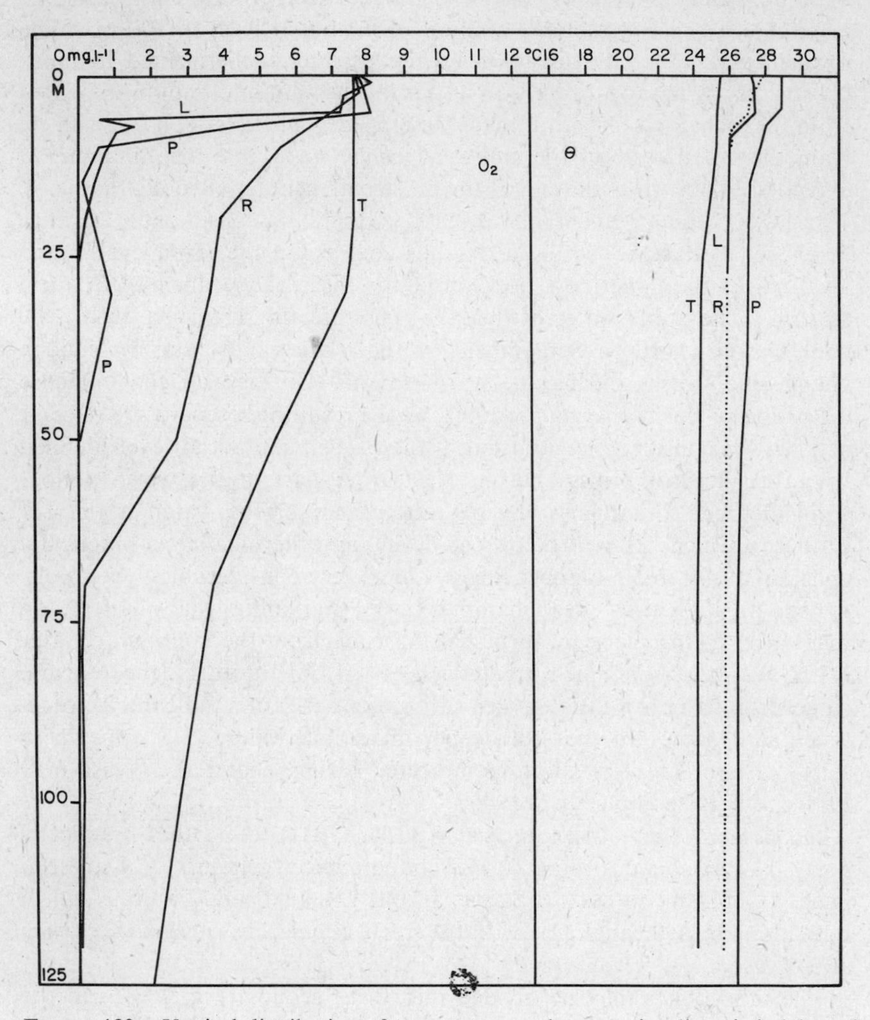

FIGURE 183. Vertical distribution of temperature and oxygen in oligomictic lakes of Indonesia. Group 1, the two least extreme examples: L, Ranu Lamongan, east Java, October 16, 1928; P, Ranu Pakis, east Java, October 24, 1928. Group 2, a typical example: R, Lake Ranau, south Sumatra, January 25, 1929. Group 3: T, south basin, Lake Toba, south Sumatra, April 1, 1929. (After Ruttner.)

In a series of large lakes, the oxygen concentration falls less rapidly, and in a more or less linear manner with depth, below the thermocline. Lake Ranau in south Sumatra provides an interesting example, because although it is a much larger lake than Ranu Lamongan, the areas being respectively 125.9 km.^{2} and 0.3396 km.^{2}, the temperature curves in the upper waters were virtually identical on certain of the days on which Ruttner studied the lakes. The considerable contrast in oxygen distribution, which is presumably due to the greater possibility and amplitude of internal seiches and the consequent greater hypolimnetic turbulence in the larger lake, is most striking.

The third type of distribution is found only in the two large basins of Lake Toba. Here, below the very feebly developed metalimnion there is an immense region of fairly low and relatively uniform oxygen concentration. In the shallower north basin, the oxygen falls from 5.35 to 4.07 mg. per liter over a depth range of from 50 to 150 m.; in the deeper southern basin, the fall is from 1.51 to 1.10 mg. per liter over a depth range of from 150 to 425 m. It is conceivable that some slow movement due to the instability caused by the heat flux from the interior of the earth plays a part in mixing the deep water of this lake.

The vertical distribution of oxygen in meromictic lakes. In meromictic lakes the stagnant monimolimnion will usually become deoxygenated even though the productivity of the lake is quite small, the long time available compensating for the very slow rate at which organic matter falls into the deepest parts of the basin. It was in fact primarily the extremely clinograde nature of the oxygen curves which led Findenegg (1932, 1933, 1934b, 1935, 1936b, 1937) to his classical study of the meromictic lakes of Carinthia and to the definition of meromixis as a widespread limnological phenomenon. In the Carinthian lakes there is by no means always a temperature inversion, but very considerable oxygen deficits and the concomitant rises in CO_2, bicarbonate, silicate, and other substances towards the bottom are quite out of harmony with the general great depth, low productivity, high transparency, and blue-green color of these lakes. The disharmony is easily resolved on the basis of seasonal studies which indicate that in such lakes full circulation is never achieved. The very striking clinograde oxygen curves were termed by Findenegg pseudo-eutrophic in his earlier work, but the term is best abandoned as it implies a usage that is unsatisfactory.

The oxygen distribution in such lakes is, in general, what would be expected (Fig. 184). The very deep water of the monimolimnion is practically or entirely devoid of the gas, but at less great depths a considerable variation in oxygen concentration may occur, dependent on the degree of circulation achieved during the cooler parts of the year. In particular, if no ice cover is developed and partial circulation can occur during the winter, a considerable ventilation of the deeper parts of the lake becomes possible. This is well shown in the Wörthersee (Findenegg 1937), in the mild winters of 1934–1935 and 1935–1936 compared with the hard winter of 1933–1934. It is, according to Findenegg, characteristic of deep meromictic lakes that when some mixing occurs, the time of the maximum oxygen content is later the greater the depth of the top of the monimolimnion. Within the latter, it seems likely that the maximum is successively later at greater depths.[25] It is remarkable in the Wörthersee, which does not circulate fully below 40 m., that the change observed in

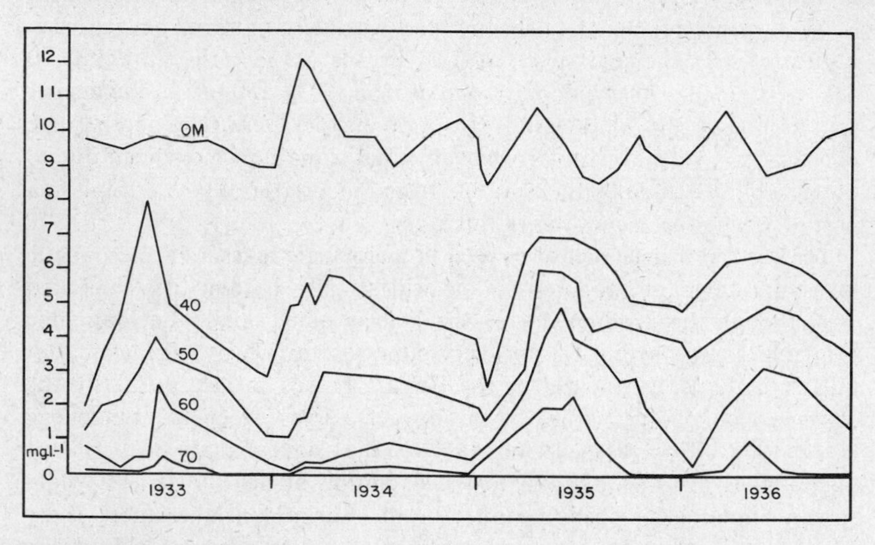

FIGURE 184. Seasonal variation of oxygen at the surface and in the monimolimnion of the Wörthersee, Austria, during four successive years. The winter 1933–1934 was the coldest, and was succeeded by a much milder winter in 1934–1935. Note the much greater oxygenation in the depths during spring and summer after the mild winter, and the tendency for the maximum oxygen content to occur later with increasing depths. (After Findenegg.)

1935 occurred with no observed change in the temperature of 4.2° C. at 60 m. between March 24 and September 14. In previous years and until January 1935 these layers lay at 4.5°, so that some mixing must have occurred to 60 m. A very similar phenomenon has been observed in the Millstättersee.

In crenogenic and ectogenic meromictic lakes in which the monimolimnion receives or has received a considerable amount of saline water containing sulfates, the disappearance of the oxygen permits reduction of

[25] This seems to be the case at least in the Wörthersee for 1935; compare the theory of heat conduction in sediments, page 505.

the sulfate and a consequent appearance of large quantities of hydrogen sulfide, as in Lake Ritom, the Hemmelsdorfersee, and Big Soda Lake. Smaller amounts of H_2S of biological origin may occur in the monimolimnion of biogenic meromictic lakes (page 774).

THE QUANTITATIVE SIGNIFICANCE OF THE OXYGEN DEFICIT

Actual, absolute and relative, apparent, and real deficits. In the earlier work on oxygen uptake, such as Thienemann (1928), the oxygen deficit was ordinarily understood to mean the difference between the concentration actually observed and the saturation concentration at the temperature of the layer from which the sample was obtained. A deficit of this kind was called by Alsterberg (1929) an *actual deficit*. In order that such a quantity should be significant, the saturation value employed should of course refer to wet air at the pressure at the surface of the lake, but this condition has not always been satisfied.

Alsterberg (1927, 1929), finding an appreciable deficit which he believed to be more or less profile bound in the Odensee, a small thermally stratified lake which he supposed to be too protected from the wind to permit any turbulent exchange, concluded that in general the hypolimnion is more isolated than had been supposed. He therefore believed that the deficit should properly be referred to the concentration present at the time of vernal circulation, which he called the *primary constant* and took to be saturation at 4° C. at the pressure of the surface of the lake. Such a deficit he termed the *absolute deficit*.

Strøm (1931) pointed out that in many lakes the water may not be exactly saturated at the circulation period, and that an empirical determination of the primary constant was always preferable. The deficit obtained by subtracting the later from the earlier value is best called the *relative deficit*.[26]

The relative deficit, considered as the change taking place in the oxygen concentration between vernal circulation and the height of summer stagnation or, more generally, between any designated dates, is a quantity of considerable value. As a measure of metabolic oxygen uptake, in default of other information the relative deficit is preferable to the actual for the following reason. The actual deficit always gives too low an estimate of oxidative metabolism, because by turbulent exchange, the existence of which is indicated by thermal data, some water rich in oxygen from higher layers is exchanged for water poor in oxygen in the lower layers.

---

[26] This is essentially Elster's (1955) usage; he calls this deficit *aktuelle relative*, but only the second word is needed. As so defined, the absolute deficits of Strøm (1931) and Hutchinson (1938) now are to be called relative.

If the surface waters were always saturated, the actual deficit would give an accurate measure of oxygen uptake in the water sample considered; but the sample is inevitably a composite, and the time of saturation of its different constituents is unknown and certainly varies with depth.

When the relative deficit is small, it is presumably always too great as an estimate of oxygen uptake, because it implies that all the water was saturated at the circulation period, whereas it is certain that the water was, as a whole, saturated at a higher mean temperature. In the case of small deficits, the difference in saturation between the circulation temperature and the observed temperature will be greater than the gain in oxygen due to turbulence. When the deficit is large, this will no longer be the case, and in lakes in which extreme clinograde oxygen curves develop, the relative deficit will always be too small, even though it is somewhat larger than the actual. In lakes exhibiting moderate hypolimnetic deficits, the early overestimate and later underestimate may balance each other, and it will subsequently be shown that in the extreme case of Lake Mendota the relative deficit is a good approximation to an acceptable estimate of hypolimnetic oxidative metabolism. In the case of less well studied lakes the relative deficit, suitably expressed, is indeed often the most illuminating quantity that can be derived from the data of chemical limnology.

As a slight additional refinement, Hutchinson (1938) has suggested that where oxidizable substances such as methane are produced anaerobically in the mud or deep water, the observed oxygen deficit should be termed the *apparent deficit*, and that the oxygen equivalent of such substances should be added to give the *real deficit*.

Intensity and capacity factors; the areal deficit. Whenever the oxygen concentration falls with depth in the hypolimnion, the form of the curve as determined by inspection is an unsatisfactory criterion of the oxidative events in the lakes, because it is so dependent on the scale of the depth axis. Green Lake, Wisconsin, exhibited the most nearly orthograde curve in the whole series studied by Birge and Juday (1911). Plotted on the same depth scale as was used for Hornindalsvatn and Tazawako in Fig. 162, the Green Lake curve would appear strongly clinograde, with a marked tendency to a negative heterograde form. Some better criterion than the shape of the oxygen curve is clearly needed.

In searching for such a criterion, it should be remembered that the oxygen concentration at any point is clearly the result of intensity factors, namely, the initial concentration of oxygen and the rate at which oxidizable substances are produced in the water above that point. A capacity factor is also involved, namely, the vertical element of volume in which the point lies and in which falling oxidizable material can be distributed, either directly or as a deficit derived from oxidation in the mud. Thienemann (1926) points this out very clearly by means of the diagrams reproduced, with a slightly modified legend, in Fig. 185.

Consider two lakes with identical epilimnetic trophogenic zones but with hypolimnetic tropholytic zones of very different volume. If the rate of organic production is taken to be identical in the two lakes, the same quantity of dead cells, faeces, and other debris will fall into both

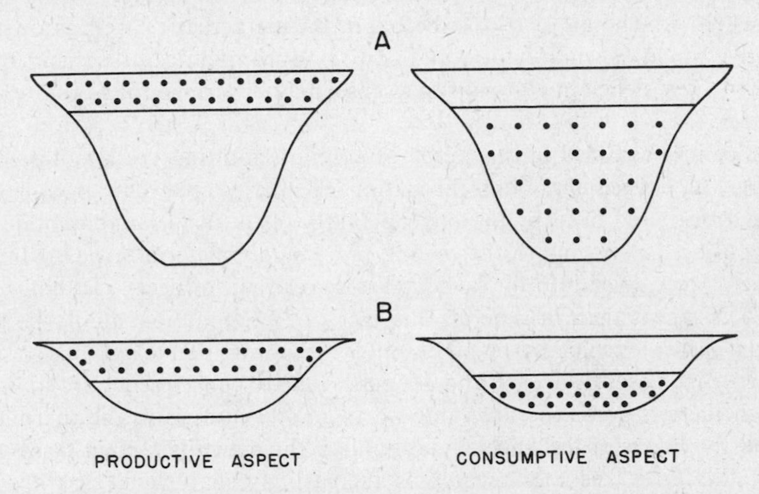

FIGURE 185. Productive aspect: diagrams showing equal production in the trophogenic layers of two lakes, *A* and *B*, of different depths. Consumptive aspect: diagrams showing the smaller concentration of material that has fallen into the large tropholytic zone of the deeper lake. (After Thienemann.)

hypolimnia in a given time. In the first lake (A), a very small uptake of oxygen per unit volume will be observed, permitting perhaps the existence of an almost unmodified orthograde oxygen curve; in the second lake (B), a very large uptake per unit volume may produce a markedly clinograde curve. Even if most of the uptake occurred when the oxidizable particles had come to rest on the bottom, the deoxygenated water produced at the interface would be distributed in a large hypolimnetic volume in lake A, in a small volume in lake B, by turbulence and horizontal water movements. The bottom waters of the deep lake would always be better oxygenated than those of the shallow lake. The very marked biological differences in the benthonic fauna found when lakes with more or less orthograde curves are compared with those with clinograde curves would, in this case, depend solely on the capacity factor provided by the volume of the hypolimnion. In two lakes of identical morphology, the same differences might be produced by variation in the nutrient supply in the epilimnion, and therefore in the rate of production of organic matter that can fall into and be oxidized in the hypolimnion. A method of calculation that permits a distinction to be made between these two cases is clearly required.

This can be done by the mode of computation introduced by Strøm (1931), who obtained from the individual absolute deficits at a series of depths the mean deficit below 1 cm.^{2} of hypolimnion surface. This type of deficit, which can of course be based on the actual as well as on the relative mode of computation and can in appropriate cases be corrected to give a real deficit in Hutchinson's sense, may be termed the *hypolimnetic areal deficit*.

Since it is assumed in this mode of computation that the hypolimnetic uptake of oxygen represents liberation of energy received through the lake surface and fixed by the phytoplankton, it is an unsuitable mode of computing deficits due to the oxidation of humic material brought in by effluents or carried into the lake by littoral erosion of peat. Hutchinson (1938) suggests that, in general, this mode of computation should be not used if the water color is over 10 Pt units. As Strøm points out, this source of error increases as deeper and deeper lakes are considered. In lakes in which there is a considerable amount of photosynthesis in the thermally stabilized layers of the upper hypolimnion, the deficit is certain to be too low; this error becomes greater as increasingly shallow lakes are considered (cf. Deevey 1940). For this reason it is advisable to restrict use of the method to lakes having maximum depths of from, say, 20 m. to 75 m. It is of course essential to exclude lakes that are not stably stratified, and meromictic lakes[27] must also be avoided. In spite of these restrictions very interesting results can be obtained.

Empirical study of the hypolimnetic deficit. Table 77 presents results from the lakes considered by Hutchinson (1938) on the basis of the data of Birge and Juday (1911), Juday and Birge (1932), Strøm (1932a, 1932b, 1933a), and Thienemann (1928). The deeper and shallower lakes are considered separately. The values are given as mg. cm.$^{-2}$ day^{-1}; the estimates of times of stagnation, which are often approximations, were derived from the original accounts by Hutchinson. The maximum depth is given after the name of each lake.

The data presented in Table 77 show very definitely that, provided lakes of moderate depth containing uncolored water in temperate regions are considered, the hypolimnetic areal relative deficit is characteristic of the region in which the lakes lie. In the least productive lakes considered, those of Norway, the deficit is minimal; in the most productive lakes,

---

[27] Hutchinson (1938) should not have included Garvin Lake, Wisconsin.

TABLE 77. *Areal relative oxygen deficits in two groups of lakes*

Lakes 50 to 75 m. Deep			Lakes 20 to 50 m. Deep		
	Depth, m.	Deficit, mg. cm.$^{-2}$ day^{-1}		Depth, m.	Deficit, mg. cm.$^{-2}$ day^{-1}
Norway and Sweden					
Feforvatn	53.5	0.0043	Steinsfjord	22	0.029
Lilla Le	53.5	0.013			
Germany (Eifel)					
Pulvermaar	74	0.014	Gemündener Maar	38	0.012
Weinfelder Maar	51	0.060	Schalkenmehrener Maar	21	0.025
			Holzmaar	21	0.036
Denmark and Baltic Germany					
Breiter Lucin	58.2	0.063	Furesø	40	0.080
Grosser Plöner See	60.5	0.070	Tollensesee	33.8	0.073
			Schöhsee	30.2	0.047
Wisconsin (soft waters)	...	...	Black Oak Lake*	26	0.043
Wisconsin (hard waters)					
Green Lake	72.2	0.14	Geneva Lake	43.3	0.090
			Okauchee Lake	28.6	0.097
			Mendota Lake	25.6	0.109

* Color, 14 Pt units. Two other shallower lakes in soft-water parts of Wisconsin for which areal absolute hypolimnetic deficits have been computed give lower results but are too shallow to be admitted to the table.

those of the Baltic region and southeastern Wisconsin, much larger values are observed. The Eifel district and the soft-water areas of Wisconsin fall between the unproductive and productive groups. To use the terminology that is so widely and often so inaccurately employed in discussing productivity, oligotrophic lakes were regarded by Hutchinson as losing hypolimnetic oxygen at the rate of 0.004 to 0.033 mg. $cm.^{-2}$ day^{-1}, eutrophic lakes at about 0.05 to 0.14 mg. $cm.^{-2}$ day^{-1}, and the lakes exhibiting values between these ranges would reasonably be termed mesotrophic. As indicating regional differences, Hutchinson set limits of 0.017 mg. $cm.^{-2}$ day^{-1} for oligotrophy, and 0.033 mg. $cm.^{-2}$ day^{-1} for eutrophy. Mortimer suggested rather different limits: 0.025 mg. $cm.^{-2}$ day^{-1} as the upper limit for oligotrophy, and 0.055 mg. $cm.^{-2}$ day^{-1} as the lower limit for eutrophy. The decision is of course purely arbitrary, but Mortimer's usage is perhaps the more convenient.

It is evident from Table 77 that within the depth range considered, there is no clear tendency for either deep or shallow lakes to exhibit greater deficits. This is entirely in line with the theoretical concepts expressed in Thienemann's diagrams (Fig. 185). In very deep lakes the error due to oxidation of even minute amounts of allochthonous organic matter probably would vitiate the use of areal deficits, as has already been indicated.

Comparison with a more rigorous treatment for Lake Mendota. In the case of Lake Mendota, for which we have a value for the mean coefficient of turbulence between 9 m. and 14 m. during the period mid-June to mid-August, it is possible to make a comparison of the relative areal deficit as computed for Table 77 with a somewhat more rigorously derived estimate of oxygen uptake. The mean rate of uptake below 9 m. during the period under consideration is from Birge and Juday's (1911) data for the years 1906 and 1907, 0.057 mg. O_2 $cm.^{-2}$ day^{-1}. The mean oxygen gradient across the 9-m. plane is -1.07×10^{-5} mg. $cm.^{-3}$ $cm.^{-1}$ The coefficient of turbulent diffusivity, derived from the observed over-all coefficient of heat conductivity by subtraction of the molecular coefficient, is 2.35×10^{-2} $cm.^{2}$ $sec.^{-1}$ The coefficient of molecular diffusivity of oxygen, 1.98×10^{-5} $cm.^{2}$ $sec.^{-1}$, is obviously negligible. The rate of passage of oxygen across the 9-m. plane is therefore 2.52×10^{-7} mg. $cm.^{-2}$ $sec.^{-1}$ or 0.0217 mg. cm^{-2} day^{-1}. If the oxygen of the part of the lake below 9 m. had remained constant during the observed period, we should be justified in assuming this rate of oxidative metabolism below 9 m. Since the oxygen content of this region actually fell at the rate of 0.057 mg. $cm.^{-2}$ day^{-1}, the total rate of uptake is 0.079 mg. $cm.^{-2}$ day^{-1}.

If we may assume that this represents the respiratory rate of the water throughout the lake, at least up to the 5-m. plane, the equivalent value for that plane, which cannot be treated directly since the coefficient of turbulence is not known, would be given by

$$0.079 \times \frac{A_9}{A_5} \times \frac{V_{5-25.6}}{V_{9-25.6}} = 0.079 \times \frac{25.0 \times 10^{10}}{29.0 \times 10^{10}} \times \frac{310 \times 10^{12}}{203 \times 10^{12}} = 0.103 \text{ mg. cm.}^{-2} \text{ day}^{-1}$$

This value is curiously close to the value given for the relative areal deficit, namely 0.109 mg. cm.^{2} day^{-1}. It is evident that the rather crude assumptions as to the ways the errors in the relative deficit cancel out are justified, at least in the present case.

If we assume that the over-all rate of oxidative metabolism in Lake Mendota is about the same at all depths up to the surface, as it appears to be from the black bottle experiments in Linsley Pond, we may compute the total rate for the lake as

$$0.079 \times \frac{A_9}{A_0} \times \frac{V_{0-25.6}}{V_{9-25.6}} = 0.118 \text{ mg. cm.}^{-2} \text{ day}^{-1}$$

This seems to be the best available estimate of the over-all metabolic rate of this lake.

Comparison of areal deficits with biological data. In view of the general occurrence of low deficits in regions of low productivity, it is reasonable to inquire into the details of the correlation between areal deficit and biological measures of productivity. Hutchinson (1938) pointed out that, considering only the four lakes of sufficient depth and of low color for which data exist—Green Lake, Lake Mendota, and Black Oak Lake in Wisconsin, and Furesø in Denmark—there is a very striking proportionality between the hypolimnetic areal relative deficit and the mean organic seston per unit area (Table 78).

TABLE 78. *Comparison of organic seston and areal relative deficit*

	Organic Seston, mg. cm.$^{-2}$	Areal Relative Deficit,* mg. cm.$^{-2}$
Green Lake	2.77	13.8
Lake Mendota	2.40	10.9
Furesø	1.56	8.0
Black Oak Lake	0.93	4.3

* After about 100 days' stagnation.

It may be objected that this comparison is arbitrary in that it is not to the productivity but to a static mass of organic matter which is distributed through both trophogenic and tropholytic zones and consists of both living photosynthetic cells and dead detritus. Nevertheless, the correlation for these four lakes is striking.

Rawson (1942), using actual deficits and various measures of productivity of even less direct significance, has come to similar conclusions. Deevey (1940), however, has found that in the shallow lakes of Connecticut the proportionality breaks down, the deficits all being smaller than would be expected.

It is possible to estimate from the Mendota and Furesø data the rate of replacement of seston implied by these figures. Taking the known caloric values of the seston of these two lakes and assuming that 1 mg. O_2 corresponds to the liberation of 3.5 cal. in oxidation of ordinary organic matter, we find that the ratio of the areal deficit per day to the seston is 0.030 for Lake Mendota, and 0.033 for Furesø. These figures would correspond to a complete replacement of the seston every month. Initially (Hutchinson 1938) this rate seemed much too low, but in view of the probability (Vol. II) that seston collected by membrane filtration or centrifugation is very largely detritus and may perhaps not contain more than 20 per cent living cells, such a value may well correspond to a replacement rate of once every six days, which seems a reasonable average figure.

Comparison of oxygen deficits and other estimates of productivity. In Linsley Pond, for which there are reasonably good figures for the mean production and consumption of oxygen in experimental bottles, the rate of production of the deficit, as calculated by Deevey (1940) below a plane equal to 1.2 times the mean summer Secchi disk transparency, is much less than the experimental values.

	Values, mg. cm.$^{-2}$ day^{-1}
Relative areal deficit	0.042
Production as O_2	0.105
Consumption of O_2	0.139

Even if we assume uniform consumption in the lake and correct the deficit accordingly, as was done in the computation for Lake Mendota, we obtain a value of only 0.050 mg. cm.$^{-2}$ day^{-1}, or about half that expected. The explanation is almost certainly to be found in the much greater proportion of anaerobic CO_2 produced in the small hypolimnion of Linsley Pond than in the hypolimnia of the larger deeper lakes set out in Table 78. Hutchinson in fact concluded that about half the CO_2 produced in the hypolimnion of Linsley Pond was the result of anaerobic metabolism, though the methane present even late in stagnation was practically negligible.

In striking contrast to this situation, Mortimer (verbal communication) has shown that the deficit development in Windermere is much greater than can be explained by the known rate of production of *Asterionella* and other algae in the trophogenic layers of the lake. He supposes that dead leaves, blowing or washing into the lake, and other organic matter in the influents play an important part in the deoxygenation of the deep water (cf. also Elster 1955).

The examples just given, the only ones at present permitting an objective estimate of the areal deficit as a measure of productivity, indicate how unreliable such an estimate may be. Nevertheless, the regional consistency of the areal deficits does indicate that they are not valueless, and in default of other information they may be quite valuable in the biological characterization of lakes.

Relation of the deficit to events at the mud surface. Grote (1934) assumed that the oxygen gradient present in the mud can be roughly obtained by dividing the concentration at the surface $[O_2]_b$ by the thickness $\overset{*}{z}_O$ of the layer in which oxygen is present. Under steady-state conditions with a constant value of $[O_2]_b$, the amount O of oxygen passing into unit area of the mud in unit time will be given by

$$(28) \qquad O = \frac{\hat{k}[O_2]_b}{\overset{*}{z}_O}$$

or

$$\overset{*}{z}_O = \frac{\hat{k}[O_2]_b}{O}$$

Mortimer points out that if $\overset{*}{z}_O$ be taken as the winter thickness of the oxidized zone of the mud, measured as the thickness of the layer of redox potential (corrected to pH 7) in excess of 0.2 volt, and if the winter rate of uptake be taken as proportional to the summer rate of uptake, we may expect from the relationship

$$(29) \qquad \log \overset{*}{z}_O = \log \hat{k}[O_2]_b - \log O$$

a linear relationship of negative slope between $\overset{*}{z}_O$ and the summer rate of generation of the deficit. This assumes that as far as the summer deficit is dependent on turbulence, the turbulence is essentially invariant from lake to lake.

Mortimer measures the summer rate of uptake as the rate of development of the actual deficit in a column of water in the deepest part of the lake. This procedure gives a value rather greater than the actual deficit computed for the whole hypolimnion, allowance being made for the volumes of the different layers. It should therefore correspond fairly closely to the relative deficits set forth in Table 77. Taking such summer deficits, it is remarkable, in view of the difficulties about variation in turbulence, how closely his values for the lakes of the English Lake District follow a straight line when plotted on double logarithmic paper against the winter values of $\mathring{z}_O$.

Mortimer continues with another semi-empirical generalization of even greater interest. The reductive power of the mud may be measured in terms of the minimum redox potential of any mud layer. Ordinarily this minimum is at about 5 cm. below the mud surface. This quantity will give, however, only a measure of intensity. The observed rate of uptake will depend not only on the reductive capacity of the mud but on the quantity of the reducing substance present. A reduced mud mixed with an equal quantity of quartz sand will have only half the oxygen-absorbing capacity of unmixed mud, though the redox potentials in such muds may be expected to be the same. If the intensity and capacity are unrelated, the rate of uptake might be expected to show a linear correlation with redox potential. But if capacity and intensity factors are interrelated, the latter would vary with the square root of the rate of oxygen uptake. This latter situation was in fact found by Mortimer among those of the English lakes for which information exists.

A final empirical relationship to the conductivity in the mud was observed by Mortimer (1941–1942). If the distribution of conductivity during winter stagnation is plotted against depth $\mathring{z}$ in the mud, the curve is, over most of its course, approximately exponential. Considering this exponential part, from about 4 cm. below to about 12 cm. below the mud surface, it follows from the form of the curve that

$$\log \frac{dC}{d\mathring{z}} \propto - \mathring{z}$$

Further it is found that

$$\left(\frac{d}{d\mathring{z}} \log \frac{dC}{d\mathring{z}}\right)^2$$

varies as the rate of uptake of oxygen as calculated above. Mortimer concludes that if we may suppose that the rate of liberation of ions into the liquid phase of the mud is proportional to the rate r_O at which oxygen is used in decomposition, we may conclude from the above observations that

$$C_{\mathring{z}} = C_L + (C_0 - C_L)e^{-\mathring{z}\sqrt{r_O/c}} \tag{30}$$

where C_L is a limiting value of the conductivity, C_0 the conductivity at the mud surface, and c a constant that involves the diffusivity of ions in the mud. This equation is of the form that expresses the steady-state temperature in a solid rod heated at one end. Although put forward very tentatively by Mortimer, his data at least suggest that a good theoretical understanding of the uptake of oxygen by the lake bottom is by no means impossible.

SUMMARY

The oxygen concentration of a lake at circulation periods, when it has time to become entirely in equilibrium with the atmosphere, will depend on the atmospheric pressure and therefore on the altitude. Where the mean atmospheric pressure at an elevated locality is not known, an approximate value can be obtained by one of the procedures described in the text. The solubility of oxygen from wet air has been the subject of much research; the most recent values have been used as far as possible in preparing the present chapter. The saturation value falls in a nonlinear manner with increasing temperature.

The process of equilibration, when the lake is circulating, depends on the rate of entry of oxygen across the lake surface. This is proportional to the difference in partial pressure in the air and in a bubble at equilibrium with water at a given moment, and to a coefficient the magnitude of which can be guessed from laboratory and oceanographic experience. It is certain that a deep lake, if its oxygen concentration departs significantly from saturation, must circulate for some weeks to come to equilibrium at all depths. This is not likely to happen in dimictic lakes in regions of continental climate. Ordinarily, saturation is considered relative to the pressure at the lake surface. At any depth, however, the quantity of a gas that can remain in solution is determined by the pressure not only of the atmosphere but of the layer of water above that depth; this defines a quantity called *absolute saturation*. If oxygen is produced well below the surface, say in excess of 4 m., the extra pressure of the water causes it to dissolve rather than to form bubbles, and supersaturation of several hundred per cent relative to the lake surface pressure may thus be occasionally recorded, though the absolute saturation is not exceeded. There is observational evidence and theoretical justification for supposing methane bubbles to form in anaerobic sediments at depths of 10 to 20 m., the absolute saturation of this gas then being exceeded.

Surface waters in very unproductive lakes are always more or less saturated. They may appear slightly supersaturated if determinations are made when the water is at its maximal diurnal temperature. In regions of high bioligical activity, supersaturation due to photosynthesis, or subsaturation due to respiration and oxidation of organic matter, are frequent. In summer in northeastern Wisconsin, the mean value, if due correction be made for altitude and for the most recent determinations of solubility, is 93 per cent saturation; the mode lies at 97 to 98 per cent saturation, the range is from 38 to 146 per cent saturation, and at least 36.4 per cent of the surface waters are supersaturated. Other less extensively studied lake districts give comparable results. Highly colored waters tend in general to be undersaturated. Dull, windless weather following the production of an algal bloom may cause catastrophic deoxygenation in shallow lakes.

The distribution of oxygen in summer in a stratified lake has been more studied than any other aspect of chemical limnology. The mere formation of an epilimnion will cause some oxygen to leave the lake as the temperature rises; every level may be saturated with respect to its temperature and the pressure at the surface, but the absolute quantity of the gas per unit volume of water increases with depth. This extreme type of distribution, found only in the deepest and most unproductive lakes, is termed *orthograde*. In contrast to this, the majority of small productive lakes during summer stratification lose oxygen from the hypolimnion, the amount lost, or oxygen deficit, increasing in most cases with depth. The resulting distribution is termed *clinograde*. It seems probable that in most lakes the clinograde curve indicates uptake of oxygen both in the free water and at the water-mud interface, the relative importance of these two processes varying from lake to lake. No case seems to have been described in which vertical turbulent movements, coupled with rapid uptake by the mud, are sufficient to explain the observed distribution. If the mud is involved, it is presumably on account of horizontal movement of the water. In several cases there is evidence that the rate of oxygen uptake, mainly by bacterial respiration, in water isolated in a bottle is sufficient to account for the deficit without the sediments being involved. In such cases, however, it is doubtful if the water would continue to lose oxygen at the rate observed in the lake without the supply of organic matter that the water *in situ* would receive from dying and sinking plankton or from diffusion out of the mud.

In some lakes, very striking maxima in oxygen concentration develop in the metalimnion during stratification. Such lakes appear in general to be transparent enough to permit considerable photosynthesis in stable layers well below the epilimnion. Oxygen would not be lost from such layers by strong turbulent exchange, and supersaturations relative to the surface pressure up to 380 per cent have been recorded. This type of distribution is termed *positive heterograde*.

Many cases are also known of marked minima developing in the metalimnion. There are two probable explanations of such *negative heterograde* distributions. One is that if any organic particle falls from the turbulent epilimnion into more or less stable metalimnetic layers, it will inevitably sink further, the turbulence now not being great enough ever to return the particle to the epilimnion. If it is in part oxidizable, the most easily oxidized part will decompose in the metalimnion at a rather higher temperature than later in its descent. In the hypolimnion, only rather resistant material will be available to bacteria in very cold water. The second possible explanation is that the minimum corresponds to a shelf in the bottom contour of the lake, and that at the depth of the shelf horizontal currents distribute water that has lost oxygen to the sediments on the shelf. Both theories are probably true, applying in greater or less degree to different lakes.

When a dimictic lake that in summer exhibited a marked clinograde curve freezes, oxygen usually tends to be lost from the deep water. The typical winter oxygen curve in such a lake is convex downward and of a form suggesting that it might be due to vertical turbulence. It seems more likely that density currents of the kind certainly responsible for winter heating in the deep water are involved in producing the winter oxygen curve. The water of such currents would tend to loose oxygen as it passed over the mud surface. In extreme cases the oxidation of ascending methane bubbles may be involved. In shallow lakes that are frozen and are carrying a load of snow, loss of oxygen may be extremely serious, causing the death of much of the fauna.

A few special aspects of the oxygen distribution are of some interest. At great altitudes the low value of the oxygen in solution at equilibrium may result in a very small concentration of oxygen at the bottom. A much larger quantity would be present in a lake of identical properties at sea level. This can have a marked biological significance.

In some warm temperate regions of variable winter temperatures, if a cold winter is succeeded by a warm winter, the winter circulation of normally warm monomictic lakes is incomplete and a rather complicated clinograde oxygen curve may develop. In the oligomictic lakes of humid tropical lowlands, very pronounced clinograde curves are almost always found whenever the lake is studied. In meromictic lakes of temperate regions, the monimolimnion is usually more or less anaerobic, and the presence of a markedly clinograde distribution in a lake which otherwise would be expected to have a very low hypolimnetic deficit is one of the commonest indications of meromixis.

The *actual oxygen deficit* of the water at any point in a lake is defined as the difference between the saturation value at the temperature of the water and at the pressure of the lake surface, and the observed value. The *absolute deficit* is the difference between the saturation value at 4° C. at the pressure of the lake surface, and the observed value. The *relative deficit* between two dates is the difference between the earlier and later determination. The relative deficit, expressed per unit of the hypolimnion surface, between vernal circulation and the height of summer stratification, or better still the relative areal deficit acquired during this period per unit time, appears to give, in holomictic lakes between 20 and 75 m. deep, a fair indication of the biological productivity of the lake. The relationship breaks down in shallow lakes with much anaerobic metabolism in the deep water and sediments. At least within the English Lake District, the relative areal deficit in summer appears to be inversely proportional to the thickness of oxidized mud in winter. The minimum redox potential in the mud of these lakes appears to vary as the square root of the rate of summer oxygen uptake. There is evidence that in using up oxygen a lake mud liberates ions which increase the conductivity of the interstitial water in the mud.

CHAPTER 10

Carbon Dioxide and the Hydrogen-Ion Concentration of Lake Waters

In Chapter 3 the nature of the dissociation of liquid water was briefly discussed. We saw that most oxygen atoms in water are associated with a pair of hydrogen atoms set about 0.99 A. from their centers and another pair set more distantly at 1.77 A. Each hydrogen, being set asymetrically between two oxygens, forms a hydrogen bridge, the nearer hydrogen of one atom being one of the further hydrogens of another. This arrangement is conventionally symbolized by H_2O. In a small proportion of the oxygen atoms there are three hydrogens in the close position, producing H_3O^+; in a negligible fraction, four such close hydrogens produce H_4O^{++}. In neutral water there will obviously be equal numbers of OH^- and H_3O^+ and of H_4O^{++} and $O^=$. It is usual to regard the molecular arrangement H_3O^+, or the hydroxonium ion, as representing a hydrated hydrogen ion $H^+ \cdot H_2O$. In the present chapter the equilibrium between H_3O^+, OH^-, and H_2O is considered, but for typographic simplicity and in order to facilitate the understanding of earlier work, the term hydrogen ion and the symbol H^+ will be used. It is important to bear in mind that no hydrogen ions exist in solution and that the apparent migration of such ions is merely the formation and destruction of a special type of hydrogen bonding in the quasi-crystalline lattice of liquid water.

CARBON DIOXIDE

A few elements, notably the halogens, and a number of nonmetallic oxides and organic substances when dissolved in water, form compounds which when dissociated increase the concentration of hydrogen ions. Such compounds are called acids. In natural waters by far the most important of such oxides is CO_2, which when it dissolves produces a small amount of H_2CO_3 which dissociates. A study of the CO_2 content of lake waters is therefore of great importance in understanding the hydrogen-ion concentration of such waters, as well as in many other ways.

Solubility of carbon dioxide in pure water. Carbon dioxide is exceedingly soluble in pure water; over the ordinary temperature range the solubility is about 200 times that of oxygen. Table 79 gives the usually accepted

TABLE 79. *Solubility of CO_2 in pure water*

Temp., ° C.	0.03%	0.033%	0.044%	Temp., °C.	0.03%	0.033%	0.044%
0	1.00	1.10	1.47	16	0.57	0.62	0.84
1	0.96	1.06	1.41	17	0.55	0.60	0.81
2	0.93	1.02	1.36	18	0.54	0.59	0.79
3	0.89	0.99	1.30	19	0.52	0.58	0.76
4	0.86	0.94	1.26	20	0.51	0.56	0.74
5	0.83	0.91	1.22	21	0.49	0.54	0.72
6	0.80	0.88	1.17	22	0.48	0.52	0.70
7	0.78	0.86	1.14	23	0.46	0.51	0.68
8	0.75	0.82	1.10	24	0.45	0.50	0.66
9	0.72	0.79	1.06	25	0.44	0.48	0.64
10	0.70	0.76	1.02	26	0.42	0.46	0.62
11	0.67	0.74	0.98	27	0.41	0.45	0.60
12	0.65	0.72	0.95	28	0.40	0.44	0.59
13	0.63	0.69	0.92	29	0.39	0.43	0.57
14	0.61	0.67	0.89	30	0.38	0.42	0.56
15	0.59	0.65	0.87				

values, ultimately due to Bohr (1899) but essentially confirmed by many later investigators, for wet atmospheres containing 0.030, 0.033, and 0.044 per cent CO_2 per unit volume of the dry components at sea level. The lowest figure is that usually employed. Callendar (1940) has adduced evidence that there has been a steady rise in the CO_2 content of the air during the present century. This he attributes to industrial production of the gas, though there are difficulties in accepting this as the whole explanation of the phenomenon (Callendar 1949, Hutchinson 1954, Suess 1955). The best data for the decade 1925–1934 give a mean value of 0.032 per cent by volume. It is quite likely that the rate of increase has accelerated since 1930. Kreutz (1941) gives a mean of 0.044 per cent for Giessen in northern Germany for 1939–1941. This figure, which Ohle accepts as the best available, seems to imply, however, so great an acceleration that it is hardly probable as a world wide average over land. The quantity in solution has therefore been also computed for 0.033 per cent as an arbitrary, but possibly more acceptable, modern figure for comparison with current limnological data. For the greater part of the older published results of limnologists, a value of 0.030 per cent is doubtless reasonable.

Throughout the range of interest in the present work, the variation of solubility with pressure obeys Henry's law.

On solution, a small part of the carbon dioxide can undergo either of two reactions

(*a*) $$CO_2 + H_2O \rightleftharpoons H_2CO_3$$

(*b*) $$CO_2 + OH^- \rightleftharpoons HCO_3^-$$

The first reaction is practically the only significant one below *p*H 8; the second reaction, above *p*H 10.

The proportion of H_2CO_3 present is very small, and has been determined by different investigators under varying conditions as 0.105 to 0.56 per cent (Quinn and Jones 1936, Wroughton 1941, Berg and Patterson 1953).

The acid H_2CO_3 is strongly dissociated

(*c*) $$H_2CO_3 \rightleftharpoons H^+ + HCO_3^-$$

(*d*) $$HCO_3^- \rightleftharpoons H^+ + CO_3^=$$

It is obvious that in view of the small quantity of H_2CO_3 in solution and the difficulty of estimating the quantity exactly, the true first dissociation constant

(1) $$K_1 = \frac{[H^+][HCO_3^-]}{[H_2CO_3]}$$

will be difficult to determine. The best value available, that of Berg and Patterson (1953), is $1.32 \pm 0.05 \times 10^{-4}$ at 25° C. Roughton (1941) obtained at 15° C., 2.5×10^{-4}. Since it is to be expected that the value would rise with temperature, Roughton's value seems not to be consistent with that of Berg and Patterson. The apparent dissociation constant at great dilution

(2) $$K_1' = \frac{[H^+][HCO_3^-]}{[CO_{2\ \mathrm{total}}]}$$

where the concentration in the numerator is that of total analytically determinable CO_2, will obviously be several hundred times smaller than the true constant K_1. Harned and Davis (1943) give K_1' the following values

°C.	K_1'	pK_1'
0	2.65×10^{-7}	6.58
5	3.04×10^{-7}	6.52
10	3.43×10^{-7}	6.46
15	3.80×10^{-7}	6.42
20	4.15×10^{-7}	6.38
25	4.45×10^{-7}	6.35

Whenever the concentration is significant, activities should be considered, not concentrations, writing for equation 1

$$(3) \qquad K_a = \frac{a_{H^+} \cdot a_{HCO_3^-}}{a_{H_2CO_3}}$$

This is called the thermodynamic dissociation coefficient.

In practice, what is generally known is the hydrogen ion activity and the concentrations of the other substances. Moreover, it is not the concentration of H_2CO_3 but rather of $CO_2 + H_2CO_3$ that is analytically available. It is therefore ordinarily necessary to deal with an imperfect apparent coefficient

$$(4) \qquad K_{a_1}' = \frac{a_{H^+}[HCO_3^-]}{a_{CO_2\ total}}$$

The activity of the total CO_2 is taken as the analytical concentration in moles multiplied by the activity of water at the concentration in question. Imperfect apparent dissociation constants increase with salinity and are of great importance in oceanography. For relatively dilute solutions we may write, if the only important ions are univalent

$$(5) \qquad pK_{a_1}' = pK_1' - 0.50\sqrt[3]{c_m}$$

where c_m is the molarity of ions present. For fresh waters, in which the more important cations are divalent, Saunders (1926), following an empirical form due to Bjerrum and Gjaldback (1919) and used by Warburg (1922), writes

$$(6) \qquad pK_{a_1}' = 6.52 - 0.53\sqrt[3]{c_m}$$

This expression gives satisfactory results with many inland waters. The second dissociation constant at great dilution is given by

$$(7) \qquad K_2 = \frac{[H^+][CO_3^=]}{[HCO_3^-]}$$

An imperfect constant is defined as

$$(8) \qquad K_{a_2}' = \frac{a_{H^+}[CO_3^=]}{[HCO_3^-]}$$

The significance of this apparent imperfect constant is greater in oceanography than in limnology, but obviously might have considerable importance in the study of saline lakes. Buch (1933) gives for fresh water K_2 as 3.2×10^{-11} at 15° C and 4.1×10^{-11} at 25° C.

From the values of the apparent dissociation constants can be found the molecular proportions of total free CO_2, HCO_3^- and $CO_3^=$ present at varying *p*H values in a very dilute solution at 15° C. (Table 80).

TABLE 80. *Proportions of* CO_2, HCO_3^-, *and* $CO_3^=$ *in water at various pH values*

*p*H	Total Free CO_2	HCO_3^-	$CO_3^=$
4	0.996	0.004	1.25×10^{-9}
5	0.962	0.038	1.20×10^{-7}
6	0.725	0.275	0.91×10^{-5}
7	0.208	0.792	2.6×10^{-4}
8	0.025	0.972	3.2×10^{-3}
9	0.003	0.966	0.031
10	0.000(2)	0.757	0.243

Below *p*H 5, only total free CO_2 is of any quantitative importance; between *p*H 7 and *p*H 9, bicarbonate is of the greatest significance; above *p*H 9.5, carbonate begins to be of importance. Figure 186, from Buch (1930), gives a graphical presentation of the same phenomenon. In view of the low solubility product of $CaCO_3$, 0.48×10^{-8}, this substance may start precipitating from a calcareous water when the *p*H is raised sufficiently to permit the existence of appreciable $CO_3^=$, but in many cases metastable

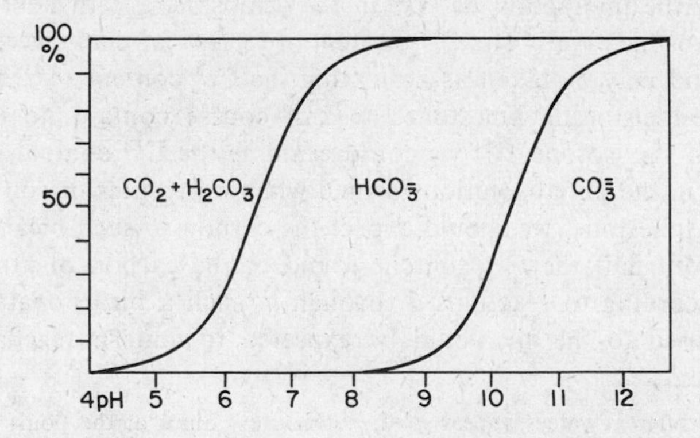

FIGURE 186. Proportions of CO_2 (and H_2CO_3), HCO_3^-, and $CO_3^=$ in solution at different *p*H values (Buch).

conditions persist for a long time (page 670). Above pH 9.0 in permanently alkaline waters the main cations appear always to be magnesium and sodium, often with appreciable potassium.

Since at pH 8.46[1] the concentrations of free CO_2 and of carbonate are identical, it is commonly said that carbonate appears in a water if the pH rises from below to above 8.4, or roughly from below to above the turning point of phenolphthalein. This is of course correct analytically; if all the CO_2 in a calcium carbonate-carbon dioxide solution at this pH is collected, it should be stoichiometrically equivalent to the CaO present. It is incorrect, however, to suppose that some $CO_3^{=}$ is not present at lower pH values and some free CO_2 at higher pH values. The free CO_2 will in fact be determined at equilibrium by the pressure in the atmosphere, though at high concentrations both its solubility and activity will be decreased.

Movement of carbon dioxide into and out from a water surface. The laboratory determinations of the invasion coefficient give values that depend greatly on the experimental conditions, the highest (Guyer and Tobler 1934) being 200 times the lowest (Becker 1924), and there are theoretical grounds (see page 587, footnote 6) for believing that no satisfactory values can be obtained for a minor very soluble component of the atmosphere. In default of some direct observations in nature, comparable to those discussed for oxygen in the previous chapter, it is obviously impossible to obtain any idea of the rapidity of the exchange.

The only evidence available is of an indirect kind, but clearly implies that a long time, of the order of days or weeks, may be needed to achieve equilibrium with the atmosphere across a normally disturbed film.

Radiocarbon content of inland waters and its significance. Atmospheric carbon and modern biological carbon contain a small quantity of the radioactive isotope C^{14}, which has a half life of 5600 years. This is formed by the interaction of N^{14} in the atmosphere with neutrons produced by cosmic rays. The C^{14} content of terrestrial plants is reasonably uniform and may be taken as indicating the C^{14} content of atmospheric CO_2. Pre-Pleistocene limestones will of course contain no detectable quantity of the isotope. If we could examine the C^{14} content of the bicarbonate in the initial solution formed when rain water or soil solutions react with limestone, we should expect the carbon of such bicarbonate to exhibit about half the C^{14} content found in the carbon of atmospheric CO_2. According to reactions *a* through *d*, such a bicarbonate solution when exposed to the air would be expected to undergo exchange with

[1] Slightly different values appear in the literature. Since at the point in question $[H^+] = \sqrt{K_1'K_2}$, the estimate of the pH will depend on the values of the dissociation constants used.

atmospheric CO_2 until an almost uniform C^{14} content resulted. Reactions *a* and *c* are not instantaneous, but proceed practically to equilibrium in a few minutes or hours under ordinary conditions. It therefore seems likely that any departures from equilibrium that may be observed in nature are to be attributed to the slowness of exchange across the water surface rather than to the time taken to establish equilibrium in solution.[2]

TABLE 81. *Radiocarbon content of carbon from various terrestrial sources and from soft-water and hard-water lakes*

Source of carbon	C^{14}, counts per min.
Modern wood, Connecticut	
Mean	5.86 ± 0.08
Range	5.78 – 6.08
Soft-water localities, emergent vegetation	
Chamaedaphne calyculata, Canaan, Conn.	6.33 ± 0.10
Chamaedaphne calyculata, Canaan, Conn.	5.99 ± 0.12
Vaccinium sp., Canaan, Conn.	6.20 ± 0.08
Soft-water localities, submerged vegetation	
Sphagnum spp., Bethany Bog, Conn.	6.03 ± 0.10
Mixed phanerogams, L. Quassapaug, Conn.	6.44 ± 0.10
Mixed phanerogams, L. Quassapaug, Conn.	5.93 ± 0.12
Mean, all soft-water plants	6.15 ± 0.08
Queechy Lake, New York	
Bicarbonate and carbonate in solution	4.60 ± 0.10
Deep-water sediments	4.68 ± 0.09
Deep-water sediments	4.97 ± 0.15
Potamogeton sp., organic matter	4.99 ± 0.11
Potamogeton sp., organic matter	5.18 ± 0.16
Potamogeton sp., calcareous incrustation	5.02 ± 0.11
Potamogeton sp., calcareous incrustation	5.09 ± 0.11
Chara sp., organic matter	5.07 ± 0.07
Chara, sp., calcareous incrustation	5.02 ± 0.10
Lake Zoar, Connecticut	
Hydrodictyon reticulatum	5.44 ± 0.11
Elodea sp.	5.35 ± 0.09

In Table 81, data are presented (Deevey, Gross, Hutchinson, and Kraybill 1954) for the C^{14} of various carbon compounds ultimately formed from carbon dioxide or bicarbonate, obtained from soft-water lakes and two hard-water lakes in eastern North America, as well as for emergent and swamp vegetation and modern wood from the same region. It will be

[2] Dr. H. C. Thomas, however, suggests verbally that the existence of inhibitors of the hydration of CO_2 or the dissociation of H_2CO_3 in natural waters cannot be excluded a priori.

observed that while the bog plants and submerged aquatic plants from the soft-water lakes give C^{14} concentrations comparable to modern wood, there is a decided deficiency in C^{14} in the plants of Lake Zoar and in all the products of Queechy Lake. Lake Zoar lies south of the Taconic marble (early Palaeozoic), but it receives its water from the marble area; Queechy Lake lies within the area of the Taconic marble. It is evident that in the course of drainage into these two lakes exchange equilibrium has not been achieved. Queechy Lake receives its water by seepage through the limestone, so that equilibrium would hardly be expected, but in the case of Lake Zoar the water must have flowed at least 24 km. after leaving the limestone area.

CALCIUM AND MAGNESIUM CARBONATES IN RELATION TO CARBON DIOXIDE

The total quantity of carbon dioxide that passes into a natural water is determined not merely by the pressure and the solubility of the gas at any given temperature, but also by the possibility of the formation of bicarbonates, mainly of the alkaline earth metals. The equilibrium involving calcium carbonate, bicarbonate, carbon dioxide, and water has been more carefully studied than that involving magnesium salts. In closed mineralized lakes, sodium and potassium bicarbonate and carbonate are of course important.

Forms of $CaCO_3$. It is important to remember that $CaCO_3$ exists under ordinary conditions in nature in two crystalline polymorphs, *calcite* and *aragonite*.[3] The latter is metastable at ordinary temperatures; it may be the stable form at − 100° C., but does not convert to calcite when heated in the absence of water until temperatures of about 400° C. are reached. In contact with water, particularly if calcium bicarbonate is present in solution, the transformation may occur at room temperature. In general, rapid crystallization from concentrated solutions at moderately high temperatures and the presence of $CaSO_4$ and certain trace elements, notably Sr, favor aragonite formation. Either form may be present in skeletal structures. The Foraminifera are calcitic, the madreporarian corals are aragonitic; most mollusks have calcite shells, but in fresh-water

[3] The deposition of hydrates, supposedly $CaCO_3 \cdot 3H_2O$, *trihydrocalcite*, and $CaCO_3 \cdot 5H_2O$, *pentahydrocalcite*, in nature has been recorded. The status of these minerals is regarded as uncertain by Palache, Berman, and Frondel (1951), though the occurrence of some natural hydrate seems very probable. Such a substance may be expected to form only at very low temperatures, near O° C. One occurrence of the supposed pentahydrocalcite was in a stream near Oslo. It is therefore conceivable that hydrated $CaCO_3$ is occasionally deposited in lakes, and may at times have been mistaken for ice.

groups aragonite is usual. A third metastable polymorph, *vaterite*, is probably formed transitorily when conchogenesis is occurring. There is evidence (e.g., Odum 1951b) that amorphous $CaCO_3$ occurs in organisms and, as will be apparent later, a colloidal $CaCO_3$ can be of considerable limnological importance.

Solubility of calcium carbonate. If calcite is suspended in pure water free of CO_2, a small quantity dissolves. It is not possible, however, to define a true solubility in the absence of CO_2, because the formation of CO_3 ions implies, according to the equilibria already discussed, a certain concentration of free CO_2 in solution. If this CO_2 is removed, the $CaCO_3$ progressively decomposes to form $Ca(OH)_2$, which is much more soluble than $CaCO_3$. An apparent solubility can be defined in the following way. If calcite suspended in pure water starts to dissolve, at any concentration a definite pressure of CO_2 will be developed. If this pressure is uncontrolled a point will be reached when equilibrium is established and no more $CaCO_3$ goes into solution. At equilibrium at 25° C.

$$(9) \qquad [Ca^{++}][CO_3^{=}] = L_{Ca^{++}\cdot CO_3^{=}} = 0.48 \times 10^{-8}$$

(Frear and Johnston 1929)

$$CO_3^{=} + H_2O \rightleftharpoons HCO_3^{-} + OH^{-}$$

$$(10) \qquad [HCO_3^{-}] = [OH^{-}]$$

$$(11) \qquad \frac{[CO_3^{=}][H^{+}]}{[HCO_3^{-}]} = K_2 = 4.1 \times 10^{-11}$$

$$(12) \qquad [OH^{-}] = \sqrt{\frac{[CO_3^{=}][H^{+}][OH^{-}][HCO_3^{-}]}{[H^{+}][CO_3^{=}]}} = \sqrt{[CO_3^{=}]\frac{K_w}{K_2}}$$

where $K_w = 10^{-14}$. Whence

$$(13) \qquad [Ca^{++}] = [CO_3^{=}] + \sqrt{\frac{[CO_3^{=}]K_w}{K_2}} = \frac{L_{Ca^{++}CO_3^{=}}}{[CO_3^{=}]}$$

$$(14) \qquad [CO_3^{=}]^2 + [CO_3^{=}]\sqrt{[CO_3^{=}]\frac{K_w}{K_2}} - L_{Ca^{++}CO_3^{=}} = 0$$

Kolthoff and Stenger (1942, p. 44) solve by a method of successive approximations, taking $L_{Ca^{++}CO_3^{=}} = 0.5 \times 10^{-8}$ and $K_2 = 5 \times 10^{-11}$, and obtain

$$(15) \qquad [Ca^{++}] = 1.27 \times 10^{-4}, \qquad [OH^{-}] = [HCO_3^{-}] = 8.83 \times 10^{-5}$$

This corresponds to 12.7 mg. $CaCO_3$ per liter, or 5.08 mg. Ca. Such a value is concordant with most of the experimental data assembled by Pia (1933) for the solubility of calcium carbonate in supposedly CO_2-free water; about 12 mg. per liter calcite or 14 mg. per liter aragonite appear to go into solution at 20° C. Johnston and Williamson (1916), in their classical study of the system $CaO-H_2O-CO_2$ found a minimum solubility at a CO_2 tension of about 3.8×10^{-7} atm. at 16° C. At this CO_2 pressure and temperature, 15.9 mg. per liter $CaCO_3$ as calcite were found to go into solution. It is probable that their results are a little too high; comparison with those of Kline (Frear and Johnston 1929) suggest that a value about 5 per cent lower would be reasonable. Such figures as are available for the minimal solubility of calcite therefore suggest that unless at least 5 mg. per liter Ca is present in a natural water, calcitic shells will tend to dissolve, whatever the CO_2 pressure may be. Aragonitic shells will be somewhat more unstable.

TABLE 82. *Solubility of* $CaCO_3$ *at 25° C. under different* CO_2 *pressures*

CO_2, vol. %	Solubility, mg. per kg.*		
(10^{-2} atm.)	$CaCO_3$	Ca	HCO_3^-
0.01	35	14	43
0.02	45	18	55
0.03	52	21	63
0.04	57	23	70
0.05	61	24	75
0.06	65	26	79
0.07	69	28	84
0.08	72	29	87
0.09	75	30	92
0.10	78	31	95
0.15	89	36	108
0.20	99	40	121
0.25	106	42	129
0.30	113	45	138

* To obtain results in millimoles Ca or $CaCO_3$, divide by 100; in millimoles HCO_3 (bicarbonate alkalinity), by 50.

TABLE 83. *Correction factor to convert Table* 82 *to solubilities at other temperatures,* °*C.*

0°	5°	10°	15°	20°	25°	30°
1.8	1.6	1.4	1.25	1.1	1.0	0.9

The solubility of $CaCO_3$ in pure water at 25° C. under various pressures of CO_2 is given in Table 82, mainly on the basis of the results of Kline (Frear and Johnston 1929). In the original publication the figures are given as millimoles per kilogram H_2O. As Pia (1933) points out in giving the same results, such a mode of presentation will not differ by more than 0.5, at any temperature, from the limnologically more convenient reference to unit volume of the solution. Values at other temperatures can be computed by multiplication by the factors given in Table 83, and in Table 84 values are set out for the atmospheric pressure limits most useful in limnology.

It should not be forgotten that at considerable altitudes the CO_2 pressure will be less than at sea level. Owing to the uncertainties at ordinary altitudes, it is hardly possible to indicate what may be expected in the higher parts of the Andes and Himalayas. Pressure of 0.00020 to 0.00025 atm. at 4500 or 5000 m. would seem reasonable.

In Fig. 187 the solubility curves for the systems $CaO\text{-}H_2O\text{-}CO_2$ and $MgO\text{-}H_2O\text{-}CO_2$ are set out. The initial effect of the addition of CO_2 to $Ca(OH)_2$ is to precipitate $CaCO_3$, the quantity of the base remaining in

TABLE 84. *Solubility of* $CaCO_3$ *at various temperatures for two important* CO_2 *pressures*

Temp., °C.	0.033%	0.044%	Temp., °C.	0.033%	0.044%
0	96	106	15	67	74
1	95	104	16	65	71
2	93	101	17	63	70
3	90	99	18	62	68
4	88	96	19	60	67
5	86	94	20	59	65
6	84	92	21	57	63
7	82	89	22	56	62
8	79	87	23	55	61
9	77	85	24	54	60
10	75	83	25	54	59
11	73	81	26	52	58
12	71	79	27	51	57
13	69	77	28	50	56
14	68	76	29	49	54

solution at equilibrium declining rapidly until a pressure of 10^{-7} to 10^{-6} atm. of CO_2 is reached, after which the solubility slowly increases to a very high value in the presence of several atmospheres of CO_2. At the extreme right, the curve breaks again and the solid phase is supposedly calcium bicarbonate, which is of course unstable at the CO_2 pressures encountered in the biosphere. The almost horizontal branch of the curve on the extreme left corresponds to the solubility of $Ca(OH)_2$ in solutions of varying OH^- concentration, regulated by the concentration of H_2CO_3.

Solubility of magnesium carbonate. The curve for the magnesium system differs from that of calcium in that all possible solid phases are

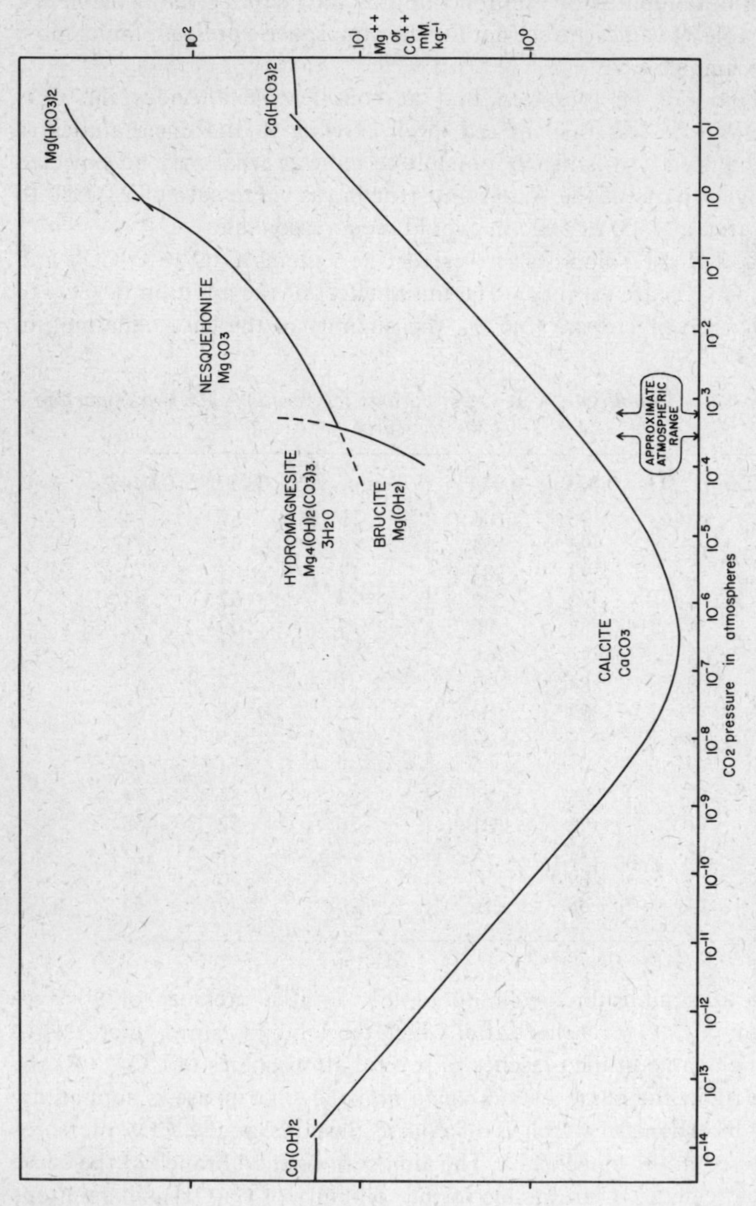

FIGURE 187. The systems $CaO-H_2O-CO_2$ and $MgO-H_2O-CO_2$. The curves indicate the solubilities of the various solid phases in millimoles per kilogram, under varying pressures of CO_2, and the nature of the stable solid phases. Atmospheric limits set at 0.029 to 0.060 per cent by volume.

much more soluble under ordinary CO_2 pressures, and in that the curve for the solubility of the hydroxide cuts the curve for the solubility of the carbonate not far to the left of the minimum, at excessively low values of the CO_2 pressure, but to the right of the hypothetical minimum at about the CO_2 pressure ordinarily present in nature. At ordinary temperatures the carbonate that precipitates when saturation is reached at from 10^{-3} to 10^{-1} atm. of CO_2 is $MgCO_3 \cdot 3H_2O$, or *nesquehonite*;[4] but at the pressures encountered in nature in well-aerated water, which are very close to that corresponding to the break of the curve, a metastable basic carbonate, $Mg_4(OH)_2(CO_3)_3 \cdot 3H_2O$ or *hydromagnesite*,[5] is evidently usually formed. The very great difference in solubility between these compounds and calcium carbonate, at all relevant pressures of carbon dioxide, implies that in ordinary hard-water lakes, in which precipitation of carbonate occurs as the result of the photosynthetic activity of plants, the precipitate consists, as far as is known, primarily[6] of $CaCO_3$. Very great enrichment in magnesium relative to calcium is obviously possible during the evaporation of a hard water. In keeping with the presence of basic carbonate or hydroxide in equilibrium with CO_2 at pressures normally encountered in the atmosphere the resulting solutions have a high *p*H of the order of 10. Britton (1942 Vol. II, pp. 101–106) finds experimentally that when 0.02 molar magnesium sulfate is titrated with sodium hydroxide or sodium carbonate, magnesium hydroxide or magnesium basic carbonate, respectively, both start to precipitate at about *p*H 10.5.

Terminology. If, starting with a suspension of $CaCO_3$ in distilled water, the system is allowed to come to equilibrium at a definite pressure of carbon dioxide in the atmosphere over the liquid phase, there will be in solution, in addition to calcium, bicarbonate, and carbonate ions and a small amount of undissociated salts, a definite quantity of CO_2. Since equilibrium has been reached, this CO_2 will not dissolve any further $CaCO_3$, and is usually referred to as *attached* CO_2. If the pressure of CO_2 in the atmosphere is now raised, more CO_2 will enter the liquid phase and will start attacking any solid $CaCO_3$ still present. Such CO_2 is therefore said to be *aggressive*.

If a bubble of air is brought into contact with any solution containing CO_2, and then analyzed, the quantity of the gas in equilibrium with the liquid phase can be determined. This is ordinarily called the CO_2

---

[4] A pentahydrate, *lansfordite*, occurs also at low temperatures in nature, but is stable only in the presence of liquid water.

[5] Another basic carbonate, *artinite*, $Mg_2(CO_3)(OH)_2 \cdot 3H_2O$, has been reported in nature, but much less commonly than hydromagnesite.

[6] As Johnston shows, at ordinary CO_2 pressures some magnesium carbonate may be expected as a contaminant, presumably as the result of diadochic substitution.

tension of the liquid. If a solution containing CO_2, bicarbonate, and carbonate ions is titrated with sodium carbonate to an end point about 8.4, what is commonly called the *free* CO_2 is determined.[7] As has already been pointed out, at $pH \simeq 8.4$, at which such a quantity may be regarded as zero, the liquid still has a definite CO_2 tension because of attached CO_2 which is present in a quantity equal to the $CO_3^=$ also present. Considering the titration of a pure CO_2 solution,

$$CO_2 + H_2O + Na_2CO_3 \rightarrow 2NaHCO_3$$

and the resulting bicarbonate comes into equilibrium with a small amount of H_2CO_2 and an equivalent amount of $CO_3^=$. The H_2CO_3 will be in equilibrium with CO_2. Therefore, starting with a solution of the gas supersaturated with respect to the atmosphere, the solution obtained at the end point will exert a tension that is not zero but is well below that of the atmosphere. The free CO_2—or as it may more precisely be termed, the *analytical free* CO_2—is thus, if the waters being studied are fairly rich in the gas, at a pH well below the end point, an approximate measure of the total CO_2 in solution, and is in fact nearly a measure of the CO_2 tension. When small amounts of CO_2 at a $pH > 7$ are present, the titration is ordinarily rather meaningless. The distinction between analytical free CO_2 and CO_2 tension has not always been made in limnological practice (cf. Juday, Birge, and Meloche 1935). Ruttner (1931) uses a graphical method based on Schäperclaus (1926) and Tillmans (1919) for estimating the attached CO_2 for any bicarbonate concentration, which he then subtracts from the analytical free CO_2 to obtain the aggressive CO_2. This may well prove satisfactory for any water $pH \leqslant 8.0$, above which both the expected analytical results and the correction for attached CO_2 will ordinarily be too small for accurate work. For small quantities of CO_2 at higher pH values, computation from the pH and bicarbonate alkalinity, or direct determination of the tension, alone give significant results.

If, instead of starting with calcium carbonate, a solution of calcium hydroxide is neutralized by passing CO_2 into it, at first $CaCO_3$ is precipitated; the precipitate then dissolves to form bicarbonate. Since on evaporation the bicarbonate decomposes, losing half its CO_2 to the air and precipitating as $CaCO_3$, one may speak of the quantity of CO_2 needed to convert $Ca(OH)_2$ to $CaCO_3$ as *firmly bound*, and the equal quantity required to form $Ca(HCO_3)_2$ from $CaCO_3$ as *half-bound*, the two fractions together constituting the *bound* CO_2. This is the usage followed by Pia (1933), Ohle, and most other European writers. Juday,

[7] Certain precautions are necessary; the technique has been discussed by Ohle (1952). Very great inaccuracies are possible, owing to variations in indicator concentration and to variations in rate of equilibration with air and in the CO_2 pressure in the latter.

Birge, and Meloche (1935) say that by bound CO_2 they mean firmly bound in Pia's sense. Actually, when an attempt was made to check by the Henderson-Hasselbalch equation the *p*H values that they calculated from the bound CO_2 and the free CO_2, used as an approximate measure of CO_2 tension, it became clear that their bound CO_2, when only bicarbonate is present, is the sum of firmly bound and half-bound fractions, and that in fact they are using the term exactly in the continental sense.

The terms alkalinity, carbonate alkalinity, alkaline reserve, titratable base, or acid-binding capacity are frequently used to express the total quantity of base, rarely free, usually in equilibrium with carbonate or bicarbonate, that can be determined by titration with a strong acid.

When this is expressed in milliequivalents per liter, it is an entirely unambiguous quantity, which will not be altered by the addition of CO_2 converting carbonate to bicarbonate. When it is desired to express alkalinity as milligrams per liter or parts per million, three conventions are possible.

(1) One may assume that the whole of the alkalinity is due to calcium carbonate and bicarbonate, and express the results as $CaCO_3$. This can be misleading in closed alkaline lakes, in which the alkalinity expressed as $CaCO_3$ may imply a quantity of calcium far in excess of the Ca actually present.

(2) One may use the CO_2 content of this calcium carbonate, i.e., express the alkalinity as firmly bound CO_2. This is the most empirical procedure, but it may lead the unwary into forgetting that there can be, and usually is, up to an equal quantity of half-bound CO_2.

(3) One can suppose all the base present to be bicarbonate, and give the alkalinity as milligrams of HCO_3^- per liter. In most moderately hard waters this gives an actual empirical quantity, which becomes conventional only when the *p*H rises over about 8.4 and appreciable carbonate appears. The present writer feels that this is the most nearly unobjectionable usage when alkalinity is reported as mass per unit volume. In the present work, the term *bicarbonate alkalinity* will be used in this sense. In oceanography the term carbonate alkalinity is used in a comparable sense, being defined as the concentration of bicarbonate plus twice the concentration of carbonate, all expressed in moles or millimoles.

The term *hardness* or degree of hardness is frequently used to express the total alkaline earth content that can produce insoluble soaps. Two kinds of hardness are usually distinguished: temporary, due to bicarbonate and carbonate; and permanent, due to sulfate and chloride. The former can be removed by boiling; the latter, usually due to magnesium sulfate, which is very soluble, remains after the temporary hardness has been removed by precipitating the $CaCO_3$. Both types can be removed with sodium carbonate. In considering waters exhibiting only or mainly temporary hardness, the measure of hardness is practically the bicarbonate alkalinity. In Germany this is expressed in parts of CaO per 100,000 parts of water; in France a degree of hardness is one part $CaCO_3$ per 100,000. An English scale in terms of grains per gallon is also in use, but will hardly appeal to the limnologist.

DISTRIBUTION OF CARBON DIOXIDE IN LAKES

Carbon dioxide in surface waters. Apart from some laboratory analyses by Saunders (1926), the only direct determinations of CO_2 tension appear to be due to Krogh (1904, Krogh and Lange 1932). In Greenland, Krogh found that the water of a spring flowing from basalt contained no CO_2. Certain other waters, from basaltic sediment at the bottom of a pond and in a waterfall carrying much glacial rock debris, were in equilibrium with air containing 0.00020 and 0.00025 atm. CO_2 respectively; these figures represent a definite undersaturation. Krogh concludes that basaltic rocks in general take up CO_2, calcium carbonate and silicic acid being formed by double decomposition.

In a well-aerated brook above a bog, the tension was 0.0003 atm.; below the bog it was 0.0014 atm., as might be expected. In the epilimnion of Furesø the tension varied from 0.00024 to 0.00060 atm., the lower values occurring in spring, the higher in autumn and winter. The mean value of 0.000395 atm. probably represents a slight supersaturation relative to the atmosphere in 1929–1930. The higher values, observed from October to January, certainly imply supersaturation.

Relationship of *p*H to CO_2 in surface water. From the Henderson-Hasselbalch equation, suitably modified to conform to activity theory if the concentration requires this, one can compute *p*H, free CO_2, or bicarbonate if one knows the other two variables and the value of pK_1'. It must, however, be emphasized that what is involved is not the analytical free CO_2, determined by titration with sodium carbonate, but the concentration (or activity) of CO_2 and H_2CO_3 actually in solution in the system. The only way to obtain this directly is to determine the CO_2 in the gaseous phase in equilibrium with the liquid phase under consideration. This has not always been realized. Thus Juday, Birge, and Meloche (1935), apparently using the unmodified Henderson-Hasselbalch equation, calculated a number of *p*H values from the analytical free CO_2 and the bicarbonate.[8] Many of the calculated values agree with observation, but the mode of computation is certainly erroneous.

Saunders, who determined in his experiments the pressure of CO_2 in

---

[8] Cf. page 655, equation 2.

equilibrium with the liquid phase, gave as the most appropriate expression to use in calculation:

$$pH = 10.7 - \sqrt[3]{c_m} + \log \frac{[HCO_3^-]}{P_{CO_2}} \quad (16)$$

where P_{CO_2} is measured in mm. Hg, and c_m is the total concentration of cations.

In most observational work, in which tolerable *p*H values are now determined with the glass electrode, one is more interested in obtaining $[CO_{2\ total}]$ from *p*H and $[HCO_3^-]$.

Juday, Birge, and Meloche give a series of mean values for analytical free CO_2 and $[HCO_3^-]$ for different *p*H classes, tabulating drainage and seepage lakes separately. Using the Henderson-Hasselbalch equation (2) and determining pK_1' according to Saunders' empirical expression (6), it is possible to compute the $[CO_2 + H_2CO_3]$ for the mean *p*H in each class, and to compare this with the analytical free CO_2. The results for drainage lakes are given in Fig. 188. As is to be expected, at high *p*H values the calculated concentrations exceed the analytical values, which are negative when $pH > 8.0$. At *p*H 7.6, the calculated and analytical values are identical. In the lower part of the range, the calculated values are again larger than the analytical, and would imply the presence of 1.047 milliequivalents or 46 mg. CO_2 per liter, in the class of mean *p*H 5.2. In the series of seepage lakes, in which the range is from *p*H 4.4 to *p*H 7.5, the calculated CO_2 is always in excess of the analytical. In the lowest *p*H class, 4.4 to 4.9, the mean calculated CO_2 is 4.3 milliequivalents or 190 mg. CO_2 per liter. It is obvious that such extreme divergence cannot be due to experimental error, and that either organic acids or a very small amount of some strong mineral acid, presumably sulfuric, must be contributing to the acidity of the more acid lakes. In the region around neutrality, the divergence of calculated and analytical CO_2 is evidently usually not great enough to cause a divergence of more than a fifth of a unit when *p*H values calculated from the analytical CO_2 are compared with the observed values. If, as seems probable, in the vicinity of neutrality the acidity is due exclusively to CO_2, it is evident that whether the analytical or the calculated values are considered the more reliable, most of the lakes with neutral water in northeastern Wisconsin are somewhat supersaturated with CO_2 relative to the atmospheric pressure of the gas. At temperatures over 10° C., at no reasonable value of the atmospheric content of the gas will the free CO_2 in equilibrium be as great as 1 mg. per liter (Table 79), even at sea level. At 500 m., the quantity in solution at 20° C., in contact with air containing 0.033 per cent by volume, would be 0.53 mg. per litre.

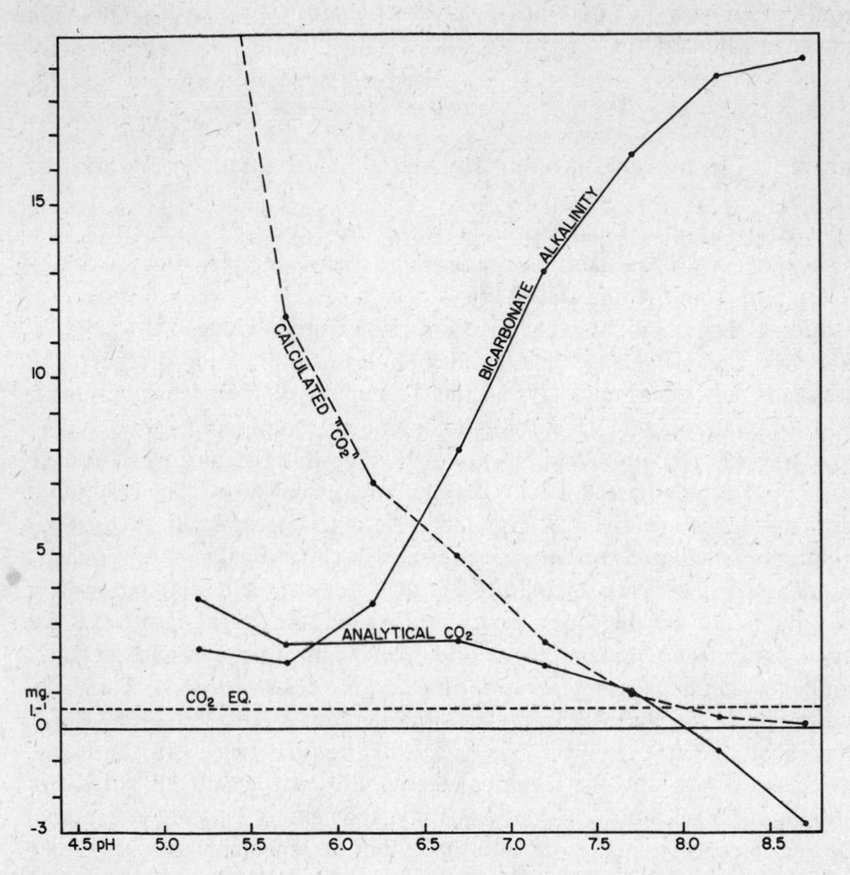

FIGURE 188. Mean bicarbonate, analytical free CO_2, and calculated CO_2 for the surface waters of drainage lakes of northeastern Wisconsin, arranged by *p*H classes. The dotted line, marked CO_2 EQ., indicates the quantity of CO_2 in solution in equilibrium at 20° C. below an atmosphere containing 0.033 per cent CO_2 by volume at an altitude of 500 m.

Apparent supersaturation with calcium carbonate. Ohle (1934a) has pointed out that in the hard-water lakes of northern Germany, the surface waters may perennially contain more calcium and apparently more bicarbonate than would be in equilibrium with the pressure of CO_2 normally exerted by the freely circulating parts of the atmosphere. At 15° C., the solubility of $CaCO_3$ varies from 65 to 74 mg. per liter over the range of pressures corresponding to CO_2 concentrations of 0.030 to 0.044 per cent by volume at sea level. Such a solubility range implies from 26 to 30 mg. Ca^{++} per liter and 79 to 90 mg. HCO_3^-. Yet, as

Ohle shows in a number of cases, significantly greater concentrations are always present. Two striking examples, analyzed at the beginning and the end of summer stagnation, are given in Table 85.

TABLE 85. *Supersaturation with calcium carbonate in two north German lakes*

Lake	Date	Temp., °C.	Ca^{++}, mg. per liter	HCO_3^-, mg. per liter
Trammersee	June 17, 1931	16.9	46.3	183.6
	Sept. 18, 1931	15.0	44.3	180.0
Grosser Plöner See	June 3, 1931	16.3	43.6	135.4
	Oct. 15, 1931	12.1	45.3	146.4

Ohle (1952) later found that the supersaturation persisted in filtered water that had stood for a month in the laboratory. Aeration with outside air and vigorous mechanical stirring for a day had no effect. Contact with precipitated calcite apparently caused a very slight drop in alkalinity, but the presence of crystallization nuclei evidently had little effect on the supersaturation. When expired air was blown through the sample, however, the conductivity rose, as if monocarbonate were going into solution as bicarbonate. Ohle supposes that excess calcium carbonate is present in a stable colloidal form; in the Grosser Plöner See, conductivity measurements before and after the solution of such $CaCO_3$ suggest that about 10 mg. per liter are present in colloidal form.

Steidtmann (1935) had previously suggested that colloidal calcium salts accounted for the apparent supersaturation of a travertine-depositing water that he examined at Lexington, Virginia, and Ruttner (1948) suspected the presence of some unrecognized soluble form of $CaCO_3$ in his experiments on the photosynthetic utilization of bicarbonate. The low C^{14} content of dissolved bicarbonate in the very hard waters of Lake Queechy, New York, when comparison is made with the products of photosynthesis and of precipitation in that lake, suggest that if the difference is real, colloidal $CaCO_3$ formed in the drainage basin or shallow littoral of the lake by photosynthesis may be present in a relatively stable form in the water (Deevey, Gross, Hutchinson, and Kraybill 1954). There is thus reason to suspect that the phenomenon observed by Ohle is very widespread in markedly hard-water lakes.

Vertical distribution of carbon dioxide during summer stagnation. In general the vertical distribution of carbon dioxide reflects in a qualitative way the vertical distribution of the oxygen deficit. Ohle (1952) has concluded that the accumulation of CO_2 in the hypolimnion gives a better measure of lake metabolism than does the oxygen deficit. Since his work is in active progress, this aspect of the matter is deferred till biological productivity is considered in Volume II.

Where there is practically no uptake of oxygen in the hypolimnion, the oxygen curve being orthograde, the CO_2 curve may likewise be expected to be orthograde. Where a clinograde oxygen curve develops, an inverse clinograde carbon dioxide curve may be expected. Where a heterograde oxygen curve is found, a heterograde carbon dioxide curve of opposite sign is usual.

Divergences from this reciprocal arrangement probably can occur. The oxygen concentration of the epilimnion of an unproductive lake with an orthograde oxygen curve is normally lower than that of the hypolimnion, owing to the lower solubility of the gas in warm water than cold. The same effect would be exhibited by CO_2 under strictly lifeless conditions, but the absolute quantity of free CO_2 is so small that there is ordinarily little chance for this effect to be exhibited undisturbed by biological complications. In hard water lakes, as soon as a small amount of photosynthesis occurs there is a production of oxygen and, ordinarily, also a production of carbonate in the epilimnion. If the oxygen production is not too great, a rather marked parellelism between the so-called negative CO_2 values and the oxygen values of the epilimnion and metalimnion may be observed. It is possible, owing to the low pressure of CO_2 in the air and the slowness with which water containing carbonate comes into equilibrium, that in some cases the epilimnion might become saturated, or in equilibrium with the air, with respect to oxygen and still remain markedly undersaturated with respect to CO_2. It is probable that the marked parallelism of the oxygen and free CO_2 curves recorded by Birge and Juday in the upper parts of Skaneateles, Cayuga, and Seneca, and to a lesser extent the other Finger Lakes except perhaps the soft-water Lake Canadice, is due to this type of phenomenon (Fig. 189). In the hypolimnia of these lakes the expected reciprocal relationship between oxygen and CO_2 is clearly demonstrated.

Although in the Finger Lakes the bicarbonate alkalinity is practically constant throughout the water column even at the height of summer stagnation, in most small productive lakes there is a striking increase not only of free CO_2 but of bicarbonate also.

Origin of hypolimnetic bicarbonate. The genesis of this increased bicarbonate concentration is not fully understood. In soft-water lakes, and in very productive lakes containing moderately hard water, with a neutral or feebly alkaline epilimnion and a hardly more acid hypolimnion, some ammonium and ferrous bicarbonates and a little manganous bicarbonate ordinarily result from reduction and anaerobic respiration in the sediments. In the moderately hard-water lakes of high productivity,

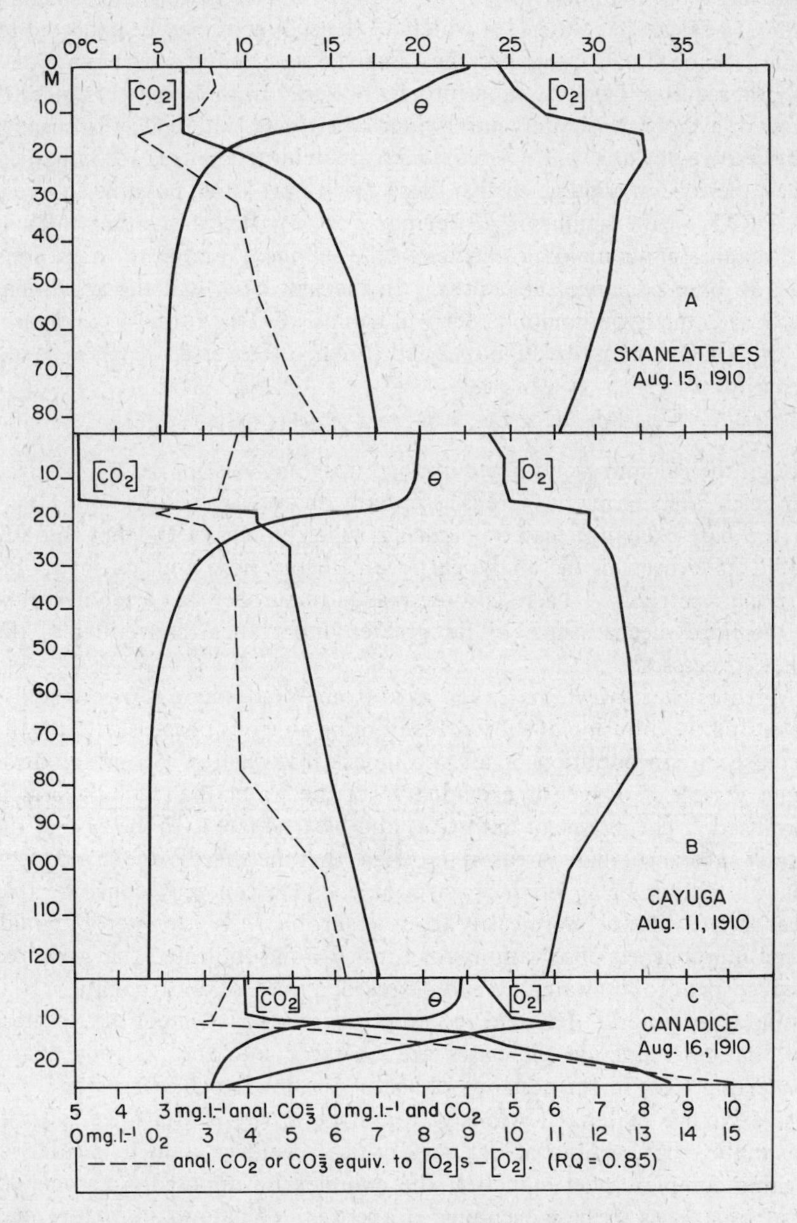

FIGURE 189. Vertical variation in $[O_2]$, oxygen and $[CO_2]$, analytical CO_2 or $CO_3^=$ in Lake Skaneateles, Lake Cayuga, and Lake Canadice. The broken line indicates the analytical CO_2 or $CO_3^=$ equivalent to the difference between the observed oxygen and oxygen saturation, a respiratory quotient of 0.85 being assumed. (From data of Birge and Juday.)

however, there is almost always an increase in magnesium and calcium also. In calcareous waters, in which all the sediment may be expected to contain some $CaCO_3$, this may be dissolved as calcium bicarbonate and may then diffuse from the mud into the water. In the sediments of soft-water or even of moderately hard-water lakes there is no reason to suspect the presence of $CaCO_3$. When the increase in bicarbonate is accompanied by an increase in alkaline earths, these could have three possible sources: (1) $CaCO_3$ in the sediments, (2) primary rock particles or other calcium and magnesium-containing silicates in the sediment, and (3) divalent ions held on base-exchange substances. In the first case, half the additional HCO_3^- in the hypolimnion is derived from $CaCO_3$. In the second case, all the HCO_3^- must be of metabolic origin, the over-all reaction being formally

$$CaSiO_3 + 2CO_2 + H_2O = Ca(HCO_3)_2 + SiO_2$$

though the calcium metasilicate implied does not exist in nature. In the third case, at its simplest, there is essentially an exchange of $2H^+$ for Ca^{++} on the base-exchange material, during which process CO_2 that initially would have been in the analytical free category now appears in the bicarbonate category. There is some reason to suspect that a modification of the third mechanism is of far greater importance than either of the other processes.

Mortimer has shown that when mud from Windermere is placed under water and the diffusion of substances into the aqueous phase is studied, the increase in concentration of all substances investigated except sulfate is much greater if oxygen is excluded from the water than if its access is permitted. This is true of both CO_2 and bicarbonate. In the case of the former substance, there is reason to believe that the effect is due to artifacts, but this cannot be so of the bicarbonate. One can only conclude that bicarbonate diffuses more easily from anaerobic than from aerobic mud. Some unpublished observations on Linsley Pond indicated that oxidized mud suspended in water actually caused a loss of bicarbonate. It is probable, in general, that oxidized mud binds Ca^{++} as might be expected, but that on reduction the bases are liberated into the aqueous phase, converting CO_2 in the aqueous phase of the mud to bicarbonate. The base-exchange material obviously cannot be a clay. Mortimer suggests reasonably, but without as yet very critical evidence, that it is a ferric-organic complex. On reduction the complex is supposedly dissociated, or at least loses its base-exchange properties. Whatever its nature may be, its importance in limnology can hardly be overestimated, as will appear in many later pages of this book.[9]

[9] Cf. also Ohle (1955b).

Influence of form of the lake basin. Hutchinson (1941) has pointed out that certain details in the vertical distribution of bicarbonate in Linsley Pond are most reasonably explained by supposing that the additional bicarbonate alkalinity that enters the hypolimnion during summer stagnation has been brought from the sediments of the lake by predominantly horizontal water movements. The bicarbonate curves, though in general inverse clinograde, exhibit points of maximum curvature and between them inflection points, as is most readily seen in the mean curve for 1937 (solid line, *A* of Fig. 190), based on weekly sampling at 1-m.

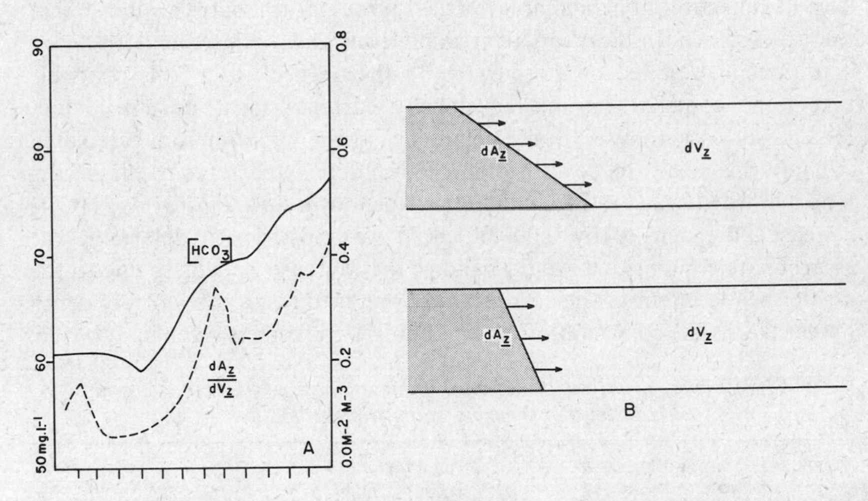

FIGURE 190. *A*, mean bicarbonate concentration (solid line) plotted against depth, Linsley Pond, summer 1937; (broken line) dA_z/dV_z plotted against depth. *B*, diagram of an element of mud surface at two different slopes, liberating material into an element of volume.

intervals. The same pattern was indicated by analysis of the less closely spaced samples of 1938, and is indeed apparent in individual series and in the differences between two such series (Table 86).

The general form of the bicarbonate curve is strikingly like that (broken line, *A* of Fig. 190) indicating dA_z/dV_z or the rate of change of mud surface with water volume, plotted against depth, except that in the bicarbonate curve no maxima are developed, presumably because if such maxima occurred they would tend to produce instability in view of the very gentle thermal gradient. The interpretation of the similarity is simply that when a horizontal element of volume is in contact with a steep, and therefore small, element of area of mud surface, the bicarbonate diffusing into unit volume of water will be less than when the element of volume is in contact with a large, gently sloping element of area (*B* of Fig. 190). Provided the horizontal component of movement, presumably to be attributed to the horizontal turbulence of imperfect internal seiches, greatly exceeds vertical turbulent transport, the type of alkalinity curve observed in Linsley Pond is the inevitable outcome of the morphology of the basin. Hutchinson showed that, at least in the zone from 4 m. to 9 m., the essentially constant coefficient of turbulence derived from thermal data could not possibly produce the observed rise in alkalinity by transport of bicarbonate vertically up from the bottom. In view of the shape of the curve, with its inflection points, simple vertical turbulent transport would in fact lead to decreases in bicarbonate at some levels and a consequent abolition of the steplike shape of the curve. In the deeper water it is probable that some contribution due to density currents must be considered. Hutchinson, assuming uniform turbulence in the hypolimnion, calculated from thermal data the contribution of such density currents to the bathylimnetic heat budget. From the experimental finding that 1 mg. HCO_3 increased the density by 0.000,001,8, it was possible to determine the bicarbonate content that would balance the decrease in density due to the greater heat content of the current and bring it to rest at the correct depth in the lake. The computation, in an abbreviated form, is given in Table 86.

TABLE 86. *Analysis of increase in bicarbonate at various levels in Linsley Pond during summer stratification*

(1) Depth, m.	(2) Observed Increase in HCO_3, mg./liter/month	(3) Increase Due to Vertical Turbulence, mg./liter/month	(4) Increase Due to Balance Density Current, mg./liter/month	(5) Increase Due to Horizontal Streaming, mg./liter/month	(6) Rate at Which Bicarbonate Leaves Mud, mg./cm.^{2}/month
4	2.42	2.41	...	0.01	0.015
5	3.03	0.25	...	2.78	3.08
6	4.00	− 1.76	...	5.76	2.24
7	4.60	− 0.42	...	5.02	1.95
8	5.09	− 0.69	...	5.78	1.88
9	5.70	+ 0.90	1.3	3.5(0)	1.42
10	7.28	+ 0.08	2.0	5.2(0)	2.03
11	7.70	− 1.53	2.3	6.9(3)	2.14
12	8.77	+ 0.83	3.3	4.6(4)	1.26

The final column is obtained by multiplying column 5 by dV/dA. Practically no bicarbonate enters the lake from the mud at 4 m.; this is presumably because the 4-m. layer is shallow enough to be in contact with water containing several milligrams of oxygen per liter throughout the summer and therefore probably retains an oxidized surface. Below this, the output rises suddenly and falls slowly to 9 m. The fall in the zone from 6 to 8 m. is obviously not rapid enough to obscure the effect of the morphometry of the lake.

It is reasonable to suppose that the density currents, if real, originate from the water in contact with mud between 6 m. and 9 m. It is quite reasonable to suppose that the density current cannot develop over oxidized mud; a sufficient diffusion outward of dissolved substances to produce such a current would be expected only from fully reduced mud. The curve for oxygen deficit shows no such detailed structure, but it is indicated in the analytical free CO_2, and of course in the metabolic CO_2 anomaly (page 679). Mortimer (1941–1942) criticises this interpretation, suggesting that at least the upper inflection is due to precipitation of iron, present initially as ferrous bicarbonate, as oxygen is brought in by turbulent exchange. Mortimer's view, however, involves the main features of the bicarbonate curve being attributed to vertical turbulent transport from below. Hutchinson's data (cf. Table 51) indicate clearly that the coefficient of turbulence derived from thermal data is quite inadequate to explain the genesis of even the rough features of the bicarbonate curve between 4 and 9 m. Although there is little detailed information about hypolimnetic water movements in small very stably stratified lakes, there is no reason to believe that internal seiches are not accompanied by horizontal irregular or turbulent movement as well as by vertical turbulence. Irregularities in internal seiches may well be able to produce the type of horizontal movement required, which alone can unify the whole picture of chemical development during the summer in Linsley Pond. That the effect has not been clearly observed elsewhere is presumably due to the fact that it is dependent on a particular kind of characteristic profile and can be detected only when the lake is productive enough to exhibit a considerable hypolimnetic metabolic CO_2 anomaly; it is likely to be completely obscured whenever photosynthetic precipitation of $CaCO_3$ causes sedimentation of carbonate from the epilimnion into the hypolimnion. Such a combination of circumstances is presumed to be uncommon, though a number of cases should ultimately be discoverable. Hooper (1954) has described a remarkable case in which adsorption of bicarbonate and cations by the mud in contact with the metaliminion is perhaps involved.

The relationship between carbon dioxide and the hypolimnetic oxygen deficit. When a quantitative comparison between loss of oxygen and gain of carbon dioxide is desired the situation becomes rather complicated.

CO_2 content at vernal circulation. There is first the problem of determining the correct base to which calculations are to be referred or, in the terminology used by Alsterberg in discussing oxygen, the primary constant for carbon dioxide. When the production is great in the hypolimnion

and no determinations at the spring circulation period have been made, in many lakes of moderate bicarbonate alkalinity no great error will be introduced by assuming saturation at the observed temperature. It has been pointed out, however, that not infrequently soft or moderately hard waters are supersaturated with respect to the air. In markedly hard waters, at the time of more than minimal photosynthesis the *p*H is usually found to be over 8.4, and more $CO_3^=$ than CO_2 is present. This excess is commonly, if rather inaccurately, regarded as a negative analytical free CO_2 concentration. Since in the lakes that exhibit such values there is often a spring phytoplankton maximum, and since we may suspect that equilibrium with the air is established less rapidly in the case of CO_2 than in that of oxygen, we may find very considerable negative values throughout the whole water column when stagnation begins. Birge and Juday (1911) indicate this, for instance, for Green Lake, Wisconsin. Here the bicarbonate alkalinity was virtually uniform throughout the lake both on February 15, 1906, when the lake was frozen, and on May 15, 1906, when it had just begun to circulate. The mean free, that is, analytical, CO_2 concentration fell during this period from -1.6 to -2.4, or 0.8 mg. per liter. Taking the mean depth as 33.1 m. (Juday 1914), the areal deficit for the whole lake would be 5.3 mg. cm.$^{-2}$ on the first day and 8.0 mg. cm.$^{-2}$ on the second. These figures represent about 1.2 and 1.8 per cent of the total CO_2 in the atmosphere above the lake. It would probably be difficult for a lake to come into equilibrium at vernal circulation with so large a mass of gas as is implied by such figures; meanwhile, photosynthesis was certainly proceeding.

Respiratory quotients of lakes. Having obtained empirically the initial value, or having decided what assumptions are legitimate in default of empirical data, there is the problem of converting the oxygen deficit into a form that can be compared with the increase in CO_2 observed. This involves a knowledge of the over-all photosynthetic or respiratory quotients, which in a closed system should be the same.

In the simple photosynthetic production of carbohydrate from CO_2 and H_2O, the quotient is unity. But aquatic plants and animals are not made of sugar, nor even starch and cellulose. If *Chlorella* is grown under optimal conditions (Myers 1949), using nitrate as a nitrogen source, and the resulting organic matter is converted again to nitrate, the over-all quotients in the two processes will be $CO_2:O_2 = 0.69$; if ammonia is the nitrogen source and final product, the quotient will be 0.91. Ohle (1952) on the basis of analytical studies on fresh-water organisms, particularly those of Krogh and Lange (1932) and Krogh and Berg (1931), believes that the best mean value is 0.85 exactly intermediate between Myers' extreme figures.

Estimates of metabolic CO_2. Water from the hypolimnion of a lake with a marked clinograde oxygen curve usually shows a considerable increase in analytical free CO_2 and an increase in bicarbonate. Part of the increase in bicarbonate can be shown to be due to the presence of $(NH_4)CO_3$, the so-called volatile alkali, which is easily removed by boiling. Usually a considerable amount of additional bicarbonate due to nonvolatile bases has also accumulated.

Assuming with Hutchinson (1941) and Ohle (1952) that the bicarbonate other than that of ammonia is derived from the solution of material in the sediments, the increment in CO_2 in the hypolimnetic water may be fractionated to (1) increment in free analytical CO_2, all of material origin; (2) ammonium bicarbonate CO_2, all of metabolic origin; (3) increment in fixed-base bicarbonate CO_2, taken to be half of metabolic origin as a lower limit.

The sum of these three quantities must usually represent a minimum estimate of the metabolic CO_2 produced in the hypolimnion. The estimate is minimal because some loss at the top of the hypolimnion by turbulent exchange is to be expected, and also because it is probable that nearly all of the increment in nonvolatile bicarbonate is due not to the solution of $CaCO_3$ but to exchange of Ca^{++} with H^+ at the surface of base-exchange materials in the sediments, and to the reduction of ferric hydroxide and its subsequent solution as ferrous bicarbonate. In any consideration of lake metabolism, rather than of the chemical equilibria obtaining at any depth and any time, it is clear that the various fractions of the metabolic CO_2 must be considered together, estimated as accurately as possible.

The CO_2 anomaly and anaerobic CO_2 production. The oxygen deficit may be converted to CO_2 by means of the respiratory quotient of 0.85. The resulting quantity can be subtracted from the free CO_2 to give the free CO_2 anomaly (ΔCO_2); or more significantly, from the total metabolic CO_2 to give[10] the metabolic CO_2 anomaly ($\Delta' CO_2$). Ohle (1934a) first pointed out that the metabolic CO_2 anomaly is often positive; in the cases where this is true, the metabolic anomaly will a fortiori be an even greater positive quantity. Such positive values imply that a very considerable quantity of CO_2 is produced anaerobically.

The data in Table 87 for Linsley Pond between April 30 and September 20, 1937, indicate the extent of this anaerobic contribution. In default of vernal analyses, the initial CO_2 has been taken as 0.9 mg. per liter,

[10] Hutchinson's (1941) original figures for Linsley Pond are defective in that a respiratory quotient of unity was employed. The figures here presented use a value of 0.85, as recommended by Ohle.

TABLE 87. *Analysis of CO_2 anomalies in Linsley Pond during summer stratification*

Depth, m.	Increment in HCO_3, mg. per liter	Volatile Alkali, mg. per liter	CO_2 of Volatile and half Nonvolatile Alkali, mg. per liter	Free CO_2, mg. per liter	Metabolic CO_2, mg. per liter	$O_2 \times 1.375 \times 0.85$, mg. per liter	ΔCO_2, mg. per liter	$\Delta' CO_2$, mg. per liter
0	9.4	0	3.4	− 0.8	2.6	1.1	− 1.9	1.5
2	11 9	0	4.3	0.6	4.9	1.2	− 0.6	2.7
5	12.6	0.8	4.9	3.0	7.9	3.1	− 0.1	4.8
7	25.1	3.6	10.4	8.6	19.0	9.6	− 1.0	9.4
9	27.0	5.4	11.7	8.8	20.3	12.1	− 3.3	8.2
11	36.6	9.3	16.6	11.8	28.4	12.8	− 1.0	15.6
13	36.2	10.2	16.8	12.0	28.8	12.6	− 0.6	16.2
14	35.1	(10.2)	16.4	12.7	29.1	11.7	1.0	17.4

or saturation at 7° C. under an atmosphere containing 0.035 per cent; a small error is likely to be introduced by such an assumption. Above the thermocline about 4 m., the changes observed are irrelevant to the present discussion and must be due to analytical error, evaporative concentration of the influents and the like. Below the thermocline about 4 m., the figures probably give a fair approximate indication of the over-all process taking place.

The details of the process are, however, not necessarily continuous and simple. In Lake Mendota (Birge and Juday 1911) it appears that early in summer stagnation the oxygen content decreases faster than seems to be implied by the increase in any CO_2 fractions. Between May 22 and July 10, 1906, the published figures imply respiratory quotients between 0.55 and 0.74 in the zone from 10 to 18 m., and in the case of the change at 15 m. between June 11 and July 10, the quite impossible value of 0.17.

Very considerable negative anomalies in the free CO_2 content have been noted by Ohle (1934a), particularly in the various basins of the Schaalsee, but in such cases they are more than balanced by the increase in the bicarbonate concentration. In the case of Lake Mendota, during the period just discussed there was practically no change in the bicarbonate alkalinity. It is possible that the observed CO_2 production, which is represented in this case by a reduction in monocarbonate rather than by an increase in free CO_2, is in part illusionary, due to technical errors. It is unlikely, however, that this is the entire explanation. Preferential oxidation of fat in sedimenting diatoms might provide a partial explanation, as this would involve a respiratory quotient of below 0.7. Later in the season, after the oxygen is exhausted in the hypolimnion, there is certainly an anaerobic production of CO_2 in the lake.

In Black Oak Lake (Juday and Birge 1932; Juday, Birge, and Meloche 1935), the only one of the lakes of northeastern Wisconsin for which both O_2 and CO_2 determinations for the same dates are recorded, the negative metabolic CO_2 anomaly, computed on the assumption that all bicarbonate in excess of the surface value is metabolic, persisted at least till August 24, 1928. The determinations for this date imply values of the RQ of 0.61 at 20 m. and 0.73 at 25 m., at which depth no oxygen was found. It is to be noted that in Lake Mendota, there is a decrease in the bicarbonate alkalinity in the epilimnion as well as a rise in this quantity in the hypolimnion. The epilimnetic decrease is presumably the result of photosynthetic precipitation of $CaCO_3$, which redissolves in the hypolimnion. Half of the CO_2 of the additional bicarbonate in the hypolimnion will of course be of metabolic origin whether the other half comes from above or below.

HYDROGEN ION CONTENTS OF NATURAL WATERS

The recorded range of *p*H appears to be from 1.7 in Katanuma, Miyagi, a volcanic lake containing 474 mg. $SO_4^=$ per liter (Yoshimura 1934, corrected by Sugawara, personal communication), to 12.0 in Lake Nakuru, an effluentless alkaline lake in the eastern Rift Valley, Kenya (Jenkin 1932).

Origin of acidity and the role of sulfuric acid. In general, very acid values, below 4.0, occur rarely; such values are known, however, in a number of volcanic lakes. Yoshimura (1934) notes, in addition to Katanuma, several Japanese lakes containing strong mineral acid at *p*H 3.2 to 4.9. Some of the crater lakes of Indonesia, such as Awoe, with 0.005*N* HCl and H_2SO_4, and Kawah Idjen, varying between 0.8 and 1.3*N* strong acid, must exhibit as low or lower *p*H values than any in Japan.[11]

Less extreme values, though well below 4.0, are sometimes observed in bog pools. Skadowsky (1923, 1926) records values of *p*H 3.2 to 3.8 in Russian bog pools, and Jewell and Brown (1929) values as low as *p*H 3.3 in pools in a sphagnum mat at the edge of Mud Lake in northern Michigan. In the Luzino bog near Moscow, studied by Skadowsky (1923), the distribution of *p*H was zonal, the margin of the bog having pools at 6.05 to 6.5, the interior pools at 3.2 to 3.8. This arrangement is no doubt usual in bogs with no large central area of open water. Wehrle (1927) noted values between 3.5 and 3.9 in pools in a sphagnum bog near Freiburg in Breisgau, Germany. In such localities, at the height of summer there

---

[11] Hydrochloric acid 0.1*N* has a *p*H of 1.04; the *p*H of Kawah Idjen water may well be under 0.7.

appears to be a tendency for the *p*H to fall about 0.1 during hot days. This is attributed to evaporative concentration of strong acids; nothing is said about a possible rise at night.

The very low *p*H values found in bogs and bog lakes have been attributed to a variety of organic acids, but are probably usually due to H_2SO_4. In some of the bay lakes of North Carolina, Frey (1949) found *p*H values between 4.34 and 4.92, which were not raised by aeration or boiling. In such waters there are about 6 to 8 mg. $SO_4^{=}$ per liter. In the absence of other dissolved matter, 7 mg. $SO_4^{=}$ per liter as sulfuric acid would give a *p*H of 3.8. A small quantity of bases and other substances are certainly in solution, however, so that the observed *p*H values are reasonable. The source of the acid in such waters is probably twofold. Rain water, which contains sulfate, percolating through peat tends to lose cations and gain H^+ by base exchange with the humus of the peat. Some peats formed under anaerobic conditions contain minute aggregates of pyrite FeS_2; if such peats are eroded and exposed to the air, the pyrite oxidizes:

$$4FeS_2 + 15O_2 + 2H_2O \rightarrow 2Fe_2(SO_4)_3 + 2H_2SO_4$$

The ferric sulfate may hydrolyze on dilution, precipitating ferric hydroxide from the dilute acid solution.

The oxidation of pyrite in the rock of the drainage basin also provides a third category of very acid water lakes. Ohle has described in great detail the Tonteich near Reinbek in northern Germany, in which a *p*H of 3.17 due to sulfuric acid is produced by weathering of an upper Miocene micaceous clay containing pyrite. The well-known acidity of water draining from mine dumps or pumped from mines has a like origin. A comparable natural case, Onumaike, north Shinano in central Japan, has been studied by Ueno. Here the main influent is a very acid stream containing up to 92 mg. H_2SO_4 per liter, derived from the oxidation of some mineral, probably pyrite, in the quartz trachyte of Mount Akaishi. The *p*H varies from 2.8 to 3.8. Both the Tonteich and Onumaike are bluish green in color, apparently due to precipitation of colored organic matter. In this they differ markedly from the bog pools at *p*H 3.2 to 4.0.

Shapiro's (1956) recent work on the acid coloring matter of lake water, a purified preparation of which can have a *p*H value of 3.6, suggests that such materials may be more important in determining acidity than most recent workers have believed.

Normal *p*H range with bicarbonate buffering. In practically every case where the water is neither very acid nor very alkaline, it may justly be assumed that the *p*H is regulated by the CO_2-bicarbonate-carbonate system. Considering equation 2 and disregarding complications arising from activity in strong solutions,

$$pK_1' = pH - \log \frac{[HCO_3^-]}{[CO_{2\ total}]} \tag{17}$$

Between 15° and 20° C. at *p*H 7, since pK_1' is about 6.4, it can be stated approximately that

$$[HCO_3^-] = 4[CO_{2\ total}] \tag{18}$$

For any given CO_2 concentration, at *p*H 6 there will be one tenth and at *p*H 8 ten times this amount of bicarbonate. Supposing that at 20° a water has a CO_2 tension corresponding to an atmosphere containing 0.035 per cent CO_2 by volume—and so, as is usual, is slightly supersaturated with respect to the contemporary atmosphere—the range of bicarbonate CO_2 corresponding to the *p*H range 6.0 to 8.0 will be 0.24 to 23.6 mg.; if the bicarbonate were all present as calcium, the range of calcium contents would be from 0.11 mg. per liter to 10.1 mg. Actually, Juday, Birge, and Meloche (1935) found in northeastern Wisconsin that waters with a mean *p*H of 6.0 had a mean bicarbonate CO_2 content of about 2.3 mg. per liter and calcium content of about 1.0 mg.; with a mean *p*H of 8.0, the bicarbonate CO_2 is 17.6 mg. per liter and the mean calcium is 8.0 mg. Some magnesium is also present, but there is evidence also of anions other than bicarbonate and of some supersaturation with CO_2. At *p*H values below 6.0, acids other than carbonic may be suspected, and in most waters at *p*H values much above 8.0, though the CO_2-bicarbonate-carbonate system is still operating, saturation deficiencies of CO_2 may be expected. Any value over 9.0 will be due either to extreme divergence from equilibrium as the result of photosynthesis or to the presence of carbonates of sodium and magnesium, more soluble than $CaCO_3$. It is in general desirable to bear in mind that when, as usually happens if the lakes of a varied but not markedly calcareous nor markedly humic terrain are considered, the *p*H values of the surface waters mostly lie between 6.0 and 8.0, this is due as much to the normal CO_2 content of the terrestrial atmosphere and to the average terrestrial abundance of calcium as it is to the fact that ideally pure water, if it could be obtained, would have a *p*H value very close to 7.0.

In the most extensive series of data available, that for the lakes of northeastern Wisconsin (Juday, Birge, and Meloche 1935), the modal value for all surface waters studied lies just on the alkaline side of neutrality. When the data are grouped to give distributions for seepage lakes and drainage lakes separately, the following results are obtained:

	Range	Mode	Median
Seepage lakes	4.4–7.5	5.8	5.95
Drainage lakes	4.8–8.9	7.7	7.45

About half of the seepage lakes and over 80 per cent of the drainage lakes have surface waters in the range from 6.0 to 8.0. The generally lower values in the seepage lakes are to be attributed to the adsorption of metallic cations, either in peaty organic matter (Williams and Thompson 1936) in the basin or in some cases by clay minerals in the soil (Gorham 1955a).

Lohammar (1938) gives data on 151 lakes in central and northern Sweden. These lakes mostly lie on noncalcareous glacial deposits or on glaciomarine sediments, though in one area there is calcareous drift. The range observed is similar to that in Wisconsin, from 4.9 to 9.0; the median lies just on the acid side of neutrality, and 95 per cent of the lakes have surface waters between *p*H 6.0 and *p*H 8.0.

Strøm (1939, 1942, 1944b), in the predominantly mountainous areas of Norway, observed a range from 5.45 to 8.7, all depths being considered. In twenty-six out of thirty-eight lakes the values at every depth, whenever the lake was studied, lay between 6.0 and 8.0. A single value below 6.0 was encountered. The higher values above 8.0 refer either to lakes in drainage basins underlaid in part by Palaeozoic limestones, or to lakes in which glacial water flows. Strøm concludes that the fine suspended rock particles of glacier milk may raise the *p*H by adsorption of hydrogen ions and liberation of small amounts of alkalies into the very dilute, unbuffered waters.

Ohle's (1934a) study of the lakes of northern Germany provides a striking example of a region dominated by calcareous lakes. The histogram of the distribution of *p*H is bimodal. A small group of four lakes have surface waters below *p*H 5.0. These are apparently deficient in bases for the same reasons as the Wisconsin seepage lakes, having received most of their water from drainage through forest soil and litter. A large group of lakes have a $pH > 6.0$; the principal mode for the whole region is between 8.0 and 9.0, and half the lakes studied have surface waters at $pH > 8.2$. Forty-five of the fifty-seven lakes contained in their surface waters more than 30 mg. Ca per liter. In all these lakes it is probable that some $CaCO_3$ is colloidal and does not influence the *p*H.

In the fens of eastern England, very considerable quantities of peat have formed in an area which receives hard-water draining from chalk or chalky boulder clay. Here the peat has become saturated with bases, and even water that has seeped through it is still slightly alkaline. Thus Saunders (1923) found in a number of determinations from lodes (drainage ditches) and artificial ponds at Wicken Fen, north of Cambridge, a range of 7.2 to 9.2. The highest values were certainly due to the photosynthetic removal of CO_2; the lowest occurred in peat cuttings or over the disturbed mud of a small ditch.

These regional examples indicate in general the ranges to be expected in the humid temperate region. They could be multiplied indefinitely.

For tropical waters in which no marked accumulation of electrolytes has occurred as the result of evaporation in closed basins, the best series is certainly that of Ruttner (1931), who found in the epilimnetic waters of Indonesian lakes values between 3.08 and 8.65. The very low values occurred in certain small bodies of water on the Dieng Plateau at an altitude around 2500 m.; they are undoubtedly due to sulfuric acid and may be attributed to the combined action of bog and volcanic waters. In more than half the surface waters of lakes, the *p*H was above 8.0. This is true of practically all the larger thermally stratified bodies of water. In all these cases there can be little doubt that the *p*H is primarily determined by photosynthetic activity.

Vertical distribution of *p*H. In general this will be determined by the utilization of CO_2 in the trophogenic layers and its liberation in the tropholytic, and by the possibility of the solution of bases as the CO_2 accumulates. To a large extent these two factors operate antagonistically. Where the geochemical possibilities of solution of bases are limited, the water will be poorly buffered and any given quantity of CO_2 liberated into the hypolimnetic water will produce a marked lowering of *p*H. In lakes in which the bicarbonate alkalinity is low, the inorganic nutrients are also usually available in small amounts; little organic matter is synthetized in the trophogenic zone, little falls into the hypolimnion, and therefore little CO_2 is produced in that region. When the bicarbonate alkalinity is high and the trophogenic zone productive, the consequent high production of CO_2 in the hypolimnion causes a relatively small lowering of the *p*H of the well-buffered water.

In the Nordfjord lakes, which include the deepest lake in Europe, a quite recognizable fall in *p*H of the very dilute water was usually noted by Strøm (1933a) in passing from the epilimnion to the hypolimnion though the oxygen curves are strikingly orthograde. Thus Strynsvatn, at the end of August, had water virtually saturated with oxygen throughout, but exhibited a fall in *p*H from 7.0 to 6.5 between 0 m. and 200 m. In contrast to this, for example, unpublished data for Mount Tom Lake, Connecticut, in September 1935, show a fall in *p*H from 7.1 to 6.85, coincident with a decline in oxygen from 7.98 to 0.88 mg. per liter. In most of the very hard-water lakes of northern Germany, Ohle found during summer stagnation differences of the order of one unit between the surface and the bottom. It is evident that while the general absolute value of *p*H is highly correlated with the broad chemical properties of any lake water, the form of the *p*H curve, though usually somewhat clinograde, varies within moderate limits in both soft-water and hard-water lakes.

Acid heterograde distributions. One rather striking special phenomenon has been recorded. In small lakes in which a considerable hypolimnetic oxygen deficit develops, not only alkaline earth but ferrous and manganous bicarbonates also may diffuse from the sediments, raising the bicarbonate alkalinity. In moderately or markedly hard-water lakes the additional bicarbonate alkalinity is small compared to the alkaline earth bicarbonate initially present, and the *p*H curve is usually clinograde in spite of the great bicarbonate alkalinity at the bottom. In very soft-water lakes the additional ferrous and manganous bicarbonate may constitute a relatively large addition to the alkalinity. The *p*H in such

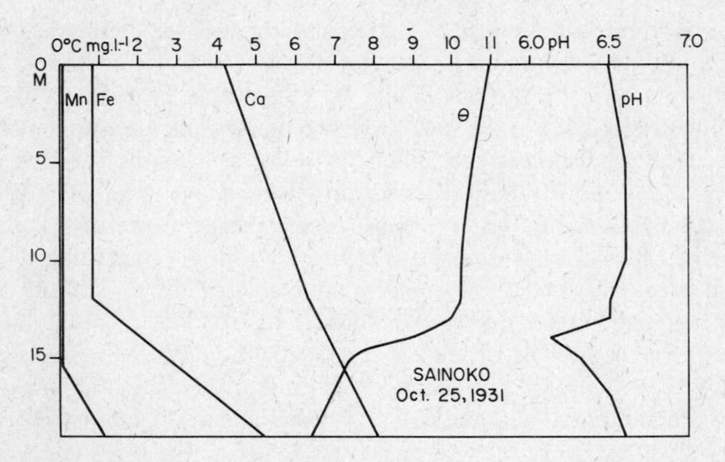

FIGURE 191. Negative heterograde distribution of *p*H in Sainoko, Japan (after Yoshimura).

lakes may not merely fall with depth less than if the additional bicarbonate were not present, it may in fact rise near the bottom, producing an acid heterograde or dichotomous curve. Yoshimura (1932d), who first called attention to this phenomenon, has noted it in a number of small soft-water lakes in Japan (Fig. 191). He predicted that it should occur elsewhere, notably in northeastern Wisconsin. Juday, Birge, and Meloche (1935) report only one clear case, however, namely Lake Mary, which is meromictic and therefore not strictly comparable with the Japanese examples. Hooper (1954) describes an American example in Weber Lake, Michigan; Åberg and Rodhe (1942) found a striking case in Skärshultsjön in south Sweden.

Alkaline heterograde distributions. Alkaline heterograde or dichotomous vertical distributions of hydrogen ion concentrations, in which the *p*H rises to a maximum in the metalimnion and then declines, are often associated with positive heterograde oxygen curves. They are obviously due to removal of sufficient CO_2 to raise the *p*H by photosynthesis.

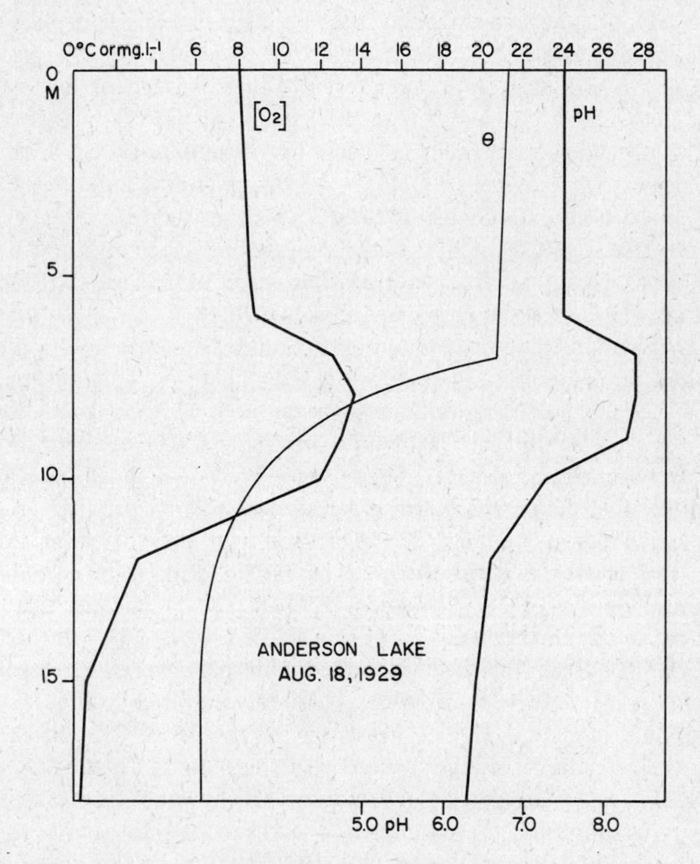

FIGURE 192. Positive heterograde distribution of *p*H and oxygen, Anderson Lake, Wisconsin, August 18, 1929 (after Juday, Birge, and Meloche).

Juday, Birge, and Meloche (1935) give, among other examples, Anderson Lake, northeastern Wisconsin (Fig. 192). It is probable that this type of distribution is limited by the general correlation of high photosynthetic productivity and high total electrolyte content, which in general will imply fairly good buffering in lakes exhibiting strong oxygen maxima.

SUMMARY

The quantity of carbon dioxide in solution in pure water under an atmosphere at sea level will depend on the quantity of the gas in the atmosphere, at present about 0.033 per cent by volume, and the temperature. At 0° C. the amount in solution is 1.10 mg. per liter; at 20° C., 0.56 mg. per liter. On solution, a small part of the CO_2 reacts with water to produce H_2CO_3, which is strongly dissociated to produce hydrogen ions, bicarbonate ions as HCO_3^-, and some $CO_3^=$. At *p*H 7.0, 20.8 per cent of the CO_2 in a dilute solution is present either as such or as H_2CO_3, 79.2 per cent as HCO_3, and 2.6×10^{-4} per cent as $CO_3^=$. At *p*H 5 almost all the CO_2 is present as such or as undissociated acid. At *p*H 8.5 nearly all is present as HCO_3^-. At greater alkalinities $CO_3^=$ becomes important, and it constitutes 24.3 per cent at *p*H 10.

The movement of CO_2 in and out of a water interface appears to be a rather slow process, though it is not possible as yet to give any satisfactory account of what is likely to happen at a lake surface.

If CO_2 dissolves in water containing carbonate, either in solution or in suspension, its solubility is greatly increased. The most significant case concerns $CaCO_3$. Starting with a suspension of $CaCO_3$ and allowing this to come to equilibrium with air containing 0.033 per cent, a very appreciable amount of $CaCO_3$, 54 mg. per liter, will go into solution. This will involve the presence in the solution of 65 mg. per liter HCO_3^- ions, of which half may be regarded as derived from the calcium carbonate and the rest from the atmosphere. The HCO_3 will be in equilibrium with a small amount of $CO_3^=$ and of $CO_2 + H_2CO_3$. Because this latter will not cause any further solution of $CaCO_3$, it is said to be *attached.* If more CO_2 is added, the amount added will be *aggressive*, dissolving more $CaCO_3$ if it is there to dissolve. The only empirical estimate of the actual amount of $CO_2 + H_2CO_3$ present is the *tension* of CO_2 or pressure of the gas in a bubble in equilibrium with the water. In the few cases in which this has been determined, natural fresh waters are sometimes slightly undersaturated, presumably as the result of photosynthesis, but more often a little supersaturated with CO_2 relative to the atmosphere. The quantity that can be determined by titration, say to *p*H 8.4 with Na_2CO_3, is not really the CO_2 present before equilibrium was disturbed. Thus at *p*H 8.4 the *analytical* CO_2 so determined will be zero, and at higher *p*H values a negative quantity, but the tension is always a definite positive quantity corresponding to the amount of $CO_2 + H_2CO_3$ in equilibrium with $CO_3^=$ and HCO_3 in the solution. Only when much aggressive CO_2 is present will the analytical CO_2 nearly correspond to the tension. In a solution of calcium bicarbonate, half the CO_2 present as

HCO_3 is easily removed by boiling and the other half remains as a precipitate of $CaCO_3$; one half of the bicarbonate CO_2 is therefore called *half bound*, the other half *firmly bound*. The sum of the two fractions is called *bound* CO_2. If the *alkalinity* is expressed in milliequivalents per liter, it does not matter whether only bicarbonate is considered or a little carbonate is present. If the results are to be expressed as milligrams per liter, the best convention would seem to be to assume conversion of all carbonate to bicarbonate and to express the results as mg. HCO_3^- per liter. Other conventions have been used and are mentioned in the text.

The pH can obviously be related to the cation, carbonate, bicarbonate, and CO_2 contents of water, though a number of features require formal presentation which must be found in the text. A good approximation is given by the equation

$$pH = 10.7 - \sqrt[3]{C_m} - \log \frac{HCO_3^-}{P_{CO_2}}$$

where c_m is the total concentration in equivalents per liter. Where pH can be accurately measured, we can obtain indirectly the CO_2 tension from this equation. In dilute acid seepage lakes the results tend to be impossibly high. In such cases the low pH must be due to acids other than carbonic; sulfuric acid is the most likely, though colored organic acids may be important.

In many hard-water lakes the quantity of Ca and bicarbonate apparently present is greatly in excess of the amount in equilibrium with air containing any likely amount of CO_2. At sea level a CO_2 content as great as 0.044 per cent by volume in the atmosphere would imply an equilibrium up to 30 mg. Ca and 90 mg. HCO_3^- per liter, yet concentrations in excess of 40 mg. Ca and of 135 mg. HCO_3^- are recorded perennially from the Grosser Plöner See and other north German lakes. It is reasonably certain that in such cases colloidal $CaCO_3$ is present.

During summer stagnation the vertical distribution of carbon dioxide, including bicarbonate, is often roughly the inverse of the oxygen distribution. If the latter is orthograde, there is little or no increase in either analytical or bicarbonate CO_2, if the oxygen curve is clinograde, there is usually an increase in analytical CO_2 and also in HCO_3^-. In many extreme cases the total quantity of CO_2, either free or bound, added in any hypolimnetic layer is much in excess of the O_2 removed, and on any reasonable assumption for the respiratory quotient, such an excess implies much anaerobic metabolism in the deep water or sediments. The additional bicarbonate that appears partly consists of ammonium bicarbonate produced in the mud by bacteria, partly of ferrous and manganous bicarbonate, of which the cations are liberated by reduction in the mud, and partly of calcium and magnesium bicarbonate. The source of the alkali earth ions is obvious in lakes with calcareous sediments, but less so in most moderately hard-water lakes. Probably the cations are held bound on oxidized iron-organic complexes and are liberated as the iron is reduced. In a few cases, notably Linsley Pond, the vertical distribution of analytical CO_2 and of bicarbonate seems to reflect the morphology of the deep part of the basin, as if these substances were moving predominantly horizontally and tended to be more concentrated at depths at which an elementary layer of water lies against a broad shelf of mud than at depths at which the mud surface in contact with the water is minimal. This pattern probably reflects horizontal imperfection in the movement of internal seiches.

The *p*H of lake waters varies from as low as 1.7 in some volcanic lakes containing free sulfuric acid, to 12.0 or more in some closed alkaline lakes rich in soda. The usual range for open lakes is between 6.0 and 9.0. Seepage lakes and all lakes in regions of acid igneous rock tend to have values below 7.0. Only very calcareous waters exhibit values well over 8.0.

There is usually a slight fall in *p*H in the hypolimnia of lakes exhibiting clinograde oxygen curves during summer stratification. In very dilute waters, liberation of ferrous and manganous bicarbonate from the sediments in contact with the deep water may produce a dichotomous *p*H curve with a minimum in the upper hypolimnion. Maxima in the *p*H curve due to photosynthesis are also known.

CHAPTER 11

Redox Potential and the Iron Cycle

The problems to be discussed in the present chapter are related, on the one hand, to the distribution of oxygen considered in Chapter 9 and on the other, to the cycles of nearly all the other elements of importance. No clear idea of the biochemical transformations in a lake can be gained without a consideration of the redox potential and the related concentration of iron. For this reason it has seemed proper to consider these topics before proceeding to a treatment of the chief elements of biological interest, such as phosphorus and nitrogen, though it might seem that the importance of these elements would justify their earlier discussion.

REDOX POTENTIAL

When two solutions of some substance which can undergo reversible oxidation and reduction are joined by a neutral salt bridge, and bright platinum electrodes connected by a wire are dipped into the two solutions, it is found that if the ratios of oxidized and reduced states of the substance differ in the two solutions, a current can be detected in the wire. The solution containing an excess of the reduced state is oxidized or, in other words, loses electrons to its platinum electrode, while that containing an excess of the oxidized state of the substance gains electrons from its platinum electrode and is reduced. If when an equal and opposite current is applied to the wire the reaction stops, and when a greater current in the opposite direction is applied the reaction is reversed, the system is said to be reversible. The current in the wire is said to be set up by the oxidation-reduction potential or redox potential of one solution relative to the other. In practice, the potential of any one solution is referred to the standard hydrogen electrode as a convenient point of reference.

The oxygen potential. Oxygen dissolved in water (cf. Cooper 1937a) generates a redox potential at a bright platinum electrode according to the reaction

$$O_2 + 2H_2O + 4e = 4OH^-$$

If the reaction were to take place reversibly, the redox potential E_h, set up at such an electrode would be given by

$$E_h = E_0 - \frac{RT}{F} \ln \frac{a_{OH^-}}{\sqrt[4]{p_O}} \quad (1)$$

where E_0 can be obtained from thermodynamic considerations, as is explained by Cooper, and where a_{OH^-} is the activity of hydroxyl ions and p_O the partial pressure of oxygen. Since the activity of hydroxyl ions depends on the pH, or more accurately the activity of hydrogen ions,

$$-\log a_{OH^-} = pK_w - pH \quad (2)$$

Equation 1 may, for any given temperature, say 18° C., at which pK_w is 14.23, be rewritten, converting to ordinary logarithms and evaluating the constants,

$$E_h = 1.234 - 0.058\, pH + 0.0145 \log p_O \quad (3)$$

At the temperatures ordinarily found in lake waters, the theoretical reversible oxygen potential determined from such an equation, when the water is in equilibrium with air at 1 atm. and at pH 7, will give results which fall from 0.86 volt at 0° C. to 0.80 volt at 25° C. It will be apparent from the equation that the potential is insensitive to changes in p_O; reducing the oxygen concentration from 100 per cent saturation to 1.0 per cent saturation (i.e., to about 0.1 mg. per liter) will lower the potential only about 0.03 volt; if the potential is determined solely by oxygen tension, the latter will become analytically undetectable long before any great fall in E_h occurs. Since a rise of one unit in pH is accompanied by a fall in potential of 0.058 volt, it is often the practice to correct potentials at different pH values to pH 7 by subtracting 0.058 volt for every unit on the acid side of neutrality, or adding 0.058 volt for every unit on the alkaline side of neutrality. A potential corrected in this way to pH 7 is usually written E_7; any other arbitrary pH can, of course, be selected, and some workers have given their results as E_5.

It is important to note that this correction is dependent on the term for the activity of hydroxyl ions in equation 1. It is therefore valid only when there is reason to believe that an oxygen potential is involved. Other redox systems may be independent of *p*H, or the *p*H may enter into the equation with a different coefficient. There is reason to believe that the potentials set up on platinum electrodes dipped into many well-oxygenated natural waters are oxygen potentials, but this does not imply that every redox potential measured in the hydrosphere can be corrected to give an E_7 value by addition of 0.058 (*p*H − 7) volt.

When the oxygen potential is actually measured by dipping a bright platinum electrode into oxygenated water, a value considerably lower than the theoretical potential of about 0.8 volt is obtained. The best value for neutral water at 25° C. is about 0.52 volt. The reason for this discrepancy appears to lie in the fact that the electrode in the presence of oxygen becomes covered with a cracked film of platinum oxide and is no longer part of a truly reversible system. It is usual to term the actual potential measured the *irreversible oxygen potential*, while the theoretical value calculated may be termed the *reversible oxygen potential.* Cooper (1937c) supposes that any reaction which proceeds to an equilibrium determined by the redox potential will proceed rapidly to a position determined by the irreversible potential and will then approach asymptotically an equilibrium position determined by the reversible potential. The reasoning involved is not entirely clear.

The redox potentials of lake waters. The first limnological studies were made by Karsinkin and Kusnetzow (1932), who were, however, concerned only with lake sediments. Kusnetzow (1935) studied lake waters, using redox indicators, but his published vertical series throw little light on the potential of well-oxygenated waters. Pearsall and Mortimer (1939) were similarly mainly concerned with the hypolimnion. A number of determinations have been published by Hutchinson, Deevey, and Wollack (1939), Deevey (1941), and Allgeier, Hafford, and Juday (1941), of the potential of the bright platinum electrode immersed in the surface water of lakes. These determinations give values that diverge little from those attributed to the irreversible oxygen potential in laboratory experiments with pure water or buffer solutions.

Hutchinson, Deevey, and Wollack, and Deevey have published determinations for samples of surface waters of eight Connecticut lakes; these determinations range from 0.456 volt to 0.479 volt. The corresponding E_7 values range from 0.461 to 0.561 volt, and the mean of these values is 0.473 volt, a little lower than that expected in laboratory experiments. Deevey added data for two lakes in New York State. One of these fell in the range of the adjacent Connecticut lakes; the other, Queechy Lake, gave a normal value of 0.491 volt for E_7 in 1938, but two very high readings, 0.635 and 0.619 volt, in August and September 1939.

Allgeier, Hafford, and Juday (1941) obtained eighteen readings *in situ* on the surface waters of twelve lakes in northern Wisconsin. They record a range for surface waters from 0.380 to 0.505 volt. Their individual readings are presented graphically; as far as can be ascertained, they correspond to a range of E_7 values from 0.37 to 0.50 volt, the mean being 0.437 volt. Two values stand out as being notably low and are discussed in detail in the subsequent paragraph; omission of these two low values raises the mean to only 0.444 volt, which is a little below the mean for Connecticut and a fortiori below the value expected from laboratory experiments. In contrast to this, Mortimer (1941–1942), in his very detailed examination of the seasonal change of redox potential at a series of depths in Esthwaite Water in the English Lake District, found the surface water to be normally at about 0.50 to 0.52 volt, though at the time of the full autumnal circulation period a very slightly lower value of E_7, 0.48 volt, was obtained.

With the exception of Queechy Lake in New York and Nebish and Helmet Lakes in Wisconsin, the available values, though often a little lower than would be expected, at first sight hardly diverge enough from the value expected for the irreversible oxygen potential to suggest that any other system is operating. The high values for Queechy are puzzling; they seem unlikely to be due to any technical error, but at present are not otherwise explicable. Mortimer obtained equally high values in water from Windermere left exposed to the air over oxidizing lake mud. The relevant data for the two Wisconsin lakes giving low surface values are entered in Table 88.

TABLE 88. *pH, oxygen concentration and redox potential of the surface water of lakes exhibiting anomalous low values of the potential*

	*p*H	O_2, mg. per liter	E_h	E_7
Nebish Lake				
July 21, 1939	7.0	7.4	0.38	0.38
August 23, 1939	7.0	...	0.45	0.45
Helmet Lake				
August 3, 1939	6.0	5.3	0.42	0.37
August 29, 1939	5.7	6.2	0.505	0.43

In Nebish Lake, which is described as eutrophic, the oxygen content on the day of the low determination is not abnormal, and no obvious explanation of the redox value is forthcoming. In Helmet Lake, which has excessively colored water (236 Pt units), the oxygen content on the day of the low potential is very low for an exposed-surface water. It seems reasonable in this case to suppose that the organic matter responsible for the high color, and presumably also for the low oxygen concentration, can under some circumstances lower the redox potential also. If this is the case, it is reasonable to inquire whether the slight difference usually to be observed when the potentials in oxygenated lake waters are compared with those obtained in the laboratory, is not also due to the presence of small amounts of organic reducing substances. In the case of Esthwaite, where there is a great deal of comparative data, Mortimer believed that the slightly lower value of the potential observable at the time of the autumnal overturn is to be attributed to reducing substances, other than ferrous iron, derived from the hypolimnion. It may be noted in passing that Müller (1938b) observed that water pressed from the sphagnum bog of the Lunzer Obersee acted as a reducing agent, forming soluble ferrous iron from ferric hydroxide. Hydrogen sulfide was absent, and he supposed an organic compound to be involved.

Vertical distribution of redox potentials. The vertical distribution of redox potentials has been considered by Kusnetzow (1935), Hutchinson, Deevey, and Wollack (1939), Allgeier, Hafford, and Juday (1941), and in great detail by Mortimer (1941–1942). In lakes in which there is, during stratification, a more or less orthograde oxygen curve, the E_h curve also tends to be orthograde, the values of the potential falling only when the mud surface is actually reached or penetrated. A clinograde oxygen curve does not of itself imply a clinograde E_h curve. This will be apparent from theory when it is remembered how little the oxygen concentration affects the oxygen potential.

A few cases of relatively orthograde E_h curves coexisting with markedly clinograde oxygen curves are actually known; Lake Quassapaug, Connecticut (Fig. 193) provides a good example. It is probable that in all such cases the mud retains an oxidized microzone in the presence of about 1 mg. O_2 per liter. In most small lakes with clinograde oxygen curves and marked hypolimnetic oxygen deficits, the E_h curve is markedly clinograde, as Kusnetzow (1935) first demonstrated. The potential may fall off at greater depths than those at which the main fall in the oxygen content occurs, but not infrequently a very large part of the hypolimnetic water may exhibit low potentials (Fig. 194). It may be assumed that in every case in which clinograde E_h curves occur, redox systems other than the oxygen-hydroxyl system are involved. In Helmet Lake, Wisconsin, the surface waters of which have already been discussed, a potential as low as $E_h = 0.375$ volt at *p*H 5.95, or $E_7 = 0.314$ volt, has been recorded in the upper hypolimnion in the absence of ferrous iron or hydrogen

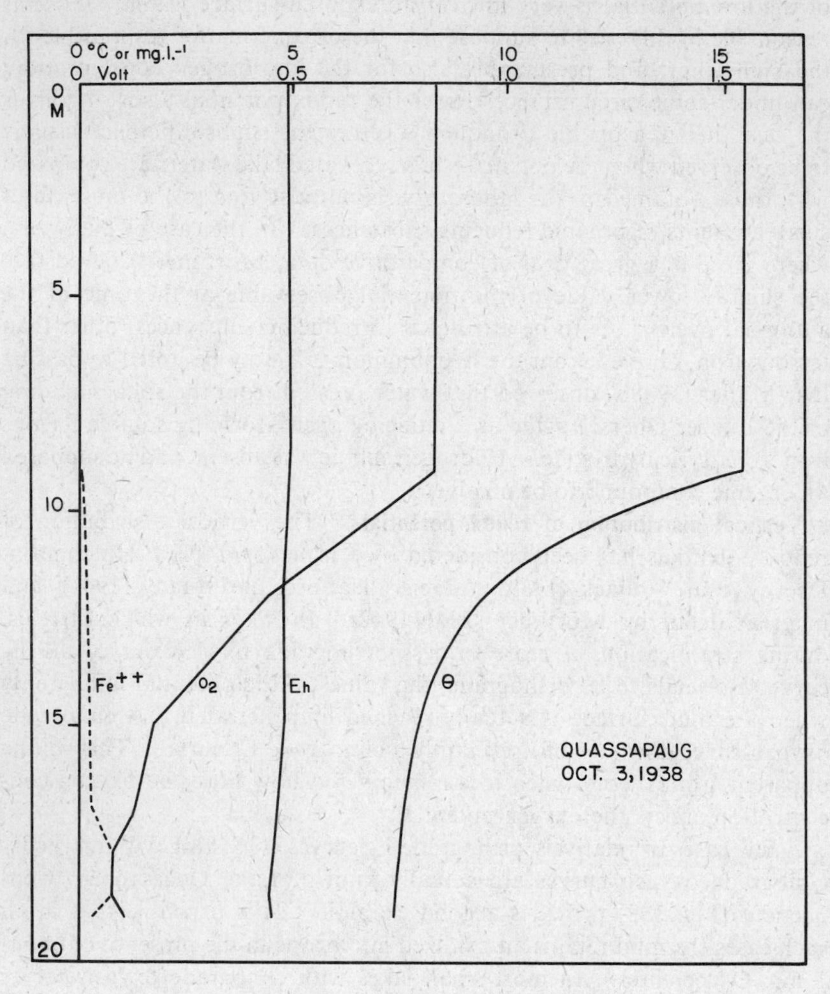

FIGURE 193. Lake Quassapaug, Connecticut. Vertical distribution of temperature, oxygen, redox potential, and ferrous iron late in summer stagnation, as an example of a lake with a clinograde oxygen curve and an orthograde redox potential curve.

sulfide and in the presence of minimal amounts of oxygen. In this case it is reasonable to suppose that some organic oxidation-reduction system is operating. In almost all the other cases that have been investigated, it seems probable that the most important system involved is the ferrous-ferric system. In the case of Lake Mary, Wisconsin, which has already been discussed as a meromictic lake, the main reducing substance in the monimolimnion appears to be hydrogen sulfide. The redox potential of the monimolimnion falls from $E_h = 0.140$ volt at 8 m. to $E_h = 0.075$ volt at 20.5 m., while the H_2S content rises from 1.2 to 2.1 mg. per liter. No ferrous iron was detected. It is probable that some other, but by no means all, meromictic lakes will be found to show a similar phenomenon.

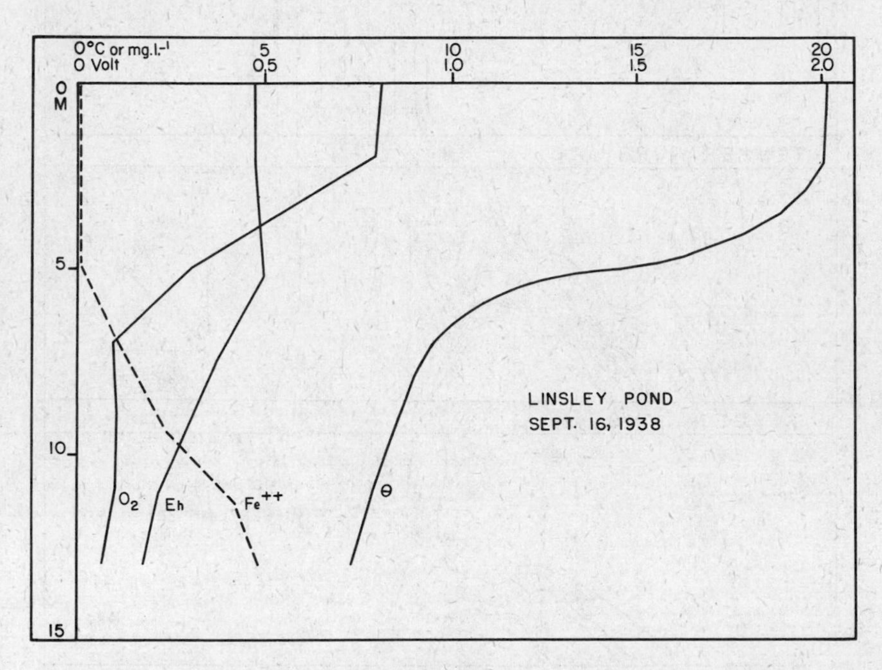

FIGURE 194. Linsley Pond, Connecticut. Vertical distribution of temperature, oxygen, redox potential, and ferrous iron, as an example of a lake with clinograde oxygen and redox potential curves.

Mortimer has made an elaborate study of the seasonal variation of redox potential in Esthwaite Water, a lake in which markedly clinograde oxygen and E_h curves are developed. His results are expressed in the form of a depth-time diagram, here reproduced (Fig. 195), and indicate graphically the rise of the isovolt surfaces from the bottom of the lake through the hypolimnion during summer stagnation. In this lake there is little vertical variation in the redox potential during winter stagnation. Ivlev's studies on the iron content of Lake Beloye suggest that the rigors of an extreme continental climate would produce marked microzonal stratification in the redox potential under the ice cover.

The redox potential of lake sediments. Except on occasion at its surface,

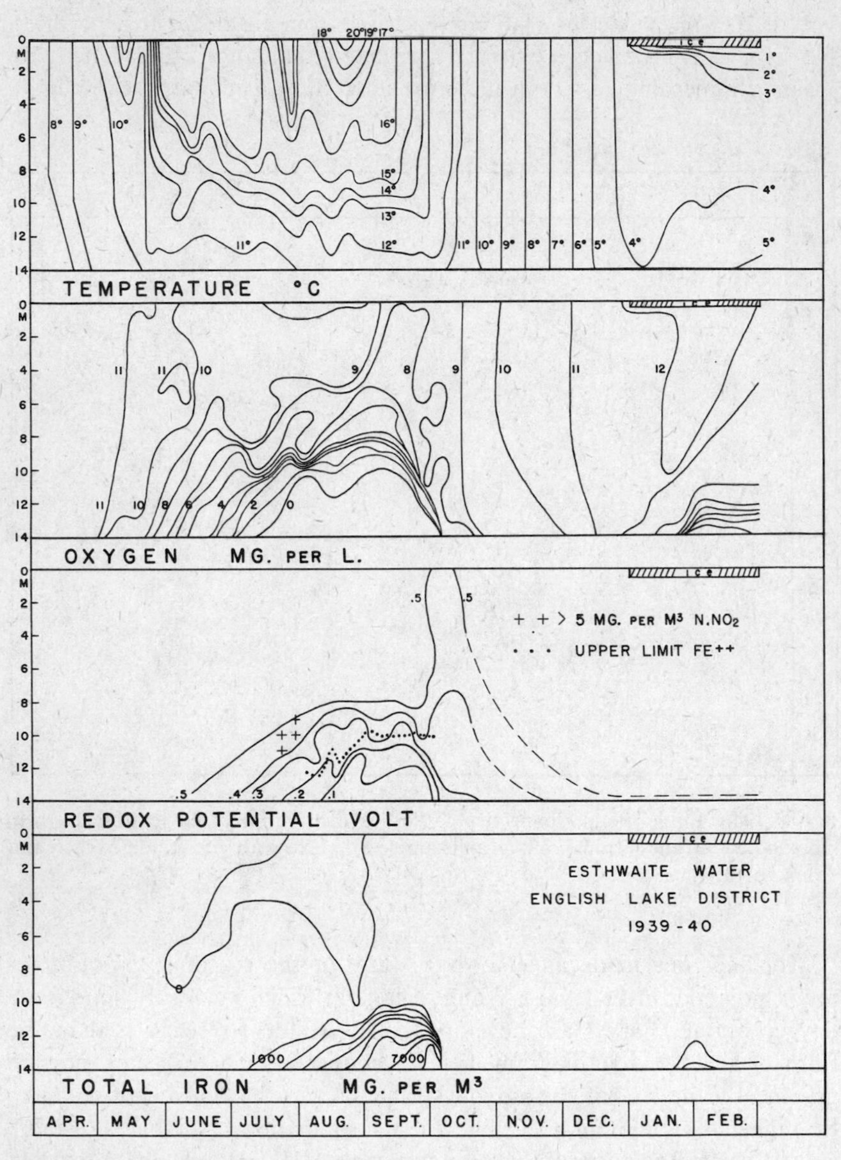

FIGURE 195. Esthwaite Water, seasonal variation in temperature, oxygen, redox potential, and iron, 1939–1940 (Mortimer).

the mud at the bottom of lakes probably always has reducing properties; the only likely cases of a thick layer of mud that did not have reducing properties would be in practically azoic lakes in which glacial rock debris is sedimenting. The redox potential of sediments generally varies around 0.0 volt. Kusnetzow (1935) gives a series of values for lakes near Moscow, from $E_h = -0.122$ to $E_h = +0.058$. The sediments of the Wisconsin lakes studied by Allgeier, Hafford, and Juday gave values between

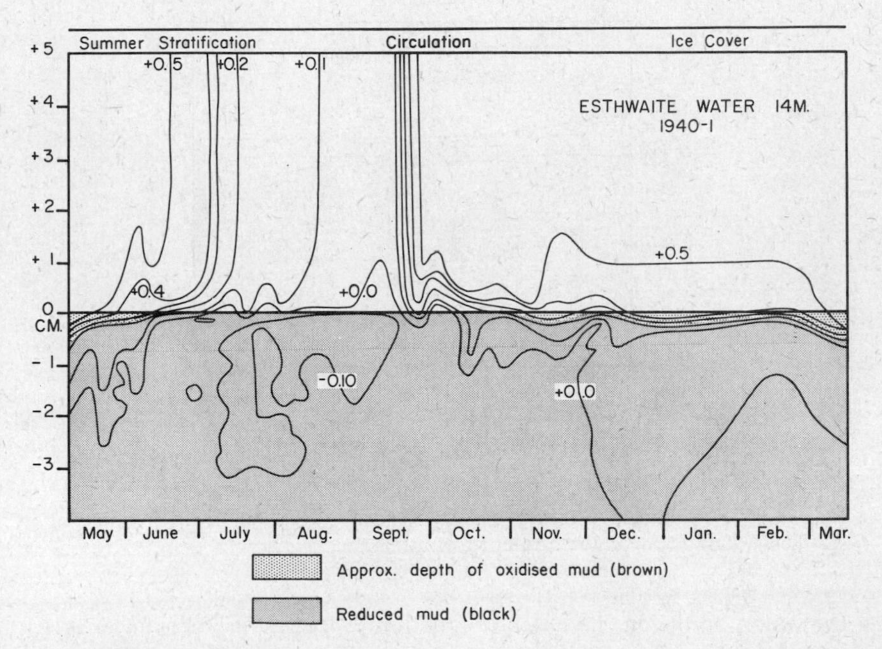

FIGURE 196. Esthwaite Water, seasonal variation in redox potential in the sediments below 14 m. of water (after Mortimer).

$E_h = -0.140$ and $E_h = +0.200$. In view of the complex nature of the vertical changes near the mud surface, these variations have little meaning.

The most complete accounts of the variation in the redox potential in the deep-water sediments of lakes are those of Mortimer for a station in 14 m. of water in Esthwaite Water from May to September 1940 (Fig. 196), and for a station in 65 m. of water in the north basin of Windermere (Fig. 197) from September 1940 to February 1941. At both stations, undisturbed cores were obtained and the potentials determined for the water and mud by means of a number of small electrodes at various depths down to 4 cm. below the mud-water interface. As in the other lake sediments so far studied, the potential at this depth below the mud lay at about 0.0 volt. At the time of vernal circulation in Esthwaite, the value at the mud-water interface corresponds to $E_7 = 0.50$ to 0.52 volt, as in the free water. The potential in the mud falls rapidly, the isovolt for 0.0 volt lying at about 2 cm. below the interface. The conditions in the top few millimeters of mud are evidently determined by diffusion of oxygen from

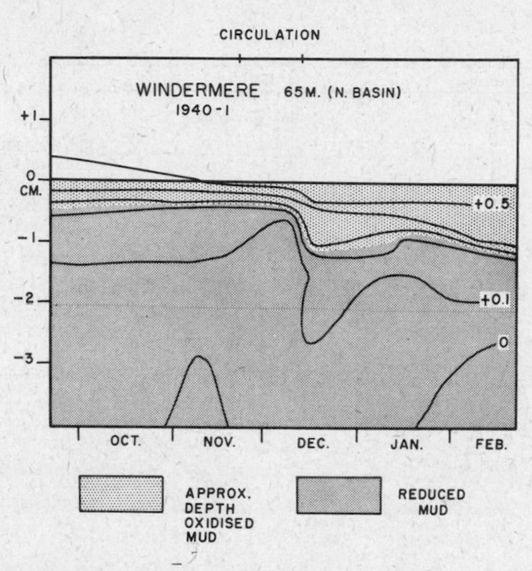

FIGURE 197. Windermere (north basin), seasonal variation of redox potential in sediments below 65 m. of water (after Mortimer).

the water, on the one hand, and reduction processes of biochemical origin in the mud, on the other. As stratification is developed and the supply of oxygen available in the contact water declines, the isovolts rise in the mud. By the middle of June the isovolt $E_7 = 0.2$ volt had crossed the mud-water interface. Prior to this time the superficial layer of mud had been marked by a brown layer, or oxidized microzone, presumably containing ferric oxide. When $E_7 < 0.2$ volt, the oxidized microzone disappears. As summer stagnation progresses, the superficial potential falls till in the middle of September a value of -0.03 is reached at the mud surface, and the bottom centimeter of water lies below the 0.0 isovolt. Oxygen was always detectable by the Winkler-Alsterberg method in the 14-m. water sample, though between mid-July and mid-September the amount present never exceeded 0.32 mg. per liter. As circulation begins again in the autumn, the process just described is reversed. Oxygen is brought into the deep water; the oxidative microzone reforms slowly as a very thin, flocculent, brown precipitate; and throughout the winter, even though an ice cover forms, the 0.2 isovolt remains below the mud surface.

The more restricted series of observations made on Windermere show that a similar movement of the isovolts occurs, but the 0.2 isovolt never rises, even at the height of stagnation, above a depth of just over half a centimeter below the interface, so that an oxidized microzone is present throughout the entire annual cycle. Samples taken on single occasions indicated that in Thirlmere, Crummock Water, Ennerdale Water, and Derwentwater, the hypolimnion is not deoxygenated enough to permit the disappearance of the microzone; in Blelham Tarn, Rydal Water, and Loweswater, a cycle similar to that found in Esthwaite occurs. The presence of an oxidized microzone throughout the entire year in some lakes, and its disappearance seasonally in the deeper parts of the basin in others, may well be the most important dichotomy in the chemical classification of lakes.

Experimental studies. Further insight into the seasonal variations in the deepwater sediments was obtained by Mortimer from experiments in which mud from 30 m. of water in Windermere was placed in aquaria and covered with a layer of lake water (Fig. 198). One aquarium (anaerobic tank) was set up with a paraffin seal over the water, the other (aerated tank) without such a seal so that free exchange with the air took place. In the aerated tank the redox potential at the mud-water interface remained high, actually increasing irregularly throughout the experiment from $E_7 = 0.45$ volt to about $E_7 = 0.64$. The conductivity of the mud rose and then fell slightly at all depths. No very great chemical changes occurred in the water over the mud, the most marked being a slight increase in sulfate at the end of the experimental period. The iron content of the water remained very low, and no ferrous iron was observed.

The events in the anaerobic tank were complex and very different from those in the aerated tank. The isovolts in the mud rose, the 0.24 isovolt reaching the surface and the oxidized microzone disappearing after about 70 days, though in the early part of the experiment the sealing of the water surface was inadequate so that the upward movement began later than it would have done if oxygen had been completely excluded from the beginning. Later in the experiment, a crack in the wax seal produced a depression of the isovolts after about 83 days, but on resealing they started to rise again. An enormous increase in the conductivity of the mud, and to a less extent in that of the water, occurred during the later part of the experimental period. Very marked chemical changes occurred in the water. Alkalinity increased irregularly throughout the entire experiment after an effective seal had been established. Ammonia

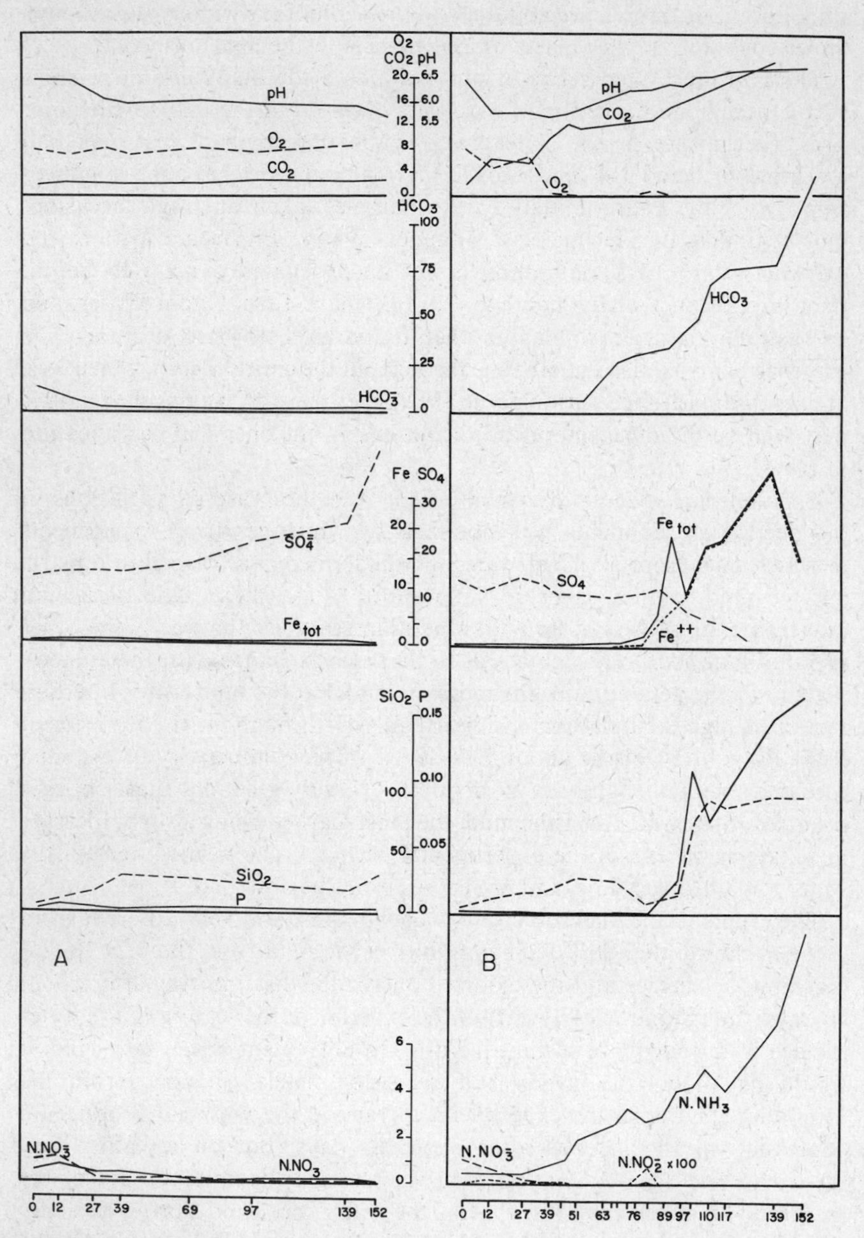

FIGURE 198. Variation in the chemical composition of water over deepwater mud from Windermere: *A*, under aerobic conditions; *B*, under anaerobic conditions. (After Mortimer.)

increased in a somewhat similar way. Substances reducing permanganate, total iron, and silicate rose suddenly at about the time of the disappearance of the oxidized microzone, but the rise did not continue throughout the experimental period. The iron content of the water fell after the crack had developed, but then rose again; the second rise involved only ferrous iron, which later declined, probably owing to precipitation as sulfide. Phosphate rose rapidly with iron, then declined and rose again. Sulfate showed a marked decline during the later part of the experiment. It will be evident that the changes in the anaerobic tank closely paralleled those in the hypolimnion of Esthwaite Water.

It is evident from these experiments that the difference between the conditions in the hypolimnia of Windermere and other lakes in which an oxidized microzone is retained throughout summer stratification, and those such as Esthwaite in which the mud surface loses such a microzone, is not necessarily dependent on the chemical nature of the mud, though in an extremely inorganic sediment reduction would not occur. The main control in ordinary cases is evidently by the mechanisms which bring a continual supply of oxygen up against the mud surface in the larger, deeper lakes with more turbulent hypolimnia. Even in the sediments on the bottom of Windermere, where the conductivity in the top 1 cm. shows little seasonal variation, there is a fall of conductivity as far into the mud as 30 cm. at the time of the autumnal circulation, when an abundant supply of oxygen is brought into the depths.

It is also evident that the presence or absence of the oxidized microzone makes an immense difference to the amount and nature of the substances leaving the mud and passing into the water. It is reasonable to conclude that two entirely different types of process are involved in the liberation of material from the mud to the water under anaerobic conditions. When the materials produced at a low redox potential are precipitated at a higher redox potential—as is the case with ferrous iron, precipitated as ferric hydroxide, phosphate in the presence of an excess of ferrous iron, precipitated as ferric phosphate, and probably manganese—it will be impossible for such substances to pass into the free water across the oxidative microzone, because at the redox potential of the latter they will tend under most conditions to precipitate. This explanation, however, does not apply to the diffusion of alkali or silicate into the water. Mortimer suggests that the very general increase in solutes passing into the water when the microzone is reduced may best be explained by assuming that such solutes, both bases and anions, were held by the ferric iron of the oxidized microzone as adsorption complexes. He further suggests that the adsorptive material in oxidized mud may well be a ferrilignoprotein complex. Reduction of the iron abolishes the adsorptive capacity and liberates bases and some anions into both the free water and the water in the mud, causing the observed changes in composition and conductivity of the mud and the water in the anaerobic tank and in small stratified lakes, such as Esthwaite and many others that have been studied. The details of such migrations, though dependent on the redox cycle, are best considered in relation to the individual substances, which will be treated in later chapters.

IRON IN LAKE WATERS

The ferrous-ferric system. The ferrous-ferric system, involving the equilibrium

$$Fe^{+++} + e \rightleftharpoons Fe^{++}$$

is a reversible system, which in acid solution is independent of *p*H. At the *p*H values normally encountered in the biosphere, the activity of ferric ions will be set as that in equilibrium with excess ferric hydroxide, which may be regarded as present in the solid phase. This concentration is very low and is dependent on the *p*H. As well as ferric ions, ferrihydroxy, ions $FeOH^{++}$ and perhaps ferrite ions $H_2FeO_3^-$ will also be present, but the concentrations are again so low that no form of trivalent iron in ionic solution is likely to be detectable analytically in normal lake waters. The equilibria involved have been exhaustively considered by Cooper (1937c, 1948a), whose treatment is followed in the present account.

The potential of a system containing ferrous and ferric ions will be given by

$$E_h = E_0 + \frac{RT}{F} \ln \frac{a_{Fe^{+++}}}{a_{Fe^{++}}} \tag{4}$$

Considering the available data for the constants involved when the activity of the ferric ions is determined by the *p*H, and for the standard potential E_0 for the ferrous-ferric system when the concentrations of the two ions are equal, Cooper derives, for the situations of ecological and geochemical interest, the equation

$$E_h = 1.011 - 0.058 \log a_{Fe^{++}} - 0.174pH \tag{5}$$

The activity coefficients are not well known, but at the great dilutions involved, the activity of the ferrous ion may be regarded as given by the concentration. Using the available data for the computation of the activities of the ferric, ferrihydroxyl, and, more doubtfully, ferrite ions, which will depend only on the *p*H in the presence of solid ferric hydroxide, and using equation 5 for computating the activity of ferrous ions, the quantities in Table 89 are obtained.

TABLE 89. *Solubility of various forms of iron*
(All computations in mg.-atom per m.³)

	*p*H 8.5	*p*H 8.0	*p*H 7.5	*p*H 7.0	*p*H 6.5	*p*H 6.0
$a_{Fe^{+++}}$	$10^{-15.4}$*	$10^{-13.9}$	$10^{-12.4}$	$10^{-10.9}$	$10^{-9.4}$	$10^{-7.9}$
$a_{FeOH^{++}}$	$10^{-9.3}$	$10^{-8.3}$	$10^{-7.3}$	$10^{-6.3}$	$10^{-5.3}$	$10^{-4.3}$
$a_{H_2FeO_3^-}$	$10^{-4.8}$	$10^{-5.3}$	$10^{-5.8}$	$10^{-6.3}$	$10^{-6.8}$	$10^{-7.3}$
$a_{Fe^{++}}$						
at E_h = 0.8	$10^{-16.0}$	$10^{-14.5}$	$10^{-13.0}$	$10^{-11.5}$	$10^{-10.0}$	$10^{-8.5}$
0.6	$10^{-12.5}$	$10^{-11.0}$	$10^{-9.5}$	$10^{-8.0}$	$10^{-6.5}$	$10^{-5.0}$
0.4	$10^{-9.0}$	$10^{-7.5}$	$10^{-6.0}$	$10^{-4.5}$	$10^{-3.0}$	$10^{-1.5}$
0.3	$10^{-7.25}$	$10^{-5.75}$	$10^{-4.25}$	$10^{-2.75}$	$10^{-1.25}$	**2**
0.25	$10^{-6.4}$	$10^{-4.9}$	$10^{-3.4}$	$10^{-1.9}$	**0.4**†	**13**
0.2	$10^{-5.5}$	$10^{-4.0}$	$10^{-2.5}$	**0.1**	**3**	**100**
0.0	10^{-2}	**0.3**	**10**	**300**	**10,000**	**300,000**

* Quantities less than 0.1 mg.-atom per m.³ or 5.6 mg. per m.³ are not likely to be analytically detectable and are entered in the table as negative powers of 10.
† The analytically detectable amounts of ferrous iron are given to one significant figure in boldface type.

As has been noted previously, Cooper supposes that if a small amount of ferrous iron in solution is added to water in equilibrium with air, the concentration of ferrous ions corresponding to the irreversible oxygen potential will be rapidly achieved, and the concentration corresponding to the reversible potential will then be slowly and asymptotically approached. Examination of Table 89 indicates that in neither case is an analytically detectable amount of either ferrous or any other ionic form of iron to be expected at equilibrium at any *p*H encountered in the surface waters of normal lakes, though at *p*H 3 appreciable quantities, even of ferric iron, will be possible in ionic solution. In the hypolimnia of small lakes and in the vicinity of deepwater sediments of many large lakes where redox potentials of less than 0.3 volt are common, ionic ferrous iron may often be expected to occur. At least in some cases of this sort, the quantitative predictions of the theory are evidently roughly confirmed, though it is probable that the theoretical values given are not very accurate, in view of the uncertainty regarding some of the quantities involved in deriving equation 4. There can be little doubt, however, that these theoretical values indicate in general the orders of magnitude of the concentrations of *ionic* iron to be expected in lake waters, and that these concentrations are usually very much less than those of *total* iron actually recorded, even in analyses of well-oxygenated surface samples.

The occurrence of iron in epilimnetic lake waters. A vast amount of analytical data exists relating to the iron content of inland waters, and still more determinations of $(Al,Fe)_2O_3$ exist. Many of these data refer to rivers and to reservoirs in which there is rapid replacement; such determinations are of little interest to the student of the dynamics of lakes.

The limnological studies of Yoshimura (1931a, 1936c), Müller (1932, 1938b), Ohle (1934a), Ruttner (1931, 1937), Juday, Birge, and Meloche (1938), Lohammar (1938), and Åberg and Rodhe (1942) have now provided a large body of systematic information on the iron contents of lake waters in Japan, Europe, and North America. These studies have been supplemented by various special investigations. The relationship of iron to brown organic matter has been the subject of study by a number of workers for many years, as has the whole question of the metabolism of the element by iron bacteria in fresh waters. These matters are dealt with in greater detail in the following paragraphs. Of the more modern work on the elucidation of the iron cycle in particular lakes, the studies of Ohle (1935, 1937), Ivlev (1937), Einsele (1936, 1937, 1938), and above all those of Mortimer (1941–1942) have proved to be of particular importance. In addition to his theoretical investigations, Cooper (1932, 1948a) has made important analytical studies on the iron content of seawater, and these as well as the experimental work of Harvey (1937a,b) are of considerable interest to the limnologist.

Bearing in mind the theoretical conclusions presented in the previous section, the extensive data of Yoshimura may first be examined. They refer to the iron contents, determined after oxidation with bromine water, of a large number of Japanese lakes. It is probable that the oxidation was imperfect, but the determinations of the series must be roughly comparable and, owing to the diversity of the lakes studied, they form the most convenient introduction to the subject. Yoshimura groups his lakes into four categories:

(1) Uncolored neutral or alkaline lakes, poor in plankton; iron generally undetectable, but occasionally as high as 50 mg. m^{-3}

(2) Moderately colored lakes, neutral or alkaline, rich in plankton; iron often undetectable, but may be as high as 260 mg. m^{-3}

(3) Lakes colored by organic material, slightly acid (*p*H 4.2 to 6.6); total iron from 30 to 2500 mg. $m.^{-3}$

(4) Lakes poor in organic matter but containing sulfuric acid of volcanic origin (*p*H 1.8 to 5.8); total iron from 90 to 19,200 mg. $m.^{-3}$

The first three categories of lakes will be familiar to limnologists over a large part of the temperate regions of the earth's surface, and in general the distribution of iron in the surface waters of such lakes follows the pattern indicated by Yoshimura's data. The fourth category is of much less general distribution and has been properly studied only in Japan. In the more acid of such lakes, the very large quantity of iron is usually in part in true solution, though suspended ferric hydroxide is visible in the water in some cases. The very large amounts of iron, above 8000 mg. $m.^{-3}$, are found only when the *p*H is below 3.3. At this *p*H there could be of the order of 1000 mg. $m.^{-3}$ of iron in ionic solution in equilibrium with the suspended ferric hydroxide. In all other cases it is evident that the problem of the form of the iron implied by the analytical data demands careful consideration.

Forms of iron in lake waters. Most of the published fractionations of the forms of iron in well-oxygenated lake waters are incomplete or have been referred to meaningless categories. Juday, Birge, and Meloche (1938) give some results in which ferric and ferrous iron were determined by the thiocyanate and ferricyanide methods, respectively, but they conclude that in general the best procedure is to estimate total and ferric iron and then to determine ferrous iron by subtraction. It is certain that such a procedure gives a quite inadequate view of the nature of the iron in an oxygenated lake water.

In his study of the use of α-α'-dipyridyl as a reagent for the ferrous iron in lake waters, Müller (1932) gives some data for the ferrous iron in the well-oxygenated water of the Lunzer Obersee during autumnal circulation. At 12 m., 160 mg. $m.^{-3}$ of ferrous iron, 300 mg. $m.^{-3}$ total iron by reduction, and 250 mg. $m.^{-3}$ total iron by oxidation are said to have occurred in the presence of 8.53 mg. O_2 per liter. It is by no means certain that the ferrous iron really represents ferrous ions in solution, as it has been shown (Harvey 1937b, Hutchinson 1941, Cooper 1948a) that in the presence of suspended ferric oxide, dipyridyl disturbs the ferrous-ferric equilibrium and gives spurious values for the ferrous iron. Hutchinson records up to 70 mg. $m.^{-3}$ of such apparent ferrous iron in the surface waters of Linsley Pond at times when the ferrous ion was only doubtfully detectable in filtrates. Müller's total iron likewise by no means certainly represents the whole of the iron in his samples. It is more likely to represent the reducible iron in Cooper's (1932) sense, which is probably largely suspended ferric hydroxide. The fact that Müller obtained rather larger amounts of reducible iron by dipyridyl than total iron by thiocyanate after oxidation cannot be completely explained, but some of both oxidizable and reducible iron was no doubt in organic combination, and a far more complex situation than is indicated by his discussion doubtless actually existed. These observations, and a number of others in the literature, indicate that a considerable amount of reducible iron is often present and that ferric iron can be liberated by oxidizing agents. Little more information was available until recently.

Rodhe (1948) has concluded from an experiment to be discussed in a later paragraph that the iron determined by ortho-phenanthroline, which is always very much less than the total iron, gives a rough measure of the iron available to the phytoplankton. It is not clear, however, in what form this *reactive iron* is actually present in the lake water. Since, in these determinations of reactive iron, hydroxylamine hydrochloride was added as a reducing agent, the determinations evidently refer to what Cooper calls reducible iron, which is largely present as ferric hydroxide adsorbed on particulate matter. Rodhe's actual determinations for surface waters are given in Table 90. It will be observed that evaporation to dryness and resolution liberates some of the previously unreactive iron of the Erken sample in a reactive form. In view of the fact that this sample had to be concentrated by evaporation tenfold before reactive iron could be determined, it is possible that some of the 10 mg. m.$^{-3}$ reactive iron recorded was liberated during this partial evaporation.

TABLE 90. *Total and reactive iron in two Swedish lake waters*

	Values, mg. m^{-3}
Erken	
Total iron (range in 1941)	90–160
Reactive iron after concentration (May 19, 1944)	10
Reactive iron after evaporation and resolution in water	30
Siggeforasjön (oligodystrophic)	
Total iron (range in 1941)	260–270
Reactive iron after evaporation and resolution (June 13, 1944)	30

Hutchinson (1941) has published a few analyses for the surface waters of Linsley Pond, a lake which would fall into the second category used in the presentation of Yoshimura's results. The total iron of the surface waters varied from 70 to 550 mg. m.$^{-3}$, though the higher values probably occur only when the melting of snow, or complete circulation, brings mineral particles from the drainage basin or material from the sediments into suspension in the water of the lake. The values in Table 91 were given by a series of analyses in March and April. These rather meager data, together with some other determinations of ferric iron in suspension made at other seasons, permit the following tentative conclusions.

(1) No significant amounts of ionic iron occur in the water. The ferrous iron recorded in solution on two occasions, on both of which particulate ferrous iron was abundant, was barely detectable and may reasonably be regarded as being in a particulate form, traces of which had passed through the membrane filter employed, which retains only material in excess of about 1μ diameter.

(2) Ferric iron is normally present in suspension in rather constant amounts. In view of the fact that the filter employed did not retain colloidal substances, there is no evidence of the existence of colloidal ferric hydroxide in any form that dissolves in the acid solution used in the determination of ferric iron with thiocyanate.

(3) Ferrous iron appears to occur in suspension, possibly in soil or sediment particles.

TABLE 91. *Distribution of various fractions of total iron in the surface water of Linsley Pond*

	Range, mg. m.$^{-3}$	Mean, mg. m.$^{-3}$
Total iron	70–300	170
Ferric iron in suspension	30–50	40
Ferric iron in solution	Not detectable	
Ferrous iron in suspension	10–40	20
Ferrous iron in solution	0–5(?)	< 5
Nonreducible iron in suspension	4–180	75
Nonreducible iron in solution	20–50	30

(4) A very variable amount of nonreducible iron occurs in suspension. The lowest values for this constituent probably refer to the iron present in plankton organisms; the highest values may well include a considerable amount of the element in iron-containing minerals in silt particles.

(5) A rather constant amount of a nonreducible iron compound occurs in colloidal or true solution. It is reasonably certain that this compound is an iron-containing organic complex, much stabler than the complexes of ferric iron and colored organic matter considered in a later section.

(6) There is evidence from the occurrence of acid-soluble sestonic phosphorus in the surface waters during the period just after the autumnal overturn, that suspended ferric phosphate is present in the surface waters for a short period at this season, but this ferric phosphate disappears in a few weeks, being either sedimented or hydrolyzed (Cooper 1948a).

It would therefore appear that at least in well-oxygenated lake waters iron compounds likely to be most constantly present are suspended ferric hydroxide in some cases associated with yellow organic matter and a nonreducible organic complex or series of complexes.

Significance of ferric hydroxide. In laboratory experiments, in the absence of organic matter, freshly formed ferric hydroxide sol slowly coagulates, and the precipitate sinks to the bottom of the experimental vessel. Such precipitation is of course much more rapid in solutions rich in electrolytes than in pure water. If the process occurred inevitably in saline water in nature, the surface waters of the sea would, as Harvey pointed out, soon become depleted of ferric hydroxide. Müller notes that a precipitate of ferric hydroxide might form in bottles of epilimnetic water from the Lunzer Obersee when hung in the hypolimnion from a buoy. In dilute lakes the ferric hydroxide sol might be stable for a long time, but this observation indicates that a slow precipitation is likely to be of importance in many lake waters. Various organic substances, either colloids, such as gum arabic, or crystalloids, such as ascorbic acid and other polyhydroxyl compounds studied by Harvey, stabilize ferric hydroxide sols. The yellow acid materials isolated from Linsley Pond water by Shapiro (1956) behave similarly.

Harvey (1939) believes that an organic substance which can be extracted from decaying seaweed or from soil and which promotes the growth of the diatom *Ditylum* in culture, actually operates by permitting iron to be retained in the medium as a stablized ferric sol. The work of various investigators, to be discussed, shows that soil extracts often do operate as growth promoting substances by permitting iron to remain in solution. The field evidence from Linsley Pond, however, strongly suggests that the whole of the iron in either colloidal or true solution was in a much more stable form than protected ferric hydroxide sol. The unequivocal ferric fraction in the Linsley samples was, moreover, in suspension and appeared to be present whenever analyses were made. It may be continually formed and continually sedimented, but its constant presence is probably explained by another observation of Harvey's, namely that diatoms from the open sea appear to be covered with a very thin film of ferric iron, which produces a faint blue color with ferrocyanide. The presence of this film explains the rather high ratios of Fe:P usually observed in analyses of such organisms (e.g., Cooper 1934), which are greatly in excess of those to be expected in living matter. Harvey has produced evidence that such films, which presumably result from the adsorption of ferric hydroxide produced from the hydrolysis of iron compounds in decomposing organisms and, as Cooper emphasizes (1948a), in faeces, can provide a source of iron for the organisms which carry them.

It is very probable that a considerable part of the ferric hydroxide found in lake water is adsorbed on plankton and on dead sestonic particles. It is also possible that the surface film itself carries an important store of iron. Cooper has shown that this is true even in the open sea of the western end of the English Channel, and he indicates, as Pearsall and Mortimer (1939) have also pointed out, that films of ferric hydroxide of considerable thickness often form on stagnant pools in bogs, presumably where oxygen is penetrating the surface of water rich in ferrous iron, at a low redox potential.

Significance of organic iron complexes. It has long been known that iron tends to be associated with the dark material both of soils and of organically stained waters (Aschan 1906, Shapiro 1956). Aschan found that it was possible to precipitate a ferric organic complex from such dark waters by the addition of 0.03 to 0.06 g. ferric chloride per liter of water. No precipitate was formed if the amount of ferric chloride was too small or too great. Aarnio (1915) found that within certain limits the so-called humus from sphagnum bogs coprecipitated with ferric hydroxide sol. In general, if the ratio of humus to Fe_2O_3 was above 3.0:1, a sol, stable for at least twenty-four hours, was formed. Decreasing the amount of humus caused precipitation until a lower limit around 0.2:1 was reached, when the sol again was stable. Much smaller amounts of Al_2O_3 were needed to produce a precipitate, and it seems likely that in nature the resultant greater mobility of ferric iron leads to a separation of iron from aluminum, the aluminum being rapidly sedimented, the iron remaining in the liquid phase.

In the long series of analyses of the surface waters of Swedish lakes made by Lohammar, there is a definite tendency for high iron contents to occur in the north Swedish lake waters, which tend to be acid and brown. In this series the highest ratios of iron to ignitible matter are of the order of 1:10. The maximum value of the ratio is given by the water of Rånträsket (lat. 65°52′ N., long. 14°59′ E.), a lake with acid yellow-brown water containing 2500 mg. Fe $m.^{-3}$ and 23.1 mg. per liter ignitible matter. The calcium content of this water is 2.5 mg. per liter, which presumably corresponds to a CO_2 loss of 2.8 mg. per liter on ignition. The ratio of iron to organic matter is thus 2.5:20.3, or approximately 1:8. Such a ratio would, according to Aarnio's experiments, be stable. Åberg and Rodhe (1942) point out that in general the amounts of iron and manganese in brown waters are less than are required to cause coprecipitation. There is, however, evidence that the organic matter may in some circumstances be partly oxidized, either bacterially or photochemically in the absence of bacteria, the iron apparently acting catalytically (Spring 1897, Åberg and Rodhe 1942). After such oxidation has proceeded, precipitation may take place; further oxidation will ultimately remove a considerable part of the organic matter, producing the deposits high in iron or manganese described as lake ochre or lake iron ore. These substances will be discussed in Volume II.

The association of high iron with brown waters is by no means constant. Although Lohammar always found some iron in the north Swedish waters of this sort and often failed to detect the element in the lakes of the south

Swedish plains, very few of the northern lakes contained any large amounts of the element. Ohle (1934a) gives a series of values of from 102 to 480 mg. Fe m.$^{-3}$ for north German lakes with brown waters, and regards these as very high, though they are certainly not richer in total iron than many less stained waters elsewhere. Although lakes with brown humic water are usually acid, it is worth noticing that in the very peculiar closed, alkaline, slightly mineralized pans of the eastern Transvaal, organic matter may be associated with high iron. Hutchinson, Pickford, and Schuurman (1932) found there:

	Total Fe	Ignitible Organic Matter
Lake Chrissie	18,600 mg. m.$^{-3}$	187.8 mg. per liter
Blaauwater Pan I	10,700 mg. m.$^{-3}$	412.3 mg. per liter

The Lake Chrissie water, which contains the greater amount of iron, is turbid gray slightly tinged with pale brown; in Blaauwater Pan I the water is deep sepia. Although these values are abnormally high for the open surface waters of lakes not containing sulfuric acid, some acid bog waters may contain even more iron, for Uspenski (1927) noted up to 50,000 mg. m.$^{-3}$ in bogs near Moscow at *p*H 4.

A good deal of work by students of the culture of microorganisms has suggested that the beneficial effects of the addition of soil extracts to culture media depend on the ability of their organic matter to retain iron in solution in such a form that the microorganisms can use the element. This has been found to be true in the culture both of bacteria (Burk, Lineweaver, and Horner 1932) and of marine and fresh-water algae (Gran 1931, Harvey 1939, Pringsheim 1936, 1946, Rodhe 1948). Rodhe, for instance, shows how the addition of soil extract may maintain in solution, after autoclaving, the whole of 0.27 mg. Fe per liter added as $FeCl_3$, whereas in a control solution without soil extract 98 per cent of the iron was precipitated.

The only direct attempt to investigate the nature of the iron-organic complexes actually found in lakes is due to Ohle (1935). In some lakes in southern Sweden, for example, in the water of Lygnen, a highly colored lake somewhat influenced by human activity, he found by the thiocyanate method 910 mg. m.$^{-3}$ after oxidation of unfiltered water; after filtration through a "Cella" filter, which retains particles in excess of 0.75, the amount of iron by the same method was 1800 mg. m.$^{-3}$ It is evident that in this and similar cases the organic particulate matter is interfering with the iron determination, even after oxidation. In Skärshultsjön, the iron determined in the filtrate was less (420 mg. m.$^{-3}$) than in the original sample (1170 mg. m.$^{-3}$). It is evident that in this case the particulate matter was retaining iron that could react to thiocyanate after oxidation. It seems unnecessary to suppose that anything more complex than ferric hydroxide was involved. The inconsistent nature of such results led to further investigation. Iron was added in solution to samples of water from the Sandkathen Moor, so that 1680 mg. m.$^{-3}$ were present. One sample was retained at its natural *p*H, which was 4.6, the other was brought to *p*H 6.3 by means of calcium bicarbonate. The results are presented graphically, but appear to indicate:

	*p*H 4.6	*p*H 6.3
Iron determinable in filtrate	1360	1260
Iron determinable in unfiltered water but not in filtrate	260	160
Iron not determinable	40	240

It is evident that in the less acid solution, part of the iron added was bound in such a way that it did not react. In view of the nature of the treatment involved in making the determination (oxidation, acidification) it seems unlikely that the differences found are due to easily reversible association of colloidal micelles. The fact that the most striking difference lay in the undeterminable fraction supports such a conclusion.

Supposed iron lignoprotein complexes. Further insight into the nature of the iron-humus complex is possibly derived from the work of Waksman and Iyer (1932a,b) and from that of the oceanographic biochemists who have used material prepared by these investigators. Waksman and Iyer, by mixing strongly alkaline solutions of lignin from straw and of protein such as casein, and then acidifying, synthesized a lignoprotein complex differing in no essential respect from the main component of naturally occurring soil humus. This is of course an entirely different substance from the yellow organic matter studied by Shapiro and probably by Aschan and by Aarnio. Waksman and Iyer prepared an iron complex of their lignoprotein, and some of their preparations were used by Gran (1933) and by Harvey (1937a) as sources of iron in the culture of diatoms. Gran speaks of his material as "ferri-ligno-protein," Harvey of his sample as "ligno-caseinate containing almost 2% of iron, similar to that used by Gran." Both Gran and Harvey found that such substances can act as iron sources for diatoms, and Harvey demonstrated that the iron so utilized in some of his experiments was in solution. It must be noted, however, that when solutions of ferrilignocaseinate are added to sea water, a great deal of the material is precipitated and only about 5 per cent of the iron added remains in solution. Although Waksman and Iyer speak of the uptake of cations by their lignoprotein preparations as base-exchange reactions, at least part of the iron is very firmly bound, for Cooper (1948a) indicates that only 41 per cent of the iron in Waksman's "ferroligno-protein" (*sic*) dissolved in sea water can be set free by boiling with bromine water in acid solution, and even less can be reduced by 10 per cent sodium sulfite at *p*H 2.8. It is clearly by no means certain that all the iron in the synthetic lignoprotein complex is in the same form, nor is it evident that the part that is unreducible and not liberated by bromine water is necessarily available to diatoms. It is easy in such cases to describe the iron that is loosely bound as being present as a sol of ferric hydroxide protected by organic matter, or as forming a base-exchange complex, or as being present as ferric lignoproteinate, according to the individual point of view of the investigator. It is also very probable that in waters rich in colored organic material, ferric hydroxide sol is stabilized by organic compounds, and that iron may exchange for ionizable hydrogen at the surface of colloids, forming base-exchange complexes or salts. Actually, the only certain information about the organic iron complexes in lake waters would seem to be that they certainly exist, and that at least under some conditions, as in the surface waters of Linsley Pond or in the Swedish lakes studied by Rodhe, they are more stable than would be expected of sols stabilized by yellow organic matter or base-exchange complexes, though not more stable than would be suggested by actual experience with synthetic iron complexes of lignoproteins. The evidence that the last named substances really occur in natural lake waters is, however, quite inadequate.

Direct determination of the availability of iron to phytoplankton. The fragmentary evidence that has been assembled in the previous paragraphs indicates that apart from particulate iron compounds in organisms and perhaps in suspended silt, the main forms of iron in lake water are ferric hydroxide in particulate form or adsorbed to particles of microscopic dimensions, and an organic iron complex or series of complexes. The experimental evidence, derived mainly from the work of Harvey, suggests that both of these forms of iron may be assimilable by diatoms and perhaps by other phytoplanktonic organisms. Goldberg (1952) finds that some marine diatoms can use only hydroxide.

It remains to examine the one study that has been made of the actual assimilability of iron naturally present in a lake water. Rodhe (1948), using surface water of Skärsjön, containing 100 mg. m.$^{-3}$ total iron, found that the quantity of the element available to iron-starved cells of *Scenedesmus quadricauda* after suitable enrichment with nitrogen and phosphorus corresponded to just over 8 mg. m.$^{-3}$, as judged by the parallel growth of cultures in media of known iron content. It would therefore appear that in this water only about 10 per cent of the total iron could be used. Rodhe found that 10 mg. m.$^{-3}$ of the iron in the water reacted to ortho-phenanthroline, and he concludes that this reagent permits the determination of the assimilable iron in a sample of lake water. As has been previously indicated, it is not entirely clear what fraction of the iron is reacting in this case.

Iron bacteria and the iron cycle in lakes. The precipitation of iron may be produced by bacteria in two ways. In the first way, only the organic matter associated with the iron is metabolized; ferric hydroxide, being left over, is precipitated. This process has already been mentioned in relation to the production of lake ochre. In the second way, iron is actually involved in the metabolism of the bacteria to a degree that greatly exceeds the normal iron requirements of living organisms. The best known iron bacteria are those such as *Leptothrix ochracea* or *Spirophyllum ferrugineum*, which can derive energy for the synthesis of organic compounds from the oxidation of ferrous salts, such as ferrous carbonate or bicarbonate. These bacteria belong to the Chlamydobacteriales and form mats of sheathed filaments. *Leptothrix* is a facultative iron organism, which can use both ferrous and manganous salts and can perhaps adsorb oxides of iron, manganese, and aluminum also. *Spirophyllum* is an obligate iron organism using only iron.

An admirable brief review of the metabolism of these forms is given by Stephenson (1939). They are mainly swamp forms; the pools in which they live doubtless show steep oxygen, redox, and ferrous iron gradients. The diffusion of oxygen from above and ferrous iron from the sediments below would permit the bacterium to utilize its position on the boundary between the reduced and the oxidized parts of the biosphere. They are also found in acid mineral springs and in more normal waters when conducted in pipes, the walls of which provide a substratum for growth. A good account of their ecology is given by Turowska (1930). In addition to these Chlamydobacteriales, stalked bacteria of the genus *Gallionella* deposit iron in their stalks; they appear to inhabit waters comparable to those occupied by the Chlamydobacteriales.

There are also a number of Eubacteriales associated with the surface films of pools and mineral springs; Dorff (1934) gives an extensive account of them, and more recent information is summarized by Huber-Pestalozzi (1938). Of greater limnological interest is the genus *Siderocapsa*. One of the species, *Siderocapsa coronata* was described from the Lunzer Obersee, but was found by Ruttner (1937) to be most characteristic of the chemocline region of meromictic lakes, at oxygen concentrations of 0.12 to 0.3 mg. per liter. In extreme cases, another iron eubacterium, *Ochrobium tectum* occurred at slightly lower levels, with oxygen concentrations less than 0.12 mg. per liter. It is reasonable to suppose that these organisms oxidize ferrous iron in the same manner as the Chlamydobacteriales of bog pools. Hardman and Henrici (1939) found that *Siderocapsa treubii* appeared on slides hung in various lakes and streams. The organism is apparently restricted to waters of *p*H 7.5 or greater. In the top 8 m. of Trout Lake, Wisconsin, it occurred abundantly in the epilimnion, where the water contained only 30 mg. Fe $m.^{-3}$ at *p*H 7.5. It is evident that no significant amounts of ionic iron were present, yet *Siderocapsa* was able to produce a sheath impregnated with ferric oxide. Hardman and Henrici conclude that *Siderocapsa* derives the iron by the oxidation of organic iron compounds in the water. If *Siderocapsa* occurs on sestonic material naturally suspended in the water, it is evident that its activity might provide a mechanism for the production of the particulate ferric iron noted in analyses.

Vertical distribution of iron. In general, in holomictic lakes with more or less orthograde oxygen curves the iron content varies little vertically. When there is any considerable microzonal oxygen deficit near the bottom, there is apt to be a considerable increase in the iron content. In the smaller shallower lakes with a markedly clinograde oxygen curve, this inverse relationship between iron and oxygen is normally well developed. The resulting inverse clinograde iron curve, however, tends to have its point of greatest curvature at a lower level than that of the oxygen curve, in the same way that the point of greatest curvature of the redox potential curve lies deeper than that of the oxygen curve.

It may be supposed that two factors underlie this observation. In the first place, the oxygen deficit is determined to a large extent by events in the free water, at least in certain lakes. Oxygen can be removed from water without increasing its iron content. The iron is mainly derived from the underlying sediments, and it is easy to imagine a zone in which much oxygen has been lost through the decomposition of falling plankton but which is too remote from the sediments to receive any great amount of iron at the time that the removal of oxygen occurs. In the second place, the iron of the deep hypolimnetic accumulation is mainly ferrous, and it appears from the observations of Mortimer and others, to be discussed in greater detail presently, that no ferrous iron is detectable until the oxygen content falls below 0.5 to 1.5 mg. $m.^{-3}$, which in most cases will be well below the region of the most rapid diminution of oxygen content with depth. When the hypolimnion is relatively small and the oxygen content reduced throughout almost its entire extent, very considerable amounts of iron may be found at almost all hypolimnetic levels, though some trace of the relationship just indicated is normally observed.

Though it has been realized, at least since 1892 when the Massachusetts State Board of Health made its classical investigation of Lake Cochituate, that the accumulation of iron in the deeper water of the hypolimnion was primarily due to ferrous iron, which oxidized and precipitated on contact with air, no very adequate data on the proportion of ferrous to total iron was published until Müller (1932) studied vertical series of samples from several Austrian lakes by means of the then newly introduced reagent α-α'-dipyridyl. Müller found that although ferrous iron was normally undetectable in the epilimnion, the large accumulations of iron in the deep water were exclusively ferrous. This ferrous iron appears to be present as bicarbonate and, as Yoshimura (1932d) has pointed out, in soft-water lakes may buffer the deep hypolimnion sufficiently to give a dichotomous *p*H curve with a minimum value in the upper hypolimnion. Subsequent work by Stangenberg (1936), and by the various authors whose work on the redox potential has been discussed, has indicated that the accumulation of iron in the ferrous form is a very general phenomenon in the hypolimnia of stratified lakes.

The most complete account of the seasonal variation of iron throughout a lake is that for Esthwaite Water by Mortimer, whose depth-time diagram is given in Fig. 195. The general changes are evidently very similar to the parallel changes in redox potential. It is to be noted, however, that early in stratification, before any detectable ferrous iron has appeared, there has been a considerable rise in the total iron of the deep hypolimnetic water. This rise is accompanied by a rise in the color of the same water, and is attributed by Mortimer to the simultaneous diffusion of ferrous iron from the already reduced mud surface and the downward turbulent movement of oxygen from above. The ferrous iron is therefore oxidized as it enters the water, producing a temporary ferric oxide sol. This may in part sediment, but in any case will later be reduced, so that at the end of stratification the whole of the iron will be present in divalent form. A very similar series of events was observed in Mortimer's experimental anaerobic tank, in which a high total iron content and an increase in water color was observed at a time when a crack in the paraffin seal was permitting a slight leak of oxygen into the water from the air.

The iron cycle was investigated in a rather less extensive way by Ivlev (1937) in Lake Beloye. The general form of the depth-time diagram is similar, though in this lake, subjected to the rigors of a far more continental climate than is Esthwaite, the rise of the concentration in the deepest water during winter stagnation under ice was more marked than the rise in the summer, though the extent of the hypolimnion involved was less great in the winter. This is probably due to the lesser hypolimnetic turbulence under an ice cover than during summer stratification. Although the whole of the data required to estimate the total iron content of the lake waters is not presented, such an estimate was made for a number of dates and plotted. The resultant graph (Fig. 199) clearly indicated the

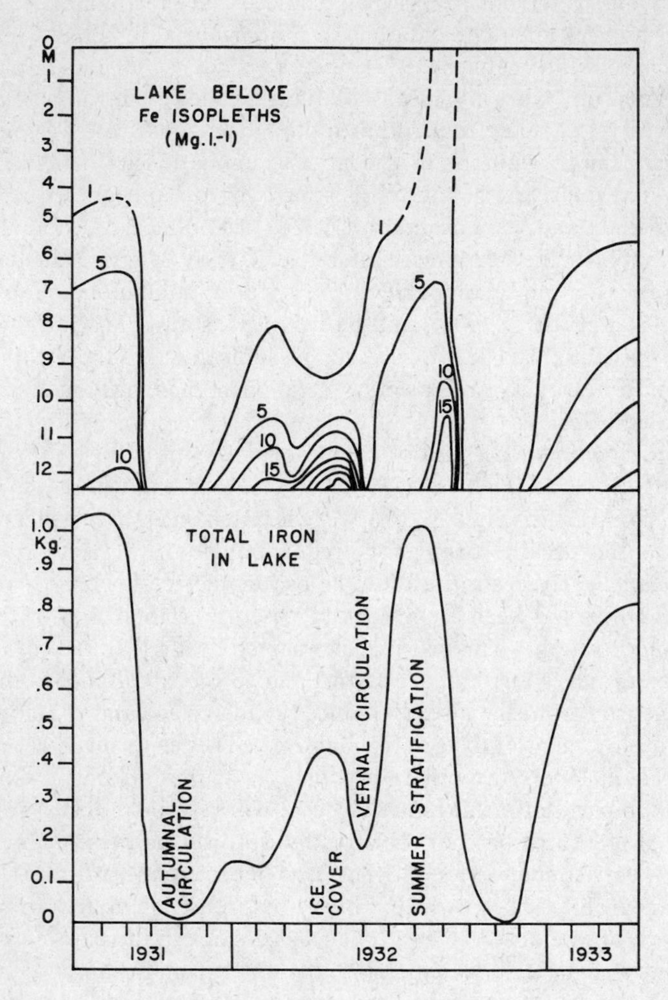

FIGURE 199. Lake Beloye, Kossino, USSR, seasonal variation in the total iron content of the water of the lake (from data of Ivlev).

great changes in the iron content of the lake, changes that are comprehensible only on the universally accepted hypothesis that most of the iron comes from the sediments.

In Esthwaite Water, Mortimer states that ferrous iron appears at a redox potential of $E_7 = 0.25$ volt. The *p*H of the water varies between 6.5 and 6.9, but the lower reading would appear to be characteristic of those times and depths at which ferrous iron is just appearing. Assuming the lower value, E_h *in situ* would be about 0.28 volt, and this at a *p*H of 6.5 would, from equation 4, permit the existence of just under 0.1 mg.-atom m.$^{-3}$ or about 5 mg. m.$^{-3}$ This is about the lower limit of detection of ferrous iron by dipyridyl as ordinarily used. Mortimer also indicates that at the end of stagnation, from 2000 to 4000 mg. m.$^{-3}$ of ferrous iron are present. From his graphs these quantities appeared to occur on August 25, at *p*H 6.6 and 6.7, respectively. At *p*H 6.6, 2000 mg. m.$^{-3}$ corresponds to a redox potential of about $E_h = 0.12$ volt, or $E_7 = 0.10$ volt; at *p*H 6.7, 4000 mg. m.$^{-3}$ corresponds to about $E_h = 0.09$ volt, or $E_7 = 0.07$ volt. These values correspond essentially to those actually observed. There would thus seem to be no reason, on the basis of the relationship developed theoretically by Cooper (1937a), for doubting that the ferrous iron in the hypolimnion of Esthwaite is in ionic form. This conclusion evidently receives confirmation from Mortimer's treatment of the conductivity of the water, for the assumption of ionic ferrous iron appears to be necessary to explain the observed values. The ferrous iron contents of the bottom waters of Anderson Lake and Helmet Lake, studied by Allgeier, Hafford, and Juday (1941) and given in the files of the Wisconsin Natural History Survey as 8100 and 3800 mg. m.$^{-3}$, are of the same order of magnitude as the values calculated from the redox potentials and *p*H values of the same layers of these lakes, namely 11,600 and 4100 mg. m.$^{-3}$, and presumably also indicate that the ferrous iron here is ionic.

In other lakes there is some evidence which may indicate that a large part of the ferrous ion exists in complex form. Such a complex, as Mortimer (1941–1942) indicates, would have to be sufficiently unstable to react to dipyridyl, but since complexes of ferrous iron behaving in the requisite way are known, the hypothesis that they occur in lake water is not unreasonable. The cases in Table 92 are recorded by Hutchinson, Deevey, and Wollack (1939) from two lakes in Connecticut, and from the unpublished files of the Wisconsin Natural History Survey for one lake

TABLE 92. *Observed and calculated ferrous iron contents in three lakes exhibiting anomalous high values in the hypolimnion*

Lake	Depth, m.	*p*H	E_h, volt	Fe^{++} calc., mg. m.$^{-3}$	Fe^{++} obs., mg. m.$^{-3}$
Job's Pond, Connecticut, Oct. 4, 1938	14	6.5	0.248	34.7	5,760
Linsley Pond, Connecticut, Sept. 16, 1938	13	7.4	0.164	1.4	4,800
Scaffold Lake, Wisconsin, 1940	11	7.95	0.123	0.16	14,000

in Wisconsin. It must be pointed out, however, that the data of the previous year indicate a *p*H in the lower hypolimnion of Scaffold Lake of only 6.6, presumably implying less discrepancy than was noted in 1940.[1]

In his discussion of the conductivity of the water maintained out of contact with air over a sample of Windermere mud, Mortimer points out that the ammonia present accounts for a very large part of the alkalinity, and that there seems to be a great excess of bases over the alkalinity or of electrolytes over the conductivity. He therefore considers possible that the ferrous iron is actually in complex and not in ionic form. He also seems to imply from his discussion that the ferrous-complex ferric system involved would set up a different potential from that given by the equivalent ionic ferrous-ferric system. This seems reasonable, though his statement that the E_0 value at *p*H 7.0 in lake muds is about 0.2 is clearly erroneous. Actually, in the anaerobic experimental tank, from which the idea of a ferrous complex is derived, the concentration of ferrous iron at the end of the experiment is apparently about 30 mg. per liter, or 540 mg.-atom $m.^{-3}$, which is certainly not excessive for *p*H 6.7 and E_7 0.07 volt. The true nature of the very high ferrous iron values supposedly found in certain American lakes at moderate redox potentials and hydrogen ion concentrations remains to be elucidated.

Variation in iron contents of lake sediments. Ivlev has studied the variation in the iron content of the mud of Lake Beloye, reporting the total iron, exchangeable iron, and iron in solution in the sediment at various depths in mud cores taken with Karsinkin and Kusnetzow's stratometer, an instrument which permits undisturbed samples of water and mud to be taken from the bottom of the lake. Unfortunately the various types of determinations seldom refer to the same dates or to dates on which vertical series of the water of the lake were taken. The iron in the sediment solution, which was obtained by centrifuging the mud under paraffin, was almost always higher than the iron in the water immediately over the mud. It was always higher in samples of mud from deep water than from samples at lesser depth. Within the mud there was considerable vertical variation in the iron content of the solution. Twenty-five centimeters below the mud surface the content was always low, from 8.5 to 14.7 mg. per liter at the deepest station. At less depths in the mud, the content oscillated markedly, the amplitude of the oscillations increasing with decreasing depth, so that at the surface of the mud the lowest and the highest values in the deep water, namely 7.5 and 88.9 mg. per liter, were recorded. In general, the high values occurred late in the stagnation

[1] There are apparently errors, either in the file cards at Madison or in Allgeier, Hafford, and Juday's Figure 9, which prevent further treatment of this case.

periods, 76.2 at the end of August, 88.9 at the end of March; the low values occurred during autumnal circulation and to a less extent during the incomplete vernal circulation.

The exchangeable iron increases likewise with increasing depth of water over the sampling station. It may be slightly less in the surface layer than 1 cm. below the mud-water interface, but from 1 cm. down it steadily declines. Practically no seasonal variation is indicated at or below 20 cm., but the upper layers show a marked seasonal change. As the hypolimnion gained iron between July 18 and August 7, 1932, the surface layers of the mud lost iron. Ivlev computes that a water column in the deepest part of the lake gained during this period 2.03 mg. cm.$^{-2}$, whereas the top cm. of the mud lost 4.63 mg. cm.$^{-2}$ During the autumnal circulation period, up to October 7, the water column lost 5.68 and the mud gained 10.83 mg. cm.$^{-2}$ The succeeding winter stagnation caused a gain of 6.12 mg. cm.$^{-2}$ by the water, a loss of 12.2 mg. cm.$^{-2}$ by the mud. Though these figures show only a rough agreement, they certainly indicate that the exchangeable iron of the mud is the major depot of iron in the lake (Fig. 200).

The total iron content varies differently from the fractions determined separately. In the deepest part of the lake the total iron at 20 cm. and below comprises about 4.55 per cent of the dry sediment. On a percentage dry basis it always increases with decreasing depth, so that the mud in contact with the water may have from 8.6 to 13.35 per cent iron. Since the water content increases towards the surface of the mud, there is a minimum in total iron per unit volume of wet mud at 10 cm. below the mud surface in all but one of the five series of analyses. Comparison with the analyses of exchangeable iron indicates that in the top centimeter practically the whole of the total iron is probably exchangeable, but that at 10 cm. only 40 to 75 per cent is exchangeable. Exact figures cannot be given, as none of the analyses refer to the same dates.

The formation of ferrous sulfide. The reduction of sulfate and the decomposition of organic sulfur compounds may both produce hydrogen sulfide. Mortimer concludes from his experience in the English Lake District that the reduction of sulfates, doubtless mediated bacterially, and the appearance of H_2S occur between $E_7 = 0.06$ and 0.10. Since this is a lower potential than that at which ferrous iron appears in quantity, namely $E_7 = 0.2$ to 0.3, it is evident that a considerable amount of ferrous iron can enter the water from the surface layer of the mud before hydrogen sulfide begins to play any important role. Since ferrous sulfide is relatively insoluble, the final stage of stagnation may be marked by a loss of ferrous iron from the deep water, as Minder (1929), Einsele (1937), Ohle (1937), and Mortimer (1941–1942) have pointed out. Einsele has shown, for

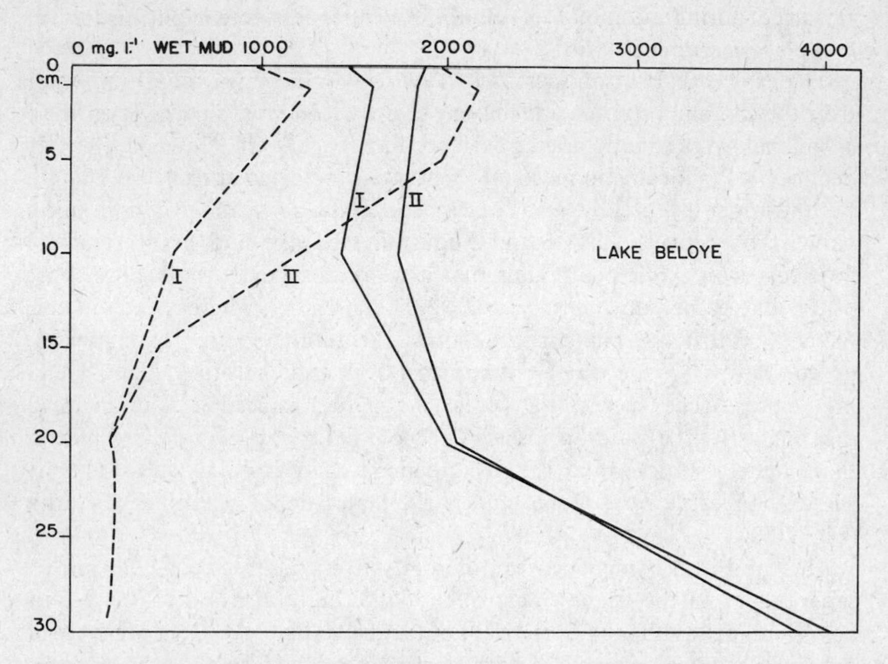

FIGURE 200. Lake Beloye, total and exchangeable iron at different depths (in cm.) in the sediments below 12.5 m. of water: I, at height of summer stratification (total, August 2, 1932; exchangeable, August 7, 1932), when the water over the mud had about 10 mg. Fe per liter; II, at autumnal circulation (total, October 7, 1932; exchangeable, October 10, 1932) when the water over the mud contained only traces of iron. The higher exchangeable than total iron in October is presumably due to the fact that different samples collected on different days are involved. It is evident that most of the superficial iron is exchangeable and that virtually no change occurs below 20 cm. of sediment. Total iron, solid lines; exchangeable iron, broken lines. (From data of Ivlev.)

example, how, under normal conditions, up to 4000 mg. Fe $m.^{-3}$ occur at the end of stagnation in the Schleinsee. When, after artificial fertilization, large amounts of H_2S were produced, the iron content started to rise as usual but reached a rather constant value, which at 11 m. was only one third of the value of earlier years.

The relationship between the *p*H, hydrogen sulfide, and iron concentration of a lake water can be considered by the elementary methods of physical chemistry. This was done by Einsele (1937), who concluded that while the precipitation of manganese sulfide was never likely to control the concentrations of either manganese or hydrogen sulfide, the precipitation of ferrous sulfide often controlled the concentrations of Fe^{++} and

H_2S. Einsele's treatment is amplified in the discussion of hydrogen sulfide given in Chapter 13 of the present work, and a series of curves is given in Fig. 204, indicating the maximum amounts of H_2S and ferrous iron that can coexist in solution at different *p*H values. These values differ somewhat from those given by Einsele, for reasons set out in detail in the discussion of the sulfur cycle. Certain cases have been described, notably by Ruttner (1931) in the Javanese lakes Telaga Pasir and Ranu Pakis, in which the iron concentration decreases with depth in the hypolimnion. In both these cases H_2S is present, and Ruttner concluded that the observed distribution was determined by the precipitation of ferrous sulfide. Einsele's calculations implied that there was not enough H_2S to account for the observed distribution of total iron. The data for Ranu Pakis is given in Table 93.

TABLE 93. *Stratification of H_2S and iron in Ranu Pakis*

Depth, m.	*p*H	H_2S,* mg. per liter	Fe obs., mg. $m.^{-3}$
50	7.49	0.5	90
70	7.40	0.7	80
100	7.35	1.1	40
150	7.31	1.2	20
154	7.23	1.3	0

* Corrected by Einsele for a supposed error making original determinations 20 per cent too low.

Using a revised and more accurate value of the solubility product for FeS than was available to Einsele, it appears from evaluation of equation 14 of Chapter 13 that not more than 4.5 mg. Fe^{++} $m.^{-3}$ can have existed in the 50 m. water, and slightly less at lower levels. It is probable that the observed iron was present as an organic compound that was decomposing slowly, the ferrous iron that would be liberated by such a process being sedimented as sulfide. Minute black balls of ferrous sulfide[2] are common in organic lake sediments (Naumann 1930, Vallentyne and Swabey 1955).

The iron content in the monimolimnia of meromictic lakes. Very high iron contents have been recorded in the monimolimnia of some meromictic

[2] The rather poorly characterized ferrous sulfide of lake sediments, supposedly FeS, has received the name *hydrotroilite*. A modern mineralogical study of this substance would be desirable. Pyrite also occurs as an autochthonous mineral in lake muds (Sugawara, Koyama, and Kozawa 1954).

lakes. Einsele (1944) found up to 41,000 mg. m.$^{-3}$ at 68 m. in the Zeller See in Austria, and Yoshimura (1936a) indicates that Takayasa found 87,000 mg. m.$^{-3}$ at 99 m. in Sikaribetuko, Hokkaido, Japan. In other meromictic lakes great increase in iron is not recorded. In some cases, as Einsele supposes, there is an inadequate supply of iron in the drainage basin to act as a source for such accumulations; in other cases, the presence of H_2S at a sufficiently high *p*H may prevent the accumulation of ferrous iron in detectable amounts, but this limitation is not always adequate to account for the observed condition.

In the monimolimnion of Lake Mary, Wisconsin, no ferrous iron occurs. The *p*H of the bottom water is given as 5.7 and the H_2S content as 2.1 mg. per liter (Fig. 208 and Table 105, Chapter 13). From Fig. 204 the maximum possible ferrous iron content would be about 1540 mg. m.$^{-3}$ It is most unlikely that the absence of ferrous iron is here controlled geochemically, as other nearby lakes do not lack the element. At *p*H 5.7, 450 mg. per liter H_2S would permit the presence of 5 mg. Fe^{++} m.$^{-3}$, which would only just be detectable; at *p*H 5.9, 275 mg. per liter would be sufficient to reduce ferrous iron concentration to the same low level. It is not likely, however, that such large amounts of H_2S would be present in the mud, in view of the rather moderate amounts in the deep water. As with certain other aspects of the chemistry of this peculiar lake, there is no obvious explanation of the observed phenomena.

The phases of stagnation. In view of the great importance of the iron cycle in the hypolimnia of lakes, it is convenient to divide the events occurring in the deep water during summer stratification into four phases. Only in the cases of lakes with extremely reductive sediments and small hypolimnia will all four phases be observed.

Phase I. Oxygen still high though falling, redox potential 0.4 to 0.5 volt, ferrous iron and H_2S absent, ionic soluble phosphorus very low. Typically retained throughout stratification by any lake with an orthograde oxygen curve.

Phase II. Oxygen much reduced, the oxygen curve being clinograde but the redox-potential curve practically orthograde. No ferrous iron or H_2S, and ionic soluble phosphorus very low. A few lakes such as Lake Quassapaug proceed no further than this.

Phase III. Oxygen much reduced, redox potential falling with depth, much ferrous iron and ionic soluble phosphorus, but no H_2S. The final condition in most small productive lakes, such as Linsley Pond.

Phase IV. Oxygen much reduced or absent, redox potential very low, much ionic soluble phosphorus, H_2S present removing ferrous iron. The final condition in the Schleinsee.

SUMMARY

The potential, relative to the standard hydrogen electrode, set up at a bright platinum electrode immersed in epilimnetic lake water usually varies from about 0.4 to 0.6 volt. This potential may be regarded as the irreversible oxygen potential; as such it can be corrected for variations in OH^- concentration. When so adjusted to *p*H 7, the observed potentials generally appear to be between 0.4 and 0.5 volt. They are thus very slightly lower than the potential of 0.52 at 25° C., which is the best value for a solution of air at a pressure of 1 atm. in pure water. A few cases are known in which E_7 is found to be above 0.60 volt; these are unexplained. The slight depression of the potential below 0.52 in many lake waters and the more marked depressions occasionally observed, in one case to a potential as low as 0.37 volt, are probably to be explained by organic reducing substances in solution in many waters.

The oxygen potential is very insensitive to changes in oxygen concentration. In the hypolimnia of some lakes the potential remains unchanged with depth from the surface to the bottom, even though little oxygen remains in the water. In such lakes the mud surface presumably remains in an oxidized condition. More frequently, at least in smaller lakes, there is a rapid fall in potential near the bottom, and not infrequently the water of the greater part of the hypolimnion exhibits potentials markedly below those prevailing in the epilimnion. This fall is due to reducing substances derived from the mud, ferrous iron being the most important.

The deepwater sediments, except at their surface when exposed to oxygenated waters, are probably always strongly reducing, showing potentials around 0.0 volt. When such mud is in contact with water to which adequate oxygen can be brought by turbulent transport, an oxidized microzone several millimeters thick develops at its surface. This microzone is usually brown in color, owing to the presence of hydrated ferric oxide. When fully developed, the microzone exhibits a potential of about 0.5 volt. As the supply of oxygen is reduced during the development of stratification, the oxidized microzone tends to become thinner and finally may disappear when the potential in the surface layer of the mud lies at about 0.2 volt. In large deep lakes with orthograde oxygen curves, the oxidized microzone may always be present. As the oxidized mud apparently has much greater adsorptive powers than the reduced mud, and as the diffusion of ferrous ions and phosphate ions into an oxygen-rich region leads to the precipitation of ferric phosphate, the oxidized microzone acts as a barrier against diffusion from the mud to the water, and its disappearance has a profound effect on the chemistry of the hypolimnetic water.

Iron is practically insoluble in ionic form at the *p*H values and redox potentials normally encountered in the epilimnetic waters of lakes. The available evidence indicates that the iron in such waters, usually occurring in small amounts of the order of 50 to 200 mg. $m.^{-3}$ is present partly as ferric hydroxide, in suspension or adsorbed on seston particles, partly as sestonic iron in various other forms, and partly as a soluble or colloidal iron organic complex, which appears to be surprisingly stable. Only a small part of the iron is available to phytoplanktonic organisms. In humic waters there is often considerable enrichment of iron, and coprecipitation of ferric hydroxide and organic matter may occur if the ratio of iron to humic material reaches a critical value. A number of bacteria precipitate ferric hydroxide. Among limnetic species, *Siderocapsa coronata*, occurring at oxygen concentrations of 0.1 to 0.3 mg. per liter in the chemocline region of European meromictic lakes, and *Ochrobium tectum*, occurring at even lower oxygen concentrations, perhaps derive energy from the oxidation of ferrous to ferric iron, as do the better known swamp forms such as *Spirophyllum ferrugineum*. *Siderocapsa treubii*, which develops on slides submerged in the epilimnia of hard-water lakes, evidently cannot have access to a significant supply of free ferrous iron for this sort of metabolic activity, but it produces a sheath impregnated with ferric hydroxide, which may be derived from soluble organic iron in the water.

The vertical distribution of iron is reflected in the vertical distribution of redox potentials. When the oxygen curve is orthograde, there is no great enrichment in iron, even of the deepest water. Whenever a microzonal oxygen deficit is well developed and the mud surface is reduced, ferrous iron tends to diffuse into the free water. If the lower layers are still receiving any appreciable amount of oxygen, ferric hydroxide may be formed and may remain dispersed for a time in the deep water. When the lower water is essentially free of oxygen, considerable amounts of ferrous iron usually accumulate in the deep water of the hypolimnion. At least in some lakes, as Esthwaite Water, the observed redox potentials, *p*H values, and ferrous iron contents of the hypolimia are concordant.

It is possible to divide the period of summer stagnation in the hypolimnion into four phases, the first being the final stage in lakes with orthograde oxygen curves, the second in a few lakes with clinograde oxygen curves but orthograde redox potential curves, the third in those small lakes in which ferrous iron accumulates in the hypolimnion, the fourth in the still more reductive cases in which H_2S begins to remove the iron.

CHAPTER 12

The Phosphorus Cycle in Lakes

Of all the elements present in living organisms, phosphorus is likely to be the most important ecologically, because the ratio of phosphorus to other elements in organisms tends to be considerably greater than the ratio in the primary sources of the biological elements. A deficiency of phosphorus is therefore more likely to limit the productivity of any region of the earth's surface than is a deficiency of any other material except water.

Phosphorus occurs in the biosphere almost exclusively in a fully oxidized state. Bacteria that can reduce phosphate to phosphite, hypophosphite, and phosphine have been recorded (Barrenscheen and Beckh-Widmanstetter 1923, Rudakov 1927), but little is known of them. Phosphine has been found in polluted springs (Lüning and Brohm 1933) and might occur in the hypolimnion under strongly reducing conditions. The phenomenon of the will-o'-the-wisp or *ignis fatuus*, supposed to be observed over marshes, has been explained as the spontaneous combustion of phosphine, containing a little H_2S or P_2H_4, produced by bacterial action, but no positive evidence for this view seems to exist, and some of the recorded occurrences of this phenomenon render the interpretation somewhat doubtful.[1] Apart from this possibility, only phosphates need be con-

[1] The subject is reviewed at greater length by Hutchinson (1952) and by Minnaert (1954).

sidered by the biogeochemist, and though pyrophosphate plays an important role inside the organism, it is easily hydrolyzed and only orthophosphate is likely to be of importance in the environment.

The quantity and forms of phosphorus in epilimnetic waters. Many old analyses indicate small quantities of phosphorus in inland waters, though it is probable that they are not very reliable. A number of such analyses, collected by Clarke (1924), indicate that the average river water of the world contains about 70 mg. P $m.^{-3}$ Though limnological experience would suggest that this figure is excessive, it is perhaps not really inconsistent with the numerous more modern analyses of lakes, for the river waters analyzed may have had a greater opportunity to take up the element than have the waters of lakes.

Real progress in the study of the phosphorus cycle of inland waters was not made until Atkins (1923) introduced into oceanography and limnology the colorimetric method for the determination of phosphate elaborated by Denigès (1920). It soon became evident, however, that more phosphorus was usually present in inland waters than could be determined as phosphate ions. Juday, Birge, Kemmerer, and Robinson (1928) introduced the concept of total and soluble phosphorus, the former being determined after, the latter before, appropriate oxidation. The difference between the two determinations was termed by them the organic phosphorus. Ohle (1939) pointed out that this fraction consisted both of sestonic and dissolved phosphorus, though the dissolved phosphorus may be in colloidal rather than in true solution. He also pointed out that part of the sestonic phosphorus might be a suspended inorganic phosphate, which, under the conditions of determination of soluble phosphate phosphorus, would go into solution. Actually, this probably happens only when sestonic ferric phosphate is present in the water at the autumnal overturn. It is also possible that colloidal or sestonic calcium phosphate may sometimes form in hard-water lakes (Gessner 1939). The categories that are useful in the present discussion are therefore: (1) soluble phosphate phosphorus, (2) acid-soluble sestonic phosphorus, (3) organic soluble (and colloidal) phosphorus, and (4) organic sestonic phosphorus. The sum or total phosphorus is also frequently of greater interest than any of the individual fractions.

The total phosphorus in lake waters is known to vary from an undetectable amount (< 1 mg. $m.^{-3}$), as in the surface waters of the Traunsee (Ruttner 1937), to immense quantities (78 g. $m.^{-3}$ in Owens Lake, California, or 208 g. $m.^{-3}$ in Goodenough Lake, British Columbia) in saline lakes in closed basins in semi-arid regions.

Apart from the tendency for concentration of phosphorus under semi-arid conditions, there is certainly a good deal of regional variation in total phosphorus content that arises from the operation of geochemical rather than climatic causes. The lake districts that are best known appear to be relatively poor in total phosphorus, though such restricted variation as occurs is of undoubted biological significance.

TABLE 94. *Regional variation in total phosphorus in surface waters*

Region	Mean, mg. m.$^{-3}$	Extremes, mg. m.$^{-3}$
Northeastern Wisconsin (Juday and Birge, 1931)	23.0	8–140
Connecticut, E. highland (Deevey, 1940)	10.8	4–21
Connecticut, W. highland (Deevey, 1940)	13.0	7–31
Connecticut, C. lowland* (Deevey, 1940)	20.0	10–31
Japan (Yoshimura, 1932a)	14.8	4.4–43.5
Austrian Alps (Ruttner, 1937)	20.0	0–46
Prov. Uppland, S. Sweden (Lohammar, 1938)*	38.3	2–162
Prov. Dalarna, S. Sweden (Lohammar, 1938)	26.0	4–92
N. Sweden (Lohammar, 1938)	21.1	7–64

* Omitting one undoubtedly contaminated locality.

The determinations of Table 94 would suggest a rather uniform low concentration of total phosphorus in the surface waters of the lakes of humid temperate regions. There is, however, clear evidence from certain other groups of lakes, not more subject to human contamination than some of the regions listed above, that much higher regional values may ultimately sometimes be expected. Ohle (1934a) found that the *soluble phosphate* phosphorus in lakes of Baltic Germany varied from 5 to 600 mg. m.$^{-3}$, the mean value being 77 mg. m.$^{-3}$; there was a clear tendency for the waters of brown lakes rich in peaty material to contain more phosphate than the clear-water lakes, but even omitting the lakes with a large quantity of humic material in their waters, the range was from 5 to 200 mg. m.$^{-3}$ and the mean 47 mg. These means are higher than any given above, even though it is certain that they represent only a fraction of the phosphorus present. It is therefore obvious that the lakes of the plain of northern Germany are richer in the element than are those of any other region for which total phosphorus analyses are available, though approached by the lakes of the Baltic coast of Sweden. The unfiltered water of the catchment area of the London water supply contains similar quantities of soluble phosphorus, the mean value for seven reservoirs analyzed in 1938 being 66 mg. m.$^{-3}$ ([Gardiner] 1939). It would, moreover, seem likely that some of the Cheshire meres, such as Rostherne and Budworth, where the soluble phosphorus can vary from 0 to 127 mg. m.$^{-3}$ (Lind 1945), also have considerable total phosphorus contents. These figures are in striking contrast to those of the English Lake District, where the soluble phosphorus is always very low, usually well below 10 mg. m.$^{-3}$ Very rarely, the occurrence of sedimentary phosphates may markedly increase the phosphorus content of surface waters. Odum (1953) finds in Florida a mean total phosphorus content of 290 mg. m.$^{-3}$ in eight lakes draining areas of phosphatic rock; the mean content in thirty-one lakes not so influenced was 38 mg. m.$^{-3}$

In lakes in which sewage or much agricultural drainage enters the basin, higher values for the total phosphorus are recorded than are usual in uncontaminated lakes. Åberg and Rodhe found 107 mg. m.$^{-3}$ in filtered water from Växjösjön, a lake that is known to be contaminated by sewage, Lohammar found up to 290 mg. m.$^{-3}$ in another south Swedish lake, and Deevey found 109 mg. m.$^{-3}$ in Dooley Pond, a small farm dam in the central Connecticut lowland; values which compare strikingly with those already given for the less modified lakes of the regions in which they lie. In the effluent of Lake Mendota in 1943 and 1944, Sawyer, Lackey, and Lenz ([1945]) record from 40 to 160 mg. m.$^{-3}$ total phosphorus; but in the outlet of Lake Kegonsa farther down the Yahara River, they record over the same period 300 to 780 mg. m.$^{-3}$ total phosphorus. The great increase between the upper and lower lake is certainly due to agricultural and domestic drainage. Tressler and Domogalla (1931), however, record a mean content of 87 mg. m.$^{-3}$ in Lake Wingra, which lies on a tributary to the main chain of the Madison lakes and was probably less contaminated when they studied it than Mendota is today. It is therefore uncertain to what extent the very high values recorded in Mendota are due to human interference, but the data of Sawyer (1946) strongly suggest that all the Madison lakes, and some others in southern Wisconsin in which the mean soluble phosphate phosphorus is in excess of 10 mg. m.$^{-3}$, owe their high phosphorus mainly to urban or agricultural drainage. It is indeed probable that the general less striking differences between the lowland and highland lakes in Connecticut and between the lakes of Uppland and Dalarna in south Sweden may also be in part of cultural origin. Small changes in the phosphorus income owing to artificial disturbance are, however, probably not always reflected in the mean phosphorus content, though they may be apparent in other ways that depend on an over-all acceleration of the metabolic cycle. It is at any rate certain that qualitative changes of cultural origin have occurred comparatively recently in the biota of Linsley Pond, though its over-all phosphorus content is no greater than that of the average northeastern Wisconsin lake.

The relationships of the mean concentrations of the various fractions into which the total phosphorus can be divided can be ascertained from some of the investigations that have already been discussed. In the very large series of lakes in northern Wisconsin discussed by Juday and Birge, the ratio of the soluble phosphate phosphorus to the other forms of phosphorus is said to vary from about 1:1 to at least 1:89. Though these authors made no attempt to separate the sestonic from the organic soluble phosphorus, the mean seston content of their series of lakes is known and five determinations of phosphorus in centrifuge seston are given, so that it is possible to estimate roughly that the mean surface lake water of the region contained more soluble than sestonic combined phosphorus. The mean values so computed, and the mean values obtained for Linsley Pond throughout 1937–1939, are given in Table 95.

TABLE 95. *Fractionation of total phosphorus in Wisconsin lakes and in Linsley Pond*

	Mean, N. Wisconsin, mg. m.$^{-3}$	Mean, Linsley Pond, mg. m.$^{-3}$
Soluble phosphate P	3	2
Organic soluble P	14	6
Sestonic P	6	13
Total P	23	21

It is possible that the difference in the distributions shown in the table is due to the inclusion among the Wisconsin lakes of a number with rather high contents of brown organic matter, with which phosphorus is associated. Einsele (1936) regards all the organic soluble phosphorus of the Schleinsee as colloidal, and records from 4 to 10 mg. m.$^{-3}$ of such phosphorus and a mean of 10 mg. m.$^{-3}$ of sestonic phosphorus in the trophogenic layers of this lake, in which soluble phosphate is detectable only at times of circulation. It is evident that Einsele's values, though not published in full, indicate magnitudes of the same order as those in the North American lakes that have been studied.

Phosphorus requirements of phytoplanktonic organisms. Although the technique of cultivation of benthic, littoral, and other algae not living in the plankton has been greatly developed during the past fifty years, and though some work on the marine planktonic diatoms has been prosecuted for about half that time, little progress has been made until recently in elucidating the requirements of single species of fresh-water phytoplankton organisms. The first important work specifically directed towards the solution of the limnological problems raised by nutritive requirements of planktonic algae appears to have been done by Frantzev (1932), whose contributions mainly established the possibility of such researches. His work was followed in the U.S.S.R. by a series of papers by Guseva, not all of which are concerned with the phosphorus requirements of the algae investigated.

Guseva (1935) found that *Synura petersenii*, in an artificial medium in which nitrogen was supplied both as nitrate and ammonia, exhibited a very marked optimum at 220 to 440 mg. P $m.^{-3}$ Very little growth occurred at 87 mg. $m.^{-3}$ or below, and the rate of increase was reduced to 5 per cent of the maximum at 1135 mg. $m.^{-3}$ In view of Rodhe's later work on two other chrysomonad genera, the high level of phosphorus tolerance, which contrasts with Rodhe's findings, and the sharp optimum, which apparently is in accord with his observations on *Dinobryon* and *Uroglena*, are noteworthy.

Guseva (1937b) later investigated two blue-green algae and the diatom *Asterionella formosa*. All three species studied raise special problems. *Anabaena lemmermanii* requires, according to Guseva, 870 mg. $m.^{-3}$ phosphorus for optimal development, whereas *Aphanizomenon flos-aquae* requires but 260; as the phosphorus is exhausted by the first species, the second displaces it. *A. lemmermannii* probably fixes nitrogen and *A. flos-aquae* does not; the iron requirements of the former are said to be greater than those of the latter, but it is uncertain what Guseva's experiments with iron really mean. In addition to these complexities, it is to be noted that the quantities of phosphorus required are high and that Rodhe failed absolutely to culture *A. flos-aquae* in artificial media. Guseva found *A. formosa* to require 650 mg. P $m.^{-3}$ for optimal growth; this is in accord with later work on this species in artificial media, but does not necessarily throw any light on its requirements in nature.

Chu (1943) made a number of experiments on the culture of lake phytoplankton and was able to obtain values (Table 96) of the ionic phosphate phosphorus below which a limitation due to phosphorus deficiency operated. In general these figures may be interpreted to mean that if the concentration given above is halved, a marked decrease in the rate of growth of the culture takes place; raising the concentration has no effect in increasing the growth rate, though in extreme cases it may cause inhibition. It will be observed that even half the concentration required by the least demanding of these species, the diatoms *Nitzschia palea* and

TABLE 96. *Minimal concentrations of phosphate phosphorus just permitting optimal growth of various algae*

	Values, mg. m.$^{-3}$
Pediastrum boryanum	89
Staurastrum paradoxum	89
Botryococcus braunii	89
Nitzschia palea	18
Tabellaria flocculosa	18
Asterionella gracillima	45

Tabellaria flocculosa, would be 9 mg. m.$^{-3}$, which is very high for the soluble phosphate phosphorus of an unpolluted lake. Chu's results in fact seem to indicate that in the majority of waters these phytoplanktonic species are living under conditions of chronic phosphorus deficiency. Such a conclusion is apparently not justified as a general conclusion, however, as is shown by the more recent and very remarkable findings of Rodhe (1948).

There appears usually to be a rather wide range of concentrations at which further increase of the phosphorus has no effect in increasing the rate of growth of the population, but in all cases that have been studied, inhibition occurs when a sufficiently large amount of phosphorus has been added to the medium. Chu found that although the nature of the nitrogen source had no effect in determining the concentration that defined the lower end of this optimal plateau, there was usually clear phosphorus inhibition at a lower concentration when ammonium salts were used as a source of nitrogen than when nitrate was employed. Chu concluded that phosphorus inhibition occurred at concentrations in excess of the values in Table 97.

Apart from the interesting differences due to the nature of the nitrogen source, it is clear that *Tabellaria* is definitely less tolerant of excess phosphate than the other species. Chu observed that in *Tabellaria* grown in the

TABLE 97. *Maximal concentrations of phosphate phosphorus just permitting optimal growth of various algae, in the presence of nitrate or of ammonia*

	$N \cdot NO_3$, mg. m.$^{-3}$	$N \cdot NH_4$, mg. m.$^{-3}$
Pediastrum boryanum	17,800	1,780
Staurastrum paradoxum	17,800	1,780
Nitzschia palea	8,900	1,780
Tabellaria flocculosa	8,900 (or less)	...
Asterionella gracillima	17,800	...

higher phosphorus concentrations there was a marked yellowing, or in extreme cases depigmentation, as the cultures aged. A similar effect was observed in the case of *Botryococcus* cultures, for which no upper limit of the optimal plateau was defined. It appears in the case of this alga that the maintenance of a deep green color is dependent on adequate intake of nitrogen, and since the experiments were run for many weeks without renewal of the medium, it is believed that excessive initial nitrogen uptake in the high phosphorus cultures may have been responsible for the pale color that ultimately developed.

Although the very high phosphorus concentrations that Chu found to be inhibitory are most unlikely ever to occur in nature, except perhaps under conditions quite unsuitable for the development of an autotrophic algal flora, the recent work of Rodhe (1948) indicates that phosphorus inhibition actually does play a significant role in the determination of the phytoplanktonic flora and its seasonal variation in a great many lakes. Rodhe found that the rate of growth of *Scenedesmus quadricauda* increased with increasing phosphorus concentration, at least up to 1000 mg. m.$^{-3}$, in a medium containing nitrate as a nitrogen source. Up to about 500 mg. m.$^{-3}$ the growth rate was proportional to the phosphorus concentration, and very slight growth was detected in a medium containing 10 mg. m.$^{-3}$ *Asterionella formosa* in the same artificial medium behaved in the same way except that no growth was detectable at a concentration of 10 mg. m.$^{-3}$ However, when experiments with *A. formosa* were performed in Erken Lake water, which certainly contains only a few milligrams of ionic phosphate phosphorus per cubic meter, the addition of 1 mg. m.$^{-3}$ caused a significant increase in the rate of reproduction, and the addition of more than 50 mg. caused no further increase in the reproductive rate and in fact caused a slight decline. It is evident, therefore, that *A. formosa* behaves quite differently in lake waters and in culture solutions, being able to utilize levels of phosphate in the former that are totally inadequate for growth in the latter. These results are in accord with the experience of other investigators, notably of Chu (1943, 1945) and of Lund (1949) at Wray Castle. It may perhaps be suggested that some material, possibly a peptide influencing the ease with which phosphorus is assimilated (Fogg and Westlake 1955), is present in lake water and lacking in culture media. Whatever the nature of the difference between lake water and culture media, it is conveniently termed, in a noncommittal way, the phosphorus-sparing factor.

Equally remarkable and illuminating results were obtained by Rodhe in his study of the Chrysophyceae *Dinobryon divergens* and *Uroglena americana*. Both species are evidently able to reproduce at a maximal rate when the phosphate phosphorus is barely detectable, and both were inhibited by the addition of as little as 5 mg. $m.^{-3}$ phosphate phosphorus to the Erken Lake water in which they were cultured.[2] Some fragmentary evidence suggests that the blue-green alga *Gloeotrichia echinulata* is also inhibited by such small amounts of phosphorus, but the problem is here complicated by the difficulty of culturing this species. It will appear subsequently, in Volume II, that these results of Rodhe go far to explain some of the more puzzling features of the seasonal periodicity of the fresh-water plankton.

It is finally to be noted that Chu (1946) found that *Nitzschia palea* could use phytin (magnesium inositol hexaphosphate) and glycerophosphoric acid as phosphorus sources, and that Rodhe likewise found that *Scenedesmus quadricauda* could utilize the phosphorus of adenylic acid. It is unfortunate that there is no information available as to the fraction constituting the soluble or colloidal organic phosphorus of lake waters. Whatever the nature of the fraction may be, it is probably more stable than is glyceromonophosphate; Margalef (1951) found that the latter material was easily hydrolyzed by living Cladocera.

The vertical distribution of phosphorus in stratified lakes. The vertical distribution of the various phosphorus fractions is very variable. In general there is a tendency for the lakes with orthograde oxygen curves to show little variation in the phosphorus content with depth, but lakes with markedly clinograde oxygen curves show a considerable increase in the phosphorus content of the lower hypolimnion in the later phases of stagnation. Usually the hypolimnetic increase in phosphorus is due mainly to the soluble component, though this is apparently not always the case; in lakes in which very irregular variations in the total phosphorus occur with depth, it is no doubt generally due to variations in the sestonic component. Examples of phosphorus distributions demonstrating these aspects of the vertical variation are given in Fig. 201.

The irregular variation due to the sestonic fraction is probably mainly dependent on the sedimentation of plankton and requires no further elucidation. The increase in soluble phosphorus at the bottom of stratified lakes is a more complex phenomenon. Though it is certainly in part due to the decomposition of the sedimenting seston and subsequent liberation of soluble phosphorus, as is shown clearly by the recovery of radiophosphorus in a soluble form in the bottom waters of Linsley Pond after the isotope had been added at the surface when the lake was in a very stable condition, there is clear evidence that in many lakes this is not the entire explanation. Einsele (1936, 1938) first pointed out that the great

---

[2] Rosenberg (1938), however, found *Dinobryon* spp. needed 440 mg. $m.^{-3}$ in artificial solutions. A case like that of *Asterionella* may perhaps be indicated.

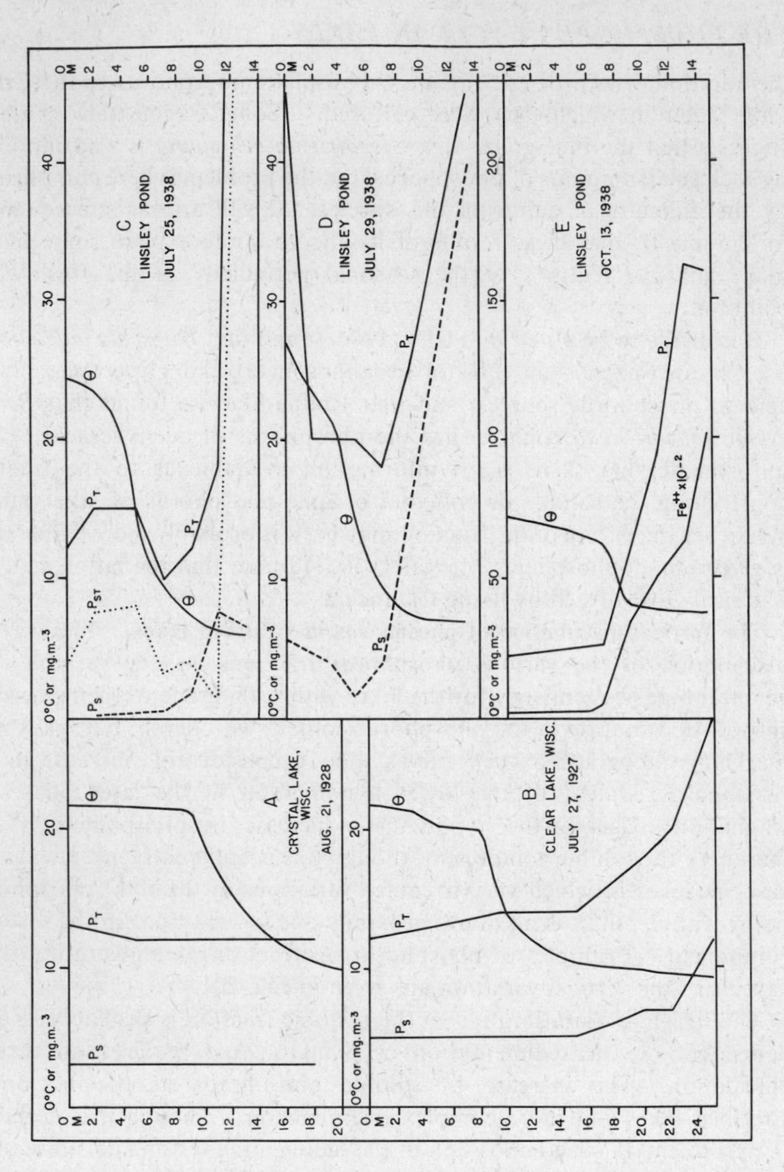

FIGURE 201. Vertical distribution of temperature and phosphorus during summer stratification: *A*, Crystal Lake, Wisconsin, soluble ionic (P_S) and total (P_T) phosphorus, showing essentially orthograde distributions; *B*, Clear Lake, Wisconsin, showing marked increase of phosphorus near the bottom, largely in the nonionic fractions; *C*, Linsley Pond, Connecticut, extreme increase in the phosphorus near the bottom water, largely in the sestonic (P_{ST}) fraction; *D*, Linsley Pond four days later than *C*, showing great increase in both soluble ionic and total phosphorus in the epilimnion after a marked rise in surface temperature; *E*, Linsley Pond, showing late autumnal condition with large quantities of phosphorus and of ferrous iron (Fe^{++}) in the deep deoxygenated water

rise in phosphorus in the deep water of small lakes concerns the soluble phosphate fraction and is a concomitant of the fall in the oxygen concentration and a rise in that of ferrous iron. During autumnal circulation the ferrous iron is oxidized to some form of ferric phosphate, which appears to be quantitatively precipitated. It is clear from experience in Linsley Pond, however, that such suspended acid-soluble sestonic phosphorus can be present in small amounts of the order of 2 mg. $m.^{-3}$ for some days or perhaps weeks in the autumn, and that it may therefore play a small part in the nutrition of phytoplankton at this time, just as ferric hydroxide is known to do in cultures of marine diatoms (Harvey 1937a, Goldberg 1952).

The nature of this ferric phosphate, which Einsele supposes to be less soluble even than the hydroxide, is not fully established. The fact that it is not present during most of the year in Linsley Pond, in spite of the continual existence of ferric hydroxide suspended in the water, indicates that it is not formed by the adsorption of phosphate from very dilute solution at about neutrality in well-aerated waters. Cooper (1948a) indeed points out that ferric phosphate suspended in water slowly hydrolyzes, the hydrolysis being more rapid at a high than at a low *p*H. The data are inadequate to permit quantitative treatment, but there is evidence that particulate ferric phosphate can occur in the sea and that it is possibly formed in the alimentary tracts of animals and so incorporated into faecal pellets. The mechanism involved is certainly different from that producing the ferric phosphate detected in Linsley Pond after autumnal circulation, but in both cases the material evidently has but a transitory existence.

Einsele finds that, in general, the ratio of soluble phosphate to iron increases with depth. This is to be expected if the quantity of iron is limited by turbulent diffusion downward of oxygen, and if the phosphate, normally present in stoichiometric amounts far smaller than the iron, tended to be precipitated as a ferric phosphate before any ferric hydroxide is formed, as Einsele supposes. The appearance of large amounts of phosphate in the deep water therefore is dependent on the disappearance of the oxidized microzone at the mud surface, which, while it is in existence, prevents free passage of phosphate ions accompanied by excess of ferrous ions into the water.

In lakes in which a considerable amount of H_2S is formed, some ferrous sulfide may be precipitated from the hypolimnion late in stagnation, removing enough iron to permit a certain amount of the accumulated

---

(note difference in the scale of this figure and others). The difference in the deep-water total phosphorus observed between *C* and *D* may be due to errors of sampling water in which the concentration gradient is very great, but it is probable that more of the phosphorus at 13 m. was in soluble form on the later date.

phosphate to remain in true solution during autumnal circulation. Hasler and Einsele (1948) have in fact suggested the addition of sulfate as a method of fertilizing lakes in which an excessive part of the circulating phosphorus is held in the hypolimnion in the way that has just been described. If a sufficient amount of the added sulfate were reduced to H_2S in the hypolimnion at the end of stagnation, all the iron but none of the phosphate might be precipitated before autumnal circulation began, and the phosphate store of the hypolimnion would then be distributed freely throughout the lake. At the same time, the precipitated ferrous sulfide would be oxidized as an oxidized microzone redeveloped and the resulting ferric sulfate would hydrolyze to ferric hydroxide and sulfate ions in equilibrium with the bases of the lake water. Hasler and Einsele also think that increased oxidation of organic matter in the lake sediments, resulting from the bacterial reduction of the sulfate, would accelerate the general metabolic cycle of the lake.

The control of the process of liberation of phosphate into the water from the mud is illustrated by Mortimer's observations on Esthwaite Water (Fig. 202) and in his anaerobic experimental tank (Fig. 198). As has been indicated above, ferrous iron appears in the free water when the redox potential falls below 0.2 volt. Some combined iron appears both in the experimental tank and in the extreme bottom water before this, but it is unaccompanied by phosphate. In the tank, the phosphate and ferrous iron rise simultaneously. In the experimental tank, the phosphate continues to rise, though less rapidly, after the ferrous iron starts to fall, owing to the formation of ferrous sulfide. In the lake, at 14 m. considerable fluctuation in phosphate occurred, not parallel with equally large variations in iron. Mortimer notes that the hypolimnion of Esthwaite was probably less stable in 1940, when these observations were made, than in the previous year, but adduces evidence against this being the explanation of the observed fluctuations, which he leaves unexplained. In the aerated tank of water over mud, no significant changes in the phosphate concentration occurred. The contrast between the aerobic and anaerobic tanks provides a model of the differences exhibited when the phosphorus cycles of lakes with orthograde and extreme clinograde oxygen curves are compared, and emphasizes the fundamental difference between lakes which retain a perennial oxidized microzone and those which during stagnation lose such a microzone where the sediments are in contact with the deep water.

It would appear probable that the main forms of manganese likely to be present in lake waters are MnO_2 in suspension and, under more or less reducing conditions, Mn^{++} in solution. Therefore, no manganic phosphate is to be expected playing a role of the kind described for ferric

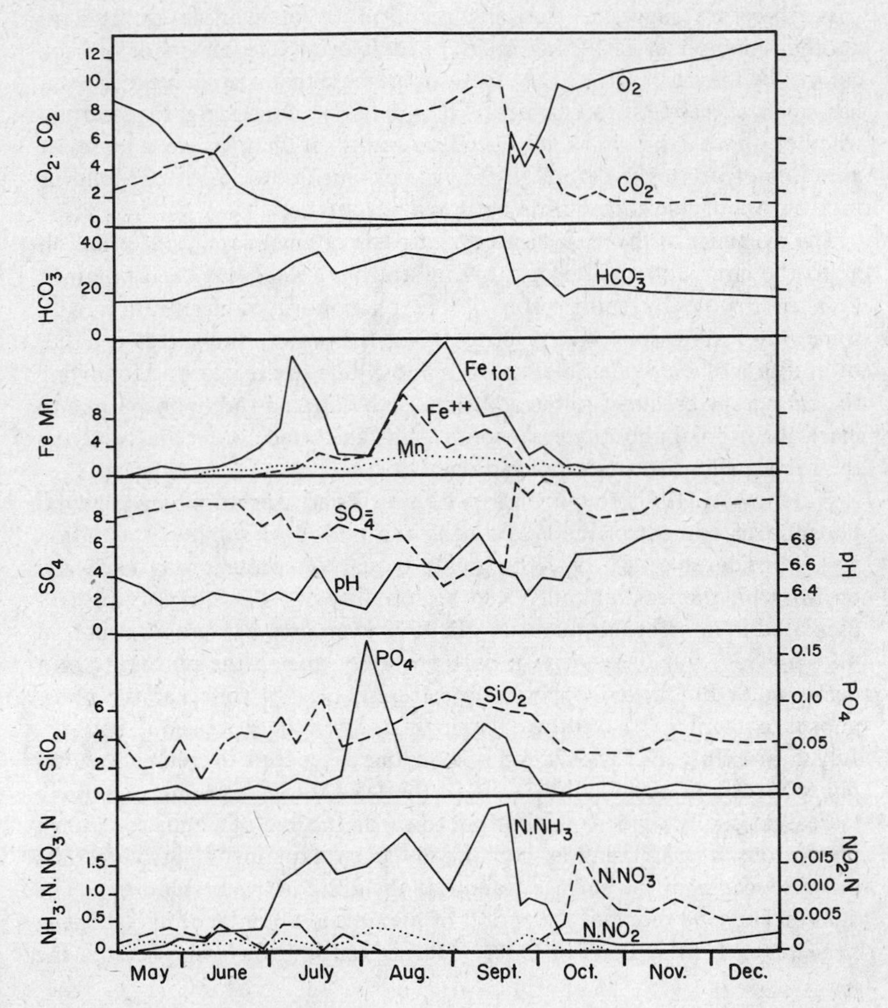

FIGURE 202. Esthwaite Water, seasonal variation in composition of water at 14 m., just over the deepest mud (after Mortimer).

phosphate. Since MnO_2 sedimenting into water containing ferrous ions is reduced and the ferrous iron is simultaneously oxidized to ferric hydroxide, addition of manganese to a lake should remove the ferrous iron from the hypolimnion. If sufficient of the element is added and is incorporated in the sediments, since manganese is reduced more easily than iron, it will enter the hypolimnion before the latter element and might even wholly displace it from the cycle in the water. Hasler and Einsele (1948) have therefore suggested that the addition of manganese be used as another method of fertilizing lakes by delivering the phosphorus from control by the iron cycle. They seem to indicate that one such experiment has been successfully performed. It would be interesting to ascertain whether those lakes in which large amounts of manganese and lesser amounts of iron exist naturally, receive copious additions of phosphorus to the trophogenic zone during autumnal circulation.

The dynamics of the phosphorus cycle: observational data. Information as to the circulation of phosphorus in stratified lakes has been obtained both indirectly, by comparison of the phosphorus concentration with some other more conservative property of the water, and directly, either by addition of a considerable excess of phosphate to the lake and following the changes subsequent to the addition, or by adding radiophosphorus to mark the normal phosphate phosphorus in the surface water and studying the subsequent distribution of radioactivity.

Hutchinson (1941) found that in Linsley Pond, variations in the total phosphorus content of the lake during the period of summer stagnation were considerable. In 1937 the greatest quantity present was 11.79 kg. on May 17, the least quantity 5.86 kg. on July 6. The quantity present oscillated in an irregular manner. In 1938 even greater changes occurred than in 1937, but these were in part probably due to the poisoning of a dense water bloom by copper sulfate treatment. In this year the phosphorus content of the entire lake increased on one occasion, between July 5 and July 12, by 9.78 kg. Both increases and decreases in total phosphorus can occur at all levels in the lake, though the most striking increases usually occur at the bottom towards the end of stagnation, when the ferrous iron is rising as has already been explained. Striking rises can also occur in the surface waters at the time of water blooms. The data in Table 98 relate to the period of maximum abundance of *Anabaena circinalis* in Linsley Pond in 1938. During the four days in question, the

TABLE 98. *Variation in various fractions of phosphorus in the epilimnion of Linsley Pond during the height of an algal bloom*

Depth,	Temp., °C.		Sol. P·PO_4, mg. m.$^{-3}$		Sol. Org. P, mg. m.$^{-3}$		Seston P, mg. m.$^{-3}$	
m.	25 July	29	25 July	29	25 July	29	25 July	29
0	24.4	27.7	0.2	5.5	7	13	4	26
2	23.4	23.4	0.2	4.0	7	6	8	10
3.5	21.9	21.3	...	...	...	...	...	...
5	15.2	14.9	0.4	1.5	7	6	8	12

number of *Anabaena* cells remained practically constant, being 10,100 per $cm.^3$ on July 25 and 10,330 on July 29; but a considerable increase in *Fragilaria* occurred, from 284 cells per $cm.^3$ on the first day to 930 on the second. The organic content of the seston in the surface water likewise rose from 3.7 to 5.6 mg. per liter, and the chlorophyll from 23 per liter to 29 per liter.

It is evident that the great rise in temperature during the four days in question has increased the rate at which phosphorus enters the superficial layers of the lake, and that this phosphorus can only have come from the marginal sediments. Such a sudden access of the element is doubtless due to an increased rate of organic decomposition in the shallowest mud and in the organic debris in the weed beds of the lake. It cannot have determined the incidence of the *Anabaena* bloom, which was developed before the rise in phosphorus began, but it may have permitted other forms to develop after the *Anabaena* had reached its maximum.

Einsele has likewise noted an increase of from 50 to 180 mg. $m.^{-3}$ over a period of fourteen days in the surface water of the Schleinsee. Another comparable case is provided by observations (Gardiner 1941) on one of the reservoirs, Barn Elms No. 8, of the London Metropolitan Water Board. This reservoir was known to produce rather heavy crops of phytoplankton and was believed to have bottom deposits rich in nutrients. The whole area of these deposits seems to have been available throughout the summer as a source of nutrient materials, for as far as can be ascertained from the published account, the reservoir does not stratify thermally. In August 1938, nineteen months after the inlet and outlet of the reservoir had been closed, an immense population of *Aphanizomenon* developed, the numbers of the alga and the concentration of soluble phosphorus varying in the following way:

	Filaments, per ml.	Sol. P, mg. $m.^{-3}$
August 2	3,183	29
August 9	2,217	26
August 16	133	83
August 23	188,000	259
August 30	0	388

Phosphate production from the mud in this case may have been the cause of the rise in the population of the alga. More phosphate was liberated than could have been assimilated by the alga, but it is to be noted that the maximum in the phosphorus concentration did not occur until after the *Aphanizomenon* population had reached its peak, had died, and had begun to decay. It is quite likely that part of the rise in total phosphorus in cases of this sort is due to dying algae adding organic matter to the mud and promoting bacterial activity.

There is abundant evidence, as has already been indicated, that all layers of Linsley Pond can receive bicarbonate from the sediments, even when the stability is at a maximum. This is clearly due to a predominantly horizontal current system; the resultant vertical pattern of distribution in the case of bicarbonate reflects the form of the lake basin; comparable passage of other materials, including phosphate, into the water from the mud is therefore to be expected. McCarter, Hayes, Jodney, and Cameron (1952) have demonstrated directly the possibility of this type of movement by introducing radiophosphorus into the hypolimnion of a small stratified lake, the Punch Bowl near Halifax, Nova Scotia (Fig. 203). Their

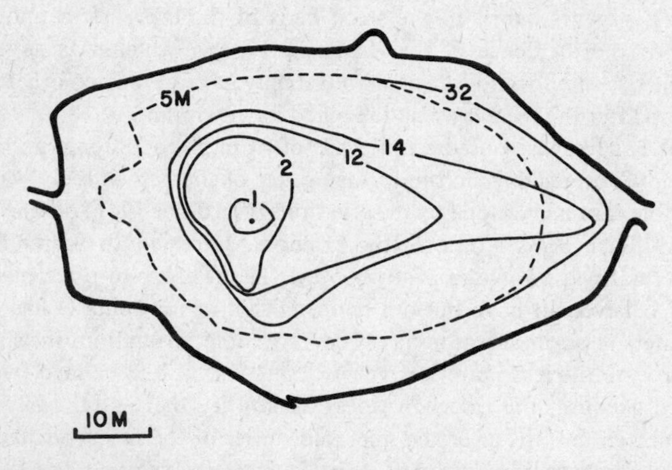

FIGURE 203. Horizontal spread of radiophosphorus introduced into the hypolimnion of the Punch Bowl near Halifax, Nova Scotia. Solid contours indicate limits of detectability of radiophosphorus at 1, 2, 12, 14 and 32 days after its introduction at the point in the center of the innermost contour. The 5-m. bathymetric contour is indicated by a broken line. (After McCarter, Hayes, Jodney, and Cameron.)

experiment showed that the radiophosphorus moved at a mean rate of 3 m. day^{-1} for a horizontal distance of 42 m., though the vertical movement was not more than 2 or 3 m. during the entire period of observation.

No trace of a morphometrically determined pattern in the total phosphorus content is apparent, however, in the mean values for Linsley Pond. It is therefore necessary to suppose that phosphorus leaving the mud is sedimented from all levels of the free water. The bicarbonate, once in the water, behaves in a conservative way; the phosphorus, always entering the seston, does not, except perhaps in the deepest water during the last phase of stagnation. The irregular changes in the total phosphorus content of the entire lake may therefore be explained on the basis of the relative rates of diffusion from the mud and sedimentation from the water. The largest single increase ever observed, 9.78 kg. in a week, corresponds to an average loss of 7 γ cm.$^{-2}$ of mud surface. Since the wet superficial mud of the deep water contains 140 γ P cm.$^{-3}$ (Hutchinson and Wollack 1940), the observed maximum output would correspond to the passage into the water, within a week, of the phosphorus in a layer of mud 0.5 mm. thick; this would seem not to place an excessive demand on the mechanism postulated.

The single rise of rather over 9 kg. in a week represents a rate of replacement of the phosphorus in the lake of about once a week. This, however, is certainly exceptionally high, as later experiments with radioactive phosphorus indicate. Even if the mean rate of increase, during those periods in which only increases occurred, were taken as representing the average rate of output from the mud, the replacement rate would be about once every two months. This figure is obviously an underestimate, but does at least serve to indicate the extremely active nature of the phosphorus cycle in the lake, The rises in plankton at times when the phosphorus content is very low, or the increases in phosphorus when the phytoplankton was not declining, have puzzled some limnologists, notably Juday and Birge (1931), but they cease to be paradoxical when the phosphorus cycle is regarded as a rapid diffusion and sedimentation of the kind just described. It is, of course, obvious that the magnitude of such cyclical changes, at least when measured in terms of rates of replacement, will tend to be much greater in small productive lakes than in large unproductive lakes. This conclusion is a particular dynamic expression of Strøm's (1933b) generalization that the relationship of volume of water to area of mud surface is of fundamental importance as a determinant of productivity.

The type of phosphorus cycle that has been outlined above is incomplete. Though the movement of phosphorus from mud to water will be compensated by sedimentation, some of the phosphorus that leaves the mud in the shallows will undoubtedly be sedimented in deep water, but except at the periods of circulation there is no mechanism available for the return of phosphorus from the deep mud to the shallow mud. The latter therefore would appear to be continually losing phosphorus that is not replaced. The very great rises that sometimes occur in the epilimnion during extreme stratification indicate that the shallow mud is by no means depleted of mobile phosphorus and suggest that a return mechanism actually exists. The most reasonable suggestion as to this mechanism would seem to be that at the vernal circulation, in which a considerable amount of free ionic phosphate is usually present at all depths, the rooted littoral vegetation can take enough phosphorus from the circulating water to replace the element in the sediments of the shallow zones of the lake when the higher plants ultimately decay.

Quantitative experimental studies. Hayes, McCarter, Cameron, and Livingstone (1952) added radiophosphorus to Bluff Lake, a small seepage lake (A = 0.4 km.^{2}, z_m = 7 m.) near Halifax, Nova Scotia, with little vegetation, which remained thermally unstratified. The material was distributed throughout the top 5 m. within a day, but took four or five days to become of uniform concentration in the small deep area below 5 m., a result of some hydrodynamic interest. The absolute concentration (corrected for decay) fell rapidly, reaching a plateau level which represents about 10 per cent of the original amount added. The investigators conclude that this represents an exchange equilibrium between the water and all forms of solid material, living or nonliving, in contact with the water.

A comparable experiment in the Punch Bowl near Halifax (Coffin, Hayes, Jodrey, and Whiteway 1949; Hayes and Coffin 1951), a bog lake with much sphagnum, showed the same kind of phenomenon in the epilimnion. Practically no phosphorus reached the hypolimnion, however, a result of interest in view of Hutchinson and Bowen's studies in the much more productive Linsley Pond. The radiophosphorus was immediately taken up by sphagnum and plankton but did not appear in rooted vegetation till two weeks after the addition.

By assuming that the uptake by, and liberation from, the solid phase is proportional to the quantity present in water and solid phase respectively, the fall in radiophosphorus to a steady-state level can be analyzed as the sum of two exponential processes, and the quantities given in Table 99

TABLE 99. *Turnover time and mobile phosphorus in three lakes studied experimentally*

Lake	A, km.^{2}	z_m	Turnover Time, Water, days	Turnover Time, Sediment, days	Total Mobile P: Phosphorus in Lake
Bluff	0.4	7	5.4	39	6.4
Punch Bowl	0.3	6.2	7.6	37	4.7
Crecy	2.04	3.8	17	176	8.7

can be determined. In this table the turnover time is the time taken to absorb from the water or liberate from the solid phase under steady-state conditions a mass of phosphorus equal to the stationary mass present, respectively, in the water or the solid phase.

The quantity in the last column represents the ratio of the total phosphorus taking part in exchange, either in sediments or in water, to that in the water alone.

Hayes *et al.* (1952) also have been able to obtain rough values of the same quantities from the decline of phosphate added to Crecy Lake, New Brunswick (Smith 1948), but it is known that these results are less likely to be accurate than in the experiments with P^{32}.

Einsele (1941) has examined the phosphorus cycle in the Schleinsee by the artificial enrichment of the lake with superphosphate. In its ordinary condition, the trophogenic layer contains detectable soluble phosphorus only at the time of circulation. The total phosphorus of the trophogenic layers is normally about 20 mg. $m.^{-3}$, occasionally rising to 25 mg. On July 3, 1937, 14 kg. phosphorus were added as superphosphate. After two days, only 7.1 kg. of soluble phosphate phosphorus remained in the circulating part of the trophogenic zone. Such phosphate as was present on this and subsequent dates tended to be distributed over an increasingly great range of depths. The total organic (sestonic and soluble) phosphate increased almost as much as the inorganic phosphorus decreased, and at the time of the maximum content of organic phosphorus the greatest quantity lay between 3 and 6 m. It is evident that phosphorus was being taken up very rapidly and sedimented as sestonic phosphorus. At the end of the summer the phosphorus content of the lake had returned to a normal figure, though it had been artificially doubled by the addition made early in July.

In 1938 the experiment was repeated on a larger scale, 94 kg. of phosphorus being added as superphosphate in two equal fractions on May 30 and June 30. The results were similar to those in the previous experiment, save that the fall in the inorganic fraction was not quite so rapid. After the second addition the total phosphorus began to fall, and then rose so that at the maximum on August 1 it surpassed the quantity that must have been present just after the addition on June 30, though it was less than the sum of both additions and the original content on May 30. The rise was due entirely to sestonic phosphorus. Such a rise must be attributed to the passage into the seston of phosphorus previously sedimented in the littoral zone, and doubtless represents an artificial enhancement of the phenomenon of littoral liberation of the element at the height of summer, at the time of the development of cyanophycean water blooms in some lakes in late July and early August.

Analysis of both net plankton and total seston indicate an enormous uptake of phosphorus relative to nitrogen during the days subsequent to fertilization. In the net plankton, the ratio N:P was normally 10–20:1; some days after fertilization it had dropped to 4:1 or even 3:1, and it is possible that immediately after fertilization even lower values would be obtained. The total plankton seems normally to have given ratios of 8–9:1, but after fertilization this fell to 1.9:1. A rapid uptake of phosphorus by the seston, and its subsequent sedimentation, is clearly the major mechanism determining the events subsequent to fertilization in the Schleinsee.

At no time did any considerable amount of the added phosphorus appear in the hypolimnion. Between August 1, 1938, when the lake contained 120.3 kg. P, and September 7, when it contained 64 kg., about 56 kg. appear to have been lost. The normal total content in September in previous years was about 40 kg., so that actually, of the 94 kg. added in 1938, only 24 kg. remained in the water in September. In May 1939, the amount present, 27 kg., was slightly greater than normal for the time of year, but the content in the following September, 37.1 kg., was entirely in accord with the quantities present at the end of summer stratification in the years prior to fertilization. It appears, in fact, that the artificial addition of phosphorus in a single or even two consecutive years has no permanent effect on the biochemistry of the lake, which evidently acts as a self-regulatory system returning to a steady-state condition determined by all the variables that operate within it, after one of these variables has been temporarily disturbed. This discovery is one of the most important that have ever been made in limnology.

The addition of a large quantity of phosphorus in 1938 undoubtedly led to a considerable increase in the standing crop of plankton. In terms of the total sestonic nitrogen, the mean summer standing crop in 1938 was 2560 mg. $m.^{-2}$ The sestonic nitrogen in the tropholytic zone showed no such increase, but the increased nitrogen of the seston of the trophogenic zone was later reflected in a great increase in the ammonia content of the tropholytic zone. Normally the ammonia nitrogen of the entire lake below 6 m. increased from about 75 kg. at the end of June to 125 kg. in the middle of September, a rise of about 50 kg. In 1938 the rise was about 270 kg., so that the extra ammonia represented about four times the normal production. The total combined nitrogen in the lake reached a maximum of 970 kg. on September 7, 1938; this figure is about 170 kg. more than was previously recorded, and apparently represents nitrogen fixed in the lake. Einsele supposes that this fixation is bacterial and occurs in the littoral and sublittoral sediments as the result of the addition of dead plankton enriched in phosphorus.

The total productivity of the lake was estimated from the loss of calcium carbonate from the trophogenic zone, incident to the removal of CO_2 by photosynthesis; from the ammonia production in the tropholytic zone; from the CO_2 production of the tropholytic zone; and from the last, assuming that the metabolic CO_2 anomaly represents only half the carbon undergoing anaerobic decomposition. The last estimate is believed to be the most reliable. The results for the entire lake (Table 100) are reported in kg. of dry organic matter produced in the five months from May to September.

TABLE 100. *Estimates of productivity of the Schleinsee before and after addition of phosphorus*

	1935–1937	1938
Biogenic ppt. $CaCO_3$	1,900	5,000
$N \cdot NH_4$ produced	2,100	5,900
CO_2 produced	3,800	9,600
CO_2 produced (corrected)	5,600	17,200
Mean biomass	1,350	2,800

Assuming that the corrected production of CO_2 gives the best estimate of the productivity, the rate of replacement is found to have been 0.83 times per month prior to fertilization, and 1.2 times per month in 1938. The mean biomass was doubled by the fertilization, the mean productivity tripled.

The mechanism of the phosphorus cycle has been further studied in Linsley Pond by Hutchinson and Bowen (1947, 1950) who marked the epilimnetic phosphate with radiophosphorus introduced into the surface layers of the lake. Their first experiment, done in 1946, merely indicated that radiophosphorus so introduced into the lake could appear in the hypolimnion under conditions of extreme stratification, within a week after the introduction. In the second paper they were able to follow the radiophosphorus in both solution and suspension for a period of four weeks. They found that though the radiophosphorus per square meter in the water column in the deepest part of the lake was always greatly in excess (140 to 190 per cent) of the quantity introduced per square meter, there was good agreement (90 to 110 per cent) if the quantity actually present in the entire lake was compared with the total quantity introduced. This implies that not merely is radiophosphorus sedimented in falling seston into the hypolimnion, but it undergoes a continual process of regeneration. Such seston as reaches the mud surface at intermediate depths must give up its phosphorus rapidly to the free water at such depths, and the phosphorus must then be distributed by the horizontal current system postulated to explain the distribution of bicarbonate in the lake.

This phenomenon is not in agreement with Einsele's results, in which his massive quantities of introduced phosphate were permanently removed from the water by the falling seston. Some radiophosphorus is undoubtedly immediately taken up by littoral flowering plants, and later is slowly liberated into the epilimnion. The radiophosphorus content of the hypolimnion was found to rise rapidly during the first week, and then much less rapidly but quite uniformly during the remaining three weeks of the experiment. This implies that initially the sedimenting seston was much richer in radiophosphorus than at any subsequent time. As a considerable amount of radiophosphorus remained in solution in the epilimnion throughout the experiment, it seems clear that an organic phosphorus compound less easily assimilated than phosphate was produced by the phytoplanktonic organisms after they had first taken up the entire supply of radiophosphorus added, and that this organic phosphorus compound contained the greater part of the soluble radiophosphorus after the first few days of the experiment. After equilibrium between the seston and organic soluble radiophosphorus may be supposed to be established, it is possible to compute, from the rate of increase of radiophosphorus in the hypolimnion and the mean specific activity of the epilimnetic seston, the mean rate of delivery of phosphorus into the hypolimnion. For the period during the first half of August during which the best data are available, the balance sheet of Table 101 can be constructed.

TABLE 101. *Phosphorus metabolism of Linsley Pond*

	Mean Rate, kg. per week
Observed increase in epilimnion	0.26
Loss to hypolimnion	1.55
Total entry into epilimnion	1.81
Observed increase in hypolimnion	3.75
Gain from epilimnion	1.55
Gain from hypolimnetic sediments	2.20

Since the mean phosphorus content of the epilimnion during this period was 5.82 kg., the rate of entry of phosphorus into the epilimnion implies one replacement about every three weeks. But because radiophosphorus is evidently lost from the plankton before it is sedimented and some of this radiophosphorus will be later utilized again, such a figure gives too low an estimate for the rate of replacement of the plankton itself.

Detailed examination of the uppermost layer of the hypolimnion (3.5 to 5.2 m.) indicates that it is very unlikely that phosphorus enters the water of this layer from the mud; the variations in its phosphorus content are explicable solely on the assumption of falling seston. Below 5.3 m. there is clear evidence of the entry of phosphorus from the lake sediments as well as from above. The process was evidently less well marked after August 15 than before that date. The localization of the diffusion of phosphorus into the lake water from the sediments to the region below

5.3 m. is presumably due to the fact that phosphorus in reduced mud in the presence of ferrous iron cannot pass through the oxidized microzone of the mud where the latter is in contact with water which contains appreciable oxygen.

The passage of phosphorus into the epilimnion from the vegetation and littoral sediments evidently involves a process that is quite different from that occurring in the hypolimnion. It is reasonable to suppose that the littoral vegetation and the organic material formed from it on decomposition constitute the major places of storage for phosphorus in the epilimnion. The radioactivity data clearly show that one week after the introduction of the P^{32} there was less radioactive material in the lake than there was at the end of the experiment. The radioactivity of the epilimnion declines irregularly and less rapidly than that of the hypolimnion increases. This can only mean that part, probably from 10 to perhaps 20 per cent of the introduced P^{32}, is immediately taken up by the littoral plants and more slowly liberated to the water. It was found, however, that the specific activity of the phosphorus in the littoral plants was much less than that of the phosphorus in the water, so that the exchanges involve only a part of the phosphorus in the littoral plants. The mobile fractions conceivably may be largely in epiphytic diatoms and the like.

The phosphorus cycle exhibited by these experiments during extreme summer stratification may therefore be considered to involve the following processes.

(1) Liberation of phosphorus into the epilimnion from the littoral, largely from the decay of littoral vegetation.

(2) Uptake of phosphorus from water by littoral vegetation.

(3) Uptake of the liberated phosphorus by phytoplankton.

(4) Loss of phosphorus as a soluble compound, less assimilable than ionic phosphate, from the phytoplankton, probably followed by slow regeneration of ionic phosphate.

(5) Sedimentation of phytoplankton and other phosphorus-containing seston, perhaps largely faecal pellets, into the hypolimnion.

(6) Liberation of phosphorus from the sedimenting seston in the hypolimnion or when it arrives at the mud-water interface.

(7) Diffusion of phosphorus from the sediments into the water at those depths at which the superficial layer of the mud lacks an oxidized microzone.

The first process is likely to be most significant after the littoral vegetation has attained its full growth, though it is reasonable to suppose that at all times there will be some partially decomposed material that can liberate phosphorus to the water. During periods of rapid growth of the littoral plants, the second process is likely to proceed far more rapidly than the first. The fact that water plants immediately take up radiophosphorus and more slowly lose it is well-established and shows that both the first and second processes are in fact going on all the time, though during the period of the experiment the first was certainly more rapid than the second. The supposed greater rapidity of the second process when the littoral vegetation is growing rapidly in the spring presumably provides a mechanism by which the cyclical nature of the migration of phosphorus in the lake can be maintained.

All the processes listed except the second and seventh either have no effect on the direction of movement or they result in a concentration of phosphorus in the deep water. Since in Linsley Pond and in many other small lakes the ferrous iron content of the deep water is at the end of summer stagnation certainly greatly in stoichiometric excess of the phosphate, it is probable that a good deal of the phosphorus derived from sedimenting seston, as well as that derived from the mud, will be precipitated at the time of circulation. After winter stagnation, however, during the spring circulation period there is usually an appreciable amount of free phosphate in the water. It is very reasonable to suppose that the uptake of some of this phosphorus, whatever its origin may be, by the littoral vegetation represents the laying up of a supply of the element at the margins of the epilimnion. As soon as the main growth of the rooted plants has come to an end and the decay of some of the older leaves begins, the overall process is reversed.

Such a sequence of events is a very plausible explanation of the rather general low phytoplankton populations encountered in early summer, followed by immense blooms of blue-green algae when the superficial layers of the lake reach their highest temperature in July and August. On this hypothesis the presence of such blooms is primarily determined by the onset of rapid decomposition at high temperatures in the littoral, and a subsequent rapid output of phosphorus into the epilimnion. As has been indicated, both Einsele (1941) and Hutchinson (1941) have noted superficial increases in total phosphorus at the time of great blue-green blooms. Though utilization of phosphate would be so rapid that the intermediate ionic stage could not be detected, there can be no question but that the phosphorus in the algae must have come from the margins of the lake.

SUMMARY

Phosphorus is in many ways the element most important to the ecologist, since it is more likely to be deficient, and therefore to limit the biological productivity of any region of the earth's surface, than are the other major biological elements.

Apart from the possible rare and transitory occurrence of phosphine under strongly reducing conditions, only orthophosphate and various undifferentiated organic phosphates are likely to be important in limnology. In ordinary cases, it is convenient to distinguish *soluble phosphate* phosphorus, sestonic *acid-soluble phosphate* phosphorus (mainly ferric and perhaps calcium phosphate), *organic soluble* (and colloidal) phosphorus, and *organic sestonic* phosphorus.

The total phosphate in natural waters varies from less than 1 mg. $m.^{-3}$ to immense quantities in a very few closed saline lakes. The total quantity depends largely on geochemical considerations, usually being greater in waters derived from sedimentary rock in lowland regions than in waters draining the crystalline rocks of mountain ranges. Most relatively uncontaminated lake districts have surface waters containing 10 to 30 mg. P $m.^{-3}$, but in some waters that are not obviously grossly polluted, higher values appear to be normal. The soluble phosphate usually is but a small fraction, of the order of 10 per cent of the total. The ratio of organic soluble to sestonic phosphorus is variable; the general orders of magnitude of the two fractions are the same. Acid-soluble phosphate may be found during autumnal circulation of small lakes of which the hypolimnia acquire much ferrous iron and phosphate during stratification.

Phosphorus is removed from water by all phytoplanktonic organisms. Some organisms, such as *Asterionella*, in natural lake water are extraordinarily efficient in the utilization of the element and can take it up from water containing less than 1 mg. $m.^{-3}$ In artificial media the efficiency of utilization appears to be less. In some cases quite moderate amounts of phosphate, suboptimal to one species, appear to be inhibitory to others.

The vertical distribution of phosphate during stratification is very variable and may be complicated. In unproductive lakes, there is usually little vertical variation in any fraction. In productive lakes with clinograde oxygen curves, there is an increase in soluble phosphate in the oxygen-deficient part of the hypolimnion. This is in part due to decomposition of sinking plankton, but in most cases is primarily caused by liberation of phosphate from sediments on reduction. An oxidized mud surface not merely holds phosphate but prevents diffusion of phosphate and ferrous ions from deeper layers in the mud, as the ferrous iron is always in excess and when oxidized precipitates all the phosphate. In lakes which proceed to phase III of stagnation, large amounts of ferric phosphate will be formed during autumnal circulation. This probably slowly hydrolyzes, and it may restore phosphate to the lake water and littoral vegetation. Where much H_2S is formed in phase IV of stagnation, the phosphate will be more mobile at circulation, and better regeneration is possible.

There is a good deal of evidence of rapid movement of phosphate from the sediments to the water, at all levels, in small lakes. At the height of summer there may be a great increase in total phosphorus in the surface waters at times of algal blooms, though soluble phosphate is undetectable. One condition for the maximum development of such blooms may well be rapid decomposition and consequent liberation of phosphate in the littoral sediments during spells of very warm weather. The phosphate would be taken up so fast by the growing algae that it never would be detectable.

Experiments with radiophosphorus have demonstrated rapid horizontal movement of the element in the hypolimnion, and cases of great increase in phosphorus at all levels of a stratified lake also imply such horizontal movement from the mud to the free water. The total phosphorus concentration in Linsley Pond does not show the steplike morphometrically determined distribution exhibited by bicarbonate. The phosphate must be engaged in too rapid a cycle to permit such a pattern to develop. When massive amounts of soluble phosphate are added to a lake, it is rapidly taken up by the phytoplankton and then sedimented. The productivity of the lake is increased, but only for a time. The chemical relations of phosphorus in mud and water evidently constitute a self-regulating system, at least within certain limits. Studies with radiophosphorus indicate rapid uptake by littoral plants and sediments, and by the plankton.

The rate of delivery of radiophosphorus into the hypolimnion permits an estimate of the fraction of epilimnetic phosphorus delivered in unit time to the hypolimnion. In the summer in Linsley Pond it appears that the phosphorus of the epilimnion is replaced from the littoral about once every three weeks. Other small lakes apparently exhibit equal or more rapid turnovers. During the period of observation, falling of seston into the hypolimnion accounted for rather less than half of the gain in phosphorus in the hypolimnion. The gain due to movement from the sediments was limited to the deep water, in which the mud surface may be regarded as reduced.

CHAPTER 13

The Sulfur Cycle in Lake Waters

In most fresh waters the sulfate ion is the second or third most abundant anion, being exceeded only by bicarbonate and in some cases by silicate. Acid lakes and some mineralized lakes in closed basins are known in which sulfate is present in higher concentration than are any of the other anions. In uncontaminated continental rain, sulfate is present in considerably greater quantity than is chloride (page 547).

THE BIOGEOCHEMICAL CYCLE OF SULFUR

Sources of Sulfate. Whatever the source of the sulfate in rain water, its presence ensures that lake waters contain the anion. There are, moreover, other possible sources from which $SO_4^=$ may enter lakes. Some sedimentary rocks contain calcium sulfate, which is moderately soluble in water; many waters characteristically rich in sulfate have derived it from such a source. The oxydation of sedimentary pyrite

$$4FeS_2 + 15O_2 + 2H_2O \rightarrow 2Fe_2(SO_4)_3 + 2H_2SO_4$$

is an equally or possibly more important source of sulfate. In dilute solution, particularly if some of the sulfuric acid formed is neutralized, the ferric sulfate hydrolyzes, producing ferric hydroxide that is precipitated and more sulfuric acid. If calcareous rocks are present in the vicinity, the sulfuric acid will produce calcium sulfate, but cases are known where very acid lakes contain appreciable amounts of sulfuric acid formed in this way.

In addition to these widespread sources of sulfate, the existence of streams of dilute sulfuric acid flowing into lakes in volcanic areas may be of local limnological significance, notably in Japan (Yoshimura, 1934). Mineral springs may also be rich in sulfate. Considerable quantities of sulfate may collect in closed basins in semi-arid regions undergoing concentration through evaporation. Moreover, though the ratio of sulfate to chloride is low in the ocean, sea water, being very concentrated by limnological standards, does in effect contain a considerable amount of sulfate in solution, so that the brackish lakes of coastal area may appear to be somewhat enriched in the anion, though they are still more enriched in chloride. All these possible local sources of sulfate will be found to provide situations of special limnological interest.

The sulfur cycle in the hydrosphere. Autotrophic organisms and many heterotrophic bacteria normally obtain their sulfur from sulfate in the medium. Since most of this sulfur is used to form amino acids such as cystine and methionine, containing –SH groups, it must undergo reduction during the course of assimilation. Most of the sulfur excreted by aerobic organisms is in oxidized form, but practically all heterotrophic bacteria can liberate sulfur as hydrogen sulfide from proteins and their decomposition products. Though this hydrogen sulfide may be largely oxidized under aerobic conditions, in lake sediments, in which little or no oxygen is present, the production of a considerable amount of hydrogen sulfide in decomposition is easily understood.

Sulfate-reducing bacteria. In addition to this general if indirect path by which a considerable part of the circulating sulfur of the biosphere is converted from sulfate to hydrogen sulfide, a more direct type of conversion, by sulfate-reducing bacteria, also occurs. The sulfate-reducing bacteria are anaerobic and heterotrophic, using sulfate as a hydrogen acceptor in metabolic oxidation in the same way that nitrate- and nitrite-reducing bacteria, and the rarer phosphate- and selenate-reducing species, use these oxyacid anions in the absence of free oxygen. Sulfite, hyposulfite, thiosulfate and elementary sulfur can be reduced as well as sulfate (Baars 1930). It is nearly always possible to find some evidence of such reduction wherever sulfate and organic matter coexist in the biosphere in the absence of oxygen. The best known sulfate-reducing bacteria belong to the genus *Desulfovibrio*; the fresh-water species *D. desulphuricans*, originally isolated from canals in Holland, is known to have been responsible for the production of hydrogen sulfide in the monimolimnion of Lake Ritom (Düggeli 1924). A closely allied species or strain, *D. aestuarii*, has been isolated from marine environments. At least one member of the group, from river mud, has been shown to effect the reduction of sulfate by the enzymatic activation of molecular hydrogen. It is not impossible, in view of the obvious capacity to reduce sulfate intracellularly possessed by all organisms that can use the sulfate ion as their sole sulfur source, that some extracellular production of hydrogen sulfide, due to a variety of forms other than *Desulfovibrio*, may occur in nature. Chle (1955a) finds that the rate of bacterial H_2S production is linearly dependent on the $SO_4^=$ concentration.

The production of hydrogen sulfide takes place primarily in deoxygenated regions of low redox potential, and in lakes such regions are mainly provided by the sediments. A considerable amount of the gas formed, at least in the earlier phases of stagnation, is likely to be fixed as ferrous sulfide, though in acid mud this compound will be appreciably soluble in the aqueous phase. Whenever the production of hydrogen sulfide is in excess of the amount of iron present in the sediments, as when much organic matter is sedimented into the hypolimnion during summer stratification in very productive lakes or when stratification is very prolonged, as in meromictic lakes, or when a large amount of sulfate is present and can undergo reduction, as in mineralized and brackish lakes, appreciable amounts of hydrogen sulfide may be found in the deep water. Often this will all have been produced from the mud, but in some lakes it appears that sulfate reduction can occur in the free water also. In general, such hydrogen sulfide produced in the anaerobic regions of the lake or its sediments is ultimately reoxidized to sulfate, completing the cycle. This can happen in two ways involving bacterial metabolism, though under some conditions H_2S can be oxidized to sulfur spontaneously by molecular oxygen.

Colorless sulfur bacteria. The first way is the metabolic oxidation of hydrogen sulfide by oxygen or by some equivalent substance, such as nitrate. Such processes yield free energy which can be used by chemautotrophic bacteria in the reduction of CO_2. The bacterial oxidation of hydrogen sulfide occurs in stages. *Beggiatoa*, one of the best known sulfur organisms, oxidizes H_2S to elementary sulfur, which is retained in the cell and further oxidized when the external supply of hydrogen sulfide is exhausted. Other species produce elementary sulfur extracellularly; this may happen in the oxidation of thiosulphate

$$5S_2O_3^= + H_2O + 4O_2 \rightarrow 5SO_4^= + H_2SO_4 + 4S$$

as is indicated by Starkey (1935) for *Thiobacillus thioparus*, though other species, such as *T. novellus*, oxidize thiosulfate quantitatively to sulfate. Elementary sulfur is an ordinary minor constituent of organic reduced lake muds (Vallentyne 1952; Sugawara, Koyama, and Kozawa 1953).

Other organisms are known which can dissolve and oxidize elementary sulfur. It is probable that many heterotrophs can, under certain conditions, oxidize thiosulfate. The oxidation of reduced sulfur compounds does not necessarily require the presence of free oxygen. *Thiobacillus denitrificans* in alkaline waters can use nitrate, liberating molecular nitrogen

$$6NO_3^- + 5S + 2CaCO_3 \rightarrow 3SO_4^= + 2CaSO_4 + 2CO_2 + N_2$$

and similar forms are known that oxidize thiosulfate at the expense of nitrate.

A good concise review of the metabolism of the bacteria oxidizing sulfur compounds is given by Stephenson (1939). Most of the well-studied species have been isolated from soils, pond and canal water, and aquatic sediments. It is quite possible, however, that the species best studied in the laboratory are not always the major agents in transforming sulfur compounds in nature. In lakes a number of morphologically recognizable sulfur bacteria have been recorded, and of these forms *Macromonas* and *Achromatium* may be abundant enough to suggest that they play a great part in the sulfur cycle in the localities that they inhabit but the details of their metabolism appear to be unknown. Much of the laboratory work has been done on species that oxidize thiosulfate, and at least some of the bacteria isolated from lakes fall into this category; but apart from a single reference by Baas-Becking (1925) to the occurrence of thiosulfate in a black mud, and a few determinations in mineral springs (Clarke 1924), the anions of the oxyacids of sulfur, other than sulfate, are quite unknown in inland waters. Sugawara, Koyama, and Kozawa (1953) could find no sulfite or thiosulfate in the muds of brackish and fresh-water lakes rich in the element. Sulfide, free sulfur, and a little sulfate can coexist in a single mud sample. Koyama and Sugawara (1953a) think that in some such cases manganese inhibits reduction of sulfate.

Photosynthetic sulfur bacteria. The second way in which hydrogen sulfide is oxidized in the hydrosphere is by the photosynthetic green and purple bacteria, of which the purple sulfur bacteria are best known. These organisms can use hydrogen sulfide as an oxygen acceptor in the photosynthetic reduction of carbon dioxide. The purple forms apparently can use thiosulfate and can carry the oxidation of hydrogen sulfide as far as sulfate; the green forms, which are much less significant quantitatively, oxidize H_2S only as far as elementary sulfur. Recent work has indicated that substances other than sulfur compounds can be used as oxygen acceptors by the purple photosynthetic sulfur bacteria, and the so-called nonsulfur bacteria can use sulfur compounds. This rather confusing aspect of the problem has been admirably treated by Van Niel (1944). In theory such organisms can complete the sulfur cycle even if they live below a barrier to molecular or turbulent diffusion, provided only that a source of radiation, which may be short infrared, is present, but the colorless sulfur bacteria can oxidize H_2S only when a continual supply of oxygen or its equivalent as nitrate can diffuse from above, though they can use such substances in complete darkness. The chemocline in a sufficiently stable meromictic lake constitutes an almost complete barrier to turbulent diffusion, and molecular diffusion is always very slow. A barrier of this sort is transparent, however, and it is therefore reasonable to find that in meromictic lakes in which much sulfate has entered the monimolimnion and much hydrogen sulfide has formed by reduction, a layer of purple bacteria frequently develops in the chemocline region, forming the so-called bacterial plate.

Since the purple bacteria, in theory, have no need of oxygen or its equivalent to oxidize hydrogen sulfide, they are often regarded as strict anaerobes, as some species probably are. Such species might be expected to occupy the surface of mud in which large quantities of hydrogen sulfide are being generated, so long as this surface did not lie so deep that the rather feeble photic requirements of the purple bacteria could not be satisfied. Colorless sulfur bacteria could exist in the free water above the purple forms, in layers receiving a little oxygen. There is some evidence of this kind of distribution in nature, but more generally it would seem that the limnetic purple bacteria are microaerophil rather than anaerobic, and in some of the best studied cases the spacial and temporal distribution and the implied environmental requirements of a colorless form such as *Macromonas* and a purple form such as *Thiopedia* do not seem to differ in any recognizable way.

Calcium sulfate. The foregoing introductory discussion has indicated that in the present state of knowledge an account of the inorganic sulfur compounds of lakes reduces to an account of the distribution of sulfate and of hydrogen sulfide. The occurrence of sulfate raises no special physicochemical problems. Of the sulfates of the common cations, calcium sulfate is the least soluble. The solubility product $[Ca^{++}][SO_4^{=}]$ for $CaSO_4 \cdot 2H_2O$ is

$$1.67 \times 10^{-4} \text{ at } 0^\circ \text{ C.}$$
$$2.36 \times 10^{-4} \text{ at } 25^\circ \text{ C.}$$

Since under an atmosphere of normal composition, containing 0.03 per cent CO_2 by volume, the concentration of ionic calcium (page 662) is unlikely to be much above 10^{-3} molar, the upper limit of sulfate to be expected before gypsum crystals form will be of the order of 0.1 to 0.2

molar, or 5 to 10 g. $SO_4^{=}$ per liter. Such concentrations occur only in saline lakes.

The solubility and ionization of hydrogen sulfide. The physical chemistry of the other principal sulfur compound to be considered, hydrogen sulfide, raises certain more complex problems of great limnological interest. The importance of these problems has already been indicated in the discussion of the limitation of the concentration of ferrous iron by hydrogen sulfide in Chapter 11. Similar problems will arise in the treatment of copper and manganese. It is therefore desirable to consider the properties of solutions of hydrogen sulfide in some detail. This is conveniently done before passing to the study of the distribution of sulfate and sulfide in individual lakes.

Hydrogen sulfide is freely soluble in water. Under a pressure of 1 atm. of the gas at 25° C. a liter of water dissolves 3.47 or 0.102 mole of the gas. In solution it behaves as a weak acid, the ionization constants being given by

$$k_1 = \frac{[OH_3^+][HS^-]}{[H_2S]} = 0.91 \times 10^{-7} \text{ at } 18^\circ \text{ C.}^{1} \tag{1}$$

$$k_2 = \frac{[OH_3^+][S^=]}{[HS^-]} \simeq 10^{-15} \tag{2}$$

Except in strongly alkaline solutions, the concentration of the sulfide ion $S^=$ will obviously be very small; even at *p*H 11, the concentration of the hydrosulfide ion HS^- will be 10,000 times greater than that of $S^=$. Since, as will appear later, in many cases the concentration of H_2S will be regulated by the solubility of ferrous sulfide, the minute quantity of $S^=$ is by no means unimportant limnologically, even though for the purposes of the present paragraph it may be disregarded. At *p*H 7 the concentrations of undissociated H_2S and of hydrosulfide ion will be approximately equal, the ratio $H_2S:HS^-$ increasing tenfold for unit fall in *p*H. If, as Baas-Becking (1925) suggests, undissociated H_2S is highly toxic even to sulfur bacteria and HS^- is much less so, the variation in the ratio of the undissociated compound and the hydrosulfide ion at different *p*H values may be of considerable ecological interest.

The relation of hydrogen sulfide concentration and base content to *p*H may be considered in the following elementary way (Baas-Becking 1925), which, since it takes no account of activity theory, will give but roughly approximate results when applied to concentrated solutions,

---

[1] Considered by Kolthoff (1931) to be the best value. Wright and Maass (1932) found 1.08×10^{-7} at 25° C. The determination at the lower temperature seems preferable in the present work.

particularly in the case of highly mineralized lake waters. In a solution containing B+ moles of base per liter

(3) $$[B^+] + [OH_3^+] = [OH^-] + [HS^-] + 2[S^=]$$

Taking

(4) $$k_w = [OH_3^+][OH^-] \simeq 10^{-14}$$

and combining equations 1, 2, and 3

(5) $$[H_2S] = \frac{([B^+] + [OH_3^+])[OH_3^+]^2 - k_w[OH_3^+]}{k_1[OH_3^+] + 2k_1k_2}$$

Within the range (pH < 11) of interest in the present work, $2k_1k_2$ may be neglected, as its value 1.82×10^{-22} is small compared to the minimum value 0.91×10^{-18} of the other term in the denominator. With no bases present and a saturated solution of H_2S under 1 atm. of the gas, the equation becomes

(6) $$[OH_3^+]^2 = 10^{-14} + 0.91 \times 10^{-7} \times 0.102$$

or approximately

(7) $$OH_3^+ = \sqrt{92.8 \times 10^{-10}} = 9.6 \times 10^{-5}$$

corresponding to a pH of about 4.0 (Baas-Becking gives a slightly different result, based on a solution of the original cubic without approximation, which corresponds to pH 4.1). At this pH, H_2S will be about 0.1 per cent ionized, so that the assumption implied in equation 6, that the whole of the quantity in solution is undissociated, is sufficiently justified.

With increasing basicity, the concentration of base will determine the relationship of H_2S concentration and pH. With a basicity of $> 10^{-4}$ moles per liter, corresponding to a bicarbonate alkalinity of at least 6.1 mg. HCO_3 per liter, and with pH > 6.0, equation 5 practically reduces to

(8) $$H_2S = \frac{[B^+][OH_3^+]}{0.91 \times 10^{-7}}$$

whence

(9) $$[HS^-] = [B^+]$$

and

(10) $$[S^=] = \frac{B \times 10^{-15}}{[OH_3^+]}$$

at all pH values likely to be of limnological interest.

These results indicate that in ordinary inland lakes, in which a small or moderate amount of alkaline earth bicarbonate constitutes the chief electrolyte, the presence of the few milligrams of hydrogen sulfide not infrequently encountered in solution in the deep water during phase IV of summer stagnation is reasonable, and that if, as is usually the case, the *p*H is a little below neutrality, this hydrogen sulfide will be present partly in an undissociated condition. In the waters of mineralized and often meromictic lakes, in which great quantities of the order of hundreds of milligrams per liter of H_2S are formed by the bacterial reduction of sulfates, very large quantities of the gas may be expected in solution at quite high *p*H values, provided that the alkalinity is great. This hydrogen sulfide will be present almost exclusively as HS^-, and such alkaline waters will probably provide the optimum conditions for the development of sulfur bacteria.

Relationship of hydrogen sulfide to iron and other heavy metals. The sulfides of the heavy metals are notoriously insoluble, and many workers have realized that the presence of large quantities of hydrogen sulfide must be inconsistent with the presence of appreciable amounts of ferrous iron in solution in hypolimnetic waters. Einsele (1937) has examined some of the available limnological data in the light of the known physical chemistry of the metallic sulfides. The present treatment, though receiving its initial impetus from Einsele's account, differs in using a more recent value of the solubility product of ferrous sulfide, and in allowing for all forms of H_2S in solution instead of only the undissociated molecule.

The best value for the solubility product for ferrous sulfide appears to be that calculated by Kapustinsky (1940) from thermodynamic data, namely 3.8×10^{-20}. Therefore, at any concentration of ferrous iron $[Fe^{++}]$

$$[S^{=}] = \frac{3.8 \times 10^{-20}}{[Fe^{++}]} \tag{11}$$

$$[HS^{-}] = \frac{[OH_3^{+}][S^{=}]}{k_2} = \frac{3.8[OH_3^{+}]10^{-5}}{[Fe^{++}]} \tag{12}$$

$$[H_2S] = \frac{[OH_3^{+}][SH^{-}]}{k_1} = \frac{4.2[OH_3^{+}]^2 10^2}{[Fe^{++}]} \tag{13}$$

Neglecting the concentration of $S^{=}$ as analytically insignificant at *p*H values of limnological interest, the total concentration of hydrogen sulfide $[H_2S]_T$ is related to the ferrous ion concentration at any *p*H by

$$[H_2S]_T = \frac{1}{[Fe^{++}]}(4.2[OH_3]^2 10^2 + 3.8[OH_3]10^{-5}) \tag{14}$$

The resulting hyperbolas have been plotted for various relevant values of *p*H in Fig. 204. These curves must be regarded as approximate only, in view of uncertainties in the values of the two ionization constants.

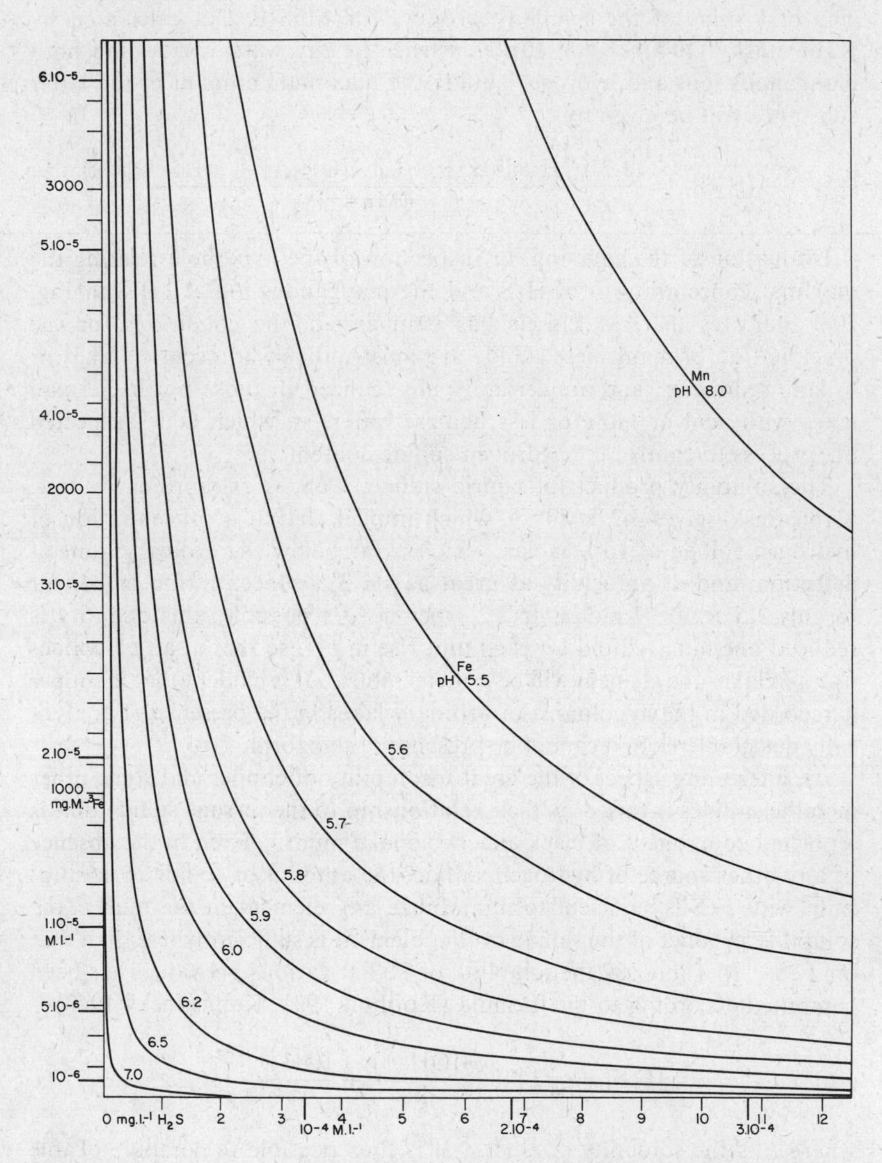

FIGURE 204. Solubility curves of FeS calculated for different *p*H values. Abscissa H_2S, ordinate Fe^{++}; both graduated in milligrams per liter and millimoles per liter. The outermost line gives the comparable curve for MnS at *p*H 8; at lower *p*H values this substance would be more soluble. The comparable curves for copper would not be distinguishable from the axes at the scale employed.

The available limnological data suggest that the relationships of both manganese and copper to hydrogen sulfide are also worth investigation. The best value of the solubility product for MnS is that calculated by Kapustinsky (1940), 1.1×10^{-15}, whence in any water containing both manganous ions and hydrogen sulfide the maximum amount of the latter substance will be given by

$$(15) \qquad [H_2S]_T = \frac{(1.21[OH_3^+]^2 \times 10^7 + 1.1[OH_3^+] + 1.1 \times 10^{-15})}{[Mn^{++}]}$$

Evaluation of the equation, or inspection of the hyperbola relating the maximal concentrations of H_2S and the manganous ion at *p*H 8 in Fig. 204, indicates that, as Einsele has emphasized, the conditions for the precipitation of manganese sulfide are most unlikely to occur in lakes or in lake sediments, and are certainly not realized in those north German lakes with acid or more or less neutral waters in which Ohle suspected the process to control the hydrogen sulfide content.

The solubility product for cupric sulfide, CuS, is extraordinarily low. Kapustinsky gives 3.2×10^{-38}, which implies that at a concentration of hydrogen sulfide of 10^{-9} molar, which is far below the ordinary limit of detection, and at an acidity as great as *p*H 3 a concentration of copper of only 3.5×10^{-11} moles, or 2.2 γ per m.^{3}, is possible; this quantity is reduced one hundredfold for each unit rise in *p*H, so that at all *p*H values usual in lakes the element will be undetectable. It is evident that if copper is recorded in the hypolimnia of ordinary lakes in the presence of analytically detectable H_2S, it cannot be present in ionic form.

An interesting aspect of the great insolubility of copper and some other metallic sulfides is raised by their relationship to the ferrous sulfide that is a normal component of black anaerobic lake muds. Even in the absence of any other source of hydrogen sulfide, the amount of sulfide in equilibrium with FeS is sufficient to immobilize any element in the mud if the solubility product of the sulfide of that element is sufficiently less than that for FeS. In Table 102 the solubility of FeS at various *p*H values has been computed according to the formula (Kolthoff 1931, Kapustinsky 1940)

$$(16) \qquad [FeS] = \sqrt{L\left(1 + \frac{[OH_3^+]}{k_2} + \frac{[OH_3^+]^2}{k_1 k_2}\right)}$$

where L is the solubility product. It is thus possible to compute (Table 102) the sulfide ion concentration and the upper limits of concentration of a number of other ions in water, in the presence of solid FeS.

Any increase in the concentration of ferrous ions will, by decreasing the concentration of sulfide, increase that of the other metals also. It is

TABLE 102. *Solubility of various heavy metals in the presence of excess ferrous sulfide*
(All computations in gram-atoms per liter)

*p*H	Fe^{++}	$S^{=}$	Cu^{++}	Ag^{+}	Pb^{++}	Cd^{++}	Sn^{++}	Zn^{++}	Co^{++}
3	2.05×10^{-2}	1.85×10^{-18}	1.7×10^{-20}	1.8×10^{-17}	3.7×10^{-11}	6.5×10^{-11}	5.4×10^{-10}	4.0×10^{-9}	1.7×10^{-5}
4	2.05×10^{-3}	1.85×10^{-17}	1.7×10^{-21}	5.6×10^{-18}	3.7×10^{-12}	6.5×10^{-12}	5.4×10^{-11}	4.0×10^{-10}	1.7×10^{-6}
5	2.06×10^{-4}	1.84×10^{-16}	1.7×10^{-22}	1.8×10^{-18}	3.7×10^{-13}	6.5×10^{-13}	5.4×10^{-12}	4.0×10^{-11}	1.7×10^{-7}
6	2.14×10^{-5}	1.78×10^{-15}	1.8×10^{-23}	5.8×10^{-19}	3.8×10^{-14}	6.7×10^{-14}	5.6×10^{-13}	4.1×10^{-12}	1.7×10^{-8}
7	2.83×10^{-6}	1.34×10^{-14}	2.4×10^{-24}	2.1×10^{-19}	5.1×10^{-15}	9.0×10^{-15}	7.5×10^{-14}	5.5×10^{-13}	2.3×10^{-9}
8	6.50×10^{-7}	5.86×10^{-14}	5.4×10^{-25}	1.0×10^{-19}	1.2×10^{-15}	2.0×10^{-15}	1.7×10^{-14}	1.2×10^{-13}	5.3×10^{-10}

evident, however, that no limnologically possible concentrations of ionic ferrous iron could ever permit the presence of analytically detectable amounts of copper or silver in the presence of FeS. For the divalent elements, the concentrations, in so far as they are determined by the existence of ferrous sulfide in the mud, will be proportional to the solubility products; therefore, per milligram of Fe^{++}, values up to the following may be expected:

Co	Zn	Sn	Cd	Pb	Cu
0.87×10^{-3}	2.3×10^{-7}	5.6×10^{-8}	6.4×10^{-9}	6.5×10^{-9}	9.4×10^{-19}

These figures may be instructively compared with the mean ratios of these elements to iron in the accessible lithosphere (Goldschmidt 1938), in which, per milligram Fe, there are:

Co	Zn	Sn	Cd	Pb	Cu
0.5×10^{-3}	8×10^{-4}	8×10^{-5}	10^{-5}	16×10^{-5}	10^{-2}

It is evident that, of all these elements, only cobalt is likely to be able to diffuse from reduced muds containing FeS in quantities, relative to iron, approaching those prevalent in the original source material of the lithosphere. Cobalt may therefore be expected to show a stratification comparable to that of iron during summer stagnation. So far as ionic solutions are concerned, none of the other metals discussed here are likely to be much enriched near the bottom of the hypolimnion of any lake with a clinograde oxygen curve, for the amounts that can leave the mud are likely to be negligible in comparison with the concentrations already present in various forms in the water. This statement may need some qualification,[2] since the decomposition of falling seston in a limited hypolimnetic volume of water may produce some enrichment, but this is quite independent of the chemistry of the mud.

The treatment of the copper cycle, to be presented in Chapter 15, will indicate that the principles that have just been discussed operate in a most striking way to produce distributions in both time and space that are quite different from those exhibited by the other substances of which the limnetic cycles have been investigated. It seems reasonable to suppose that when the other divalent heavy metals are investigated, the same principles will be found to underlie their migrations.

---

[2] Complexes of amino acids and metallic sulfides are known and are soluble in water (Neuberg and Mandl 1948). They may prove ultimately to be of some significance in the chemistry of hypolimnetic waters.

DISTRIBUTION OF SULFUR IN LAKE WATERS

The sulfate content of the aerated regions of lakes. The sulfate content of lake waters varies from extremely small amount usually reported as traces, to quantities corresponding to saturated sodium sulfate in a few closed and usually dry lakes in which this salt is the predominant electrolyte present in the basin. The smallest recorded quantity of sulfate is probably <0.5 mg. per liter, the lowest value found in the open lakes of Africa; the largest is presumably the 60.3 g. per liter recorded in Red Lake, Wyoming, in 1888, when this usually dry lake contained water (Pemberton and Tucker 1893).

Omitting the extreme type of saline lake (page 570) in a closed basin, a few series of analyses taken from the very large number available will indicate the normal range in sulfate to be expected.

In northeastern Wisconsin, where the lakes in general drain glacial deposits derived from ancient crystalline and metamorphic rocks, Juday, Birge, and Meloche (1938) studied 234 lakes and recorded from 0.75 to 7.86 mg. SO_4 per liter, the mean content being 4.0 mg. per liter and the modal class containing 3.0 to 3.9 mg. per liter. The mean chloride content of these lakes is 1.02 mg. per liter, so that the ratio SO_4:Cl for the region is 3.93:1. As far as chloride and sulfate are concerned, these lakes contain slightly concentrated rain water.

In southeastern Wisconsin, considerably higher sulfate and chloride contents were encountered (Birge and Juday 1911). In the eleven lakes analyzed, the sulfate of their surface waters varied from 3.6 to 18.4 mg. per liter, the entire range being covered by the seasonal and annual variations of Lake Mendota. It is possible, however, that the lowest figure is a misprint for 13.6 mg. per liter, in which case the range is from 3.7 to 18.4 mg. per liter. The mean value is 10.6 mg. per liter, and correcting the supposed misprint only raises this mean to 10.7 mg. per liter. The mean chloride content is 4.8 mg. per liter, so that the ratio of the means SO_4:Cl is 2.22. Four lakes in northeastern Wisconsin, analyzed at the same time, gave a range of from 1.4 to 7.9 mg. SO_4 per liter, the mean being 5.2 mg., which is in fair concordance with the later work.

The values of the sulfate contents encountered in the surface waters of the north German lakes studied by Ohle (1934a) differ little from those of the Wisconsin lakes. The range was found by Ohle to be from 1.4 to 14.9 mg. per liter. As in southeastern Wisconsin, almost the entire range is covered by a single lake, the Trammersee, in which from 3.5 to 14.6 mg. per liter are recorded on different dates. The mean sulfate content was 7.6 mg. per liter. Owing to the high regional chloride content, the ratio SO_4:Cl was 0.3:1. Ohle found slight indications that when an almost isolated part of a lake was more productive than the main basin the sulfate content was rather higher, but the evidence for this is not very satisfactory.

In the large subalpine lakes of Switzerland (Bourcart 1906, Collet 1925) the sulfate content varies from a trace in the Lac d'Annecy to 44.8 mg. SO_4 per liter in Lake Geneva. The lakes containing the greatest amounts of sulfate drain from watersheds which include Triassic rocks containing gypsum. In seven subalpine lakes in Bavaria, Schwager (1894, 1897) found from 6.5 mg. per liter in the Königsee to 33.3 mg. per liter in the Starnbergersee, the mean being 19.7 mg. per liter. In these lakes the mean chloride was 1.5 mg. per liter, so that the ratio SO_4:Cl was 12.8:1. Most of the variation is doubtless to be explained by the lithology of the drainage basin.

Beauchamp (1953b) records from less than 0.5 to 2.9 mg. per liter in the open lakes of central Africa, and believes that in some cases the quantity of sulfate present is so low that it may limit algal production. The small quantities of sulfate in these waters is attributed, at least partly, to lack of sulfate-containing sedimentary rocks in the region.

Vertical distribution of sulfur compounds in stratified lakes. Mortimer (1941–1942), in his experiments with Windermere mud, found that sulfate increased in concentration in the aerated water over mud which retained an oxidized microzone at its upper surface. The increase was very slight during the first forty days of the experiment, but then became progressively more and more rapid. The meaning of this change was not elucidated. In his anaerobic tank, sulfate varied slightly and irregularly in concentration till the eighty-third day of the experiment. Then, as the ferrous iron rose, the sulfate fell, none being left on the one hundred and thirty-ninth day of the experiment. Hydrogen sulfide was not determined, but since at the time of the disappearance of the sulfate, when at least 10 mg. SO_4 per liter had been used up, about 32,000 mg. Fe^{++} $m.^{-3}$ were present at *p*H 6.6, reference to Fig. 204 will indicate that no analytically detectable amount of the gas was to be expected.

These two experiments suggest that an increase in sulfate with depth would be likely to occur during stagnation, at least if stratification was maintained for a long enough period and the oxidized microzone persisted; a decrease in sulfate would be expected if the oxidized microzone disappeared and the redox potential at the mud surface fell low enough.

Mortimer found that sulfate declined from about 8 mg. per liter to about 4 mg. per liter in the deep water over the mud in Esthwaite Water during summer stratification, but the high spring values persisted at the surface. He indicates that a more elaborate treatment of his data, as yet unpublished, demonstrated that the reduction occurred almost exclusively at the water-mud interface and not in the free water. Mortimer did not detect H_2S chemically in any of the Esthwaite samples, but sometimes the odor of the gas was apparent. This is in line with the high iron content of the waters, about 5000 mg. Fe m.$^{-3}$ at a relatively high *p*H of 6.7. The main fall in sulfate, in the bottom water of Esthwaite, did not occur until the middle of August, when ferrous iron had reached a maximum value and oxygen had entirely disappeared. Mortimer concluded that the bacterial production of hydrogen sulfide from sulfate does not occur until the redox potential has fallen at least to 0.10 volt. This is in accord with observations on the stratification of sulfate in Linsley Pond (Hutchinson 1941), in which on September 20, 1937, there was a slight increase in sulfate in the hypolimnion:

Depth, m.	SO_4, mg. per liter
0	0.75
2	0.68
5	1.47
9	1.64
13	1.70

Experience in the subsequent year, when redox potential measurements were made, suggests that the lowest values likely to have been encountered would have been well above 0.10 volt. Certain observations, to be presented in the next paragraph, on the northeastern Wisconsin lakes, notably Adelaide, are not in accord with Mortimer's results, and the matter clearly requires further research.

In the lakes of northeastern Wisconsin, Juday, Birge, and Meloche (1938) made no analyses of complete vertical series of samples. They did however determine sulfate in twelve lakes on pairs of samples from the surface and bottom during summer stagnation. In nine of the twelve pairs, sulfate was present in slightly greater amounts at the bottom than at the surface; in the other three pairs, the reverse was true. The greatest decrease at the bottom was shown by Ike Walton Lake, which contained 4.0 mg. per liter at the surface and 3.6 mg. per liter at the bottom. Though no hydrogen sulfide determinations were made, it is evident that sulfate reduction does not greatly affect the sulfate concentration of the bottom water in any of the lakes on which deep water analyses were made. A few of the lakes of this region, studied by Allgeier, Hafford, and Juday (1941) in their investigation of redox potentials, are found to have hydrogen sulfide in the deep water at the end of summer stratification. The data of these investigations, in so far as they can be read from the graphs, are given in Table 103.

It is possible that neither the data nor the constants employed in constructing Fig. 194 are accurate enough to suggest that there is any significant

TABLE 103. *Hydrogen sulfide and ferrous iron in certain Wisconsin lakes*

Lake	Depth, m.	H_2S, mg./l.	Fe^{++}, mg./m.^{3}	*p*H	Max. H_2S from Fig. 204, mg./l.
Scaffold Lake, Aug. 10, 1939	10	2.0	8000	6.5	0.015
Adelaide Lake,	4	0.0	0	6.25	No limitation
Aug. 12, 1939	5	0.25	0	6.2	No limitation
	8	0.25	0	6.0	No limitation
	12	0.25	0	5.8	No limitation
	17	0.40	800	5.8	2.7
	21	0.90	1500	5.7	2.2
Helmet Lake	6	0.0	0	5.7	No limitation
Aug. 29, 1939	8	1.2	2000	5.6	2.7
	10	1.6	4000	5.6	1.3

discrepancy between the maximal calculated H_2S and the observed quantity of the gas at the bottom of Helmet Lake. Presumably the water at 10 m. in this lake may be regarded as saturated with respect to FeS. In Scaffold Lake a more curious situation arises, for here the highest H_2S content recorded in the Wisconsin lake district appears to coincide with one of the highest ferrous iron concentrations observed in the region, at a *p*H which makes the coincidence of the observed values appear to be impossible. It is not quite certain that the recorded data all refer to the same water sample, as the account is a little vague. If the analytical methods were valid and only one sample were involved, it must have contained finely dispersed ferrous sulfide. It is most unfortunate that it is not possible to pursue this paradoxical case further.

Birge and Juday (1911) give several vertical series of sulfate determinations for the lakes of southeastern Wisconsin; these show only small irregular variations. Hydrogen sulfide was evidently determined on several lakes, but the results of such analyses are presented only for Lake Mendota, in which lake on September 22, 1908, the quantity of H_2S rose from 0.006 mg. per liter at 15 m. to 0.22 mg. at 21.5 m.

Sugawara, Koyama, and Noda (1952) have obtained evidence that when ferrous iron is oxidized and a precipitate of ferric hydroxide is formed, this precipitate may adsorb sulfate ions, removing them from the water and greatly enriching the mud. In some Japanese lakes the water in the sediments contains up to ten times as much sulfate as is present in the free water of the lake.

The most extensive study of the vertical distribution of inorganic sulfur compounds is undoubtedly that of Ohle (1934a) on the lakes of Holstein. Ohle used, in his study of sulfate, the Kuhlmann and Grossfeld technique, in which the sulfate is precipitated by the addition of a known amount of $BaCl_2$. An amount of K_2CrO_4 equivalent to the barium chloride is then added; this precipitates the excess barium. The excess chromate, equivalent to the original sulfate, is determined iodometrically. Ohle felt that the method left much to be desired, particularly in analyzing waters rich in organic colloidal matter.

In the lakes in which marked oxygen deficits developed during summer stagnation, hydrogen sulfide generally appeared in the bottom water unless, as in the case of the Hertasee, the alkalinity was too great to permit measurable amounts of the gas to occur in the presence of ferrous iron. In such a case the odor of the mud gave evidence of the formation of the gas. In recent years (Ohle 1955a) a considerable increase in H_2S production has been noted in the north German lakes. This is at least in part due to the accession of sulfate in the influents from recently drained land, sulfide in the previously reduced soil oxidizing and dissolving as sulfate.

The simplest case of marked stratification in sulfur compounds is provided by the Grosses Sager Meer, at the end of summer stagnation, in which there is a moderate increase in sulfate at the top of the hypolimnion and a very marked microzonal increase near the bottom (Fig. 205). The only irregularity, an increase at 14.3 m., is exhibited by a

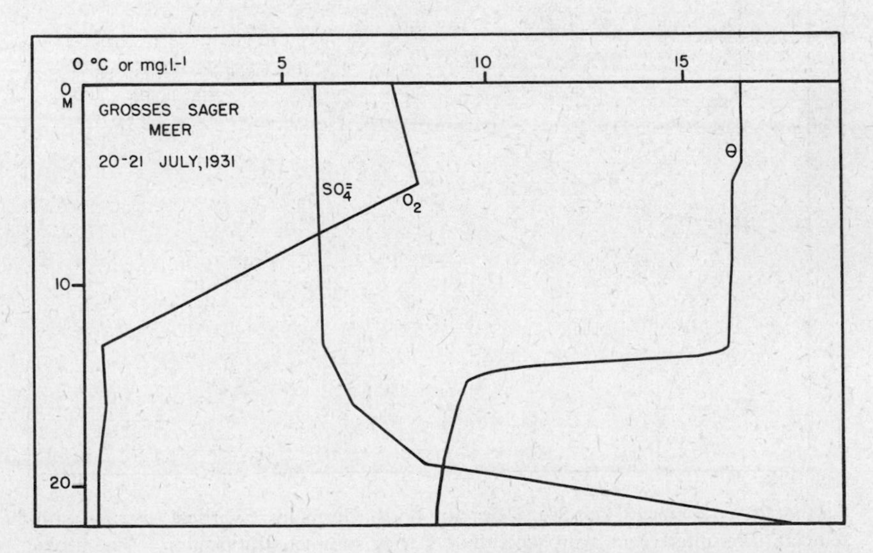

FIGURE 205. Grosses Sager Meer, Holstein, north Germany. Vertical distribution of temperature, oxygen, and sulfate during summer stratification. (After Ohle.)

sample taken at a marginal station. Sulfate reduction, as had just been indicated, apparently does not occur even in the sediments of this lake. It is reasonably certain that the increase near the bottom represents sulfate that has diffused from the sediments. In the more acid lakes in which sulfate reduction occurs and in which the resulting H_2S is not precipitated as FeS, a fall in sulfate may occur in the hypolimnion, but if the SO_4 equivalent of the H_2S produced be added to the sulfate actually present, a great increase in total sulfur in the lower part of the hypolimnion becomes apparent. In some cases, notably the Kleiner Uklei See, the rate of change of reduction with depth does not keep pace with the rate of increase in total sulfur, so that a marked hypolimnetic minimum in the sulfate content results in spite of a regular increase in total sulfur (Fig. 206). This type of distribution suggests reduction of sulfate to H_2S in the free water, but if a system of horizontal currents of the kind postulated to explain the bicarbonate distribution in Linsley Pond be accepted, it is conceivable that the reduction occurs at the mud-water interface.

When the reduction occurs in the hypolimnia of somewhat alkaline lakes, as in the Hertasee in which the *p*H at the bottom is 7.2, and there is

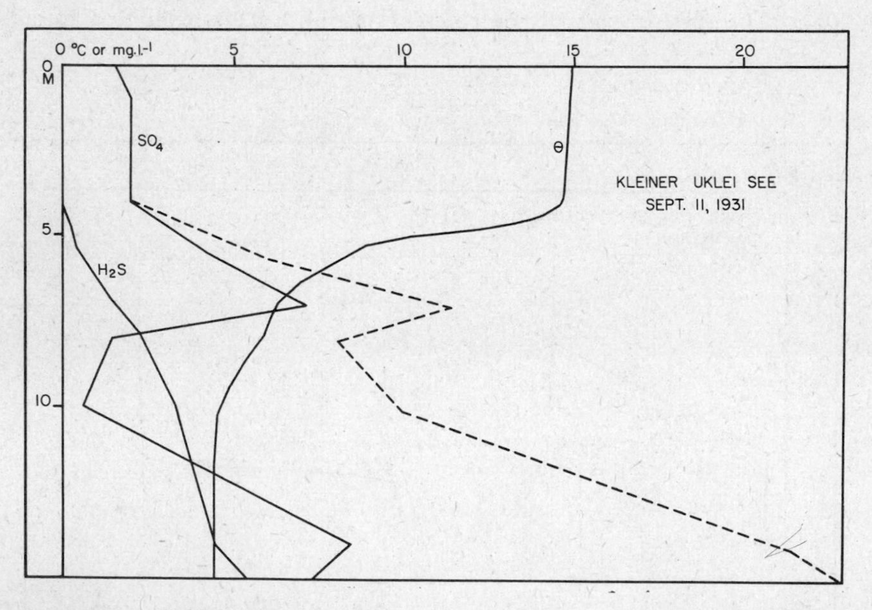

FIGURE 206. Kleiner Uklei See, Holstein, north Germany. Vertical distribution of temperature, sulfate, and hydrogen sulfide during summer stratification. The broken line gives the sum of the sulfate found and that equivalent to the hydrogen sulfide present. (After Ohle.)

considerable enrichment in iron also, no H_2S can remain in solution, and the sulfate curve may fall with depth in spite of the lack of H_2S *in situ* to balance the decline.

Both of Ohle's lakes with more or less orthograde oxygen curves give evidence of a fall in sulfate in the hypolimnion, the resulting distributions somewhat resembling that of the Hertasee. In the case of the Breiter Lucin, the more strictly orthograde of the two, the fall in sulfate looks like a case of sulfate reduction. In the case of the Drewitzer See, the greater part of the hypolimnion is of uniform sulfate content but lower in the anion than is the metalimnion. This case is probably comparable to the cases of epilimnetic stratification to be discussed below.

The most curious aspect of Ohle's sulfate analyses is a seemingly persistent stratification of sulfate in the epilimnion. In view of Ohle's remarks about the accuracy of his method of analysis, too much stress should not be placed on his single observations; but when the same pattern is repeated on a subsequent visit to the lake, it is hard to dismiss the vertical variations involved as due entirely to the experimental error.

That the pattern can be repeated is shown by the data for both the Grosser Plöner See and the Edebergsee (Fig. 207). In the Trammersee, another lake for which two sets of analyses are available, the earlier one suggests, if the determinations are taken at face value, that a very complex microstratification of sulfate existed in the epilimnion and metalimnion. It is to be noted that in the Edebergsee and less strikingly in other lakes also, very marked stratification of sulfate can occur in a thick layer of water of very low stability. In the case of the October series for the Edebergsee, the sulfate increases from 5.2 to 12.4 mg. per liter in passing from the surface to 6 m. This surely must be well outside the experimental error, but the accompanying fall in temperature is only 0.28° C. As the observations were made on a day that at first was heavily overcast but later cleared to bright sunshine, it would seem possible that this gentle temperature gradient had been established since the previous night. A thick autumnal epilimnion of the kind indicated by Ohle's observations would ordinarily imply fairly thorough nocturnal mixing. It is extremely difficult, however, to imagine any mechanism by which the sulfate distribution apparently observed could be developed rapidly each day. Ohle is noncommittal as to the meaning of the peculiar epilimnetic and metalimnetic sulfate maxima exhibited by these and some others of the lakes that he studied, though he apparently considered them as real. His remarks about technique indicate that some reserve in accepting the observations may be necessary, but if they prove to be valid, it is evident that some sort of a sulfur cycle must be in operation under fully aerobic conditions, as is known to be the case with nitrogen.

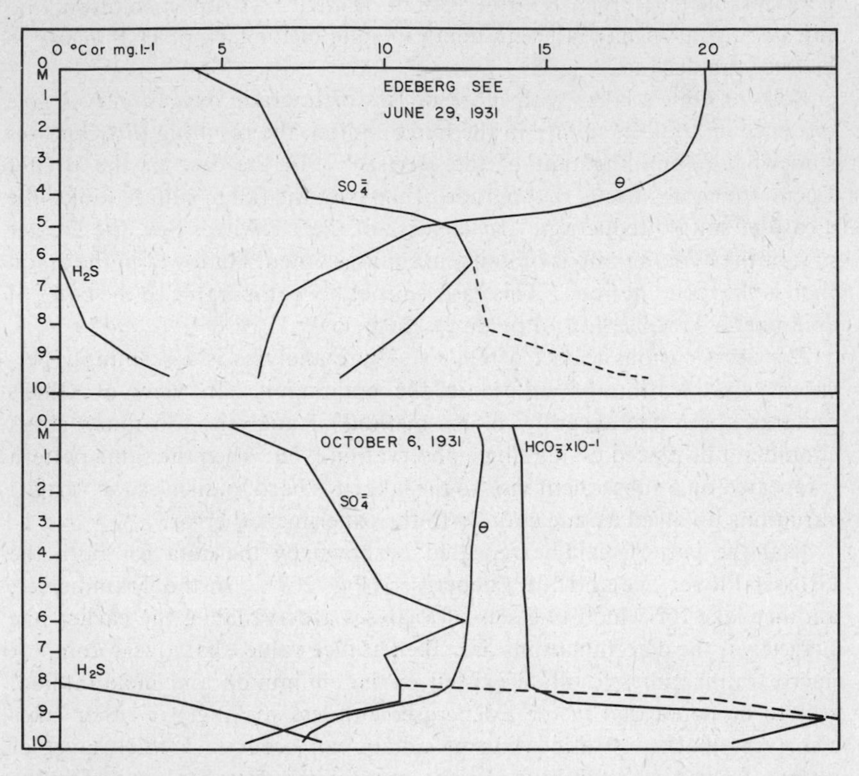

FIGURE 207. Edeberg See, north Germany. *Upper panel*, vertical distribution of temperature, sulfate, and hydrogen sulfide, June 29, 1931; *lower panel*, the same substances and bicarbonate alkalinity, October 6, 1931. The broken line gives the sum of the sulfate and the sulfate equivalent of H_2S. Note the over-all similarity of the epilimnetic sulfate stratification on the two dates, in spite of the negligible stability and uniform bicarbonate concentration in the upper 7 m. in the October series. The bicarbonate curve is plotted on a scale one tenth of that of the sulfur compounds. (After Ohle.)

In the tropical lakes of Indonesia studied by Ruttner (1931), the odor of H_2S frequently indicated the presence of small amounts of the gas in the deep oxygen-deficient waters. In general, these lakes contain extremely little sulfate, but in a few cases Ruttner believed the hydrogen sulfide of the deep water to be generated by sulfate reduction. This was supposedly the case in the shallow Telaga Pasir in east Java, in which H_2S increased from 1.2 mg. per liter at 10 m. to 3.1 mg. at 17 m. Ruttner also considered sulfate reduction to have contributed to the H_2S in Lake Ranau in south Sumatra, where the 90-m. water contained 0.58 mg. per liter and the 200-m.

water 2.04 mg. It is probable that the other lakes of the region did not contain as much H_2S as these two. Circulation in all the Indonesian lakes takes place irregularly and rarely. When it occurs, the mixing of considerable amounts of oxygen-free water into the tropholytic zone may cause fish deaths, and an odor of H_2S is said to be perceptible on such occasions. These oligomictic lakes are in fact intermediate between ordinary holomictic lakes and the very deep meromictic lakes of the tropics, such as Tanganyika and Nyasa.

An extreme case which exhibits certain puzzling features is provided by the Son-sakesar-kahar, a lake at the western end of the Son plateau in the Salt Range, Punjab, Pakistan (LaTouche 1910, Hutchinson 1937a and unpublished). The lake, normally about 5 km. long and 1.6 km. wide, had in 1932 a depth of about 9 m. over a large part of its area. There is no outlet and the area is variable. A sample of water collected March 10, 1932, at a depth of 6 m. contained the following inorganic constituents (in milligrams per liter):

Na	25,644
K	1,783
Ca	223
Mg	2,233
Fe	2.4
Al	10
HCO_3	1,667
SO_4	17,176
Cl	34,400
SiO_2	326

A vertical series of analyses of bicarbonate and H_2S is given in Table 104. No oxygen could be detected at any level even after appropriate steps had been made to remove the H_2S.

TABLE 104. *Vertical variation temperature of bicarbonate and hydrogen sulfide in Son-sakesar-kahar*

Depth, m.	Temp., °C.	HCO_3, mg. per liter	H_2S, mg. per liter
0	14.7	1580	3.4
1	15.3		
2	15.6	1580	
3	12.6		5.1
4	12.5	1580	
5	12.3		
6	11.4	1710	68.1
7	11.5		
8	11.7	1710	79.9

It is not certain if this lake is meromictic or, as is more probable, circulates irregularly during exceptionally cold spells in winter. In 1932 the entire lake was dull pink, owing to a dense bloom of the purple sulfur bacterium *Lamprocystis roseopersicina*. The colonies in the surface and in 3-m. samples which contained small amounts of H_2S were pink and presumably living, the colonies in the deeper water were white and probably dead. Information received from Dr. Hem Singh Pruthi, who visited the lake, indicates that the *Lamprocystis* bloom is not always present. The same organism has been recorded coloring the Schliersee, Bavaria, under ice in 1886. Presumably in this case enough H_2S was produced during prolonged winter stagnation to permit the growth of *Lamprocystis* in massive amounts similar to the bloom in the very different Son-sakesar-kahar.

Hydrogen sulfide in the monimolimnion of meromictic lakes. In some meromictic lakes, notably those of the Austrian Alps, very little if any hydrogen sulfide occurs, even in the deepest water. This condition is not, however, universally met with in lakes showing biogenic meromixis. In general, where much iron occurs, H_2S would not be expected: in the best known case of a biogenic meromictic lake containing appreciable H_2S in the monimolimnion, Lake Mary, Wisconsin (Fig. 208), no ferrous iron occurs but it is not altogether clear why this should be the case (page 724). The data given for this lake on August 12, 1939, by Allgeier, Hafford, and Juday (1941) are set out in Table 105. These data indicate the existence of hydrogen sulfide at redox potentials well above those regarded by Mortimer as setting the upper limit for the presence of the gas. Lake Mary resembles the nearby holomictic Lake Adelaide in this respect.

TABLE 105. *Vertical distribution of temperature, oxygen, pH, redox potential and hydrogen sulfide in Lake Mary*

Depth, m.	Temp., °C.	O_2, mg. per liter	*p*H	E_h, volt	H_2S, mg. per liter
0	19.7	6.6	6.25	0.450	0.0
2	15.8	5.5	6.2	0.485	0.0
3	8.4	0.7	...	0.405	0.6
4	4.4	...	6.1	0.388	...
5	...	0.0	6.1	0.265	0.6
8	4.0	...	5.9	0.145	...
10	4.0	0.0	5.9	0.130	1.5
12	...	...	...	0.125	...
13	4.1	...	...	...	...
15	4.2	0.0	5.8	0.110	1.9
18	...	...	...	0.100	...
20.5	4.3	0.0	5.7	0.075	2.1

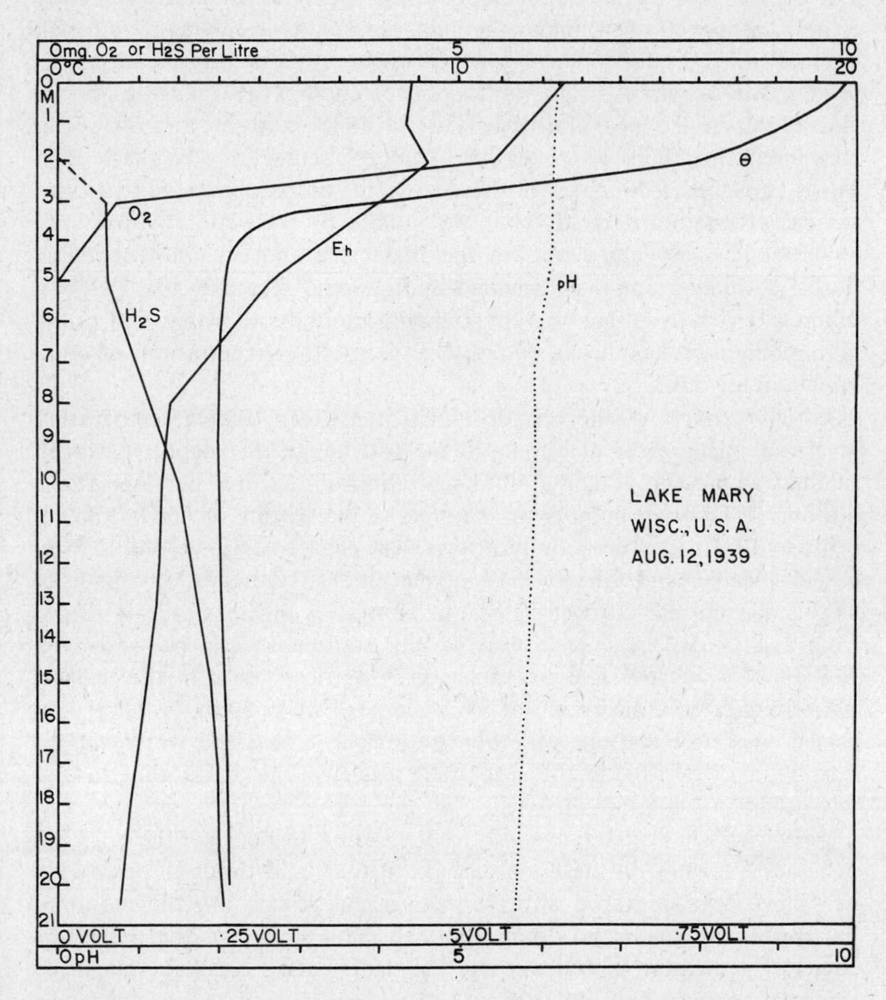

FIGURE 208. Lake Mary, Wisconsin. Vertical distribution of oxygen, redox potential, *p*H, and hydrogen sulfide on August 12, 1939 (after Allgeier, Hafford and Juday).

In ectogenic and crenogenic meromictic lakes, large quantities of sulfate may be present in the monimolimnion, and part of this sulfate, on undergoing reduction in the sediments in contact with the monimolimnion if not in the water itself, may give rise to amounts of hydrogen sulfide far in excess of those present in other types of lake. This phenomenon has been recorded in the Lac de la Girotte, Haute Savoie, France (Delebecque 1898), and in Lake Ritom, Switzerland (Bourcart 1906; Collet, Mellet, and Ghezzi 1918; Düggeli 1924; Collet 1925), both of which lakes apparently exhibit crenogenic meromixis, the monimolimnion receiving water rich in sulfates from mineral springs which apparently discharge from gypsiferous Triassic strata. The same phenomenon is also known in the Hemmelsdorfersee near Lübeck (Griesel 1935) and in the two deep meromictic members of the chain of lakes opening into Wakasa Bay, Japan (Yoshimura 1932b). In these cases the monimolimnetic hydrogen sulfide is certainly derived from the sulfide of sea water by bacterial reduction. Analogous cases are the Black Sea and certain Norwegian fjords. A final example is provided by Big Soda Lake, Nevada. Here a saline lake rich in sulfate was covered with more dilute water, and in the monimolimnion has accumulated the greatest concentration of H_2S known in any lake.

Little is known of the Lac de la Girotte, except that a temperature inversion was present at about 25 m. and below this depth hydrogen sulfide was present, reaching a concentration of 15.5 mg. per liter at the bottom. The other crenogenic example, Lake Ritom, is better known. Prior to 1917 the monimolimnion was very sharply delimited at 12.5 m. (Collet, Mellet, and Ghezzi 1918; Collet 1925). Observations made in 1913 indicated no hydrogen sulfide at 12 m., 6.1 mg. per liter at 13 m., from 11.8 to 19.4 mg. per liter at various stations at 13.5 m., and from 20.9 to 24.0 mg. per liter at 15 m. The vertical series in the deepest station indicated a maximum of 31.2 mg. per liter at 30 m., a value confirmed by determinations at other stations at which the water was at least 30 m. deep. Below this level there was evidence of a decline to 23.2 mg. per liter at 40 to 45 m.

As has been indicated (page 487), there is direct evidence of the existence of mineral springs on the floor of Lake Ritom, and though a little H_2S may have been delivered directly from such springs into the monimolimnion, there can be no doubt that the main source of the gas is the bacterial reduction of sulfates in the sediments. Analyses given by Bourcart are given in Table 106.

Analyses made in 1913 (Collet, Mellet, and Ghezzi 1918) indicate an increase in total solids at the surface to 304.0 mg. per liter, and a slight increment in the H_2S content of the bottom water, which rose to 23.2 mg. per liter in 1913 and 24.0 mg. in 1916. Collet thinks that this implies a continual increase in the hydrogen sulfide content of the monimolimnion, and such an idea is obviously reasonable a priori. The rate of increase from 1903 to 1913, however, would not seem to have been maintained from 1913 to 1916, and taken at its face value implies that the accumulation of the gas did not begin until a couple of decades before Bourcart examined the deep water in 1903.

TABLE 106. *Analysis of water from surface, monimolimnion and outlet of Lake Ritom*

(All computations in milligrams per liter)

	At 0 m.	At 44 m.	At Outlet
Total residue at 170° C.	122.7	2365.3	128.8
SiO_2	2.8	10.0	2.2
$(Al,Fe)_2O_3$	Trace	1.2	Trace
CaO	37.9	737.0	38.8
MgO	11.1	196.2	12.1
K_2O	2.4	4.2	2.0
Na_2O	2.0	2.7	1.7
So_3	48.1	1376.7	53.3
H_2S	...	15.9	...

The chemocline region in 1913 was rose colored, due to the presence of *Chromatium okenii*, the color beginning at 13.0 m. and having its maximum intensity at 13.5 m. At least the 13-m. sample is said also to have been opalescent, owing to the presence of elementary sulfur. A similar colored bacterial layer existed in 1916 at about 12.7 m. Most of the deepwater samples were apparently bacteriologically sterile, though in one case a small population of *Desulfovibrio desulphuricans* was noted by Düggeli. This bacterium was extremely abundant in the anaerobic mud, and it is probable that most of the sulfate reduction took place in the sediments rather than in the free water. The decline in H_2S below 30 m. is not explained.

Disturbance of the water level in 1917, when the lake began to be used as a power source for a hydroelectric installation, led to some mixing of the monimolimnion and a decline in the H_2S content. A great lowering of the level of the lake, by about 30 m., in 1918 produced full circulation and complete disappearance of the hydrogen sulfide.

The two coastal meromictic lakes studied by Yoshimura (1932b) in Japan are of interest because detailed information exists relating to the vertical distribution of sulfate as well as of H_2S. These two lakes, Hirugako and Suigetuko form part of a chain, the shallow fresh-water lake Mikatako flowing into Suigetuko, and the latter, by an artificial connection made in 1801, draining into Hirugako, which was joined to the sea in 1630. The water of Hirugako is slightly diluted sea water, but the dilution normally affects the top 5 m. more than the rest of the lake, which is cut off as a fairly permanent monimolimnion, though circulation occasionally takes place during very stormy winters. The monimolimnion of Suigetuko is about one fifth the concentration of Hirugako, and the mixolimnion is very much more dilute. The data in Table 107 are for Hirugako on November 11, 1927.

TABLE 107. *Vertical variation of temperature, oxygen, chloride and sulfur compounds in Hirugako*

Depth, m.	Temp., °C.	O_2, mg./l.	H_2S, mg./l.	SO_4, g./l.	Cl, g./l.	SO_4 calc., g./l.	H_2S calc., mg./l.
0	17.5	7.30	0.00	2.34	16.73	2.33	...
5	17.5	7.75	0.00	2.33	16.82	2.32	...
10	18.0	6.75	0.00	2.42	17.02	2.37	...
15	18.2	5.90	0.00	2.42	17.12	2.38	...
20	16.7	3.88	0.00	2.38	17.12	2.38	...
25	13.7	1.23	0.00	1.21	16.97	2.37	417
30	13.4	0.00	0.49	...	17.07	...	...
32.5	13.3	0.00	2.39	...	...	...	...
35	13.0	0.00	5.4	0.98	17.07	2.37	490

In the top twenty meters the sulfate content (SO_4 calc.) is essentially that to be expected in sea water diluted to the chlorinity of the lake. In the deeper part of the monimolimnion the sulfate concentration falls off, but very much less hydrogen sulfide (H_2S calc.) is present than can account for the difference between the expected and observed SO_4. It is possible that all the missing sulfate has been reduced to sulfide, and much of this sulfide has been precipitated as FeS. Yoshimura states that on the rare occasions that the lake circulates completely, the water becomes laden with a white suspension of elementary sulfur.

The data for the more inland meromictic Suigetuko, on November 12–14, 1927, are given in Table 108. The slight deficiency of sulfate in the

TABLE 108. *Vertical variation of temperature, oxygen, chloride and sulfur compounds in Suigetuko*

Depth, m.	Temp., °C.	O_2, mg./l.	H_2S, mg./l.	SO_4, mg./l.	Cl, mg./l.	SO_4 calc., mg./.	H_2S calc., mg./l.
0	13.2	9.34	0.00	80	920	128	17
5	15.3	9.20	0.00	67	920	128	22
10	16.7	1.67	0.00	200	1580	220	7
13	15.7	1.56	0.00	280	2350	327	17
14	15.3	0.00	0.37	360	...	...	...
15	14.7	0.00	1.60	369	2400	334	12
20	12.3	0.00	23.89	356	2960	413	20.5
25	12.6	0.00	47.52	322	3280	445	44
30	12.9	0.00	54.5	250	...	456	73
32	12.7	0.00	53.8	252	3280	456	72

mixolimnion is considered by Yoshimura to be of no significance. In the monimolimnion, in striking contrast to the condition in Hirugako, the hydrogen sulfide produced accounts for over 72 per cent of the sulfate reduced.

The Hemmelsdorfersee differs from the two preceding lakes mainly in becoming filled with salt water on the rare occasions of catastrophic marine floods, after which the chemocline slowly descends. In 1914, after forty-two years' stagnation, the chemocline lay at 35.35 m. and the monimolimnion contained the following quantities of hydrogen sulfide:

Depth, m.	H_2S, mg. per liter
32 (and above)	0
33	234
35	291
40	281
43	304

A well-developed bacterial zone occupied by *Thiopedia rosea* occurred in the lake in September 1921, when the chemocline lay at about 39.8 m. Utermöhl (1925) found few *Thiopedia* at 28 m. and 30 m., large numbers between 32 m. and 35 m., with a maximum at the latter depth, and none at 36 m. or below. It is evident that any H_2S used by this organism must have passed, either by molecular diffusion or very limited turbulent diffusion, from a depth of several meters below the maximum.

In Big Soda Lake, Nevada, water originally containing 11.9 g. SO_4 per liter has been covered, owing to an artificial rise in ground water consequent on irrigation, by a much more dilute mixolimnion. Hutchinson found that after not more than twenty-eight years of stagnation, the monimolimnion contained 456 mg. per liter H_2S except near the bottom, where 786 mg. per liter were present. These values seem to be the highest yet recorded.

The formation of bacterial zones in relation to sulfur compounds. In the Hemmelsdorfersee and in Lake Ritom, very marked coloration of the water at or just above the level of the chemocline indicated the presence of massive populations of purple sulfur bacteria. Similar bacterial zones or plates have been observed in the chemocline region of the Black Sea, and they are also known in some holomictic lakes. In the last named type of locality there is often no evidence of the presence of hydrogen sulfide in or immediately below the bacterial zone, so that the occurrences at first sight appear rather anomalous. Modern work, however, indicates that the requirements of the photosynthetic bacteria are not as rigorously defined as was formerly believed.

The organisms which were formerly classed as Thiorhodaceae or true purple sulfur bacteria are known to be able, at least in some cases, to utilize not only H_2S and thiosulphate but also simple organic compounds and molecular hydrogen as oxygen acceptors in photosynthesis; the Athiorhodaceae or nonsulfur purple bacteria can use, at least in some cases, hydrogen sulfide, molecular hydrogen, and thiosulfate, as well as simple organic compounds. As the matter at present stands, the Thiorhodaceae are auto-auxotrophic, the Athiorhodaceae require an external source of at least some of the vitamins.

The belief that these organisms are strictly anaerobic also requires modification. The capacity to tolerate moderate amounts of oxygen is exhibited by some of them, and it appears that in some cases microaerophil strains can be developed in a culture from anaerobic strains. The major problems involved are admirably treated by Van Niel (1944), who gives full references to the biochemical studies on which the foregoing statements are based.

Utermöhl (1925) found, in the north German lakes, that the large colorless sulfur bacterium *Macromonas bipunctatus* often formed populations in narrowly defined layers in the hypolimnion. He concluded that the optimum concentration of oxygen for this form lay between 2.0 and 3.0 mg. per liter. No hydrogen sulfide was detected, either by odor or by the lead acetate test, in the layer in which the bacterium formed a massive population in the Krummensee at the height of summer stagnation. In general, sulfur droplets were visible in the cells of *Macromonas* only when the organism was collected near the bottom at the end of the stagnation period.

In a number of the lakes studied by Utermöhl, *Thiopedia rosea* was also present. This form was generally found at or near the bottom. In the Krummensee, in which on July 13, 1922, the *Macromonas bipunctatus* maximum lay at 9 m., the maximum population of *Thiopedia* was found at 11 m., though it is possible that even more cells of the purple form may have been present between the 11-m. layer and the bottom at 12 m.

Very similar observations have been made by Ruttner (1937) in the lakes of the Austrian Alps. Ruttner found no analytically detectable H_2S in any of the lakes that he studied, yet he noted the presence of species supposed to be sulfur bacteria in the following cases. In the Lunzer Obersee, a holomictic lake, *Chromatium weissei* formed a rose-colored layer scarcely 50 cm. thick, at the limits of detectable oxygen. In the meromictic Krottensee a very marked stratification of purple bacteria, of *Macromonas mobilis*, and of the iron bacterium *Ochrobium tectum* occurred on June 24, 1933, in the vicinity of the chemocline. The data are given in Table 109.

The organism designated as *Chromatium* sp. is smaller than that referred

TABLE 109. *Vertical distribution of supposed sulfur organisms and of* Ochrobium tectum *in the Krottensee*

Depth, m.	Temp., °C.	O_2, mg./l.	*p*H	Fe, mg./m.^{3}	*C. weissei*	*Chromatium* sp.	*Lamprocystis*	*M. mobilis*	*O. tectum*
20	4.6	0.79	7.46	0	0	0	0	60	0
21	...	...	...	...	0	5800	8300	0	0
22	...	...	...	...	140	90	...	0	...
22.5	...	...	...	...	...	10	...	0	...
25	4.7	0.0	7.38	30	20	0	100	0	2360
30	4.9	0.0	7.1	13,600	0	0	10	0	0

rather doubtfully to *C. weissei*, it is also said to lack sulfur droplets, but it is not clear from this statement that the species referred to *C. weissei* actually contained them. It is evident from the iron content and *p*H that any significant upward movement of H_2S from the sediments at the bottom of the monimolimnion is out of the question.

The data of Table 110 refer to another meromictic lake, the Toplitzsee, on June 18, 1933. On a previous occasion, *Siderocapsa* was absent but

TABLE 110. *Vertical distribution of sulfur and iron organisms in the Toplitzsee*

Depth, m.	Temp., °C.	O_2, mg./l.	*p*H	Fe, mg./m.^{3}	*Lamprocystis*	*M. mobilis*	*S. coronata*
10	6.1	7.46	7.62	0	0	0	0
12.5	5.6	3.01	...	...	...	...	...
15	5.6	0.76	...	...	0	0	0
17.5	...	0.24	...	...	...	...	...
20	5.65	0.25	7.28	50	0	20	10
22.5	5.8	0.14	...	...	...	...	...
25	5.9	0.15	...	...	80	70	190
30	5.9	0.11	7.28	40	650	170	0
50	5.9	0.11	7.3	10	240	50	0

Ochrobium tectum occurred, its distribution resembling that of *Lamprocystis*, with maxima for both species at 50 m.

In the Schleinsee, Vetter (1937) found little difference in the vertical distribution of *Macromonas bipunctatus* and that of *Thiopedia rosea*. Both appeared in massive populations in the deep water when the oxygen fell below about 2.5 mg. per liter, and were found at greater concentrations only when the oxygen rose after a previous low value. During late June and throughout July, when the populations were developing, H_2S was always undetectable at or above 10 m. and was usually absent from the 11-m. level, just above the deepest mud. Early in August, when H_2S began to appear in quantitatively detectable amounts, the maximal populations of both forms lay just above the layer containing H_2S, but later very large numbers of both *Macromonas* and *Thiopedia* were observed at 9 m. and 10 m., where 0.5 to 0.9 mg. H_2S per liter was present. The maximum population of *Thiopedia*, however, appears to have developed at 10 m. at the end of October, when the oxygen was rising from 0.8 to 2.3 mg. per liter; the maximal *Macromonas* population developed during the same period at 11 m., and the oxygen rose from 0.2 to 3.1 mg. per liter. No H_2S was present at either level at this time. A later rise of oxygen to 7.5 mg. per liter, as full circulation intervened, was accompanied by a great decline in both forms. *Lamprocystis roseopersicina* and *Thiodictyon elegans* were both observed in small numbers in the lake, but no indication is given of their distribution.

It will be observed that in all the lakes discussed in the present section, there is a strong suggestion that organisms which have frequently been regarded as typical sulfur bacteria flourish in regions in which not only is H_2S absent, but which are in some cases remote from any possible significant source of the gas. The most recent studies of the physiology of these organisms indicate that such distributions are not entirely unexpected, but the observations raise the limnological problem of the nature of the substances used by the colored forms as oxygen acceptors in photosynthesis. The production of molecular hydrogen in the deepwater sediments of highly productive lakes is not unknown, but is unlikely to be of great quantitative significance. Simple organic compounds may be present, but the information available (page 895), fragmentary as it is, does not suggest that even in the anaerobic hypolimnion of a lake such as Mendota, their concentration will be very great. The presence of microaerophil sulfur bacteria in well-defined bacterial zones emphasizes one of the most unfortunate gaps in our present knowledge of lake chemistry.

The effect of artificial contamination with sulfur compounds. Some industrial wastes, notably solutions used in the treatment of wood pulp in paper manufacture, may contain large amounts of sulfite. If these wastes enter a lake, they may lead to immense production of hydrogen sulfide. A remarkable case has been described by Liebmann (1938). The Bleiloch Dam on the River Saale, near Saalburg, is the largest artificial lake in Germany. Its upper end received the discharge from a large paper mill. This discharge consisted of a solution of calcium bisulfite, free sulfurous acid, carbohydrates, pectins, and other organic compounds, including lignin, in part sulfonated, in suspension. During the colder months this material appears to have undergone oxidation throughout the circulating waters of the dam, or to have sedimented. During summer stratification a very great uptake of oxygen occurred throughout the hypolimnion, and hydrogen sulfide was produced in considerable amounts.

Unfortunately, Liebmann gives no information of any consequence on temperature. His data indicate that at the end of summer, at the deeper lower end of the dam, hydrogen sulfide is present both at the bottom and in a layer extending from about 6 m. to 16 m., supposedly at the top of the hypolimnion; the middle of the hypolimnion, from about 16 to 36 m., contained water very poor in oxygen but with little or no H_2S. The upper layer containing hydrogen sulfide is interpreted as a density current, rich in organic matter and sulfur compounds undergoing reduction. The lower layer containing H_2S clearly derives its gas from the black sulfide-rich sediment, formed from the settling of the particulate matter in the wastes from the paper mill. Liebmann obtained some experimental evidence that different organisms were responsible for the hydrogen sulfide production in the two layers. In experiments with water from the upper layer, H_2S production occurred only at or above 16° C., apparently the temperature of the layer. In samples of the fluid mud from the bottom, H_2S production occurred at 5° C. The water in contact with the mud contained a remarkable ciliate fauna, consisting largely of species of the genera *Metopus* and *Chaenia*. These ciliates must be anaerobes capable of tolerating up to at least 8 mg. H_2S per liter. They are said to be quite characteristic of this type of locality, and to feed on the bacteria of the mud, which bacteria are supposed to live for a time symbiotically in the vacuoles in the cytoplasm of the ciliates.

Sulfuric acid as a cause of extreme acidity in lakes. As has already been indicated in the discussion of hydrogen ion concentration, some lakes in volcanic regions may be extremely acid, owing to the presence of free sulfuric acid. The most remarkable of such lakes is certainly Lake Kata-numa, Miyagi Prefecture, Japan, a small lake in a crater, in which the water contains 474 mg. SO_4 per liter. This sulfuric acid is apparently produced by the oxidation of sulfur which occurs in suspension in elementary form on the crater wall and in the turbid, whitish water. The water of the lake was found by Yoshimura (1934) to have a *p*H of 1.4, and although it was homoiothermal when visited on November 13, 1932, no layer was more than 61.5 per cent saturated with oxygen. Yoshimura supposes that the sulfur is continually oxidized to produce the observed concentration of sulfuric acid.

More usually, sulfuric acid is produced by the oxidation of pyrite in the immediate surroundings of the lake. The most remarkable case is the Tonteich near Reinbek, described by Ohle (1936) and already discussed. The pond occupies an abandoned clay pit in upper Miocene micaceous clay. Ohle gives a complete mineral analysis of the water, and also three analyses of watery extracts of 20 g. of clay in 1000 cc. of water, made from samples collected at depths of 5 cm., 25 cm., and 50 cm. in a profile in the clay about 100 m. from the pond. The analyses of Table 111 indicate the changes that take place as the clay weathers and gives up some of its constituents to the water running into the pond.

TABLE 111. *Analysis of water of the Tonteich near Reinbek, and of aqueous extracts of the clay in which the pond lies*

Constituents	Water	Extract 5 cm.	Extract 25 cm.	Extract 50 cm.
Na	12.2	2.0	2.6	2.2
K	1.2	0.5	0.6	1.1
$N \cdot NH_4$	0.14	0.12	0.16	0.14
Ca	82.4	47.7	64.5	26.0
Mg	20.7	2.9	8.2	5.4
Fe	6.60	1.05	5.55	0.04
Mn	0.60	1.05	0.29	0.08
Al	4.40	1.72	2.36	1.58
Si	19.3	3.39	12.3	3.36
$N \cdot NO_2$	0.002	...	0.001	...
$N \cdot NO_3$	1.56	0.38	0.21	0.09
$P \cdot PO_4$	0.0037	0.012	0.025	0.026
Cl	16.6	0.4	0.00	0.00
SO_4	351	102	283	57.5

It will be clear from the analyses of the clay extracts that there is a maximum in the soluble constituents in the middle of the profile. At 50 cm. weathering evidently has not proceeded far enough to liberate all the material that can go into solution, and at the top of the profile most of the soluble material has already been leached out by rain water and has run into the pond. In all the extracts, sulfate is the only anion of any importance; the chloride in the pond, and most of the sodium also, may be attributed to human contamination. Almost all of the sulfate in the extract from the bottom of the profile must represent gypsum that has gone into solution; very little free sulfuric acid is present at this depth. At 25 cm. pyrite, which is easily recognizable in the clay, has undergone oxidation

$$2FeS_2 + 7O_2 + 2H_2O \rightarrow 2FeSO_4 + 2H_2SO_4$$

Further oxidation of the ferrous sulfate would produce ferric sulfate that would tend to hydrolyze, though at pH 3, 0.7 mg. of ferric iron could be present in true solution. The free sulfuric acid would tend to liberate alkalies and alkaline earths from minerals other than gypsum and pyrite,

The most striking unexplained difference between the extract of clay from the middle of the profile and the water of the pond is the high magnesium content of the latter; no indication of the probable origin of this magnesium is given. Ohle concluded, from a comparison of the ionic composition and the conductivity of the pond water, that the material present in the water was from 3.3 to 9.4 per cent undissociated, and suspected that the undissociated material was probably colloidal iron and alumina. In view of the fact that the iron is likely to have been largely ferric this seems very probable. Metallic iron immersed in the water is corroded, presumably forming ferrous sulfate first, which then would be oxidized and in part hydrolyzed.

With this remarkable case before him, Ohle reviews the literature of the occurrence of free sulfuric acid in soils and ground waters. He concludes that, in general, wherever *p*H values lower than *p*H 5 are recorded in nature, it is reasonable to expect that sulfuric acid is present, and that the frequent uncritical attribution of such acidities in bog waters to organic acids is erroneous. This has been confirmed for certain of the bay lakes in North Carolina in Frey (1949).

SUMMARY

Rain water normally contains one or two milligrams of sulfate per liter, but in inland rain there is much less chloride than this. Lake waters in regions where there are no soluble sedimentary deposits of sulfates or chlorides normally retain the high ratio of SO_4:Cl found in uncontaminated rain, the sulfate ion ordinarily being the second or third most abundant anion in lake water. The concentration of sulfate in lake water is known to vary from rather under 1 mg. SO_4 per liter in some lakes in basins deriving their water from ancient crystalline rocks, to 60.3 g. per liter in a closed inland basin depositing sodium sulfate crystals. The quantities ordinarily to be expected in open lakes lie between 3 and 30 mg. per liter, and depend largely on the lithology of the drainage basin. Marked temporal variations in sulfate content in the surface waters of lakes have been recorded, but as no systematic study of them has been made, the meaning is unknown.

Under anaerobic conditions sulfate can be used as a hydrogen acceptor in the metabolic oxidation of organic matter by certain bacteria, notably *Desulfovibrio desulphuricans*, which produces hydrogen sulfide. This gas is also produced in the bacterial decay of any organic matter, such as protein, containing reduced sulfur.

In the hypolimnetic mud of all lakes in which the oxygen of the deep water is exhausted during summer stagnation, some production of H_2S may be expected. Owing to the insolubility of ferrous sulfide in neutral or alkaline water, hydrogen sulfide appears in the free water of the deep hypolimnion only when this water is acid or when enough decomposition or sulfate reduction has occurred to precipitate all the iron present as FeS. Theoretical treatment indicates that a mud containing ferrous sulfide cannot lose the ions of certain rarer metals, such as copper, silver, lead, and cadmium, to the free water in analytically detectable quantities at any *p*H value normally encountered in lakes. This result will appear of some importance in consideration of the copper cycle in Chapter 15.

Under aerated water, lake mud will probably generally be found to lose sulfur to the water as sulfate. Under entirely anaerobic conditions the sulfate initially present in the water will be reduced at the surface of the mud, though the H_2S so produced is often precipitated as FeS, so that the water loses sulfur compounds. Lakes with moderately orthograde to moderately clinograde oxygen curves may therefore be expected to exhibit an increase in sulfate with depth during the period of stratification, whereas lakes with extreme clinograde oxygen curves will show a decrease, though the equivalent H_2S may not appear in the free water. Mortimer thinks that H_2S of bacterial origin does not appear till the redox potential falls below 0.10 volt, but data exist in apparent conflict with this conclusion.

Possible intermediate products of sulfate reduction, such as thiosulfate, seem to be absent in lake waters. In some north German lakes the rate of increase of total sulfur with depth is much greater than the rate of increase of sulfate reduction with depth. In such lakes the sulfate curve shows a marked minimum in the intermediate layers of the hypolimnion, though the total inorganic sulfur, present as SO_4 and H_2S, may increase regularly with depth. Curious unexplained vertical variations have been observed in the epilimnion under conditions of extreme instability, the sulfate content at the surface being sometimes much less than that at 1 m. even though the temperature gradient suggests that complete nocturnal mixing of the top few meters of the lake can take place. The meaning of this is unknown.

Hydrogen sulfide can be oxidized by sulfur bacteria that use the oxidation as a source of energy. It can also be oxidized by photosynthetic colored bacteria which use the gas as an oxygen acceptor in photosynthesis. Both types of bacteria are known to form bacterial zones at or above the the upper limit at which hydrogen sulfide can be detected. In at least certain cases, it is clear that these organisms are not metabolizing H_2S in the lake, but it is uncertain if they are using nonsulfur-containing organic matter.

In meromictic lakes, particularly those exhibiting crenogenic or ectogenic meromixis, in which an external source of water rich in sulfate delivers this anion into the anaerobic monimolimnion, enormous quantities of hydrogen sulfide may be generated, the greatest amount being 786 mg. per liter at the bottom of Big Soda Lake, Nevada. A rather similar condition may occur in waters contaminated with sulfur-containing wastes, such as sulfite residues from paper mills. A few cases are known in which appreciable quantities of free sulfuric acid occurs in lakes. Such acid, if not of industrial origin, is formed by the oxidation of volcanic sulfur or sedimentary pyrite.

CHAPTER 14

The Silica Cycle in Lake Waters

As has been indicated in Chapter 8, silica is often the most abundant acidic substance other than bicarbonate in inland waters, and may in fact be present in greater concentration than that substance in some waters. Apart from its quantitative importance, silica is of immense significance as a major nutrient for diatoms, the dominant phytoplanktonic organisms of most lakes. It is also possible that the details of the silica cycle in fresh waters will prove of considerable interest in relation to the general geochemistry of silicon at the earth's surface.

FORMS OF SILICA IN INLAND WATERS

Silicon is almost universally present in some more or less reactive form in all natural waters. A priori one might expect the element always to occur in an oxidized state and in the following forms: (1) in true solution as undissociated silicic acids, orthosilicate ions of the form $H_nSiO_4^{(4-n)-}$, or as complex silicate ions; (2) as colloidal silica; (3) as sestonic mineral particles. It seems likely that these categories grade imperceptibly into one another, and not impossible that in acid or neutral water no truly stable solution of silica exists. In the lakes of Japan, Tanaka (1953b) suspects that sometimes the concentration of the total silica content may be two or three times that of the soluble silica.

Nature of soluble silica. What is commonly called soluble silica can be determined by the method of Diénert and Wandenbulcke (1923, Robinson and Kemmerer 1930) in which a yellow color, due to a complex silicomolybdate, is generated by adding ammonium molybdate to the water acidified with sulfuric acid. It appears (Strickland 1952), however, that two yellow silicomolybdic acids exist, the proportions depending on the conditions of the experimental procedures. All the limnological work on silica is likely to suffer from inaccuracies due to this situation.

Diénert and Wandenbulcke claim that their reaction is given only by crystalloid silicic acid or silicate, a view which receives confirmation from the work of Harman (1927). The last named author made an extended study of the physical chemistry of sodium silicates in aqueous solution and concluded that at the concentrations that normally occur in natural waters, practically the whole of the silicate is crystalloid, in true solution. Harman further believed that he had obtained approximate values for the ionization constants for the reactions

$$[H_2O] + [H_2SiO_3] \rightleftharpoons [OH_3^+] + [HSiO_3^-] \qquad k_1 = 4.2 \times 10^{-10}$$

$$[H_2O] + [HSiO_3^-] \rightleftharpoons [OH_3^+] + [SiO_3^=] \qquad k_2 = 0.5 \times 10^{-10}$$

These values would indicate that at the hydrogen ion activity ordinarily encountered in lake waters, the undissociated acid will be several tens to several thousands of times more abundant than the monovalent ion $HSiO_3^-$, and that the concentration of the divalent $SiO_3^=$ will be negligible except under conditions of extreme alkalinity. Okura (1954), who finds some colloidal silica in natural waters, concludes that it easily passes into the molecular or ionic form, but according to Carman (1940, Cooper 1952) the only ions likely to be present are of the form $H_nSiO_4^{(4-n)-}$. The tetrahedral groups SiO_4^{4-} have a strong tendency to aggregate in a three-dimensional network except in alkaline solutions, and a neutral silica sol is evidently easily formed in neutral or acid water. The phenomena may be more complex in lake water than in the laboratory, owing to the probable presence of colloidal ferric and aluminium hydroxides, either freely dispersed or adsorbed to suspended particles. It is not unlikely that silica or silicate ions would become associated with such hydroxides. Moreover, there is a possibility, at least in acid waters, of the subsequent formation of aluminosilicate ions, perhaps of the form $Al(OH)(HSiO)^+$ or $Al(HSiO_3)^{++}$; Mattson (1933) has obtained evidence of the existence of such complex ions in experiments on the electrodialysis of soils.

Krauskopf (1956), in a very important paper received after this book went to press, concludes that amorphous silica forms stable solutions of H_4SiO_4 containing up to about 100 mg. per liter over the temperature range of limnological interest.

Sources of silica in natural waters. The major source of silica in natural waters is certainly from the decomposition of aluminosilicate minerals in the drainage basin from which the waters flow. In the presence of water, CO_2 reacts with silicates to produce carbonates and silica, the equilibrium pressure of CO_2 being well below that of the CO_2 of the atmosphere (Urey 1952). It is also commonly supposed that organic acids may play a part in the decomposition. Under acid conditions, however, the resulting silica appears to be relatively immobile and easily aggregated, whereas in alkaline waters some silicate ions are formed and there is considerably more movement of the silica. In the drainage of tropical regions, in which almost complete decomposition of organic matter occurs and very intense leaching takes place, the silica is likely to be more mobile than in the acid waters of temperate and subarctic lands.

As with most other constituents, the silica content of water draining igneous rocks is likely to be less than that draining sediments. Conway (1942) concludes that the mean igneous water contains 10.3 mg. SiO_2 per liter, the mean sedimentary water 19.4 mg. Since the total residue is much less in the former case, the proportion of silica, 37.7 per cent, is greater than the 11.9 per cent for sedimentary water.

Possible biogeochemical decomposition of silicate minerals. There is a distinct possibility, of considerable limnological importance, that the purely inorganic liberation of silica may be supplemented by specific biogeochemical processes. Several investigators have concluded that benthonic diatoms can attack the aluminosilicate clay minerals and obtain silica directly from them, greatly accelerating inorganic decomposition. This process appears to have been discoverd by Murray and Irving (1891). Vernadsky (1922) and Coupin (1922) found that *Nitzschia* spp. can attack kaolin, and Vinogradov and Boichenko (1942) have made a more detailed study of the process by which *Nitzschia palea* and *Navicula minuscula* decompose nacrite, a mineral closely related to kaolinite, which though rare can be obtained in a very pure crystalline condition. It was found that nacrite was an adequate source of silica for the diatoms in pure culture. The decomposition of individual crystals by single diatom cells could be followed under the microscope. The crystal begins to decompose when it comes in contact with the slime sheath secreted by the diatom. Decomposition is marked first by an increase in opacity; the crystal appears to swell and loses its birefringence. It then starts exfoliating along the cleavage plane. Later the entire crystal becomes an amorphous mass, alumina is liberated into the medium during the decomposition. There is evidence that the lattice of the nacrite is changed to give an intermediate crystalline substance during the breakdown of the crystal. It is not clear if the silica of the nacrite is assimilated quantitatively, but since the digestion of the mineral is external to the cell, it would seem likely that some soluble silicate would diffuse into the medium. If this supposition is confirmed, it may well turn out that benthonic diatoms play a considerable part in liberating to the water silica present in the clay minerals of the sediments.

King and Davidson (1933) find that autolyzing diatoms liberate silica, and suppose an enzymatic reaction to be involved. The evidence for this, though suggestive, is not unequivocal.

SILICA IN LAKE WATERS

The silicate content of surface waters. The limnological data relating to surface samples from a number of well-studied regions is summarized in Table 112.

TABLE 112. *The silica content of surface waters*

	Range, mg. per liter	Mean, mg. per liter
Austrian Alps (Ruttner 1937)	0.0–3.0	1.11
English Lake District (Pearsall 1930)	0.1–2.4	1.16
Carinthia (Findenegg 1935, from graph)	0.2–2.6	1.26
Northern Germany (Ohle 1934a)	0.02–3.8	1.35
Northeastern Wisconsin (Juday, Birge, and Melohe 1938)	0.2–25.6	2.05*
Connecticut (Hutchinson, unpublished)	0.6–11.1	2.95
Japan (Yoshimura 1930)	2.3–27.3	10.9
Central Africa (Beadle 1932, Beauchamp 1939)	Trace–33	13.6
Texas and Mexico (Deevey 1957)	4–50	20.9
Indonesia (Ruttner 1931)	1–71	31.3
Guatemala and El Salvador (Deevey 1957)	25–77.5	48.1

* Mode 0.0 to 0.9 mg. per liter.

The table indicates the very considerable variation that can be observed even within a restricted region. Much of this variation is certainly due to utilization by diatoms, and perhaps in some cases by other types of precipitation in the lake. The data, so far as they go, suggest that greater silicate concentrations can occur in tropical than in temperate regions. This is in line with theoretical expectations and with the averages for the rivers of the world (Clarke 1924, Mattson 1941). The high Central American values are in part probably due to the presence in the drainage basins of easily eroded volcanic ash. Löffler (verbal communication) finds that in the lakes of the Chilean Andes the silica concentration depends mainly on the size of the drainage basin.

Vertical distribution and seasonal cycle of silica. The vertical distribution of silica tends to resemble that of other substances that can be derived from the sediments. In view of the fact that at first sight it would seem unlikely that silica would be involved in an oxidation-reduction cycle, the similarity of the distribution to those of some of the other substances that have been considered in the preceding few chapters might be surprising. It is probable that the movement of silica, like that of phosphate but to a lesser degree, is determined by the state of oxidation of the iron present at the mud-water interface, though very little is known definitely about this.

In his experimental tanks set up over mud from Windermere, Mortimer (1941–1942) found that under aerobic conditions the silica in the water rose from a few to 20 or 30 mg. per liter, but in the anaerobic tank the rise was much greater, reaching 100 mg. per liter in the course of 110 days, after which time there was no further increase. Comparison with the events in the bottom water of Esthwaite indicates that during summer stratification the rises in phosphate and in iron are of the same order as those observed in the tank, but the rise in silica in the tank was disproportionately great. It is not clear if this is due to differences in the composition of Windermere and Esthwaite sediments. In nature, values of the same order as those obtained in the anaerobic tank have not been reported, and it is possible that the conditions provided by the tank do not, in the case of silica, duplicate those found naturally. Mortimer believed that the whole of the available silica passed from the mud during the 100 days of the experiment, and that the lack of further rise indicated the establishment of equilibrium concentrations in the two phases. A slight decline in silica was noted on the eighty-third day, when some oxygen was accidentally introduced through a crack in the paraffin seal of the tank. This fall is attributed by Mortimer to the precipitation of a ferric silicate complex.

The general progression of the seasonal variations in silica in a small and productive lake is beautifully demonstrated in Yoshimura's (1930) study of the silica cycle in Takasuka-numa (Fig. 209), an investigation that initiated the serious study of the seasonal variation of silica in lakes. In this more or less monomictic lake, all layers contained 12 to 13 mg. SiO_2 per liter during the long winter circulation period. A slight fall at all levels occurred during the spring. During the first half of summer stratification, the silicate of the trophogenic layer continues to decrease; in the tropholytic zone there is a marked increase, producing an inverse clinograde depth distribution. At the height of summer, 6.3 mg. SiO_2 per liter are recorded at the surface, and 12 to 15 mg. per liter at 6 m. Subsequently, during the latter part of summer stratification, the silica begins to rise at all levels, so that in the middle of September, when the lake is still partly stratified, 10.9 to 11.5 mg. per liter are present in the trophogenic zone.

Yoshimura computed the total mass of silicate present in the lake. The observed variations are evidently due largely to the development of diatom populations in the late winter and spring. The silica content of the lake as a whole, moreover, is evidently largely controlled by the inflow of

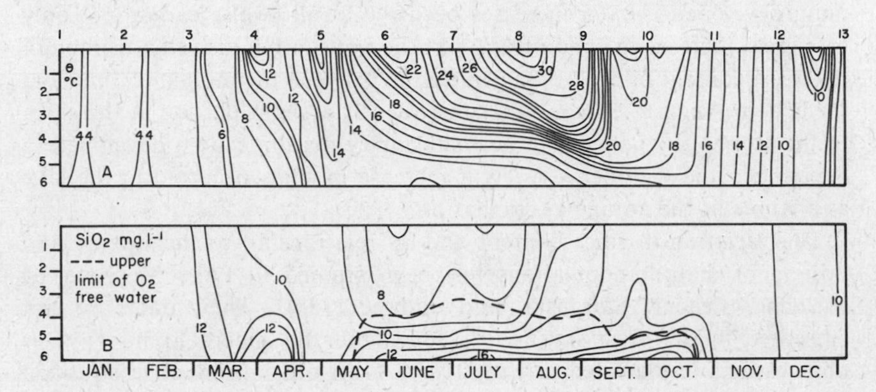

FIGURE 209. Time-depth diagram of variation in *A* temperature, and *B* silica, in Takasuka-numa, Japan (after Yoshimura).

water, which is artificially regulated. Water entering in April evaporates in the lake, and practically its entire mineral load must remain in the basin. During the period of the rise in water level in April, about 210 kg. SiO_2 are brought into the lake, but the SiO_2 content of the water of the lake as a whole rises by only 40 kg. It appears, therefore, that 170 kg. are taken up by diatoms. This implies a mean daily uptake of 7.1 kg. During the rise in the silica content in the late summer, when abstraction by diatoms is certainly very low, the total content of the lake increases by 260 kg.; of this, 180 kg. can certainly be attributed to inflowing water. Since this last estimate is based on relatively few analyses, it is not very accurate but does indicate that a large part, though probably not the whole, of the rise can be attributed to the influent. In other lakes, to be discussed in the next few paragraphs, in which a similar late summer rise occurs, it is reasonably certain that most of the silica is derived from the sediments. It is therefore not surprising that the estimate of the influent silica falls short of the observed rise. Yoshimura considers it possible, but rather unlikely, that underwater springs may also be responsible for part of the increment.

In the Schleinsee, in which well-defined winter stagnation occurs under ice, Einsele and Vetter (1938) found high values during autumnal circulation, followed by the establishment of inverse clinograde stratification after the lake froze. This stratification involved a fall in the silica content of the trophogenic zone, which was not counterbalanced by the rise in the tropholytic zone. At vernal circulation the SiO_2 content of the entire lake rose slightly. Early in summer stratification, a very pronounced fall in the silica of the trophogenic zone took place as the main diatom maximum developed. Later increases occurred at all levels, leading not only to a great increase in concentration at the bottom but also to a tripling of the silica content of the trophogenic zone between the middle of June and the end of August. There was subsequently a slight decline in the silica of the trophogenic zone as a new subsidiary diatom bloom developed in September and October. A slight increase in silica occurred in the lake as a whole at the autumnal circulation.

The variation in silica content and its relationship to the waxing and waning of diatom populations has been studied in Lake Mendota by Meloche, Leader, Safranski, and Juday (1938). Their data are not presented in the most satisfactory form, but the general trend of their observations is indicated in Fig. 210. As in other productive stratified lakes, an inverse clinograde distribution develops during both stratification periods. There is a marked increase in silica in all layers except the deepest at the time of the vernal overturn. This is followed by a fall in early summer, it being evident that the minimum amount of silica is present in the lake in May. There is then a sudden rise in both silica and diatoms in the trophogenic zone early in June, and this rise in silicate evidently continues throughout the summer, after the diatom population has declined. During the early part of autumnal circulation, a fall in the silica content of the trophogenic zone is accompanied by a moderate rise in the diatom population.

These cases and the very similar cycle reported by Mortimer for Esthwaite water indicate that a general rise in the silica of the trophogenic layer during the later part of summer stratification is usual. In some cases the incidence of this rise can be explained by influent water and by the cessation of growth of diatom populations, but neither explanation will fit Lake Mendota, where the influent water cannot possibly have been responsible for the very rapid increase in June, with a rising diatom population. Since it has been demonstrated that phosphate must leave the epilimnetic mud surface in small productive lakes during summer stratification, and since there is less reason to doubt that silica can enter oxygenated water from mud than that phosphate can, as is clearly shown in Mortimer's experimental tanks of water over Windermere mud, the late summer increase in silicate can be most reasonably explained by the diffusion of

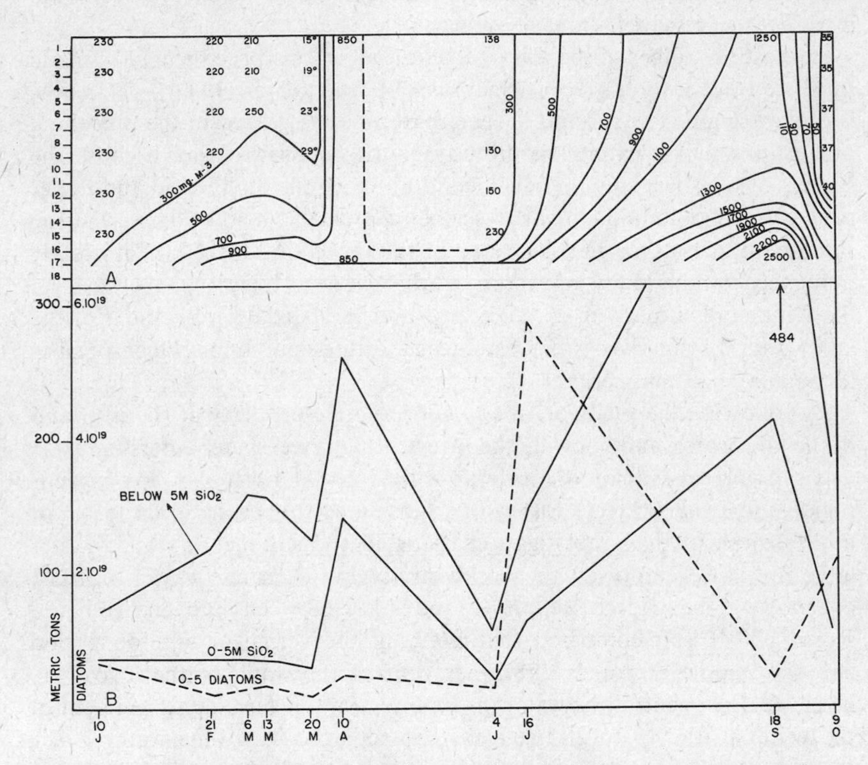

FIGURE 210. Silica cycle in Lake Mendota: *A*, time-depth diagram of concentration of SiO_2; *B*, total silica dissolved in the top 5 m. (0–5 m.) and in the rest of the lake, and total diatom population in the top 5 m. (From data of Meloche, Leader, Safranski, and Juday.)

silica from the shallow-water sediments. The Mendota observations do, however, raise the very interesting question as to why this process should be limited to the later part of the stagnation period, for it is quite clear that it does not take place during May, when stratification is just developing. It is reasonable to suppose that in this lake, and probably in the others, the movement of silica from the mud to the oxygenated water of the epilimnion is under the control of temperature. It is indeed possible that the fall in silica that accompanies the rather poorly developed autumnal diatom bloom in Mendota is due to the upper waters having cooled off sufficiently to prevent the input of silica from keeping pace with its utilization, and it must be supposed that far more silica was liberated early in June than could be used even by the great population then developing. It is most unfortunate that in the work of Meloche, Leader, Safranski, and Juday the thermal condition of the lake can only be inferred from the most general remarks about stratification.

Interesting studies of the silica cycle in Lake Constance have been made by Elster and Einsele (1937), Grim (1939), and Elster (1939). This lake is of course much larger and deeper than those discussed in the preceding paragraphs, but it cannot be directly contrasted with them, because the studies of the silica content have been limited to the epilimnion and upper part of the hypolimnion, no analyses being available below 50 m. During summer stratification, the chemistry of the top 20 m. of the lake is greatly influenced by the inflowing waters of the Rhine. In spring, the water of the Rhine falls rapidly in conductivity, and the partial replacement of the lake water by this river water produces a dilute epilimnion which persists throughout the summer.

By following the variation in any conservative property of the river and of the lake water, and knowing the rate at which river water enters the lake, it is possible to estimate the rate at which the lake water of any layer is replaced by river water. Elster and Einsele used carbonate hardness for this purpose in their investigation made in the summer of 1935; Elster later, in 1937, employed the conductivity in the same way. In 1937, considering the widest section of the lake, between Langenargen and Rorschach, Elster computed that between the vernal circulation period and the middle of August, 78.2 per cent of the water in the 0 to 5 m. layer, 72.4 per cent of the water in the 5 to 10 m. layer, and 41 per cent of the water in the 10 to 20 m. layer was replaced by Rhine water. The silica content of the river falls far less than does the total inorganic concentration in May, and later rises slightly. In the lake, the silica content continues to fall until the middle of August at 0 m. and 5 m., and to a rather less degree at 10 m. The change at 20 m. is small and irregular. Knowing the rate of replacement of the epilimnetic water and the silica content of the Rhine, it is possible to calculate the amount of silica incorporated into diatom shells during the period of summer stagnation. For 1935 the data of Elster and Einsele, as integrated by Grim, indicate that between May 1 and August 26 1.2 mg. SiO_2 cm.$^{-2}$ were removed biologically. Elster calculated from the data of 1937 that 2.19 mg. SiO_2 cm.$^{-2}$ were removed between vernal circulation and August 17. The difference is evidently significant and indicates a greater diatom production in 1937 than in 1935. The over-all rate for 1937 corresponds to a removal of 0.0157 mg. SiO_2 cm.$^{-2}$ day^{-1}, but it is evident that the rate of removal was variable, being greatest in August, during the first seventeen days of which month it averaged 0.046 mg. SiO_2 cm.$^{-2}$ day^{-1}. The time of the maximum evidently corresponds to the development of *Cyclotella* to produce the major phytoplankton bloom.

Grim has compared the 1935 figures derived from the chemical study of the lake with results of his own obtained from the silicon content of diatom shells, and estimates the rate of production of diatoms in the lake. He concludes that between the middle of March and the middle of December at least 3 mg. SiO_2 cm.$^{-2}$, and perhaps twice that amount, were incorporated into diatom shells. Since the two major diatom blooms, *Synedra* in June and *Cyclotella* in late July and August, both occurred within the period covered by the chemical studies, the results of the latter should give a figure for total silica consumption little below that derived from the planktological studies. Since there is an evident discrepancy between the two methods of estimation, it would seem very probable that not only the original silica present at vernal circulation and that brought in by the Rhine, but also a considerable quantity derived from the sediments in contact with the fairly warm water of the epilimnion, contributes to diatom growth and the silica cycle of the lake.

Heterograde distribution of silica. The development of large populations of diatoms in stable strata may produce markedly heterograde silicate distributions. A most remarkable example is given by Birge and Juday (1911), who found in the thermocline of Beasley Lake, Wisconsin a great maximum of *Fragilaria* and *Asterionella* at 5 m. The population had produced great supersaturation of oxygen, and a marked minimum in the total silica content (Fig. 172). Since the diatoms were apparently included in the samples analyzed, it is evident that the population had existed for some time in the 4 to 6 m. stratum, and that many dead diatoms must have fallen into the hypolimnion, transferring to the deeper part of the lake silica initially in the densely populated metalimnetic zone. There was also evidence of precipitation of calcium carbonate by the photosynthetic activities of the diatoms.

BIOCHEMICAL ASPECTS OF SILICA IN LAKES

Biological utilization of silica. There can be no doubt that the main mechanism of loss of silica from lake waters is its utilization by diatoms, though a few other quantitatively unimportant groups of organisms also require silicon.

Diatoms, particularly planktonic species, are capable of using silica at quite low concentrations, as is to be expected from the data on the occurrence of the material in lake waters. Lund (1954) finds *Asterionella* and *Tabellaria* to be limited in nature by concentrations of the order of 0.5 mg. SiO_2 per liter, and *Melosira italica subarctica* by about 0.8 mg.

In culture, Chu (1942) found that *Fragilaria crotonensis*, *Nitzschia palea*, and *N. acicularis* grow best at about 30 mg. SiO_2 per liter, which concentration is markedly harmful not only to certain green algae but also to *Asterionella.* Lewin (1955a) found a similar high optimum in experiments with *Navicula pelliculosa.* In the last named species, Lewin (1955b) demonstrated a clear inverse relationship between nutrient concentration, including silica, in the medium and the production of sheath material by the organism. The sheath appears to be composed of a polyglucuronide formed when, owing to nutrient deficiency, the products of photosynthesis can no longer be used for growth. It is doubtful that the sheath has any relationship to silica uptake in this species, but the observations of Vinogradov and Boichenko (1942), already mentioned, suggest that in other species this may be the case. Lewin (1954) further found that *N. pelliculosa,* after being starved of silica, takes up the material rapidly when soluble silica is supplied. The uptake depends on an aerobic respiratory mechanism, probably involving high-energy phosphate (Lewin 1955c). It is probable that *N. pelliculosa* uses only orthosilicate in true solution, for which colloidal SiO_2 and sodium methyl siliconate are ineffective as substitutes (Lewin 1955a). It seems certain that other species are capable of using a wider variety of silica sources.

The newly absorbed silica is immediately deposited as a thickening of the existing frustule, and there is no evidence of storage of a significant amount in the cell. Uptake can be specifically inhibited by the sulfhydryl inhibitor $CdCl_2$ at concentrations which do not affect respiration. Washing cells also can prevent silica absorbtion, and the material that is washed out can be replaced to some extent by $SO_4^{=}$ ions or by ascorbic acid, and completely by Na_2S, $Na_2S_2O_3$, glutathione, 1-cysteine,*dl*-methionine, or a mixture of sulfate and ascorbic acid. It appears that the uptake of silica must depend on water-soluble sulfhydryl groups, presumably in the cell membrane. Silica appears to stimulate endogenous, but not exogenous, respiratory metabolism in *N. pelliculosa* (Lewin 1955c). It may also be noted that there is evidence that silica is involved catalytically in organic synthesis and growth in *Chilomonas* (Mast and Pace 1937) and *Amoeba* (Pace 1933). It is clear that extension of this extremely interesting field of silica metabolism will produce results of great geochemical as well as biochemical interest.

Utilization by other organisms. Apart from diatoms, a few phytoplankters use silica. The most important cases in fresh waters are no doubt the yellow-brown algae that form siliceous cysts. None of these organisms are likely to have any significant quantitative effect on the silica cycle.

Among animals, the siliceous sponges may be abundant enough to have a small effect on the cycle of silicon, as Votintsev (1948) finds in Lake Baikal. Jørgensen (1944) has examined the capacity of the fresh-water sponge *Spongilla lacustris* to utilize the silica dissolved in fresh waters. Down to

0.02 mM. SiO_2, or about 1.2 mg. per liter, a normal production of spicules occurred. Below this concentration, very few thin microscleres were produced; the formation of megascleres was much less inhibited than that of microscleres. Sponges grown in an artificial medium in paraffin-coated vessels grew much better spicules if powdered mica, orthoclase, or spicules from the sponge *Iophon* were present. In general, mica and *Iophon* spicules were a much better source of silica than was orthoclase. There was no evidence that contact between such solid siliceous material and the sponge promoted the uptake of the mineral silica, and the solution of the latter appears to have been an ordinary inorganic process. Though it is possible that biochemical processes in lake sediments may greatly promote the diffusion of silica into the water from the mud, as presumably happened in Mortimer's anaerobic tank, the specific biochemical decomposition of aluminosilicates appears, as far as is known, to be solely a function of diatoms.

SUMMARY

Silica occurs at the *p*H values of natural waters mainly as orthosilicate in an undissociated condition, though there is evidence for the existence of colloidal silica and possibly of complex aluminosilicate ions such as $Al(OH)(HSiO_3)^+$ or $Al(HSiO_3)^{++}$. The most obvious source of silica is the solution of rocks in the drainage basin under the influence of CO_2. The quantities present in surface waters range up at least to 77.5 mg. per liter in open lakes, and in some cases silica is evidently more abundant than are the inorganic acids or oxyanions in solution. Tropical fresh waters probably tend to have more than do temperate waters.

Though much silica must be brought into lakes by the influents, there is evidence from quantitative studies of the silica in the entire water mass of certain lakes that silica must enter the water from the lake sediments; there is some indication that this occurs most easily from the epilimnetic mud at high temperatures. The silica concentration is usually increased in summer stratification in the anaerobic deep water of lakes having clinograde oxygen curves. Experimental studies indicate that this is regulated by redox potential, possibly by the reduction of ferric silicate previously formed in the superficial layers of mud when in contact with oxygenated water.

The development of diatom blooms constitutes the most important mechanism by which silica is removed from lake waters. Some benthonic diatoms can apparently use silica in aluminosilicates with which their slime sheaths come in contact. There is evidence in *Navicula pelliculosa* that the development of the polyglucuronide slime sheath follows nutrient depletion, including that of silica, but in this species apparently only dissolved orthosilicate can be utilized. Uptake of dissolved silica by diatoms involves an aerobic process and the presence of water-soluble sulfhydryl groups on the cell membrance. The frustules of dying diatoms appear to lose silica to the water, but it is not known if this involves au enzymatic autolysis or is merely dependent on simple decomposition processes.

CHAPTER 15

Minor Metallic Elements in Lake Waters

The metallic elements to be considered in the present chapter are all likely to be less concentrated in lake water than is iron. Some play a considerable part, as trace elements, in metabolism; others, though of no known biological importance, have geochemical cycles of considerable interest. The final group of strongly radioactive elements may provide important chronological information in certain cases.

MANGANESE

Manganese exists at low redox potentials and low *p*H values in bivalent, ionic, freely soluble, manganous form, which generally accompanies ferrous iron. For both of the oxidative reactions likely to be of importance in the manganese cycle, the E_0 value is higher than in the case of the ferrous-ferric system:

$$Mn^{++} \leftrightharpoons Mn^{+++} + e \qquad E_0 = 1.51$$

$$Mn^{++} + 2H_2O \leftrightharpoons MnO_2 + 4H^+ + 3e \qquad E_0 = 1.28$$

$$Fe^{++} \leftrightharpoons Fe^{+++} + e \qquad E_0 = 0.77$$

Disregarding all other considerations, we should expect to find manganous ions in detectable quantities at redox potentials too high to permit the presence of detectable ferrous ions. The observational data to be given hereafter are in accord with this expectation.

Forms of manganese in the biosphere. On oxidation, relatively insoluble and more or less hydrated oxides are produced, but it is not possible to derive at present a simple theoretical treatment parallel to that given for the ferrous-ferric system, because while at least two and probably more different solid oxidized forms of manganese occur freely in soils and perhaps elsewhere in the biosphere, nothing whatever is known of the identity of the solid oxidized components in lakes.

The formation of MnO_2 should be expected as the final oxidation product, and it is not impossible that subsequent studies will demonstrate that this oxide, in suspension or adsorbed on particulate matter, is the only important form of oxidized manganese present. It must be remembered, however, that in soils there is clear evidence (Mann and Quastel 1946) of continual bacterial oxidation of Mn^{++} to manganic hydroxide, conventionally to be regarded as $Mn(OH)_3$. This material dismutes, the reaction being equivalent to

$$Mn_2O_3 \rightarrow MnO_2 + MnO$$

The rate of dismutation is dependent on the *p*H, being rapid in acid solutions, slow around neutrality, and quite negligible in alkaline solutions. The manganous ions formed can of course undergo further biological oxidation, so that finally all the manganese will be converted into MnO_2. It would be reasonable to suppose that in hard-water lakes, particularly at the time of autumnal circulation, some manganic manganese as well as manganese dioxide might occur temporarily in suspension.

The natural oxides of manganese occurring as minerals are:

Manganosite	MnO
Pyrochroite	$MnO \cdot H_2O$
Hausmannite	Mn_3O_4 (i.e., $Mn \cdot Mn_2O_4$)
Pyrolusite	MnO_2
Manganite	$Mn_2O_3 \cdot H_2O$

Various indefinite hydrated materials of the general form $m MnO \cdot n MnO_2 \cdot 2H_2O$ are known.

Dion and Mann (1946) have concluded that in more or less neutral soils a hydrated manganic oxide $Mn_2O_3 \cdot nH_2O$ is often present, and that hausmannite may be expected in alkaline soils. Fujimoto and Sherman (1948) find that an indefinite hydrate $m MnO \cdot n MnO_2 \cdot 2H_2O$ is present in some Hawaiian soils, and has different properties from Dion and Mann's preparations of manganous manganite. Heintz and Mann (1947) found that pyrophosphate and various hydroxyacids, such as citric, tartaric, and malonic, could form stable complexes with manganese extracted from hydrated manganic hydroxide and from hausmannite, but not from manganese dioxide. There is evidence (Jones and Leeper 1950) that under some conditions, however, pyrolusite gives up manganese more easily to roots and to reducing agents than does hausmannite. The specific surface of the preparation is doubtless important. None of these results, perhaps, have any direct limnological importance at the moment, but they at least direct attention to the complexity of the situation likely to await anyone who starts investigating the manganese cycle in a lake.

Total manganese in epilimnetic lake waters. In the surface waters of the north German lakes, Ohle (1934a) finds from less than 5 mg. m.$^{-3}$ to 200 mg. This last figure, for one date for the surface water of the Kleiner Uklei See, may be due to a misprint, as the 1-m. sample is given as 25 mg. m.$^{-3}$ and no mention is made of the high value in the text. Omitting this value, the highest is 133 mg. m.$^{-3}$, and the mean is 23 mg. The inclusion in the mean data of the determination for the Kleiner Uklei See only raises this value to 25 mg. m.$^{-3}$ If only those lakes in which both iron and manganese are recorded are considered, the mean values are Mn 31 mg. m.$^{-3}$, and Fe 40 mg. m.$^{-3}$, giving a ratio Fe:Mn of 1.29:1. The individual values of the ratio vary from 0.78:1 to 81:1.

There are six determinations for the Trammersee, the lake that gave the highest unequivocal values; these cover the whole range:

	Mn. Values, mg. m.$^{-3}$
May 5, 1931	10
June 17, 1931	5
September 18, 1931	35
January 27, 1932	8
March 22, 1932	< 5
April 26, 1932	133

The maximal value is clearly associated with spring circulation, and it seems likely that sedimentation occurs under the ice. The fairly high September value is quite superficial; at 1 m. only 9 mg. m.$^{-3}$ were present.

Even if the high superficial value of the Kleiner Uklei See be rejected, Ohle gives several other cases of higher manganese contents at 0 m. than at 1 m. This suggests the association of manganese, perhaps as MnO_2, with the surface film.

Juday, Birge, and Meloche (1938) have given analyses for eight lakes in northern Wisconsin. Their values tend to be lower than those of Ohle, ranging from 3 mg. m.$^{-3}$ to 23 mg. m.$^{-3}$, the mean being 8.4 mg. The mean value of the Fe:Mn ratio is 4.8:1, the individual values varying from 1.3:1 to 10:1.

Three unpublished determinations of Mn and Fe in the surface waters of Linsley Pond, August–October 1937, give a range from 50 to 250 mg.

Mn m.$^{-3}$, and 1.8:1 to 4.2:1 for the Fe:Mn ratio. The mean value is 140 mg. m.$^{-3}$, with a ratio of 2.5:1.

Similar high values, from 80 to 120 mg. m.$^{-3}$ had been recorded earlier in North America by Wiebe (1930), in the Mississippi at Fairport, Iowa. These figures refer to the manganese of filtered water; unfortunately, no iron contents are given. Turekian and Kleinkopf (1956) give the very high mean of 400 mg. m.$^{-3}$ for superficial waters in Maine.

By far the most important body of data relate to the surface waters of the Swedish lakes studied by Lohammar (1938). These may be arranged in two groups, those of the lowlands of south Sweden (Uppland and Dalarna) and those in north Sweden. In many lakes of the former group, iron could not be detected, and in obtaining a mean, two values have been computed, one giving the sum of the mean iron contents of each lake divided by the number of lakes containing iron; the second, the same sum divided by the total number of lakes. The true mean obviously lies between the two figures. The ratio of the mean contents (Table 113) has been given, on the assumption that the low iron figure is comparable to the mean manganese figure for all lakes and the high iron figure to the manganese content (56 mg. m.$^{-3}$) of the lakes with detectable iron.

TABLE 113. *Iron and manganese contents of Swedish lake waters*

	Fe		Mn		
	Range*	Mean	Range	Mean	Fe:Mn
N. Sweden	< 200–3100	960	Tr. < 10–460	33	29.1:1
S. Sweden	Tr. < 200–2500	220–370	Tr. < 10–850	44	5.0–6.6:1

* In all discussion of Lohammar's and similar long series of analyses, the term range refers to the range of individual determinations; mean refers to the mean of means for individual lakes. Where a few entries of the form < *n* are mixed with a number of determinations, they are taken as *n*/2 in computing means.

The problem of the Fe:Mn ratio is of considerable interest. The mean geochemical ratio for the accessible lithosphere is 50:1 (Goldschmidt 1938). In Northern Sweden, 26 out of 66 lakes, or 39.4 per cent contain waters in which this ratio is exceeded. In south Sweden, it is exceeded in one certain case out of 47 lakes, and perhaps in another where the data are imperfect, i.e., in only 2 to 4 per cent of the available cases. It is evident that except in some of the humic lake water of north Sweden, manganese tends to remain in the water in relatively greater amounts than does iron.[1] Sometimes this phenomenon can be very striking, as in the

[1] Data given by Yoshimura (1931b) are vitiated by a known error (Yoshimura 1936c) in the iron contents and do not imply the excessive ratios that they would appear to do if taken at their face value.

few cases in which more manganese than iron is present. It is difficult to gain any idea of the cause of these extreme variations, but Ohle suggests that drainage from the litter of coniferous forests may be enriched in manganese, as conifers often contain great accumulations of the element. As is indicated in Volume II the sediments laid down in Linsley Pond during the time of the postglacial coniferous forest are abnormally high in manganese.

Acid-soluble manganese. The only attempt to determine any fraction of the total manganese that might represent the amount of the element available to the phytoplankton was made by Harvey (1949). The method used by Harvey depends on the catalytic action of manganese ions on the oxidation of tetramethyldiaminodiphenylmethane, commonly called tetrabase, in the presence of periodate. A transitory blue color is developed, and the intensity of this color, within certain limits, is proportional to the quantity of manganese in solution. Full instructions for the performance of the estimation, which requires critical control of the *p*H, concentration, and time, are given in Harvey's paper. Since the tetrabase is insoluble in alkaline or neutral media, the method gives, not the quantity of ionic manganese initially present in the sample, but the amount that goes into solution at *p*H 4.6 to 4.8, at which the determinations are best performed. Harvey found for various fresh waters, after standing for 24 hours at *p*H 4.7, the values given in Table 114.

The five lakes are arranged in the order of increasing productivity, and it will be observed that the manganese content rises in almost the same order. The quantity of easily acid-soluble manganese in these waters appears to be about the same as that of the total manganese in some of

TABLE 114. *Acid-soluble manganese in various English waters*

	Mn, mg. m.$^{-3}$
River Yealm, Devon	
On leaving moorland, Wisdome	1.15
Yealm Bridge, after passing for 11 km. through agricultural land	10
English Lake District	
Ennerdale Water	6.5, 8.0
Loweswater	11
Windermere, north basin	10
Blelham Tarn	20
Esthwaite Water	> 40
Black Beck, main influent to Esthwaite	12

the waters previously discussed. It is evident that determination of total acid-soluble manganese on the same whole waters and on membrane filtrates is very desirable.

Manganese requirements of the phytoplankton. It has been known (Hopkins 1930) for some time that *Chlorella* requires manganese. Later Pirson (1937) and Emerson and Lewis (1939) demonstrated that manganese deficiency greatly reduced the rate of photosynthesis in a dim light and less markedly in a bright light. The addition of manganese to a manganese-starved culture may raise the photosynthetic rate with great rapidity (Pirson 1937, Noack and Pirson 1939). Pirson and Bergmann (1955) have found that, in *Chorella*, glucose can be utilized in the dark by cells growing heterotrophically in a manganese-deficient medium, but that the element is required when such cultures are strongly illuminated. Guseva (1937b) found that *Scenedesmus*, *Chlamydomonas*, *Zygnema*, *Eudorina*, *Synura*, and *Aphanizomenon* all required the element, though in different quantities. The requirement by fresh-water algae is probably universal; in view of the importance of manganese as an activator of enzyme systems, this is to be expected. Harvey has shown, moreover, that the marine diatom *Ditylum brightwelli*, certain marine species of *Chlorella* and *Chlamydomonas*, two undetermined brown Chrysomonads, and a red *Cryptomonas* are all severely limited by manganese deficiency, while Levring (1946) has shown that the green seaweed *Ulva* requires the element.

There is, however, a clear indication of considerable variation in the threshold concentration at which deficiency begins to be apparent, for the diatom *Coscinodiscus excentricus* showed little reduction in growth in media markedly deficient for *Chlamydomonas*, and evidently requires much less manganese than do the other species studied (Harvey 1947). In view of the considerable differences in manganese requirements that can be exhibited by different organisms, and of the differences in content from water to water from season to season, it is quite likely that differences in manganese content may be found to play a part in regulating the qualitative composition of the phytoplankton.

In the case of *Chlamydomonas*, the addition of from 0.5 to 2.0 mg. Mn $m.^{-3}$ produced large if not maximal crops in water which otherwise did not support the organism. The effect of addition of 0.1 mg. $m.^{-3}$ was apparent in the early phases of growth of the population, but the resulting final crop was less than one fifth of that obtained by addition of 0.5 mg. m^{-3} Harvey considered that at such very low concentrations the element is not fully utilized. Addition of the element either as divalent manganous ions or as heptavelent permanganate was equally effective.

Guseva (1937a, 1939) found some evidence that both inorganic nitrogen and manganese in concentrations within the natural range inhibited *Aphanizomenon flos-aquae*; she therefore explained the decline of this alga in the Ucha Reservoir on the Moscow-Volga canal not only by phosphorus depletion but also by a concomitant increase in inorganic nitrogen and manganese concentration. This work, though interesting, requires confirmation in view of the many pitfalls inherent in the experimental study of the Myxophyceae.

Suspended manganese. As has been indicated, a general consideration of the redox potentials permitting the formation of appreciable amounts of the more oxidized forms of the element, and the detailed study of the behavior of manganese in soils, would suggest that manganese dioxide, which as pyrolusite is a very common mineral in the more oxidized sedimentary rocks, is likely to be formed as a solid precipitate in lake waters in which manganous ions are being oxidized, as at the top of the hypolimnion and probably throughout the whole lake during the circulation periods. Although the compound has not actually been identified in lake seston, there is evidence of suspended insoluble manganese often being present.

Braidech and Emery (1935) found no (i.e., less than 1 mg. $m.^{-3}$) manganese in certain municipal water supplies, and comment that sometimes sand filtration can remove the whole of the spectrographically detectable manganese from a water. Moreover, an unpublished determination from Linsley Pond (November 16, 1938) indicated that at the time of autumnal circulation 9.6 mg. per liter of organic seston and 1.7 mg. per liter of seston ash were present. Of this latter quantity, 0.2 mg. or 12 per cent was manganese and 0.12 mg. or 7 per cent was iron. The corresponding content of oxides would be 18 per cent MnO_2 and 10 per cent Fe_2O_3. A little of the iron was probably present as ferric phosphate; there was practically no silica in the sample. It is evident that during circulation, part of the seston ash had been formed by the oxidation of manganous manganese and ferrous iron previously in solution in the hypolimnion; probably part of the ferric iron had already sedimented. Suspended manganese may of course be included in the easily acid-soluble manganese determined by Harvey's method, and as in the case of suspended ferric iron, it may well be available to the phytoplankton.

Apart from the precipitation of MnO_2, it is possible that suspended manganese may in some cases be present owing to the adsorption of manganous ions on organic seston. Harvey finds that not only living *Chlorella* but also dead marine organic detritus rapidly takes up the element, though it is not adsorbed on powdered pumice, volcanic ash, or kaolin.

Vertical distribution of manganese. The vertical distribution of manganese has been studied by Yoshimura (1931a,b, 1932d); Ohle (1934a); Ruttner (1931, 1937); Juday, Birge, and Meloche (1938); and notably by Einsele (1937, 1940, 1944). The results of such studies indicate that manganese, though it behaves in a general way like iron, tends to have a slightly different pattern of distribution when examined in detail. These differences are not always apparent in lakes in which very small hypolimnetic accumulation occurs, as in most of the German lakes studied by Ohle. Yoshimura noted that in some of the Japanese lakes with more or less orthograde oxygen curves, a slight microzonal increase in manganese might occur without an increase in iron. This observation is concordant with those to be discussed in the following paragraphs.

The simplest kind of manganese stratification in lakes with clinograde oxygen curves is an inversely clinograde distribution, but one in which the concentration of manganese increases most rapidly at a slightly higher level than the maximum rate of increase of iron. The ratio of iron to manganese therefore increases with depth. Such a distribution is recorded by Yoshimura, at least in Takasuka-numa and Nakatuna-ko; by Einsele in the Schleinsee in Bavaria, and in a very well-developed form in the meromictic Hüttensteinersee or Krottensee and the Zeller See in the Austrian Alps. The last named lake contains no less than 13,000 mg. m.$^{-3}$ in the deep water of the monimolimnion. A comparable distribution has been observed in Linsley Pond in September (Fig. 211) and in some other lakes in eastern North America (Deevey 1941).

Einsele points out that since manganese is more easily reducible than iron, a significant amount of manganous ion can accumulate at a redox potential slightly greater than that at which ferrous iron appears; the observed distribution is thus easily explained. Iron and manganese both leave the mud and enter the water in a divalent state. At a certain critical depth, enough oxygen is delivered by turbulent exchange from above to oxidize the iron but not the manganese in solution. Iron is then removed from the upper layers of the deep hypolimnion and is sedimented. This process of impoverishment of iron in the upper part of the deeper layers of the hypolimnion, and of enrichment in iron as the precipitated hydroxide is reduced in the lowest part of the hypolimnion, is believed to explain the observed pattern of distribution. As stagnation progresses and lower and lower redox potentials appear at higher and higher levels, the process becomes progressively less significant and the Fe:Mn ratio is found to rise, at least in the hypolimnion of the Schleinsee, extensively studied by Einsele. The ratio is apparently quite variable in the hypolimnion and may well depend here on the degree of reduction of the sediment-water interface. Manganese, as is to be expected, leaves the mud more easily than iron, so that in the Schleinsee a ratio of Fe:Mn of 1.5:1 in the hypolimnetic water corresponds to a ratio of 17:1 in the top layer of the sediments.

A second type of vertical distribution of manganese is found most strongly developed in certain tropical lakes, notably those of Java studied by Ruttner. Here there is a marked tendency for the curve to be positively heterograde, with a strong maximum in the upper part of the

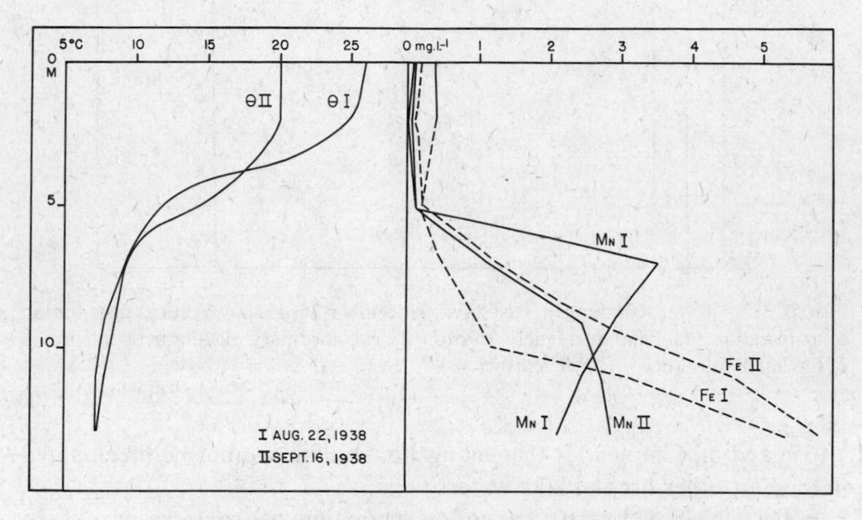

FIGURE 211. Linsley Pond, Connecticut. Vertical distribution of temperature, total iron, and manganese: I, August 22, 1938; II, September 16, 1938. Note the transformation of the heterograde manganese curve of the earlier date into an inverse clinograde curve in September, with loss of manganese at the upper hypolimnetic levels, while the iron increases throughout the hypolimnion.

epilimnion. The most extreme case is that of Ranu Klindungan (Fig. 212), in which 15,000 mg. Mn $m.^{-3}$ were present at 9 m. At this level the Fe:Mn ratio was 0.002:1. Below 12 m., iron rises and manganese falls, so that at 125 m., near the bottom of the lake, the ratio is near equality. Ruttner thinks a manganiferous mineral spring is involved in this case. Einsele considers the distribution in Ranu Klindungan as an extreme example of a pattern known elsewhere. Ohle has noticed markedly heterograde manganese curves in some of the moderately humic lakes of Baltic Germany, and finds in general an increase in H_2S below the manganese maximum. It is evident, however, from the calculations of Einsele

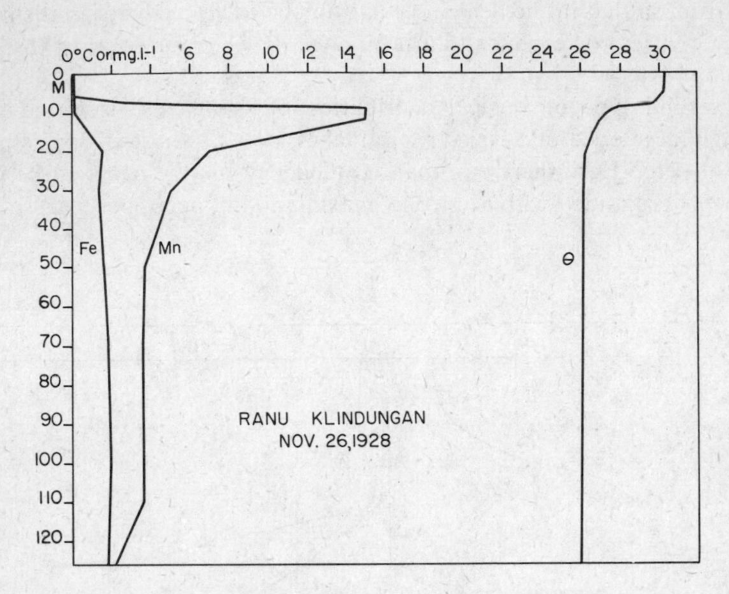

FIGURE 212. Ranu Klindungan, east Java, November 26, 1928. Vertical distribution of temperature, iron, and manganese, to show the extraordinary metalimnetic maximum of the latter substance. (After Ruttner.)

(1937) and of Chapter 13, that manganous sulfide cannot be precipitated in these or other normal lake waters.

In the Schleinsee towards the end of stagnation, as the upper part of the hypolimnion is oxygenated, manganese is certainly precipitated from the newly oxygenated water, to collect quantitatively, presumably after reduction, in the upper part of the still unmixed layers, producing on a small scale a distribution like that found in the Javanese lakes. Hasler and Einsele (1948) believe that by adding manganese to a lake this process could be accentuated, and as the MnO_2 that is precipitated is reduced, an equivalent amount of ferrous iron would be oxidized. Ultimately, all the iron might be replaced by manganese. As phosphate is not precipitated on oxidation of a manganous solution, this treatment might permit much more phosphorus to be regenerated from the mud. Unpublished data from Linsley Pond (Fig. 211), in which a positive heterograde distribution during August has been replaced during September by an inverse clinograde distribution of the kind already described, casts some doubt on the universality of the explanation put forward by Einsele. It is evident that much remains to be learned about the genesis of the more complex patterns exhibited by the vertical distribution of the element.

COPPER

In spite of the little attention that has been given to the occurrence of copper in lakes, the one adequate study that has been made has shown that the copper cycle in inland waters is complex and interesting. Atkins (1933) found in several English rivers remote from industrial contamination, copper contents of from 0 to 36 mg. Cu $m.^{-3}$ In a spectrographic study of samples from a number of city water supplies in the United States, Braidech and Emery (1935) found from 5 to 600 mg. Cu $m.^{-3}$, the mean being 91 mg.; a sample from Lake Michigan, the only untreated lake water analyzed, contained 70 mg. Cu. $m.^{-3}$

In the one extended limnological study on the metal yet carried out, Riley (1939) found that the copper contents of the surface waters of three lakes in Connecticut varied from 9 to 215 mg. Cu $m.^{-3}$ Considering the limited geographical extent of his main study, it is evident that Riley's range of values is concordant with that of Braidech and Emery. Turekian and Kleinkopf (1956) record the very high mean value of 1160 mg. $m.^{-3}$ for lakes and streams in Maine. Riley (unpublished) has also made analyses of a few surface samples from the Old World, finding 15 mg. $m.^{-3}$ in a summer sample from Windermere, 14 and 17 mg. $m.^{-3}$ in the waters of two brown-water tarns in Westmorland, England, and 10 mg. $m.^{-3}$ in the alkaline and saline water of Pang-gong Tso on the western borders of Tibet.

Forms of copper in lake waters. In his study of the Connecticut lakes, Riley separated the total copper into three fractions, *soluble* or *ionic*, *organic*, and *sestonic*. Two determinations of the copper content of ultra filtrates indicate that about 80 per cent of the so-called organic copper is associated with colloidal material. The term organic may therefore be misleading, as there is evidence (page 887) that in some lakes inorganic colloids are more abundant than organic. Experiments indicated that the addition of ionic copper to natural waters was followed by adsorption of part of the added copper. Later, after the water had stood some days in the laboratory, some of the adsorbed copper reappeared in ionic form, presumably owing to the bacterial decomposition of organic matter to which it had been attached. Riley obtained evidence that water rich in brown humic matter, when in contact with lake mud, takes up copper more easily than does water poor in such matter. The sestonic fraction, moreover, represents a proportion of the total seston that is considerably greater than the copper likely to be present in the living organisms of the plankton, and therefore probably represents copper adsorbed on the surface of organic detritus (Benoit 1956).

The surface waters of the three Connecticut lakes studied by Riley gave the mean and extreme values of the various fractions of the total copper content recorded in Table 115.

TABLE 115. *Fractionation of copper compounds in surface waters of Connecticut lakes*

	Ionic Cu, mg. m^{-3}	Organic Cu, mg. $m.^{-3}$	Sestonic Cu, mg. $m.^{-3}$	Total Cu, mg. $m.^{-3}$
Linsley Pond	5–35	4–93	0–67	11–210
Mean	11.4	27.1	14.5	53.0
Quonnapaug	4–18	4–66	0–30	9–109
Mean	10.2	21.6	9.0	40.8
Quassapaug	5–24	1–177	1–15	16–203
Mean	10.2	24.9	5.0	40.1

Relationship to *p*H and to hardness. It is a common experience (Ellis 1937, Hasler 1949) in treating reservoirs and other waters with copper sulfate to poison algal blooms, that more of the salt is needed to destroy the algae in hard than in soft water. This is usually and presumably correctly attributed to the formation of a basic copper carbonate in hard waters. Such a salt, being very insoluble, removes as a precipitate part of the ionic copper added. Nichols, Henkel, and McNall (1946) found that addition of 200 mg. per liter of $CuSO_4 \cdot 5H_2O$ to a sample of Lake Mendota water produced a bluish-white precipitate and a fall in *p*H from 8.0 to 6.8. They considered the precipitate to be a basic carbonate of somewhat variable composition. Two well-defined basic copper carbonates, malachite $CuCO_3 \cdot Cu(OH)_2$ and azurite $2CuCO_3 \cdot Cu(OH)_2$ are known in nature, and the physical chemistry of the production of basic carbonates of copper is likely to prove complicated. It is possible, however, to obtain some empirical information as to the likelihood of the precipitation of such basic carbonates controlling the copper content of hard waters not artificially enriched with copper.

A number of determinations of the copper content of the surface waters of lakes in northeastern Wisconsin were made in 1938 by Herbert C. Stecker. These analyses have not been published, but were made available for use in preparing the present work through the kindness of Professor V. W. Meloche and Professor A. D. Hasler of the University of Wisconsin. They were made on water concentrated by evaporation to one fifth of its original volume, the actual determination being done colorimetrically with sodium diethyldithiocarbamate. Presumably the results obtained relate not only to ionic copper but to some part of the organic copper as defined by Riley, for coagulation of organic colloidal material during concentration is likely to have liberated copper. This supposition is in accordance with the rather high values, from 10 to 120 mg. $m.^{-3}$

found in the 136 lakes studied. The mean value was 29 mg. m.$^{-3}$, and the mode lies in the 20 to 29 mg. m.$^{-3}$ class. There is no indication of a systematic decrease in copper content with increasing pH or increasing bicarbonate content. The mean copper content of waters at $p\text{H} \leqslant 6.4$ is 29 mg. m.$^{-3}$, and of waters at $p\text{H} \geqslant 8.0$ is slightly higher, 32 mg. m.$^{-3}$ The slightly greater copper content of the more alkaline waters, which is also apparent when the relationship of copper content to bicarbonate concentration is considered, is probably due to a tendency for waters with more total solids in solution to contain more copper than do dilute soft waters. It is, at any rate, quite certain that the figures give no indication of the copper content of the harder waters being regulated by precipitation of ionic copper as basic carbonate.

Vertical distribution of copper during stagnation periods. During stratification there is a slight tendency for the ionic copper, and sometimes for the other fractions, to increase near the bottom of the lake (Fig. 213). In Riley's Connecticut data this is most noticeable in certain vertical series from Lake Quonnapaug. The tendency is feeble, however, compared to the increments in iron (Fig. 215), manganese, and phosphorus observed under like conditions.

Several series of vertical analyses of copper, presumably mainly ionic, are included in the unpublished material of the Wisconsin Geological and Natural History Survey. Through the kindness of Professor A. D. Hasler it has been possible to consult this material during the preparation of the present work. In general, these analyses show little variation from the surface to the bottom, even at the height of summer stagnation. As one of the few extreme cases, Scaffold Lake is of interest. In July 1939 the redox potential fell from 0.380 volt at 3 m. to 0.008 volt at 9.5 m. In the same depth range, the iron rose from 130 mg. m.$^{-3}$ to 2170 mg. m.$^{-3}$, the manganese from 140 to 580 mg. m.$^{-3}$, and the copper only from 63 to 78 mg. m.$^{-3}$ In the summer of 1940 a fall in redox potential from 0.327 volt at 6 m. to 0.123 volt at 11 m. was accompanied by a rise in ferrous iron from 0 to 14,000 mg. m.$^{-3}$, and in copper from 53 only to 58 mg. m.$^{-3}$ In Nebish Lake, the hypolimnion of which was unusually rich in copper, the redox potential fell from 0.352 volt at 7 m. to 0.03 volt at 16 m. The bottom water contained some H_2S, and the ferrous iron was low though it showed a fivefold rise from 20 to 110 mg. m.$^{-3}$ going from 7 m. to 16 m. In the same depth range, copper rose from 67 to 109 mg. m.$^{-3}$ It is possible that some of the ferrous iron and copper in this case represented sulfide in suspension or colloidal solution, derived from the decomposition of falling seston under highly reducing conditions. It is evident that the unpublished Wisconsin data confirm the conclusions on vertical distribution to be derived from Riley's investigation.

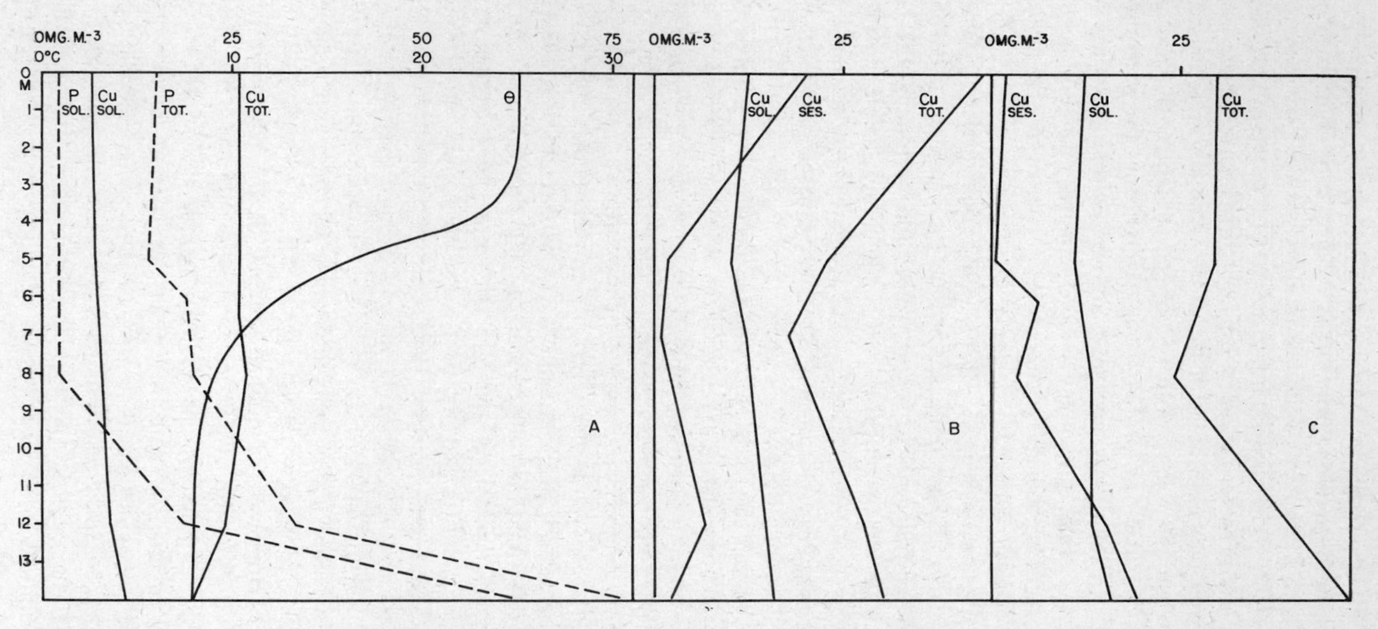

FIGURE 213. Linsley Pond, Connecticut. *A*, vertical distribution of temperature, total and ionic copper (solid lines), and total and ionic phosphorus (broken lines), at height of summer stratification, phase III, showing the great difference in behavior between copper and a substance such as phosphorus that is controlled by the redox cycle at the mud surface; *B* and *C*, total sestonic and ionic copper on two dates fourteen days apart (*B*, September 1, 1935; *C*, September 15, 1935), showing apparent control of the distribution by falling seston. (After Riley.)

The manifest difference between the behavior of copper and of those substances such as manganese, iron, phosphorus, and ammonia, the concentration of which rises with falling redox potential, is easily explained by the virtual immobility of copper, unless in complex form with an amino compound, in the presence of ferrous sulfide. Where a small rise in copper does take place at the bottom of the hypolimnion, it must be explained either by the sedimentation and subsequent decomposition, in the hypolimnetic water, of copper-containing seston, or by diffusion of copper-containing organic compounds, possibly copper sulfide amino complexes, from the mud.

The seasonal cycle of copper in a lake. The most striking feature of the copper cycle as it is described by Riley is the tendency for the content of the metal to increase at all levels during the later part of summer stratification and in the subsequent autumnal circulation period. The intensity of this seasonal variation evidently differs from year to year. It was very well marked in all three lakes in 1935, but was not exhibited by Quassapaug in 1936 and in that year was also less striking in Linsley Pond than in the preceding season. Only in Quonnapaug was a really well-marked maximum observed in the second year of the investigation (Fig. 214). In general, the high autumnal copper contents appear to involve striking maxima in the organic fraction, but in some cases, notably in Linsley Pond in 1935, there were also large amounts of ionic and sestonic copper in the surface waters at the season of the maximum.

Statistical studies by Riley indicate that in a given lake the total copper content is directly correlated with the bicarbonate content and with the rate of fall of the epilimnetic temperature, and inversely correlated with the precipitation of the previous three weeks. The correlation with alkalinity suggests that part of the copper, like the alkaline earth bicarbonate, is derived from the sediments of the basin. The temperature effect is considered by Riley to be due to the uptake of copper by littoral plants as the temperature is rising in the spring, and a liberation of the element into the water as the temperatures fall and the plants decay in the autumn. It would not be unreasonable to suppose that the correlation with falling autumnal temperatures, which is largely responsible for the autumnal copper maximum, is really a correlation with the amount of mud surface that has become oxidized at any given stage of the descent of the thermocline. Copper, largely immobilized as CuS at the surface of the mud under conditions of reduction, may well be able to diffuse into the circulating water as soon as the superficial layer of the sediments has become oxidized. The only difficulty in this argument relates to Lake Quassapaug, where it is doubtful that the mud surface ever became reduced during the years of Riley's investigation. But it is quite possible that

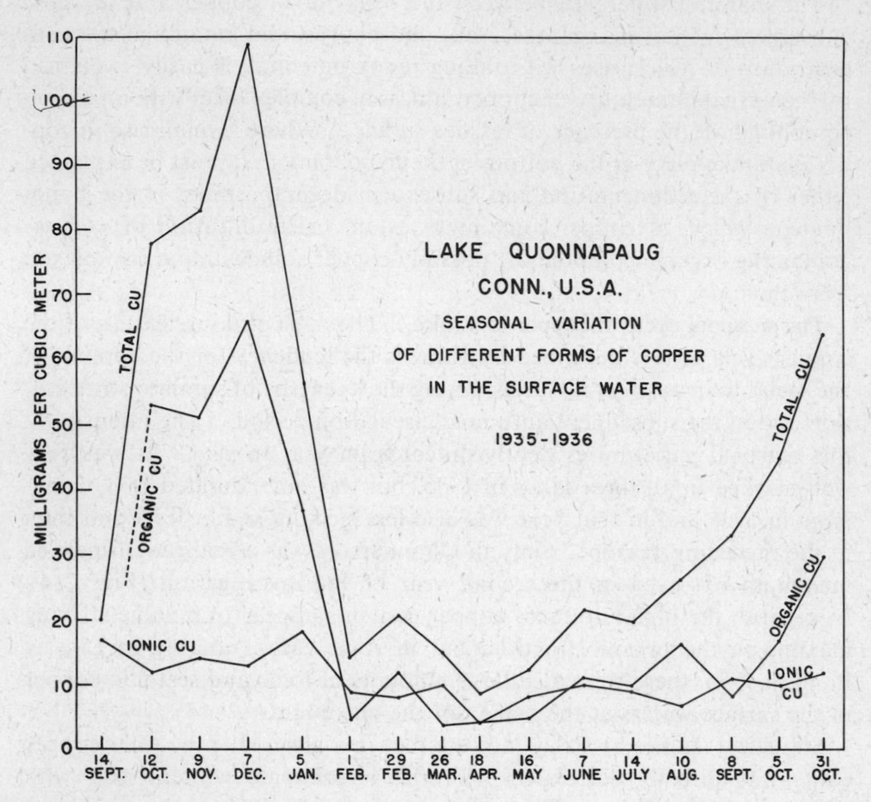

FIGURE 214. Lake Quonnapaug, Connecticut. Seasonal variation in the various fractions of copper in the surface water (after Riley).

enough change in redox potential and sulfide content occurred just below the oxidized microzone to permit enhanced outward diffusion of copper ions during the autumnal circulation period, producing the observed result.

Copper toxicity as a possible ecological factor. The presence of ionic copper in solution in lake waters raises several interesting biological problems. In general, the concentration of ionic copper by weight is somewhat greater than is the concentration of phosphorus as ionic phosphate. On an atomic basis the two elements are clearly present in amounts of the same order of magnitude. During the *Dinobryon* maximum in Linsley Pond in 1936, when the soluble phosphate phosphorus is recorded as 2 mg. m.$^{-3}$, the ionic copper content was 6 mg. m.$^{-3}$ Since Rodhe's (1948) studies indicate that the phosphorus content was little below the maximum concentration tolerated by *Dinobryon divergens*, it is evident that even on a molecular basis copper cannot be much more and is probably somewhat less toxic to this organism than is phosphate.

At certain seasons, as in the autumn of 1935, ionic copper contents in excess of 30 mg. Cu $m.^{-3}$ have been recorded in the hypolimnion of Quonnapaug and at all depths in Linsley Pond. Riley found that concentrations of this magnitude were toxic to *Hydra* spp., *Daphnia* spp., and probably to many other species of small fresh-water animals. It is therefore not inconceivable that on occasions when the soluble copper rises rapidly at the end of summer stratification, this rise may contribute to the disappearance of certain members of the microfauna and microflora from the lake.

The treatment of reservoirs with copper sulfate has long been used as a method for destruction of excessive algal blooms. Hale (1922) has given a considerable quantity of data on the amount of copper needed to poison the various phytoplanktonic algae that may become nuisances in water supplies. From these data it appears probable that certain species of *Coelastrum*, *Navicula*, and *Uroglenopsis* may be sensitive to amounts of copper, of the order of 30 mg. $m.^{-3}$, that can occur in ionic form in the trophogenic zones of certain lakes during autumnal circulation. *Anabaena*, *Aphanizomenon*, *Tabellaria*, and *Synura*, though relatively susceptible, can apparently tolerate about 50 mg. Cu $m.^{-3}$, which is probably more ionic copper than would occur in the water of the trophogenic zone of an untreated and uncontaminated lake. Pillai (1954), working with blue-green algae tolerant of high salinities, found that an amount of copper inhibiting *Phormidium tenue* stimulated *Spirulina* and *Oscillatoria* spp.

Churchill (in Hasler 1949) thought that unicellular algae are usually more susceptible than are filamentous algae. He observed considerable changes in the qualitative composition of cultures of natural plankton when treated with copper, owing to this difference in susceptibility. Churchill also found that certain multicellular plants, such as *Riccia* and *Lemna*, are injured by exposure to concentrations as low as 30 mg. $m.^{-3}$

It is often noted that after an algal population has been greatly reduced by copper sulfate treatment, a secondary maximum in total phytoplankton, of a rather transitory nature, may develop. This phenomenon has been neatly studied by Tressler (1937) in the Muscoot-Croton Reservoir, 60 km. north of New York City. The water is treated at one end of the reservoir as it enters the long narrow basin. Somewhat below the place of treatment, a well-defined maximum in phytoplankton can be observed. As the water moves on down the reservoir, the plankton steadily declines. It is probable that the main factor in producing the secondary maximum is the rapid regeneration of nutrient elements from the dying cells of the more susceptible forms, which thus become available for the nutrition of the more resistant species.

Riley made some experiments in which he compared the copper tolerances of organisms living in the epilimnion and hypolimnion of the Connecticut lakes. The planktonic copepod *Diaptomus spathulocrenatus* was found to be unable to live at a concentration of 50 mg. m.$^{-3}$ ionic Cu, though it survived, as might be expected, at one-fifth this concentration. The copepod *Mesocyclops edax*, which is common in the hypolimnia of the lakes studied by Riley, was able to survive indefinitely at 50 mg. m.$^{-3}$ ionic copper, and almost as well at twice that concentration. Similar differences were observed when the littoral larvae of a species of *Tanypus*, which can survive at 50 mg. m.$^{-3}$ but not at 500 mg. m.$^{-3}$ ionic copper, were compared with the larvae of *Chaoborus punctipennis*, which can tolerate at least 10,000 mg. Cu m.$^{-3}$, though it is perhaps rather less active than is normal at concentrations above 500 mg. m.$^{-3}$ These indications of a greater resistance of hypolimnetic arthropods over that of epilimnetic organisms of the same major taxonomic groups are doubtless merely nonspecific expressions of general differences in cuticular permeability, and also possibly in the ability of the animal to exclude unneeded fluid from the lumen of the alimentary tract. However unspecific such differences may be, they are doubtless of the greatest ecological importance in enabling the animals of the profundal benthos to live not merely under conditions of low oxygen and high CO_2 tension, but also in the presence of a variety of more or less toxic substances in concentrations seldom or never encountered in the well-oxygenated illuminated waters of the epilimnion.

Copper in lake sediments. The addition of copper sulfate to lakes causes a marked increase in the copper content of the sediments. In Lake Mendota, to which copper sulfate has not been added, the sediments contain 85 mg. Cu. kg.$^{-1}$ dry, while in the other lakes of the Yahara Basin, to which copper sulfate has been added, the sediments formed in recent years contain from 223 to 605 mg. Cu. kg.$^{-1}$ dry (Sawyer, Lackey, and Lenz 1945; Nichols, Henkel, and McNall 1946). Hasler (1949), on the basis of unpublished studies, concludes that this quantity of copper in the sediments has reduced the bemthonic fauna of the treated lakes; in Lake Monona, of which the sediments contain the greatest amount of copper, *Pisidium* is entirely absent and the number of bottom organisms was determined by Jones to be but 800 m.$^{-2}$ as against 9000 m.$^{-2}$ in Lake Mendota. Hasler also quotes work by Frey showing that both the molluscan fauna and the higher vegetation have been reduced by the accumulation of the element, which is clearly having a generally deleterious effect on the whole biocoenosis of the lakes in which it is enriched.

In the mud of Lake Mendota, remote from small areas that have been treated with copper sulfate, Nichols, Henkel, and McNall (1946) found from 25 to 68 mg. Cu kg.$^{-1}$ dry sediment. These determinations refer to copper liberated by prolonged digestion with concentrated sulfuric acid, as in the Kjeldahl determination of total nitrogen. It is possible that a small amount of copper present in undecomposed rock in the silt fraction of the sediment would not be set free by this treatment. From 3.3 to 62 per cent of the copper was found to dissolve after one minute of boiling with decinormal hydrochloric acid. The sediments of Lake Wingra contained from 18 to 33 mg. Cu kg.$^{-1}$, of which 11.0 to 39.4 per cent was easily soluble in hot dilute HCl. In the sediment of Allequash Lake in northeastern Wisconsin, 25 mg. Cu kg.$^{-1}$ dry were found.

The lower part of the range of values for total copper is extended somewhat if older strata sampled by a corer are considered. The full range would appear to be from 7 to 68 mg. Cu kg.$^{-1}$ dry. The copper content of igneous rocks varies from a mean figure of 15 to 16 mg. kg.$^{-1}$ for acidic rocks to values perhaps as high as 150 mg. kg.$^{-1}$ for basic rocks, and the majority of sedimentary rocks have very similar copper contents (summary in Rankama and Sahama 1950). The data for lake muds obtained by Nichols, Henkel, and McNall therefore seem inherently reasonable.

There is considerable variation, not only horizontally but also vertically, in core samples. In Mendota a vertical profile gave

Depth, m.	Cu, mg. kg.$^{-1}$
0.31–0.61	25
0.61–0.92	20
0.92–1.22	7
1.22–2.44	26

The meaning of the minimum at about 1 m. below the modern mud surface is by no means clear.

Treatment of a lake with copper sulfate produces an immense increase in the copper of the sediment. In the Madison lakes that have been treated, the amount of copper present is usually well over 100 mg. kg.$^{-1}$ dry sediment, and in the case of one sample from Lake Monona reached the immense value of 1093 mg. kg.$^{-1}$ dry. Usually from about one third to rather over one half of this excess copper is easily soluble in dilute hot HCl, and is supposed by Nichols, Henkel, and McNall to represent a precipitate of basic carbonate. The rest of the copper they regard as being in organic combination.

There is no direct evidence as to the nature of the copper compounds in sediments from untreated lakes. A basic carbonate may occur naturally in the sediments of hard-water lakes as well as under conditions of artificial enrichment. As has already been pointed out, however, there is very little evidence that precipitation of basic carbonate occurs in untreated lakes. The sulfide would of course be included as easily soluble copper by Nichols, Henkel, and McNall, and presumably always occurs in sufficiently reduced mud. The most interesting possibility is the occurrence of native metallic copper as a constituent of lake mud. The free metal can form at redox potentials in excess of those required to produce the sulfide. Lovering (1927) has described a swamp deposit near Cooke, Montana, in the headwaters of Clarke's Fork of the Yellowstone River. The deposit consisted of a black claylike material full of blackened blades of grass and twigs. In it, native copper occurred in loose spongy masses, usually 0.5 to 1.0 cm. in diameter but occasionally up to more than 2.5 cm. across. The metal is probably derived from chalcopyrite, which occurs in small amounts associated with pyrite higher up the valley. Oxidation of chalcopyrite produces ferric hydroxide and a weakly acid solution containing copper sulfate. Lovering found that ordinary bacteria such as *Escherichia coli* could lower the redox potential of a culture medium to which 0.01 to 0.08 per cent Cu had been added, sufficiently to permit the formation of microscopic particles of metallic copper. He considers that the same process occurred in the Cooke swamp deposit, and it seems likely that it must also occur in the less reduced parts of lake sediment profiles.

ZINC

The occurrence of zinc in inland waters has been even less studied than has the occurrence of copper, though from the available knowledge of the presence of the two elements in the ocean it might be supposed that they would be present in amounts of the same order of magnitude in fresh waters as is the case in the sea. The spectrographic studies of Braidech and Emery tend to confirm this supposition; they found from 5 to 300 mg. m.$^{-3}$ of zinc in the series of samples which they examined, the mean content being 62 mg. m.$^{-3}$ Samples from Lake Michigan contained 200 to 300 mg. m.$^{-3}$ The Lake Michigan values, treated as 250 mg. m.$^{-3}$ in obtaining the mean, are the highest encountered by Braidech and Emery, but 200 mg. m.$^{-3}$ occurred in water from a mountain stream at Helena, Montana, and in water from deep artesian wells at Jacksonville, Florida.

Morita (1955) found that the zinc extractable by dithizone from the waters of Japanese mountain lakes varied from 1.3 to 5 mg. m.$^{-3}$, whereas from the lakes and ponds of lowlands 5.6 to 18 mg. m.$^{-3}$ were recorded. The ratio of the extractable copper to extractable zinc varies from 0.11 to 0.38, being usually higher in the unproductive mountain lakes. The very acid water of the volcanic lake Kata-numa contained 79 mg. Zn m.$^{-3}$ It is probable that these figures are too low as estimates of total zinc, but they tend to confirm the hypothesis that the element is usually present in quantities at least as great as those of copper.

Zinc has been recorded in considerable amounts in Bear Lake, Idaho, by Kemmerer, Bovard, and Boorman (1923), who found 650 mg. m.$^{-3}$ in the surface waters. Hutchinson (1932) found that this amount of zinc added as bicarbonate to pond water, is markedly toxic to various species of Cladocera. Though it is possible that the rather restricted fauna of Bear Lake may be explained by its apparent richness in zinc, it seems likely that if the analysis is reliable, much of the zinc is likely to have been present in suspension, as a basic carbonate.

ALUMINUM AND GALLIUM

Braidech and Emery (1935) found from 100 to 1000 mg. m.$^{-3}$ of aluminum in the water supplies that they studied; their mean value was 400 mg. m.$^{-3}$ The untreated sample from Lake Michigan contained 600 mg. m.$^{-3}$ Such figures appear to be consistent with the older and probably not very reliable analyses of lakes and rivers which have been summarized by Clarke (1924). Moreover, Lehmann (1931), using more accurate methods than were available to earlier workers, found 120 mg. m.$^{-3}$ in Würzburg tap water.

Vertical variation in aluminum content of lake water. Several limnological investigations in which aluminum was determined have been made. Mulley and Wittmann (1914) record from 270 to 2900 mg. m.$^{-3}$ in the Lunz lakes. Some indications of stratification are suggested by the following figures for the Lunzer Untersee in Table 116. Thermal stratification can hardly have existed on any of the dates given in the table, and the meaning of the vertical variations is completely obscure.

TABLE 116. *Vertical distribution of aluminum in the Lunzer Untersee*

	Aluminum, mg. m.$^{-3}$		
Depth, m.	May 8, 1909	Feb. 18,1910	April 28, 1910
1	320	...	...
5	...	1330	800
10	370	...	...
25	...	800	270
30	690	...	...

Ohle (1934a) has published without comment several analyses from various levels in the basin of the Schaalsee, Lauenberg, which indicate about 200 mg. Al m.$^{-3}$ at all levels during summer stagnation, with very little evidence of any stratification in the vertical distribution of the element.

A considerable number of analyses of the acid-soluble aluminum present in lake waters from northeastern Wisconsin were made in 1938 by H. C. Stecker. Like the copper analyses made at the same time, they have been made available for the present work by Professor V. W. Meloche and Professor A. D. Hasler. The determinations were done colorimetrically after concentration with HCl to one-tenth the original volume. The material determined presumably includes all ionic aluminum and colloidal and suspended hydrated alumina, but may not include all the alumina of suspended clay and other aluminosilicate material. The range of aluminum contents varied from a trace (or in one case 1 mg. m.$^{-3}$) to 70 mg. m.$^{-3}$ The mean content is 16 mg. m.$^{-3}$ if entries of a trace are regarded as referring, on an average, to 5 mg. m.$^{-3}$, and 13.5 mg. m.$^{-3}$ if such entries are regarded as zero. The modal acid-soluble aluminum content must be lower, however, since just half or 65 out of 128 of the analyses are entered as trace in Stecker's table.

A few vertical series are given; in most of these the aluminum content was so low that the series showed nothing of interest, but in Scaffold Lake the water contained 61 mg. m.$^{-3}$ at all depths from the surface to 7 m., a depth well in the hypolimnion and at a temperature of 6.2° C. on July 29. At the bottom of the lake there was a marked fall, so that the 9-m. sample contained only 12 mg. m.$^{-3}$ An even more marked fall in the soluble copper content occurred in the same stratum; since this part of the lake is anaerobic and rich in ferrous iron, it is reasonable to suppose that the copper has been largely removed as sulfide, but in view of the instability of Al_2S_3 in the presence of water, this explanation would hardly apply in the case of aluminum.

It is to be noted that all of Stecker's values are lower than those obtained by other workers. It is obviously not yet possible to know if this is because Stecker did not determine the whole of the aluminum present, or whether it is due to technical errors or to a paucity of aluminum in the waters of Wisconsin.

Tanaka (1953b) has recently studied the vertical distribution of aluminum in several Japanese lakes. The quantity in the epilimnetic waters varied from 220 to 1200 mg. m.$^{-3}$, in line with most previous investigations. In the lakes studied, in all of which the oxygen curves were clinograde and both iron and manganese were strongly enriched in the hypolimnion, the aluminum content was found to undergo an increase up to tenfold in the bottom few centimeters over the hypolimnetic mud. In passing from the epilimnion to the bottom, a maximum rate of increase with depth is first found for manganese, then for iron, and lastly for aluminum. Tanaka thinks that aluminum is liberated from alumino-silicate mineral particles sinking through the hypolimnion or incorporated in the mud, as the ferric iron usually found in such particles is reduced to ferrous. The aluminum freed is supposed to be stabilized in solution by forming complexes with organic matter in the mud.

Aluminum content in relation to *p*H. The solubility of aluminum varies greatly with the hydrogen ion concentration and is minimal at about neutrality. Below *p*H 4.5 appreciable quantities of aluminum go into solution, apparently mostly as oxyaluminum $Al_2O_3H^{++}$ and as alumino-hydroxyl $Al(OH)^{++}$ ions. Above *p*H 8 appreciable amounts of the aluminate ion $HAl_2O_4^-$ will be present. The data obtained by Stecker show no tendency for the lake waters with a *p*H below 6.5 or above 7.5 to contain more aluminum than do the more frequently occurring waters between these *p*H values. There is also no systematic change in distribution with bicarbonate content. Unfortunately, the number of extremely acid lakes is very small in this series, but it is quite clear from the data for lakes whose waters have a *p*H value of 8.0 or higher that there is no greater tendency for aluminum to accumulate in alkaline than in neutral lakes. In the case of lakes in closed basins, however, there is some evidence of the accumulation of aluminate.

In a series of analyses of waters brought back from the high-altitude lakes of Indian Tibet published by Hutchinson (1937b), data are given on *p*H, total solids and aluminum, as set out in Table 117. It will be ob-

TABLE 117. *Aluminum content, p*H, *and total solids of lake waters from Indian Tibet*

	At Depth, m.	*p*H	Total Solids, mg. per liter	Aluminum, mg. m.$^{-3}$
Open lakes				
Ororotse Tso	1.25 and 6.0*	7.1	78	700
Yaye Tso	0	8.7	138	600
Closed lakes				
Tso Moriri	0 and 50*	9.0	1,368	300
Pang-gong Tso	0	9.35	12,872	600
Mitpal Tso	0	9.1	1,528	6,300
Khyagar Tso	0	9.5	5,294	4,800
Pangur Tso	9	9.6 +	6,736	22,000
Tso Kar	0	8.9	79,266	5,200

* Mean of composite sample.

served that the open lakes and the first two closed lakes, which are very large, 50 m. or more deep, with clear blue or blue-green water, contain an amount of aluminum which, though perhaps rather high, is not outside the range already given. The four last closed lakes, which are not only salt and alkaline but relatively shallow and with green water, contain very considerable amounts of aluminum. If, as seems likely, this is the result of the solution of aluminum from the mud in the alkaline water, it is evident that there must also be processes precipitating aluminum; otherwise, it would not be only the small lakes, with a high ratio of bottom area to water volume, that would contain an excessive amount of the element. An even more extreme case of a shallow closed alkaline lake containing large quantities of aluminum is provided by Goodenough Lake in British Columbia (Wait 1900), in which the water contains no less than 97,000 mg. m.$^{-3}$ of aluminum.

Gallium in lake waters. Gallium, which normally accompanies aluminum in its geochemical migration, has been detected spectrographically in the surface waters of Linsley Pond by Hutchinson (1944a), who concluded that between 0.1 and 1.0 mg. m.$^{-3}$ were probably present.

MOLYBDENUM

The presence and distribution of molybdenum in lake waters is likely to prove of considerable interest, in view of the fact that the element apparently plays a part both in nitrate assimilation and in the fixation of molecular nitrogen. The amount of molybdenum normally present in inland waters is evidently extremely small. Braidech and Emery reported but traces in seven out of twenty-four samples, and no molybdenum in the others. Ter Meulen (1932) found 0.9 mg. m.$^{-3}$ in canal water from near Delft, Holland.

It appears that the nitrogen-fixing Nostocaceae, like the nitrogen-fixing bacteria, require molybdenum, though it can be replaced rather inefficiently by vanadium or by tungsten (Bortels 1940). Ter Meulen found that *Azolla* in which a species of symbiotic *Anabaena* lives, concentrated molybdenum from the Delft canal water, the content of the dry *Azolla* containing symbiotic *Anabaena azollae* being 1.1 mg. kg.$^{-1}$ This probably corresponds to about 100 mg. per ton of the living plant, and so to a concentration factor of about 100.

In view of the indications that heavy *Anabaena* blooms may add significant amounts of combined nitrogen to lakes in which they develop, and in view also of the very low concentration in which molybdenum apparently occurs in inland waters, it is not unlikely that detailed study of the cycle of the element in lakes will provide information of considerable interest.

NICKEL AND COBALT

Braidech and Emery (1935) detected nickel in sixteen out of the twenty-four samples from water supplies which they examined. The greatest amount was in a specimen from Indianapolis, containing 300 mg. m.$^{-3}$

This sample did not contain detectable cobalt and is presumably contaminated. The normal range would appear to be from 0 to 10 mg. Ni $m.^{-3}$, with a mean of 5 mg. Lake Michigan water contained 2 mg. Ni $m.^{-3}$ Only three of the waters that they studied contained any trace of cobalt, and one of these was contaminated by mine water; no cobalt was detected in the Lake Michigan sample. Turekian and Kleinkopf (1956) who did not determine cobalt, found a mean content of 20 mg. Ni $m.^{-3}$ in the lakes and streams of Maine.

Maliuga (1946), using a polarographic method, found 1.2 to 19.0 mg. Ni $m.^{-3}$ and up to 6.7 mg. Co $m.^{-3}$ in various fresh waters. The saline waters of the Sea of Aral and Lake Karakul contained markedly less cobalt. Of the three fresh-water lakes studied, the water of Lake Baikal contained 5 mg. Ni $m.^{-3}$ and 2.3 mg. Co $m.^{-3}$; samples from two smaller lakes near Moscow contained 13 to 19 mg. Ni $m.^{-3}$ and 5.7 to 6.6 mg. Co $m.^{-3}$ The means of the three lake waters correspond to a Co:Ni ratio of 1:2.5.

Benoit (1956) determined the cobalt content of various Connecticut waters by spectrophotometric determination of the cobalt nitroso-R salt. His figures are in general two orders of magnitude less than those of Maliuga. Under conditions of extreme thermal stratification in August, the epilimnion of Linsley Pond contained 0.02 to 0.04 mg. Co. $m.^{-3}$, all apparently sestonic; the hypolimnion, 0.05 to 0.105 mg. Co. $m.^{-3}$, of which 0.02 to 0.04 mg. $m.^{-3}$ appeared to be in solution. In spite of some irregularities it is reasonable to suppose that cobalt is behaving more or less like iron. The distribution of soluble copper determined at the same time shows no unequivocal hypolimnetic enrichment (Fig. 215). Presumably the soluble copper at all levels is derived from copper organic complexes, mostly in the seston, whereas cobalt, like ferrous iron, can also be added to the hypolimnion from the reduced mud. This difference in behavior is in accord with the mobility of Co and Cu in the presence of ferrous sulfide, as has already been indicated. There is no clear explanation of the discrepancy between Benoit's and Maliuga's figures.

Benoit made some determinations of vitamin B_{12} in Linsley Pond water collected two months later than that analyzed for cobalt but presumably containing about the same amount of the element. The vitamin B_{12} content varied from 60 to 150 γ $m.^{-3}$, the corresponding cobalt content being 0.0022 to 0.0056 mg. $m.^{-3}$ It is evident that only a small part of the cobalt is likely to be present as a part of the molecule of the vitamin.

Though the cobalt and vitamin B_{12} or cyanocobalamin requirements of higher plants and invertebrate animals are still inadequately understood, it is certain that a number of microorganisms require the vitamin in their environment and that others can synthesize it if provided with cobalt.

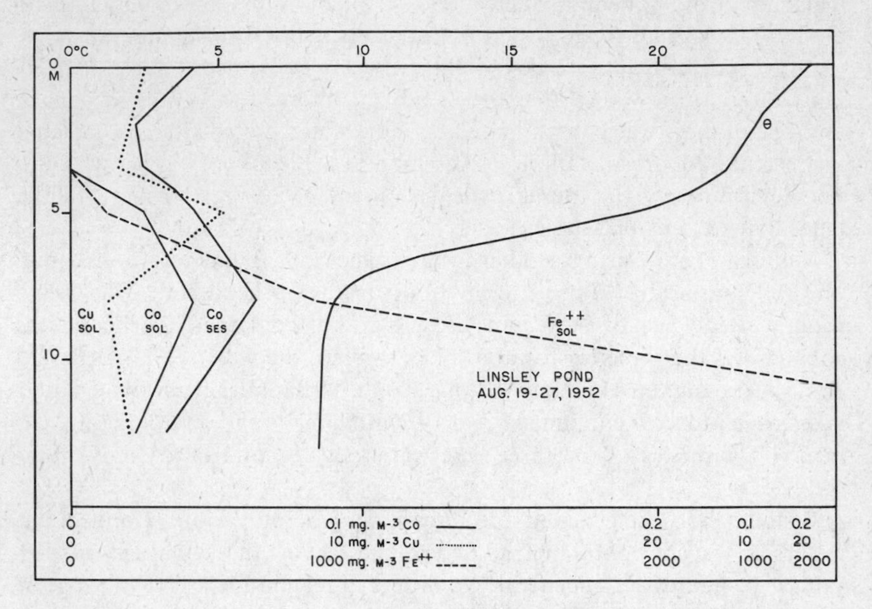

FIGURE 215. Linsley Pond, Connecticut. Vertical distribution of temperature, total and soluble cobalt, ferrous iron, and soluble ionic copper during summer stagnation. (After Benoit.)

Among the phytoplanktonic algae, experiments with the chelating agent ethylene-diamine-tetracetic acid (EDTA) have demonstrated that cobalt is required by *Synura* (Provasoli and Pinnter 1953) and by various blue-green algae, including *Diplocystis aeruginosa* (Holm-Hansen, Gerloff, and Skoog 1954). Owing to the very small amounts of cobalt required, it has usually been easier to demonstrate a need for the vitamin as such than to prove the essentiality of cobalt for species that can presumably synthesize B_{12}. The Euglenineae, notably *Euglena gracilis*, require the vitamin (Hutner, Provasoli, Stokstad, Hoffman, Belt, Franklin, and Jukes 1949; Hutner, Provasoli, Schatz, and Haskins 1950; Robbins, Hervey, and Stebbins 1950, 1951), as do at least some strains of *Ochromonas*, *Poteriochromonas*, and of the marine diatom *Amphora perpusilla* (Hutner and Provasoli 1951, 1953; Hine and Dawbarn 1954). Even more important ecologically is Droop's (1955) finding that the very important marine plankton diatom *Skeletonema costata* requires cyanocobalamin or some of the less well understood related compounds. Among the green alga which appear to require B_{12} are a strain of *Chlamydomonas chlamydogama* and a marine species of *Stichococcus* (Lewin, 1954). The subject is at present under active study in a number of laboratories.

The quantity of cobalt in the B_{12} required for optimal growth of *Euglena gracilis* is 0.6 γ m.$^{-3}$, the analytical detection of which would at present be quite impractical. Even if Benoit's rather than Maliuga's figures are taken as indicating the general range of the cobalt concentration of inland waters, it is obvious that cobalt deficiency is very unlikely. It is possible, however, that some waters might be B_{12} deficient from the standpoint of the Euglenineae or of *Ochromonas*, and that spacial and temporal variation in the concentration of this substance will prove of considerable ecological importance.

Robbins, Hervey, and Stebbins (1950) found in a small pond in which *Euglena* occurs regularly, a quantity of vitamin B_{12} as great as 2 mg. m.$^{-3}$ in late November, but in the spring and summer the amount varied irregularly between 0.1 and 0.6 mg. m.$^{-3}$ These figures are in general higher than those given for Linsley Pond water by Benoit. It is quite likely that the algal associations of small ponds with a great ratio of mud surface to water volume are determined in part by the vitamin B_{12} that is evidently produced by a number of microorganisms in the mud.

OTHER MINOR METALLIC CONSTITUENTS

Chromium, lead, and tin have been recorded in most, and silver in all, of the city water supplies studied by Braidech and Emery (1935). Some of the waters had undergone filtration and other kinds of treatment, and it seems likely that they varied in the extent to which they had been exposed to metal pipes. As has been pointed out, however, the data given by these authors for the copper contents of their waters are inherently reasonable when considered in relation to the limited limnological studies of the element, and their records of zinc and aluminum are concordant with the known distribution of these elements in the hydrosphere. It would therefore seem probable that the data for the other metals that they determined can be used to gain some idea of the distribution of these elements in the lake waters. To the limnologist, their Chicago sample, consisting of untreated water from Lake Michigan, is of course the most interesting of Braidech and Emery's samples, but it has seemed desirable also to present the range and mean values throughout the whole series for the four metals under discussion (Table 118).

Turekian and Kleinkopf (1956) give for the lakes and streams of Maine mean values of 260 mg. Pb m.$^{-3}$ and 20 mg. Cr m.$^{-3}$, so that the values of Table 118 do not seem excessive.

The high regular occurrence of silver in all Braidech and Emery's samples is most curious. It is unlikely, in view of the diversity of origin of the samples, that the silver constantly present in them is due to contamination by industrial wastes or by the destructive cleansing of domestic

TABLE 118. *Minor metallic elements in North American waters*

Metal	Range, mg. m.$^{-3}$	Mean, mg. m^{-3}.	Lake Michigan, mg. m.$^{-3}$	No. Samples Showing None
Chromium	0–40	5	2	2 of 24
Lead	1–180	26	2	0* of 24
Silver	10–200	28	20	0 of 24
Tin	0–40 (100)†	13 (17†)	40	2 of 24

* In the text, lead is said to be present in most samples, but in the table it is recorded in all.
† Including a single very high value which possibly implies contamination.

silverware. The cycle of silver in the biosphere might be supposed to resemble that of copper, being largely determined by the insolubility of the sulfide and the ease with which the element is likely to be adsorbed on organic matter. Unfortunately, practically nothing is known about this. Relative to the copper content, Braidech and Emery's mean value represents a great enrichment of silver, for the ratio of Cu:Ag derived from the means of their determination is 3.2:1, whereas for the accessible lithosphere the ratio is about 1000:1 (Goldschmidt 1938). Braidech and Emery comment on this high silver content, pointing out that some of their samples contained more silver than copper, but no explanation of the phenomenon was suggested. It seems not unlikely that the results are due to technical errors, but it is obviously a matter that should be reinvestigated. The extraordinary nature of the recorded silver contents will be realized when it is borne in mind that they imply a quantity of silver in many natural waters of the same order as the quantity of phosphorus. The data given for tin raise a similar problem, but they are somewhat less regular than are the silver determinations.

Apart from these four metals and those elements that have been discussed here at greater length, Braidech and Emery record antimony in five of their twenty-four samples. One from Bismarck, North Dakota, derived from the upper part of the Missouri River, contained enough of the element for a quantitative estimate of 40 mg. Sb m.$^{-3}$ This is very surprising, as the element is hardly known in the biosphere. The Lake Michigan sample is not one of those containing the element in traces. No less than 20 mg. m.$^{-3}$ of vanadium is reported from St. Louis, Missouri, but no other sample contained the element. It has, however, been recorded qualitatively in water from Brookline, Massachusetts (Hayes 1875), and more recently Sugawara, Tanaka, and Naitô 1953), in a preliminary paper on technique, note 0.2 to 0.4 mg. V m.$^{-3}$ in the surface waters of Japanese lakes. Lake Michigan contained 70 mg. m.$^{-3}$ titanium, five samples from other localities contained 20 mg. Ti m.$^{-3}$, and six more showed traces of the element (Braidech and Emery 1935). Elsewhere, 200 mg. $m.^{-3}$ are recorded as a mean value for Maine (Turekian and Kleinkopf 1956); 50 mg. Ti. $m.^{-3}$ are reported (Hutchinson 1941) in hypolimnetic water from Linsley Pond. Such records may all refer to sestonic titanium.[2]

RADIUM AND OTHER STRONGLY RADIOACTIVE ELEMENTS

The radioactive elements which have been studied in natural waters belong to the radium and thorium series; practically nothing is known about the geochemical migration of the actinium series, which accordingly need not be discussed. The radioactive disintegrations of the radium and thorium series may be conveniently presented as in Fig. 216.

The further disintegrations beyond the gases of atomic number 86, are of no significance in the present work. Starting with any of the elements of very long half life, the daughter elements will accumulate until the rates at which they decay just balance the rates at which they are produced. At such radioactive equilibrium, the ratio of uranium to radium is $3 \times 10^6:1$. In the course of the history of an atom of uranium undergoing ordinary radioactive disintegration, it will twice pass through the stage of atomic number 90, which is equivalent to saying that it will twice be an isotope of thorium. Moreover, one of the thorium isotopes involved, ionium, has a quite long half life of 6.9×10^4 years. Similarly, in the history of its disintegration a thorium atom will twice go through the stage of atomic number 88 or become an isotope of radium; one of these isotopes, mesothorium 1, has a half life of 6.7 years, which is long enough for it to undergo some geochemical migration with radium, independently of the parent element thorium. These isotopic relationships are of great importance in the geochemical cycles of the radioactive elements in the biosphere.

Before proceeding to the very inadequate data relating to inland waters, it is desirable to examine the far better known history of the migration of the elements of the radium series in the sea. One of the few investigations of the behavior of these elements in an inland water is fully comprehensible only in the light of the results of such oceanographic investigations.

The uranium content of sea water varies from about 1.1 to 1.8 mg. $m.^{-3}$ The best mean value would appear to be 1.3 mg. $m.^{-3}$, or 1.3×10^{-9} g. per g. (Föyn, Karlik, Petterson, and Rona, 1939). Such a quantity would

---

[2] Just as this book goes to press, Dr. Löffler informs the author of the existence of minute traces of gold in the water of Lake Titicaca. The whole subject of minor metallic constituents is under intense investigation, and the present chapter is certain to be out of date before it appears.

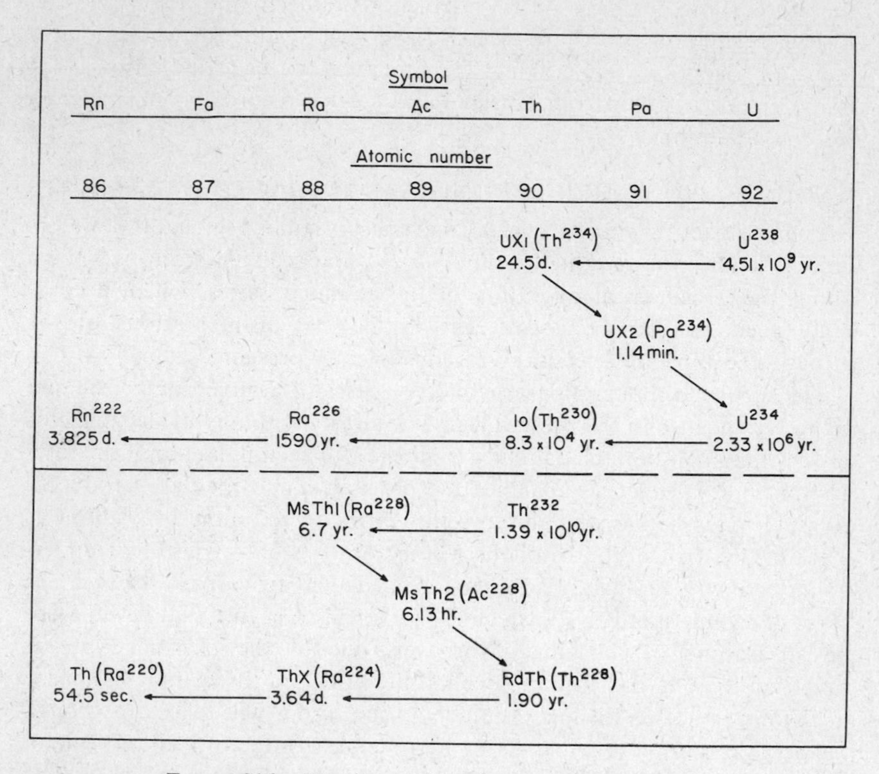

FIGURE 216. Uranium and thorium disintegration series.

be in equilibrium with 4.3×10^{-16} g. Ra per g. Actually, the best determinations of the radium content of sea water give 0.8 to 0.9×10^{-16} g. per g., corresponding to about 0.25×10^{-9} g. U per g. It is therefore evident that radium is being continually lost from the sea. In superficial marine sediments these relationships are reversed, for such sediments contain so-called unsupported radium, or radium in excess of the amount required for equilibrium. Wherever sedimentation has proceeded in deep water at a moderate velocity, there is a well-defined maximum in radium content below the surface of the sediment. This implies that radium is lost to the sediments from the water, not as radium itself but as some precursor of moderate or long half life. Examination of the diagram of the disintegrations of the radium series (Fig. 216) indicates that ionium is the only possible precursor. This conclusion is in accord with the properties of ionium, for it is an isotope of thorium and has chemical properties practically identical to those of the latter element. It is known that thorium can be removed very completely from aqueous solution by coprecipitation on ferric hydroxide, and it is reasonably certain that this is continually happening in the sea. In its sedimentation thorium would take with it such ionium as is formed from the uranium dissolved in the ocean. The position of the maximum of unsupported radium, and more generally the form of the rising and falling curve of radium concentration, in marine sedimentary cores has been made the basis of a chronometric technique by Piggot and Urry (1942a,b, Urry and Piggot 1942).

Uranium in inland waters and in aquatic organisms. Hoffmann (1942) has published a number of determinations of the uranium content of various Austrian spring, well, and surface waters. None of them are from lakes, but several river waters are included. The range, from 0.16 to 47 mg. $m.^{-3}$, probably indicates the quantities to be expected in lake waters, and shows that the quantity of uranium in fresh waters is not much less than the quantities of most of the commoner heavy metals. The highest value recorded by Hoffmann is from the Danube at Vienna. Vienna tap-water gave a very low value of about 0.17 mg. $m.^{-3}$ This quantity would be in equilibrium with 0.57×10^{-16} g. Ra per g., which is practically identical with the value given for the radium content of Vienna tap water by Pertz (1937). It is conceivable that this identity is a mere coincidence, the samples having been taken several years apart and not necessarily drawn from the same source. It would appear, however, to be the only empirical information bearing on the question as to whether equilibrium is ever established in fresh waters. More recently, Judson and Osmond (1955) have recorded 0.13 to 3.5 mg. U $m.^{-3}$ in surface waters from Wisconsin, Illinois, and Texas, and much higher values from ground waters in uraniferous localities; Lake Mendota water contained 0.4 mg. U $m.^{-3}$ All surface waters are more radioactive than their uranium content would imply. At least part of the additional radioactivity is due to elements of relatively short half life, possibly ThX (Ra^{224}).

Hoffmann found 9.1×10^{-4} per cent uranium in the ash of green algae from the River March at Angern. The water in which the algae were growing contained 4 mg. U $m.^{-3}$ Assuming that the ash represents about 5 per cent of the dry weight, and the dry weight about 10 per cent of the wet weight, the uranium content of the living algae would be about 45 mg. per ton and the enrichment factor therefore of the order of 10.

Radium in inland waters and in aquatic organisms. A good many analyses of the radium content of inland waters have been published. Many of these relate to mineral springs and other localities of little limnological interest, and most of the earlier determinations are probably quite unreliable. Two analyses of direct interest in the present work may be mentioned: Pertz (1937) found 0.6×10^{-16} g. per g. in Vienna tap water, and Brunowsky and Kunasheva (1935) found from 1.6 to 5.1×10^{-16} g. per g. in the water of the pond at the Peterhof Orangery near Leningrad.

The first of these analyses, which evidently refers to water derived from rocks from which little uranium dissolves, may well be at the lower end of the expected range. In view of the transitory nature of the majority of inland waters, it is unlikely that equilibrium between uranium and radium is often established in them, though in this one case, in which analyses of both elements have been made from the same source, radioactive equilibrium is apparently indicated. There is a good deal of evidence that organisms concentrate radium considerably. Marine diatoms contain about 2×10^{-12} g. Ra per g. dry, which probably corresponds to about 10^{-13} g. per g. wet, implying a concentration factor of about 1000. Wiesner (1938), analyzing the larger marine algae, obtained a mean value of 0.92×10^{-13} g. per g. dry, which would correspond to a concentration factor of the order of 100.

No data appear to be available for the fresh-water phytoplankton, but in the filamentous algae Wiesner found from 1.4×10^{-13} g. per g. dry in *Cladophora alpina* to 173×10^{-13} g. per g. dry in *Zygnema* spp. from the Lunzer See, the mean being 59.8×10^{-13}. The mean radium in the living algae may well be 6.10^{-13} g. per g., and if, as Wiesner supposes, the lake water contains about as much radium as Vienna tap water, which drains from a geologically similar area, the mean concentration factor would be of the order of 10,000, and in the case of the *Zygnema* may well be several fold more. As might be expected from the known geochemical distribution of uranium, *Spirogyra* growing in waters draining off granite contained more radium than did specimens growing in waters draining off other rocks.

The differences noted by Brunowsky and Kunasheva (1935) in the radium content of the Peterhof Orangery pond apparently indicate a seasonal biological cycle, radium being removed from the water by *Lemna* as it developed during the summer and liberated into the water when the plant decayed in the autumn.

Vernadsky, Brunowsky, and Kunasheva (1937), working with a mixture of *Lemna minor* and *Lemna polyrhiza*, found that the living plant contained 10.3×10^{-17} g. mesothorium 1 per g., which represents about one hundredfold concentration over the amount found in the medium, which was 8.25×10^{-19} g. per g. Such a concentration is to be expected in view of the fact that MsTh1 is an isotope of radium, which also is concentrated by *Lemna* about one hundred fold; the two isotopes would be treated identically by the plant. Vernadsky, Brunowsky, and Kunasheva further concluded that their *Lemna* did not contain more than about 1 per cent of the thorium required for equilibrium with the MsTh1 content; thorium, unlike the other metallic elements, therefore appears not to be concentrated at all by the plant.

Possibility of age determinations of lake sediments by radiochemical methods. It is clear from the preceding paragraphs that much work remains to be done on the relationships of the various members of both the radium and thorium families in inland waters. That such studies may yield remarkable results is indicated by Urry's work on the varved clays deposited in the lake that existed in late Glacial times in the vicinity of Hartford, Connecticut.

Urry (1949) studied only the radium content of the varves. He found that while the mean radium content of the sediment remained practically constant at 1.36×10^{-12} g. cm.$^{-3}$ throughout a period of 700 years, the radium content of the summer and winter components differed considerably and showed systematic and reciprocal variation with time. The radium content of the winter clay tends to fall with time over a range of 1.7 to 1.5×10^{-12} g. per g. dry clay, while that of the summer clay tends to rise from 0.96 to 1.04×10^{-12} g. per g. dry clay. Urry thinks that the most reasonable explanation of the difference is that at least part of the radioactive material was sedimented in nonequilibrium concentrations, though minute particles of old unaltered minerals containing equilibrium quantities of the radioactive elements may also be included in the sediments.

To produce the observed effect, the ratio of ionium to uranium must be in excess of the equilibrium quantity in the summer varves, so that those laid down late in the sequence now contain more unsupported radium than do those that were laid down early. In the winter varves, the ionium to uranium ratio would have to be less than that of equilibrium, so that those laid down late have had less time to produce radium than those laid down early. The observed data suggest that about three-quarters of the uranium goes into the winter varves, accompanied by little over half the ionium. Since the ionium is certainly far less soluble than the uranium, the greater seasonal variation of the latter element is reasonable. The main difference between the winter and summer layers is the much greater amount of colloidal matter in the former, and it may be supposed that the additional uranium sedimented in the winter is adsorbed to the colloidal material.

Assuming that this explanation is correct, it is possible to estimate the age of the varves from the systematic variations in the radium contents of the summer and winter series. The two series give concordant results, which would indicate that the receding ice front passed Hartford about 19,500 years ago. Since radiocarbon dating indicates that the lake discharged ten or eleven thousand years ago (Preston, Person, and Deevey 1955), this estimate appears to be excessive.

SUMMARY

Manganese is present in detectable quantities in nearly all surface waters. The quantity present is usually less than that of iron, but in most cases the ration Fe:Mn is well below the mean geochemical ratio, which is 50:1. Nothing is known of the forms of manganese in surface waters, save that sestonic manganese may be present and that at least part is soluble at *p*H 4.6 to 4.8. The concentrations at the surface undergo irregular seasonal changes.

There is often a considerable enrichment of the element, presumably as Mn^{++}, in the hypolimnia of small productive lakes. This enrichment begins at a higher level and a higher redox potential than in the case of iron. There is often a maximum in the manganese curve in the metalimnion or upper part of the hypolimnion. This may be due to re-solution of manganese precipitated when oxygen is mixed into the previously reduced layers just above the depth of the maximum, but not all cases can be explained in this way. The quantities required by phytoplankton appear to vary from species to species, and it is not impossible that the element plays some part in regulating phytoplankton succession.

Copper has almost always been detected wherever it has been sought. The quantities present range from an undetectable amount to several hundred milligrams per cubic meter. The element is apparently present in ionic form, in organic or colloidal form, and in suspension, the organic fraction being usually the most important. The suspended copper represents a fraction of the seston far greater than the proportion to be expected in living cells, so that the detritus of the seston is probably enriched in copper. Little enrichment of copper occurs in the hypolimnetic water; this is to be expected, since the element is insoluble in the presence of ferrous sulfide. Very considerable increases in copper content have been recorded at autumnal circulation, perhaps due to oxidation of sulfide in the mud as oxygen reaches the deeper layers. It is just possible that quantities of copper toxic to some fresh-water organisms may be present occasionally at the time of such maxima. Copper added as sulfate to lake waters, as an algicide, is precipitated, supposedly in part as basic carbonate. In reduced lake muds the copper is likely to be present as sulfide, though under some circumstances the occurrence of free metallic copper may be possible.

Zinc has been very little studied; it is probably present in concentrations as great or greater than that of copper.

Aluminum is evidently normally present. The forms in lake waters are not known, though the very large amounts found in some shallow, closed alkaline lakes presumably represent aluminate ion. The element may be unchanged, depauperated, or enriched in the hypolimnion. The most careful work suggests enrichment just over the bottom, possibly as aluminum organic complexes.

Gallium has been once detected; it is likely to be universally present with aluminum, in quantities of the order of 0.1 to 1.0 mg. $m.^{-3}$

Molybdenum has been once detected in fresh water, just under 1 mg. $m.^{-3}$ being present. The cycle of this element in relation to biological nitrogen fixation may prove important.

Nickel and cobalt have both been detected. The quantity of the former element is not certainly known; the best data for cobalt suggests about 0.02 to 0.04 mg. $m.^{-3}$ Cobalt is important biochemically as a constituent of vitamin B_{12}, but only a very small amount of the cobalt in lake water is present as the vitamin. Cobalt appears to be somewhat enriched in the hypolimnion along with iron, in accordance with theoretical speculation.

Chromium, lead, silver, and tin have been recorded in Lake Michigan water and in various other city water supplies. The first two elements occur in lake waters in Maine, but all require further investigation. Titanium is also known from several waters, but is almost certainly entirely in suspension. Small amounts of vanadium, 0.2 to 0.4 mg. $m.^{-3}$, have been found in some Japanese lakes.

Uranium has been detected in all surface waters in which it has been sought; the amount to be expected in regions not specially uraniferous is evidently 0.1 to 1.0 mg. $m.^{-3}$ The radioactivity of such waters appears always to be greater than that due to uranium alone. The radium (Ra^{226}) content may be expected to be of the order of 10^{-7} mg. $m.^{-3}$; it is possible that MsTh1 (Ra^{228}) and ThX (Ra^{224}), in the thorium disintegration series, also contribute significantly to the radioactivity of lake waters. Thorium itself and its isotope ionium are probably never present except in suspension. Variation in the adsorption of uranium on clay particles may give seasonally different ratios of uranium to ionium, and as the latter decays, variations in the U:Ra ratio in sediments. This effect has been observed in the varved clays deposited in an ice lake of Wisconsin age near Hartford, Connecticut.

CHAPTER 16

The Nitrogen Cycle in Lake Waters

The ultimate source of the nitrogen, which plays a fundamental part in the metabolism of organisms, is certainly the molecular nitrogen of the atmosphere. A very little juvenile combined nitrogen may be continually added to the biosphere from volcanoes. Practically all the nitrogen compounds of the biosphere are, however, clearly the result of fixation of atmospheric nitrogen. This fixation may occur electrically or photochemically in the atmosphere, but there seems to be little doubt that bacterial fixation in soils is of far greater quantitative significance.

Molecular nitrogen is ordinarily and rightly regarded as an extremely stable substance, but in some ways this insistence is misleading. Though the formation of NO, NO_2, or N_2O_4 from O_2 and N_2 involves a gain in free energy, the production of nitric acid from its elements involves a loss of free energy under standard conditions. Moreover, if water and the two main gases of the atmosphere could react to reach the equilibrium indicated thermodynamically, a solution of 0.1 molar nitric acid would result. On a geochemical scale, the whole of the oxygen of the atmosphere would be exhausted before the equilibrium concentration was reached. Lewis and Randall (1923), in discussing the thermodynamics of nitrogen compounds, express a hope that nature will not discover a catalyst for this reaction between N_2, O_2, and H_2O. Actually, the system consisting of nitrogen-fixing and nitrifying organisms does effectively constitute the required activator, and the entire biosphere would accumulate nitrate, if it were not for nitrate assimilation and denitrification.

Biochemical changes in the concentration of molecular nitrogen thus involve nitrogen fixation, assimilation, and denitrification. Nitrogen, either newly fixed or assimilated as nitrate or ammonia, is incorporated into proteins or other compounds in organisms. On death or after excretion, a variety of compounds may be liberated; of these, ammonia is certainly the most important. Most heterotrophic bacteria take part in the process of ammonification of organic compounds, a process which is largely the result of the deamination which occurs when proteins or the products of their hydrolysis are used as energy sources. In the presence of oxygen and under conditions which seem to be satisfied in most lakes, a considerable part of the ammonia so formed is nitrified, apparently in two stages, nitrite being first formed and then nitrate. The production of ammonia, nitrite, and nitrate, in this regular order, from dead diatoms suspended in sea water in the dark, has been elaborately studied by Von Brand and Rakestraw (1940, 1941; Von Brand, Rakestraw, and Renn 1937, 1939; Von Brand, Rakestraw, and Zabor 1942). In acid waters, the purely chemical reaction between ammonia and nitrous acid produced during the decomposition might liberate molecular nitrogen; this process is believed to occur in some acid soils and possibly also in the interior of plant cells.

The oxidation of ammonia to nitrite and of nitrite to nitrate is accompanied by a fall in free energy, and is therefore a possible energy source utilized by nitrifying bacteria. Ammonia, nitrite, and nitrate are all possible nitrogen sources for most green plants, and the accumulation of nitrate in the biosphere is prevented partly by the uptake of nitrate by such organisms. Moreover, a very large number of bacteria can use nitrate or nitrite as hydrogen acceptors in the place of oxygen, so that if enough organic matter is present, nitrate reduction also is important in the removal of nitrate from the biosphere. This process of reduction may stop with the production of nitrite, which may be accompanied by hydroxylamine, or may proceed to molecular nitrogen or perhaps to N_2O or to ammonia. Nitrate reduction proceeding to N_2 is generally spoken of as *denitrification*. It is a process which locally appears wasteful, but one which evidently maintains the atmosphere in the condition in which it is known.

The forms of nitrogen present in lake waters may be roughly, though conveniently, grouped as:

(1) Molecular nitrogen N_2 in solution.

(2) Organic nitrogen compounds, including a great variety of decomposition products ranging from proteins to such simple substances as amino acids, urea, and the methylamines.

(3) Ammonia, mainly as NH_4^+ and NH_4OH.

(4) Nitrite, mainly as NO_2^-, though in acid waters some HNO_2 will be present; the single recorded case of hydroxylamine NH_2OH is conveniently considered along with nitrite.

(5) Nitrate, mainly as NO_3.

In considering the nitrogen cycle in lakes, it is convenient to concentrate on the inorganic compounds, reserving a detailed discussion of what is known about the organic compounds to Chapter 17, in which all the organic compounds of lake waters are considered.

OCCURRENCE, FIXATION, AND FORMATION OF ELEMENTARY NITROGEN

Molecular nitrogen in lake waters. The available data were mainly collected at a time when only gasometric methods were available for the determination of dissolved oxygen, and many of the reported determinations are probably not very accurate. The first analysis was that of Hoppe-Seyler (1896), who used the nitrogen content to demonstrate that oxygen had been lost from deepwater samples of Lake Constance. Other analyses were given by Delebecque (1898) and by Voigt (1905). The best series of determinations appears to be that of Birge and Juday (1911), though unfortunately they early gave up the estimation of molecular nitrogen in their samples, after having obtained results which appear to be of some interest.

It is usually supposed that the molecular nitrogen of lake waters represents atmospheric nitrogen which has dissolved at the surface of the lake or of its influents. If the water comes into equilibrium with air at a period of circulation, the entire lake will be saturated at the temperature of circulation. Later, as heating occurs, nitrogen will be lost from the surface waters which, though they contain less of the gas than formerly, will be still saturated at their new and higher temperature. If all the heating of the deeper water is by wind-driven currents and turbulent mixing, the various layers of the lake will be approximately saturated, though the warmer ones will contain less nitrogen. In a stratified lake, a slight inverse clinograde curve relating the absolute concentration to the depth is therefore to be expected, even though the water is saturated with respect to its temperature throughout.

Birge and Juday concluded that at the surface of the lakes which they studied, the water was always saturated with nitrogen, but that in the deeper layers of stratified lakes slight supersaturation was of frequent occurrence. They do not give any indication of the source of the standard values used in estimating the percentage saturation. If the tables given by Fox (1909) for dry air are employed in such a calculation, and if corrections are made for the vapor pressure of water and for altitude, it would seem that supersaturation is usual in the Wisconsin lakes. It is known, moreover, from the work of Rakestraw and Emmel (1938), that Fox's determinations, at least for sea water, are slightly too high,[1] so that the supersaturations apparently implied by Birge and Juday's analyses cannot be due to errors in Fox's table. It appears almost certain, therefore, that there is a systematic error in the determinations of molecular nitrogen dissolved in the waters of the Wisconsin lakes.

An attempt has been made to obtain a criterion of saturation applicable to Birge and Juday's data, from these data themselves. In Fig. 217, curve *A* represents Fox's data for dry air at sea level, curve *B* gives these data corrected for an atmosphere saturated with water vapor, at the altitude of Green Lake, Wisconsin. Curve *C* is derived from curve *B* by applying a further correction, derived from a comparison of Fox's determinations for chlorinity 16 and 20 per cent with those for the same chlorinity given by Rakestraw and Emmel; this treatment should correct for supposed supersaturation in Fox's experiments. Curve *C*, if the correction for supposed supersaturation is independent of salinity and of small changes of pressure, should give the expected saturation values for the Wisconsin lakes. The points entered on the graph indicate the observed nitrogen contents and temperatures of all the *surface* samples taken by Birge and Juday. Curve *D* has been drawn by eye, parallel to curve *C*, on the assumption that the latter curve has the correct slope, to fit as well as possible these data for the surface samples, which are assumed to be saturated. Curve *D* has subsequently been used to estimate the percentage saturation values of the determinations in the *vertical* series given by Birge and Juday.

In Fig. 218 the vertical distribution of molecular nitrogen in Green Lake during summer stratification is indicated. It is quite obvious, in this as in all but one of the other cases, that no hypolimnetic supersaturation is indicated by the present mode of treating the data. Since Alsterberg (1930a) drew far-reaching conclusions from the supposed supersaturation in the hypolimnia of Green Lake and of other lakes in Wisconsin, the apparent illusionary nature of this supersaturation is of some importance. In a single case, Otter Lake, Birge and Juday record what appears to be a valid supersaturation of molecular nitrogen. The full results have never been published, but through the kindness of Dr. A. D. Hasler it has been

[1] Van Slyke, Dillon, and Margaria (1934) give values of the solubility coefficients at 25° C. and 38° C. hardly different from those of Fox, and quote work by other biochemists which appears also to be in agreement with Fox's and their own determinations. A renewed study of the solubility of nitrogen in pure water over the entire range of biological temperatures is evidently needed.

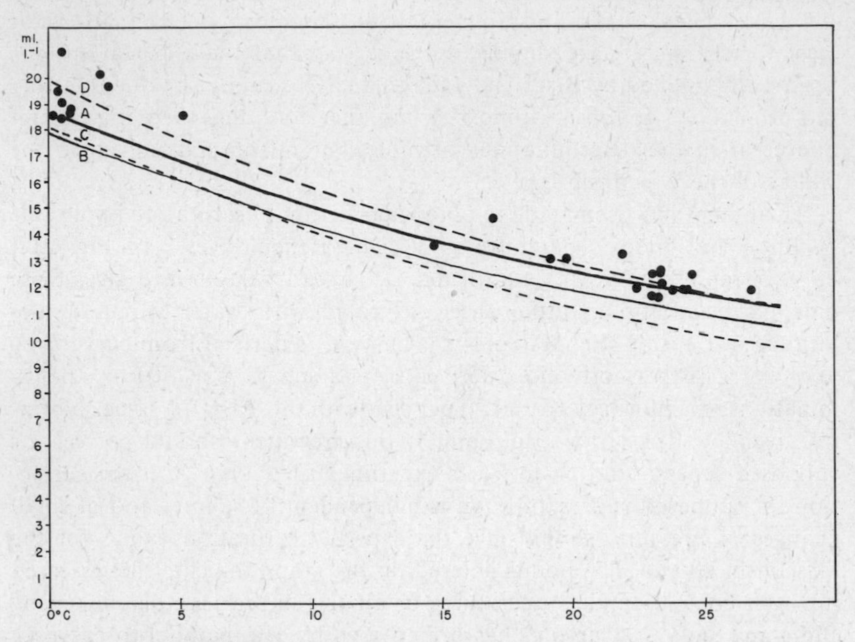

FIGURE 217. The solubility of molecular nitrogen (ml. per liter) in waters at various temperatures. A, solubility computed for dry air at sea level from the determinations of Fox; B, solubility computed for wet air at an elevation of 278 m.; C, curve B corrected for the apparent supersaturation effect in Fox's experiments; D, curve drawn by eye, parallel to C, to fit the determinations made by Birge and Juday on surface waters of Wisconsin Lakes, and presumably a suitable working standard for estimation of saturation in vertical series of analyses by these investigators.

possible to consult the original tables of analyses (Table 119). It is to be noted that the general accuracy of the gasometric method employed is to some extent confirmed by the fair concordance between the oxygen concentrations determined by boiling out the gas and those determined by the Winkler titration.

It is evident that in both 1905 and 1906 the upper hypolimnion of Otter Lake was supersaturated with molecular nitrogen. This supersaturation, however, exists only relative to the pressure at the surface of the lake. At the pressures of 1.35 to 1.5 atm. prevailing in the 3.5 to 5-m. zone, the observed nitrogen contents are very slightly undersaturated, so that there can have been no tendency for the gas to escape in the form of bubbles.

The quantity of nitrogen in excess of saturation at the pressure of the

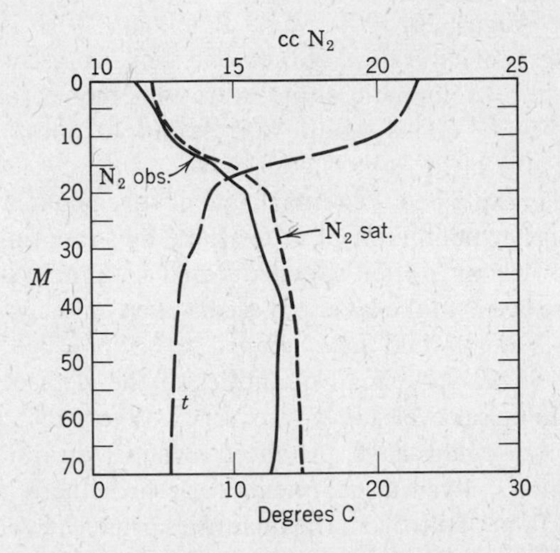

FIGURE 218. Green Lake, Wisconsin. Vertical distribution of molecular nitrogen during summer stratification. The broken line indicates saturation as defined by curve *D* of Fig. 217. Note that the water is slightly, but probably not significantly, undersaturated.

TABLE 119. *Molecular nitrogen in Otter Lake, Wisconsin*

Depth, m.	Temp., °C.	O_2 (gasometric), ml.	O_2 (Winkler), ml.	N_2, ml.	N_2, % sat.*	Implied Temp. of Solution, °C.
August 17, 1905						
3.0	19.5	7.0	...	12.2	93.9	28.0
3.5	19.2	10.3	...	16.5	125.9	8.2
6.0	9.5	1.47	...	17.7	110.7	5.1
10.0	8.5	1.78	...	17.3	103.5	6.2
August 20, 1906						
0.0	26.7	6.1	6.0	11.9	104.2	24.2
3.0†	23.3	7.3	7.1	12.9	106.5	19.9
3.5	21.8	15.5	14.7	15.9	128.2	9.8
4.0	18.8	14.2	13.3	17.7‡	134.0	5.1
5.0	13.4	7.4	7.2	19.2	130.5	1.4
10.5	8.8	2.9	3.1	18.7	114.8	2.5

* From Fig. 217.
† August 27, 1906.
‡ Reanalysis on August 26 gave 17.6 ml. N_2 at 4.0 m.

lake surface and at the observed temperatures, when summed from 3.0 m. to 10.5 m. on August 20, 1906, was 2.36 ml. $cm.^{-2}$ It is possible that the 0.7 ml. in excess of saturation at 0 m. and 3 m. represents a systematic error, so that the hypolimnetic supersaturation may perhaps be better set at 1.84 ml. $cm.^{-2}$ This would correspond to about 1 ml. $cm.^{-2}$ when the whole area of the lake is considered.

Birge and Juday explained the remarkable nitrogen content of the metalimnion and upper hypolimnion of Otter Lake by supposing that ground water, which they claim is often supersaturated with molecular nitrogen,[2] had entered these layers of the lake. It seems most unlikely, however, that enough ground water would have seeped in to produce the observed effect. In view of the very small quantities of the element normally involved in the biological cycle, it is also very unlikely that biological denitrification *in situ* could have produced a supersaturation of several milligrams per liter. Even in sediments for which there is unequivocal evidence of liberation of nitrogen, the quantities produced seem inadequate to explain the observations in Otter Lake. Allgeier, Peterson, Juday, and Birge (1932) found not more than about 0.5 ml. N_2 produced from a liter of Mendota mud in 100 days of anaerobic decomposition at 7° C. When it is remembered that a column of area of cross section 1 $cm.^2$ and volume of 1 liter has a length of 10 m., this source of nitrogen seems unlikely.

Dalton (in Otley 1819, 1831) found that the gas liberated from the floating island of Derwentwater (page 175) contained equal volumes of methane and nitrogen with a little carbon dioxide. Russell (in Symons [1889]) found the proportions of nitrogen in the gas from the same source to vary from 15.62 to 91.31 per cent (presumably by volume), most of the rest of the gas being methane. Düggeli (1936) has found up to 52.9 per cent nitrogen in the gas, otherwise mainly methane, ascending from the mud of the Rotsee, a somewhat polluted lake near Lucerne. Most of

---

[2] Matsue, Egusa, and Saeki (1953) record up to 160 per cent saturation with nitrogen in artesian waters. According to Sugawara and Tochikubo (1955), in an important paper received too late for detailed consideration, some ground waters are supersaturated with both nitrogen and argon, others only with nitrogen. The former class is explained by bubbles being carried by the ground water to a depth at which a considerable pressure is achieved. The bubbles go into solution, producing supersaturation when the water is brought to the earth's surface. The waters supersaturated with respect to nitrogen alone are supposed to be due to organic decomposition. Sugawara and Tochikubo record supersaturation, relative to the prevailing temperature, of both argon and nitrogen at the bottoms of Kizakiko and Nakatsunako. They suggest a mechanism comparable to that of Alsterberg, apparently assuming heating without mixing in the bottom layers of these lakes. To the writer this seems improbable.

this nitrogen is doubtless of atmospheric origin and has diffused into the methane bubble as it formed. Neglecting water vapor and gases other than methane and atmospheric nitrogen, the percentage by volume of nitrogen in such a bubble formed at depth z, will be, from equations 7 and 11 of Chapter 9, equal to

$$\frac{79.1}{1 + 0.0967z}$$

so that at moderate depths a considerable amount of nitrogen may be expected in the bubble. As the bubble ascends, it will enter regions containing no methane, and all such regions will initially also contain less nitrogen than the absolute saturation value. Nitrogen and methane will therefore pass into the intermediate water from the bubble, though of course this process will become increasingly slow as absolute saturation with nitrogen is approached. In order for such an explanation to hold, it would be necessary to suppose that this mechanism could remove all the nitrogen from a layer of water or mud 0.5 to 1.0 m. thick on the bottom of the lake. Though this explanation of the phenomenon is perhaps the most reasonable, it is curious that Birge and Juday, who were quite aware of the necessity for distinguishing methane from nitrogen in their analyses, indicate nothing about the former gas. In view of this difficulty, the supersaturation of molecular nitrogen in Otter Lake must be left unexplained.

Birge and Juday point out that a condition of the kind found in this lake would be very favorable for the development of the so-called gas disease in fish, which appears to be due to the body fluids coming into equilibrium with nitrogen, either at a low temperature or under other conditions leading to the presence of more nitrogen in a lower than in a more superficial layer. If the fish now moves upward, there will be a tendency for nitrogen to come out of solution, producing a condition comparable to decompression sickness, bends, or caisson sickness in man. No evidence of the occurrence of such a condition in the fish of Otter Lake was noted.

The nitrogen balance of a lake and the problem of nitrogen fixation. Theoretically, there are three possible sources of nitrogen compounds: (1) influents to the lake, including ground water, (2) precipitation on the lake surface, and (3) fixation in the lake and its sediments.

In considering these sources, it is necessary to bear in mind that the lake, theoretically, can lose nitrogen compounds (*a*) in the effluents, (*b*) by diffusion of volatile nitrogen compounds from its surface (probably usually insignificant), (*c*) by denitrification in the lake, and (*d*) in the formation of permanent sediments.

The available information on the relative importance of these processes is extremely meager. It is reasonably certain that in most cases the greater part of the nitrogen compounds is derived from the soil and enters by the influents, and that the major loss is by the effluent. Most of the available data merely indicate this in a rough way.

The presence of nitrogen compounds in rain is well known, though their origin is by no means as well established as is commonly believed (Hutchinson 1954). Both the total quantity and the qualitative distribution of these compounds is very variable. In general, in the temperate zone ammonia is more abundant than nitrate, whereas in the tropics the partition between the two forms of nitrogen is usually more equal or may be in favor of nitrate. These observations strongly suggest that most of the nitrate is produced by nitrification, doubtless of ammonia associated with dust particles. The quantities of both nitrate and ammonia present compare favorably with those to be detected in the surface waters of lakes, as can be seen from Table 67, but very few lakes would receive their entire water supply from unmodified rain, and the rate of replacement of water by rain in any such case is likely to be small compared to the rate of change of the form of nitrogen under biological influences. Though it is possible, particularly in sterile mountainous regions, that the nitrogen in precipitated water is of considerable importance, it is much more likely that in most lakes the origin of the fixed nitrogen is to be sought in the soil of the drainage basin or in the lake itself.

Quantitative estimates of nitrogen balance and evidence of denitrification. The first attempt to construct a quantitative balance sheet which would throw light on this problem is that for Windermere prepared by Mortimer (1939). Unfortunately, more than half of the data used by Mortimer for organic nitrogen are derived, by means of a conversion factor, from the oxygen consumed, and direct Kjeldahl determinations were employed in only the last five months of the year investigated. Mortimer estimated the total combined nitrogen entering the lake in a year as 326 metric tons, and the total leaving the lake as 318 metric tons. About 5 tons have been lost to the sediments annually during postglacial time, so that the loss of 323 tons balances the income of 326 tons very accurately. Mortimer, however, believes that the estimate of the income is about 20 to 50 tons too small, as certain subsidiary sources of nitrogen, such as dead leaves, are not considered, and that the estimate for the contribution of the small streams is probably too low. He therefore concludes that an over-all denitrification to molecular nitrogen takes place in the lake.

Rohlich and Lea (1949) have examined the problem of the nitrogen balance of Lake Mendota, and have concluded that between October 1, 1948, and October 1, 1949, the balance sheet was as follows:

	Metric Tons
Income	
Organic nitrogen	38.0
Inorganic nitrogen	117.8
Total nitrogen	155.8
Loss by effluent	
Organic nitrogen	23.8
Inorganic nitrogen	17.6
Total nitrogen	41.4
Retained or lost in lake	114.4

Some of the nitrogen retained or lost in the lake must have entered the sediments, but it would seem reasonable to suppose that a considerable part underwent bacterial denitrification to molecular nitrogen.

Mortimer has also noted that in the case of the less productive lakes in the English Lake District, the mean concentration of nitrate and of total nitrogen in the influents was about the same as or a little lower than the concentrations in the effluents, but in the more productive lakes the concentration in the effluents was definitely less than in the influents, just as in Windermere and Mendota (Table 120). He considered, therefore, that in the early less productive stages of the history of a lake, denitrification,

TABLE 120. *Forms of nitrogen in influents and effluents of lakes in the English Lake District*

Lake	Total N in Mud, % dry wt.	Influent			Effluent		
		$N \cdot NO_3$	Org. N	Tot. N	$N \cdot NO_3$	Org. N	Tot. N
LESS PRODUCTIVE							
Wastwater	0.45	0.10	0.13	0.23	0.11	0.10	0.21
Ennerdale	0.37	0.06	0.12	0.18	0.05	0.16	0.21
Buttermere	0.34	0.07	0.17	0.24	0.06	0.20	0.26
Crummock	0.40	0.06	0.20	0.26	0.08	0.22	0.30
Hawes Water	...	0.08	0.13	0.21	0.08	0.19	0.27
Thirlmere	...	0.07	0.44	0.51	0.15	0.38	0.53
Mean	0.39	0.07	0.20	0.27	0.09	0.21	0.30
MORE PRODUCTIVE							
Loweswater	0.88	0.36	0.26	0.62	0.09	0.26	0.35
Esthwaite	0.85	0.29	0.87	1.16	0.19	0.55	0.74
Blelham Tarn	1.03	0.35	0.40	0.75	0.16	0.30	0.46
Mean	0.92	0.33	0.51	0.84	0.15	0.37	0.52

if it occurs, does so at a rate not in excess of nitrogen fixation, and in the more mature productive lakes denitrification tends to take place more rapidly than do processes of fixation. It is evident that in no case is it necessary to postulate very intense fixation in the lake.

Nitrogen-fixing organisms in lakes. It is becoming apparent that a considerable number of species of ordinary bacteria can fix molecular nitrogen. This is true not only of the well known aerobic species of *Azotobacter* and the anaerobic *Clostridium pasteurianum*, which have long been recognized as nitrogen-fixing organisms, but also of species of *Azotomonas*, *Aerobacter*, *Methanomonas*, and *Pseudomonas*. The most recent work on such organisms in soils, that of Anderson (1955), may be consulted for a brief review of current knowledge of the distribution of the capacity to use molecular N_2.

It has been known since the classical investigations of Beijerinck (1901) that *Azotobacter* occurs frequently in water as well as in soils. More recently, Kluyver and Van Reenen (1933) have shown that *A. agile* is to be regarded as specifically an aquatic form. Lantzsch (1925), studying ponds at Wielenbach, discovered no evidence of nitrogen-fixing organisms in the free water, but found small numbers of a supposed nitrogen fixer (referred, no doubt erroneously, to *Bacillus turcosum*) on the leaves of *Elodea* and *Potamogeton*. Sarles (personal communication) has found *A. agile* abundant in such situations. In the sediments of the ponds that he studied, Lantzsch found both aerobic and anaerobic nitrogen-fixing bacteria, the former tending to occupy the more superficial layers. Klein and Steiner (1929) found marked evidence of the existence of nitrogen fixers in the Lunzer See. The numbers are said to have been minimal at the time of the circulation. The most marked activity occurred in the mud, in the water lying over the mud, and at the surface. Klein and Steiner conclude, however, that the lower activity observed in the free water would, when integrated over the entire lake, be considerable.

It is impossible to know how far the results reported by investigators determining bacterial activity by ordinary bacteriological methods throw any light on the problem as to whether bacterial fixation does occur significantly in lakes. All that such investigations really show is that the bacterial flora is such that fixation could occur if an adequate energy source were available. In the free water at least, it does not seem likely that such a condition is often realized. Moreover, when a considerable amount of organic matter suitable as an energy source is present, as in organic sediments or in polluted waters, the concomitant presence of inorganic nitrogen compounds is likely to depress fixation. This is probably the case in the sediments of Mortimer's three more productive lakes. Baier (1936) found but feeble indications of nitrogen fixation in the ponds which he studied in Kiel; all these localities are probably somewhat polluted and have high combined nitrogen contents. Nümann (1941) thought, however, that significant fixation occurs at times in the Schleinsee, but his evidence is not critical.

It has recently become apparent that certain blue-green algae are important nitrogen-fixing organisms, and in the case of the genus *Anabaena* there is good reason for believing that such algal fixation is of significance in those lakes in which the genus occurs abundantly. The energy required for fixation by such organisms is of course provided by the products of photosynthesis. As far as is known, the faculty is most generally developed among the Myxophyceae, in the family Nostocaceae. Odintsova (1941), however, reports fixation by *Gloeocapsa minor* of the Chroococcaceae, and Watanabe, Nishigaki, and Konishi (1951) find that *Tolypothrix tenuis* (fam. Scytonemataceae) and *Calothrix brevissima* (fam. Rivulariaceae) are effective nitrogen fixers; like the species of *Anabaena* studied by De (1939), they may be of economic importance in rice fields. In the Nostocaceae the fixation of molecular nitrogen has been demonstrated critically, using N^{15}, in *Nostoc muscorum* by Burris, Eppling, Wahlin, and Wilson (1943); by less critical, though seemingly quite adequate methods, in various species of *Anabaena* by De (1939, Fritsch and De 1938) and by Fogg (1942); in *Cylindrospermum* as well as in *Nostoc* and *Anabaena* by Bortels (1940); and in *Anabaenopsis* spp. by Watanabe, Nishigaki, and Konishi (1951). Though the species of *Anabaena* used in the investigations of De and of Fogg are apparently not planktonic, there is evidence (Guseva 1937a) of considerable rises in the total combined nitrogen of cultures of *Anabaena lemmermanni*, though in this case bacteria were present.

Hutchinson (1941) likewise recorded increases in total combined nitrogen, of the order of 460 mg. $m.^{-3}$, in the surface waters of Linsley Pond during a ten-day period when the epilimnetic water was soupy with *Anabaena circinalis*, and even greater increases when some of this water was enclosed in a bottle suspended from a buoy. Aleev and Mudretsova (1937) had earlier indicated nitrogen fixation, supposedly by algae, during a water bloom, but did not indicate what species were involved. It is also not without interest to note that Utterback, Phifer, and Robinson (1942) record *Anabaena* as the main phytoplanktonic organism in Crater Lake, Oregon, the water of which is rich in ionic phosphate but very poor in combined nitrogen. None of the observations in nature are critical, but they all point to *Anabaena* being a rather important nitrogen-fixing organism when it occurs in massive quantities in the plankton.

The discovery that the purple bacteria *Rhodospirillum rubrum*, *Rhodopseudomonas* spp., *Rhodomicrobium vannielii*, and in all probability

Chromatium spp., can fix molecular nitrogen (Kamen and Gest 1949; Lindstrom, Lewis, and Pinsky 1951; Bregoff and Kamen 1951) suggests that the photosynthetic bacteria may ultimately be found to play a part like that of *Anabaena* in the few lakes in which massive populations of these organisms develop.[3]

AMMONIA IN LAKE WATERS

The production of ammonia. Ammonia, under which term NH_3, NH_4^+, and NH_4OH will be included, is the major nitrogenous end product of the bacterial decomposition of organic matter, and is an important excretory product of invertebrate animals also. It seems probable that a considerable amount of ammonia produced in the decomposition of planktonic organisms may be liberated by direct bacterial action, without the formation of soluble intermediate products. This point of view has been adopted by Von Brand, Rakestraw, and their associates in a long series of studies of the decomposition of marine diatoms in sea water in the dark (Von Brand and Rakestraw 1940, 1941; Von Brand, Rakestraw, and Renn 1937, 1939; Von Brand, Rakestraw, and Zabor 1942). Cooper (1937b), however, points out that the rate of formation of intermediates is likely to be less rapid than their ammonification, so that analytically detectable amounts of such intermediates might never be present. Von Brand, Rakestraw, and their co-workers, moreover, often found that such inorganic nitrogen compounds accumulated in their experiments, all of which had passed through an ammonia stage, represented the whole of the nitrogen lost from the introduced particulate matter, mainly dying diatoms, and that the organic nitrogen already present in the sea water might remain entirely outside the process of decomposition. In lake water, in which polypeptides, if not amino acids, are supposed to occur in small amounts in solution, it is possible that the presence of intermediate compounds is of greater significance than in the sea, but it is probable that the concentrations of such intermediates are usually too low for effective bacterial decomposition until they are adsorbed on solid surfaces.

Apart from the direct production of ammonia from proteins and its indirect production by the bacterial deamination of amino acids produced

[3] From time to time, claims of nitrogen fixation by organisms other than the bacteria and the algae discussed above have been made; none of these claims are fully established, and most of them have been forgotten. A recent and not yet fully studied case of alleged fixation by a fresh-water organism is given by Seaman (1949), who concludes that the ciliate referred to *Colpidium campylum*, but actually probably *Tetrahymena*, can use molecular nitrogen.

in decomposition and liberated into the water, Cooper (1937b) points out that certain other types of compounds may be expected to result from the decomposition of aquatic organisms. In the sea, trimethylamine and its oxide may be quite important, since both compounds occur in marine fishes and invertebrates. In fresh water animals, these compounds are rare or absent, but some trimethylamine oxide may be formed from choline and other betaines, and then decompose to give dimethylamine and formaldehyde, as Cooper indicates. It has also been supposed that methylamines may be formed by algae in fresh waters. Such methylamines as may be produced are likely to be decomposed by bacterial action, to give ammonia.

Urea is another widespread decomposition and excretory product. Cooper points out that the small amount of amide nitrogen (page 895) which has been recorded from lake waters is probably mainly urea. Urea in aqueous solution is in equilibrium with ammonium cyanate, and the cyanate ion is itself hydrolyzed to ammonium and bicarbonate ions. Cooper therefore believes that although urea-splitting bacteria have commonly been isolated from the sea, as they have also from lakes and lake sediments, the decomposition in the biosphere is not dependent on them and may be purely chemical.

Apart from the production of ammonia by the direct or indirect decomposition of proteins and other nitrogenous organic matter, production by bacterial reduction of nitrate is also of considerable importance. Moreover, in the utilization of nitrate as a nitrogen source by green plants, it is clear that reduction must occur, and there is evidence that in some algae the reduction process can lead to extracellular accumulation of less oxidized nitrogen compounds derived from nitrate. Beckwith (1933) found that *Chlorella* produced small amounts of nitrite; ZoBell (1935) has confirmed this and has extended the observation to several marine diatoms. Warburg and Negelein (1920) found that *Chlorella* could, under some conditions, produce ammonia from nitrate, and this has also been noted by Pearsall and Loose (1937).

Against these processes which lead to the production of ammonia may be set the reverse processes which lead to removal of ammonia, namely, assimilation of ammonia as a nitrogen source by bacteria and, above all, by algae, and nitrification or oxidation of ammonia by bacteria as an energy source. Both of these processes are of great importance and will be discussed in later paragraphs, when the problem of the forms of ammonia present in lake waters have been examined.

Forms of ammonia present in lake waters. Ammonia in aqueous solution is present mainly as NH_4^+ and as undissociated NH_4OH. The proportions of these two forms will depend greatly on the *p*H, and this variation may be of considerable ecological importance, as Cooper (1937b) and other workers have emphasized. At 18° C.

$$(1) \qquad \frac{a_{NH_4^+} \cdot a_{OH^-}}{a_{NH_4OH}} = 17.15 \times 10^{-6}$$

Since at this temperature

$$(2) \qquad p_{a_{OH}} = 14.24 - p_{a_H}$$

the ratio of ammonium ion to undissociated hydroxide will be given by

$$(3) \qquad \frac{a_{NH_4^+}}{a_{NH_4OH}} = 2.98 \times 10^{(9-pH)}$$

the term written p_{a_H} being more commonly though less correctly written pH. Thus at pH 6.0 the ratio will be about 3000:1, at pH 7.0 about 300:1, and at pH 8.0 about 30:1. As will be indicated, the formation of any considerable amount of ammonia at a pH of, say, 9.5 when the concentrations of the two forms are about equal, or a rise in pH to this value or above it in the presence of a considerable quantity of ammonia, may imply the production of a toxic concentration of the undissociated hydroxide.

There can be little doubt that part of the ammonia present in lake waters can be bound to colloidal particles, particularly in neutral or alkaline brown humic waters. Ohle (1934a) finds that in northern Germany the ammonia content of highly colored waters tends to be greater than that of less colored waters, though this is not true to the lakes of northern Wisconsin (Juday, Birge, and Meloche 1938). Ohle (1935) found, moreover, that a humic water artificially enriched in ammonia at pH 5.6 contained 780 mg. free $N \cdot NH_3$ m.$^{-3}$, but when the same water was brought to pH 7.1 by the addition of a calcium bicarbonate buffer, only 703 mg. free $N \cdot NH_3$ m.$^{-3}$ were detected.

Cooper (1948b) has called attention to the probability that much of the ammonia in the sea is adsorbed to particulate matter of such dimensions and rarity that the ordinary volumes of water used in analysis are too small to be statistically adequate samples. He suspects that many cases of irregular series of ammonia determinations, formerly referred to contamination of an unexplained nature, are due to the irregular distribution of particulate ammonia, which he takes to consist of adsorbed NH_3 or NH_4^+, or possibly of weakly hydrolyzable amide groups ($CO \cdot NH_2$). This particulate ammonia is presumably associated with microscopic seston or with the surface film or with matter adsorbed thereon. As will be later discussed, ammonia is probably associated with the surface film in some lakes. In view of the probability that nitrifying bacteria in soils act primarily on adsorbed ammonia, Cooper believes that the occurrence of particulate ammonia is of considerable importance in the nitrogen cycle in the sea. As the quantity of colloidal and particulate matter is likely to be much greater in most lakes than in the open ocean, Cooper's argument as to the significance of particulate ammonia is likely to apply even more strongly to inland waters. It is desirable to point out that the so-called albuminoid ammonia of water analysts is present in the water as organic matter which can lose ammonia on alkaline hydrolysis, and is therefore not to be regarded as ammonia in any form, ionized or undissociated NH_4OH, free or bound.

Assimilation of ammonia. In view of the necessity of reduction in the process of assimilation of nitrate, it is unlikely on general grounds that ammonia would not be as good or a better source of nitrogen than is nitrate. Many higher terrestrial plants can in fact use both ammonia and nitrate, but it is known that the *p*H of the nutrient medium plays a part in the availability of ammonia. Tiedjens and Robbins (1931) found that several crop plants assimilated ammonia preferentially above *p*H 7, and nitrate preferentially below *p*H 7. At *p*H 4.0 no ammonia was used. It is important, in considering such results, to remember the variation in the ratio of ammonium ion to the undissociated hydroxide. The results on terrestrial phanerogams seem to suggest that the ion is much less available than the undissociated compound. The evidence relating to the unicellular algae is far from uniform, and permits no clear and easily interpreted conclusions. It is certain that most algae can use ammonia, and many forms can employ as nitrogen sources a variety of simple organic nitrogen compounds. A rather full bibliography, mostly relating to forms of little interest to the limnologist, is given by Ludwig (1938).

Carteria is known to use glycine (Schreiber 1927), *Chlamydomonas* glycine, alanine, and asparagine (Braarud and Föyn 1930). *Chlorella* can use not only ammonia, nitrite, and nitrate, but also acetamine, guanidine carbonate, uric acid, glycine, alanine, calcium aspartate, calcium glutamate, and less easily valine, leucine, phenylalanine, tyrosine, *l*-cystine, and perhaps urea; hydroxylamine, glucosamine, diphenylamine, and various azo compounds, proved to be unsuitable as nitrogen sources, or toxic (Ludwig 1938). Harvey (1940) found that marine phytoplankton could not use trimethylamine oxide. Pearsall and Loose (1937) found that *C. vulgaris* and Myers (1949) that *C. pyrenoidosa* used ammonia in preference to nitrate, and the same preference has been observed in a number of marine diatoms by ZoBell (1935) and by Harvey (1940).

Other investigators, however, working mainly with fresh-water species, have found that they are able to utilize nitrate under a wider range of conditions and at greater dilutions than they can use ammonia, the latter substance never being a better source of nitrogen than the former. Chu (1943) concluded that at all concentrations nitrate is a better nitrogen source than ammonia for *Botryococcus braunii.* Over a considerable range of concentrations, *Pediastrum boryanum*, *Staurastrum paradoxum*, and *Nitzschia palea* used either source equally well; but at considerable dilution, of the order of 200 mg. $m.^{-3}$, which would be normal or even excessive in nature, nitrate was definitely much superior to ammonia. ZoBell, it is to be noted, had obtained exactly the opposite result with the marine *N. closterium.*

Until careful studies of planktonic forms supposedly preferring nitrate and of those supposedly preferring ammonia have been made side by side by the same worker, it is probably premature to draw any definite conclusion, but it is hard to avoid the suspicion that real differences do exist and that, in view of the great variability in the ratio of ammonia nitrogen to nitrate nitrogen dissolved in lake waters, such differences may play a part in determining the constitution of phytoplanktonic populations.[4]

At very high *p*H values, effects due to the high concentration of undissociated ammonium hydroxide may be expected. Österlind (1947) found that in a medium containing 53,000 mg. N $m.^{-3}$, near the upper limit of concentration studied by Chu, nitrate was superior to ammonia at all *p*H values permitting growth of *Scenedesmus quadricauda.* Between *p*H 7.0 and 9.0, growth could take place with both nitrate and ammonia as nitrogen sources, though nitrate was always better. The optimal *p*H, about 8.0, was well defined and appeared to be identical in the two media. It is therefore unlikely that the availability of ammonia below *p*H 9.0 depends on the ratio of NH_4^+ to undissociated NH_4OH. Above *p*H 9.5 no growth whatever occurred with ammonia, though with nitrate some growth could take place in media as alkaline as *p*H 11. As Rodhe (1948) points out, very high *p*H values can be reached by photosynthesis in rich algal cultures, and the toxic effect of undissociated ammonium hydroxide under such conditions may have misled some workers into thinking that ammonia is not a satisfactory nitrogen source. This, however, is unlikely to have happened in Chu's experiments.

Distribution of ammonia in lake waters. In the aerated waters of the trophogenic layers of lakes, the concentration of ammonia is very variable. In some of the well-studied localities, such as the large lakes of the English Lake District, no ammonia is detectable in the surface layers at any season (Pearsall 1930). This is also true, at least in the summer, of a consider-

---

[4] This point of view receives considerable support from the recent and still unpublished observations of Proctor.

able number of the lakes of northeastern Wisconsin studied by Juday, Birge, and Meloche (1938), who found in 438 surface samples from 276 lakes:

$N \cdot NH_3$, mg. m.$^{-3}$	No. Samples	Per Cent
0	174	39.8
$<$ 10 (trace)	150	34.3
10	99	22.6
20	14	3.1
30	1	0.2

Ratio of nitrate to ammonia in surface waters. In the Austrian Alps, Ruttner (1937) found from 0 to 30 mg. $N \cdot NH_3$ m.$^{-3}$ in the trophogenic layers of the lakes which he studied, the over-all mean being 9 mg. m.$^{-3}$; in the same lakes the over-all mean for nitrate nitrogen was 238 mg. $N \cdot NO_3$ m.$^{-3}$, so that the ratio $N \cdot NO_3 : N \cdot NH_3$ was 26.5:1. In marked contrast to these observations are the data from the work of Domogalla and Fred (1926), who found in the lakes of the Yahara Basin, near Madison in southeastern Wisconsin, the quantities of ammonia and nitrate nitrogen given in Table 121.

TABLE 121. *Ammonia and nitrate in the Madison lakes, Wisconsin*

	$N \cdot NH_3$, mg. m.$^{-3}$		$N \cdot NO_3$, mg. m.$^{-3}$		Mean
Lake	Range	Mean	Range	Mean	$N \cdot NO_3 : N \cdot NH_3$
Mendota	68–205	132	10–72	32	0.24:1
Monona	168–544	407	48–133	86	0.21:1
Waubesa	49–325	158	22–125.5	64	0.34:1
Kegonsa	80–269	144	21–131.5	50	0.35:1
Wingra	52–192	91	39–175	90	0.99:1

The very high ammonia concentrations recorded in Lake Monona are certainly due to sewage contamination, and all the lakes are somewhat modified by cultural influences. Lake Wingra, which is least modified, shows the highest ratio of nitrate to ammonia, but even in this lake the ammonia nitrogen constitutes, on an average, half the inorganic combined nitrogen. Moreover, it is quite evident from the work of Karcher (1939) on the forest lakes of Masuria, that waters much less artificially disturbed than those in the vicinity of Madison, can have a distribution of inorganic combined nitrogen greatly in favor of ammonia. In the surface layers of nine lakes, Karcher found in the late winter and spring (February to May), when nitrate is most plentiful, 90 to 720 mg. $N \cdot NH_3$ m.$^{-3}$ and 8 to 103 mg. $N \cdot NO_3$ m.$^{-3}$ The mean quantities for these surface waters during this period of the year are 238 mg. $N \cdot NH_3$ m.$^{-3}$ and 36 mg. $N \cdot NO_3$ m.$^{-3}$, corresponding to a ratio of 0.13:1. In no single case is the ratio more in favor of nitrate than 0.51:1, and at all other seasons even less nitrate relative to ammonia occurred. Nümann similarly found that in the Schleinsee, near Langenargen, ammonia is the main form of inorganic combined nitrogen. As was pointed out in the previous section, the great variations observed in the ratio are likely to be of some significance if it ultimately appears that certain of the phytoplanktonic algae really assimilate ammonia better than nitrate, and others nitrate better than ammonia.

In unstratified lakes, where full circulation is maintained throughout the summer, ammonia produced by decomposition in the sediments throughout the entire area of the lake can always be added to the water, and in some cases, as in the Mummelsee in Masuria (Karcher 1939), this results in a summer maximum in ammonia content, the increased summer temperature apparently increasing ammonification more rapidly than assimilation. This temporal pattern is far from universal, however, and in the small ponds in Denmark studied by Nygaard (1938) the maximum content of ammonia was normally found in winter, either in February if no nitrification took place, or rather earlier, in November, December or January, if nitrification occurred.

Seasonal variation in surface waters. In the relatively few stratified lakes which have been studied on a sufficient number of occasions throughout a complete annual cycle and at several depths, the ammonia content of the trophogenic zone is usually minimal during the period of summer stratification, and there may be in dimictic lakes indications of a second minimum under the ice. Of the lakes which have been well studied, the Untersee below Constance (Elster and Einsele 1938) departs from this pattern except in the more or less isolated Gnaden Basin; in most of the Untersee the lowest values, 15 mg. $N \cdot NH_3$ m.$^{-3}$, are at the time of autumnal circulation. In most lakes, maximal ammonia concentrations appear in the trophogenic zone at the periods of full circulation; such maxima may be best developed in the autumn, but this is possibly due primarily to the nature of the circulation periods, being observed most strikingly in the monomictic Takasuka-numa (Yoshimura 1932a) and in many of the Masurian forest lakes, in which the spring circulation tends to be incomplete. In other cases, such as that of Lake Waubesa, the vernal ammonia maximum at the surface is more marked than the autumnal; in yet other lakes, such as Lake Mendota and Esthwaite Water in which the surface ammonia is always very low, the two maxima differ little in development.

The minima during stratification in the trophogenic zone are clearly related to the biochemical utilization of ammonia, either as a nitrogen source or in nitrification. As will appear later, the main fall in ammonia in the summer is usually accompanied by a fall in nitrate, and so must be due directly or indirectly to assimilation. Unequivocal evidence of the assimilation of ammonia without a rapid passage through a nitrate stage is difficult to obtain in nature, but the culture experiments which have already been mentioned indicate that such a process must frequently occur. The very sudden fall of ammonia, accompanied by a great increase of phytoplankton but without much change in nitrate concentration, observed in Lake Monona by Domogalla and Fred can be most easily explained as direct assimilation. A direct assimilation presumably occurs in localities such as the acid ponds studied by Nygaard (1938), in which nitrate is never found, but his data do not throw any clear light on the process.

Ammonia apparently associated with surface films. Within the trophogenic zone, one rather interesting aspect of vertical distribution is apparent in some of Karcher's data. In several of his Masurian lakes there appears to be a significantly greater amount of ammonia at the surface than at a depth of 1 m. This was observed, for instance, in series taken in May in three lakes (Table 122).

TABLE 122. *Superficial concentration of ammonia in Masurian Forest Lakes*

Depth, m.		Gelguhnersee (May 12, 1937), mg. m.$^{-3}$	Otzkosee (May 10, 1937), mg. m.$^{-3}$	Eupotecksee (May 8, 1937), mg. m.$^{-3}$
0	$N \cdot NH_3$	180	310	530
1	$N \cdot NH_3$	100	130	480
Difference		+ 80	+ 180	+ 50
0	$N \cdot NO_3$	21	12	8
1	$N \cdot NO_3$	73	27	15
Difference		− 52	− 15	− 7

Karcher concludes that the excess of ammonia is due to the inhibition of nitrification by ultraviolet light. This explanation is imperfect, for the excess of ammonia is never balanced by a deficiency of nitrate at the surface, and the total inorganic combined nitrogen is therefore always greater at 0 m. than at 1 m. The most reasonable explanation of the observations is that ammonia is held at the surface film or adsorbed to minute particles associated with the film, much as Cooper has supposed may occur in the sea. Whatever the origin of the observed distribution, it must be established very rapidly, for the top layer at this season must be very unstable, and is presumably mixed completely every night. It would seem probable that under such conditions material associated with the surface film might escape mixing more easily than material in solution in the free water.

Seasonal variation in deep water. In the tropholytic zone the seasonal cycle is roughly the inverse of that of the trophogenic zone. The onset of stratification leads to accumulation of ammonia in the deeper layers of the hypolimnion. In large deep unproductive lakes such an accumulation may be absent or limited to the stratum directly over the mud, but in many small productive lakes a large part of the hypolimnion is greatly enriched at the end of summer stratification, and not inconsiderable amounts of ammonia may also accumulate in the deep water of ice-covered lakes. At times of circulation the water of the tropholytic zone, enriched in ammonia, is mixed into the water of the superficial layers, increasing its ammonia content. The vernal and autumnal maxima in the ammonia concentration of the trophogenic zone could be produced entirely in this simple way; and in many cases such an interpretation is no doubt essentially correct. Karcher points out that in the surface layers of the temporarily meromictic Jegodschinsee no autumnal maximum in ammonia occurred, at least in 1937, because not enough of the deep water, rich in ammonia, was mixed into the freely circulating water mass.

In the Schleinsee, a small moderately productive lake, Einsele and Vetter (1938) found about 80 to 100 mg. $N \cdot NH_3$ m.$^{-3}$ in the trophogenic zone (0 to 8 m.) in summer. Nümann (1941), in a later account referring to 1937–1938, found only 30 to 60 mg. $N \cdot NH_3$, but concluded that ammonia is the most important inorganic nitrogen compound present. The nitrate data are inadequate, though it seems certain that no more nitrate than ammonia is present. At the time of the autumnal circulation all layers contain about 200 mg. $N \cdot NH_3$ m.$^{-3}$ During the winter Einsele and Vetter (1938) found a slow, irregular decline to a minimum just after the breaking of the ice, and then a sudden rise to 300 mg. m.$^{-3}$ at the time of vernal circulation. Comparable but less marked changes are indicated by Nümann. The deeper layers of the tropholytic zone accumulate massive amounts of ammonia during summer stagnation and less amounts in the winter, under the ice. The rate of increase in ammonia at the bottom of the lake was most rapid between August 6 and September 8, or about one month later than the maximum rate of increase of phosphorus and of iron. The time of the maximum rate of ammonia production is believed by Einsele and Vetter to be determined by the falling of large amounts of dead plankton from the trophogenic into the tropholytic zone between the middle of July and the end of August. Nümann accepted the same interpretation, but it is probably unnecessary as a general explanation of the high hypolimnetic ammonia concentration. Nümann found a marked inverse correlation between organic soluble and colloidal nitrogen and ammonia nitrogen in the tropholytic zone during summer stagnation. At the time of minimum oxygen content, organic soluble nitrogen disappeared completely from the deepest water of the lake. The anaerobic deamination implied is interesting and should be looked for in other localities.

Estimation of the total mass of ammonia nitrogen in the Schleinsee can be made from Einsele and Vetter's data, and from the morphometric map which they give. Such an estimation cannot be very accurate, for the chemical determinations are all presented graphically, but the curves introduced below the depth-time diagram (Fig. 219), based on their figure,

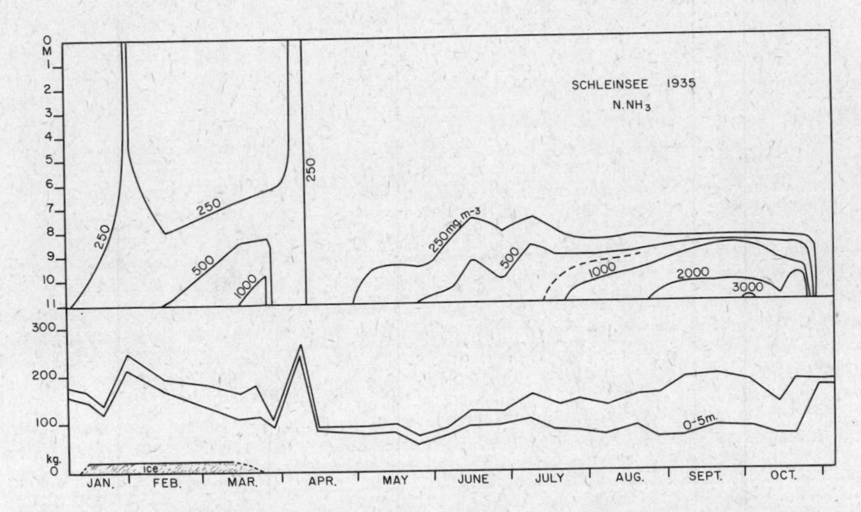

FIGURE 219. Schleinsee, Bavaria. *Upper panel*, time-depth diagram for ammonia nitrogen; *lower panel*, seasonal variation of ammonia nitrogen in the entire lake and in the top 5 m. (From data of Einsele and Vetter.)

presumably give the general nature of the variation of the total ammonia content. The epilimnetic rise at the time of autumnal circulation is evidently due to the mixing of the deep hypolimnetic water with the rest of the lake, but the maximum, preceded by a minimum at the spring overturn must have some other explanation, though it is not clear what that explanation can be.

A very similar study was made by Mortimer (1941–1942) in the course of his work on Esthwaite Water (Fig. 220). In this lake, as in other lakes of the English Lake District, the ammonia content of the surface waters is always low, not exceeding about 50 mg. $N \cdot NH_3$ m.$^{-3}$ at either the vernal or autumnal circulation period. As summer stratification is

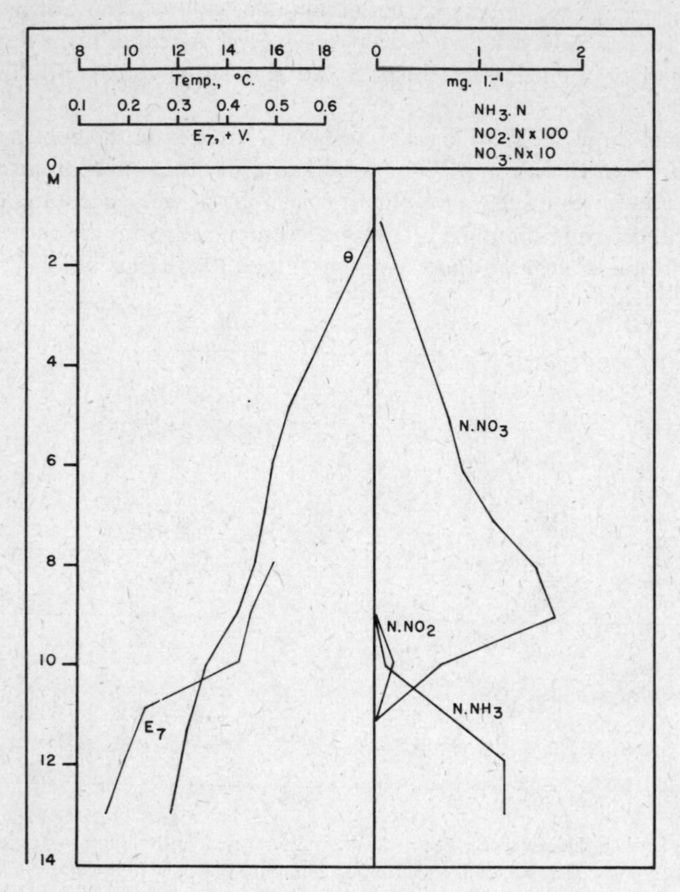

FIGURE 220. Esthwaite Water, vertical distribution of nitrate, nitrite, and ammonia nitrogen, and of redox potential (approximate), August 25, 1939 (after Mortimer).

developed, the ammonia disappears entirely from the surface waters. At the end of July very large quantities begin to accumulate in the deeper part of the hypolimnion; this accumulation continues through August, and more slowly through September. On September 22, 1939, 2600 mg. $N \cdot NH_3$ m.$^{-3}$ were present in the water just over the mud. In general, the rise of hypolimnetic ammonia follows a course similar to that of the rise of iron in the same layers. Mortimer points out that although all the nitrate initially present in the hypolimnion is reduced during the summer, this reduction accounts for only about one tenth of the ammonia production. The other nine-tenths must have come from the mud, as Mortimer believes, or from decomposition of falling plankton and faeces in the free water, as Einsele and Vetter consider to have happened in the Schleinsee, or from deamination of organic nitrogen compounds, as is certainly in part true of the Schleinsee. The known dual origin of the hypolimnetic phosphorus in Linsley Pond suggests that in the case of ammonia all possible sources make contributions.

The timing of the major increase indicates that early in stagnation ammonia does not enter the hypolimnion in appreciable quantities. Only after the oxidized microzone of the mud has disappeared does the great and rapid increase occur. Since ammonium salts are freely soluble, the only explanation of the observed changes, if the ammonia comes from the mud, would seem to be that while the oxidized microzone exists, it is capable of adsorbing all the ammonia which might have diffused from the mud into the free water. Part of the ammonia so adsorbed would be nitrified, but since nitrate is also freely mobile, there should be a marked increase in total combined nitrogen at this time unless the oxidized microzone really forms some sort of barrier. When the oxidized microzone is reduced, it appears to lose its adsorbtive power; the ammonia which it has held is liberated, and no further barrier to the free passage of ammonia produced in the mud is interposed between the site of production and the water of the hypolimnion.

When the thermal stratification of Esthwaite is destroyed in October, there appears to be a sudden fall in the ammonia content of the lake, as there is in the Schleinsee in the spring but not in the autumn. Mortimer's data, presented only in graphical form, are related to the water column over 1 $m.^2$ in the deep (13 m.) part of the lake, in which the main collecting station was situated. As in the case of the Schleinsee, it is possible to estimate roughly the total quantity of ammonia in Esthwaite. If this is done, it appears that at the time of the autumnal overturn there is a very considerable drop in the total ammonia content, for at the end of September, when the lower layers are still stratified, the lake contains about 700 kg. $N \cdot NH_3$, whereas during autumnal circulation little more than 50 mg. $m.^{-3}$ appear to be present in all layers, and this corresponds to 260 kg. for the whole lake.

As far as can be judged from Mortimer's data, this decrease in the total ammonia content was not paralleled by an increase in nitrate, though rather later, after the autumnal circulation, nitrate increased slowly throughout the lake without a corresponding fall in ammonia. The most reasonable, though not entirely satisfactory, explanation of the phenomenon apparently indicated by Mortimer's results is that at the time of circulation, when the water is oxygenated throughout the lake, the formation of an oxidized microzone permits adsorption of ammonia from the water, and that later this adsorbed ammonia is nitrified and liberated again into the lake.

A third and most interesting example of an ammonia cycle which can be treated quantitatively is provided by Yoshimura's (1932a) study of Takasuka-numa. In this lakelet or pond, maximum depth 6 m., the thermal cycle is essentially monomictic, though the lowest water temperatures may descend below 4° C. during the long winter circulation period. The maximum ammonia concentrations, which may reach 200 to 250 mg. $N \cdot NH_3$ m.$^{-3}$ in the trophogenic zone, appear after full autumnal circulation has been established; there is a subsequent decline of an irregular kind during the winter, when from 50 to 100 mg. m.$^{-3}$ are present throughout the circulating water mass. During summer stratification, ammonia starts to rise in May in the deeper layers of the tropholytic zone, shortly after oxygen becomes undetectable, just as happens in Esthwaite Water or in the Schleinsee. But after the end of July this conventional behavior ceases, for although the lake is very stably stratified, a marked decline in ammonia occurs in all the hypolimnetic layers.

This phenomenon was observed in both 1929 and 1930 by Yoshimura, who regarded it as characteristic of the lake, though he was at a loss to explain it. The fall in ammonia is not compensated by any rise in nitrate, which would hardly be expected in deoxygenated water, nor by any increase in albuminoid nitrogen. When full circulation is established, there is a great increase in ammonia throughout the lake, so that the total quantity of nitrogen in this form, which had fallen from 12.3 kg., all in the tropholytic zone, to 3.8 kg. during the month July 16 to August 15, 1930, and had remained low during the succeeding two months of stagnation, now rose suddenly to 20.5 kg. in the circulating water on November 11. This increase represents a very great increment in the total inorganic combined nitrogen, which was 24.6 kg. on October 2 and 40.2 kg. on November 11. During this same period, the albuminoid nitrogen fell irregularly from only 18.0 to 14.3 kg. It is conceivable that a compensating fall in nonalbuminoid organic nitrogen occurred, either in solution or in the seston, or that the results are due to denitrification and nitrogen fixation. Neither hypothesis, however, seems very likely.

It is also possible that in the hypolimnion of Takasuka-numa the ammonia is taken up by the sediments, possibly by adsorption on falling seston, during the later phases of stagnation, and is slowly liberated to the circulating water during the early phases of the long winter overturn. Such an hypothesis, which is the converse of that put forward to explain the observations made in Esthwaite, is not supported by any independent evidence. It is clear, however, that the well-established differences in the details of the cycles in Esthwaite and Takasuka-numa demonstrate that the ammonia cycles in a small lake may be much more complex than would at first appear.

It has been indicated in this discussion that the time of the most rapid rise of ammonia in the deeper water of the hypolimnion has been regarded as being determined either by the degree of oxidation of the mud-water interface or by the incidence of large quantities of falling plankton which are undergoing decomposition in the tropholytic layer. It is quite possible that in some lakes one, and in other lakes the other event, is in fact the controlling factor, and it has already been pointed out that in the case of phosphorus both the analogous processes are known to occur in Linsley Pond simultaneously. Mortimer, however, in his experiments with aerated and anaerobic tanks (page 701), found that ammonia does leave the mud in great quantities as the oxygen disappears from the anaerobic tank. The rise in ammonia in the anaerobic tank began before that of iron when an oxidized microzone still covered the mud surface, but the most rapid passage of ammonia into the water occurred late in the experiment, when the redox potential (E_7) of the mud surface lay at about $+ 0.1$ volt. This extremely rapid rise in ammonia thus occurred at a time subsequent to the most rapid increase in phosphorus and when the ferrous iron content of the water in the tank was actually falling, presumably owing to sulfide production. In the general sequence, the events of the tank, in so far as they diverge in details from the sequence in Esthwaite, appear to approach those in the Schleinsee, in which the period of most rapid rise in ammonia follows that of the most rapid increase in iron. While the explanation of the rise in ammonia in the hypolimnion of the Schleinsee as being due to falling plankton may be correct, Mortimer's experiment suggests that such an hypothesis is not necessary to explain the phenomena which have been observed.

NITRITE AND HYDROXYLAMINE IN LAKE WATERS

The distribution of nitrite. Minute amounts of nitrite are sometimes found even in unpolluted oxygenated surface waters of lakes, though any appreciable nitrite content in surface waters has long been regarded as a warning of sewage contamination. Juday, Birge, and Meloche (1938), examining 504 surface samples collected during the summer months from 307 lakes in northeastern Wisconsin, made the following determinations:

No. Samples	$N \cdot NO_2$, mg. m.$^{-3}$
369	None
125	Trace (apparently < 1 mg.)
5	1
4	2
1	4 (sewage pollution)

The small quantities of nitrite present in surface waters show marked seasonal variation. In Lake Mendota, Domogalla, Juday, and Peterson (1925) found nitrite almost always to be present. The minimal quantities occurred in the latter part of summer stagnation, the absolute minimum of zero occurring on September 21, 1923, which was also the date of the absolute nitrate minimum of 8.3 mg. $N \cdot NO_3$ $m.^{-3}$ The maximum amounts of nitrite occurred in the winter under ice, 17.9 mg. $N \cdot NO_2$ $m.^{-3}$ being recorded on January 9, 1923. In 1923 nitrite fell to 6.2 mg. $m.^{-3}$ early in June, but in the succeeding year the January maximum was less well marked and high values of 10 to 12 mg. $m.^{-3}$ persisted until the investigation terminated in June. It seems likely that the nitrite present in summer, which tends to vary with the nitrate, is produced by nitrate reduction, but the meaning of the increase during the early part of winter stagnation, when both ammonia and nitrate are low, is not apparent.

A not dissimilar nitrite cycle is recorded by Einsele and Vetter (1938) in the Schleinsee. From 1 to 5 mg. $N \cdot NO_2$ $m.^{-3}$ were present in the winter and spring, the absolute maximum occurring in November. After a minimum in April, the nitrite rose again to 4 mg. $m.^{-3}$ in May. After the middle of June, little or no nitrite was detectable in the surface waters during summer stratification and the earlier part of the long autumn circulation period. Einsele and Vetter attributed the presence of nitrite to nitrate reduction, and believed that the seasonal variations reflected variations in nitrate, for which substance they did not have satisfactory data.

Mortimer also found nitrite at the surface of Esthwaite Water on various occasions, and concluded that it was probably produced as the result of the activity of the phytoplankton. Since reduction of nitrate to nitrite is well known in cultures of diatoms and of *Chlorella* in the laboratory, such an explanation of the minute amounts of nitrite often observed in unpolluted and well-oxygenated surface waters is very reasonable.

Relation of nitrite and hydroxylamine to oxidation and reduction. The most striking feature of the vertical distribution of nitrite is the development of very marked maxima just at or above the part of the hypolimnion at which oxygen concentration declines most rapidly. Ideally, the nitrite maximum may be expected to lie between a well-oxygenated region rich in nitrate and a practically anaerobic region rich in ammonia. This type of distribution was first clearly indicated by Müller (1934), though his best examples (Toplitzsee, Hüttensteinersee) are perhaps a little overdramatic, the great accumulation of ammonia in the anaerobic depths being due to meromixis. Esthwaite Water late in stagnation shows the ideal distribution beautifully, as does Blelham Tarn (Pearsall and Mortimer 1939). very marked nitrite maxima at 8 to 9 m. in the Schleinsee, above the level

(9 to 10.6 m.) of the high ammonia concentration, doubtless indicates the same pattern, though nitrate data are not available in this case.

A narrow band of high nitrite concentration would be expected whereever a markedly dichotomic or direct clinograde nitrate curve is found. Certain lakes, however, fail to show the expected distribution. In Kizakiko in September, Tanaka (1953a) found evidence of striking nitrate

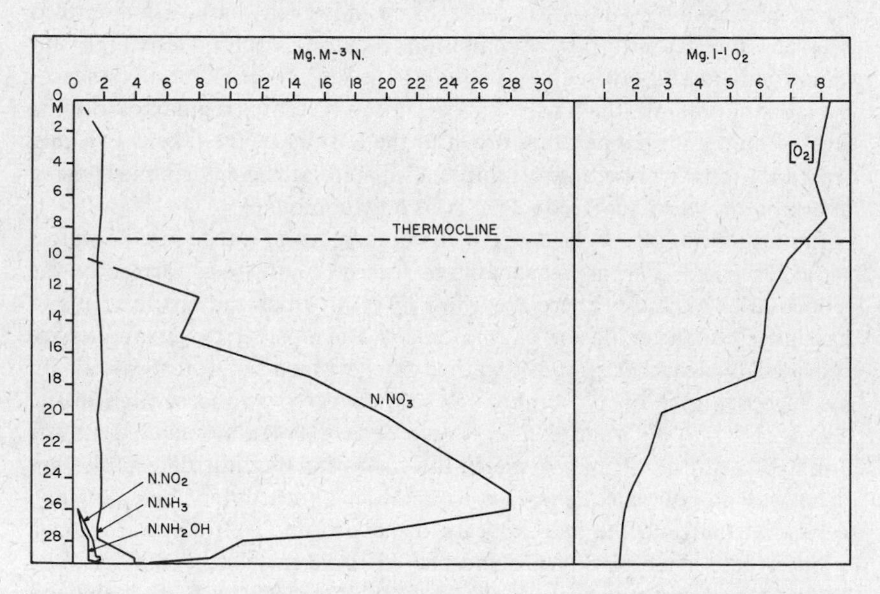

FIGURE 221. Kizakiko, Japan. Vertical distribution of nitrate, nitrite, and hydroxylamine, September 5, 1952. (From data of Tanaka.)

reduction with appearance of both nitrite and hydroxylamine in the bottom 3.5 m. in the presence of 0.4 to 0.5 mg. O_2 per liter, but no evidence of ammonia production in spite of the fact that the nitrate nitrogen lost was not balanced by the nitrogen in the nitrite and in the hydroxylamine formed (Fig. 221). In the lakes of northeastern Wisconsin, nitrate has usually been removed by assimilation in the trophogenic layer in July and August, producing an inverse clinograde nitrate curve. Some reduction has meanwhile occurred at the bottom, so that there is often a small accumulation of ammonia at the level of the maximum nitrate content. In some series of this sort, nitrite, the intermediate stage in reduction, is also present in small amounts at the bottom. In general, lakes such as Presque Isle Lake and Trout Lake show a pattern during August not unlike that developed in Esthwaite early in summer stagnation. In the single case of Adelaide Lake, which develops a very perfect dichotomic nitrate curve, there is no indication of nitrite at any level, though not enough depths were studied to reveal a narrow band of nitrite-rich water between 10 and 15 m. Lake Mary, which is meromictic and contains massive amounts of ammonia in the monimolimnion, would also be expected to exhibit the same kind of nitrite distribution which Müller found in the meromictic lakes of the Austrian Alps; again a narrow zone of nitrite-rich water may have been missed. Lake Mary, however, does raise a curious problem in relation to nitrate reduction, because in spite of the high ammonia content and low redox potential (page 774) of its essentially anaerobic monimolimnion, the nitrate concentration increases regularly from the level of minimum temperature down to the bottom of the lake. It would seem that nitrate reduction is inhibited, in spite of the environment being of a kind of which such reduction is usually complete.

In most, if not all, of the cases of the occurrence of nitrite in lake waters so far discussed, it seems reasonable to regard the nitrite as formed by the reduction of nitrate. There are a few lakes in which the nitrite seems to have arisen by the oxidation of ammonia. The most striking examples are provided by Karcher's (1939) studies of the Masurian forest lakes. In the Jegodschinsee on November 18, 1937, a narrow zone of high nitrite (14 mg. $N \cdot NO_2$ m.$^{-3}$) at 12.5 m. is recorded. The ammonia content was 9 mg. $N \cdot NH_3$ m.$^{-3}$; half a meter lower, nitrite was not detectable, but there was no equivalent increase in ammonia, the content being 10 mg. m.$^{-3}$ At the time the lake appears to have been freely circulating at a temperature of 6.8 to 6.9° C., and the 10 to 13-m. zone contained over 8 mg. O_2 per liter. No nitrate was detectable in any layer. It is difficult to interpret this case in terms of nitrate reduction, but it is also not at all clear why an excess of ammonia should have been present at 12.5 m. and should have undergone oxidation to nitrite. Whatever process took place must have occurred rapidly. A more spectacular case is provided by the Omulefsee at the end of winter stagnation under ice. Oxygen occurred abundantly down to the bottom at 16 m. An extraordinary nitrate and nitrite maximum occurred 1 m. from the bottom. Ammonia was abundant in the top 5 m., but declined between 5 m. and 10 m. above the rise in oxidized nitrogen. There was thus a very marked minimum in inorganic combined nitrogen at 10 m. The fact that ammonia fell at a higher level than nitrate and nitrite rose, precludes an unequivocal interpretation of this case as one of oxidation of ammonia in the free water, but it is hard to give any other reasonable explanation.

A better case is probably that of the Grosser Scewssee on March 9, 1937, evidently at the end of winter stagnation under ice and with rather over 5 mg. O_2 per liter from 2 m. to the bottom at 6 m. Here, there is a small deficiency in the ammonia curve at 5 m., corresponding to a decrease of about 8 to 10 mg. m.$^{-3}$ when compared with the 3 and 6 m. levels, and a marked maximum in nitrate of 60 to 65 mg. m.$^{-3}$ and in nitrite of 4.6 mg. m.$^{-3}$ It is reasonable to attribute the deepwater maxima of oxidized nitrogen here to the oxidation of ammonia, but it must be admitted that a slight minimum in the ammonia curve immediately under the surface is not reflected in the presence of nitrite or an equivalent increase in nitrate. A fourth case of a nitrite maximum, given by Karcher, in the Otzkosee at 5 m. early in summer stratification, may also possibly imply nitrification of ammonia.

In examining both Karcher's data and those of Einsele and Vetter, one is struck by the irregularity of the distribution of nitrite, which, as in the Jegodschinsee, may be present at a single depth in a lake of quite low stability. Similarly, Müller found in the Hüttensteinersee on September 21, 1933, traces of nitrite at 10 and 11 m., and then none until 17 m. when the maximum at the top of the deoxygenated region began. It seems most reasonable to conclude that nitrite formation in the hypolimnion depends not only on the chemical conditions but also on the existence of rather a special bacterial flora which may develop transitorily over a limited depth range. It would seem that in such cases nitrifying bacteria perhaps associated with detritus may be distributed in narrow zones of the kind which are implied by Whitney's measurements of transparency.

NITRATE IN LAKE WATERS

Bacterial nitrification and nitrate reduction. Ammonia is oxidized according to the reactions

$$NH_4^+ + OH^- + \tfrac{3}{2}O_2 = H^+ + NO_2^- + 2H_2O$$
$$\Delta F^\circ = -\ 59{,}400 \text{ cal.}$$

Cooper (1937b) has considered the likelihood of this reaction occurring unmediated by biological systems in the sea. He thinks it is possible that a chemical oxidation may occur at the interfaces provided by bubbles in the surface layers. There is clear evidence that photochemical activation of the reaction can also occur at the surface of the sea. According to Rao and Dhar (1931), in addition to ultraviolet light certain oxides must be present as photosensitizers. ZoBell (1933) and Rakestraw and Hollaender (1936) found that the reaction with sunlight or an artificial ultraviolet source proceeded more easily in natural sea water than in artificial sea water or distilled water. Natural sea water loses this property when autoclaved at 120° C. Cooper suggests that colloidal silica, which would be peptized at such a temperature, may act as the photosensitizer. If this is so, the reaction may be of some significance in the surface waters of lakes, but it is probable that biochemical oxidation is always of far greater importance in inland waters than are the possible photochemical or purely chemical mechanisms. In the sea, bacterial nitrification evidently is very difficult to demonstrate in the free water away from the bottom, and the existence of nonbiological mechanisms of oxidation of ammonia are therefore of considerable interest. In lakes, in which bacterial nitrification appears always to occur in the free water, at least at some season of the year, the nonbiological mechanisms are likely to be of less quantitative importance.

The oxidation of ammonia to nitrite involves certain intermediate steps. According to Corbet (1935) the sequence is

$$NH_3 \rightarrow NH_2OH \rightarrow H_2N_2O_2 \rightarrow HNO_2$$

Recent work (Lees and Quastel 1946a,b; Lees 1948), however, indicates that in soils hydroxylamine cannot be nitrified unless it is first converted into pyruvic oxime. Of the various compounds which Lees and Quastel have studied, pyruvic oxime was the only one of which the nitrogen could be oxidized directly by soil bacteria. Cooper (1937b) suggested that the production and accumulation of hyponitrite might explain certain cases in which a fall in the ammonia content in sea water stored in the dark is not balanced by a rise in nitrite and nitrate concentrations. There are suggestions in the oceanographic literature of comparable events occurring *in situ* in the sea, and if they are known in the ocean, like phenomena may also occur in lakes. However, in view of the work of Lees and Quastel, conclusions as to what may happen in nature, based on Corbet's scheme, are rather uncertain. In small lakes in which interaction with sediments or with large quantities of falling seston is possible, the sudden disappearance and reappearance of considerable amounts of ammonia in the free water is more reasonably attributed to adsorption on and release from particulate matter in the sediments than to the formation of hypothetical intermediates such as hyponitrite.

The further oxidation of nitrite proceeds according to the reaction

$$NO_2^- + \tfrac{1}{2}O_2 \text{ (gas)} = NO_3^- \qquad \Delta F^\circ = -18{,}000 \text{ cal.}$$

Cooper points out that at equilibrium and with an oxygen activity of 0.2 atm.

$$a_{NO_3^-} = 3.1 \times 10^{12} a_{NO_2^-} \tag{4}$$

The equilibrium concentration of nitrite in aerated water with an ordinary concentration of nitrate will therefore be analytically undetectable.

Since the oxidation both of ammonia to nitrite and of nitrite to nitrate is accompanied by a fall in free energy, these reactions are available as energy sources to any organisms which can activate them. Such a capacity is possessed by a limited number of bacteria found in soils, lake waters, and marine and fresh-water sediments. The commonest form oxidizing ammonia is *Nitrosomonas* and that oxidizing nitrite is *Nitrobacter*, but nothing is known about the specific composition of the limnetic flora. The chemical physiology of these organisms is complex; good recent reviews are given by Stephenson (1939) and by Porter (1946). In general, oxidation of ammonia proceeds in pure culture only in the virtual absence of ordinary organic matter, any appreciable amount of methionine being inhibitory (Quastel and Scholefield 1950), in the presence of calcium, and over a relatively restricted *p*H range, which, however, can differ greatly with the organism used; copper (Lees 1948) may be required as a trace element. The energy liberated by the oxidation is partly used to reduce CO_2 and therefore to form organic matter. The great difficulty of culturing the nitrifying bacteria in the presence of many kinds of organic matter appears inconsistent with their frequent occurrence in organic soils and sediments. It appears, however, from the work of Lees and Quastel (1946a,b; Quastel and Scholefield 1951), that the oxidation of ammonia takes place only in association with the surfaces of fine particles in soils. The bacteria associated with such particles may often be likewise associated with loci of base-exchange adsorption of NH_4^+. In the sea, though there is analytical evidence that nitrification occurs in the free water, nitrifying bacteria are only known to occur in the sediments, and most of the nitrification which does occur no doubt takes place in the water in the immediate vicinity of the bottom. In lakes, on the other hand, the Wisconsin school has provided plenty of evidence of intense nitrification in the free water, provided that oxygen is present, and the process is not to be expected in the anaerobic muds of small lakes with extreme clinograde oxygen curves. It is not unlikely that the general prevalence of nitrification at all oxygen-rich levels in lakes is due to the presence of more particulate matter in suspension, and therefore of more adsorbed ammonia, than would be expected in the ocean.

The observed concentration of nitrate will naturally depend on the balance of biochemical production and destruction. Though the activities of nitrifying bacteria alone account for the biochemical production of nitrate in lakes, the destruction of nitrate is accomplished in two major ways. One of these, the assimilation of nitrate by green plant cells, has already been mentioned. It always involves reduction, but there is little evidence of the extent to which the reduction can affect the extracellular concentration of unassimilated nitrate in plankton-rich natural waters, though it certainly can do so in certain *Chlorella* cultures. The other process removing nitrate is bacterial denitrification. This is one of a class of events in bacterial metabolism by which the oxygen of highly oxidized anions is used in the oxidation of organic matter. Analogous processes occur, leading to the reduction of sulfate and of various organic substances and, more rarely, of phosphate and selenate, by certain bacteria.

In all these events, the more oxidized zone of the biosphere is reduced and the more reduced is oxidized without the direct utilization of molecular oxygen. In the case of nitrate, the reaction may be expressed

$$C_6H_{12}O_6(aq.) + 12NO_3^- = 12NO_2^- + 6CO_2(aq.) + 6H_2O(liq.)$$
$$\Delta F^\circ = -460{,}000 \text{ cal.}$$

and for nitrite, reduced to molecular nitrogen,

$$C_6H_{12}O_6(aq.) + 8NO_2^- = 4N_2(aq.) + 2CO_2(aq.) + 4CO_3^= + 6H_2O(liq.)$$
$$\Delta F^\circ = -728{,}000 \text{ cal.}$$

The free energy change compares favorably with the fall of 699,000 cal. in the oxidation of a gram molecule of glucose in aqueous solution by dissolved molecular oxygen. In the presence of organic matter, therefore, both the reduction of nitrate to nitrite and the reduction of nitrite to hyponitrite (Cooper 1937b) or free nitrogen are processes which are capable of providing energy to organisms.

A rather large variety of bacteria can reduce nitrate, including the common *Escherischia coli* and *Serratia marcescens*, and there are known cases in which the process can take place easily in mixed culture of species not capable of this kind of metabolism in pure culture. ZoBell and Upham (1944, ZoBell 1946) note that thirty-four of the sixty species of marine bacteria which they studied can reduce nitrate in media enriched in organic matter to which nitrate has been added, but after a detailed survey of the available evidence ZoBell concludes, as others have done before him, that denitrification to molecular nitrogen is a rare and unimportant process in the ocean; it must, however, be an integral part of the large-scale geochemical cycle of the element (Hutchinson 1944b).

In sewage, bacteria causing the reduction of nitrate and loss of nitrogen in gaseous form evidently occur (Wooldridge and Corbet 1940), and under some conditions the process is probably of significance in aerobic soils (Corbet and Wooldridge 1940). The evidence from lakes is scanty, but perhaps suggests that in inland waters denitrification to N_2 is somewhat more important than in the ocean, and is of increasing significance in small organically polluted bodies of water. A slightly less extreme process ending in the production of N_2O, and probably accounting for the small stationary concentration of that gas in the atmosphere, is now well known in soils but has as yet not been detected in lake waters or lacustrine sediments. The process presumably involves formation of hyponitrous acid; Cooper suspects the occurrence of hyponitrite in the sea. The reduction of nitrate as far as ammonia is less well studied than is denitrification, restricting that term to the production of N_2, as is usually done. The production of ammonia by reduction, however, is clearly of much greater importance than has often been believed (Woods 1938). It is certainly performed by anaerobic soil bacteria of various kinds and by some strains of *Escherischia coli*. It is likely to be an energetically wasteful process, the adaptive significance of which is not entirely clear, except in so far as reduced nitrogen is needed as a source of amino groups. In the best studied cases, the process appears to involve the reduction of nitrate, through nitrite and probably hydroxylamine, by molecular hydrogen. It is evident that the process occurs extensively in lake waters, under conditions roughly the inverse of those promoting nitrification.

Nitrification and nitrate reduction in lake waters. Domogalla, Fred, and Peterson (1926) found that excess of ammonia added to samples of the surface waters of Lake Mendota collected with sterile apparatus was always nitrified, though the rate varied seasonally, being maximal in April and May and in October, and minimal in January and August. The rate of nitrification thus appears to follow the changes in the concentration of ammonia, but the concordance is not exact. The autumnal ammonia maximum is greater than the vernal, but for the rate of nitrification the opposite is true. An essentially similar pattern in the variation of rates of nitrification was observed by Domogalla and Fred (1926) in the surface waters of the lakes of the Yahara Basin in the vicinity of Madison. The concentration of nitrate at first rises with the increase in rate of nitrification in the spring, but begins to fall before the rate of nitrification is maximal, and remains low throughout the autumnal period of high ammonia and rapid nitrification.

The variation in the rate of nitrate reduction was studied in the Wisconsin lakes by Domogalla, Fred, and Peterson, and by Domogalla and Fred, at the same time as and by comparable methods to those used in the investigation of nitrification, excess nitrate being added and the time required to reduce it ascertained. Nitrate-reducing bacteria were found always to be present. In general, the rate of reduction is minimal in the winter and spring and maximal during July and August. In Domogalla and Fred's paper, evidence is presented graphically for a secondary maximum in the rate of nitrate reduction at the time of the autumnal overturn, which presumably partly explains the low nitrate which accompanies the high ammonia and rapid nitrification of this season.

In the deep water of Lake Mendota, the rate of nitrification and the nitrate content were maximal at the time of the spring ammonia maximum in March, but as thermal stratification set in and the hypolimnion became oxygen-free, no new increase in nitrification could accompany the great rise in ammonia, and the seasonal curve for nitrification, as well as that for nitrate concentration, is practically unimodal.

The investigations of Klein and Steiner (1929) on the Lunzer Untersee indicate that, as in Wisconsin, denitrification proceeds most rapidly in the summer. The Austrian investigators found that nitrification in the free water was most marked in winter, but the number of observations which they made is not great enough to permit any detailed conclusions as to the period of maxima.

Mortimer (1941–1942), discussing certain experiments not reported in detail, found that significant nitrification occurred in the laboratory in the waters which he examined, only if some sediment were present. He concluded, therefore, that the production of nitrate cannot occur to any great extent in the free water. While this conclusion is concordant with what is known about nitrification in soils, it can hardly be of general validity in lakes, and is in fact not concordant with some of Mortimer's own observations in Esthwaite Water.

Very little information exists as to the products of nitrate reduction in inland waters. The Wisconsin investigators did not study the fate of the nitrate lost from their cultures. Klein and Steiner never found ammonia, and concluded that free nitrogen was not often produced in significant quantities, the main product of the bacterial reduction of nitrate being nitrite. In the oxygen-free hypolimnion of many lakes, it would seem certain that reduction to ammonia is usual. Baier (1936), examining various pond and canal waters in the vicinity of Kiel, found that ammonia was in general produced almost quantitatively from nitrate, though in two out of six experiments some free nitrogen was liberated and in half the samples there was evidence of the production of nitrite. It is probable that in the Wisconsin work in which the fall in nitrate content appears to have been used as the criterion of nitrate reduction, the nitrite produced would be estimated as nitrate, and would therefore go undetected, so that in this investigation ammonia and perhaps some free nitrogen may well have been the main product. In one of Baier's samples, in which free nitrogen was produced, a bacterium comparable to *Thiobacillus denitrificans* was responsible for the reaction, and Klein and Steiner found denitrification associated with thiosulfate oxidation in the mud of the Lunzer Untersee.

Seasonal and vertical distribution of nitrate in lakes. It will be convenient to begin with Lake Mendota, because although it is unlikely to be the simplest, it certainly provides the best known case. In the surface waters, the net effect of the opposing processes of nitrification and nitrate reduction and assimilation permit a rise in nitrate from January until about the middle of April, when a maximum concentration of 72 mg. $N \cdot NO_3^-$ $m.^{-3}$ is reached (Fig. 222). Laboratory experiments (Fig. 223) suggest that the rate of nitrification continues to increase slightly until the early part of May, but that little increase in the rate of nitrate reduction is apparent at this time. The decline of nitrate concentration after the spring maximum is therefore reasonably attributed to nitrate assimilation by the phytoplankton. During the summer months, the nitrate concentration remains low, not exceeding 20 mg. $N \cdot NO_3^-$ $m.^{-3}$, and these low values apparently continue into the winter, for though the rate of nitrification as determined in the laboratory increases in the autumn and that of nitrate reduction decreases, the reductive processes appear always to equal or exceed the oxidative, except in the spring.

In some of the other lakes of the Yahara Valley, notably Lake Wingra and less markedly Lakes Kegonsa and Waubesa, the nitrate rises in the autumn, from October at least until December. There is not enough data to indicate if this rise continues through the late winter and early spring until the maximum in March or April is reached. Such a temporal pattern is very probable, however, and may quite likely be the most characteristic type of seasonal variation in the nitrate content in the trophogenic zone. Pearsall's (1930) data for the surface waters of the lakes of the English Lake District show just such a series of variations, the maximum nitrate concentrations occurring in the winter and spring and the minimum concentrations in July and August. The nitrate concentration in the lakes was found to rise in the autumn and winter before any marked increase in the phosphorus content had occurred, and Pearsall attributed considerable significance to the resulting variations in the nitrate:phosphate ratio. Karcher (1939) concluded that in the Masurian forest lakes, nitrate increased throughout the winter, reaching its maximum value in the spring, and the same kind of variation is certainly indicated by the large number of nitrate determinations on the surface waters of Connecticut lakes made by Deevey (1940). Nygaard's (1938) studies show that this type of seasonal variation is also frequent in ponds, in which an ammonia maximum in the early winter may be succeeded by a nitrate maximum which, though marked, is usually not stoichiometrically equivalent to the decline which simultaneously takes place in ammonia.

There is a little evidence that although acidophilous nitrifying bacteria may occur in peat bogs, there is a general tendency for nitrification to take place much less rapidly in acid than in neutral or alkaline waters. Nygaard found that nitrate never occurred in four of a series of fifteen carefully studied ponds. One of these four ponds was a culturally eutrophic

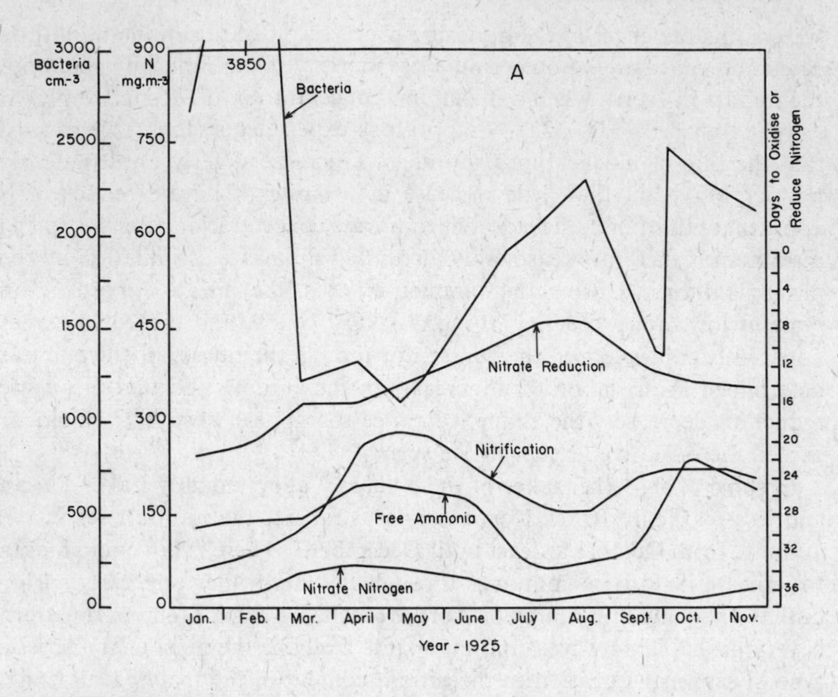

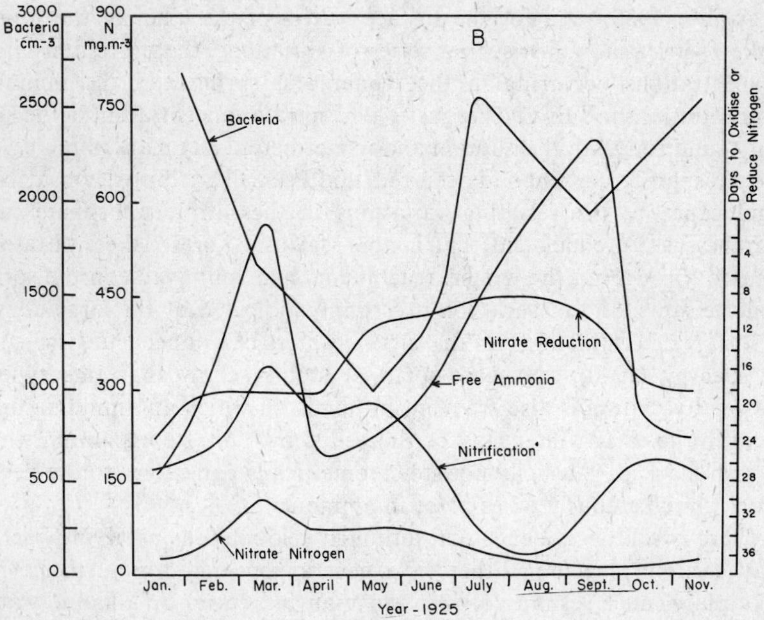

FIGURE 222. Lake Mendota, 1925. Bacterial counts, ammonia ($N \cdot NH_3$), nitrate nitrogen ($N \cdot NO_3$), and rate of nitrate reduction and of nitrification: *A*, in surface-water samples; *B*, in bottom-water samples. (After Domogalla, Fred, and Peterson.)

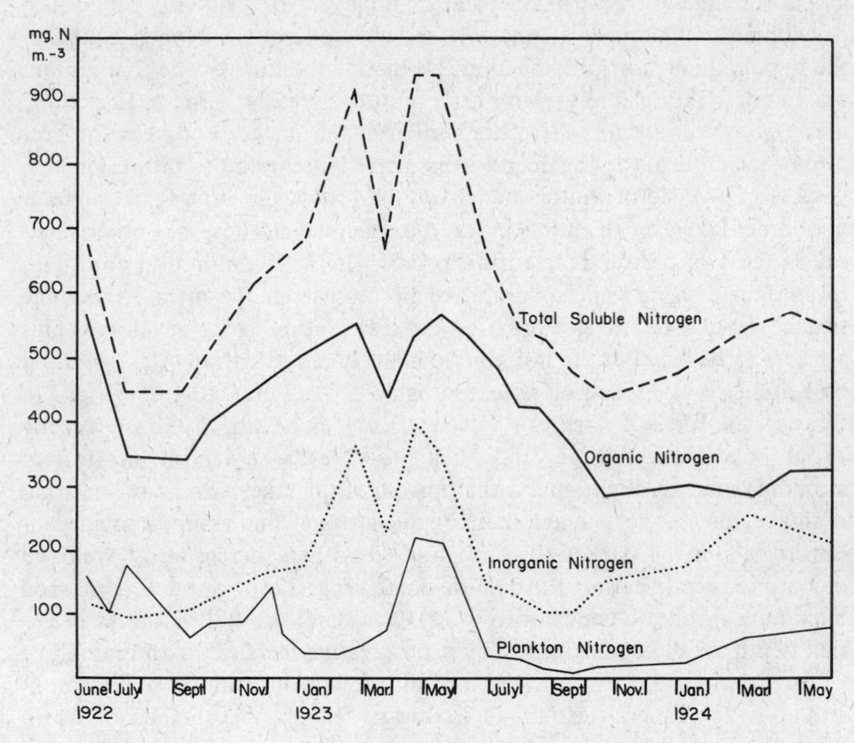

FIGURE 223. Seasonal distribution of the various forms of nitrogen in Lake Mendota, 1922–1924 (after Domogalla, Juday, and Peterson).

pond containing much *Elodea*, which may have removed nitrate as fast as it was formed. The other three localities were very acid (pH $\leqslant$ 5.0), and nitrification clearly did not occur in them. In these very acid ponds an ammonia maximum occurred late in the winter, when the other ponds exhibited nitrate maxima. Karcher usually found no more than a trace of nitrate in the trophogenic layers of acid lakes during the summer, but always obtained detectable quantities in comparable neutral and alkaline waters. Karcher concludes that nitrification proceeds so much more slowly in the acid waters than in those with a pH value of over 7.0, that in the former the process never keeps pace with the processes tending to remove nitrate, whereas in the latter a small steady-state concentration can always be maintained. Some acid lakes rich in supposedly colloidal humic material did, however, contain detectable nitrate, which is supposed to have been held by adsorption.

In the deep water of Lake Mendota, nitrate accumulates in the spring, and at the time of the vernal circulation all layers are evidently rich. The development of summer stagnation and the consequent oxygen deficit of the hypolimnion stops nitrification, so that at the time of the ammonium maximum in the deep water very little nitrate is present. In broad outline, the seasonal variation is therefore similar at the surface and near the bottom of the lake, but in the deep waters there is rather more nitrate present just before the end of winter stagnation than is ever present at the surface.

In other lakes, in which less marked oxygen deficits develop, other vertical patterns of nitrate distribution occur. If the hypolimnion continues to contain oxygen while assimilative processes remove nitrate from the surface layers, an inverse clinograde distribution may develop. This appears to be usual in the holomictic lakes of the Austrian Alps, is noted by Karcher in the Jegodschinsee, and is even more markedly developed in some of the lakes of northern Wisconsin. It is beautifully illustrated by Trout Lake (Juday, Birge, and Meloche 1938), where both nitrate and ammonia increase slightly near the bottom of the lake. Some of the lakes of this region are too deficient in inorganic nitrogen and assimilating organisms to exhibit any marked stratification. In others there is well-developed dichotomic stratification. Adelaide Lake (Fig. 224) exhibited a marked maximum in nitrate concentrati n at 10 m., and the fall in values below that depth is fully compensated by a progressive increase in ammonia.

The development of a dichotomic nitrate stratification is well demonstrated in Mortimer's account of Esthwaite Water. At the end of April, when the investigation began, the lake was practically homoiothermal at 10° C., and was well oxygenated throughout. The nitrate content at all depths was about 250 to 300 mg. $N \cdot NO_3$ m.$^{-3}$ Throughout May, the concentration fell at all levels, so that at the beginning of June only 100 mg. m.$^{-3}$ were present. Thereafter, as the hypolimnion became deoxygenated, the nitrate in the deeper water of the lake disappeared, becoming reduced to ammonia, and the nitrate at the surface declined, presumably owing to continued assimilation by phytoplankton at the height of summer stratification. Thus, though the vertical distribution of the ammonia is inversely clinograde, that of nitrate is dichotomic, with a marked maximum of about 200 mg. m.$^{-3}$ at 9 m. This dichotomic vertical pattern is apparently characteristic of moderately productive stably stratified lakes during summer stagnation. With the breakdown of stratification the nitrate content of Esthwaite Water increased at all levels. When the lake first became homoiothermal at 10° C. at the end of October, all layers contained about 100 mg. $N \cdot NO_3$ m.$^{-3}$, but by the end of December, when the temperature had fallen to 6.6° C., all layers contained 250 mg. m.$^{-3}$ Later, under the ice cover a further increase took place, so that at

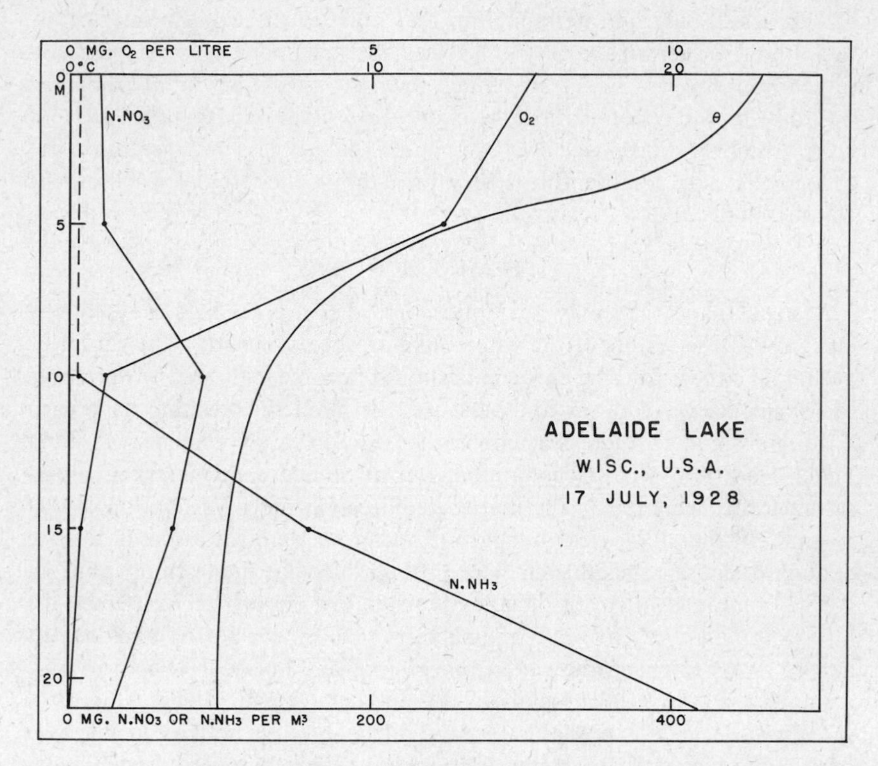

FIGURE 224. Vertical distribution of the forms of nitrogen in Adelaide Lake, Wisconsin, during the height of summer stratification (from data of Juday, Birge, and Meloche).

the beginning of February the surface waters below the ice contained 360 mg. $N \cdot NO_3$ m.$^{-3}$; at 9 m., 440 mg. m.$^{-3}$ were present. Below this depth, the concentration fell off to about 150 mg. m.$^{-3}$ at the bottom, at 14 m.

As has been indicated, Mortimer thinks it unlikely that nitrification occurs in the free water of lakes. In support of this view, he cites not only some unpublished experiments already touched upon but also the seasonal observations in Esthwaite Water. He believes that if nitrification of ammonia which has passed into the water explains the winter rise in nitrate, then ammonia should begin to accumulate in the hypolimnion as soon as the concentration of the deep water has fallen so low that nitrification no longer can take place. Actually, there appears to be a lag between the time when nitrate no longer increases in the spring and ammonia begins to increase. Mortimer concludes, therefore, that ammonia, as such, cannot leave the mud until the oxidative microzone disappears, though nitrate reduction may start in the water prior to this. In spite of the reasonableness of Mortimer's argument, it is very difficult to see how the rise in nitrate under the ice, when horizontal transport would be minimal, could affect the entire water column, except the bottom layer which has been accumulating ammonia, unless nitrification were occurring on a considerable scale throughout the greater part of the volume of the lake.

SUMMARY

Molecular nitrogen may be supposed to dissolve at the surface of the lake, and if the waters of the whole lake are being mixed, as at vernal or autumnal circulation, the entire lake should become saturated with respect to the surface temperature and pressure. In the few lakes that have been adequately studied, there is usually rather unsatisfactory evidence for slight undersaturation in the hypolimnion, but in one case very marked supersaturation is recorded. The matter requires further investigation, but it is unlikely that biological nitrogen fixation or denitrification is intense enough to cause detectable changes in the molecular nitrogen in solution.

The combined nitrogen of lakes is probably mainly derived from the inflowing water. Nitrogen-fixing bacteria certainly occur in the sediments and to a less extent in the water, but so far have not been shown to play a significant part in lake metabolism. In lakes in which large blooms of *Anabaena* occur, it is probable that the nitrogen fixing activity of this alga may make significant contributions to the nitrogen undergoing biogeochemical transformation in the lake.

Ammonia is produced by practically all heterotrophic bacteria in the course of organic decomposition. The free NH_4^+ ion and the undissociated base have very different solubilities and physiological properties. A change in the *p*H in the presence of ammonia may produce much more striking effects, owing to the toxicity of NH_4OH, than an identical change when ammonia is absent. It is probable that part of the ammonia in lake water is adsorbed to seston, and some observations suggest that appreciable quantities may become associated with the surface film. During periods of stratification, ammonia accumulates in the hypolimnion in the same general way as ferrous iron, phosphorus, and silicate. It is probable that apart from the cessation of nitrification when the redox potential falls below about 0.4 volt, this accumulation is also determined by the fall in the adsorptive power of the mud when the oxidized microzone disappears. During the autumnal circulation period this ammonia is distributed throughout the lake. The available data, however, indicate that rather large amounts of ammonia can disappear from the water either before or after autumnal circulation without any compensating increase in the other forms of nitrogen.

At least in some lakes, nitrification can take place in the free water whenever oxygen is present, whereas nitrate reduction occurs at all times. Nitrification appears to proceed most rapidly in winter, and nitrate reduction in summer. The existence of denitrifying bacteria that liberate molecular nitrogen is known, but it seems probable that most of the nitrate reduced forms ammonia. It is probable that the presence of surfaces is needed for the nitrification of ammonia, only adsorbed ammonia being oxidized. The apparent difference between at least some lakes and the sea, where it is very hard to demonstrate nitrification in the free water, may be accounted for by the greater amount of suspended matter in inland waters.

The nitrification process is sensitive to *p*H and occurs less rapidly, or in extreme cases not at all, in acid waters. Nitrite and hydroxylamine are known to occur in lakes, as intermediates in the oxidation or reduction processes. In general, maximal amounts of nitrate tend to be present at the end of winter or at the vernal circulation period. In some unproductive lakes with orthograde oxygen curves, little stratification develops. More usually in such lakes, nitrate is removed from the trophogenic zone in the summer, but little increase occurs in the hypolimnion. In more productive lakes with clinograde oxygen curves, nitrate is usually removed by assimilation in the trophogenic layer and by reduction near the bottom of the lake, producing a marked dichotomic distribution with a nitrate maximum in the middle water. In such cases it is usual to find a nitrite maximum between the nitrate maximum in the middle water and the ammonia maximum at the bottom. In these cases the nitrite is clearly produced by reduction of nitrate. Minute amounts of nitrite present in the surface waters of some unpolluted lakes may perhaps be due to reduction of nitrate by the phytoplankton. A few cases have been described in which thin layers of water rich in nitrite are associated with vertically restricted low ammonia contents, suggesting the formation of nitrite by oxidation of ammonia.

CHAPTER 17

Organic Matter in Lake Waters

The organic compounds that are present in either true or colloidal solution in lake waters, like nearly all the organic substances found free in the biosphere, have proved difficult to investigate. This difficulty will be appreciated when it is realized that there is no known way by which the total quantity of organic matter present in a natural water can be estimated with the degree of accuracy usual in routine quantitative inorganic analysis.

TOTAL ORGANIC CONTENT

Methods of estimation and interpretation of results. Significant results have been obtained by the following methods.

Determination of the loss on ignition of the dry residue after evaporation. Correction can be made for the CO_2 lost from alkaline earth carbonates, and in theory a reanalysis would permit an estimation of the error due to the loss of small quantities of chloride. The main error in the method is due to the retention of water by the dry residue, which water cannot be removed at temperatures below those at which charring occurs. Drying *in vacuo* should give reasonably reliable results but does not seem to have been used in limnological work.

Determination of the quantity of some oxidizing agent required to oxidize the organic matter under standard conditions. Potassium permanganate has been most usually employed. Oxidation is usually performed, under either alkaline or acid conditions, at 100° C. for a stated arbitrary time. Oxalic acid is then added and the excess back-titrated with standard permanganate. It appears from the work of Juday and Birge (1932) that under acid conditions, oxidation for thirty minutes destroys organic matter equivalent to 40 per cent of the organic carbon present. Wet oxidation with other strong oxidizing agents, such as chromic acid, has been made the basis of techniques widely used in pedology but apparently so far not applied to lake waters.

Determination of organic carbon by combustion and of organic nitrogen by the Kjeldahl or any other suitable method. The sample should have first been extracted with ether and the ether extract weighed. The nitrogen is usually computed as crude protein containing 52 per cent carbon, and the balance of the carbon is supposed to be present as carbohydrate, including lignin, etc., containing 45 per cent carbon. The total organic matter is considered to be the sum of the ether extract, the crude protein, and the carbohydrate estimated in this way. This method has been widely used by the Wisconsin investigators (Birge and Juday 1934). A technique involving characteristically ingenious microgasometry, but based essentially on the same general principles, has been used by Krogh and Lange (1932) in their study of Lake Furesø. The validity of the estimates of total organic matter obtained by methods of this sort obviously depends on the assumptions made as to the distribution of carbon and nitrogen, which assumptions are certainly only fairly close approximations.

Interrelation of various types of analytical result. Juday and Birge (1932) investigated the relationship of the quantity of permanganate reduced—or the oxygen consumed, as it is usually called—to the organic carbon. There is considerable variation in the oxygen consumed by waters within a narrowly defined carbon class, but the mean value for each class, when plotted against carbon content, gives a nearly straight line that cuts the axis corresponding to zero organic carbon at a point corresponding to about 2 mg. per liter permanganate oxygen consumed. It is hard to believe, however, that this represents the regular occurrence of a small amount of noncarbonaceous oxidizable matter. From the slope of the line, it appears that 9 mg. per liter oxygen consumed is equivalent to 8.7 mg. carbon per liter. Since 32 mg. of oxygen are needed to oxidize 12 mg. carbon to CO_2, it would appear that the mode of procedure used by Juday and Birge oxidizes about 40 per cent of the organic carbon present.

The relationship of permanganate oxygen consumed to water color has been considered by Juday and Birge (1932), Ohle (1934a), Thunmark (1937), and Åberg and Rodhe (1942). Juday and Birge give a mean curve which tends to diverge from rectilinearity toward the color axis, implying that increase in the oxygen consumed at high concentrations corresponds to a greater increase in color than at low concentrations. Thunmark's curve is very similar in form to that of Juday and Birge, but is displaced about 8 mg. per liter along the oxygen-consumed axis (Fig. 225). Prac-

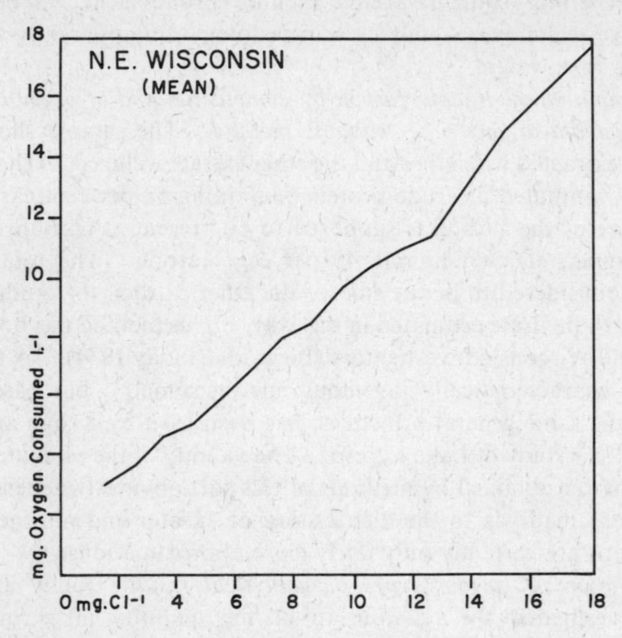

FIGURE 225. Mean permanganate oxygen consumed for waters of varying organic carbon content.

tically all Åberg and Rodhe's points fall along Thunmark's line, the chief exception being that for Växjösjön, a lake which, as the result of cultural eutrophy, is evidently more productive of phytoplankton than the other lakes Åberg and Rodhe examined. The results of all these investigations indicate that about 10 to 20 mg. oxygen consumed per liter correspond to a material not appreciably contributing to the water color. The plot of water color against carbon content (Fig. 226) indicates the same sort of phenomenon (Juday and Birge 1933), and it would seem to be rather generally recognized that in the clearest lakes, which received no water from peat deposits and which support a limited plankton population, a small amount of relatively uncolored organic matter is usually present in the water. Ohle's results are rather irregular, but they show the same

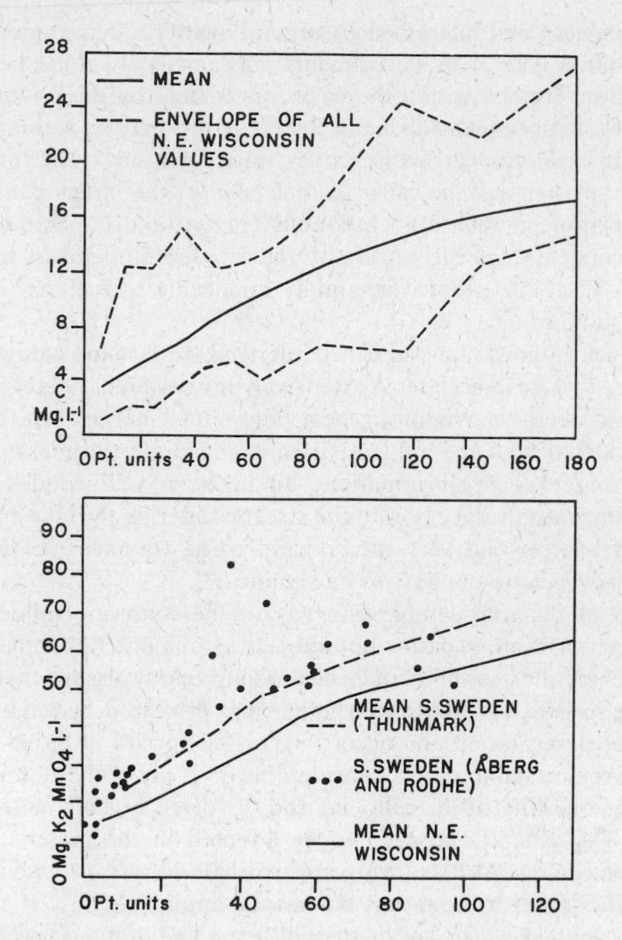

FIGURE 226. Mean color (Pt units) of waters of varying organic carbon content, or oxygen consumed.

tendency for the rate of increase of color with respect to oxygen consumed to increase as the latter quantity increases, and the same indication of the existence of organic matter that can be oxidized with permanganate but which imparts almost no appreciable color to the water.

In the next section the possibility will therefore be explored of considering the waters of lakes as containing two proximate organic fractions, one relatively uncolored and formed from plant material, mainly planktonic, that has developed in the lake, the other highly colored and essentially consisting of the extractives of peat from the margins of the lake or from peat bogs and soils in the catchment area of the lake.

Autochthonous and allochthonous organic matter. The organic matter found within a lake or in its sediments may obviously either be material that has been formed within the basin, ultimately by the photosynthetic activity of the green plants living in the lake, or it may be material formed outside the basin and carried into it by water or wind. The former type of organic matter may be called *autochthonous*, the latter *allochthonous*. The distinction between autochthonous (*gyttja*) and allochthonous (*dy*) organic sediments was early made by the Swedish students of sedimentation (Von Post 1862), but it is equally applicable to material in true or colloidal solution.

The allochthonous material can be derived from many sources, but in the regions that have been most extensively investigated, notably northern Europe and northern Wisconsin, peat bogs either marginal to the lake or in the headwaters of the influent streams constitute the most important external source of organic matter. In lakes in well-wooded districts, leaves falling into the lake or into the streams entering the lake are another important source, and very rarely wind-borne fragments of grasses or herbaceous vegetation appear to be significant.

In view of the considerable diversity of the sources of allochthonous organic matter even in basins not subject to human contamination, and also in view of the possibility that the autochthonous organic matter varies from lake to lake, according to whether it is produced by phytoplankton or by rooted vegetation and according to the species of either involved, there is every reason to expect considerable diversity in the composition as well as the quantity of the colloidal and dissolved organic matter of lake waters. The general approach to be adopted in the present section is therefore to be regarded as giving approximate results, probably of widespread validity but by no means universally applicable; it is at least likely that wherever lake waters have received their allochthonous organic matter from marginal peat or from streams draining peat bogs, results of the same general kind could be obtained.

By far the most valuable data on the proximate composition of the organic matter dissolved in lake waters are those given by Birge (Birge and Juday 1934). Having at his disposal analyses from several hundred lakes, Birge has presented his data as averages for classes of lakes grouped according to the total organic carbon content of the water prior to the removal of the seston. Only a selection of such classes is considered in the published report (Table 123). For each class the total organic seston, the total organic matter in colloidal or true solution, and the composition of this organic matter are given. The composition is presented in terms of three arbitrary categories. The ether extract is directly determined. The crude protein, which certainly contains a large number of substances other than proteins, is obtained by multiplication of the total nitrogen content by 6.25 in the usual way; the carbohydrate is obtained from the residual carbon not accounted for by the ether extract or the crude protein, which is supposed to contain 52 per cent carbon. The carbohydrate is computed on the assumption that it contains 45 per cent carbon.

TABLE 123. *Proximate composition of organic matter from northeastern Wisconsin waters containing varying amounts of total organic carbon*

Carbon Content mg. per liter	Organic Seston, mg. per liter	Organic Matter in Solution, mg. per liter	Crude Protein, per cent	Ether Extract, per cent	Carbo-hydrate, per cent	C:N Ratio
1.0–1.9	0.62	3.09	24.3	2.3	73.6	12.2
5.0–5.9	1.27	10.33	19.4	1.3	79.0	15.1
10.0–10.9	1.89	20.48	14.4	0.4	85.2	20.1
15.0–15.9	2.32	31.30	12.9	0.2	86.9	22.4
20.0–25.9	2.22	48.12	9.9	0.2	89.9	29.0

The figures of Table 123 clearly indicate that an increase in total organic carbon is accompanied by a decline in the crude protein fraction and in the C:N ratio. The ether extract content of the dissolved organic matter falls at first with increasing carbon, but evidently tends to a low constant value. The seston increases at first, but evidently tends to a high constant value in the upper part of the range. The water color is known to increase with increasing carbon content.

Since the darkest waters of high carbon content are known to derive a considerable part of their organic coloring matter from peat bogs, it is natural to attempt to explain Birge's distributions in terms of the progressive addition of allochthonous organic matter low in nitrogen to autochthonous organic matter having a composition comparable to that of the organic matter of the 1.0 to 1.9 mg. C per liter class. Birge himself attempted to do this, though he used, in order to obtain a proportionality factor between seston and autochthonous dissolved organic matter and to obtain the proximate composition of this autochthonous organic matter, a wider range of lakes than the three included in the lowest class of carbon contents. It is reasonably certain that some of the lakes included by Birge in his group supposedly free from allochthonous organic matter did in fact contain a significant amount, but since his ratio of dissolved organic matter to sestonic organic matter is 6:1 rather than 5:1 as in the calculations to follow, and his crude protein content in the autochthonous organic matter is 20 per cent rather than 24.3 per cent, the differences counterbalance each other and the results indicated by Birge are essentially the same as those to be set out in the following paragraphs. The assumption of a proportionality between plankton and autochthonous organic matter would seem to be reasonable, but it introduces an hypothetical element into the calculation, which is emphasized by the fact that such an assumption is certainly invalid when highly productive lakes are considered.

The class of lakes containing least carbon supported 0.62 mg. per liter organic seston, which is assumed to be in equilibrium with the 3.09 mg. per liter dissolved organic matter. The mean organic seston of the region contains

	Per Cent
Crude protein	37.0
Ether extract	4.0
Carbohydrate	59.0

It is evident, when these figures are compared with those for the composition of the dissolved organic matter of the lakes poorest in carbon, that in forming this organic matter nitrogen has been lost more rapidly than carbon. Assuming that in each of the other classes the seston is responsible for autochthonous organic matter in a ratio of 0.62:3.09, or as 1:4.98, and that this organic matter contains.

	Per Cent
Crude protein	24.35
Ether extract	2.3
Carbohydrate	73.6

the composition of the allochthonous organic matter can be obtained. For the class of organic carbon contents between 5.0 and 5.9 mg. per liter, the 10.33 mg. per liter total soluble organic matter will consist of

	Mg. per Liter
Crude protein	2.00
Ether extract	0.11
Carbohydrate	8.18

Subtracting 1.54 mg. per liter crude protein, 0.14 mg. per liter ether extract, and 4.64 mg. per liter carbohydrate, due to the 1.27 mg. per liter organic seston, the corrected values become

	Mg. per Liter
Crude protein	0.46
Ether extract	− 0.03
Carbohydrate	3.54

Neglecting the very small, meaningless negative ether extract, the water appears to contain 4.00 mg. per liter allochthonous organic matter; of this 11.55 per cent is crude protein and the rest carbohydrate, according to the analytical conventions here adopted.

When all classes of carbon contents are considered in this way, the values in Table 124 are obtained.

TABLE 124. *Estimates of allochthonous organic matter and its properties in the waters of Table* 123

Carbon Content, mg. per liter	Allochthonous Dissolved Organic Matter		Crude Protein, per cent	Carbo-hydrate, per cent	Water Color, Pt Units		
	mg. per liter	% total			In lake	Per mg. total org. matter	Per mg. alloch. sol. org. matter
5.0– 5.9	4.0	38.8	11.5	88.5	25	2.4	6.3
10.0–10.9	11.2	54.8	5.9	94.1	67	3.2	6.0
15.0–15.9	17.9	56.0	7.0	93.0	135	4.3	7.5
20.0–25.9	36.2	75.2	5.7	94.3	...	...	...

At least in the upper part of the range of carbon contents, above 10.0 mg. C per liter, the nonplanktogenic organic matter calculated in this way appears to consist rather constantly of a material containing about 6 per cent crude protein, with a C:N ratio of 47:1. By inspection of Fig. 226, indicating the relationship of water color to carbon content given by Juday and Birge (1933), it is possible to estimate approximately the mean water color for three of the classes of water in Table 124. It will be apparent that whereas there is a steady rise in the ratio of color units to the total soluble organic matter, there is no such systematic rise in the ratio to the allochthonous organic matter. It is reasonable, therefore, to attribute most or all of the water color to the latter fraction, a conclusion already reached on qualitative grounds by Birge and probably held by all previous investigators of the matter.

Birge, as has been indicated, performed calculations similar to those presented here, but rejected the results because of the very low nitrogen contents of the calculated allochthonous fractions. These he regarded as suspect, since they were so much lower than the mean nitrogen content of the waters of forty-one bogs in northern Wisconsin. In such bog waters the total organic matter had a mean crude protein content of 13.3 per cent. This figure apparently represents the organic matter including seston; it must include a little protein derived from plankton, since there is no reason to believe that all the bogs would be entirely free of organisms. The dissolved and colloidal organic matter, which alone is of interest here, would have had a somewhat lower crude protein content. It is apparent, moreover, that the mean bog water contains organic matter actually richer in nitrogen than does the water of the highest class of lakes arranged according to carbon contents.

There is therefore, right from the beginning of the calculation, no possibility of reconstructing the series of lake waters by mixing water from the lakes of the lowest class of carbon contents with mean bog water. It is possible that some of the bogs would have contained waters in which the soluble organic matter was so low in nitrogen that they could have served as a standard of pure allochthonous organic matter. Acid peat containing only 2 per cent ash and as little as 0.65 per cent N is in fact known (Waksman and Stevens 1928); the aqueous extract of such a peat, rather than the mean of a series of bog waters in some of which there is likely to be a good deal of planktonic and other life, would provide a more critical standard of comparison. Until the composition deduced for the allochthonous fraction is shown to be impossible by some such comparison, the treatment given here, even if it is somewhat artificial, is illuminating and does permit the recognition of two theoretical categories of organic matter in solution:

(1) *Autochthonous organic matter*, containing about 24 per cent crude protein, with a C:N ratio of about 12:1, probably not imparting a very strong brown color to the water, largely derived from the decomposition of plankton.

(2) *Allochthonous organic matter*, containing about 6 per cent crude protein, with a C:N ratio of 45–50:1, imparting a strong brown color to the water, derived from peaty material either in the lake sediments or marginal bog, or in the catchment area of the influent streams. Much of the material derived from marginal rooted vegetation may belong in this category.

Colloidal organic matter in lake waters. One fundamental but practically unsolved problem is that of the nature of the solutions, whether true or colloidal, that are involved. The only attempt at separation by ultrafiltration appears to be that of Krogh and Lange. Their analytical technique was somewhat different from that of the Wisconsin workers, but it is probable that their mean results on Lake Furesø in Denmark actually do indicate differences in the composition of the seston and dissolved organic contents of that lake and those of northern Wisconsin. Krogh and Lange obtained the mean contents indicated in Table 125.

The soluble organic matter contained much more nitrogen than the supposedly pure autochthonous organic matter of the least colored Wisconsin lakes, but the same loss of protein in passing from the particulate to the dissolved matter is apparent in both the Danish and American analyses. The ratio of the soluble to the sestonic organic matter is 5.6 :1,

TABLE 125. *Proximate composition of seston, colloidal and soluble organic matter of Furesø*

Constituent	Fraction, mg. per liter	Protein, per cent	Fat, per cent	Carbohydrate, per cent
Sestonic organic matter	1.56	53.2	9.9	36.9
Colloidal organic matter	0.67	41.5	11.7	46.8
Soluble organic matter	8.8	37.5	...	62.5

in agreement with experience in Wisconsin. The most unexpected feature of the results is the very small quantity of colloidal organic matter present. The smallness of this organic colloidal fraction is the more remarkable when it is compared with the mean total colloid, which is 3.5 mg. per liter; 81 per cent of this amount is evidently inorganic, presumably consisting of hydrated ferric oxide and alumina and perhaps of silicates also. Krogh and Lange found the dissolved organic matter of Furesø to be extremely stable, hardly decomposed after months of storage in either the presence or absence of bacteria. In this respect the material resembles not only lignin complexes but also the 5 mg. per liter organic matter found by Krogh (1934) normally to be present in sea water. It is possible, however, for the material in the sea to act as a nitrogen source for bacterial metabolism when a pure carbohydrate is added as an energy and carbon source (Waksman and Carey 1935).

Ohle (1934b) has examined the colloidal content of lake water by indirect methods. The total ionic content of the lake water is first determined by analysis, and the conductivity of the water is calculated. The apparent conductivity of the water is then measured, and the degree of dissociation $\hat{\alpha}$ calculated as the ratio of the observed to the theoretical reading. Table 126 gives the values obtained in a series of north German lakes.

TABLE 126. *Indirect estimate of colloidal matter in North German lakes*

Lake	*p*H	Color, Ohle units	C, 10^{-4} ohm cm.$^{-1}$	$\hat{\alpha}$	$\mu_{CA} - C_{CA}$
Schwarzsee	4.32	38	0.856	1.529	19.3
Garrensee	6.97	3.8	0.604	1.235	0.73
Schwarze Kuhle	5.02	23.6	0.475	1.049	...
Grundloser Kolk	4.48	4.0	0.300	1.028	1.50
Drewitzer See	8.34	3.8	2.459	0.960	18.0
Sarnekower See	9.15	45.0	3.220	0.838	60.6
Grosser Ukleisee	7.6	8.0	4.430	0.871	50.3
Hertasee	8.4	70	2.332	0.726	60.5
Klinkerteich	8.08	19	8.560	0.879	111.0

It is apparent that the acid waters, poor in electrolytes, tend to give apparent dissociation coefficients greater than unity. Ohle considers that in such cases the observed conductivity is due not only to ionized electrolytes but also to the cataphoresis of highly dispersed colloidal particles. In the waters rich in electrolytes, from the alkaline lakes, the apparent dissociation coefficient is less than unity, and this is interpreted as due to the adsorption of alkaline earth bicarbonate by the less dispersed colloidal particles.

A very similar result is obtained when the refractive index μ_{CA} is compared with the conductivity C_{CA}, both quantities being expressed in terms of the calcium content (mg. per liter), of $CaCl_2$ solutions having the same conductivity or refractive index as the sample. Part of the refractive index will be due to the undissociated organic matter, but where this consists of small charged particles, these will contribute to the apparent conductivity and the difference $\mu_{CA} - C_{CA}$ will be small. In waters in which the electrolyte content, though high, behaves in part as if undissociated, owing to the adsorption of shells of ions around the colloidal particles, the difference $\mu_{CA} - C_{CA}$ will be greater. Ohle gives data for nine lakes in which the difference is less than 19.3; in these the *p*H varies from 4.32 to 7.20. He also gives data for eight lakes in which the difference is greater than 38.5; in these the *p*H varies from 7.16 to 9.15. Highly colored as well as relatively uncoloured waters occur in each group. Such results indicate very clearly the presence of colloidal material and its effects on certain of the physical properties of lake water, but they do not throw any light on its chemical nature.

THE COLORED ORGANIC MATTER OF LAKE WATERS

Two views have been held as to the nature of the colored material, often of largely allochthonous origin, occurring in lake waters. One view is that the material is colloidal and related to the protein-lignin complexes supposed to constitute the true humus of soils; the other view is that the brown material is comparable to the humic acids, of uncertain composition though low in nitrogen, that have long been recognized in soil extracts.

Supposed relationship to lignin-protein complexes. It is often believed that the humus of soil consists to a large if variable degree of protein-lignin complexes. Such complexes, dark in color and having solubilities comparable to those classically ascribed to humus, have been synthesized by Waksman and Iyer (1932a,b,c, 1933a,b). Not all the organic matter that would have been regarded as humified by the older workers can be considered as proteolignin. In particular, the main part of the humus of acid peat is not a lignin derivative but apparently consists of hemicelluloses or polyuronides (Waksman and Stevens 1928). Steiner and Meloche (1935) found that lignin-like materials occur in both lake plankton and in lake sediments, so that a possible source of proteolignin complexes is available in lakes. Practically no information is forthcoming about the presence of lignin derivatives in the water.

Ohle (1933) has determined the methoxyl content of the residue left after evaporation of the water of the Kleiner Uklei See near Plön. He found 0.12 mg. per liter CH_3O to be present; this corresponds to about 1 per cent of the residue. The methoxyl content of the so-called lignin of lake plankton and of lacustrine mosses and *Isoetes* was found by Steiner and Meloche to lie between 0.5 and 1.05 per cent. The Kleiner Uklei See, which has colored water (12.8 on Ohle's scale, or about 40 Pt units) undoubtedly receives a good deal of its organic matter from the leaves of trees and from forest litter. Since the lignin of higher plants may contain as much as 16 per cent CH_3O, it is possible that at least 6 per cent, and possibly much more, of the organic matter dissolved in the waters of this lake is present as a lignin-like material. Ohle, however, supposes that the methoxyl is present in alkali humate in true solution.

Although some oceanographers, as has been indicated in Chapter 11, have been favorably impressed with the idea that protein-lignin complexes are important stabilizers of ferric hydroxide in natural waters, the evidence that such complexes actually exist in lake or sea water is, to say the least, unimpressive.

Chemical studies on the yellow coloring matter of lake waters. The idea that the brown or yellow stain of lake waters is due to a loosely defined humic acid, or group of such acids, is an old one. The first important study founded on such an hypothesis is that of Aschan (1908). Very recently Shapiro (1956) has concluded that the comparison of the coloring matter of lakes with what would have been termed humic acid, rather than with a lignin-protein complex, is correct, though the yellow acids in lakes seem to have a molecular weight that is considerably less than that ordinarily ascribed to humic acids.

Shapiro (1956) finds that at least a considerable part of the yellow coloring matter characterizing all but the clearest lakes is due to a group of yellow carboxylic acids, soluble in ether and in ethyl acetate, which pass through a dialysis membrane. The unresolved mixture of such substances has a mean molecular weight of about 450, very much less than that of the brown materials derived from soils and ordinarily termed humic acids in a broad sense. Like the humic acids, the solid yellow material derived from lakes is not clearly crystalline.

The mixture isolated by Shapiro forms salts highly soluble in water and is strongly fluorescent in solution. In spite of a very high carbon content, about 55 per cent, there is no certain evidence from the infrared spectrum of any phenolic structure in the molecule. The nitrogen content is very low and may represent nitrogenous impurities. Chromatographic separation indicates that about five individual compounds of this sort are ordinarily present in lake waters. The visual color of aqueous solutions of the material per unit mass of carbon is comparable to that found by Juday and Birge (1933) for untreated lake waters. The color depends on the *p*H, however, as was noted for lake water by Ohle (1934a). As in the observations of Hutchinson, Vallentyne, and Townsley given in Fig. 117, there is apparently no definite turning point from yellow to brown when the solution is made alkaline. The yellow materials apparently stabilize colloidal ferric hydroxide, as would be expected from the experience of earlier workers on brown-water lakes.

It is reasonable to suppose that the highly colored allochthonous material postulated in discussing Birge's observations represents a mixture of substances comparable to those isolated by Shapiro. The more nitrogenous substances to be discussed in the next section may be the less colored fraction deduced from Birge's results. Ohle's (1940) observations that the color and colloid content do not increase proportionately in the lakes of south Sweden would be explained if the colored substances are not colloidal and the ratio of such material to nitrogenous colloidal compounds is greater in the colored than in the relatively uncolored lakes.

Goryunova (1954), who claims to have identified half the organic matter present in the water of Lake Beloye, finds that it is mostly polysaccharide; starch is very low, cellulosic material occurs at the surface, and there is also a saturated fatty acid of high molecular weight. The last named substance may perhaps be identical with one of Shapiro's acids.

NITROGENOUS ORGANIC MATTER

Another method of approach to the problem of the organic content of lake waters was employed by Domogalla, Juday, and Peterson (1925; Peterson, Fred, and Domogalla 1925) in a very important investigation of the organic nitrogen compounds in Lake Mendota. Using large quantities of water from which the seston had been removed by centrifugation, and which had been evaporated under reduced pressure, they determined the ordinary forms of inorganic nitrogen, the free amino acid content, the amino acid content after hydrolysis, and the amino acid content of the phosphotungstic acid precipitate. The last named fraction, in general, was of the same order of magnitude as the difference between free and total amino acid contents. The seasonal variations of these fractions is shown in Fig. 227.

In general the organic nitrogen in the surface waters accounts for from

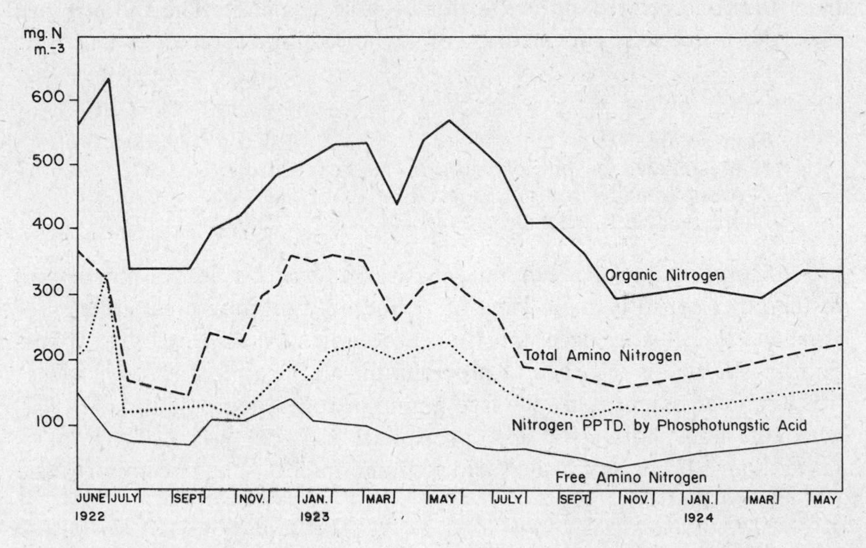

FIGURE 227. Seasonal distribution of organic nitrogen, total amino nitrogen, and free amino nitrogen in surface water of Lake Mendota 1922–1924. (After Peterson, Fred, and Domogalla.)

just over 50 per cent to about 75 per cent of the total soluble nitrogen, and rather over half of this organic nitrogen is present in amino groups. Of this amino nitrogen about one-third is present as free amino acids, the other two-thirds as a variety of polypeptides, presumably ranging from dipeptides to proteins. Evidence presented in Table 127 indicates that

TABLE 127. *Proximate fractionate of organic nitrogen in Lake Mendota*

	At 0 m., mg. m.$^{-3}$		At 20 m., mg. m.$^{-3}$	
N Constituent	Mean	Extremes	Mean	Extremes
Plankton	103.4	39.4–242.0	86.4	44.9–187.5
Total soluble	593.3	377.1–969.0	842.0	380.4–1739.3
Phosphotungstic acid ppt	169.7	113.0–336.5	177.1	89.2–319.6
Free amino	88.2	35.5–236.0	88.0	35.5–135.8
Peptide	180.5	77.0–436.0	173.4	117.3–266.0
Organic nonamino	186.6	95.5–278.3	181.3	66.3–276.0

much of the nonamino nitrogen is present as the nonamino groups of amino acids such as arginine, tryptophane, and histidine.

In addition to the routine precipitation with phosphotungstic acid, data are given for the soluble nitrogen precipitated by a variety of reagents from the concentrated surface water of June 3, 1922. The following of these determinations appear to be of particular importance:

	Fraction, mg. m.$^{-3}$	Per Cent
Total soluble N	558.0	100
N precipitated by phosphotungstic acid	207.0	37.1
N precipitated by sodium tungstate	127.1	22.8
N precipitated by trichloroacetic acid	70.0	12.6

Trichloroacetic acid precipitates native proteins but leaves most of all of the other constituents in solution. Sodium tungstate precipitates proteins and their higher decomposition products, such as proteoses. Phosphotungstic acid is regarded as precipitating all the amino nitrogen except free amino acids. Taking the free amino nitrogen figure of June 7, 1922, when the phosphotungstic acid precipitate nitrogen was 216 mg. m.$^{-3}$ rather than 207 mg. m.$^{-3}$, the various amino fractions may be apportioned roughly as follows:

	Fraction, mg. m.$^{-3}$
Protein N	70.0
Proteose, peptone, etc.	57.1
Lower polypeptide N	79.9
Free amino acid N	140.0

Early in June 1922, therefore, about 20 per cent of the amino nitrogen or 12.5 per cent of the total soluble nitrogen was present as proteins, and about 40 per cent of the amino nitrogen or 25 per cent of the total soluble nitrogen as free amino acids. It must be borne in mind, however, that the proportion of free amino nitrogen is at this time rather above the mean figure of 32.8 per cent of the total amino nitrogen. At other times the amount of organic colloidal nitrogen may be proportionately somewhat greater.

Data from certain other Wisconsin lakes are presented in Table 128.

The relationship of the seston nitrogen to the total organic nitrogen in the surface waters of the Wisconsin lakes is shown in Fig. 228. It will be observed that both the individual Mendota data and the data relating to the other lakes of low seston content are distributed about the dotted line that represents the same relationship for the five classes of lakes arranged according to the carbon content in Table 123. In the range of sestonic nitrogen up to about 300 mg. m.$^{-3}$, there is no apparent reason to suspect large systematic departures from simple proportionality; relatively little of the sestonic nitrogen is contributed by the allochthonous fraction, and this is increasingly true as the latter increases. There is no reason to doubt that in the lakes of low and moderate productivity the

TABLE 128. *Fractionation of organic nitrogen (mg. m.$^{-3}$) in various lakes in Wisconsin*

Lake	Plankton N	Total Sol. N	Phospho-tung. Ppt. N	Free Amino N	Peptide N	Non-amino N
Michigan Feb. 28, 1924	20.5	383.4	58.2	40.5	56.9	45.9
Devils						
Oct. 27, 1922	38.3	337.7	136.1	21.8	103.6	176.4
July 11, 1922	15.2	281.4	132.4	74.9	72.4	71.3
Oct. 5, 1923	14.3	314.3	110.4	69.5	67.7	39.3
Turtle Jan. 18, 1923	30.5	650.6	196.6	82.5	192.5	211.6
Bass Dec. 20, 1923	33.6	600.6	176.6	63.3	177.7	185.3
Green						
July 18, 1923	42.3	424.0	130.3	93.3	96.0	120.5
Oct. 17, 1923	48.7	460.8	136.8	70.9	122.7	102.8
Geneva July 16, 1923	50.9	457.7	135.5	74.5	135.9	126.8
Rock July 6, 1923	73.1	683.0	143.0	120.5	176.7	252.7
Madeline Dec. 12, 1923	95.6	616.5	150.5	82.2	151.1	221.3
Waubesa July 20, 1923	299.0	842.6	367.2	115.5	294.0	334.4
Monona July 20, 1923	388.8	885.8	326.6	49.0	315.8	347.3
Kegonsa July 20, 1923	696.6	839.4	379.0	99.7	299.3	331.3
Wingra July 6, 1923	882.0	896.0	298.0	78.9	299.5	343.6

nitrogen in solution is largely dependent on that in suspension in organisms. In the case of the most productive lakes, it is evident that any relationship of this sort tends to break down. Inspection of the data presented suggests that the different constituents of the soluble organic nitrogen depend in different ways on the seston nitrogen. The free amino nitrogen appears to reach a limiting value of 80 to 100 mg. m.$^{-3}$ at very low seston nitrogen figures; both the peptide and the nonamino nitrogen appear to go on increasing until the seston nitrogen is about 300 mg. m.$^{-3}$ This is well shown by comparing mean values for samples all taken at the same season (Table 129).

Peterson, Fred, and Domogalla (1925) made determinations of some individual amino acids and other organic compounds in the hydrolyzed

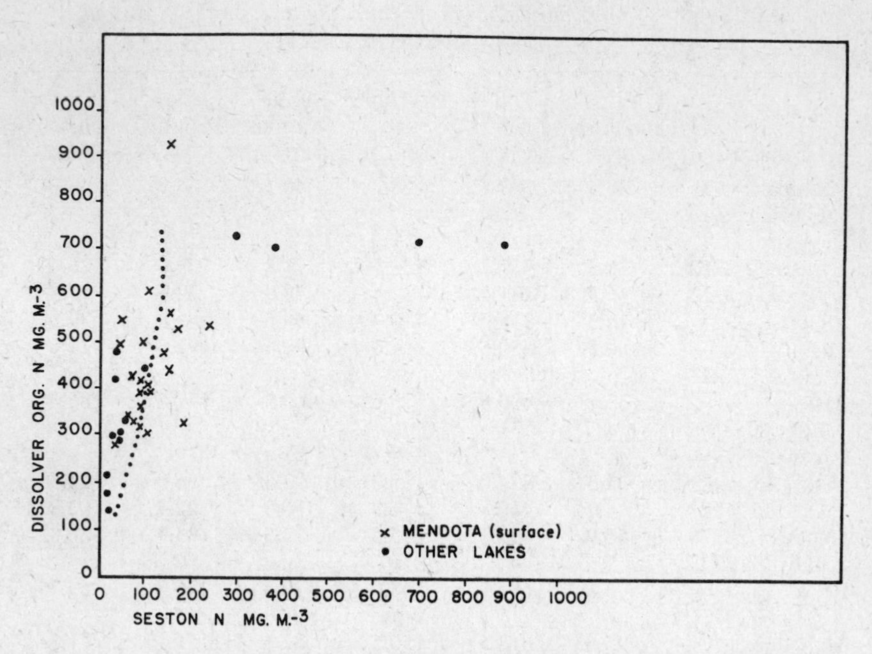

FIGURE 228. Organic soluble nitrogen plotted against sestonic nitrogen. Crosses and circles indicate various determinations for Lake Mendota and other lakes in Wisconsin, from the earlier work of Domogalla, Juday, and Peterson; dotted line, northeastern Wisconsin, mean values for Birge's various classes of organic carbon content. Note a regular proportionate increase, in both northern and southern Wisconsin, of soluble nitrogen with increasing sestonic nitrogen, up to about 200 mg. sestonic nitrogen, above which the increasing planktonic populations of lakes of excessive culturally determined productivity seem to add no more soluble nitrogen to the water.

TABLE 129. *Comparison of nitrogen fractions (mg. m.$^{-3}$) in lakes with high and low seston nitrogen*

	Seston N	Free Amino N	Peptide N	Nonamino N
Lakes with less than 100 mg. m.$^{-3}$ seston N	45.4	90.8	120.3	142.8
Mendota (June–July)	140.3	78.7	178.2	218.4
Lakes with more than 200 mg. m.$^{-3}$ seston N	506.6	85.8	302.1	339.2

soluble organic matter of certain samples. The most complete of these analyses refer to Lake Mendota water collected at the height of summer (Table 130).

TABLE 130. *Detailed analysis of organic nitrogen compounds in Lake Mendota*

N Constituent	At 0 m. (June 18, 1924), mg. m.$^{-3}$	At 20 m. (June 25, 1924), mg. m.$^{-3}$	N Constituent	At 0 m. (June 18, 1924), mg. m.$^{-3}$	At 20 m. (June 25, 1924), mg. m.$^{-3}$
Total soluble	515.6	766.9	Cystine	1.5	6.1
Organic soluble	324.2	357.3	Arginine	41.5	46.3
Total amino	189.0	221.4	Purine	8.4	9.5
Free amino	54.0	81.0	Amide	12.4	19.3
Tryptophane	10.3	13.3	Amine	14.2	16.0
Tyrosine	10.4	12.5	Unidentified amino	159.5	180.8
Histidine	5.7	10.2	nonamino	60.0	41.9

The amide nitrogen is, as Cooper (1937b) suggests, probably mainly urea. Table 131 gives analyses for other lakes.

TABLE 131. *Amino acids (mg. m.$^{-3}$) in lake waters*

N Constituent	Devils Lake, Oct. 10, 1923	Green Lake, July 18, 1923	Turtle Lake, Jan. 18, 1924	Lake Michigan, Feb. 28, 1924
Tryptophane	12.6	15.6	9.6	6.5
Tyrosine	17.6	9.6	16.7	8.3
Histidine	14.8	19.2	22.7	6.7
Cystine	3.3	4.4	7.5	2.1
Free amino	69.5*	93.3	82.5	40.5
Total amino	137.2*	189.3	275.0	97.4

* October 5, 1923.

The data for the individual amino acids can be recomputed as percentages of a hypothetical protein containing 16 per cent N, the breakdown of which might be supposed to produce the entire amino fraction (Table 132). This method of presentation allows comparison with the very large

TABLE 132. *Amino acids as per cent of total amino nitrogen*

N Constituent	Mendota 0 m.	Mendota 20 m.	Devils	Green	Michigan	Turtle
Tyrosine	11.4	11.66	26.4	10.43	17.6	12.58
Tryptophane	6.35	7.02	10.67	9.61	7.77	3.85
Histidine	1.78	3.19	6.38	6.00	4.07	4.86
Cystine	0.79	2.75	2.40	2.32	2.16	2.72
Arginine	10.73	10.33	...	...	...	...

number of amino acid analyses collected by Block and Bolling (1947). Such comparison indicates that an amino acid composition as high in both tyrosine and tryptophane is very improbable, for it can be paralleled by no known protein and is not likely to be simulated by a mixture of known proteins. In the two fresh-water blue-green algae that have been studied, *Phormidium valderianum* and *Gloeotrichia echinulata.* Mazur and Clarke (1938, 1942) found little tyrosine and still less tryptophane; in view of the large amount of humin nitrogen recorded, the results are probably not very accurate but evidently differ greatly from those given in Table 132. In diatoms of unspecified origin they found 6.2 per cent of the nitrogen to be present as tryptophane but only 0.2 per cent as tyrosine. The available information therefore gives no hint of source proteins of peculiar composition, and the most reasonable explanation of the observations, if they are not due to analytical error, is that certain amino acids are much more rapidly removed by bacterial metabolism than are tyrosine and tryptophane.

Fogg and Westlake (1955) have recently determined the amino and peptide nitrogen in nine lake waters from England. Their range for amino nitrogen, from 6 mg. m.$^{-3}$ to 25 mg. m.$^{-3}$, overlaps the lower end of the Wisconsin range; but their peptide nitrogen values, from 16 mg. m.$^{-3}$ to 43 mg. m.$^{-3}$, are all lower than those of the American investigators. Nevertheless, their data provide welcome confirmation of the occurrence of amino nitrogen dissolved in lake water. They have also found a remarkable polypeptide, rich in serine and threonine with smaller amounts of glycine, glutamic acid, and tyrosine, to be excreted into culture medium by *Anabaena cylindrica.* This compound forms complexes with iron, copper, and phosphate, and is likely to be of great ecological importance. Other fresh-water algae form other polypeptides.

THE VITAMIN CONTENTS OF LAKE WATERS

Although a considerable amount of work has been done on the vitamins and accessory growth substances in soils and soil extracts, very little is known about the distribution of such substances in the hydrosphere. In some of the early work on diatom culture it was found that growth was greatly stimulated by the addition of small amounts of natural sea water to an artificial inorganic medium (Allen 1914). More recently Harvey (1939) has studied the requirements of the diatom *Ditylum brightwelli* and finds that two organic substances must be present in the medium if normal growth is to take place. These substances appear to be absent in the waters of the English Channel in the summer, and their temporal variation may explain the appearance of the species only in the winter plankton.

Although soil extracts are often used effectively in algal cultures, Chu (1942) finds that most of the common fresh-water species of phytoplankton are able to develop on purely inorganic media; his work does not exclude the possibility that some forms are hetero-auxotrophic. The genus *Synedra* would appear to offer interesting possibilities in this direction, as a personal communication from Chu indicates that he had difficulties with members of the genus, of which some species are apparently indicators of strong eutrophication. *Asterionella formosa* and *Fragilaria crotonensis*, which also appear to enter lakes when the productivity is increased by human interference in the drainage basin (Patrick 1943, Hasler 1947), are, according to Chu, auto-auxotrophic. The possibility that the substance controlling the phosphorus requirements of *Asterionella*, discovered by Rodhe (1948) and already discussed, is an organic vitamin-like compound must be taken seriously. Rodhe also suspected *Synura uvella*, *Cryptomonas ovata*, *Melosira granulata*, and *Stephanodiscus astraea* to be stimulated by an organic substance in soil extract. Such a substance, which is extremely thermolabile, appeared to be necessary for *Gloeotrichia echinulata*.

Thiamin. A few determinations of the vitamin contents of lake waters have been made. Hutchinson (1943) found from 0.11 to 0.29 mg. m.$^{-3}$ of thiamin in unfiltered water from Linsley Pond, and from 0.008 to 0.077 mg. m.$^{-3}$ of the vitamin in filtered water. Other lakes in Connecticut contained from 0.029 to 1.9 mg. m.$^{-3}$ in the unfiltered water. The highest value occurred in water very rich in *Polycystis aeruginosa*. Samples of seston contained:

Linsley Pond, phytoplankton mainly *Oscillatoria*

December 1, 1941	16 γ per gram
February 2, 1942	24 γ per gram
April 29, 1942	49 γ per gram
May 25, 1942	49 γ per gram

Smith Brothers Pond, phytoplankton mainly *Polycystis*

July 10, 1942	12 γ per gram

Crustacean plankton contained 11 to 37 γ per gram, but rotifers appeared to be somewhat richer in thiamin than the crustaceans. Benthonic insect larvae fell within the crustacean range. The organic matter of the mud of Linsley Pond contained about 10 γ per gram. It would seem, therefore, that the phytoplankton, or possibly bacteria associated with organic tripton, produce a good supply of thiamin, adequate for the organisms that feed directly or indirectly upon the seston.

Von Witsch (1946) found, however, that the thiamin content of *Chlorella* depends greatly on the environment and the age of the culture. Strong light and abundant CO_2, though favoring rapid growth, decreased the thiamin content. The amount of the vitamin also decreased markedly after the completion of the initial period of rapid growth. Young cultures under suboptimal conditions contained 18 γ per gram; old cultures grown under conditions of high CO_2 content and high light intensity contained but one tenth of this amount. The loss of thiamin on aging was also observed in cultures of *Nitschia palea*. In young cultures 4 to 5.5 γ per gram were present; in old, 0.25 to 0.38 γ per gram. A number of filamentous green algae collected in nature showed low contents, none in excess of 3.3 γ per gram. It is possible that the rather high values observed in the lake plankton indicate that the samples in question were in the physiological state of a young culture under suboptimal conditions. This would not be unreasonable in the case of the spring samples, which contained the greatest amount of the vitamin, though the absolute amounts recorded are much greater than those given by Von Witsch.

No considerable amount of thiamin collects in solution in the water. The mean value of the thiamin content of filtered water is 0.048 mg. $m.^{-3}$, or 24 per cent of the total thiamin in the unfiltered water. This is based on too few samples to have much significance, but probably indicates the order of magnitude of the quantity of soluble thiamin to be expected in the waters of biologically productive lakes.

Niacin and biotin. Hutchinson and Setlow (1946) have investigated the niacin (nicotinic acid amide) content of Linsley Pond. In the surface water of the lake, the dissolved niacin varied from 0.15 to 0.89 mg. $m.^{-3}$, but the very high values, in excess of 0.7 mg. $m.^{-3}$, are found only during winter under an ice cover. The mean concentration was 0.4 mg. $m.^{-3}$ The sestonic niacin varies from 0.05 to 0.87 mg. $m.^{-3}$, 21 to 85 per cent of the total niacin being in solution. It appears that the sestonic niacin per unit volume of water depends as much on the concentration of the vitamin in the seston as on the concentration of the latter in the water. In living phytoplankton the niacin content is of the order of 50 to 100 γ per gram, but when little plankton and much detritus is present in the seston, the niacin content is much lower. No vertical stratification of any significance could be found when the soluble niacin was determined at different depths during the height of summer stagnation. The vertical distribution of the sestonic niacin reflects that of the phytoplankton. It is evident that ordinarily the bacterial destruction of the vitamin maintains a concentration in the water at all depths of between 0.15 and 0.35 mg. $m.^{-3}$ Similar values were encountered in the water of Bantam Lake, another eutrophic lake in Connecticut.

Such concentrations probably represent a limit below which bacterial decomposition of niacin proceeds with extreme slowness or is balanced by bacterial synthesis. At very low temperatures, near 0° C., in the presence of photosynthetic organisms, bacterial decomposition is apparently slowed less than is production, mainly no doubt by the phytoplankton; under these circumstances higher concentrations of the vitamin can occur than in summer. In the hypolimnion, where synthesis by autotrophic organisms is not to be expected, no such effect is observed in spite of the low temperature.

Somewhat unsatisfactory determinations of biotin by Hutchinson and Setlow indicate from 0.1 to 4.0 γ per $m.^3$ biotin in solution in the surface waters of Linsley Pond, and a like range for the sestonic biotin. The seasonal cycle of biotin is probably like that of niacin, with some accumulation under the ice.

Significance of vitamins in lake waters. Although the quantities of both niacin and thiamin in solution, 0.15 to 0.89 mg. $m.^{-3}$ and 0.008 to 0.077 mg. $m.^{-3}$ respectively, may seem so small as to appear quite insignificant biologically, a little reflection will show that this is probably not the case. The phytoplankton of many lakes is capable of flourishing in nature in water containing less than 5 mg. $m.^{-3}$ phosphate phosphorus in solution. In setting up a medium for some hetero-auxotrophic organism, no microbiologist is likely to be so generous with vitamins as to use them in concentrations of the order of 10 per cent or even 1 per cent of his phosphorus source. Yet in Linsley Pond and the few other Connecticut lakes that have been examined, this is just the sort of proportionality that prevails between the more abundant vitamins and the soluble phosphate phosphorus. If the phytoplankton can use phosphorus at the low concentrations that are known to support increase in nature, then any species that may require thiamin or niacin is likely to be able to obtain these vitamins from the environment. In view of the greater quantity of niacin in solution in winter than in summer, and the probability that such a temporal distribution will prove general for the water-soluble vitamins, it is particularly interesting that the only autotrophic truly open-water plankter so far encountered that requires accessory organic substances other than vitamin B_{12}, the marine diatom *Ditylum*, should be a winter form.

The presence of vitamin B_{12} or cyanocobalamin, which at the present time seems likely to prove of greater ecological importance than any other accessory organic nutrient, has been discussed in Chapter 15.

INDIRECT EVIDENCE OF ORGANIC MATTER

Surface-active materials. Foam is often observed on fresh waters, but very little work appears to have been done on its occurrence and properties. According to Miyake and Abe (1948), just over two thirds of the foaming power of sea water is due to its inorganic salts, and the rest to unidentified organic matter. Some foam can form even on Lake Baikal; Votintsev (1953) has determined that it is lower in organic matter than is sea foam; it does contain largely colloidal surface-active organic compounds. The subject would probably repay further investigation.

Adam (1937) introduced a method for determining the surface tension of water *in situ* by placing on it drops of solutions of *n*-dodecyl alcohol in pure white mineral oil. The solutions are of graded concentration, from 0.2 to 1.5 per cent. The capacity of such drops to spread on water is related to the surface tension, and by finding what concentration will just spread, the tension can be estimated. Adam found some depression in slow-flowing streams as compared with less stagnant waters. Hardman (1941) applied the method to the surfaces of about forty lakes in Wisconsin. In general, uncolored waters poor in plankton exhibited virtually no depression of surface tension below pure water of surface tension 72.7 at 20° C. Colored bog lakes usually have lower surface tensions than pure water, the depression ranging from 0 to 20 dynes $cm.^{-1}$, with an average value of 6 or 7 dynes $cm.^{-1}$ Massive plankton production also can produce marked depressions; on Lake Monona during a bloom of blue-green algae a value of less than 52 dynes $cm.^{-1}$ was recorded. Locally in shallow water the presence of certain flowering plants lowers surface tension. Water near *Potamogeton*, *Utricularia*, and *Castalia* often had its surface tension depressed to 52 to 56 dynes $cm.^{-1}$ If a drop of the test oil was allowed to float past a group of *Castalia* leaves, the drop contracted as it approached each leaf and expanded as it left the zone of influence of the leaf.

When wind has been blowing over a lake which previously showed a depression of the surface tension, the tension may be raised to normal except downwind, where the depression can be detected. Foaming was occasionally observed and was accompanied by a depression of 2 to 9 dynes $cm.^{-1}$

Fluorescence. Merker (1931) observed fluorescence in water in which plants and animals had been kept, and Thomas (1948) has found that lake waters are frequently fluorescent. Shapiro (1956) has found that the yellow organic substances of lake waters are a major source of the fluorescence.

Odors of lake waters. The "unforgettable unforgotten river smell" can easily be appreciated in water derived from the weedy margins of lakes. Fish, moreover, appear from the work of Hasler and Wisby (1951) to discriminate between the odors of different streams. The odors appear to be due to thermolabile volatile organic compounds. Rinses of various aquatic plants were shown by Walker and Hasler (1949, Hasler 1954) to have specific odors that could be detected by the blunt-nosed minnow *Hyporhynchus notatus*.

The odors of various species of phytoplankton organisms have long been a source of difficulty to the waterworks engineer. Aromatic odors are mainly produced by diatoms (Whipple 1899), *Asterionella* being said to smell like rose geranium. Blue-green algae are described as having a grassy odor, and some green and yellow algae a fishy odor. All of these organisms are presumably adding minute amounts of specific organic compounds to the water.

Antibiotic effects. In the succeeding volume it will be necessary to consider the production of antibiotic substances of various sorts by algae, as there is growing evidence that part of the phenomenon of seasonal succession in the phytoplankton is due to such materials. It is also quite possible that higher plants may produce substances that inhibit plankton (Langhans, in Schreiter 1928), a view which receives some support from the studies of Hasler and Jones (1949).

SUMMARY

Organic matter is present in all lake waters but is extremely hard to investigate. The simplest methods of estimation are (1) as ignition loss corrected for loss of CO_2 from calcium carbonate, (2) as the material oxidized under standard conditions by an oxidizing agent such as permanganate, or (3) as organic carbon. The carbon of the organic matter, oxidized under standard conditions, corresponds to about 40 per cent of the organic carbon. All modes of estimation give a quantity correlated with water color, but there is evidence of the existence of an oxidizable fraction of the organic matter that does not contribute to the color.

Very roughly the organic matter of lakes can be considered either as autochthonous or allochthonous; the latter consisting mainly of extractives from soils or peaty material. The organic content of a series of waters of increasing carbon content has been studied in northeastern Wisconsin. The C:N ratio tends to rise from about 12 to about 30 in passing from waters with 1 mg. organic carbon per liter to waters containing 30 mg. Assuming that the autochthonous organic matter is proportional to the organic seston in the lake, it is possible to obtain some idea of the allochthonous component. This material appears to contain very little nitrogen, the C:N ratio being about 47:1, and to have an effect on water color of about 7 Pt units per milligram per liter. In Furesø, the only lake studied by ultrafiltration, most of the nonsestonic organic matter was not colloidal, and most of the colloid was inorganic. In acid water there is evidence of colloidal material that can undergo cataphoresis, and of adsorption of alkaline earth ions by colloids in neutral or alkaline waters.

The yellow or brown coloring matter is mainly a mixture of dibasic organic acids, probably free of nitrogen and having a lower molecular weight than the humic acids of soils.

There is a large body of data suggesting that the organic nitrogen present in lake water is either protein or protein degradation products, as would be expected.

A considerable quantity of free amino nitrogen, of the order of 100 mg. $m.^{-3}$, has been recorded in some of the lakes of southeastern Wisconsin. In these lakes, as in the less productive waters of northeastern Wisconsin, the total nitrogen increases up to a certain point with increasing seston nitrogen. When the latter reaches 300 mg. $m.^{-3}$, the relationship breaks down. The free amino nitrogen likewise increases with seston nitrogen, but only as long as the latter is below 100 mg. $m.^{-3}$ Presumably, in very productive lakes a level of dissolved organic matter is achieved, permitting so much bacterial activity that the causal relationship between producing organisms and product is obscured. The total amino nitrogen in these lakes has been partly referred to specific amino acids. Of this, 10 to 25 per cent appears to be present as tyrosine, and up to 10 per cent as tryptophane. If these analyses are reliable, they probably indicate a differential breakdown, acting more vigorously on the aliphatic amino acids.

Certain specific compounds are known by their biological activity. Thiamin, niacin, biotin, and cyanocobalamin have all been identified in lake water. The greatest quantity of any of these vitamins in solution is 0.89 mg. $m.^{-3}$ recorded for niacin. The latter substance appears to be more abundant in winter under ice than at other seasons; the same may be true of biotin. The quantities present are probably biologically significant.

Surface-active materials are certainly produced by the vegetation of lakes and may lower the surface tension from 73 to 52 dynes $cm.^{-1}$ Their nature is unknown. It is probable that substances producing foam are often present.

Lake water is usually fluorescent. Several substances appear to be involved. All appear to be yellow in ordinary light and to contribute significantly to the color of the water.

The waters of shallow lakes often have a distinctive odor, and many specific phytoplankton odors have been described by waterworks engineers. There is a considerable possibility that odor is of fundamental importance in directing fish migration.

Antibiotic substances are produced by phytoplankton and perhaps by higher plants; their occurrence will be of some importance in the discussion of phytoplankton periodicity in Volume II.

List of Symbols

An effort has been made to avoid any duplication of symbols; for this reason a number of changes have been necessary in adapting the mathematical expressions of other workers for use in the present work. The ordinary mathematical conventions as to e, π, etc., are of course adopted. In a few cases elaboration of conventional usage is necessary. Thus μ being the dynamic coefficient of molecular viscosity, μ_R is used for the refractive index. Only symbols used in mathematical expressions are listed, since there is inevitable duplication with chemical symbols.

A	Area of lake.
$\bar{\mathrm{A}}$	Mean area of a lake of variable level.
A'	Area of drainage basin (including lake).
$\mathrm{A_e}$	Area at depth z_e.
$\mathrm{A}_n^{(i)}$, $\mathrm{B}_n^{(i)}$	Coefficients in Fourier expansions used in theory of internal seiches.
$\mathring{\mathrm{A}}_n$	Constants (phase angles) in Fourier expansion in theory of heat conduction in sediments; $n = 1, 2, 3$, etc.
A_z	Area enclosed by contour of depth z ($\mathrm{A}_0 = \mathrm{A}$).
A, A_x, A_y, A_z	Coefficients of turbulence, direction unspecified or in direction of the x, y, and z axes (ordinarily vertical with $A_{sz} = A_{\theta z} = A_z$, which latter is used except in special contexts).
A_F	Mixing coefficient, analogous to $A_{\theta z}$. Applied to Ertel's source function F.
A_s, A_{sx}, A_{sy}, A_{sz}	Coefficients of eddy diffusivity, direction unspecified or along the designated axes.
A_θ, $A_{\theta x}$, $A_{\theta y}$, $A_{\theta z}$	Coefficients of eddy conductivity, direction unspecified or along the designated axes.

A_v, A_{vx}, A_{vy}, A_{vz}	Coefficients of eddy viscosity, direction unspecified or along the designated axes.
A_0	Coefficient of turbulence in water of indifferent stability, when $A_{vz} = A_{sz} = A_0$.
A, B	Points designating positions of stations.
B	Work done by wind in distributing heat throughout lake, per unit area of lake surface.
B	Cloud cover of sky (in tenths).
C	Conductivity.
C_{CA}	Conductivity expressed as concentration of $CaCl_2$ solution.
C_L	Limiting value of conductivity of mud.
$C_{\overset{*}{z}}$	Conductivity at depth $\overset{*}{z}$ in mud.
C_0	Conductivity at surface of mud.
C	Virtual temperature of isothermal circulation in McEwen's theory of turbulent heating.
C_1	Virtual temperature at surface in McEwen's theory.
D_L	Development of shore line or ratio of length of shore line (L) to length of circumference of a circle of area equal to that of lake.
D_V	Development of volume, or ratio of volume of lake (V) to volume of a right cone of base equal to area of lake and height equal to maximum depth of lake.
D	Depth of frictional resistance.
E, E_z	Stability of any layer, specifically at depth z; where $E_z = (1/\rho_z)(d\rho_z/dz)$.
E_h	Oxidation-reduction potential.
E_0	Constant in equation relating E_h to activities of reacting substances in oxidations or reductions, giving redox potential when activities of the oxidized and reduced substances are equal.
E	Energy.
E_f	Rate of discharge by effluent per unit area of lake surface.
E_k	Kinetic energy of a seiche.
${}^nE_m$	Energy of the mth wave in a train of n waves.
E_p	Potential energy of a seiche.
E_s	Total energy of a seiche.
E_v	Mean rate of evaporation from lake surface per unit area of lake surface.
E_v'	Mean rate of evaporation from drainage basin per unit area of basin.
F, F_z	Functions.
$F_s(b)$	Defined as $F_s(b) = 2b/[b - \ln(1 + b)]$, a function of b, the optical depth per meter, used in scattering theory.
F	Ertel's source function.
G	Work per unit area required to distribute observed heat in a lake uniformly throughout all depths (G = B + S; for a third-class lake, S = 0, B = G).

$[G_1], [G_2], \cdots [O_2]$ etc. — Concentrations of gases, $[G_1]_b$ concentration at bottom of lake, $[G_1]_z$ at depth z. The quantity of solute is here expressed in whatever units are being employed in context. $[G_1]_S$, $[G_1]_s$ concentrations of G_1 at saturation, as defined in text. $[G_1]_c$ critical concentration for bubble formation. $[G_1]_0$, $[G_1]_t$, $[G_1]_\infty$ concentrations at times 0, t, and ∞.

H — Defined by $H = -\phi_s^2/g\kappa_s^2$ in theory of internal seiche.

$H^{(i)}$ — Roots of cubic equation giving H for i = 1, 2, 3.

H — Area of cross section of lake.

$\bar{H}$ — Mean area of cross section.

H_e — Area of cross section of epilimnion.

$H_e(x)$, $H_h(x)$ — H_e and H_h as functions of x.

H_h — Area of cross section of hypolimnion.

H_x — Area of cross section at point x on long axis of lake.

I — Intensity of radiation (I is used in contexts in which only radiant energy is considered, Q where other forms of energy may be involved).

I_c — Solar constant.

I_H — Illumination from sky.

I_m — Maximum illumination on a plane set in such an angle as to receive the greatest amount of radiant energy possible.

I_R — Intensity of light reflected from lake surface.

I_t — Total illumination received on a point recorder from all directions.

I_u — Intensity of radiation scattered upward from below a Secchi disk.

I_u' — Intensity of radiation scattered upward from between a Secchi disk and the surface.

I_v — Vertical illumination, on a horizontally placed plane.

I_w — Illumination on a horizontal unit facing downward just below water surface.

I_z — Illumination at depth z.

I_0 — Intensity of radiation at lake surface, or in general at zero distance.

I_n — Rate of flow of influents per unit area of lake surface.

$J^{(n)}$ — Defined by $J^{(n)}\,\mathrm{erfc}\,(z) = \int_0^\infty J^{(n-1)}\,\mathrm{erfc}\,(\xi)d\xi$ etc., $J^{(0)}\,\mathrm{erfc}\,(z) = \mathrm{erfc}\,(z)$.

K — Molecular diffusion coefficient in water.

K_w — Dissociation constant of water.

K_1 — Lunisolar diurnal tide.

K_2 — Lunisolar semidiurnal tide.

K_a, K_{a_1}, K_{a_1}' — Thermodynamic dissociation constants (these in the logarithmic form are prefaced with a p).

K_1, K_2 — First and second dissociation constants, specifically of carbonic acid.

K_1', K_2' — Apparent dissociation constants.

L — Length of shore line.

[L]	Dimension of length in dimensional analysis.
L_z	Length of contour of depth z.
L	Solubility product.
L_e	Latent heat of evaporation.
[M]	Dimension of mass in dimensional analysis.
M_2	Lunar semidiurnal tide.
M	Length of path of radiation in atmosphere.
N_2	Lunar elliptical semidiurnal tide.
O_1	Lunar diurnal tide.
P	Partial pressure of a gas in atmosphere.
P_a	Defined as $P_a = 100 - P_s$.
P_h	Atmospheric pressure at altitude h.
P_s	Percentage scattering of light.
P_w	Defined as $P_w = \log K_w^{-1}$.
P_z	Pressure in a bubble of gas at depth z.
P_0	Atmospheric pressure at sea level.
P_1	Solar diurnal tide.
P_r	Precipitation over lake per unit area.
P_r'	Mean precipitation over drainage basin per unit area.
Q	Energy flux per unit area, nature indicated by subscript.
Q_A	Long-wave radiant energy from atmosphere.
Q_B	Net radiation surplus.
Q_E	Energy lost in evaporation as latent heat.
Q_e	Energy lost in heating evaporated water vapor.
Q_H	Scattered and reflected radiant energy from sky and clouds.
Q_i	Net energy flux by advection.
Q_M	Long-wave radiation from mountains around lake.
Q_R	Radiant energy reflected from lake surface.
Q_s	Net energy flux as sensible heat.
Q_S	Direct radiant energy from sun.
Q_S'	Net radiation flux; $Q_S' = (Q_S + Q_H - Q_R - Q_U)$.
Q_U	Radiant energy scattered upward from water.
Q_W	Long-wave radiant energy emitted by water.
Q_θ	Rate of storage of heat in lake.
R′	Fraction of light of sky reflected from water surface.
R_b	Bowen ratio, $R_b = Q_s/Q_E$.
R_e	Reynold's number, $R_e = \rho v l \mu^{-1}$.
R_f	Reflectivity as fraction of incident light.
R_i	Richardson's number, $R_i = gE(v')^2$; the derivatives R_i' and R_i'' are also used.
R_p	Ratio of I_w to I_0.
R_t	Total fraction of light returned by reflection and upward scattering.
R_x, R_y, R_z	Component of frictional forces in directions of the x, y, and z axes, per unit volume.
R	Ratio of maximum value of ordinates of normal curves for ordinary and internal seiches.

S — Stability of a lake per unit area, or amount of work per unit area needed to mix a lake in any observed condition to a uniform vertical density.
S_s — Scattering in a beam of light of unit energy traveling unit distance.
S_2 — Solar semidiurnal tide.
S — Quantity of any substance of concentration s in solution, that is transferred across a plane.
S_h — Set-up, defined by $S_h = |h_0| + |h_l|$.
T — Period of a seiche.
[T] — Dimension of time in dimensional analysis.
T_i — Period of internal seiche.
T_m — Period of a seiche computed by Merian's formula.
T_n — Period of the nth nodal seiche (T_1, T_2, T_3, etc., periods of uninodal, binodal, trinodal, etc., seiches).
T_p — Period of a half pendulum day; at latitude ϕ, $T_p = \pi(\Omega \sin \phi)^{-1}$.
T_s — Period of transverse seiche.
T_t — Computed period of seiche, allowing for turbulent dissipation of energy.
T — Surface tension of water.
U — Defined as $U = z/2\sqrt{Kt}$.
V — Volume of lake.
$V_{z_1-z_2}$ — Volume between the planes of depth z_1 and z_2. $V_{z_1} = V_{0-z_1}$.
W — Wind velocity.
W_c — Critical wind velocity.
W — Defined as $W = (\pi/2z_h)^2 At$.
a, b, c, d — Numbers; where used alone, a is an area much smaller than the area A of a lake.
a_n — (n = 1, 2, 3, etc.) coefficients in Fourier expansion in the theory of heat conduction in sediments.
a — Quantity defining the temperature gradient in McEwen's theory of turbulent heating.
a_{H^+} — Activity of "hydrogen ion" and similarly with all other solutes.
b — Optical path per meter.
b, b_z — Breadth of lake, in general or at point x.
b_e — Breadth of base of epilimnion.
b_i — Breadth of epilimnion at various depths (i = 1, 2, 3, etc.).
b_j — Breadth of hypolimnion at various depths (j = 1, 2, 3, etc.).
$b(x)$, $b_e(x)$, $b_i(x)$, $b_j(x)$ — Last four quantities as functions of x measured along long axis of lake.
b_s — Breadth of littoral shelf.
$\bar{b}$ — Mean breadth.
c — Chézy coefficient in $v = c\sqrt{\bar{z} \cdot dz/dx}$.
$\overset{*}{c}$ — Specific heat of sediments.
c_m — Concentrations of cations.

c_p	Specific heat of water at constant pressure.
c	Constant in any unenumerated expression.
c_s	Defined as $c_s = 1 + (\tau_b/\tau_a)$.
d_1	Loss in intensity of light in passing from surface of water to Secchi disk.
d_2	Loss in intensity of light in passing from Secchi disk to eye.
f	Function.
g_1, g_2, etc.	Proportions of gases in bubbles.
g	Acceleration due to gravity.
h, h_1, h_2, etc.	Altitudes above sea level.
h, h_x, $h(x)$	Height of a denivellated water surface above its mean position, in general, or at a point x or as a function of x.
h_m	Maximum value of h.
h_w	Height of a wave from trough to crest.
h_{wm}	Maximum possible height of a wave under specified conditions.
h_0	Depth below mud surface at which oxygen is not detectable.
$h_1^{(i)}$, $h_2^{(i)}$, $h_3^{(i)}$	Displacements of surface (h_1) and of upper (h_2) and lower (h_3) pycnoclines in a three-layered system, due to the tilting of the surface (i = 1), the upper (i = 2), or the lower (i = 3) pynocline.
i, j	Sets of numbers.
i	Element of illumination.
i_x	Slope of water surface at x.
j_1, j_2, etc.	Roots of Bessel functions.
k	In the Munk-Anderson theory of the thermocline, defined as $k = -E(\theta')^{-1}$ and taken as a virtual constant over moderate ranges of θ.
$\overset{*}{k}$	Molecular thermal diffusivity in sediments.
$\hat{k}$	Molecular diffusivity of oxygen in sediments.
k_c	Coefficient, dependent on the nature of cloud cover.
k_q	Defined as $k_q = \ln \left(\frac{1 + \sqrt{1 - (z_q/z_m)}}{1 - \sqrt{1 - (z_q/z_m)}} \right)^2$
k_1, k_2, etc.	Dissociation constants.
l	Mixing length.
l	Length of lake, or in general hydromechanical treatment any length, specified as l_0, l_1, etc., or l_n.
l_q	Length to end in a truncated quartic lake.
l_s	Wave length of a seiche considered as a stationary wave.
l_w	Wave length, in hydrodynamics.
m, n	Unspecified numbers.
n	Nodality of a seiche.
p, q, r, s	Numbers used as indices in dimensional analysis.
p_w	Pressure of water vapor in air.
p_w'	Pressure of water vapor in air saturated with respect to a water surface.
p, p_z	Hydrostatic pressure, unspecified or at depth z.

p_m — Pressure at depth at which no current is flowing, or is assumed not to be flowing, between two stations **A** and **B**.
p_t — Pressure of gas in equilibrium with water at time t.
p_O — Pressure of oxygen in equilibrium with a specified concentration in solution at a given temperature.
q — Rate of biochemical oxygen consumption.
$\mathrm{r_b}$ — Rate of bacterial respiration.
r — Radius
r_d — Reflectivity of a Secchi disk.
r_i — Radius of an inertia current.
r_z — Radius of a circle of area A_z.
r_O — Rate of oxygen uptake in mud.
s — Sheltering coefficient.
$\bar{\mathbf{s}}$ — Defined by $\bar{\mathbf{s}}^2 = \lim_{t^* \to \infty} \frac{1}{t^*} \int_0^{t^*} \mathbf{s}^2(t) \cdot \mathrm{d}t$.
$\mathbf{s}(t)$ — Defined by $\mathbf{s}(t) = s(t) - \bar{s}$.
s — Concentration of any substance in solution or suspension.
$s(t)$ — s as a function of time.
$\bar{s}$ — Mean value of $s(t)$.
t — Time.
$\overset{*}{t}$ — A specified time interval.
t_p — Time measured in half pendulum days.
u — Defined as $\mathrm{u} = (z_\mathrm{m} - z)/2\sqrt{At}$.
u' — Shear, defined as $u' = -\,\mathrm{d}u/\mathrm{d}z$.
u, v, w — Velocities along the x, y, and z axes, respectively.
$\bar{\mathbf{u}}, \bar{\mathbf{v}}, \bar{\mathbf{w}}$ — Mean turbulent velocities.
$\bar{u}, \bar{v}, \bar{w}$ — Mean velocities along three axes.
$u_\mathrm{m}, v_\mathrm{m}, w_\mathrm{m}$ — Maximum velocities along the three axes.
$\mathbf{u}(t), \mathbf{v}(t), \mathbf{w}(t)$ — Turbulent velocities at time t, along the three axes.
$u(t), v(t), w(t)$ — Velocity along the three axes as functions of time.
u_x, v_x, w_x — Velocity along the three axis at any value of x.
u_z, v_z, w_z — Velocities along the three axes at depth z.
u_0, v_0, w_0 — Velocities along the three axes at surface.
$\mathrm{v_b}$ — Volume of bacteria.
v — Used in general case for velocity in unspecified directions, with or without subscripts.
v' — Shear, defined as $v' = \mathrm{d}v/\mathrm{d}z$.
$\overset{*}{v}$ — Velocity of water due to wave of finite amplitude.
$\dot{v}$ — Acceleration or $\mathrm{d}v/\mathrm{d}t$.
$\dot{v}'$ — Acceleration of a longshore current.
$v_\mathrm{A,\,B}$ — Velocity between stations **A** and **B**.
v_E — Eastward horizontal component of velocity.
v_w — Wave velocity.
w — Defined as 2π times reciprocal of period of wave.
x — Distance along horizontal coordinate, normally along the long axis of lake.
y — Distance along horizontal coordinate at right angles to x axis.

z Depth or distance along vertical coordinate (positive values downward).

$\bar{z}$ Mean depth.

$\dot{z}$ Mean rate of increase of z.

$\overset{*}{z}$ Depth of sediment below sediment-water interface.

$\overset{*}{z}_{O}$ Depth of sediment in which oxygen is present.

z_b, z_{bx} Depth to bottom, specifically at x, from actual water surface.

z_b', z_{bx}' Depth to bottom, specifically at x, from mean undisturbed water surface; $z_b' = z_b - h$.

z_c Depth of cryptodepression, below sea level.

z_d Depth of open water (cf. z_s).

z_e Depth of epilimnion.

z_g Depth to center of gravity of lake.

z_h Depth of hypolimnion; in theoretical two-layered systems, $z_h = z_m - z_e$.

z_m Maximum depth.

z_q Depth at end of a truncated lake.

z_r Relative depth or maximum depth as percentage of mean diameter, $z_r = 50 z_m \sqrt{\pi}(\sqrt{A})^{-1}$.

z_s Depth of littoral shelf.

z_u Depth of lake of uniform depth (often z_m can reasonably be used).

z_x Depth at x.

z_1, z_2, etc. Thickness of layers, arbitrarily numbered 1, 2, etc., from above downward.

$z_{1,x}$, $z_{2,x}$, etc. Thickness of layers 1 and 2 at x, etc.

Δ Increment.

ΔCO_2, $\Delta' CO_2$ Free and metabolic CO_2 anomalies.

Θ Quantity of heat below unit area in deepest part of lake (Forelian heat budget).

Θ_b Quantity of heat below unit area, averaged over entire area of lake (Birgean heat budget).

Θ_{ba} Annual heat budget per unit area, averaged over entire surface.

Θ_{bs} Summer heat income, averaged over entire suface.

Θ_{bw} Winter heat income, per unit area, averaged over entire surface.

Θ_m Annual heat budget of sediments, per unit area, averaged over entire mud surface.

Θ_z Heat flux through unit area at depth z.

Ξ Solid angle in scattering theory.

O Quantity of oxygen passing through unit area of a plane.

Π Period length.

$\Phi(\ )$ Error function.

X Defined as

$$X = \frac{8}{\pi^2}\left[(1 - e^{-W}) - \frac{1}{15}(1 - e^{-9W}) + \frac{1}{45}(1 - e^{-25W}) - \cdots\right].$$

Ω Angular velocity of rotation of earth.

α, β Molecular entrance and exit coefficients.

α' Proportionality factor relating θ_0 to some power of $\sqrt{t}$.

$\hat{\alpha}$ Degree of dissociation, ratio of observed to calculated conductivity.

α_s Specific volume of water.

α_z, α_{zA} Specific volume of water at depth z, at depth z and station A, etc.

α_1, β_1 Convective entrance and exit coefficients.

β_s Defined as $\beta_s = A_v{}^2 l^2 / g z_u{}^5 \pi^2$.

β_v, β_θ Coefficients of R_i in expressions giving A_v/A_0 and A/A_0.

γ As γ^2, coefficient of resistance (always positive) in expressions of the form $\tau_a = \gamma^2 \rho_a (W - v_0)^n$.

γ_q Defined as $\gamma_q = 2\sqrt{1 - (z_q/z_m)}$.

ϵ Mean slope.

ζ Arbitrary or specially defined variable in integration.

η Extinction coefficient.

η', η_t' Zenith extinction coefficient in lake, fractionated in the same way as η_t, ($\eta_t' = \eta_w' + \eta_c' + \eta_p'$).

η'' Vertical extinction coefficient in lake.

η_a Extinction coefficient of air.

η_c Fraction of extinction coefficient due to dissolved coloring matter.

η_p Fraction of extinction coefficient due to suspended matter.

η_t Extinction coefficient due to all forms of absorbtion in laboratory experiment; $\eta_t = \eta_w + \eta_c + \eta_p$.

η_w Fraction of extinction coefficient due to water.

θ Temperature (ordinarily °C.).

$\bar{\theta}$ Mean temperature.

θ_A, θ_W Temperature (°C.) of air and water respectively (used only for a water surface).

θ_K, θ_{KA}, θ_{KW} Temperature measured from absolute zero (°K.); in general, of air or of water, respectively.

θ_{Kh} Absolute temperature of surface of lake at height h above sea level.

θ_{Ks} Absolute temperature of air at a station for which pressure data exist.

θ_m Mean temperature of sediments, averaged over time and depth.

θ_z Temperature at depth z.

$\theta_{\overset{*}{z}}$ Temperature in sediments at distance $\overset{*}{z}$ below sediment water interface.

θ_0 Temperature at surface.

θ', θ_z', etc. Defined as $\theta_z' = d\theta_z/dz$.

θ'', θ_z'', etc. Defined as $\theta_z'' = d^2\theta_z/dz^2$.

κ Defined as $\kappa = 2\pi/l_w$.

κ_s Defined in theory of internal seiche as $\kappa_s = 2\pi/l_s = 2\kappa_0$.

κ_0 Defined in theory of internal seiche by $\kappa_0 = \pi/l_s$.

μ Dynamic coefficient of molecular viscosity.

μ_{CA} Refractive index expressed as concentration of equivalent $CaCl_2$ solution.

μ_{R}	Refractive index.
ν	Kinematic coefficient of molecular viscosity.
ξ	Arbitrary variable in integration.
ρ, ρ_z	Density of water, unspecified or at depth z.
$\overset{*}{\rho}$	Density of sediments.
ρ_{a}	Density of air.
ρ_{e}	Density of water of epilimnion.
ρ_{h}	Density of water of hypolimnion.
$\rho_1, \rho_2, \cdots \rho_m$ etc.	Density of water of designated layers $1, 2, \cdots m$, specifically above a plane of oscillation.
$\rho_1', \rho_2', \cdots \rho_n'$ etc.	Density of water of designated layers $1, 2, \cdots n$, below a plane of oscillation.
τ	Stress.
$\tau_{\mathrm{a}}, \tau_{\mathrm{a}x}, \tau_{\mathrm{a}y}$	Stress at air-water interface, components along x and y axes.
$\tau_{\mathrm{b}}, \tau_{\mathrm{b}x}, \tau_{\mathrm{b}y}$	Stress at water-sediment interface, and its components along x and y axes.
τ_x', τ_y'	Derivatives of τ_x, τ_y with respect to z.
ϕ	Geographical latitude.
ϕ_{s}	Defined in theory of internal seiche as $\phi_{\mathrm{s}} = 2\pi/l_{\mathrm{s}}$.
ϕ_0	Defined in theory of internal seiche as $\phi_0 = (\pi/l_{\mathrm{s}})\sqrt{-g\mathrm{H}^{\mathrm{i}}}$.
ψ	Angular distance of light source from zenith.
ψ'	Angular height of any light source.
ψ_{i}	Angle of incidence (for sunlight on a plane lake surface).
ψ_{r}	Angle of refraction.
ψ_{r}'	Defined as $\psi_{\mathrm{r}}' = \pi/2 - \psi_{\mathrm{r}}$.
ψ_{s}	Angular distance of sun from zenith.
ψ_{s}'	Angular height of sun.
ψ_{w}	Apparent angular distance of light source seen from a point below lake surface.
ω_n	A quantity defined in Chapter 7, used in Ertel's theory of the thermocline.
[ ]	Used to express concentration. In the cases of all substances other than permanent gases, the ordinary chemical convention of expressing solute in moles is employed unless the contrary is indicated, when such brackets are used; for permanent gases, the symbol refers to concentration in the unit employed in context.
$>$	Greater than.
$<$	Less than.
$\gg$	Much greater than.
$\ll$	Much less than.
$\geqslant$	Equal to or greater than (not less than).
$\leqslant$	Equal to or less than (not greater than).
$\sim$	About, approximately.
$\simeq$	Approximately equal to.
log	Logarithm to base 10.
ln	Logarithm to base e.

Bibliography and Index of Authors

It is impossible but that offences will come: but woe unto him, through whom they come. Luke 17: 1.

The standard abbreviations of the *World List of Scientific Periodicals*, 3rd ed., London, Butterworth Scientific Publications, 1950, are used throughout.

The asterisk (*) indicates works of exceptional limnological importance, either on account of the discoveries reported or because of extensive lists of references.

Aarnio, B., 1915. Über die Ausfallung des Eisenoxyds und der Tonerde in finnländischen Sand- und Grusböden. *Geotekn. Medd.*, 16:1–76. **711**†

* Åberg, B., and Rodhe, W., 1942. Über die Milieufaktorenin ei nigen südschwedischen Seen. *Symb. bot. upsaliens*, 5, No. 3, 256 pp. **383, 392, 402, 408, 413, 602, 686, 706, 711, 879**

Adam, N. K., 1937. A rapid method for determining the lowering of tension of exposed water surfaces, with some observations on the surface tension of the sea and of inland water. *Proc. roy. Soc.*, ser. B, 122:134–149. **900**

[Addison, J.] 1705. Remarks on several parts of Italy, etc., in the years 1701, 1702, 1703. London, J. Tonson, 534 pp. The later editions, at least from the 5th, were not anonymous; failure to recognize the existence of the first edition has led some authors to imply that the lake of Geneva was observed by Addison post mortem. **300**

Adeney, W. E., 1905. Unrecognized factors in the transmission of gases through water. *Sci. Trans. R. Dublin Soc*, 8:161–168. Also *Phil. Mag.*, ser. 6, 9:360–369. **589**

Adeney, W. E., 1926. On the rate and mechanism of the aeration of water under open-air conditions. *Sci. Proc. R. Dublin Soc.*, n.s. 18:211–217. **589**

* Adeney, W. E., and Becker, H. G., 1919. On the rate of solution of atmospheric nitrogen and oxygen by water, Part II. *Sci. Proc. R. Dublin Soc.*, n.s. 15:609–628. **588**

Adeney, W. E., and Becker, H. G., 1920. On the rate of solution of atmospheric nitrogen and oxygen by water, Part III. *Sci. Proc. R. Dublin Soc.*, n.s. 16:143–152. **589**

† Numbers in bold type are references to pages of this volume.

Adeney, W. E., Leonard, A. G. G., and Richardson, A. M., 1922. On the aeration of quiescent columns of distilled water and of solutions of sodium chloride. *Sci. Proc. R. Dublin Soc.*, n.s. 17:19–28. Also *Phil. Mag.*, ser. 6, 45:837–845. **457, 589**

Adlercreutz, E., 1928. Über das Vorkommen von Jod in verschiedenartigen Wässern in Finnland. *Acta med. scand.*, 69:325–391. **562**

Ahlmann, H. W., 1919. Geomorphological studies in Norway. *Geogr. Ann.*, 1:1–252. **43, 57**

Aitken, J., 1911. The sun as fog producer. *Proc. roy. Soc. Edinb.*, 32:183–215. **548**

Albrecht, F., 1933. Ein Strahlungsbilanzmesser zur Messung des Strahlungshaushaltes von Oberflächen. *Met. Z.*, 50:62–65. **378**

Alden, W. C., 1928. Landslide and flood at Gros Ventre, Wyoming. *Trans. Amer. Instit. min. (metall.) Engrs*, 76:347–360. **43**

Aleev, B. S., and Mudretsova, K. A., 1937. The role of the phytoplankton in the nitrogen dynamics in the water of a "blossoming" pond. *Microbiology, Moscow*, 6:329–338 (not seen: ref. *Chem Abstr.*, 33:5446). **847**

Alison, M. S., 1899. On the origin and formation of pans. *Trans. geol. Soc. S. Afr.*, 4:159–161. **132**

Allen, E. J., 1914. On the culture of the plankton diatom *Thalassiosira gravida*, Cleve, in artificial sea water. *J. Mar. biol. Ass. U.K.*, 10:417–439. **896**

Allen, J. A., 1955. Solubility of oxygen in water. *Nature, Lond.*, 175:83. **581**

Allgeier, R. J., Hafford, B. C., and Juday, C., 1941. Oxidation-reduction potentials and *p*H of lake waters and of lake sediments. *Trans. Wis. Acad. Sci. Arts Lett.*, 33:115–133. **693, 694, 695, 719, 767, 774**

Allgeier, R. J., Peterson, W. H., Juday, C., and Birge, E. A., 1932. The anaerobic fermentation of lake deposits. *Int. Rev. Hydrobiol.*, 26:444–461. **584, 842**

Alsterberg, G., 1922. Die respiratorischen Mechanismen der Tubificiden. *Acta Univ. lund. (Lunds Univ. Arsskr.)*, NF Avd. 2, 18, No. 1, 175 pp. **609**

Alsterberg, G. 1927. Die Sauerstoffschichtung der Seen. *Bot. Notiser.*, 25:255–274. **282, 605, 618, 625, 639**

Alsterberg, G., 1929. Über das aktuelle und absolute O_2-Defizit der Seen im Sommer. *Bot. Notiser.*, 1929:354–376. **610, 639**

Alsterberg, G., 1930a. Die O_2-Primärkonstante in den verschiedenen Seenbereichen während des Jahres. *Bot. Notiser.*, 1930:251–304. **576, 839**

* Alsterberg, G., 1930b. Die thermischen und chemischen Ausgleiche in den Seen zwischen Boden- und Wasserkontakt, sowie ihre biologische Bedeutung. *Int. Rev. Hydrobiol.*, 24:290–327. **454, 476, 613, 629**

Alsterberg, G., 1931. Die Ausgleichströme in den Seen im Sommerhalbjahr bei Abwesenheit der Windwirkung. *Int. Rev. Hydrobiol.*, 25:1–32. **454, 476, 613**

American Public Health Association, 1917. *Standard Method for the Examination of Water and Sewage.* Boston, 1917, 3rd ed., xvi, 115 pp. Later ed. to 10th, 1955, xix, 522 pp. **579**

Anderson, C. A., 1941. Volcanoes of the Medicine Lake Highland, California. *Bull. Dept. Geol. Sci. Univ. Calif.*, 25:347–422. **34**

Anderson, E. N., and Walker, E. R., 1920. An ecological study of algae of some sandhill lakes. *Trans. Amer. micro. Soc.*, 39:51–85. **127, 128**

* Anderson, E. R., 1952. Energy-budget studies. Water-loss investigations: Vol. 1, Lake Hefner Studies. Tech. Rep., *U.S. Geol. Surv. Circ.*, 229:71–119. **375, 376, 377, 513, 514**

* Anderson, E. R., and Pritchard, D. W., 1951. *Physical Limnology of Lake Mead.* U. S. Navy Electronics Laboratory, San Diego, California, Report 258, Problem NEL 2J1, 152 pp. **297, 381, 507, 514**

Anderson, G. R., 1955. Nitrogen fixation by Pseudomonas-like soil bacteria. *J. Bact.*, 70:129–133. **846**

Anderson, T., and Flett, J. S., 1903. Report on the eruptions of the Soufrière, in St. Vincent in 1902, and on a visit to Montagne Pelée, in Martinique. *Phil. Trans.*, ser. A, 200:353–553. **30**

Andrews, E., 1870. The North American lakes considered as chronometers of post-glacial time. *Trans. Chicago Acad. Sci.*, 2:5–9. **178**

Ångstrom, A., 1925. On the albedo of various surfaces of ground. *Geogr. Ann.*, 7:323–342. **375, 376**

Antonescu, C. S., 1931. Über das Vorkommen eines ausgeprägten metalimnischen Sauerstoffminimums in einem norddeutschen See (Sakrower See bei Potsdam). *Arch. Hydrobiol.*, 22:580–596. **621**

Arakawa, H., 1954. Fujiwhara on five centuries of freezing dates of Lake Suwa in the Central Japan. *Arch. Met. Wien.* ser. B., *Allgem. biol. Klimat.*, 6:152–166. **531**

Arnold, J. R., and Libby, W. F., 1950. *Radiocarbon Dates* (*September 1, 1950*). Univ. Chicago, Institute for Nuclear Studies, 15 pp.; also *Science*, 113:111–120. **32**

Aschan, O., 1906. Humusämnena i de nordiska inlandsvattnen och deras betydelse, särskildt vid sjömalmernas daning. *Bidrag Finl. Nat. Folk*, 66:1–176. **711**

Aschan, O., 1908. Die wasserlöslichen Humusstoffe (Humussole) der nordischen Güssgewässer. *J. prakt. Chem.*, NF 77:172–188. **889**

Atkins, W. R. G., 1923. The phosphate content of fresh and salt waters in its relationship to the growth of algal plankton. *J. Mar. biol. Ass. U.K.*, 13:119–150. **728**

Atkins, W. R. G., 1933. The rapid estimation of the copper content of sea water. *J. Mar. biol. Ass. U.K.*, 19:63–66. **811**

Atkins, W. R. G., 1947. Electrical conductivity of river, rain, and snow water. *Nature, Lond.*, 159:674. **549**

Auer, V., 1930. Peat bogs in southeastern Canada. *Mem. geol. Surv. Can.*, 162, 32 pp. **143**

Auerbach, M., Maerker, W., and Schmalz, J., 1924. Hydrographisch-biologische Bodensee—Untersuchungen I. *Arch. Hydrobiol. Suppl.*, 3:597–738. **451, 596**

Auerbach, M., Maerker, W., and Schmalz, J., 1926. Hydrographisch-biologische Bodensee—Untersuchungen II. *Verh. naturw. Ver. Karlsruhe*, 30 (not seen: ref. Thienemann 1928). **625**

Auerbach, M., and Ritzi, M., 1937. Oberflächen- und Tiefenströme des Bodensees, IV. *Arch. Hydrobiol.*, 32:409–433. **292**

* Aufsess, O. von, 1904. Die Farbe der Seen. *Ann. Phys. Lpz.*, ser. 4, 13:678–711. **383, 401**

Aurén, T. E., 1939. Radiation climate in Scandinavian Peninsula. *Ark. Mat. Ast. Fys.*, 26A, No. 20, 50 pp. **372**

Ayers, J. C., 1956. A dynamic height method for the determination of currents in deep lakes. *Limnol. Oceanogr.*, 1:150–161. **264**

Baars, J. K., 1930. Over sulfaat reductie door bacterien. Diss. Delft, 164 pp. **754**

Baas-Becking, L. G. M., 1925. Studies on the sulfur bacteria. *Ann. Bot. Lond.*, 39:613–650. **756, 758**

Baldi, E., 1949. Relazione del direttore sull' attività scientifica dell' istituto negli anni 1947–1948. *Mem. Ist. ital. Idrobiol. de Marchi*, 5:1–29. **115**

Baldi, E., Pirocchi, L., and Tonolli, V., 1949. Relazione preliminare sulle ricerche idrobiologiche condotte sul Lago di Lugano. *Pallanza, Ist. ital. Idrobiol. de Marchi* (Pubbl. a cura dell' Ispettorato Federate Svizzero per la Pesca, Berna), mimeogr., 35 pp. **488**

Baldwin, R. B., 1949. *The Face of the Moon.* Chicago, Univ. Chicago Press, 239 pp. **37, 38, 151**

Baier, C. R., 1936. Studien zur Hydrobakteriologie stehender Binnengewässer. *Arch. Hydrobiol.*, 29:183–264. **846, 870**

Barnes, T. C., 1932. The physiological effect of trihydrol in water. *Proc. nat. Acad. Sci., Wash.*, 18:136–137. **202**

Barnes, T. C., 1933. A possible physiological effect of the heavy isotope of H in water. *J. Amer. chem. Soc.*, 55:4332–4333. **202**

Barnes, T. C., 1937. *A Textbook of General Physiology.* Philadelphia, Blakiston, xxii, 554 pp. **218**

* Barnes, T. C., and Jahn, T. L., 1934. Properties of water of biological interest. *Quart. Rev. Biol.*, 9:292–341. **197, 202, 218**

Barnes, T. C., and Larson, E. J., 1934. The influence of heavy water of low concentration on Spirogyra, Planaria, and on enzyme action. *Protoplasma, Leipzig*, 22:431–443. **217**

Barrell, J., 1912. Criteria for the recognition of ancient delta deposits. *Bull. geol. Soc. Amer.*, 23:397–446. **186**

Barrenscheen, H. K., and Beckh-Widmanstetter, H. A., 1923. Über bakterielle Reduktion organisch gebundener Phosphorsäure. *Biochem. Z.*, 140:279–283. **727**

Barringer, D. M., 1905. Coon Mountain and its crater. *Proc. Acad. nat. Sci. Philad.*, 57:861–886. **149**

* Beadle, L. C., 1932. Scientific results of the Cambridge expedition to the East African lakes, 1930–1. 4. The waters of some East African lakes in relation to their fauna and flora. *J. Linn. Soc. (Zool.)*, 38:157–211. **462, 463, 634, 791**

Beadnell, H. J. L., 1905. Topography and Geology of the Fayum Province of Egypt. *Mem. geol. Surv. Egypt*, Cairo, 101 pp. **136**

Beath, O. A., Eppson, H. F., and Gilbert, C. S., 1935. Selenium and other toxic minerals in soils and vegetation. *Bull. Wyo. agric. Exp. Sta.*, 206:1–55. **563**

* Beauchamp, R. S.A., 1939. Hydrology of Lake Tanganyika. *Int. Rev. Hydrobiol.*, 39:316–353. **463, 465, 634, 791**

Beauchamp, R. S. A., 1940. Chemistry and hydrography of Lakes Tanganyika and Nyasa. *Nature, Lond.*, 146:253–256. **634, 635**

Beauchamp, R. S. A., 1953a. Hydrological data from Lake Nyasa. *J. Ecol.*, 41:226–239. **463, 635**

Beauchamp, R. S. A., 1953b. Sulphates in African inland waters. *Nature, Lond.*, 171:769. **766**

Beauchamp, R. S. A., 1954. *East African Fisheries Research Organization, Annual Report 1953.* Jinja, Uganda, East African High Commission, 44 pp. **339**

Becker, H. G., 1924. Mechanism of absorption of moderately soluble gases in water. *Industr. Engng. Chem.*, 16:1220–1224. **658**

Becker, H. G., and Pearson, E. F., 1923. Irregularities in the rate of solution of oxygen by water. *Sci. Proc. R. Dublin Soc.*, n.s. 17:197–200. **589**

Beckwith, T. D., 1933. Metabolic studies upon certain Chlorellas and allied forms. *Univ. Calif. at Los Angeles Publ. Biol. Sci.*, 1:1–34. **849**

Beijerinck, M. W., 1901. Ueber oligonitrophile Mikroben. *Zbl. Bakt.*, II, 7:561–582. **846**

van Bemmelen, R. W., 1930. The origin of Lake Toba. *Proc. Fourth Pacif. Sci. Congr.*, Java, 1929, IIA(*Physical papers*):115–124. **35**

van Bemmelen, R. W., 1940. *The Geology of Indonesia:* Vol. 1A, *General Geology of Indonesia and Adjacent Archipelagoes.* Gov. Printing Office, Hague, 732 pp. **14, 22, 25, 30, 33**

Benoit, R. J., 1956. Studies on the biogeochemistry of cobalt and related elements. Unpublished thesis, Yale University. **811, 825**

Benton, G. S., Blackburn, R. T., and Snead, V. O., 1950. The role of the atmosphere in the hydrologic cycle. *Trans. Amer. geophys. Un.*, 31:61–73. **229**

Berg, D., and Patterson, A., 1953. The high field conductance of aqueous solution of carbon dioxide at 25°. The true ionization constant of carbonic acid. *J. Amer. chem. Soc.*, 75:5197–5200. **655**

Berg, L. S., 1908. Aral'skoe more (Die Aral See. Versuch einer physisch-geographischen Monographie). *Bull. Turkestan Sect. Russ. Geogr. Soc.*, 5, xxiii, 580 pp. (Russian). **316**

Berg, L. S., 1934. Niveau de la mer Caspienne dans les temps historiques. *Probl. phys. Geogr., Moscow* (*Akad. nauk. Instit. Geogr.*), 1:Russian text, 11–58, 61–64 (bibl.); French summary, 58–61. **7, 239**

Berg, L. S., (1938) 1950. *Natural Regions of U.S.S.R.* (trans. G. A. Titelbaum from the 2nd Russian ed., 1938; ed. T. A. Morrison and C. C. Nikiforoff). New York, Macmillan, xxxi, 436 pp. **7, 8, 44, 239**

Bergsten, F., 1926. The sieches of Lake Vetter. *Geogr. Ann.*, 8:1–73. **307, 308, 311, 325**

Bergsten, F., 1928. On periods in the water heights of Lake Vänern. *Geogr. Ann.*, 10:140–144. **247**

Bergsten, F., 1949. The duration of the water stages of the Swedish lakes Vänern and Storsjön. *Gen. Ass. int. Un. Geod.*, Oslo, 1948, *Ass. int. Hydrol. Sci., 1. Trav. Comm. Potamol. Limnol.* **236**

Bernal, J. D., and Fowler, R. F., 1933. A theory of water and ionic solution with particular reference to H and OH ions. *J. chem. Phys.*, 1:515–548. **197, 210**

Bertrand, G., 1943a. Sur le magnésium de l'eau de pluie récoltée à Grignon. *C.R. Acad. Sci., Paris*, 216:364–367. **550**

Bertrand, G., 1943b. Sur le magnésium et le calcium de l'eau de pluie récoltée à Paris. *C.R. Acad. Sci., Paris*, 216:701–709. **550**

Bertrand, G., 1944. Origin multiple du magnésium de l'eau de pluie. *C.R. Acad. Sci., Paris*, 219:14–16. **550**

Bertrand, G., 1945. Le potassium dans l'eau de pluie. *C.R. Acad. Sci., Paris*, 220:865–868. **550**

Birge, E. A., 1897. Plankton studies on Lake Mendota. II. The crustacea from the plankton from July, 1894, to December, 1896. *Trans. Wis. Acad. Sci. Arts Lett.*, 11:274–448. **428**

Birge, E. A., 1906. The oxygen dissolved in the waters of Wisconsin lakes. *Trans. Amer. Fish. Soc.*, 35:142–163. **600**

Birge, E. A., 1907. The respiration of an inland lake. *Trans. Amer. Fish Soc.*, 36:223–241. **600**

Birge, E. A., 1908. Gases dissolved in the waters of Wisconsin lakes. *Bull. U.S. Bur. Fish.*, 28:1275–1295 (*Proc. Fourth Intern. Fish. Cong.*). **600**

Birge, E. A., 1910a. An unregarded factor in lake temperatures. *Trans. Wis. Acad. Sci. Arts Lett.*, 16:989–1004. **509**

Birge, E. A., 1910b. On the evidence for temperature seiches. *Trans. Wis. Acad. Sci. Arts Lett.*, 16:1005–1016. **428**

Birge, E. A., 1910c. The apparent sinking of ice in lakes. *Science*, 32:81–82. **456, 530, 531**

* Birge, E. A., 1915. The heat budgets of American and European lakes. *Trans. Wis. Acad. Sci. Arts Lett.*, 18:166–213. **439, 492, 495, 501** fn., **504, 511**

* Birge, E. A., 1916. The work of the wind in warming a lake. *Trans. Wis. Acad. Sci. Arts Lett.*, 18:341–391. **429, 495, 508**

* Birge, E. A., and Juday, C., 1911. The inland lakes of Wisconsin. The dissolved gases and their biological significance. *Bull. Wis. geol. nat. Hist. Surv.*, 22, 259 pp. **503, 554, 579, 587, 596** fn., **600, 601, 605** fn., **623, 640, 642, 644, 678, 680, 765, 768, 797, 838**

* Birge, E. A., and Juday, C., 1914. A limnological study of the Finger Lakes of New York. *Bull. U.S. Bur. Fish.*, 32:523–609. **90, 428** fn., **579, 596** fn.

Birge, E. A., and Juday, C., 1921. Further limnological observations on the Finger Lakes of New York. *Bull. U.S. Bur. Fish.*, 37:210–252. **508**

* Birge, E. A., and Juday, C., 1929. Transmission of solar radiation by the waters of the inland lakes. *Trans. Wis. Acad. Sci. Arts Lett.*, 24:509–580. **391, 392**

Birge, E. A., and Juday, C., 1930. A second report on solar radiation and inland lakes. *Trans. Wis Acad. Sci. Arts Lett.*, 25:285–335. **391, 392**

Birge, E. A., and Juday, C., 1931. A third report on solar radiation and inland lakes. *Trans. Wis. Acad. Sci. Arts Lett.*, 26:383–425. **391, 392, 395**

Birge, E. A., and Juday, C., 1932. Solar radiation and inland lakes, fourth report. Observations of 1931. *Trans. Wis. Acad. Sci. Arts Lett.*, 27:523–562. **391, 392**

* Birge, E. A., and Juday, C., 1934. Particulate and dissolved organic matter in inland lakes. *Ecol. Monogr.*, 4:440–474. **231, 879, 882**

* Birge, E. A., Juday, C., and March, H. W., 1928. The temperature of the bottom deposits of Lake Mendota; a chapter in the heat exchanges in the lake. *Trans. Wis. Acad. Sci. Arts Lett.*, 23:187–231. **454, 505**

Bjerrum, N., and Gjaldback, J. K., 1919. Undersøgelser over de faktorer, som bestemmer jorbundens reaktion. *K. VetHøjsk. Aarsskr.*, 1919:48–91. (not seen: ref. Warburg 1922, Buch 1930). **656**

Black, R. F., and Barksdale, W. L., 1949. Oriented lakes of Northern Alaska. *J. Geol.*, 57:105–118. **100**

Blackwelder, E., 1948. The geological background in a symposium on the Great Basin, with emphasis on glacial and postglacial times. *Bull. Univ. Utah biol. Ser.*, 107:1–16. **15**

Bloch, R. J., and Bolling, D., 1947. *The Amino Acid Composition of Proteins and Foods*, rev. ed. Springfield, Ill., Charles C Thomas, 414 pp. **896**

Bogatschew, W. W., 1928. Mytilaster im Kaspischen Meere. *Russk. gidrobiol. Zh.* (*Russ. hydrobiol. Z.*), Saratov., 7:189. **7**

Bohr, C., 1899. Definition und Methode zur Bestimmung der Invasions und Evasionscoefficienten bei der Auflösung von Gasen in Flüssigkeiten. Werthe der genannter Constanten sowie der Absorptionscoefficienten der Kohlensäure bei Auflösung in Wasser und in Chlornatium-lösungen. *Ann. Phys., Lpz.*, 304(NF 68):500–525. **654**

Bohr, C., and Bock, J., 1891. Bestimmung der Absorption einiger Gase in Wasser bei den Temperaturen zwischen 0 und 100°. *Ann. Phys., Lpz.*, 280(NF 44):318–343. **579**

Bolz, H. M., and Fritz, H., 1950. Tabellen und Diagramme zur Berechnung der Gegenstrahlung und Ausstrahlung. *Z. Met.*, 4:314–317. **377**

Bond, R. M., 1935. Investigations of some Hispaniolan lakes (Dr. R. M. Bond's expedition). II. Hydrology and hydrography. *Arch. Hydrobiol.*, 28:137–161. **571**

Bonython, C. W., and Mason, B., 1953. The filling and drying of Lake Eyre. *Geogr. J.*, 119:321–330. **228, 244, 245**

Boone, E., and Baas-Becking, L. G. M., 1931. Salt effects on eggs and nauplii of *Artemia salina* L. *J. gen. Physiol.*, 14:753–763. **571**

Bortels, H., 1940. Über die Bedeutung des Molybdäns für stickstoffbindende Nostocaceen. *Arch. Mikrobiol.*, 11:155–186. **824, 847**

du Boys, P. 1891. Essai théorique sur les seiches. *Arch. Sci. phys. nat.*, Ser. 3, 25:628–652. **302**

Boule, M., Glangeaud, P., Rouchon, G., and Vernière, A., 1901. *Le Puy-de-Dôme et Vichy; guide du touriste, du naturaliste et de l'archéologue.* Paris, Masson et Cie, iii, 378 pp. **38, 39**

Bourcart, F. E., 1906. *Les lacs alpins suisses, étude chimique et physique.* Genève, Georg et Cie, 130 pp. **486, 766, 775**

Bourgin, A., 1945. Lacs d'altitude des Alpes Françaises, I. *Rev. Géogr. alp.*, 35:739–745. **110**

Boussinesq, J., 1877. Essai sur la théorie des eaux courantes. *Mem. prés. Acad. Sci. Paris*, 23:1–680. **251**

Bowen, V. T., 1948. Studies of the mineral metabolism of some manganese accumulator organisms. Unpublished thesis, Yale University. **560**

Bowman, I., 1935. Our expanding and contracting "desert." *Geogr. Rev.*, 25:43–61. **244**

Braarud, T., and Føyn, B., 1930. Beiträge zur Kenntnis des Stoffwechsels im Meere. *Avh. norske VidenskAkad.*, Mat.-nat. Kl., 14:1–24. **851**

Bradtke, F., 1910. Stehende Seespiegelschwankungen, beobachtet am Lagower See und Tschetschsee. Diss. Halle (not seen: ref. Halbfass 1922). **314**

Braidech, M. M., and Emery, F. H., 1935. The spectrographic determination of minor chemical constituents in various water supplies in the United States. *J. Amer. Wat. Wks Ass.*, 27:557–580. **807, 811, 821, 824, 827, 829**

Branca, W., and Fraas, E., 1905. Das kryptovulkanische Becken von Steinheim. *Abh. preuss. Akad. Wiss.*, I., 10, 64 pp. **151**

von Brand, T., and Rakestraw, N. W., 1940. Decomposition and regeneration of nitrogenous organic matter in sea water. III. Influence of temperature and source and condition of water. *Biol. Bull., Woods Hole*, 79:231–236. **837, 848**

von Brand, T., and Rakestraw, N. W., 1941. Decomposition and regeneration of nitrogenous organic matter in sea water. IV. Interrelationship of various stages: influence of concentration and nature of particulate matter. *Biol. Bull., Woods Hole*, 81:63–69. **837, 848**

von Brand, T., Rakestraw, N. W., and Renn, C. E., 1937. The experimental decomposition and regeneration of nitrogenous organic matter in sea water. *Biol. Bull., Woods Hole*, 72:165–175. **837, 848**

von Brand, T., Rakestraw, N. W., and Renn, C. E., 1939. Further experiments on the decomposition and regeneration of nitrogenous organic matter in sea water. *Biol. Bull., Woods Hole*, 79:285–296. **837, 848**

von Brand, T., Rakestraw, N. W., and Zabor, J. W., 1942. Decomposition and regeneration of nitrogenous organic matter in sea water. V. Factors influencing the length of the cycle: observations upon the gaseous and dissolved organic nitrogen. *Biol. Bull., Woods Hole*, 83:273–282. **837, 848**

Bregoff, H. M., and Kamen, M. D., 1951. Photohydrogen production in *Chromatium. J. Bact.*, 63:147–149. **848**

Bretschneider, C. L., 1952. The generation and decay of wind waves in deep water. *Trans. Amer. geophys. Un.*, 33:381–389. **353**

Bretz, J H., 1932. The Grand Coulee. *Spec. Publ. Amer. geogr. Soc.*, No. 15, 86 pp. **111**

Bretz, J H., 1951. The stages of Lake Chicago: their causes and conditions. *Amer. J. Sci.*, 249:409–429. **82**

Britton, H. T. S., 1942. *Hydrogen Ions. Their Determination and Importance in Pure and Industrial Chemistry*, 3rd ed. London, Chapman and Hall. I, xix, 420 pp; II, xix, 443 pp. **665**

Brochu, M., 1954. Lacs d'érosion différentielle glaciaire sur le bouclier canadien. *Rev. Géomorph. dynam.*, 5:274–279. **72** fn.

Brönsted, J. N., and Wesenberg-Lund, C., 1911. Chemisch-physikalische Untersuchungen der dänischen Gewässer nebst Bemerkungen über ihre Bedeutung für unsere Auffassung der Temporalvariationen. *Int. Rev. Hydrobiol.*, 4:251–290, 437–492. **428, 452**

Brooks, C. E. P., 1919. The secular variation of rainfall. *Quart. J. Roy. met. Soc.*, 45:233–245. **242**

Brooks, C. E. P., 1923. Variations in the levels of the Central African lakes, Victoria and Albert. *Geophys. Mem.* (Great Britain: Meteorological Office), 20:337–344. **242**

Brooks, J. L., 1947. Turbulence as an environmental determinant of relative growth in Daphnia. *Proc. nat. Acad. Sci., Wash.*, 33:141–148. **259**

* Brooks, J. L., 1950. Speciation in ancient lakes. *Quart. Rev. Biol.*, 25:131–176. **24, 464, 635**

Broughton, W. A., 1941. The geology, ground water, and lake basin seal of the region south of the Muskellunge Moraine, Vilas County, Wisconsin. *Trans. Wis. Acad. Sci. Arts Lett.*, 33:5–20. **96, 232**

Brown, C. A., 1943. Vegetation and lake level correlations at Catahoula Lake, Louisiana. *Geogr. Rev.*, 33:435–445. **125**

Brown, E., 1669. An account from the same Dr. Brown concerning an uncommon lake, called the Zirchnitzer, in Carniola. *Phil. Trans.* 4(No. 54):1083–1085. **107**

Brown, E., 1674. Some queries and answers, relating to an account given in No. 54 by Dr. Edw. Brown of a strange lake in Carniola, called the Zirchnitz-Sea. *Phil. Trans.*, 9(No. 109):194–197. **107**

Brown, H., 1949a. A table of relative abundances of nuclear species. *Rev. mod. Phys.*, 21:625–634. **222**

Brown, S., 1949b. Rare gases and the formation of the earth's atmosphere. In *The Atmosphere of the Earth and Planets*, ed. G. P. Kuiper. Chicago, Univ. Chicago Press, Chap. 9:260–268. **221**

* Brückner, E., 1890. Klimaschwankungen seit 1700 nebst Bemerkungen über die Klimaschwankungen der Diluvialzeit. *Geogr. Abh.* (herausgeg. A. Penck) IV, No. 2, 324 pp. Wien und Olmütz, ed. Hölzel. **7, 238, 239**

Bruno, B., 1938. Die Wasserstandsschwankungen des Wenersees. *Geogr. Ann.*, 20:308–315. **247**

Brunowsky, B. K., and Kunasheva, K. D., 1935. Beiträge zum Radiumgehalt von Pflanzen und Gewässern. *Trav. Lab. Biogéochim. USSR*, 3:31–41. **832**

Brunt, D., 1946. Patterns in ice and cloud. *Weather*, 1:184–185. **531**

Bruyant, C., 1903. Les seiches du lac Pavin. *Rev. d'Auvergne* (Clermont-Ferrand), 20:81 (not seen: ref. Halbfass 1923). **314**

Bryan, K., 1925. The Papago county, Arizona. *Wat. Supp. Pap., Wash.*, 499, 436 pp. **128**

Bryson, R. A., and Bunge, W. W., 1956. The stress-drop jet in Lake Mendota. *Limnol. Oceanogr.*, 1:42–46. **341**

Bryson, R. A., and Kuhn, P. M., 1952. *On Certain Oscillatory Motions of Lakes.* Report to the University of Wisconsin Lake Investigations Committee, No. 5, mimeogr., 10 pp. **308, 309, 310, 314, 315**

Bryson, R. A., and Suomi, V. E., 1951. Midsummer renewal of oxygen within the hypolimnion. *J. Mar. Res.*, 10:263–269. **296**

Bryson, R. A., and Suomi, V. E., 1952. The circulation of Lake Mendota. *Trans. Amer. geophys. Un.*, 33:707–712. **288**

Bryson, R. A., Suomi, V. E., and Stearns, C. R., 1952. *The Stress of the Wind on Lake Mendota.* Report to the University of Wisconsin Lake Investigations Committee, No. 6, mimeogr., 6 pp. **274**

Buch, K., 1930. Die Kohlensaürefaktoren des Meerwassers, I. *Rapp. Cons. Explor. Mer.*, 67:5–88. 1933. **657**

Buch, K., 1933. On boric acid in the sea and its influence on the carbonic acid equilibrium. *J. Cons. Perm. int. Explor. Mer*, 8:309–325. **657**

Buchan, A., 1871. Remarks on the deep-water temperature of Lochs Lomond, Katrine, and Tay. *Proc. roy. Soc. Edinb.*, 7:791–795. **426**

Buchanan, J. Y., 1886. Distribution of temperature in Loch Lomond during the autumn of 1885. *Proc. roy. Soc. Edinb.*, 13:403–428. **427**

Bucher, W. H., 1933. Cryptovolcanic structures in the United States. *Rep. 16th int. geol. Cong.*, 2:1055–1084. **38**

Buckley, E. R., 1900. Ice ramparts. *Trans. Wis. Acad. Sci. Arts Lett.*, 13:141–157. **532**

Büdel, A., 1926. Transkaukasien. Eine technische Geographie. *Petermanns Mitt.*, Erg. Heft. No. 189, 152 pp. **170**

von Bülow, B. Fr., and Otto, K., 1931. Der Arsengehalt von Wasser, Grund und Umgebung des "Roten Sees" sowie der Werra und einiger ihrer Zuflüsse nahe Witzenhausen. *Arch. Hydrobiol.*, 22:129–133. **417, 563**

Buffle, J. P., Jung, C., and Rossier, P., 1938. Observations d'un phénomène d'optique lacustre: l'iris du 8 Mars 1938 sur le lac de Genève. *C.R. Soc. Phys. Hist. nat. Genève*, 55:71–73 (suppl. to *Arch. Sci. phys. nat.*, 5th ser., 20). **419**

Bullard, S. E., 1954. The Interior of the Earth. *The Solar System* (ed. G. P. Kuiper): Vol. II, *The Earth as a Planet.* Chicago, Univ. Chicago Press, Chap. 3:57–137. **460**

Burk, D., Lineweaver H., and Horner, C. K., 1932. Iron in relation to the stimulation of growth by humic acids. *Soil Sci.*, 33:413–453. **712**

Burris, R. H., Eppling, F. J., Wahlin, H. B., and Wilson, P. W., 1943. Detection of nitrogen-fixation with isotopic nitrogen. *J. biol Chem.*, 148:349–357. **847**

Burt, W. V., 1955. Interpretation of spectrophotometer readings on Chesapeake Bay Water. *J. Mar. Res.*, 14:33–46. **385** fn.

Buxtorf, A., 1922. Das Längenprofil des schweizerisch-französischen Doubs zwischen dem Lac des Brenets und Soubey. *Ecl. geol. Helv.*, 16:527–539. **44**

Byrstein, J. A., and Beliaev, G. M., 1946. Deĭstvie vody Ozera Balkhash na Volgo-Caspiiskikh bezpozvonochnykh (The action of the water of Lake Balkash on the Volga-Caspian invertebrates.) *Zool. Zh.*, 25:225–236 (not seen: ref. *Biol. Abstr.*, 23, No. 22965), **571**

Cabot, E. C., 1946. Dual-drainage anomalies in the far north. *Geogr. Rev.*, 36:474–482. **232**

Caldenius, C. C., 1932. Las glaciaciones cuaternarias en la Patagonia y Tierra del Fuego. *Geogr. Ann.*, 14:1–164. **80**

Callendar, G. S., 1940. Variations of the amount of carbon dioxide in different air currents. *Quart. J. R. met. Soc.*, 66:395–400. **654**

Callendar, G. S., 1949. Can carbon dioxide influence climate? *Weather*, 4:310–314. **654**

Campbell, M. R., 1914. *Origin of the Scenic Features of the Glacier National Park.* Washington, U.S. Dept. of the Interior, 42 pp. **84**

* Capart, A., 1949. Sondages et carte bathymétrique. *Explor. hydrobiol. Lac Tanganyika* (*Inst. roy. Sci. nat. Belg.*), 2, fasc. 2, 16 pp. **23, 189**

Capart, A., 1952. Le milieu géographique et géophysique. *Explor. hydrobiol. Lac Tanganyika* (*Inst. roy. Sci. nat. Belg.*), 1:3–27. **403, 462, 463, 465**

Carman, P. C., 1940. Constitution of colloidal silica. *Trans. Faraday Soc.*, 36:964–973. **789**

Caton-Thompson, G., and Gardiner, E. W., 1929. Recent work on the problem of Lake Moeris. *Geogr. J.*, 73:20–60. **136**

Chamisso, A. von, 1821. In O. von Kotzebue, *Entdeckungs—Reise in die Süd-See und nach der Berings-Strasse.* Weimar, Geb. Hoffmann, 3 vol., 240 pp. **145**

Charlesworth, J. K., 1939. Some observations on the glaciation of northeast Ireland. *Proc. R. Irish Acad.*, 45B:255–295. **86, 189**

Charlesworth, J. K., 1953. *The Geology of Ireland: An Introduction.* London, Oliver and Boyd, xvi, 276 pp. **57, 173**

Chézy, A., 1775. Mémoire sur la vitesse de l'eau conduite dans une rigole donnée. (The original statement of the Chézy formula was never published and is apparently lost; a work with the above title was, however, left in manuscript by Chézy and has been printed by G. Mouret in *Ann. Ponts Chauss.*, 91st ann., 11th ser., tome 61, 1921, II:241–246. An English translation had appeared earlier, published by C. Herschel, *J. Ass. Engng Soc., Philad.*, 18:363–369, 1897. Both Herschel and Mouret give interesting historic notes. The formula had of course been widely disseminated and used long before its belated publication). **262**

Cholnoky, E. von, 1897. Limnologie des Plattensees. *Resultate der wissenschaftlichen Erforschung des Plattensees*, Bd. 1, *Physikalische Geographie*, Teil 3, Wien, ed. Hölzel, 118 pp. **316, 324, 329**

Christophersen, E., 1927. Vegetation of Pacific equatorial islands. *Bull. Bishop Mus., Honolulu*, No. 44, 79 pp. **145**

Chrystal, G., 1904. Some results in the mathematical theory of seiches. *Proc. roy. Soc. Edinb.*, 25:328–337. **302**

Chrystal, G., 1905a. Some further results in the mathematical theory of seiches. *Proc. roy. Soc. Edinb.*, 25:637–647. **302, 305**

* Chrystal, G., 1905b. On the hydrodynamical theory of seiches. With a bibliographical sketch. *Trans. roy. Soc. Edinb.*, 41:599–649. **302, 307**

Chrystal, G., 1908. An investigation of the seiches of Loch Earn by the Scottish Lake Survey. Part III. Observations to determine the periods and nodes. Part IV. Effect of meteorological conditions upon the denivellation of lakes. Part V. Mathematical appendix on the effect of pressure disturbances upon the seiches in a symmetrical parabolic lake. *Trans. roy. Soc. Edinb.*, 46:455–517. **302, 307, 325**

Chrystal, G., 1908. Seiches in the lakes of Scotland. *Proc. roy. Inst. G.B.*, 18:657–676. **302**

* Chrystal, G., 1910. Seiches and other oscillations of lake surfaces, observed by the Scottish Lake Survey. In J. Murray and L. Pullar, *Bathymetrical Survey of the Scottish Fresh-water Lochs*, Vol. 1. Edinburgh, Challenger Office, pp. 29–90. **302, 306, 307, 311, 312, 314, 315, 316, 326, 328**

Chrystal, G., and Maclagan-Wedderburn, E., 1905. Calculation of the periods and nodes of Lakes Earn and Treig, from the bathymetric data of the Scottish Lake Survey. *Trans. roy. Soc. Edinb.*, 41:823–850. **302, 307**

* Chu, S. P., 1942. The influence of the mineral composition of the medium on the growth of planktonic algae. Part I. Methods and culture media. *J. Ecol.*, 30:284–325. **797, 897**

* Chu, S. P., 1943. The influence of the mineral composition of the medium on the growth of planktonic algae. Part II. The influence of the concentration of inorganic nitrogen and phosphate phosphorus. *J. Ecol.*, 31:109–148. **732, 734, 852**

Chu, S. P., 1945. Phytoplankton. *Rep. Freshw. biol. Ass. Brit. Emp.*, 13:20–23. **734**

Chu, S. P., 1946. The utilization of organic phosphorus by phytoplankton. *J. Mar. biol. Assoc. U.K.*, 26:285–295. **735**

* Chumley, J., 1910. Bibliography of limnological literature. In J. Murray, and L. Pullar, *Bathymetrical Survey of the Scottish Fresh-water Lochs*, Vol. 1. Edinburgh, Challenger Office, pp. 659–753. (*See preface*)

Church, P. E., 1942. The annual temperature cycle of Lake Michigan. I. Cooling from late autumn to the terminal point, 1941–42. *Institute of Meteorology, University of Chicago, Miscellaneous Reports*, No. 4, 48 pp. **294, 454** fn.

Church, P. E., 1945. The annual temperature cycle of Lake Michigan. II. Spring warming and summer stationary periods, 1942. *Department of Meteorology, University of Chicago, Miscellaneous Reports*, No. 18, 100 pp. **295, 495, 500**

Cialdi, A., 1866. *Sul moto ondoso del mare*, seconda ediz. Roma. xxviii, 693 pp. (not seen). **401**

* Clarke, F. W., 1924. The data of geochemistry. Fifth ed. *Bull. U.S. geol. Surv.* 770, 841 pp. **229, 550, 553, 554, 555, 565, 566, 571, 728, 756, 791, 821**

Clay, P. H., Gemert, A. V., and Clay, J., 1939. The penetrating cosmic radiation in water and rock down to 450 m. water. *Physica, 'sGrav.*, 6:184–204. **422**

Close, M., 1871. On some corries and their rock basins in Kerry. *J. roy. geol. Soc. Ireland*, 12:236–248. **62**

Coffin, C. C., Hayes, F. R., Jodrey, L. H., and Whiteway, S. G., 1949. Exchange of materials in a lake as studied by the addition of radioactive phosphorus. *Canad. J. Res.*, D27:207–222. **744**

Cohn, E. J., and Edsall, J. T., 1943. *Proteins, Amino Acids and Peptides as Ions and Dipolar Ions.* New York, Reinhold Publishing Corp., xviii, 686 pp. **209** fn.

Coldring, A., 1876. Fremstilling af Resultaterne af nogle Undersøgelser over de ved Vindens Kraft fremkaldte Strømninger i Havet. *Danske vidensk. Selsk Kjøbenhavn, 5 Raekke naturv. og math.*, Avd. 9, No. 3. **274**

Cole, G. A., 1954. Studies on a Kentucky Knobs lake. I. Some environmental factors. *Trans. Ky. Acad. Sci.*, 15:31–47. **441**

Collet, L. W., 1922. Alpine lakes. *Scot. geogr. Mag.*, 38:73–101. **57**

* Collet, L. W., 1925. *Les lacs. Leur mode de formation—leurs eaux—leur destin—éléments d'hydrogéologie.* Paris, xii, 320 pp. **13, 30, 44, 50, 51, 53, 56, 57, 96, 105, 106, 114, 487, 766, 776**

Collet, L. W., Mellet, R., and Ghezzi, C., 1918. Il Lago Ritom. *Commun. Serv. Eaux*, No. 13, Berne, 101 pp. **486, 776**

Collins, J. R., 1925. Changes in the infra-red absorption spectrum of water with temperature. *Phys. Rev.*, 26:771–779. **383**

Collins, M. I., 1923. Studies in vegetation of arid and semi-arid New South Wales. Part I. The plant ecology of the Barrier District. *Proc. Linn. Soc. N.S.W.*, 48:229–266. **135**

Collins, W. D., and Williams, K. T., 1933. Chloride and sulfate in rain water *Ind. Engng. Chem.*, 25:944–945. **544, 547**

Comstock, F. N., 1900. An example of wave-formed cusp at Lake George, N. Y. *Amer. Geol.*, 25:192–194. **180**

* Conway, E. J., 1942. Mean geochemical data in relation to oceanic evolution. *Proc. R. Irish Acad.*, 48B:119–159. **229, 545, 548, 553, 554** fn, **555, 790**

* Conway, E. J., 1943. The chemical evolution of the ocean. *Proc. R. Irish Acad.*, 48B:161–212. **229**

Cooke, C. W., 1933. Origin of the so-called meteorite scars of South Carolina. *J. Wash. Acad. Sci.*, 23:569–570. **155**

Cooke, C. W., 1934. Discussion of the origin of the supposed meteor scars of South Carolina. *J. Geol.*, 42:88–96. **155**

Cooke, C. W., 1939. Scenery of Florida. *Bull. Fla geol. Surv.*, 17, 118 pp. **102, 104**

Cooke, C. W., 1940. Elliptical bays in South Carolina and the shape of eddies. *J. Geol.*, 48:205–211. **155**

Cooper, L. H. N., 1932. Iron in the sea and in marine plankton. *Proc. roy. Soc.*, ser. B, 118:419–438. **706, 707**

Cooper, L. H. N., 1934. The determination of phosphorus and nitrogen in plankton. *J. Mar. biol. Ass. U.K.*, 19:755–759. **710**

Cooper, L. H. N., 1937a. Oxidation-reduction potentials in sea water. *J. Mar. biol. Ass. U.K.*, 22:167–176. **692, 719**

* Cooper, L. N. H., 1937b. The nitrogen cycle in the sea. *J. Mar. biol. Ass. U.K.*, 22:183–204. **848, 849, 850, 865, 866, 868, 895**

* Cooper, L. H. N., 1937c. Some conditions governing the solubility of iron. *Proc. roy. Soc.*, ser. B, 124:299–307. **693, 704**

Cooper, L. H. N., 1948a. Some chemical considerations on the distribution of iron in sea water. *J. Mar. biol. Ass. U.K.*, 27:314–321. **704, 706, 707, 709, 710, 713, 737**

Cooper, L. H. N., 1948b. Particulate ammonia in sea water. *J. Mar. biol. Ass. U.K.*, 27:322–325. **850**

Cooper, L. H. N., 1952. Factors affecting the distribution of silicate in the North Atlantic Ocean and the formation of North Atlantic deep water. *J. Mar. biol. Ass. U.K.*, 30:511–526. **789**

Corbet, A. S., 1935. The formation of hyponitrous acid as an intermediate compund in the biological or photochemical oxidation of ammonia to nitrous acid. II. Microbiological oxidation. *Biochem. J.*, 29:1088–1096. **866**

Corbet A. S., and Wooldridge, W. R., 1940. The nitrogen cycle in biological system. 3. Aerobic denitrification in soils. *Biochem. J.*, 34:1036–1040. **868**

Cornish, V., 1934. *Ocean waves and kindred geophysical phenomena* and additional notes by Harold Jeffreys. Cambridge Univ. Press, xv, 164 pp. **356, 357**

Cornish, V., 1935. *Scenery and the Sense of Sight.* Cambridge Univ. Press, xii, 110 pp. **77** fn.

Coste, J. H., and Wright, H. L., 1935. The nature of the nucleus in hygroscopic droplets. *Phil. Mag.*, ser. 7, 20:209–234. **542, 548**

Coster, H. P., and Gerrard, J. A. F., 1947. A seismic investigation of the outflow of Windermere. *Geol. Mag.*, 84:224–228. **76**

* Cotton, C. A., 1941. *Landscape as Developed by the Processes of Normal Erosion.* Cambridge Univ. Press, xviii, 291 pp. **3, 44, 183**

* Cotton, C. A., 1944. *Volcanoes as Landscape Forms.* Christchurch, N.Z., and London, Whitcombe and Tombs, 416 pp. **25, 27, 28, 29, 37, 41, 149**

* Cotton, C. A., 1948. *Climatic Accidents in Landscape Making.* New York, John Wiley & Sons, xx, 354 pp. **60, 63, 64, 69**

Coupin, H., 1922. Sur l'origine de la carapace siliceuse des diatomées. *C.R. Acad. Sci., Paris*, 175:1226–1229. **790**

Cox, C., and Munk, W., 1954. Statistics of the sea surface. *J. Mar. Res.*, 13:198–227. **419**

Cox, J., 1904. Note on an apparently accidental formation of frazil ice in a cryophorus. *Proc. & Trans. roy. Soc. Can.*, 2 ser., 10(Sect. III):3–4. **209**

Craig, H., and Boato, G., 1955. Isotopes. *Ann. Rev. phys. Chem.*, 6:403–432. **213, 214, 215, 216**

Craig, H., Boato, G., and White, D. E., 1954. Isotope geochemistry of thermal waters. *Bull. geol. Soc. Amer.*, 65:1243. **215**

Craig, W., Brant, I., Hite, M. P., and Davis, W. M., 1925. The "undertow." *Science*, 62:30–33. **358**

Crary, A. P., Cotell, R. D., and Sexton, T. F., 1952. Preliminary report on scientific work on "Fletcher's Ice Island," T3. *Arctic*, 5:211–223. **50**

Cross, P. C., Burnham, J., and Leighton, P. A., 1937. Raman spectrum and the structure of water. *J. Amer. chem. Soc.*, 59:1134–1147. **200**

Cvijić, J., 1901. Morphologische und glaciale Studien aus Bosnien, der Hercegovina und Montenegro. II. Die Karstpoljen. *Abh. geogr. Ges., Wien*, 3, No. 2, 85 pp. **106**

Cvijić, J., 1902. Les crypto-dépressions de l'Europe. *Géographie*, 5:247–254. **107**

Cvijić, J., 1903. *Atlas der Seen von Makedonien, Altserbien und Epirus.* Belgrad. (Not seen: ref. in Peucker 1903.) **166, 167**

Dalton, J., 1819. Note to J. Otley (1819). Account of the floating island in Derwent Lake, Keswick. *Mem. Manchr. lit. pil. Soc.*, 2nd ser., 3:69. **842**

Damas, H., 1937. La stratification thermique et chimique des lacs Kivu, Edouard et Ndalga (Congo Belge). *Verh. int. Ver. Limnol.*, 8(III):51–68. **463, 488**

Damas, H., 1955. Recherches limnologiques dans quelques lacs du Ruanda. *Verh. int. Ver. Limnol.*, 12:335–341. **463**

Darby, William, 1818. *The Emigrant's Guide to the Western and Southwestern States and Territories.* New York, Kirk and Merceing, 311 pp. **115**

Darton, N. H., 1905. The Zuni Salt Lake. *J. Geol.*, 13:185–193. **33**

Darwin, C. R., 1839. Observations on the Parallel Roads of Glen Roy and of other parts of Lochaber in Scotland, with an attempt to prove that they are of marine origin. *Phil. Trans.*, 39:39–82. **54** fn.

David, Sir W. T. E., and Browne, W. R., 1950. *The Geology of the Commonwealth of Australia:* Vol. II, *Physiography and Economic Geology.* London, Edward Arnold, iii, 618 pp. **11, 38, 63, 64, 109, 116, 122, 134, 135**

David, T. W. E., and Priestley, R. E., 1909. Geological observations in Antarctica by the British Antarctic Expedition. In Shackleton (1909), *The Heart of the Antarctic*, Vol. 2, Appendix II, pp. 268–307. **462**

Davis, F. J., 1941. Surface loss of solar and sky radiation by inland lakes. *Trans. Wis. Acad. Sci. Arts Lett.*, 33:83–93. **376, 410**

Davis, R., and Schaeffer, O. A., 1955. *Chlorine-36 in Nature.* Brookhaven National Laboratory, Upton, N. Y., 14 pp. **571**

* Davis, W. M., 1882. On the classification of lake basins. *Proc. Boston Soc. nat. Hist.*, 21:315–381. **2, 115**

Davis, W. M., 1887. On the classification of lake basins. *Science*, 10:142. **2**

Davis, W. M., 1925. The undertow myth. *Science*, 61:207–208. **358**

Davis, W. M., 1931. Undertow and rip tides. *Science*, 73:526–527. **358**

* Davis, W. M., 1933. The lakes of California. *Calif. J. Min.*, 29:175–236. **3, 14, 19, 34, 46, 47, 113, 114**

Davison, C., 1921. *A Manual of Seismology.* Cambridge Univ. Press, xi, 256 pp. **362**

Day, P. C., 1926. Precipitation in the drainage area of the Great Lakes 1875–1924, with discussion on the levels of the separate lakes and relation to the annual precipitation. *Mon. Weath. Rev., Wash.*, 54:85–101. **234**

* De, P. K., 1939. The role of the blue-green algae in nitrogen fixation in rice fields. *Proc. roy. Soc.*, ser. B, 127:121–139. **847**

Deane, R. E., 1950. Pleistocene geology of the Lake Simcoe District, Ontario. *Mem. geol. Surv. Can.*, 256, vii, 108 pp. **86**

Deevey, E. S., 1940. Limnological studies in Connecticut. V. A contribution to regional limnology. *Amer. J. Sci.*, 238:717–741. **642, 646, 729, 871**

* Deevey, E. S., 1941. Limnological studies in Connecticut. VI. The quantity and composition of the bottom fauna in thirty-six Connecticut and New York lakes. *Ecol. Monogr.*, 11:413–455. **693, 808**

Deevey, E. S., 1942. A re-examination of Thoreau's "Walden." *Quart. Rev. Biol.*, 17:1–11. **96, 427**

Deevey, E. S., 1953. Paleolimnology and climate. In *Climatic Change: Evidence, Causes, and Effects*, ed. H. Shapley. Cambridge, Harvard Univ. Press, Chap. 22:273–318. **130**

* Deevey, E. S., 1957. Limnologic studies in Middle America, with a chapter on Aztec limnology. *Trans. Conn. Acad. Arts Sci.*, 39:213–328. **299, 464, 494, 495, 508, 511, 791**

Deevey, E. S., Jr., Gross, M. S., Hutchinson, G. E., and Kraybill, H. L., 1954. The natural C^{14} contents of materials from hard-water lakes. *Proc. nat. Acad. Sci., Wash.*, 40:285–288. **659, 671**

Deevey, E. S., and Flint, R. F., 1957. Postglacial hypsithermal interval. *Science*, 125:182–184. **49**

Defant, A., 1908. Über die stehenden Seespiegelschwankungen (Seiches) in Riva am Gardasee. *S.B. Akad. Wiss. Wien*, Math.-Nat. Kl., 117, Abt. IIa: 697–780. **307, 312**

Defant, A., 1932. Beiträge zur theoretischen Limnologie. *Beitr. freien Atmos.* (Bjerknes-Festband), 29:143–150. **282, 324**

De Geer, G., 1909. Dal's Ed., Some stationary ice borders of the last glaciation. *Geol. Fören. Stockh. Förh.*, 31:511–556. **90**

De la Beche, H. T., 1819. Sur la profondeur et la température du lac de Genève. *Bibliogr. Univ. Sci. Arts, Genève*, 12:118–126. **426**

De la Beche, H. T., 1820. Sur la température des lacs de Thun et de Zug, en Suisse. *Bibliogr. Univ. Sci. Arts, Genève*, 14:144–145. **426**

* Delebecque, A., 1898. *Les Lacs Français.* Paris, Chamerot et Renouard, xi, 436 pp. **3, 44, 47, 50, 84, 109, 110, 122, 167, 190, 416, 487, 600, 775, 838**

Delebecque, A., and Duparc, C., 1894. Sur les changements survenues au glacier de la Tête Rousse depuis la catastrophe de Saint Gervais du juillet 1892. *C.R. Acad. Sci., Paris*, 117:333-334. See also Delebecque 1898. **53**

Delebecque, A., and Ritter, E., 1892. Sur les lacs du plateau central de la France. *C.R. Acad. Sci., Paris*, 115:74-75. **27, 30**

Demoll, R., 1922. Temperaturwellen (Seiches) und Planktonwellen. *Arch. Hydrobiol.*, 13:313–320. **288, 340, 341**

Denigès, G., 1920. Réaction de coloration extrêmement sensible des phosphates et des arséniates. *C.R. Acad. Sci., Paris*, 171:802–804. **728**

Denison, F. N., 1897. The Great Lakes as a sensitive barometer. *Rep. Brit. Ass.*, London, 1897, (67 Toronto) Meeting 2:567–568. **318**

Denison, F. N., 1898. The Great Lakes as a sensitive barometer. *Weath. Rev., Wash.*, 26:261–262. **316, 318**

de Terra, H., 1934. Physiographic results of a recent survey in Little Tibet. *Geogr. Rev.*, 24:12–41. **22** de Terra, H., and Hutchinson, G. E., 1934. Evidence of recent climatic changes shown by Tibetan highland lakes. *Geogr. J.*, 84:311–320. **183, 241**

Dhar, N. R., and Ram, A., 1933. Variations in the amounts of ammoniacal and nitric nitrogen in rain water of different countries, and the origin of nitric nitrogen in the atmosphere. *J. Indian chem. Soc.*, 10:125–133. **550**

Diénert, F., and Guillerd, A., 1949. Mouvement de l'eau dans les lacs. *Gen. Ass. int. Un. Geod.*, Oslo, 1948, *I. Trav. Comm. Potamol. Limnol.*, 378–385. **278**

Diénert, F., and Wandenbulcke, F., 1923. Sur la dosage de la silice dans les eaux. *C.R. Acad. Sci., Paris*, 176:1478–1480. **789**

Dietz, R. S., and LaFond, E. C., 1950. Natural slicks on the ocean. *J. Mar. Res.*, 9:69–76. **361**

Dion, H. G., and Mann, P. J. G., 1946. Three-valent manganese in soils. *J. agric. Sci.*, 36:239–245. **802**

Dixey, F., 1924. Lake level in relation to rainfall and sunspots. *Nature, Lond.*, 114:659–661. **242, 244**

Dixey, F., 1939. Early Cretaceous valley-floor peneplain of the Lake Nyasa region and its relation to tertiary rift structure. *Quart. J. geol. Soc. Lond.*, 95:75–108. **24**

Dixey, F., 1941. The Nyasa Rift Valley. *S. Afr. geogr. J.*, 23:21–25. **24**

Dixon, W. A., 1878. Notes on the meteorology and natural history of a guano island. *J. roy. Soc., N.S.W.* (for 1877), 11:165–175. **144**

Dole, M., 1936. Relative atomic weights of oxygen in water and in air. *J. Amer. chem. Soc.*, 57:2731. **214**

Domogalla, B. P., and Fred, E. B., 1926. Ammonia and nitrate studies of lakes near Madison, Wisc. *J. Amer. Soc. Agron.*, 18:897–911. **853, 869**

Domogalla, B. P., Fred, E. B., and Peterson, W. H., 1926. Seasonal variations in the ammonia and nitrate content of lake waters. *J. Amer. Wat. Wks Ass.*, 15:369–395. **869**

Domogalla, B. P., Juday, C., and Peterson, W. H., 1925. The forms of nitrogen found in certain lake waters. *J. biol. Chem.*, 63:269–285. **862, 890**

Domratschev, P. F., 1933. Materialien zu einer physikalisch-geographischen Charakteristik des Sees. *Serv. hydrometeorol. USSR, Inst. Hydrol. Explorations des lacs de l'USSR*, 4 (Balkasch Exped.), Russian text, 31–53; German summ., 54–56. **239**

Doodson, A. T., Carey, R. M., and Baldwin, R., 1920. Theoretical determination of the longitudinal seiches of Lake Geneva. *Trans. roy. Soc. Edinb.*, 52:629–642. **307**

Dorff, P., 1934. Die Eisenorganismen. Systematik und Morphologie (Inaug. Diss. Berlin). *Pflanzenforschung*, 16, 62 pp. **715**

* Dorsey, N. E., 1940. *Properties of Ordinary Water-Substance in All Its Phases: Water-Vapor, Water, and All the Ices.* Amer. chem. Soc. Monogr. Ser., New York, Reinhold Publishing Corp., 673 pp. **197, 201, 202, 207, 209** fn., **211, 588**

Downing, A. L., and Truesdale, G. A., Aeration in Aquaria. *Zoologica, N.Y.*, 41:129–143. **587**

Drischel, H., 1940. Chlorid-, Sulfat-, und Nitratgehalt der atmosphärischen Niederschlage in Bad Reinerz und Oberschreiberhau im Vergleich zu bisher bekannten Werten anderer Orte. *Balneologe*, 7:321–334. **544, 548, 549**

Droop, M. R., 1955. A pelagic marine diatom requiring cobalamin. *J. Mar. biol. Ass. U.K.*, 34:229–231. **826**

Drower, H. S., 1954. Water supply, irrigation and agriculture. In *A History of Technology*, ed. M. S. Singer, E. J. Holmyard, and A. R. Hall. Oxford, Clarendon Press, Vol. 1, pp. 526–557. **147**

Düggeli, M., 1924. Hydrobiologische Untersuchungen in Pioragebiet. Bakteriologische Untersuchungen am Ritomsee. *Z. Hydrol.*, 2:65–206. **754, 776**

Düggeli, M., 1936. Die Bakterienflora im Schlamm des Rotsees. *Z. Hydrol.*, 7:205–365. **842**

Dufour, L., 1873. Recherches sur la réflexion de la chaleur solaire a la surface du Lac Léman. *Bull. Soc. vaud. Sci. nat.*, 12:1–108. **375**

* Dussart, B., 1948. Recherches hydrographiques sur le lac Léman. *Ann. Sta. centr. Hydrobiol. appl.*, 2:187–206. **294, 296**

Dussart, B., 1954. Température et mouvements des eaux dans les lacs. *Sta. centr. Hydrobiol. appl.*, 5:7–128. **336, 338**

du Toit, A. L., 1907. Geological survey of the eastern portion of Griqualand West. *Rep. geol. Comm. C.G.H.*, 11:87–176. **132**

du Toit, A. L., 1908. Geological survey of portions of Mafeking and Vryburg. *Rep. geol. Comm. C.G.H.*, 12:123–192. **132**

Dyson, J. L., 1948a. Glaciers and glaciation in Glacier National Park. *Spec. Bull. Glacier nat. Hist. Ass.*, No. 2, 24 pp. **61**

Dyson, J. L., 1948b. Shrinkage of Sperry and Grinnell Glaciers, Glacier National Park. *Geogr. Rev.*, 38:94–103. **61**

Ebert, H., 1900. Periodische Seespiegelschwankungen (Seiches), beobachtet am Starnberger See. *S.B. bayer. Akad. Wiss.*, Math.-phys. Kl., 30:435–462. **315**

Edmondson, W. T., 1955. Measurement of conductivity of lake water *in situ*. *Ecology*, 37:201–204 (appeared in 1956). **486**

Eggleton, F. E., 1931. A limnological study of the profundal bottom fauna of certain fresh-water lakes. *Ecol. Monogr.*, 1:231–332. **488**

Eggleton, F. E., 1956. Limnology of a meromictic, interglacial, plunge-basin lake. *Trans. Amer. micr. Soc.*, 75:334–378. **111, 488**

Ehmert, A., 1937. Die Absorptionskurve der harten Komponente die kosmischen Ultrastrahlung. *Z. Phys.*, 106:751–773. **422**

* Einsele, W., 1936. Ueber die Beziehungen des Eisenkreislaufs zum Phosphatkreislauf im eutrophen See. *Arch. Hydrobiol.*, 29:664–686. **706, 731, 735**

Einsele, W., 1937. Physikalisch-chemische Betrachtung einiger Probleme des limnischen Mangan-und Eisenkreislaufs. *Verh. int. Ver. Limnol.*, 8:69–84. **706, 721, 722, 760, 808, 810**

Einsele, W., 1938. Über chemische und kolloidchemische Vorgänge in Eisenphosphat-systemen unter limnochemischen und limnogeologischen Gesichtspunkten. *Arch. Hydrobiol.*, 33:361–387. **706, 735**

Einsele, W., 1940. Versuch einer Theorie der Dynamik der Mangan-und Eisenschichtung im eutrophen Seen. *Naturwissenschaften*, 28:257–264, 280–285. **808**

* Einsele, W., 1941. Die Umsetzung von zugeführtem, anorganischen Phosphat im eutrophen See und ihre Rückwirkung auf seinen Gesamthaushalt. *Z. Fisch.*, 39:407–488. **745, 750**

Einsele, W., 1944. Der Zeller See, ein lehrreicher Fall entremer limnochemischer Verhältnisse. *Z. Fisch.*. 42:151–168. **488, 724, 808**

* Einsele, W., and Vetter, H., 1938. Untersuchungen über die Entwicklung der physikalischen und chemischen Verhältnisse im Jahreszyklus in einem mässig eutrophen See (Schleinsee bei Langenargen). *Int. Rev. Hydrobiol.*, 36:285–324. **794, 856, 862**

Ekblaw, W. E., 1918. On unknown shores; the traverse of Grant and Ellesmere Lands. Appendix II of D. B. MacMillan, *Four years in the White North.* New York, Harpers, pp. 333–370. **416**

Ekman, S., 1904. Die Phyllopoden, Cladoceren und freilebenden Copepoden der nord-schwedischen Hochgebirge. *Zool. Jb.*, Abt. Syst. Geogr. Biol. Tiere, 21:1–170. **461**

* Ekman, V. W., 1905. On the influence of the earth's rotation on ocean currents. *Ark. Mat. Astr. Fys.*, 2, No. 11, 52 pp. **256, 261, 263, 266, 270**

Ekwall, E., 1947. *The Concise Oxford Dictionary of English Place Names*, 3rd ed. Oxford, Clarendon Press, xlvii, 530 pp. **123** fn.

Ellis, M. M., 1937. Detection and measurement of stream pollution. *Bull. U.S. Bur. Fish.*, 48:365–437. **812**

Elrod, M. J., 1912. *Some Lakes of Glacier National Park*. Washington, U.S. Dept. Interior, 29 pp. **64**

Elschner, C., 1915. The Leeward Islands of the Hawaiian group. Honolulu, reprinted from the *Sunday Advertiser*, 68 pp. **145**

Elschner, C., 1922. Kolloide Phosphate. *Kolloidzschr.*, 31:94–96. **145**

* Elster, H. J., 1939. Beobachtungen über das Verhalten der Schichtgrenzen nebst einigen Bemerkungen über die Austauschverhältnisse im Bodensee (Obersee). *Arch. Hydrobiol.*, 35:286–346. **264, 294, 796**

Elster, H. J., 1955. Limnologische Untersuchungen im Hypolimnion verschiedener Seetypen. *Mem. Ist. ital. Idrobiol. de Marchi*, Suppl. 8:83–119. **639, 647**

Elster, H. J., and Einsele, W., 1937. Beiträge zur Hydrographie des Bodensees (Obersee). *Int. Rev. Hydrobiol.*, 35:522–585. **796**

Elster, H. J., and Einsele, W., 1938. Beiträge zur Kenntnis der Hydrographie des Untersees (Bodensee). *Int. Rev. Hydrobiol.*, 36:241–284. **854**

Eméleus, H. J., James, F. W., King, A., Pearson, T. G., Purcell, R. H., and Briscoe, H. V. A., 1934. The Isotopic ratio in hydrogen: a general survey by precise density comparisons. *J. chem. Soc.*, 137:1207–1219. **214**

Emerson, R., and Lewis, C. M., 1939. Factors influencing the efficiency of photosynthesis. *Amer. J. Bot.*, 26:808–822. **806**

Emory, K. P., 1934. Archaeology of the Pacific Equatorial islands. *Bull. Bishop Mus., Honolulu*, No. 123, 43 pp. **145**

Endrös, A., 1903. Seeschwankungen (Seiches) beobachtet am Chiemsee. Diss. K. Technischen Hochschule zu München, Traunstein, 117 pp. **306, 315, 319, 320, 326**

Endrös, A., 1904. Seiches kleiner Wasserbecken. *Petermanns Mitt.*, 50:294–295. **312**

Endrös, A., 1905. Die Seiches des Waginger-Tachingersees. *S. B. bayer. Akad. Wiss.*, Math.-Phys. Kl., 35:447–476. **314, 324**

Endrös, A., 1906a. Die Seiches des Waginger-Tachingersees. *Petermanns Mitt.*, 52:94–95. **314**

* Endrös, A., 1906b. Die Seeschwankungen (Seiches) des Chiemsees. *S.B. bayer. Akad. Wiss.*, Math.-Phys. Kl., 36:297–350. **315, 319, 320**

Endrös, A., 1906c. Seiches-Beobachtungen an den grösseren Seen des Salzkammergutes. *Petermanns Mitt.*, 52:252–258. **314, 315**

* Endrös, A., 1908. Vergleichende Zusammenstellung der Hauptseichesperioden der bis jetzt untersuchten Seen mit Anwendung auf verwandte Probleme. *Petermanns Mitt.*, 54:39–47, 60–68, 86–88. **306, 311, 316, 318, 319, 331**

Endrös, A., 1913. Der Simssee und seine Seeschwankungen. *Progr. Gymn. Freising*, 1913 (not seen: ref. Halbfass 1923). **314**

Enge, J., 1931. Der Anstieg des Toten Meeres 1800–1900, und seine Erklärung. Inaug. Diss. Leipzig., Borna-Leipzig, 42 pp. **242**

* Epstein, S., and Mayeda, T., 1953. Variation of O^{18} content of waters from natural sources. *Geochim. et cosmochi. Acta.*, 4:213–224. **214, 215**

Erikson, H. A., 1933. Light intensity at different depths in lake water. *J. opt. Soc. Amer.*, 23:170–177. **383**

Eriksson, E., 1952a. Composition of atmospheric precipitation, I. *Tellus*, 4:215–232. **544, 545, 546**

Eriksson, E., 1952b. Composition of atmospheric precipitation, II. *Tellus*, 4:280–303. **544, 545, 546**

Ertel, H. E., 1954. Theorie der thermischen Sprungschicht in Seen. *Acta. Hydrophys.*, 1:151–171. **434**

Evans, G. L., and Meade, G. E., 1945. Quaternary of the Texas High Plains. *Univ. Tex. Publ.*, 4401:485–507. **128**

Ewan, T., 1894. On the absorption spectra of dilute solutions. *Proc. roy. Soc.*, 57:117–161. **382**

Ewell, R. H., and Eyring, H., 1937. Theory of the viscosity of liquids as a function of temperature and pressure. *J. chem. Phys.*, 5:726-736. **200**

Ewing, G. T., 1950. Relation between band slicks at the surface and internal waves in the sea. *Science*, 111:91–94. **361**

Ewing, M., Press, F., and Donn, W. L., 1954. An explanation of the Lake Michigan wave of 26 June 1954. *Science*, 120:684–686. **299** fn.

Exner, F. M., 1908a. Über eigentümliche Temperaturschwankungen von eintägiger Periode in Wolfgangsee. *S.B. Akad. Wiss. Wien*, Math.-Nat. Kl., 117, Abt. IIa:9–26. **338**

Exner, F. M., 1908b. Ergebnisse einiger Temperaturregistrierungen im Wolfgangsee. *S.B. Akad. Wiss. Wien*, Math.-Nat. Kl., 117, Abt. IIa:1295–1315. **338**

Exner, F. M., 1928. Temperaturseiches im Lunzer See. *Ann. Hydrogr., Berl.*, 56:14–20, 142. **338**

Fairbridge, R. W., 1949. Geology of the country around Waddamana, Central Tasmania. *Pap. roy. Soc. Tasm.*, 1948:111–149. **65**

Fairchild, H. L., 1918. Postglacial uplift of northeastern America. *Bull. geol. Soc. Amer.*, 29:187–238. **11**

Fairchild, H. L., 1919. Pleistocene marine submergence of the Hudson, Champlain, and St. Lawrence valleys. *Bull. N.Y. St. Mus.*, 209–210, 76 pp. **11**

Fairchild, H. L., 1926. Geological romance of the Finger Lakes. *Sci. Mon., N.Y.*, 23:161–173. **86**

Fairchild, H. L., 1934. Silencing the "guns" of Seneca Lake. *Science*, 79:340–341. **89, 362**

Falckenberg and Bolz, *see* Bolz and Fritz, 1950.

Falkenberg, G., 1928. Absorptionskonstanten einiger meteorologisch wichtiger Körper für infrarote Wellen. *Met. Z.*, 45:334–337. **377**

Fantoli, G., 1897. *Sul regime idraulico dei laghi.* Milano, U. Hoepli, xiv, 349 pp. **235**

Fatio de Duillier, J. C., 1730. Remarques . . . sur l'histoire naturelle des environs du lac de Genève. In J. Spon, *Histoire de Genève . . . rectifiée et considerablement augmentée par amples notes*. Genève, Fabri and Barrillot, Vol. 4, pp. 289–330. This is a revised edition, edited by J. A. Gautier and F. Abauzit, of a much earlier *Histoire* by Spon published in France. **299, 300**

Fenneman, N. M., 1902. On the lakes of southeastern Wisconsin. *Bull. Wis. geol. nat. Hist. Surv.*, 8, xv, 178 pp. **142, 179, 358**

Fenner, C., 1918. The physiography of the Glenelg River. *Proc. roy. Soc. Vict.*, n.s., 30:99–120. **38, 42**

Fenner, C., 1921. Craters and lakes of Mt. Gambier. *Trans. roy. Soc. S. Aust.*, 45:169–205. **26, 29**

Finbak, C., and Viervoll, H., 1943. The structure of liquids. II. The structure of liquid water. *Tidsskr. Kemi Bergv.*, 3:36–40. **201**

Finch, R. H., 1937. A tree-ring calendar for dating volcanic events, Cinder Cone, Lassen National Park, California. *Amer. J. Sci.*, 5th ser., 33:140–146. **40**

Findeisen, W., 1937. Entstehen die Kondensationskerne an der Meeresoberfläche? *Met. Z.*, 54:377–379. **542**

Findeisen, W., 1939. Das Verdampfen der Wolken-und Regentropfen. *Met. Z.*, 56:453–460. **543**

Findenegg, I., 1932. Die Schichtungsverhältnisse im Wörthersee. *Arch. Hydrobiol.*, 24:253–262. **488, 637**

Findenegg, I., 1933. Alpenseen ohne Vollzirkulation. *Int. Rev. Hydrobiol.* 28:295–311. **488. 637**

Findenegg, I., 1934a. Umschichtungsvorgänge im Millstätter und Weissensee in Karnten. *Int. Rev. Hydrobiol.*, 31:88–98. **488**

Findenegg, I., 1934b. Zur Frage der Entstehung pseudoeutropher Schichtungsverhältnisse in den See. *Arch. Hydrobiol.*, 27:621–625. **488, 637**

Findenegg, I., 1934c. Die Entstehung sommerlicher Temperaturinversionen in Ostalpenseen. *Bioklim. Beibl.*, 1934:160–165. **488, 491**

* Findenegg, I., 1935. Limnologische Untersuchungen im Kärntner Seengebiete. Ein Beitrag zur Kenntnis des Stoffhaushaltes in Alpenseen. *Int. Rev. Hydrobiol.*, 32:369–423. **439, 480, 488, 637, 791**

Findenegg, I., 1936a. Die Bedeutung des Klimas für die Entstehung biologische Seetypen. *Bioklim. Beibl.*, 1936:57–63. **488**

Findenegg, I., 1936b. Über den Sauerstoffgehalt tiefer Seen und seine indikatonische Bedeutung für ihren Trophiezustand. *Arch. Hydrobiol.*, 30:337–344. **488, 489, 637**

Findenegg, I., 1937. Holomiktische und meromiktische Seen. *Int. Rev. Hydrobiol.*, 35:586–610. **481, 488, 637**

* Findenegg, I., 1938. Sechs Jahre Temperaturlotungen in den Kärntner Seen. *Int. Rev. Hydrobiol.*, 37:364–384. **488**

Firbas, F., 1949. *Spät und nacheiszeitliche Waldgeschichte Mitteleuropas nördlich der Alpen. I. Allgemeine Waldgeschichte.* G. Fischer, Jena, 480 pp. **26**

Fireman, E. L., 1953. Measurement of the (n, H^3) cross section in nitrogen and its relationship to tritium production in the atmosphere. *Phys. Rev.*, 91:922–926. **212**

de Fischer-Foster, C., and Brunner, C., 1849. Recherches sur la température du lac de Thoune. *Mém. Soc. Phys. Genève*, 12:255–276. See also *Arch. Sci. phys. nat.*, 12:20–39, where the first author's name is given as Fischer-Ooster, which form has been widely quoted. **426**

Fisher, C., 1936. The meteor craters in Esthonia. *Nat. Hist., N.Y.*, 38:292–299. **149**

Fitzgerald, D., 1895. The temperature of lakes. *Trans. Amer. Soc. civ. Engrs*, 34:67–114. **427, 453**

Fjeldstad, J. E. 1929. Ein Beitrag zur Theorie der winderzeugten Meereströmungen, *Beitr. Geophys.* 23:237–247. **267**

Flint, R. F., 1947. *Glacial Geology and the Pleistocene Epoch.* New York, John Wiley & Sons (London, Chapman and Hall), xviii, 589 pp. **48, 49, 82, 86, 89, 95**

* Flint, R. F., and Deevey, E. S., 1951. Radiocarbon dating of late Pleistocene events. *Amer. J. Sci.*, 249:257–300. **8**

Flint, R. F., and Dorsey, H. G., 1945. Glaciation of Siberia. *Bull. geol. Soc. Amer.* 56:89–106. **56**

Flint, R. F., and Irwin, W. H., 1939. Glacial geology of Grand Coulee Dam, Washington. *Bull. geol. Soc. Amer.*, 50:661–680. **11**

Fogg, G. E., 1942. Studies on nitrogen fixation by blue-green algae. I. Nitrogen fixation by *Anabaena cylindrica* Lemm. *J. exp. Biol.*, 19:78–87. **847**

* Fogg, G. E., and Westlake, D. F., 1955. The importance of extracellular products of algae in fresh waters. *Verh. int. Ver. Limnol.*, 12:219–231. **734, 896**

Folse, J. A., 1929. *A New Method of Estimating Stream-Flow Based on a New Evaporation Formula.* Carnegie Institute of Washington, Pub. 400. xi, 237 pp. **523**

Forbes, S. T., 1887. The lake as a microcosm. *Bull. Peoria (Illinois) Scientific Association*, 1887:77–87. Reprinted in *Bull. Ill. nat. Hist. Surv.*, 15:537–550 (1925) and in *Selected Readings in Biology*, II (2nd ed.), Chicago Univ. Press, 1956, pp. 1–22. **2**

Forel, F. A., 1873. Les taches d'huile connues sous le nom de fontaines et chemins du lac Léman. *Bull. Soc. vaud. Sci. nat.* 12:148–155. **360**

Forel, F. A., 1874. Une variété nouvelle ou peu connue de gloire étudiée sur la lac Léman. *Bull. Soc. vaud. Sci. nat.*, 13:357–370. **422**

Forel, F. A., 1876. Le formule des seiches. *C.R. Acad. Sci., Paris*, 83:712–714. Also *Arch. Sci. phys. nat.*, 57:278. **301**

Forel, F. A., 1880. Températures lacustres: recherches sur la température du la Léman et d'autres lacs d'eau douce. *Arch. Sci. phys. nat.*, ser. 3, 3:501–515. **492**

* Forel, F. A., 1885. Faune profonde des lacs suisses. *N. Denkschr. schweiz. Ges. Naturw.*, 29 (4), 234 pp. **600**

Forel, F. A., 1889. Ricerche fisiche sui laghi d'Insubria. *Rend. R. Ist. Lombardo*, ser. 2, 22:739, 742. **415**

Forel, F. A., 1892a. La thermique des lacs l'eau douce. *Verh. schweiz. naturf. Ges.*, 75:5–8. **437**

* Forel, F. A., 1892b. *Le Léman: monographie limnologique.* Tome 1, *Géographie, Hydrographie, Géologie, Climatologie, Hydrologie.* Lausanne, F. Rouge, xiii, 543 pp. **190**

Forel, F. A., 1893. Die Schwankungen der Bodensees. *Schr. Ver. Gesch. Bodenseen* (Lindau) 22 (not seen: ref. Halbfass, 1923). **300, 315**

* Forel, F. A., 1895. *Le Léman: monographie limnologique.* Tome 2, *Mécanique, Chimie, Thermique, Optique, Acoustique.* Lausanne, F. Rouge, 651 pp. **296, 299, 300, 310, 311, 325, 328, 417, 420, 421, 437, 492**

Forel, F. A., 1897. Réfractions et mirages. Passage d'un type à l'autre, sur le lac Léman. *Bull. Soc. vaud, Sci. nat.*, 4 ser., 32:271–277. **314, 421**

Forel, F. A., 1898a. Quelques études sur les lacs de Joux. *Bull. Soc. Sci. nat.*, 4 ser., 33:79–100. **314**

Forel, F. A., 1898b. Les flaques d'eau libres dans la glace des lacs gelés. *Bull. Soc. vaud. Sci. nat.*, 4 ser., 34:272–278. **231**

Forel, F. A., 1901a. Handbuch der Seenkunde: allgemeine Limnologie. Stuttgart, 249 pp. **3, 114**

Forel, F. A., 1901b. Études thermique des lacs du nord de l'Europe. *Arch. Sci. phys. nat.*, ser. 4, 12:35–55. **492**

Forel, F. A., 1912. The Fata Morgana. *Proc. roy. Soc. Edinb.*, 32:175–182. **421**

Forsh, F., 1944. Isparenie i kondensatsiia s poverkhnosti ozera Baikal. *Akad. Nauk, SSSR, Referaty Nauch no -issle dovabli'shikh Rabot*, 145 (not seen: ref. *Met. Abstr.* 1, 1950:751). **525**

Föyn, E., Karlik, B., Petterson, H., and Rona, E., 1939. The radioactivity of sea water. *Göteborgs VetenskSamh. Handl.*, B., 6, No. 12. **829**

Fox, C. J. J., 1907. On the coefficients of absorption of the atmospheric gases in distilled water and in sea water. *Publ. Circ. Cons. Explor. Mer.*, 41, 23 pp. **579**

Fox, C. J. J., 1909. On the coefficients of absorption of nitrogen and oxygen in distilled water and sea water, and of atmospheric carbonic acid in sea water. *Trans. Faraday Soc.*, 5:68–87. **579, 838**

Francis, J. R. D., 1951. The aerodynamic drag of a free water surface. *Proc. roy. Soc.*, A, 206:387–406. **273, 274, 352**

Franssila, M., 1940. Zür Frage des Wärme—und Feuchtaustauches über Binnenseen (Mitt. Meteorolog. Inst. Univ. Helsinki, No. 42). *Comment. phys.-math. Helsingf.*, 10, No. 14, 36 pp. **378**

Frantzev, A. W., 1932. Ein Versuch der physiologischen Erforschung der Produktionsfähigkeit des Moskauflusswassers. (In Russian, German summary.) *Microbiology, Moscow*, 1:112–130. **732**

Frazer, J. G., 1898. *Pausanias' description of Greece*, translated with a commentary, Vol. IV. London, Macmillan, 447 pp. (Discharge of Lake Pheneos, pp. 262–263.) **109**

Frear, G. L., and Johnston, J., 1929. Solubilities of calcium carbonate (calcite) in certain aqueous solutions at 25°. *J. Amer. chem. Soc.*, 51:2082–2093. **661, 662**

* Frey, D. G., 1949. Morphometry and hydrography of some natural lakes of the North Carolina coastal plain: the bay lake as a morphometric type. *J. Elisha Mitchell sci. Soc.*, 65:1–37. **152, 682, 785**

Frey, D. G., 1950. Carolina bays in relation to the North Carolina coastal plain. *J. Elisha Mitchell sci. Soc.*, 66:44–52. **151**

Frey, D. G., 1953. Regional aspects of the late-glacial and postglacial pollen succession of southeastern North Carolina. *Ecol. Monogr.*, 23:289–313. **152**

Frey, D. G., 1954. Evidence for the recent enlargement of the bay lakes of North Carolina. *Ecology*, 35:78–88. **155**

Frey, D. G., 1955a. Stages in the ontogeny of the Carolina bays. *Verh. int. Ver. Limnol.*, 12:660–668. **152**

* Frey, D. G., 1955b. Längsee: a history of meromixis. *Mem. Ist. ital. Idrobiol. de Marchi*, suppl., 8:141–161. **482, 489**

Friedman, J., 1953. Deuterium content of natural waters and other substances. *Geochim. et cosmoch. Acta.*, 4:89–103. **211, 214, 215**

Fries, C., 1938. Geology and ground water of the Trout Lake region, Vilas County, Wisconsin. *Trans. Wis. Acad. Sci. Arts Lett.*, 31:305–322. **96**

Fries, C., 1943. *Biberland. Ein Buch über den Biber und sein Werk* (trans. by H. Willecke from Swedish: *Bäverland En bok om bavern ochshans verk*). Neudamm, Verlag J. Neumann, 122 pp. **146**

Fritsch, F. E., and De, P. K., 1938. Nitrogen fixation by blue-green algae. *Nature, Lond.*, 142:878. **847**

Fritz, S., 1951. Solar radiant energy and its modification by the earth and its atmosphere. In *Compendium of Meteorology*. Boston, American Society of Meteorologists, 1951, pp. 12–13. **368** fn., **369**

Fritz, S., and MacDonald, T. H., 1949. Average solar radiation in the United States. *Heating and Ventilating*, 46, No. 7, 61-64. **374**

Fuchs, K., 1876. *Les volcans et les tremblements du terre*. Paris. (Not seen: ref. Forel 1895.) **328**

Fuchs, V. E., 1939. The geological history of the Lake Rudolf Basin, Kenya Colony. *Phil. Trans.*, ser. B., 229:219–274. **12, 23, 24**

Fuchs, V. E., 1950. Pleistocene events in the Baringo Basin. *Geol. Mag.*, 87:149–174. **24**

Fujimoto, C. K., and Sherman, G. D., 1948. Behavior of manganese in the soil and the manganese cycle. *Soil Sci.*, 66:131–145. **802**

Fujiwhara, S., 1914. The horizontal rainbow. *Mon. Weath. Rev., Wash.*, 42:426–430. **419**

Fülleborn, F., 1906. *Das deutsche Nyasa—und Ruvuma-Gebiet.* Berlin. (Not seen: ref. Halbfass 1922.) **312**

Fuller, M. L., 1914. The geology of Long Island, New York. *Prof. Pap. U.S. geol. Surv.*, 82, 231 pp. **95, 96**

Gabbard, J. L., and Dole, M., 1937. A redetermination of the deuterium ratio in normal water. *J. Amer. chem. Soc.*, 59:181–185. **212**

Gaillard, D., DuB., 1904. Wave action in relation to engineering structures. *Prof. Pap. Cps. Engrs, U.S.*, 31, 232 pp. **360**

Gajl, K., 1924. Über zwei faunistiche Typen aus der Umgebung von Warschau auf Grund von Untersuchungen an Phyllopoda und Copepoda (exkl. Harpacticidae). *Bull. int. Acad. Cracovie* (*Acad. pol. Sci.*), ser. B., 1924:13–55. **228**

Garbini, A., 1897. Alcune notizie fisiche sulle acque del Benaco. *Riv. geogr. ital.*, 4:23–29. **416**

Gardiner, A. C., 1939. In E. F. W., Mackenzie, Report on the results of the bacteriological, chemical, and biological examination of the London waters for the twelve months ending 31 December, 1938. *Rep. metrop. Wat. Bd.*, 33 (all biological and biochemical sections are by Gardiner). **730**

Gardiner, A. C., 1941. Silicon and phosphorus as factors limiting development of diatoms. *J. Soc. chem. Ind., Lond.*, 60:73–78. **741**

Gardiner, E. W., 1929. The origin of the Fayum Depression: a critical commentary on a new view of its origin. *Geogr. J.*, 74:371–383. **136**

Garretty, M. D., 1937. Some notes on the physiography of the Lake George region, with special reference to the origin of Lake George. *J. roy. Soc. N.S.W.*, 70:285–293. **11**

Gauthier, H., 1928. *Recherches sur la faune des eaux continentales de l'Algerie et de la Tunisie.* Alger., Imp. Minerva, 419 pp., 3 pl., 6 maps. **136**

* Gavazzi, A., 1904. Die Seen des Karstes. Erster Teil: Morphologisches Material *Abh. der k.k. geogr. Ges. Wien*, 5, No. 2, 136 pp. **107**

Gay, C., 1833. Aperçu sur les recherches d'histoire naturelles faites dans l'Amérique du Sud, et princepalement dans le Chili, pendant les années 1830 et 1831. *Ann. Sci. nat.*, ser. 1, 28:369–393. **176**

Gemsege, P., 1755. The agitation of Pibley dam, with remarks. *Scots Magazine*, 17:594–595. **300**

* Geological Society of America, 1945. Glacial Map of North America. *Spec. Pap. geol. Soc. Amer.*, 60; viii, 37 pp. (2nd ed. 1948). **54**

Geren, H. O., 1941. Climate of New York. In *Climate and Man.* U.S. Dept. of Agriculture (House Document 27, 77th Congr., 1st Sess.), pp. 1025–1034. **526**

Gessner, F., 1939. Die Phosphorarmut der Gewässer und ihre Beziehungen zum Kalkgehalt. *Int. Rev. Hydrobiol.*, 38:203–211. **728**

Gessner, F., 1955. Die limnologischen Verhältnisse in den Seen und Flüssen von Venezuela. *Verh. int. Ver. Limnol.*, 12:284–294. **464**

* Gilbert, G. K., 1885. The topographic features of lake shores. *Rep. U.S. geol. Surv.*, 5:69–123. **139, 142, 143**

Gilbert, G. K., 1895. Lake basins created by wind erosion. *J. Geol.*, 3:47–49. **128**

Glangeaud, P., 1913. Les régions volcaniques du Puy-de-Dôme. II. La chaîne des Puys. *Bull. Carte géol. Fr.*, No. 135, Paris. (Not seen: ref. Collet 1925.) **39, 41**

Glangeaud, P., 1919. *Le Massif Central de la France. Clermont-Ferrand.* (Not seen: ref. Collet 1923.) **30, 38**

Götzinger, G., 1909. Studien über das Eis der Lunzer Ober- und Unter- sees. *Int. Rev. Hydrobiol.*, 2:386–396. **529**

Goldberg, E. D., 1952. Iron assimilation by marine diatoms. *Biol. Bull., Wood's Hole*, 102:243–248. **714, 737**

Goldring, W., 1922. The Champlain Sea. *Bull. N.Y. St. Mus.* (17th Rep. of Director, 1920–21), 239–240:153–187. **11**

Goldschmidt, V. M., 1938. Geochemische Verteilungsgesetze der Elemente. IX. Die Mengenverhältnisse der Elemente und der Atom-Arten. *Skr. norske VidenskAkad.*, Mat.-nat. Kl., 1937, No. 4., 148 pp. **764, 804, 828**

Goldstein, S., 1931. On the stability of superposed streams of fluids of different densities. *Proc. roy. Soc.*, ser. A., 132:524–548. **346**

Gorham, E., 1955a. On some factors affecting the chemical composition of Swedish fresh waters. *Geochim. et cosmoch. Acta*, 7:129–150. **556, 684**

Gorham, E., 1955b. On the acidity and salinity of rain. *Geochim. et cosmoch. Acta*, 7:231–239. **549, 551** fn.

Goryunova, S. V., 1954. Characterization of dissolved organic substances in water of Lake Beloie. *Trudy. Inst. Mikrobiol. Akad. Nauk. SSSR*, 3:185–193 (not seen: ref. *Chem. Abstr.*, 49:9841 ff.). **890**

Goto, K., and Okabe, K., 1940. Comparison of the density of water from various places on the earth. *Bull. chem. Soc. Japan.*, 15:76–81. **214**

Grahmann, R., 1937. Die Entwicklungsgeschichte des Kaspisees und des Schwarzen Meeres. *Mitt. Ges. Erdk. Lpz.*, 54:24–47. **6**

Gran, H. H., 1931. On the conditions for the production of plankton in the sea. *Rapp. Cons. Explor. Mer.*, 75:37–46. **712**

Gran, H. H., 1933. Studies on the biology and chemistry of the Gulf of Maine. III. Distribution of phytoplankton in August 1932. *Biol. Bull., Wood's Hole*, 64:159–182. **713**

Gran, H. H., and Ruud, B., 1927. Über die Planktonproduktion in Hurdals-See. *Avh. norske VidenskAkad.*, Mat.–nat. Kl., 1927, No. 1, 33 pp. **594**

Grant, C., 1945. A biological explanation of the Carolina bays. *Sci. Mon., N.Y.*, 61:443–450. **153**

Greeley, A. W., 1888. *International Polar Expedition. Report on the Proceedings of the United States Expedition to Lady Franklin Bay, Grinnell Land*, Vol. 1. Washington, Govt. Printing Office, viii, 545 pp. **461**

Greenbank, J. T., 1945. Limnological conditions in ice-covered lakes, especially as related to winter-kill of fish. *Ecol. Monogr.*, 15:343–392. **630, 632**

Greene, C. H., and Voskuyl, R. J., 1939. The deuterium-protium ratio. I. The densities of natural waters from various sources. *J. Amer. chem. Soc.*, 61:1342–1349. **212, 214**

Gregory, J. W., 1914. Is the earth drying up? *Geogr. J.*, 43:148–172, 293–313. **239** fn.

Gregory, J. W., 1921. *The rift valleys and geology of East Africa.* London, Seeley, Service, 479 pp. **22**

Griesel, R., 1920. Physikalische und chemische Eigenschaften des Hemmelsdorfer Sees bei Lübeck. Inaug. Dissert., Rostock. **315, 482**

Griesel, R., 1935. Die Aussüssung des Hemmelsdorfer Sees. *Mitt. Geogr. Ges. Lübeck*, 2 Reihe, 38:77–83. **482, 776**

Griesseier, H., 1953. Zur Reflection der direkten Sonnenstrahlung an einer bewegten Wasseroberfläche. *Acta Hydrophys.*, 1:107–133. **418**

Grim, J., 1939. Beobachtungen am Phytoplankton des Bodensees (Obersee) sowie deren rechnerische Auswertung. *Int. Rev. Hydrobiol.*, 39:193–315. **796**

Gross, P., Steiner, H., and Suess, H., 1936. The inversion of cane sugar in mixtures of light and heavy water. *Trans. Faraday Soc.*, 32:883–889. **217**

Grosse, A. V., Johnston, W. M., Wolfgang, R. L., and Libby, W. F., 1951. Tritium in nature. *Science*, 113:1–2. **212**

Grote, A., 1934. *Der Sauerstoffhaushalt der Seen. Die Binnengewässer*, 14. Stuttgart, E. Schweizerbartsche Verlagsbuchhandlung, 217 pp. **257** fn., **447, 474, 586, 609, 647**

Gruissmayr, F. B., 1931. Der Einfluss der grossen kanadischen Seen auf die Fruhlingstemperature der Union. *Geogr. Ann.*, 13:95–104. **526**

Gunn, J. P., Todd., H. J., and Mason, K., 1930. The Shyok Flood, 1929. *Himalayan J.*, 2:35–47. **51**

Guseva, K. A., 1935. *Microbiology, Moscow*, 4 (not seen: ref. Lastochkin 1945). **732**

Guseva, K. A., 1937a. Deistvie mangantsa na razvitie vodoroslei. (Effect of manganese on the development of algae.) *Microbiology, Moscow*, 6:292–307. **807, 847**

Guseva, K. A., 1937b. Gidrobiologii i mikrobiologii uchinskugo vodokhranchsheh a kanala Moskva- Volga. *Microbiology, Moscow*, 6:449–464. (Hydro- and microbiological studies of the Ucha reservoir of the Moscow Volga canal. II. The development of *Anabaena lemmermanni* Richter, *Aphanizomenan flos-aquae*, and *Astenonella formosa* Hassall in the reservoir during the first summer of its existence.) **732, 806**

Guseva, K. A., 1939. The blooming of the Uchinskii reservoir. *Biol. Soc. nat. Moscow*, Sect. biol., 48:30–32 (not seen: ref. *Chem. Abstr.*, 34:6327). **807**

Gustafson, T., and Kullenberg, B., 1936. Untersuchungen von Trägheitsströmungen in der Ostsee. *Svenska hydrogr. biol. Komm. Skr.*, Ny ser. Hydro., No. 13, 28 pp. **261**

Guyer, A., and Tobler, B., 1934. Zur Kenntnis der Geschwindigkeit der Gasexsorption von Flüssigkeiten. *Helv. chim. acta.*, 17:257–271. **658**

Hacker, W., 1933. Sichttiefe, Wärmegang und Durchlüftung in Hochgebirgsseen *Geogr. Jber, aus Öst.*, 16:88–105. **416, 441**

Halbfass, W., 1902. Über Seespiegelschwankungen im Madüsee. *Z. Gewässerk.*, 5:15 (not seen: ref. Chumley 1910, Halbfass 1922). **315**

Halbfass, W., 1903. Über Seespiegelschwankungen in Madüsee. *Z. Gewässerk.*, 6:65 (not seen: ref. Chumley 1910, Halbfass 1922.) **315**

Halbfass, W., 1905. Die Thermik der Binnenseen und das Klima. *Petermanns Mitt.*, 51:219–233. **492**

Halbfass, W., 1910a. Ergebnisse neuerer simultaner Temperaturmessungen in einigen tiefen Seen Europas. *Petermanns Mitt.*, 56(II):59–64. **492**

Halbfass, W., 1910b. Gibt es im Madüsee Temperaturseiches? *Int. Rev. Hydrobiol.*, Hydrog. Suppl. 1, 3:1–40. **336, 338**

* Halbfass, W., 1922. Die Seen der Erde. *Petermanns Mitt.*, Erganzungsheft 185, vi, 169 pp. **41** fn., **93, 170, 173, 314**

* Halbfass, W., 1923. *Grundzüge einer vergleichenden Seenkunde.* Berlin, Borntraeger, viii, 354 pp. **3, 109** fn., **175, 231, 234, 299, 312, 314, 316, 362, 403, 416, 420, 440, 441, 445, 479** fn., **492** fn., **495, 524**

* Halbfass, W., 1933. Seen. In *Handbuch der Geophysik.* Berlin, Borntraeger, Bd. 7., lf.T:122–182. **223**

* Halbfass, W., 1934. Der Jahreswasserhaushalt der Erde. Ist er quantitative eine konstante Grösse? *Petermanns Mitt.*, 80:137–140, 177–179. **223, 229**

* Halbfass, W., 1937. Nachträge zu meinem Buche: Die Seen der Erde. *Int. Rev. Hydrobiol.*, 35:246–294. **19, 170**

Hale, F. E., 1922. Tastes and odors in New York water supply. *J. Amer. Wat. Wks Ass.*, 9:829–837. **817**

Hand, I. F., 1941. A summary of total solar and sky radiation measurements in the United States. *Mon. Weath. Rev., Wash.*, 69:95–125. **372**

Hand, I. F., 1949. Weekly mean values of daily total solar and sky radition. *Tech. Pap. U.S. Weath. Bur.*, 11, 17 pp. **525, 528**

Harbeck, G. E., and Kennon, F. W., 1952. The water-budget water-loss investigations: Vol. 1, Lake Hefner Studies. Tech. Rep., *U.S. geol. Surv. Circ.* 229, pp. 17–34. **518**

Harding, S. T., 1949. Statistics of lake levels (with special reference to the magnitude and duration of fluctuations). *Gen. Ass. Int. Un. Geod.*, Oslo, 1948. *Ass. Int. Hydrol. sci., I. Trav. Comm. Potamol. Limnol.*, 349–357. **244** fn., **245**

Hardman, Y., 1941. The surface tension of Wisconsin lake waters. *Trans. Wis. Acad. Sci. Arts Lett.*, 33:395–404. **900**

Hardman, Y., and Henrici, A. T., 1939. Studies of fresh water bacteria. V. The distribution of *Siderocapsa treubii* in some lakes and streams. *J. Bact.*, 37:97–104. **716**

Harman, R. W., 1927. Aqueous solutions of sodium silicates. VII. Silicate ions. Electrometric titrations, Diffusion, Colorimetric Estimation. *J. phys. Chem.*, 31:616–625. **789**

Harned, H. S., and Davis, R., 1943. The ionization constant of H_2CO_3 in water, and the solubility of CO_2 in water and aqueous solutions from 0 to 50°. *J. Amer. chem. Soc.*, 65:2030–2037. **656**

Harrington, M. W., 1894. The currents of the Great Lakes. *Bull. U.S. Weath. Bur.*, B, 5 pp. **289**

* Harris, R. A., 1909. *Manual of Tides:* Part V, *Currents, Shallow-Water Tides, Meteorological Tides, and Miscellaneous Matters.* Report Superintendent of the Coast and Geodetic Survey, showing the progress of work from July 1, 1906, to June 30, 1907. App. 6, 231–545. (See Chap. X, Tides in lakes and wells, pp. 483–487.) **330, 333**

Harris, S. W., 1954. An ecological study of the waterfowl of the Pot-Holes area, Grant County, Washington. *Amer. Midl. Nat.*, 52:403–432. **127**

Harrison, J. M., 1954. Ungava (Chubb) Crater Lake and Glaciation. *J. roy. astro. Soc. Can.*, 48:16–20. **149**

Harvey, E. N., Barnes, D. K., McElroy, W. D., Whiteley, A. M., Pease, D. C., and Cooper, K. W., 1944. Bubble formation in animals. I. Physical factors. *J. cell. comp. Physiol.*, 24:1–22. **583** fn.

* Harvey, H. W., 1933. On the rate of diatom growth. *J. mar. biol. Ass. U.K.*, 19:253–276. **202**

Harvey, H. W., 1937a. The supply of iron to diatoms. *J. mar. biol. Ass. U.K.*, 22:205–219. **706, 713, 737**

Harvey, H. W., 1937b. Note of colloidal ferric hydroxide in sea water. *J. mar. biol. Ass. U.K.*, 22:221–225. **706, 707**

* Harvey, H. W., 1939. Substances controlling the growth of a diatom. *J. mar. biol. Assoc. U.K.*, 23:499–520. **710, 712, 896**

Harvey, H. W., 1940. Nitrogen and phosphorus required for the growth of phytoplankton. *J. mar. biol. Ass. U.K.*, 24:115–123. **851**

Harvey, H. W., 1947. Manganese and the growth of phytoplankton. *J. mar. biol. Ass. U.K.*, 26:562–579. **806**

Harvey, H. W., 1949. On manganese in sea and fresh waters. *J. mar. biol. Ass. U.K.*, 28:155–164. **805**

Hasler, A. D., 1938. Fish biology and limnology of Crater Lake, Oregon. *J. Wildlife Mgmt*, 2:94–103. **403**

* Hasler, A. D., 1947. Eutrophication of lakes by domestic drainage. *Ecology*, 28:383–395. **897**

Hasler, A. D., 1949. Antibiotic aspects of copper treatment of lakes. *Trans. Wis. Acad. Sci. Arts Lett.*, 39:97–103. **812, 817, 818**

Hasler, A. D., 1954. Odour perception and orientation in fishes. *J. Fish. Res. Rd. Can.*, 11:107–129. **900**

Hasler, A. D., and Einsele, W. G., 1948. Fertilization for increasing productivity of natural inland waters. *Trans. 13th. N. Amer. Wild L. Conf.*, March 8, 9, and 10, 1948, pp. 527–554. **738, 740, 810**

Hasler, A. D., and Jones, E., 1949. Demonstration of the antagonistic action of large aquatic plants on algae and rotifers. *Ecology*, 30:359–364. **901**

Hasler, A. D., and Wisby, W. J., 1951. Discrimination of stream odors by fishes and its relation to parent stream behaviour. *Amer. Nat.*, 85:223–238. **900**

Hasluck, M., 1936a. A historical sketch of the fluctuations of Lake Ostrovo in West Macedonia. *Geogr. J.*, 87:338–347. **107**

Hasluck, M., 1936b. The archeaological history of Lake Ostrovo in West Macedonia. *Geogr. J.*, 88:448–465. **107**

Hasluck, M., 1937. Causes of the fluctuations in level of Lake Ostrovo, West Macedonia. *Geogr. J.*, 90:446–457. **107**

Haurwitz, B., 1951. The slope of lake surfaces under variable wind stresses. *Tech. Memor. U.S. Army Eros. Bd.*, 25, 23 pp. **328**

Hawes, G. H., 1881. On liquid carbon dioxide in smoky quartz. *Amer. J. Sci.*, 3rd ser., 21:203–209. **202**

Hayes, A. A., 1875. On the wide diffusion of vanadium and its association with phosphorus in many rocks. *Proc. Amer. Acad. Arts Sci.*, 10:294–299. **828**

Hayes, F. R., and Coffin, C. C., 1951. Radioactive phosphorus and exchange of lake nutrients. *Endeavour*, 10:78–81. 744

* Hayes, F. R., McCarter, J. A., Cameron, M. L., and Livingstone, D. A., 1952. On the kinetics of phosphorus exchange in lakes. *J. Ecol.*, 40:202–216. **744, 745**

Hayford, J. F., 1922. *Effects of Winds and of Barometric Pressures on the Great Lakes.* Carnegie Inst., Washington., publ. No. 317, v, 133 pp. **274, 308, 316, 318, 319, 333**

Hazen, A., 1892. A new color standard for natural waters. *Amer. chem. J.*, 14:300–310. **413**

Hearne, V., 1705. Memorabilia nonnulla lacus Vetteri. *Phil. Trans.*, 24:1938–1946. **300, 401** fn.

Hedin, S. A., 1904. *Scientific Results of a Journey in Central Asia, 1899 to 1902*: Vol. 1, *The Tarim River.* Stockholm, Lithographic Inst., Swedish Army, 523 pp. **127, 242**

Hedin, S. A., 1917. *Southern Tibet*: Vol. II, *Lake Manasarovar and the Sources of the Great Indian Rivers—from the End of the Eighteenth Century to 1913.* Stockholm, Lithographic Inst., Swedish Army, xi, 330 pp. **240**

Heim, A., 1885. *Handbuch der Gletscherkunde.* Stuttgart, J. Engelhorn, 560 pp. **50**

Heim, A., 1894. Die Enstehung der alpinen Rand-Seen. *Vjschr. naturf. Ges. Zürich*, 39:66–84. **77**

Heim, A., 1905. Das Säntisgebirge. *Beitr. geol. Karte Schweiz*, NF, Lief. 16 (not seen: ref. Collet 1925). **12**

Heintze, S. G., and Mann, P. J. G., 1947. Soluble complexes of manganic manganese. *J. agri. Sci.*, 37:23–26. **802**

* Hellström, B., 1941. Wind effect on lakes and rivers. *IngenVetenskAcad. Handl.*, 158, 191 pp. **263, 273, 274, 277, 281, 282, 284**

Henderson, L. J., 1913. *The Fitness of the Environment; an Inquiry into the Biological Significance of the Properties of Matter.* New York, Macmillan, xv, 317 pp. **195**

Henkel, J., 1912. Zusammenstellung von Zahlen für die Wasserführung der Flüsse. *Geogr. Anz.*, 13:266 (not seen: ref. Kalle 1945). **229**

Henry, A. J., 1902. Wind velocity and fluctuations of water level of Lake Erie. *Bull. U.S. Weath. Bur.*, 262, 22 pp. **318, 333**

Herbette, F., 1914. Le problème du desséchement de l'Asia intérieure. *Ann. Geogr.*, 23:1–30. **239** fn.

Herre, A. W. C. T., 1933. The fishes of Lake Lanao: a problem in evolution. *Amer. Nat.*, 67:154–162. **41**

Hess, H., 1933. Das Eis der Erde. In *Handbuch der Geophysik*. Berlin, Borntraeger, Bd. 7, lf.1: 1–121. **222**

Hills, E. S., 1939. The lunette, a new land form of aeolian origin. *Austr. Geogr.*, 3:15–21. **127**

Hine, D. C., and Dawbarn, M. C., 1954. The determination of vitamin B_{12} activity in the organs and excreta of sheep. II. The influence of cobalt on the production of factors possessing vitamin B_{12} activity in the rumen contents of sheep. *Nature, Lond.*, 170:794. **826**

Hobbs, W. H., 1911. Requisite conditions for the formation of ice ramparts. *J. Geol.*, 19:157–160. **532**

Hobbs, W. H., 1912. *Earth Features and Their Meaning*. New York, Macmillan, xxxix, 506 pp. **3, 8, 14, 65, 114, 125, 126**

Höll, K., 1951. Über den Kaliumgehalt der Gewässer. *Verh. int. Ver. Limnol.*, 11:137–143. **556**

Hoffmann, J., 1942. Über in Süsswässern gelöste und von Sedimenten mitgerissene Uranmengen. *Chem. d. Erde*, 14:239–252. **831**

Holland, T. H., 1894. Report on the Gohma Landslip, Garhwal. *Rec. geol. Surv. India.*, 27:55–64. **43**

Holland, T. H., and Christie, W. A. K., 1909. The origin of the salt deposits of Rajputana. *Rec. geol. Surv. India*, 38:154–186. **546**

Holm-Hansen, O., Gerloff, G. C., and Skoog, F., 1954. Cobalt as an essential element for blue-green algae. *Physiol. Plant.*, 7:665–675. **826**

Holroyd, A., and Parker, H. B., 1949. Investigations on the dynamics of aeration (the aeration of clean water). *J. Inst. Sew. Purif.*, 1949:292–312. **587**

Holsinger, E. C. T., 1955. The distribution and periodicity of the phytoplankton of three Ceylon lakes. *Hydrobiologia*, 7:25–35. **442** fn.

Holtedahl, O., 1924. Studier over Isrand-terrassene syd for de store østlandske sjøer. *Skr. norske VidenskAcad.*, 1924, No. 14, 110 pp. **84, 96**

Honda, K., Terada, T., Yoshida, Y., and Isitani, D., 1909. Secondary undulations of oceanic tides. *J. Coll. Sci. Tokyo*, 24:1–113. **330**

Hooper, F. F., 1954. Limnological features of Weber Lake, Cheboygan County, Michigan. *Pap. Mich. Acad. Sci.*, 39:229–240. **677, 686**

Hopkins, D. M., 1949. Thaw lakes and sinks in the Imuruk Lake area, Seward Peninsula, Alaska. *J. Geol.*, **57**:119–131. **99**

Hopkins, E. F., 1930. The necessity and function of manganese in the growth of Chlorella. *Science*, 72:609–610. **806**

Hoppe-Seyler, F., 1896. Ueber die Vertheilung absorbierter Gase im Wasser des Bodensees und ihre Beziehungen zu den in ihm lebenden Thierens und Pflanzen. *Schr. Ver. Gesch. Bodensees*, 24:29-48. **600, 838**

Horie, S., 1956. Morphometry of Japanese lakes. *Jap. J. Limnol.*, 18:1–28. **170**

Horner (*sic*, ? Hörner), J. C., 1821. Instruction für die astronomischen und physikalischen Arbeiten auf dieser Reise. In O. von Kotzebue, *Entdeckungs—Reise in die Süd-*

See und nach der Berings-Strasse, 3 vols., 240 pp. Weimar, Hoffmann, Vol. I, pp. 73–91. **401**

* Hough, J. L., 1953. *Final Report on the Project Pleistocene Chronology of the Great Lakes Region.* Office of Naval Research Contract No. N6ori–07133, Project NR–018–122. Univ. Illinois, 108 pp. (mimeogr.). **82**

Hough, J. L., 1955. Lake Chippewa, a low stage of Lake Michigan indicated by bottom sediments. *Bull. geol. Soc. Amer.*, 66:957–968. **82, 178** fn.

Houghton, H. G., *see* Rankama and Sahama 1950.

* Hubbs, C. L., and Miller, R. R., 1948. The zoological evidence. In: A symposium on the Great Basin with emphasis on glacial and postglacial times. *Bull. Univ. Utah Biol. Serv.*, 107:18–166. **15, 16, 17**

Huber-Pestalozzi, G., 1936. Beobachtungen an einem "Blutsee" im Samnaun (K. K. Graubünden, Schweiz). *Arch. Hydrobiol.*, 29:265–273. **417**

* Huber-Pestalozzi, G., 1938. *Das Phytoplankton des Süsswassers Systematik und Biologie 1.* Teil: *Allgem. Teil, Blaualgen, Bakterien, Pilze. Die Binnengewässer*, 16 pt.1. Stuttgart, E. Schweizerbartsche Verlagsbuchhandlung, 342 pp. **715**

Hulburt, E. O., 1934. The polarization of light at sea. *J. opt. Soc. Amer.*, 24:35–42. **418**

Humphreys, W. J., 1929. The horizontal rainbow. *J. Franklin Inst.*, 207:661–664. **419**

Humphreys, W. J., 1935. The "sinking" of lake and river ice. *Mon. Weath. Rev., Wash.*, 62:133–134. **530**

Huntington, E., 1907. *The Pulse of Asia.* Boston and New York, Houghton Mifflin, xxi, 415 pp. **7, 238**

Hurst, H. E., and Phillips, P., 1931. The Nile Basin: Vol. 1, General description of the basin. Topography of the White Nile. *Physical Dept. Bull.* No. 21, Cairo. **242**

Hutchinson, G. E., 1932. Experimental studies in ecology. I. The magnesium tolerance of Daphnidae and its ecological significance. *Int. Rev. Hydrobiol.*, 28:90–108. **821**

Hutchinson, G. E., 1936. Alkali deficiency and fish mortality. *Science*, 84:18. **599**

Hutchinson, G. E., 1937a. *The Clear Mirror: A Pattern of Life in Goa and in Indian Tibet.* Cambridge Univ. Press, xi, 171 pp. **773**

Hutchinson, G. E., 1937b. Limnological studies in Indian Tibet. *Int. Rev. Hydrobiol.*, 35:134–177. **63, 65, 441, 495, 504, 508, 531, 564, 565, 578, 823**

Hutchinson, G. E., 1937c. A contribution to the limnology of arid regions primarily founded on observations made in the Lahontan Basin. *Trans. Conn. Acad. Arts Sci.* 33:47–132. **26, 136, 226** fn., **227** fn., **480, 481, 483** fn., **484, 487, 495, 508, 511, 512**

Hutchinson, G. E., 1938. On the relation between the oxygen deficit and the productivity and typology of lakes. *Int. Rev. Hydrobiol.*, 36:336–355. **639, 640, 642, 645, 646**

Hutchinson, G. E., 1941. Limnological studies in Connecticut. IV. The mechanism of intermediary metabolism in stratified lakes. *Ecol. Monogr.*, 11:21–60. **341, 454, 466** fn., **469** fn., **488, 614, 618, 675, 679, 707, 708, 740, 750, 767, 829, 847**

Hutchinson, G. E., 1943. Thiamin in lake waters and aquatic organisms. *Arch. Biochem.*, 2:143–150. **897**

Hutchinson, G. E., 1944a. Limnological studies in Connecticut. VII. A critical examination of the supposed relationship between phytoplankton periodicity and chemical changes in lake waters. *Ecology*, 25:3–26. **824**

Hutchinson, G. E., 1944b. Nitrogen in the biogeochemistry of the atmosphere. *Amer. Scient.*, 32:178–195. **551, 868**

Hutchinson, G. E., 1950. Survey of contemporary knowledge of biogeochemistry. III. The biogeochemistry of vertebrate excretion. *Bull. Amer. Mus. nat. Hist.*, 96, 554 pp. **144**

Hutchinson, G. E., 1952. The biogeochemistry of phosphorus. In *The Biology of Phosphorus*, pp. 1–35, ed. L. F. Wolterink. Michigan State College Press, vii, 147 pp. **727** fn.

Hutchinson, G. E., 1954. The biochemistry of the terrestrial atmosphere. In *The solar system*: II, *The Earth as a Planet*, pp. 371–433, ed. G. P. Kuiper. Chicago, Univ. Chicago Press. **543, 551, 654, 844**

Hutchinson, G. E., and Bowen, V. T., 1947. A direct demonstration of the phosphorus cycle in a small lake. *Proc. nat. Acad. Sci., Wash.*, 33:148–153. **747**

Hutchinson, G. E., and Bowen, V. T., 1950. Limnological studies in Connecticut. IX. A quantitative radiochemical study of the phosphorus cycle in Linsley Pond. *Ecology*, 31:194–203. **747**

Hutchinson, G. E., Deevey, E. S., and Wollack, A., 1939. The oxidation-reduction potential of lake waters and its ecological significance. *Proc. nat. Acad. Sci., Wash.*, 25:87–90. **693, 695, 719**

Hutchinson, G. E., and Löffler, H., 1956. The thermal classification of lakes. *Proc. nat. Acad. Sci., Wash.*, 42:84–86. **462**

Hutchinson, G. E., and Pickford, G. E., 1932. Limnological observations on Mountain Lake, Virginia. *Int. Rev. Hydrobiol.*, 27:252–264. **46**

Hutchnison, G. E., Pickford, G. E., and Schuurman, J. F. M., 1932. A contribution to the hydrobiology of pans and other inland waters of South Africa. *Arch. Hydrobiol.*, 24:1–136. **132, 712**

Hutchinson, G. E., and Setlow, J. K., 1946. Limnological studies in Connecticut. VIII. The niacin cycle in a small inland lake. *Ecology*, 27:13–22. **898**

Hutchinson, G. E., and Wollack, A. C., 1940. Studies on Connecticut lake sediments. II. Chemical analyses of a core from Linsley Pond. *Amer. J. Sci.*, 238:493–517. **743**

Hutner, S. H., and Provasoli, L., 1951. The phytoflagellates. In *Biochemistry and Physiology of Protozoa*, Vol. 1, ed. A. Lwoff. New York, Academic Press, pp. 29–129. **826**

Hutner, S. H., and Provasoli, L., 1953. A pigmented marine diatom requiring vitamin B_{12} and uracil. *News Bull. phycol. Soc. Amer.*, 6(No. 18):7–8. **826**

Hutner, S. H., Provasoli, L., Schatz, A., and Haskins, C. P., 1950. Some approaches to the study of the role of metals in the metabolism of microorganisms. *Proc. Amer. phil. Soc.*, 94:152–170. **826**

Hutner, S. H., Provasoli, L., Stokstad, E. L. R., Hoffmann, C. E., Belt, M., Franklin. A. L., and Jukes, T. H., 1949. The assay of anti-pernicious anemia factor with Euglena, *Proc. Soc. exp. Biol., N.Y.*, 70:118–120. **826**

Hyman, L. H., 1937. Studies on the morphology, taxonomy, and distribution of North American triclad turbellaria. VII. The two species confused under the name *Phagocata gracilis*, the validity of the generic name *Phagocata* Leidy 1847, and its priority over *Fonticola* Komarek, 1926. *Trans. Amer. micro. Soc.*, 56:298–310. **217** fn.

Huzayyin, S. A., 1941. The place of Egypt in prehistory. A correlated study of climates and cultures in the Old World. *Mém. Inst. Égypte*, 43, xxxiv, 474 pp. **136**

Ingalls, A. G., 1934. "Guns" of Seneca Lake. *Science*, 79:479–480. **362**

Ingham, G., 1950. Effect of materials absorbed from the atmosphere in maintaining soil fertility. *Soil Sci.*, 70:205–212. **549**

Iversen, H. W., 1952. Laboratory study of breakers. In: Gravity waves. *Circ. U.S. Bur. Stand.*, 521:9–32. **357**

* Ivlev, V. S., 1937. Material zur Studium der Stoffbilanz im See. Die Eisenbilanz. *Arb. limnol. Sta. Kossino*, 21:Russian text, 21–53; German Summ., 54–59. **706, 717**

Jackson, D. D., 1905. The normal distribution of chlorine in the natural waters of New York and New England. *Wat. Supp. Pap., Wash.*, 144, 31 pp. **545**

Jacobs, W. C., 1951. The energy exchange between sea and atmosphere and some of its consequences. *Bull. Scripps. Instn. Oceanogr.*, 6:27–122. **229**

Jacobsen, J. P., 1905. Die Löslichkeit von Sauerstoff im Meerwasser durch Winklers Titriermethode bestimmt. *Medd. Komm. Havundersøg., Kbh.*, (c) Hydrogr., No. 8, 13 pp. **579**

* Jaeger, F., 1939. Die Trockenseen der Erde. *Petermanns Mitt.*, Ergänzungsh. 236, 159 pp. **109, 127, 131, 132, 133, 136**

Järnefelt, H., 1932. Zur Limnologie einiger Gewässer Finnlands, IX. *Ann. Soc. Zool.-bot. fenn. Vanamo*, 12:145–283. **622**

Järnefelt, H., 1938. Die Entstehungs und Entwichlungsgeschichte der finnischen Seen. *Geol. Meere*, 2:199–223. **57**

Jallabert, J., 1745. Seiche ou flux et reflux du lac de Genève. *Hist. Acad. roy. des Sciences*, 1742:36–43. **300**

James, H. R., 1941. Beer's law and the properties of organic matter in lake waters. *Trans. Wis. Acad. Sci. Arts Lett.*, 33:73–82. **413**

* James, H. R., and Birge, E. A., 1938. A laboratory study of the absorption of light by lake waters. *Trans. Wis. Acad. Sci. Arts Lett.*, 31:1–154. **381, 383**

Jamieson, T. T., 1863. On the Parallel Roads of Glen Roy and their place in the history of the Glacial Period. *Quart. J. geol. Soc. Lond.*, 19:235–259. **54**

Jardine, J., *see* Leslie 1838 and Buchan 1872.

Jarvis, N. D., Clough, R. W., and Clarke, E. D., 1926. Iodine content of Pacific Coast salmon. *Univ. Wash. Publ. Fisheries*, 1:109–140. **563**

Jeffreys, H., 1919. The periods of seiches in long lakes. *Observatory*, 42:123. **306**

Jeffreys, H., 1923. The free oscillations of water in an elliptical lake. *Proc. Lond. math. Soc.*, ser. 2, 23:455–476. **307**

Jeffreys, H., 1925a. The flow of water in an inclined channel of rectangular section. *Phil. Mag.*, ser. 6, 49:793–807. **252**

* Jeffreys, H., 1925b. On the formation of water waves by wind. *Proc. roy. Soc.*, ser. A, 107:189–206. **350**

Jeffreys, H., 1926. On the formation of water waves by wind (second paper). *Proc. roy. Soc.*, ser. A, 110:241–247. **350**

Jeffreys, H., 1928. The more rapid longitudinal seiches of a narrow lake. *Mon. Not. R. astr. Soc. geophys. Suppl.*, 1:495–500. **307**

Jekimov, A. P., and Krawitz, T. P., 1926. In: Der Baikal. *Arb. Magnet Meteorol. Observatorium Irkutsk*, 1 (not seen: ref. Sterneck 1928). **334**

Jenkin, P. M., 1930. A preliminary limnological survey of Loch Awe (Argyllshire). *Int. Rev. Hydrobiol.*, 24:24–46. **444**

Jenkin, P. M., 1932. Report of the Percy Sladen expedition to some Rift Valley Lakes in Kenya in 1929. VII. Summary of the ecological results, with special reference to the alkaline lakes. *Ann. Mag. nat. Hist.*, ser. 10, 18:133–181. **681**

Jenkin, P. M., 1942. Seasonal changes in the temperature of Windermere (English Lake District). *J. Anim. Ecol.*, 11:248–269. **428** fn., **443** fn., **444, 503**

Jennings, J. N., 1950. The origin of the Fenland meres. *Geol. Mag.*, 87:217–225. **119, 121**

Jennings, J. N., and Lambert, J. M., 1953. The origin of the broads. *Geogr. J.*, 119:91. **121**

Jentzsch, A., 1912. Über einige Seen in der Gegend von Meseritz und Birnbaum. *Abh. Preuss. geol. Landesanst.*, NF., 48 (not seen: ref. Halbfass 1923). **236**

Jerlov, N. G., 1951. Optical studies of ocean waters. *Rept. Swedish Deep-Sea Exped.*, 1947–1948, 3:1–59. **408, 416**

Jermakoff, N. W., 1926. Über ein dispersoidologisches Klassifikationsprinzip für die Ursachen der Färbung natürlicher Gewässer. *Russ. hydrobiol. Zh.*, 5:Russian text, 233–239, German summ. 239–240. **417**

Jewell, M. E., and Brown, H. W., 1929. Studies on northern Michigan bog lakes. *Ecology*, 10:427–475. **681**

* Johnson, D. W., 1919. Shore processes and shoreline development. New York, John Wiley & Sons, 584 pp. **139, 142, 177, 178, 179, 184, 186**

Johnson, D. W., 1936. Origin of the supposed meteoric scars of Carolina. *Science*, 84:15–18. **156**

Johnson, D. W., 1937. Role of artesian water in forming the Carolina bays. *Science*, 86:255–258. **156**

* Johnson, D. W., 1942. *Origin of the Carolina Bays.* New York, Columbia Univ. Press, 341 pp. **153, 156**

Johnson, D. W., 1944. Mysterious craters of the Carolina coast. *Amer. Scient.*, 32:1–22. **156**

Johnson, J. W., 1948. The characteristics of wind waves on lakes and protected bays. *Trans. Amer. geophys. Un.*, 29:671–681. **350**

Johnson, J. W., 1950. Relationships between wind and waves, Abbotts Lagoon, California. *Trans. Amer. geophys. Un.*, 31:386–392. **353**

Johnson, J. W., and Rice, E. K., 1952. A laboratory investigation of wind-generated waves. *Trans. Amer. geophys. Un.*, 33:845–854. **353**

Johnson, Willard, 1899. An unrecognized process in glacial erosion. *Science*, 9:106. **60**

* Johnsson, H., 1946. Termisk-hydrologiska Studier i Sjön Klämmingen. *Geogr. Ann., Stockh.*, 28:1–154. **283, 376, 378, 381, 495, 514, 518, 529**

Johnsson, H., 1949. Wind currents in Lake Klämmingen. *Gen. Ass. int. Un. Geod.*, Oslo, 1948, *Ass. int. Hydrol. Sci.*, *I. Trav. Comm. Potamol. Limnol.*, 369–372. **283**

Johnston, J., 1915. The solubility-product constant of calcium and magnesium carbonates. *J. Amer. chem. Soc.*, 37(II):2001–2020. **665** fn.

* Johnston, J., and Williamson, E. D., 1916. The complete solubility curve of calcium carbonate. *J. Amer. chem. Soc.*, 38(I):975–983. **661**

Jones, L. H. P., and Leeper, G. W., 1950. The availability of various manganese oxides to plants. *Science*, 111:463–464. **802**

Jones, W. C., 1925. The undertow myth. *Science*, 61:444. **358, 440**

Jørgensen, C. B., 1944. On the spicule-formation of *Spongilla lacustris* L. 1. The dependence of the spicule-formation on the content of dissolved and solid silicic acid of the milieu. *Biol. Medd. Kbh.*, 19, No. 7, 45 pp. **798**

* Juday, C., 1914. The inland lakes of Wisconsin. The hydrography and morphometry of the lakes. *Bull. Wis. geol. nat. Hist. Surv.*, 27, Sci. Ser. No. 9, xv, 137 pp. **90, 92, 678**

Juday, C., 1916. Horizontal rainbows on Lake Mendota. *Mon. Weath. Rev., Wash.*, 44:65–67. **419, 420**

* Juday, C., and Birge, E. A., 1931. A second report on the phosphorus content of Wisconsin lake waters. *Trans. Wis. Acad. Sci. Arts Lett.*, 26:353–382. **729, 743**

* Juday, C., and Birge, E. A., 1932. Dissolved oxygen and oxygen consumed in the lake waters of northeastern Wisconsin. *Trans. Wis. Acad. Sci. Arts Lett.*, 27:415–486. **488, 489, 576, 598, 642, 681, 879**

* Juday, C., and Birge, E. A., 1933. The transparency, the color and the specific conductance of the lake waters of northeastern Wisconsin. *Trans. Wis. Acad. Sci. Arts Lett.*, 28:205–259. **401, 403, 415, 880, 885, 890**

* Juday, C., Birge, E. A., Kemmerer, G. I., and Robinson, R. J., 1928. Phosphorus content of lake waters of northeastern Wisconsin. *Trans. Wis. Acad. Sci. Arts Lett.*, 23:233–248. **728**

* Juday, C., Birge, E. A., and Meloche, V. W., 1935. The carbon dioxide and hydrogen ion concentration of the lake waters of northeastern Wisconsin. *Trans. Wis. Acad Sci. Arts Lett.*, 29:1–82. **666, 667, 668, 681, 683, 686, 687**

* Juday, C., Birge, E. A., and Meloche, V. W., 1938. Mineral content of the lake waters of northeastern Wisconsin. *Trans. Wis. Acad. Sci. Arts Lett.*, 31:223–276. **488, 552, 554, 555, 557, 561, 562, 706, 707, 765, 767, 791, 803, 808, 850, 853, 861, 874**

Judson, S., 1950. Depressions of the northern portion of the southern High Plains of New Mexico. *Bull. geol. Soc. Amer.*, 61:253–274. **129**

Judson, S., and Osmond, J. K., 1955. Radioactivity in ground and surface water. *Amer. J. Sci.*, 253:104–116. **831**

Junge, C., 1956. Recent investigations in air chemistry. *Tellus*, 8:127–139. **543**

Junner, N. R., 1937. The geology of the Bosumtwi caldera and surrounding country. *Bull. Gold Coast geol. Surv.*, 8:5–38 (not seen: ref. *Bibliogr. geol. Excl. N. Amer.*, 6:132). **37**

Jutson, J. T., 1934. The physiography (geomorphology) of western Australia (2nd ed.), *Bull. geol. Surv. W. Aust.*, 95, xvi, 366 pp. **134, 135**

Kaleczinsky, A. von, 1901. Über die unganischen warmen und keissen Kochsalzseen. *Földt. Közl.*, 31 (not seen: ref. Halbfass 1923). **479**

Kalle, K., 1938. Zur Problem des Meereswasserfarbe. *Ann. Hydrogr., Berl.*, 66:1–13. **415, 416**

Kalle, K., 1939. Die Farbe des Meeres. *Rapp. Cons. Explor. Mer.*, 109:98–105. **416**

* Kalle, K., 1945. *Der Stoffhaushalt der Meeres. Probleme der kosmichen Physik*, 23. Leipzig, Akad. Verlagsgesellschaft, Becker & Erler, 263 pp. (originally published 1943). **223, 228, 229**

Kamen, M. D., and Gest, H., 1949. Evidence for a nitrogenase system in the photosynthetic bacterium *Rhodospirillum rubrum*. *Science*, 109:560. **848**

Kanthack, F. E., 1941. The fluctuations of Lake Nyasa. *Geogr. J.*, 98:20–33. **242, 244**

Kapustinsky, A. F., 1940. Solubility product and the solubility of metal sulphides in water. *C.R. Acad. Sci. URSS*, 28:144–147. **760, 762**

Karcher, F. H., 1939. Untersuchungen über den Stickstoffhaushalt in ostpreussischen Waldseen. *Arch. Hydrobiol.*, 35:177–266. **488, 853, 854, 864, 871**

Karsinkin, G. S., and Kusnetzow, S. I., 1932. Neue Methoden in der Limnologie. *Arb. Limnol. Sta. Kossino*, 13–14: Russian text, 47–63; German summ., 63–68. **693**

Kaufman, S., and Libby, W. F., 1954. The natural distribution of tritium. *Phys. Rev.*, 93:1337–1344. **212**

Kawada, S., 1943. Untersuchung des neuen Sees gebildet infolge der Erdhebens vom Jahre 1941 in Taiwan (Formosa). *Bull. Earthq. Res. Inst. Tokyo*, 21:317–325 (not seen: ref. *Bibliogr. Geol. Excl. N. Amer.*, 14:129). **46**

Kemmerer, G., Bovard, J. F., Boorman, W. R., 1923. Northwestern lakes of the United States: biological and chemical studies with reference to possibilities in production of fish. *Bull. U.S. Bur. Fish.*, 39:51–140. **168, 440, 821**

Kendall, P. F., 1902. A system of glacier lakes in the Cleveland Hills. *Quart. J. geol. Soc. Lond.*, 58:471–571. **56**

Keulegan, G. H., 1951. Wind tides in small closed channels. *J. Res. nat. Bur. Stand.*, 46:358–381. **273, 275, 276, 278, 282, 351**

Kikuchi, K., 1937. Studies on the vertical distribution of the plankton crustacea. *Rec. Oceanogr. Wks. Jap.*, 9:61–85. **403**

Kimball, H. H., 1924. Records of total solar radiation intensity and their relation to daylight intensity. *Mon. Weath. Rev., Wash.*, 52:473–479. **370**

King, E. J., and Davidson, V., 1933. The biochemistry of silicic acid. IV. The relation of silica to the growth of phytoplankton. *Biochem. J.*, 27:1015–1021. **791**

King, L. C., 1942. *South African Scenery*. London, Oliver and Boyd, xxiv, 340 pp. **46**

Kirshenbaum, I., 1951. *Physical Properties and Analysis of Heavy Water.* New York, Toronto, London, McGraw-Hill Book Co., xv, 438 pp. **211, 213**

Kirshenbaum, I., Graff, J., and Forstat, H., 1951, *see* Kirshenbaum, 1951.

Klausener, Carl, 1908. Die Blutseen der Hochalpen. Eine biologische Studie auf hydrographischer Grundlage. *Int. Rev. Hydrobiol.*, 1:359–424. **417**

* Klein, G., and Steiner, M., 1929. Bakteriologisch-chemische Untersuchungen am Lunzer-Untersee. I. Die bakteriellen Grundlagen des Stickstoff- und Schwefelumsatzes im See. *Öst. botn. Z.*, 78:289–324. **846, 870**

Kluyver, A. J., and Van Reenen, W. J., 1933. Über *Azotobacter agilis* Beijerinck. *Arch. Mikrobiol.*, 4:280–300. **846**

Knauthe, K., 1899. Beobachtungen über den Gasgehalt der Gewässer im Winter. *Biol. Zbl.*, 19:783–799. **632**

Knipowitsch, N., 1922. Hydrobiologische Untersuchungen im Kaspischen Meere in den Jahren 1914–1915. *Int. Rev. Hydrobiol.*, 10:394–440, 561–602. **313**

Kobayashi, S., 1954. Distribution of flourine in various states in terrestrial waters. *Bull. chem. Soc. Japan*, 27:314–317. **562**

Köhler, H., 1937. Studien über Nebelfrost und Schneebildung und über den Chlorgehalt des Nebelfrostes, des Schnees und das Seewassers im Halddegebiet. *Bull. geol. Inst. Univ. Upsala*, 26:279–308. **542, 543**

Koenen, A. von, 1913. Die Entstehung einer Insel im Seeburger See. *Jb. preuss. geol. Landesanst.*, 32, T1.II:485–486. **175**

Kohler, M. A., 1952. Lake and pan evaporation. Water-loss investigations: Vol. 1, Lake Hefner studies Tech. Rep., *U.S. geol. Surv. Circ.*, 229:127–148. **522**

Kolden, K., and Strøm, K. M., 1939. Bathygraphy of Fyresvatn. *Norsk geogr. Tidssbr.*, 7:228–232. **71**

Kolesnikov, V. P., 1950. *Paleontologiia SSSR*. Tom. X, Part III.12, Akchagilskie i Apsheronskie Molliuski. Moscow, Izdatelstvo Akadem ii Nauk, 259 pp. **5**

Kolthoff, I. M., 1931. The solubilities and solubility products of metallic sulfides in water. *J. phys. Chem.*, 35:2711–2721. **758** fn., **762**

Kolthoff, I. M., and Stenger, V. A., 1942. *Volumetric Analysis:* Vol. I, *Theoretical Fundamentals*, 2nd. ed New York, Interscience Publishers, xv, 309 pp. **661**

Kosack, H. P., 1952. Die Verbreitung der Karst- und Pseudokarsterscheinungen über die Erde. *Petermanns. Mitt.*, 6:16–21. **109**

Kovalevsky, S. A., 1933. *The Face of the Caspian Sea* (*Paleogeography of the Caspian during the Quaternary*). Azneft, Baku, USSR, 129 pp., 22 maps and diagrams (not seen: abst. in Zavoico 1935). **6, 7**

Koyama, T., 1955. Gaseous metabolism in lake muds and paddy soils. *J. Earth Sci. Nagoya Univ.* 3:65–76. **585** fn.

Koyama, T., and Sugawara, K., 1953a. Sulphur metabolism in bottom muds and related problems. *J. Earth Sci. Nagoya Univ.*, 1:24–34. **756**

Koyama, T., and Sugawara, K., 1953b. Separation of the components of atmospheric salt and their distribution (continued). *Bull. chem. Soc. Japan*, 26:123–126. **548**

Koźmiński, Z., and Wisniewski, J., 1935. Über die Vorfrühlingthermik der Wigry-Seen. *Arch. Hydrobiol.*, 28:198–235. **456**

Kranz, W., 1934. Fünfte Fortsetzung der Beiträge zum nördlinger Ries Problem. *Zbl. Miner. Geol. Paläont.*, 1934B:262–271. **151**

Kraus, E., Meyer, R., and Wegener, A., 1928. Untersuchungen über den Krater von Sall auf Ösel. *Beitr. Geophys.*, 20:312–378. **149**

Krauskopf, K. B., 1956. Dissolution and precipitation of silica at low temperatures. *Geochim. et Cosmochim. Acta*, 10:1–26. **789**

Krecker, F. H., 1928. Periodic oscillations in Lake Erie. *Contr. Stone Lab. Ohio Univ.*, 1, 22 pp. **318**

Krecker, F. H., 1931. Vertical oscillations or seiches in lakes as a factor in the aquatic environment. *Ecology*, 12:156–163. **139, 318, 329**

Kreutz, W., 1941. Kohlensäuregehalt der unteren Luftschichten in Abhändigkeit von Witterungsfaktoren. *Angew. Bot.*, 23:89–117. **655**

Krogh, A., 1904. On the tension of carbonic acid in natural waters and especially in the sea. *Medd. Grønland*, 26:331–405. **668**

Krogh, A., 1910. Some experiments on the invasion of oxygen and carbonic oxide into water. *Skand. Arch. Physiol.*, 23:224–235. **587**

Krogh, A., 1934. Conditions of life at great depths in the ocean. *Ecol. Monogr.*, 4:430–439. **887**

Krogh, A., and Berg, K., 1931. Über die chemische Zusammensetzung des Phytoplanktons aus dem Frederiksborg-Schlosssee und ihre Bedeutung für die Maxima der Cladoceren. *Int. Rev. Hydrobiol.*, 25:204–218. **678**

Krogh, A., and Lange, E., 1932. Quantitative Untersuchungen über Plankton, Kolloide und gelöste organische und anorganische Substanzen in dem Füresee. *Int. Rev. Hydrobiol.*, 26:20–53. **579, 615, 668, 678, 879**

Kuchař, K., 1947. Chomutovské kamencové jezero. Le lac d'alum de Chomutov. *Sborn. čsl. Státní Úst. Hydrol.*, 1947:22–28 (not seen: ref. *Bibliogr. Geol. Excl. N. Amer.*, 14:142). **148**

Kuenen, P. H., 1941. Geochemical calculations concerning the total mass of sediments in the earth. *Amer. J. Sci.*, 239:161–190. **554, 555**

Kufferath, J., 1952. Le milieu biochimique. *Exploration hydrobiologique du lac Tanganika* (1946–1947), *Inst. roy. Sci. nat. Belg. Bruxelles*, 1:31–47. **462, 465**

Kuiper, G. P., 1952. Planetary atmosphere and their origin. In *The Atmospheres of the Earth and Planets*, 2nd ed., ed. G. P. Kuiper. Chicago, Univ. Chicago Press, Chap. 12, pp. 306–405. **222** fn., **223**

Kulp, L. J., 1951. Origin of the hydrosphere. *Bull. goel. Soc. Amer.*, 62:326–330. **222**

Kusnetzow, S. I., 1935. Microbiological researches in the study of the oxygenous regimen of lakes. *Verh. int. Limn. Ver.*, 7:562–582. **693, 695, 699**

Kusnetzow, S. I., and Karsinkin, G. S., 1931. Direct method for the quantitative study of bacteria in water, and some considerations on the causes which produce a zone of oxygen-minimum in Lake Glubokoye. *Zbl. Bakt.*, II, 83:169–174. **614**

Lahn, E., 1945. Contribution à l'étude géomorphologique des lacs du Toros occidental (Bati Toros göllerinin jeomorfolojisi). *Meden Tetkik ve Arama*, Ankara, ser. 10:387–400 (not seen: ref. *Bibliogr. Geol. Excl. N. Amer.*, 12:134). **19**

* Lambert, J. M., 1953. The past, present and future of the Norfolk Broads. *Trans. Norfolk Norw. Nat. Soc.*, 17:223–258. **121**

Langbein, W. B., 1951. Research on evaporation from lakes and reservoirs. *Gen. Ass. int. Un. Geod.*, Bruxelles, *Ass. int. Hydrol.*, 3:409–425. **525**

Langmuir, I., 1938. Surface motion of water induced by wind. *Science*, 87:119–123. **275, 279, 452, 454, 476**

Langmuir, I., 1948. The production of rain by a chain reaction in cumulus clouds at temperatures above freezing. *J. Meteorol.*, 5:175–192. **543**

Lantzsch, C., 1925. Der Boden der Wielenbacher Teiche (Kolloidgehalt, Bakterientätigheit). *Arch. Hydrobiol.*, 15:406–411. **846**

* Lastochkin, D., 1945. Achievements in Soviet hydrobiology of continental waters. *Ecology*, 26:320–331. **732**

La Touche, T. D., 1910. Lakes of the Salt Range in the Punjab. *Rec. geol. Surv. India,*. 40:36–51. **773**

La Touche, T. H. D., 1912. The geology of the Lonar Lake, with a note on the Lonar soda deposit by W. A. K. Christie. *Rec. geol. Surv. India*, 41:266–285. **37**

Lauscher, F., 1941. Lichtmessungen in einigen Hochgebirgsgewässern der Ostalpen. *Arch. Hydrobiol.*, 37:583–597. **391**

* Lauscher, F., 1955. Sonnen und Himmelsstrahlung im Meer und in Gewässern. In *Handbuch der Geophysik*. Berlin, Bomtraeger, Bd 8, Lf4:723–768. See Preface.

Lawson, A. C., 1904. The geomorphogeny of the upper Kern Basin. *Bull. Dept. Geol. Univ. Calif.*, 3:291–376. **43**

Leakey, L. B. S., 1951. *Olduvai Gorge*, with chapters on the geology and fauna by Hans Reck and A. T. Hopwood. Cambridge Univ. Press, xvi, 163 pp. **133**

Le Conte, J., 1883. Physical studies of Lake Tahoe. *Overland Monthly*, ser. 2, 2:506–516, 595–612. **168, 403**

Leeflang, K. W. A., 1938. De chemische Samenstellung van den Neerslag in Nederland. *Chem. Weekbl.*, 35:658–664. **550**

Lees, H., 1948. The effects of zinc and copper on soil nitrification. *Biochem. J.*, 42:534–538. **866, 867**

Lees, H., and Quastel, J. H., 1946a. Biochemistry of nitrification in soil. 2. The site of soil nitrification. *Biochem. J.*, 40:815–823. **866, 867**

Lees, H., and Quastel, J. H., 1946b. Biochemistry of nitrification in soil. 3. Nitrification of various organic nitrogen compounds. *Biochem. J.*, 40:824–828. **866, 867**

Lehmann, K. B., 1931. Aluminiumbestimmung und Aluminiumgehalt von Nährungsmitteln. *Arch. Hyg., Berl.*, 106:309–335. **821**

Leighly, J., 1942. Effects of the Great Lakes on the annual march of air temperatures in their vicinity. *Pap. Mich. Acad. Sci.*, 27:377–414. **527**

Leslie, J., 1838 (*fide* Chumley; printing consulted dated 1842). Climate. In Encyclopaedia Britannica, Edinburgh, 7th ed., Vol. 6, pp. 743–764. **426**

Leuchs, K., 1937. *Geologie von Asiens:* Bd. 1. T.2, *Zentralasien.* In *Geologie der Erde*, ed. E. Krenkel. Berlin, Bomtraeger, 317 pp. **19**

Leutelt-Kipke, S., 1934. Ein Beitrag zur Kenntnis der hydrographischen und hydrochemischen Verhältnisse einiger Tiroler Hoch- und Mittelgebirgsseen. *Arch. Hydrobiol.*, 27:286–352. **453, 454**

Leutelt-Kipke, S., 1935. Hydrographische und hydrochemische Untersuchungen an Hochgebirgsseen des Bulgarischen Rilo Dag. *Arch. Hydrobiol.*, 38:415–436. **441**

Leverett, F., and Taylor, F. B., 1915. *Pleistocene of Indiana and Michigan.* U.S. Geological Survey, liii, 523 pp. **82**

Levring, T., 1946. Some culture experiments with Ulva and artificial sea water. *K. fysiogr. Sällsk. Lund Förh.*, 16:45–56. **806**

Lewin, J. C., 1954. Silicon metabolism in diatoms. I. Evidence for the role of reduced sulfur compounds in silicon utilization. *J. gen. Physiol.*, 37:589–599. **798, 826**

Lewin, J. C., 1955a. Silicon metabolism in diatoms. II. Sources of silicon for growth of *Navicula pelliculosa.* *Plant Physiol.*, 30:129–134. **798**

Lewin, J. C., 1955b. The capsule of the diatom *Navicula pelliculosa.* *J. gen. Microbiol.*, 13:162–169. **798**

Lewin, J. C., 1955c. Silicon metabolism in diatoms. III. Respiration and silicon uptake in *Navicula pelliculosa.* *J. gen. Physiol.*, 39:1–10. **798**

Lewin, R. A., 1954. A marine *Stichococcus* which requires vitamin B_{12} (cobalamin). *J. gen. Microbiol.*, 10:93–96. **826**

Lewis, A. N., 1933. Note on the origin of the Great Lake and other lakes on the Central Plateau. *Pap. roy. Soc. Tasm.*, 1932:15–38. **65**

Lewis, G. N., 1933. The biochemistry of water containing hydrogen isotope. *J. Amer. chem. Soc.*, 55:3503–3504. **217**

Lewis, G. N., and Randall, M., 1923. *Thermodynamics and the Free Energy of Chemical Substances.* New York and London, McGraw-Hill Book Co., 653 pp. **836**

Libby, W. F., 1946. Atmospheric helium and radiocarbon from cosmic radiation. *Phys. Rev.*, 69:671–672. **212**

Liebmann, Hans, 1938. Biologie und Chemismus der Bleilochsperre. Zugleich ein Beitrag zur Frage der Wirkung von Abwässern aus Sulfitzellulose-fabriken auf stehende Gewässer. *Arch. Hydrobiol.*, 33:1–81. **782**

Liljequist, G., 1941. Winter temperatures and ice conditions on Lake Vetter with special regard to the 1939/40. *Geogr. Ann.*, 23:24–52. **452, 454**

Lind, E., 1945. The phytoplankton of some Cheshire meers. *Mem. Manchr. lit. phil. Soc.*, 86:83–105. **730**

Lindberg, A. M., and Lindberg, C. A., 1934. Flying around the North Atlantic. *Nat. geogr. Mag.*, 66:259–337. **55**

Lindstrom, E. S., Lewis, S. M., and Pinsky, J. M., 1951. Nitrogen fixation and hydrogenase in various bacterial species. *J. Bact.*, 61:481–487. **848**

Lipp, H., 1929. *Met. Jahrb. Bayern*, 1928 (not seen: ref. Sauberer and Ruttner 1941). **370**

Litynaki, A., 1926. Limnological studies of Lake Wigri. *Arch. Hydrob. Rybact.*, 1:1–78. **622**

Livingstone, D. A., 1954. On the orientation of lake basins. *Amer. J. Sci.*, 252:547–554 **100, 129, 179, 285**

Livingstone, D. A., 1957. In Deevey, E. S. (ed.), *Biostratonomy of Arctic Lakes.* Final Rep. to the Office of Naval Research, Contract Nonr-340(00), mimeogr. **495, 502, 508**

Lönnerblad, G., 1930. Über die Sauerstoffabsorption des Bodensubstrats in einigen Seentypen. *Bot. Notiser*, 1930:53–60. **609**

Lönnerblad, G., 1931. Über den Sauerstoffhaushalt der dystrophen Seen. *Acta. Univ. Lund.*, NF Avd. 2.27, No. 14, 53 pp. **598, 617, 629**

Löwl, F., 1888. Der Lüner See. *Z. dtsch. öst. Alpenver.*, 19:25–34. **104**

* Lohammar, G., 1938. Wasserchemie und höhere Vegetation schwedischer Seen. *Symb. Bot. Upsaliens.*, 3:1–252. **554, 555, 560, 684, 706, 729, 804**

Lohuis, D., Meloche, V. W., and Juday, C., 1938. Sodium and potassium content of Wisconsin lake waters and their residues. *Trans. Wis. Acad. Sci. Arts Lett.*, 31:285–304. **552, 554, 555, 559**

Lombardini, E., 1845. Della natura dei laghi, e delle opere intese a regolarne l'effluso. *Mem. Ist. Lombardo*, 2:393–528. **234** fn.

Longuet-Higgins, M. S., 1952, *see* Mortimer 1952a.

Longuet-Higgins, M. S., 1953. Mass transport in water waves. *Phil. Trans.*, ser. A, 245:535–581. **348**

Lorenz, J. R., 1859. Der Vrana-See auf Cherso. *Petermanns Mitt.*, 1859:510–512. **106**

Lovering, T. S., 1927. Organic precipitation of metallic copper. *Bull. U.S. geol. Surv.*, 795C:45–52. **820**

Low, E. M., and Hutchinson, G. E. (in press). Survey of contemporary knowledge of biogeochemistry. V. The biogeochemistry of bromine and iodine. *Bull. Amer. Mus. nat. Hist.* **546, 563** fn.

Lubbock, G., 1894. The Gohna Lake. *Geogr. J.*, 4:457. **43**

Lucas, K., 1904. A bathymetric survey of the lakes of New Zealand. *Geogr. J.*, 23:645–660, 744–760. **69**

Ludlow, F., 1929. The Shyok Dam in 1928. *Himalayan J.*, 1:4–10. **51**

Ludwig, C. A., 1938. The availability of different forms of nitrogen to a green alga. *Amer. J. Bot.*, 25:448–458. **851**

Lüning, O., and Brohm, K., 1933. Über des Vorkommen von Phosphorwasserstoff in Brunnenwässern. *Z. Untersuch. Lebensmitt.*, 66:460. **727**

Lütschg, O., 1915. Der Märjelensee und seine Abflussverhältnisse. *Ann. schweiz. Landeshydrographie*, I (not seen: ref. Collet 1925). **51**

Lugeon, M., and Jérémine, E., 1911. Les bassins fermés des Alpes Suisses. *Bull. Soc. vaud. Sci. nat.*, 47:461–650. **104**

Lund, J. W. G., 1949. Studies on *Asterionella formosa.* I. The origin and nature of the cells producing seasonal maxima. *J. Ecol.*, 37:389–419. **734**

* Lund, J. W. G., 1954. The seasonal cycle of the plankton diatom, *Melosira italica* (Ehr.) Kütz, subsp. *subarctica* O. Müll. *J. Ecol.*, 42:151–179. **734, 797**

Lunelund, H., 1924. Über die Wärme und Lichtstrahlung in Finland. *Comment. Phys. Math., Helsingf.*, 2, No. 11, 147 pp. **375**

Lyall-Grant, I. H., and Mason, K., 1940. The upper Shyok Glaciers, 1939. *Himalayan J.*, 12:52–63. **51**

Lyell, C., 1837. *Principles of Geology*, 1st American ed. (from 5th English ed.). Philadelphia and Pittsburgh, Kay, Vol. I, 546 pp.; Vol. II, 532 pp. **12, 53**

MacCarthy, G. E., 1937. The Carolina bays. *Bull. geol. Soc. Amer.*, 48:1211–1226. **153**

Mackereth, F. J. H., and Heron, J., 1954. In *Rep. Freshw. biol. Ass. Brit. Emp.*, 22:20–21. **561, 562**

MacIntire, W. H., and Young, J. B., 1923. Sulphur, calcium, magnesium and potassium content and reaction of rainfall at different points in Tennessee. *Soil Sci.*, 15:205–227. **548**

Maclaren, M., 1931. Lake Bosumtwi, Ashanti. *Geogr. J.*, 78:270. **37**

Magrini, G. P., 1903. I recenti studi sulle sesse e le sesse nei laghi italiani. *Riv. geogr. ital.*, 12:127–135, 216–225, 291, 299. **314**

Makkaweev, W. M., 1936. Bemerkungen über die Theorie der langen internen Wellen. *V. Hydrologische Konferenz der Baltischen Staaten, Helsinki*, June 1936. *Ber.* 15C, pp. 13–19 (not seen: ref. in Mortimer 1952). **341**

Maliuga, D. P., 1946. Sur la géochimie du nickel et du cobalt disperse dans la biosphère. *Trav. Lab. Biogéochim. URSS*, 8:73–143. **825**

Mann, P. J. G., and Quastel, J. H., 1946. Manganese metabolism in soils. *Nature, Lond.*, 158:154. **802**

Marchetti, E. A. A., 1949. The Nahuel Huapi Lake. *Gen. Ass. int. Un. Geol.*, Oslo, 1948, *Ass. int. Hydrol. Sci., I. Trav. Comm. Potamol. Limnol.*, 379. **170**

de Marchi, G., 1949. Caractéristiques hydrologiques et hydrauliques des grands lacs subalpins. *Gen. Ass. int. Un. Geol.*, Oslo, 1948, *Ass. int. Hydrol. Sci., I. Trav. Comm. Potamol. Limnol.*, 327–331. **235**

Marciano, J. J., and Harbeck, G. E., 1952. Mass-transfer studies. Water-loss investigations: Vol. I, Lake Hefner Studies. *Tech. Rep. U.S. geol. Surv.*, 229:46–70. **514**

Margalef, R., 1951. Rôle des entomostracés dans la régeneration des phosphates. *Verh. int. Ver. Linmol.*, 11:246–247. **735**

Marquardsen, H., 1916. Die Seen Tanganjika, Moero, Bangweolo. *Mitt. dtsch. Schütgeb.*, 29:97–98. **465**

Marr, J. E., 1895. The tarns of Lakeland. *Geol. Mag.*, Dec. IV, 2:44. **60, 61**

Marr, J. E., 1897. The origin of lakes. *Sci. Progr. Lond.*, n.s 1:218 (not seen: ref. Chumley 1910.) **76**

Marr, J. E., 1916. *The Geology of the Lake District and the Scenery as Influenced by Geological Structure.* Cambridge Univ. Press, xii, 220 pp. **47, 61, 73, 76**

Marr, J. E., and Adie, R. H., 1898. The lakes of Snowdon. *Geol. Mag.*, Dec. IV, 5:51–61. **61**

Martel, E. A., 1892. Les katavothres du Péloponèse. *Rev. Géogr.*, 30:241–251, 336–346. **109**

Martin, N. V., 1955. Limnological and biological observations in the region of the Ungava or Chubb Crater, Province of Quebec. *J. Fish. Res. Bd. Can.*, 12:487–496. **149, 403**

Martineau, Harriet, 1855. *A Complete Guide to the English Lakes.* Windermere, J. Garnett; London, Whittaker. 1st ed. not seen; 2nd ed. ? 1859, front matter, 210 pp. and advert. **76**

Martonne, E. de, and Aufrère, L., 1928. L'extension des régions privées d'écoulement vers l'océan. *Publ. Un. Géogr. int.*, 3, Paris, 194 pp. **226**

Mason, K., 1929. Indus floods and Shyok glaciers. *Himalyan J.*, 1:10–29. **42** fn., **51**

Mason, M. A., 1952. Some observations of breaking waves. In: Gravity waves. *Circ. U.S. Bur. Stand.*, 512:215–220. **358**

Massachusetts State Board of Health, 1892. On the amount of dissolved oxygen contained in waters of ponds and reservoirs at different depths (correctly attributable to T. M. Drown). *Rep. Mass. Bd publ. Hlth*, 23:373–381. **716**

Mast, S. O., and Pace, D. M., 1937. The effect of silicon on growth and respiration in *Chilomonas paramecium. J. cell. comp. Physiol.*, 10:1–13. **798**

Matsue, Y., Egusa, S., and Saeki, A., 1953. Nitrogen gas contents in flowing water of artesian wells and springs. In relation to high supersaturation inducing the so-called gas disease of fish. *Bull. Japan Soc. sci. Fish.*, 19:439–444. **842** fn.

Mattson, S., 1933. The laws of soil colloidal behaviour. XI. Electrodialysis in relation to soil processes. *Soil Sci.*, 36:149–163. **789**

Mattson, S., 1941. The laws of soil colloidal behavior. XXIII. The constitution of the pedosphere and soil classification. *Soil Sci.*, 51:407–425. **791**

Maucha, R., 1932. *Hydrochemische Methoden in der Limnologie. Die Binnengewässer*, 12. Stuttgart, Schweizerbartsche Verlagsbuchhandlung, 173 pp. **557, 558**

Maurer, J., 1913. Über die Grösse der jährlichen Verdunstung auf Schweizer Seen am nord alpinen Fuss. *Met. Z.*, 30:209–213. **524**

Maurer, J., 1924. Die strengen Winter Suddeutschlands und der Schweiz, bewertet nach den grossen Seegefrörnen seit 1400. *Met. Z.*, 41:85–86. (See also trans. by W. W. Reed, Severe winters in southern Germany and Switzerland since the year 1400 determined from severe lake freezes. *Mon. Weath. Rev., Wash.*, 52:222.) **531**

Mawson, Sir D., 1929. South Australian algal limestones in the process of formation. *Quart. J. geol. Soc. Lond.*, 85:613–623. **137**

Mawson, Sir D., 1950. Occurrence of water in Lake Eyre, South Australia. *Nature, Lond.*, 166:667–668. **136, 228**

Mazur, A., and Clarke, H. T., 1938. The amino acids of certain marine algae. *J. biol. Chem.*, 118:631–634. **896**

Mazur, A., and Clarke, H. T., 1942. Chemical components of some autotrophic organisms. *J. biol. Chem.*, 143:39–42. **896**

* McCarter, J. A., Hayes, F. E., Jodney, L. H., and Cameron, M. L., 1952. Movement of materials in the hypolimnion of a lake as studied by the addition of radioactive phosphorus. *Can. J. Zool.*, 30:128–133. **341, 742**

McClendon, J. F., and Hathaway, J. C., 1924. Inverse relation between iodine in food and drink and goiter, simple and exophthalmic. *J. Amer. med. Ass.*, 82:1668–1672. **563**

* McEwen, G F., 1929. A mathematical theory of the vertical distribution of temperature and salinity in water under the action of radiation, conduction, evaporation, and mixing due to the resultant convection. *Bull. Scripps Instn Oceanogr. tech.*, 2:197–306. **457, 466, 469** fn.

McGee, W. J., 1893. A fossile earthquake. *Bull. geol. Soc. Amer.*, 4:411–415. **12**

McNown, J. S., 1952. Waves and seiche in idealized ports. In: Gravity waves. *Circ. U.S. Bur. Stand.*, 521:153–164. **308**

Meen, V. B., 1950. Chubb Crater, Ungava, Quebec. *Bull. geol. Soc. Amer.*, 61:1485. **149**

Meen, V. B., 1952. Solving the riddle of Chubb Crater. *Nat. geogr. Mag.*, 101:1–32. **149**

Meinertzhagen, R., 1927. Ladakh, with special reference to its natural history. *Geogr. J.*, 70:129–163. **142, 183**

Meinhardus, W., 1928. *Das Kreislauf des Wassers.* Göttingen. (Not seen: ref. Kalle 1945.) **223**

Meinzer, O. E., 1922. Map of the Pleistocene lakes of the Basin-and-Range Province and its significance. *Bull. geol. Soc. Amer.*, 33:541–552. **17**

Mellor, J. W., 1922. *Higher Mathematics for Students of Physics and Chemistry*, 4th ed. London, Longmans, Green, xv, 641 pp. **612**

Meloche, V. M., Leader, G., Safranski, L., and Juday, C., 1938. The silica and diatom content of Lake Mendota water. *Trans. Wis. Acad. Sci Arts Lett.*, 31:363–376. **794**

Melton, F. A., and Schriever, W., 1933. The Carolina bays—are they meteorite scars? *J. Geol.*, 41:52–66. **153**

Mercanton, P. L., 1932. Étude de la circulation des eaux du lac Léman. *Mém. Soc. vaud. Sci. nat.*, 4:225–241. **294**

Merian, J. R., 1828. *Über die Bewegung tropfbarer Flüssigkeiten in Gefässen.* Basle. (Not seen: ref. Forel 1876.) **301**

Merker, E., 1931. Die Fluoreszens und die Lichtdurchlässigkeit der bewohnten Gewässer. *Zool. Jb.*, Abt. Allgem. Zool. Physiol. Tiere, 49:69–104. **900**

Merrill, G. P., 1928. The Siberian meteorite. *Science*, 67:489–490. **149**

Middlemiss, C. S., 1890. Geological sketch of Naini-Tal. *Rec. geol. Surv. India*, 23:213–234. **109**

Mill, R. H., 1895. Bathymetrical survey of the English lakes. *Geogr. J.*, 6:46–73, 135–166. **185**

Millikan, R. A., 1926. High-frequency rays of cosmic origin. *Proc. nat. Acad. Sci., Wash.*, 12:48–55. **422**

Miller, N. H. J., 1914. The composition of rain-water collected in the Hebrides and in Iceland. With special reference to the amount of nitrogen as ammonia and as nitrates. *J. Scot. met. Soc.*, 16:141–158. **544, 546, 550**

Mills, E. A., 1913. *In Beaver World.* Boston and New York, Houghton Mifflin, ix, 228 pp. **146**

Minder, L., 1926. Biologische-chemische Untersuchungen im Zürichsee. *Z. Hydrol.*, 3:1–70. **595**

Minder, L., 1929. Chemische Untersuchungen am Stausee Waggital. *Verh. int. Ver. Limnol.*, 4:454–461. **721**

Minder, Leo, 1941. Über die Löslichkeit des Sauerstoffs in Gebirgsgewässern. Rechenhilfer für limnologische Untersuchungen. *Vjschf. naturf. Ges. Zürich*, 86:157–183. **576, 577, 579**

Minnaert, M., 1954. *The Nature of Light and Colour in the Open air.* Trans. H. M. Kremer-Priest, revised by K. E. Brian Jay. New York, Dover, 362 pp. **419, 421, 727** fn.

* Mitchell, G. F., 1951. Studies in Irish quaternary deposits, No. 7. *Proc. R. Irish Acad.*, 53B:111–206. **96**

Miyake, Y., 1944. The possibility and the available limit of the formation of air-bubbles in sea water. *Umi to Sora* (Sea and Sky), 24:121–134 (Jap., not seen: ref. *Biol. Abstr.*, 24:3311). **581** fn.

Miyake, Y., and Abe, T., 1948. A study on the foaming of sea water. *J. Mar. Res.*, 7:67–73. **899**

Möller, Lotte, 1928. Hydrographische Arbeiten am Sakrower See bie Potsdam. *Z. ges. Erdk. Berl.*, Sonderband 1828–1928:535–551. **287**

Möller, Lotte, 1933. Der Sakrower See bei Potsdam. Ein Beitrag zur Frage der Wirkung der Windzirkulation auf die physikalische und chemische Schichtung in einem See. *Verh. int. Ver. Limnol.*, 6:201–216. **287**

Möller, Lotte, 1955. Geographische Verteilung der Konzentration gelöster Substanzen von Grund- und Oberflächengewässern Sudwestdeutschlands in limnologisches Sicht. *Verh. int. Ver. Limnol.*, 12:351–359. **559**

Moon, H. P., 1939. The geology and physiography of the Altiplano of Peru and Bolivia. *Tr. Limn. Soc. Lond.*, ser. III, 1:27–43. **14**

Moore, W. G., 1942. Field studies of the oxygen requirements of certain freshwater fishes. *Ecology*, 23:319–329. **599**

Morgan, J., and Warren, B. E., 1938. X-ray analysis of the structure of water. *J. chem. Phys.*, 6:666–673. **199, 200**

Morgan, L., 1868. *The American Beaver and His Workers.* Philadelphia, J. B. Lippincott, xv, 330 pp. **146, 147**

Morandini, G., 1940. *Missione di Studio al Lago Tana:* 3. *Ricerche limnologiche*, parte prima, Geografia-Fisica. Rome, R. accademia d'Italia, 315 pp. **522**

Morandini, G., 1949. Richerche limnologiche sugli alti laghi alpini della Venezia tridentina (secondo contributo), Sella, Lagorai, Val delle Pozze. *Boll. Pesca, Piscic. Idrobiol.*; n.s. 4:75–115. **529**

Morita, Y., 1955. Distribution of copper and zinc in various phases of the earth's materials. *J. Earth Sci., Nagoya Univ.*, 3:33–57. **820**

Mortimer, C. H., 1939. The work of the Freshwater Biological Assoc. of Great Britain in regard to water supplies. The nitrogen balance of large bodies of water. *Off. Circ. Brit. Waterw. Ass.*, 21:1–10. **844**

* Mortimer, C. H., 1941–42. The exchange of dissolved substances between mud and water in lakes. *J. Ecol.*, 29:280–329; 30:147–201. **336, 466, 648, 677, 694, 695, 706, 719, 721, 766, 792, 857, 870**

Mortimer, C. H., 1951. The use of models in the study of water movements in lakes. *Verh. int. Ver. Limnol.*, 11:254–260. **285, 336**

* Mortimer, C. H., 1952a. Water movements in lakes during summer stratification; evidence from the distribution of temperature in Windermere, with an appendix by M. S. Longuet-Higgins. *Phil. Trans.*, ser. B., 236:355–404. **250, 336, 341, 449**

Mortimer, C. H., 1952b. Water movements in stratified lakes, deduced from observations in Windermere and model experiments. *Gen. Ass. int. Un. Geod. phys.*, Bruxelles, *Ass. int. Hydrol. Sci.*, *C.R. Rapp*, 3:335–349. **285, 336**

* Mortimer, C. H., 1953a. The resonant response of stratified lakes to wind. *Schweiz. Z. Hydrol.*, 15:94–151. **250, 287, 336, 337, 338**

Mortimer, C. H., 1953b. A review of temperature measurement in limnology. *Mitt. int. Ver. Limnol.*, 1, 25 pp. **427**

Mortimer, C. H., 1955a. Some effects of the earth's rotation on water movement in stratified lakes. *Verh. int. Ver. Limnol.*, 12:66–77. **339**

Mortimer, C. H., 1955b. Saturation of atmospheric oxygen in fresh waters. 7 pp. mimeogr. advance copy, to appear in *Mitt. int. Ver. Limnol.* **579, 581**

Mortimer, C. H., and Moore, W. H., 1953. The use of thermistors for the measurement of lake temperatures. *Mitt. int. Ver. Limnol.*, 2, 42 pp. **427**

Müller, H., 1932. Die Verwendung von α–α′-Dipyridyl zur Bestimmung von Ferro- und Gesamteisen in natürlichen Wässern. *Mikrochemie*, 12:307–314. **706, 707, 717**

Müller, H., 1934. Über das Auftreten von Nitrit in einigen Seen der österreichischen Alpen. *Int. Rev. Hydrobiol.*, 30:428–439. **862**

Müller, H., 1938a. Über die Auswirkungen des Schneedruckes auf die Schwingrasen und die biochemische Schichtung des Lunzer Obersees. *Int. Rev. Hydrobiol.*, 36:362–370. **706**

Müller, H., 1938b. Beiträge zur Frage der biochemischen Schichtung im Lunzer Ober- und Untersee. *Int. Rev. Hydrobiol.*, 36:433–500. **695, 706**

Mulley, G., and Wittmann, J., 1914. Analysen des Wassers der Lunzer Seen. *Int. Rev. Hydrobiol.*, Hydrogr. Supp. 3 (to Bd 5.), No. 4, 11 pp. **821**

* Munk, W. H., and Anderson, E. R., 1948. Notes on a theory of the thermocline. *J. Mar. Res.*, 7:276–295. **256, 275, 430, 443**

Murphey, B. F., 1941. Relative abundance of the oxygen isotopes. *Phys. Rev.*, 59:320. **213**

Murray, James, 1909. Physics, Additional notes. In Shackleton 1909, Appendix III, pp. 339-343. **462**

Murray, John, 1888. On the effect of winds on the distribution of temperature in the sea-and-freshwater lochs of the west of Scotland. *Scot. geogr. Mag.*, 4:345–365. **229, 282**

* Murray, Sir John, 1910. The characteristics of lakes in general, and their distribution over the surface of the globe. In Murray, J., and Pullar, L., *Bathymetric Survey of the Scottish Fresh-Water Lochs*. Edinburgh, Challenger Office, Vol. I, 514–658. **3, 37** fn., **143**

Murray, John, and Irving, R., 1891. On the silica and the siliceous remains of organisms in modern seas. *Proc. roy. Soc. Edinb.*, 18:229–250. **790**

* Murray, John, and Pullar, L., 1910. *Bathymetric Survey of the Scottish Fresh-Water Lochs.* Vol. I, lviii, 785 pp. Vol. II, lviii, pt. 1, 435 pp.; pt. 2, 281 pp. Vol. III, lviii. Pl. I–LI. Vol. IV, Pl. LII–CV. Vol. V, Pl. I–LXVII. Vol. VI, Pl. LXVIII–CXXXIV, Edinburgh, Challenger Office.

Myers, J., 1949. The pattern of photosynthesis in Chlorella. In *Photosynthesis in Plants,* ed T. Franck and W. E. Loomis. Ames, Iowa State Coll. Press, pp. 349–364. **678, 851**

Nagéra, J. J., 1926. Los hoyos del Campo del Ciele y el meteorito. *Publ. Direcc. Min., B. Aires,* 19, 9 pp. **149**

* Nakamura, S., and Honda, K., 1911. Seiches in some lakes of Japan. *J. Coll. Sci. Tokyo,* 28, No. 5, 95 pp. **306, 312, 314, 315, 324**

Naumann, E., 1922. Die Sestonfärbung des Süsswassers. *Arch. Hydrobiol.,* 13:647–692. **417, 418**

Naumann, E., 1930. *Einführung in die Bodenkunde der Seen. Die Binnengewässer, 9.* Stuttgart, Schweizerbartsche Verlagsbuchhandlung, 126 pp. **723**

Nelson, K. H., and Thompson, T. G., 1954. Deposition of salts from sea water by frigid concentration. *J. Mar. Res.,* 13:166–182. **570**

Neuberg, C., and Mandl, I., 1948. An unknown effect of amino acids. *Arch. Biochem.,* 19:149–161. **764** fn.

Neumann, G., 1952. On the complex nature of ocean waves and the growth of the sea under the action of wind. In: Gravity waves. *Circ. U.S. Bur. Stand.,* 521:61–68. **354**

* Neumann, J., 1953. Energy balance and evaporation from sweet-water lakes of the Jordan Rift. *Bull. Res. Coun. Israel,* 2:337–357. **378, 381, 495, 505, 507, 513, 514**

Neumann, J., 1954. On the annual variation in evaporation from lakes in the middle latitudes. *Arch. Met. Wien.,* ser. B, Allgeim. biol. Klimat., 5:297–306. **525**

Newcombe, C. L., 1950. Environmental factors of Sodon Lake—a dichothermic lake in southern Michigan. *Ecol. Monogr.,* 20:207–227. **488**

Newcombe, C. L., and Dwyer, P. S., 1949. An analysis of the vertical distribution of temperature in a dichothermic lake of southeastern Michigan. *Ecology,* 30:443–449. **470**

Newcombe, C. L., and Slater, J. V., 1949. Temperature characteristics of Sodon Lake—a dichothermic lake in southeastern Michigan. *Hydrobiologia,* 1:346–378. **488**

Newcombe, C. L., and Slater, J. V. 1950. Environmental factors of Sodon Lake—a dichothermic lake in southeastern Michigan. *Ecol. Monogr.,* 20:207–227. **488**

Newell, N. D., 1949. Geology of the Lake Titicaca region, Peru and Bolivia. *Mem. geol. Soc. Amer.,* 36, ix, 111 pp. **14**

Nichols, M. S., Henkel, T., and McNall, D., 1946. Copper in lake muds of the Madison area. *Trans. Wis. Acad. Sci. Arts Lett.,* 38:333–350. **812, 818, 819**

Nielsen, N., 1937. A volcano under an ice cap. Vatnajökull, Iceland. *Geogr. J.,* 90:6–23. **31**

* Nilsson, E., 1940. Ancient changes of climate in British East Africa and Abyssinia; a study of ancient lakes and glaciers. *Geogr. Ann.,* 22:1–79. **49**

Nininger, H. H., and Figgins, J. D., 1933. The excavation of a meteorite crater near Haviland, Kiowa County, Kansas. *Proc. Colo. Mus. nat. Hist.,* 12:9–15. **149**

Noack, K., and Pirson, A., 1939. Die Wirkung von Eisen und Mangan auf die Stickstoffassimilation von Chlorella. *Ber. dtsch. bot. Ges.,* 57:442–452. **803**

Norman, G. W. H., 1938. The last Pleistocene ice front in Chibougamau District. *Proc. roy. Soc. Can.,* ser. 3, 32(sect. 4):69–86. **95**

Nümann, V., 1938. Die Verbreitung des Rheinwassers im Bodensee. *Int. Rev. Hydrobiol.*, 36:501–530. **292**

Nümann, W., 1941. Der Stickstoffhaushalt eines mässig eutrophen Sees (Schleinsee). *Z. Fisch.*, 39:387–405. **847, 856**

* Nygaard, G., 1938. Hydrobiologische Studien über dänische Teiche und Seen. I. Teil.: Chemisch-physikalische Untersuchungen und Planktonwägungen. *Arch. Hydrobiol.*, 32:523–692. **854, 855, 871**

Oana, S., 1948. Distribution of heavy water in natural waters. *Chemical Res. Univ. Nagoya*, 3: Japanese text, 71–89; English summ., 90. **214**

Oberholzer, J., 1900. Monographie einiger prähistorischer Bergstürze in den Glarneralpen. *Beitr. geol. Karte Schweiz*, NF 9, whole 39, 109 pp. **44**

O'Brien, T. P., 1939. *The Prehistoric Uganda Protectorate.* Cambridge Univ. Press. xi, 318 pp. **11, 22**

Odintsova, S. V., 1941. Niter formation in deserts. *C.R. Acad. Sci. URSS*, 32:578–580. **847**

Odum, H. T., 1950. The biogeochemistry of strontium. Unpublished thesis, Yale University. **550, 560**

Odum, H. T., 1951a. Strontium in Florida waters. *Wat. Supp. Resour. Pap., Florida*, 6:20–21. **560**

Odum, H. T., 1951b. Nudibranch spicules made of amorphous calcium carbonate. *Science*, 114:395. **661**

Odum, H. T., 1952. The Carolina bays and a Pleistocene weather map. *Amer. J. Sci.*, 250:263–270. **155**

Odum, H. T., 1953. Dissolved phosphorus in Florida waters. *Rep. Fla geol. Surv.*, 9:1–40. **730**

Odum, H. T., 1956. Primary production in flowing waters. *Limnol. Oceanogr.*, 1:102–117. **587**

Odum, H. T., and Parrish, B., 1954. Boron in Florida waters. *Quart. J. Fla Acad. Sci.*, 17:105–109. **549, 563, 571**

Ohle, W., 1933. Chemisch-stratigraphische Untersuchungen der Sedimentmetamorphose eines Waldsees. *Biochem. Z.*, 258:420–428. **889**

* Ohle, W., 1934a. Chemische und physikalische Untersuchungen norddeutscher Seen. *Arch. Hydrobiol.*, 26:386–464, 584–658. **388, 413, 561, 598, 670, 679, 680, 684, 706, 712, 729, 765, 768, 791, 803, 808, 821, 850, 879, 890**

Ohle, W., 1934b. Über organische Stoffe in Binnenseen. *Verh. int. Ver. Limnol.*, 6:249–262. **887**

Ohle, W., 1935. Organische Kolloide in ihrer Wirking auf den Stoffhaushalt der Gewässer. *Naturwissenschaften*, 23:480–484. **706, 712, 850**

* Ohle, W., 1936. Der schwefelsaure Tonteich bei Reinbeck, Monographie eines idiotrophen Weihers. *Arch. Hydrobiol.*, 30:604–662. **783**

Ohle, W., 1937. Kolliodgele als Nährstoffregulatoren der Gewässer. *Naturwissenschaften*, 25:471–474. **706, 721**

Ohle, W., 1939. Zur Vervollkommnung der hydrochemischen Analyse. III. Die Phosphorbestimmung. *Angew. Chem.*, 51:906–911. **728**

Ohle, W., 1940. Vergleichend-chemische Erforschung der südschwedischen Braunwasserseen. *Forsch. u. Fortschr.*, 16:54–55. **890**

* Ohle, W., 1952. Die hypolimnische Kohlendioxyde-Akkumulation als productionsbiologischer Indikator. *Arch. Hydrobiol.*, 46:153–285. **579, 666** fn., **671, 678, 679**

Ohle, W., 1955a. Die Ursachen der rasanten Seeneutrophierung. *Verh. int. Ver. Limnol.*, 11:373–382. **755, 769**

Ohle, W., 1955b. Ionenaustausch der Gewässersedimente. *Mem. Ist. ital. Idrobiol. de Marchi.*, suppl. vol. 8:221–248. **554, 555, 557** fn., **674** fn.

Okura, T., 1954. The state of silica in natural waters. *J. chem. Soc. Japan, Pure Chem. Sect.*, 75:872–876. **789**

Olson, F. C. W., 1950. The currents of western lake Erie. Unpublished thesis, Ohio State University. **268, 291, 333**

Olson, F. C. W., 1952. The currents of western Lake Erie. *Abstr. Thes. Ohio St. Univ.*, 62:419–424. **268, 278, 291, 318**

Olson, T. A., 1932. Some observations on the interrelationships of sunlight, aquatic plant life and fishes. *Trans. Amer. Fish. Soc.*, 62:278–289. **599**

Ormerod, G. W., 1859. Rock basins in the granite of the Dartmoor District. *J. geol. Soc.*, 15:16–29. **286** fn.

Ostenfeld, C. H., and Wesenburg-Lund, C., 1905. A regular fortnightly exploration of the plankton of the two Icelandic lakes Thingvallavatn and Myvatn. *Proc. roy. Soc. Edinb.*, 25:1092–1167. **38**

Österlind, Sven, 1947. Growth of a planktonic green algae at various carbonic acid and hydrogen-ion concentrations. *Nature, Lond.*, 159:199–200. **852**

Otley, J., 1819. Account of the floating island in Derwent Lake, Keswick (1814). *Mem. Manchr lit. phil. Soc.*, 2nd ser., 3:64–69. **175, 842**

Otley, J., 1831. Further observations on the floating island of Derwent Lake; with remarks on certain other phenomena (1825). *Mem. Manchr lit. phil. Soc.*, 2nd ser., 5:19–24. **175, 842**

Ousley, R., 1788. An account of the moving of a bog, and the formation of a lake in the county of Galway, Ireland. *Trans. R. Irish Acad.*, 2:3–5. **46**

Owston, P. G., and Lonsdale, K., 1948. The crystalline structure of ice. *J. Glaciol.*, 1:118–123. **209** fn.

Pace, D. M., 1933. The relation of inorganic salts to growth and reproduction in *A. proteus*. *Arch. Protistenk.*, 79:133–145. **798**

Palache, C., Berman, H., and Frondel, C., 1951. *The System of Mineralogy* of James Dwight Dana and Edward Salisbury Dana. New York, John Wiley & Sons, Vol. II, xi, 1124 pp. **568, 570, 660**

Palazzo, L., 1905. Studi limnologichi sul lago de Bolsena. *Atti V Congr. geogr. ital.*, II, Napoli (not seen: ref. Halbfass 1923). **314**

Palmén, E., and Laurila, E., 1938. Über die Einwirkung eines Sturmes auf den hydrographischen Zustand im nördlichen Ostseegebiet. *Comment. phys.-math., Helsingf.*, 10, No. 1, 53 pp. **274**

Panov, B. P., 1932. Variations du niveau de lac Balkhash. *Serv. Hydrometeorol. URSS Inst. Hydrol. Explorations der lacs de l'URSS*, 1:Russian text, 34–40; French summ., 40–41. **240**

Parker, G. G., and Cooke, C. W., 1944. Late Cenozoic geology of southern Florida with a discussion of ground water. *Bull. Fla geol. Surv.*, 27, 119 pp. **8, 102**

Parmenter, R., 1929. Hydrography of Lake Erie. *Bull. Buffalo Soc. nat. Sci.*, 14:25–50. **336, 339**

Passarge, S., 1904. *Die Kalahari*. Berlin, Reimer, xvi, 822 pp. **132**

* Passarge, S., 1911. Die pfannenförmigen Hohlformen der südafrikanischen Steppen. *Petermanns Mit.*, 57:57–130. **132**

Patrick, R., 1943. The diatoms of Linsley Pond, Connecticut. *Proc. Acad. nat. Sci. Philad.*, 95:53–110. **897**

Pauling, L., 1939. *The Nature of the Chemical Bond and the Structure of Molecules and Crystals*. Ithaca, Cornell Univ. Press, xiv, 429 pp. **196** fn., **197**

* Peach, B. N., and Horne, J., 1910. The Scottish lakes in relation to the geological features of the country. In Murray and Pullar, *Bathymetric Survey of Scottish Fresh-Water Lochs*, Vol. 1, pp. 439–513. **57, 62, 68, 69, 84, 96, 114, 115**

* Pearsall, W. H., 1921. The development of vegetation in the English lakes, considered in relation to the general evolution of glacial lakes in rock basins. *Proc. roy. Soc.*, ser. B, 92:259–248. **554**

Pearsall, W. H., 1930. Phytoplankton in the English Lakes. I. The proportions in the waters of some dissolved substances of biological importance. *J. Ecol.*, 18:306–320. **791, 852** fn., **871**

Pearsall, W. H., and Hewitt, T., 1933. Light penetration into fresh waters. II. Light penetration and changes in vegetation limits in Windermere. *J. exp. Biol.*, 10:306–312. **395**

Pearsall, W. H., and Loose, L., 1937. The growth of *Chorella vulgaris* in pure culture. *Proc. roy. Soc.*, ser. B, 121:451–501. **849, 851**

Pearsall, W. H., and Mortimer, C. H., 1939. Oxidation-reduction potentials in waterlogged soils, natural waters and muds. *J. Ecol.*, 27:483–501. **693, 710, 862**

Pearsall, W. H., and Ullyott, P., 1934. Light penetration into fresh water. III. Seasonal variations in the light conditions in Windermere in relation to vegetation. *J. exp. Biol.*, 11:89–93. **394**

Pemberton, H., and Tucker, G. P., 1893. The deposits of native soda near Laramie, Wyoming. *J. Franklin Inst.*, 135:52–57. **765**

Penck, A., 1882. *Die Vergletscherung der deutschen Alpen, ihre Ursachen, periodische Wiederkekr und ihr Einfluss auf die Bodengestaltung.* Leipzig, J. A. Barth, viii, 483 pp. **2**

Penck, A., 1894. *Morphologie der Erdoberfläche.* Stuttgart, Engelhorn, Vol. I, xiv, 471 pp.; Vol. II, 696 pp. **3, 44, 109, 223**

Penck, A., 1905. Glacial features in the surface of the Alps. *J. Geol.*, 13:1–19. **77**

Penck, A., and Brückner, E., 1909. *Die Alpen im Eiszeitalter.* Leipzig, Tauchnitz, Vol. 3, xii, 717–1176. **77, 84**

Pennak, R. W., 1944. Diurnal movements of zooplankton organisms in some Colorado lakes. *Ecology*, 25:387–403. **442**

Perkins, E. A., 1893. The seiche in America. *Amer. met. J.*, 10:251–263. **316, 318**

Perl, G., 1935. Zur Kenntnis der wahren Sonnenstrahlung in verschiedenen geographischen Breiten. *Met. Z.*, 52:85–89. **370**

Perntner, J. M., and Exner, F. M., 1910. *Meteorologische Optik.* Wien and Leipzig, W. Braumüller, xvii, 799 pp. **421**

Pertz, H., 1937. Über die Radioaktivität von Quellwässern. *S.B. Acad. Wiss. Wien.*, Abt. IIa, 146:611–622. **831, 832**

* Pesta, O., 1929. *Der Hochgebirgssee der Alpen. Die Binnengewässer*, 8. Stuttgart, Schweizerbartsche Verlagsbuchhandlung, xi, 156 pp. **362, 416, 441**

Peterson, W. H., Fred, E. B., and Domogalla, B. P., 1925. The occurrence of amino acids and other organic nitrogen compounds in lake water. *J. biol. Chem.*, 23:287–295. **890, 893**

Pettit, B., 1936. On the color of Crater Lake water. *Proc. nat. Acad. Sci., Wash.*, 22:139–146. **408**

Peucker, K., 1903. The lakes of the Balkan Peninsula. *Geogr. J.*, 21:530–533. **167**

Philippson, A., 1892. *Der Peloponnes.* Berlin, R. Friedlander, 642 pp. **109, 137**

* Pia, J., 1933. *Kohlensäure und Kalk. Die Binnengewässer*, 13. Stuttgart, Schweizerbartsche Verlagsbuchhandlung, vii, 183 pp. **661, 662, 666**

Pietenpol, W. B., 1918. Selective absorption in the visible spectrum of Wisconsin lake waters. *Trans. Wis. Acad. Sci. Arts Lett.*, 19:562–593. **383**

Piggot, C. S., and Urry, Wm. D., 1942a. Radioactivity of ocean sediments. IV. The radium content of sediments of the Cayman Trough. *Amer. J. Sci.*, 240:1–12. **831**

Piggot, C. S., and Urry, Wm. D., 1942b. Time relations in ocean sediments. *Bull. geol. Soc. Amer.*, 53:1187–1210. **831**

Pillai, V. K., 1954. Some factors controlling algal production in salt-water lagoons. *Proc. Indo-Pac. Fish Counc.*, Sect. II, 5:78–81. **817**

Pirson, A., 1937. Ernährungs- und stoffwechsel-physiologische Untersuchungen an *Fontinalis* und *Chlorella*. *Z. Bot.*, 31:193–267. **806**

Pirson, A., and Bergmann, L., 1955. Manganese requirement and carbon source in *Chlorella*. *Nature, Lond.*, 176:209–210. **806**

Poole, H. H., and Atkins, W. R. G., 1926. On the penetration of light into sea-water. *J. Mar. biol. Ass. U.K.*, 14:177–198. **375, 389, 391, 392**

Porter, J. R., 1946. *Bacterial Chemistry and Physiology*. New York, John Wiley & Sons, 1073 pp. **867**

Post, H. von, 1862. Studier öfver nutidens koprogena jordbildningar, gyttja, dy, torf och mylla. *K. svenska VetenskAkad. Handl.*, 4, No. 1, 59 pp. **882**

Potonié, H., 1913. Eine im Ögelsee (Prov. Brandenburg) plötzlich neu entstandene Insel. *J. Preuss. Geol. Landesanst.*, 32, I(für 1911):187–218. **175**

Potter, M. C., 1930. Hydrion concentration of rain and potable water. *Nature, Lond.*, 126:434–5. **549**

Powell, W. M., and Clarke, G. L., 1936. The reflection and absorption of daylight at the surface of the ocean. *J. opt. Soc. Amer.*, 26:111–120. **375, 393**

Power, F. D., 1905. Phosphate deposits of Ocean and Pleasant Islands. *Trans. Aust. Inst. Min. Engrs*, 10:213–232. **144**

Prandtl, L., 1925. Bericht über Untersuchungen zur ausgebilteten Turbulenz. *Z. angew. Math. Mech.*, 5:136–139. **255**

Preston, R. S., Person, E., and Deevey, E. S., 1955. Yale natural radiocarbon measurements, II. *Science*, 122:954–960. **145, 833**

Pringal, E., 1908. Über den westlichen Einfluss von Spuren nitroser Gase auf die Kondensation von Wasserdampf. *Ann. Phys. Lpz.*, 4 folge, 26:727–750. **542**

Pringsheim, E. G., 1936. Das Rätsel der Erdabkochung. *Bot. Zbl. Beih.*, 55(A): 100–121. **712**

Pringsheim, E. G., 1946. *Pure Cultures of Algae, Their Preparation and Maintenance.* Cambridge Univ. Press, xii, 119 pp. **712**

Proudman, J., 1915. Free and forced longitudinal tidal motion in a lake. *Proc. Lond. Math. Soc.*, 2nd ser., 14:240–250. **302**

Proudman, J., 1925. On tidal features of local coastal origin and on sea-seiches. *Mon. Not. R. astr. Soc. geophys. Suppl.*, 1:247–270. **330, 331**

* Proudman, J., 1953. *Dynamical Oceanography*. London, Metheun; New York, John Wiley & Sons, xii, 409 pp. **250, 260** fn., **267, 270, 309, 310, 332, 334**

Prout, W., 1833. *On Chemistry, Meteorology and the Function of Digestion, Considered with Reference to Natural Theology*. Bridgewater treatises on the power, wisdom, and goodness of God as manifested in the creation, Treatise VIII (1st ed. not seen), 1st Amer. ed. 1834, Philadelphia, Carey, Lea, and Blanchard, xiv, 306 pp. **195**

Prouty, W. F., 1935. The Carolina bays and elliptical lake basins. *J. Geol.*, 43:200–207. **153**

* Prouty, W. F., 1952. Carolina bays and their origin. *Bull. geol. Soc. Amer.*, 63:167–224. **151, 152, 153**

* Provasoli, L., and Pintner, I. J., 1953. Ecological implications of *in vitro* nutritional requirements of algae flagellates. *Ann. N.Y. Acad. Sci.*, 56:839–851. **826**

Pushin, N. A., and Grebenshchikov, E. V., 1923. The adiabatic cooling of water and the temperature of its maximum density as a function of pressure. *J. chem. Soc.*, 123(pt. 2):2717–2725. **457**

Quastel, J. H., and Scholefield, P. G., 1950. Nitrification of amino acids. *Nature, Lond.*, 166:239. **867**

* Quastel, J. H., and Scholefield, P. G., 1951. Biochemistry of nitrification in soil. *Bact. Rev.*, 15:1–53. **867**

Quensel, P. D., 1910. On the influence of the ice age on the continental watershed of Patagonia. *Bull. geol. Instn, Univ. Upsala*, 19:60–92. **80**

Quinn, E. L., and Jones, C. L., 1936. Carbon dioxide. *Amer. chem. Soc. Monogr. Ser.*, New York, Reinhold Publ. Corp., 294 pp. **655**

Quirke, T. T., 1925. The undertow. *Science*, 61:468. **358**

Raisz, E., 1934. Rounded lakes and lagoons of the coastal plains of Massachusetts. *J. Geol.*, 42:839–848. **139, 156, 171**

Rakestraw, N. W., and Emmel, V. M., 1938. The solubility of nitrogen and argon in sea water. *J. phys. Chem.*, 42:1211–1215. **839**

Rakestraw, N. W., and Hollaender, A., 1936. Photochemical oxidation of ammonia in sea water. *Science*, 84:442–443. **865**

Ramsay, A. C., 1862. *A Physical Geology and Geography of Great Britain*, 3rd ed. London, Stanford, xii, 349 pp. **77**

Ramsay, W., 1931. Changes of sea-level resulting from the increase and decrease of glaciations. *Fennia*, 52, No. 5, 62 pp. **6**

Rankama, K., and Sahama, T. G., 1950. *Geochemistry*. Chicago, Univ. Chicago Press, xvi, 912 pp. **544, 570, 819**

Rao, G. G., and Dhar, N. R., 1931. Photosynthesized oxidation of ammonia and ammonium salts and the problem of nitrification in soils. *Soil Sci.*, 31:379–384. **865**

* Rawson, D. S., 1942. A comparison of some large alpine lakes in western Canada. *Ecology*, 23:143–161. **442, 495, 646**

Rawson, D. S., 1944. The calculation of oxygen saturation values and their correction for altitude. *Spec. Publ. Amer. Soc. Limnol. Oceanogr.*, 15, 4 pp. **579**

* Rawson, D. S., 1950. The physical limnology of Great Slave Lake. *J. Fish. Res. Bd Can.*, 8(1)1950:3–66. **32** fn., **79, 495**

Rawson, D. S., 1953. The limnology of Amethyst Lake, a high Alpine type near Jasper, Alberta. *Canad. J. Zool.*, 31:193–210. **442, 495**

Rawson, D. S., and Moore, J. E., 1944. The saline lakes of Saskatchewan. *Canad. J. Res.*, D, 22:141–201. **564, 570**

Redfield, A. C., 1949. The exchange of oxygen across the sea surface. *J. Mar. Res.*, 7:347–361. **589, 590**

Regener, E., 1932. Über das Spektrum der Ultrastrahlung. I. Die Messungen im Herbst 1928. *Z. Phys.*, 74:433–454. **422**

Reinwaldt, I., and Luha, A., 1928. Bericht über geologische Untersuchungen an Kaalijärv (Krater von Sall) auf Ösel. *S.B. naturf. Ges. Tartu* (Tartu Ülikodi juures oleva Loodusuurijate Seltsi Arvanded), 35:30–70. **149**

Reinwaldt, J. A., 1938. Der Krater von Sall (Kaali Järv), ein Meteorkraterfeld in Estland. *Natur u. Volk*, 68:16–24. **149**

Reynolds, O., 1883. An experimental investigation of the circumstances which determine whether the motion of water shall be direct or sinuous, and of the law of resistance in parallel channels. *Phil. Trans.*, 174:935–982. **251**

Reynolds, O., 1894. On the dynamical theory of incompressible viscous fluids and the determination of the criterion. *Phil. Trans.*, ser. A., 186:123–164. **251**

Richards, O. W., 1934. The effect of deuterium on the growth of yeast. *J. Bact.*, 28:289–294. **218**

Richardson, L. F., 1926. Atmospheric diffusion shown on a distance-neighbour graph. *Proc. roy. Soc.*, ser. A., 110:709–737. **256, 257, 258**

Richardson, L. F. and Stommel, H., 1948. Note on eddy diffusion in the sea. *J. Met.*, 5:238–240. **258**

Richter, E., 1892. Die Temperaturverhaltnisse der Alpenseen. 9 *dtsch Geogrtags*, Wien, 189 (not seen: ref. Halbfass 1923, etc.). **427, 428, 487**

Ricker, W. E., 1934. A critical discussion of various measures of oxygen saturation in lakes. *Ecology*, 15:348–363. **576, 579, 581**

* Ricker, W. E., 1937. Physical and chemical characteristics of Cultus Lake, British Columbia. *J. Biol. Bd. Can.*, 3:363–402. **440, 470, 478, 495**

Riesenfeld, E. H., and Chang, T. L., 1936. Über den Gehalt an HDO und H_2O^{18} in Regen und Schnee. *Ber. dtsch. chem. ges.*, 69B:1305–1307. **214**

Riffenburg, H. B., 1925. Chemical character of ground waters of the northern Great Plains. *Wat. Supp. Pap., Wash.*, 560B:31–52. **544, 549**

Riikoja, H., 1929. Über die sommerliche Sauerstoffschichtung in den eutrophen Seen. *S.B. naturf. Ges. Tartu.*, 35: Esthonian text, xx–xxxi; German summ., xxxi–xxxv. **622**

* Riley, G. A., 1939. Limnological studies in Connecticut. Part I. General limnological survey. Part. II. The copper cycle. *Ecol. Monogr.*, 9:66–94 **403, 811**

Robbins, W. T., Hervey, A., and Stebbins, M. E., 1950. Studies on Euglena and vitamin B_{12}. *Bull. Torrey bot. Cl.*, 77:423–441. **826, 827**

Robbins, W. T., Hervey, A., and Stebbins, M. E., 1951. Further observations on Euglena and B_{12}. *Bull. Torrey bot. Cl.*, 78:363–375. **826**

Robinson, R. J., and Kemmerer, G., 1930. Determination of silica in mineral waters. *Trans. Wis. Acad. Sci. Arts Lett.*, 25:129–134. **789**

Robinson, R. A., and Stokes, R. H., 1955. *Electrolytic Solutions.* New York, Academic Press, 512 pp. **209**

* Rodhe, W., 1948. Environmental requirements of fresh-water plankton algae. *Symbolae Botan. Upsalienses*, 10, No. 1, 149 pp. **707, 712, 714, 733, 734, 816, 852, 897**

Rodhe, W., 1949. The ionic composition of lake waters. *Verh. int. Ver. Limnol.*, 10:377–386. **554, 555, 556, 557**

Röntgen, W. C., 1892. Über die Consitution des flüssigen Wassers. *Ann. Phys., Lpz.*, NF 45, (whole 281):91–97. **196**

Rogers, A. W., 1907. Geological survey of parts of Bechuanaland and Griqualand West. *Rep. geol. Comm. C.G.H.*, 1906, 11:9–85. **132**

Rogers, A. W., 1908. Geological Survey of parts of Vryburg, Kuruman, Hay, and Gordonia. *Rep. geol. Comm. C.G.H.*, 1907, 12:9–122. **132**

Rogers, A. W., 1922. Post-Cretaceous climates of South Africa. *S. Afr. J. Sci.*, 19:1–31. **131, 132, 133**

Rohleder, H. P. T., 1933. Steinheim Basin and the Pretoria Salt Pan, volcanic or meteoric origin? *Geol. Mag.*, 70:489–498. **149, 151**

Rohleder, H. P. T., 1936. Lake Bosumtwi, Ashanti. *Geogr. J.*, 87:51–65. **37**

Rohlich, G. A., and Lea, W. L., 1949. *The Origin of Plant Nutrients in Lake Mendota.* Report to Univ. Wisconsin Lake Investigations Committee (mimeogr.) 7(+1) pp. **844**

Roscoe, H. E., and Lund, J., 1889. On Schutzenberger's process for the estimation of dissolved oxygen in water. *J. chem. Soc.*, 55:552–576. **579**

Rosenberg, M., 1938. The culture of algae and its applications. *Rep. Freshw. biol. Ass. Brit. Emp.*, 6:43–46. **735** fn.

Ross, F. W., and Utterback, C. L., 1939. Intensity fluctuations in components of solar radiation with atmospheric conditions. *J. Cons. int. Explor. Mer.*, 14:193–200. **368**

Rossby, C.-G., 1952. On the nature of the general circulation of the lower atmosphere. In *The Atmospheres of the Earth and Planets*, ed. G. P. Kruper. Chicago, Univ. Chicago Press, Ch. II, 16–48. **225**

Rossby, C.-G., and Egnér, H., 1955. On the chemical climate and its variation with the atmospheric circulation pattern. *Tellus*, 7:118–133. **546**

Rossby, C.-G., and Montgomery, R. B., 1935. The layer of frictional influence in wind and ocean currents. *Pap. phys. Oceanogr.*, 3, No. 3, 101 pp. **267**

Rossolimo, L., 1932a. Die Thermik der Bodenablagerungen des Beloje-Sees zu Kossino. *Arb. limnol. Sta. Kossino.*, 15:Russian text, 44–62; German summ., 63–66. **454, 495, 507**

Rossolimo, L., 1932b. Über die Gasausscheidung im Beloje-See zu Kossino. *Arb. limnol. Sta. Kossino*, 15:Russian text, 67–81; German summ., 82–84.

Rossolimo, L., 1935. Die Boden-Gasausscheidung und das Sauerstoffregime der Seen. *Verh. int. Ver. Limnol.*, 7:539–561. **630**

Roughton, F. J. W., 1941. The kinetics and rapid thermochemistry of carbonic acid. *J. Am. chem. Soc.*, 63:2930–2934. **655**

Rowland, H. A., 1880. On the mechanical equivalent of heat, with subsidiary researches on the variation of the mercurial from the air thermometer, and on the variation of the specific heat of water. *Proc. Amer. Acad. Arts Sci.*, 15:75–200. **206**

Roy, C. J., 1945. Silica in natural waters. *Amer. J. Sci.*, 243:393–403.

* Rubey, W. W., 1951. The geological history of sea water: an attempt to state the problem. *Bull. geol. Soc. Amer.*, 62:1111–1148. **222, 223**

Rubin, M., and Suess, H., 1955. U.S. Geological survey radio carbon dates, II. *Science*, 121:481–488. **152**

Rudakov, K. I., 1927. Die Reduktion der mineralischen Phosphate auf biologischen Wegen. *Zentr. Bakt.*, II, 70:202–214. **727**

Russell, E. J., and Richards, E. H., 1919. The amounts and composition of rain and snow falling at Rothamsted. *J. agric. Sci.*, 9:309–337. **546**

Russell, I. C., 1885. Geological history of Lake Lahontan. *Monogr. U.S. geol. Surv.*, 11, xiv, 287 pp. **16, 484**

Russell, I. C., 1893. Geological reconnaisance in central Washington. *Bull. U.S. geol. Surv.*, 108. 168 pp. **47, 126**

* Russell, I. C., 1895. *Lakes of North America; a Reading Lesson for Students of Geography and Geology*. Boston and London, Ginn, 125 pp. (reprint 1900). **3, 15** fn.

Russell, I. C., 1897. *Volcanoes of North America; a Reading Lesson for Students of Geography and Geology*. New York and London, Macmillan, xiv, 346 pp. **30**

Russell, R. J., 1927. The land forms of Surprise Valley, northwestern Great Basin. *Univ. Calif. Publ. Geogr.*, 2:323–358. **15, 43**

Russell, R. J., 1928. Basin-range structure and stratigraphy of the Warner range, Northeastern California. *Bull. Dept. geol. Sci., Univ. Calif.*, 17:387–496. **15**

Russell, R. J., 1933. Alpine land forms of western United States. *Bull. geol. Soc. Amer.*, 44:927–950. **59**

* Russell, R. J., 1942. Geomorphology of the Rhône delta. *Ann. Ass. Amer. Geogr.*, 32:147–254. **119**

Russell, W. J., *see* Symons 1889

* Ruttner, F., 1931. Hydrographische und hydrochemische Beobachtungen auf Java, Sumatra und Bali. *Arch. Hydrobiol. Supp.*, 8:197–454. **28, 29, 37, 464, 635, 666, 685, 706, 723, 772, 791, 808**

* Ruttner, F., 1937. Limnologische Studien an einigen Seen der Ostalpen. *Arch. Hydrobiol.*, 32:167–319. **488, 706, 715, 728, 729, 780, 791, 808, 853**

Ruttner, F., 1948. Die Veränderungen des Äquivalentleitvermögens als Mass der Karbonatassimilation der Wasserpflanzen. *Schweiz. Z. Hydrol.*, 11:72–89. **671**

Ruttner, F., and Sauberer, F., 1938. Durchsichtigkeit des Wassers und Planktonschichtung. *Inter. Rev. Hydrobiol.*, 37:405–419. **396**

Saderra Masó, M., 1911. *The Eruption of Taal Volcano, January 30, 1911.* Dept. Interior, Weather Bureau, Manila, 45 pp. **33**

Sandström, J. W., 1908. Dynamische Versuche mit Meerwasser. *Ann. Hydrogr., Berl.*, 36:6–23. **282**

Sanford, K. S., and Arkell, W. J., 1928. The relation of Nile and Faiyum in Pliocene and Pleistocene times. *Nature, Lond.*, 121:670–671. **136**

Sapper, K., 1902. Beiträge zur physischen Geographie von Honduras. *Z. Ges. Erdk. Berl.*, 1902:33–56, 143–164, 231–241. **109**

Sapper, K., 1913. Entwurf von Höhenschichtlinien der mittleren Vulkanregion Nicaraguas. *Petermanns Mit.*, 54(1):310–311. **191**

Sarasin, E., 1886. Limnimètre enregistreur au bord du lac de Zurich. *Arch. Sci. phys. nat.*, ser. 3, 16:210–212. **315**

Sarasin, E., 1895. Sur les seiches du lac de Thoune. *Arch. Sci. phys. nat.*, ser. 3, 34:368–371. **314**

Sarasin, E., 1897–1900. Les seiches du lac des Quatres Cantons. *Arch. Sci. phys. nat.*, ser. 4, 4:458–460; 5:389–390; 6:382–384; 8:382–383; 10:600. See also *Mitt. naturf. Ges. Luzern*, 4 (not seen: ref. Halbfass 1922). **315**

Sarasin, E., and Pasquier, L. du, 1895. Les seiches du lac de Neuchâtel. *Bull. Soc. neuchâtel. Sci. nat.*, 23:1–9. **315**

Sauberer, F., 1937a. Messung des Strahlungshaushaltes horizontaler Oberflächen bei heiterem Wetter. *Met. Z.*, 54:213–221. **380**

Sauberer, F., 1937b. Messungen des Strahlungshaushaltes horizontaler Flächen bei Bewölkung 4–10. *Met. Z.*, 54:273–278. **380**

Sauberer, F., 1937c. Einiges über Erfahrungen mit dem Strahlungsbilanzmesser nach F. Albrecht. *Met. Z.*, 54:329–333. **378**

Sauberer, F., 1938. Zur Methodik der Durchsichtigkeitsmessung im Wasser. *Arch. Hydrobiol.*, 33:343–360. **400**

* Sauberer, F., 1939. Beiträge zur Kenntnis des Lichtklimas einiger Alpenseen. *Int. Rev. Hydrobiol.*, 39:20–55. **392, 402, 408**

*Sauberer, F., 1945. Beiträge zur Kenntnis der optischen Eigenschaften der Kärtner Seen. *Arch. Hydrobiol.*, 41:259–314. **392, 408**

Sauberer, F., and Eckel, O., 1938. Zur Methodik der Strahlungsmessungen unter Wasser. *Int. Rev. Hydrobiol.*, 37:257–289. **375**

* Sauberer, F., and Ruttner, F., 1941. *Die Strahlungsverhältnisse der Binnengewässer. Probleme der kosmischen Physik*, Bd. 21. Leipzig, Akademische Verlagsgesellschaft, x, 240 pp. **371, 376, 377, 399, 408, 409**

Saunders, J. T., 1923. The hydrogen ion concentration of the waters of Wichen Fen. In *The Natural History of Wichen Fen*, ed. J. Stanley Gardiner. Cambridge, England, Bowes, pp. 162–171. **684**

* Saunders, J. T., 1926. The hydrogen ion concentration of natural waters. I. The

relation of *p*H to the pressure of carbon dioxide. *Brit. J. exp. Biol.*, 4:46–72. **656, 668**

Sauramo, M., 1939. The mode of the land upheaval in Fennoscandia during late-Quaternary time. *Bull. Comm. géol. Finl.*, No. 125, 26 pp.

de Saussure, H. B., 1779. *Voyages dans les Alpes, précédés d'un essai sur l'histoire naturelle des environs de Genève.* Tome 1, Chap. 1, Le Lac de Genève; Chap. 2, De la profondeur de la température des eaux du lac. Neuchatel, S. Fauche (quarto; octavo reprint 1803). **300, 426**

de Saussure, H. B., 1796. *Voyages dans les Alpes*, etc. Tome 5, Chap. 18, Recherche sur la température de la mer, des lacs, de la terre à différentes profondeurs. **426**

Sawyer, C. N., 1946. Fertilization of lakes by agricultural and urban drainage. *J. New Engl. Wat. Wks Ass.*, 61:109–127. **730**

Sawyer, C. N., Lackey, J. B., and Lenz, R. T., 1945. *An Investigation of the Odor Nuisance Occurring in the Madison Lakes, Particularly Monona, Waubesa and Kegonsa, from July 1943–July 1944.* Rept. Governor's Committee, Madison, Wis., 92 pp., 25 fig., 27 tables. **730, 818**

Sawyer, W. R., 1931. The spectral absorption of light by pure water and Bay of Fundy water. *Contrib. Canad. Biol.*, 7:75–89. **382**

Schaaf, S. A., and Sauer, F. M., 1950. A note on the tangential transfer of energy between wind and tide. *Trans. Amer. geophys. Un.*, 31:867–869. **351**

Schäperclaus, W., 1926. Die örtlichen Schwankungen der Alkalinität und der *p*H, ihre Ursachen, ihre Beziehungen zueinander und ihre Bedeutung. *Z. Fisch.*, 24:71 (not seen: ref. Ruttner 1931). **666**

Scheffer, V. B., and Robinson, R. J., 1939. A limnological study of Lake Washington. *Ecol. Monogr.*, 9:95–143. **495**

Schiøtz, O. E., 1887. Resultatet af en del undersøgelser over temperaturforholdene i dybet af Vøsen. *Forh. skand. naturf. Møte*, 13 (not seen: ref. Strøm 1945). **457**

Schmassmann, H., 1948. Die chemisch-technologische Beurteilung natürlicher Wässer hinsichlich ihrer Agressivität auf Eisen und Mangan. *Schweiz. Arch. angew. Wiss.*, 14:6–13 (not seen: ref. Mortimer 1956). **579**

Schmassmann, H., 1949. Die Sauerstoffsättigung naturlicher Wasser, ihre Ermittlung und ihre Bedeutung in der Hydrologie. *Schweiz. Z. Hydrol.*, 11:430–462.

Schmidt, C., 1882. Hydrologische Unersuchungen. *Mélang. chem. phys. Acad. St. Petersburg*, 11:595–647. **559**

Schmidt, W., 1915. Über den Energie-gehalt der Seen. Mit Beispielen vom Lunzer Untersee nach Messungen mit einem einfachen Temperaturlot. *Int. Rev., Hydrobiol. Suppl.*, 6, Leipzig. **510**

* Schmidt, W., 1925. *Der Massenaustausch in freier Luft und verwandte Erscheinungen. Probleme der kosmischen Physik*, Bd. 7. Hamburg, H. Grand, viii, 118 pp. **466, 624** fn.

* Schmidt, W., 1928. Über Temperatur und Stabilitätsverhältnisse von Seen. *Geogr. Ann.*, 10:145–177. **503, 508, 510, 511**

Schofield, S. J., 1943. The origin of Okanagan Lake. *Trans. roy. Soc. Can.*, 3rd ser., 37(Sect. 4):89–92. **72**

Schofield, S. J., 1946. The origin of Kootenay Lake. *Trans. roy. Soc. Can.*, 3rd ser., 40(Sect. 4):93–98. **73**

Schokalsky, J. de, 1914. Une dénivellation recentre et brusque du niveau de la mer Caspienne. *Ann. Géogr.*, 23:151–159. **239**

Schostokowitsch, W. B., 1926. Thermische Verhaltnisse des Baikalsees. *Verh. magnet. met. Observat. Irkutsk*, 1:1–30 (Russian, German summ.; not seen: ref. Mortimer 1953a). **338**

Schreiber, E., 1927. Die Reinkultur von marinem Phytoplankton und deren

Bedeutung für die Erforschung der Produktionsfähigkeit des Meerwasser. *Wiss. Meeresuntersuch.*, Abt. Helgoland, NF 16, No. 10, 33 pp. **851**

Schreiter, T., 1928. Untersuchungen über den Einfluss einer Helodea-wucherung auf das Netzplankton des Hirschberger Grossteiches in Böhmen in den Jahren 1921 bis 1925 incl. *Rec. Inst. Rech. agron. Rep. tschechosl.*, 5, 98 pp. (not seen: ref. Hasler and Jones 1949). **901**

Schriever, W., 1955. Were the Carolina bays oriented by gyroscopic action? *Science*, n.s. 120:806. **156**

Schultz, K., 1899. *Beiträge zur Kenntnis des Gmundener See.* Gmunden (not seen: ref. Halbfass 1922). **314**

Schwager, A., 1894. *Geogr. Jh.*, 1894:91 (not seen: ref. Clarke 1924). **766**

Schwager, A., 1897. *Geogn. Jh.*, 1897:65 (not seen: ref. Clarke 1924). **766**

Schweitzer, A., 1909. Die Seiches des Walensees. *Mitt. phys. Ges. Zürich* (not seen: ref. Halbfass 1922). **314**

Scott, I. D., 1926. Ice-push on lake shores. *Pap. Mich. Acad. Sci.*, 7:107–123. **532**

Scott, W., 1910. The fauna of a solution pond. *Proc. Ind. Acad. Sci.*, 26:395. **101**

Scott, W., 1916. Report on the Lakes of the Tippecanoe Basin. *Ind. Univ. Stud.*, No. 31, 39 pp. **92, 495**

Scrope, G. P., 1858. *The Geology and Extinct Volcanoes of Central France.* London, J. Murray, 258 pp. **27**

Seaman, G. R., 1949. Utilization of atmospheric nitrogen by the animal microorganism *Colpidium campylum. Anat. Rec.*, 105:522. **848** fn.

Sears, P. B., 1936. Crisis under water. *Science*, 83:81. **599**

Secchi, A., *see* Cialdi 1866

Selivanov, L. S., 1939a. La biogéochimie du brome dispersé. I. Le brome dans les plantes et les eaux douces. *Trav. Lab. Biogéochim. URSS*, 5:113–122 (Russian text, French summ.). **562**

Selivanov, L. S., 1939b. La biogéochimie du brome dispersé. II. Le brome dans les sols, les ares et les tourbes. *Trav. Lab. Biogéochim. URSS*, 5:123–144 (Russian text, French summ.). **562**

Selivanov, L. S., 1944. Géochimie et biogéochimie du brome dispersé. *Trav. Lab. Biogéochim. URSS*, 7:55–75 (Russian text). **562**

Selivanov, L. S., 1946. Géochimie et biogéochimie du brome dispersé. *Trav. Lab. Biogéochim. URSS*, 8:5–72 (Russian text). **546, 562, 563**

Seljakov, N., 1936. To what class of symmetry does ordinary ice belong? *C.R. Acad. Sci. URSS*, 10:293–294. **209** fn.

Seljakov, N., 1937. The nature of ordinary ice. *C.R. Acad. Sci. URSS*, 11:183–186. **209** fn.

Sellards, E. H., 1914. Some Florida lakes and lake basins. *Rep. Fla geol. Surv.*, 6:115–159. **104**

Shackleton, E. H., 1909. *The Heart of the Antarctic.* London, W. Heinemann, Vol. 1, xlvii, 371 pp.; Vol. 2, xv, 419 pp. **462**

Shand, S. J., 1946. Dust devils? Parallelism between the South African salt pans and the Carolina bays. *Sci. Mon., N.Y.*, 62:95. **155**

Shapiro, J., 1956. The coloring matter of natural waters. Unpublished thesis, Yale University. **682, 710, 711, 889, 900**

Sharpe, C. F. S., 1938. *Landslides and Related Phenomena; a Study of Mass Movements of Soil and Rock.* New York, Columbia University Press, xii, 137 pp. **42** fn.

Shaw, A. N., 1924. The formation of heavy ice in a cryophorus. *Trans. roy. Soc. Can.*, 18(Sect. III):187–190. **209**

Shepard, F. P., 1936. Undertow, rip tide or "rip current." *Science*, 84:181–182. **358**

Shepard, F. P., 1937. Origin of the Great Lakes basins. *J. Geol.*, 45:76–88. **81**

Shorygin, A. A., 1943. Changes in the quantity and composition of the benthos in the northern part of the Caspian Sea in the course of the years 1935–1940. *Zool. Zh.*, 24:Russian text, 148–159; English summ., 159–160. **239**

Shuman, J. W., 1931. Notes on lake levels. *Mon. Weath. Rev., Wash.*, 59:97–105. **246**

Simojoki, H., 1940. Über die Eisverhältnisse der Binnenseen Finnlands. *Ann. Acad. Sci. Fenn.*, A 62, No. 6, 194 pp. **532** fn.

Simony, F., 1850. Die Seen des Salzkammergutes. *S.B. Akad. Wiss. Wien*, Math.-Nat. Kl., 4:542–566. **426**

* Simpson, Sir G. C., 1941a. On the formation of cloud and rain. *Quart. J. R. met. Soc.*, 67:99–133. **542**

Simpson, Sir G. C., 1941b. Sea-salt and condensation nuclei. *Quart. J. R. met. Soc.*, 67:163–169. **542**

Skadowsky, S., 1923. Hydrophysiologische und hydrobiologische beobachtungen über die Bedeutung der Reaction des Mediums für die Süsswasserorganismen. *Verh. int. Ver. Limnol.*, 1:341–358. **681**

Skadowsky, S., 1926. Über die aktuelle Reaktion der Süsswasserbecken und ihre biologische Bedeutung. *Verh. int. Ver. Limnol.*, 3:109–144. **681**

Skeats, E. W., and James, A. U. G., 1937. Basaltic barriers and other surface features of the newer basalts of western Victoria. *Proc. roy. Soc. Vict.*, n.s. 49:245–278. **38, 42**

Sloane, H., 1694. A letter from Hans Sloane, M.D. and S.R.S., with several accounts of the earthquakes in Peru, October 20, 1687. And at Jamaica, February 19, 1687, and June 7, 1692. *Phil. Trans.*, 18, No. 209, 78–100. **12**

Smith, L. B., and Keyes, F. G., 1934. Steam research. III. The volumes of unit mass of liquid water and their correlation as a function of pressure and temperature. *Proc. Amer. Acad. Arts Sci.*, 69:285–312. **206**

Smith, L. L., 1931. Solution depressions in sandy sediments of the coastal plain in South Carolina. *J. Geol.*, 39:641–652. **110**

Smith, L. L., 1941. A limnological investigation of a permanently stratified lake in the Huron Mountain region of northern Michigan. *Pap. Mich. Acad. Sci.*, 26:281–296. **488, 491**

Smith, M. W., 1948. Preliminary observations upon the fertilization of Crecy Lake, New Brunswick. *Trans. Amer. Fish. Soc.*, 75(for 1945):165–174. **745**

Smith, R. A., 1872. *Air and Rain. The Beginners of a Chemical Climatology.* London, Longmans, Green, xiii, 600 pp. **574, 549**

Smuts, J. C., 1938. Past climates and pre-Stellenbosch stone implements of Rietvlei (Pretoria) and Benoni. *Trans. roy. Soc. S. Afr.*, 25:367–388. **133**

Smuts, J. C., Jr., 1945. Climates and pre-Paleolithic artifacts of the Witwatersrand. *Trans. roy. Soc. S. Afr.*, 31:39–67. **133**

Snodgrass, R. E., and Heller, E., 1902. The buds of Clipperton and Cocos Islands. *Proc. Wash. Acad. Sci.*, 4:501–520. **145**

Solomon, J. D., 1939, *see* O'Brien 1939.

Spencer, J. W., 1891. Post-pleistocene subsidence versus glacial dams. *Bull. geol. Soc. Amer.*, 2:465–476. **81**

Spencer, L. J., 1933. Meteorite craters as topographical features of the earth's surface. *Geogr. J.*, 81:226–242. **37, 149**

Spitzer, L., 1949. The terrestrial atmosphere above 300 km. In *The Atmospheres of the Earth and Planets*, ed. G. P. Knifer. Chicago, Univ. Chicago Press, Chap. VII. **368**

Spring, W., 1897 (1898). Sur le rôle des composés ferriques et de matières humiques dans le phénomène de la coloration des eaux, et sur l'élimination de ces substances sous l'influence de la lumière solaire. *Bull. Acad. Belg. Cl. Sci.*, 34:578–600. Also *Arch. Sci. phys. Nat.*, ser. 4, 5:5–26. **413, 417, 711**

Stahl, A. F. von, 1923. *Handbuch der Regionalen Geologie:* Bd. V, Abt. 5, Kaukasus. 80 pp. **4**

Stangenberg, Marjan, 1936. Eisenverteilung in den Seen des Suwalki Gebiets wahrend des Sommers. *Arch. Hydrobiol. Rybact.*, 10:48–75. **717**

* Stanković, S., 1932. Die Fauna des Ohrid—Sees und ihre Herkunft. *Arch. Hydrobiol.*, 23:557–617. **19**

Stappers, L., 1915. Composition chimique des eaux de surface des lacs Moëro et Tanganika. In *Composition, analyse et étude des produits de la Colonie* (not seen: ref. Kupperath 1952). **465**

Starkey, R. L., 1935. Products of the oxidation of thiosulphate by bacteria in mineral media. *J. gen. Physiol.*, 18:325–349. **755**

Steinböck, O., 1938. Arbeiten über die Limnologie der Hochgebirgsgewässer. *Int. Rev. Hydrobiol.*, 37:467–509. **441**

Steidtmann, E., 1935. Travertine-depositing waters near Lexington, Virginia. *Science*, 82:333–334. **671**

Steiner, J. F., and Meloche, V. W., 1935. A study of ligneous substances in lacustrine materials. *Trans. Wis. Acad. Sci. Arts Lett.*, 29:389–402. **889**

Steinhauser, F., 1939. Die Zunahme der Intensität der direkten Sonnenstrahlung mit der Höhe im Alpengebiet und die Verteilung der "Trübung" in den unteren Luftschichten. *Met. Z.*, 56:172–181. **371**

Steinmayer, R. A., 1939. Bottom sediments of Lake Pontchartrain, Louisiana. *Bull. Amer. Ass. Petrol. Geol.*, 23:1–23. **119**

Stephenson, M., 1939. *Bacterial Metabolism*, 2nd ed. London, Longmans, Green, xiv, 391 pp. **715, 756, 867**

Sterneck, R., 1928. Die Gezeiten des Baikalsees. *Ann. Hydrogr. Berl.*, 56:221–225. **334**

Stevenson, T., 1852. Observations on the relation between the height of waves and their distance from the windward shore. *Edinb. New* (Jameson's) *Phil. J.*, 53:358–359. **355**

Stommel, H., 1949. Horizontal diffusion due to oceanic turbulence. *J. Mar. Res.*, 8:199–225. **252** fn., **257, 258**

Stott, V., and Bigg, P. H., 1928. Density and specific volume of water. In *International Critical Tables*, National Research Council. New York, McGraw-Hill Book Co., Vol. 3, pp. 24–26. **206**

Strachey, R., 1894. The landslip at Gohna, in British Garwhal. *Geogr. J.*, 4:162–170. **43**

Strickland, C., 1940. *Deltaic Formation with Special Reference to the Hydrographic Process of the Ganges and the Brahmaputra.* Calcutta and Bombay, Longmans, Green, xiii, 157 pp. **118, 119**

Strickland, J. D. H., 1952. The preparation and properties of silicomolybdic acid. *J. Amer. chem. Soc.*, 74:862–876. **789**

* Strøm, K. M., 1931. Feforvatn. A physiographic and biological study of a mountain lake. *Arch. Hydrobiol.*, 22:491–536. **443** fn., **444, 601, 617, 639, 642**

Strøm, K. M., 1932a. Tyrifjord. A limnological study. *Skr. norske VidenskAkad.*, Mat.-nat. Kl., 1932, No. 3, 84 pp. **595, 597, 642**

Strøm, K. M., 1932b. Lilla Le. A preliminary survey of a remarkable lake. *Geogr. Ann.*, 1932:259–272. **642**

* Strøm, K. M., 1933a. Nordfjord lakes. A limnological survey. *Skr. norske VidenskAkad.*, Mat.-nat. Kl., 1932 No. 8, 56 pp. **69, 187, 446, 594, 642, 685**

Strøm, K. M., 1933b. The nutrition of algae. *Arch. Hydrobiol.*, 25:38–47. **743**

Strøm, K. M., 1934. Flakevatn. A semi-arctic lake of central Norway. *Skr. norske VidenskAkad,*. Mat.-nat. Kl., 1934, No. 5, 28 pp. **49, 63, 441**

Strøm, K. M., 1935a. Bessvatn and other lakes of eastern Jotunheim. *Skr. norske VidenskAkad.*, Mat.-nat. Kl., 1935, No. 4, 29 pp. **49, 57, 63, 90, 446**

Strøm, K. M., 1935b. Sammensatte votsjøer og litt om geomorfologiske sjøtyper i Norge. *Norsk geogr. Tidsskr.*, 5:334–341. **49, 63, 90, 93**

Strøm, K. M., 1937. Eikesdalsvatn. A limnological reconnaissance. *Skr. norske VidenskAkad.*, Mat.-nat. Kl., 1937, No. 8, 11 pp. **71, 84**

Strøm, K. M., 1938a. The catastrophic emptying of a glacier-dammed lake in Norway 1937. *Geol. Meere*, 2:443–444. **51, 52**

* Strøm, K. M., 1938b. Moskenesøy. A study in high-latitude cirque lakes. *Skr. norske VidenskAkad.*, Mat.-nat. Kl., 1938, No. 1, 32 pp. **61, 597**

Strøm, K. M., 1939. Conductivity and reaction in Norwegian lake waters. *Int. Rev. Hydrobiol.*, 38:250–258. **684**

Strøm, K. M., 1940. Tyrifjord geomorphology. Relation of a lake basin and its surroundings to the geological structure of the region. *Norsk geogr. Tidsskr.*, 8:85–93 pp. **72**

Strøm, K. M., 1942. Hadeland Lakes. A limnological outline. *Skr. norske VidenskAkad.*, Mat.-nat. Kl., 1941, No. 7, 42 pp. **684**

* Strøm, K. M., 1944a. Heat in a south Norwegian lake. Studies on Lake Eikeren during the years 1934 and 1935. *Geofys. Publ.*, 16, No. 3, 23 pp. **494, 495, 503**

Strøm, K. M., 1944b. High mountain limnology. Some observations on stagnant and running waters of the Rondane area. *Avh. norske VidenskAkad.*, Mat.-nat. Kl., 1944, No. 8, 24 pp. **684**

* Strøm, K. M., 1945. The temperature of maximum density in fresh waters. *Geofys. Publ.*, 16, No. 8, 14 pp. **208, 457**

Strøm, K. M., 1955. Waters and sediments in the deep of lakes. *Mem. Inst. ital. Idrobiol. de Marchi*, suppl., 8:345–356. **486**

Strøm, K. M., and Østtveit, H., 1948. Blankvatn, a meromictic lake near Oslo. *Skr. norske VidenskAkad.*, Mat.-nat. Kl., 1948, No. 1, 41 pp. **98**

Stromsten, F. A., 1927. Lake Okoboji as a type of aquatic environment. *Stud. nat. Hist. Ia Univ.*, 12, No. 5, 52 pp. **504**

Stumpf, T., 1923. Rheindelta im Bodensee. Aufnahme im Frühlung 1921. *Mitt. Amt. Wasserw., Bern*, No. 15 (not seen: ref. Collet 1925). **183**

Suess, E., 1888. *Das Antlitz der Erde.* Wien, Tempsky, Vol. II, 703 pp. **27**

Suess, E., 1891. Beiträge zur geologischen Kenntnis der östlichen Afrika. IV. Die Brücke der ostlichen Afrika. *Denkscht Akad. Wiss. Wien*, Math.-Nat. Kl., 58:555–558. **22**

Suess, H., 1955. Radiocarbon concentration in modern wood. *Science*, 122:415–417. **654**

Sugawara, K., 1948. Distribution of fine particles of salts in the air (Kagaku no Ryoiki). *J. Jap. Chem.*, 82:341 (not seen: ref. *Chem. Abstr.*, 45:4499). **548, 549**

Sugawara, K., in press. To appear in *Verh. int. Ver. Lim.*, 13. **563**

Sugawara, K., Koyama, T., and Kozawa, A., 1953. Distribution of various forms of sulphur in lake, river, and sea muds. *J. Earth Sci., Nagoya Univ.*, 1:17–23. **755, 756**

Sugawara, K., Koyama, T., and Kozawa, A., 1954. Distribution of various forms of sulphur in lake, river, and sea muds, (II). *J. Earth Sci., Nagoya Univ.*, 2:1–4. **723** fn.

Sugawara, K., Koyama, T., and Noda, B., 1952. A radiometric investigation of sulphate co-precipitation in lake waters. *Ann. Rep. Res. Comm. Appl. artif. Isotopes Japan*, 2:36–41. **768**

Sugawara, K., Oana, S., and Koyama, T., 1949. Separation of the components of atmospheric salt and their distribution. *Bull. chem. Soc. Japan*, 22:47–52. **548, 549, 550**

Sugawara, K., Tanaka, M., and Naitô, H., 1953. New colorimetric microdetermination of vanadium in natural waters. *Bull. chem. Soc. Japan*, 26:417–420. **828**

Sugawara, K., and Tochikubo, I., 1955. The determination of argon in natural waters with special reference to the metabolisms of oxygen and nitrogen. *J. Earth Sci., Nagoya Univ.*, 3:77–84. **842** fn.

Supan, A. G., 1896. *Grundzüge der physischen Erdkunde*, 2nd ed. Leipzig, Veit, ix, 706 pp. **2**

Suslov, S. P., 1947. *Fizicheskaia geografiia SSSR:* Zapadniia Sibir', vostochniia Sibir', Dal'ni Vostok, Sredniaia Asia (in Russian). Leningrad and Moscow, 544 pp. **44, 110, 170, 173, 183, 186, 239, 242, 451, 460, 501, 525, 533**

Sutherland, W., 1900. The molecular constitution of water. *Phil. Mag.*, ser. 5, 50:460–489. **196**

Sverdrup, H. U., 1942. *Oceanography for Meteorologists.* New York, Prentice-Hall, 121 pp. **272**

Sverdrup, H. U., 1952. Evaporation from the oceans. In *Compendium of meteorology.* Boston, American Meteorological Society, pp. 1071–1981. **228**

Sverdrup, H. U., Johnson, M. W., and Fleming, R. H., 1942. *The Oceans: Their Physics, Chemistry, and General Biology.* New York, Prentice-Hall, xii, 1087 pp. **266** fn., **274** fn.

Sverdrup, H. U., and Munk, W. H., 1947. Wind, sea and swell: Theory of relations for forecasting. *U.S. Navy Dept. Hydrographic Office Pub. No.* 601 (*U.S. Hydrogr. Off. Tech. Rep. No.* 1), v, 44 pp. **348, 351, 352**

Swartout, J. A., and Dole, M., 1939. The protium-deuterium ratio and the atomic weight of hydrogen. *J. Amer. chem. Soc.*, 61:2025–2029. **212**

Symons, G. J., [?1889]. *The Floating Island in Derwentwater, Its History, and Mystery, with Notes of Other Dissimilar Islands.* London, E. Stanford; Simpkins Marshall and Co., 64 pp. **175, 176, 842**

Tanaka, M., 1953a. Occurrence of hydroxylamine in lake waters as an intermediate in bacterial reduction of nitrate. *Nature, Lond.*, 171:1160–1161. **863**

Tanaka, M., 1953b. Étude chimique sur le métabolisme minéral dans les lacs. *J. Earth Sci., Nagoya Univ.*, 1:119–134. **788, 822**

Tanakadate, H., 1930. The problem of caldera in the Pacific region. *Proc. Fourth Pacif. Sci. Congr.*, Java, 1929, IIB:729–744. **32, 33**

Tarr, R. S., 1897. Evidence of glaciation in Labrador and Baffin Land. *Amer. Geol.*, 19:191–197. **59**

Tarr, R. S., 1898. Wave-formed cuspate forelands. *Amer. Geol.*, 22:1–12. **180**

Tarr, R. S., and Martin, L., 1914. *Alaskan Glacier Studies.* Washington, National Geographic Society, xxvii, 498 pp. **53**

Taylor, G. I., 1916. Skin friction of the wind on the earth's surface. *Proc. roy. Soc.*, ser. A, 92:196–199. **273**

Taylor, G. I., 1931a. Effect of variation in density on the stability of superposed streams of fluid. *Proc. roy. Soc.*, ser. A, 132:499–523. **346**

Taylor, G. I., 1931b. Internal waves and turbulence in a fluid of variable density. *Rapp. Cons. Explor. Mer*, 76:35–43. **255, 346**

Taylor, H. S., Swingle, W. W., Eyring, H., and Frost, A. H., 1933. The effect of water containing the isotope of hydrogen upon fresh-water organisms. *J. chem. Phys.* 1:751. **217**

Taylor, T. G., 1907. The Lake George Senkungfeld: A Study of the Evolution of Lakes George and Bathurst, N.S.W. *Proc. Linn. Soc. N.S.W.*, 32:325–345.

Taylor, W. R., 1939. Algae collected on the presidential cruise of 1938. *Smithson. misc. Coll.*, 98, No. 9, 18 pp. **145**

Teis, R. V., 1939. Isotopic composition of water from some rivers and lakes of the U.S.S.R. *C.R. Acad. Sci. URSS*, 24:779–782. **214**

Teis, R. V., 1946. Variation in deuterium concentration in the process of melting of ice. *C.R. Acad. Sci. URSS*, 53:529–532. **214**

Teis, R. V., and Florenskii, K. P., 1940. Isotopic composition of snow. *C.R. Acad. Sci. URSS*, 28:70–74. **214**

Terada, T., 1906. Notes on seiches. *Proc. Tokyo math.-phys. Soc.*, 3:174–181. **302**

ter Meulen, H., 1932. Distribution of molybdenum. *Nature, Lond.*, 130:966. **824**

Thienemann, A., 1913a. Der Zusammenhang zwischen dem Sauerstoffgehalt des Tiefenwassers und der Zusammensetzung der Tierfauna unserer Seen. Vorlaüfige Mitteilung. *Int. Rev. Hydrobiol.*, 6:243–249. **600**

* Thienemann, A., 1913b. Physikalische und chemische Untersuchungen in den Maaren der Eifel. *Verh. naturh. Ver. preuss. Rheinl.*, 70:249–302. **487, 600**

* Thienemann, A., 1915. Physikalische und chemische Untersuchungen der Maaren der Eifel. *Verh. naturh. Ver. preuss. Rheinl.*, 71:281–389. **487, 600**

* Thienemann, A., 1920. Untersuchungen über die Beziehungen zwischen Sauerstoffgehalt des Wassers und der Zusammensetzung der Fauna in norddeutschen Seen. *Arch. Hydrobiol.*, 12:1–65. **600**

* Thienemann, A., 1925a. *Die Binnengewässer Mitteleuopas. Die Binnengewässer*, 1. Stuttgart, Schweizerbartsche Verlagsbuchhandlung, 255 pp. See Preface.

* Thienemann, A., 1925b. Mysis relicta. *Z. Morph. Ökol. Tiere*, 3:389–440. **605**

* Theinemann, A., 1926. Der Nahrungskrieslauf im Wasser. *Verh. dtsch. zool. Ges.* (*Zool. Anz. Suppl.*, Bd. 2), 31:29–79. **641**

* Thienemann, A., 1928. *Der Sauerstoff im eutrophen und oligotrophen Seen. Die Binnengewässer*, 4. Stuttgart, Schweizerbartsche Verlagsbuchhandlung, 175 pp. **595, 600, 607, 624, 639, 642**

Thijsse, J. Th., 1952. Growth of wind-generated waves and energy transfer. In: Gravity waves. *Circ. U.S. Bur. Stand.*, 521:281–287. **358**

Thomas, E. A., 1948. Limnologische Untersuchungen am Türlersee. *Schweiz. Z. Hydrol.*, 11:90–177. **488, 900**

Thoreau, H. D., 1854. *Walden; or Life in the Woods.* Boston, Ticknor and Fields, 357 pp. **96**

Thoreau, H. D., 1906. *The Writings of Henry David Thoreau*, ed. Bradford Tomey: Vol. 20, *Journal.* Boston and New York, Houghton Mifflin, pp. 60–66. **427**

Thorade, H., 1914. Die Geschwindigkeit von Triftströmung und die Ekmansche Theorie. *Ann. Hydrogr., Berl.*, 42:379–390. **256, 267**

Thoroddsen, T., 1906. Island. Grundriss der Geographie und Geologie. *Petermanns Mitt.*, 32, Ergänzunghefte 152 u., 153, 358 pp. **38**

Thoulet, J., 1894. Contribution à l'étude des lacs des Vosges. *Géographie*, 15 (not seen: ref. Mortimer 1952). **334**

Thunmark, S., 1937. Über die regionale Limnologie von Südschweden. *Årsb. Sverig. geol. Unders.*, 31(C. 410):1–160. **879**

Thwaites, F. T., 1929. Glacial geology of part of Vilas County, Wisconsin. *Trans. Wis. Acad. Sci. Arts Lett.*, 24:109–125. **95**

Thwaites, F. T., 1947. Geomorphology of the basin of Lake Michigan. *Pap. Mich. Acad. Sci.*, 33:243–251. **81**

Tiedjens, V. A , and Robbins, W. R,. 1931. The use of ammonia and nitrate nitrogen by certain crop plants. *Bull. N.J. agric. Exp. Sta.*, 526:1–46. **851**

Tilho, J., 1928. Variation et disparition possible du Tchad. *Ann. Géogr.*, 37:238–260. **244**

Tillmans, J., 1919. Über die quantitative Bestimmung der Reaktion in naturliche Wässern. *Z. Untersuch. Nahr- u. Genussm.*, 38:1–16. **666**

Tiselius, D., 1722. Utförlig Beskrifning ofver Den store Swea och Giotha Sjön Wättern. (Not seen: ref. Hellström 1941). **300**

Tison, L. J., 1949. Variations des niveaux du lac Tanganika. *Gen. Ass. int. Un. Geod.*, Oslo, 1948, *Ass. int. Hydro. Sci.*, I. *Trav. Comm. Potamol. Limnol.*, 360–362. **242**

Tomlinson, D., 1935. Rare aquatic phenomena. *Science*, 82:418. **599**

Tonolli, V., 1947a. Differenziamento microgeografico in popolazioni planctiche d'alta montagna. *Mem. Ist. ital. Idrobiol. de Marchi.*, 3:273–305. **63, 64**

Tonolli, V., 1947b. Abaco per la determinazione grafica dei valori di saturazione dell'ossigeno disciolto. *Mem. Ist. ital. Idrobiol. de Marchi*, 3:463–466. **576**

Trelease, S. F., and Beath, O. A., 1949. *Selenium, Its Geological Occurrence and Its Biological Effects in Relation to Botany, Chemistry, Agriculture, Nutrition, and Medicine.* New York, published by the authors, x, 292 pp. **563**

Tressler, W. L., 1937. The effects of copper sulphate treatment on certain genera of freshwater plankton organisms. *Int. Rev. Hydrobiol.*, 35:178–185. **817**

Tressler, W. L., and Domogalla, B. P., 1931. Limnological studies of Lake Wingra. *Trans. Wis. Acad. Sci. Arts Lett.*, 26:331–351. **730**

Tressler, W. L., Wagner, L. G., and Bere, R., 1940. A limnological study of Chautauqua Lake. II. Seasonal variations. *Trans. Amer. micr. Soc.*, 59:12–30. **403**

Trolle, A., 1913. Hydrographical observations from the Danmark expedition. *Medd. Grønland*, 41:271–426. **461, 484, 495**

Trueman, A. E., 1938. *The Scenery of England and Wales.* London, Gollancz, 351 pp. **44, 73**

Truesdale, G. A., and Downing, A. L., 1954. Solubility of oxygen in water. *Nature, Lond.*, 173:1236. **579, 581, 587**

Truesdale, G. A., Downing, A. L., and Lowden, G. F., 1955. The solubility of oxygen in pure water and sea-water. *J. appl. Chem., Lond.*, 5:53–63. **579, 580, 598** fn.

Turekian, K. K., and Kleinkopf, M. D., 1956. Estimates of the average abundance of Cu, Mn, Pb, Ti, Ni, and Cr in surface waters of Maine. *Bull. geol. Soc. Amer.*, 67:1129–1131. **804, 811, 825, 827, 829**

Turowska, I., 1930. Études sur les conditions vitales des bactéries ferrugineuses. *Bul. int. Cracovie* (*Acad. pol. Sci.*), Cl. Sci. Math., ser. B, 1928:225–282. **715**

Uéno, Masuzo, 1934. Acid water lakes in north Shinano. *Arch. Hydrobiol.*, 27:571–584.

Ule, W., 1892. Die Bestimmung der Wasserfarbe in den Seen. *Petermanns Mitt.*, 38:70–71. **415**

Ule, W., 1893. Über die Beziehungen zwischen den Mansfelder Seen und dem Mansfelder Bergbau. *Z. prak. Geol.*, 1:339–346. **110**

Ule, W., 1901. Der Würmsee (Starnbergersee) in Oberbayern. *Wiss. Veröff. Ver., Erdk. Lpz.*, 5, 210 pp. **338**

Urey, H. C., 1952. *The Planets. Their Origin and Development.* New Haven, Yale University Press, xvii, 245 pp. **221, 790**

Urey, H. C., Brickwedde, F. G., and Murphy, G. M., 1932. A hydrogen isotope of mass 2 and its concentration. *Phys. Rev.*, 40:1–15. **213**

Urry, W. D., 1949. The radium content of varved clay and a possible age of the Hartford, Connecticut, deposits. *Amer. J. Sci.*, 246:689–700. **833**

Urry, W. D., and Piggot, C. S., 1942. Concentration of the radio elements and their significance in red clay. *Amer. J. Sci.*, 240:93–103. **831**

Uspenski, E. E., 1927. *Eisen als Faktor für die Verbreitung niederer Wasserpflanzen. Pflanzenforschung*, 9. Jena, Fischer, vi, 104 pp. **712**

Utermöhl, H., 1925. Limnologische Phytoplanktonstudien. Die Besiedelung ostholsteinischer Seen mit Schwebpflanzen. *Arch. Hydrobiol. Suppl.*, 5:1–527. **779, 780**

Utterback, C. L., Phifer, L. D., and Robinson, R. J., 1942. Some chemical, physical and optical characteristics of Crater Lake. *Ecology*, 23:97–103. **393, 408, 847**

Vallentyne, J. R., 1952. Sulfur in ether extracts of lake sediments. *Science*, 116:667. **755**

Vallentyne, J. R., and Swabey, Y. S., 1955. A reinvestigation of the history of Lower Linsley Pond, Connecticut. *Amer. J. Sci.*, 253:313–340. **723**

Vallot, J., and Delebecque, A., 1892. Sur les causes de la catastrophe survenue à Saint-Gervais (Haute-Savoie), le 12 juillet 1892. *C.R.*, 115:264–266. **50**

Valvasor, J. W., 1688. Extract of a letter written to the Royal Society out of Carniola by Mr. John Weichard Valvasor, R. Soc. S., being a full and accurate description of the wonderful Lake of Zirknitz in that country. *Phil. Trans.*, 16, No. 191, 411–426. (Transactions for December 1687, presumably issued later; title page dated 1688). **107**

Van Dorn, W. G., 1953. Wind stress on an artificial pond. *J. Mar. Res.*, 12:249–276. **273, 275, 277, 278, 312**

Van Neil, C. B., 1944. The culture, general physiology, morphology and classification of the non-sulphur purple and brown bacteria. *Bact. Rev.*, 8:1–118. **757, 780**

Van Slyke, D. D., Dillon, R. T., and Margaria, R., 1934., xviii. Solubility and physical state of atmospheric nitrogen in blood cells and plasma. *J. biol. Chem.*, 105:571–596. **839** fn.

Vaucher, J. P. E., 1833. Mémoire sur les seiches du lac de Genève. *Mém. Soc. Phys. Genève*, 6:35–94. **300**

Veatch, A. C., 1906. Geology and underground resources of northern Louisiana and southern Arkansas. *Prof. Pap. U.S. geol. Surv.*, 46, 422 pp. **115**

Verber, J. L., 1955. Rotational water movements in western Lake Erie. *Verh. int. Ver. Limnol.*, 12:97–104. **291**

Verber, J. L., Bryson, R. A., Suomi, V. E., 1953. Currents in Lake Mendota, Wisconsin. *Ohio J. Sci.*, 53:221–225. **268, 288, 341**

* Vereščagin, G. J., 1937. Études du lac Baikal. Quelques problèmes limnologiques. *Verh. int. Ver. Limnol.*, 8:189–207. **460, 495**

Vereščagin, G., Gorbov, A. I., Mendelejev, I., 1934. Sur l'existence de l'eau de densité anormale dans des conditions naturelles. *C.R. Acad. Sci. URSS*, 1939, 2:134–137. **214**

Vernadsky, W., 1922. Sur la problème de la décomposition du kaolin par les organismes. *C.R. Acad. Sci., Paris*, 175:450–452. **790**

Vernadsky, W. I., Brunowsky, B. K., and Kunasheva, C. G., 1937. Concentration of mesothorium I by duckweed (Lemna). *Nature, Lond.*, 140:317–318. **832**

Vernon, R. O., 1942. Tributary valley lakes of western Florida. *J. Geomorph.*, 5:302–311. **115**

Vernon, R. O., 1951. Geology of Citrus and Levy Counties, Florida. *Bull. Fla geol. Surv.*, 33, 256 pp. **102, 104**

Vetter, H., 1937. Limnologische Untersuchungen über das Phytoplankton und seine Beziehungen zur Ernährung des Zooplanktons im Schleinese bei Langenargen am Bodensee. *Int. Rev. Hydrobiol.*, 34:499–561. **781**

Vinberg, G., 1940. On the measurement of the rate of exchange of oxygen between a water basin and the atmosphere. *C.R. Acad. Sci. URSS*, 26:666–669. **591**

Vinogradov, A. P., and Boichenko, E. A., 1942. Decomposition of kaolin by diatoms. *C.R. Acad. Sci. URSS*, 37:135–138. **790, 798**

Vivenzio, G., 1788. *Isoria de' tremuoti avvenuti nella provincia della Calabria Ulteriore, e nella cittá di Messina nell'anno 1783.* Napoli, 2 vols. **12, 46**

Voigt, M., 1905. Die vertikale Verteilung des Planktons im Grossen Plöner See und ihre Beziehungen zum Gasgehalt dieses Gewässers. *ForschBer. biol. Sta. Plön*, 12:115–144. **838**

Voisey, A. H., 1949a. The geology of the country around the Great Lake, Tasmania. *Pap. roy. Soc. Tasm.*, 1947:95–103. **65, 68**

Voisey, A. H., 1949b. The geology of the country between Arthur's Lakes and the Lake River. *Pap. roy. Soc. Tasm.*, 1948:105–110. **65**

Volker, A., 1949a. Bilan d'eau du lac IJssel. *Gen. Ass. int. Un. Geol.*, Oslo, 1948, *Ass. int. Hydrol. Sci., I. Trav. Comm. Potamol Limnol.*, 320–326. **523**

Volker, A., 1949b. Movements des eaux dans le lac IJssel. *Gen. Ass. int. Un. Geol.*, Oslo, 1948, *Ass. int. Hydrol. Sci., I. Trav. Comm. Potamol. Limnol.*, 386–391. **280**

Von Arx, W. S., 1948. The circulation systems of Bikini and Rongelap Lagoons. *Trans. Amer. geophys. Un.*, 29:861–870. **286**

Von Engeln, O. D., 1933. Palisade Glacier of the High Sierra of California. *Bull. geol. Soc. Amer.*, 44:575–600. **64, 65**

Voskuyl, R. J., Ingraham, M. G., and Rustad, B. M., 1943. *S.A.M. Report*, Oct. 1, 1943, *see* Kirschenbaum 1951. **213**

Votintsev, K. K., 1948. Role of sponges in the dynamics of silicic acid in Lake Baikal. *C.R. Acad. Sci. URSS*, 62:661–663 (not seen: ref. Krauskopf 1956). **798**

Votintsev, K. K., 1953. Organicheskoe veschestvo v pene Baĭkala (Organic matter in Baikal foam). *C.R. Acad. Sci. URSS*, 42:425–427. **900**

Wadell, Hakon, 1920. Vatnajökull. Some studies and observations from the greatest glacial area in Iceland. *Geogr. Ann.*, 2:300–323. **31**

Wagner, G., 1922, *see* Kalle 1945.

Wagner, P. A., 1922. The Pretoria Salt Pan, a soda caldera. *Mem. geol. Surv. S. Afr.*, 20, 136 pp. **37**

Wait, F. G., 1900. Natural waters. *A.R. geol. Surv. Can.*, n.s. 11:41R–55R. **824**

Waksman, S. A., and Carey, C. C., 1935. Decomposition of organic matter in sea water by bacteria. *J. Bact.*, 29:531–543, 545–561. **887**

Waksman, S. A., and Iyer, K. R. N., 1932a. Synthesis of a humus nucleus, an important constituent of humus in soils, peats, and composts. *J. Wash. Acad. Sci.*, 22:41–50. **713, 888**

Waksman, S. A., and Iyer, K. R. N., 1932b. Contribution to our knowledge of the chemical nature and origin of humus. I. On the synthesis of the "humus nucleus." *Soil Sci.*, 34:43–69. **713, 888**

Waksman, S. A., and Iyer, K. R. N., 1932c. Contribution to our knowledge of the chemical nature and origin of humus. II. The influence of "synthesized" humus compounds and of "natural humus" upon soil microbiological processes. *Soil Sci.*, 34:71–79. **888**

Waksman, S. A., and Iyer, K. R. N., 1933a. Contribution to our knowledge of the chemical nature and origin of humus. III. The base-exchange capacity of "synthesized humus" (ligno-protein) and of "natural humus" complexes. *Soil Sci.*, 36:57–67. **888**

Waksman, S. A., and Iyer, K. R. N., 1933b. Contribution to our knowledge of the chemical nature and origin of humus. IV. Fixation of proteins by lignins resistant to microbial decomposition. *Soil Sci.*, 36:69–82. **888**

Waksman, S. A., and Stevens, K. R., 1928. Contribution to the chemical composition of peat. I. Chemical nature of organic complexes in peat and methods of analysis. *Soil Sci.*, 26:113–137, 239–252. **886, 889**

Walker, G. T., 1936. The variations of level in lakes; their relations with each other and with sunspot numbers. *Quart. J. R. met. Soc.*, 62:451–454. **242, 244**

Walker, T. J., and Hasler, A. D., 1949. Detection and discrimination of odors of aquatic plants by the bluntnose minnow (*Hyborhynchus notatus*). *Physiol. Zoöl.*, 22:45–63. **900**

Wallace, A. R., 1893. The ice age and its work. *Fortnightly Rev.*, 60:616–663, 750–774. **78**

Wallace, R. E., 1948. Cave-in lakes in the Nabesna, Chisana, and Tanana river valleys, eastern Alaska. *J. Geol.*, 56:171–181. **98**

Wallén, A., 1914. Om afdunstnings bestämningar. *Tekn. Tidskr., Stockh.*, 1914. Väg-och vatten byggnadskonst. (Not seen: ref. Johnsson 1946.) **523**

Warburg, E. J., 1922. Carbonic acid compounds and hydrogen ion activities in blood and salt solutions. A contribution to the theory of the equation of Lawrence J. Henderson and K. S. Hasselbach. *Biochem. J.*, 16:153–340. **656**

Warburg, O., and Negelein, E., 1920. Über die Reduktion der Salpetersäure in grünen Zellen. *Biochem. Z.*, 110:66–115. **849**

Warren, E. R., 1927. *The Beaver. Its Works and Its Ways.* Baltimore, Williams and Wilkins Co. (*Monogr. Amer. Soc. Mammal.*, No. 2), xx, 177 pp. **146**

Washburn, E. W., and Smith, E. R., 1934. An examination of water from various natural sources for variations in isotopic composition. *J. Res. nat. Bur. Stand.*, 12:305–311. **214**

* Wasmund, E., 1927–28. Die Strömungen in Bodensee, verglichen mit bisher in Binnenseen bekannten Strömungen. *Int. Rev. Hydrobiol.*, 18:84–114, 231–260; 19:21–155. **292**

Watanabe, A., Nishigaki, S., and Konishi, C., 1951. Effect on nitrogen-fixing blue-green algae on the growth of rice plants. *Nature, Lond.*, 168:748–749. **847**

Waterman, T. H., 1954. Polarization patterns in submarine illumination. *Science*, 120:927–932. **410**

Waterman, T. H., 1955. Polarization of scattered sunlight in deep water. *Deep Sea Res.*, 3(suppl.):426–434. **410**

Watson, E. R., 1903. Internal oscillation in the waters of Loch Ness. *Nature, Lond.*, 69:174. **334**

Watson, E. R., 1904a. Movements of the waters of Loch Ness, as indicated by temperature observations. *Geogr. J.*, 24:430. **334**

Watson, E. R., 1904b. On the ionization of air in vessels immersed in deep water. *Geogr. J.*, 24:437–440. **422** fn.

Watson, T. L., 1897. Lakes with more than one outlet. *Amer. Geol.*, 19:267–270. **232**

Watson, T. L., 1899. Some notes on the lakes and valleys of the Upper Nugsuak Peninsula, North Greenland. *J. Geol.*, 7:655–666. **59**

Wayland, E. J., 1934. Rifts, rivers, rains and early man in Uganda. *J. R. Anthropol. Inst.*, 64:333–352. **11**

Wedderburn, E. M., 1907. The temperature of fresh-water lochs of Scotland, with special reference to Loch Ness. *Trans. roy. Soc. Edinb.*, 45:407–489. **334, 338, 428, 444**

Wedderburn, E. M., 1908. Temperature observations in Loch Garry (Invernessshire). With notes on currents and seiches. *Proc. roy. Soc. Edinb.*, 29:98–113. **286, 314**

Wedderburn, E. M., 1910. Current observations in Loch Garry. *Proc. roy. Soc. Edinb.*, 30:312–323. **286**

* Wedderburn, E. M., 1911. The temperature seiche. Part 1. Temperature observations in the Madüsee, Pomerania. *Trans. roy. Soc. Edinb.*, 47:619–628. Part 2. Theory of temperature oscillations, *ibid.*, 628–634. Part 3. Computation of the period of the Madüsee, *ibid.*, 634–636. **334, 336, 338**

* Wedderburn, E. M., 1912. Temperature observations in Loch Earn, with a further contribution to the hydrodynamical theory of the temperature seiche. *Trans. roy. Soc. Edinb.*, 48:629–695. **334, 338**

Wedderburn, E. M., and Williams, A. M., 1911. The temperature seiche. Part 4. Experimental verification of the temperature seiche theory. *Trans. roy. Soc. Edinb.*, 47:636–642. **334**

Wedderburn, E. M., and Watson, W., 1909. Current and temperature observations on Loch Ness. *Proc. roy. Soc. Edinb.*, 29:619–647. **286**

Wedderburn, E. M., and Young, A. W., 1915. Temperature observations in Loch Earn, Part. II. *Trans. roy. Soc. Edinb.*, 50:741–767. **334**

Wehrle, E., 1927. Studien über Wasserstoffionenkonzentrationsverhältnisse und Besiedung an Algenstandorten in der Umgebung von Freiburg im Breisgau. *Z. Bot.*, 19:207–287. **681**

* Welch, P. S., 1935. *Limnology.* New York and London, McGraw-Hill Book Co. 471 pp. (2nd ed., ix, 536 pp., 1952). **187**

* Welch, P. S., 1948. *Limnological Methods.* Philadelphia, Blakiston, 381 pp. **164, 427**

Wellington, J. H., 1943. The Lake Chrissie problem. *S. Afr. Geogr. J.*, 25:50–64. **131**

Wells, B. W., and Boyce, S. G., 1953. Carolina bays: additional data on their origin age, and history. *J. Elisha Mitchell Sci. Soc.*, 69:119–141. **152**

Wentworth, C. K., 1931. Geology of the Pacific equatorial islands. *Occ. Pap. Bishop Mus.*, 8, No. 15, 25 pp. **144**

Wenz, W., 1925. *Gastropoda extramarina tertiaria, Hydrobiidae. Fossilium Catalogus 1*: *Animalia*, Pars 32, pp. 1863–2230. **5** fn.

Wharton, W. J., 1898. Notes on Clipperton Atoll (northern Pacific). *Quart. J. geol. Soc. Lond.*, 54:228–229. **145**

Whewell, W., 1833. *Astromomy and General Physics Considered with Reference to Natural Theology.* Bridgewater treatises on the power, wisdom, and goodness of God as manifested in the creation, Treatise III. Philadelphia, Carey, Lea, and Blanchard, 284 pp. **195**

Whipple, G. C., 1898. Classifications of lakes according to temperature. *Amer. Nat.*, 32:25–33. **439**

* Whipple, G. C., 1899. *The Microscopy of Drinking Water.* New York, John Wiley & Sons, xxi, 300 pp. (4th ed., 1927, rev. G. M. Fair). **413, 901**

Whipple, G. C., and Parker, H. N., 1902. On the amount of oxygen and carbonic acid dissolved in natural waters and the effect of these gases upon the occurrence of microscopic organisms. *Trans. Amer. micro. Soc.*, 23:103–144. **579**

Whipple, G. C., and Whipple, M. C., 1911. The solubility of oxygen in sea water. *J. Amer. chem. Soc.*, 33:362–365. **579**

White, P., and Watson, W., 1906. Some experimental results in connection with the hydrodynamical theory of seiches. *Proc. roy. Soc. Edinb.*, 26:142–156. **306**

Whitney, L. V., 1937. Microstratification of the waters of inland lakes in summer. *Science*, 85:224–225. **396**

Whitney, L. V., 1938a. Microstratification of inland lakes. *Trans. Wis. Acad. Sci. Arts Lett.*, 31:155–173. **396**

Whitney, L. V., 1938b. Continuous solar radiation measurements in Wisconsin lakes. *Trans. Wis. Acad. Sci. Arts Lett.*, 31:175–200. **375, 391**

Whitney, L. V., 1938c. Transmission of solar energy and the scattering produced by suspensoids in lake waters. *Trans. Wis. Acad. Sci. Arts Lett.*, 31:201–221. **404**

Wiebe, A. H., 1930. The manganese content of the Mississippi River water at Fairport, Iowa. *Science*, 71:248. **804**

Wiebe, A. H., 1939a. Dissolved oxygen profiles at Norris Dam and in the Big Creek sector of Norris Reservoir (1937) with a note on the oxygen demand of the water (1938). *Ohio J. Sci.*, 39:27–36. **296**

Wiebe, A. H., 1939b. Density currents in Norris Reservoir. *Ecology*, 20:446–450. **296**

Wiebe, A. H., 1940. The effect of density currents upon the vertical distribution of temperature and dissolved oxygen in Norris Reservoir. *J. Tenn. Acad. Sci.*, 15:301–308. **296**

Wiebe, A. H., 1941. Density currents in impounded waters—their significance from the standpoint of fisheries management. *Trans. Sixth N. Amer. Wildl. Conf.*, 6:256–264. **296**

Wiesner, R., 1939. Bestimmung des Radiumgehaltes von Algen. *S.B. Akad. Wiss. Wein.*, Math-Nat. Kl., 147(Abt. 2a):521–528. **832**

Wilhelmy, H., 1943. Die "Pods" der südrussischen steppen. *Petermanns Mitt.*, 89:129–141. **127**

Williams, H., 1932. Geology of the Lassen Volcanic National Park, California. *Bull. Dept. Geol. Univ. Calif.*, 21:195–385. **26, 43**

* Williams, H., 1941. Calderas and their origin. *Bull. Dept. Geol. Univ. Calif.*, 25:239–346. **26, 31, 33, 37, 151**

* Williams, H., 1942. The geology of Crater Lake National Park, Oregon. *Bull. Carneg. Inst. Wash.*, 540, 162 pp. **32**

Williams, K. T., and Thompson, T. G., 1936. Effect of sphagnum on the *p*H of salt solutions. *Int. Rev. Hydrobiol.*, 33:271–275. **684**

* Willis, B., 1936. East African plateaus and rift valleys. *Studies in Comparative Seismology*. Carnegie Institution of Washington, Publ. 470, x, 358 pp. **12, 22, 23, 24, 28**

Wills, H. M., 1941. Climate of Michigan. In *Climate and Man*. U.S. Dept. Agriculture, *Year Book of Agriculture* (House Document 27, 77th Congr., 1st Sess.), pp. 914–924. **527**

Wilson, A. W. G., 1908. Shoreline studies on Lakes Ontario and Erie. *Bull. geol. Soc. Amer.*, 19:493. **173**

Wilson, I. T., 1936. A study of sedimentation of Winona Lake. *Proc. Ind. Acad. Sci.*, 45:295–304. **187**

Wilson, I. T., 1938. The accumulated sediment in Tippecanoe Lake and a comparison with Winona Lake. *Proc. Ind. Acad. Sci.*, 47:234–253. **173, 187** fn., **422**

Wilson, I. T., 1945. A study of the sediment in Douglas Lake, Cheboygan County, Michigan. *Pap. Mich. Acad. Sci.*, 30:391–419. **187, 188**

Wilson, I. T., and Opdyke, D. F., 1941. The distribution of the chemical constituents in the accumulated sediment of Tippecanoe Lake. *Invest. Ind. Lakes*, 2:16–42. **187** fn.

Wilson, J. T., Zumberge, J. H., and Marshall, E. W., 1954. *A Study of Ice on an Inland Lake*. Project 2030, Snow, Ice, and Permafrost Establishment, Corps of Engineers, U.S. Army, Wilmette, Illinois, Report 5, pt. 1, 78 pp. **528, 533**

Wilson, V. C., 1938. Cosmic-ray intensities at great depths. *Phys. Rev.*, 53:337–343. **422**

Wing, L., 1943. Freezing and thawing of lakes and rivers as phenological indicators. *Mon. Weath. Rev., Wash.*, 71:149–155. **531**

Winkler, L. W., 1889. Die Löslichkeit des Sauerstoffs im Wasser. *Ber. dtsch. Chem. Ges.*, 22: 1764–1774. **579**

Wiseman, J. D. H., and Sewell, R. B. S., 1937. The floor of the Arabian Sea. *Geol. Mag.*, 74:219–230. **23**

Witsch, H. Von, 1946. Wachstum und Vitamin–B_1–Gehalt von Süsswasseralgen unter verschiedenen Aussenbedingungen. *Naturwissenschaften*, 33:221–222. **897**

Witting, R., 1909. Zur Kenntnis des vom Winde erzeugten Oberflächenstromes. *Ann. Hydrogr., Berl.*, 37:193–203. **268**

Witting, R., 1914. Bericht, abgegeben von dem Arbeitsausschuss zur Untersuchung des Wassers und des Planktons der finnischen Binnengewässer. *Fennia*, 35, No. 6, 41 pp.; No. 7, 21 pp. **383**

Wittstein, G. C., 1860. Beobachtungen und Betrachtungen über die Farbe des Wassers. *S.B. bayer. Akad. Wiss.*, Math.-phys. Kl., 1:603–624. **417**

Wojeikoff, J. A., 1902. Der jahrliche Warmeaustausch in den nordeuropaischen Seen. *Z. Gewässerk.*, 5:193–199. **492**

Woldstedt, P., 1921. Studien an Rinnen und Sanderflachen in Norddeutschland. *Jb. preuss. geol. Landesanst.*, 42:780–820. **97**

* Woldstedt, P., 1926. Probleme der Seenbildung in Norddeutschland. *Z. Ges. Erdk. Berl.*, 1926:103–124. **93, 97**

Woltereck, R., Tressler, W. S., and Buñag, D. M., 1941. Insehn und Seen der Philippinen. *Int. Rev. Hydrobiol.*, 41:37–176. **483**

Wood, R. W., 1911. *Physical Optics*, rev. ed. New York, Macmillan Co., xvi, 705 pp. **421**

Woodcock, A. H., 1941. Surface cooling and streaming in shallow fresh and salt waters. *J. Mar. Res.*, 4:153–161. **457, 589**

Woodcock, A. H., 1944. A theory of surface water motion deduced from the wind-induced motion of the Physalia. *J. Mar. Res.*, 5:196–205. **280**

Woodcock, A. H., 1950. Condensation nuclei and precipitation. *J. Met.*, 7:161–162. **542**

Woodcock, A. H., 1953. Salt nuclei in marine air as a function of altitude and wind force. *J. Met.*, 10:362–371. **542**

Woodcock, A. H., and Riley, G. A., 1947. Patterns in pond ice. *J. Met.*, 4:100–101. **531**

Woods, D. D., 1938. The reduction of nitrate to ammonia by *Clostridium welchii*. *Biochem. J.*, 32:2000–2012. **869**

Wooldridge, W. R., and Corbet, A. S., 1940. The nitrogen cycle in biological systems. 1. Some conditions affecting the distribution of nitrogenous compounds during treatment of sewage by the activated sludge process. *Biochem. J.*, 34:1015–1025. **868**

[Wordsworth, William] 1810. *Select Views in Cumberland, Westmoreland, and Lancashire*, by the Rev. Joseph Wilkinson, Rector of East and West Wretham in the county of Norfolk, and Chaplain to the Marquis of Huntley. London, published for the Rev.

Joseph Wilkinson by R. Ackerman, at his Repository of Arts, 101 Strand, 6, xxxiv, 46 pp., 48 pl. **73**

Wordsworth, William, 1820. *The River Duddon, a Series of Sonnets: Vaudracour and Julia; and Other Poems.* To which is annexed a topographical description of the country of the lakes, in the north of England. London, Longman, Hurst, Rees, Orme and Broun, viii, 321 pp. [The topographical description of pp. 213–309, i.e., the Introduction (i–xxxiv) of the previous entry, here first appearing under Wordsworth's name. Many later editions and reprints exist.] **73**

Worthington, E. B., 1930. Observations on the temperature, hydrogen-ion concentration, and other physical conditions of the Victoria and Albert Nyanzas. *Int. Rev. Hydrobiol.*, 24:328–357. **463**

Worthington, E. B., 1932. The lakes of Kenya and Uganda. *Geogr. J.*, 79:275–297. **41**

Wright, A. W., 1881. On the gaseous substances contained in the smoky quartz of Branchville, Conn. *Amer. J. Sci.*, 3rd ser., 21:209–216. **202**

Wright, H. L., 1940. Atmospheric opacity at Valentia. *Quart. J. R. met. Soc.*, 66:66–77. **542**

Wright, R. H., and Maass, O., 1932. Electrical conductivity of aqueous solutions of hydrogen sulfide and the state of the dissolved gas. *Canad. J. Res.*, 6:588–595. **758** fn.

Wroughton, F. J. W., 1941. The kinetics and rapid thermochemistry of carbonic acid. *J. Amer. chem. Soc.*, 63(2):2930–2934. **655**

Wüst, G., 1936. Oberflächen Salzgehalt, Verdunstung und Niederschlag auf dem Weltmeere. *Landerkundliche Forschung: Festschrift . . . Norbert Krebs*, ed. H. Lovis and W. Panzer. Stuttgart, J. Engelhorns nachf., pp. 347–359. **228**

Wyberg, W., 1918. The limestone resources of the Union. Vol. 1, *Mem. geol. Surv. S. Afr.*, 11, vii, 122 pp. **132**

Wyman, J., Scholander, P. F., Edwards, G. A., and Irving, L., 1952. On the stability of gas bubbles in sea water. *J. Mar. Res.*, 11:47–62. **583** fn.

* Yoshimura, S., 1930. Seasonal variation of silica in Takasuka-numa, Saitama. *Jap. J. Geol. Geogr.*, 7:101–113. **791, 792**

Yoshimura, S., 1931a. Seasonal variation of iron and manganese in the water of Takasuka-numa, Saitama. *Jap. J. Geol. Geogr.*, 8:269–279. **706, 808**

Yoshimura, S., 1931b. Contributions to the knowledge of the stratification of iron and manganese in the lake water of Japan. *Jap. J. Geol. Geogr.*, 9:61–69. **804** fn., **808**

Yoshimura, S., 1932a. Seasonal variations in content of nitrogenous compounds and phosphate in the water of Takasuka Pond, Saitama, Japan. *Arch. Hydrobiol.*, 24:155–172. **729, 854, 860**

Yoshimura, S., 1932b. Vertical distribution of the amount of sulphate dissolved in the water of Lakes Suigesu and Hiruga with reference to the origin of hydrogen sulphide in their bottom water. *Geophys. Mag., Tokyo*, 69:315–321. **776, 777**

Yoshimura, S., 1932c. Limnological reconnaissance of Lake Busyu, Hukui, Japan. *Sci. Rep. Tokyo Bunrika Daig.*, Sect. C., No. 1, pp. 1–27. **44**

Yoshimura, S., 1932d. On the dichotomous stratification of hydrogen ion concentration of some Japanese lake waters. *Jap. J. Geol. Geogr.*, 9:155–185. **686, 717, 808**

Yoshimura, S., 1932e. Contributions to the knowledge of nitrogenous compounds and phosphate in the lake waters of Japan. *Proc. Imp. Acad. Japan*, 8:94–97. **729**

Yoshimura, S., 1934. Kata-numa, a very strong acid-water lake of Volcano, Katanuma, Miyagi Prefecture, Japan. *Arch. Hydrobiol.*, 26:197–202. **681, 754, 783**

* Yoshimura, S., 1936a. A contribution to the knowledge of deep water temperatures of Japanese lakes. Part I. Summer temperatures. *Jap. J. Astr. Geophys.*, 13:61–120. **437, 724**

Yoshimura, S., 1936b. A contribution to the knowledge of deep water temperatures of Japanese lakes. Part II. Winter temperatures. *Jap. J. Astr. Geophys.*, 14:57–83. **428** fn., **441**, **446**, **458** fn., **474**, **479**

Yoshimura, S., 1936c. Contributions to the knowledge of iron dissolved in the lake waters of Japan. Second report. *Jap. J. Geol. Geogr.*, 13:39–56. **441**, **452**, **453**, **454**, **706**, **804** fn.

Yoshimura, S., 1936d. The effect of salt-breeze on the chemical composition of freshwater lakes near the sea. *Arch. Hydrobiol.*, 30:345–351. **561**

Yoshimura, S., 1937. Abnormal thermal stratifications on inland lakes. *Proc. Imp. Acad. Japan*, 13:316–319. **479**, **481**, **487**

Yoshimura, S., 1938a. Limnology of two lakes on the oceanic island, Kita-Daito-Zima. *Proc. Imp. Acad. Japan*, 14:12–15. **145**

* Yoshimura, S., 1938b. Dissolved oxygen of the lake waters of Japan. *Sci. Rep. Tokyo Bunrika Daig.*, Sect. C, No. 8, pp. 63–277. **32**, **41**, **191**, **403**, **483**, **576**, **598**, **601**, **604**, **620**, **622**, **627**

Yoshimura, S., and Miyadi, D., 1936. Limnological observations of two crater lakes of Miyaki Island. Western north Pacific. *Jap. J. Geol. Geogr.*, 13:339–352. **487**

Young, R. K., 1929. Tides and seiches on Lake Huron. *J. R. astr. Soc. Can.*, 23:445–455. **319**, **333**

Zavoico, B. B., 1935. Review: Kovalevsky, S. A. The face of the Caspian Sea. *Bull. Amer. Ass. Petrol. Geol.*, 19:120–125. **7** fn.

Zerbe, W. B., 1947. The tide in the David Taylor Model Basin. *Trans. Amer. Geophys. Un.*, 30:357–368. **332**

Zerbe, W. B., 1952. The tide in an enclosed basin. In: Gravity waves. *Circ. U.S. Bur. Stand.*, 521:267–277. **332**, **334**

ZoBell, C. E., 1933. Photochemical nitrification of sea water. *Science*, 77:27–28. **865**

ZoBell, C. E., 1935. The assimilation of ammonium nitrogen by *Nitschia closterium* and other marine phytoplankton. *Proc. nat. Acad. Sci., Wash.*, 21:517–522. **849**, **851**

ZoBell, C. E., 1946. *Marine Microbiology*. Waltham, Mass., Chronica Botanica, xv, 240 pp. **868**

ZoBell, C. E., Sisler, F. D., and Oppenheimer, C. H., 1953. Evidence of biochemical heating in Lake Mead mud. *J. sediment. Petrol.*, 23:13–17. **507**

ZoBell, C. E., and Stadler, J., 1940. The effect of oxygen tension on the oxygen uptake of lake bacteria. *J. Bact.*, 39:307–322. **614**

ZoBell, C. E., and Upham, H. C., 1944. A list of marine bacteria including descriptions of sixty new species. *Bull. Scripps. Instn. Oceanogr.*, 5:239–292. **868**

Zotov, V. D., 1941. Pot-holing of limestones by development of solution cups. *J. Geomorph.*, 4:71–73. **101**

* Zumberge, J. H., 1952. *The Lakes of Minnesota, Their Origin and Classification*. Minneapolis, University of Minnesota Press, xiii, 99 pp. **3**, **93**, **114**, **118**, **123**, **125**, **139**, **176**

Zumberge, J. H., 1955. Glacial erosion in tilted rock layers. *J. Geol.*, 63:149–158. **59**, **72** fn.

Zumberge, J. H., and Potzger, J. E., 1955. Pollen profiles, radiocarbon dating, and geologic chronology of the Lake Michigan Basin. *Science*, 121:309–311. **82**

Zumberge, J. H., and Wilson, J. T., 1953. Quantitative studies on thermal expansion and contraction of lake ice. *J. Geol.*, 61:374–383. **532**

Index of Lakes

The symbol † indicates a lake of the geological past. If the basin of such a lake is occupied by a modern lake, reference to the existing lake follows.

In the case of a large lake, the bearings refer to the approximate center. Where bearings are not given, the exact position of the lake is not available.

Index of Organisms

'he names of families, genera, and species of organisms mentioned
'he text are listed here in a separate index in order that, where
sible, the authors of specific names, ordinarily omitted from the
:, can be given and to indicate the general taxonomic position of
.ain forms that may be unfamiliar to readers not specialists in the
vant groups. The grade of major category to which each taxon is
rred differs from group to group, according to the probable famili-
y of the category in question. Although in general an attempt
been made to give correct names in the text, in a few cases there is
ie doubt as to the identification of the organism involved. In such
es the name cited by the original author has been used, and its
bable correct identification is given in parentheses after the index
.tion.

'wo particularly important organisms require special comment.
.ch if not all of the experimental work supposedly performed on
zschia, referred to *N. closterium*, probably was actually done on
ieodactylum tricornutum Bohlin, the systematic position of which
incertain. The common blue-green bloom-forming alga, which is
inarily referred to *Polycystis aeruginosa* or *Microcystis aeruginosa*
tz., is now apparently to be called *Anacystis cyanea* (Kütz.) Dr.

and Daily, though a full discussion of this case has not yet been p lished.

The nomenclature of the higher groups is that of the sixth edi of Bergey's *Manual of Determinative Bacteriology* for the bacteria Fritch's *Structure and Reproduction of Algae* for the lower plants, of Hyman's *The Invertebrata* for the lower animals.

General Index

(Italic paging refers to figures.)